BIG IDEAS MATH®

COURSE 2 ACCELERATED

A Bridge to Success

TEACHING EDITION

Ron Larson
Laurie Boswell

BIG IDEAS LEARNING®
Erie, Pennsylvania
BigIdeasLearning.com

Big Ideas Learning, LLC
1762 Norcross Road
Erie, PA 16510-3838
USA

For product information and customer support, contact Big Ideas Learning at **1-877-552-7766** or visit us at ***BigIdeasLearning.com***.

Cover
Songquan Deng/Shutterstock.com

Printed in the U.S.A.

ISBN 13: 978-1-68033-125-7
ISBN 10: 1-68033-125-6

2 3 4 5 6 7 8 9 10 WEB 17 16 15

AUTHORS

Ron Larson is a professor of mathematics at Penn State Erie, The Behrend College, where he has taught since receiving his Ph.D. in mathematics from the University of Colorado. Dr. Larson is well known as the lead author of a comprehensive program for mathematics that spans middle school, high school, and college courses. His high school and Advanced Placement books are published by Holt McDougal. Ron's numerous professional activities keep him in constant touch with the needs of students, teachers, and supervisors. Ron and Laurie Boswell began writing together in 1992. Since that time, they have authored over two dozen textbooks. In their collaboration, Ron is primarily responsible for the pupil edition and Laurie is primarily responsible for the teaching edition of the text.

Laurie Boswell is the Head of School and a mathematics teacher at the Riverside School in Lyndonville, Vermont. Dr. Boswell received her Ed.D. from the University of Vermont in 2010. She is a recipient of the Presidential Award for Excellence in Mathematics Teaching. Laurie has taught math to students at all levels, elementary through college. In addition, Laurie was a Tandy Technology Scholar, and served on the NCTM Board of Directors from 2002 to 2005. She currently serves on the board of NCSM, and is a popular national speaker. Along with Ron, Laurie has co-authored numerous math programs.

ABOUT THE BOOK

Big Ideas Math: A Bridge to Success was developed using the same research-based strategy of a balanced approach that has become synonymous with the *Big Ideas Math* series. This approach opens doors to abstract thought, reasoning, and inquiry as students persevere to answer the Essential Questions that introduce each section.

From start to finish, this program was designed with the student in mind. Students are subtly introduced to "Habits of Mind" that help them internalize concepts for a greater depth of understanding. These habits serve students well not only in mathematics, but across all curricula throughout their academic careers. *Big Ideas Math: A Bridge to Success* exposes students to highly motivating and relevant problems. Woven throughout the program are the depth and rigor students need to prepare for career-readiness and other college-level courses.

We consider the Big Ideas Math series to be the crowning jewel of 30 years of achievement in writing educational materials.

Ron Larson   Laurie Boswell

TEACHER REVIEWERS

- Lisa Amspacher
  Milton Hershey School
  Hershey, PA
- Mary Ballerina
  Orange County Public Schools
  Orlando, FL
- Lisa Bubello
  School District of Palm Beach County
  Lake Worth, FL
- Sam Coffman
  North East School District
  North East, PA
- Kristen Karbon
  Troy School District
  Rochester Hills, MI
- Laurie Mallis
  Westglades Middle School
  Coral Springs, FL
- Dave Morris
  Union City Area School District
  Union City, PA
- Bonnie Pendergast
  Tolleson Union High School District
  Tolleson, AZ
- Valerie Sullivan
  Lamoille South Supervisory Union
  Morrisville, VT
- Becky Walker
  Appleton Area School District
  Appleton, WI
- Zena Wiltshire
  Dade County Public Schools
  Miami, FL

STUDENT REVIEWERS

- Mike Carter
- Matthew Cauley
- Amelia Davis
- Wisdom Dowds
- John Flatley
- Nick Ganger
- Hannah Iadeluca
- Paige Lavine
- Emma Louie
- David Nichols
- Mikala Parnell
- Jordan Pashupathi
- Stephen Piglowski
- Robby Quinn
- Michael Rawlings
- Garrett Sample
- Andrew Samuels
- Addie Sedelmyer
- Tyler Steffy
- Erin Taylor
- Reid Wilson

CONSULTANTS

- Patsy Davis
  Educational Consultant
  Knoxville, Tennessee

- Bob Fulenwider
  Mathematics Consultant
  Bakersfield, California

- Linda Hall
  Mathematics Assessment Consultant
  Norman, Oklahoma

- Ryan Keating
  Special Education Advisor
  Gilbert, Arizona

- Michael McDowell
  Project-Based Instruction Specialist
  Fairfax, California

- Sean McKeighan
  Interdisciplinary Advisor
  Norman, Oklahoma

- Bonnie Spence
  Differentiated Instruction Consultant
  Missoula, Montana

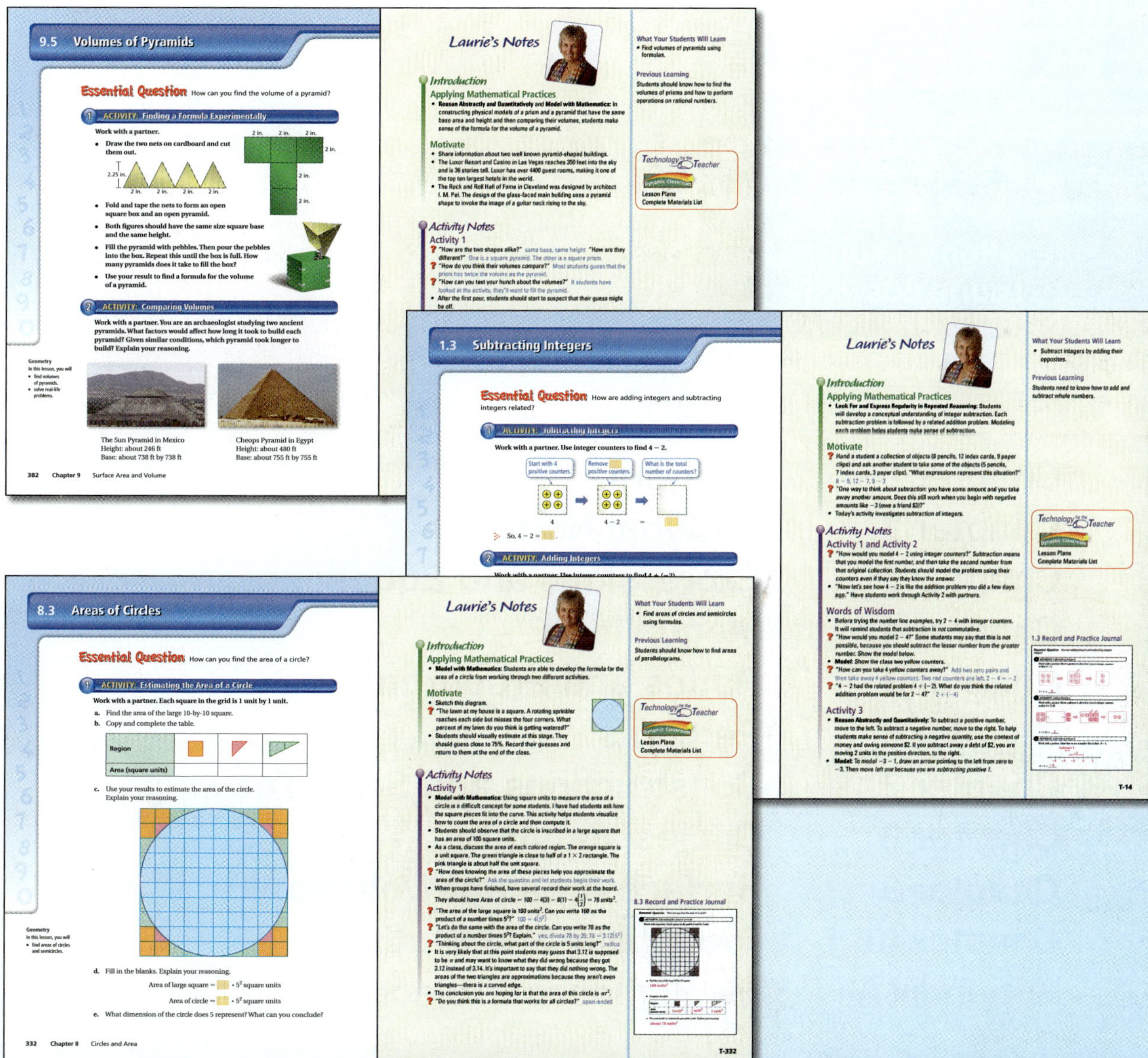

BIG IDEAS MATH

Big Ideas Math Course 2 Accelerated allows students to follow the Accelerated, or "Compacted," Traditional Pathway. Students complete 7th grade, 8th grade, and Algebra 1 in two years without skipping any learning objectives..

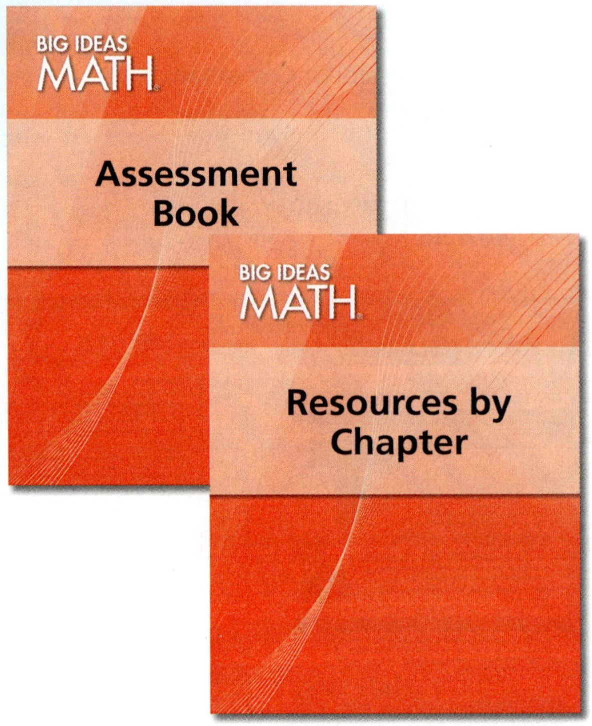

COURSE 2 ACCELERATED

Using *Big Ideas Math Course 2 Accelerated* and *Big Ideas Math Algebra 1*, students can complete the Compacted Pathway and have the opportunity for conceptual understanding, procedural fluency, and application through the use of focus, coherence, and rigor.

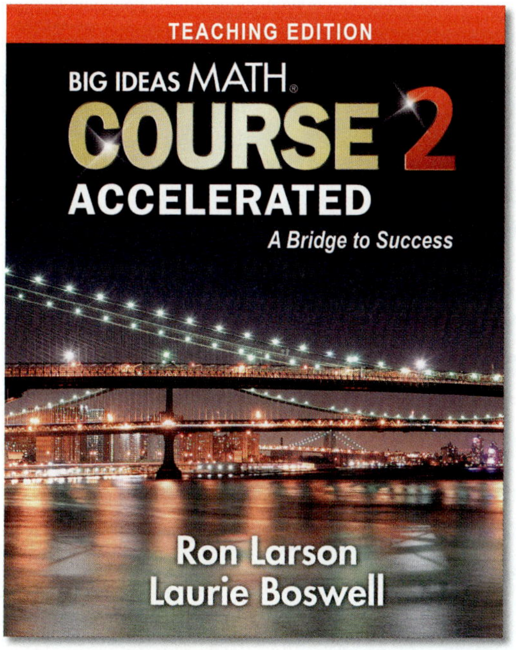

Chapter 11 Transformations
Chapter 12 Angles and Triangles
Chapter 13 Graphing and Writing Linear Equations
Chapter 14 Real Numbers and the Pythagorean Theorem
Chapter 15 Volume and Similar Solids
Chapter 16 Exponents and Scientific Notation
Additional Topics

11 Transformations

“Before my school had Big Ideas Math I would always lose test points because I left units off my answers. Now I see why they are so important.”

Angles and Triangles

“Word problems used to confuse me. Now I understand how to look for patterns and what the question is asking!”

13

Graphing and Writing Linear Equations

“I like the Big Ideas Math Tutorials because they help explain the math when I am at home.”

Real Numbers and the Pythagorean Theorem

14

“I really like the Graphic Organizers because they show me another way to take notes.”

15 Volume and Similar Solids

"*Using the Interactive Manipulatives from the Dynamic Student Edition helps me to see the mathematics that I am learning.*"

Exponents and Scientific Notation

"The Skills Review Handbook helps me review topics that I learned before."

Additional Topics

PROGRAM OVERVIEW

Print *Also available online*

- ### Pupil Edition

- ### Teaching Edition
  - **Laurie's Notes**

- ### Record and Practice Journal
  Available in English and Spanish

- ### Resources by Chapter and Assessment Book
  - **Resources by Chapter**
    - **Family and Community Involvement: English and Spanish**
    - **Start Thinking! and Warm Up**
    - **Extra Practice**
    - **Enrichment and Extension**
    - **Puzzle Time**
    - **Technology Connection**
  - **Assessment Book**
    - **Quizzes**
    - **Chapter Tests**
    - **Cumulative Assessment**
    - **Alternative Assessment**
    - **End-of-Course Tests**

Technology

Student Edition

With complete English and Spanish Audio

- Dynamic eBook App
  - Lesson Tutorial Videos
  - Dynamic Investigations
- Online Home Edition
  - Multi-Language Glossary
  - Basic Skills Handbook
  - Skills Review Handbook
  - Game Closet

Dynamic Classroom

- Interactive Manipulatives
- Answer Presentation Tool
- Mini-Assessment
- Closure Activity

Dynamic Assessment and Progress Monitoring Tool

- Homework and Assessment Creation
- Progress Monitoring
- Direct Ties to Remediation
- All-In-One Reporting
- Online Chat Tutor

Dynamic Teaching Tools

- Interactive Whiteboard Lesson Library
  - Includes standard and customizable lessons
  - Compatible with SMART®, Promethean®, and Mimeo® technology
- ExamView® Assessment Suite
- Real-Life STEM Videos
- Editable Online Resources
  - Lesson Plans
  - Pacing Guides
  - Assessment Book
  - Resources by Chapter

SCOPE AND

Regular Pathway

Grade	Domain	Focus
Grade 6	**Ratios and Proportional Relationships**	– Understand Ratio Concepts; Use Ratio Reasoning
	The Number System	– Perform Fraction and Decimal Operations; Understand Rational Numbers
	Expressions and Equations	– Write, Interpret, and Use Expressions, Equations, and Inequalities
	Geometry	– Solve Problems Involving Area, Surface Area, and Volume
	Statistics and Probability	– Summarize and Describe Distributions; Understand Variability
Grade 7	**Ratios and Proportional Relationships**	– Analyze Proportional Relationships
	The Number System	– Perform Rational Number Operations
	Expressions and Equations	– Generate Equivalent Expressions; Solve Problems Using Linear Equations and Inequalities
	Geometry	– Understand Geometric Relationships; Solve Problems Involving Angles, Surface Area, and Volume
	Statistics and Probability	– Analyze and Compare Populations; Find Probabilities of Events
Grade 8	**The Number System**	– Approximate Real Numbers; Perform Real Number Operations
	Expressions and Equations	– Use Radicals and Integer Exponents; Connect Proportional Relationships and Lines; Solve Systems of Linear Equations
	Functions	– Define, Evaluate, and Compare Functions; Model Relationships
	Geometry	– Understand Congruence and Similarity; Apply the Pythagorean Theorem; Apply Volume Formulas
	Statistics and Probability	– Analyze Bivariate Data

SEQUENCE

Compacted Pathway

Grade 6

Ratios and Proportional Relationships	– Understand Ratio Concepts; Use Ratio Reasoning
The Number System	– Perform Fraction and Decimal Operations; Understand Rational Numbers
Expressions and Equations	– Write, Interpret, and Use Expressions, Equations, and Inequalities
Geometry	– Solve Problems Involving Area, Surface Area, and Volume
Statistics and Probability	– Summarize and Describe Distributions; Understand Variability

Grade 7 Accelerated

Number and Quantity	– Analyze Proportional Relationships; Perform Real Number Operations; Use Radicals and Integer Exponents
Algebra	– Generate Equivalent Expressions; Connect Proportional Relationships and Lines; Solve Problems Using Linear Equations and Inequalities
Geometry	– Understand Geometric Relationships and Similarity; Solve Problems Involving Angles, Surface Area, and Volume
Statistics and Probability	– Analyze and Compare Populations; Find Probabilities of Events

Grade 8 Algebra 1

Number and Quantity	– Use Rational Exponents; Perform Real Number Operations
Algebra	– Solve Linear and Quadratic Equations; Solve Inequalities and Systems of Equations
Functions	– Define, Evaluate, and Compare Functions; Write Sequences; Model Relationships
Geometry	– Apply the Pythagorean Theorem
Statistics and Probability	– Represent and Interpret Data; Analyze Bivariate Data

LEARNING PROGRESSION

Compacted Pathway

Domain	Grade 6	Grade 7 Accelerated	Algebra 1
Number and Quantity			
Ratios and Proportional Relationships	Use ratios to solve problems. *Chapter 5*	Use proportional relationships to solve problems. *Chapters 5, 6*	
The Number System	Perform operations with multi-digit numbers and find common factors and multiples. *Chapter 1* Divide fractions by fractions. *Chapter 2* Extend understanding of numbers to the rational number system. *Chapter 6*	Perform operations with rational numbers. *Chapter 2* Extend understanding of numbers to the real number system. *Chapter 14*	Use properties of rational exponents, rational numbers, and irrational numbers. *Chapters 6, 9, 10*
Algebra			
Expressions and Equations	Perform arithmetic with algebraic expressions. *Chapter 3* Solve one-variable equations and inequalities. *Chapter 7* Analyze relationships between dependent and independent variables. *Chapter 7*	Write equivalent expressions. *Chapter 3* Use numerical and algebraic expressions, equations, and inequalities to solve problems. *Chapters 3, 4* Understand the connections between proportional relationships, lines, and linear equations. *Chapter 13* Work with radicals and integer exponents. *Chapters 14, 16*	Perform arithmetic on linear and quadratic polynomials. *Chapters 7–9* Solve linear and absolute value equations and inequalities, and quadratic and exponential equations. *Chapters 1–4, 6–10* Solve linear systems of equations and inequalities, and nonlinear systems of equations. *Chapters 5, 9*
Functions			
Functions			Interpret and analyze linear, quadratic, exponential, absolute value, step, and piecewise functions. *Chapters 3, 4, 6, 8–10* Build linear, quadratic, exponential, and absolute value functions from existing functions. *Chapters 3, 6, 8, 10*

MIDDLE SCHOOL MATH

Domain	Grade 6	Grade 7 Accelerated	Algebra 1
Geometry			
Geometry	Solve real-world and mathematical problems involving area, surface area, and volume. *Chapters 4, 8*	Draw, construct, and describe geometrical figures and describe the relationships between them. *Chapters 7–9* Solve problems involving angle measure, area, surface area, and volume. *Chapters 7–9* Understand congruence and similarity. *Chapters 11, 12* Use the Pythagorean Theorem. *Chapter 14* Solve problems involving volume of cylinders, cones, and spheres. *Chapter 15*	
Statistics and Probability			
Statistics and Probability	Develop understanding of statistical variability and summarize and describe distributions. *Chapters 9, 10*	Make inferences about a population, compare two populations, and use probability models. *Chapter 10*	Summarize, represent, and interpret data using measures of center and spread. *Chapter 11*

PACING GUIDE

Regular:

Chapters 1–10	154 Days

Accelerated:

Chapters 1–16:	152 Days

	Regular	Accelerated
Scavenger Hunt	**1 Day**	**1 Day**
Chapter 1	**14 Days**	**4 Days**
Chapter Opener	1 Day	1 Day
Section 1.1	2 Days	1 Day
Section 1.2	2 Days	0.5 Day
Section 1.3	2 Days	0.5 Day
Study Help/Quiz	1 Day	0 Days
Section 1.4	2 Days	0.5 Day
Section 1.5	2 Days	0.5 Day
Chapter Review/Chapter Tests	2 Days	0 Days
Chapter 2	**12 Days**	**5 Days**
Chapter Opener	1 Day	0.5 Day
Section 2.1	2 Days	0.5 Day
Section 2.2	2 Days	0.5 Day
Study Help/Quiz	1 Day	0 Days
Section 2.3	2 Days	0.5 Day
Section 2.4	2 Days	1 Day
Chapter Review/Chapter Tests	2 Days	2 Days
Chapter 3	**15 Days**	**12 Days**
Chapter Opener	1 Day	1 Day
Section 3.1	2 Days	0.5 Day
Section 3.2	3 Days	1.5 Days
Study Help/Quiz	1 Day	0 Days
Section 3.3	2 Days	1 Day
Section 3.4	2 Days	1 Day
Section 3.5	2 Days	1 Day
Topic 1		1 Day
Topic 2		2 Days
Topic 3		1 Day
Chapter Review/Chapter Tests	2 Days	2 Days

	Regular	Accelerated
Chapter 4	**12 Days**	**7 Days**
Chapter Opener	1 Day	1 Day
Section 4.1	2 Days	1 Day
Section 4.2	2 Days	0 Days
Study Help/Quiz	1 Day	1 Day
Section 4.3	2 Days	1 Day
Section 4.4	2 Days	1 Day
Chapter Review/Chapter Tests	2 Days	2 Days
Chapter 5	**17 Days**	**11 Days**
Chapter Opener	1 Day	1 Day
Section 5.1	2 Days	1 Day
Section 5.2	2 Days	1 Day
Section 5.3	3 Days	2 Days
Study Help/Quiz	1 Day	1 Day
Section 5.4	2 Days	1 Day
Section 5.5	2 Days	1 Day
Section 5.6	2 Days	1 Day
Chapter Review/Chapter Tests	2 Days	2 Days
Chapter 6	**18 Days**	**11 Days**
Chapter Opener	1 Day	1 Day
Section 6.1	2 Days	1 Day
Section 6.2	2 Days	1 Day
Section 6.3	2 Days	1 Day
Section 6.4	2 Days	1 Day
Study Help/Quiz	1 Day	1 Day
Section 6.5	2 Days	1 Day
Section 6.6	2 Days	1 Day
Section 6.7	2 Days	1 Day
Chapter Review/Chapter Tests	2 Days	2 Days
Chapter 7	**17 Days**	**11 Days**
Chapter Opener	1 Day	1 Day
Section 7.1	2 Days	1 Day
Section 7.2	2 Days	1 Day
Section 7.3	4 Days	3 Days
Study Help/Quiz	1 Day	1 Day
Section 7.4	3 Days	1 Day
Section 7.5	2 Days	1 Day
Chapter Review/Chapter Tests	2 Days	2 Days

PACING GUIDE (CONTINUED)

	Regular	Accelerated
Chapter 8	**12 Days**	**7 Days**
Chapter Opener	1 Day	1 Day
Section 8.1	2 Days	1 Day
Section 8.2	2 Days	1 Day
Study Help/Quiz	1 Day	0 Days
Section 8.3	2 Days	1 Day
Section 8.4	2 Days	1 Day
Chapter Review/Chapter Tests	2 Days	2 Days
Chapter 9	**16 Days**	**10 Days**
Chapter Opener	1 Day	1 Day
Section 9.1	2 Days	1 Day
Section 9.2	3 Days	2 Days
Section 9.3	2 Days	1 Day
Study Help/Quiz	1 Day	0 Day
Section 9.4	2 Days	1 Day
Section 9.5	3 Days	2 Days
Chapter Review/Chapter Tests	2 Days	2 Days
Chapter 10	**20 Days**	**13 Days**
Chapter Opener	1 Day	1 Day
Section 10.1	2 Days	1 Day
Section 10.2	2 Days	2 Days
Section 10.3	2 Days	2 Days
Section 10.4	2 Days	1 Day
Section 10.5	3 Days	1 Day
Study Help/Quiz	1 Day	1 Day
Section 10.6	3 Days	1 Day
Section 10.7	2 Days	1 Day
Chapter Review/Chapter Tests	2 Days	2 Days
Chapter 11	**0 Days**	**10 Days**
Chapter Opener		1 Day
Section 11.1		1 Day
Section 11.2		1 Day
Section 11.3		1 Day
Section 11.4		1 Day
Study Help/Quiz		1 Day
Section 11.5		1 Day
Section 11.6		1 Day
Section 11.7		2 Days
Chapter Review/Chapter Tests		2 Days

	Regular	Accelerated
Chapter 12	**0 Days**	**8 Days**
Chapter Opener		1 Day
Section 12.1		1 Day
Section 12.2		1 Day
Section 12.3		2 Days
Section 12.4		1 Day
Chapter Review/Chapter Tests		2 Days
Chapter 13	**0 Days**	**12 Days**
Chapter Opener		1 Day
Section 13.1		1 Day
Section 13.2		2 Days
Section 13.3		1 Day
Study Help/Quiz		1 Day
Section 13.4		1 Day
Section 13.5		1 Day
Section 13.6		1 Day
Section 13.7		1 Day
Chapter Review/Chapter Tests		2 Days
Chapter 14	**0 Days**	**9 Days**
Chapter Opener		1 Day
Section 14.1		1 Day
Section 14.2		1 Day
Section 14.3		1 Day
Section 14.4		2 Days
Section 14.5		1 Day
Chapter Review/Chapter Tests		2 Days
Chapter 15	**0 Days**	**8 Days**
Chapter Opener		1 Day
Section 15.1		1 Day
Section 15.2		1 Day
Section 15.3		1 Day
Section 15.4		2 Days
Chapter Review/Chapter Tests		2 Days
Chapter 16	**0 Days**	**11 Days**
Chapter Opener		1 Day
Section 16.1		1 Day
Section 16.2		1 Day
Section 16.3		1 Day
Section 16.4		1 Day
Study Help/Quiz		1 Day
Section 16.5		1 Day
Section 16.6		1 Day
Section 16.7		1 Day
Chapter Review/Chapter Tests		2 Days

Mathematical Practices

Make sense of problems and persevere in solving them.

- Multiple representations are presented to help students move from concrete to representative and into abstract thinking
- *Essential Questions* help students focus and analyze
- *In Your Own Words* provide opportunities for students to look for meaning and entry points to a problem

Reason abstractly and quantitatively.

- Visual problem solving models help students create a coherent representation of the problem
- Opportunities for students to decontextualize and contextualize problems are presented in every lesson

Construct viable arguments and critique the reasoning of others.

- *Error Analysis*; *Different Words, Same Question*; and *Which One Doesn't Belong* features provide students the opportunity to construct arguments and critique the reasoning of others
- *Inductive Reasoning* activities help students make conjectures and build a logical progression of statements to explore their conjecture

Model with mathematics.

- Real-life situations are translated into diagrams, tables, equations, and graphs to help students analyze relations and to draw conclusions
- Real-life problems are provided to help students learn to apply the mathematics that they are learning to everyday life

Use appropriate tools strategically.

- *Graphic Organizers* support the thought process of what, when, and how to solve problems
- A variety of tool papers, such as graph paper, number lines, and manipulatives, are available as students consider how to approach a problem
- Opportunities to use the web, graphing calculators, and spreadsheets support student learning

Attend to precision.

- *On Your Own* questions encourage students to formulate consistent and appropriate reasoning
- Cooperative learning opportunities support precise communication

Look for and make use of structure.

- *Inductive Reasoning* activities provide students the opportunity to see patterns and structure in mathematics
- Real-world problems help students use the structure of mathematics to break down and solve more difficult problems

Look for and express regularity in repeated reasoning.

- Opportunities are provided to help students make generalizations
- Students are continually encouraged to check for reasonableness in their solutions

Go to *BigIdeasMath.com* for more information on the Mathematical Practices.

Mathematical Content

Chapter Coverage

1 · **2** · 3 · 4 · **5** · **6** · 7 · 8 · 9 · 10 · 11 · 12 · 13 · **14** · 15 · **16** · AT

Number and Quantity

- The Real Number System
- Quantities

1 · 2 · **3** · **4** · 5 · **6** · 7 · 8 · 9 · 10 · 11 · 12 · **13** · 14 · 15 · 16 · **AT**

Algebra

- Seeing Structure in Expressions
- Reasoning with Equations and Inequalities

1 · 2 · 3 · 4 · 5 · 6 · **7** · **8** · **9** · 10 · **11** · **12** · 13 · 14 · **15** · 16 · AT

Geometry

- Congruence
- Similarity, Right Triangles, and Trigonometry
- Geometric Measurement and Dimension

1 · 2 · 3 · 4 · 5 · 6 · 7 · 8 · 9 · **10** · 11 · 12 · 13 · 14 · 15 · 16 · AT

Statistics and Probability

- Making Inferences and Justifying Conclusions
- Conditional Probability and the Rules of Probability

11 Transformations

"Just 2 more minutes. I'm almost done with my 'cat tessellation' painting."

"If you hold perfectly still..." "...each frame becomes a horizontal..." "...translation of the previous frame..."

What Your Students Have Learned

- Draw polygons in the coordinate plane given vertices and find lengths of sides whose endpoints have the same *x*- or *y*-coordinate.
- Find areas of triangles and special quadrilaterals.
- Understand ratios and describe ratio relationships.
- Draw geometric shapes with given conditions.
- Represent proportional relationships with equations.
- Find unit rates associated with ratios of perimeters and areas.
- Use proportionality to solve ratio problems.
- Reproduce a scale drawing at a different scale.

What Your Students Will Learn

- Identify and use congruent figures by finding corresponding angles and corresponding sides.
- Verify the properties of translations, reflections, and rotations.
- Describe translations, reflections, rotations, and dilations using coordinates.
- Understand that figures are congruent (or similar) when they can be related by a sequence of translations, reflections, and rotations (and dilations).
- Describe a sequence that exhibits congruence or similarity between two figures.

Pacing Guide for Chapter 11

Chapter Opener Accelerated	1 Day
Section 1 Accelerated	1 Day
Section 2 Accelerated	1 Day
Section 3 Accelerated	1 Day
Section 4 Accelerated	1 Day
Study Help / Quiz Accelerated	1 Day
Section 5 Accelerated	1 Day
Section 6 Accelerated	1 Day
Section 7 Accelerated	2 Days
Chapter Review/ Chapter Tests Accelerated	2 Days
Total Chapter 11 Accelerated	12 Days
Year-to-Date Accelerated	104 Days

Technology for the Teacher

BigIdeasMath.com
Chapter at a Glance
Complete Materials List
Parent Letters: English and Spanish

What Your Students Have Learned

- Reflect points in the x- and y-axis.
- Draw polygons in a coordinate plane using given vertices.

Additional Topics for Review

- Geoboard
- Congruent Angles
- Congruent Sides
- Right Angle
- Perimeter
- Area
- Vertical
- Horizontal
- Quadrants
- Clockwise
- Counterclockwise
- Ratio
- Proportion

Try It Yourself

1. a. $(7, -3)$ **b.** $(-7, 3)$

2. a. $(-4, -6)$ **b.** $(4, 6)$

3. a. $(5, 5)$ **b.** $(-5, -5)$

4. a. $(-8, 3)$ **b.** $(8, -3)$

5–10. See Additional Answers.

Record and Practice Journal Fair Game Review

1. $(1, -1); (-1, 1)$

2. $(-2, 4); (2, -4)$

3. $(-3, -3); (3, 3)$

4. $(4, 3); (-4, -3)$

5–12. See Additional Answers.

Math Background Notes

Vocabulary Review

- Reflection
- Coordinate Plane
- Ordered Pair
- x-Axis
- y-Axis
- x-Coordinate
- y-Coordinate
- Opposite
- Polygon
- Quadrilateral
- Triangle
- Vertices

Reflecting Points

- Students should have studied reflecting points in a previous course. If you sense that a review is necessary, then go over Example 1 in detail.
- You may also want to expand upon Example 1 and reflect the point in the y-axis as well as in both axes. When reflecting a second time, however, be sure that students use the reflected point and not the original point.
- If you choose to review reflection in both axes, then an excellent follow-up question would be whether reflection in the x-axis followed by reflection in the y-axis produces a different result than reflection in the y-axis followed by reflection in the x-axis. (The answer is no; the end result is the same.)
- **Common Error:** Students may confuse reflection in the x-axis with reflection in the y-axis (and vice versa). This is typically because algebraically, the coordinate substitutions necessary to come up with a reflected point do not seem natural to students. To reflect in the x-axis, you replace the y-coordinate with its opposite (and vice versa).
- To overcome confusion, have students actually plot the point as well as its reflection. Discourage rote memorization of algebraic procedures.

Drawing a Polygon in a Coordinate Plane

- Students have been studying polygons since elementary school. Also, students should be familiar with drawing a polygon in a coordinate plane from a previous course.
- You may want to expand upon Example 2 and the exercises that follow and ask students to classify the polygons. In a previous course, they should have classified triangles and reviewed classifying quadrilaterals.
- Students may mistakenly classify the quadrilateral in Example 2 as a kite, because it may appear to them to have two pairs of congruent adjacent sides. In fact, there is only one pair of congruent adjacent sides, so the polygon cannot be classified any further than "quadrilateral." Students do not learn the distance formula until Chapter 14, so the best that students can do at this point is to use a ruler to measure the side lengths. The take-away lesson here is that students should not "eyeball" a polygon to come up with a classification.
- **FYI:** The triangle in Exercise 9 is obtuse scalene. The quadrilateral in Exercise 10 is a square.

Reteaching and Enrichment Strategies

If students need help...	If students got it...
Record and Practice Journal • Fair Game Review Skills Review Handbook Lesson Tutorials	Game Closet at *BigIdeasMath.com* Start the next section

What You Learned Before

Reflecting Points

Example 1 **Reflect (3, −4) in the *x*-axis.**

Plot (3, −4).

To reflect (3, −4) in the x-axis, use the same x-coordinate, 3, and take the opposite of the y-coordinate. The opposite of −4 is 4.

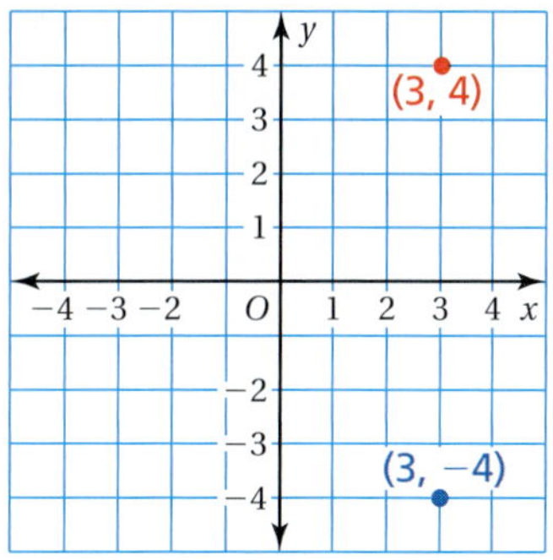

So, the reflection of (3, −4) in the x-axis is (3, 4).

Try It Yourself

Reflect the point in (a) the *x*-axis and (b) the *y*-axis.

1. (7, 3) **2.** (−4, 6) **3.** (5, −5) **4.** (−8, −3)

5. (0, 1) **6.** (−5, 0) **7.** (4, −6.5) **8.** $\left(-3\frac{1}{2}, -4\right)$

Drawing a Polygon in a Coordinate Plane

Example 2 **The vertices of a quadrilateral are *A*(1, 5), *B*(2, 9), *C*(6, 8), and *D*(8, 1). Draw the quadrilateral in a coordinate plane.**

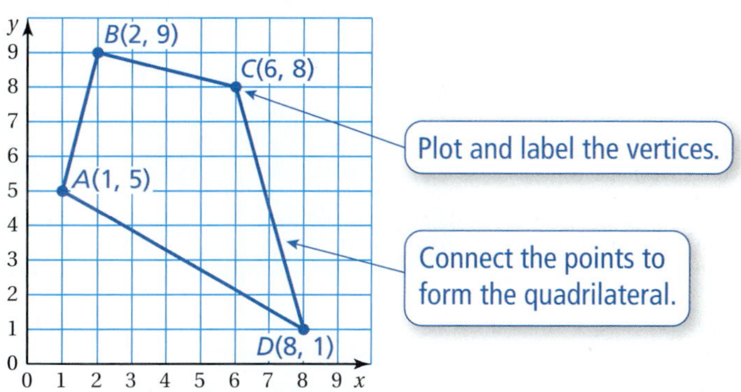

Try It Yourself

Draw the polygon with the given vertices in a coordinate plane.

9. $J(1, 1)$, $K(5, 6)$, $M(9, 3)$ **10.** $Q(2, 3)$, $R(2, 8)$, $S(7, 8)$, $T(7, 3)$

11.1 Congruent Figures

Essential Question How can you identify congruent triangles?

Two figures are congruent when they have the same size and the same shape.

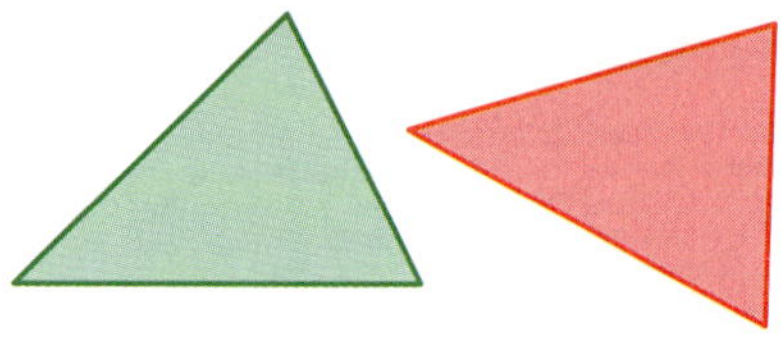

Congruent
Same size *and* shape

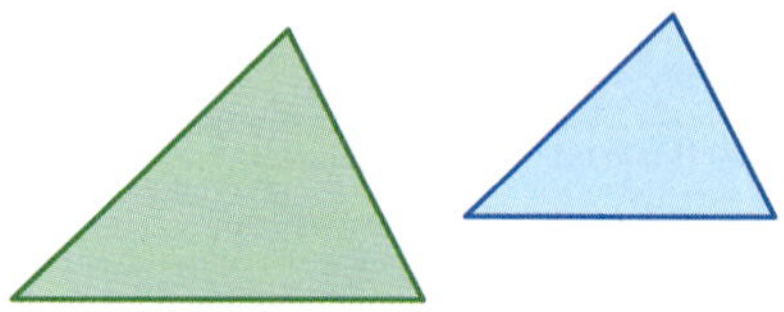

Not Congruent
Same shape, but not same size

1 ACTIVITY: Identifying Congruent Triangles

Work with a partner.

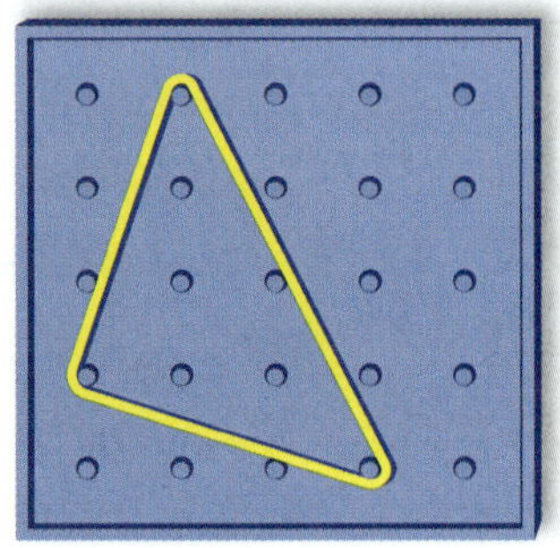

- **Which of the geoboard triangles below are congruent to the geoboard triangle at the right?**
- **Form each triangle on a geoboard.**
- **Measure each side with a ruler. Record your results in a table.**
- **Write a conclusion about the side lengths of triangles that are congruent.**

a.

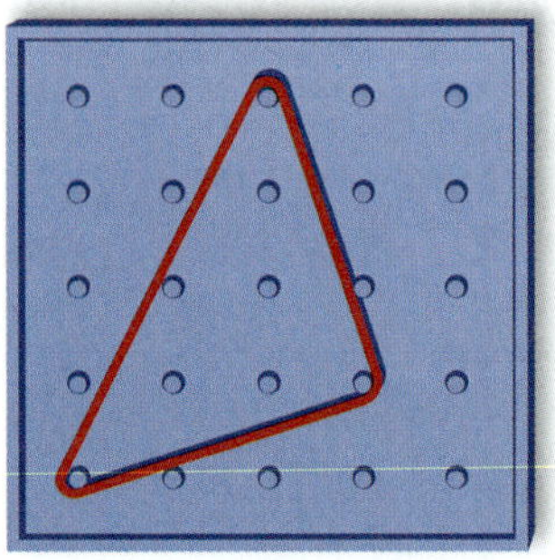

b.

c.

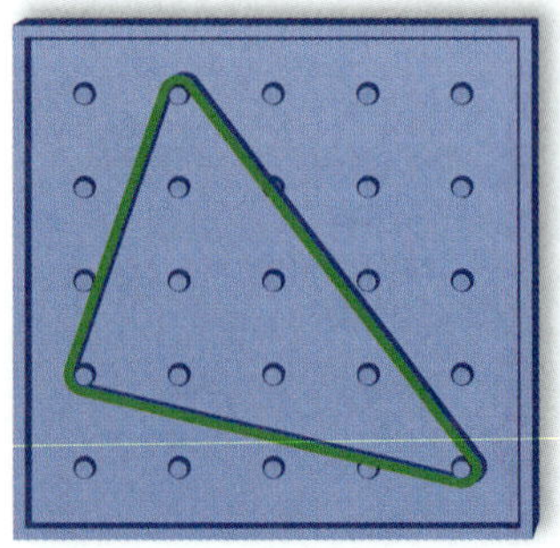

d.

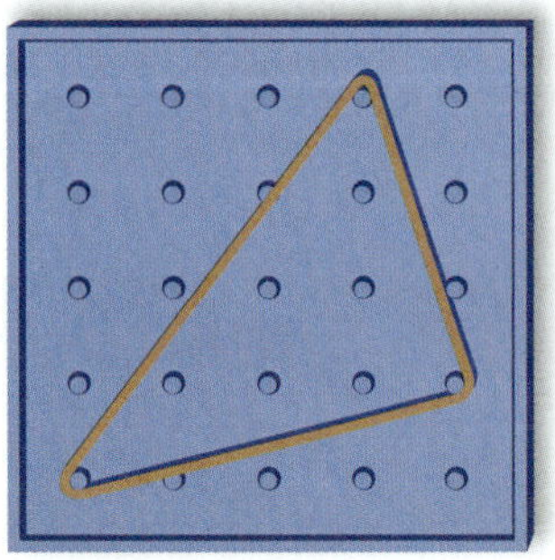

e.

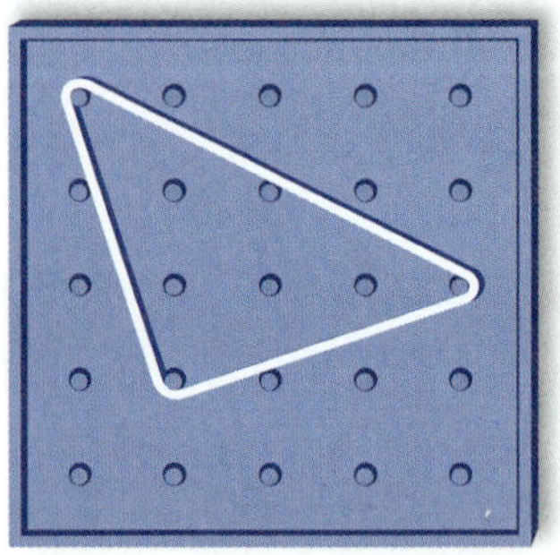

f.

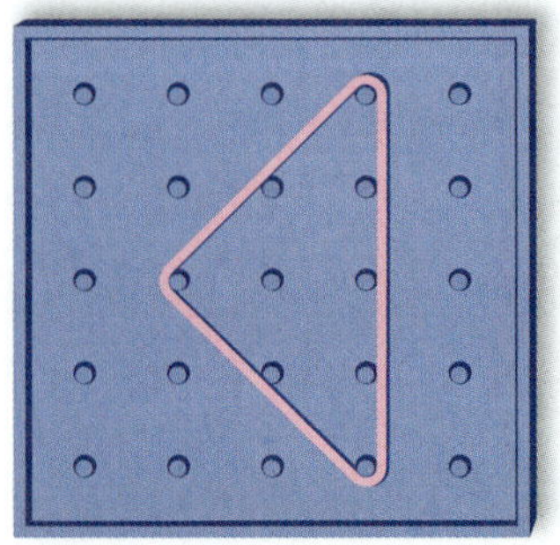

Geometry
In this lesson, you will
- name corresponding angles and corresponding sides of congruent figures.
- identify congruent figures.

Laurie's Notes

Introduction

Applying Mathematical Practices

- **Use Appropriate Tools Strategically:** Geoboards allow students to change the shape or orientation of a figure with little effort. There is no erasing to do! When you change the location of a vertex, the two sides meeting at that vertex change automatically to meet at the new vertex.

Motivate

- ? "What word is used to describe figures that have the same size and shape?" congruent
- **Use Appropriate Tools Strategically:** Display a variety of objects such as templates, stencils, rubber stamps, and cookie cutters.
- ? "What do all of these objects have in common?" They can be used to create figures that are congruent to one another.
- ? "Can you think of other things that can be used to create congruent figures?" *Sample answer:* computer graphics program, clay molds
- Use one of the objects to create a few congruent figures.
- "Today you will use geoboards to decide whether triangles are congruent."

Activity Notes

Activity 1

- **Classroom Management:** If this is the first time your students have used geoboards, give them time to explore and play. I hand out the geoboards with two rubber bands arranged vertically and two arranged horizontally, and students are to return them the same way. If your geoboards are a different size than the 5×5 shown, don't worry. Your students will adjust. Also, you can use geoboard dot paper in place of geoboards.
- **Common Misconception:** Measuring between two pins horizontally or vertically is *not* the same as measuring across a diagonal. The distance between two diagonal pins is longer.
- Have students use their rulers to measure each side in millimeters. Recommend that they try to measure from the center of one pin to the center of another pin.
- Students with good spatial skills may be able to rotate or reflect some triangles, so they may not want to measure each side. Encourage students to leave the triangles in the orientations given and measure each side.
- Depending upon the accuracy of their measurements, students may believe that all of the triangles are congruent.
- The yellow triangle has two congruent sides. The triangles in parts (b) and (f) also have two congruent sides but they are not congruent to the yellow triangle.
- Have students share their results and observations. If they use vocabulary such as *isosceles* and *scalene* to describe the triangles, ask them to explain what these words mean. None of the triangles are equilateral.

What Your Students Will Learn

- Name corresponding angles and corresponding sides of congruent figures.
- Identify and use congruent figures to find lengths or measurements.

Previous Learning

Students should know how to analyze and compare angles and segments.

Technology for the Teacher

Dynamic Classroom

Lesson Plans
Complete Materials List

11.1 Record and Practice Journal

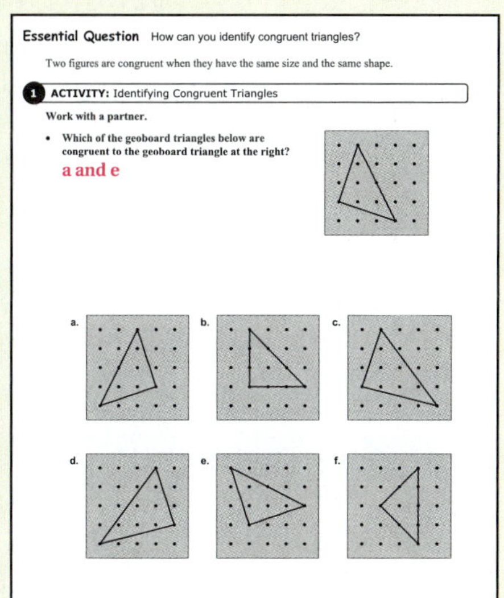

Essential Question How can you identify congruent triangles?

Two figures are congruent when they have the same size and the same shape.

1 ACTIVITY: Identifying Congruent Triangles

Work with a partner.

- **Which of the geoboard triangles below are congruent to the geoboard triangle at the right?**
  a and e

a. b. c. d. e. f.

Differentiated Instruction

Kinesthetic

Provide students with cutouts of congruent figures. Let them turn and flip the cutouts to discover that figures can be congruent even though they may have different orientations.

11.1 Record and Practice Journal

- Form each triangle on a geoboard.

  Check students' work.

- Measure each side with a ruler. Record your results in the table.

Sample answer:

	Side 1	Side 2	Side 3
Given Triangle	126 mm	126 mm	180 mm
a.	126 mm	126 mm	180 mm
b.	120 mm	120 mm	170 mm
c.	165 mm	126 mm	200 mm
d.	165 mm	126 mm	200 mm
e.	126 mm	126 mm	180 mm
f.	115 mm	115 mm	160 mm

- Write a conclusion about the side lengths of triangles that are congruent.

  Sample answer: The corresponding sides of triangles that are congruent have the same length.

2 ACTIVITY: Forming Congruent Triangles

Work with a partner.

a. Form the given triangle in Activity 1 on your geoboard. Record the triangle on geoboard dot paper. See Additional Answers.

b. Move each vertex of the triangle one peg to the right. Is the new triangle congruent to the original triangle? How can you tell?

yes; The new triangle is the same shape and size as the original triangle.

c. On a 5-by-5 geoboard, make as many different triangles as possible, each of which is congruent to the given triangle in Activity 1. Record each triangle on geoboard dot paper. See Additional Answers.

What Is Your Answer?

3. **IN YOUR OWN WORDS** How can you identify congruent triangles? Use the conclusion you wrote in Activity 1 as part of your answer.

Congruent triangles are the same size and shape. The corresponding sides of congruent triangles have the same length.

4. Can you form a triangle on your geoboard whose side lengths are 3, 4, and 5 units? If so, draw such a triangle on geoboard dot paper.

yes;

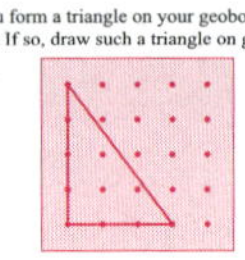

Laurie's Notes

Activity 2

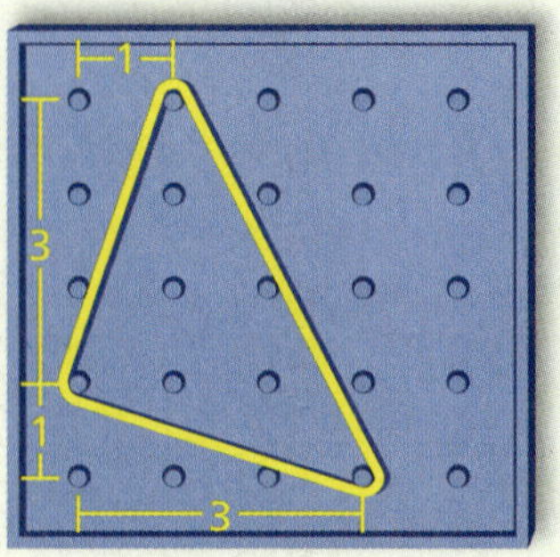

- In Activity 1, students concluded that the triangles in parts (a) and (e) are congruent to the yellow triangle and that each triangle has two congruent sides (the three triangles are isosceles).
- Point out that for each of these triangles, you can move three units (vertically or horizontally) and then one unit (horizontally or vertically) to get from one end of the congruent side to the other, as shown in the figure at the right.
- **Make Sense of Problems and Persevere in Solving Them:** To sketch additional triangles congruent to the yellow triangle, encourage students to consider different orientations.

What Is Your Answer?

- In answering Question 4, the units are not inches or centimeters, but *geoboard units*. This means that the vertical or horizontal distance between two pins is one geoboard unit. Determining the 3- and 4-unit side lengths is straightforward. Convincing yourself that the third side has a length of 5 units is less obvious. The triangle should be a 3-4-5 right triangle.
- Ask students to describe how they determined that the side lengths are 3, 4, and 5 units.

Closure

- Name 3 different examples of congruent figures or objects in the classroom. *Sample answer:* eraser, marker, chalkboard, bulletin board

Math Practice

Recognize Usefulness of Tools

What are some advantages and disadvantages of using a geoboard to construct congruent triangles?

The geoboard at the right shows three congruent triangles.

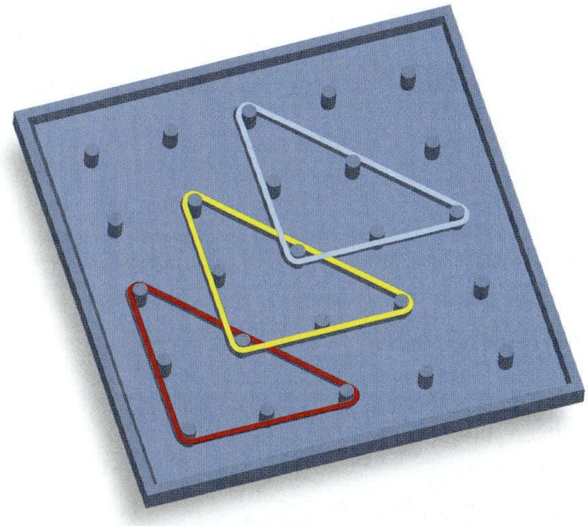

2 ACTIVITY: Forming Congruent Triangles

Work with a partner.

a. Form the yellow triangle in Activity 1 on your geoboard. Record the triangle on geoboard dot paper.

b. Move each vertex of the triangle one peg to the right. Is the new triangle congruent to the original triangle? How can you tell?

c. On a 5-by-5 geoboard, make as many different triangles as possible, each of which is congruent to the yellow triangle in Activity 1. Record each triangle on geoboard dot paper.

What Is Your Answer?

3. **IN YOUR OWN WORDS** How can you identify congruent triangles? Use the conclusion you wrote in Activity 1 as part of your answer.

4. Can you form a triangle on your geoboard whose side lengths are 3, 4, and 5 units? If so, draw such a triangle on geoboard dot paper.

Practice Use what you learned about congruent triangles to complete Exercises 4 and 5 on page 470.

11.1 Lesson

Check It Out
Lesson Tutorials
BigIdeasMath.com

Key Vocabulary
congruent figures, *p. 468*
corresponding angles, *p. 468*
corresponding sides, *p. 468*

Key Idea

Congruent Figures

Figures that have the same size and the same shape are called **congruent figures**. The triangles below are congruent.

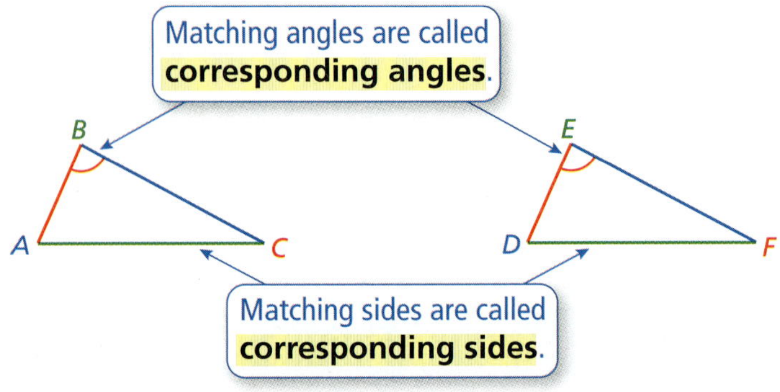

EXAMPLE 1 Naming Corresponding Parts

The figures are congruent. Name the corresponding angles and the corresponding sides.

Corresponding Angles	*Corresponding Sides*
$\angle A$ and $\angle W$	Side AB and Side WX
$\angle B$ and $\angle X$	Side BC and Side XY
$\angle C$ and $\angle Y$	Side CD and Side YZ
$\angle D$ and $\angle Z$	Side AD and Side WZ

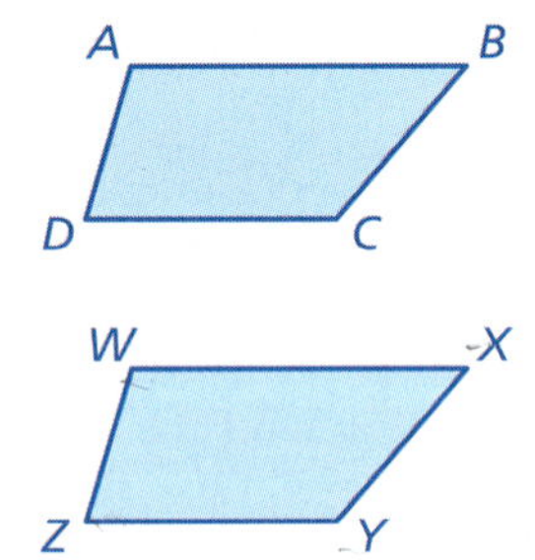

On Your Own

Now You're Ready
Exercises 6 and 7

1. The figures are congruent. Name the corresponding angles and the corresponding sides.

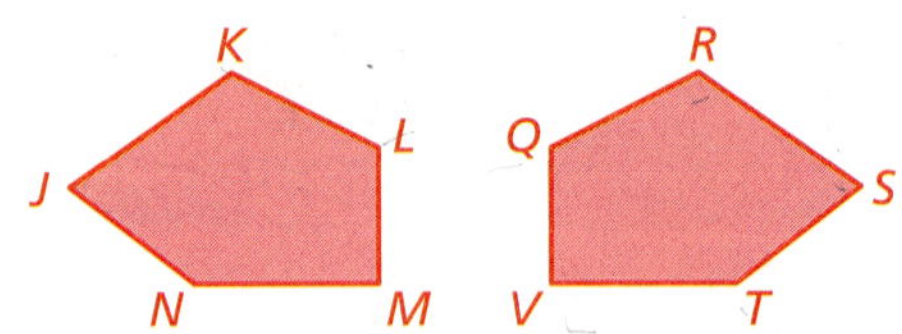

Key Idea

Identifying Congruent Figures

Two figures are congruent when corresponding angles and corresponding sides are congruent.

Triangle ABC is congruent to Triangle DEF.

$$\triangle ABC \cong \triangle DEF$$

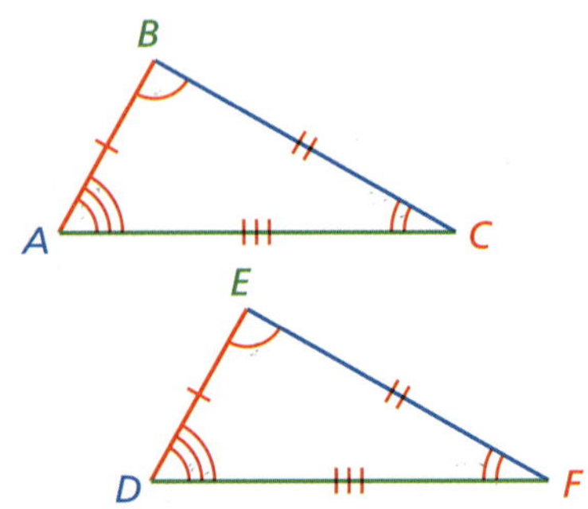

Reading

The symbol $\cong$ means *is congruent to*.

Multi-Language Glossary at BigIdeasMath.com

Laurie's Notes

Introduction

Connect

- **Yesterday:** Students explored congruent triangles by measuring side lengths.
- **Today:** Students will use the definition of congruent figures to determine whether two figures (polygons) are congruent.

Motivate

- **Preparation:** Cut coffee stirrers or uncooked spaghetti into lengths of 3 inches and 5 inches.
- Ask students to pick up 3 pieces (either 2 short and 1 long or 2 long and 1 short). Have them make a triangle using the 3 pieces.
- ? "Is anyone's triangle congruent to yours? Explain." Students with the same number of long and short pieces will have congruent triangles.
- Ask two students with congruent triangles to make their triangles on the overhead or document camera. Trace the triangles on the board. Show how congruent angles and segments are conveyed by marks without using words.

Lesson Notes

Key Idea

- Use 3 different colors to draw the sides of the triangles that correspond (match).
- Students often think about corresponding parts in terms of the greatest/least angle measures and longest/shortest sides.

Example 1

- **Representation:** Have students use color-coding to show corresponding parts.
- **Notation:** Remind students that the angle symbol ($\angle$) is needed when talking about corresponding angles; they should not write the letter only.
- ? "Is Side *AB* the same as Side *BA*?" yes

On Your Own

- When students have finished, ask them to share their answers. Students should use proper terminology such as "Angle *K*" and "Side *JK*."

Key Idea

- **Attend to Precision:** Discuss the use of arcs and tick marks to indicate congruent angles and sides.
- This definition of congruent figures involves 2 parts, corresponding angles being congruent and corresponding sides being congruent. For a triangle, 6 congruencies must be established, for a quadrilateral, 8 congruencies.
- **FYI:** You can use the congruent symbol "$\cong$" to write angle and side congruency statements. For example, write the 6 congruency statements for the pair of triangles shown.
  Congruent angles: $\angle A \cong \angle D$, $\angle B \cong \angle E$, $\angle C \cong \angle F$
  Congruent sides: Side $AB \cong$ Side DE, Side $BC \cong$ Side EF, Side $AC \cong$ Side DF

Goal Today's lesson is defining **congruent figures** and determining if two figures are congruent.

Technology for the Teacher

Dynamic Classroom

Lesson Tutorials
Lesson Plans
Answer Presentation Tool

Extra Example 1

The figures are congruent. Name the corresponding angles and corresponding sides.

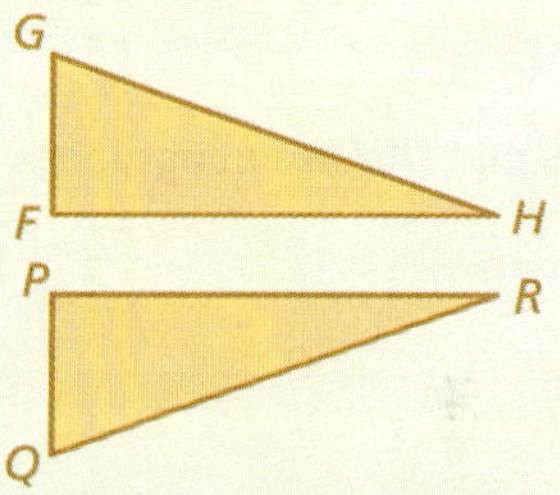

Corresponding angles: $\angle F$ and $\angle P$, $\angle G$ and $\angle Q$, $\angle H$ and $\angle R$;
Corresponding sides:
Side *FG* and Side *PQ*,
Side *GH* and Side *QR*,
Side *HF* and Side *RP*

On Your Own

1. Corresponding angles: $\angle J$ and $\angle S$, $\angle K$ and $\angle R$, $\angle L$ and $\angle Q$, $\angle M$ and $\angle V$, $\angle N$ and $\angle T$;
   Corresponding sides:
   Side *JK* and Side *SR*,
   Side *KL* and Side *RQ*,
   Side *LM* and Side *QV*,
   Side *MN* and Side *VT*,
   Side *NJ* and Side *TS*

English Language Learners

Symbols

Make sure that students understand the markings on congruent angles and sides. Use the figures in the *Key Ideas* to show which angles and sides are congruent.

Extra Example 2

Tell whether the two figures are congruent. Explain your reasoning.

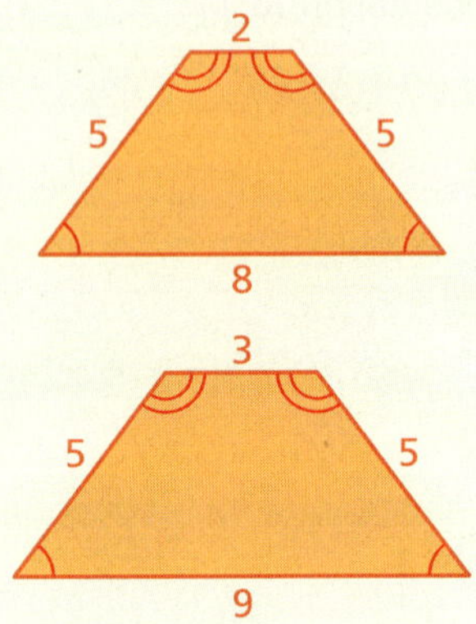

not congruent; The side lengths of the corresponding bases are not congruent.

Extra Example 3

Triangles *ABC* and *DEF* are congruent.

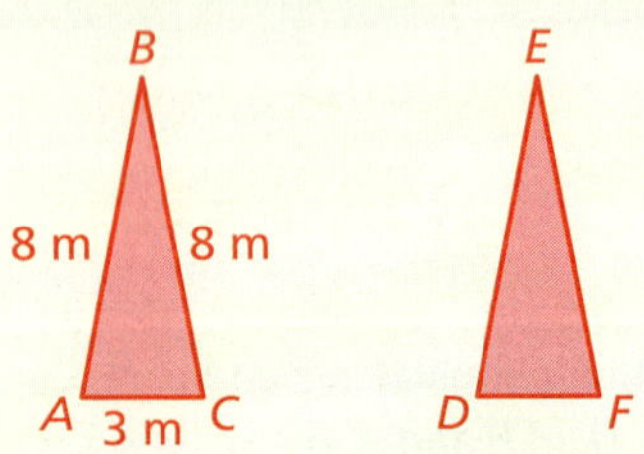

a. What is the length of Side *EF*? 8 m

b. What is the perimeter of *DEF*? 19 m

On Your Own

2. Square B

3. $\angle L$; 8 ft

Laurie's Notes

Example 2

- Work through the example. Point out that the orientation of the figure does not influence whether the two figures are congruent.
- **Common Error:** Students often refer to Square C as a diamond versus square.

Example 3

- Draw two congruent scalene triangles with different orientations. Label the triangles $\triangle PQR$ and $\triangle STU$.

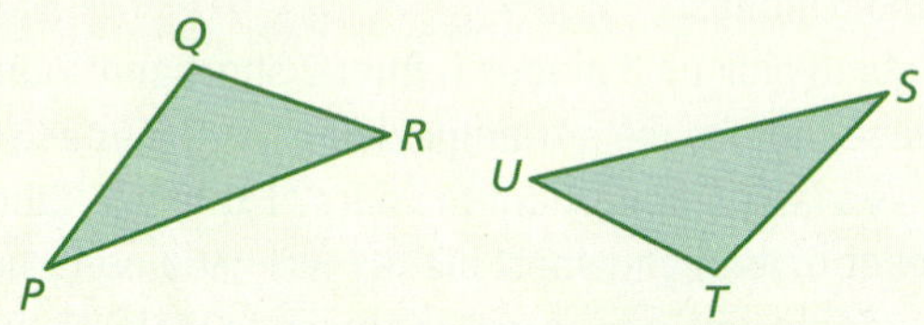

? "Which pairs of sides appear to be the same length?" Side *PQ* and Side *ST*, Side *QR* and Side *TU*, Side *RP* and Side *US* "Which pairs of angles appear to be the same measure?" $\angle P$ and $\angle S$, $\angle Q$ and $\angle T$, $\angle R$ and $\angle U$

- Students should always look at the figures and first determine if they are in the same orientation.

? "Are the two trapezoids in the same orientation?" yes

- Work through the example, as shown.

Discuss

- To preview transformations, ask students these questions.

? "How could you move Quadrilateral *ABCD* to get Quadrilateral *WXYZ* in Example 1?" Slide (move) the quadrilateral down.

? "What can you do to Pentagon *JKLMN* to get Pentagon *SRQVT* in On Your Own Question 1?" Reflect (flip) the pentagon in (or across) a line at the right of the pentagon to form a mirror image of the pentagon.

? "What can you do to Square A to get Square C in Example 2?" Rotate (turn) the square and slide it to the right.

On Your Own

- **Think-Pair-Share:** Students should read each question independently and then work in pairs to answer the questions. When they have answered the questions, the pair should compare their answers with another group and discuss any discrepancies.

Closure

- Draw a triangle and a quadrilateral. Exchange papers with your partner. Draw a figure congruent to your partner's figure. Then mark the corresponding sides and the corresponding angles of the congruent figures.

EXAMPLE 2 Identifying Congruent Figures

Which square is congruent to Square A?

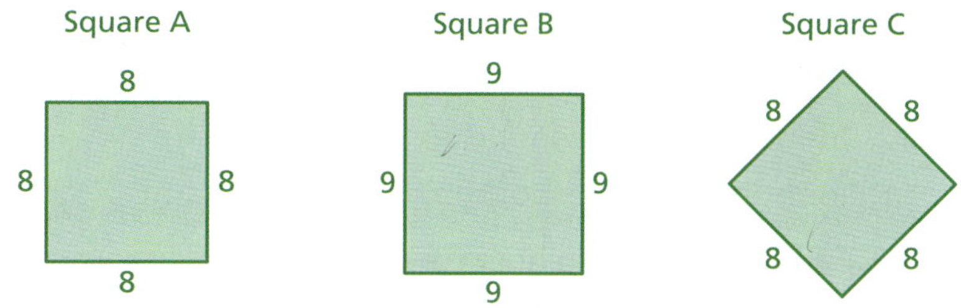

Each square has four right angles. So, corresponding angles are congruent. Check to see if corresponding sides are congruent.

Square A and Square B

Each side length of Square A is 8, and each side length of Square B is 9. So, corresponding sides are not congruent.

Square A and Square C

Each side length of Square A and Square C is 8. So, corresponding sides are congruent.

∴ So, Square C is congruent to Square A.

EXAMPLE 3 Using Congruent Figures

Trapezoids *ABCD* and *JKLM* are congruent.

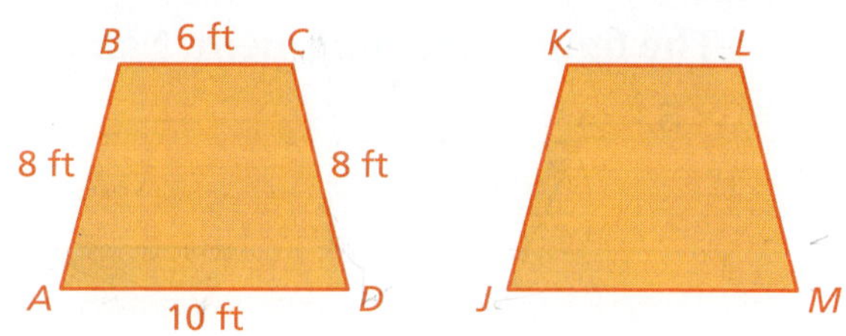

a. What is the length of side *JM*?

Side *JM* corresponds to side *AD*.

∴ So, the length of side *JM* is 10 feet.

b. What is the perimeter of *JKLM*?

The perimeter of *ABCD* is $10 + 8 + 6 + 8 = 32$ feet. Because the trapezoids are congruent, their corresponding sides are congruent.

∴ So, the perimeter of *JKLM* is also 32 feet.

On Your Own

Now You're Ready
Exercises 8, 9, and 12

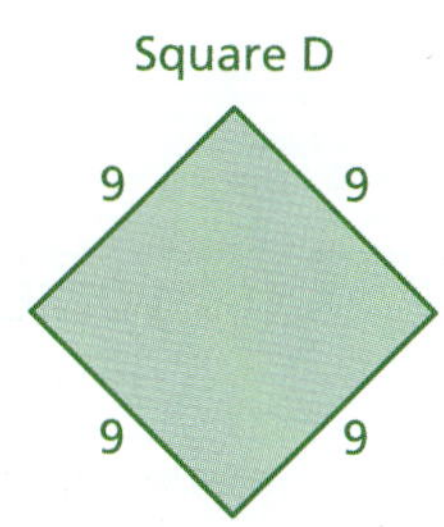

2. Which square in Example 2 is congruent to Square D?
3. In Example 3, which angle of *JKLM* corresponds to $\angle C$? What is the length of side *KJ*?

11.1 Exercises

Check It Out
Help with Homework
BigIdeasMath.com

Vocabulary and Concept Check

1. **VOCABULARY** $\triangle ABC$ is congruent to $\triangle DEF$.
   a. Identify the corresponding angles.
   b. Identify the corresponding sides.

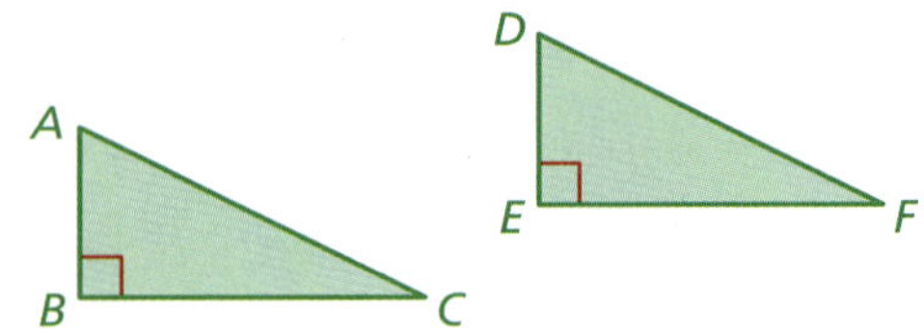

2. **VOCABULARY** Explain how you can tell that two figures are congruent.

3. **WHICH ONE DOESN'T BELONG?** Which one does *not* belong with the other three? Explain your reasoning.

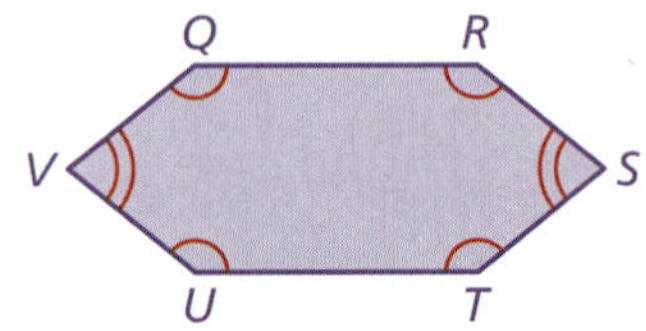

$\angle R$	$\angle U$	$\angle V$	$\angle Q$

Practice and Problem Solving

Tell whether the triangles are *congruent* or *not congruent*.

4.
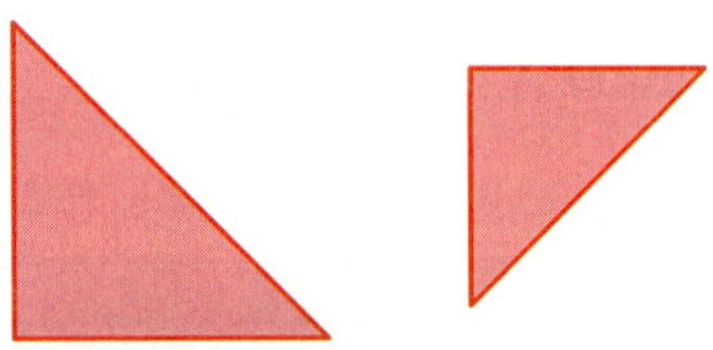

5.

The figures are congruent. Name the corresponding angles and the corresponding sides.

1 6.
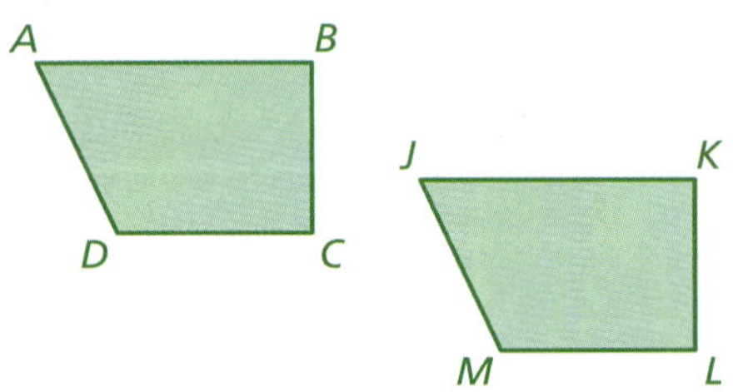

7.
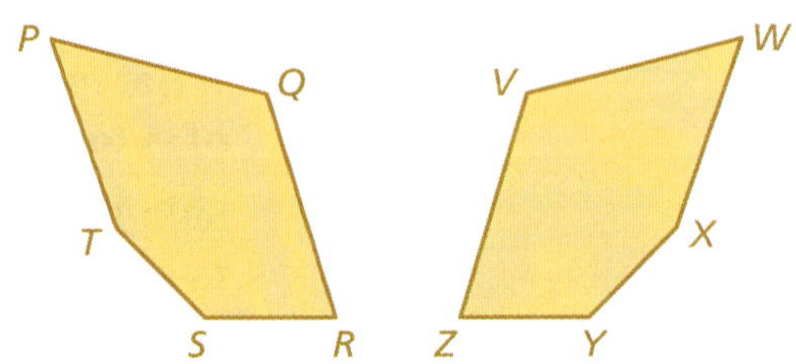

Tell whether the two figures are congruent. Explain your reasoning.

2 8.
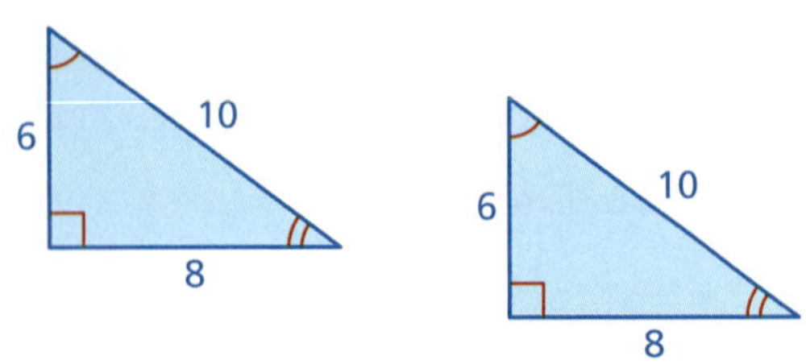

9.
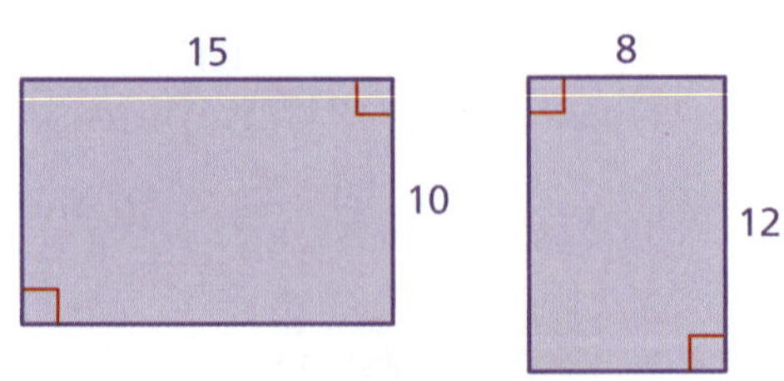

10. **PUZZLE** Describe the relationship between the unfinished puzzle and the missing piece.

Assignment Guide and Homework Check

Level	Assignment	Homework Check
Accelerated	1–6, 8, 11–20	6, 8, 11, 14

For Your Information

- **Exercise 1** Students may not be familiar with the symbol "△" used in this exercise. Mention that this symbol is used to represent a triangle.

Common Errors

- **Exercise 5** Students may think that the figures are not congruent because they do not have the same orientation. Remind them that as long as the figures have the same size and the same shape, they are congruent.
- **Exercises 6 and 7** Students may forget to write the angle symbol with the name. Remind them that A is a point and $\angle A$ is the angle.
- **Exercises 8, 9, and 11** Students may think that the figures are congruent because corresponding angles are congruent *or* corresponding sides are congruent. Remind them that there are two parts for determining that two figures are congruent, corresponding angles *and* corresponding sides.
- **Exercise 12** In part (b), students may have difficulty determining the corresponding angle. Point out that the fronts of the houses form congruent figures and remind them that corresponding angles in congruent figures have matching positions.

11.1 Record and Practice Journal

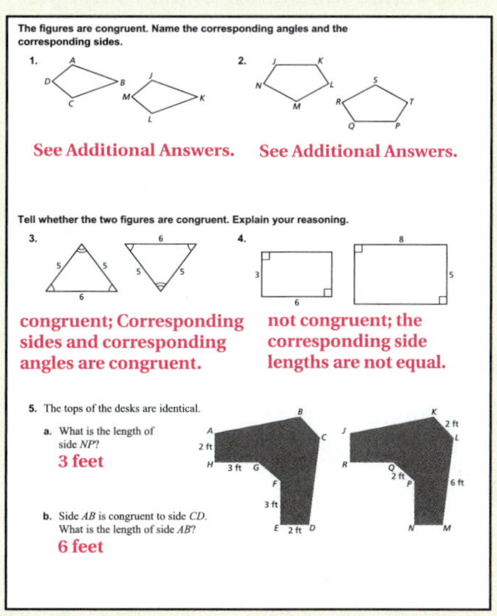

The figures are congruent. Name the corresponding angles and the corresponding sides.

1. See Additional Answers.

2. See Additional Answers.

Tell whether the two figures are congruent. Explain your reasoning.

3. congruent; Corresponding sides and corresponding angles are congruent.

4. not congruent; the corresponding side lengths are not equal.

5. The tops of the desks are identical.

   a. What is the length of side *NP*? **3 feet**

   b. Side *AB* is congruent to side *CD*. What is the length of side *AB*? **6 feet**

Vocabulary and Concept Check

1. a. $\angle A$ and $\angle D$, $\angle B$ and $\angle E$, $\angle C$ and $\angle F$

   b. Side AB and Side DE, Side BC and Side EF, Side AC and Side DF

2. Two figures are congruent if they have the same size and the same shape.

3. $\angle V$ does not belong. The other three angles are congruent to each other, but not to $\angle V$.

Practice and Problem Solving

4. not congruent

5. congruent

6. $\angle A$ and $\angle J$, $\angle B$ and $\angle K$, $\angle C$ and $\angle L$, $\angle D$ and $\angle M$; Side AB and Side JK, Side BC and Side KL, Side CD and Side LM, Side DA and Side MJ

7. $\angle P$ and $\angle W$, $\angle Q$ and $\angle V$, $\angle R$ and $\angle Z$, $\angle S$ and $\angle Y$, $\angle T$ and $\angle X$; Side PQ and Side WV, Side QR and Side VZ, Side RS and Side ZY, Side ST and Side YX, Side TP and Side XW

8. congruent; Corresponding sides and corresponding angles are congruent.

9. not congruent; Corresponding sides are not congruent.

10. The unfinished portion of the puzzle and the missing piece are congruent.

11. The corresponding angles are not congruent, so the two figures are not congruent.

Practice and Problem Solving

12. See Additional Answers.

13. See *Taking Math Deeper*.

14–15. See Additional Answers.

Fair Game Review

16–20. See Additional Answers.

Mini-Assessment

1. The figures are congruent. Name the corresponding angles and the corresponding sides.

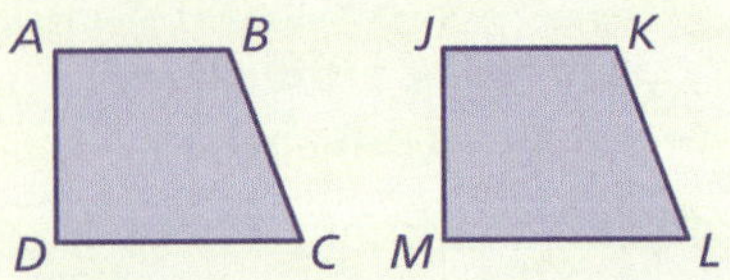

$\angle A$ and $\angle J$, $\angle B$ and $\angle K$, $\angle C$ and $\angle L$, $\angle D$ and $\angle M$; Side *AB* and Side *JK*, Side *BC* and Side *KL*, Side *CD* and Side *LM*, Side *AD* and Side *JM*

Tell whether the two figures are congruent. Explain your reasoning.

2.

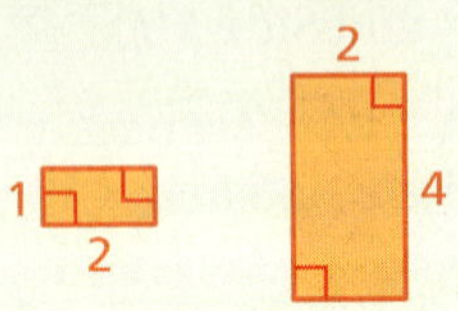

not congruent; Corresponding sides are not congruent.

3.

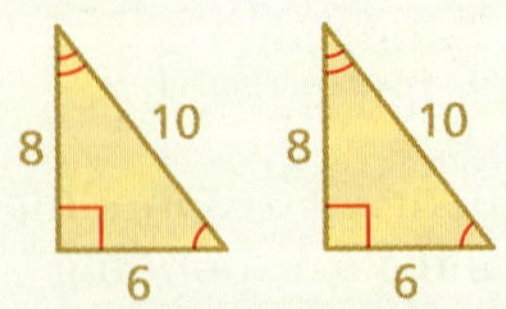

congruent; Corresponding angles and corresponding sides are congruent.

4. Trapezoids *ABCD* and *EFGH* are congruent.

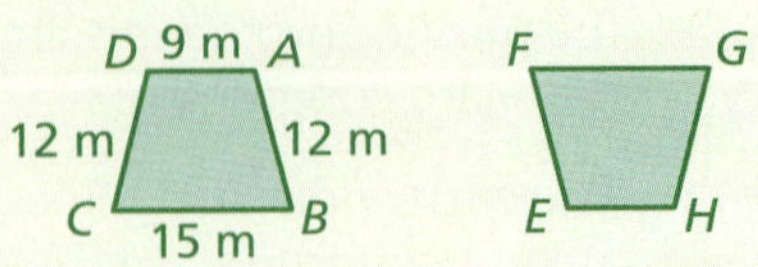

   a. What is the length of Side *FG*? 15 m

   b. What is the perimeter of Trapezoid *EFGH*? 48 m

Taking Math Deeper

Exercise 13

This problem requires students to visualize congruent shapes and requires "thinking outside the box."

1. The two basic ways to use a single line to divide a rectangle into two congruent figures are given in the problem.

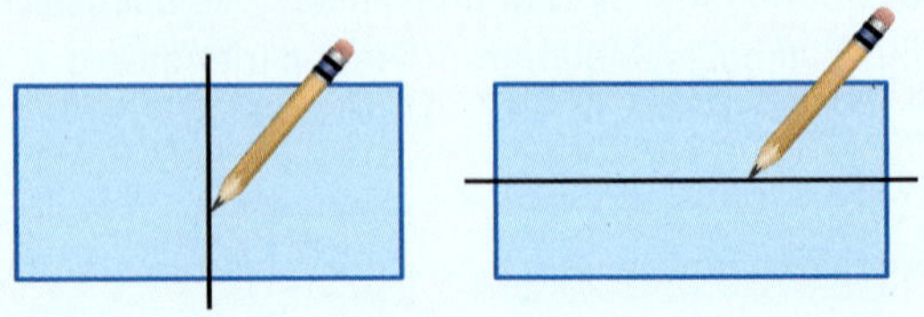

2. Two other ways are to draw the diagonal lines between opposite corners.

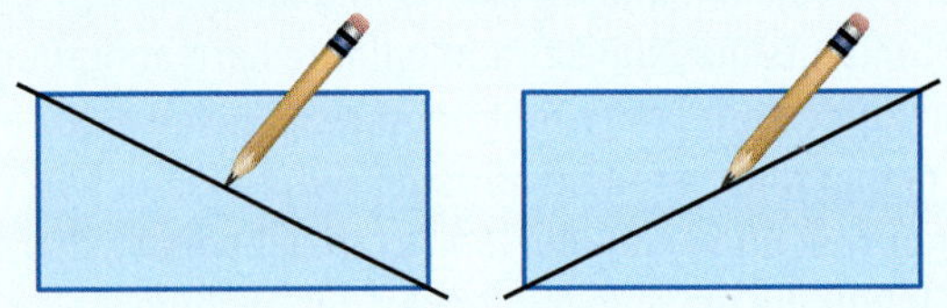

3. Some students may have trouble thinking of a third way to create two congruent figures. As it turns out, it is fun to realize that there are infinitely many ways to do this. Here are two of them.

For the trapezoids to be congruent, the shorter bases must be congruent and the longer bases must be congruent.

Note that in each case the figures are congruent when the line goes through the center of the rectangle.

Reteaching and Enrichment Strategies

If students need help. . .	If students got it. . .
Resources by Chapter • Practice A and Practice B • Puzzle Time Record and Practice Journal Practice Differentiating the Lesson Lesson Tutorials Skills Review Handbook	Resources by Chapter • Enrichment and Extension • Technology Connection Start the next section

11. ERROR ANALYSIS Describe and correct the error in telling whether the two figures are congruent.

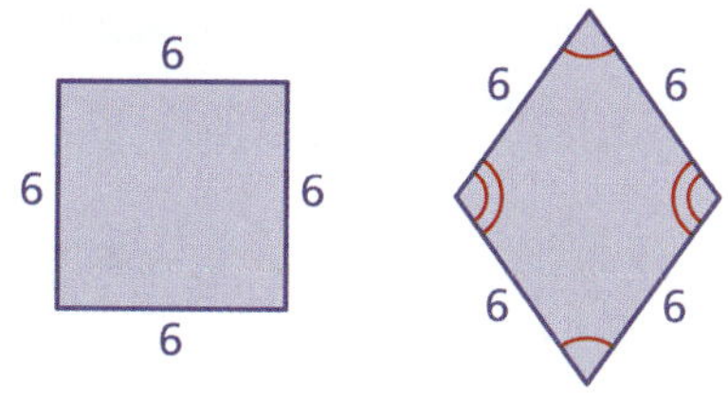

> ✗ Both figures have four sides, and the corresponding side lengths are equal. So, they are congruent.

③ **12. HOUSES** The fronts of the houses are identical.

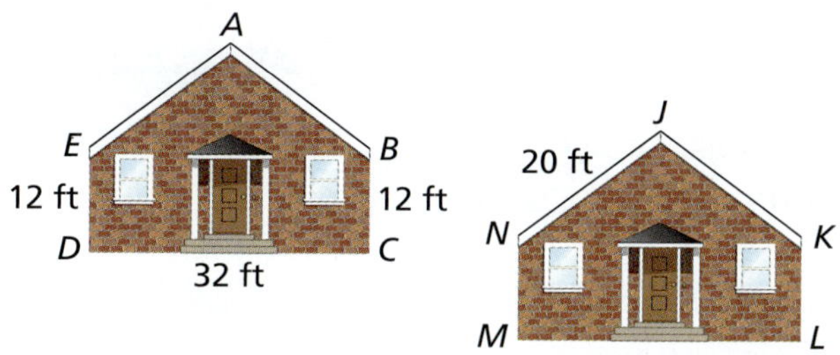

a. What is the length of side LM?

b. Which angle of $JKLMN$ corresponds to $\angle D$?

c. Side AB is congruent to side AE. What is the length of side AB?

d. What is the perimeter of $ABCDE$?

13. REASONING Here are two ways to draw *one* line to divide a rectangle into two congruent figures. Draw three other ways.

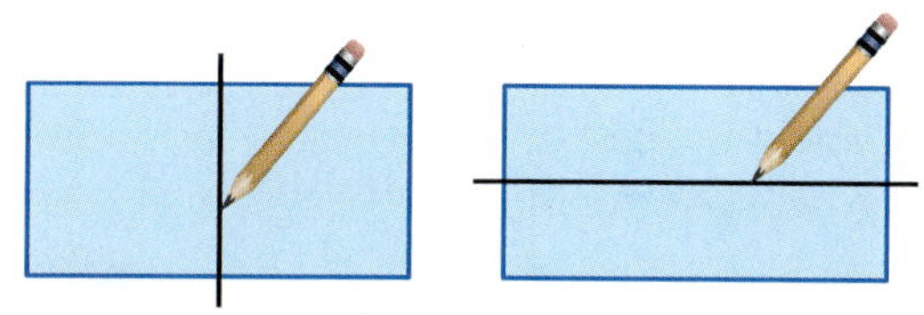

14. CRITICAL THINKING Are the areas of two congruent figures equal? Explain. Draw a diagram to support your answer.

15. True or False? The trapezoids are congruent. Determine whether the statement is *true* or *false*. Explain your reasoning.

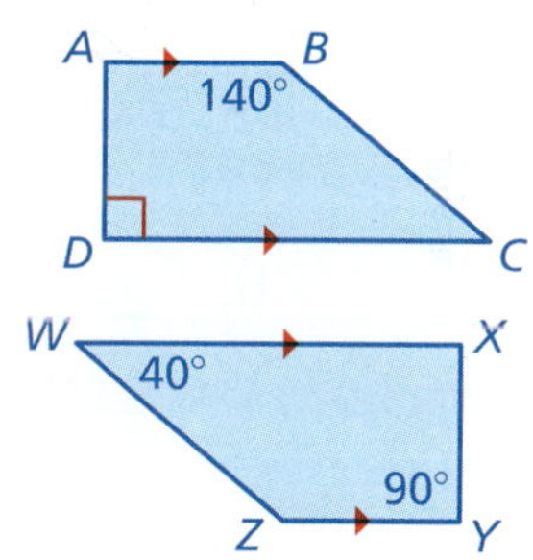

a. Side AB is congruent to side YZ.

b. $\angle A$ is congruent to $\angle X$.

c. $\angle A$ corresponds to $\angle X$.

d. The sum of the angle measures of $ABCD$ is 360°.

Fair Game Review What you learned in previous grades & lessons

Plot and label the ordered pair in a coordinate plane. *(Skills Review Handbook)*

16. $A(5, 3)$ **17.** $B(4, -1)$ **18.** $C(-2, 6)$ **19.** $D(-4, -2)$

20. MULTIPLE CHOICE You have 2 quarters and 5 dimes in your pocket. Write the ratio of quarters to the total number of coins. *(Skills Review Handbook)*

Ⓐ $\frac{2}{5}$ Ⓑ 2 : 7 Ⓒ 5 to 7 Ⓓ $\frac{7}{2}$

11.2 Translations

Essential Question How can you arrange tiles to make a tessellation?

The Meaning of a Word ● Translate

When you **translate** a tile, you slide it from one place to another.

When tiles cover a floor with no empty spaces, the collection of tiles is called a *tessellation*.

1 ACTIVITY: Describing Tessellations

Work with a partner. Can you make the tessellation by translating single tiles that are all of the same shape and design? If so, show how.

a. Sample:

Tile Pattern

Single Tiles

b.

c.

Geometry

In this lesson, you will
- identify translations.
- translate figures in the coordinate plane.

2 ACTIVITY: Tessellations and Basic Shapes

Work with a partner.

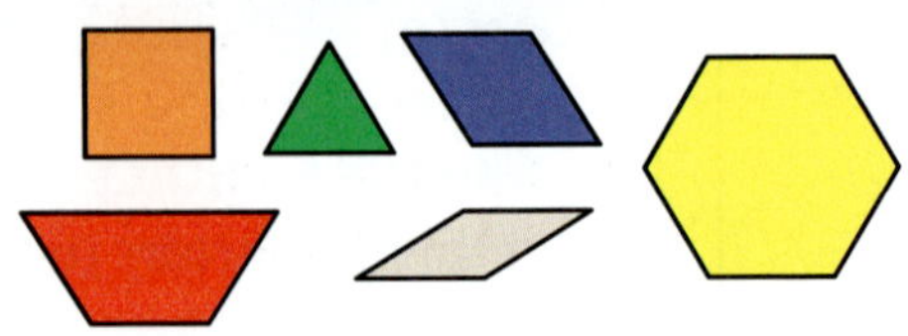

a. Which pattern blocks can you use to make a tessellation? For each one that works, draw the tessellation.

b. Can you make the tessellation by translating? Or do you have to rotate or flip the pattern blocks?

Laurie's Notes

Introduction

Applying Mathematical Practices

- **Construct Viable Arguments and Critique the Reasoning of Others:** As students manipulate geometric shapes, they develop spatial reasoning and make conjectures. Mathematically proficient students explore the truth of their conjectures by considering all cases.

Motivate

- **Whole Class Activity:** Model translations by having all students stand in an open area facing the same direction. Give directions such as: two steps right; three steps backward, etc. They should step backward or sideways as needed without turning their torso to keep their orientation.
- **The Meaning of a Word:** To **translate** a figure, you slide it to a new location in the plane. All points in the figure move the same distance and direction. The size, shape, and orientation of the figure do not change.
- Model a translation on the overhead (or document camera) by sliding a transparency with a shape or design on it.
- To help students develop spatial skills, have them explore transformations by moving and manipulating pattern blocks and designs traced on tracing paper or transparencies.

Activity Notes

Discuss

- Point out examples of tessellations. Tessellations may be found in the tile patterns on a floor, wall, or ceiling. They are also in wallpaper books.
- **FYI:** The word *tessellation* comes from the Latin word *tessellae*, which is what ancient Romans called small tiles used for pavements and walls.

Activity 1

- Have students trace on tracing paper the single tile they feel they can translate to form the tessellation.
- Remind students they are looking for a single tile that can be traced repeatedly to cover the surface without gaps or overlaps. Compare this to buying a box of tiles of the same size and shape to cover a floor.
- Make transparencies of each tessellation. When students share their solutions, have them trace the single tile on a clear transparency and then slide it over the tessellation transparency to show that the tile works.

Activity 2

- **Management Tip:** Place at least 10 triangles and 6 of the remaining pattern block shapes into a plastic bag for each pair of students.
- **Review:** "What is the name of each pattern block shape?" square, triangle, rhombus or parallelogram (tan and blue), hexagon, trapezoid
- **Discuss:** Four of the six shapes can be used to make a tessellation using translations only. The other two shapes (triangles and trapezoids) require both translations and rotations (or flips) to make a tesselation.

What Your Students Will Learn

- Identify whether two figures are translations of one another.
- Translate figures and draw images after translations in the coordinate plane.

Previous Learning

Students should know how to plot points in the coordinate plane.

Technology for the Teacher

Dynamic Classroom

Lesson Plans
Complete Materials List

11.2 Record and Practice Journal

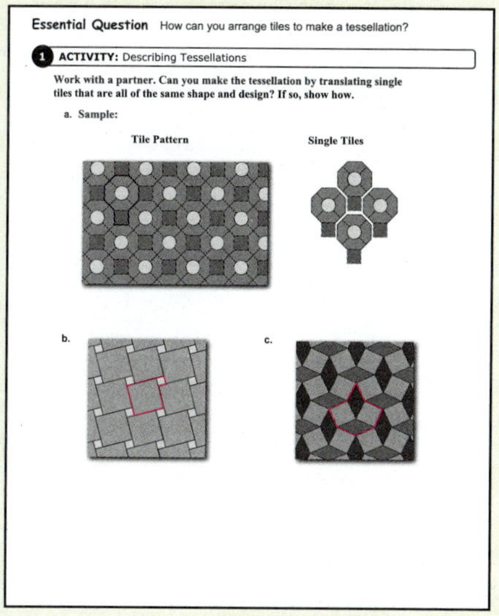
Essential Question How can you arrange tiles to make a tessellation?

1 ACTIVITY: Describing Tessellations

Work with a partner. Can you make the tessellation by translating single tiles that are all of the same shape and design? If so, show how.

a. Sample:

Tile Pattern Single Tiles

b.

c.

Differentiated Instruction

Auditory

To help students remember what each type of transformation is, use the words *slide* for translation, *flip* for reflection, and *turn* for rotation.

11.2 Record and Practice Journal

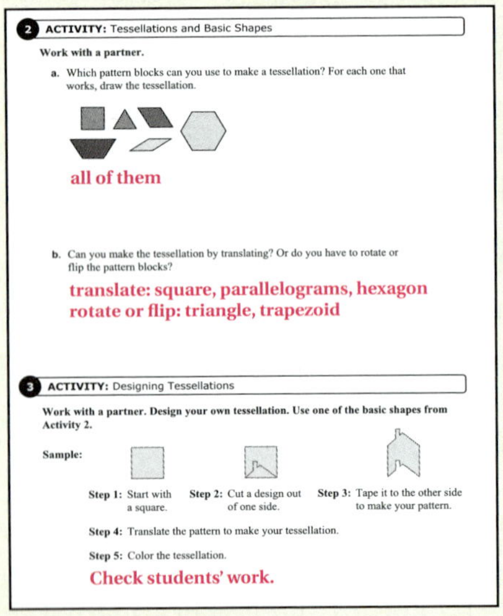

2 ACTIVITY: Tessellations and Basic Shapes

Work with a partner.

a. Which pattern blocks can you use to make a tessellation? For each one that works, draw the tessellation.

all of them

b. Can you make the tessellation by translating? Or do you have to rotate or flip the pattern blocks?

translate: square, parallelograms, hexagon
rotate or flip: triangle, trapezoid

3 ACTIVITY: Designing Tessellations

Work with a partner. Design your own tessellation. Use one of the basic shapes from Activity 2.

Sample:

Step 1: Start with a square.

Step 2: Cut a design out of one side.

Step 3: Tape it to the other side to make your pattern.

Step 4: Translate the pattern to make your tessellation.

Step 5: Color the tessellation.

Check students' work.

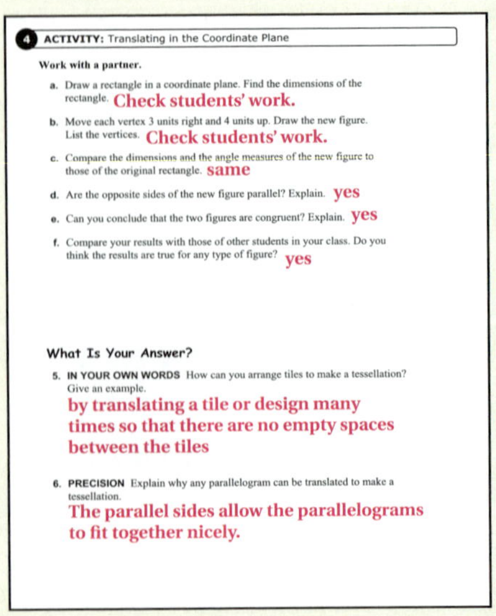

4 ACTIVITY: Translating in the Coordinate Plane

Work with a partner.

a. Draw a rectangle in a coordinate plane. Find the dimensions of the rectangle. Check students' work.

b. Move each vertex 3 units right and 4 units up. Draw the new figure. List the vertices. Check students' work.

c. Compare the dimensions and the angle measures of the new figure to those of the original rectangle. same

d. Are the opposite sides of the new figure parallel? Explain. yes

e. Can you conclude that the two figures are congruent? Explain. yes

f. Compare your results with those of other students in your class. Do you think the results are true for any type of figure? yes

What Is Your Answer?

5. **IN YOUR OWN WORDS** How can you arrange tiles to make a tessellation? Give an example.
by translating a tile or design many times so that there are no empty spaces between the tiles

6. **PRECISION** Explain why any parallelogram can be translated to make a tessellation.
The parallel sides allow the parallelograms to fit together nicely.

Laurie's Notes

Activity 3

- This activity could be started in class, and students could finish their designs for homework.
- Explain to students that they need to start with one of the basic shapes from Activity 2.
- Use the *cut and bump* method to create the tessellation. Whatever shape is *cut* from one edge of the shape must *bump out* on the edge to where the shape slides. After the initial shape is created, additional artwork (such as windows and roof coloring) can be added to provide additional details.
- **Management Tip:** Some students find it helpful to draw their designs on grid paper. The grid helps the students to be more accurate with their designs.
- This activity could be used as a project that students work on throughout the chapter. They may find other transformations in later sections that they want to incorporate into their tessellations.

Activity 4

- The *size* of the original rectangle does not matter, but encourage students to draw a rectangle that is not a square. A square is a special case of a rectangle, and their observations may only be true for the square.

? When students finish, ask a few questions.
  - "Did the second rectangle overlap with the original rectangle?" Answers will vary.
  - "Was the second rectangle in the same quadrant as the original rectangle?" Answers will vary.
  - "When you were locating the vertices of the second rectangle did you count 3 units right and 4 units up for each vertex?" Some may not have; they may have guessed at the location of the fourth vertex after locating the first three vertices.

? **Construct Viable Arguments and Critique the Reasoning of Others:** "What conjectures can you make about the two rectangles?" Students should share many conjectures, such as the rectangles are congruent; they have the same shape, size, perimeter, and area; and their opposite sides are parallel. They should also give evidence for their conjectures. Students may also make conjectures about the vertices when a rectangle is translated.

- **Discuss:** A translation of a line is a line. A translation of a line segment is a line segment of the same length as the original.

What Is Your Answer?

- **Think-Pair-Share:** Students should read each question independently and then work in pairs to answer the questions. When they have answered the questions, the pair should compare their answers with another group and discuss any discrepancies.

Closure

- Would *any* rectangle cover or tessellate a flat surface? Explain. Yes, because the sides are parallel.

3 ACTIVITY: Designing Tessellations

Work with a partner. Design your own tessellation. Use one of the basic shapes from Activity 2.

Sample:

Step 1: Start with a square.

Step 2: Cut a design out of one side.

Step 3: Tape it to the other side to make your pattern.

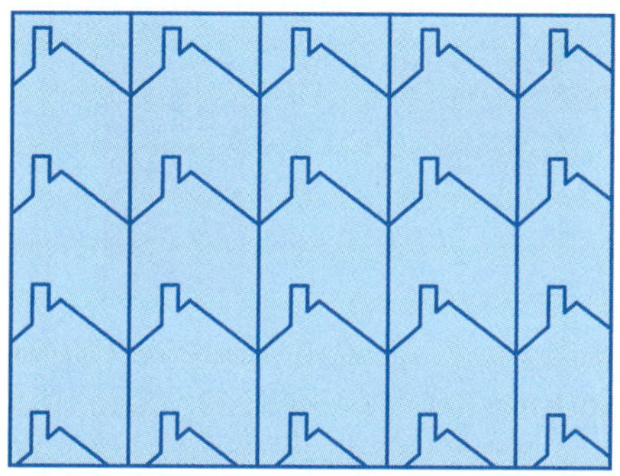

Step 4: Translate the pattern to make your tessellation.

Step 5: Color the tessellation.

4 ACTIVITY: Translating in the Coordinate Plane

Work with a partner.

a. Draw a rectangle in a coordinate plane. Find the dimensions of the rectangle.

b. Move each vertex 3 units right and 4 units up. Draw the new figure. List the vertices.

c. Compare the dimensions and the angle measures of the new figure to those of the original rectangle.

d. Are the opposite sides of the new figure still parallel? Explain.

e. Can you conclude that the two figures are congruent? Explain.

f. Compare your results with those of other students in your class. Do you think the results are true for any type of figure?

Math Practice

Justify Conclusions

What information do you need to conclude that two figures are congruent?

What Is Your Answer?

5. **IN YOUR OWN WORDS** How can you arrange tiles to make a tessellation? Give an example.

6. **PRECISION** Explain why any parallelogram can be translated to make a tessellation.

Practice Use what you learned about translations to complete Exercises 4–6 on page 476.

11.2 Lesson

Check It Out
Lesson Tutorials
BigIdeasMath.com

Key Vocabulary
transformation, *p. 474*
image, *p. 474*
translation, *p. 474*

A **transformation** changes a figure into another figure. The new figure is called the **image**.

A **translation** is a transformation in which a figure *slides* but does not turn. Every point of the figure moves the same distance and in the same direction.

EXAMPLE 1 Identifying a Translation

Tell whether the blue figure is a translation of the red figure.

a.

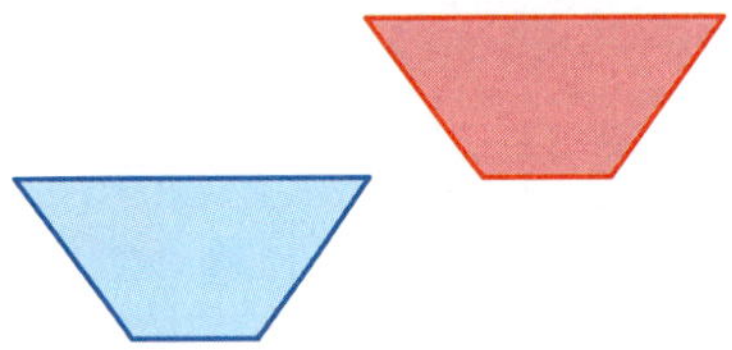

The red figure *slides* to form the blue figure.

So, the blue figure is a translation of the red figure.

b.

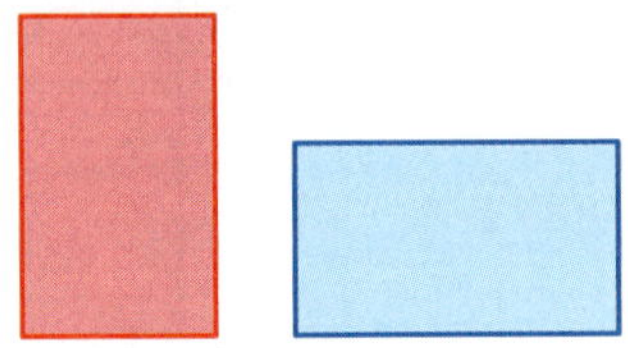

The red figure *turns* to form the blue figure.

So, the blue figure is *not* a translation of the red figure.

On Your Own

Now You're Ready
Exercises 4–9

Tell whether the blue figure is a translation of the red figure. Explain.

1.

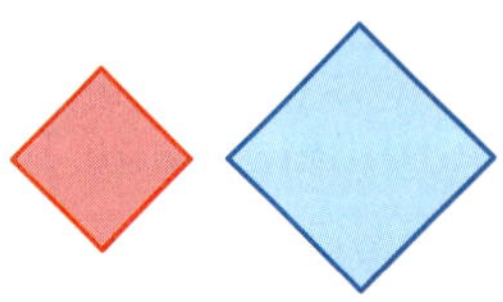

2.

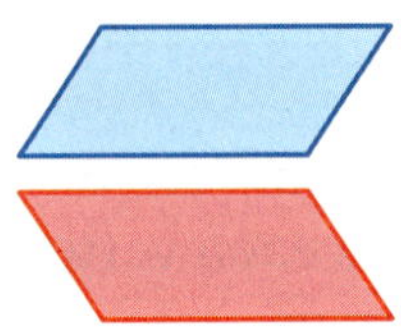

3.

Key Idea

Translations in the Coordinate Plane

Words To translate a figure a units horizontally and b units vertically in a coordinate plane, add a to the x-coordinates and b to the y-coordinates of the vertices.

Positive values of a and b represent translations up and right. Negative values of a and b represent translations down and left.

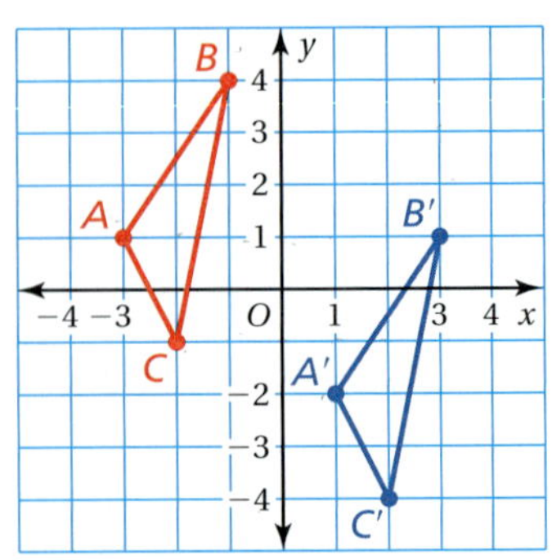

Algebra $(x, y) \rightarrow (x + a, y + b)$

In a translation, the original figure and its image are congruent.

Reading

A′ is read "*A* prime." Use *prime* symbols when naming an image.

$A \rightarrow A'$
$B \rightarrow B'$
$C \rightarrow C'$

Laurie's Notes

Introduction

Connect

- **Yesterday:** Students explored translations by manipulating pattern blocks and sketching translations.
- **Today:** Students will use their visual skills to draw translations in the coordinate plane.

Motivate

- Share a quick story about movie animation. Perhaps some of your students have made flipbooks where images are drawn at a slightly different location on each card, so that as you flip through the cards the image appears to move. If the image were a baseball, it would be translated to a new location on each card and appear to be moving as you flip the cards.
- Make a flipbook in advance to share with the class.

Lesson Notes

Discuss

- Discuss the introductory vocabulary: transformation, image, and translation.
- Relate a "translation" to the Motivate activity yesterday when each student would move (slide) the same distance in the same direction. Images in the flipbook are translations.

Example 1

- **Common Misconception:** The translation does not need to be in a horizontal or vertical direction. It can also be in a diagonal direction.
- Students generally have little difficulty identifying translations.

On Your Own

- **Think-Pair-Share:** Students should read each question independently and then work in pairs to answer the questions. When they have answered the questions, the pair should compare their answers with another group and discuss any discrepancies.
- For any questions that are translations, describe the direction of the translation.

Key Idea

- Write the Key Idea, using the language of A and A' (A prime) as you identify the coordinates of the original figure (A, B, C) and its image (A', B', C').
- In this example, red triangle ABC is the original figure and blue triangle $A'B'C'$ is the translated image.
- Use a third color to draw the translation arrow from A to A'.

? "How was vertex A translated to vertex A'?" It moved 4 units right and 3 units down.

Goal Today's lesson is identifying **translations**.

Technology for the Teacher

Dynamic Classroom

Lesson Tutorials
Lesson Plans
Answer Presentation Tool

Extra Example 1

Tell whether the blue figure is a translation of the red figure.

a.

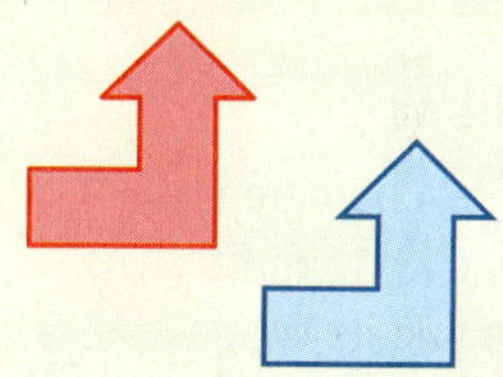

a translation

b.

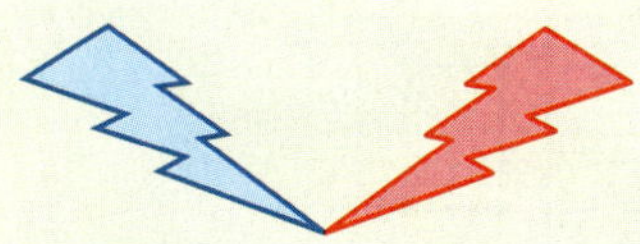

not a translation

On Your Own

1. no; The blue figure is larger than the red figure.
2. no; The red figure flips to form the blue figure.
3. yes; The red figure slides up and to the left to form the blue figure.

Extra Example 2

Translate the red triangle 2 units right and 5 units up. What are the coordinates of the image?

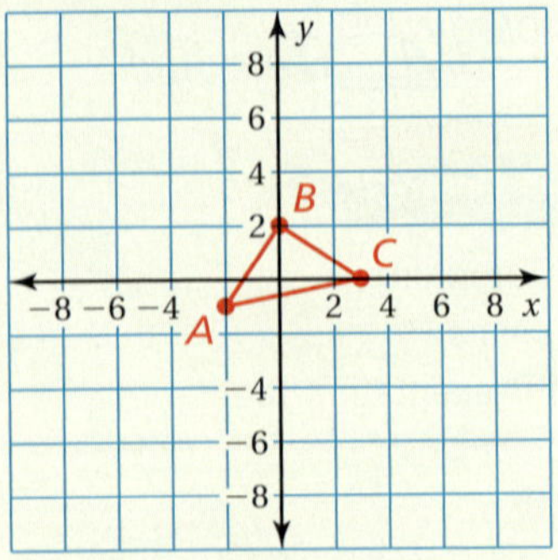

$A'(0, 4)$, $B'(2, 7)$, $C'(5, 5)$

On Your Own

4. $A'(-6, 3)$, $B'(-2, 7)$, $C'(-3, 4)$

Extra Example 3

The vertices of a rectangle are $A(1, 4)$, $B(3, 4)$, $C(3, 1)$ and $D(1, 1)$. Draw the figure and its image after a translation 3 units left and 4 units down.

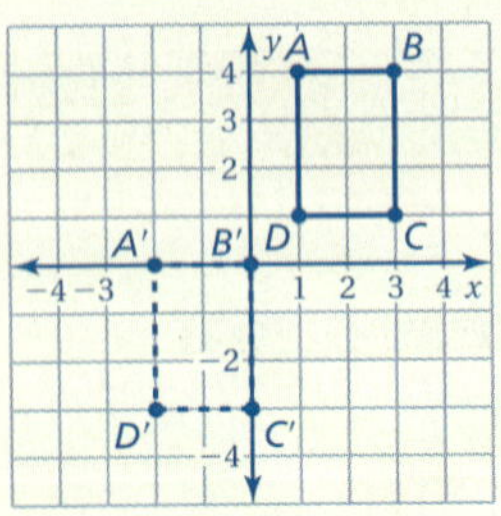

On Your Own

5. See Additional Answers.

English Language Learners

Visual

Make a poster in the classroom to illustrate the movement of a point in a coordinate plane based on the coordinate notation.

$x + h$	→
$x - h$	←
$y + k$	↑
$y - k$	↓

Laurie's Notes

Example 2

- Draw $\triangle ABC$ and label the vertices on a transparency. Slide the transparency 3 units to the right and 3 units down.
- Alternatively, you can model the translation on an interactive white board. Draw $\triangle ABC$. Copy $\triangle ABC$ and slide the copy to the new position.
- **Representation:** The **image** of a transformation (translation, reflection, rotation, or dilation) is written with the prime symbol. This helps to distinguish the image from the original figure, often referred to as the pre-image.
- After the result of the translation has been drawn, you can draw arrows from A to A', B to B', and C to C'. The resulting figure appears to be a 3-D diagram of a triangular prism.
- Explain that translating the triangle on a diagonal is equivalent to translating the triangle horizontally and then vertically. The two steps focus on what happens to each of the coordinates in an ordered pair.

? "Is the blue triangle the same size and shape as the red triangle?" yes

- Reinforce the concept of same size and shape by talking about the lengths of corresponding sides, the measures of the corresponding angles, and the perimeters and areas of the two triangles.

On Your Own

- Students should work in pairs.

Example 3

- Plot the four ordered pairs.
- **Common Error**: Students may interchange x- and y-directions in plotting the ordered pairs.

? Ask questions about the translation.

  - "In what quadrant is the original square?" IV
  - "If a figure is translated in the coordinate plane 4 units left, what will change, the x-coordinate or the y-coordinate?" x-coordinate
  - "If a figure is translated in the coordinate plane 6 units up, what will change, the x-coordinate or the y-coordinate?" y-coordinate

- **Look for and Express Regularity in Repeated Reasoning:** Explain the notation in the table. Use an alternate color to draw attention to the repeated pattern (subtracting 4 and adding 6) that occurs with each ordered pair.
- Draw the new image.

? "In what quadrant is the image?" II

? "Is the blue square the same size and shape as the red square?" yes

On Your Own

- **Neighbor Check:** Have students work independently and then have their neighbors check their work. Have students discuss any discrepancies.

Closure

- Draw a right triangle in Quadrant II. Translate the triangle so that the image is in Quadrant IV. Describe the translation.

EXAMPLE 2 Translating a Figure in the Coordinate Plane

Translate the red triangle 3 units right and 3 units down. What are the coordinates of the image?

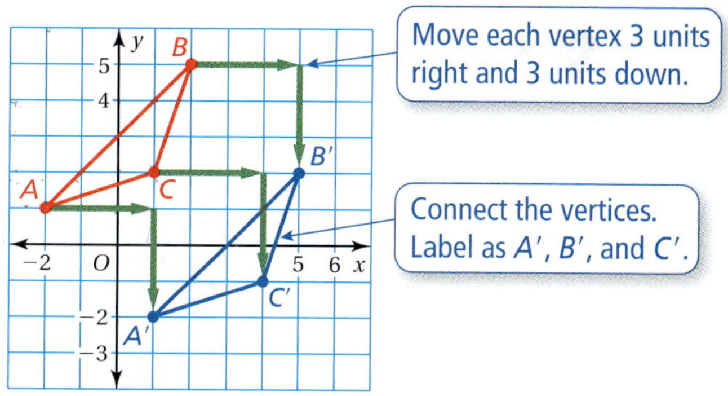

∴ The coordinates of the image are $A'(1, -2)$, $B'(5, 2)$, and $C'(4, -1)$.

On Your Own

Now You're Ready
Exercises 10 and 11

4. **WHAT IF?** The red triangle is translated 4 units left and 2 units up. What are the coordinates of the image?

EXAMPLE 3 Translating a Figure Using Coordinates

The vertices of a square are $A(1, -2)$, $B(3, -2)$, $C(3, -4)$, and $D(1, -4)$. Draw the figure and its image after a translation 4 units left and 6 units up.

Add -4 to each x-coordinate. So, subtract 4 from each x-coordinate.

Add 6 to each y-coordinate.

Vertices of $ABCD$	$(x - 4, y + 6)$	Vertices of $A'B'C'D'$
$A(1, -2)$	$(1 - 4, -2 + 6)$	$A'(-3, 4)$
$B(3, -2)$	$(3 - 4, -2 + 6)$	$B'(-1, 4)$
$C(3, -4)$	$(3 - 4, -4 + 6)$	$C'(-1, 2)$
$D(1, -4)$	$(1 - 4, -4 + 6)$	$D'(-3, 2)$

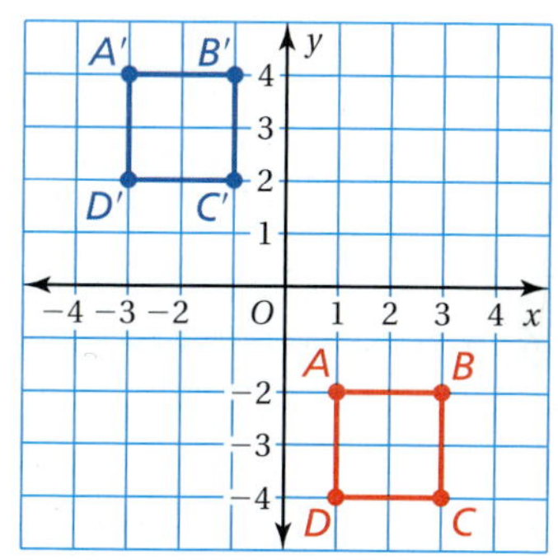

∴ The figure and its image are shown at the above right.

On Your Own

Now You're Ready
Exercises 12–15

5. The vertices of a triangle are $A(-2, -2)$, $B(0, 2)$, and $C(3, 0)$. Draw the figure and its image after a translation 1 unit left and 2 units up.

11.2 Exercises

Check It Out
Help with Homework
BigIdeasMath.com

Vocabulary and Concept Check

1. **VOCABULARY** Which figure is the image?

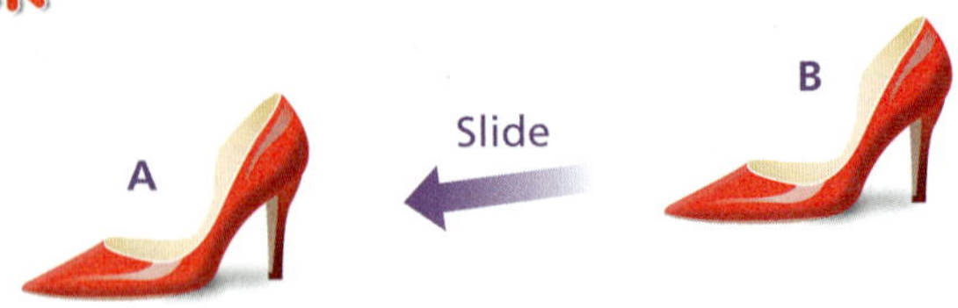

2. **VOCABULARY** How do you translate a figure in a coordinate plane?

3. **WRITING** Can you translate the letters in the word TOKYO to form the word KYOTO? Explain.

Practice and Problem Solving

Tell whether the blue figure is a translation of the red figure.

① 4.
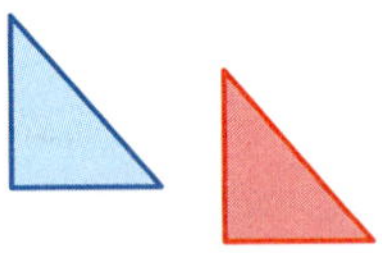

5.
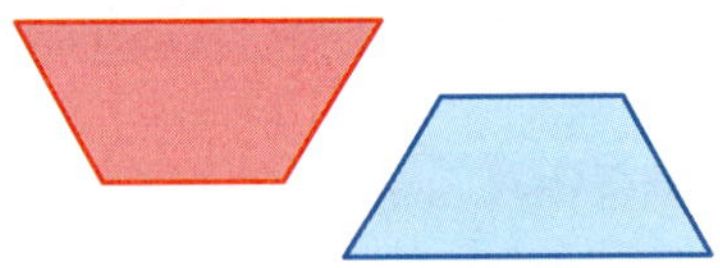

6.
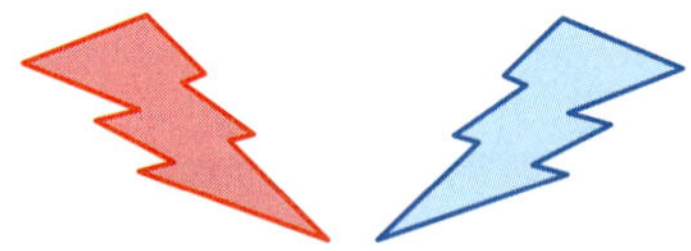

7.
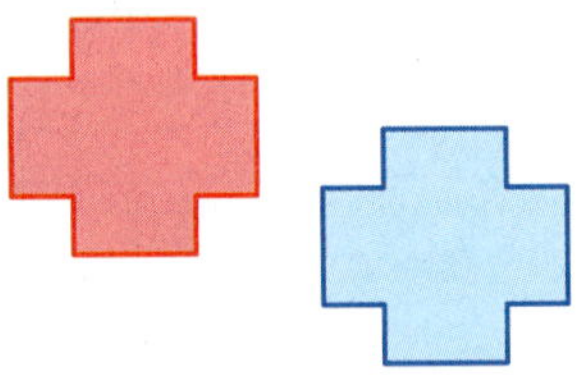

8.
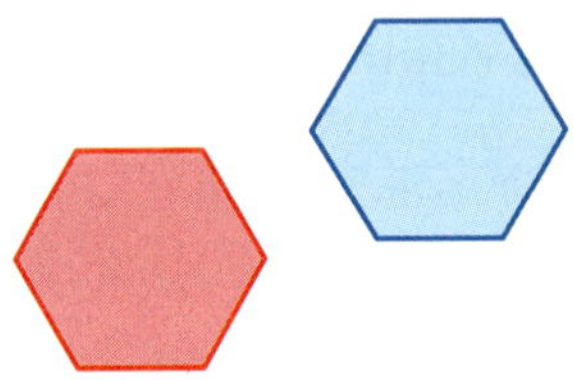

9.
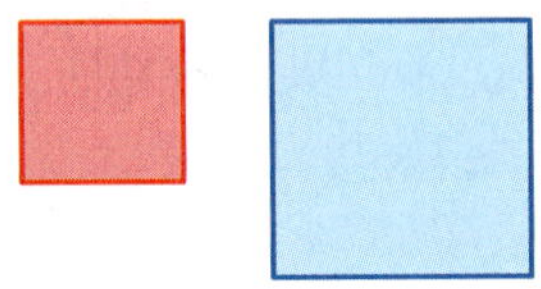

② 10. Translate the triangle 4 units right and 3 units down. What are the coordinates of the image?

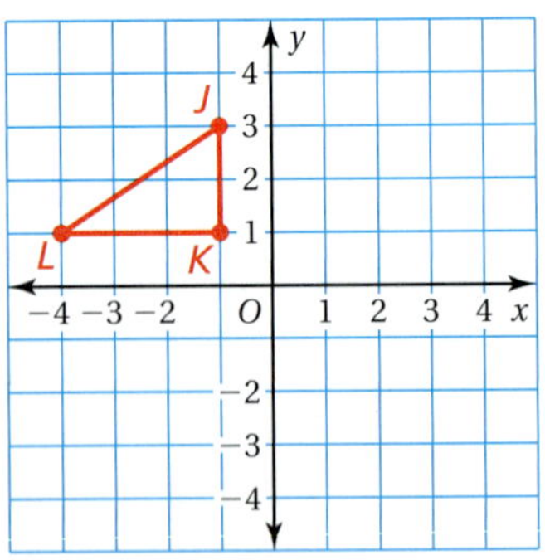

11. Translate the figure 2 units left and 4 units down. What are the coordinates of the image?

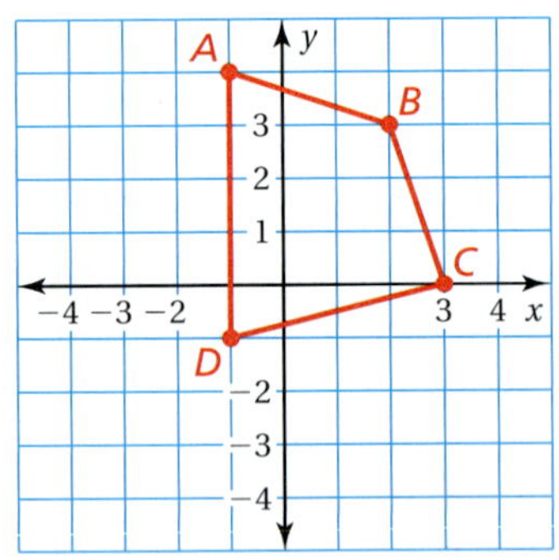

The vertices of a triangle are $L(0, 1)$, $M(1, -2)$, and $N(-2, 1)$. Draw the figure and its image after the translation.

③ 12. 1 unit left and 6 units up

13. 5 units right

14. $(x + 2, y + 3)$

15. $(x - 3, y - 4)$

16. **ICONS** You can click and drag an icon on a computer screen. Is this an example of a translation? Explain.

Assignment Guide and Homework Check

Level	Assignment	Homework Check
Accelerated	1–6, 12–20 even, 21–28	12, 18, 20, 22

For Your Information

- **Exercise 3** The Japanese language is composed of symbols, not letters. KYO means *capitol* and TO means *new*. So, KYO TO was the ancient capitol of Japan, and TO KYO is the modern day capitol of Japan.

Common Errors

- **Exercises 4–9** Students may forget that the objects must be the same size to be a translation. Remind them that the size stays the same. Tell students that when the size is different, it is a scale drawing.
- **Exercises 10–15** Students may translate the shape the wrong direction or mix up the units for the translation. Tell them to redraw the original on graph paper. Also, tell students to write the direction of the translation using arrows to show the movement left, right, up, or down.
- **Exercises 17 and 18** Students may struggle finding the translation. Encourage students to plot the points in a coordinate plane and count the change left, right, up, or down.
- **Exercises 19 and 20** Students may count the translation to the wrong point. Ask them to label the red figure with points *A*, *B*, etc. and the corresponding points on the blue figure as *A*′, *B*′, etc.

11.2 Record and Practice Journal

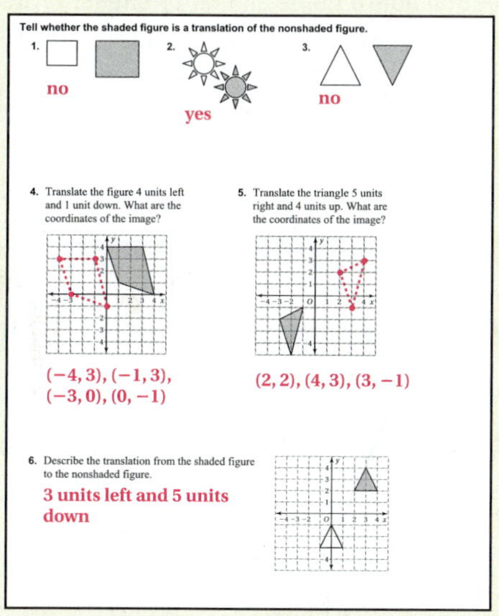
Tell whether the shaded figure is a translation of the nonshaded figure.

1. no
2. yes
3. no

4. Translate the figure 4 units left and 1 unit down. What are the coordinates of the image?
(−4, 3), (−1, 3), (−3, 0), (0, −1)

5. Translate the triangle 5 units right and 4 units up. What are the coordinates of the image?
(2, 2), (4, 3), (3, −1)

6. Describe the translation from the shaded figure to the nonshaded figure.
3 units left and 5 units down

Vocabulary and Concept Check

1. A
2. Move each vertex according to the translation.
3. yes; Translate the letters T and O to the end.

Practice and Problem Solving

4. yes
5. no
6. no
7. yes
8. yes
9. no
10. $J'(3, 0), K'(3, -2), L'(0, -2)$
11. $A'(-3, 0), B'(0, -1), C'(1, -4), D'(-3, -5)$
12.

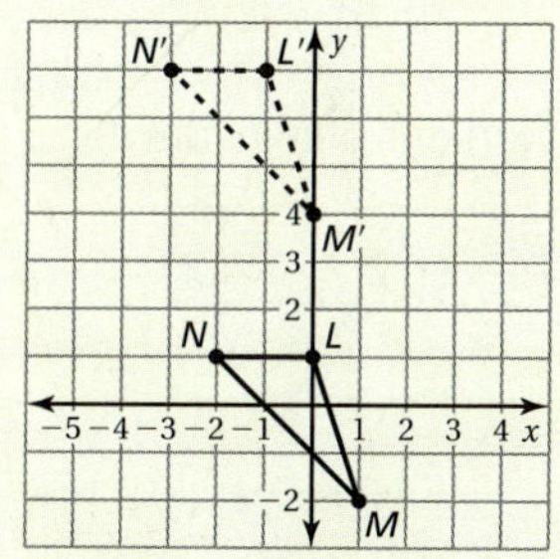

13.

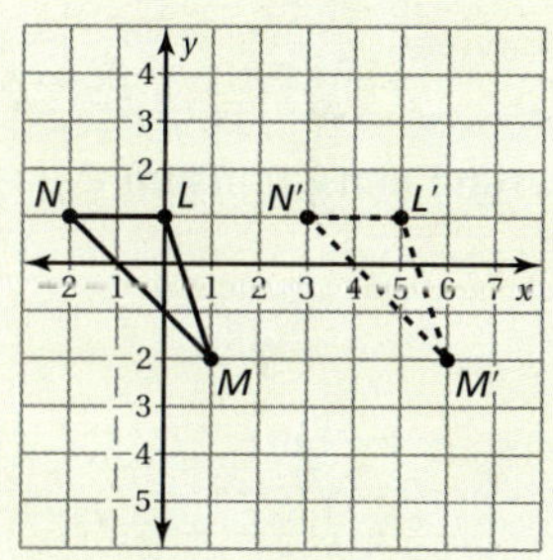

14.

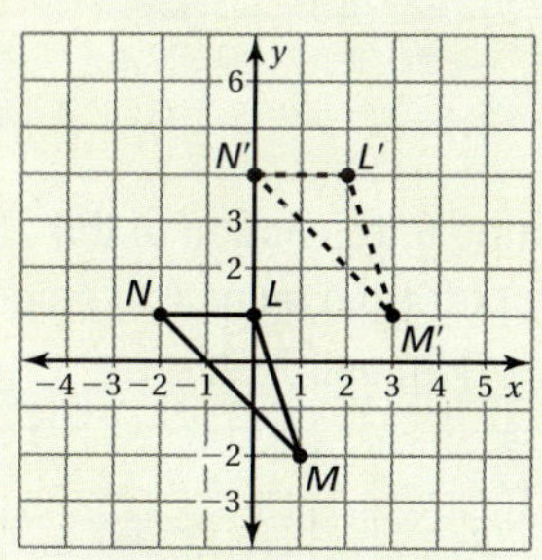

15. See Additional Answers.
16. Yes, because the figure slides.
17. 2 units left and 2 units up
18. 5 units right and 9 units up

19–22. See Additional Answers.

Practice and Problem Solving

23. See *Taking Math Deeper.*

Fair Game Review

24–28. See Additional Answers.

Mini-Assessment

The vertices of a triangle are *A* (1, 3), *B* (4, 3), and *C* (3, 0). Draw the figure and its image after the translation.

1. 2 units left and 3 units down

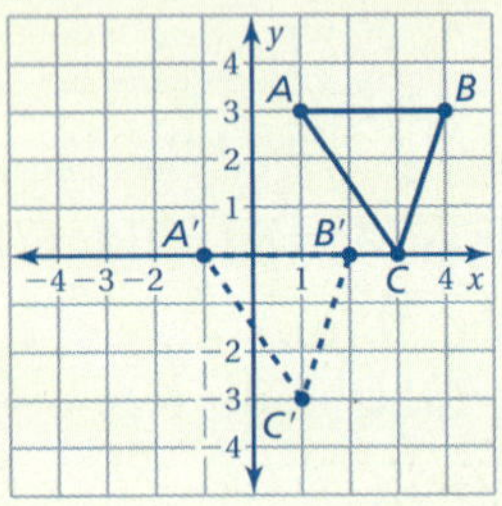

2. 1 unit left and 4 units down

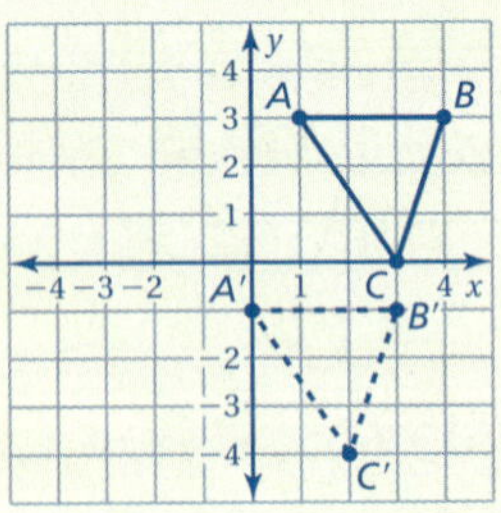

3. 1 unit right and 2 units up

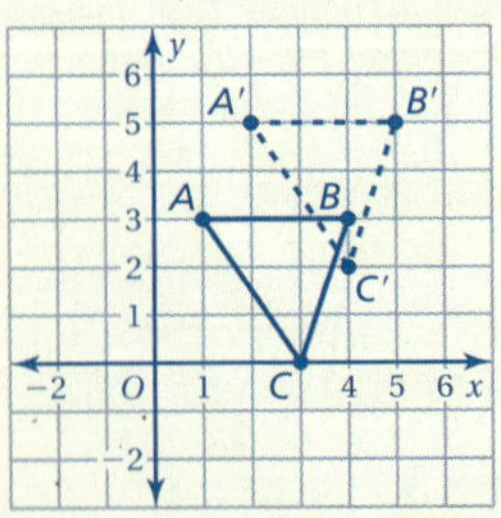

4. Describe a translation of the helicopter from point *A* to point *B*.
5 units right and 7 units up

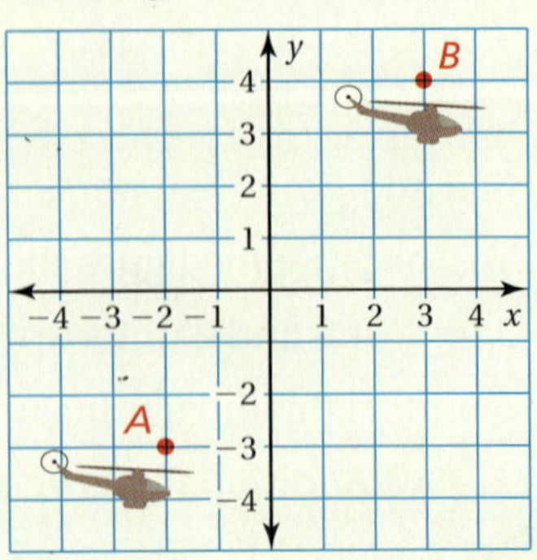

Taking Math Deeper

Exercise 23

There are thousands of correct answers to this question. This could be a nice discussion question for pairs or groups of students.

1. Here is one translation that takes 5 moves.

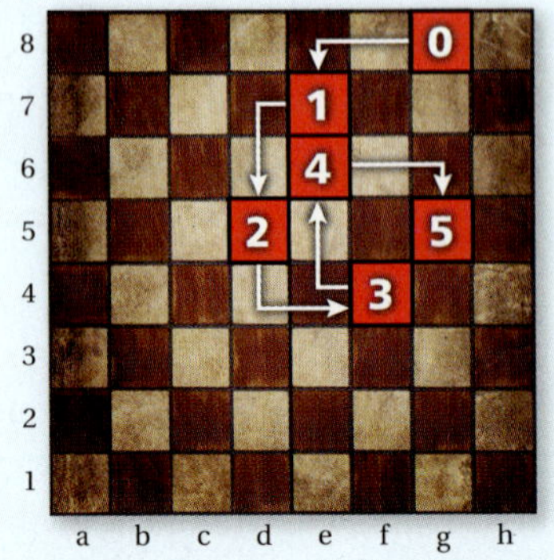

2. Here is one translation that takes only 3 moves.

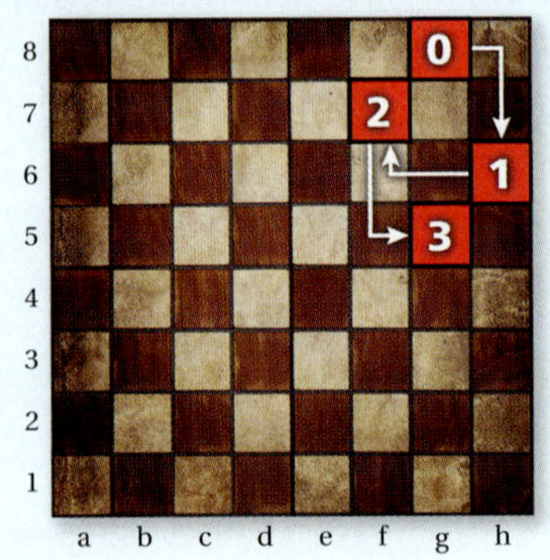

3. It is not possible to move from g8 to g5 in less than 3 moves.

Project

Create a board game similar to chess. Write the rules for your game and play your game with another student.

Reteaching and Enrichment Strategies

If students need help. . .	If students got it. . .
Resources by Chapter • Practice A and Practice B • Puzzle Time Record and Practice Journal Practice Differentiating the Lesson Lesson Tutorials Skills Review Handbook	Resources by Chapter • Enrichment and Extension • Technology Connection Start the next section

Describe the translation of the point to its image.

17. $(3, -2) \rightarrow (1, 0)$

18. $(-8, -4) \rightarrow (-3, 5)$

Describe the translation from the red figure to the blue figure.

19.

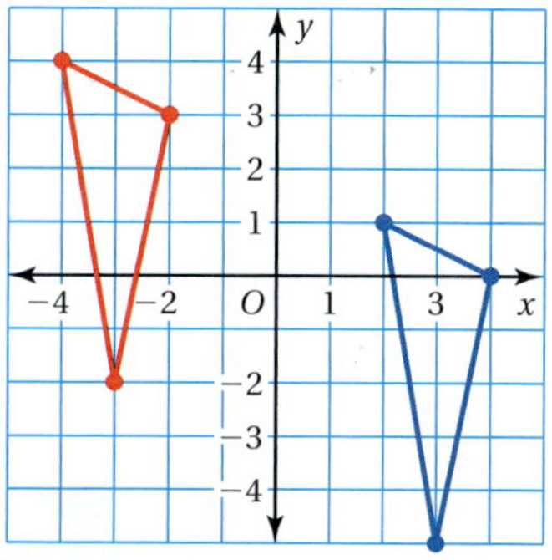

20.

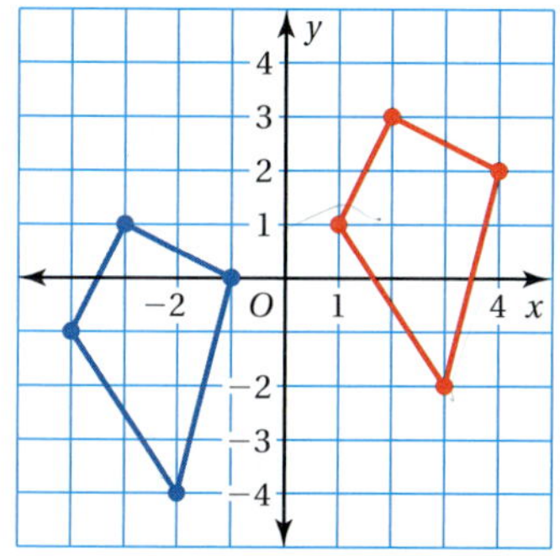

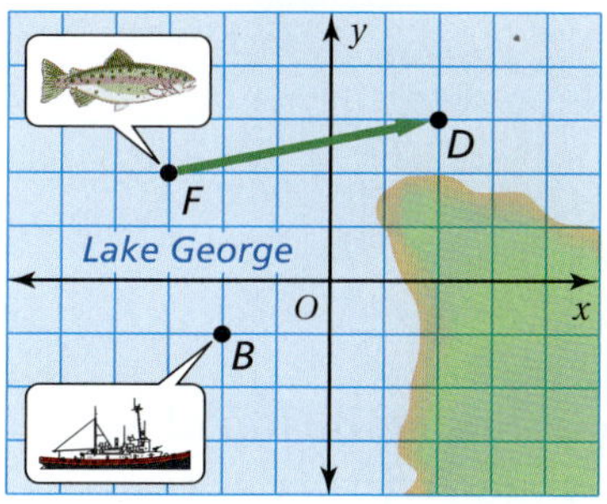

21. FISHING A school of fish translates from point F to point D.

a. Describe the translation of the school of fish.

b. Can the fishing boat make the same translation? Explain.

c. Describe a translation the fishing boat could make to get to point D.

22. REASONING The vertices of a triangle are $A(0, -3)$, $B(2, -1)$, and $C(3, -3)$. You translate the triangle 5 units right and 2 units down. Then you translate the image 3 units left and 8 units down. Is the original triangle congruent to the final image? If so, give two ways to show that they are congruent.

23. Problem Solving In chess, a knight can move only in an L-shaped pattern:

- *two* vertical squares, then *one* horizontal square;
- *two* horizontal squares, then *one* vertical square;
- *one* vertical square, then *two* horizontal squares; or
- *one* horizontal square, then *two* vertical squares.

Write a series of translations to move the knight from g8 to g5.

Fair Game Review What you learned in previous grades & lessons

Tell whether you can fold the figure in half so that one side matches the other. *(Skills Review Handbook)*

24. 

25. 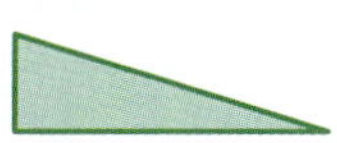

26. 

27. 

28. MULTIPLE CHOICE You put \$550 in an account that earns 4.4% simple interest per year. How much interest do you earn in 6 months? *(Section 6.7)*

Ⓐ \$1.21 Ⓑ \$12.10 Ⓒ \$121.00 Ⓓ \$145.20

11.3 Reflections

Essential Question How can you use reflections to classify a frieze pattern?

The Meaning of a Word ● Reflection

When you look at a mountain by a lake, you can see the **reflection**, or mirror image, of the mountain in the lake.

If you fold the photo on its axis, the mountain and its reflection will align.

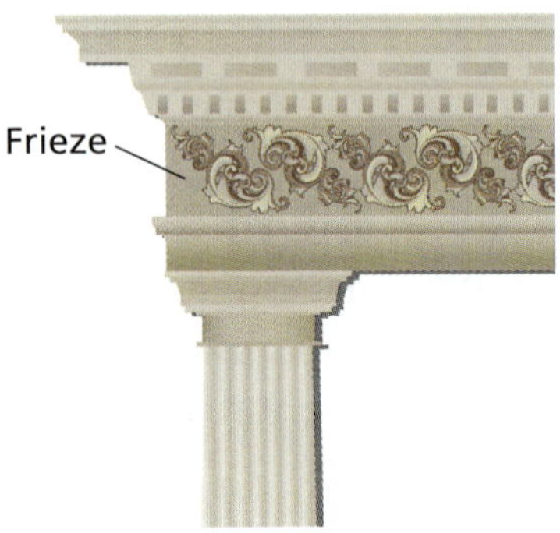

A *frieze* is a horizontal band that runs at the top of a building. A frieze is often decorated with a design that repeats.

- All frieze patterns are translations of themselves.
- Some frieze patterns are reflections of themselves.

1 ACTIVITY: Frieze Patterns and Reflections

Work with a partner. Consider the frieze pattern shown.

a. Is the frieze pattern a reflection of itself when folded horizontally? Explain.

b. Is the frieze pattern a reflection of itself when folded vertically? Explain.

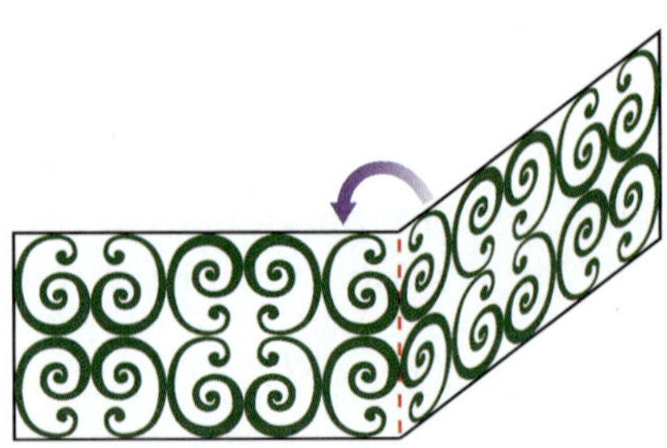

Geometry

In this lesson, you will

- identify reflections.
- reflect figures in the *x*-axis or the *y*-axis of the coordinate plane.

Laurie's Notes

Introduction

Applying Mathematical Practices

- **Construct Viable Arguments and Critique the Reasoning of Others:** As students manipulate geometric shapes they make conjectures and develop spatial reasoning. Mathematically proficient students explore the truth of their conjectures by considering all cases.

Motivate

- Before class, practice folding a long strip of scrap paper. Cut it to make a frieze pattern. A common design is the stick figure. Practice various folds so you can create reflections.

Vertical line of symmetry

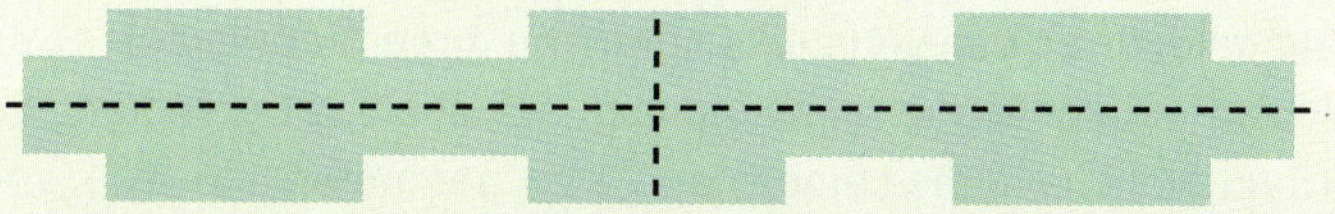

Horizontal and vertical lines of symmetry

- **Teaching Tip:** Lay the cut-out, still folded, on the overhead projector. Ask students what it will look like when it is unfolded.

Activity Notes

Discuss

- Today's investigation involves working with a pattern known as a frieze. The text shows an example of an architectural frieze on a building. Friezes also occur on wallpaper borders, designs on pottery, ironwork railings, and the headbands and belts of the indigenous people of North America, to name a few.
- A frieze is a pattern which repeats in one direction and can always be translated onto itself. Friezes may also contain reflections, and that is the focus of this investigation.

Activity 1

- Provide students with tracing paper so that they can sketch the original design. Some students will not need the tracing paper, but it should be offered to all students.
- When students finish, listen to the evidence they give as they explain whether the pattern reflects onto itself horizontally or vertically.

What Your Students Will Learn

- Identify whether two figures are reflections of one another.
- Reflect figures in the x-axis or the y-axis of the coordinate plane.

Previous Learning

Students should know how to plot points in the coordinate plane.

Technology for the Teacher

Dynamic Classroom

Lesson Plans
Complete Materials List

11.3 Record and Practice Journal

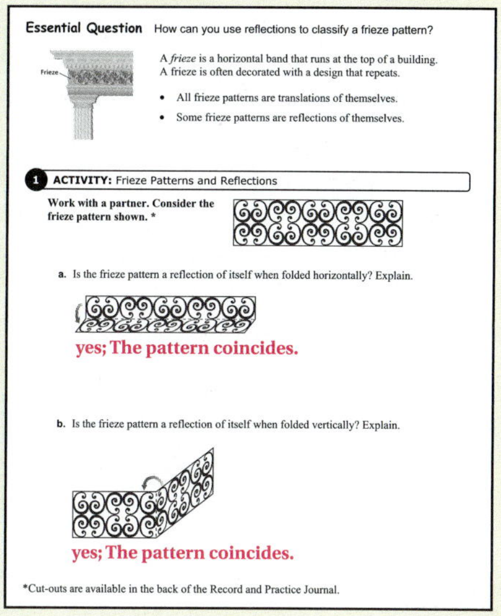

Essential Question How can you use reflections to classify a frieze pattern?

A *frieze* is a horizontal band that runs at the top of a building. A frieze is often decorated with a design that repeats.

- All frieze patterns are translations of themselves.
- Some frieze patterns are reflections of themselves.

1 ACTIVITY: Frieze Patterns and Reflections

Work with a partner. Consider the frieze pattern shown. *

a. Is the frieze pattern a reflection of itself when folded horizontally? Explain.

yes; The pattern coincides.

b. Is the frieze pattern a reflection of itself when folded vertically? Explain.

yes; The pattern coincides.

*Cut-outs are available in the back of the Record and Practice Journal.

Differentiated Instruction

Kinesthetic

Have students fold paper in half or in quarters and use scissors to cut out various shapes. Open the paper and find the lines of symmetry. Depending on the cut out, there may be more than one line of symmetry.

11.3 Record and Practice Journal

2 ACTIVITY: Frieze Patterns and Reflections

Work with a partner. Is the frieze pattern a reflection of itself when folded *horizontally*, *vertically*, or *neither*?

a. 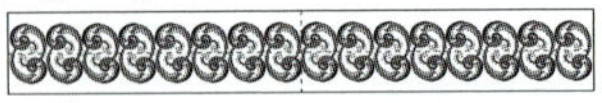

neither

b. 

horizontally

3 ACTIVITY: Reflecting in the Coordinate Plane

Work with a partner.

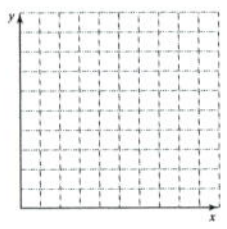

a. Draw a rectangle in Quadrant I of a coordinate plane. Find the dimensions of the rectangle.
Check students' work.

b. Copy the axes and the rectangle onto a piece of transparent paper.

Flip the transparent paper once so that the rectangle is in Quadrant IV. Then align the origin and the axes with the coordinate plane.

Draw the new figure in the coordinate plane. List the vertices.
Check students' work.

c. Compare the dimensions and the angle measures of the new figure to those of the original rectangle.
same

d. Are the opposite sides of the new figure still parallel? Explain.
yes

e. Can you conclude that the two figures are congruent? Explain.
yes

f. Flip the transparent paper so that the original rectangle is in Quadrant II. Draw the new figure in the coordinate plane. List the vertices. Then repeat parts (c)–(e).
Check students' work; same; yes; yes

g. Compare your results with those of other students in your class. Do you think the results are true for any type of figure?
yes

What Is Your Answer?

4. **IN YOUR OWN WORDS** How can you use reflections to classify a frieze pattern?
Fold a frieze pattern horizontally or vertically. If it coincides, then it is a reflection of itself.

Laurie's Notes

Activity 2

- Students should work with partners.
- Students may wish to have tracing paper to test their thinking about the patterns shown.
- Make an overhead transparency of the designs to help facilitate discussion. Have clear transparencies available for students to trace their answers.
- Remind students that a reflection that folds onto itself in a frieze must be horizontal or vertical.
- **Common Error:** Students will see the rotation in the pattern and identify it as a reflection.

Activity 3

- The original rectangle can be any size as long as it is in Quadrant I. Encourage students to draw a rectangle that is not a square. A square is a special case of a rectangle and if a square is used, any conclusions may only be true for a square.
- If transparent paper is not available, you can present this activity on the overhead projector or document camera. Flip the coordinate plane and have students list vertices and discuss relationships between the vertices.
- In part (f), start from the original position of the rectangle in Quadrant I.

? **Construct Viable Arguments and Critique the Reasoning of Others:** "What conjectures can you make about the three rectangles?" Students should share and support many conjectures, such as the rectangles are congruent; they have the same shape, size, perimeter, and area; and opposite sides are parallel. Students may also talk about the vertices when a rectangle is reflected about the *x*- or *y*-axis.

- **Extension:** Draw a frieze pattern for the following cases.
  - The pattern is a reflection of itself when folded horizontally.
  - The pattern is a reflection of itself when folded vertically.
  - The pattern is not a reflection of itself when folded horizontally or vertically.

What Is Your Answer?

- Students should consider their work in all three activities to answer the question.

Closure

- Imagine footprints in sand left by someone walking normally. Are the footprints a reflection? no
- Imagine footprints in mud left by a rabbit hopping normally. Are the footprints a reflection? yes

2 ACTIVITY: Frieze Patterns and Reflections

Work with a partner. Is the frieze pattern a reflection of itself when folded *horizontally*, *vertically*, or *neither*?

a.

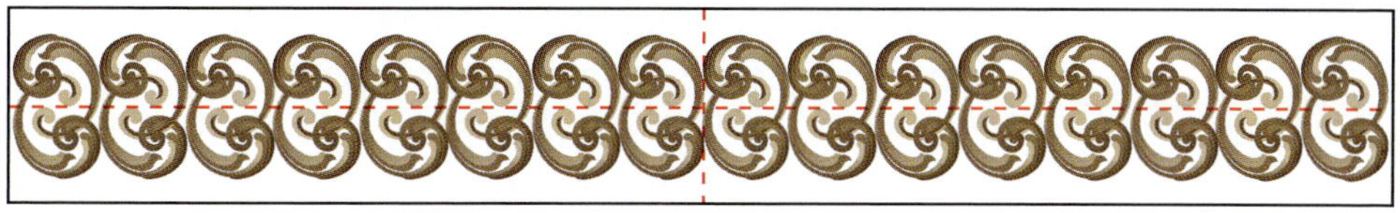

b.

3 ACTIVITY: Reflecting in the Coordinate Plane

Work with a partner.

a. Draw a rectangle in Quadrant I of a coordinate plane. Find the dimensions of the rectangle.

b. Copy the axes and the rectangle onto a piece of transparent paper.

Flip the transparent paper once so that the rectangle is in Quadrant IV. Then align the origin and the axes with the coordinate plane.

Draw the new figure in the coordinate plane. List the vertices.

c. Compare the dimensions and the angle measures of the new figure to those of the original rectangle.

d. Are the opposite sides of the new figure still parallel? Explain.

e. Can you conclude that the two figures are congruent? Explain.

f. Flip the transparent paper so that the original rectangle is in Quadrant II. Draw the new figure in the coordinate plane. List the vertices. Then repeat parts (c) – (e).

g. Compare your results with those of other students in your class. Do you think the results are true for any type of figure?

Math Practice

Look for Patterns

What do you notice about the vertices of the original figure and the image? How does this help you determine whether the figures are congruent?

What Is Your Answer?

4. **IN YOUR OWN WORDS** How can you use reflections to classify a frieze pattern?

Practice Use what you learned about reflections to complete Exercises 4–6 on page 482.

11.3 Lesson

Check It Out
Lesson Tutorials
BigIdeasMath.com

Key Vocabulary
reflection, *p. 480*
line of reflection, *p. 480*

A **reflection**, or *flip*, is a transformation in which a figure is reflected in a line called the **line of reflection**. A reflection creates a mirror image of the original figure.

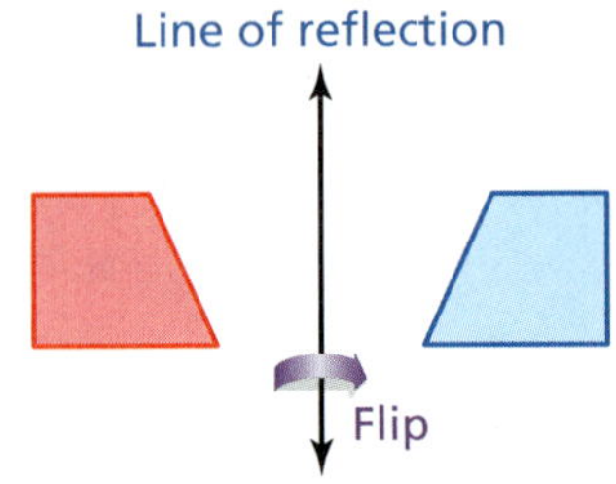

EXAMPLE 1 Identifying a Reflection

Tell whether the blue figure is a reflection of the red figure.

a.

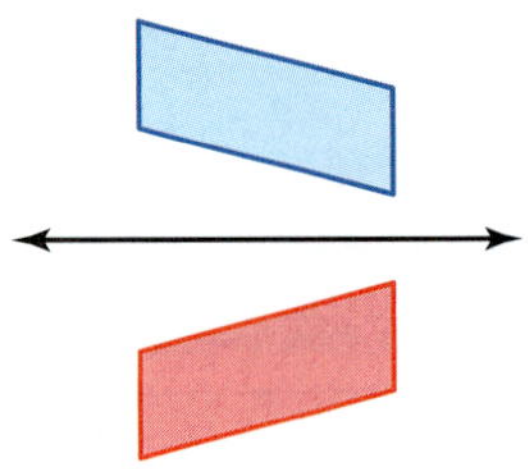

The red figure can be *flipped* to form the blue figure.

∴ So, the blue figure is a reflection of the red figure.

b.

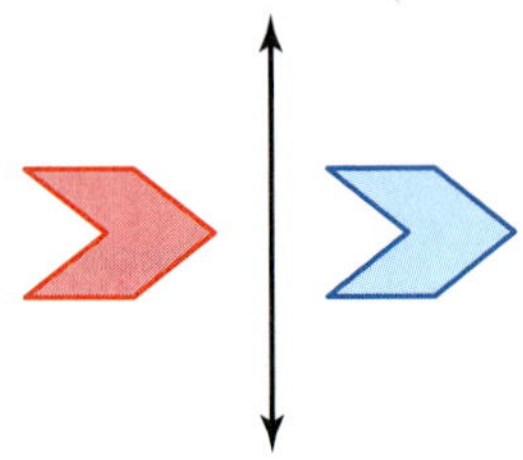

If the red figure were *flipped*, it would point to the left.

∴ So, the blue figure is *not* a reflection of the red figure.

On Your Own

Now You're Ready
Exercises 4–9

Tell whether the blue figure is a reflection of the red figure. Explain.

1.

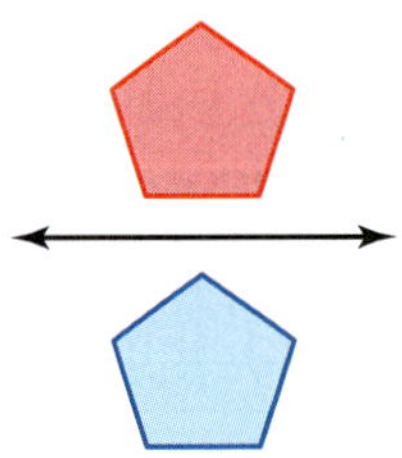

2.

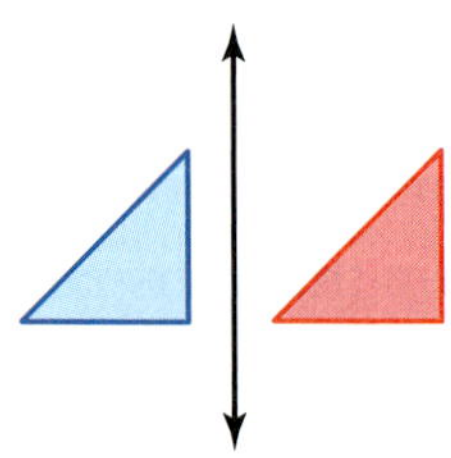

3.

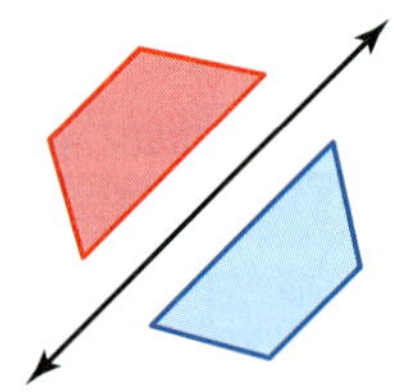

Key Idea

Reflections in the Coordinate Plane

Words To reflect a figure in the x-axis, take the opposite of the y-coordinate.

To reflect a figure in the y-axis, take the opposite of the x-coordinate.

Algebra Reflection in x-axis: $(x, y) \rightarrow (x, -y)$
Reflection in y-axis: $(x, y) \rightarrow (-x, y)$

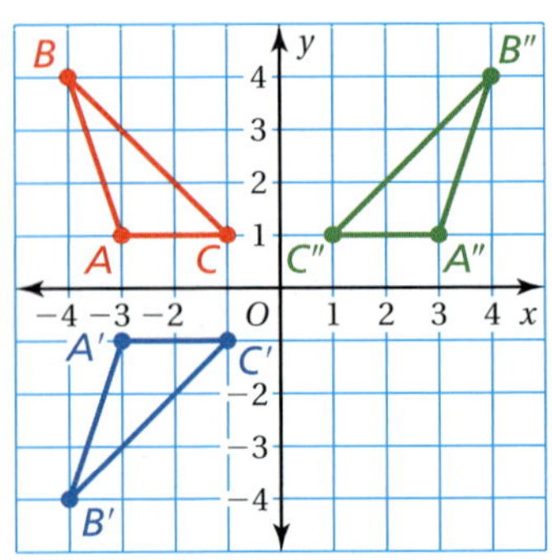

In a reflection, the original figure and its image are congruent.

Laurie's Notes

Introduction

Connect

- **Yesterday:** Students explored reflections in frieze patterns.
- **Today:** Students will use their visual skills to draw reflections in the coordinate plane.

Motivate

? Write the word **MOM** on a transparency, and ask a few questions.

- "What is special about this word?" Listen for ideas about reflection.
- "Describe the result when the word is reflected in the red line." MOM
- "Describe the result when the word is reflected in the green line." WOW

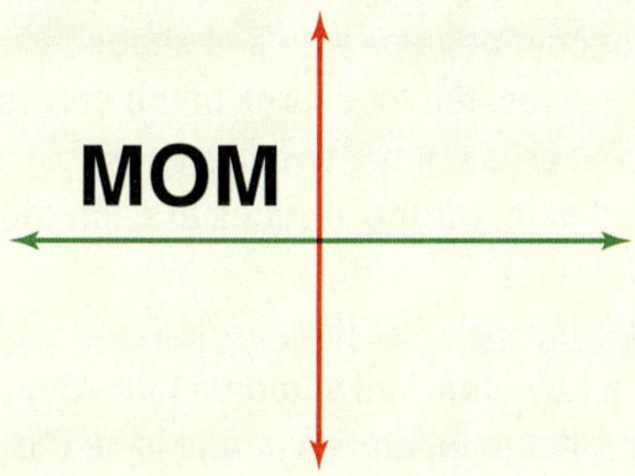

? "Can you think of other words that behave in a similar fashion?"

Lesson Notes

Discuss

- Discuss the introductory vocabulary: reflection (flip) and line of reflection.
- Relate lines of reflection to the red and green lines in the Motivate activity.

Example 1

- **Common Error:** Students may call part (b) a reflection because the shapes remain the same size and the orientation is the same. It is actually a translation.
- Offer tracing paper to students who struggle with spatial reasoning.

On Your Own

- **Neighbor Check:** Have students work independently and then have their neighbors check their work. Have students discuss any discrepancies.

Key Idea

- Write the Key Idea, using the language of A and A' as you identify the coordinates of the original figure (A, B, C) and its image (A', B', C') reflected in the x-axis. The reflection of (A, B, C) in the y-axis is the image (A'', B'', C''). Note that you read A'' as *A double prime*.
- Discuss how the coordinates change when you reflect a figure in each axis.
- **Attend to Precision:** Students may think $(x, -y)$ means there is a positive x-coordinate and a negative y-coordinate. Read the ordered pair as (x, the opposite of y) and explain that when y is negative, $-y$ is positive. Have students read the ordered pair this way also.

Goal Today's lesson is identifying and drawing **reflections**.

Technology for the Teacher

Dynamic Classroom

Lesson Tutorials
Lesson Plans
Answer Presentation Tool

Extra Example 1

Tell whether the blue figure is a reflection of the red figure.

a.

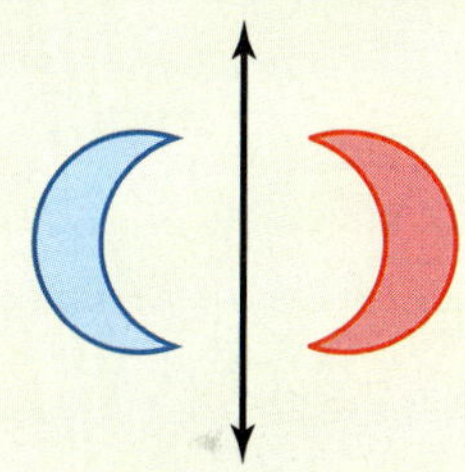

a reflection

b.

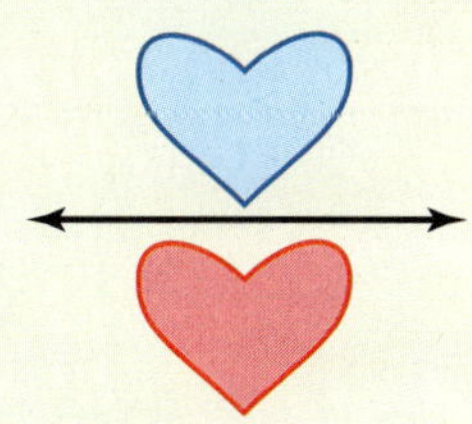

not a reflection

On Your Own

1. no; It is a translation.
2. no; It is a translation.
3. yes; The red figure can be flipped to form the blue figure.

Extra Example 2

The vertices of a parallelogram are $A(-1, -1)$, $B(2, -1)$, $C(4, -3)$, and $D(1, -3)$. Draw the figure and its reflection in the x-axis. What are the coordinates of the image?

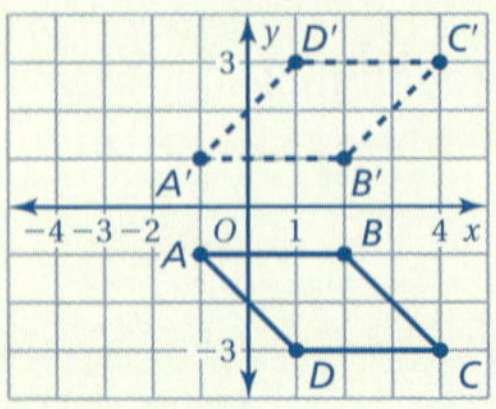

$A'(-1, 1)$, $B'(2, 1)$, $C'(4, 3)$, $D'(1, 3)$

Extra Example 3

The vertices of a triangle are $A(1, -2)$, $B(4, -2)$, and $C(1, 4)$. Draw the figure and its reflection in the y-axis.

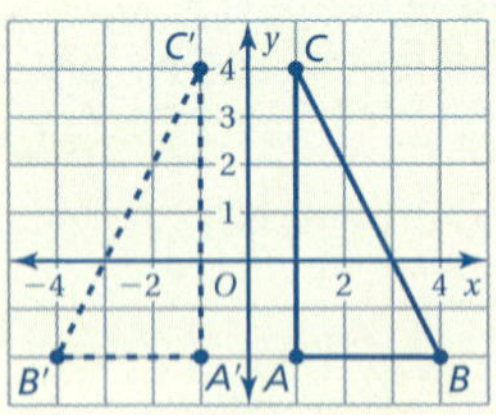

$A'(-1, -2)$, $B'(-4, -2)$, $C'(-1, 4)$

On Your Own

4. See Additional Answers.

English Language Learners

Vocabulary

Tell students that different words may be used to describe a reflection in a coordinate plane. For example, a figure is a reflection *in* the x-axis, *about* the x-axis, *across* the x-axis, or *over* the x-axis. The same words can be used to describe a reflection in the y-axis as well.

Laurie's Notes

Example 2

- Draw $\triangle ABC$ and label the vertices.
- ? "Which is the x-axis?" the horizontal axis
- We want to reflect the triangle from above the x-axis to below the x-axis.
- Note the suggestion boxes on the graph. Start with point A. Say, "Because A is 1 unit above the x-axis, it will be reflected to 1 unit below the x-axis." Repeat using similar language for points B and C. When students don't use this approach, they can easily translate the triangle instead of reflecting it.
- **Common Error:** The numbers written horizontally along the x-axis may cause students to be off by one number when they find the coordinates of each point in the blue triangle.
- ? "Is the blue triangle the same size and shape as the red triangle?" yes
- Reinforce the concept of same size and shape by talking about the lengths of corresponding sides, the measures of the corresponding angles, and the perimeters and areas of the two triangles.
- Write the ordered pairs for the vertices of each triangle.

  $A(-1, 1)$ $B(-1, 3)$ $C(6, 3)$

  $A'(-1, -1)$ $B'(-1, -3)$ $C'(6, -3)$

  Refer back to the Key Idea. Tell students that when you reflect a point across the x-axis, the point and its image have the same x-coordinates, and the y-coordinates are opposites.

Example 3

- This problem is similar to Example 2 except the original figure is a quadrilateral and it is reflected in the y-axis.
- **Model with Mathematics:** Set up a table to show the change in the x-coordinate when you reflect a point in the y-axis.
- ? "Is the blue quadrilateral the same size and shape as the red quadrilateral?" yes
- ? Ask questions like the following about each vertex and its image.
  - "How many units is P from the y-axis?" 2 units
  - "How many units is P' from the y-axis?" 2 units

On Your Own

- **Think-Pair-Share:** Students should read each question independently and then work in pairs to answer the questions. When they have answered the questions, the pair should compare their answers with another group and discuss any discrepancies.

Closure

- Draw a right triangle in Quadrant II. Reflect the triangle in the x-axis. Reflect the original triangle in the y-axis.

EXAMPLE 2 Reflecting a Figure in the *x*-axis

The vertices of a triangle are $A(-1, 1)$, $B(-1, 3)$, and $C(6, 3)$. Draw the figure and its reflection in the x-axis. What are the coordinates of the image?

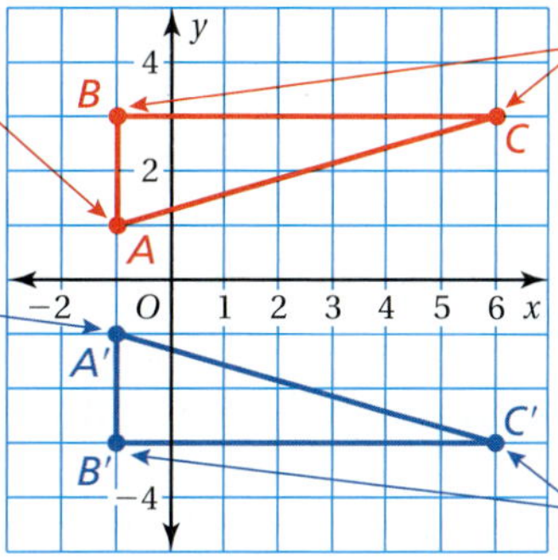

The coordinates of the image are $A'(-1, -1)$, $B'(-1, -3)$, and $C'(6, -3)$.

EXAMPLE 3 Reflecting a Figure in the *y*-axis

The vertices of a quadrilateral are $P(-2, 5)$, $Q(-1, -1)$, $R(-4, 2)$, and $S(-4, 4)$. Draw the figure and its reflection in the y-axis.

Take the opposite of the *x*-coordinate.

The *y*-coordinate does not change.

Vertices of $PQRS$	$(-x, y)$	Vertices of $P'Q'R'S'$
$P(-2, 5)$	$(-(-2), 5)$	$P'(2, 5)$
$Q(-1, -1)$	$(-(-1), -1)$	$Q'(1, -1)$
$R(-4, 2)$	$(-(-4), 2)$	$R'(4, 2)$
$S(-4, 4)$	$(-(-4), 4)$	$S'(4, 4)$

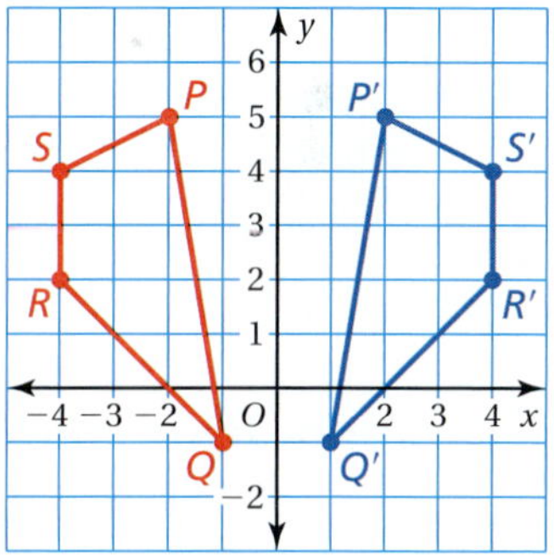

The figure and its image are shown at the above right.

On Your Own

Now You're Ready
Exercises 10–17

4. The vertices of a rectangle are $A(-4, -3)$, $B(-4, -1)$, $C(-1, -1)$, and $D(-1, -3)$.
   a. Draw the figure and its reflection in the x-axis.
   b. Draw the figure and its reflection in the y-axis.
   c. Are the images in parts (a) and (b) congruent? Explain.

11.3 Exercises

Check It Out
Help with Homework
BigIdeasMath.com

Vocabulary and Concept Check

1. **WHICH ONE DOESN'T BELONG?** Which transformation does *not* belong with the other three? Explain your reasoning.

2. **WRITING** How can you tell when one figure is a reflection of another figure?

3. **REASONING** A figure lies entirely in Quadrant I. The figure is reflected in the x-axis. In which quadrant is the image?

Practice and Problem Solving

Tell whether the blue figure is a reflection of the red figure.

(1) 4.

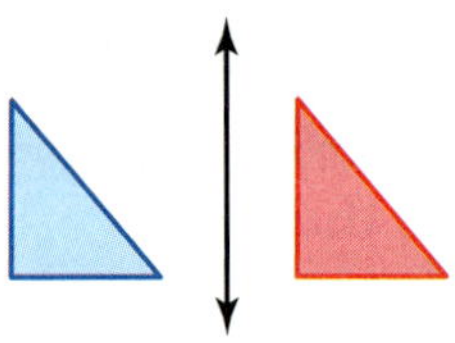

5.

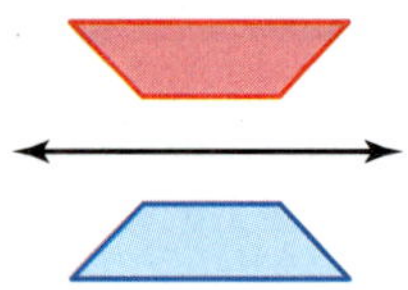

6.

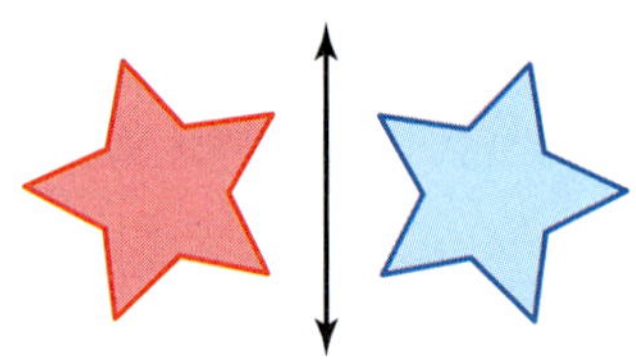

7.

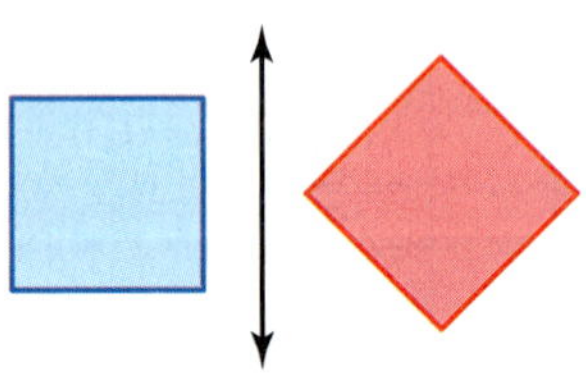

8.

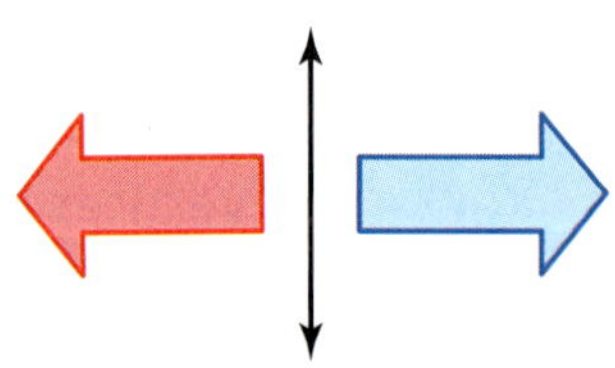

9.

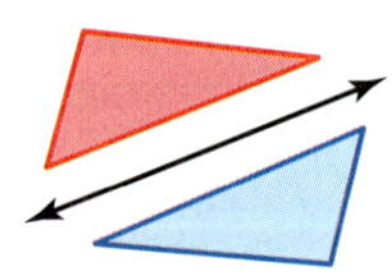

Draw the figure and its reflection in the x-axis. Identify the coordinates of the image.

(2) 10. $A(3, 2)$, $B(4, 4)$, $C(1, 3)$

11. $M(-2, 1)$, $N(0, 3)$, $P(2, 2)$

12. $H(2, -2)$, $J(4, -1)$, $K(6, -3)$, $L(5, -4)$

13. $D(-2, -1)$, $E(0, -1)$, $F(0, -5)$, $G(-2, -5)$

Draw the figure and its reflection in the y-axis. Identify the coordinates of the image.

(3) 14. $Q(-4, 2)$, $R(-2, 4)$, $S(-1, 1)$

15. $T(4, -2)$, $U(4, 2)$, $V(6, -2)$

16. $W(2, -1)$, $X(5, -2)$, $Y(5, -5)$, $Z(2, -4)$

17. $J(2, 2)$, $K(7, 4)$, $L(9, -2)$, $M(3, -1)$

18. **ALPHABET** Which letters look the same when reflected in the line?

A B C D E F G H I J K L M N O P Q R S T U V W X Y Z

Assignment Guide and Homework Check

Level	Assignment	Homework Check
Accelerated	1–6, 10–22 even, 24–33	10, 14, 20, 27

Common Errors

- **Exercises 4–9** Some students may struggle with the visual and think that a translation is actually a reflection. Give students tracing paper to trace the objects, and then fold the paper to see if the vertices line up.
- **Exercises 10–17** Students may reflect in the incorrect axis. Refer them back to Examples 2 and 3.
- **Exercise 18** Students may need to copy the alphabet and fold their paper on the line to see which letters look the same.

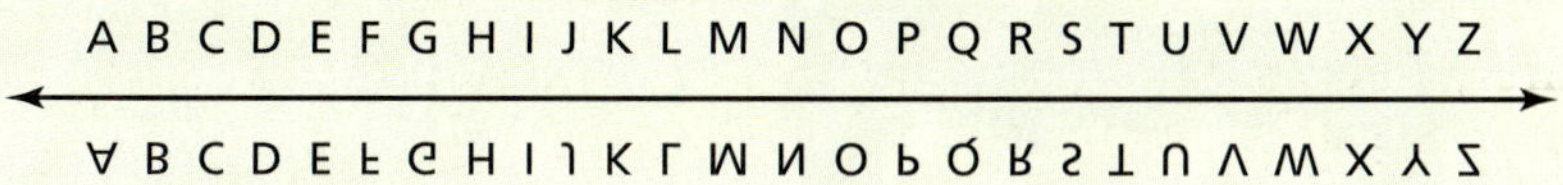

11.3 Record and Practice Journal

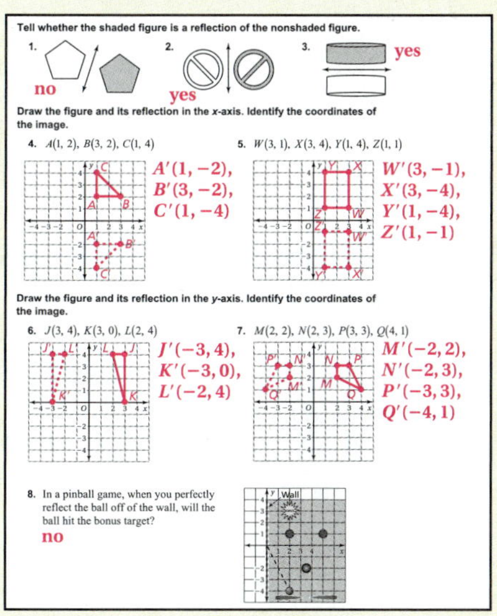
Tell whether the shaded figure is a reflection of the nonshaded figure.

1. no 2. yes 3. yes

Draw the figure and its reflection in the x-axis. Identify the coordinates of the image.

4. $A(1, 2)$, $B(3, 2)$, $C(1, 4)$ — $A'(1, -2)$, $B'(3, -2)$, $C'(1, -4)$
5. $W(3, 1)$, $X(3, 4)$, $Y(1, 4)$, $Z(1, 1)$ — $W'(3, -1)$, $X'(3, -4)$, $Y'(1, -4)$, $Z'(1, -1)$

Draw the figure and its reflection in the y-axis. Identify the coordinates of the image.

6. $J(3, 4)$, $K(3, 0)$, $L(2, 4)$ — $J'(-3, 4)$, $K'(-3, 0)$, $L'(-2, 4)$
7. $M(2, 2)$, $N(2, 3)$, $P(3, 3)$, $Q(4, 1)$ — $M'(-2, 2)$, $N'(-2, 3)$, $P'(-3, 3)$, $Q'(-4, 1)$

8. In a pinball game, when you perfectly reflect the ball off of the wall, will the ball hit the bonus target? no

Vocabulary and Concept Check

1. The third one because it is not a reflection.
2. A figure is a reflection of another figure if one is the mirror image of the other.
3. Quadrant IV

Practice and Problem Solving

4. no
5. yes
6. yes
7. no
8. yes
9. no

10. 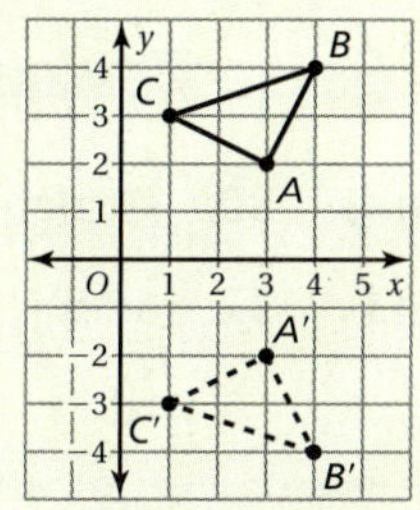

$A'(3, -2)$, $B'(4, -4)$, $C'(1, -3)$

11. 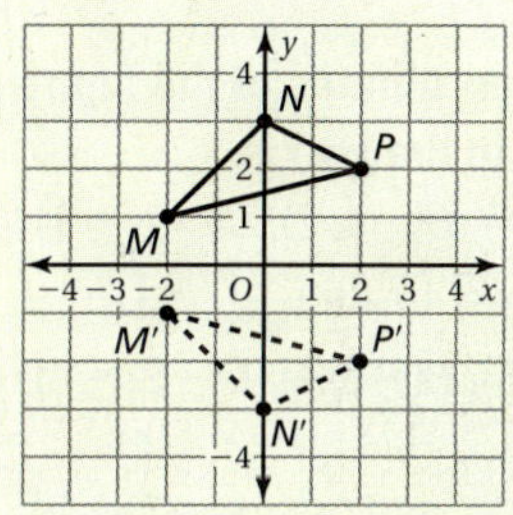

$M'(-2, -1)$, $N'(0, -3)$, $P'(2, -2)$

12. 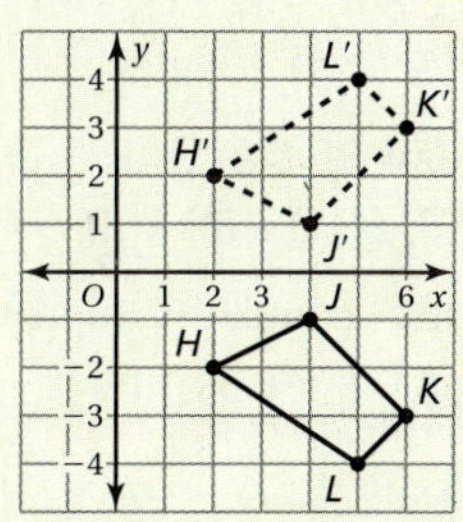

$H'(2, 2)$, $J'(4, 1)$, $K'(6, 3)$, $L'(5, 4)$

13–17. See Additional Answers.

18. B, C, D, E, H, I, K, O, X

Practice and Problem Solving

19. x-axis **20.** y-axis

21. y-axis **22.** x-axis

23. $R'(3, -4), S'(3, -1), T'(1, -4)$

24. $W'(-4, 5), X'(-4, 2), Y'(0, 2), Z'(2, 5)$

25. yes; Translations and reflections produce images that are congruent to the original figure.

26. $(-x, -y)$

27. See *Taking Math Deeper*.

28. See Additional Answers.

Fair Game Review

29. obtuse **30.** straight

31. right **32.** acute

33. B

Mini-Assessment

Find the coordinates of the figure after reflecting in the y-axis.

1. $A(-2, 4), B(-4, 2), C(-1, -1)$
   $A'(2, 4), B'(4, 2), C'(1, -1)$
2. $A(-2, 5), B(-5, 1), C(-3, -4)$
   $A'(2, 5), B'(5, 1), C'(3, -4)$

Find the coordinates of the figure after reflecting in the x-axis.

3. $A(-4, -2), B(4, -1), C(1, -6)$
   $A'(-4, 2), B'(4, 1), C'(1, 6)$
4. $A(-2, 5), B(4, 8), C(5, 1)$
   $A'(-2, -5), B'(4, -8), C'(5, -1)$
5. Will the letter E look the same when reflected in the y-axis? no

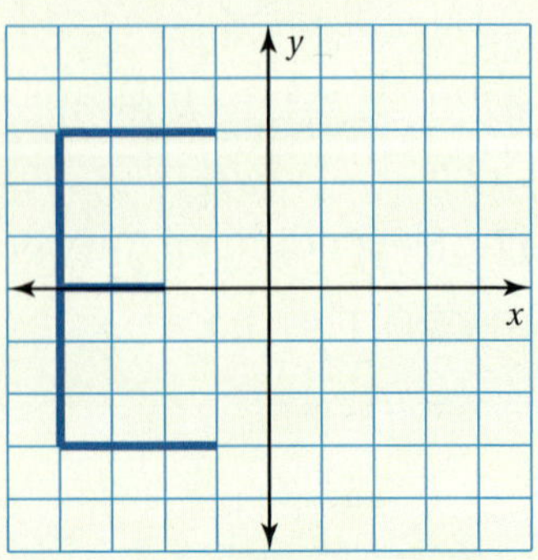

Taking Math Deeper

Exercise 27

Students need a mirror to see this one.

① Looking straight on, this is the ambulance.

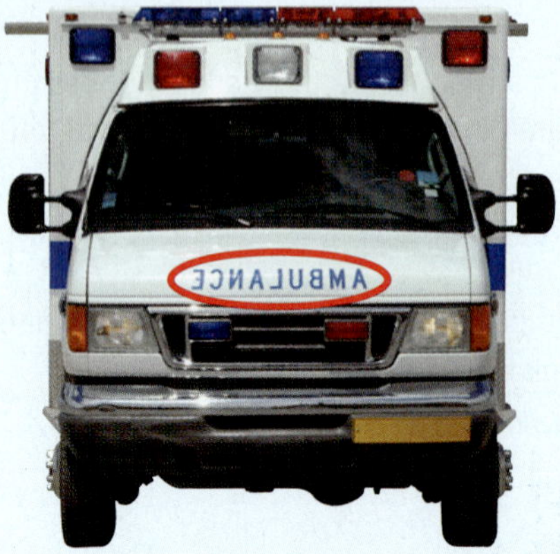

② **a.** Looking in a mirror, this is what you see.

③ **b.** The word "AMBULANCE" is printed backwards so that when the ambulance comes up behind a car, the word will look correct in the rear-view mirror.

Reteaching and Enrichment Strategies

If students need help. . .	If students got it. . .
Resources by Chapter • Practice A and Practice B • Puzzle Time Record and Practice Journal Practice Differentiating the Lesson Lesson Tutorials Skills Review Handbook	Resources by Chapter • Enrichment and Extension • Technology Connection Start the next section

The coordinates of a point and its image are given. Is the reflection in the *x-axis* or *y-axis*?

19. $(2, -2) \rightarrow (2, 2)$

20. $(-4, 1) \rightarrow (4, 1)$

21. $(-2, -5) \rightarrow (2, -5)$

22. $(-3, -4) \rightarrow (-3, 4)$

Find the coordinates of the figure after the transformations.

23. Translate the triangle 1 unit right and 5 units down. Then reflect the image in the y-axis.

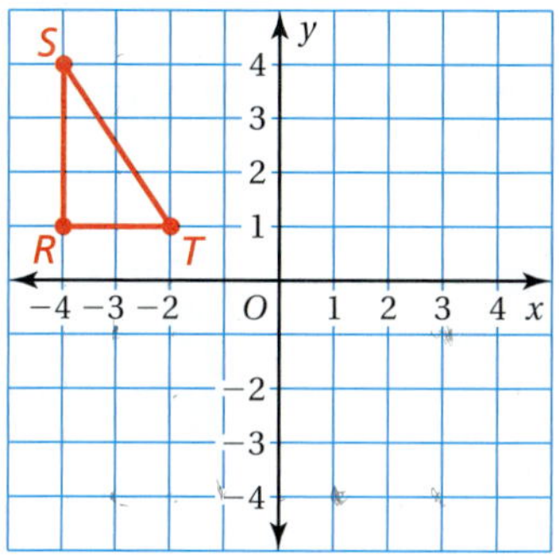

24. Reflect the trapezoid in the x-axis. Then translate the trapezoid 2 units left and 3 units up.

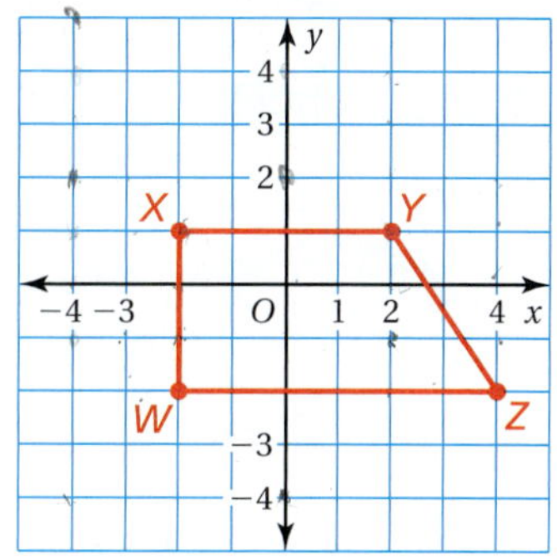

25. REASONING In Exercises 23 and 24, is the original figure congruent to the final image? Explain.

26. NUMBER SENSE You reflect a point (x, y) in the x-axis, and then in the y-axis. What are the coordinates of the final image?

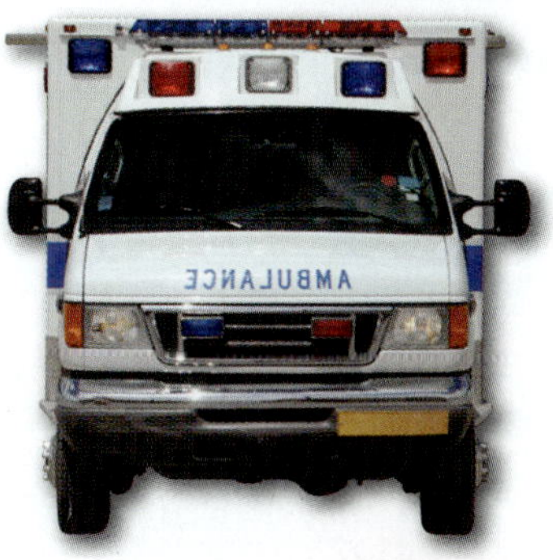

27. EMERGENCY VEHICLE Hold a mirror to the left side of the photo of the vehicle.

a. What word do you see in the mirror?

b. Why do you think it is written that way on the front of the vehicle?

28. Critical Thinking Reflect the triangle in the line $y = x$. How are the x- and y-coordinates of the image related to the x- and y-coordinates of the original triangle?

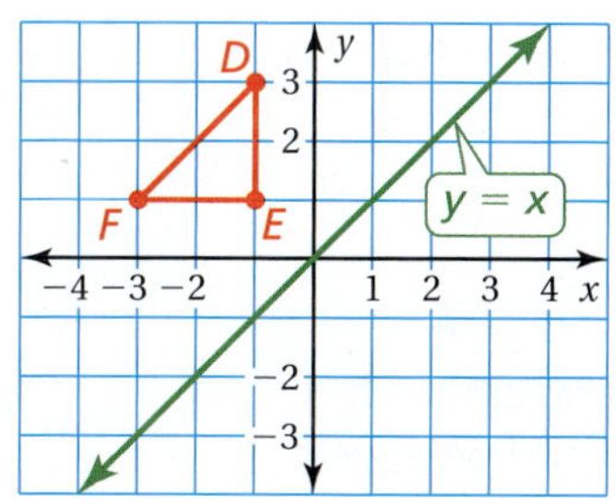

Fair Game Review What you learned in previous grades & lessons

Classify the angle as *acute, right, obtuse,* or *straight.* *(Skills Review Handbook)*

29. 

30. 

31. 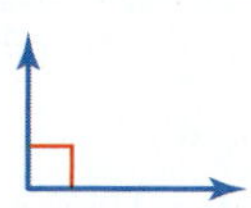

32. 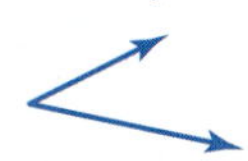

33. MULTIPLE CHOICE 36 is 75% of what number? *(Section 6.3 and Section 6.4)*

Ⓐ 27 Ⓑ 48 Ⓒ 54 Ⓓ 63

11.4 Rotations

Essential Question What are the three basic ways to move an object in a plane?

The Meaning of a Word ● Rotate

A bicycle wheel can **rotate** clockwise or counterclockwise.

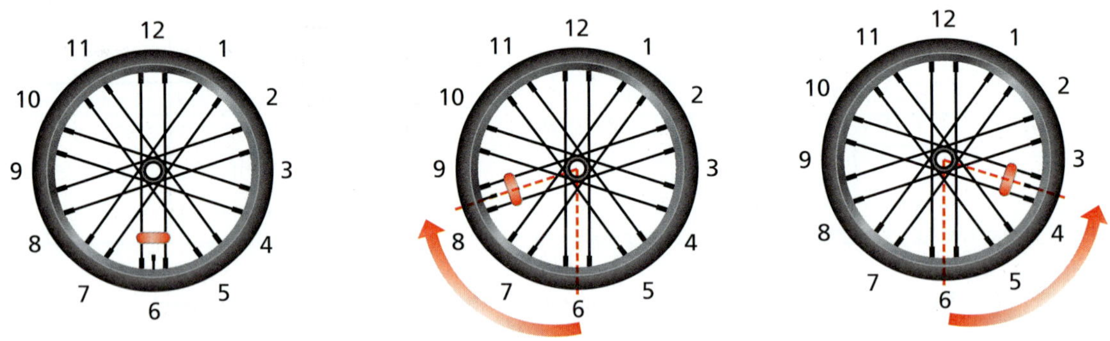

1 ACTIVITY: Three Basic Ways to Move Things

There are three basic ways to move objects on a flat surface.

_____ the object. _____ the object. _____ the object.

Work with a partner.

a. What type of triangle is the blue triangle? Is it congruent to the red triangles? Explain.

b. Decide how you can move the blue triangle to obtain each red triangle.

c. Is each move a *translation*, a *reflection*, or a *rotation*?

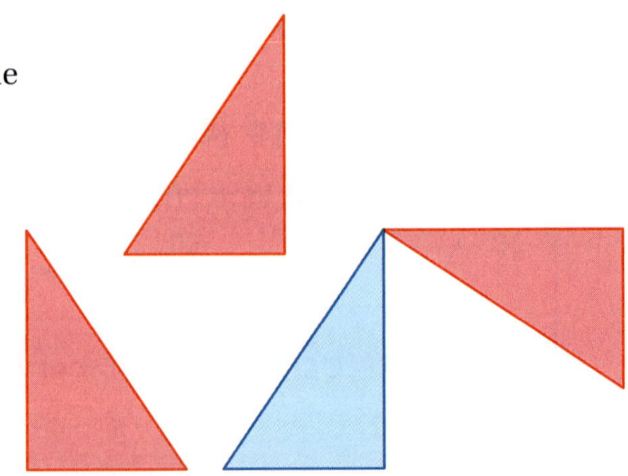

Geometry

In this lesson, you will

- identify rotations.
- rotate figures in the coordinate plane.
- use more than one transformation to find images of figures.

Laurie's Notes

Introduction

Applying Mathematical Practices

- **Construct Viable Arguments and Critique the Reasoning of Others:** As students manipulate geometric shapes, they develop spatial reasoning and make conjectures. Mathematically proficient students explore the truth of their conjectures by considering all possible cases.

Motivate

- **Time to Play:** *Name Five Twice.* In this game, students will name things that rotate: the first five objects rotate about a point in the center of the object (like a wheel) and the next five objects rotate about a point not in the center of the object (like a windshield wiper). Give students time to work with partners to generate two lists of five.

  Example 1: car tire, Ferris wheel, merry-go-round, dial on a combination lock

  Example 2: windshield wiper, lever—as on a mechanical arm or wrench
- **FYI:** Rotation is generally the most challenging transformation for students to visualize.

Activity Notes

Activity 1

- To help students visualize, offer them tracing paper so that they can sketch and transform the blue triangle. If tracing paper is not readily available, ahead of time ask a local doughnut shop or bakery for a donation of a box of tissue paper.
- For students with higher spatial skills, sketching the blue triangle on tissue paper will not be necessary.
- After students have finished the activity, ask them to share their results. If necessary, prompt them to explain their reasoning.

What Your Students Will Learn

- Identify rotations of figures.
- Rotate figures in the coordinate plane.
- Use more than one transformation in the coordinate plane to find images of figures.
- Describe sequences of transformations of congruent figures.

Previous Learning

Students should know how to plot points in the coordinate plane.

Technology for the Teacher

Dynamic Classroom

Lesson Plans

Complete Materials List

11.4 Record and Practice Journal

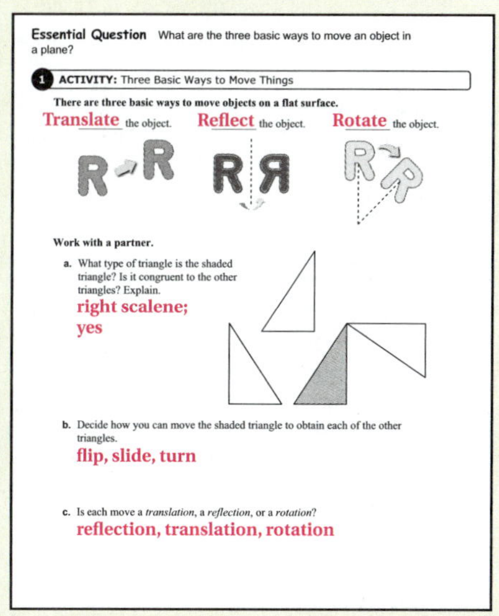

Essential Question What are the three basic ways to move an object in a plane?

1 ACTIVITY: Three Basic Ways to Move Things

There are three basic ways to move objects on a flat surface.

Translate the object. Reflect the object. Rotate the object.

Work with a partner.

a. What type of triangle is the shaded triangle? Is it congruent to the other triangles? Explain.

right scalene; yes

b. Decide how you can move the shaded triangle to obtain each of the other triangles.

flip, slide, turn

c. Is each move a *translation*, a *reflection*, or a *rotation*?

reflection, translation, rotation

Differentiated Instruction

Kinesthetic

Project a coordinate plane using an overhead projector. Give one student a geometric shape to place in one quadrant of the coordinate plane. Give another student a duplicate shape to place in another quadrant of the coordinate plane so that the shape is a rotation of the first shape. Discuss whether the duplicate shape could also be a translation or reflection of the first shape.

11.4 Record and Practice Journal

2 ACTIVITY: Rotating in the Coordinate Plane

Work with a partner.

a. Draw a rectangle in Quadrant II of a coordinate plane. Find the dimensions of the rectangle.
Check students' work.

b. Copy the axes and the rectangle onto a piece of transparent paper.
Align the origin and the vertices of the rectangle on the transparent paper with the coordinate plane. Turn the transparent paper so that the rectangle is in Quadrant I and the axes align.
Draw the new figure in the coordinate plane. List the vertices.
Check students' work.

c. Compare the dimensions and the angle measures of the new figure to those of the original rectangle.
same

d. Are the opposite sides of the new figure still parallel? Explain.
yes

e. Can you conclude that the two figures are congruent? Explain.
yes

f. Turn the transparent paper so that the original rectangle is in Quadrant IV. Draw the new figure in the coordinate plane. List the vertices. Then repeat parts (c)–(e).
Check students' work; same; yes; yes

g. Compare your results with those of other students in your class. Do you think the results are true for any type of figure?
yes

What Is Your Answer?

3. **IN YOUR OWN WORDS** What are the three basic ways to move an object in a plane? Draw an example of each.
translation, reflection, rotation

4. **PRECISION** Use the results of Activity 2(b).

a. Draw four angles using the conditions below.
- The origin is the vertex of each angle.
- One side of each angle passes through a vertex of the original rectangle.
- The other side of each angle passes through the corresponding vertex of the rotated rectangle.

Check students' work.

b. Measure each angle in part (a). For each angle, measure the distances between the origin and the vertices of the rectangles. What do you notice?
All angle measures are 90°. The distances are equal.

c. How can the results of part (b) help you rotate a figure?
***Sample answer:* You can use a ruler and a protractor to rotate each vertex of a figure.**

5. **PRECISION** Repeat the procedure in Question 4 using the results of Activity 2(f).
b. All angle measures are 180°. The distances are equal.

Laurie's Notes

Activity 2

- The size of the rectangle students draw is irrelevant, as long as they draw it in Quadrant II. Also, encourage students not to draw a square. Their observations with a square may not apply to a rectangle that is not a square.
- **Teaching Tip:** Before students turn the transparent paper, suggest that they hold it in place with the tip of a sharp pencil located at the origin. Then tell them to rotate the transparent paper a quarter turn clockwise. To make sure each student performs the rotation correctly, I tell them to rotate the paper so that the top edge (vertical) becomes the right edge (horizontal).
- If you are using an interactive white board and you have access to the Internet in the classroom, ahead of time find an applet online that demonstrates a 90° clockwise rotation. Share it with students during class.
- **Construct Viable Arguments and Critique the Reasoning of Others:** Ask students what conjectures they can make about the three rectangles. Make sure they provide evidence for their conjectures. Students may suggest, for instance, that the rectangles are congruent, have the same perimeter or area, all have opposite sides that are parallel, etc. They may also make conjectures about the coordinates of the vertices after rotation and their relation to those of the original rectangle.

What Is Your Answer?

- In Questions 4 and 5, make sure students know that the "original rectangle" is the rectangle they drew in part (a) of Activity 2.
- You may need to demonstrate part (a) of Question 4. After students have finished Question 4, ask a volunteer to share his or her results.

Closure

- Draw a right triangle in Quadrant I with the right angle at (0, 0). Rotate the triangle 90° clockwise about the origin.

2 ACTIVITY: Rotating in the Coordinate Plane

Work with a partner.

a. Draw a rectangle in Quadrant II of a coordinate plane. Find the dimensions of the rectangle.

b. Copy the axes and the rectangle onto a piece of transparent paper.

Align the origin and the vertices of the rectangle on the transparent paper with the coordinate plane. Turn the transparent paper so that the rectangle is in Quadrant I and the axes align.

Draw the new figure in the coordinate plane. List the vertices.

c. Compare the dimensions and the angle measures of the new figure to those of the original rectangle.

d. Are the opposite sides of the new figure still parallel? Explain.

e. Can you conclude that the two figures are congruent? Explain.

f. Turn the transparent paper so that the original rectangle is in Quadrant IV. Draw the new figure in the coordinate plane. List the vertices. Then repeat parts (c)–(e).

g. Compare your results with those of other students in your class. Do you think the results are true for any type of figure?

Math Practice

Calculate Accurately

What must you do to rotate the figure correctly?

What Is Your Answer?

3. **IN YOUR OWN WORDS** What are the three basic ways to move an object in a plane? Draw an example of each.

4. **PRECISION** Use the results of Activity 2(b).

   a. Draw four angles using the conditions below.
   - The origin is the vertex of each angle.
   - One side of each angle passes through a vertex of the original rectangle.
   - The other side of each angle passes through the corresponding vertex of the rotated rectangle.

   b. Measure each angle in part (a). For each angle, measure the distances between the origin and the vertices of the rectangles. What do you notice?

   c. How can the results of part (b) help you rotate a figure?

5. **PRECISION** Repeat the procedure in Question 4 using the results of Activity 2(f).

Practice Use what you learned about transformations to complete Exercises 7–9 on page 489.

11.4 Lesson

Check It Out
Lesson Tutorials
BigIdeasMath.com

Key Vocabulary

rotation, *p. 486*
center of rotation, *p. 486*
angle of rotation, *p. 486*

Key Idea

Rotations

A **rotation**, or *turn*, is a transformation in which a figure is rotated about a point called the **center of rotation**. The number of degrees a figure rotates is the **angle of rotation**.

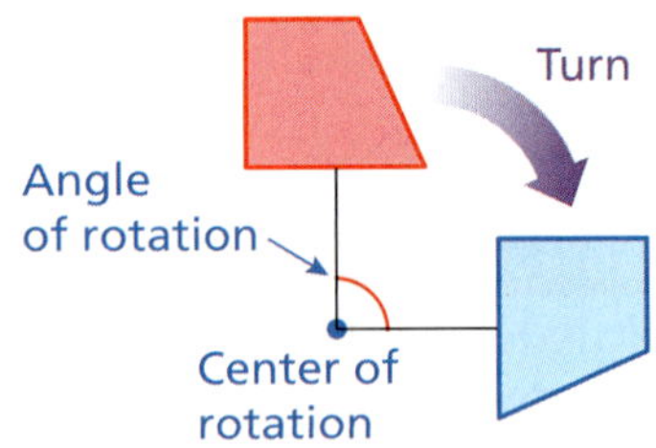

In a rotation, the original figure and its image are congruent.

EXAMPLE 1 Identifying a Rotation

You must rotate the puzzle piece 270° clockwise about point *P* to fit it into a puzzle. Which piece fits in the puzzle as shown?

Ⓐ  Ⓑ  Ⓒ  Ⓓ 

Rotate the puzzle piece 270° clockwise about point *P*.

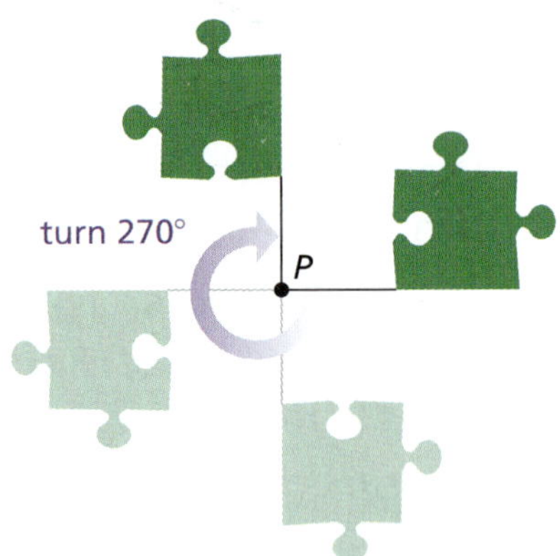

So, the correct answer is Ⓒ.

Study Tip

When rotating figures, it may help to sketch the rotation in several steps, as shown in Example 1.

On Your Own

Now You're Ready
Exercises 10–12

1. Which piece is a 90° counterclockwise rotation about point *P*?
2. Is Choice D a rotation of the original puzzle piece? If not, what kind of transformation does the image show?

Multi-Language Glossary at BigIdeasMath.com

Laurie's Notes

Introduction

Connect

- **Yesterday:** Students explored and sketched rotations.
- **Today:** Students will use their visual skills to draw rotations in the coordinate plane.

Motivate

- Use a marker to make two sizeable dots, one at the tip of your middle finger and one at the base of your palm. Anchor your elbow on a level surface. Wave at the class so that your elbow is the pivot.
- ? Do a "wave" of 90°, by starting in the horizontal position and "waving" to the vertical position. Ask the following questions.
  - "Through how many degrees did I wave my hand?" 90°
  - "Did my elbow move?" no
  - "Did the two points move the same distance?" no "If not, which point moved farther?" The point on the tip of the middle finger moved farther.
- Relate this motion to that of a windshield wiper. The farther a point on the wiper is from the point of rotation, the farther it travels.

Lesson Notes

Key Idea

- The rotation is hard to visualize because the center of rotation is generally not attached to the shape being rotated. My hand is connected to my forearm, which is connected to my elbow, so the "wave" is easier to see as a rotation. When a diagram only shows the original figure and the image, it is harder to see the angle of rotation.
- **Construct Viable Arguments and Critique the Reasoning of Others:** Note that the relationships between the coordinates of the vertices of a figure before and after rotation are not given. If you feel that your students are ready, explore this with them. Have students make conjectures about the relationships and explain their reasoning.

Example 1

- Model a rotation of 270° clockwise using a transparency with an arrow pointing to the right. Lightly place your finger on the middle to act as the center of rotation. Turn the transparency 90° clockwise, 3 times, stopping each time for students to see where the arrow is pointing. After 270° the arrow will be pointing up.

On Your Own

- If you have a puzzle piece that you can use to model these two questions, then it would help those students who have difficulty visualizing the movement of the pieces.

Goal Today's lesson is identifying and drawing **rotations**.

Technology for the Teacher

Dynamic Classroom

Lesson Tutorials
Lesson Plans
Answer Presentation Tool

Extra Example 1

Tell whether the blue figure is a 180° clockwise rotation of the red figure.

a.

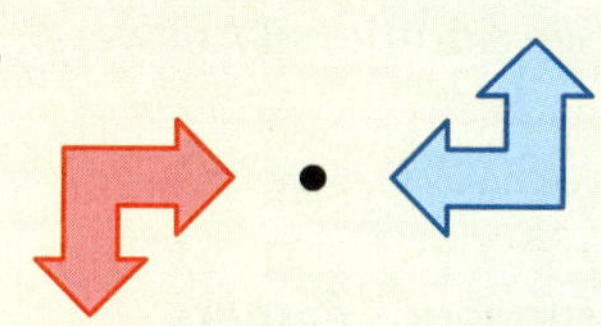

Is a 180° clockwise rotation.

b.

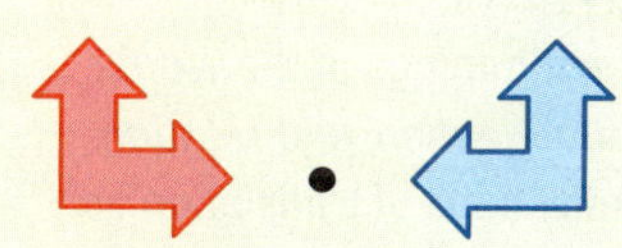

Is not a 180° clockwise rotation.

On Your Own

1. C
2. no; reflection

Laurie's Notes

Example 2

- Tracing paper or a transparency will be needed by many students for this example. Students need to *see* where the trapezoid rotates, before they can plot the ordered pairs.
- Draw the trapezoid and label the vertices.
- **Common Error:** Students will rotate the trapezoid about vertex Z instead of rotating about the origin.
- **Teaching Strategy:** Remind students that when a figure is rotated 180°, what was on the top will rotate to the bottom, and vice versa. Model this by holding a sheet of paper and rotating it 180°.
- **Extension:** List the coordinates of the original trapezoid and the image.

  $W(-4, 2) \longrightarrow W'(4, -2)$
  $X(-3, 4) \longrightarrow X'(3, -4)$
  $Y(-1, 4) \longrightarrow Y'(1, -4)$
  $Z(-1, 2) \longrightarrow Z'(1, -2)$

? **Look for and Express Regularity in Repeated Reasoning:** "What happens to the coordinates of the point (x, y) when you rotate the point 180° about the origin?" The coordinates become $(-x, -y)$.

Example 3

- In this example, a triangle is rotated 90° counterclockwise and the center of rotation is one of the vertices instead of the origin.
- Hold a sheet of paper facing the students.

? "When I rotate the sheet of paper 90° counterclockwise, which way will the top of the paper rotate?" to the left

- Work through the problem as shown.

? "Did all of the points on the triangle move? Explain." no; Vertex L did not move because it is the center of rotation.

- **Big Idea:** The lengths of the sides of a triangle do not change when you rotate the triangle.

On Your Own

- **Think-Pair-Share:** Students should read each question independently and then work in pairs to answer the questions. When they have answered the questions, the pair should compare their answers with another group and discuss any discrepancies.

Extra Example 2

The vertices of a triangle are $A(-4, 1)$, $B(-1, 6)$, and $C(-1, 1)$. Rotate the triangle 90° clockwise about the origin. What are the coordinates of the image?

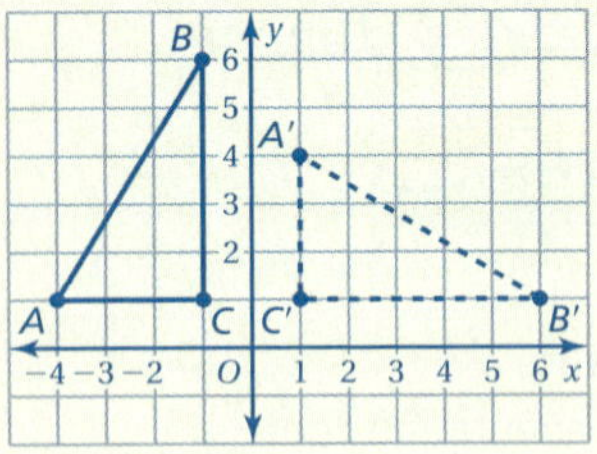

$A'(1, 4)$, $B'(6, 1)$, and $C'(1, 1)$

Extra Example 3

The vertices of a trapezoid are $A(1, 2)$, $B(4, 4)$, $C(4, 6)$, and $D(1, 6)$. Rotate the trapezoid 270° counterclockwise about vertex A. What are the coordinates of the image?

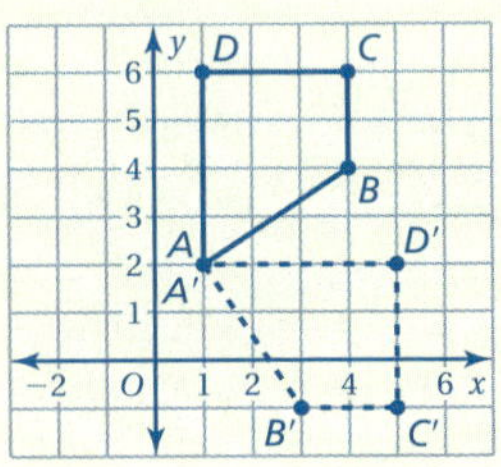

$A'(1, 2)$, $B'(3, -1)$, $C'(5, -1)$, $D'(5, 2)$

On Your Own

3. See Additional Answers.

English Language Learners

Vocabulary

Discuss the meanings of the words *translation*, *reflection*, and *rotation*. Students may think of translation as a process of writing text in another (parallel) language. Mathematically, a translation is when all points of a figure move along parallel lines. Have students visualize a sun setting on the horizon of the ocean. At the point where half of the sun has set, its reflection in the water gives the appearance of a full-circled sun. A rotation about a point in the plane is similar to a nail rotating around the wheel of a tire.

EXAMPLE 2 Rotating a Figure

The vertices of a trapezoid are $W(-4, 2)$, $X(-3, 4)$, $Y(-1, 4)$, and $Z(-1, 2)$. Rotate the trapezoid 180° about the origin. What are the coordinates of the image?

Study Tip

A 180° clockwise rotation and a 180° counterclockwise rotation have the same image. So, you do not need to specify direction when rotating a figure 180°.

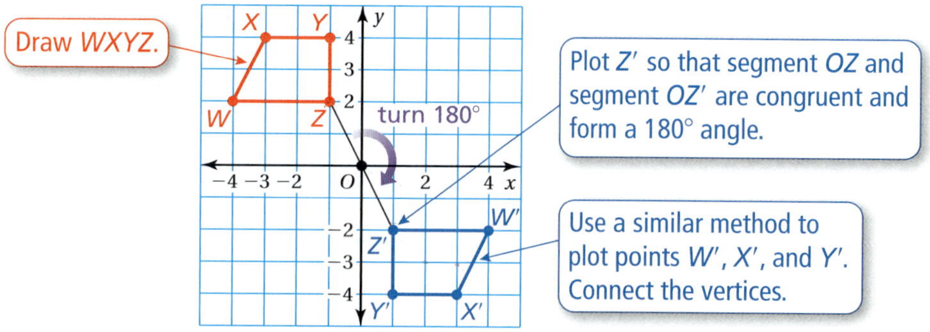

The coordinates of the image are $W'(4, -2)$, $X'(3, -4)$, $Y'(1, -4)$, and $Z'(1, -2)$.

EXAMPLE 3 Rotating a Figure

The vertices of a triangle are $J(1, 2)$, $K(4, 2)$, and $L(1, -3)$. Rotate the triangle 90° counterclockwise about vertex L. What are the coordinates of the image?

Common Error

Be sure to pay attention to whether a rotation is clockwise or counterclockwise.

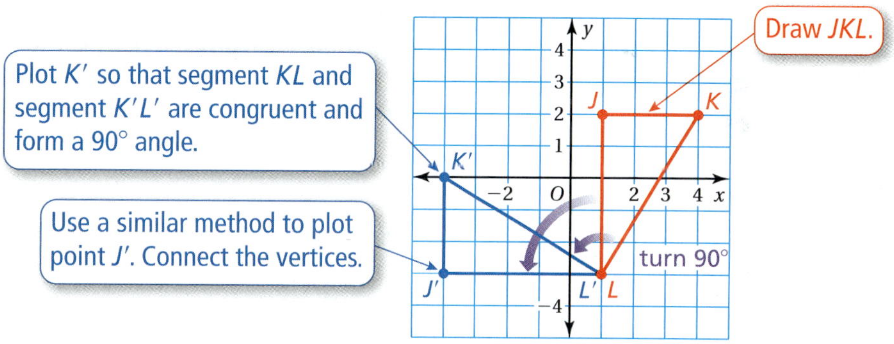

The coordinates of the image are $J'(-4, -3)$, $K'(-4, 0)$, and $L'(1, -3)$.

On Your Own

Now You're Ready
Exercises 13–18

3. A triangle has vertices $Q(4, 5)$, $R(4, 0)$, and $S(1, 0)$.
   - **a.** Rotate the triangle 90° counterclockwise about the origin.
   - **b.** Rotate the triangle 180° about vertex S.
   - **c.** Are the images in parts (a) and (b) congruent? Explain.

EXAMPLE 4 Using More than One Transformation

The vertices of a rectangle are $A(-3, -3)$, $B(1, -3)$, $C(1, -5)$, and $D(-3, -5)$. Rotate the rectangle 90° clockwise about the origin, and then reflect it in the y-axis. What are the coordinates of the image?

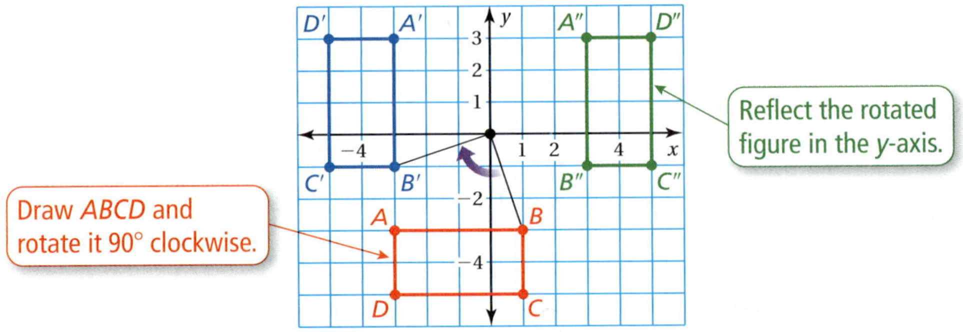

The coordinates of the image are $A''(3, 3)$, $B''(3, -1)$, C'' $(5, -1)$ and D'' $(5, 3)$.

The image of a translation, reflection, or rotation is congruent to the original figure. So, two figures are congruent when one can be obtained from the other by a sequence of translations, reflections, and rotations.

EXAMPLE 5 Describing a Sequence of Transformations

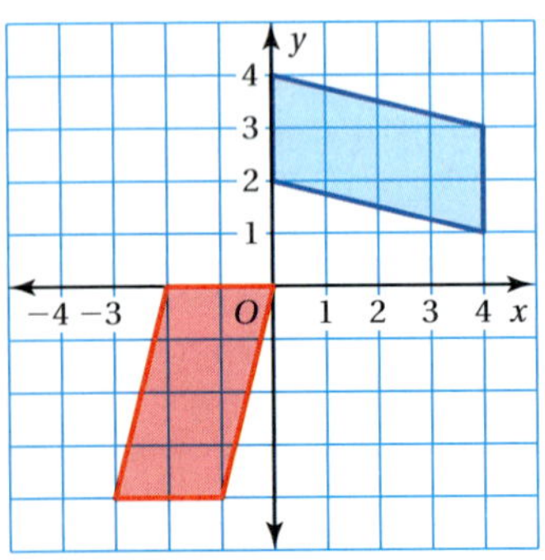

The red figure is congruent to the blue figure. Describe a sequence of transformations in which the blue figure is the image of the red figure.

You can turn the red figure 90° so that it has the same orientation as the blue figure. So, begin with a rotation.

After rotating, you need to slide the figure up.

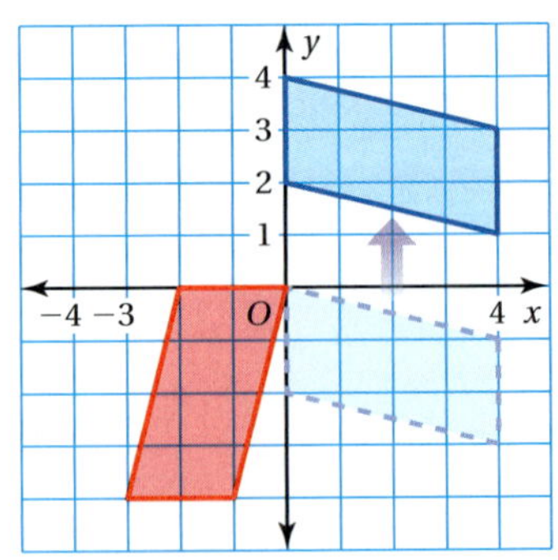

So, one possible sequence of transformations is a 90° counterclockwise rotation about the origin followed by a translation 4 units up.

On Your Own

Now You're Ready
Exercises 22–25

4. The vertices of a triangle are $P(-1, 2)$, $Q(-1, 0)$, and $R(2, 0)$. Rotate the triangle 180° about vertex R, and then reflect it in the x-axis. What are the coordinates of the image?

5. In Example 5, describe a different sequence of transformations in which the blue figure is the image of the red figure.

Laurie's Notes

Example 4

- This example involves two transformations. First the rectangle is rotated and then it is reflected. Work slowly and carefully through this example.
- ? "Can you visualize where the image will be after the two transformations?" Answers will vary.
- Graph the original rectangle *ABCD*.
- ? "Does it matter whether you rotate the rectangle 90° clockwise or 90° counterclockwise?" yes
- To construct the image after the rotation, draw the segment *OB*, where *O* represents a point at the origin. Next locate vertex *B′* by drawing *OB′* so that ∠*BOB′* is 90° and *OB* and *OB′* are the same length. After you locate vertex *B′*, the remaining vertices should be relatively easy to locate.
- Reflect rectangle *A′B′C′D′* in the *y*-axis to obtain rectangle *A″B″C″D″*.
- **Teaching Tip:** Use transparent paper to help visualize the transformations.
- **Common Error:** Because the reflection resembles a translation, students may label the vertices of rectangle *A″B″C″D″* incorrectly. Watch for this.
- ? **Construct Viable Arguments and Critique the Reasoning of Others:** "Are rectangles *ABCD* and *A″B″C″D″* congruent? Explain." yes; Rectangles *ABCD* and *A′B′C′D′* are congruent because in a rotation, the original figure and its image are congruent. Rectangles *A′B′C′D′* and *A″B″C″D″* are congruent because in a reflection, the original figure and its image are congruent. Because rectangles *ABCD* and *A″B″C″D″* are both congruent to rectangle *A′B′C′D′*, they are congruent to each other.
- **Construct Viable Arguments and Critique the Reasoning of Others** and **Attend to Precision:** If time allows, ask students whether the order in which you perform the transformations matters. (The answer is yes.) Have them thoroughly explain their reasoning.

Example 5

- Ask a volunteer to read the problem. Give students time to come up with an answer. Ask volunteers to share their answers with the class.
- Use transparent paper to help visualize the transformations.
- ? **Extension:** "Is there is a single transformation to get from the red figure to the blue figure?" 90° counterclockwise rotation about the point (−2, 2). Students may need a hint to consider moving the center of rotation.

On Your Own

- **Think-Pair-Share:** Students should read each question independently and then work in pairs to answer the questions. When they have answered the questions, the pair should compare their answers with another group and discuss any discrepancies.

Closure

- Draw a right triangle in Quadrant II. Reflect the triangle in the *x*-axis. Rotate the original triangle about the origin 90° clockwise.

Extra Example 4

The vertices of a rectangle are *A*(−1, 1), *B*(−4, 1), *C*(−4, 5), and *D*(−1, 5). Rotate the rectangle 90° clockwise about the origin, and then reflect it in the *y*-axis. What are the coordinates of the image?

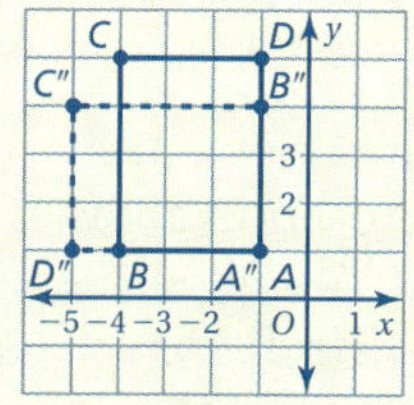

A″(−1, 1), *B″*(−1, 4), *C″*(−5, 4), *D″*(−5, 1)

Extra Example 5

The red figure is congruent to the blue figure. Describe a sequence of transformations in which the blue figure is the image of the red figure.

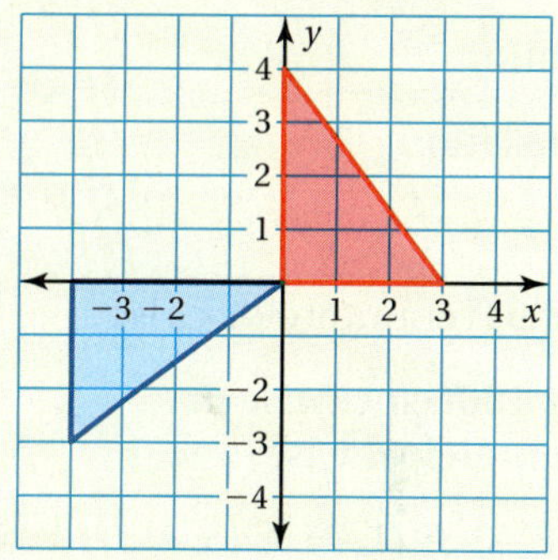

Sample answer: 90° clockwise rotation about the origin followed by a translation 4 units left

On Your Own

4. 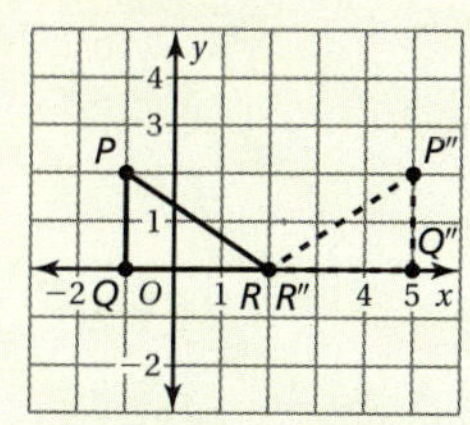

   P″(5, 2), *Q″*(5, 0), *R″*(2, 0)

5. *Sample answer:* 90° clockwise rotation about the origin followed by a translation 4 units right and 1 unit up

Vocabulary and Concept Check

1. $(0, 0)$; $(1, -3)$
2. Quadrant I
3. Quadrant IV
4. Quadrant III
5. Quadrant II
6. What are the coordinates of the figure after a 270° clockwise rotation about the origin?; $A'(-4, 2)$, $B'(-4, 4)$, $C'(-1, 4)$, $D'(-1, 2)$; $A'(4, -2)$, $B'(4, -4)$, $C'(1, -4)$, $D'(1, -2)$

Practice and Problem Solving

7. reflection
8. rotation
9. translation
10. no
11. yes; 90° counterclockwise
12. yes; 180° clockwise or counterclockwise
13. $A'(2, 2)$, $B'(1, 4)$, $C'(3, 4)$, $D'(4, 2)$
14. $F'(-1, -2)$, $G'(-3, -5)$, $H'(-3, -2)$
15. $J'(0, -3)$, $K'(0, -5)$, $L'(-4, -3)$

Assignment Guide and Homework Check

Level	Assignment	Homework Check
Accelerated	1–9, 16–26 even, 27–34	18, 20, 22, 26, 28

Common Errors

- **Exercises 7–12** Students with minimal spatial skills may not be able to tell whether a figure is rotated. Give them tracing paper and have them copy the red figure and rotate it.

11.4 Record and Practice Journal

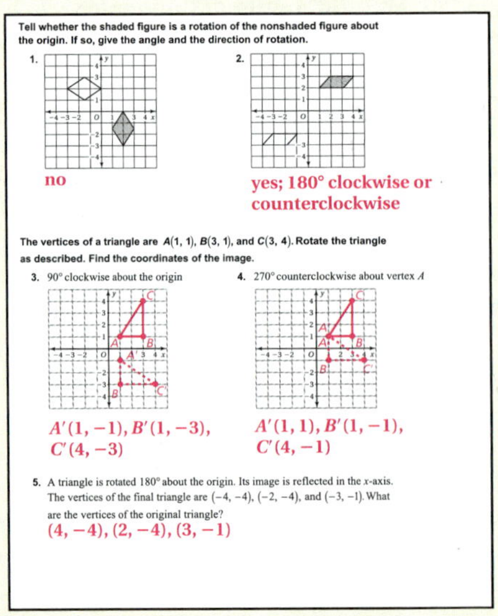
Tell whether the shaded figure is a rotation of the nonshaded figure about the origin. If so, give the angle and the direction of rotation.

1. no

2. yes; 180° clockwise or counterclockwise

The vertices of a triangle are $A(1, 1)$, $B(3, 1)$, and $C(3, 4)$. Rotate the triangle as described. Find the coordinates of the image.

3. 90° clockwise about the origin
$A'(1, -1)$, $B'(1, -3)$, $C'(4, -3)$

4. 270° counterclockwise about vertex A
$A'(1, 1)$, $B'(1, -1)$, $C'(4, -1)$

5. A triangle is rotated 180° about the origin. Its image is reflected in the x-axis. The vertices of the final triangle are $(-4, -4)$, $(-2, -4)$, and $(-3, -1)$. What are the vertices of the original triangle?
$(4, -4)$, $(2, -4)$, $(3, -1)$

11.4 Exercises

Check It Out
Help with Homework
BigIdeasMath.com

Vocabulary and Concept Check

1. **VOCABULARY** What are the coordinates of the center of rotation in Example 2? Example 3?

MENTAL MATH **A figure lies entirely in Quadrant II. In which quadrant will the figure lie after the given clockwise rotation about the origin?**

2. 90°
3. 180°
4. 270°
5. 360°

6. **DIFFERENT WORDS, SAME QUESTION** Which is different? Find "both" answers.

What are the coordinates of the figure after a 90° clockwise rotation about the origin?

What are the coordinates of the figure after a 270° clockwise rotation about the origin?

What are the coordinates of the figure after turning the figure 90° to the right about the origin?

What are the coordinates of the figure after a 270° counterclockwise rotation about the origin?

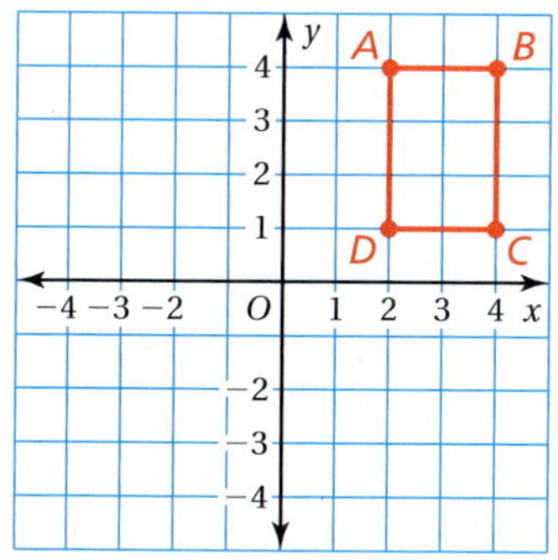

Practice and Problem Solving

Identify the transformation.

7.

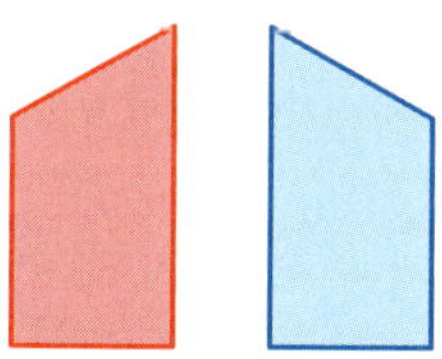

8.

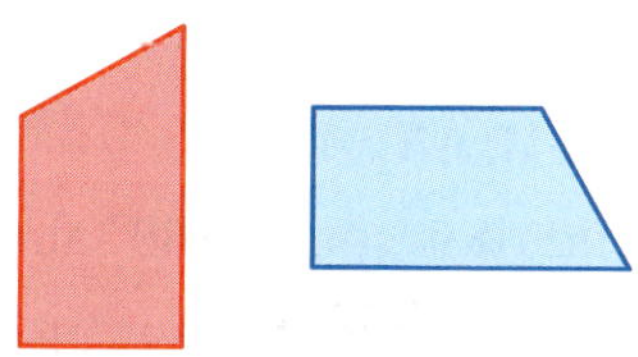

9.

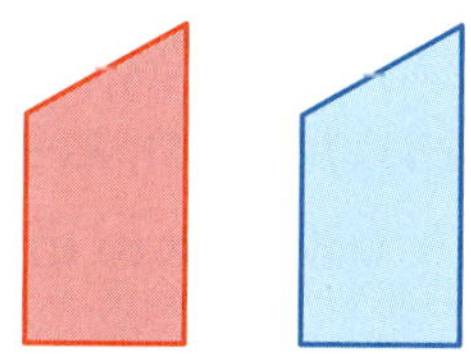

Tell whether the blue figure is a rotation of the red figure about the origin. If so, give the angle and direction of rotation.

1 10.

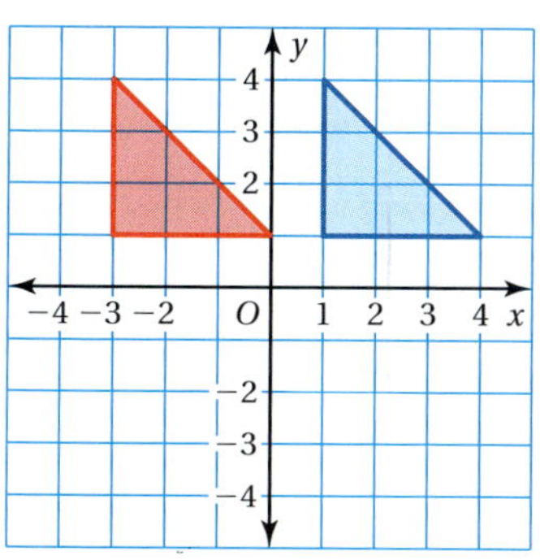

11.

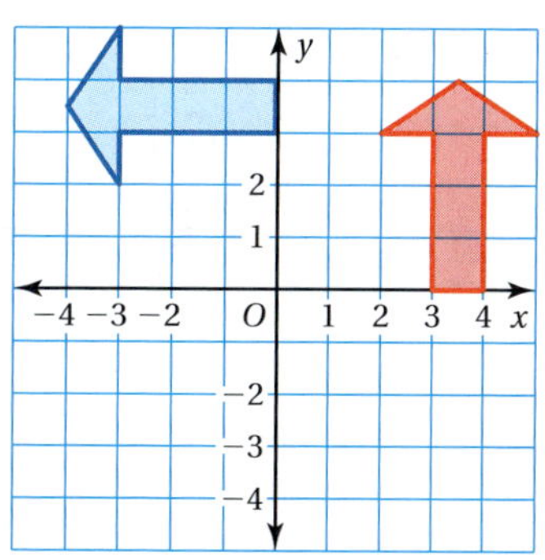

12.

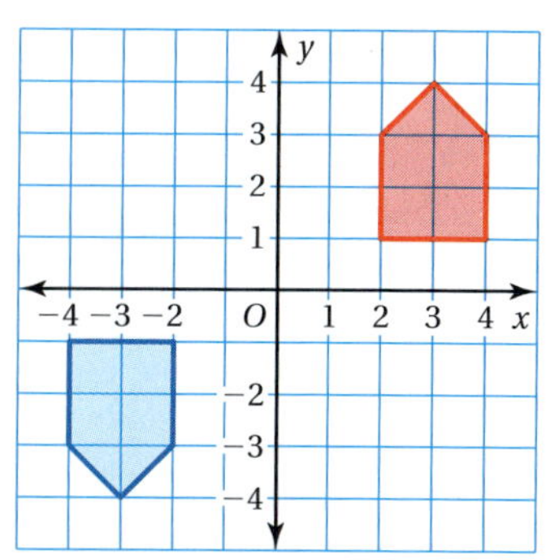

The vertices of a figure are given. Rotate the figure as described. Find the coordinates of the image.

2 3 **13.** $A(2, -2)$, $B(4, -1)$, $C(4, -3)$, $D(2, -4)$
90° counterclockwise about the origin

14. $F(1, 2)$, $G(3, 5)$, $H(3, 2)$
180° about the origin

15. $J(-4, 1)$, $K(-2, 1)$, $L(-4, -3)$
90° clockwise about vertex L

16. $P(-3, 4)$, $Q(-1, 4)$, $R(-2, 1)$, $S(-4, 1)$
180° about vertex R

17. $W(-6, -2)$, $X(-2, -2)$, $Y(-2, -6)$, $Z(-5, -6)$
270° counterclockwise about the origin

18. $A(1, -1)$, $B(5, -6)$, $C(1, -6)$
90° counterclockwise about vertex A

A figure has *rotational symmetry* if a rotation of 180° or less produces an image that fits exactly on the original figure. Explain why the figure has rotational symmetry.

19.

20.

21.

The vertices of a figure are given. Find the coordinates of the figure after the transformations given.

4 **22.** $R(-7, -5)$, $S(-1, -2)$, $T(-1, -5)$

Rotate 90° counterclockwise about the origin. Then translate 3 units left and 8 units up.

23. $J(-4, 4)$, $K(-3, 4)$, $L(-1, 1)$, $M(-4, 1)$

Reflect in the x-axis, and then rotate 180° about the origin.

The red figure is congruent to the blue figure. Describe two different sequences of transformations in which the blue figure is the image of the red figure.

5 **24.**

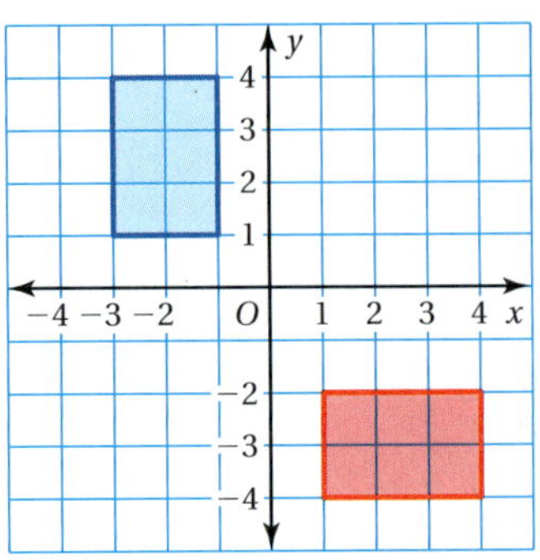

25.

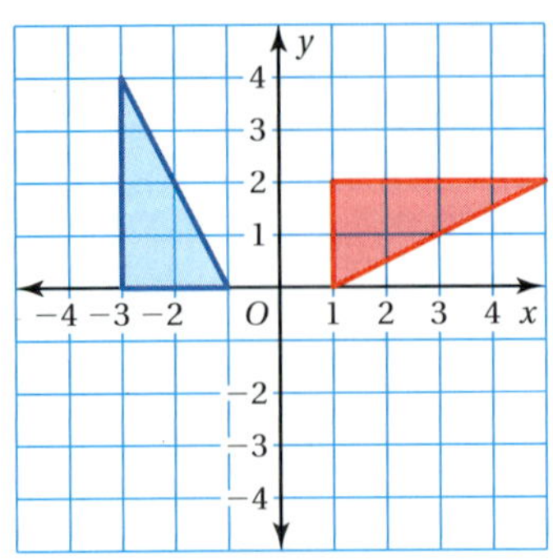

Common Errors

- **Exercises 13–18** Students may rotate the figure in the wrong direction. Remind them what clockwise and counterclockwise mean. It may be helpful for students to draw an arrow on a graph of the figure in the direction of rotation.

Practice and Problem Solving

16. $P'(-1, -2)$, $Q'(-3, -2)$, $R'(-2, 1)$, $S'(0, 1)$
17. $W'(-2, 6)$, $X'(-2, 2)$, $Y'(-6, 2)$, $Z'(-6, 5)$
18. $A'(1, -1)$, $B'(6, 3)$, $C'(6, -1)$
19. It only needs to rotate 120° to produce an identical image.
20. It only needs to rotate 90° to produce an identical image.
21. It only needs to rotate 180° to produce an identical image.
22. $R''(2, 1)$, $S''(-1, 7)$, $T''(2, 7)$
23. $J''(4, 4)$, $K''(3, 4)$, $L''(1, 1)$, $M''(4, 1)$
24. *Sample answer:* Rotate 90° counterclockwise about the origin and then translate 5 units left; Rotate 90° clockwise about the origin and then translate 1 unit right and 5 units up.
25. *Sample answer:* Rotate 180° about the origin and then rotate 90° clockwise about vertex $(-1, 0)$; Rotate 90° counterclockwise about the origin and then translate 1 unit left and 1 unit down.
26. **a.** $A'(6, 2)$, $B'(3, 2)$, $C'(1, 4)$, $D'(6, 4)$

    b. Reflect the trapezoid in the x-axis and then in the y-axis, or reflect the trapezoid in the y-axis and then in the x-axis.

English Language Learners

Kinesthetic

Give students examples of translations, reflections, and rotations in the classroom. You can move a chair to show a translation, use a mirror to show a reflection, and spin the chair around to show a rotation.

Practice and Problem Solving

27. See *Taking Math Deeper*.

28. See Additional Answers.

29. (2, 4), (4, 1), (1, 1)

Fair Game Review

30. no **31.** yes

32. yes **33.** no

34. B

Mini-Assessment

Tell whether the blue figure is a rotation of the red figure about the origin. If so, give the angle and direction of rotation.

1. yes; 90° clockwise rotation

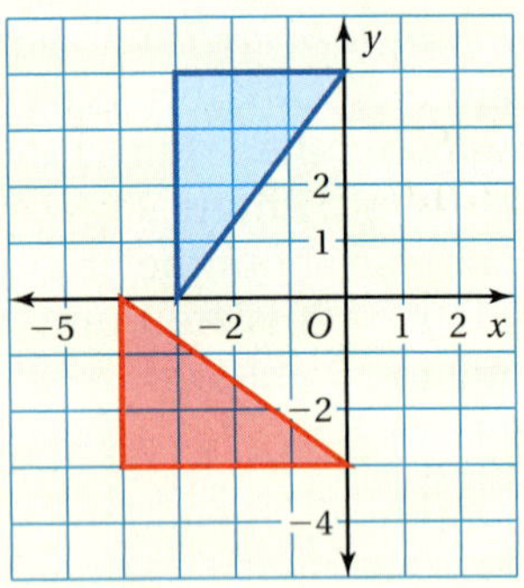

2. yes; 90° counterclockwise rotation

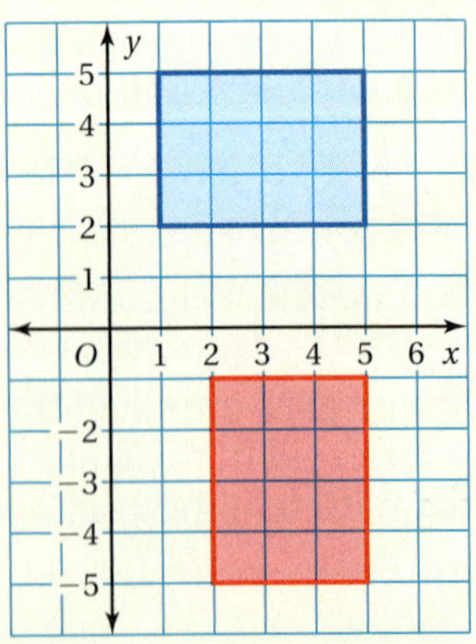

The vertices of a figure are given. Rotate the figure as described. Find the coordinates of the image.

3. $L(3, 2)$, $M(1, 1)$, $N(1, 5)$; 90° counterclockwise about the origin $L'(-2, 3)$, $M'(-1, 1)$, $N'(-5, 1)$

4. $T(2, 5)$, $U(5, 4)$, $V(6, 1)$, $W(2, 1)$; 270° clockwise about the origin $T'(-5, 2)$, $U'(-4, 5)$, $V'(-1, 6)$, $W'(-1, 2)$

Taking Math Deeper

Exercise 27

Students can use *Guess, Check, and Revise* to find a correct sequence of transformations. This is a good question for students to discuss in pairs or groups.

1 Do the rotations.

Original position

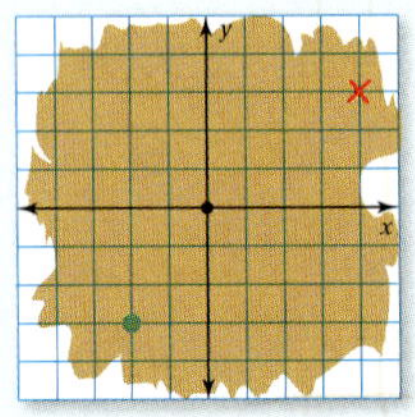

Rotate 180° about the origin.

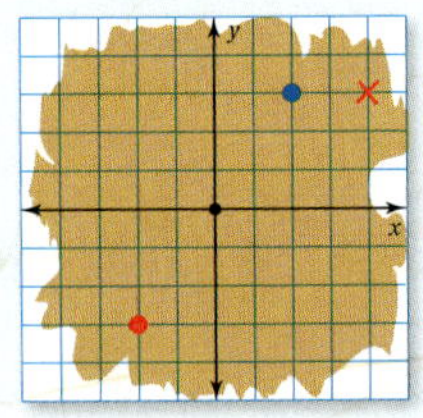

Rotate 90° counterclockwise about the origin.

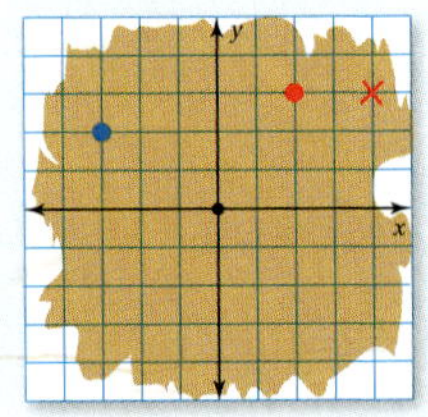

2 Do the reflection.

Position after rotations.

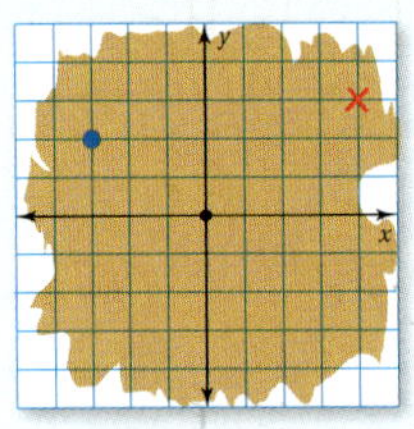

Reflect in *y*-axis.

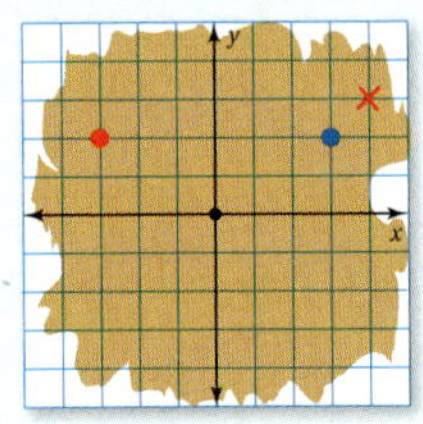

3 Do the translation.

Position after rotations and reflection.

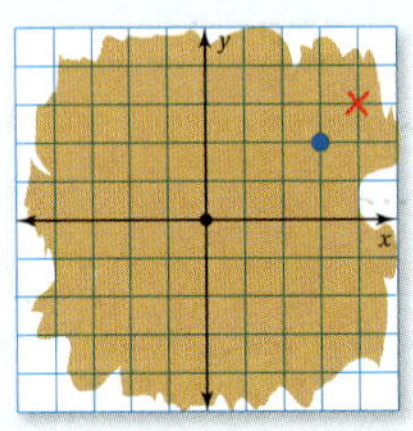

Translate 1 unit right and 1 unit up.

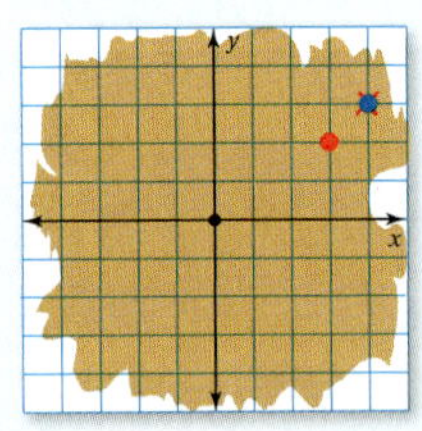

Reteaching and Enrichment Strategies

If students need help...	If students got it...
Resources by Chapter • Practice A and Practice B • Puzzle Time Record and Practice Journal Practice Differentiating the Lesson Lesson Tutorials Skills Review Handbook	Resources by Chapter • Enrichment and Extension • Technology Connection Start the next section

26. **REASONING** A trapezoid has vertices $A(-6, -2)$, $B(-3, -2)$, $C(-1, -4)$, and $D(-6, -4)$.

  a. Rotate the trapezoid 180° about the origin. What are the coordinates of the image?

  b. Describe a way to obtain the same image without using rotations.

27. **TREASURE MAP** You want to find the treasure located on the map at ×. You are located at ●. The following transformations will lead you to the treasure, but they are not in the correct order. Find the correct order. Use each transformation exactly once.

- Rotate 180° about the origin.
- Reflect in the y-axis.
- Rotate 90° counterclockwise about the origin.
- Translate 1 unit right and 1 unit up.

28. **CRITICAL THINKING** Consider $\triangle JKL$.

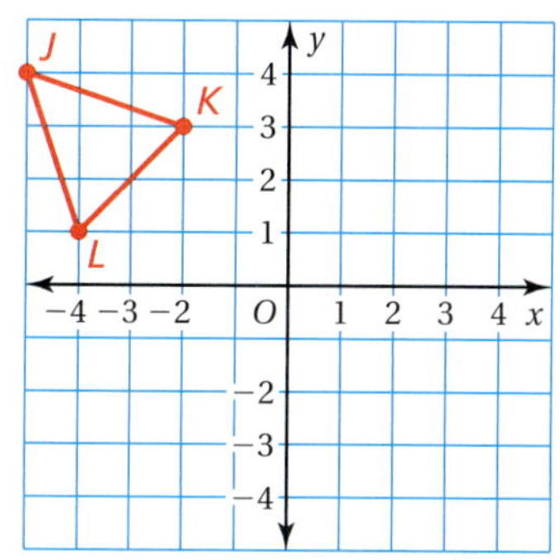

  a. Rotate $\triangle JKL$ 90° clockwise about the origin. How are the x- and y-coordinates of $\triangle J'K'L'$ related to the x- and y-coordinates of $\triangle JKL$?

  b. Rotate $\triangle JKL$ 180° about the origin. How are the x- and y-coordinates of $\triangle J'K'L'$ related to the x- and y-coordinates of $\triangle JKL$?

  c. Do you think your answers to parts (a) and (b) hold true for any figure? Explain.

29. **Reasoning** You rotate a triangle 90° counterclockwise about the origin. Then you translate its image 1 unit left and 2 units down. The vertices of the final image are $(-5, 0)$, $(-2, 2)$, and $(-2, -1)$. What are the vertices of the original triangle?

Fair Game Review What you learned in previous grades & lessons

Tell whether the ratios form a proportion. *(Section 5.2)*

30. $\frac{3}{5}, \frac{15}{20}$ 31. $\frac{2}{3}, \frac{12}{18}$ 32. $\frac{7}{28}, \frac{12}{48}$ 33. $\frac{54}{72}, \frac{36}{45}$

34. **MULTIPLE CHOICE** What is the solution of the equation $x + 6 \div 2 = 5$? *(Section 3.3)*

  (A) $x = -16$ (B) $x = 2$ (C) $x = 4$ (D) $x = 16$

11 Study Help

Check It Out
Graphic Organizer
BigIdeasMath.com

You can use a **summary triangle** to explain a concept. Here is an example of a summary triangle for translating a figure.

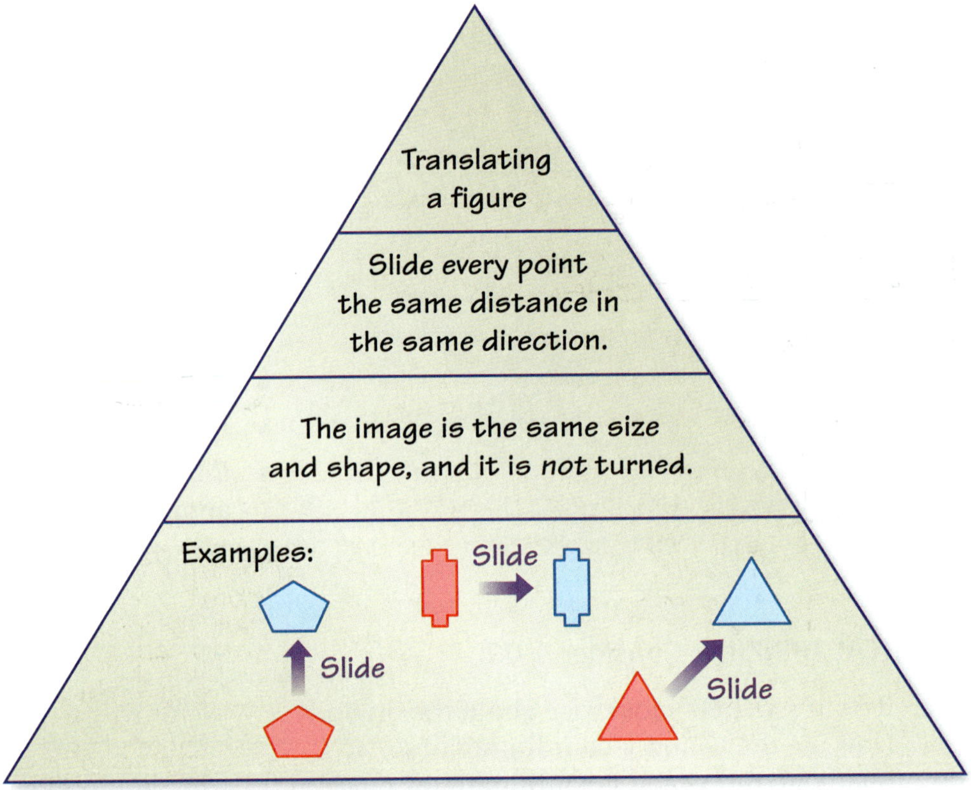

On Your Own

Make summary triangles to help you study these topics.

1. congruent figures
2. reflecting a figure
3. rotating a figure

After you complete this chapter, make summary triangles for the following topics.

4. similar figures
5. perimeters of similar figures
6. areas of similar figures
7. dilating a figure
8. transforming a figure

"I hope my owner sees my summary triangle. I just can't seem to learn 'roll over.' "

Sample Answers

1.

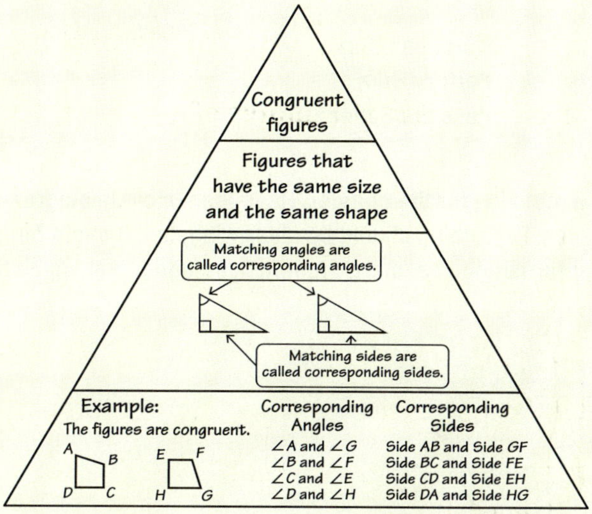

2.

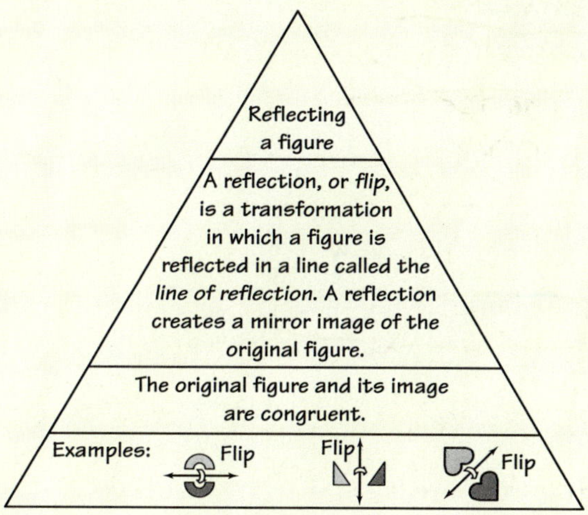

3.

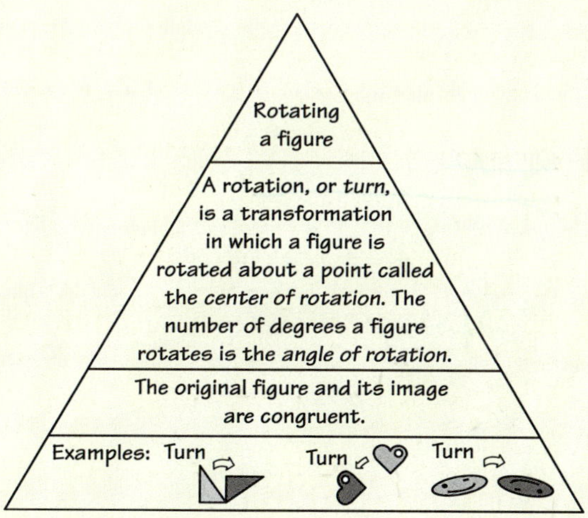

List of Organizers

Available at *BigIdeasMath.com*

Comparison Chart
Concept Circle
Example and Non-Example Chart
Formula Triangle
Four Square
Idea (Definition) and Examples Chart
Information Frame
Information Wheel
Notetaking Organizer
Process Diagram
Summary Triangle
Word Magnet
Y Chart

About this Organizer

A **Summary Triangle** can be used to explain a concept. Typically, the summary triangle is divided into 3 or 4 parts. In the top part, students write the concept being explained. In the middle part(s), students write any procedure, explanation, description, definition, theorem, and/or formula(s). In the bottom part, students write an example to illustrate the concept. A summary triangle can be used as an assessment tool, in which blanks are left for students to complete. Also, students can place their summary triangles on note cards to use as a quick study reference.

Technology for the Teacher

Editable Graphic Organizer

Answers

1. not congruent; Corresponding side lengths are not congruent.
2. congruent; Corresponding angles and side lengths are congruent.
3. no
4. yes
5. yes
6. no
7. *Sample answer:* rotate 90° clockwise about vertex $(-1, 1)$, translate 1 unit right and 1 unit down; rotate 270° counterclockwise about vertex $(-1, 1)$, translate 1 unit down and 1 unit right
8. *Sample answer:* rotate 180° clockwise about the origin, translate 1 unit right and 1 unit down; translate 1 unit left and 1 unit up, reflect in x-axis, reflect in y-axis
9. 6 units right and 4 units down
10. no; It will be 1 unit to the right of the hole.

Alternative Quiz Ideas

100% Quiz	**Math Log**
Error Notebook	Notebook Quiz
Group Quiz	Partner Quiz
Homework Quiz	Pass the Paper

Math Log

Ask students to keep a math log for the chapter. Have them include diagrams, definitions, and examples. Everything should be clearly labeled. It might be helpful if they put the information in a chart. Students can add to the log as they are introduced to new topics.

Reteaching and Enrichment Strategies

If students need help. . .	If students got it. . .
Resources by Chapter • Practice A and Practice B • Puzzle Time Lesson Tutorials *BigIdeasMath.com*	Resources by Chapter • Enrichment and Extension • Technology Connection Game Closet at *BigIdeasMath.com* Start the next section

Technology for the Teacher

Online Assessment
Assessment Book
ExamView® Assessment Suite

11.1–11.4 Quiz

Tell whether the two figures are congruent. Explain your reasoning. *(Section 11.1)*

1.

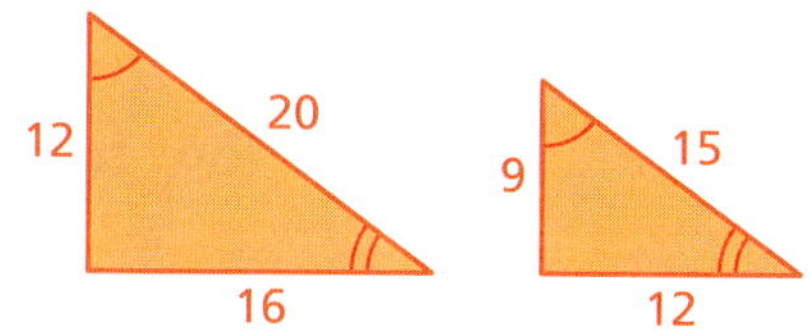

2.

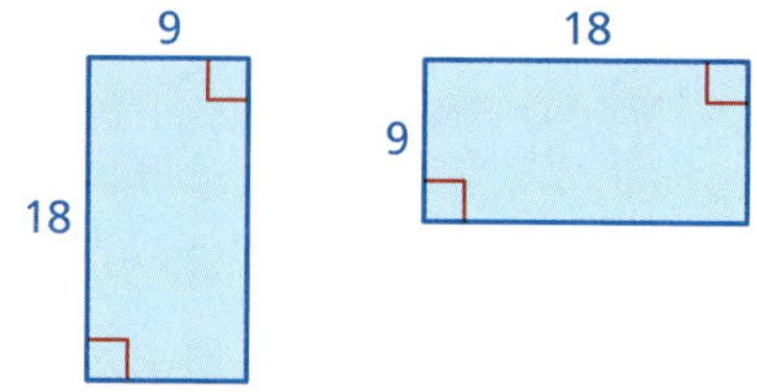

Tell whether the blue figure is a translation of the red figure. *(Section 11.2)*

3.

4.

Tell whether the blue figure is a reflection of the red figure. *(Section 11.3)*

5.

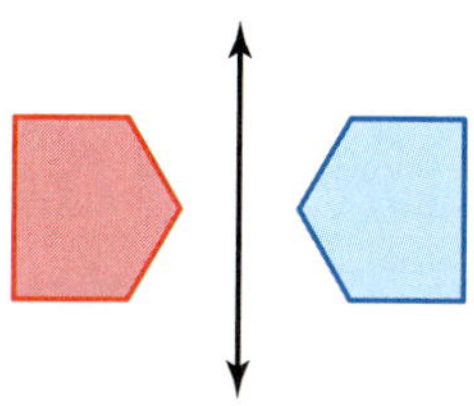

6.

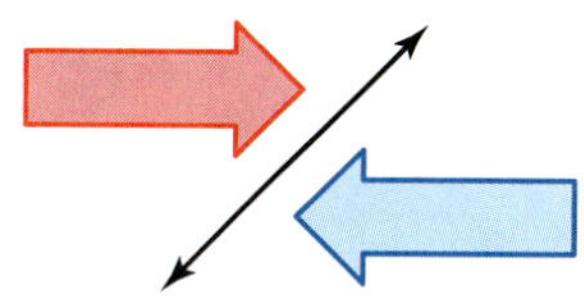

The red figure is congruent to the blue figure. Describe two different sequences of transformations in which the blue figure is the image of the red figure. *(Section 11.4)*

7.

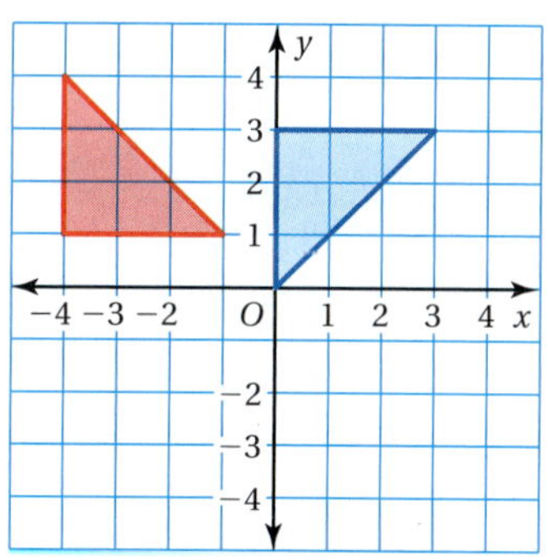

8.

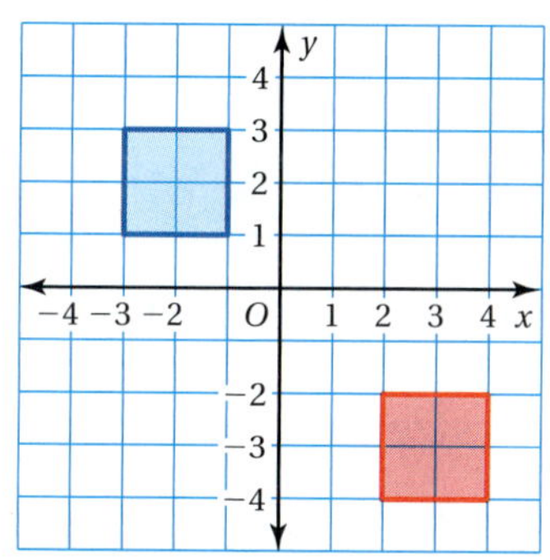

9. AIRPLANE Describe a translation of the airplane from point A to point B. *(Section 11.2)*

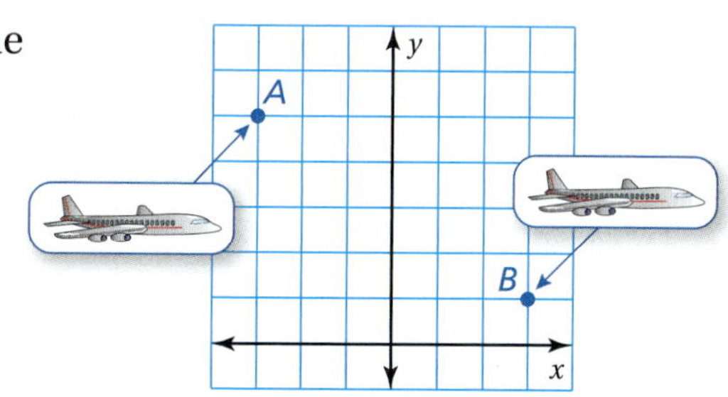

10. MINIGOLF You hit the golf ball along the red path so that its image will be a reflection in the y-axis. Does the golf ball land in the hole? Explain. *(Section 11.3)*

11.5 Similar Figures

Essential Question How can you use proportions to help make decisions in art, design, and magazine layouts?

Original photograph

In a computer art program, when you click and drag on a side of a photograph, you distort it.

But when you click and drag on a corner of the photograph, the dimensions remain proportional to the original.

Distorted

Distorted

Proportional

1 ACTIVITY: Reducing Photographs

Work with a partner. You are trying to reduce the photograph to the indicated size for a nature magazine. Can you reduce the photograph to the indicated size without distorting or cropping? Explain your reasoning.

a.

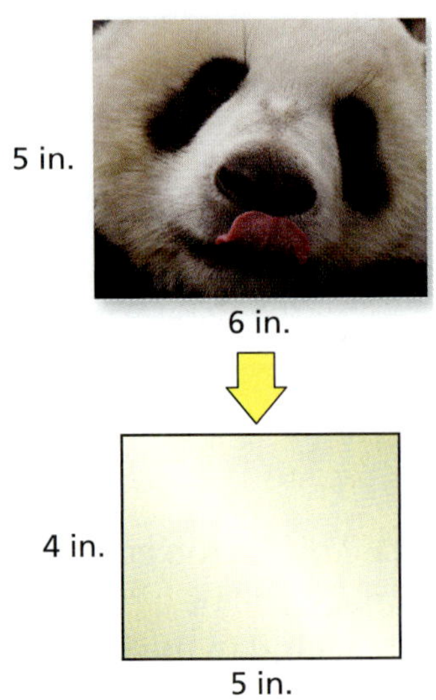

b.

Geometry

In this lesson, you will

- name corresponding angles and corresponding sides of similar figures.
- identify similar figures.
- find unknown measures of similar figures.

Laurie's Notes

Introduction

Applying Mathematical Practices

- **Attend to Precision:** Students should use precise vocabulary to communicate their thinking to others. In working with similar figures, they should be able to state which ratios are in proportion and why.

Motivate

- Draw a simple stick figure or other image on a stretchable surface, such as a balloon, physical therapy elastic, or play putty.
- ? Ask students what they think will happen to the figure when you pull the picture to the right. Students should recognize that the image will be distorted. Pull one of the sides of the picture to confirm.
- Pull the top of the picture so that students see this result as the same.
- ? Ask students what they think will happen if the stretchable surface is pulled in both directions (right and up). Students should recognize that the image will enlarge proportionally.
- **Alternative:** If you can display a computer image to the class, you can drag a side to distort the image, or drag a corner to change the size of the image proportionally.

Activity Notes

Activity 1

- Photography is a good context to use to examine similarity. A common misconception is that standard photo sizes are proportional and, in fact, most are not. A 5″ × 7″ photo and a 4″ × 6″ photo are not proportional.
- **Common Misconception:** Students often believe that if you subtract the same amount from each dimension, the resulting ratio will be proportional to the first. For example, $\frac{5}{7} \neq \frac{5-1}{7-1} = \frac{4}{6}$.
- ? Ask questions about proportions and ratios:
  - "What is a proportion?" two equal ratios
  - "Are the two ratios 2 : 3 and 4 : 6 equal?" yes
  - "Are the ratios 2 : 3 and 8 : 9 equal? Explain." No, listen for students to get at the idea that $2 \times 4 = 8$, but 3×4 is 12, not 9.
- Remember, students have *not* learned a formal definition for similar figures. Remind students that the task is to decide if the photograph can be reduced to the new dimensions without distorting it. Therefore, students must use the information about keeping the side lengths proportional.

Words of Wisdom

- **Attend to Precision:** Listen to how students describe their proportions. There are many correct ways to set up a proportion and some students might hear one way and incorrectly think their way is wrong.
- ? Ask students, "Did anyone set up their proportions differently?" Here are two possibilities. The key is to make sure *like things* are being compared.

$$\frac{\text{length (original)}}{\text{width (original)}} = \frac{\text{length (new size)}}{\text{width (new size)}};\ \frac{\text{length (original)}}{\text{length (new size)}} = \frac{\text{width (original)}}{\text{width (new size)}}$$

What Your Students Will Learn

- Name corresponding angles and corresponding sides of similar figures.
- Identify similar figures using corresponding angles or corresponding sides.
- Find unknown measures of similar figures by writing proportions and using the Cross Products Property.

Previous Learning

Students should know how to write ratios and have a basic understanding of proportions.

Technology for the Teacher

Dynamic Classroom

Lesson Plans
Complete Materials List

11.5 Record and Practice Journal

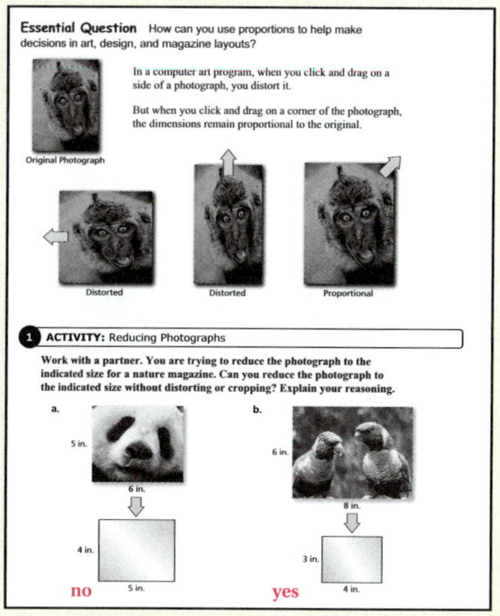

Essential Question How can you use proportions to help make decisions in art, design, and magazine layouts?

In a computer art program, when you click and drag on a side of a photograph, you distort it.

But when you click and drag on a corner of the photograph, the dimensions remain proportional to the original.

Original Photograph

Distorted | Distorted | Proportional

1 ACTIVITY: Reducing Photographs

Work with a partner. You are trying to reduce the photograph to the indicated size for a nature magazine. Can you reduce the photograph to the indicated size without distorting or cropping? Explain your reasoning.

a. 5 in. × 6 in. → 4 in. × 5 in. no

b. 6 in. × 8 in. → 3 in. × 4 in. yes

English Language Learners

Vocabulary

Ask students what *similar* means. Ask them if *similar* things are exactly alike. Explain that *similar figures* are not exactly alike. Similar figures have the same shape, but not necessarily the same size.

11.5 Record and Practice Journal

2 ACTIVITY: Creating Designs

Work with a partner.

a. Tell whether the dimensions of the new designs are proportional to the dimensions of the original design. Explain your reasoning.

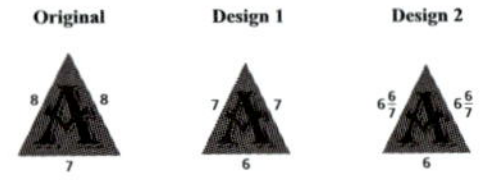

Design 2 is proportional.
Design 1 is not proportional.

b. Draw two designs whose dimensions are proportional to the given design. Make one bigger and one smaller. Label the sides of the designs with their lengths.

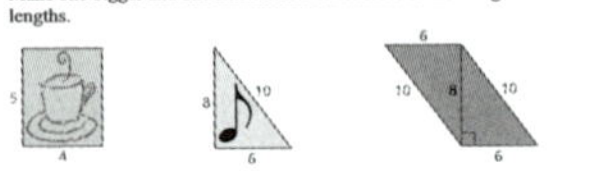

Check students' work.

What Is Your Answer?

3. **IN YOUR OWN WORDS** How can you use proportions to help make decisions in art, design, and magazine layouts? Give two examples.

Check students' work.

4. a. Use a computer art program to draw two rectangles whose dimensions are proportional to each other.

b. Print the two rectangles on the same piece of paper.

c. Use a centimeter ruler to measure the length and the width of each rectangle. Record your measurements here.

"I love this statue. It seems similar to a big statue I saw in New York."

a–c. Check students' work.

d. Find the following ratios. What can you conclude?

$\frac{\text{Length of larger}}{\text{Length of smaller}}$ $\frac{\text{Width of larger}}{\text{Width of smaller}}$

They are proportional.

Laurie's Notes

Activity 2

- This activity is similar to the first activity. In Activity 1, students were asked whether the original figure would become distorted. In this activity, students are asked whether the dimensions of the designs are proportional.

? "What type of triangle is the original design?" Isosceles

? "Triangles don't have a *length* and *width* as rectangles do. What dimensions will you compare to decide whether the triangular designs are proportional?" Listen for language such as base and sides or base and legs.

- **Common Error:** Students think that subtracting 1 from each side length of the triangle produces a new triangle proportional to the original triangle.
- In completing part (b), students are not expected to make a scale drawing. They are drawing a figure that should look similar to the original. The rectangle should not look like a square. The right scalene triangle should not look equilateral.
- **Construct Viable Arguments and Critique the Reasoning of Others** and **Attend to Precision:** Have students share the dimensions of their new figures. They should explain how they came up with the new dimensions. Listen for methods that use multiplication, not addition.

What Is Your Answer?

- **Technology:** Question 4 provides a great opportunity to have students work with a computer art program to enhance their understanding of similar figures.

Closure

- Are all three of these triangles proportional? Yes.

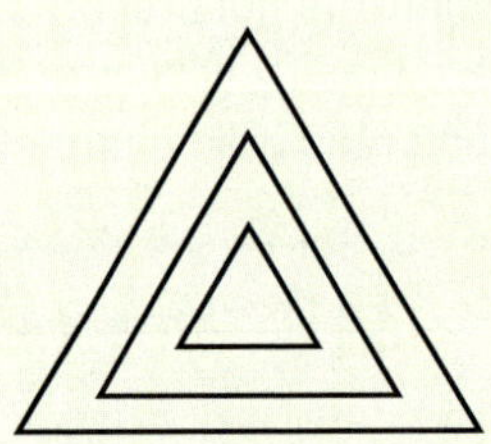

2 ACTIVITY: Creating Designs

Work with a partner.

a. Tell whether the dimensions of the new designs are proportional to the dimensions of the original design. Explain your reasoning.

b. Draw two designs whose dimensions are proportional to the given design. Make one bigger and one smaller. Label the sides of the designs with their lengths.

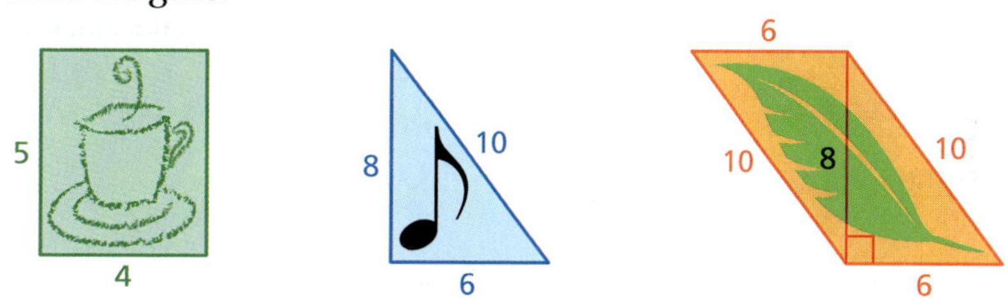

> **Math Practice**
>
> **Analyze Relationships**
>
> How can you use mathematics to determine whether the dimensions are proportional?

What Is Your Answer?

3. **IN YOUR OWN WORDS** How can you use proportions to help make decisions in art, design, and magazine layouts? Give two examples.

4. **a.** Use a computer art program to draw two rectangles whose dimensions are proportional to each other.

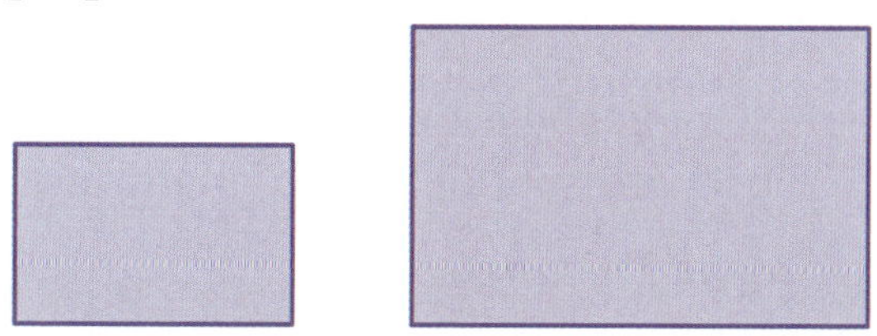

b. Print the two rectangles on the same piece of paper.

c. Use a centimeter ruler to measure the length and the width of each rectangle.

d. Find the following ratios. What can you conclude?

$$\frac{\text{Length of larger}}{\text{Length of smaller}} \qquad \frac{\text{Width of larger}}{\text{Width of smaller}}$$

"I love this statue. It seems similar to a big statue I saw in New York."

Practice Use what you learned about similar figures to complete Exercises 4 and 5 on page 498.

11.5 Lesson

Check It Out
Lesson Tutorials
BigIdeasMath.com

Key Vocabulary
similar figures, *p. 496*

Key Idea

Similar Figures

Figures that have the same shape but not necessarily the same size are called **similar figures**.

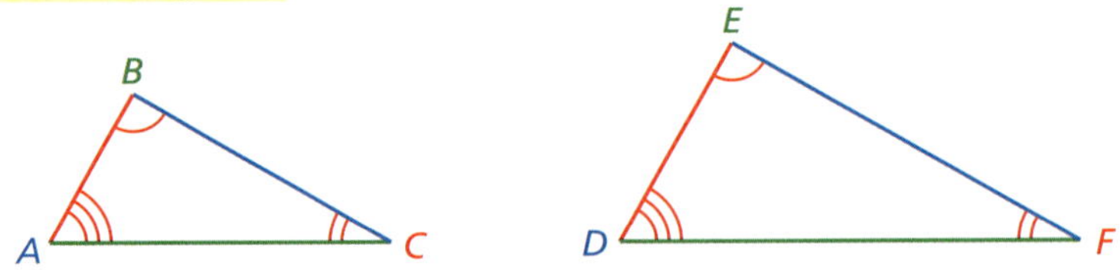

Triangle *ABC* is similar to Triangle *DEF*.

Words Two figures are similar when

- corresponding side lengths are proportional and
- corresponding angles are congruent.

Symbols

Side Lengths	*Angles*	*Figures*
$\frac{AB}{DE} = \frac{BC}{EF} = \frac{AC}{DF}$	$\angle A \cong \angle D$ $\angle B \cong \angle E$ $\angle C \cong \angle F$	$\triangle ABC \sim \triangle DEF$

Reading

The symbol ~ means *is similar to*.

Common Error

When writing a similarity statement, make sure to list the vertices of the figures in the correct order.

EXAMPLE 1 Identifying Similar Figures

Which rectangle is similar to Rectangle A?

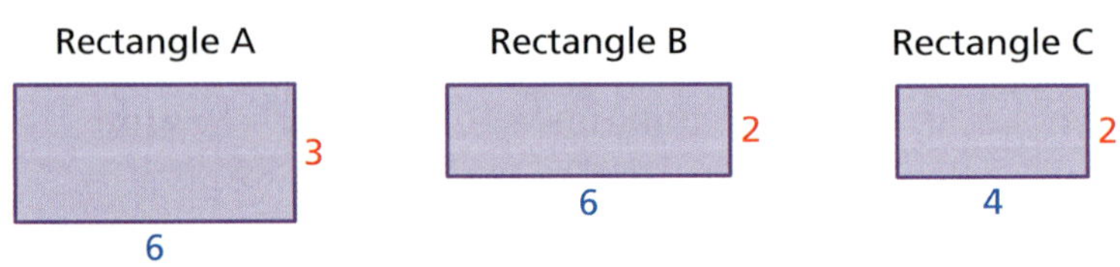

Each figure is a rectangle. So, corresponding angles are congruent. Check to see if corresponding side lengths are proportional.

Rectangle A and Rectangle B

$\frac{\text{Length of A}}{\text{Length of B}} = \frac{6}{6} = 1$ $\quad$ $\frac{\text{Width of A}}{\text{Width of B}} = \frac{3}{2}$ $\quad$ Not proportional

Rectangle A and Rectangle C

$\frac{\text{Length of A}}{\text{Length of C}} = \frac{6}{4} = \frac{3}{2}$ $\quad$ $\frac{\text{Width of A}}{\text{Width of C}} = \frac{3}{2}$ $\quad$ Proportional

So, Rectangle C is similar to Rectangle A.

On Your Own

Now You're Ready
Exercises 4–7

1. Rectangle D is 3 units long and 1 unit wide. Which rectangle is similar to Rectangle D?

Laurie's Notes

Introduction

Connect

- **Yesterday:** Students developed an intuitive understanding about proportional polygons.
- **Today:** Students will use the formal definition of similar figures.

Motivate

- Place an item on an overhead projector, such as an index card, school ID, or other rectangular item. Ask questions about the actual item and its projected image.

? "How does the actual item compare to its projection?" Listen for: "They look alike," "They have the same shape," or "They're similar." It is unlikely that they know the mathematical definition of similar.

- Place a different-shaped item on the overhead.

? "There are two items and two projected images. Which projection goes with which item? How do you know?" Listen for students to say the items are the same shape but different sizes.

Lesson Notes

Key Idea

- Discuss the tilde symbol ~ that denotes similarity. Explain that the order in which the vertices of the triangle are written identifies how the sides and angles correspond.
- Remind students that *congruent* angles have the same measure.
- **Big Idea:** Discuss the need for two conditions to be met for two figures to be similar: corresponding side lengths are proportional *and* corresponding angles are congruent.
- **Representation:** Point out the color-coding, which should help students see the corresponding parts.
- Take your time in this section. There is a great deal of vocabulary, symbols, representations, *and* the fundamental concept of similarity. Give students time to ask questions and think about all that is being presented.

Example 1

? "What do you know about the angles of a rectangle?" 4 right angles

? "Are the corresponding angles congruent?" Yes.

? "What else must you check to know that the rectangles are similar?" corresponding side lengths are proportional

- Note that the problem has students focus on the dimensions of the rectangles, using the words *length* and *width,* without using the side names that can confuse students.

On Your Own

- Check that students correctly identify Rectangle B.

Goal Today's lesson is using proportions to determine if two figures are similar.

Technology for the Teacher

Dynamic Classroom

Lesson Tutorials
Lesson Plans
Answer Presentation Tool

Extra Example 1

Which parallelogram is similar to Parallelogram A?

Parallelogram A

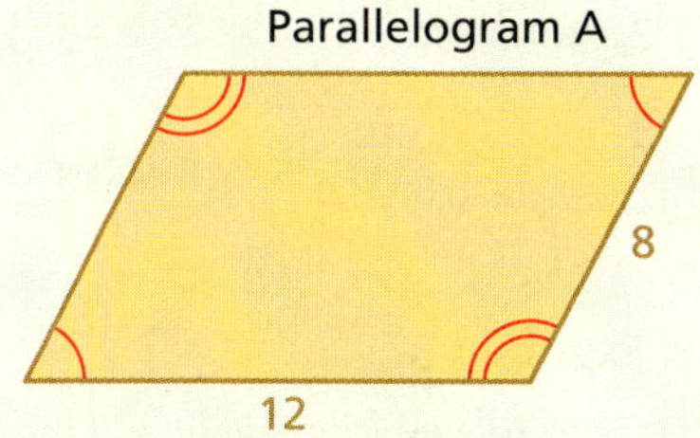

Parallelogram B

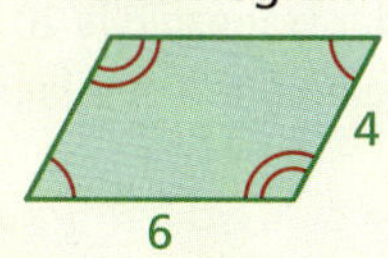

Parallelogram C

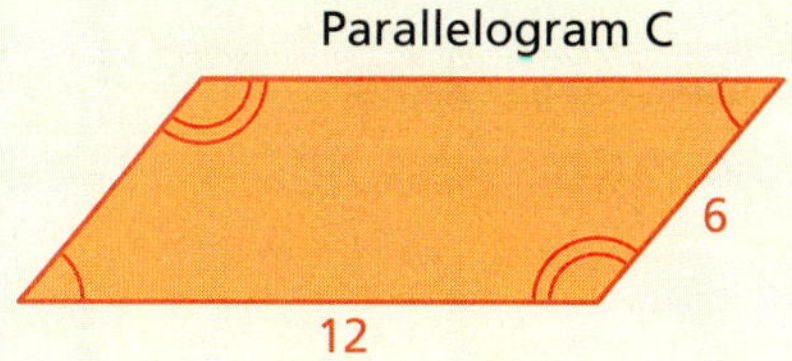

Parallelogram B

On Your Own

1. Rectangle B

Differentiated Instruction

Visual

Bring in examples of figures that are similar and figures that have the same size and shape. Ask students to identify the figures that have the same size and shape (congruent). Then ask students to identify the figures that have the same shape (similar).

Extra Example 2

The triangles are similar. Find *x*.

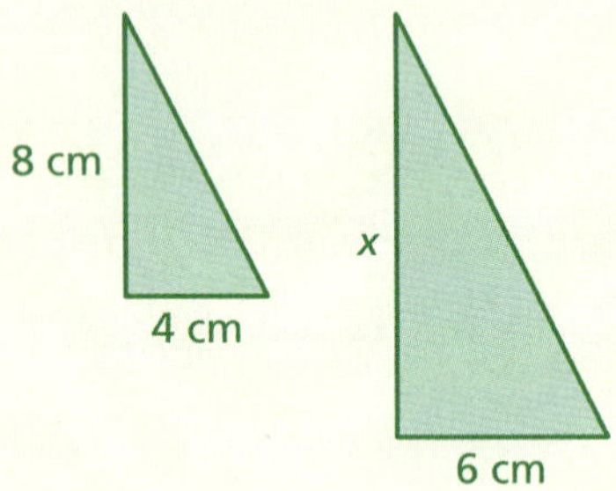

12 cm

On Your Own

2. $x = 4$ ft
3. $x = 24$ cm

Extra Example 3

The artist draws a larger replica of the painting in Example 3. The shorter base of the similar trapezoid is 10 inches. What is the height *h* of this trapezoid?
8 inches

On Your Own

4. 18 in.

Laurie's Notes

Example 2

- Draw the two triangles and state that they are similar.
- ? "Because the triangles are similar, what do you know?" Corresponding side lengths are proportional and corresponding angles are congruent.
- ? "The small triangle has a side that is 6 meters long. What is the length of the corresponding side in the large triangle?" 9 meters
- ? "The small triangle has a side that is 8 meters long. What is the labeled length of the corresponding side in the large triangle?" *x* meters
- Set up and solve the proportion.
- **Attend to Precision:** State the solution with the correct units.
- ? "Did anyone solve the proportion differently?" Some may say that because you add half of 6 to 6 to get 9, you can add half of 8 to 8 to get 12. Others may say that you can simplify the ratio $\frac{6}{8}$ as $\frac{3}{4}$ and then use mental math to solve for the missing side.
- **Connection:** Consider the larger triangle as a scale model (enlarged) of the smaller triangle. Ask students to find the scale factor. 1.5 : 1 or 3 : 2

On Your Own

- **Neighbor Check:** Have students work independently and then have their neighbors check their work. Have students discuss any discrepancies.

Example 3

- Share some information about the Berlin Wall.
  - It was built to keep East Germans from escaping to West Germany.
  - East German crews began tearing up streets and spreading barbed wire at midnight on August 13, 1961. By morning, the border with West Germany was closed. This is how the Berlin Wall began.
  - The wall was over a hundred miles long and went through four major changes. The final version of the Berlin Wall was 12 feet tall.
  - The Berlin Wall symbolized the boundary between communism and democracy until it was torn down in 1989.
  - Thierry Noir is one of the artists who turned the Berlin wall into the world's longest painting canvas. He wanted to "demystify" the wall with his bright-colored paintings.
- Draw the two trapezoids, one representing the painting and one representing the replica. Label the known dimensions.
- ? "The trapezoid in the actual painting has a side of length 15 inches. What is the length of the corresponding side in the replica?" 3.75 inches
- ? "You are to find the height *h* of the trapezoid in the replica. What is the length of the corresponding side in the actual painting?" 12 inches
- Set up and solve the proportion.
- **Attend to Precision:** State the solution with the correct units.

Closure

- Sketch two figures that look similar. Describe how you would determine whether your figures are actually similar.

EXAMPLE 2 Finding an Unknown Measure in Similar Figures

The triangles are similar. Find x.

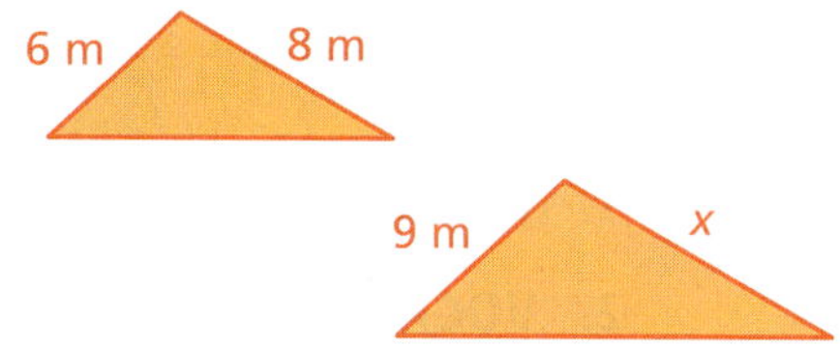

Because the triangles are similar, corresponding side lengths are proportional. So, write and solve a proportion to find x.

$$\frac{6}{9} = \frac{8}{x} \qquad \text{Write a proportion.}$$

$$6x = 72 \qquad \text{Cross Products Property}$$

$$x = 12 \qquad \text{Divide each side by 6.}$$

So, x is 12 meters.

On Your Own

Now You're Ready
Exercises 8–11

The figures are similar. Find x.

2.

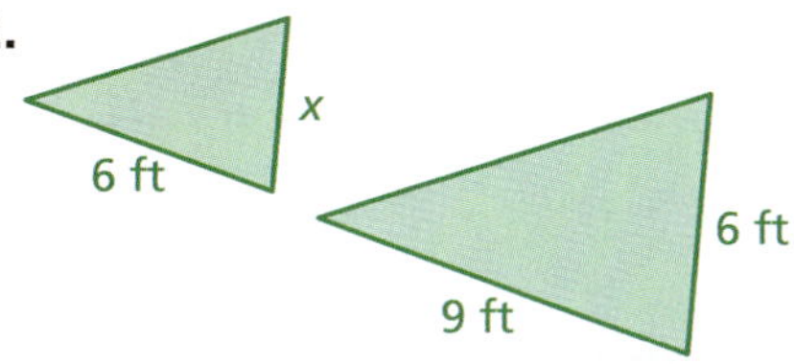

3.

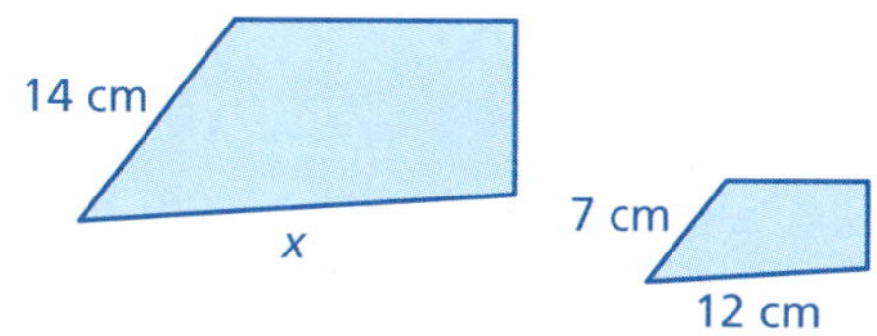

EXAMPLE 3 Real-Life Application

An artist draws a replica of a painting that is on the Berlin Wall. The painting includes a red trapezoid. The shorter base of the similar trapezoid in the replica is 3.75 inches. What is the height h of the trapezoid in the replica?

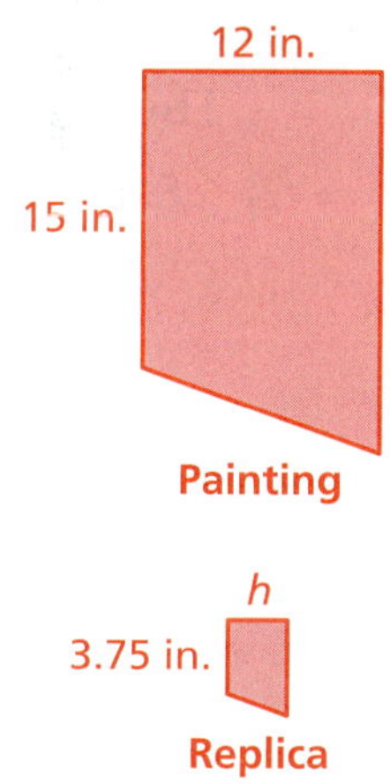

Because the trapezoids are similar, corresponding side lengths are proportional. So, write and solve a proportion to find h.

$$\frac{3.75}{15} = \frac{h}{12} \qquad \text{Write a proportion.}$$

$$12 \cdot \frac{3.75}{15} = 12 \cdot \frac{h}{12} \qquad \text{Multiplication Property of Equality}$$

$$3 = h \qquad \text{Simplify.}$$

So, the height of the trapezoid in the replica is 3 inches.

On Your Own

4. **WHAT IF?** The longer base in the replica is 4.5 inches. What is the length of the longer base in the painting?

11.5 Exercises

Check It Out
Help with Homework
BigIdeasMath.com

Vocabulary and Concept Check

1. **VOCABULARY** How are corresponding angles of two similar figures related?
2. **VOCABULARY** How are corresponding side lengths of two similar figures related?
3. **CRITICAL THINKING** Are two figures that have the same size and shape similar? Explain.

Practice and Problem Solving

Tell whether the two figures are similar. Explain your reasoning.

1 4.

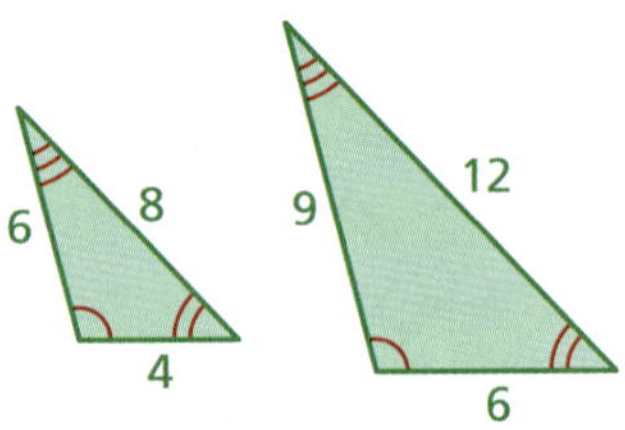

5.

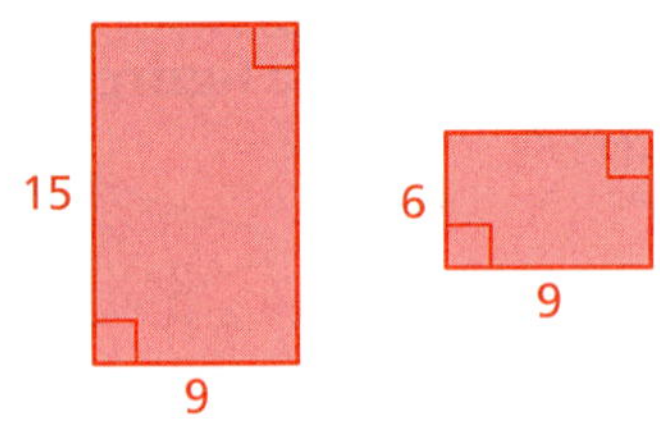

In a coordinate plane, draw the figures with the given vertices. Which figures are similar? Explain your reasoning.

6. Rectangle A: (0, 0), (4, 0), (4, 2), (0, 2)
   Rectangle B: (0, 0), (−6, 0), (−6, 3), (0, 3)
   Rectangle C: (0, 0), (4, 0), (4, 2), (0, 2)

7. Figure A: (−4, 2), (−2, 2), (−2, 0), (−4, 0)
   Figure B: (1, 4), (4, 4), (4, 1), (1, 1)
   Figure C: (2, −1), (5, −1), (5, −3), (2, −3)

The figures are similar. Find x.

2 8.

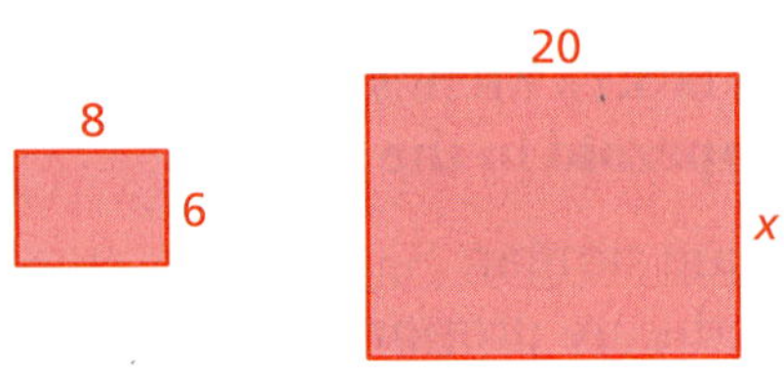

9.

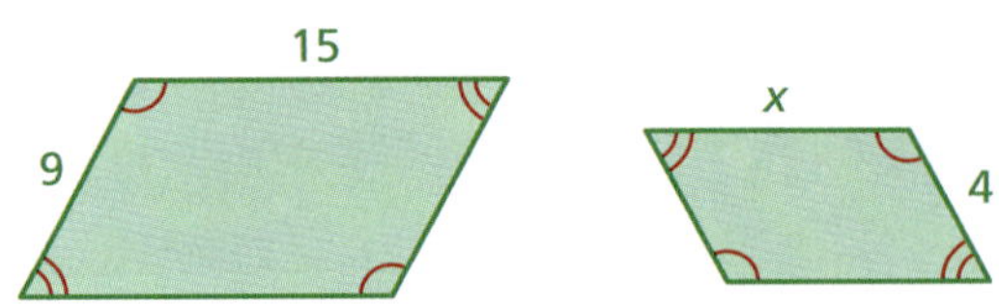

10.

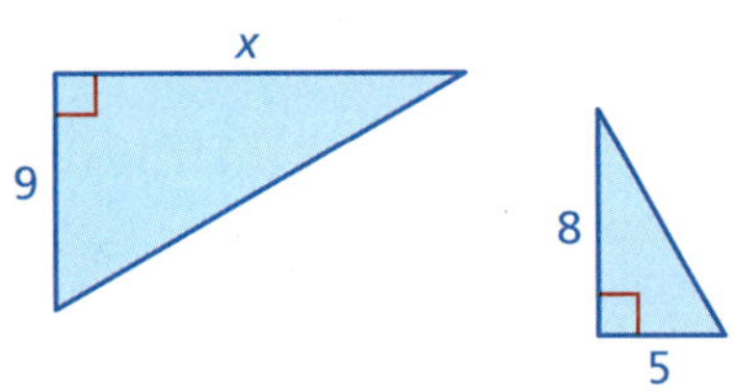

11.

12. **MEXICO** A Mexican flag is 63 inches long and 36 inches wide. Is the drawing at the right similar to the Mexican flag?

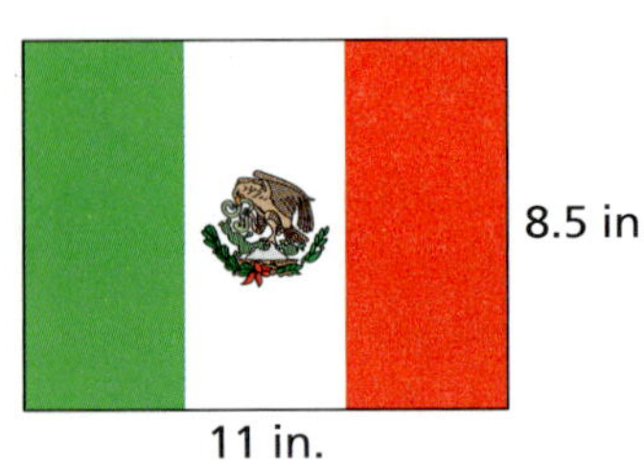

13. **DESKS** A student's rectangular desk is 30 inches long and 18 inches wide. The teacher's desk is similar to the student's desk and has a length of 50 inches. What is the width of the teacher's desk?

Assignment Guide and Homework Check

Level	Assignment	Homework Check
Accelerated	1–5, 6–12 even, 13–25	6, 10, 16, 18, 20

Common Errors

- **Exercise 4** Students may think that the triangles are not similar because an optical illusion makes them appear to have different angles. Remind them to pay attention to the congruency markings on the figure.
- **Exercise 5** Students may be confused because the orientations of the rectangles are different. Tell them it is helpful to redraw the figures with the same orientations.
- **Exercises 8–11** Students may form the wrong proportions. Remind them to make sure like things are being compared.

11.5 Record and Practice Journal

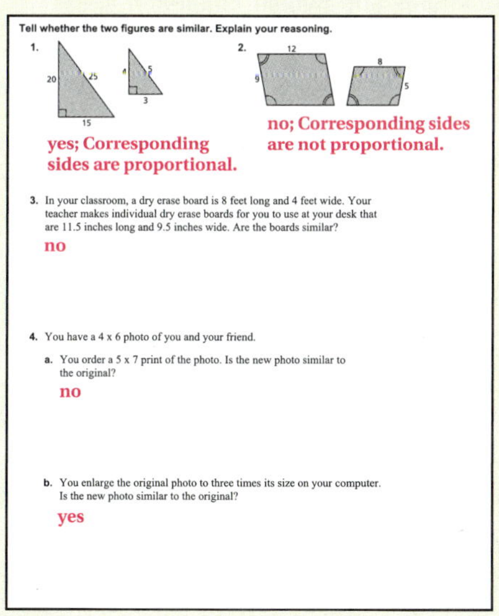
Tell whether the two figures are similar. Explain your reasoning.

1. yes; Corresponding sides are proportional.

2. no; Corresponding sides are not proportional.

3. In your classroom, a dry erase board is 8 feet long and 4 feet wide. Your teacher makes individual dry erase boards for you to use at your desk that are 11.5 inches long and 9.5 inches wide. Are the boards similar?
   no

4. You have a 4 x 6 photo of you and your friend.
   a. You order a 5 x 7 print of the photo. Is the new photo similar to the original?
   no
   b. You enlarge the original photo to three times its size on your computer. Is the new photo similar to the original?
   yes

Vocabulary and Concept Check

1. They are congruent.
2. They are proportional.
3. Yes, because the angles are congruent and the side lengths are proportional.

Practice and Problem Solving

4. similar; Corresponding angles are congruent. Because $\frac{4}{6} = \frac{6}{9} = \frac{8}{12}$, the corresponding side lengths are proportional.
5. not similar; Corresponding side lengths are not proportional.

6–7. See Additional Answers.

8. 15 **9.** $6\frac{2}{3}$

10. 14.4 **11.** 14

12. no **13.** 30 in.

14. **a.** sometimes; They are similar only when corresponding side lengths are proportional and corresponding angles are congruent.

b. always; All angles are congruent and all sides are proportional.

c. sometimes; Corresponding angles are always congruent, but corresponding side lengths are not always proportional.

d. never; They do not have the same shape.

15. See *Taking Math Deeper.*

16. **a.** yes **b.** no

17. 3 times

Practice and Problem Solving

18. yes; A scale drawing is a proportional drawing of an object, so corresponding angles are congruent and corresponding side lengths are proportional.

19–20. See Additional Answers.

Fair Game Review

21. $\frac{16}{81}$ **22.** $\frac{9}{64}$

23. $\frac{49}{16}$ **24.** $\frac{169}{16}$

25. C

Mini-Assessment

1.

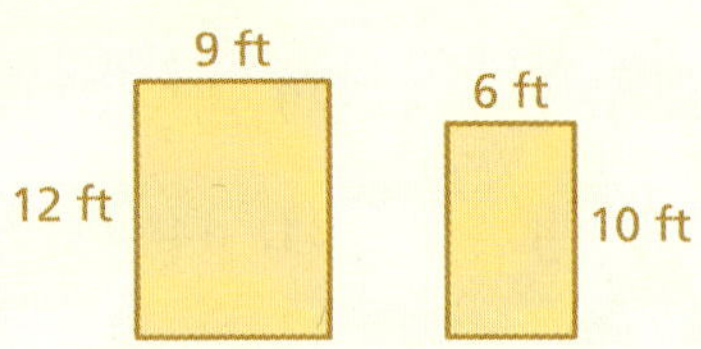

no; corresponding side lengths are not proportional

2. Are the two triangular stickers similar? Explain your reasoning.

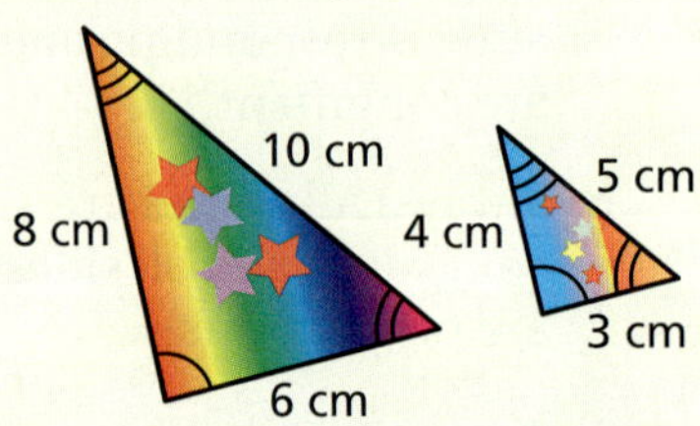

yes; corresponding side lengths are proportional and corresponding angles are congruent

3. The figures are similar. Find *x*.

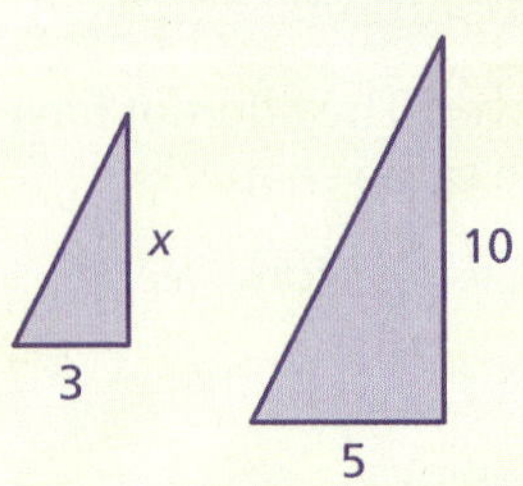

6

Taking Math Deeper

Exercise 15

Students may have difficulty finding an entry point when solving this problem. They could begin by reasoning about properties of different quadrilaterals.

1 Decide which type(s) of quadrilateral(s) to draw.

The quadrilaterals must each have two 130° angles and two 50° angles. So, you can eliminate rectangles because they can only have 90° angles.

One way to be sure that two quadrilaterals will not be similar is to use two quadrilaterals that have different shapes, such as trapezoids and parallelograms.

2 Construct a trapezoid and a parallelogram with the four given interior angle measures.

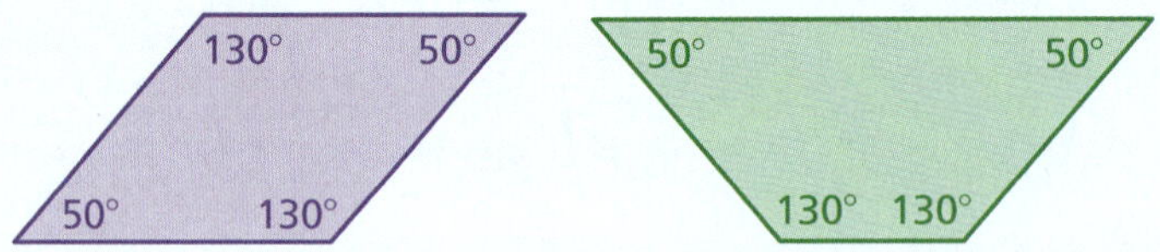

3 Students should pay close attention to orientation and size in problems like this. They may think that the two parallelograms below would justify their reasoning. However, the parallelograms are similar.

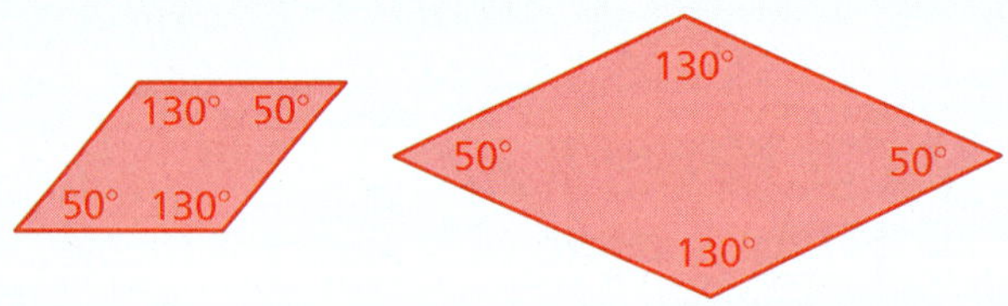

Project

Ask students if they can draw *three* different quadrilaterals each having two 100° angles and two 80° angles.

Reteaching and Enrichment Strategies

If students need help...	If students got it...
Resources by Chapter • Practice A and Practice B • Puzzle Time Record and Practice Journal Practice Differentiating the Lesson Lesson Tutorials Skills Review Handbook	Resources by Chapter • Enrichment and Extension • Technology Connection Start the next section

14. **LOGIC** Are the following figures *always, sometimes,* or *never* similar? Explain.

   a. two triangles
   b. two squares
   c. two rectangles
   d. a square and a triangle

15. **CRITICAL THINKING** Can you draw two quadrilaterals each having two 130° angles and two 50° angles that are *not* similar? Justify your answer.

16. **SIGN** All the angle measures in the sign are 90°.

   a. You increase each side length by 20%. Is the new sign similar to the original?
   b. You increase each side length by 6 inches. Is the new sign similar to the original?

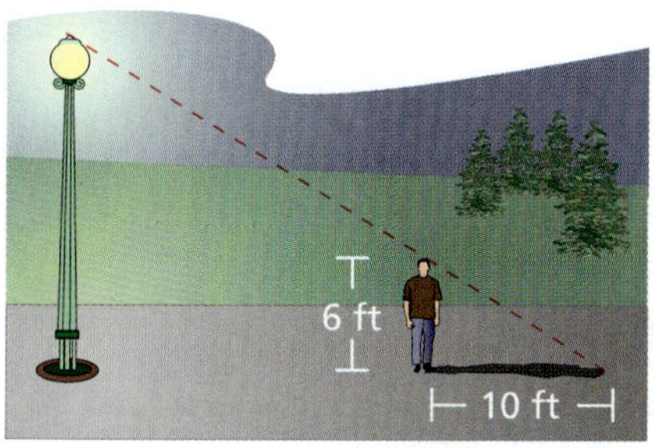

17. **STREETLIGHT** A person standing 20 feet from a streetlight casts a shadow as shown. How many times taller is the streetlight than the person? Assume the triangles are similar.

18. **REASONING** Is an object similar to a scale drawing of the object? Explain.

19. **GEOMETRY** Use a ruler to draw two different isosceles triangles similar to the one shown. Measure the heights of each triangle to the nearest centimeter.

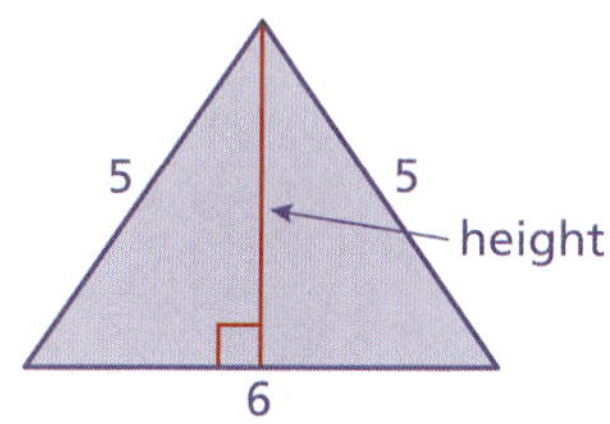

   a. Is the ratio of the corresponding heights proportional to the ratio of the corresponding side lengths?
   b. Do you think this is true for all similar triangles? Explain.

20. Critical Thinking Given $\triangle ABC \sim \triangle DEF$ and $\triangle DEF \sim \triangle JKL$, is $\triangle ABC \sim \triangle JKL$? Give an example or a non-example.

Fair Game Review What you learned in previous grades & lessons

Simplify. *(Skills Review Handbook)*

21. $\left(\frac{4}{9}\right)^2$
22. $\left(\frac{3}{8}\right)^2$
23. $\left(\frac{7}{4}\right)^2$
24. $\left(\frac{6.5}{2}\right)^2$

25. **MULTIPLE CHOICE** You solve the equation $S = \ell w + 2wh$ for w. Which equation is correct? *(Topic 3)*

   (A) $w = \frac{S - \ell}{2h}$
   (B) $w = \frac{S - 2h}{\ell}$
   (C) $w = \frac{S}{\ell + 2h}$
   (D) $w = S - \ell - 2h$

11.6 Perimeters and Areas of Similar Figures

Essential Question How do changes in dimensions of similar geometric figures affect the perimeters and the areas of the figures?

1 ACTIVITY: Creating Similar Figures

Work with a partner. Use pattern blocks to make a figure whose dimensions are 2, 3, and 4 times greater than those of the original figure.

a. Square

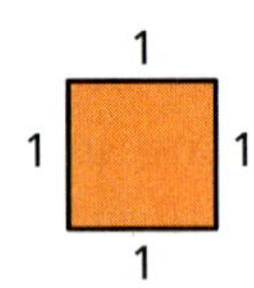

b. Rectangle

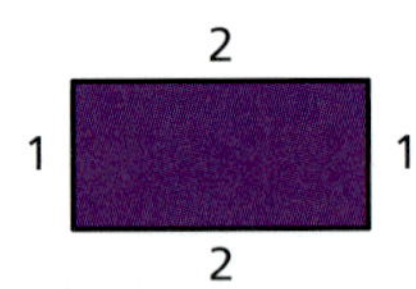

2 ACTIVITY: Finding Patterns for Perimeters

Work with a partner. Copy and complete the table for the perimeter *P* of each figure in Activity 1. Describe the pattern.

Figure	Original Side Lengths	Double Side Lengths	Triple Side Lengths	Quadruple Side Lengths
	$P =$ ▢			
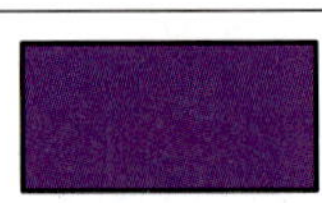	$P =$ ▢			

3 ACTIVITY: Finding Patterns for Areas

Work with a partner. Copy and complete the table for the area *A* of each figure in Activity 1. Describe the pattern.

Figure	Original Side Lengths	Double Side Lengths	Triple Side Lengths	Quadruple Side Lengths
	$A =$ ▢			
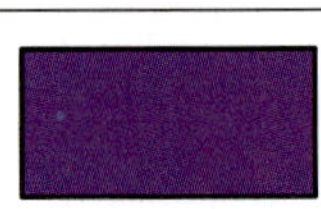	$A =$ ▢			

Geometry

In this lesson, you will

- understand the relationship between perimeters of similar figures.
- understand the relationship between areas of similar figures.
- find ratios of perimeters and areas for similar figures.

Laurie's Notes

Introduction

Applying Mathematical Practices

- **Look for and Express Regularity in Repeated Reasoning:** Students investigate how perimeters and areas of similar figures are related by finding a pattern.

Motivate

- Show various stages of the fractal known as The Sierpinski Triangle, which can be found in many places online. Ask students how many triangles (of various sizes) they see in each stage.

Activity Notes

Activity 1

- Use the pattern blocks provided in the Record and Practice Journal. Cut them out, store them in a reclosable bag, and reuse them each year.
- Circulate to make sure students are constructing the figures correctly.

Activities 2 and 3

? Ask questions to review perimeter and area.

- "What does perimeter mean?" the distance around a figure (you want an understanding of what perimeter *means*, not a formula)
- "How is it found?" Add the lengths of the sides of the figure.
- "What does area mean?" the amount of surface that a figure covers (you want an understanding of what area *means*, not a formula)
- "How is it found?" The type of figure determines which area formula you use.

? "When you make a scale drawing of a figure whose dimensions are twice the original dimensions, is the scale drawing similar to the figure? Explain." Yes, the corresponding sides are proportional and the corresponding angles are congruent.

? **Look for and Express Regularity in Repeated Reasoning:** Ask questions to compare the perimeter of each new figure with the perimeter of the original figure.

- "When the dimensions of a figure are doubled, what happens to the perimeter?" The perimeter also doubles.
- "When the dimensions of a figure are tripled, what happens to the perimeter?" The perimeter also triples.
- "When the dimensions of a figure are quadrupled, what happens to the perimeter?" The perimeter also quadruples.

- Repeat the above questions for area. The correct answers are different!
- **Suggestion:** When students look at a table entry, they should consider statements such as the following:
  "I double the original side lengths and the perimeter is _____." double
  "I double the original side lengths and the area is _____." 4 times greater
- **Look for and Express Regularity in Repeated Reasoning:** Students should recognize a pattern as they work through these activities.

What Your Students Will Learn

- Understand the relationship between perimeters of similar figures.
- Understand the relationship between areas of similar figures.
- Find ratios of perimeters and areas for similar figures.
- Use proportions of similar figures to find perimeters and areas.

Previous Learning

Students should know how to plot ordered pairs. Students also need to remember how to solve a proportion.

Technology for the Teacher

Dynamic Classroom

Lesson Plans
Complete Materials List

11.6 Record and Practice Journal

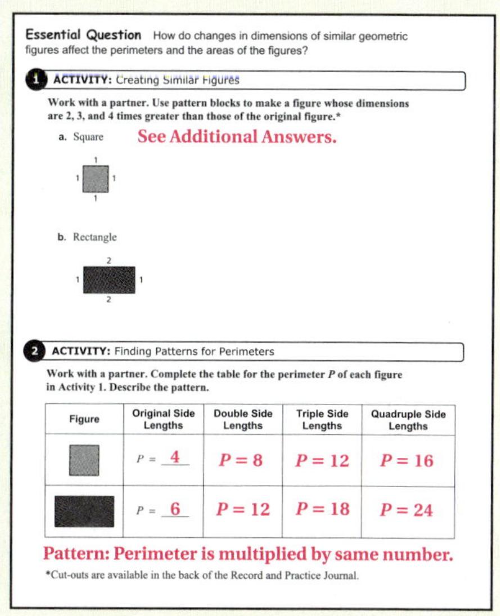

Essential Question How do changes in dimensions of similar geometric figures affect the perimeters and the areas of the figures?

1 ACTIVITY: Creating Similar Figures

Work with a partner. Use pattern blocks to make a figure whose dimensions are 2, 3, and 4 times greater than those of the original figure.*

a. Square See Additional Answers.

b. Rectangle

2 ACTIVITY: Finding Patterns for Perimeters

Work with a partner. Complete the table for the perimeter P of each figure in Activity 1. Describe the pattern.

Figure	Original Side Lengths	Double Side Lengths	Triple Side Lengths	Quadruple Side Lengths
(square)	$P = 4$	$P = 8$	$P = 12$	$P = 16$
(rectangle)	$P = 6$	$P = 12$	$P = 18$	$P = 24$

Pattern: Perimeter is multiplied by same number.

*Cut-outs are available in the back of the Record and Practice Journal.

English Language Learners

Vocabulary

Have students select objects around the classroom and use tape to mark the perimeters of the objects. Label the objects "perimeter." Have students select other objects in the classroom and cover them with square sheets of paper. Label these objects "area." Have students identify which units are best for measuring perimeter and area. Add this information to the labels. Keep the objects in the classroom until the students understand the concepts of perimeter and area.

11.6 Record and Practice Journal

3 ACTIVITY: Finding Patterns for Areas

Work with a partner. Complete the table for the area A of each figure in Activity 1. Describe a pattern.

Figure	Original Side Lengths	Double Side Lengths	Triple Side Lengths	Quadruple Side Lengths
(square)	$A =$ 1	$A = 4$	$A = 9$	$A = 16$
(rectangle)	$A =$ 2	$A = 8$	$A = 18$	$A = 32$

Pattern: Area is multiplied by the square of the number.

4 ACTIVITY: Drawing and Labeling Similar Figures

Work with a partner.

a. Find another rectangle that is similar and has one side from $(-1, -6)$ to $(5, -6)$. Label the vertices.

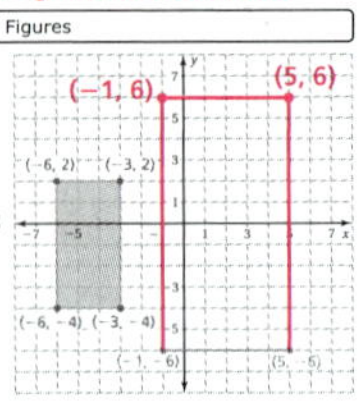

Check that the two rectangles are similar by showing that the ratios of corresponding sides are equal.

$$\frac{\text{Shaded Length}}{\text{Unshaded Length}} \stackrel{?}{=} \frac{\text{Shaded Width}}{\text{Unshaded Width}}$$

$$\frac{\text{change in } y}{\text{change in } y} \stackrel{?}{=} \frac{\text{change in } x}{\text{change in } x}$$

$$\frac{6}{12} \stackrel{?}{=} \frac{3}{6}$$

$$\frac{1}{2} \stackrel{?}{=} \frac{1}{2}$$

The ratios are equal. So, the rectangles are similar.

b. Compare the perimeters and the areas of the figures. Are the results the same as your results from Activities 2 and 3? Explain.
See Additional Answers.

c. There are three other rectangles that are similar to the shaded rectangle and have the given side.

- Draw each one. Label the vertices of each.
  See Additional Answers.
- Show that each is similar to the original shaded rectangle.
  See Additional Answers.

What Is Your Answer?

5. **IN YOUR OWN WORDS** How do changes in dimensions of similar geometric figures affect the perimeters and the areas of the figures?
   Perimeter is multiplied by the number. Area is multiplied by the square of the number.

6. What information do you need to know to find the dimensions of a figure that is similar to another figure? Give examples to support your explanation.
   lengths of a pair of corresponding sides and the side length that corresponds to the unknown length

Laurie's Notes

Activity 4

? "What are the dimensions of the rectangle?" 3 units by 6 units

- It is okay for students to put their fingers on the sides and count units.
- **Common Error:** Students count the lattice points beginning with the vertex and end up with dimensions 4 by 7 instead of 3 by 6.
- **Connection:** Notice in part (a) the use of language that is also used for slope. To compute the "change in y" or the "change in x," students should just look at the diagram and count.
- Do not skip the last step of showing that the blue rectangles are similar to the original red rectangle. Students need the practice of writing proportions. Note that while the terms length and width are usually interchangeable, the length here is the longer of the two sides. The color reference is also easier for students to understand rather than saying "corresponding side in the left rectangle to the corresponding side in the right rectangle."
- Keep the language simple so students focus on the concept.
- Have students share their three solutions. Check to see that the coordinates are correctly labeled, for instance, $(5, -6)$ and not $(-6, 5)$.

? "What is the ratio of corresponding sides when the rectangles are the same size?" 1 : 1

What Is Your Answer?

- **Big Idea:** When the dimensions are doubled, tripled, or quadrupled, the resulting figure is similar to the original figure. Students should see a pattern in the perimeters and in the areas. The Key Ideas in the lesson will define these patterns further.

Closure

- **Exit Ticket:** Find the perimeter and area of a rectangle with dimensions 3 inches by 4 inches. If you double the dimensions, what will be the new perimeter and area? Original dimensions: perimeter is 14 in., area is 12 in.^{2} New dimensions: perimeter is $2(14) = 28$ in., area is $4(12) = 48$ in.^{2}

4 ACTIVITY: Drawing and Labeling Similar Figures

Work with a partner.

a. Find a blue rectangle that is similar to the red rectangle and has one side from $(-1, -6)$ to $(5, -6)$. Label the vertices.

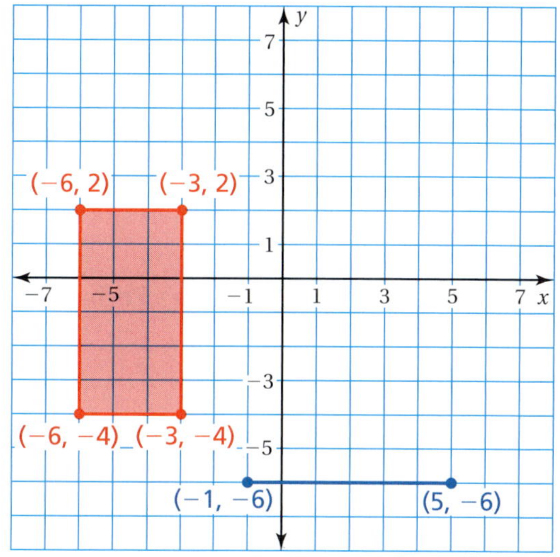

Check that the two rectangles are similar by showing that the ratios of corresponding sides are equal.

$$\frac{\text{Red Length}}{\text{Blue Length}} \stackrel{?}{=} \frac{\text{Red Width}}{\text{Blue Width}}$$

$$\frac{\text{change in } y}{\text{change in } y} \stackrel{?}{=} \frac{\text{change in } x}{\text{change in } x}$$

$$\frac{\square}{\square} \stackrel{?}{=} \frac{\square}{\square}$$

$$\frac{\square}{\square} \stackrel{?}{=} \frac{\square}{\square}$$

The ratios are equal. So, the rectangles are similar.

b. Compare the perimeters and the areas of the figures. Are the results the same as your results from Activities 2 and 3? Explain.

c. There are three other blue rectangles that are similar to the red rectangle and have the given side.

- Draw each one. Label the vertices of each.
- Show that each is similar to the original red rectangle.

Math Practice

Analyze Givens

What values should you use to fill in the proportion? Does it matter where each value goes? Explain.

What Is Your Answer?

5. **IN YOUR OWN WORDS** How do changes in dimensions of similar geometric figures affect the perimeters and the areas of the figures?

6. What information do you need to know to find the dimensions of a figure that is similar to another figure? Give examples to support your explanation.

Practice Use what you learned about perimeters and areas of similar figures to complete Exercises 8 and 9 on page 504.

11.6 Lesson

Check It Out
Lesson Tutorials
BigIdeasMath.com

Key Idea

Perimeters of Similar Figures

When two figures are similar, the ratio of their perimeters is equal to the ratio of their corresponding side lengths.

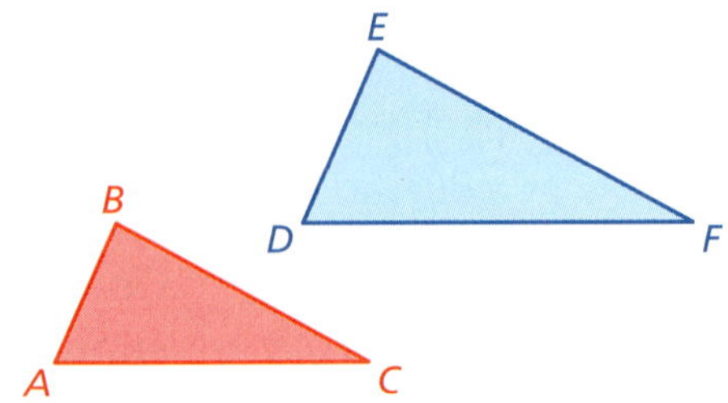

$$\frac{\text{Perimeter of } \triangle ABC}{\text{Perimeter of } \triangle DEF} = \frac{AB}{DE} = \frac{BC}{EF} = \frac{AC}{DF}$$

EXAMPLE 1 Finding Ratios of Perimeters

Find the ratio (red to blue) of the perimeters of the similar rectangles.

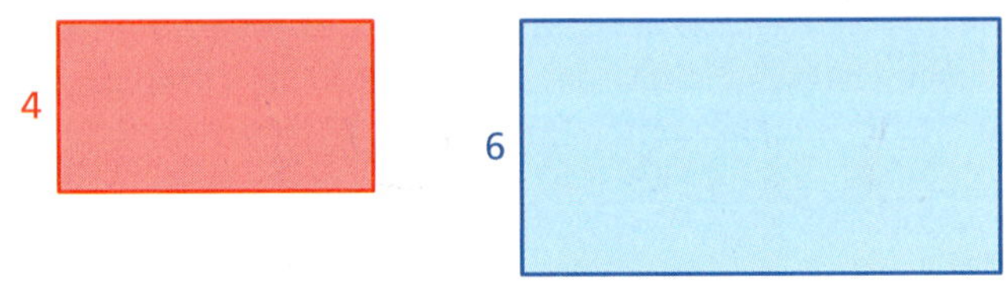

$$\frac{\text{Perimeter of red rectangle}}{\text{Perimeter of blue rectangle}} = \frac{4}{6} = \frac{2}{3}$$

∴ The ratio of the perimeters is $\frac{2}{3}$.

On Your Own

1. The height of Figure A is 9 feet. The height of a similar Figure B is 15 feet. What is the ratio of the perimeter of A to the perimeter of B?

Key Idea

Areas of Similar Figures

When two figures are similar, the ratio of their areas is equal to the *square* of the ratio of their corresponding side lengths.

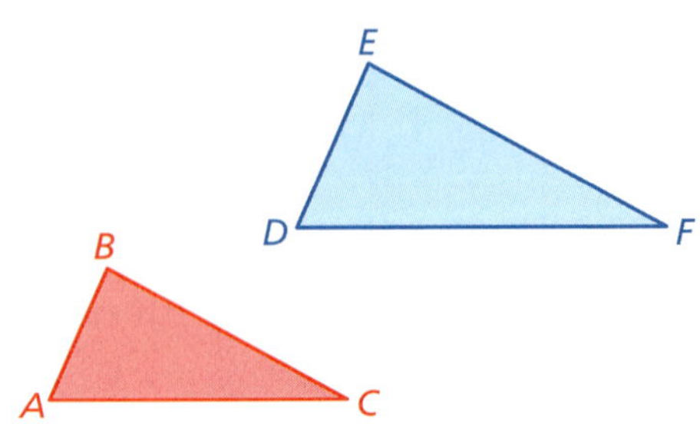

$$\frac{\text{Area of } \triangle ABC}{\text{Area of } \triangle DEF} = \left(\frac{AB}{DE}\right)^2 = \left(\frac{BC}{EF}\right)^2 = \left(\frac{AC}{DF}\right)^2$$

Laurie's Notes

Introduction

Connect

- **Yesterday:** Students used pattern blocks to investigate how changes in the dimensions of similar figures affect the perimeter and area of the figures.
- **Today:** Students will use the stated relationships to solve problems.

Motivate

- **Story Time:** Tell students that your neighbor's lawn is twice the size of your lawn. In other words, it is twice as long and twice as wide. If it takes you one-half hour to mow your lawn, about how long does it take your neighbor to mow his lawn? You will answer this in the Closure activity.

Lesson Notes

Key Idea

? "How do you identify similar triangles?" Corresponding sides are proportional and corresponding angles are congruent.

- Write some side lengths on the two triangles, such as 3-4-5 and 6-8-10. Then find the two perimeters, 12 and 24, to show the same ratio of 1 : 2.

Example 1

? "What is the ratio of the corresponding sides?" 4 : 6 or 2 : 3 The ratio of the perimeters is also 2 : 3.

? **Make Sense of Problems and Persevere in Solving Them:** "Why do you not need to know both dimensions of one of the rectangles to find the ratio of the perimeters?" Listen for students to explain that because the rectangles are similar, only one pair of corresponding sides is necessary.

On Your Own

- Check that students have correctly written the ratio.

Key Idea

- **Representation:** Draw two triangles whose corresponding sides appear to have a ratio of 1 : 2.

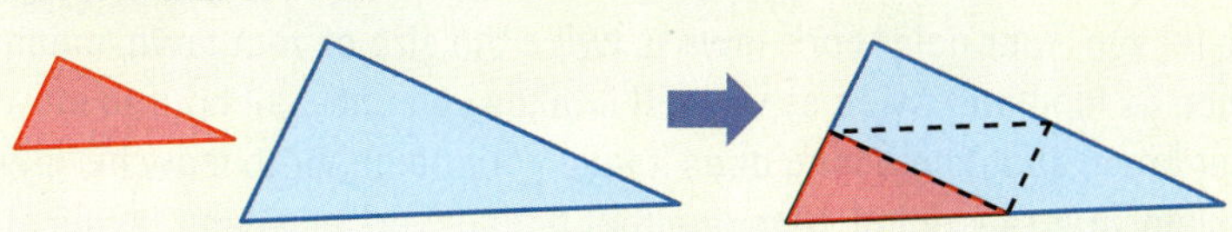

? "If the corresponding sides have a ratio of 1 : 2, then what will the ratio of the areas be?" 1 : 4; You can also use pattern blocks to show this relationship.

- **Model with Mathematics:** There are 4 copies of the smaller triangle inside the larger. Another way to state this relationship is that the larger triangle has an area 4 times greater than the area of the smaller triangle.

Goal Today's lesson is finding ratios of perimeters and areas of similar figures.

Technology for the Teacher

Dynamic Classroom

Lesson Tutorials
Lesson Plans
Answer Presentation Tool

Extra Example 1

Find the ratio (red to blue) of the perimeters of the similar trapezoids. $\frac{7}{5}$

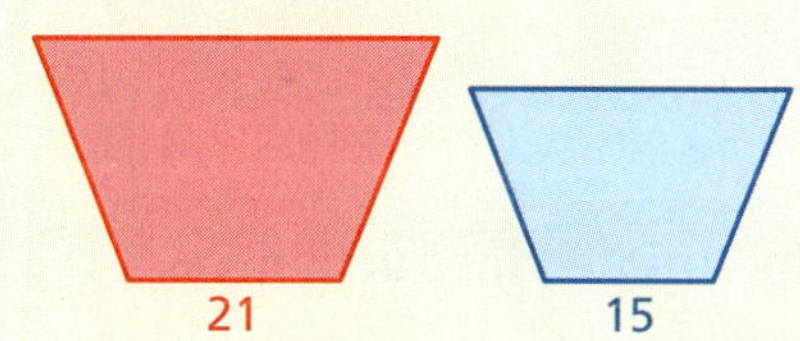

On Your Own

1. $\frac{3}{5}$

Extra Example 2

Find the ratio (red to blue) of the areas of the similar parallelograms. $\frac{81}{36} = \frac{9}{4}$

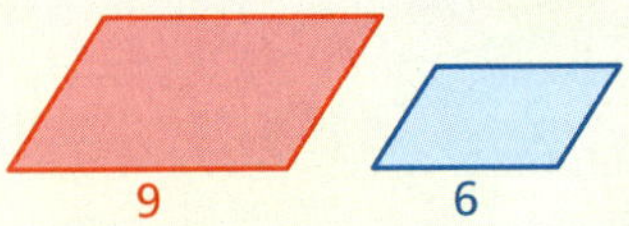

On Your Own

2. $\frac{64}{49}$

Extra Example 3

In Example 3, the width of the pool is 22 yards. Find the perimeter P and the area A of the pool. 132 yd; 968 yd^2

On Your Own

3. 96 yd; 512 yd^2

Differentiated Instruction

Kinesthetic

Provide students with large pieces of construction paper, rulers, and protractors. Have students work in pairs to draw a large right triangle on the paper. Record the lengths of the sides and the measures of the angles in a table. Connect the midpoints of each side of the large triangle and record the side lengths and angle measures of the second triangle in the table. Connect the midpoints of the sides of the second triangle to form a third triangle. Record the side lengths and angle measures of the third triangle in the table. Have students determine if the triangles are similar. If they are similar, find the ratios of the perimeters and areas of each pair of triangles.

Laurie's Notes

Example 2

- As with Example 1, only one pair of corresponding sides is given. The actual areas cannot be computed. Students must use the relationship stated in the Key Idea.
- Remind students that $\left(\frac{3}{5}\right)^2$ means $\left(\frac{3}{5}\right) \cdot \left(\frac{3}{5}\right) = \frac{9}{25}$.

On Your Own

- Have students work in pairs on Question 2.

Example 3

- Ask a volunteer to read the problem.
- ? "What information do you know?" one side length in each object, area and perimeter of the volleyball court, and the fact that the objects are similar
- ? "Are the sides whose lengths are given corresponding sides?" yes
- **Make Sense of Problems and Persevere in Solving Them:** Give students time to think about solution strategies. There are many different strategies, and it is important for students to come up with an entry point that makes sense to them.
- One strategy, which makes use of the Key Ideas in the section, is given in the solution.
- Another strategy is to find the length ℓ of the volleyball court by solving $2(10) + 2\ell = 60$ for ℓ and then setting up a proportion to find the length of the pool. From there, you can find the area and perimeter of the pool.
- Discuss any other strategies that students might suggest. Ask students which strategy they prefer.
- **Attend to Precision:** Make sure students include the correct units in their answers.

On Your Own

- Have students work in pairs on Question 3.

Closure

- **Exit Ticket:** Return to the question used to motivate at the beginning of the lesson. Your neighbor's lawn is twice the size of your lawn, meaning twice as long and twice as wide. If it takes you one-half hour to mow your lawn, about how long does it take your neighbor to mow his lawn? Explain your reasoning. (Assume that both of you mow your lawns at the same rate.)
  2 hours; because the dimensions of your neighbor's lawn are double the dimensions of your lawn, you have the ratio neighbor : you = 2 : 1. The area of your neighbor's lawn is $\left(\frac{2}{1}\right)^2 = \frac{4}{1}$ times the area of your lawn, so it should take your neighbor 4 times longer to mow his lawn, $4\left(\frac{1}{2}\right) = 2$ hours.

EXAMPLE 2 Finding Ratios of Areas

Find the ratio (red to blue) of the areas of the similar triangles.

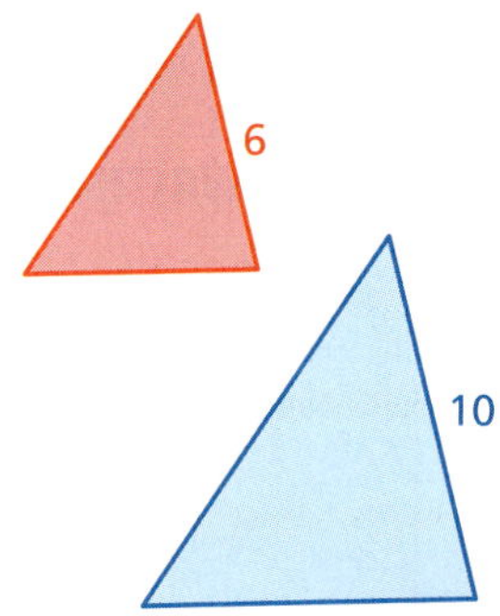

$$\frac{\text{Area of red triangle}}{\text{Area of blue triangle}} = \left(\frac{6}{10}\right)^2$$

$$= \left(\frac{3}{5}\right)^2 = \frac{9}{25}$$

∴ The ratio of the areas is $\frac{9}{25}$.

On Your Own

Now You're Ready
Exercises 4–7

2. The base of Triangle P is 8 meters. The base of a similar Triangle Q is 7 meters. What is the ratio of the area of P to the area of Q?

EXAMPLE 3 Using Proportions to Find Perimeters and Areas

A swimming pool is similar in shape to a volleyball court. Find the perimeter *P* and the area *A* of the pool.

18 yd

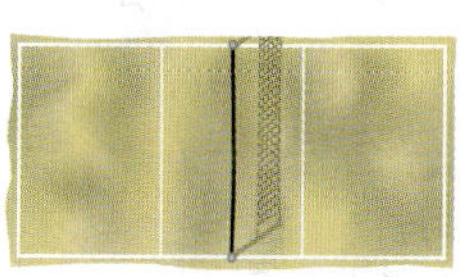

10 yd

Area = 200 yd²
Perimeter = 60 yd

The rectangular pool and the court are similar. So, use the ratio of corresponding side lengths to write and solve proportions to find the perimeter and the area of the pool.

Perimeter

$$\frac{\text{Perimeter of court}}{\text{Perimeter of pool}} = \frac{\text{Width of court}}{\text{Width of pool}}$$

$$\frac{60}{P} = \frac{10}{18}$$

$$1080 = 10P$$

$$108 = P$$

Area

$$\frac{\text{Area of court}}{\text{Area of pool}} = \left(\frac{\text{Width of court}}{\text{Width of pool}}\right)^2$$

$$\frac{200}{A} = \left(\frac{10}{18}\right)^2$$

$$\frac{200}{A} = \frac{100}{324}$$

$$64{,}800 = 100A$$

$$648 = A$$

∴ So, the perimeter of the pool is 108 yards, and the area is 648 square yards.

On Your Own

3. **WHAT IF?** The width of the pool is 16 yards. Find the perimeter P and the area A of the pool.

11.6 Exercises

Check It Out
Help with Homework
BigIdeasMath.com

Vocabulary and Concept Check

1. **WRITING** How are the perimeters of two similar figures related?

2. **WRITING** How are the areas of two similar figures related?

3. **NUMBER SENSE** Rectangle *ABCD* is similar to Rectangle *WXYZ*. The area of *ABCD* is 30 square inches. Explain how to find the area of *WXYZ*.

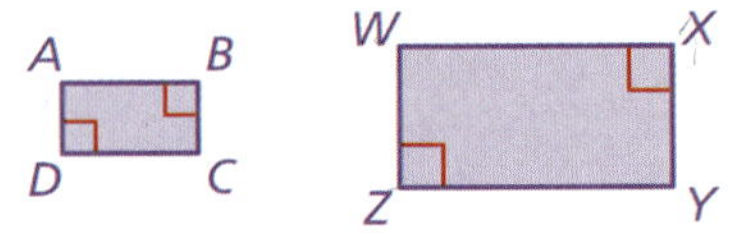

$\frac{AD}{WZ} = \frac{1}{2}$ $\quad \frac{AB}{WX} = \frac{1}{2}$

Practice and Problem Solving

The two figures are similar. Find the ratios (red to blue) of the perimeters and of the areas.

4.

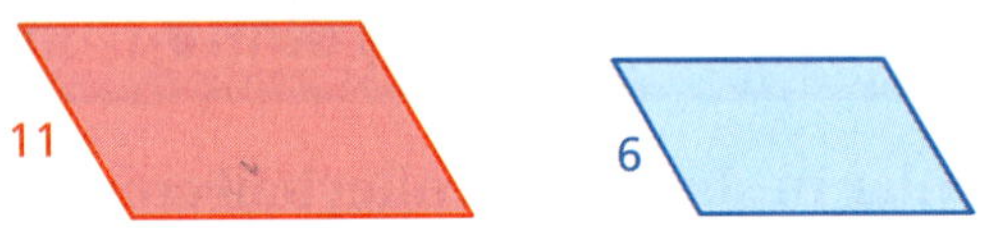

5.

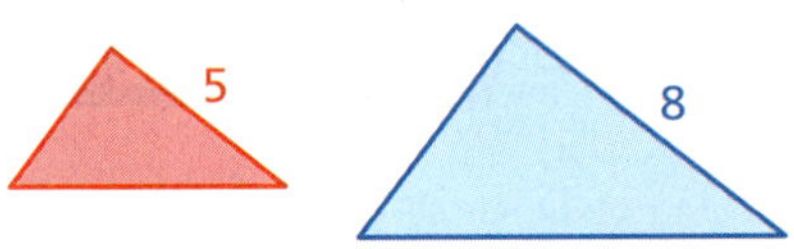

6.

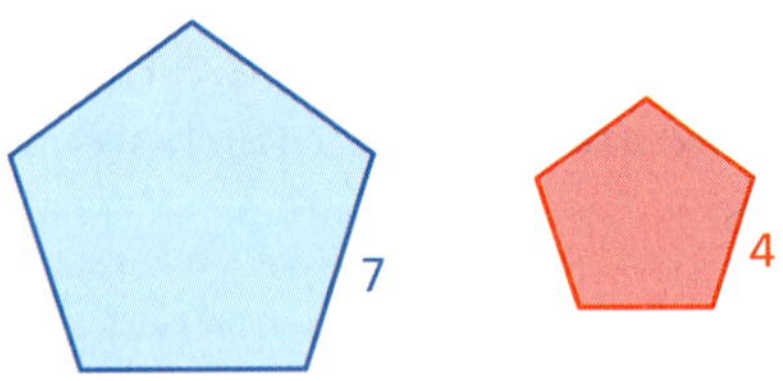

7.

8. **PERIMETER** How does doubling the side lengths of a right triangle affect its perimeter?

9. **AREA** How does tripling the side lengths of a right triangle affect its area?

The figures are similar. Find *x*.

10. The ratio of the perimeters is 7 : 10.

11. The ratio of the perimeters is 8 : 5.

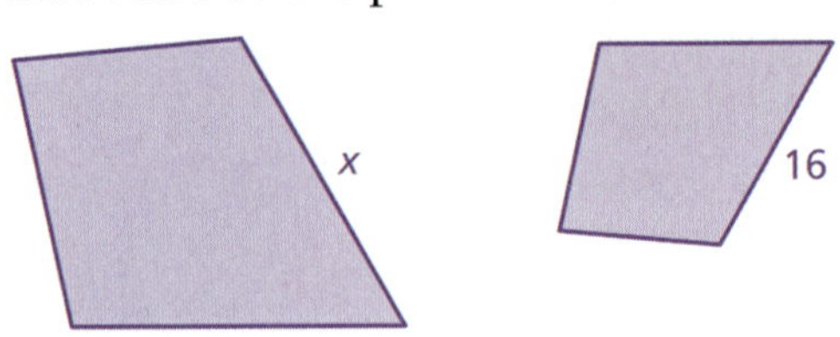

12. **FOOSBALL** The playing surfaces of two foosball tables are similar. The ratio of the corresponding side lengths is 10 : 7. What is the ratio of the areas?

13. **CHEERLEADING** A rectangular school banner has a length of 44 inches, a perimeter of 156 inches, and an area of 1496 square inches. The cheerleaders make signs similar to the banner. The length of a sign is 11 inches. What is its perimeter and its area?

Assignment Guide and Homework Check

Level	Assignment	Homework Check
Accelerated	1–4, 6, 8, 9, 10–14 even, 15–24	6, 10, 16, 17, 19

Common Errors

- **Exercises 4–7** Students may find the reciprocal of the ratio. For example, they may find the ratio of blue to red instead of red to blue. Remind students to read the directions carefully.
- **Exercises 4–7** When finding the ratio of the areas, students may forget to square the ratio of the corresponding side lengths. Remind them of the Key Idea at the bottom of page 78.

11.6 Record and Practice Journal

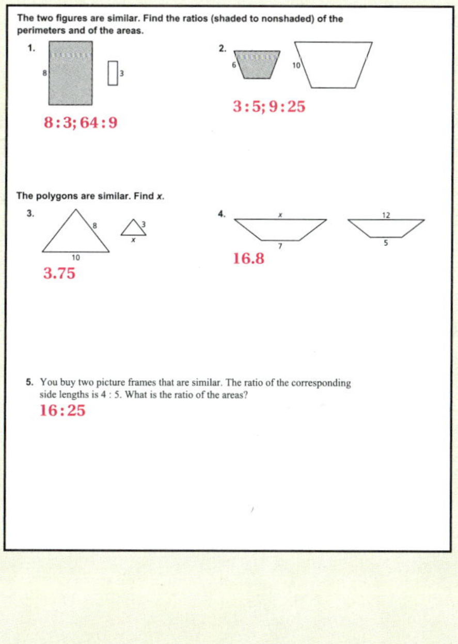
The two figures are similar. Find the ratios (shaded to nonshaded) of the perimeters and of the areas.

1. 8:3; 64:9
2. 3:5; 9:25

The polygons are similar. Find x.

3. 3.75
4. 16.8

5. You buy two picture frames that are similar. The ratio of the corresponding side lengths is 4 : 5. What is the ratio of the areas?
16:25

Vocabulary and Concept Check

1. The ratio of the perimeters is equal to the ratio of the corresponding side lengths.
2. The ratio of the areas is equal to the square of the ratio of the corresponding side lengths.
3. Because the ratio of the corresponding side lengths is $\frac{1}{2}$, the ratio of the areas is equal to $\left(\frac{1}{2}\right)^2$. To find the area, solve the proportion $\frac{30}{x} = \frac{1}{4}$ to get $x = 120$ square inches.

Practice and Problem Solving

4. $\frac{11}{6}$; $\frac{121}{36}$
5. $\frac{5}{8}$; $\frac{25}{64}$
6. $\frac{4}{7}$; $\frac{16}{49}$
7. $\frac{14}{9}$; $\frac{196}{81}$
8. The perimeter doubles.
9. The area is 9 times larger.
10. 8.4
11. 25.6
12. 100 : 49
13. 39 in.; 93.5 in.^{2}

Practice and Problem Solving

14. *ABCD*: $P = 14, A = 12$; *WXYZ*: $P = 28, A = 48$; yes; Corresponding side lengths are proportional and corresponding angles are congruent.

15. 108 yd

16. See *Taking Math Deeper*.

17. **a.** 400 times greater; The ratio of the corresponding lengths is $\frac{120 \text{ in.}}{6 \text{ in.}} = \frac{20}{1}$. So, the ratio of the areas is $\left(\frac{20}{1}\right)^2 = \frac{400}{1}$.

b. 1250 ft^2

18. See Additional Answers.

19. 15 m **20.** 12.5 bottles

Fair Game Review

21. $x = -2$ **22.** $b = 1.6$

23. $n = -4$ **24.** B

Mini-Assessment

The two figures are similar. Find the ratio (red to blue) of the perimeters and of the areas.

1.

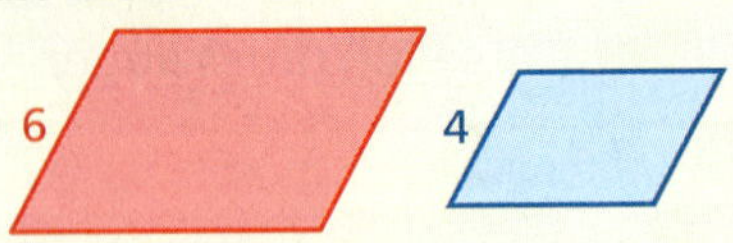

$\frac{3}{2}, \frac{9}{4}$

2.

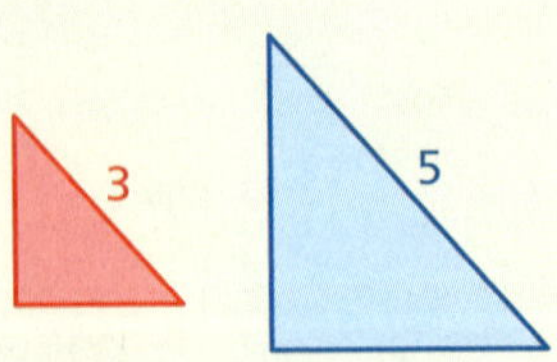

$\frac{3}{5}, \frac{9}{25}$

3. The ratio of the corresponding side lengths of two similar cellular phones is 3 : 4. The perimeter of the smaller phone is 9 inches. What is the perimeter of the larger phone? 12 in.

Taking Math Deeper

Exercise 16

There are several very different ways to solve this problem. This would be a good problem to encourage students to "think outside the box." You might have students work in pairs to see how many different ways they can solve the problem.

1 Recognize that the smaller piece is one-fourth the size of the larger piece.

2 Use unit prices.

$$\frac{\text{Cost}}{\text{Area}} = \frac{1.31}{9 \times 21} \approx \$0.007 \text{ per in.}^2$$

Area of the 18×42 piece $= 18 \cdot 42 = 756 \text{ in.}^2$

Cost of the 18×42 piece $\approx 756 \cdot 0.007 \approx \$5.29$

3 To find a more precise solution, use a proportion.

$$\frac{\text{Area}}{\text{Area}} = \frac{\text{Cost}}{\text{Cost}}$$

$$\frac{756}{189} = \frac{x}{1.31}$$

$$\$5.24 = x$$

Reteaching and Enrichment Strategies

If students need help. . .	If students got it. . .
Resources by Chapter • Practice A and Practice B • Puzzle Time Record and Practice Journal Practice Differentiating the Lesson Lesson Tutorials Skills Review Handbook	Resources by Chapter • Enrichment and Extension • Technology Connection Start the next section

14. **REASONING** The vertices of two rectangles are $A(-5, -1)$, $B(-1, -1)$, $C(-1, -4)$, $D(-5, -4)$ and $W(1, 6)$, $X(7, 6)$, $Y(7, -2)$, $Z(1, -2)$. Compare the perimeters and the areas of the rectangles. Are the rectangles similar? Explain.

15. **SQUARE** The ratio of the side length of Square A to the side length of Square B is 4 : 9. The side length of Square A is 12 yards. What is the perimeter of Square B?

16. **FABRIC** The cost of the fabric is \$1.31. What would you expect to pay for a similar piece of fabric that is 18 inches by 42 inches?

17. **AMUSEMENT PARK** A scale model of a merry-go-round and the actual merry-go-round are similar.

    a. How many times greater is the base area of the actual merry-go-round than the base area of the scale model? Explain.

    b. What is the base area of the actual merry-go-round in square feet?

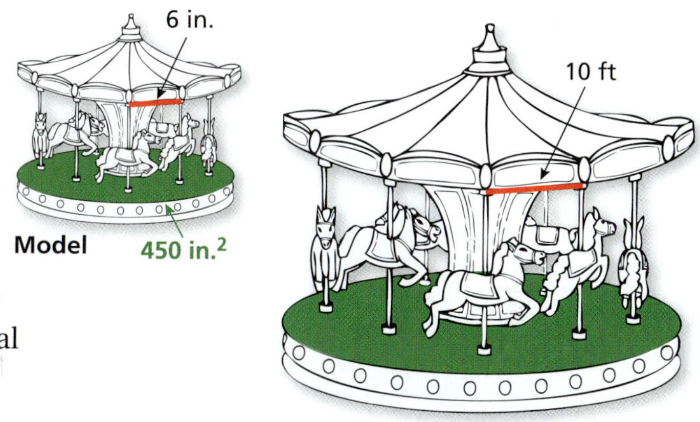

18. **STRUCTURE** The circumference of Circle K is π. The circumference of Circle L is 4π.

    a. What is the ratio of their circumferences? of their radii? of their areas?

    b. What do you notice?

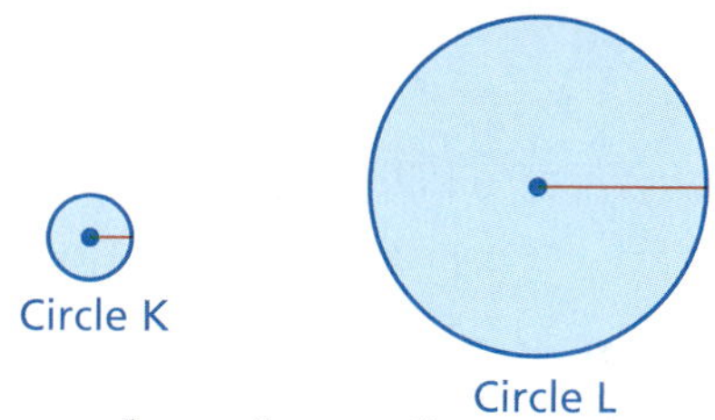

19. **GEOMETRY** A triangle with an area of 10 square meters has a base of 4 meters. A similar triangle has an area of 90 square meters. What is the *height* of the larger triangle?

20. **Problem Solving** You need two bottles of fertilizer to treat the flower garden shown. How many bottles do you need to treat a similar garden with a perimeter of 105 feet?

Fair Game Review What you learned in previous grades & lessons

Solve the equation. Check your solution. *(Topic 2)*

21. $4x + 12 = -2x$

22. $2b + 6 = 7b - 2$

23. $8(4n + 13) = 6n$

24. **MULTIPLE CHOICE** Last week, you collected 20 pounds of cans for recycling. This week, you collect 25 pounds of cans for recycling. What is the percent of increase? *(Section 6.5)*

    Ⓐ 20% Ⓑ 25% Ⓒ 80% Ⓓ 125%

11.7 Dilations

Essential Question How can you enlarge or reduce a figure in the coordinate plane?

The Meaning of a Word ● Dilate

When you have your eyes checked, the optometrist sometimes **dilates** one or both of the pupils of your eyes.

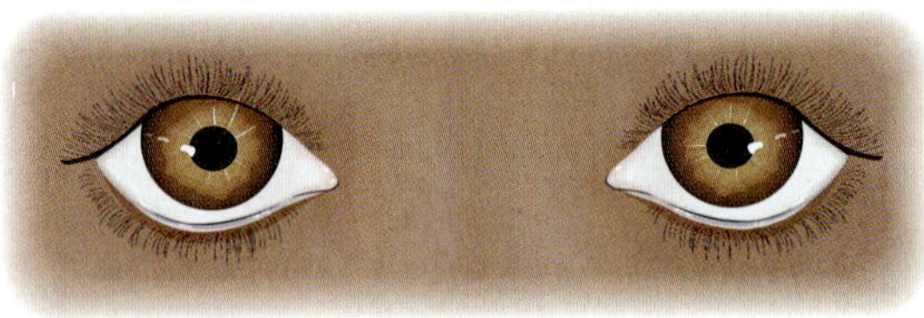

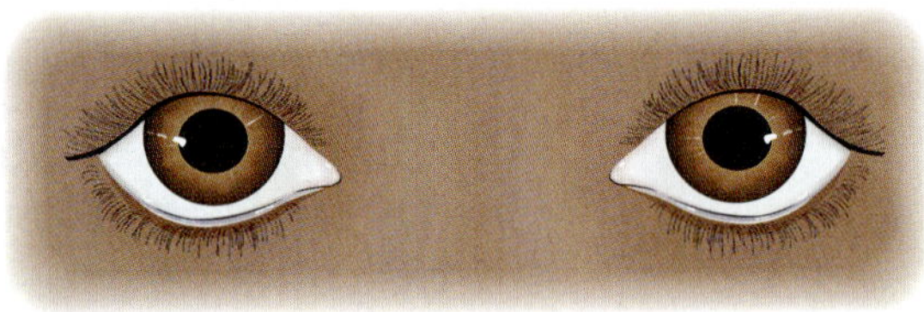

1 ACTIVITY: Comparing Triangles in a Coordinate Plane

Work with a partner. Write the coordinates of the vertices of the blue triangle. Then write the coordinates of the vertices of the red triangle.

a. How are the two sets of coordinates related?

b. How are the two triangles related? Explain your reasoning.

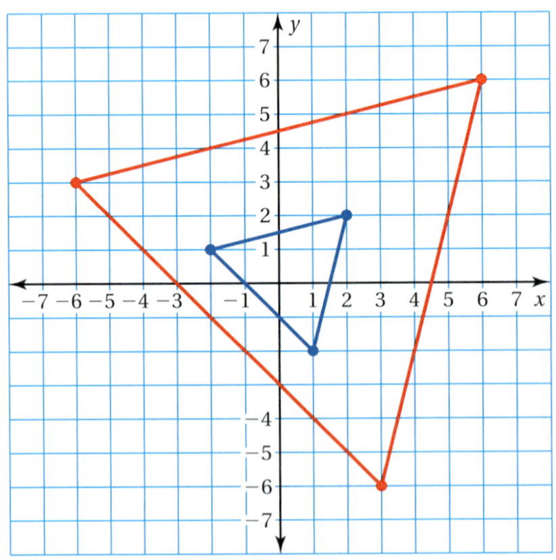

c. Draw a green triangle whose coordinates are twice the values of the corresponding coordinates of the blue triangle. How are the green and blue triangles related? Explain your reasoning.

d. How are the coordinates of the red and green triangles related? How are the two triangles related? Explain your reasoning.

Geometry

In this lesson, you will

- identify dilations.
- dilate figures in the coordinate plane.
- use more than one transformation to find images of figures.

Laurie's Notes

Introduction

Applying Mathematical Practices

- **Model with Mathematics:** Although students are not able to calculate the side lengths of the triangles in Activity 1, they analyze the relationships between the corresponding segments (slope) and between the corresponding vertices and conjecture that the triangles are similar.

Motivate

- Draw a figure on a transparency and display with the overhead projector. Trace over the projection on the board. Move the projector to enlarge the figure and also to reduce the figure. Each time trace over the projection.
- ? "What do you notice about the three figures?" Students should describe the figures as being the same shape, but different sizes.
- Alternatively, you can use a document camera.

Discuss

- ? "Did the size or shape of the figures change when you performed a translation, reflection, or rotation?" no
- "There is a fourth type of transformation. M.C. Escher used this type in creating his graphic art work."
- ? "At an eye appointment, the optometrist dilates your pupils. Who knows what it means to dilate your pupils?" Listen for enlarge your pupils.

Activity Notes

Activity 1

- Check that students record the coordinates correctly. It is common to interchange the *x*- and *y*-coordinates.
- When describing the relationship between the coordinates of the blue and red triangles, check that students state that both the *x*- and *y*-coordinates have been multiplied by a factor of 3.
- The red and blue triangles are similar— the same shape, but different size. Students are not able to determine side lengths or angle measures, but *may* say the red triangle is enlarged proportionally from the blue triangle. They may also say that the corresponding sides have the same slope.
- ? For part (c), ask, "How will you determine the coordinates of the green triangle?" Multiply the coordinates of the blue triangle by 2.
- **Model with Mathematics:** Once the triangles are drawn, the students may observe that there is a line that can be drawn from the origin through 3 corresponding vertices. For example, the origin and the vertices (2, 2), (4, 4), and (6, 6) lie on a line.
- Students may also correctly observe that corresponding sides are parallel.
- Point out that there is an order implied when asked how the green and blue triangles are related. Green to blue is the ratio 2 : 1, while blue to green is the ratio 1 : 2.
- **Attend to Precision:** Students may incorrectly say that the red triangle is 3 times larger than the blue triangle. The ratios of the side lengths of the red triangle to the blue triangle are 3 : 1. The areas are not in the ratio of 3 : 1.

What Your Students Will Learn

- Identify whether two figures are dilations of one another.
- Dilate figures in the coordinate plane using one or more than one transformation.
- Describe sequences of transformations in the coordinate plane.

Previous Learning

Students should know how to multiply integers and plot points in the coordinate plane.

Technology for the Teacher

Dynamic Classroom

Lesson Plans
Complete Materials List

11.7 Record and Practice Journal

Essential Question How can you enlarge or reduce a figure in the coordinate plane?

1 ACTIVITY: Comparing Triangles in a Coordinate Plane

Work with a partner. Write the coordinates of the vertices of the shaded triangle. Then write the coordinates of the vertices of the nonshaded triangle.

shaded: (−2, 1), (2, 2), (1, −2)

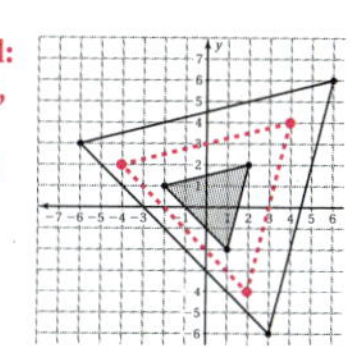

nonshaded: (−6, 3), (6, 6), (3, −6)

a. How are the two sets of coordinates related?
The shaded coordinates are multiplied by 3 to get the nonshaded coordinates.

b. How are the two triangles related? Explain your reasoning.
similar triangles

c. Draw a dashed triangle whose coordinates are twice the values of the corresponding coordinates of the shaded triangle. How are the dashed and shaded triangles related? Explain your reasoning.
similar triangles

Differentiated Instruction

Visual

Have students use two different colored pencils when drawing a figure and its image after a dilation. This will make the figures easily identifiable.

11.7 Record and Practice Journal

d. How are the coordinates of the nonshaded and dashed triangles related? How are the two triangles related? Explain your reasoning.

Multiply coordinates of nonshaded by $\frac{2}{3}$ to get the coordinates of dashed; similar triangles

2 ACTIVITY: Drawing Triangles in a Coordinate Plane

Work with a partner.

a–c.

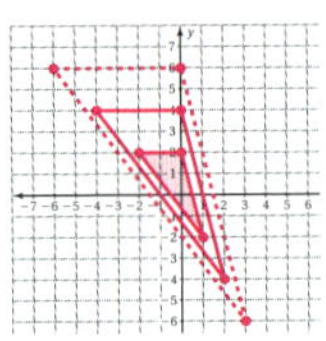

a. Draw the triangle whose vertices are $(0, 2)$, $(-2, 2)$, and $(1, -2)$.

b. Multiply each coordinate of the vertices by 2 to obtain three new vertices. Draw the triangle given by the three new vertices. How are the two triangles related?

similar triangles; scale factor of 2

c. Repeat part (b) by multiplying by 3 instead of 2.

similar triangles; scale factor of 3

3 ACTIVITY: Summarizing Transformations

Work with a partner. Make a table that summarizes the relationships between the original figure and its image for the four types of transformations you studied in this chapter.

Translation	same size and shape, slides left, right, up and/or down
Reflection	same size and shape, mirror image of original
Rotation	same size and shape, rotated about a point
Dilation	different size, same shape, the image is an enlargement or reduction of the original

What Is Your Answer?

4. **IN YOUR OWN WORDS** How can you enlarge or reduce a figure in the coordinate plane?

Multiply each coordinate by the same amount. To enlarge, multiply by a number greater than 1. To reduce, multiply by a number between 0 and 1.

5. Describe how knowing how to enlarge or reduce figures in a technical drawing is important in a career such as drafting.

Sample answer: to show details of an object

Laurie's Notes

Activity 2

- The triangle formed is an obtuse scalene triangle. Given the distinctive shape of the triangle, the two new triangles formed will also be distinctive. They will be the same shape but not the same size.
- **Construct Viable Arguments and Critique the Reasoning of Others** and **Attend to Precision:** As you listen to students describing how the triangles are related, you should hear them mention previously learned concepts such as proportional dimensions, similar figures, and scale factor.

? **Extension:** "What do you think the triangle would look like if you had multiplied each coordinate by $\frac{1}{2}$?" Students should say the triangle would be smaller or reduced from the original triangle.

Activity 3

- You may need to give suggestions to students about what aspects or categories to write about. Students should be thinking about sides, angles, orientation, size, and shape.

What Is Your Answer?

- Students should work independently and then share their results with the class.

Closure

- **Writing Prompt:** To enlarge a trapezoid that is drawn in the coordinate plane you . . .

2 ACTIVITY: Drawing Triangles in a Coordinate Plane

Work with a partner.

a. Draw the triangle whose vertices are (0, 2), (−2, 2), and (1, −2).

b. Multiply each coordinate of the vertices by 2 to obtain three new vertices. Draw the triangle given by the three new vertices. How are the two triangles related?

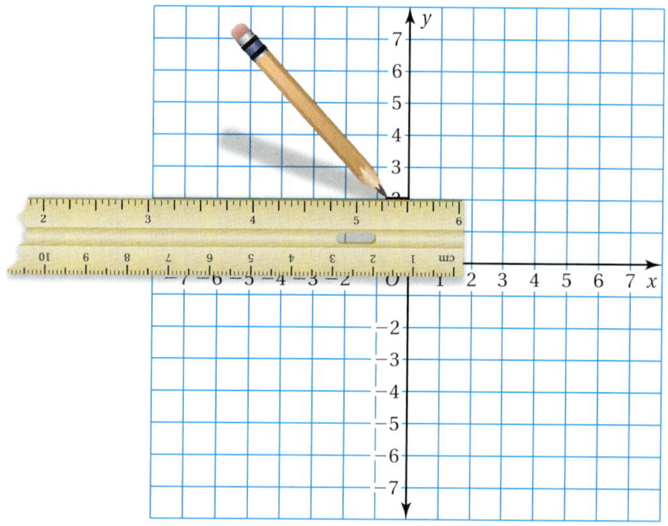

c. Repeat part (b) by multiplying by 3 instead of 2.

Math Practice

Use Prior Results

What are the four types of transformations you studied in this chapter? What information can you use to fill in your table?

3 ACTIVITY: Summarizing Transformations

Work with a partner. Make a table that summarizes the relationships between the original figure and its image for the four types of transformations you studied in this chapter.

What Is Your Answer?

4. **IN YOUR OWN WORDS** How can you enlarge or reduce a figure in the coordinate plane?

5. Describe how knowing how to enlarge or reduce figures in a technical drawing is important in a career such as drafting.

Practice Use what you learned about dilations to complete Exercises 4–6 on page 511.

11.7 Lesson

Check It Out
Lesson Tutorials
BigIdeasMath.com

A **dilation** is a transformation in which a figure is made larger or smaller with respect to a point called the **center of dilation**.

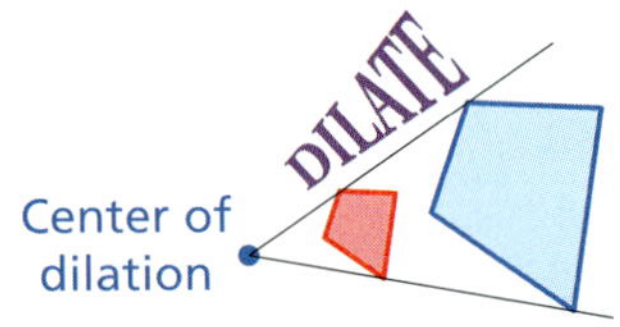

Key Vocabulary
dilation, *p. 508*
center of dilation, *p. 508*
scale factor, *p. 508*

EXAMPLE 1 Identifying a Dilation

Tell whether the blue figure is a dilation of the red figure.

a.

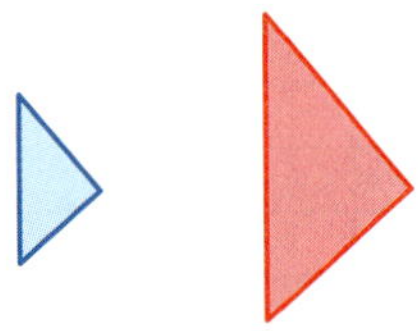

Lines connecting corresponding vertices meet at a point.

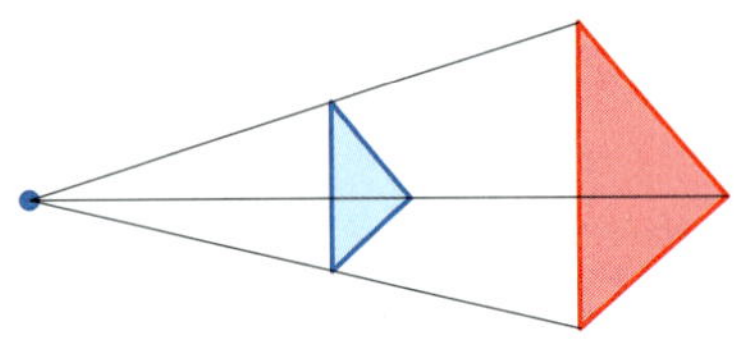

So, the blue figure is a dilation of the red figure.

b.

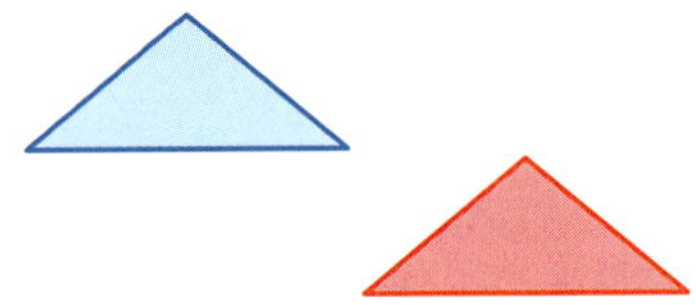

The figures have the same size and shape. The red figure *slides* to form the blue figure.

So, the blue figure is *not* a dilation of the red figure. It is a translation.

On Your Own

Now You're Ready
Exercises 7–12

Tell whether the blue figure is a dilation of the red figure. Explain.

1.

2.

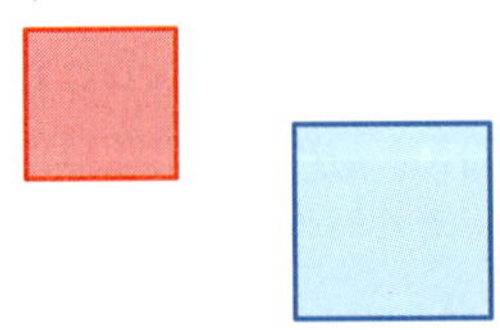

In a dilation, the original figure and its image are similar. The ratio of the side lengths of the image to the corresponding side lengths of the original figure is the **scale factor** of the dilation.

Key Idea

Dilations in the Coordinate Plane

Words To dilate a figure with respect to the origin, multiply the coordinates of each vertex by the scale factor k.

Algebra $(x, y) \rightarrow (kx, ky)$

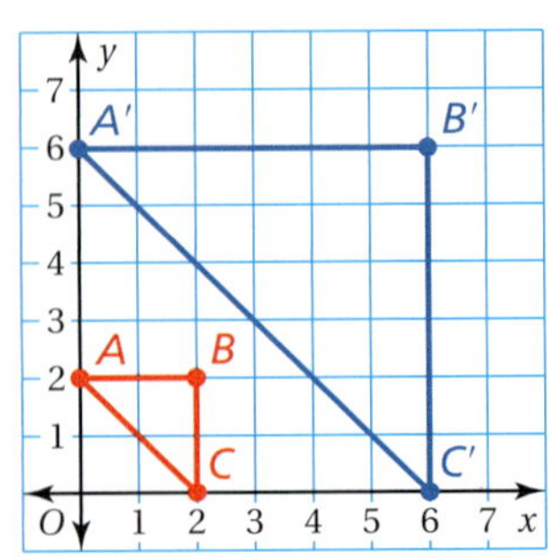

- When $k > 1$, the dilation is an enlargement.
- When $k > 0$ and $k < 1$, the dilation is a reduction.

Multi-Language Glossary at BigIdeasMath.com

Laurie's Notes

Introduction

Connect

- **Yesterday:** Students explored enlarging and reducing triangles in the coordinate plane.
- **Today:** Students will identify dilations and use scale factors to enlarge and reduce polygons in the coordinate plane.

Motivate

- Cut a rectangle out of heavier card stock. Use a flashlight to cast a shadow of the rectangle onto the wall.
- ? "What do you notice about the shadow?" similar to the rectangle
- Vary the distance between the bulb of the flashlight and the rectangle. Discuss how this changes the shadow.
- ? "Is the shadow always similar to the original figure?" It should be if the figure is held parallel to the wall and the flashlight is perpendicular to the wall.

Discuss

- Explain what a dilation and the center of a dilation are, based on the Motivate activity.

Lesson Notes

Example 1

- It is important to draw the line segments connecting the corresponding vertices. It enables you to locate the center of dilation.
- **Extension:** In part (a), if the blue triangle were rotated, you would not have a dilation. The triangles would still be similar, but there would not be a center of dilation.

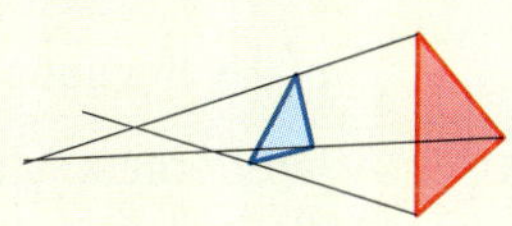

- In part (b), segments connecting corresponding vertices are parallel.

On Your Own

- Ask students to explain their answers.

Key Idea

- Write the Key Idea, including the example in the coordinate plane.

Discuss

- **Write:** Scale factor = side length of image : side length of original
- Put numbers on the vertical sides of the triangles in Example 1(a) to approximate the lengths, such as 4 (blue) and 8 (red). The blue triangle is reduced by an approximate scale factor of 4 : 8 or $\frac{1}{2}$.
- ? "Can the red triangle in Example 1(a) be a dilation of the blue triangle?" yes; It is an enlargement of the blue triangle.
- Discuss the difference between a scale factor for an enlargement, $k > 1$, and a scale factor for a reduction, $k > 0$ and $k < 1$.
- ? "What do you think the image would look like if $k = 1$?" The two figures would be congruent and there would be no center of dilation.

Goal Today's lesson is drawing **dilations** in the coordinate plane.

Technology for the Teacher

Dynamic Classroom

Lesson Tutorials
Lesson Plans
Answer Presentation Tool

English Language Learners

Vocabulary

Discussion about dilations brings up a lot of vocabulary words from previous lessons. Make sure that students are familiar with the definitions and uses of *ratio*, *similar figures*, and *scale factor*.

Extra Example 1

Tell whether the blue figure is a dilation of the red figure. Explain.

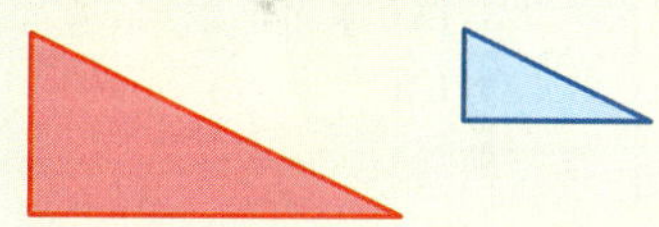

yes; Lines connecting corresponding vertices meet at a point.

On Your Own

1. no; It is a reflection.
2. yes; Lines connecting corresponding vertices meet at a point.

Laurie's Notes

Extra Example 2

The vertices of a triangle are $D(1, 4)$, $E(1, 1)$, and $F(3, 1)$. Draw the triangle and its image after a dilation with a scale factor of 2. Identify the type of dilation.

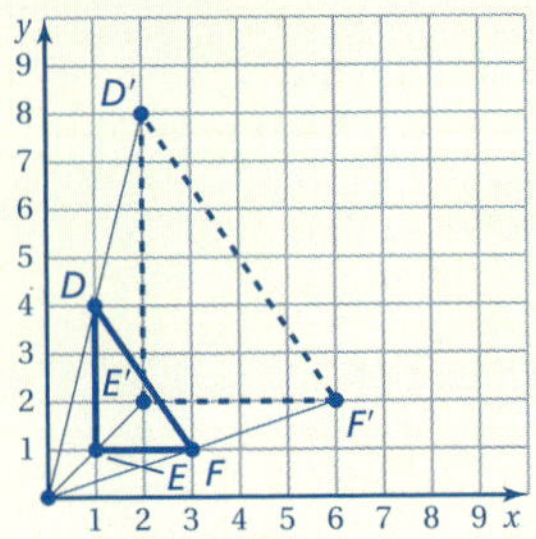

enlargement

Extra Example 3

The vertices of a rectangle are $J(-4, 2)$, $K(4, 2)$, $L(4, -2)$ and $M(-4, -2)$. Draw the rectangle and its image after a dilation with a scale factor of 0.5. Identify the type of dilation.

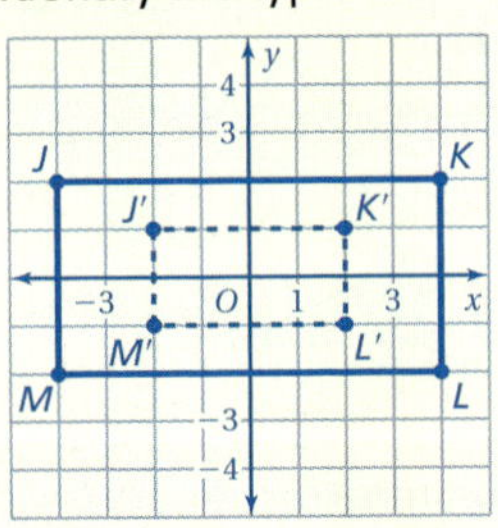

reduction

On Your Own

3. $A'(2, 6), B'(4, 6), C'(4, 2)$
4. $W'(-1, -1.5), X'(-1, 2), Y'(1, 2), Z'(1, -1.5)$

Example 2

- Work through the example.
- Use color to identify the two different triangles and their vertices.
- Note that the two triangles are similar. Side AB is horizontal and its image side $A'B'$ is horizontal. Side BC is vertical and its image side $B'C'$ is vertical. Side AC and its image side $A'C'$ are parallel.
- Refer to the Study Tip.
- **Construct Viable Arguments and Critique the Reasoning of Others** and **Attend to Precision:** Ask the following questions.
  - ? "How do you think the perimeters of the two triangles compare? Explain." Because the two triangles are similar with a scale factor of 3, the perimeter of the larger triangle is 3 times the perimeter of the smaller triangle.
  - ? "How do you think the areas of the two triangles compare? Explain." The area of the larger triangle is the square of the scale factor, or $3^2 = 9$ times greater than the area of the smaller triangle.

Example 3

- Work through the example.
- Use color to identify the two different rectangles and their vertices.

? "What is the product of a positive integer and a negative integer?" negative integer

? "Where is the center of dilation?" the origin (0, 0)

- **Construct Viable Arguments and Critique the Reasoning of Others** and **Attend to Precision:** Ask the following questions.
  - ? "How do you think the perimeters of the two rectangles compare? Explain." Because the two rectangles are similar with a scale factor of 0.5, the perimeter of the smaller rectangle is $\frac{1}{2}$ the perimeter of the larger rectangle.
  - ? "How do you think the areas of the two rectangles compare? Explain." The area of the smaller rectangle is the square of the scale factor, or $\left(\frac{1}{2}\right)^2 = \frac{1}{4}$ times the area of the larger rectangle.

On Your Own

- Students should make a table to record the original coordinates and the coordinates of the dilation.
- Ask students to draw each dilation.

EXAMPLE 2 Dilating a Figure

Draw the image of Triangle *ABC* after a dilation with a scale factor of 3. Identify the type of dilation.

Multiply each x- and y-coordinate by the scale factor 3.

Vertices of ABC	$(3x, 3y)$	Vertices of $A'B'C'$
$A(1, 3)$	$(3 \cdot 1, 3 \cdot 3)$	$A'(3, 9)$
$B(2, 3)$	$(3 \cdot 2, 3 \cdot 3)$	$B'(6, 9)$
$C(2, 1)$	$(3 \cdot 2, 3 \cdot 1)$	$C'(6, 3)$

Study Tip

You can check your answer by drawing a line from the origin through each vertex of the original figure. The vertices of the image should lie on these lines.

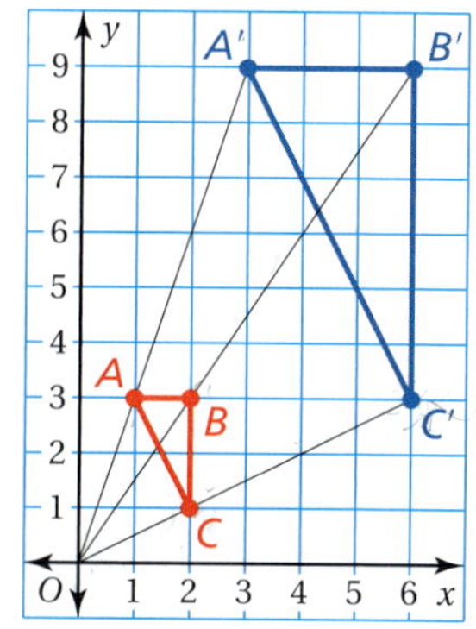

∴ The image is shown at the right. The dilation is an *enlargement* because the scale factor is greater than 1.

EXAMPLE 3 Dilating a Figure

Draw the image of Rectangle *WXYZ* after a dilation with a scale factor of 0.5. Identify the type of dilation.

Multiply each x- and y-coordinate by the scale factor 0.5.

Vertices of $WXYZ$	$(0.5x, 0.5y)$	Vertices of $W'X'Y'Z'$
$W(-4, -6)$	$(0.5 \cdot (-4), 0.5 \cdot (-6))$	$W'(-2, -3)$
$X(-4, 8)$	$(0.5 \cdot (-4), 0.5 \cdot 8)$	$X'(-2, 4)$
$Y(4, 8)$	$(0.5 \cdot 4, 0.5 \cdot 8)$	$Y'(2, 4)$
$Z(4, -6)$	$(0.5 \cdot 4, 0.5 \cdot (-6))$	$Z'(2, -3)$

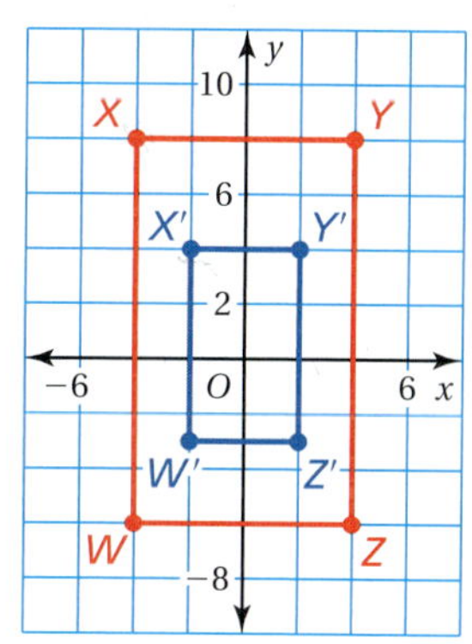

∴ The image is shown at the right. The dilation is a *reduction* because the scale factor is greater than 0 and less than 1.

On Your Own

Now You're Ready
Exercises 13–18

3. **WHAT IF?** Triangle ABC in Example 2 is dilated by a scale factor of 2. What are the coordinates of the image?
4. **WHAT IF?** Rectangle $WXYZ$ in Example 3 is dilated by a scale factor of $\frac{1}{4}$. What are the coordinates of the image?

EXAMPLE 4 Using More than One Transformation

The vertices of a trapezoid are $A(-2, -1)$, $B(-1, 1)$, $C(0, 1)$, and $D(0, -1)$. Dilate the trapezoid with respect to the origin using a scale factor of 2. Then translate it 6 units right and 2 units up. What are the coordinates of the image?

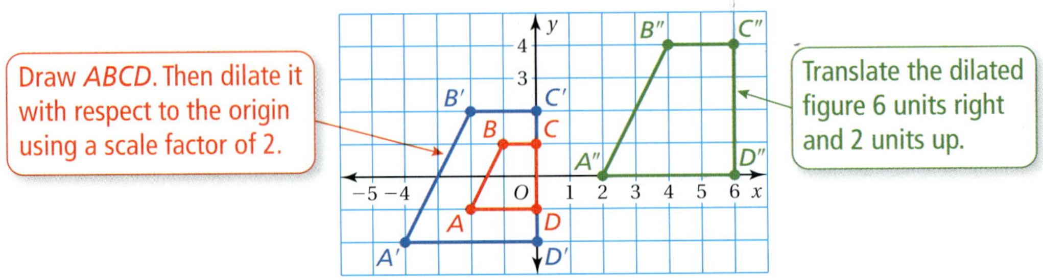

The coordinates of the image are $A''(2, 0)$, $B''(4, 4)$, $C''(6, 4)$, and $D''(6, 0)$.

The image of a translation, reflection, or rotation is congruent to the original figure, and the image of a dilation is similar to the original figure. So, two figures are similar when one can be obtained from the other by a sequence of translations, reflections, rotations, and dilations.

EXAMPLE 5 Describing a Sequence of Transformations

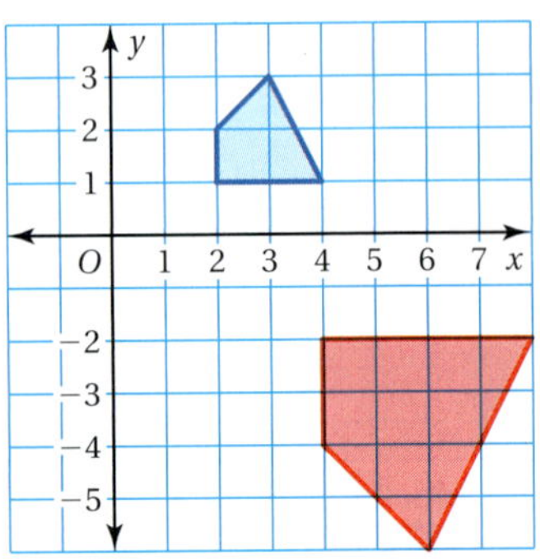

The red figure is similar to the blue figure. Describe a sequence of transformations in which the blue figure is the image of the red figure.

From the graph, you can see that the blue figure is one-half the size of the red figure. So, begin with a dilation with respect to the origin using a scale factor of $\frac{1}{2}$.

After dilating, you need to flip the figure in the x-axis.

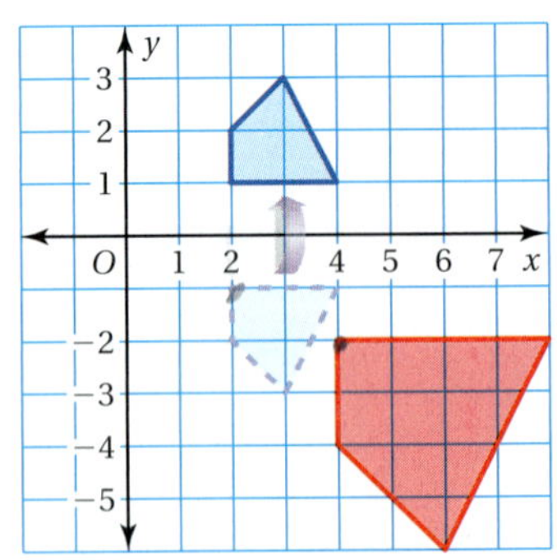

So, one possible sequence of transformations is a dilation with respect to the origin using a scale factor of $\frac{1}{2}$ followed by a reflection in the x-axis.

On Your Own

Now You're Ready
Exercises 23–28

5. In Example 4, use a scale factor of 3 in the dilation. Then rotate the figure 180° about the image of vertex C. What are the coordinates of the image?

6. In Example 5, can you reflect the red figure first, and then perform the dilation to obtain the blue figure? Explain.

Laurie's Notes

Example 4

- The legs of the original trapezoid are not congruent, and there is a right angle. Both of these attributes will help students with orientation questions.
- Plot the original trapezoid. Say that you will perform two transformations.

? "How will a dilation with a scale factor of 2 change the trapezoid?" It will double the size of the trapezoid.

? "What will a translation do to the dilated trapezoid?" It will shift the trapezoid to another location.

- Set up a table to organize the vertices, then draw each trapezoid.

Vertices of *ABCD*	Vertices of *A′B′C′D′*	Vertices of *A″B″C″D″*
$A(-2, -1)$	$A'(-4, -2)$	$A''(2, 0)$
$B(-1, 1)$	$B'(-2, 2)$	$B''(4, 4)$
$C(0, 1)$	$C'(0, 2)$	$C''(6, 4)$
$D(0, -1)$	$D'(0, -2)$	$D''(6, 0)$

? "How is each pair of trapezoids related?" *ABCD* and *A′B′C′D′* are similar; *A′B′C′D′* and *A″B″C″D″* are congruent; *ABCD* and *A″B″C″D″* are similar.

? **Construct Viable Arguments and Critique the Reasoning of Others** and **Attend to Precision:** "When is a figure and its image after a sequence of transformations congruent? Similar?" The figure and its image are congruent when there are translation(s), reflection(s), or rotation(s) with no dilations. They are similar when a dilation is involved.

Example 5

- Students should recognize that a dilation is involved.

? **Construct Viable Arguments and Critique the Reasoning of Others** and **Attend to Precision:** "What is the scale factor of the dilation and how do you know?" 0.5; The horizontal side of the blue trapezoid is 2 and the corresponding side of the red trapezoid is 4. So, the red figure must be reduced by a scale factor of $\frac{1}{2}$ to obtain the blue figure.

- Once the dilation is performed, students should recognize that the trapezoid must be reflected about the *x*-axis.
- **Discuss:** Make the connection between scale drawings, similarity, and dilation. Discuss scale factor in similar figures and how it relates to area and perimeter.
- **Extension:** Look back at Example 4 and ask students how the red figure can be transformed into the green figure using a single transformation. Use a dilation from a point other than the origin Then ask them what the center of dilation is and how to find it. $(-6, -2)$; Draw lines through vertices and find the point where the lines meet (intersect).

On Your Own

- Students can answer Question 6 without doing the transformations.

Closure

- **Exit Ticket:** The vertices of a triangle are $T(-2, 3)$, $R(3, 2)$, and $S(3, 1)$. Draw the triangle and its image after a dilation with a scale factor of 2. What are the coordinates of Triangle $T'R'S'$? $T'(-4, 6)$, $R'(6, 4)$, and $S'(6, 2)$

Extra Example 4

The vertices of a trapezoid are $A(-4, 0)$, $B(-2, 4)$, $C(2, 4)$, and $D(6, 0)$. Dilate the trapezoid with respect to the origin using a scale factor of 0.5. Then translate it 2 units right and 3 units down. What are the coordinates of the image?

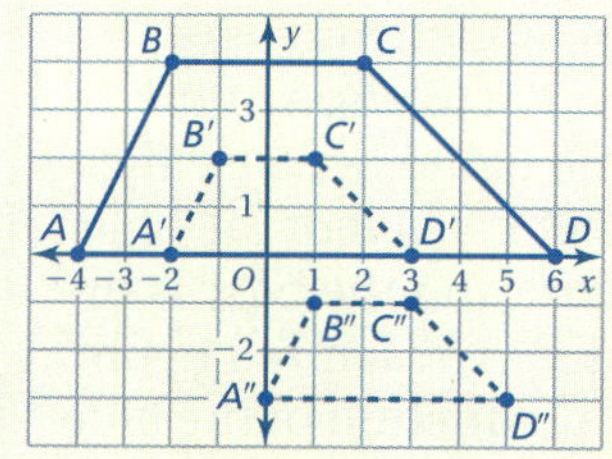

$A''(0, -3)$, $B''(1, -1)$, $C''(3, -1)$, $D''(5, -3)$

Extra Example 5

The red figure is similar to the blue figure. Describe a sequence of transformations in which the blue figure is the image of the red figure.

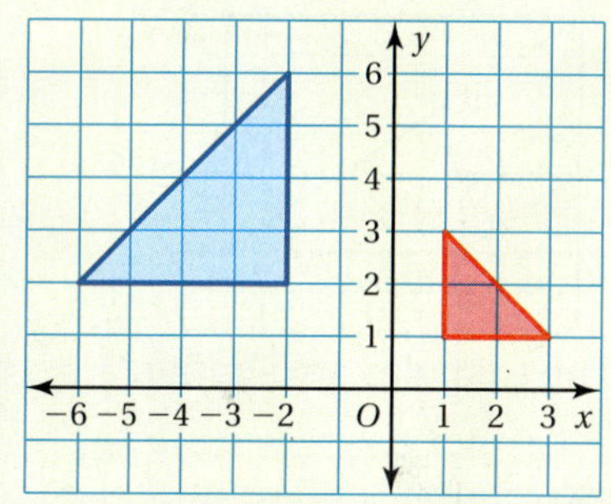

Sample answer: One possible sequence of transformations is a dilation with respect to the origin using a scale factor of 2 followed by a reflection in the *y*-axis.

On Your Own

5. 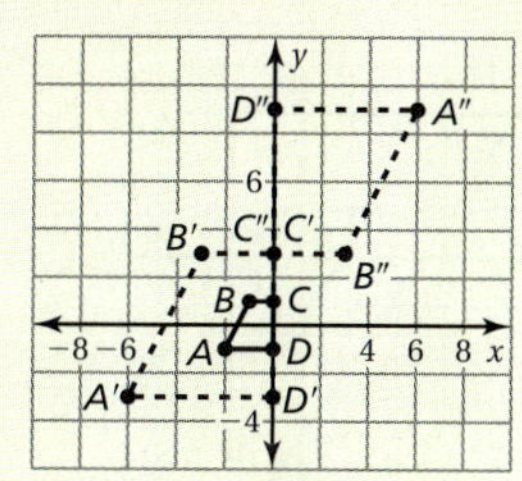

$A''(6, 9)$, $B''(3, 3)$, $C''(0, 3)$, $D''(0, 9)$

6. yes

A′(−6,−3), B′(−3,3), C′(0,3), D′(0,−3)

Vocabulary and Concept Check

1. A dilation changes the size of a figure. The image is similar, not congruent, to the original figure.
2. $k > 1$; $k > 0$ and $k < 1$
3. The middle red figure is not a dilation of the blue figure because the height is half of the blue figure and the base is the same. The left red figure is a reduction of the blue figure and the right red figure is an enlargement of the blue figure.

Practice and Problem Solving

4. 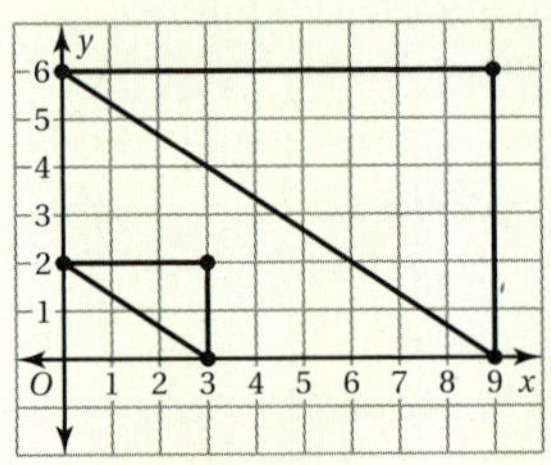

The triangles are similar.

5. 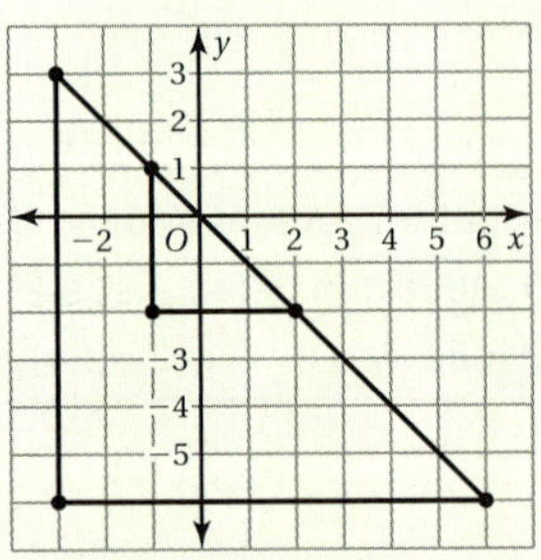

The triangles are similar.

6. 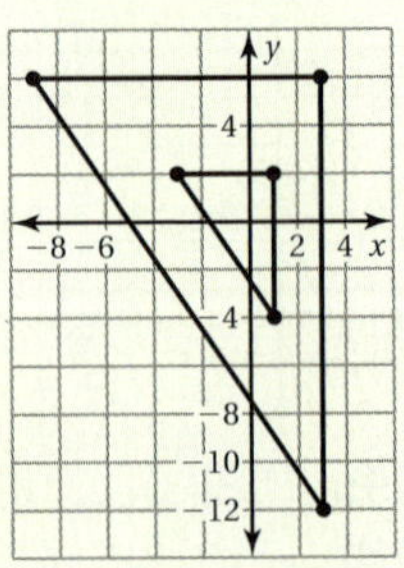

The triangles are similar.

7–18. See Additional Answers.

Assignment Guide and Homework Check

Level	Assignment	Homework Check
Accelerated	1–6, 10–18 even, 19, 20–32 even, 38–41	10, 14, 18, 26, 30

Common Errors

- **Exercises 7–12** Students may think that a figure with the same shape as another is a dilation. Remind them that in order for a transformation to be a dilation, lines connecting corresponding vertices must meet at a point.
- **Exercises 13–18** Students may confuse the image of a dilation with the original polygon when both are drawn in the same coordinate plane. Remind them to pay attention to the prime notation when determining the type of dilation.

11.7 Record and Practice Journal

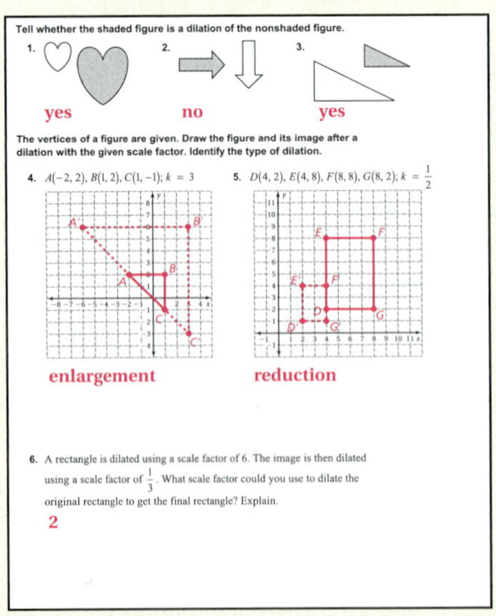

Tell whether the shaded figure is a dilation of the nonshaded figure.

1. yes
2. no
3. yes

The vertices of a figure are given. Draw the figure and its image after a dilation with the given scale factor. Identify the type of dilation.

4. $A(-2, 2), B(1, 2), C(1, -1); k = 3$

enlargement

5. $D(4, 2), E(4, 8), F(8, 8), G(8, 2); k = \frac{1}{2}$

reduction

6. A rectangle is dilated using a scale factor of 6. The image is then dilated using a scale factor of $\frac{1}{3}$. What scale factor could you use to dilate the original rectangle to get the final rectangle? Explain.

2

11.7 Exercises

Check It Out
Help with Homework
BigIdeasMath.com

Vocabulary and Concept Check

1. **VOCABULARY** How is a dilation different from other transformations?

2. **VOCABULARY** For what values of scale factor k is a dilation called an *enlargement*? a *reduction*?

3. **REASONING** Which figure is *not* a dilation of the blue figure? Explain.

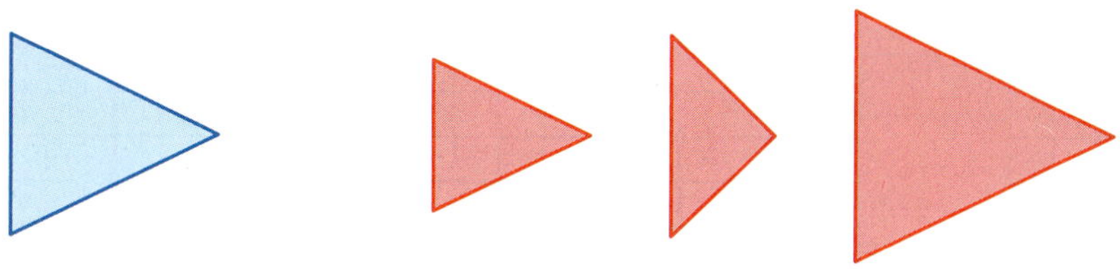

Practice and Problem Solving

Draw the triangle with the given vertices. Multiply each coordinate of the vertices by 3, and then draw the new triangle. How are the two triangles related?

4. (0, 2), (3, 2), (3, 0)

5. (−1, 1), (−1, −2), (2, −2)

6. (−3, 2), (1, 2), (1, −4)

Tell whether the blue figure is a dilation of the red figure.

1 7.

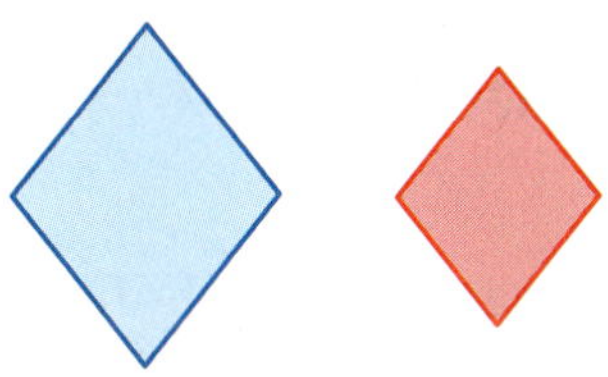

8.

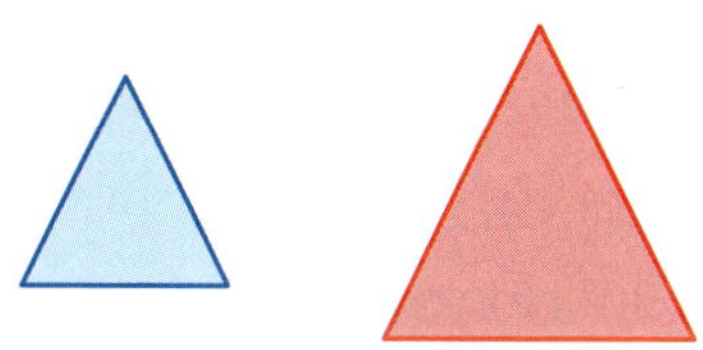

9.

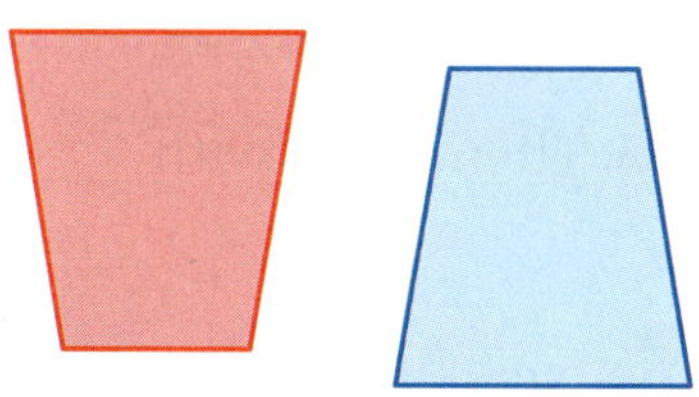

10.

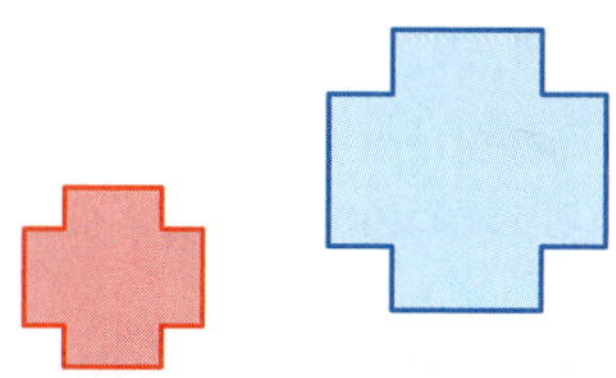

11.

12.

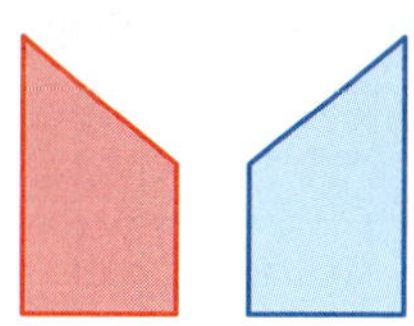

The vertices of a figure are given. Draw the figure and its image after a dilation with the given scale factor. Identify the type of dilation.

2 3 13. $A(1, 1)$, $B(1, 4)$, $C(3, 1)$; $k = 4$

14. $D(0, 2)$, $E(6, 2)$, $F(6, 4)$; $k = 0.5$

15. $G(-2, -2)$, $H(-2, 6)$, $J(2, 6)$; $k = 0.25$

16. $M(2, 3)$, $N(5, 3)$, $P(5, 1)$; $k = 3$

17. $Q(-3, 0)$, $R(-3, 6)$, $T(4, 6)$, $U(4, 0)$; $k = \frac{1}{3}$

18. $V(-2, -2)$, $W(-2, 3)$, $X(5, 3)$, $Y(5, -2)$; $k = 5$

19. ERROR ANALYSIS Describe and correct the error in listing the coordinates of the image after a dilation with a scale factor of $\frac{1}{2}$.

Vertices of ABC	$(2x, 2y)$	Vertices of $A'B'C'$
$A(2, 5)$	$(2 \cdot 2, 2 \cdot 5)$	$A'(4, 10)$
$B(2, 0)$	$(2 \cdot 2, 2 \cdot 0)$	$B'(4, 0)$
$C(4, 0)$	$(2 \cdot 4, 2 \cdot 0)$	$C'(8, 0)$

The blue figure is a dilation of the red figure. Identify the type of dilation and find the scale factor.

20.

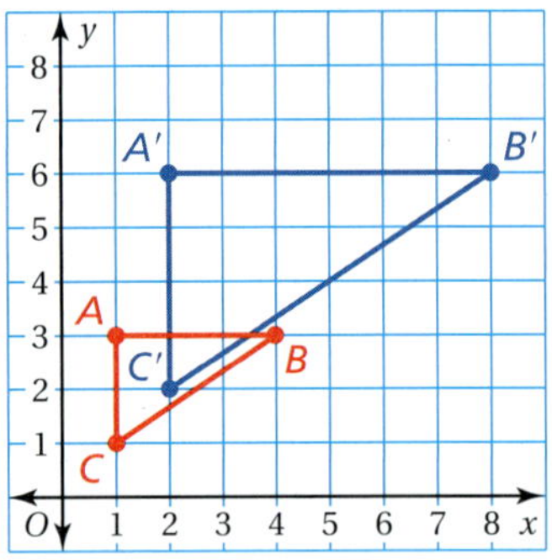

21.

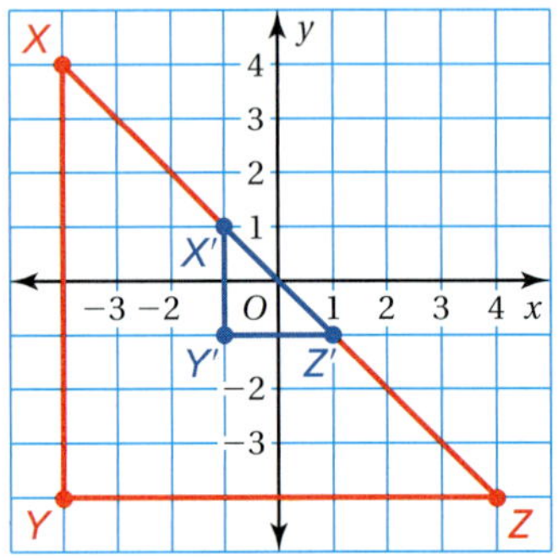

22.

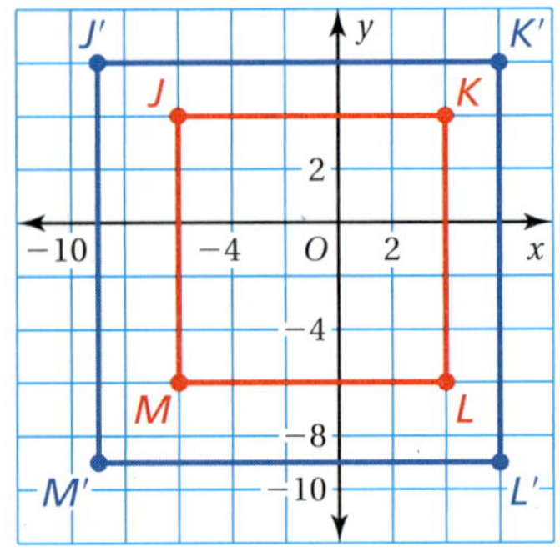

The vertices of a figure are given. Find the coordinates of the figure after the transformations given.

4 **23.** $A(-5, 3)$, $B(-2, 3)$, $C(-2, 1)$, $D(-5, 1)$

Reflect in the y-axis. Then dilate with respect to the origin using a scale factor of 2.

24. $F(-9, -9)$, $G(-3, -6)$, $H(-3, -9)$

Dilate with respect to the origin using a scale factor of $\frac{2}{3}$. Then translate 6 units up.

25. $J(1, 1)$, $K(3, 4)$, $L(5, 1)$

Rotate 90° clockwise about the origin. Then dilate with respect to the origin using a scale factor of 3.

26. $P(-2, 2)$, $Q(4, 2)$, $R(2, -6)$, $S(-4, -6)$

Dilate with respect to the origin using a scale factor of 5. Then dilate with respect to the origin using a scale factor of 0.5.

The red figure is similar to the blue figure. Describe a sequence of transformations in which the blue figure is the image of the red figure.

5 **27.**

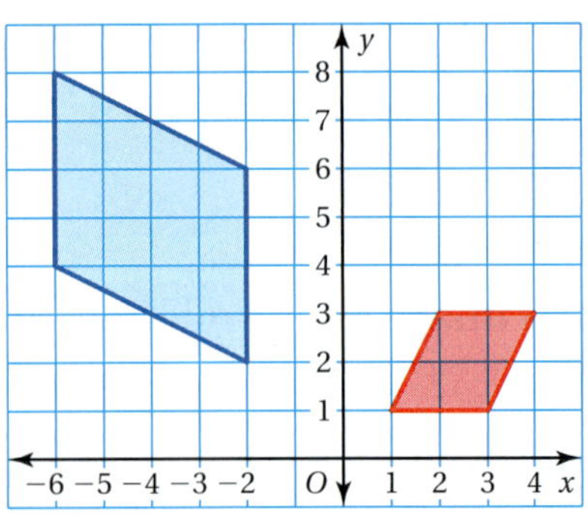

28.

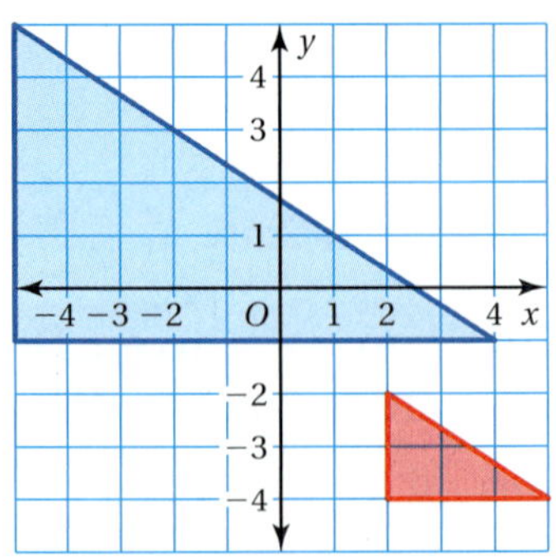

29. STRUCTURE In Exercises 27 and 28, is the blue figure still the image of the red figure when you perform the sequence in the opposite order? Explain.

Common Errors

- **Exercises 20–22** Students may indicate the wrong type of dilation and give the reciprocal of the scale factor. Remind students that you multiply each coordinate of the original figure by the scale factor to obtain the corresponding image coordinate.
- **Exercises 27 and 28** Students may give the sequence of transformations in reverse order. Encourage students to perform the sequence of transformations on the original figure to check their answers

Practice and Problem Solving

19. Each coordinate was multiplied by 2 instead of divided by 2. The coordinates should be $A'(1, 2.5)$, $B'(1, 0)$, and $C'(2, 0)$.

20. enlargement; 2

21. reduction; $\frac{1}{4}$

22. enlargement; $\frac{3}{2}$

23. $A''(10, 6)$, $B''(4, 6)$, $C''(4, 2)$, $D''(10, 2)$

24. $F''(-6, 0)$, $G''(-2, 2)$, $H''(-2, 0)$

25. $J''(3, -3)$, $K''(12, -9)$, $L''(3, -15)$

26. $P''(-5, 5)$, $Q''(10, 5)$, $R''(5, -15)$, $S''(-10, -15)$

27. *Sample answer:* Rotate 90° counterclockwise about the origin and then dilate with respect to the origin using a scale factor of 2

28. *Sample answer:* Dilate with respect to the origin using a scale factor of 3 and then translate 11 units left and 11 units up

29. Exercise 27: yes; Exercise 28: no; Explanations will vary based on sequences chosen in Exercises 27 and 28.

English Language Learners

Vocabulary

Help students remember the mathematical meaning of dilation. The everyday meaning is to enlarge or expand, such as the dilation of a pupil of an eye. In mathematics, a dilation makes a figure either larger or smaller, depending on the scale factor of the dilation.

Practice and Problem Solving

30. See *Taking Math Deeper.*

31–37. See Additional Answers.

Fair Game Review

38. complementary; $x = 50$

39. supplementary; $x = 16$

40. complementary; $x = 9$

41. B

Mini-Assessment

1. Tell whether the blue figure is a dilation of the red figure.

no

2. The vertices of a triangle are $A(1, 2)$, $B(2, 4)$, $C(3, 1)$. Draw the triangle and its image after a dilation with a scale factor of 2. Identify the type of dilation.

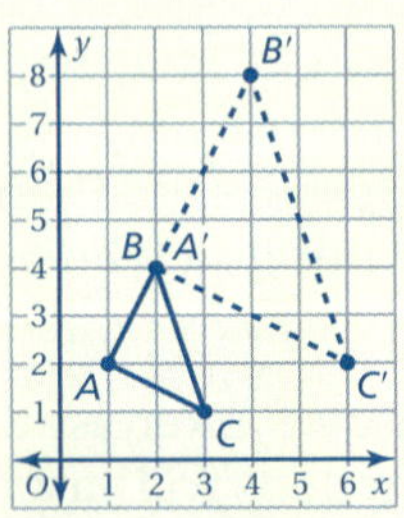

enlargement

3. The red figure is similar to the blue figure. Describe a sequence of transformations in which the blue figure is the image of the red figure.

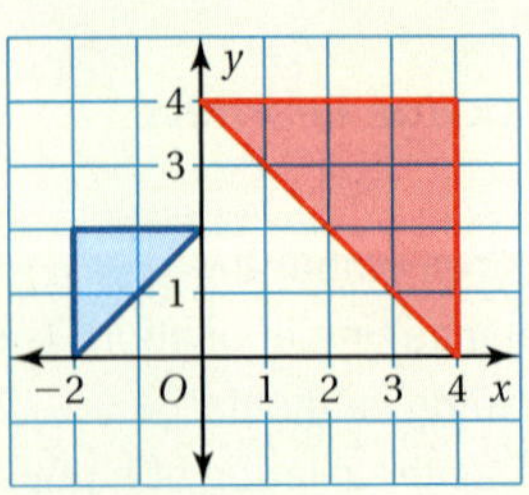

Sample Answer: A dilation with respect to the origin using a scale factor of 0.5 followed by a reflection in the y-axis.

Taking Math Deeper

Exercise 30

The concept seems to occur frequently on state tests. When you enlarge a figure by a scale factor of k:

1. the perimeter increases by a scale factor of k, and
2. the area increases by a scale factor of k^2.

1 Draw a diagram.

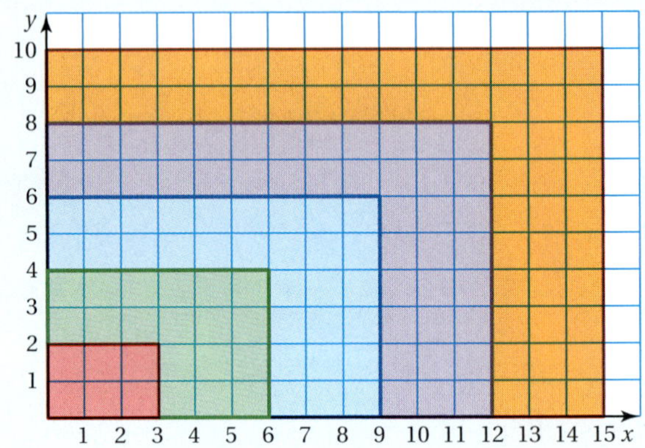

2 Compute the areas.

Area $= 3 \times 2 = 6$

$k = 2$: Area $= 6 \times 4 = 24$

$k = 3$: Area $= 9 \times 6 = 54$

$k = 4$: Area $= 12 \times 8 = 96$

$k = 5$: Area $= 15 \times 10 = 150$

3 How many times greater is the area?

$k = 2$: $\frac{24}{6} = 4$ times greater

$k = 3$: $\frac{54}{6} = 9$ times greater

$k = 4$: $\frac{96}{6} = 16$ times greater

$k = 5$: $\frac{150}{6} = 25$ times greater

Reteaching and Enrichment Strategies

If students need help. . .	If students got it. . .
Resources by Chapter • Practice A and Practice B • Puzzle Time Record and Practice Journal Practice Differentiating the Lesson Lesson Tutorials Skills Review Handbook	Resources by Chapter • Enrichment and Extension • Technology Connection Start the next section

30. **OPEN-ENDED** Draw a rectangle on a coordinate plane. Choose a scale factor of 2, 3, 4, or 5, and then dilate the rectangle. How many times greater is the area of the image than the area of the original rectangle?

31. **SHADOW PUPPET** You can use a flashlight and a shadow puppet (your hands) to project shadows on the wall.

 a. Identify the type of dilation.

 b. What does the flashlight represent?

 c. The length of the ears on the shadow puppet is 3 inches. The length of the ears on the shadow is 4 inches. What is the scale factor?

 d. Describe what happens as the shadow puppet moves closer to the flashlight. How does this affect the scale factor?

32. **REASONING** A triangle is dilated using a scale factor of 3. The image is then dilated using a scale factor of $\frac{1}{2}$. What scale factor could you use to dilate the original triangle to get the final image? Explain.

CRITICAL THINKING The coordinate notation shows how the coordinates of a figure are related to the coordinates of its image after transformations. What are the transformations? Are the figure and its image similar or congruent? Explain.

33. $(x, y) \rightarrow (2x + 4, 2y - 3)$

34. $(x, y) \rightarrow (-x - 1, y - 2)$

35. $(x, y) \rightarrow \left(\frac{1}{3}x, -\frac{1}{3}y\right)$

36. **STRUCTURE** How are the transformations $(2x + 3, 2y - 1)$ and $(2(x + 3), 2(y + 1))$ different?

37. **Problem Solving** The vertices of a trapezoid are $A(-2, 3)$, $B(2, 3)$, $C(5, -2)$, and $D(-2, -2)$. Dilate the trapezoid with respect to vertex A using a scale factor of 2. What are the coordinates of the image? Explain the method you used.

Fair Game Review What you learned in previous grades & lessons

Tell whether the angles are *complementary* or *supplementary*. Then find the value of *x*. *(Section 7.2)*

38.

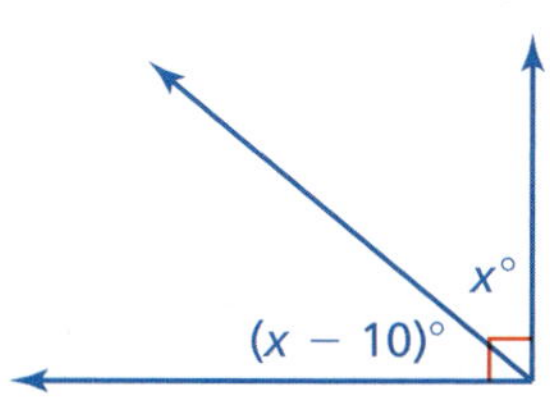

39.

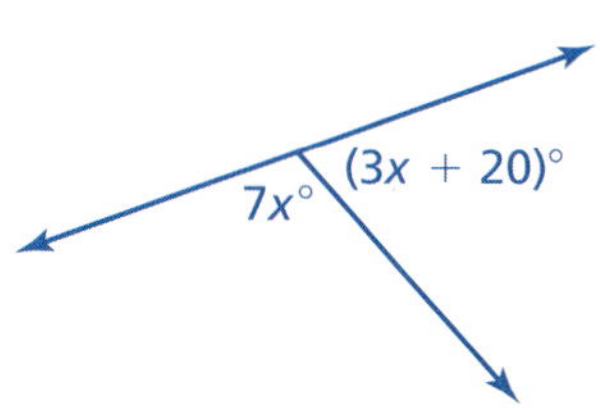

40.

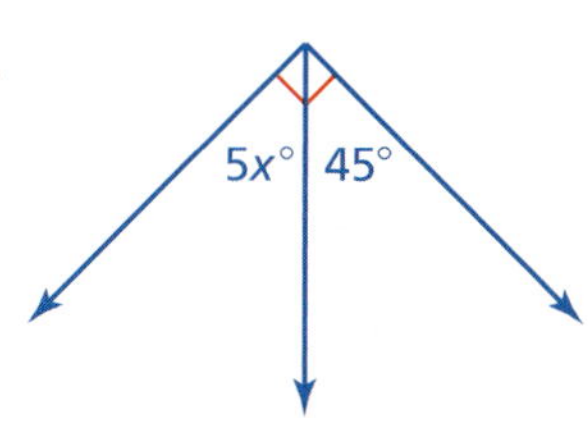

41. **MULTIPLE CHOICE** Which quadrilateral is *not* a parallelogram? *(Section 7.4)*

 (A) rhombus (B) trapezoid (C) square (D) rectangle

11.5–11.7 Quiz

Check It Out
Progress Check
BigIdeasMath.com

1. Tell whether the two rectangles are similar. Explain your reasoning. *(Section 11.5)*

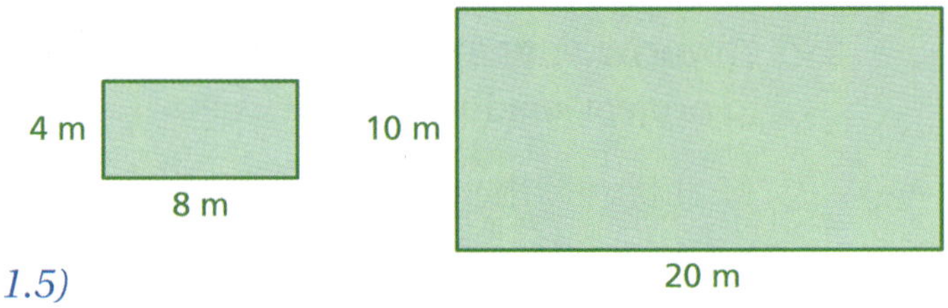

The figures are similar. Find *x*. *(Section 11.5)*

2.

3.

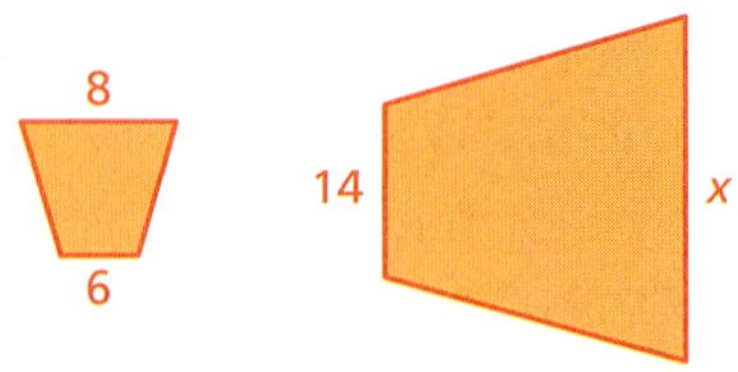

The two figures are similar. Find the ratios (red to blue) of the perimeters and of the areas. *(Section 11.6)*

4.

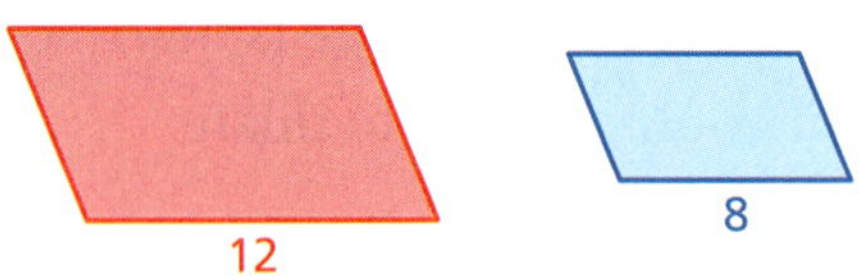

5.

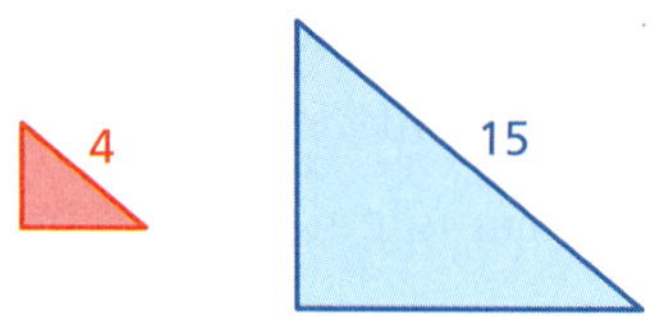

Tell whether the blue figure is a dilation of the red figure. *(Section 11.7)*

6.

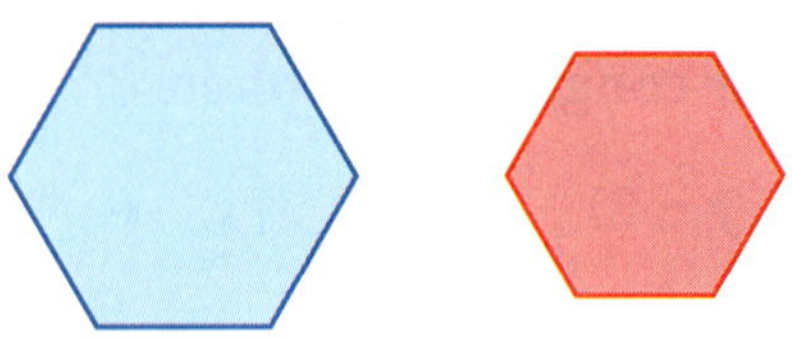

7.

8. SCREENS The TV screen is similar to the computer screen. What is the area of the TV screen? *(Section 11.6)*

9. GEOMETRY The vertices of a rectangle are $A(2, 4)$, $B(5, 4)$, $C(5, -1)$, and $D(2, -1)$. Dilate the rectangle with respect to the origin using a scale factor of $\frac{1}{2}$. Then translate it 4 units left and 3 units down. What are the coordinates of the image? *(Section 11.7)*

10. TENNIS COURT The tennis courts for singles and doubles matches are different sizes. Are the courts similar? Explain. *(Section 11.5)*

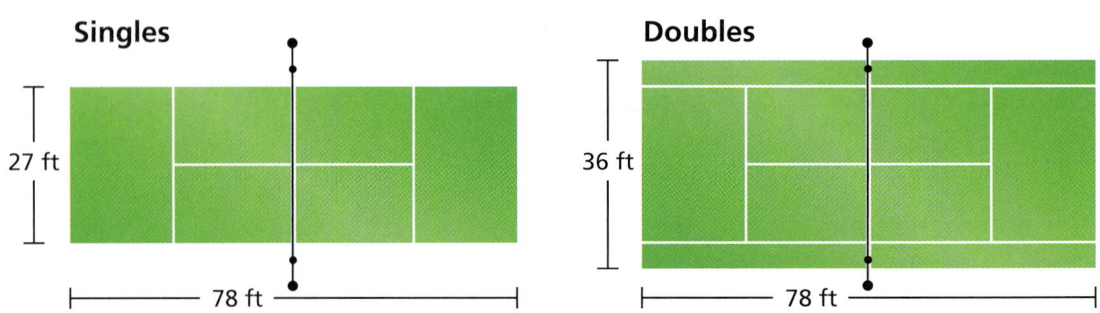

Alternative Assessment Options

Math Chat | Student Reflective Focus Question
Structured Interview | **Writing Prompt**

Writing Prompt

Ask students to write a few paragraphs about how transformations of figures are used in real life. They should include real-life examples of translations, reflections, rotations, and dilations. Then have students share what they have written with the class.

Study Help Sample Answers

Remind students to complete Graphic Organizers for the rest of the chapter.

4.

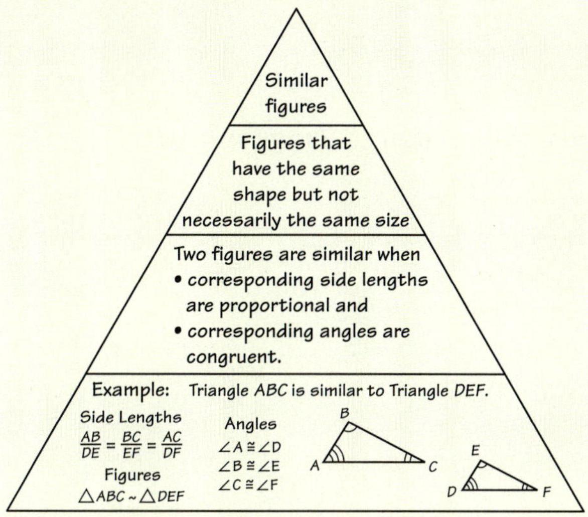

5–8. Available at *BigIdeasMath.com.*

Reteaching and Enrichment Strategies

If students need help. . .	If students got it. . .
Resources by Chapter • Practice A and Practice B • Puzzle Time Lesson Tutorials *BigIdeasMath.com*	Resources by Chapter • Enrichment and Extension • Technology Connection Game Closet at *BigIdeasMath.com* Start the Chapter Review

Answers

1. yes; Corresponding angles are congruent. Because $\frac{4}{10} = \frac{8}{20}$, the corresponding side lengths are proportional.
2. 16.5
3. $18\frac{2}{3}$
4. $\frac{3}{2}, \frac{9}{4}$
5. $\frac{4}{15}, \frac{16}{225}$
6. yes
7. no
8. 300 in.^{2}
9. $A''(-3, -1), B''(-1.5, -1), C''(-1.5, -3.5), D''(-3, -3.5)$
10. no; Corresponding side lengths are not proportional.

Technology for the Teacher

Online Assessment
Assessment Book
ExamView® Assessment Suite

For the Teacher
Additional Review Options

- *BigIdeasMath.com*
- Online Assessment
- Game Closet at *BigIdeasMath.com*
- Vocabulary Help
- Resources by Chapter

Answers

1. 3 ft
2. 20 ft
3. $\angle A$ and $\angle K$, $\angle B$ and $\angle L$, $\angle C$ and $\angle M$; Side *AB* and Side *KL*, Side *BC* and Side *LM*, Side *AC* and Side *KM*
4. $\angle R$ and $\angle W$, $\angle Q$ and $\angle X$, $\angle T$ and $\angle Y$, $\angle S$ and $\angle Z$; Side *RQ* and Side *WX*, Side *QT* and Side *XY*, Side *TS* and Side *YZ*, Side *SR* and Side *WZ*

Review of Common Errors

Exercises 1 and 2

- Students may forget to include the units in their answers. Remind them that the units must be included to make their answers complete.

Exercises 3 and 4

- Students may forget to write the angle symbol with the name. Remind them that *A* is a point and $\angle A$ is the angle.

11 Chapter Review

Check It Out
Vocabulary Help
BigIdeasMath.com

Review Key Vocabulary

congruent figures, *p. 468*
corresponding angles, *p. 468*
corresponding sides, *p. 468*
transformation, *p. 474*
image, *p. 474*
translation, *p. 474*
reflection, *p. 480*
line of reflection, *p. 480*
rotation, *p. 486*
center of rotation, *p. 486*
angle of rotation, *p. 486*
similar figures, *p. 496*
dilation, *p. 508*
center of dilation, *p. 508*
scale factor, *p. 508*

Review Examples and Exercises

11.1 Congruent Figures *(pp. 466–471)*

Trapezoids *EFGH* and *QRST* are congruent.

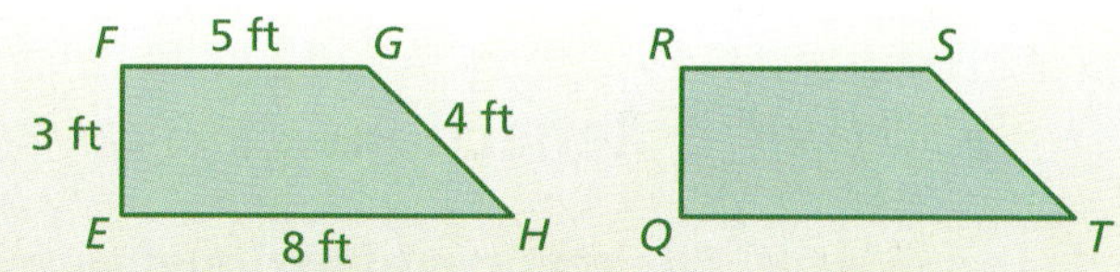

a. What is the length of side *QT*?

Side *QT* corresponds to side *EH*.

So, the length of side *QT* is 8 feet.

b. Which angle of *QRST* corresponds to ∠*H*?

∠*T* corresponds to ∠*H*.

Exercises

Use the figures above.

1. What is the length of side *QR*?
2. What is the perimeter of *QRST*?

The figures are congruent. Name the corresponding angles and the corresponding sides.

3. 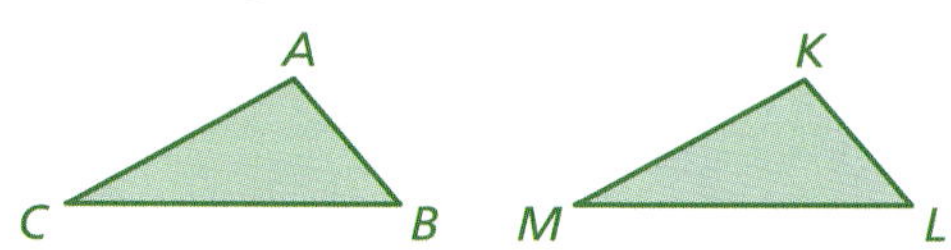

4. 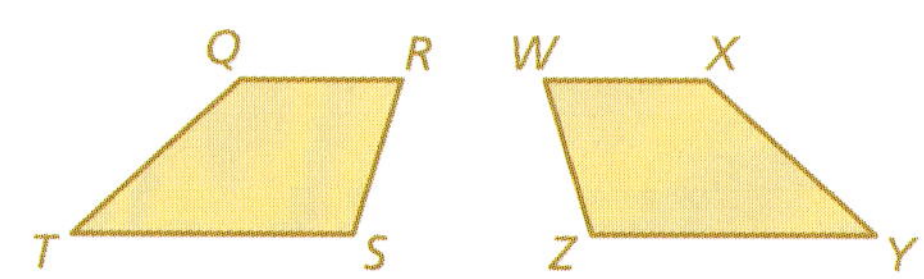

11.2 Translations *(pp. 472–477)*

Translate the red triangle 4 units left and 1 unit down. What are the coordinates of the image?

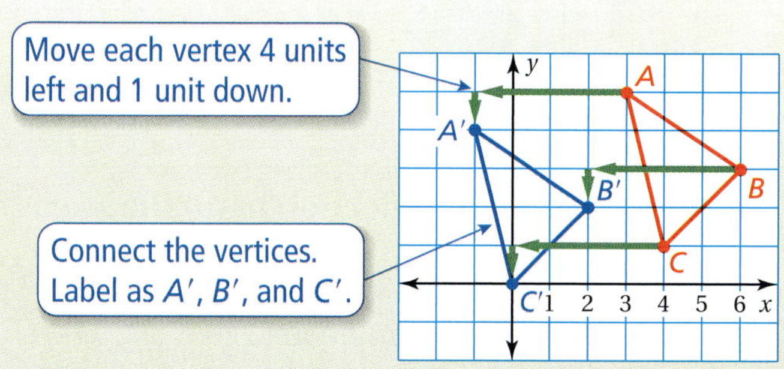

The coordinates of the image are $A'(-1, 4)$, $B'(2, 2)$, and $C'(0, 0)$.

Exercises

Tell whether the blue figure is a translation of the red figure.

5.

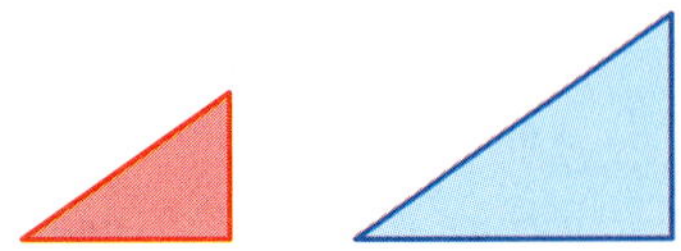

6.

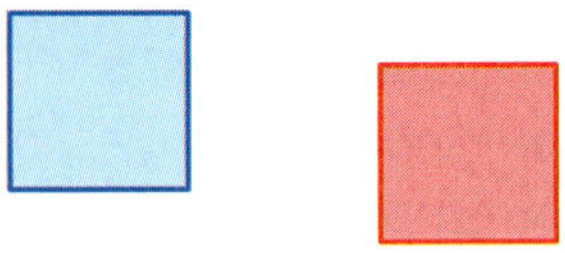

7. The vertices of a quadrilateral are $W(1, 2)$, $X(1, 4)$, $Y(4, 4)$, and $Z(4, 2)$. Draw the figure and its image after a translation 3 units left and 2 units down.

8. The vertices of a triangle are $A(-1, -2)$, $B(-2, 2)$, and $C(-3, 0)$. Draw the figure and its image after a translation 5 units right and 1 unit up.

11.3 Reflections (pp. 478–483)

The vertices of a triangle are $A(-2, 1)$, $B(4, 1)$, and $C(4, 4)$. Draw the figure and its reflection in the x-axis. What are the coordinates of the image?

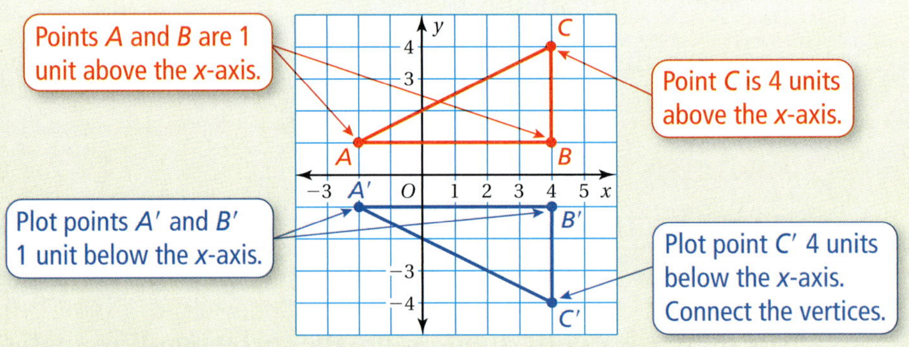

∴ The coordinates of the image are $A'(-2, -1)$, $B'(4, -1)$, and $C'(4, -4)$.

Exercises

Tell whether the blue figure is a reflection of the red figure.

9.

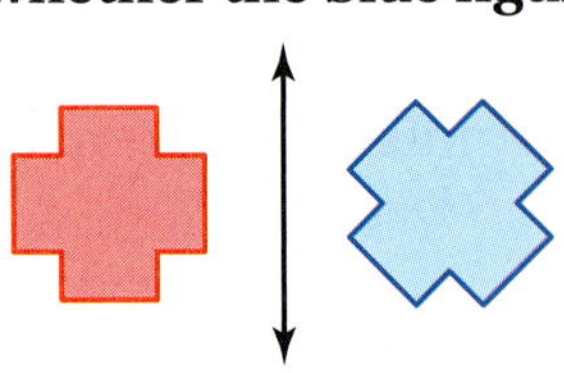

10.

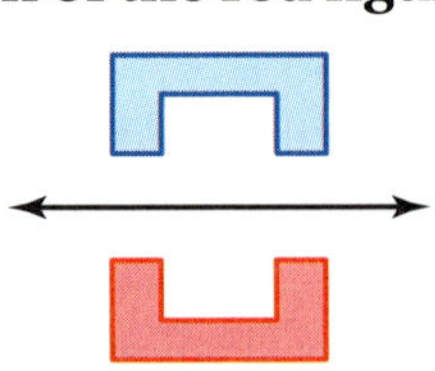

Draw the figure and its reflection in (a) the x-axis and (b) the y-axis.

11. $A(2, 0)$, $B(1, 5)$, $C(4, 3)$

12. $D(-5, -5)$, $E(-5, -1)$, $F(-2, -2)$, $G(-2, -5)$

13. The vertices of a rectangle are $E(-1, 1)$, $F(-1, 3)$, $G(-5, 3)$, and $H(-5, 1)$. Find the coordinates of the figure after reflecting in the x-axis, and then translating 3 units right.

Review of Common Errors (continued)

Exercises 5 and 6

- Students may forget that the objects must be the same size to be a translation. Remind them that the size stays the same. Tell students that when the size is different, it is a scale drawing.

Exercises 7 and 8

- Students may translate the shape the wrong direction or mix up the units for the translation. Tell them to write the direction of the translation on their drawings, using arrows to show the movement left, right, up, or down.

Exercises 9 and 10

- Some students may not be able to visualize whether the blue figure is a reflection of the red figure. Give them tracing paper and have them trace the objects. Then have them fold the paper to determine whether corresponding vertices of each figure align.

Exercises 11 and 12

- Some students may reflect the image obtained in part (a) to answer part (b). Point out that the directions are asking for a reflection of *the original figure* in each part of the exercise.

Exercises 11–13

- Students may confuse reflection in the *x*-axis with reflection in the *y*-axis (and vice versa). Have students label the axes in their drawings so it is clearer which way to reflect.

Answers

5. no **6.** yes

7.

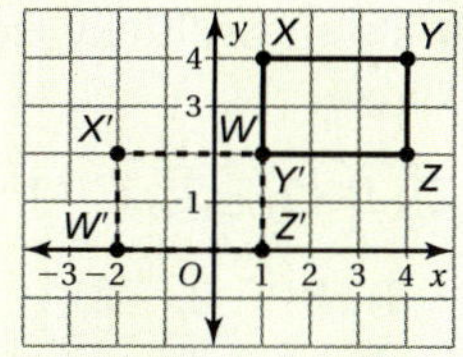

8.

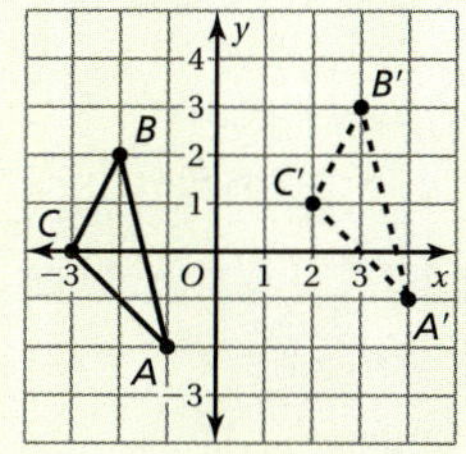

9. no **10.** yes

11. a.

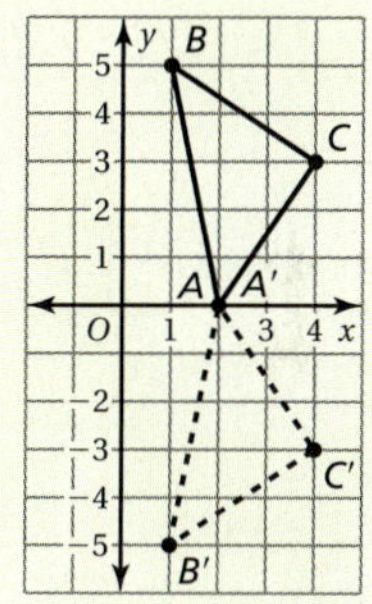

b.

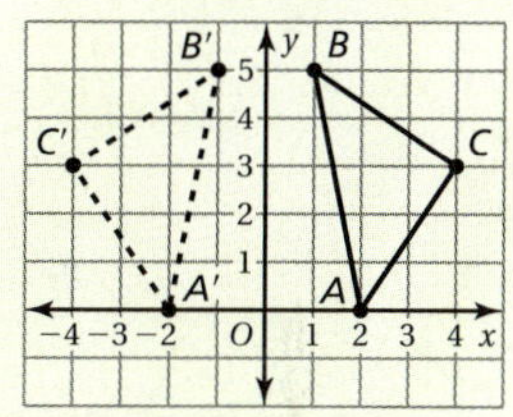

12. a.

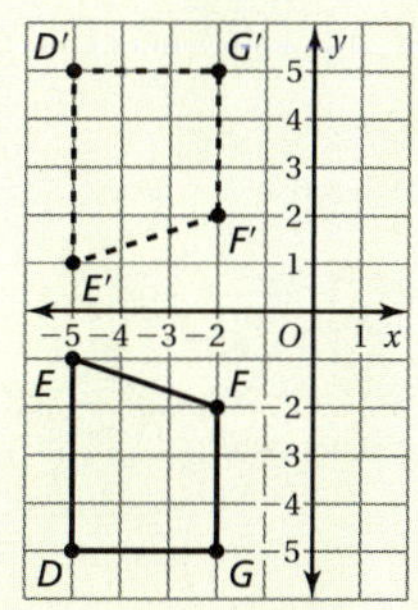

b.

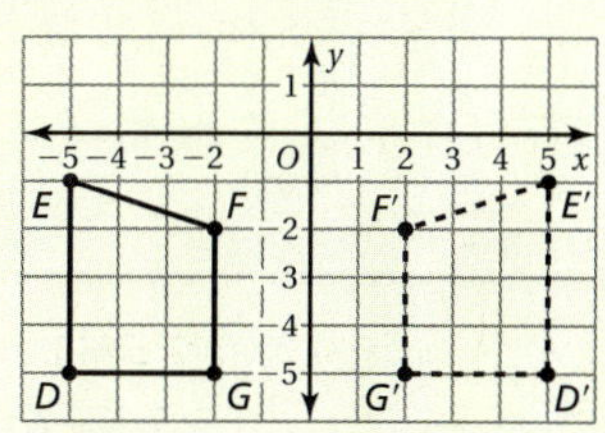

13. $E'(2, -1), F'(2, -3),$
$G'(-2, -3), H'(-2, -1)$

Answers

14. no

15. yes; 180° counterclockwise or clockwise

16. $A'(4, -2)$, $B'(2, -2)$, $C'(3, -4)$

17. $A'(-2, -4)$, $B'(-2, -2)$, $C'(-4, -3)$

Review of Common Errors (continued)

Exercises 14–17

- Students with minimal spatial skills may not be able to tell whether a figure is rotated or how to rotate a figure. Give them tracing paper and have them copy the figure and rotate it.

Exercises 18 and 19

- Students may think that because corresponding side lengths may have changed by the same amount, the figures are similar. Remind them that the *ratios* of corresponding side lengths must be the same for the two objects to be similar.

Exercises 20 and 21

- Students may write the proportion incorrectly. For example, for Exercise 20 they may write $\frac{14}{20} = \frac{x}{7}$ instead of $\frac{14}{20} = \frac{7}{x}$. Remind them that the lengths of corresponding sides should both be in the numerator or the denominator OR the side lengths of the larger shape or smaller shape should be in the numerators, and the corresponding side lengths of the other shape should be in the denominators.

Exercises 22–24

- Students may find the reciprocal of the ratio. For example, they may find the ratio of blue to red instead of red to blue. Remind students to read the directions carefully.
- When finding the ratio of the areas, students may forget to square the ratio of the corresponding side lengths. Remind them of the Key Idea at the bottom of page 78.

Exercises 25 and 26

- Students may think that a figure with the same shape as another is a dilation. Remind students that in order for a transformation to be a dilation, lines connecting corresponding vertices must meet at a point.

Exercises 27–29

- Students may confuse the image of a dilation with the original polygon when both are drawn in the same coordinate plane. Remind students to use and pay attention to the prime notation when drawing the image and determining the type of dilation.

11.4 Rotations (pp. 484–491)

The vertices of a triangle are $A(1, 1)$, $B(3, 2)$, and $C(2, 4)$. Rotate the triangle 90° counterclockwise about the origin. What are the coordinates of the image?

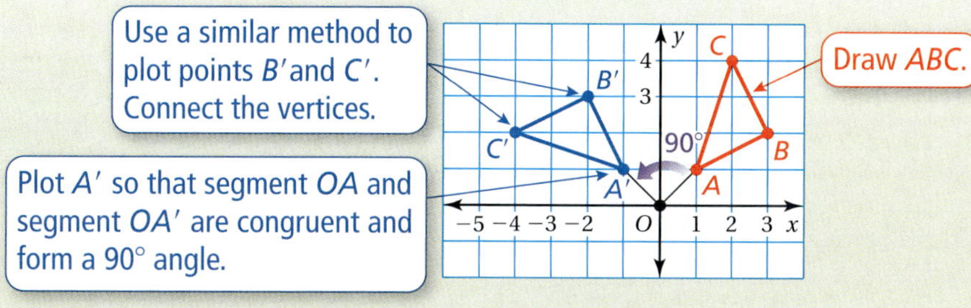

∴ The coordinates of the image are $A'(-1, 1)$, $B'(-2, 3)$, and $C'(-4, 2)$.

Exercises

Tell whether the blue figure is a rotation of the red figure about the origin. If so, give the angle and the direction of rotation.

14.

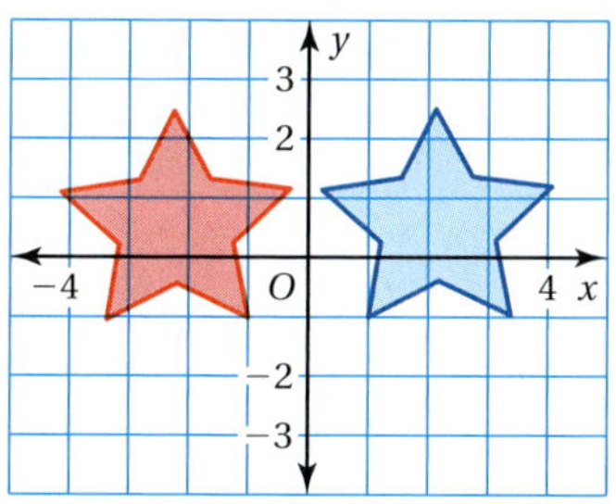

15.

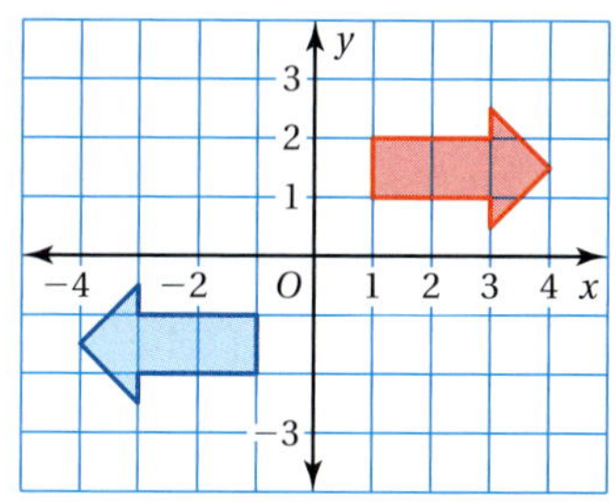

The vertices of a triangle are $A(-4, 2)$, $B(-2, 2)$, and $C(-3, 4)$. Rotate the triangle about the origin as described. Find the coordinates of the image.

16. 180°

17. 270° clockwise

11.5 Similar Figures (pp. 494–499)

a. Is Rectangle A similar to Rectangle B?

Each figure is a rectangle. So, corresponding angles are congruent. Check to see if corresponding side lengths are proportional.

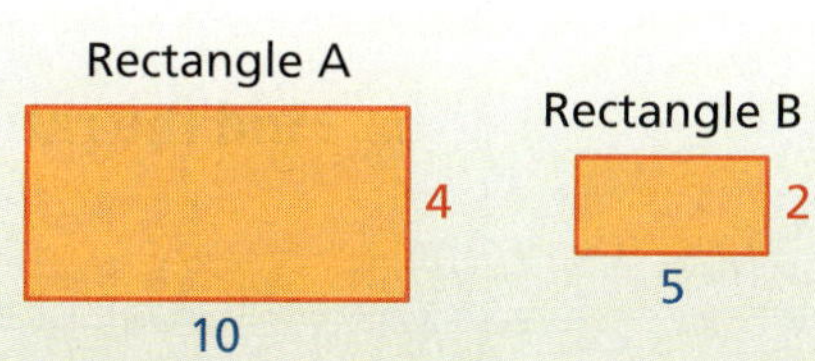

$$\frac{\text{Length of A}}{\text{Length of B}} = \frac{10}{5} = 2 \qquad \frac{\text{Width of A}}{\text{Width of B}} = \frac{4}{2} = 2$$

Proportional

∴ So, Rectangle A is similar to Rectangle B.

b. The two rectangles are similar. Find x.

Because the rectangles are similar, corresponding side lengths are proportional. So, write and solve a proportion to find x.

$\frac{10}{24} = \frac{4}{x}$	Write a proportion.
$10x = 96$	Cross Products Property
$x = 9.6$	Divide each side by 10.

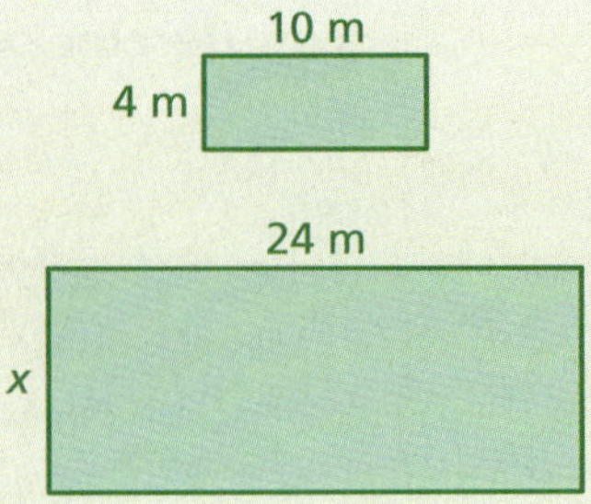

So, x is 9.6 meters.

Exercises

Tell whether the two figures are similar. Explain your reasoning.

18.

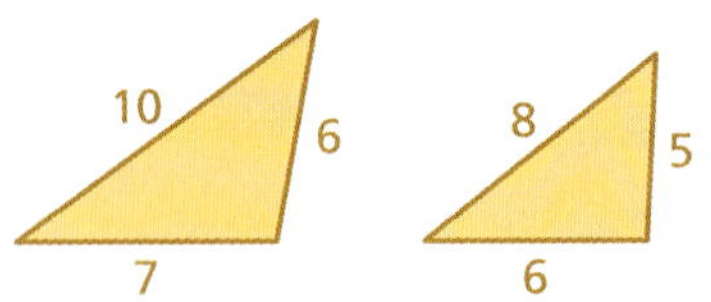

19.

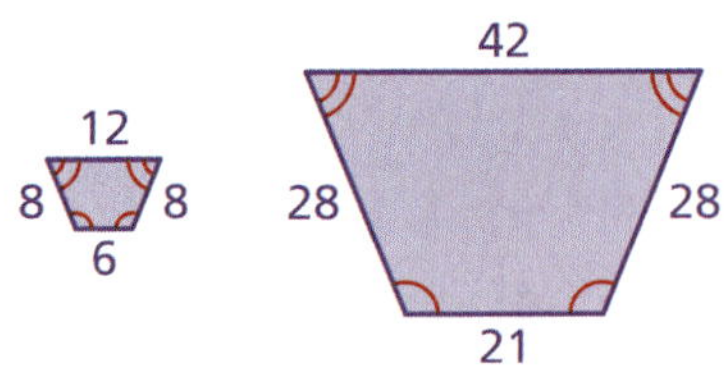

The figures are similar. Find x.

20.

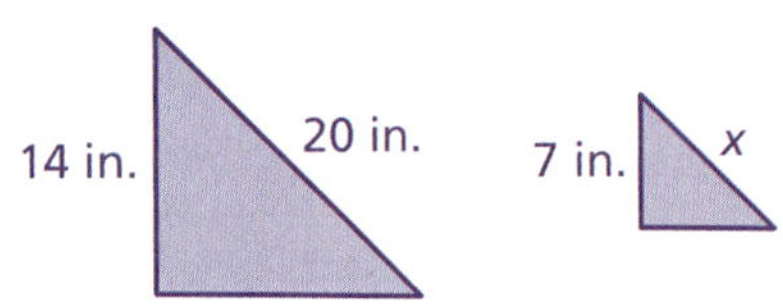

21.

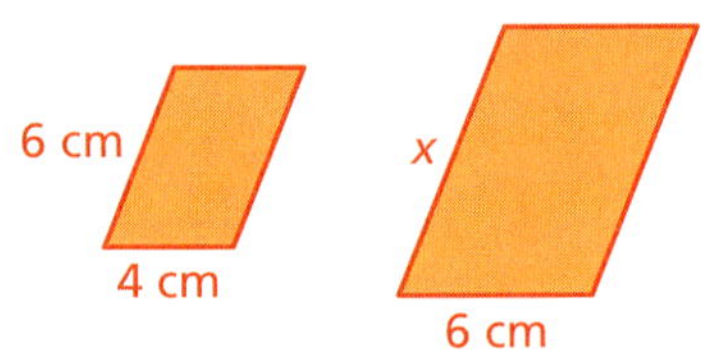

11.6 Perimeters and Areas of Similar Figures (pp. 500–505)

a. Find the ratio (red to blue) of the perimeters of the similar parallelograms.

$$\frac{\text{Perimeter of red parallelogram}}{\text{Perimeter of blue parallelogram}} = \frac{15}{9}$$

$$= \frac{5}{3}$$

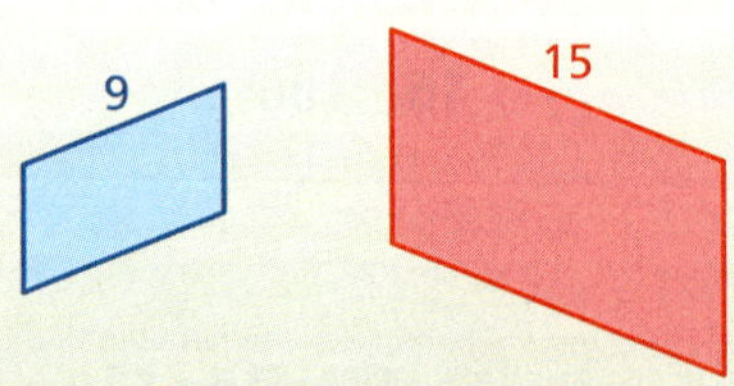

The ratio of the perimeters is $\frac{5}{3}$.

b. Find the ratio (red to blue) of the areas of the similar figures.

$$\frac{\text{Area of red figure}}{\text{Area of blue figure}} = \left(\frac{3}{4}\right)^2$$

$$= \frac{9}{16}$$

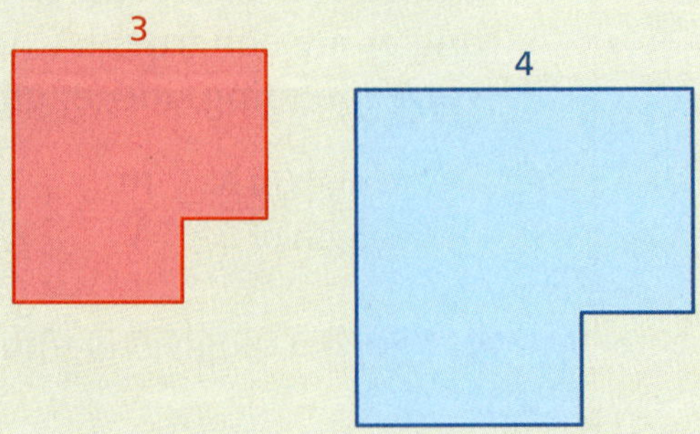

The ratio of the areas is $\frac{9}{16}$.

Review Game

Transformations

Materials

- masking tape

Players: 2

Directions

- Each player chooses an object in the classroom. The object must be something that can easily be moved by the student.
- Players use tape to create a coordinate plane beside the object.
- One player moves the object to a different position on the coordinate plane.
- The other player tries to guess whether the transformation is a translation, reflection, or rotation.
- The players switch spots and repeat the process.

Who Wins?

Each time a player guesses the correct transformation, they receive one point. The player with the most points at the end of the game wins.

For the Student Additional Practice

- Lesson Tutorials
- Multi-Language Glossary
- Self-Grading Progress Check
- *BigIdeasMath.com*
  Dynamic Student Edition
  Student Resources

Answers

18. no; The lengths of corresponding sides are not proportional.

19. yes; The lengths of corresponding sides are proportional and corresponding angles are congruent.

20. 10 in.

21. 9 cm

22. $\frac{3}{4}$; $\frac{9}{16}$

23. $\frac{7}{4}$; $\frac{49}{16}$

24. 9 : 16

25. no

26. yes

27. 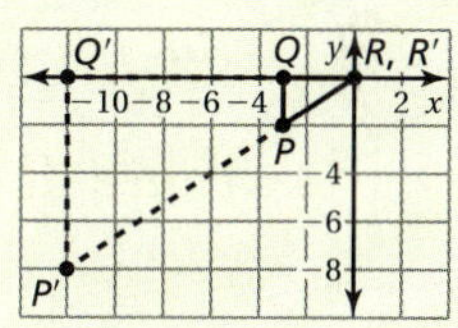

enlargement

28. 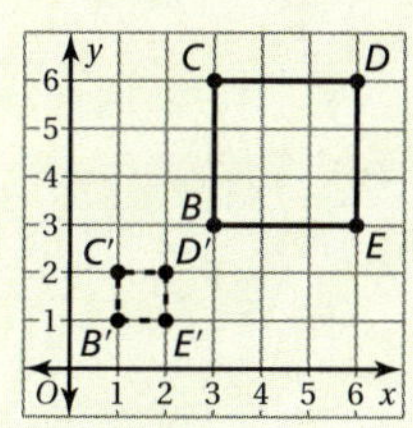

reduction

29. $Q''(-4, 2)$, $R''(14, 2)$, $S''(14, -7)$, $T''(-4, -7)$

My Thoughts on the Chapter

Teacher Tip

Not allowed to write in your teaching edition? Use sticky notes to record your thoughts.

What worked. . .

What did not work. . .

What I would do differently. . .

Exercises

The two figures are similar. Find the ratios (red to blue) of the perimeters and of the areas.

22.

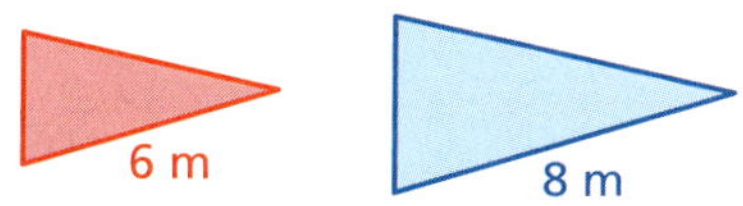

23.

24. **PHOTOS** Two photos are similar. The ratio of the corresponding side lengths is 3 : 4. What is the ratio of the areas?

11.7 Dilations (pp. 506–513)

Draw the image of Triangle *ABC* after a dilation with a scale factor of 2. Identify the type of dilation.

Multiply each *x*- and *y*-coordinate by the scale factor 2.

Vertices of *ABC*	(2*x*, 2*y*)	Vertices of *A′B′C′*
A(1, 1)	(2 • 1, 2 • 1)	*A′*(2, 2)
B(1, 2)	(2 • 1, 2 • 2)	*B′*(2, 4)
C(3, 2)	(2 • 3, 2 • 2)	*C′*(6, 4)

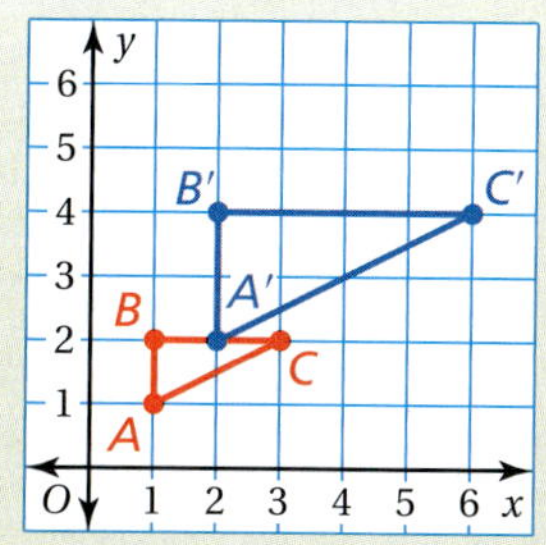

The image is shown at the above right. The dilation is an *enlargement* because the scale factor is greater than 1.

Exercises

Tell whether the blue figure is a dilation of the red figure.

25.

26.

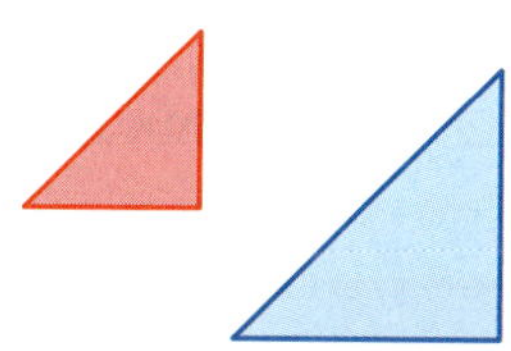

The vertices of a figure are given. Draw the figure and its image after a dilation with the given scale factor. Identify the type of dilation.

27. $P(-3, -2)$, $Q(-3, 0)$, $R(0, 0)$; $k = 4$

28. $B(3, 3)$, $C(3, 6)$, $D(6, 6)$, $E(6, 3)$; $k = \frac{1}{3}$

29. The vertices of a rectangle are $Q(-6, 2)$, $R(6, 2)$, $S(6, -4)$, and $T(-6, -4)$. Dilate the rectangle with respect to the origin using a scale factor of $\frac{3}{2}$. Then translate it 5 units right and 1 unit down. What are the coordinates of the image?

11 Chapter Test

Check It Out
Test Practice
BigIdeasMath.com

Triangles *ABC* and *DEF* are congruent.

1. Which angle of *DEF* corresponds to $\angle C$?

2. What is the perimeter of *DEF*?

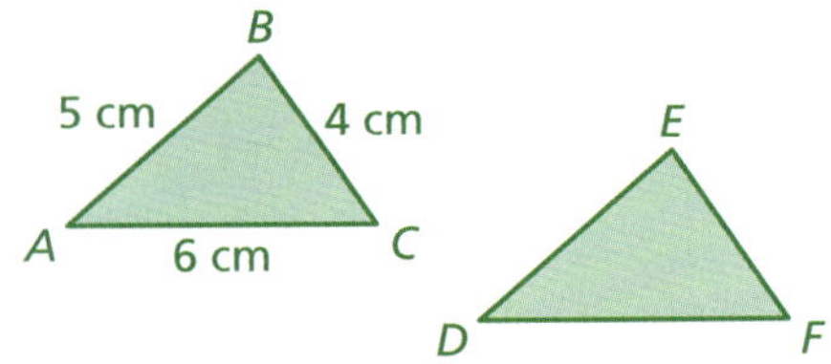

Tell whether the blue figure is a *translation, reflection, rotation,* or *dilation* of the red figure.

3. 

4. 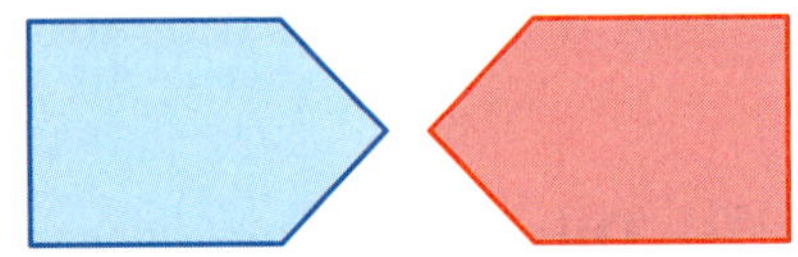

5. 

6. 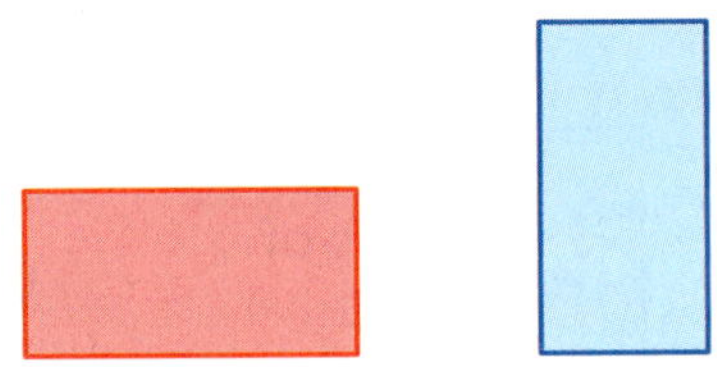

7. The vertices of a triangle are $A(2, 5)$, $B(1, 2)$, and $C(3, 1)$. Reflect the triangle in the x-axis, and then rotate the triangle 90° counterclockwise about the origin. What are the coordinates of the image?

8. The vertices of a triangle are $A(2, 4)$, $B(2, 1)$, and $C(5, 1)$. Dilate the triangle with respect to the origin using a scale factor of 2. Then translate the triangle 2 units left and 1 unit up. What are the coordinates of the image?

9. Tell whether the parallelograms are similar. Explain your reasoning.

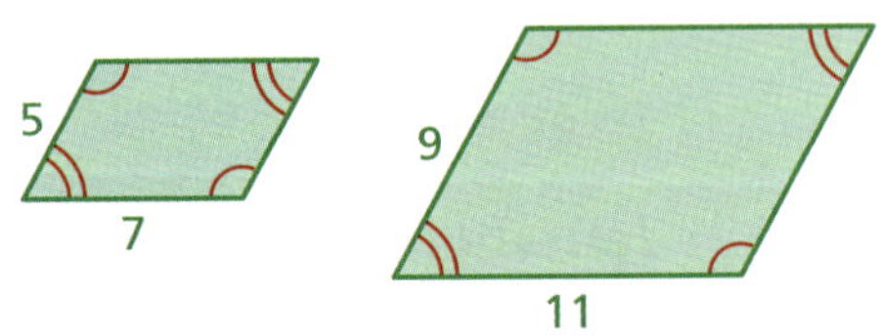

The two figures are similar. Find the ratios (red to blue) of the perimeters and of the areas.

10. 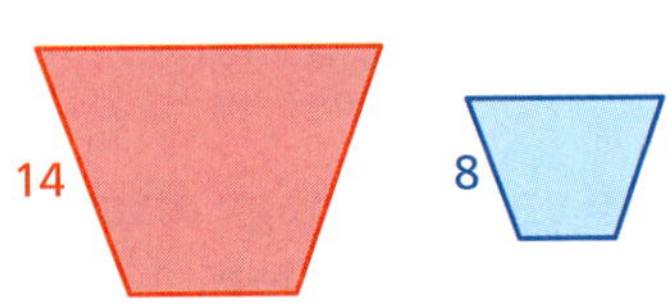

11. 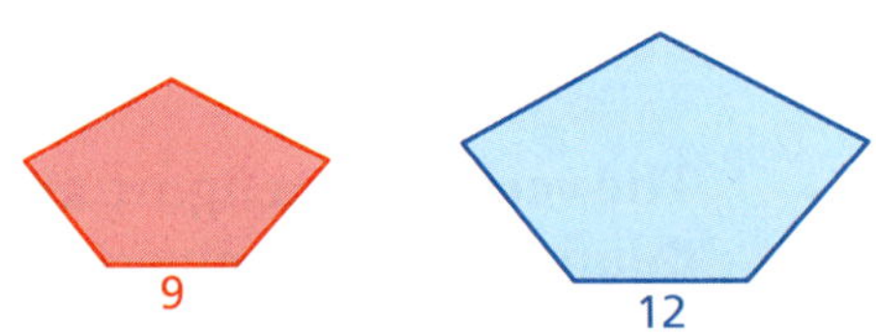

12. **SCREENS** A wide-screen television measures 36 inches by 54 inches. A movie theater screen measures 42 feet by 63 feet. Are the screens similar? Explain.

13. **CURTAINS** You want to use the rectangular piece of fabric shown to make a set of curtains for your window. Name the types of congruent shapes you can make with one straight cut. Draw an example of each type.

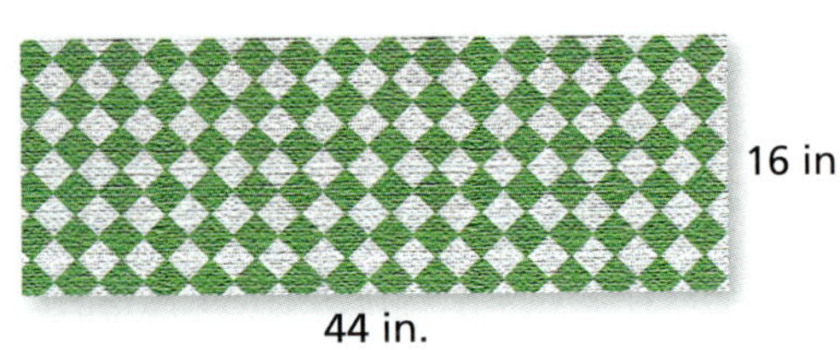

Test Item References

Chapter Test Questions	Section to Review
1, 2, 13	11.1
5	11.2
4	11.3
6, 7	11.4
9, 12	11.5
10, 11	11.6
3, 8	11.7

Test-Taking Strategies

Remind students to quickly look over the entire test before they start so that they can budget their time. On this test it is very important for students to **Stop** and **Think**. When students hurry on a test dealing with transformations, they may end up with incorrect coordinates of transformed figures. Encourage students to work carefully and deliberately.

Common Errors

- **Exercise 1** Students may have difficulty determining the corresponding angle. Point out that the triangles are congruent figures and remind students that corresponding angles in congruent figures have matching positions.
- **Exercises 3–6** Students may confuse the terms *translation*, *reflection*, *rotation*, and *dilation*. Review these terms prior to the test.
- **Exercise 7** Students may reflect the triangle in the wrong axis or rotate it in the wrong direction. Refer them to the Key Idea on page 56 to review reflections. Also, remind them what *counterclockwise* means and suggest drawing an arrow on a graph of the triangle in the direction of rotation.
- **Exercise 9** Students may think that because the side lengths change by the same amount, the objects are similar. Remind them that the *ratios* of corresponding side lengths must be the same for the objects to be similar.
- **Exercises 10 and 11** When finding the ratio of the areas, students may forget to square the ratio of the corresponding side lengths. Refer them to the Key Idea at the bottom of page 78 to review ratios of areas of similar figures.

Reteaching and Enrichment Strategies

If students need help. . .	If students got it. . .
Resources by Chapter • Practice A and Practice B • Puzzle Time Record and Practice Journal Practice Differentiating the Lesson Lesson Tutorials *BigIdeasMath.com* Skills Review Handbook	Resources by Chapter • Enrichment and Extension • Technology Connection Game Closet at *BigIdeasMath.com* Start Cumulative Assessment

Answers

1. $\angle F$
2. 15 cm
3. dilation
4. reflection
5. translation
6. rotation
7. $A''(5, 2), B''(2, 1), C''(1, 3)$
8. $A''(2, 9), B''(2, 3), C''(8, 3)$
9. no; The lengths of corresponding sides are not proportional.
10. $\frac{7}{4}; \frac{49}{16}$
11. $\frac{3}{4}; \frac{9}{16}$
12. yes; Because both screens are rectangles, the corresponding angle measures are congruent. Corresponding side lengths are proportional.
13. 2 rectangles

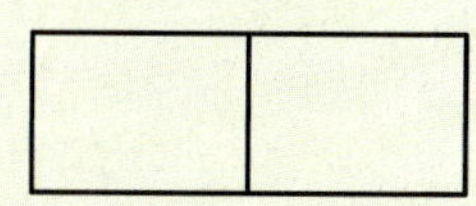

2 right triangles

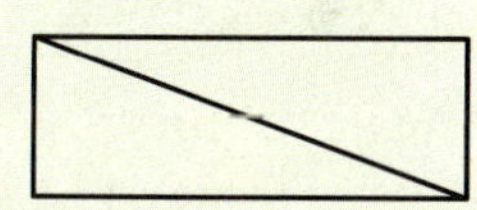

2 right trapezoids

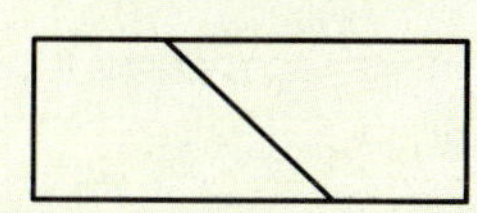

Technology for the Teacher

Online Assessment
Assessment Book
ExamView® Assessment Suite

Test-Taking Strategies
Available at *BigIdeasMath.com*

After Answering Easy Questions, Relax

Answer Easy Questions First
Estimate the Answer
Read All Choices before Answering
Read Question before Answering
Solve Directly or Eliminate Choices
Solve Problem before Looking at Choices
Use Intelligent Guessing
Work Backwards

About this Strategy

When taking a multiple choice test, be sure to read each question carefully and thoroughly. After skimming the test and answering the easy questions, stop for a few seconds, take a deep breath, and relax. Work through the remaining questions carefully, using your knowledge and test-taking strategies. Remember, you already completed many of the questions on the test!

Answers

1. 270°
2. D
3. I
4 C
5. G

Technology for the Teacher

Performance Tasks
Online Assessment
Assessment Book
ExamView® Assessment Suite

Item Analysis

1. **Gridded Response:** Correct answer: 270°

   Common Error: The student confuses a rotation of 90° with a rotation of 180° and thinks that a 90° clockwise rotation has the same result as a 90° counterclockwise rotation, getting an answer of 90°.

2. **A.** The student makes an operation error.

   B. The student makes an operation error.

   C. The student makes an operation error.

   D. Correct answer

3. **F.** The student does not apply inverses in the correct order.

   G. The student confuses inverse operations.

   H. The student confuses inverse operations.

   I. Correct answer

4. **A.** A translation *slides* a figure.

   B. A reflection *flips* a figure.

   C. Correct answer

   D. A dilation changes the size of a figure.

5. **F.** The student only translates the triangle 3 units right.

   G. Correct answer

   H. The student only translates the triangle 2 units down.

   I. The student translates the triangle 2 units right and 3 units down.

11 Cumulative Assessment

Test-Taking Strategy
After Answering Easy Questions, Relax

"After answering the easy questions, relax and try the harder ones. For this, the image is flipped. So, it's D."

1. A clockwise rotation of 90° is equivalent to a counterclockwise rotation of how many degrees?

2. The formula $K = C + 273.15$ converts temperatures from Celsius C to Kelvin K. Which of the following formulas is *not* correct?

   A. $K - C = 273.15$

   B. $C = K - 273.15$

   C. $C - K = -273.15$

   D. $C = K + 273.15$

3. Joe wants to solve the equation $-3(x + 2) = 12x$. What should he do first?

   F. Subtract 2 from each side.

   G. Add 3 to each side.

   H. Multiply each side by -3.

   I. Divide each side by -3.

4. Which transformation *turns* a figure?

   A. translation

   B. reflection

   C. rotation

   D. dilation

5. A triangle is graphed in the coordinate plane below.

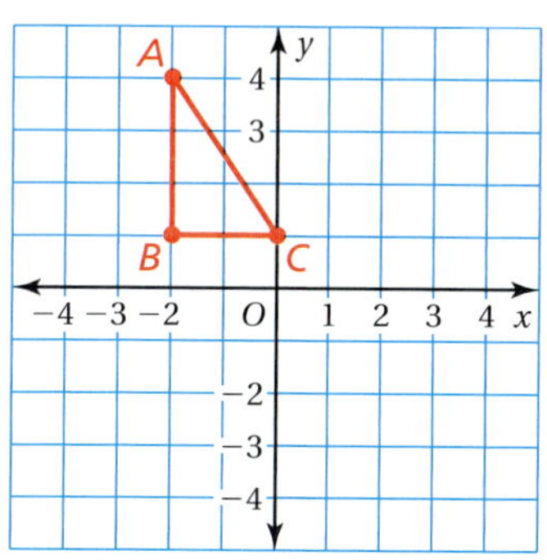

   Translate the triangle 3 units right and 2 units down. What are the coordinates of the image?

   F. $A'(1, 4)$, $B'(1, 1)$, $C'(3, 1)$

   G. $A'(1, 2)$, $B'(1, -1)$, $C'(3, -1)$

   H. $A'(-2, 2)$, $B'(-2, -1)$, $C'(0, -1)$

   I. $A'(0, 1)$, $B'(0, -2)$, $C'(2, -2)$

6. Dale solved the equation in the box shown. What should Dale do to correct the error that he made?

$$-\frac{x}{3} + \frac{2}{5} = -\frac{7}{15}$$
$$-\frac{x}{3} + \frac{2}{5} - \frac{2}{5} = -\frac{7}{15} - \frac{2}{5}$$
$$-\frac{x}{3} = -\frac{13}{15}$$
$$3 \cdot \left(-\frac{x}{3}\right) = 3 \cdot \left(-\frac{13}{15}\right)$$
$$x = -2\frac{3}{5}$$

A. Add $\frac{2}{5}$ to each side to get $-\frac{x}{3} = -\frac{1}{15}$.

B. Multiply each side by -3 to get $x + \frac{2}{5} = \frac{7}{5}$.

C. Multiply each side by -3 to get $x = 2\frac{3}{5}$.

D. Subtract $\frac{2}{5}$ from each side to get $-\frac{x}{3} = -\frac{5}{10}$.

7. Jenny dilates the rectangle below using a scale factor of $\frac{1}{2}$.

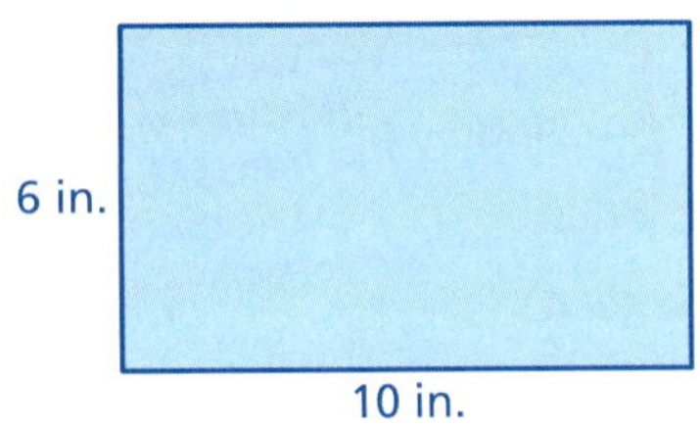

What is the area of the dilated rectangle in square inches?

8. The vertices of a rectangle are $A(-4, 2)$, $B(3, 2)$, $C(3, -5)$, and $D(-4, -5)$. If the rectangle is dilated by a scale factor of 3, what will be the coordinates of vertex C'?

F. $(9, -15)$

G. $(-12, 6)$

H. $(-12, -15)$

I. $(9, 6)$

9. In the figures, Triangle *EFG* is a dilation of Triangle *HIJ*.

Which proportion is *not* necessarily correct for Triangle *EFG* and Triangle *HIJ*?

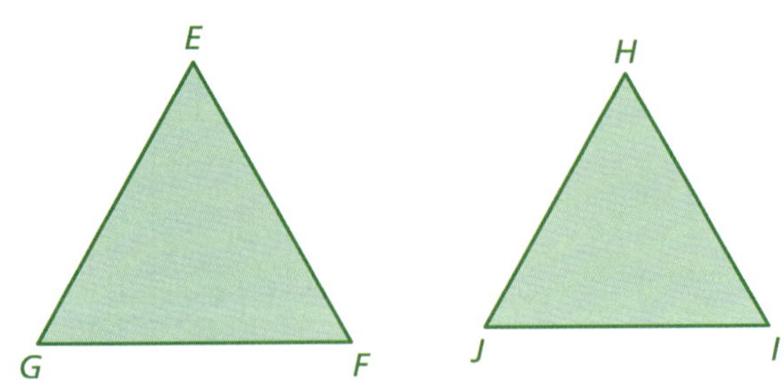

A. $\frac{EF}{FG} = \frac{HI}{IJ}$

B. $\frac{EG}{HI} = \frac{FG}{IJ}$

C. $\frac{GE}{EF} = \frac{JH}{HI}$

D. $\frac{EF}{HI} = \frac{GE}{JH}$

Item Analysis (continued)

6. **A.** The student *undoes* the addition of $\frac{2}{5}$ incorrectly by adding $\frac{2}{5}$ to each side, rather than subtracting.

   B. The student performs an order of operations error by not subtracting $\frac{2}{5}$ from each side first.

   C. Correct answer

   D. The student subtracts the numerators and denominators to get $-\frac{7-2}{15-5} = -\frac{5}{10}$.

7. **Gridded Response:** Correct answer: 15 square inches

   Common Error: The student finds half the area of the original rectangle, getting an answer of 30 square inches.

8. **F.** Correct answer

   G. The student used the coordinates of vertex A'.

   H. The student used the coordinates of vertex D'.

   I. The student used the coordinates of vertex B'.

9. **A.** The student chooses a proportion that correctly represents a relationship between pairs of corresponding sides of the triangles.

   B. Correct answer

   C. The student chooses a proportion that correctly represents a relationship between pairs of corresponding sides of the triangles.

   D. The student chooses a proportion that correctly represents a relationship between pairs of corresponding sides of the triangles.

Answers

6. C
7. 15
8. F
9. B

Answers

10. F

11. *Part A* translation up

Part B dilation (reduction)

Part C 2

12. A

Item Analysis (continued)

10. **F.** Correct answer

G. The student chooses an answer based only on visual approximation.

H. The student thinks that because 12 is 4 more than 8, x should be 4 more than 12.

I. The student thinks that because 8 is 4 less than 12, x should be 4 less than 21.

11. **2 points** The student demonstrates a thorough understanding of working with transformations. In Part A, the student correctly describes the transformation as a translation up. In Part B, the student correctly describes the transformation as a dilation (reduction). In Part C, the student correctly determines that the scale factor is 2. The student provides clear and complete work and explanations.

1 point The student demonstrates a partial understanding of working with transformations. The student provides some correct work and explanation.

0 points The student demonstrates insufficient understanding of working with transformations. The student is unable to make any meaningful progress toward a correct answer.

12. **A.** Correct answer

B. The student performs a reflection in the x-axis.

C. The student rotates the rectangle 90° clockwise.

D. The student thinks 180° causes the figure to end up where it started.

10. In the figures below, Rectangle *EFGH* is a dilation of Rectangle *IJKL*.

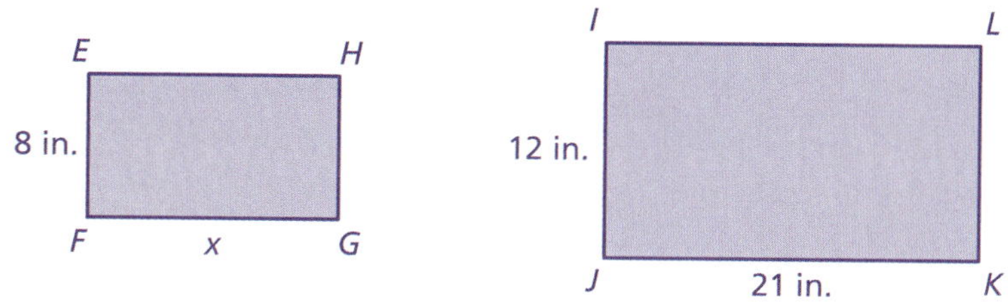

What is x?

F. 14 in.

G. 15 in.

H. 16 in.

I. 17 in.

11. Several transformations are used to create the pattern.

Think Solve Explain

Part A Describe the transformation of Triangle *GLM* to Triangle *DGH*.

Part B Describe the transformation of Triangle *ALQ* to Triangle *GLM*.

Part C Triangle *DFN* is a dilation of Triangle *GHM*. Find the scale factor.

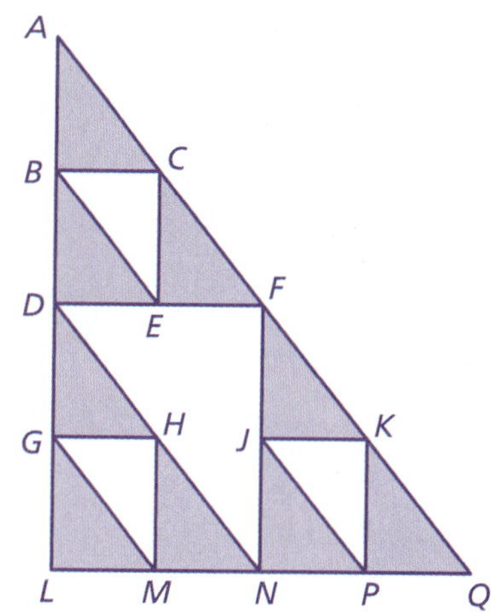

12. A rectangle is graphed in the coordinate plane below.

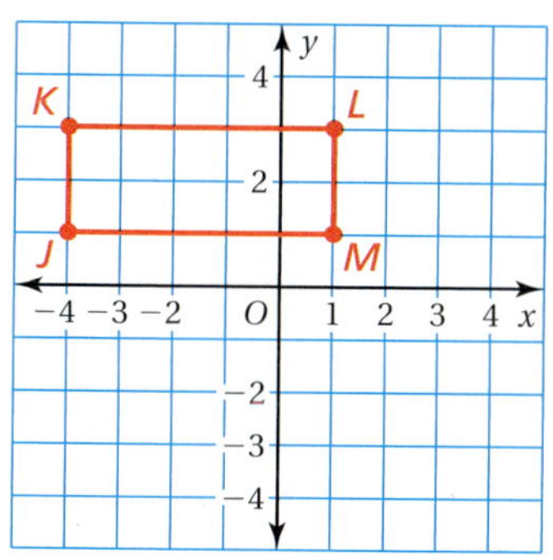

Rotate the rectangle 180° about the origin. What are the coordinates of the image?

A. $J'(4, -1), K'(4, -3), L'(-1, -3), M'(-1, -1)$

B. $J'(-4, -1), K'(-4, -3), L'(1, -3), M'(1, -1)$

C. $J'(1, 4), K'(3, 4), L'(3, -1), M'(1, -1)$

D. $J'(-4, 1), K'(-4, 3), L'(1, 3), M'(1, 1)$

12 Angles and Triangles

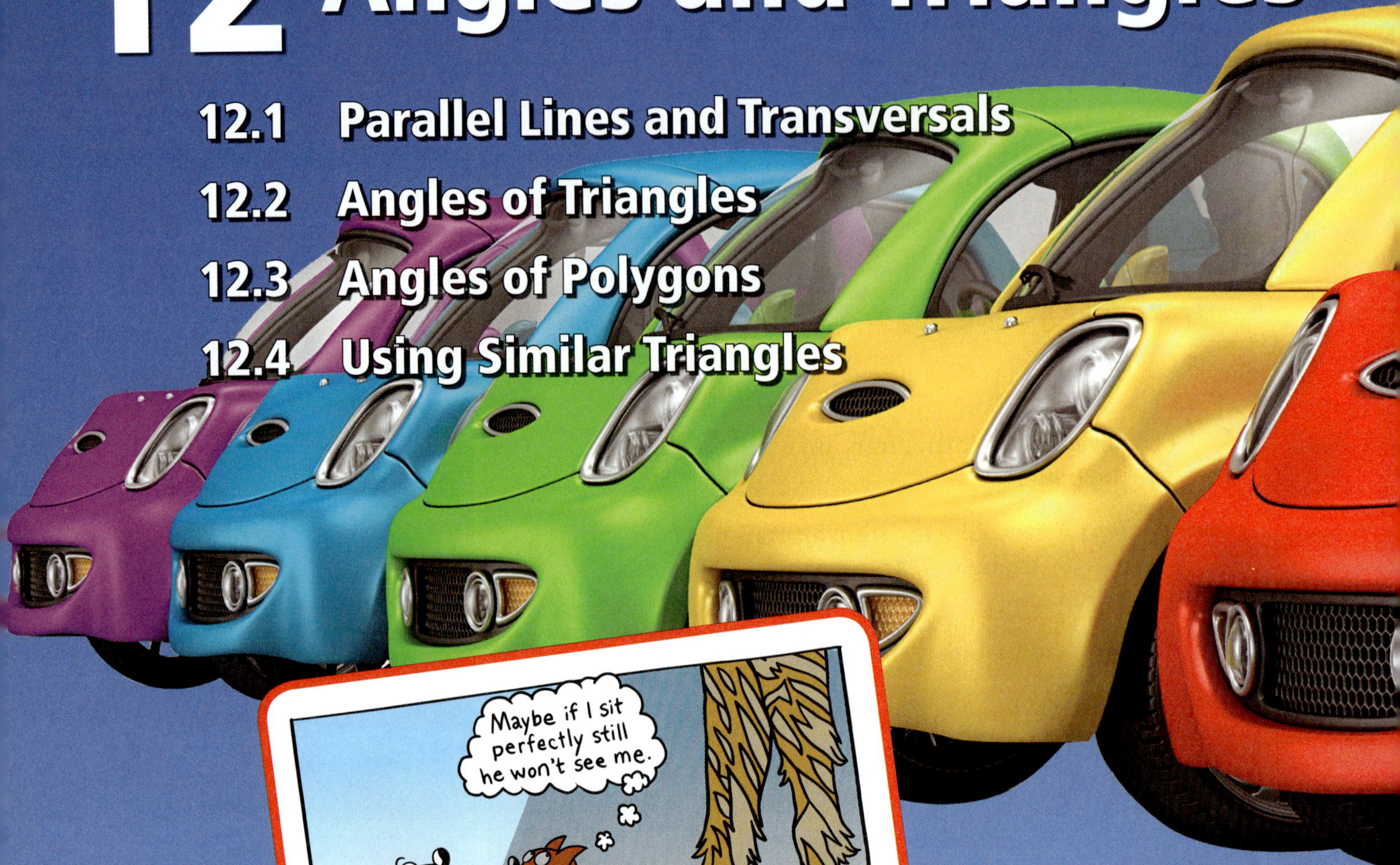

"Let's use shadows and similar triangles to indirectly measure the height of the giant hyena standing right behind you."

"Start with any triangle."

"Tear off the angles. You can always rearrange the angles so that they form a straight line."

"What does that prove?"

What Your Students Have Learned

- Use reasoning about multiplication and division to solve ratio and rate problems.
- Measure and describe relationships among vertical, adjacent, supplementary, and complementary angles.
- Use proportions to solve problems.

What Your Students Will Learn

- Classify and determine the measures of angles created when parallel lines are cut by a transversal.
- Demonstrate that the sum of the interior angle measures of a triangle is 180° and apply this fact to find the unknown measures of angles and the sum of the angles of polygons.
- Use indirect measurement and similar triangles to solve problems that include height and distance.

Pacing Guide for Chapter 12

Chapter Opener Accelerated	1 Day
Section 1 Accelerated	1 Day
Section 2 Accelerated	1 Day
Section 3 Accelerated	1 Day
Section 4 Accelerated	2 Days
Chapter Review/ Chapter Tests Accelerated	2 Days
Total Chapter 12 Accelerated	8 Days
Year-to-Date Accelerated	112 Days

Technology for the Teacher

BigIdeasMath.com
Chapter at a Glance
Complete Materials List
Parent Letters: English and Spanish

What Your Students Have Learned

- Identify and use adjacent or vertical angles to solve for an unknown value.
- Identify and use complementary or supplementary angles to solve for an unknown value.

Additional Topics for Review

- Vocabulary of Angles and Triangles
- Similar Triangles
- Solving Multi-Step Equations

Try It Yourself

1. vertical; $x = 112$
2. adjacent; $x = 44$
3. complementary; $x = 78$
4. supplementary; $x = 50$

Record and Practice Journal Fair Game Review

1. vertical; 128
2. adjacent; 55
3. vertical; 37
4. adjacent; 15
5. 76°
6. supplementary; 63
7. complementary; 21
8. complementary; 49
9. supplementary; 14
10. 53

Math Background Notes

Vocabulary Review

- Adjacent Angles
- Vertical Angles
- Congruent Angles
- Complementary Angles
- Supplementary Angles

Adjacent and Vertical Angles

- Two intersecting lines will form four angles with a common vertex.
- *Adjacent angles* share a common side and have the same vertex.
- When two adjacent angles form a larger angle, the sum of the measures of the smaller angles is equal to the measure of the larger angle.
- The opposite angles formed by the intersection of two lines are *vertical angles.*
- Vertical angles are *congruent angles*, meaning they have the same measure.
- In Question 2, remind students of the corner mark used to designate a right angle.

Complementary and Supplementary Angles

- Two angles are complementary angles when the sum of their measures is 90°.
- Two angles are supplementary angles when the sum of their measures is 180°.
- In Question 3, remind students of the corner mark used to designate a right angle.

Reteaching and Enrichment Strategies

If students need help...	If students got it...
Record and Practice Journal • Fair Game Review Skills Review Handbook Lesson Tutorials	Game Closet at *BigIdeasMath.com* Start the next section

What You Learned Before

"I just remember that C comes before S and 90 comes before 180. That makes it easy."

Adjacent and Vertical Angles

Example 1 Tell whether the angles are *adjacent* or *vertical*. Then find the value of x.

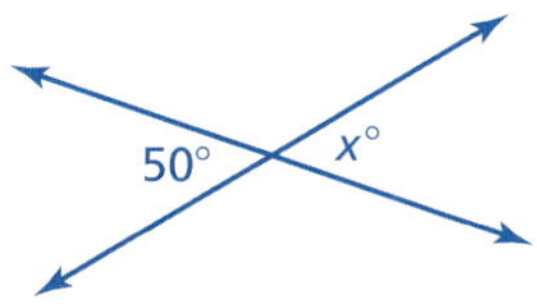

The angles are vertical angles. Because vertical angles are congruent, the angles have the same measure.

So, the value of x is 50.

Try It Yourself

Tell whether the angles are *adjacent* or *vertical*. Then find the value of x.

1.

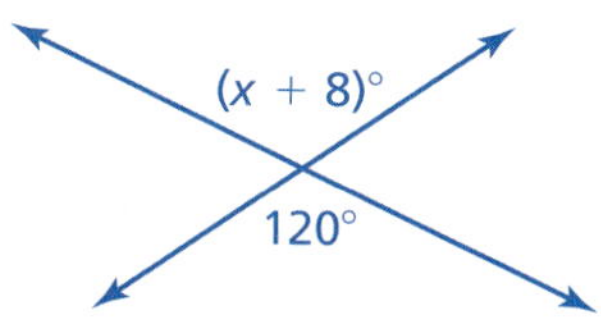

2.

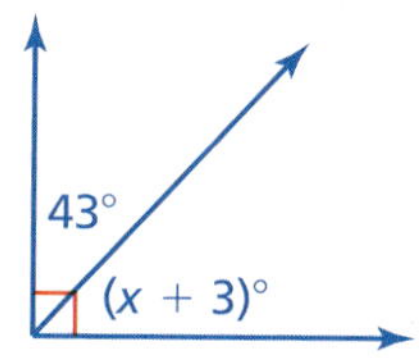

Complementary and Supplementary Angles

Example 2 Tell whether the angles are *complementary* or *supplementary*. Then find the value of x.

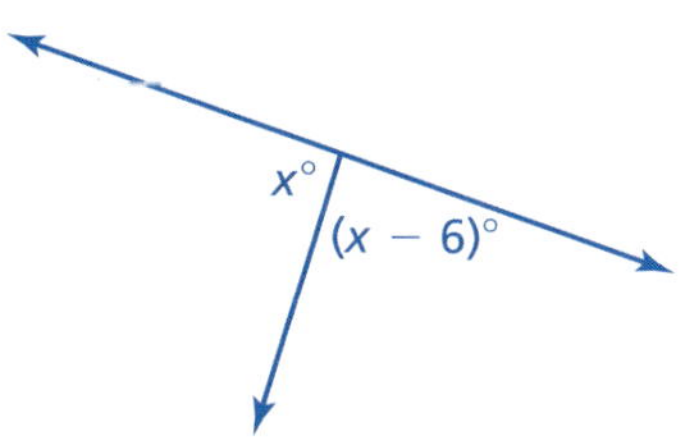

The two angles make up a straight angle. So, the angles are supplementary angles, and the sum of their measures is 180°.

$x + (x - 6) = 180$	Write equation.
$2x - 6 = 180$	Combine like terms.
$2x = 186$	Add 6 to each side.
$x = 93$	Divide each side by 2.

Try It Yourself

Tell whether the angles are *complementary* or *supplementary*. Then find the value of x.

3.

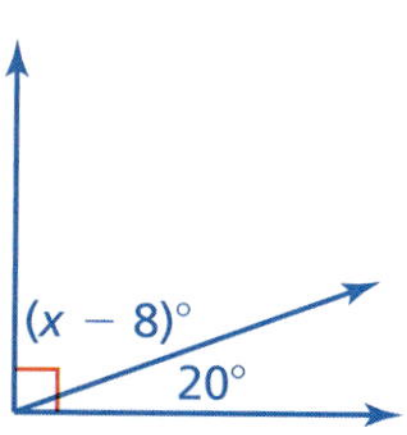

4.

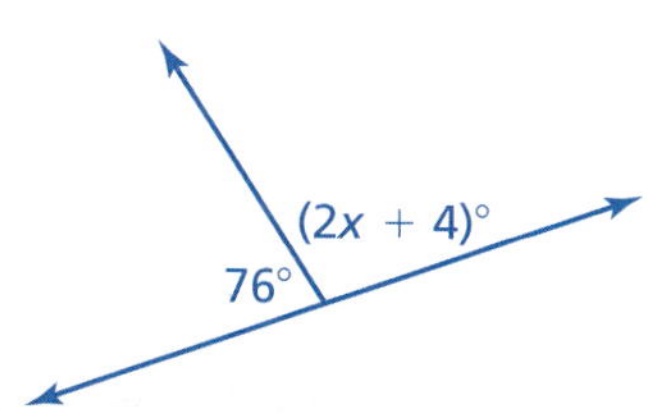

12.1 Parallel Lines and Transversals

Essential Question How can you describe angles formed by parallel lines and transversals?

The Meaning of a Word ● Transverse

When an object is **transverse**, it is lying or extending across something.

1 ACTIVITY: A Property of Parallel Lines

Work with a partner.

- **Discuss what it means for two lines to be parallel. Decide on a strategy for drawing two parallel lines. Then draw the two parallel lines.**
- **Draw a third line that intersects the two parallel lines. This line is called a transversal.**

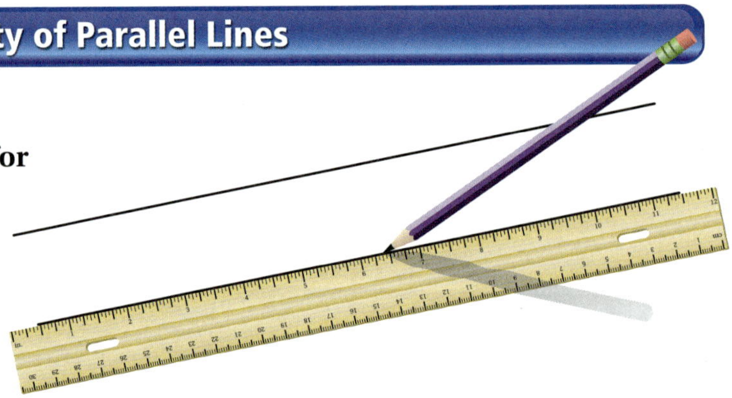

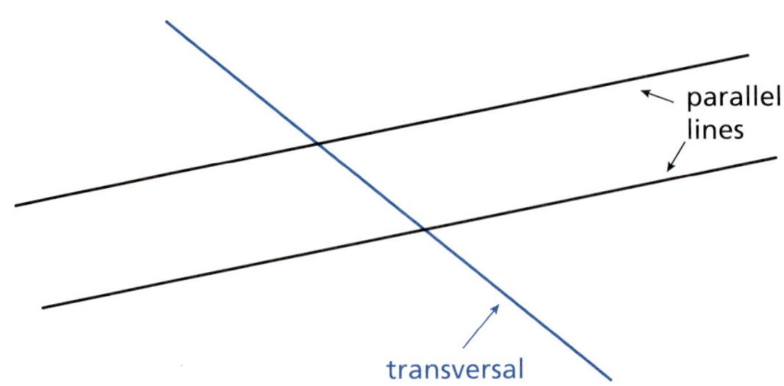

Geometry

In this lesson, you will

- identify the angles formed when parallel lines are cut by a transversal.
- find the measures of angles formed when parallel lines are cut by a transversal.

a. How many angles are formed by the parallel lines and the transversal? Label the angles.

b. Which of these angles have equal measures? Explain your reasoning.

Laurie's Notes

Introduction

Applying Mathematical Practices

- **Use Appropriate Tools Strategically:** There are many appropriate tools mathematically proficient students may use to gain an understanding of a new concept or to solve a problem. Several tools are useful in the investigations today.

Motivate

- **Preparation:** Use a transparency that has two parallel lines drawn on it on the overhead projector. Place a grid on top of the parallel lines so that the lines pass through obvious lattice points.
- ? "What appears to be true about the lines?" parallel
- Compute the slopes and conclude that the lines are parallel.
- Now remove the grid but keep the transparency with the (parallel) lines.
- ? "Are the lines still parallel? How do you know?" The point of these questions is not for students to give an answer, but for students to consider what it means for two lines to be parallel.

Discuss

- Ask students to give examples of parallel lines in real life, where the lines are
  - functional (as with railroad tracks or lanes of a highway).
  - aesthetic (on clothing).
  - coincidentally parallel (the edge of table and a pencil).
- Parallel lines are common. Often there is a reason for the parallelism.

Activity Notes

Activity 1

- Discuss the meaning of the word *transverse*. Give examples, such as a path that cuts through a field or a street that cuts through the downtown area of a city.
- **Use Appropriate Tools Strategically:** Students should have tools available such as rulers, protractors, and tracing paper.
- Students will likely create parallel lines by tracing opposite edges of a rigid object they believe has parallel edges, such as a ruler or an index card. There are other methods. Ask students to share their methods, and give explanations as to why the lines are parallel.
- Students will use protractors to measure the angles formed. If tracing paper is available, some students may simply trace the acute angle and the obtuse angle and slide the tracing paper around.

What Your Students Will Learn

- Identify and find the measures of angles formed when parallel lines are cut by a transversal, including corresponding, interior, and exterior angles.

Previous Learning

Students should know the definition of similar triangles.

Technology for the Teacher

Dynamic Classroom

Lesson Plans
Complete Materials List

12.1 Record and Practice Journal

Essential Question How can you describe angles formed by parallel lines and transversals?

1 ACTIVITY: A Property of Parallel Lines

Work with a partner.

- **Discuss what it means for two lines to be parallel. Decide on a strategy for drawing two parallel lines. Then draw the two parallel lines.**
- **Draw a third line that intersects the two parallel lines. This line is called a *transversal*.**

Sample answer:

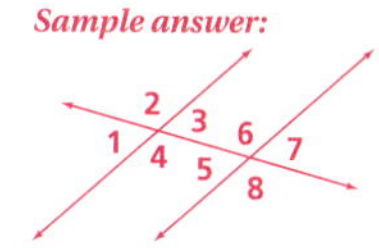

a. How many angles are formed by the parallel lines and the transversal? Label the angles.

8

b. Which of these angles have equal measures? Explain your reasoning.

$\angle 2, \angle 4, \angle 6,$ and $\angle 8$ are congruent.
$\angle 1, \angle 3, \angle 5,$ and $\angle 7$ are congruent.

Differentiated Instruction

Visual

In order to be successful in geometry, students need to be proficient with the vocabulary. Have them write vocabulary words in their math notebook glossaries. Next to the definitions students should illustrate the terms with color-coded diagrams.

12.1 Record and Practice Journal

2 ACTIVITY: Creating Parallel Lines

Work with a partner.

a. If you were building the house in the photograph, how could you make sure that the studs are parallel to each other?

Makes sure vertical angles are congruent and the angles that correspond to them are congruent.

b. Identify sets of parallel lines and transversals in the photograph.

studs are parallel; diagonal board is a transversal

3 ACTIVITY: Using Technology to Draw Parallel Lines and a Transversal

Work with a partner. Use geometry software to draw two parallel lines intersected by a transversal.

a. Find all of the angle measures.

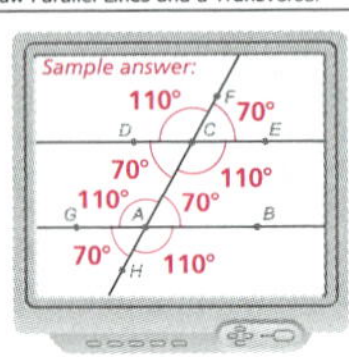

b. Adjust the figure by moving the parallel lines or the transversal to a different position. Describe how the angle measures and relationships change.

Check students' work.

What Is Your Answer?

4. **IN YOUR OWN WORDS** How can you describe angles formed by parallel lines and transversals? Give an example.

See Additional Answers.

5. Use geometry software to draw a transversal that is perpendicular to two parallel lines. What do you notice about the angles formed by the parallel lines and the transversal?

All eight angle measures = 90°.
See Additional Answers.

Laurie's Notes

Activity 2

- Students will have different thoughts about how to make sure the studs are parallel to each other. Be prepared for students who do not know much about construction.
- The transversals should be obvious, but remember that the horizontal boards perpendicular to the studs are also transversals.

Activity 3

- **Use Appropriate Tools Strategically:** Seeing and being able to manipulate parallel lines and a transversal in a dynamic environment is very powerful. There is dynamic geometry software free for download. If a computer lab is not available, a class demonstration is an alternative.
- Be sure that students *construct* parallel lines versus drawing lines that look parallel. They should draw the first line, plot a point not on that line, and construct a line through the point parallel to the first line.
- Students should quickly see that when the lines are parallel the four acute angles are congruent and the four obtuse angles are congruent.
- When the lines are not parallel, only the four pairs of vertical angles will be congruent.

What Is Your Answer?

- In Question 5, make sure that students are constructing perpendicular lines versus using eyesight to draw two lines that appear perpendicular.

Closure

- Identify five pairs of parallel lines in your classroom. Note that the definition of *line* is modified to include things such as the metal molding on either side of the white board and the side casings on the door frame.

2 ACTIVITY: Creating Parallel Lines

Math Practice

Use Clear Definitions

What do the words *parallel* and *transversal* mean? How does this help you answer the question in part (a)?

Work with a partner.

a. If you were building the house in the photograph, how could you make sure that the studs are parallel to each other?

b. Identify sets of parallel lines and transversals in the photograph.

3 ACTIVITY: Using Technology

Work with a partner. Use geometry software to draw two parallel lines intersected by a transversal.

a. Find all the angle measures.

b. Adjust the figure by moving the parallel lines or the transversal to a different position. Describe how the angle measures and relationships change.

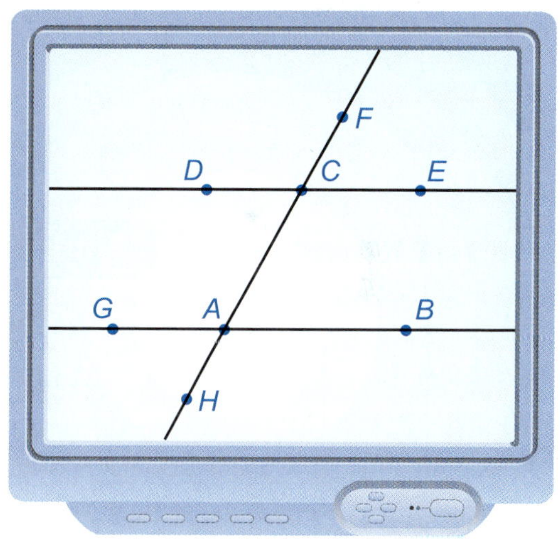

What Is Your Answer?

4. **IN YOUR OWN WORDS** How can you describe angles formed by parallel lines and transversals? Give an example.

5. Use geometry software to draw a transversal that is perpendicular to two parallel lines. What do you notice about the angles formed by the parallel lines and the transversal?

Practice Use what you learned about parallel lines and transversals to complete Exercises 3–6 on page 531.

12.1 Lesson

Check It Out
Lesson Tutorials
BigIdeasMath.com

Key Vocabulary
transversal, *p. 528*
interior angles, *p. 529*
exterior angles, *p. 529*

Lines in the same plane that do not intersect are called *parallel lines*. Lines that intersect at right angles are called *perpendicular lines*.

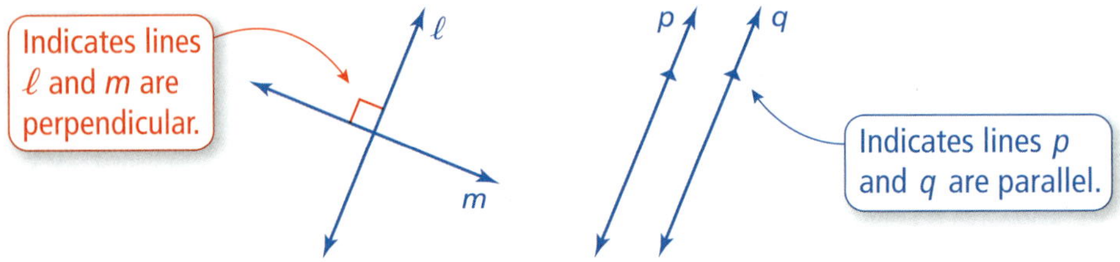

A line that intersects two or more lines is called a **transversal**. When parallel lines are cut by a transversal, several pairs of congruent angles are formed.

Key Idea

Study Tip
Corresponding angles lie on the same side of the transversal in corresponding positions.

Corresponding Angles

When a transversal intersects parallel lines, corresponding angles are congruent.

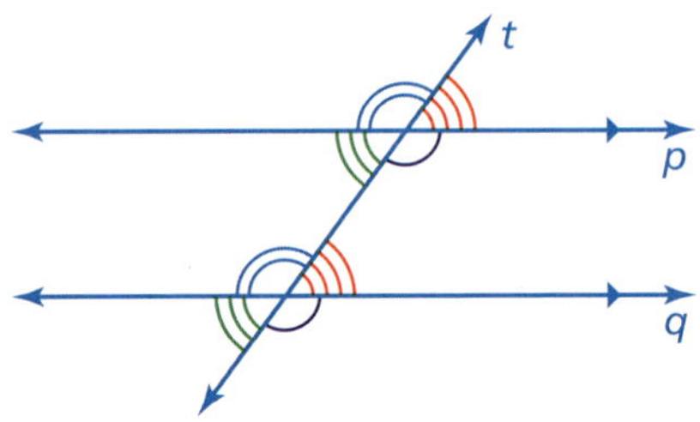

Corresponding angles

EXAMPLE 1 Finding Angle Measures

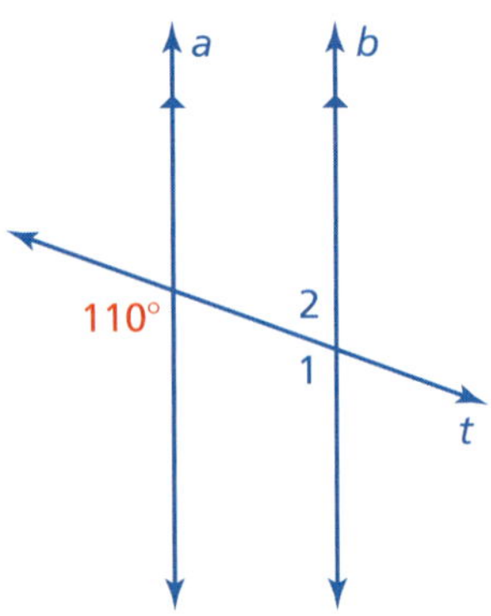

Use the figure to find the measures of (a) $\angle 1$ and (b) $\angle 2$.

a. $\angle 1$ and the 110° angle are corresponding angles. They are congruent.

So, the measure of $\angle 1$ is 110°.

b. $\angle 1$ and $\angle 2$ are supplementary.

$\angle 1 + \angle 2 = 180^\circ$	Definition of supplementary angles
$110^\circ + \angle 2 = 180^\circ$	Substitute 110° for $\angle 1$.
$\angle 2 = 70^\circ$	Subtract 110° from each side.

So, the measure of $\angle 2$ is 70°.

On Your Own

Now You're Ready
Exercises 7–9

Use the figure to find the measure of the angle. Explain your reasoning.

1. $\angle 1$

2. $\angle 2$

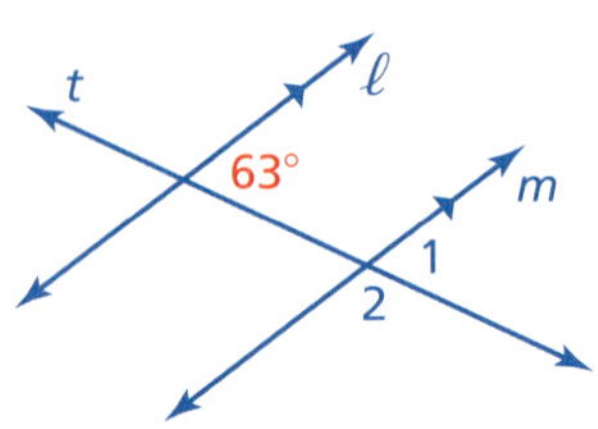

Multi-Language Glossary at BigIdeasMath.com

Laurie's Notes

Introduction

Connect

- **Yesterday:** Students explored angles formed when parallel lines are intersected by a transversal.
- **Today:** Students will find the measures of many types of angles, all formed when parallel lines are cut by a transversal.

Motivate

- **Preparation:** Make a model to help discuss the big ideas of this lesson. Cut 3 strips of card stock; punch holes in the middle of two strips and punch two holes in the third strip. Attach the strips using brass fasteners.

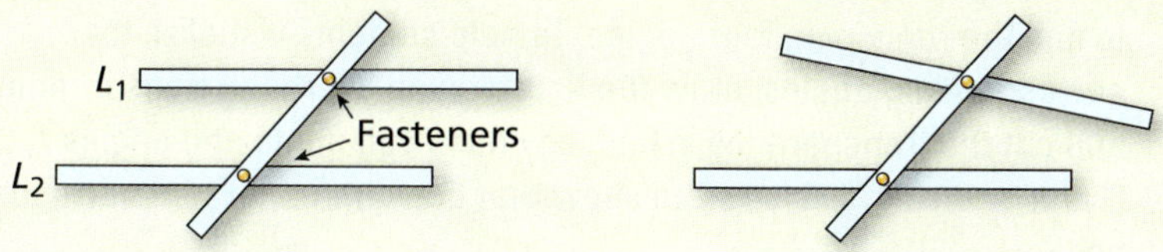

Model A Model B

- Place the model on the overhead or document camera. Demonstrate to students that the pieces are moveable, by transforming from Model A to Model B.
- Focus students' attention on the connection between the 4 angles on L_1 and the 4 angles on L_2. Pairs of vertical angles will always be congruent whether or not L_1 and L_2 are parallel.
- Place the model on the overhead or document camera, and encourage students to point to the angles that they think are congruent. Use models A and B.

Lesson Notes

Key Idea

- Write the informal definitions of parallel lines and perpendicular lines. Draw examples of each and discuss the notation used in the diagram.
- Write the definition of transversal. Explain that a line that intersects two or more lines is called a **transversal** even if the lines are *not* parallel. Only when the lines are parallel are the pairs of angles in this lesson congruent.
- Write the Key Idea. Identify corresponding angles which are color-coded in the diagram. Mention the Study Tip.
- Students will ask what is meant by *corresponding position.* The corresponding angles are both above or below the parallel lines (when in horizontal position) and on the same side of the transversal (left or right).
- **Attend to Precision:** In a formal geometry course, the rules in the Key Ideas of this lesson would be stated in if-then (conditional) form. Discuss this with students.

Example 1

? "Are the lines parallel? How do you know?" yes; blue arrow marks

- Students will need to recall the definition of supplementary angles.

Goal Today's lesson is finding measures of angles formed by parallel lines and a **transversal**.

Technology for the Teacher

Dynamic Classroom

Lesson Tutorials
Lesson Plans
Answer Presentation Tool

Extra Example 1

Use the figure to find the measures of (a) $\angle 1$ and (b) $\angle 2$.

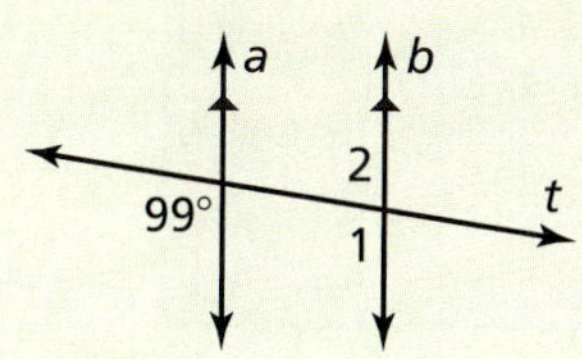

a. 99° **b.** 81°

On Your Own

1. 63°; Corresponding angles are congruent.
2. 117°; $\angle 1$ and $\angle 2$ are supplementary.

Extra Example 2

Use the figure to find the measures of the numbered angles.

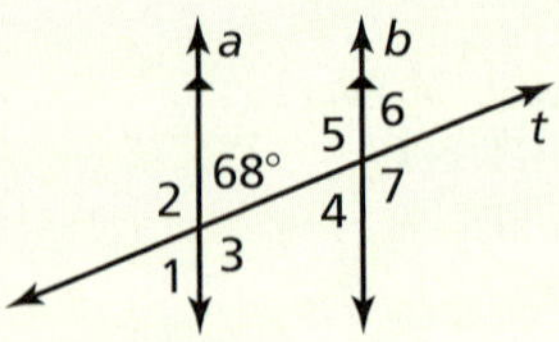

$\angle 1 = 68°$, $\angle 2 = 112°$, $\angle 3 = 112°$, $\angle 4 = 68°$, $\angle 5 = 112°$, $\angle 6 = 68°$, $\angle 7 = 112°$

On Your Own

3. $\angle 1 = 121°$, $\angle 2 = 59°$, $\angle 3 = 121°$, $\angle 4 = 121°$, $\angle 5 = 59°$, $\angle 6 = 121°$, $\angle 7 = 59°$

Extra Example 3

The painting shows several parallel lines and transversals. What is the measure of $\angle 1$?

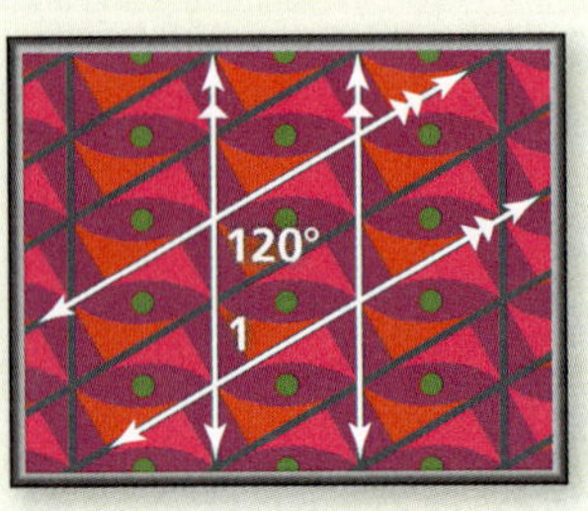

60°

English Language Learners

Vocabulary and Symbols

Make sure that students understand that the arrowhead marks *on* the lines indicate that the lines or line segments are parallel.

Laurie's Notes

Example 2

- Draw the figure on the board or overhead.
- ? "Can you find the measures of all the angles if you only know one angle?" Students may not know the answer at this point, but by the end of this example, they will see that they can.
- **Look for and Express Regularity in Repeated Reasoning** and **Big Idea:** If you know any angle when a transversal intersects two parallel lines, then you can use vertical, supplementary, and corresponding angles to find all 7 of the other measures. It is not necessary to learn other theorems about alternate exterior or interior angles. **Students should be able to do all of the homework after this example.**
- Once angles 1, 2, and 3 are found, you can use corresponding angles to find the remaining four angles. To help students visualize the corresponding angles, draw the figure on an overhead transparency and cut the transparency in half. Lay the given angle and angles 1, 2, and 3 over angles 4, 5, 6, and 7 to show that they are congruent corresponding angles.

On Your Own

- **Question 3:** Students can say $\angle 3 = 121°$ because of vertical angles *or* because it is the supplement of $\angle 2$.

Discuss

- Use the model from the beginning of class to talk about other pairs of angles. Make the lines parallel even though the angles still have the same definition.
- Identify the four angles that are interior (between the two parallel lines) and the four angles that are exterior (outside the two parallel lines).
- ? "Are there pairs of interior angles that appear congruent?" yes; 3 & 6 and 4 & 5 (in the diagram)
- ? "Are there pairs of exterior angles that appear congruent?" yes; 1 & 8 and 2 & 7 (in the diagram)

Example 3

- ? "Are the two dashed lines parallel? How do you know?" Students may not be sure that they are parallel. Explain that because all the letters are slanted at an 80° angle, the lines are parallel.
- It is helpful to label other angles around $\angle 1$. For example:

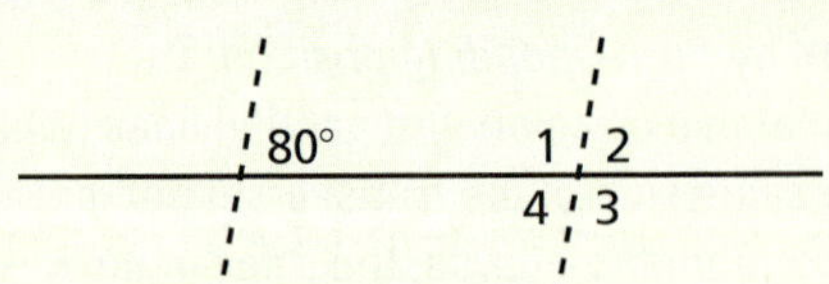

- ? "What angle is congruent to the 80° angle?" $\angle 2$
- ? "How can you find the measure of $\angle 1$?" Because $\angle 2$ and $\angle 1$ are supplementary and $\angle 2$ is congruent to the 80° angle, $\angle 1 = 180° - \angle 2 = 180° - 80° = 100°$.

EXAMPLE 2 Using Corresponding Angles

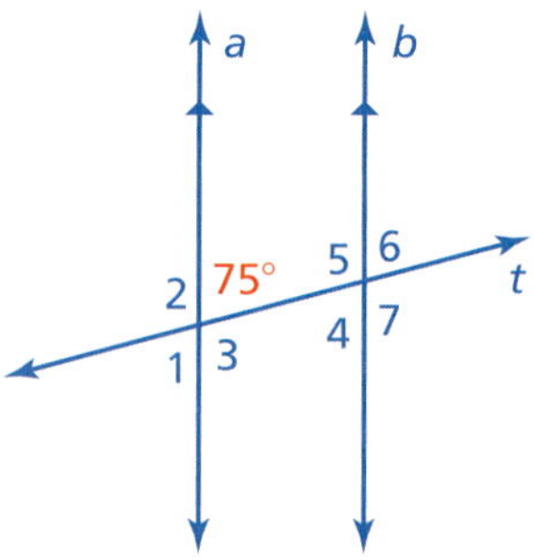

Use the figure to find the measures of the numbered angles.

∠1: ∠1 and the 75° angle are vertical angles. They are congruent.

So, the measure of ∠1 is 75°.

∠2 and ∠3: The 75° angle is supplementary to both ∠2 and ∠3.

$75° + \angle 2 = 180°$	Definition of supplementary angles
$\angle 2 = 105°$	Subtract 75° from each side.

So, the measures of ∠2 and ∠3 are 105°.

∠4, ∠5, ∠6, and ∠7: Using corresponding angles, the measures of ∠4 and ∠6 are 75°, and the measures of ∠5 and ∠7 are 105°.

On Your Own

Now You're Ready
Exercises 15–17

3. Use the figure to find the measures of the numbered angles.

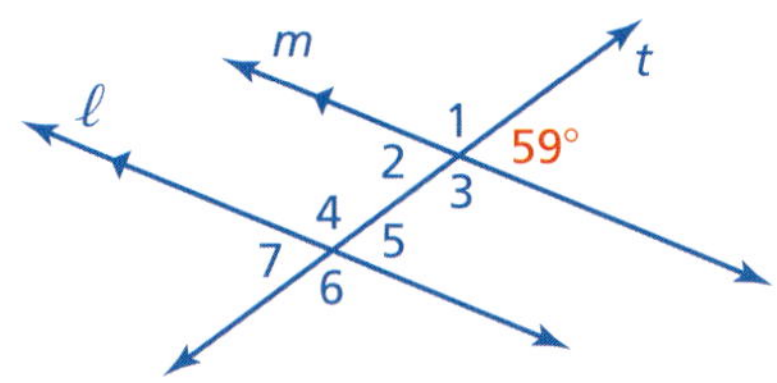

When two parallel lines are cut by a transversal, four **interior angles** are formed on the inside of the parallel lines and four **exterior angles** are formed on the outside of the parallel lines.

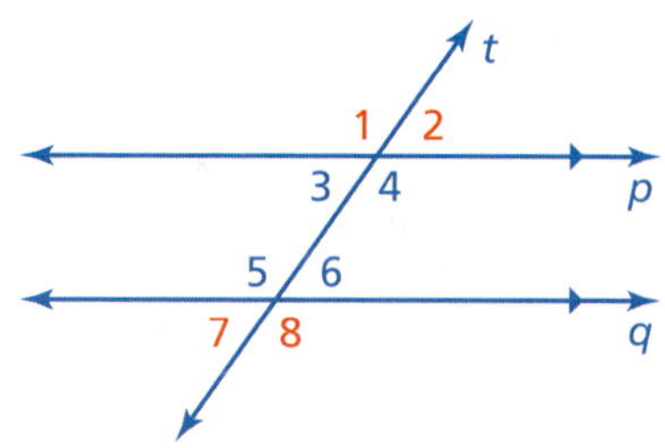

∠3, ∠4, ∠5, and ∠6 are interior angles.
∠1, ∠2, ∠7, and ∠8 are exterior angles.

EXAMPLE 3 Using Corresponding Angles

A store owner uses pieces of tape to paint a window advertisement. The letters are slanted at an 80° angle. What is the measure of ∠1?

Ⓐ 80° Ⓑ 100° Ⓒ 110° Ⓓ 120°

Because all the letters are slanted at an 80° angle, the dashed lines are parallel. The piece of tape is the transversal.

Using corresponding angles, the 80° angle is congruent to the angle that is supplementary to ∠1, as shown.

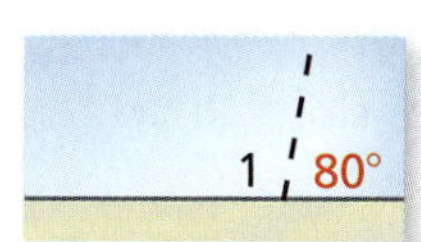

The measure of ∠1 is 180° − 80° = 100°. The correct answer is Ⓑ.

On Your Own

Now You're Ready
Exercises 18 and 19

4. **WHAT IF?** In Example 3, the letters are slanted at a 65° angle. What is the measure of ∠1?

Key Idea

Alternate Interior Angles and Alternate Exterior Angles

When a transversal intersects parallel lines, alternate interior angles are congruent and alternate exterior angles are congruent.

Study Tip

Alternate interior angles and alternate exterior angles lie on opposite sides of the transversal.

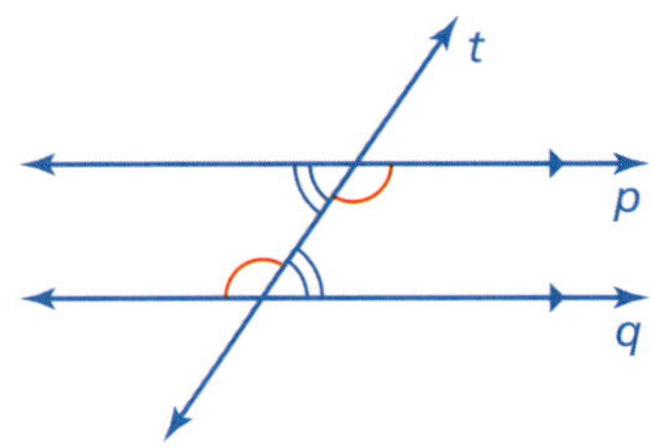

Alternate interior angles

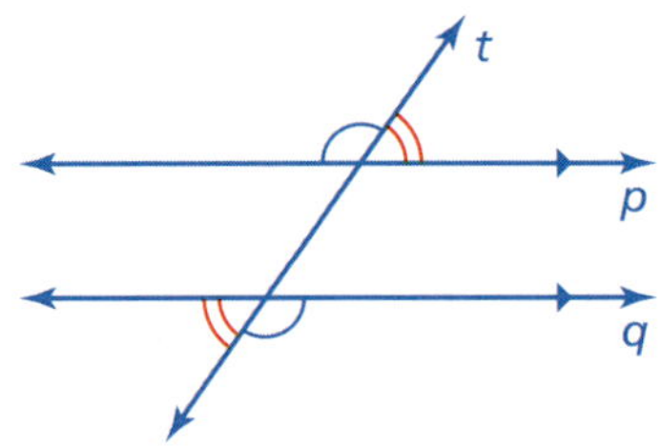

Alternate exterior angles

EXAMPLE 4 Identifying Alternate Interior and Alternate Exterior Angles

The photo shows a portion of an airport. Describe the relationship between each pair of angles.

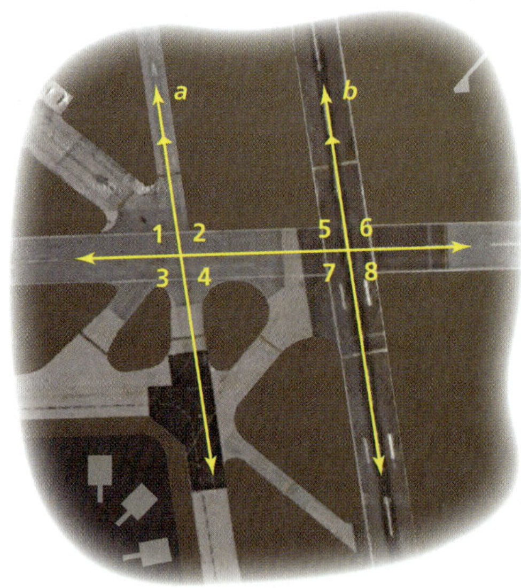

a. ∠3 and ∠6

∠3 and ∠6 are alternate exterior angles.

So, ∠3 is congruent to ∠6.

b. ∠2 and ∠7

∠2 and ∠7 are alternate interior angles.

So, ∠2 is congruent to ∠7.

On Your Own

Now You're Ready
Exercises 20 and 21

In Example 4, the measure of ∠4 is 84°. Find the measure of the angle. Explain your reasoning.

5. ∠3 6. ∠5 7. ∠6

Laurie's Notes

On Your Own

- **Think-Pair-Share:** Students should read the question independently and then work in pairs to answer the question. When they have answered the question, the pair should compare their answer with another group and discuss any discrepancies.

Key Idea

- Write the *Key Idea*. Identify the angles which are marked congruent in the diagram. Mention the *Study Tip*.

Example 4

? "Are the lines parallel? How do you know?" yes; yellow arrow marks

- Work through the explanation as shown. This example helps students identify these new angle pairs.
- **Note:** There is a great deal of vocabulary in this section, so students may need extra practice. It is also important not to draw the parallel lines in the same orientation all of the time, particularly horizontal and vertical.

On Your Own

- Draw the diagram on the board. When students have finished, ask volunteers to come to the board to record their answers and explain their reasoning.
- **Question 7:** Students can say $\angle 6 = 96°$ because of alternate exterior angles *or* because it is the supplement of $\angle 8$.

Closure

- **Exit Ticket:** Find the measure of each angle. Explain your reasoning.

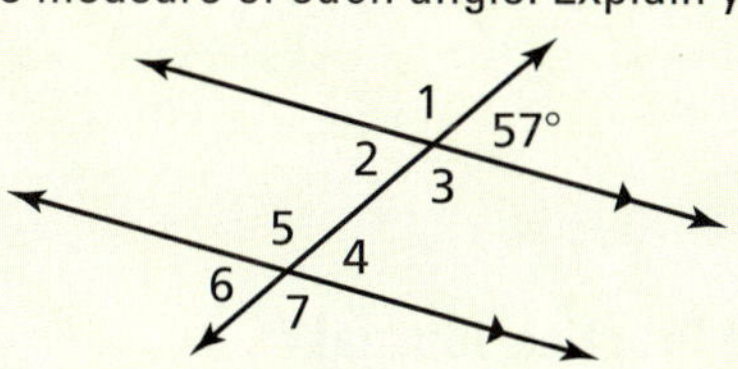

$\angle 1 = 123°$, $\angle 2 = 57°$, $\angle 3 = 123°$, $\angle 4 = 57°$, $\angle 5 = 123°$, $\angle 6 = 57°$, $\angle 7 = 123°$

On Your Own

4. 115°

Extra Example 4

Describe the relationship between each pair of angles.

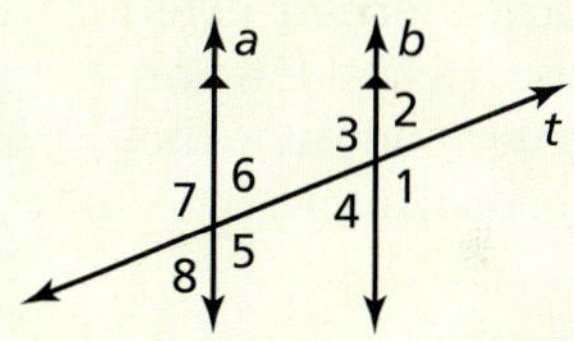

a. $\angle 1$ and $\angle 7$ $\angle 1$ and $\angle 7$ are alternate exterior angles. So, $\angle 1$ is congruent to $\angle 7$.

b. $\angle 3$ and $\angle 5$ $\angle 3$ and $\angle 5$ are alternate interior angles. So, $\angle 3$ is congruent to $\angle 5$.

On Your Own

5. 96°; $\angle 3$ and $\angle 4$ are supplementary.
6. 84°; Alternate interior angles are congruent.
7. 96°; $\angle 5$ and $\angle 6$ are supplementary.

Vocabulary and Concept Check

1. *Sample answer:*

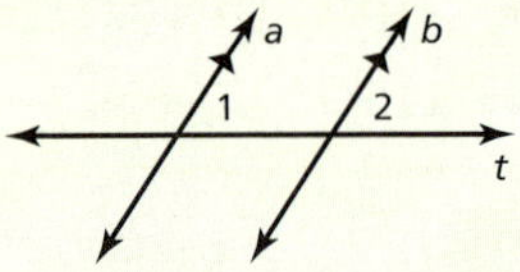

2. "The measure of $\angle 5$" doesn't belong because $\angle 2$, $\angle 6$, and $\angle 8$ are congruent and $\angle 5$ is not a corresponding, alternate interior, or alternate exterior angle with the other three angles. $\angle 2$ and $\angle 8$ are congruent because they are alternate exterior angles. $\angle 6$ and $\angle 8$ are congruent because they are vertical angles.

Practice and Problem Solving

3. m and n
4. t
5. 8
6. $\angle 5$, $\angle 7$, $\angle 1$, and $\angle 3$ are congruent. $\angle 8$, $\angle 6$, $\angle 4$, and $\angle 2$ are congruent.
7. $\angle 1 = 107°$, $\angle 2 = 73°$
8. $\angle 3 = 95°$, $\angle 4 = 85°$
9. $\angle 5 = 49°$, $\angle 6 = 131°$
10. The two lines are not parallel, so $\angle 5 \neq \angle 6$.
11. 60°; Corresponding angles are congruent.
12. *Sample answer:* Railroad tracks are parallel, and the out of bounds lines on a football field are parallel.

Assignment Guide and Homework Check

Level	Assignment	Homework Check
Accelerated	1–6, 10, 14–24 even, 25–35	14, 16, 22, 26, 28

Common Errors

- **Exercises 7–9** Students may mix up some of the definitions of congruent angles and find incorrect angle measures. Encourage them to look at the Key Ideas and color-code the figure they are given to determine what angles are congruent.
- **Exercise 11** Students may not realize that the line in front of the cars is the transversal. Remind them that lines are infinite and can be extended. Draw a diagram of the parallel parking spaces to help students visualize that $\angle 1$ and $\angle 2$ are corresponding angles.

12.1 Record and Practice Journal

Use the figure to find the measures of the numbered angles.

1.

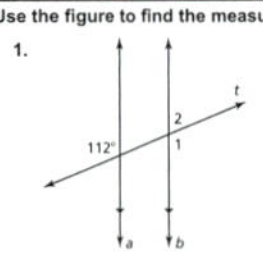

$\angle 1 = 112°$, $\angle 2 = 68°$

2.

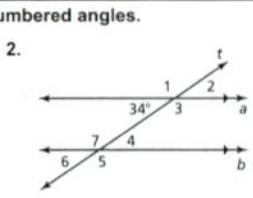

$\angle 1 = 146°$, $\angle 2 = 34°$, $\angle 3 = 146°$, $\angle 4 = 34°$, $\angle 5 = 146°$, $\angle 6 = 34°$, $\angle 7 = 146°$

Complete the statement. Explain your reasoning.

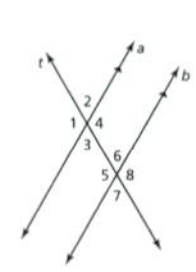

3. If the measure of $\angle 1 = 150°$, then the measure of $\angle 6 =$ **30°**.
4. If the measure of $\angle 3 = 42°$, then the measure of $\angle 5 =$ **138°**.
5. If the measure of $\angle 6 = 28°$, then the measure of $\angle 3 =$ **28°**.
6. You paint a border around the top of the walls in your room. What angle does x need to be to repeat the pattern?

26°

12.1 Exercises

Vocabulary and Concept Check

1. **VOCABULARY** Draw two parallel lines and a transversal. Label a pair of corresponding angles.

2. **WHICH ONE DOESN'T BELONG?** Which statement does *not* belong with the other three? Explain your reasoning. Refer to the figure for Exercises 3–6.

The measure of $\angle 2$	The measure of $\angle 5$
The measure of $\angle 6$	The measure of $\angle 8$

Practice and Problem Solving

In Exercises 3–6, use the figure.

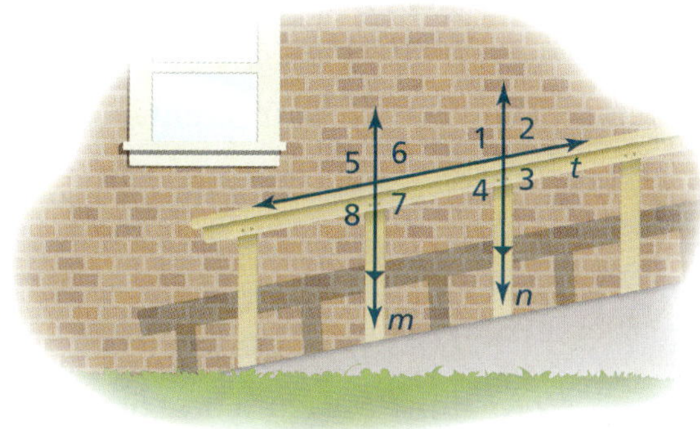

3. Identify the parallel lines.
4. Identify the transversal.
5. How many angles are formed by the transversal?
6. Which of the angles are congruent?

Use the figure to find the measures of the numbered angles.

1 7.

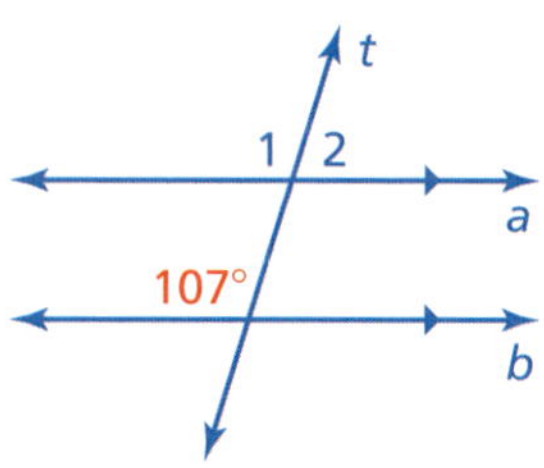

8.

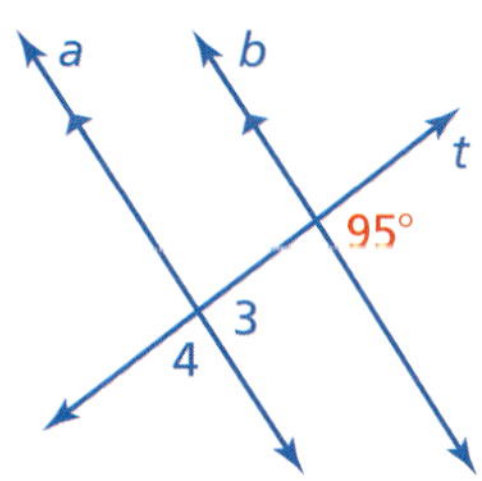

9.

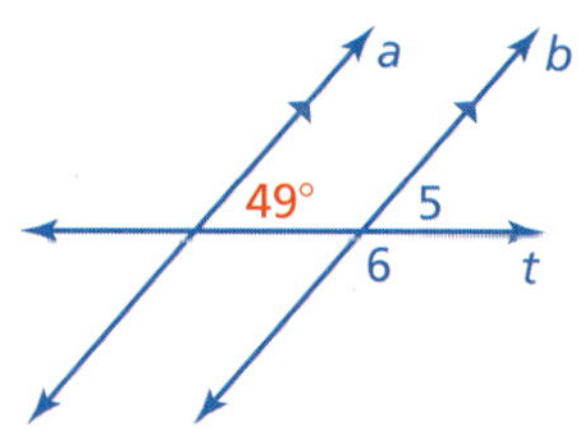

10. **ERROR ANALYSIS** Describe and correct the error in describing the relationship between the angles.

✗ $\angle 5$ is congruent to $\angle 6$.

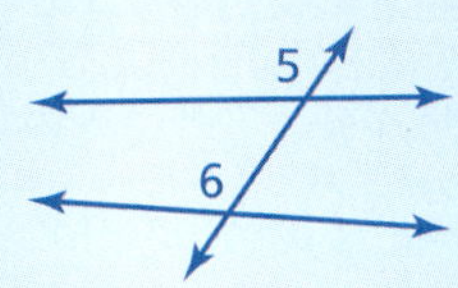

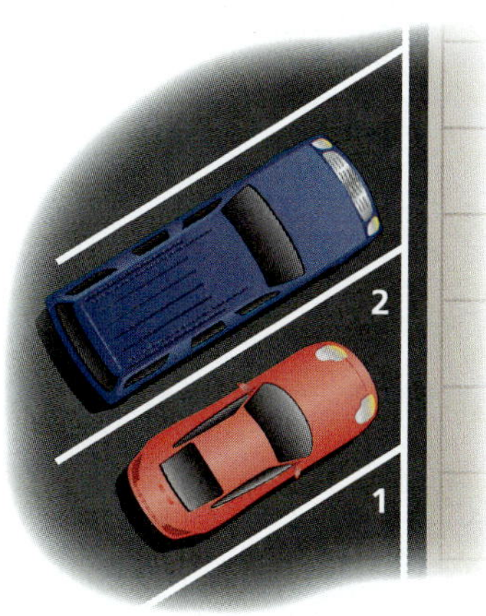

11. **PARKING** The painted lines that separate parking spaces are parallel. The measure of $\angle 1$ is 60°. What is the measure of $\angle 2$? Explain.

12. **OPEN-ENDED** Describe two real-life situations that use parallel lines.

13. **PROJECT** Trace line p and line t on a piece of paper. Label $\angle 1$. Move the paper so that $\angle 1$ aligns with $\angle 8$. Describe the transformations that you used to show that $\angle 1$ is congruent to $\angle 8$.

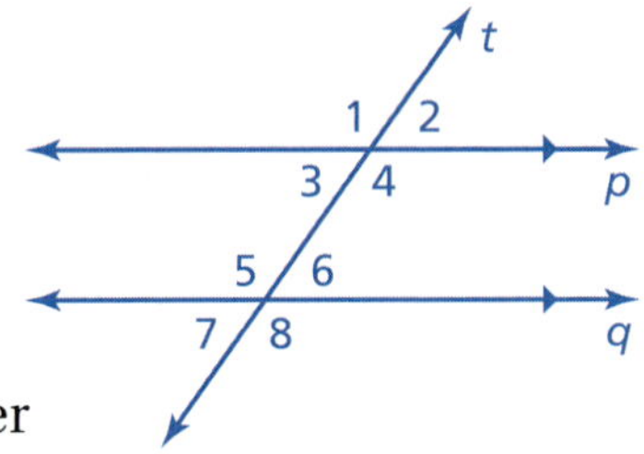

14. **REASONING** Two horizontal lines are cut by a transversal. What is the least number of angle measures you need to know in order to find the measure of every angle? Explain your reasoning.

Use the figure to find the measures of the numbered angles. Explain your reasoning.

2 15.

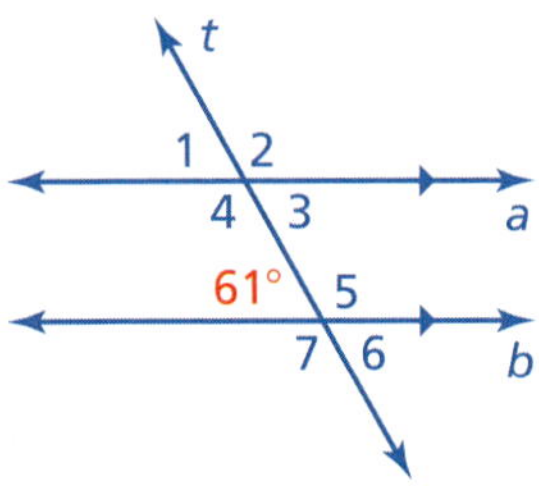

16.

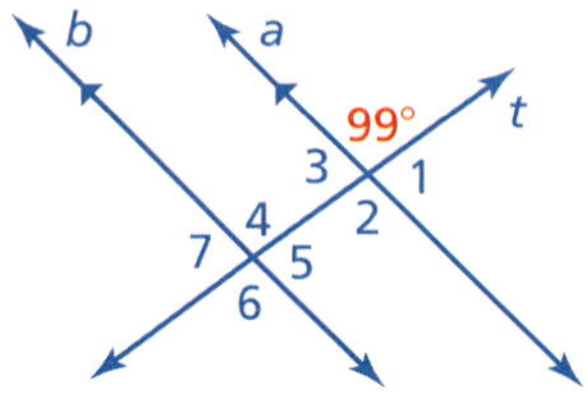

17.

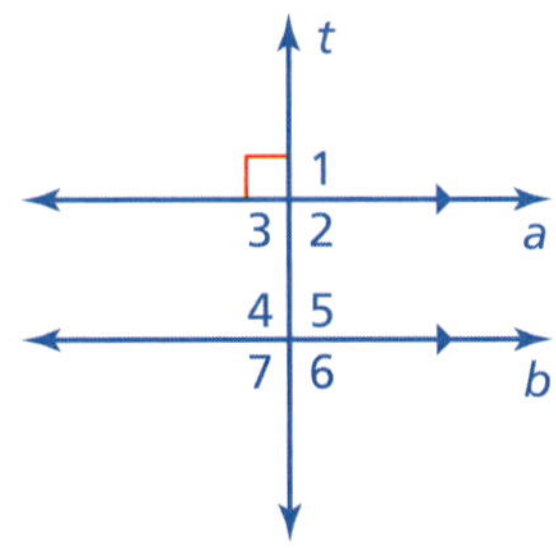

Complete the statement. Explain your reasoning.

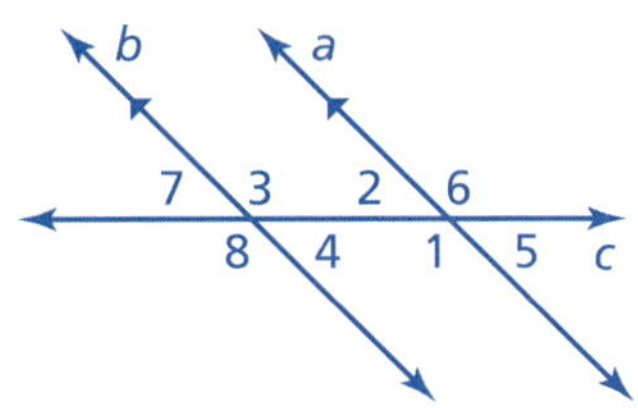

3 18. If the measure of $\angle 1 = 124°$, then the measure of $\angle 4 =$ ____.

19. If the measure of $\angle 2 = 48°$, then the measure of $\angle 3 =$ ____.

4 20. If the measure of $\angle 4 = 55°$, then the measure of $\angle 2 =$ ____.

21. If the measure of $\angle 6 = 120°$, then the measure of $\angle 8 =$ ____.

22. If the measure of $\angle 7 = 50.5°$, then the measure of $\angle 6 =$ ____.

23. If the measure of $\angle 3 = 118.7°$, then the measure of $\angle 2 =$ ____.

24. **RAINBOW** A rainbow forms when sunlight reflects off raindrops at different angles. For blue light, the measure of $\angle 2$ is 40°. What is the measure of $\angle 1$?

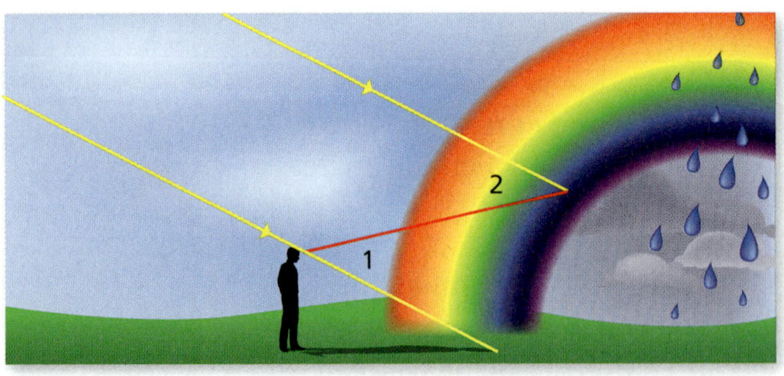

25. **REASONING** When a transversal is perpendicular to two parallel lines, all the angles formed measure 90°. Explain why.

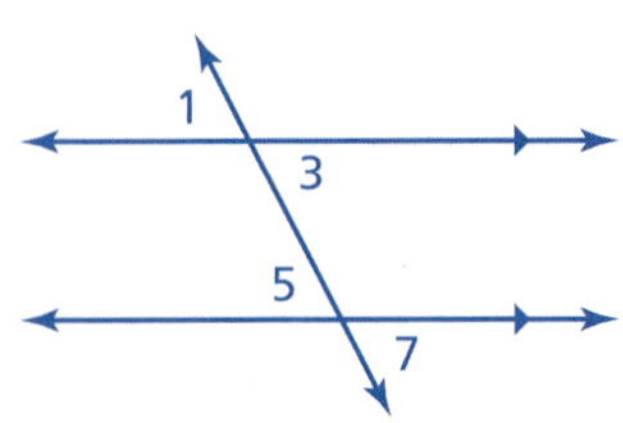

26. **LOGIC** Describe two ways you can show that $\angle 1$ is congruent to $\angle 7$.

Common Errors

- **Exercises 15–17** Students may not understand alternate interior and exterior angles and say that an exterior angle is congruent to the alternate interior angle. For example, in Exercise 15, a student may say the measure of $\angle 2$ is 61°. Use corresponding angles to show that this is not true.
- **Exercises 18–23** Students may use some of the definitions of congruent angles incorrectly in finding the angle measure of the unknown angle. Review the definitions and give an example with an adequate explanation of how to find the missing angle.
- **Exercises 27 and 28** Students may only see one set of parallel lines and think that they cannot find the measure of the missing angle. Point out the small arrows that denote that two lines are parallel. Encourage them to find the measure of an angle that is near the missing angle and then rotate the figure to help them visualize how to solve for the missing angle.

Differentiated Instruction

Auditory

Students may confuse the measures of *complementary angles* and *supplementary angles*. Show students that *complementary* comes before *supplementary* in the dictionary and that *90* comes before *180* numerically. So, the sum of complementary angles is 90° and the sum of supplementary angles is 180°.

Practice and Problem Solving

13. *Sample answer:* rotate 180° and translate down.

14. You only need one angle because half of the angles are congruent to that angle and you can find the other angles using relationships.

15. $\angle 6 = 61°$; $\angle 6$ and the given angle are vertical angles.
$\angle 5 = 119°$ and $\angle 7 = 119°$; $\angle 5$ and $\angle 7$ are supplementary to the given angle.
$\angle 1 = 61°$; $\angle 1$ and the given angle are corresponding angles.
$\angle 3 = 61°$; $\angle 1$ and $\angle 3$ are vertical angles.
$\angle 2 = 119°$ and $\angle 4 - 119°$; $\angle 2$ and $\angle 4$ are supplementary to $\angle 1$.

16. $\angle 2 = 99°$; $\angle 2$ and the given angle are vertical angles.
$\angle 1 = 81°$ and $\angle 3 = 81°$; $\angle 1$ and $\angle 3$ are supplementary to the given angle.
$\angle 4 = 99°$; $\angle 2$ and $\angle 4$ are alternate interior angles.
$\angle 5 = 81°$ and $\angle 7 = 81°$; $\angle 5$ and $\angle 7$ are supplementary to $\angle 4$.
$\angle 6 = 99°$; $\angle 6$ and the given angle are alternate exterior angles.

17–26. See Additional Answers.

Practice and Problem Solving

27. 130

28. 115

29. **a.** no; They look like they are spreading apart.

b. Check students' work.

30. See *Taking Math Deeper*.

Fair Game Review

31. 13 **32.** 14

33. 51 **34.** 3

35. B

Mini-Assessment

Use the figure to find the measures of the numbered angles.

1.

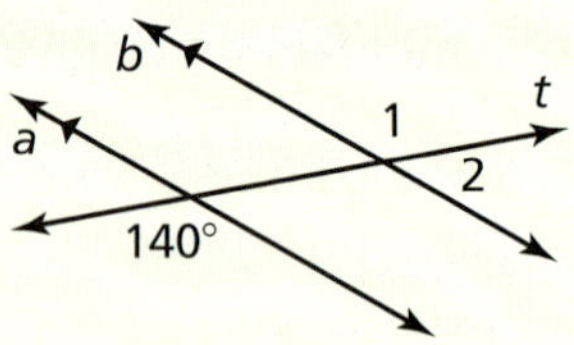

$\angle 1 = 140°$; $\angle 2 = 40°$

2.

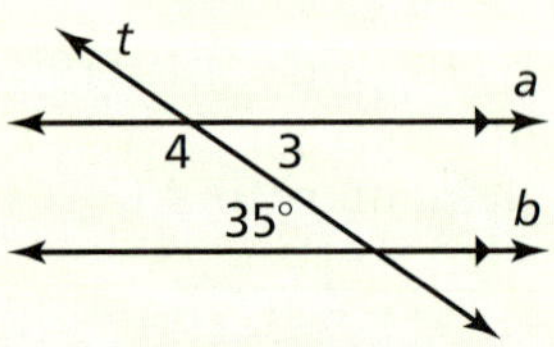

$\angle 3 = 35°$; $\angle 4 = 145°$

3.

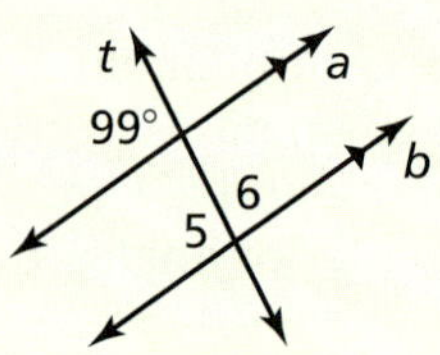

$\angle 5 = 99°$; $\angle 6 = 81°$

4.

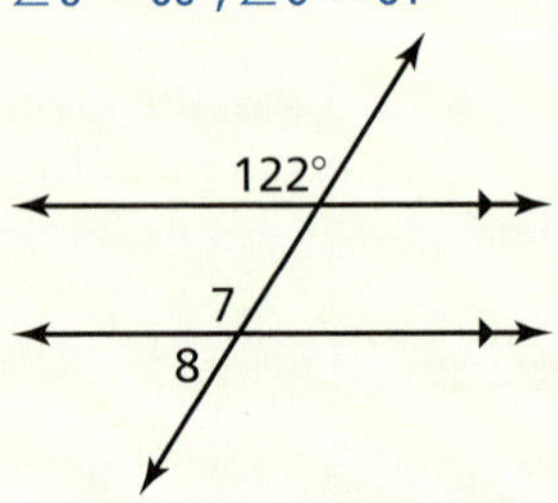

$\angle 7 = 122°$; $\angle 8 = 58°$

Taking Math Deeper

Exercise 30

This problem uses a well-known reflective property in physics. This property applies to mirrors, billiard tables, air hockey tables, and many other objects. The property states that when the hockey puck bounces off the side board, its out-going angle is equal to its in-coming angle.

1 Solve for m.

$$m + m + 64 = 180$$
$$2m = 116$$
$$m = 58$$

2 Solve for x.

a. Using the property of alternate interior angles, you can determine that $x = 64$.

3 Answer the question.

b. The goal is slightly wider than the hockey puck. So, there is some leeway allowed for the measure of x.

By studying the diagram, you can see that x cannot be much greater. However, x can be a little less and still have the hockey puck go into the goal.

Project

Write a report about at least two other games that use angles as part of their strategy.

Reteaching and Enrichment Strategies

If students need help. . .	If students got it. . .
Resources by Chapter • Practice A and Practice B • Puzzle Time Record and Practice Journal Practice Differentiating the Lesson Lesson Tutorials Skills Review Handbook	Resources by Chapter • Enrichment and Extension • Technology Connection Start the next section

CRITICAL THINKING **Find the value of x.**

27.

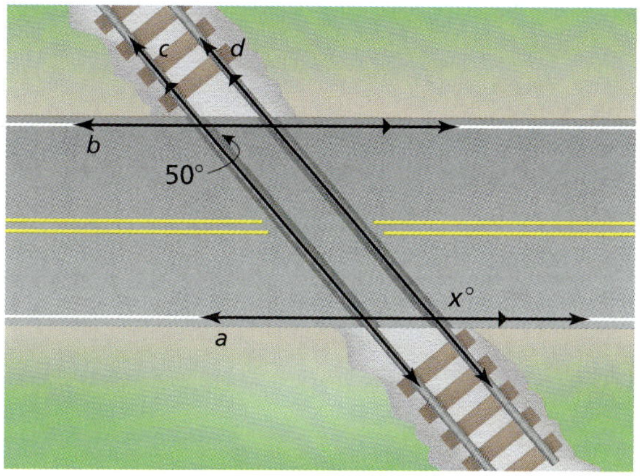

28.

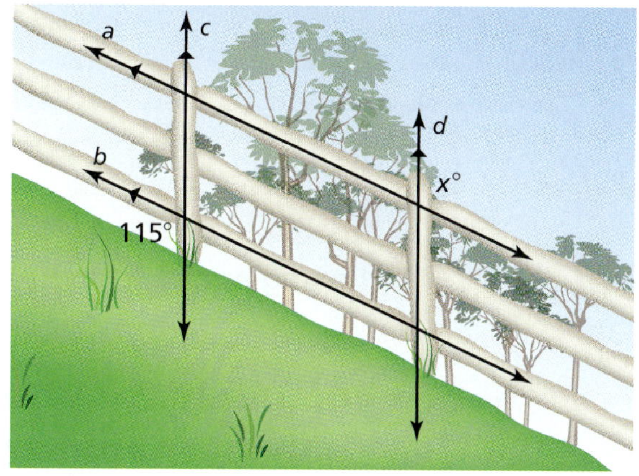

29. **OPTICAL ILLUSION** Refer to the figure.

a. Do the horizontal lines appear to be parallel? Explain.

b. Draw your own optical illusion using parallel lines.

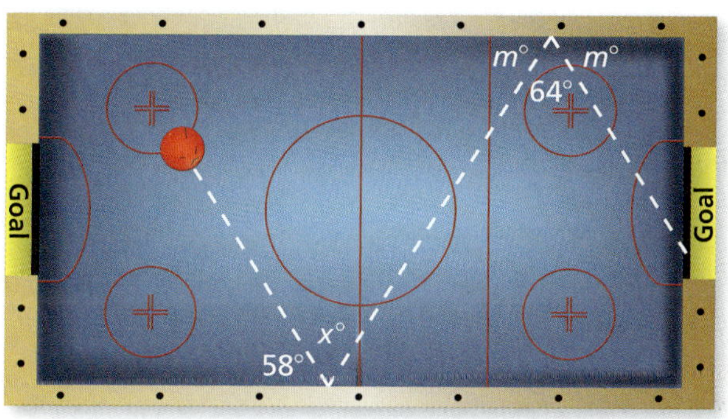

30. Geometry The figure shows the angles used to make a double bank shot in an air hockey game.

a. Find the value of x.

b. Can you still get the red puck in the goal when x is increased by a little? by a lot? Explain.

Fair Game Review What you learned in previous grades & lessons

Evaluate the expression. *(Section 1.4)*

31. $4 + 3^2$ **32.** $5(2)^2 - 6$ **33.** $11 + (-7)^2 - 9$ **34.** $8 \div 2^2 + 1$

35. **MULTIPLE CHOICE** The triangles are similar. What length does x represent? *(Section 11.5)*

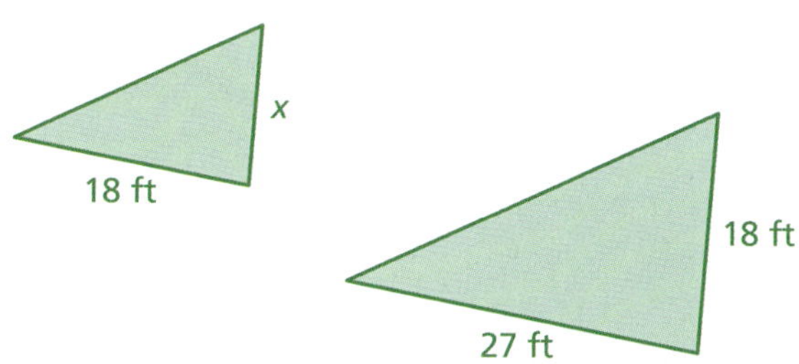

Ⓐ 2 ft Ⓑ 12 ft
Ⓒ 15 ft Ⓓ 27 ft

12.2 Angles of Triangles

Essential Question How can you describe the relationships among the angles of a triangle?

1 ACTIVITY: Exploring the Interior Angles of a Triangle

Work with a partner.

a. Draw a triangle. Label the interior angles *A*, *B*, and *C*.

b. Carefully cut out the triangle. Tear off the three corners of the triangle.

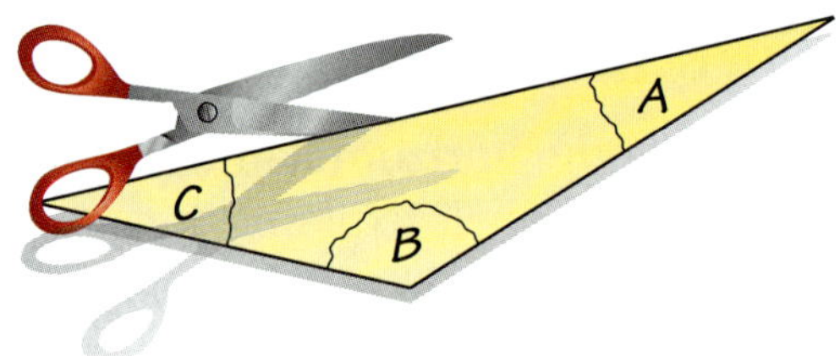

c. Arrange angles *A* and *B* so that they share a vertex and are adjacent.

d. How can you place the third angle to determine the sum of the measures of the interior angles? What is the sum?

e. Compare your results with those of others in your class.

f. **STRUCTURE** How does your result in part (d) compare to your conclusion in Lesson 7.3, Activity Question 7?

2 ACTIVITY: Exploring the Interior Angles of a Triangle

Work with a partner.

a. Describe the figure.

b. **LOGIC** Use what you know about parallel lines and transversals to justify your result in part (d) of Activity 1.

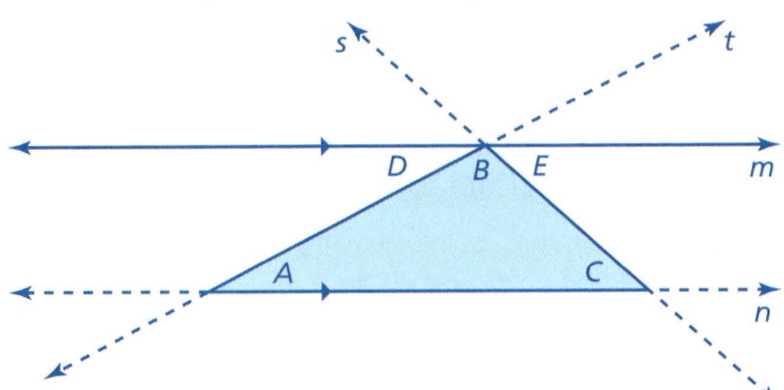

Geometry

In this lesson, you will

- understand that the sum of the interior angle measures of a triangle is 180°.
- find the measures of interior and exterior angles of triangles.

Laurie's Notes

Introduction

Applying Mathematical Practices

- **Construct Viable Arguments and Critique the Reasoning of Others:** Mathematically proficient students are able to make conjectures and construct arguments to explain their reasoning.

Motivate

- Make teams of three students. Give them 3 minutes to make a list of as many words as they can that begin with the prefix *tri-*.
- Some examples are: triangle, triathlon, tricycle, tri-fold, triangulate, triad, triaxial, trilogy, trimester, trinary, trinity, trio, trilingual, trillium.
- Provide dictionaries if necessary.
- The goal of this activity is to demonstrate that *tri-* is a common prefix.

Activity Notes

Activity 1

- Many students will have heard of the property they are investigating today. Having heard the property and internalizing it for all triangles are two different levels of knowledge.
- The sides of the triangle must be straight; otherwise the three angles will not lie adjacent to one another when placed about a point on the line.
- **Teaching Tip:** If you cannot gain access to enough pairs of scissors, you can cut out several triangles in advance using a paper cutter. It is okay to have multiple copies of the same triangle because different pairs of students will get one copy of the triangle.
- The conclusion, or rule, that students should discover is that the angle measures of any triangle will sum to 180°.
- **Management Tip:** There will be torn pieces of scrap paper resulting from this investigation. To help keep the room clean, cluster 4–6 desks together in a circle and tape a recycled paper or plastic bag to the front edge of one of the desks. Students are expected to put scraps of paper in the bags when they are finished with the investigation.

Activity 2

- **Construct Viable Arguments and Critique the Reasoning of Others** and **Attend to Precision:** In this activity, you want to hear students make statements based upon evidence they have. Instead of, "Angles C and E are the same measure because they look it," students should say, "Angles C and E are the same measure because lines m and n are parallel and angles C and E are alternate interior angles."
- If students are stuck, ask them to think about what they learned in the last lesson. They should mark the diagram with what they know.
- Students should make the connection that angles B, D, and E are the same as the three angles they placed about a point in Activity 1, where their conjecture was that the sum of the three angle measures is 180°. What is different in this activity is that students are asked to *justify* their answers.

What Your Students Will Learn

- Understand that the sum of the interior angle measures of a triangle is 180°.
- Find and use the measures of interior and exterior angles of triangles.

Previous Learning

Students should know basic vocabulary associated with angles and triangles.

Technology for the Teacher

Dynamic Classroom

Lesson Plans
Complete Materials List

12.2 Record and Practice Journal

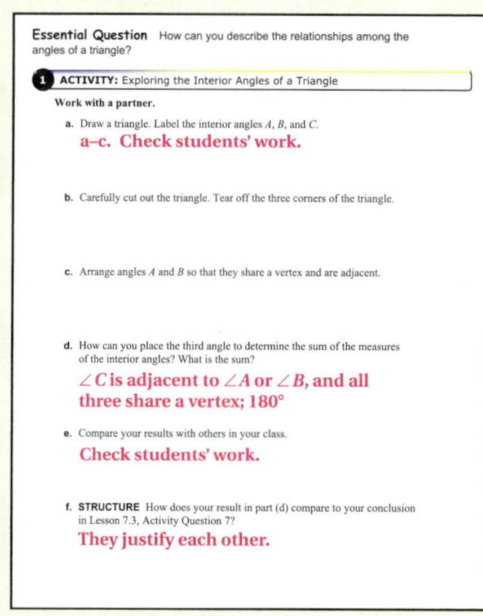

Essential Question How can you describe the relationships among the angles of a triangle?

1 ACTIVITY: Exploring the Interior Angles of a Triangle

Work with a partner.

a. Draw a triangle. Label the interior angles A, B, and C.
a–c. Check students' work.

b. Carefully cut out the triangle. Tear off the three corners of the triangle.

c. Arrange angles A and B so that they share a vertex and are adjacent.

d. How can you place the third angle to determine the sum of the measures of the interior angles? What is the sum?
$\angle C$ is adjacent to $\angle A$ or $\angle B$, and all three share a vertex; 180°

e. Compare your results with others in your class.
Check students' work.

f. **STRUCTURE** How does your result in part (d) compare to your conclusion in Lesson 7.3, Activity Question 7?
They justify each other.

Differentiated Instruction

Kinesthetic

When talking about right, acute, and obtuse angles of a triangle, ask students if it is possible to draw a triangle with 2 right angles. Students should see by drawing the two right angles with a common side that the remaining two sides of the right angles will never meet. So, no triangle can be formed with 2 right angles. Ask students if it is possible for a triangle to have 2 obtuse angles. Students should reach the same conclusion. No triangle can be formed with 2 obtuse angles.

12.2 Record and Practice Journal

2 ACTIVITY: Exploring the Interior Angles of a Triangle

Work with a partner.

a. Describe the figure.

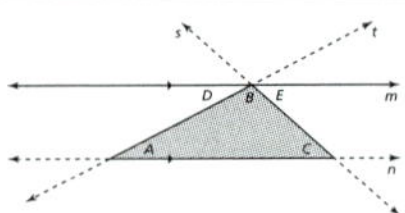

m parallel to *n*; sides of triangle extended; *s* and *t* are transversals.

b. **LOGIC** Use what you know about parallel lines and transversals to justify your result in part (d) of Activity 1.

See Additional Answers.

3 ACTIVITY: Exploring an Exterior Angle of a Triangle

Work with a partner.

a. Draw a triangle. Label the interior angles *A*, *B*, and *C*.

a–c. Check students' work.

b. Carefully cut out the triangle.

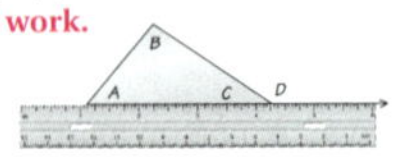

c. Place the triangle on a piece of paper and extend one side to form *exterior angle D*, as shown.

d. Tear off the corners that are not adjacent to the exterior angle. Arrange them to fill the exterior angle, as shown. What does this tell you about the measure of exterior angle *D*?

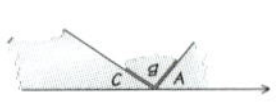

$m\angle D = m\angle A + m\angle B$

4 ACTIVITY: Measuring the Exterior Angles of a Triangle

Work with a partner.

a. Draw a triangle and label the interior and exterior angles as shown. Check students' work.

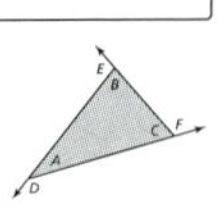

b. Use a protractor to measure all six angles of your triangle. Complete the table to organize your results. What does the table tell you about the measure of an exterior angle of a triangle? Check students' work.

Exterior Angle	$D =$ ____°	$E =$ ____°	$F =$ ____°
Interior Angle	$B =$ ____°	$A =$ ____°	$A =$ ____°
Interior Angle	$C =$ ____°	$C =$ ____°	$B =$ ____°

$m\angle D = m\angle B + m\angle C$; $m\angle E = m\angle A + m\angle C$; $m\angle F = m\angle A + m\angle B$

What Is Your Answer?

5. **REPEATED REASONING** Draw three triangles that have different shapes. Repeat parts (b)–(d) from Activity 1 for each triangle. Do you get the same results? Explain.

Check students' work.

6. **IN YOUR OWN WORDS** How can you describe the relationships among angles of a triangle?

Interior angle measures sum to 180°; Exterior angle measure = sum of nonadjacent interior angle measures

Laurie's Notes

Activity 3

- **Teaching Tip:** When asked to draw a triangle, suggest to students that they draw a scalene triangle. Draw the triangle large enough so that it is easy to see and measure.
- **Teaching Tip:** If gaining access to sufficient pairs of scissors is a problem, have pre-cut triangles available for students to use.
- If students are unclear about the direction "extend one side" in part (c), explain that they should draw a ray and place the triangle against the ray, as shown. The extended ray is one side of angle *D* and the second side of angle *D* is a side of the triangle *ABC*.
- **Construct Viable Arguments and Critique the Reasoning of Others:** Ask students to summarize their findings. Students are making a conjecture about an exterior angle of a triangle.

Activity 4

- Students should draw a large, scalene triangle.
- Students should extend the sides of the triangle in order to form one exterior angle at each vertex.
- **FYI:** In the diagram at the right, Angles 1 and 2 are exterior angles for interior angle *A*. There are two exterior angles at each vertex. These are congruent, vertical angles. It does not matter which of the two angles students draw and measure.

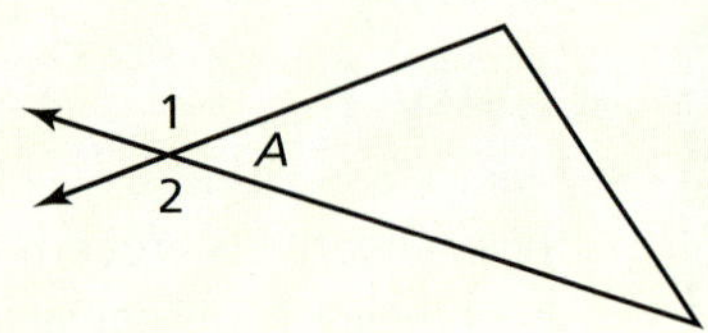

? "What are the interior angles of the triangle?" angles *A*, *B*, and *C*

? "What are the exterior angles of the triangle?" angles *D*, *E*, and *F*

? "What do you notice about the measure of an exterior angle of a triangle?" It is the same as the sum of the two non adjacent interior angles.

- **Attend to Precision:** Encourage students to be precise with their language. They may say that the exterior angle is the sum of two interior angles. Which two? How do they describe the two angles?
- **Extension:** Students may also observe that the sum of the exterior angle measures is 360°. Compare results with different groups of students. Is this true for the triangles drawn by all the students?

What Is Your Answer?

- **Construct Viable Arguments and Critique the Reasoning of Others:** Students should understand that conjectures need to be verified. Repeating an investigation on a different type of triangle gives additional evidence, or in some cases, it might become a counter-example.

Closure

- Draw the triangle shown and label the two angles. Find the measure of angles 1 and 2. 56°, 146°

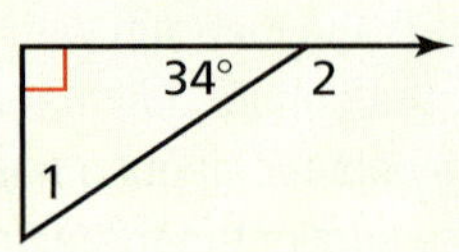

3 ACTIVITY: Exploring an Exterior Angle of a Triangle

Math Practice

Maintain Oversight

Do you think your conclusion will be true for the exterior angle of any triangle? Explain.

Work with a partner.

a. Draw a triangle. Label the interior angles *A*, *B*, and *C*.

b. Carefully cut out the triangle.

c. Place the triangle on a piece of paper and extend one side to form *exterior angle D*, as shown.

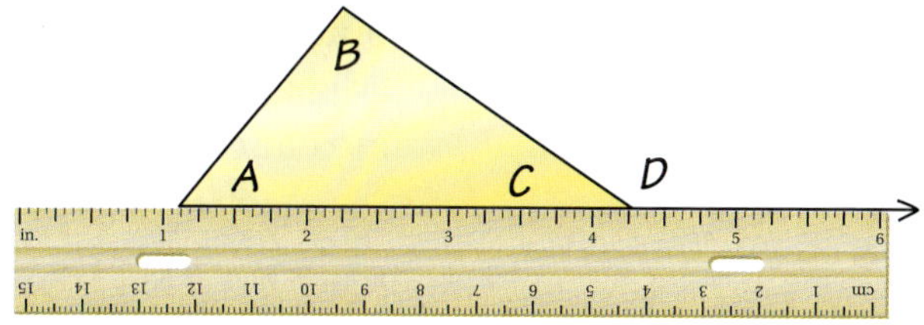

d. Tear off the corners that are not adjacent to the exterior angle. Arrange them to fill the exterior angle, as shown. What does this tell you about the measure of exterior angle *D*?

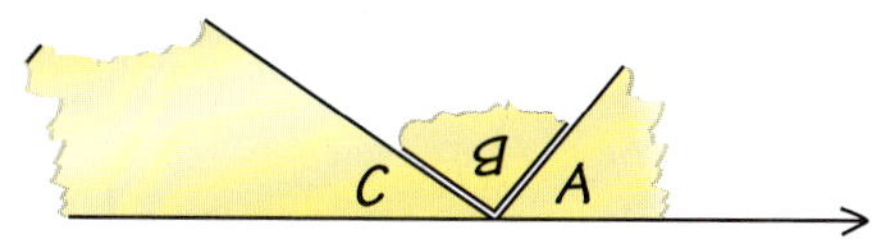

4 ACTIVITY: Measuring the Exterior Angles of a Triangle

Work with a partner.

a. Draw a triangle and label the interior and exterior angles, as shown.

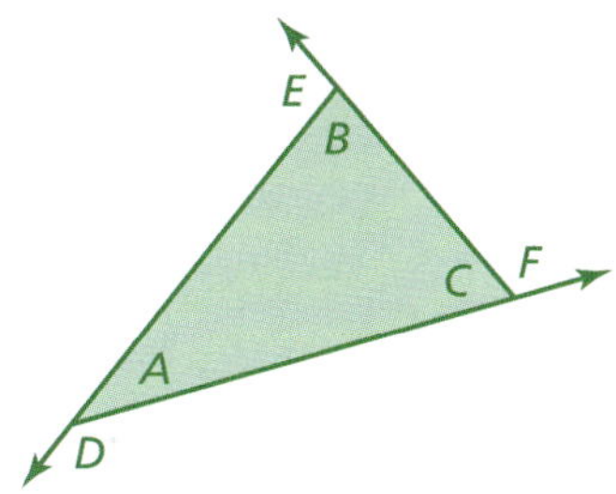

b. Use a protractor to measure all six angles. Copy and complete the table to organize your results. What does the table tell you about the measure of an exterior angle of a triangle?

Exterior Angle	$D =$ ___°	$E =$ ___°	$F =$ ___°
Interior Angle	$B =$ ___°	$A =$ ___°	$A =$ ___°
Interior Angle	$C =$ ___°	$C =$ ___°	$B =$ ___°

What Is Your Answer?

5. **REPEATED REASONING** Draw three triangles that have different shapes. Repeat parts (b)–(d) from Activity 1 for each triangle. Do you get the same results? Explain.

6. **IN YOUR OWN WORDS** How can you describe the relationships among angles of a triangle?

Practice Use what you learned about angles of a triangle to complete Exercises 4–6 on page 538.

12.2 Lesson

Check It Out
Lesson Tutorials
BigIdeasMath.com

Key Vocabulary
interior angles of a polygon, *p. 536*
exterior angles of a polygon, *p. 536*

The angles inside a polygon are called **interior angles**. When the sides of a polygon are extended, other angles are formed. The angles outside the polygon that are adjacent to the interior angles are called **exterior angles**.

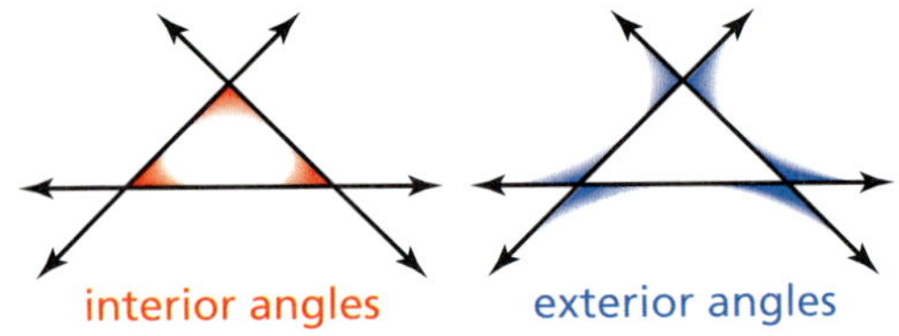

Key Idea

Interior Angle Measures of a Triangle

Words The sum of the interior angle measures of a triangle is 180°.

Algebra $x + y + z = 180$

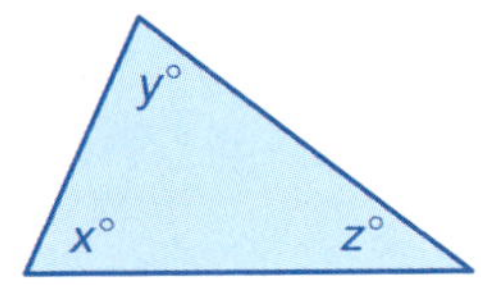

EXAMPLE 1 Using Interior Angle Measures

Find the value of *x*.

a.

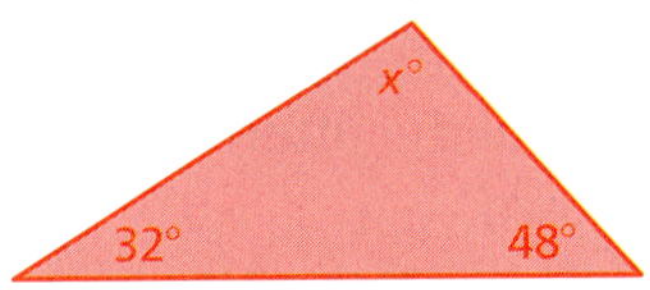

$$x + 32 + 48 = 180$$
$$x + 80 = 180$$
$$x = 100$$

b.

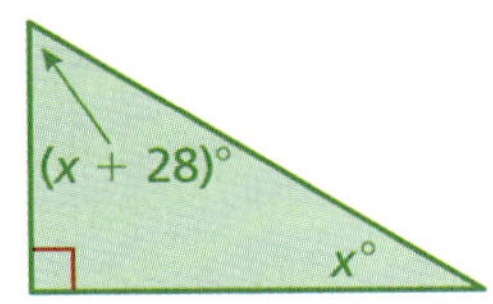

$$x + (x + 28) + 90 = 180$$
$$2x + 118 = 180$$
$$2x = 62$$
$$x = 31$$

On Your Own

Now You're Ready
Exercises 4–9

Find the value of *x*.

1.

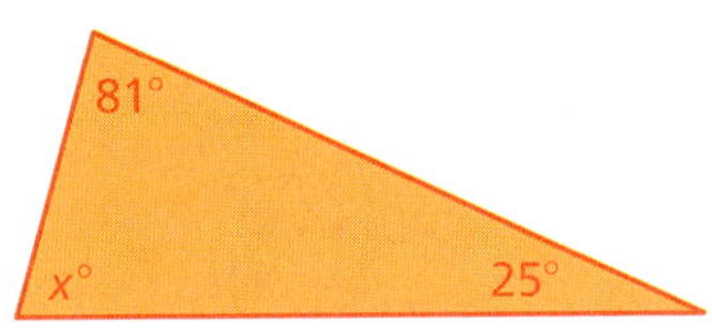

2.

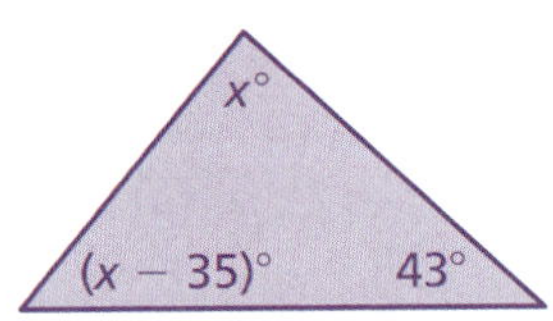

Key Idea

Exterior Angle Measures of a Triangle

Words The measure of an exterior angle of a triangle is equal to the sum of the measures of the two nonadjacent interior angles.

Algebra $z = x + y$

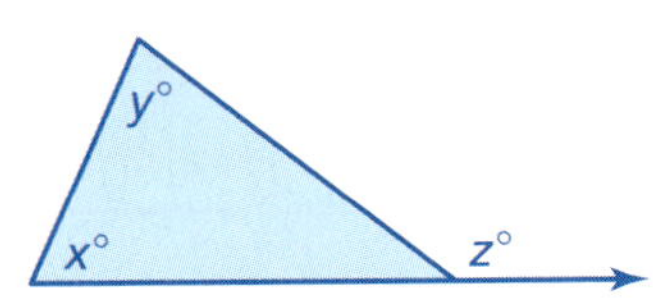

Multi-Language Glossary at BigIdeasMath.com

Laurie's Notes

Introduction

Connect

- **Yesterday:** Students explored the sum of the angle measures of a triangle and the vocabulary associated with triangles.
- **Today:** Students will find the missing angle measure of a triangle and classify the triangle.

Motivate

- Discuss the Ohio State flag.
  The blue triangle represents hills and valleys. The red and white stripes represent roads and waterways. The 13 leftmost stars represent the 13 original colonies. The 4 stars on the right bring the total to 17, representing that Ohio was the 17th state admitted to the Union.

? "Do you know any other flags that contain a triangle?" Answers will vary.

Lesson Notes

Discuss

- Discuss interior and exterior angles of a triangle. Tell students that there are six exterior angles of a triangle, 2 at each vertex.

? "What is the relationship between the two exterior angles at each vertex?" They are congruent, because they are vertical angles.

Key Idea

- The property is written with variables to suggest that you can solve an equation to find the third angle when you know the other two angles. This is also called the *Triangle Sum Theorem.*

? "What type of angles are the remaining angles of a right triangle? a triangle with an obtuse angle?" Both are acute.

? "Do you think an obtuse triangle could have a right angle? Explain." no; The sum of the angle measures would be greater than 180°.

Example 1

- Some students may argue that all they need to do is add the angle measures and subtract from 180. Remind them that they are practicing a *process*, one that works when the three angle measures are given as algebraic expressions, such as $(x + 10)°$, $(x + 20)°$, and $(x + 30)°$.

Key Idea

- This is also called the *Exterior Angle Theorem*, and the two nonadjacent interior angles are also called the *remote interior angles.*

? "If the interior angle of a triangle is acute, what do you know about the exterior angle at that vertex?" obtuse and supplement of the acute angle

? "Could an exterior angle of a triangle be acute? Explain." yes; if the interior angle is obtuse

Goal Today's lesson is finding interior and exterior angle measures of triangles.

Technology for the Teacher

Dynamic Classroom

Lesson Tutorials
Lesson Plans
Answer Presentation Tool

Extra Example 1

Find the value of *x*.

a.

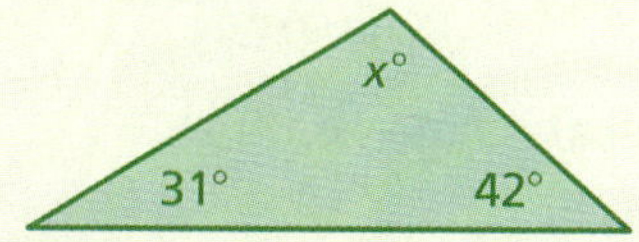

107

b.

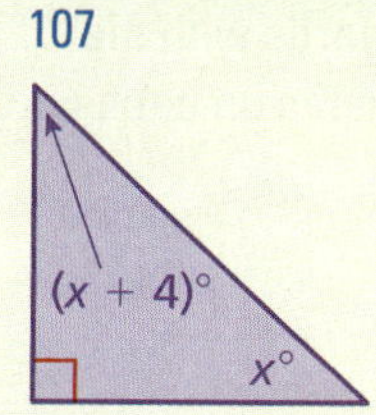

43

On Your Own

1. 74
2. 86

Laurie's Notes

Extra Example 2

Find the measure of the exterior angle.

a.

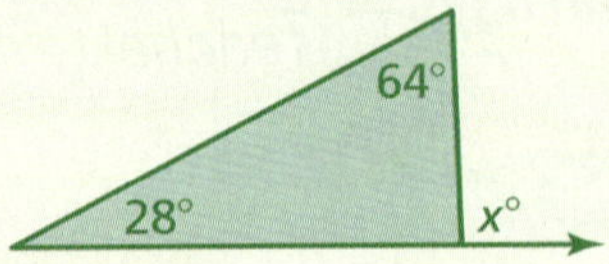

92°

b.

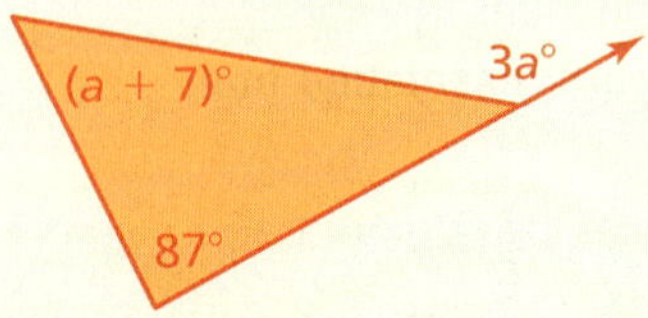

141°

Extra Example 3

A car travels around the park shown below. What is the value of *x*?

62

On Your Own

3. 70° **4.** 140°

5. 54.3

English Language Learners

Pair Activity

Create sets of index cards with the measures of three angles on each card, as shown below.

30°, 60°, 90°
40°, 50°, 60°
20°, 75°, 85°

Have students work together to determine whether the angles could be the angles of a triangle. If not, they should choose two of the angles and determine the measure of the third angle that would form a triangle.

Example 2

- Write the problem in part (a).
- ? "How do you find the measure of the exterior angle?" The exterior angle will be the sum of the two remote interior angles, so 36° + 72°.
- ? **Reason Abstractly and Quantitatively:** "Explain two different ways to find the measure of the third angle of the triangle." Use the *Triangle Sum Theorem*, or find the supplement of the exterior angle.
- The problem in part (b) involves writing an equation versus doing mental math.
- Use the *Exterior Angle Theorem* to write the equation. Solve the equation as shown.
- **Common Error:** Students will solve for the variable *a* correctly and forget to answer the question asked, meaning they forget to substitute the value of *a* into the expression for the exterior angle.

Example 3

- Add a little interest by sharing information from the Department of the Navy website, *history.navy.mil*. The *Bermuda Triangle* is an imaginary area located off the southeastern Atlantic coast of the U.S. where a supposedly high incidence of unexplained disappearances of ships and aircraft occurs. The vertices of the triangle are Bermuda; Miami, Florida; and San Juan, Puerto Rico.
- Set up the equation and work through the problem as shown.
- This is a good review of equation solving and work with decimals.

On Your Own

- **Neighbor Check:** Have students work independently and then have their neighbors check their work. Have students discuss any discrepancies.

Closure

- **Exit Ticket:** Find the value of *x*. Then classify the triangle.

a.

75; acute triangle

b.

46; obtuse triangle

EXAMPLE 2 Finding Exterior Angle Measures

Study Tip

Each vertex has a pair of congruent exterior angles. However, it is common to show only one exterior angle at each vertex.

Find the measure of the exterior angle.

a.

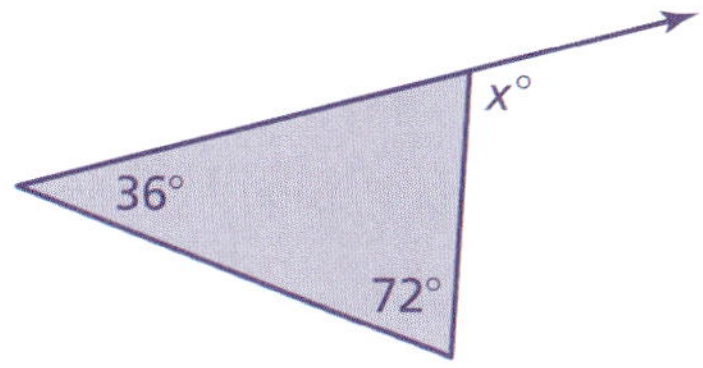

$x = 36 + 72$

$x = 108$

So, the measure of the exterior angle is 108°.

b.

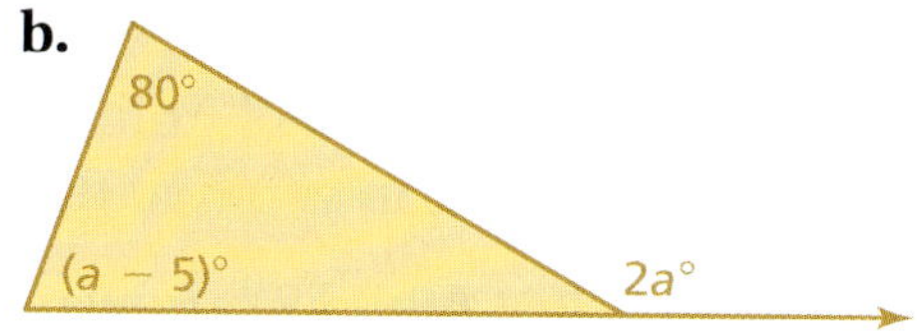

$2a = (a - 5) + 80$

$2a = a + 75$

$a = 75$

So, the measure of the exterior angle is $2(75)° = 150°$.

EXAMPLE 3 Real-Life Application

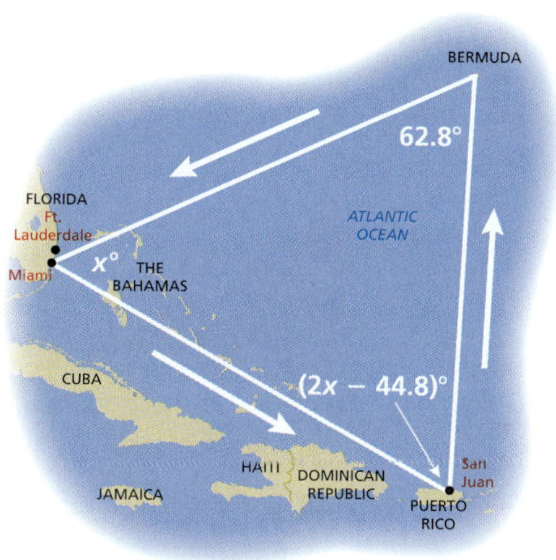

An airplane leaves from Miami and travels around the Bermuda Triangle. What is the value of *x*?

Ⓐ 26.8 Ⓑ 27.2 Ⓒ 54 Ⓓ 64

Use what you know about the interior angle measures of a triangle to write an equation.

$x + (2x - 44.8) + 62.8 = 180$	Write equation.
$3x + 18 = 180$	Combine like terms.
$3x = 162$	Subtract 18 from each side.
$x = 54$	Divide each side by 3.

The value of x is 54. The correct answer is Ⓒ.

On Your Own

Now You're Ready
Exercises 12–14

Find the measure of the exterior angle.

3.

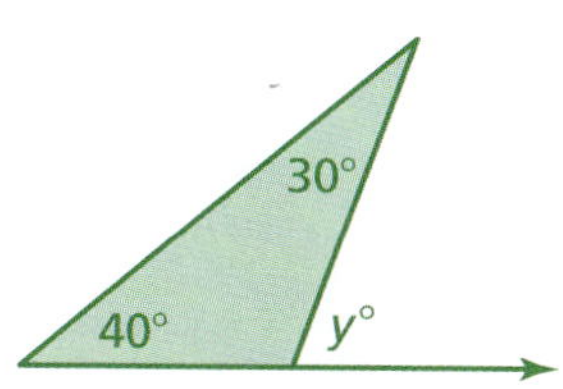

4.

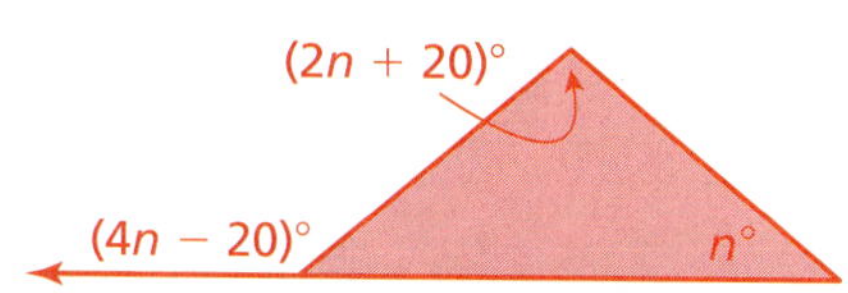

5. In Example 3, the airplane leaves from Fort Lauderdale. The interior angle measure at Bermuda is 63.9°. The interior angle measure at San Juan is $(x + 7.5)°$. Find the value of x.

12.2 Exercises

Vocabulary and Concept Check

1. **VOCABULARY** You know the measures of two interior angles of a triangle. How can you find the measure of the third interior angle?
2. **VOCABULARY** How many exterior angles does a triangle have at each vertex? Explain.
3. **NUMBER SENSE** List the measures of the exterior angles for the triangle shown at the right.

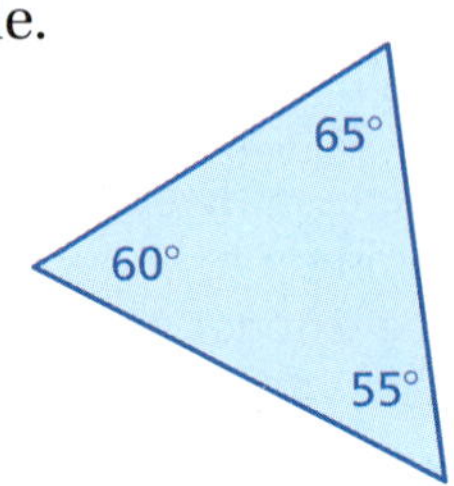

Practice and Problem Solving

Find the measures of the interior angles.

1 4.

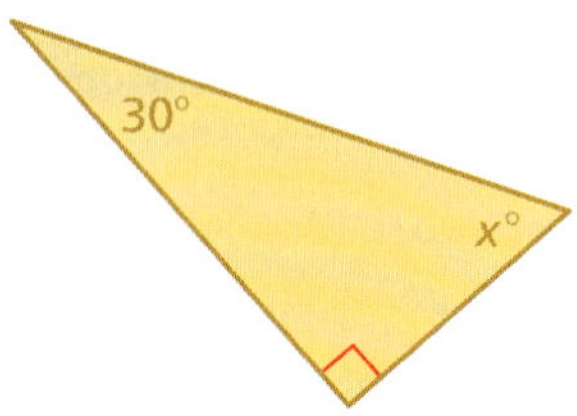

5.

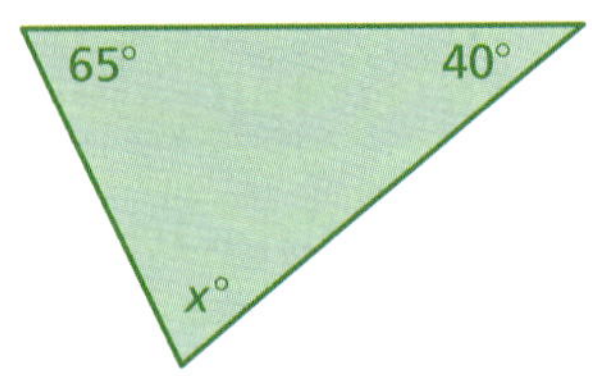

6.

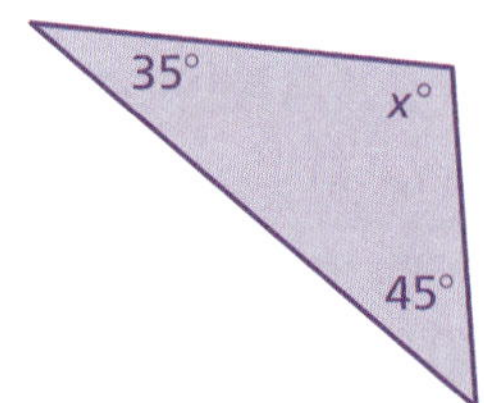

7.

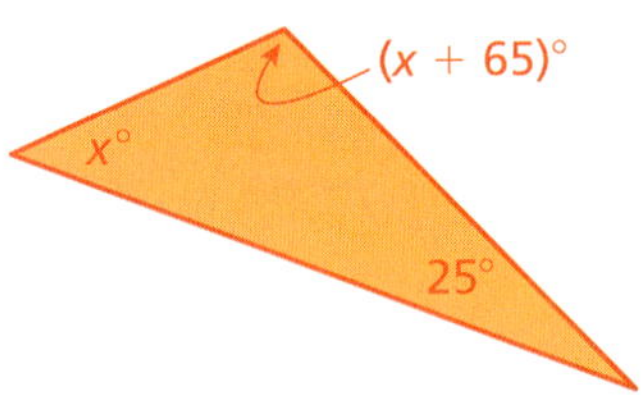

8.

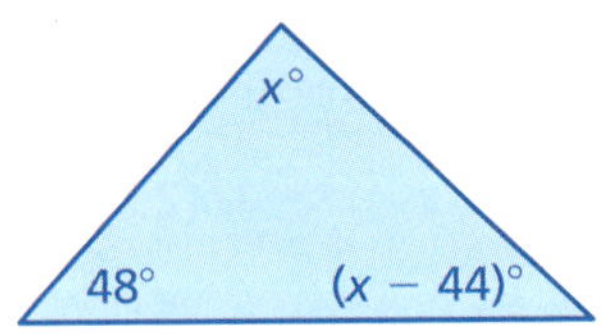

9.

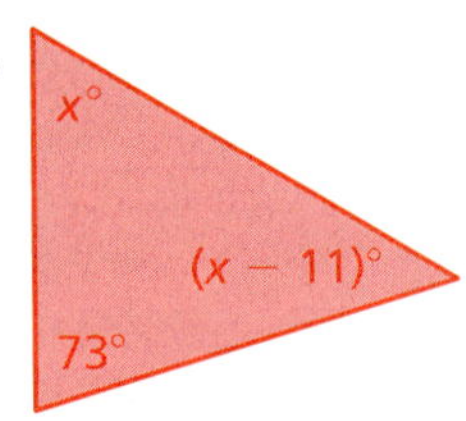

10. **BILLIARD RACK** Find the value of x in the billiard rack.

11. **NO PARKING** The triangle with lines through it designates a no parking zone. What is the value of x?

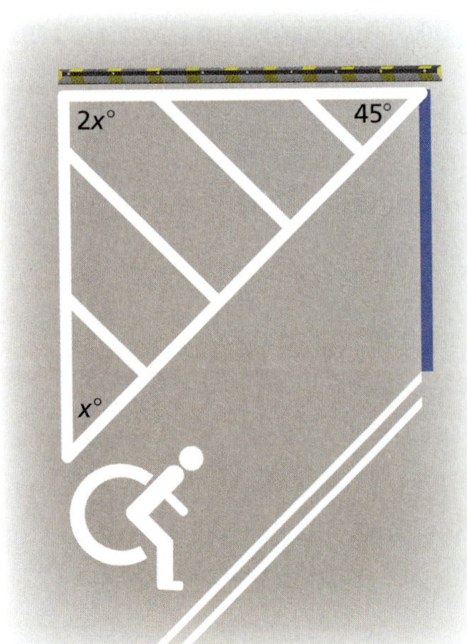

Assignment Guide and Homework Check

Level	Assignment	Homework Check
Accelerated	1–6, 8–14 even, 15–26	8, 10, 14, 17, 18

Common Errors

- **Exercises 7–11** Students may forget to combine like terms when solving for *x*. Remind them that because there are two variables on the same side of the equal sign, they should start by combining like terms.
- **Exercises 7–9, 14, 17** Students may solve for the variable, but forget to find the measures of the angles. Remind them to read the directions carefully and to answer the question.

12.2 Record and Practice Journal

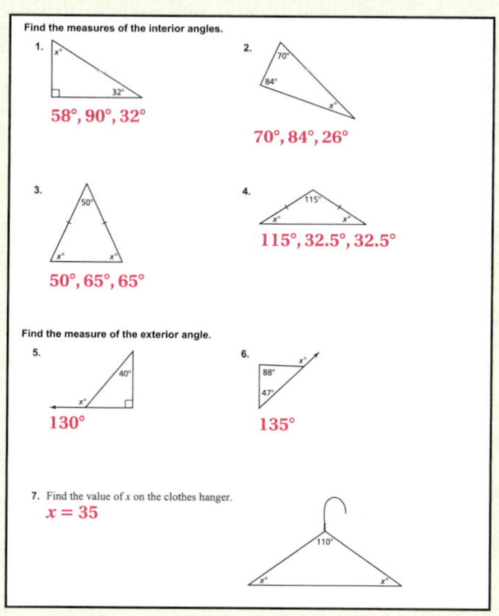
Find the measures of the interior angles.

1. 58°, 90°, 32°
2. 70°, 84°, 26°
3. 50°, 65°, 65°
4. 115°, 32.5°, 32.5°

Find the measure of the exterior angle.

5. 130°
6. 135°
7. Find the value of x on the clothes hanger. $x = 35$

Vocabulary and Concept Check

1. Subtract the sum of the given measures from 180°.
2. 2; When the sides are extended, 2 angles are formed that are adjacent to the interior angle.
3. 115°, 120°, 125°

Practice and Problem Solving

4. 30°, 60°, 90°
5. 40°, 65°, 75°
6. 35°, 45°, 100°
7. 25°, 45°, 110°
8. 44°, 48°, 88°
9. 48°, 59°, 73°
10. 60
11. 45
12. 128°
13. 140°
14. 108°
15. The measure of the exterior angle is equal to the sum of the measures of the two nonadjacent interior angles. The sum of all three angles is not 180°;

$$(2x - 12) = x + 30$$
$$x = 42$$

The exterior angle is $(2(42) - 12)° = 72°$.

Practice and Problem Solving

16. See *Taking Math Deeper.*

17. 126°

18. no; The two nonadjacent interior angles could be any two angles that sum to 120°.

19–22. See Additional Answers.

Fair Game Review

23. $x = -4$ **24.** $y = -1$

25. $n = -3$ **26.** A

Mini-Assessment

Find the value of x.

1.

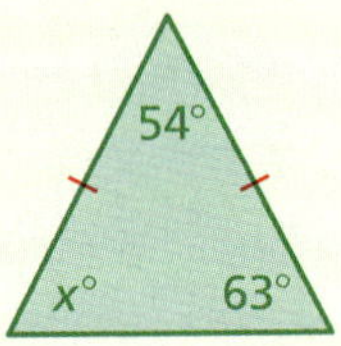

63

2.

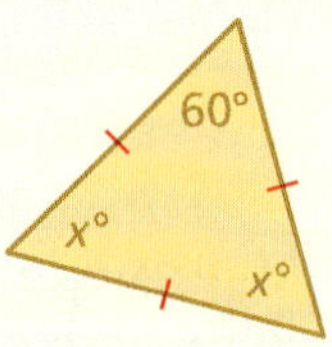

60

3.

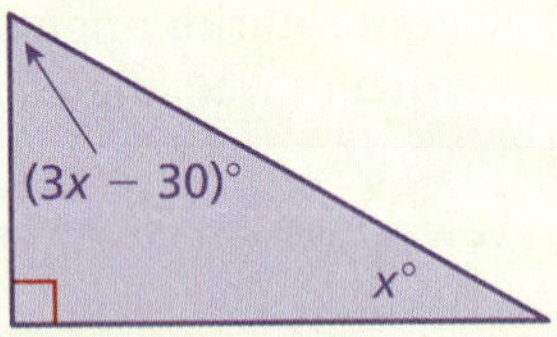

30

Find the measure of the exterior angle.

4.

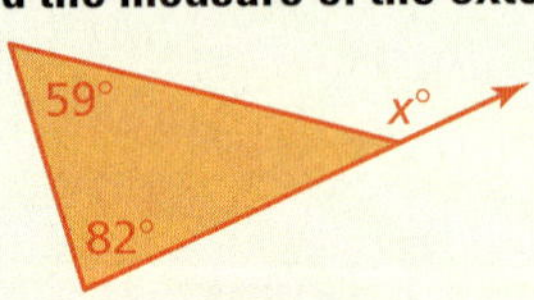

141°

5.

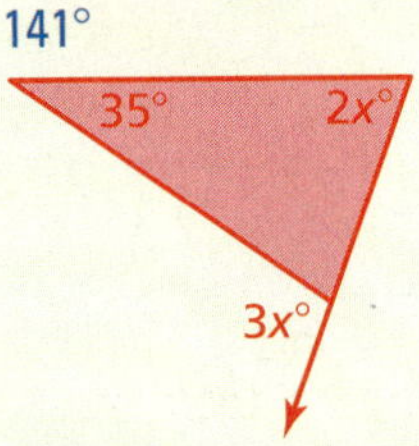

105°

Taking Math Deeper

Exercise 16

For this exercise, you can use a tape diagram to help visualize the relationship between the interior angle measures of the triangle.

1 Express the ratio 2 : 3 : 5 using a tape diagram.

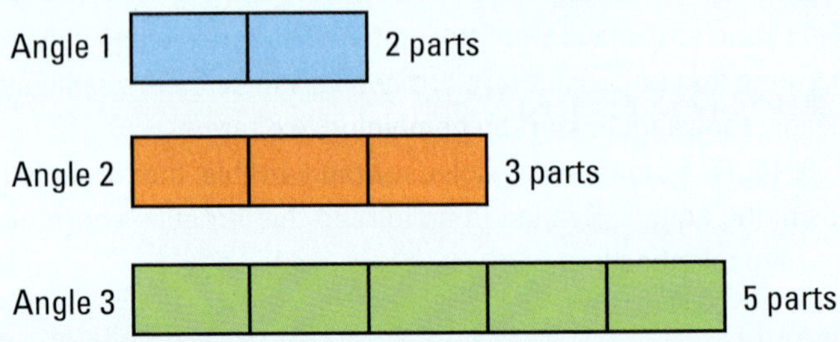

2 Interpret the tape diagram.

There are a total of $2 + 3 + 5 = 10$ parts, which represent the 180° in the triangle.

So, each part must represent $\frac{180}{10} = 18°$.

3 Answer the question.

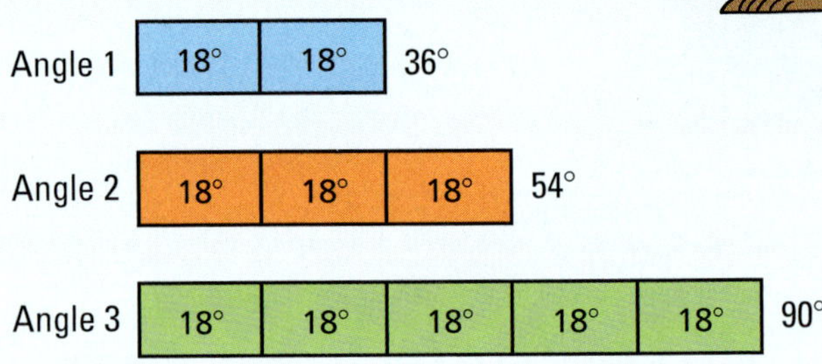

The interior angle measures of the triangle are 36°, 54°, and 90°.

Project

Tell students that the ratio of the interior angle measures of a triangle is $x: y: z$. Ask students to assign an integer between 1 and 10 to each variable, find the interior angle measures, and construct the triangle.

Reteaching and Enrichment Strategies

If students need help. . .	If students got it. . .
Resources by Chapter • Practice A and Practice B • Puzzle Time Record and Practice Journal Practice Differentiating the Lesson Lesson Tutorials Skills Review Handbook	Resources by Chapter • Enrichment and Extension • Technology Connection Start the next section

Find the measure of the exterior angle.

2 **12.**

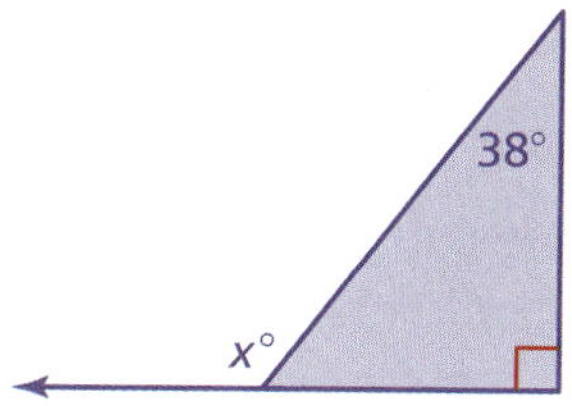

13.

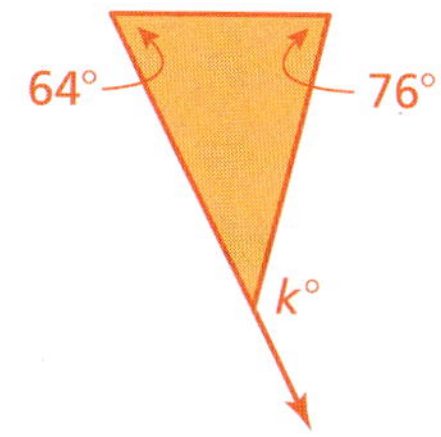

14.

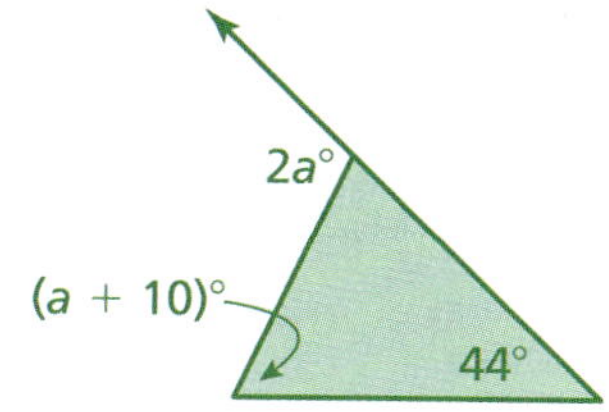

15. ERROR ANALYSIS Describe and correct the error in finding the measure of the exterior angle.

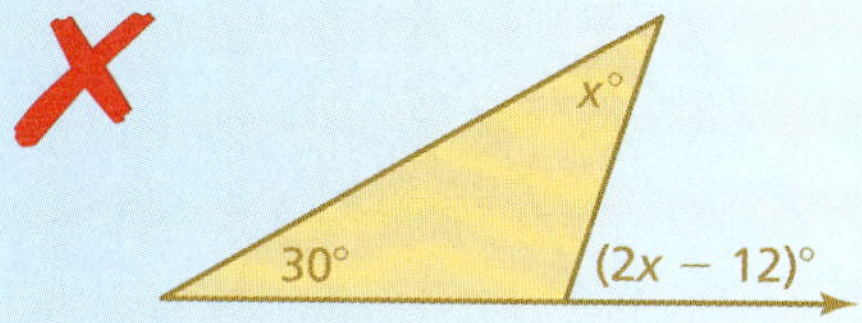

$$(2x - 12) + x + 30 = 180$$
$$3x + 18 = 180$$
$$x = 54$$

The exterior angle is $(2(54) - 12)° = 96°$.

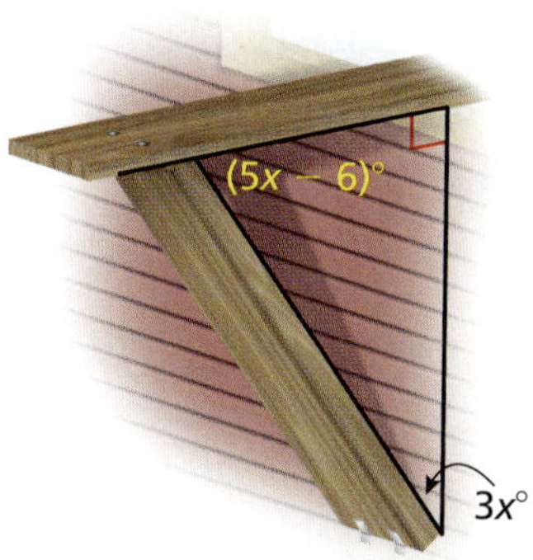

16. RATIO The ratio of the interior angle measures of a triangle is 2 : 3 : 5. What are the angle measures?

17. CONSTRUCTION The support for a window air-conditioning unit forms a triangle and an exterior angle. What is the measure of the exterior angle?

18. REASONING A triangle has an exterior angle with a measure of 120°. Can you determine the measures of the interior angles? Explain.

Determine whether the statement is *always, sometimes,* or *never* true. Explain your reasoning.

19. Given three angle measures, you can construct a triangle.

20. The acute interior angles of a right triangle are complementary.

21. A triangle has more than one vertex with an acute exterior angle.

22. Precision Using the figure at the right, show that $z = x + y$. (*Hint:* Find two equations involving w.)

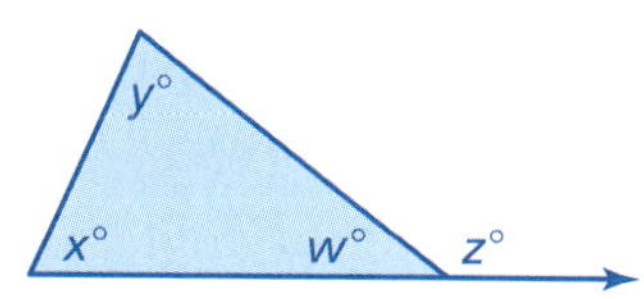

Fair Game Review What you learned in previous grades & lessons

Solve the equation. Check your solution. *(Topic 1)*

23. $-4x + 3 = 19$ **24.** $2(y - 1) + 6y = -10$ **25.** $5 + 0.5(6n + 14) = 3$

26. MULTIPLE CHOICE Which transformation moves every point of a figure the same distance and in the same direction? *(Section 11.2)*

Ⓐ translation Ⓑ reflection Ⓒ rotation Ⓓ dilation

12 Study Help

Check It Out
Graphic Organizer
BigIdeasMath.com

You can use an **example and non-example chart** to list examples and non-examples of a vocabulary word or item. Here is an example and non-example chart for transversals.

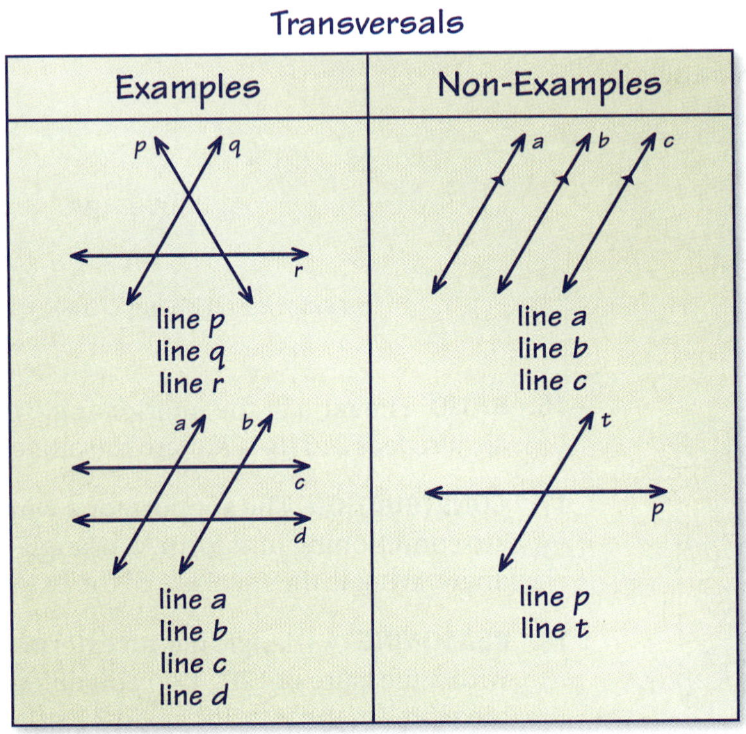

On Your Own

Make example and non-example charts to help you study these topics.

1. interior angles formed by parallel lines and a transversal
2. exterior angles formed by parallel lines and a transversal

After you complete this chapter, make example and non-example charts for the following topics.

3. interior angles of a polygon
4. exterior angles of a polygon
5. regular polygons
6. similar triangles

"What do you think of my example & non-example chart for popular cat toys?"

Sample Answers

1. Interior angles formed by parallel lines and a transversal

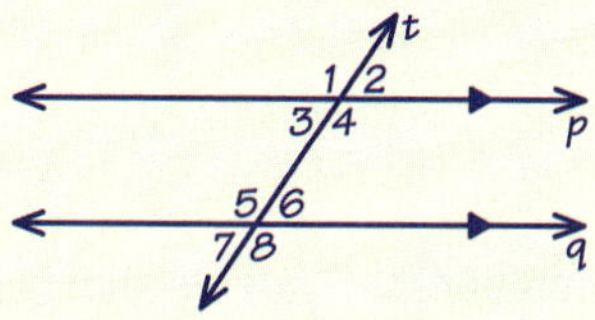

Examples	Non-Examples
∠3	∠1
∠4	∠2
∠5	∠7
∠6	∠8

2. Exterior angles formed by parallel lines and a transversal

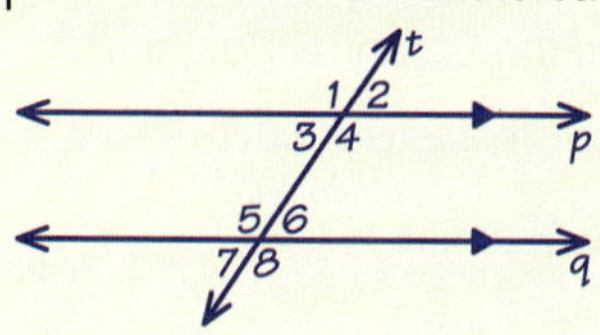

Examples	Non-Examples
∠1	∠3
∠2	∠4
∠7	∠5
∠8	∠6

List of Organizers

Available at *BigIdeasMath.com*

Comparison Chart
Concept Circle
Definition (Idea) and Example Chart
Example and Non-Example Chart
Formula Triangle
Four Square
Information Frame
Information Wheel
Notetaking Organizer
Process Diagram
Summary Triangle
Word Magnet
Y Chart

About this Organizer

An **Example and Non-Example Chart** can be used to list examples and non-examples of a vocabulary word or term. Students write examples of the word or term in the left column and non-examples in the right column. This type of organizer serves as a good tool for assessing students' knowledge of pairs of topics that have subtle but important differences, such as complementary and supplementary angles. Blank example and non-example charts can be included on tests or quizzes for this purpose.

Technology for the Teacher

Editable Graphic Organizer

Answers

1. $\angle 2 = 82°$; $\angle 2$ and the given angle are alternate exterior angles.
2. $\angle 6 = 82°$; $\angle 6$ and the given angle are vertical angles.
3. $\angle 4 = 82°$; $\angle 4$ and the given angle are corresponding angles.
4. $\angle 1 = 98°$; $\angle 4$ and $\angle 1$ are supplementary.
5. 123°; $\angle 1$ and $\angle 7$ are alternate exterior angles.
6. 122°; $\angle 2$ and $\angle 8$ are alternate interior angles and $\angle 8$ and $\angle 5$ are supplementary.
7. 119°; $\angle 5$ and $\angle 3$ are alternate interior angles.
8. 60°; $\angle 4$ and $\angle 6$ are alternate exterior angles.
9. 60°; 60°; 60°
10. 115°; 40°; 25°
11. 45°; 45°; 90°
12. 105°
13. 60°
14. $\angle 1 = 108°$, $\angle 2 = 108°$; Because of alternate interior angles, the angle below $\angle 1$ is 72°. This angle is supplementary to both $\angle 1$ and $\angle 2$.
15. Exterior angle with wall: $180 - 15 = 165°$; Exterior angle with ground: $180 - 5(15) = 105°$;
    $$x + 5x + 90 = 180$$
    $$x = 15$$

Technology for the Teacher

Online Assessment
Assessment Book
ExamView® Assessment Suite

Alternative Quiz Ideas

100% Quiz
Error Notebook
Group Quiz
Homework Quiz
Math Log
Notebook Quiz
Partner Quiz
Pass the Paper

Notebook Quiz

A notebook quiz is used to check students' notebooks. Students should be told at the beginning of the course what the expectations are for their notebooks: notes, class work, homework, date, problem number, goals, definitions, or anything else that you feel is important for your class. They also need to know that it is their responsibility to obtain the notes when they miss class.

1. On a certain day, what was the answer to the warm up question?
2. On a certain day, how was this vocabulary term defined?
3. For Section 12.1, what is the answer to On Your Own Question 1?
4. For Section 12.2, what is the answer to the Essential Question?
5. On a certain day, what was the homework assignment?

Give the students 5 minutes to answer these questions.

Reteaching and Enrichment Strategies

If students need help. . .	If students got it. . .
Resources by Chapter • Practice A and Practice B • Puzzle Time Lesson Tutorials *BigIdeasMath.com*	Resources by Chapter • Enrichment and Extension • Technology Connection Game Closet at *BigIdeasMath.com* Start the next section

12.1–12.2 Quiz

Check It Out
Progress Check
BigIdeasMath.com

Use the figure to find the measure of the angle. Explain your reasoning. *(Section 12.1)*

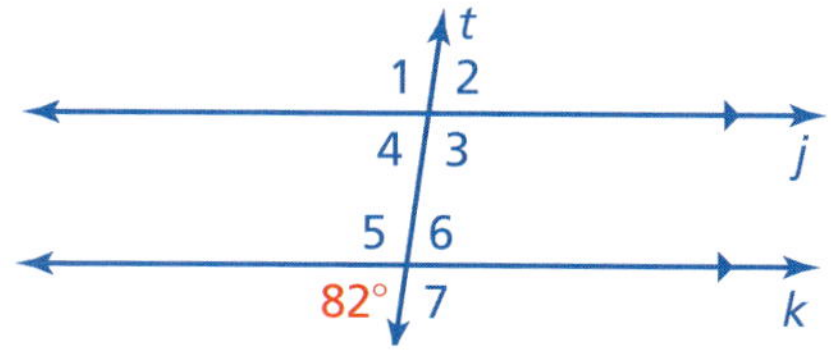

1. $\angle 2$ **2.** $\angle 6$

3. $\angle 4$ **4.** $\angle 1$

Complete the statement. Explain your reasoning. *(Section 12.1)*

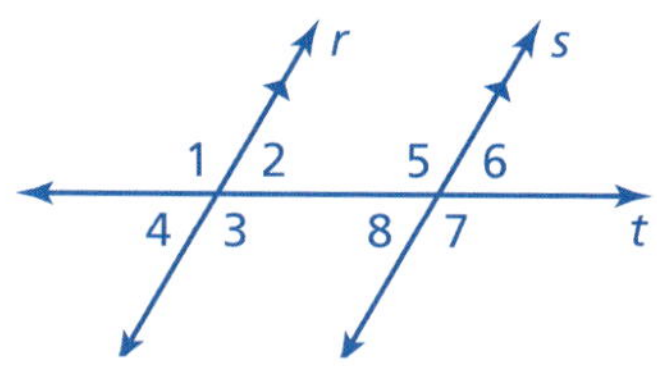

5. If the measure of $\angle 1 = 123°$, then the measure of $\angle 7 =$ ____.

6. If the measure of $\angle 2 = 58°$, then the measure of $\angle 5 =$ ____.

7. If the measure of $\angle 5 = 119°$, then the measure of $\angle 3 =$ ____.

8. If the measure of $\angle 4 = 60°$, then the measure of $\angle 6 =$ ____.

Find the measures of the interior angles. *(Section 12.2)*

9.

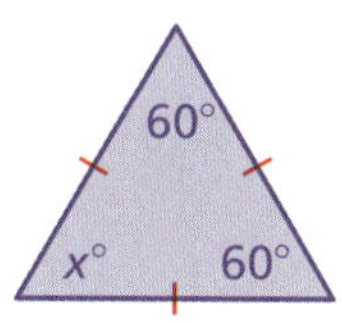

10.

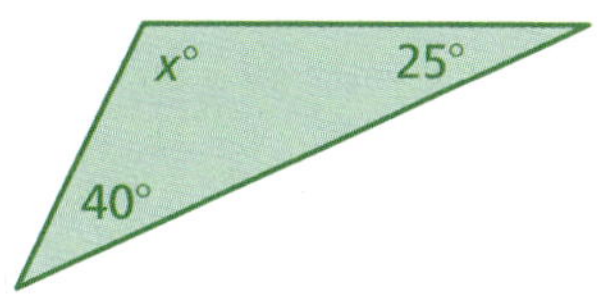

11.

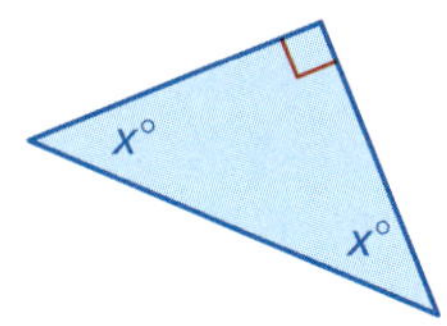

Find the measure of the exterior angle. *(Section 12.2)*

12.

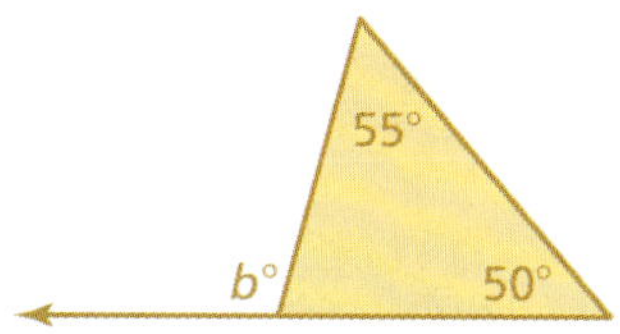

13.

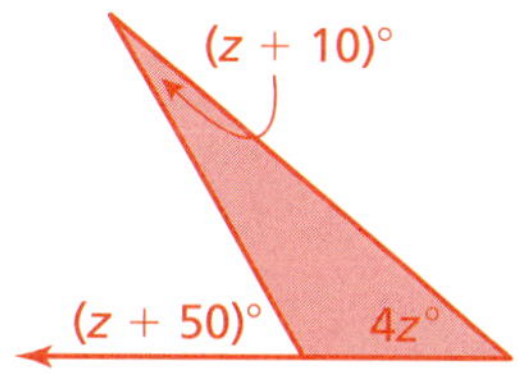

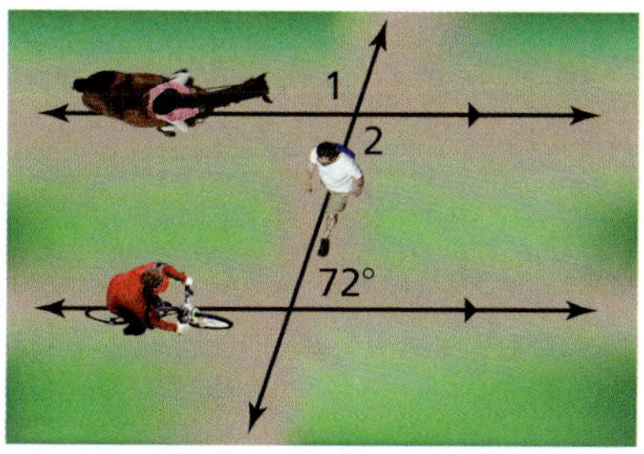

14. PARK In a park, a bike path and a horse riding path are parallel. In one part of the park, a hiking trail intersects the two paths. Find the measures of $\angle 1$ and $\angle 2$. Explain your reasoning. *(Section 12.1)*

15. LADDER A ladder leaning against a wall forms a triangle and exterior angles with the wall and the ground. What are the measures of the exterior angles? Justify your answer. *(Section 12.2)*

12.3 Angles of Polygons

Essential Question How can you find the sum of the interior angle measures and the sum of the exterior angle measures of a polygon?

1 ACTIVITY: Exploring the Interior Angles of a Polygon

Work with a partner. In parts (a) – (e), identify each polygon and the number of sides n. Then find the sum of the interior angle measures of the polygon.

a. Polygon: ______ Number of sides: $n =$ ______

Draw a line segment on the figure that divides it into two triangles. Is there more than one way to do this? Explain.

What is the sum of the interior angle measures of each triangle?

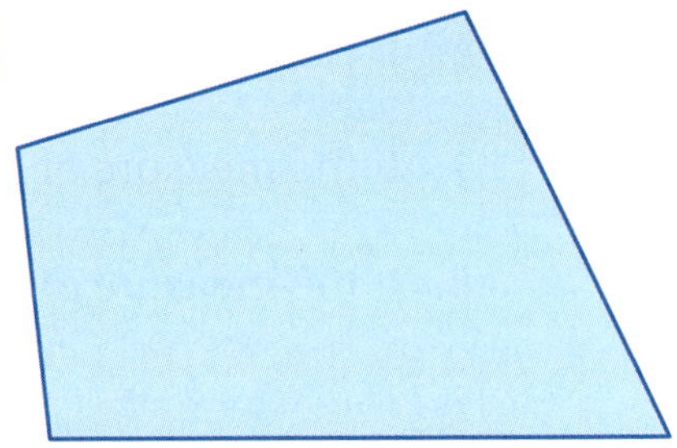

What is the sum of the interior angle measures of the figure?

b.

c.

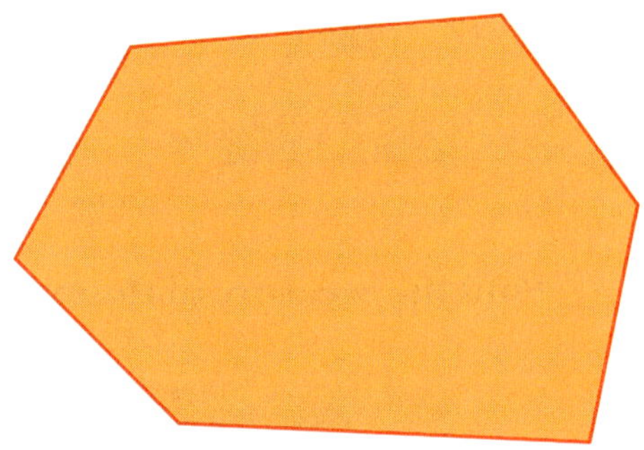

d.

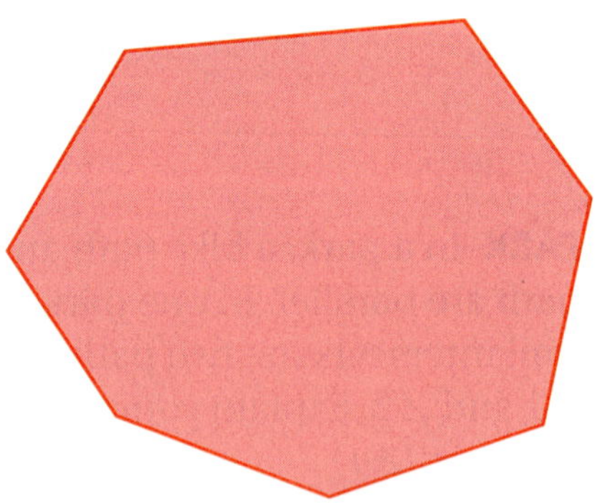

e.

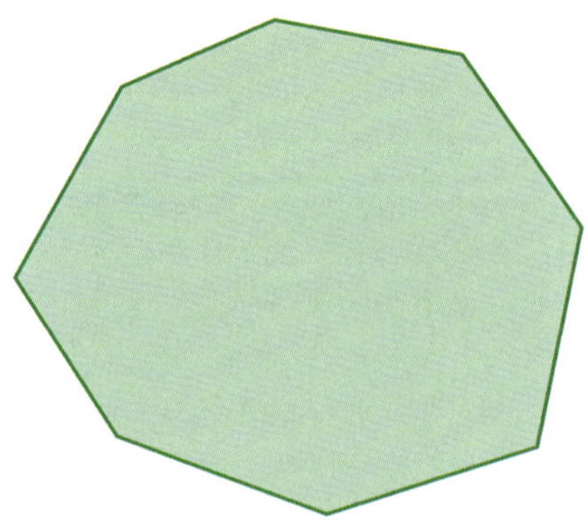

f. **REPEATED REASONING** Use your results to complete the table. Then find the sum of the interior angle measures of a polygon with 12 sides.

Number of Sides, n	3	4	5	6	7	8
Number of Triangles						
Angle Sum, S						

Geometry

In this lesson, you will
- find the sum of the interior angle measures of polygons.
- understand that the sum of the exterior angle measures of a polygon is 360°.
- find the measures of interior and exterior angles of polygons.

Laurie's Notes

Introduction

Applying Mathematical Practices

- **Look for and Express Regularity in Repeated Reasoning:** In making sense of how to find the sum of the interior angles of a polygon, students will repeat a strategy, each time with a polygon having one more side than previously.

Motivate

- ? "How many of you are looking forward to getting your driver's license?"
- Tell them that they will likely be tested on road signs.
- Draw several shapes and ask students if they know the names of the shapes and what they are used for on highway signs.

Activity Notes

Activity 1

- You may need to guide students through the part (a) of the activity with the quadrilateral.
- Students recognize that the diagonal divides the quadrilateral into two triangles and that each triangle has interior angle measures that sum to 180°. The confusion is that the two triangles have a total of 6 angles and the quadrilateral has only 4 angles. Help students recognize that the diagonal divides two angles of the quadrilateral.
- The approach taken for the remaining polygons is slightly different. All of the diagonals are drawn from one vertex, forming triangles inside the polygon.

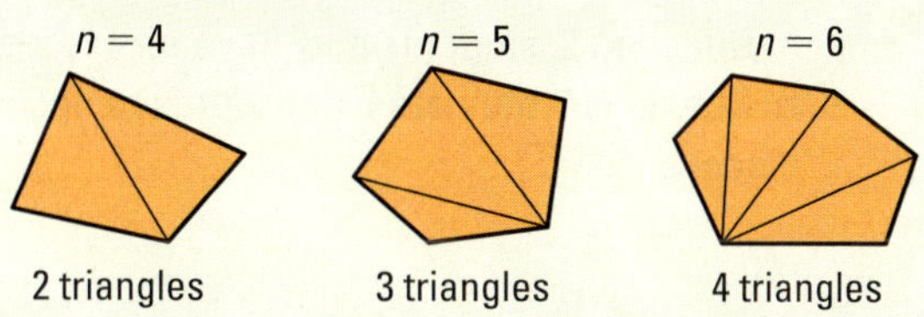

- **Look for and Express Regularity in Repeated Reasoning:** Students will use repeated reasoning as they explore the remaining polygons: the pentagon has 3 interior triangles so 3×180, the hexagon has 4 interior triangles so 4×180, and so on.
- **Construct Viable Arguments and Critique the Reasoning of Others:** Some students may conjecture that the number of triangles is two less than the number of sides in the polygon. Other students may just notice that the sum of the interior angle measures is increasing by 180° each time.
- **Connection:** Students who snowboard or skateboard will recognize the number pattern quickly. They know the multiples of 180° well.
- **Model with Mathematics:** The table helps to organize the data. This also will help students make the connection between the number of sides and the number of triangles formed.
- Ask volunteers to share how they found the interior angle sum for a polygon with 12 sides.

What Your Students Will Learn

- Find the sum of the interior angle measures of polygons using a formula.
- Understand that the sum of the exterior angle measures of a polygon is 360°.
- Find the measures of interior and exterior angles of polygons by using formulas and writing and solving equations.

Previous Learning

Students should know how to solve multi-step equations.

Technology for the Teacher

Dynamic Classroom

Lesson Plans
Complete Materials List

12.3 Record and Practice Journal

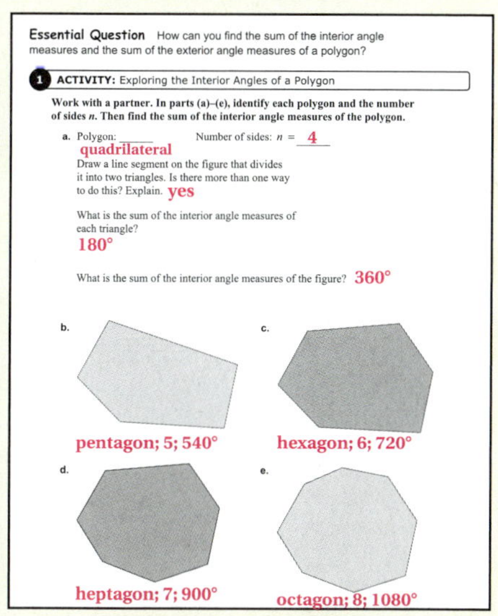

Essential Question How can you find the sum of the interior angle measures and the sum of the exterior angle measures of a polygon?

1 ACTIVITY: Exploring the Interior Angles of a Polygon

Work with a partner. In parts (a)–(e), identify each polygon and the number of sides n. Then find the sum of the interior angle measures of the polygon.

a. Polygon: quadrilateral Number of sides: n = 4

Draw a line segment on the figure that divides it into two triangles. Is there more than one way to do this? Explain. yes

What is the sum of the interior angle measures of each triangle?
180°

What is the sum of the interior angle measures of the figure? 360°

b. pentagon; 5; 540°

c. hexagon; 6; 720°

d. heptagon; 7; 900°

e. octagon; 8; 1080°

Differentiated Instruction

Kinesthetic

Another way to discover the sum of the angle measures of a polygon is to have the students cut the polygon into triangles. This can be done for convex and concave polygons.

12.3 Record and Practice Journal

f. **REPEATED REASONING** Use your results to complete the table. Then find the sum of the interior angle measures of a polygon with 12 sides.

Number of Sides, n	3	4	5	6	7	8
Number of Triangles	1	2	3	4	5	6
Angle Sum, S	180°	360°	540°	720°	900°	1080°

1800°

2 ACTIVITY: Exploring the Exterior Angles of a Polygon

Work with a partner.

a. Draw a convex pentagon. Extend the sides to form the exterior angles. Label one exterior angle at each vertex A, B, C, D, and E, as shown.

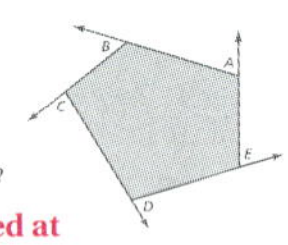

b. Cut out the exterior angles. How can you join the vertices to determine the sum of the angle measures? What do you notice?

all angles adjacent and joined at vertices; they form 360°

c. **REPEATED REASONING** Repeat the procedure in parts (a) and (b) for each figure below.

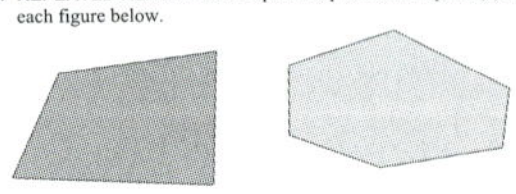

What can you conclude about the sum of the measures of the exterior angles of a convex polygon? Explain.

Sum is always 360°.

What Is Your Answer?

3. **STRUCTURE** Use your results from Activity 1 to write an expression that represents the sum of the interior angle measures of a polygon.

$(n - 2) \cdot 180°$

4. **IN YOUR OWN WORDS** How can you find the sum of the interior angle measures and the sum of the exterior angle measures of a polygon?

Interior: sum $= (n - 2) \cdot 180°$
Exterior: sum $= 360°$

Laurie's Notes

Discuss

? "Have you heard of the words convex and concave? If so, what do they mean?" Students may be familiar with the words from science class. They often say that a concave lens caves in.

- Define convex and concave polygons.
- Draw examples of polygons and have students identify them as being convex or concave.

Activity 2

- **Teaching Tip:** If sufficient pairs of scissors are not available, you could group students with 3 to 4 per group or have one student be "the cutter" at the front of the room.
- Remind students to draw their pentagon with a straightedge and large enough to be able to work with it easily. Students should also extend the sides far enough so that the lines aid the cutting process.
- **Look for and Express Regularity in Repeated Reasoning:** Students will use repeated reasoning as they explore the two remaining polygons: the quadrilateral has 4 exterior angles to cut, the hexagon has 6 exterior angles to cut. Each time, the angles are placed about a single point, a technique students used in the previous lesson when they tore the angles off their paper triangle and placed them about a point.
- **Construct Viable Arguments and Critique the Reasoning of Others:** Some students may conjecture that the sum of the exterior angle measures of any polygon is 360°.

What Is Your Answer?

- In Question 3, students may have an expression that they use.
- In Question 4, they are using words to describe the formula they have discovered.

Closure

- Use the results of the investigations to find the sum of the interior angle measures and the sum of the exterior angle measures of a decagon, a polygon with 10 sides. 1440°; 360°

A polygon is **convex** when every line segment connecting any two vertices lies entirely inside the polygon. A polygon is **concave** when at least one line segment connecting any two vertices lies outside the polygon.

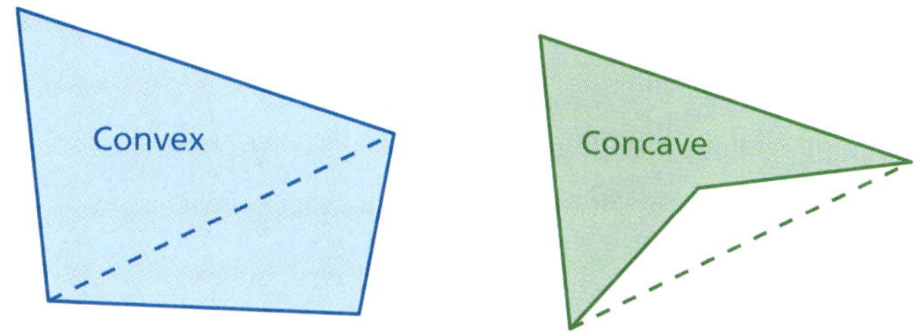

2 ACTIVITY: Exploring the Exterior Angles of a Polygon

Math Practice

Analyze Conjectures

Do your observations about the sum of the exterior angles make sense? Do you think they would hold true for any convex polygon? Explain.

Work with a partner.

a. Draw a convex pentagon. Extend the sides to form the exterior angles. Label one exterior angle at each vertex *A*, *B*, *C*, *D*, and *E*, as shown.

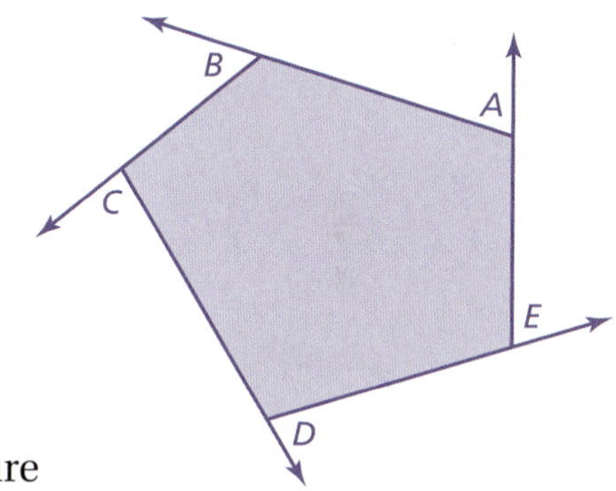

b. Cut out the exterior angles. How can you join the vertices to determine the sum of the angle measures? What do you notice?

c. **REPEATED REASONING** Repeat the procedure in parts (a) and (b) for each figure below.

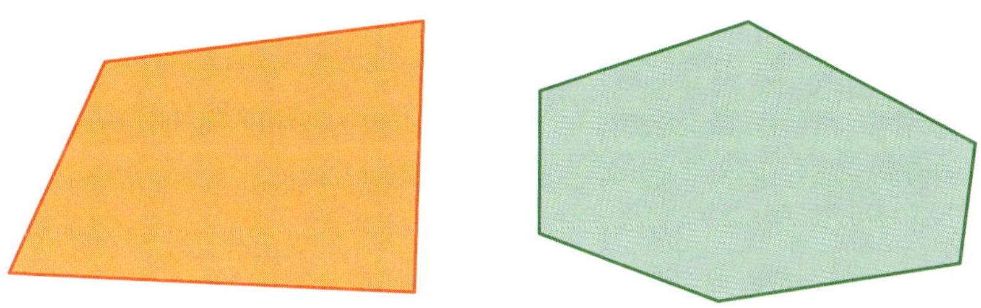

What can you conclude about the sum of the measures of the exterior angles of a convex polygon? Explain.

What Is Your Answer?

3. **STRUCTURE** Use your results from Activity 1 to write an expression that represents the sum of the interior angle measures of a polygon.

4. **IN YOUR OWN WORDS** How can you find the sum of the interior angle measures and the sum of the exterior angle measures of a polygon?

Practice Use what you learned about angles of polygons to complete Exercises 4–6 on page 547.

12.3 Lesson

Check It Out
Lesson Tutorials
BigIdeasMath.com

Key Vocabulary
convex polygon, *p. 543*
concave polygon, *p. 543*
regular polygon, *p. 545*

A *polygon* is a closed plane figure made up of three or more line segments that intersect only at their endpoints.

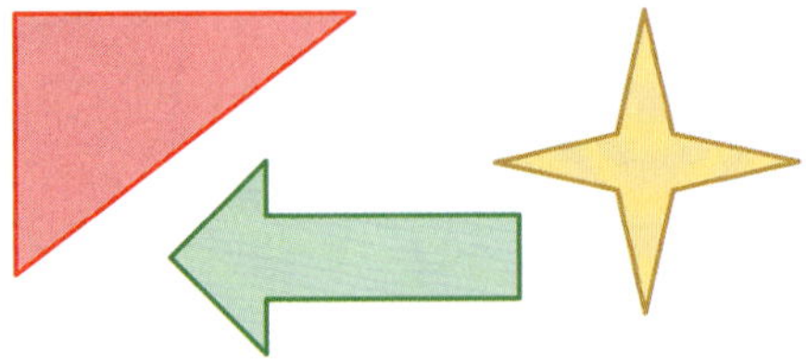

Polygons

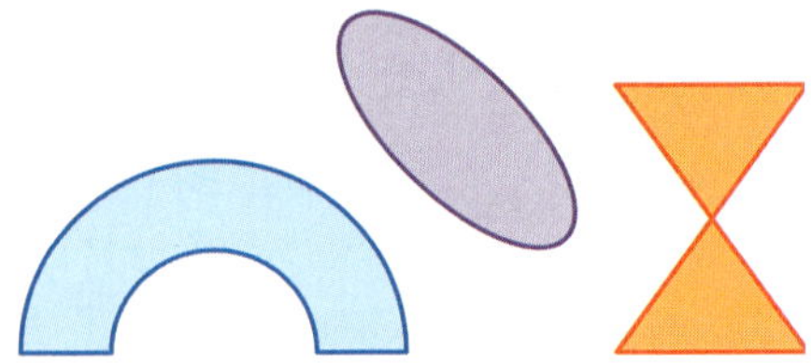

Not polygons

Key Idea

Interior Angle Measures of a Polygon

The sum S of the interior angle measures of a polygon with n sides is

$$S = (n - 2) \cdot 180^\circ.$$

EXAMPLE 1 Finding the Sum of Interior Angle Measures

Reading
For polygons whose names you have not learned, you can use the phrase "*n*-gon," where *n* is the number of sides. For example, a 15-gon is a polygon with 15 sides.

Find the sum of the interior angle measures of the school crossing sign.

The sign is in the shape of a pentagon. It has 5 sides.

$S = (n - 2) \cdot 180^\circ$	Write the formula.
$= (5 - 2) \cdot 180^\circ$	Substitute 5 for n.
$= 3 \cdot 180^\circ$	Subtract.
$= 540^\circ$	Multiply.

The sum of the interior angle measures is 540°.

On Your Own

Now You're Ready
Exercises 7–9

Find the sum of the interior angle measures of the green polygon.

1. 

2. 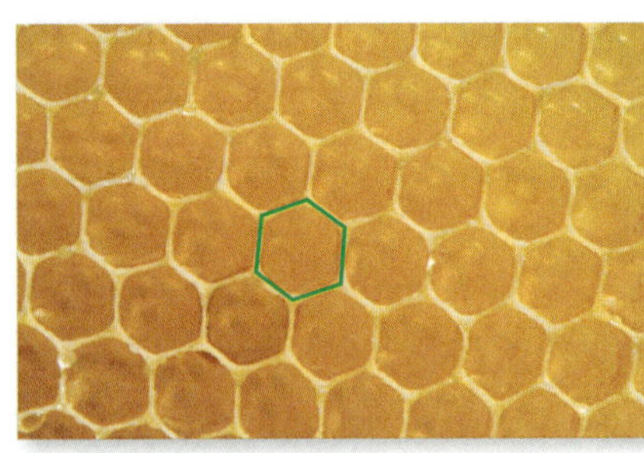

Multi-Language Glossary at BigIdeasMath.com

Laurie's Notes

Introduction

Connect

- **Yesterday:** Students explored finding the sums of interior and exterior angle measures of a polygon.
- **Today:** Students will use a formula to find the sums of interior and exterior angle measures of a polygon.

Motivate

? "Did you ever wonder why bees use a hexagonal structure for their honeycomb? Why not squares? or circles? or octagons?"

- Draw a few cells of the honeycomb.

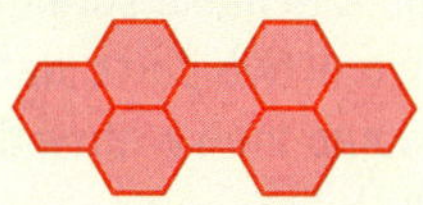

- Mathematicians have concluded that a hexagon is the most appropriate geometric form for the maximum use of a given area. This means that a hexagonal cell requires the minimum amount of wax for construction while it stores the maximum amount of honey.

Lesson Notes

Key Idea

- Write the definition of a polygon. Draw examples of shapes which are and are not polygons. Students should be able to explain why some are not polygons.
- In all of the samples shown, the interior is shaded. The polygon is the figure formed by the line segments. The polygonal region contains the interior of the polygon.
- Write the Key Idea. This is the same equation that students wrote yesterday, but in the more common form. This form highlights the fact that the sum is a multiple of 180°.

Example 1

- Review the names of common polygons: Triangle (3), Quadrilateral (4), Pentagon (5), Hexagon (6), Octagon (8), and Decagon (10). It is also common to say n-gon and replace n with 9 to talk about a 9-sided polygon.
- Read the problem. The polygon is a pentagon.
- Write the equation, substitute 5 for n, and solve.

On Your Own

- **Think-Pair-Share:** Students should read each question independently and then work in pairs to answer the questions. When they have answered the questions, the pair should compare their answers with another group and discuss any discrepancies.

? "What are the names of the polygons in Questions 1 and 2?"
7-gon or heptagon; hexagon

Goal Today's lesson is finding the interior and exterior angle measures of a polygon.

Technology for the Teacher

Dynamic Classroom

Lesson Tutorials
Lesson Plans
Answer Presentation Tool

Extra Example 1

Find the sum of the interior angle measures of the polygon.

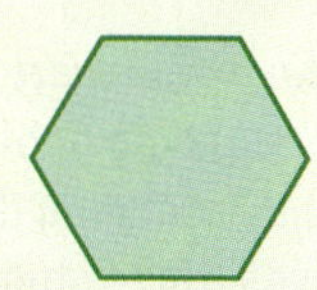

720°

On Your Own

1. 900°
2. 720°

Extra Example 2

Find the value of *x*.

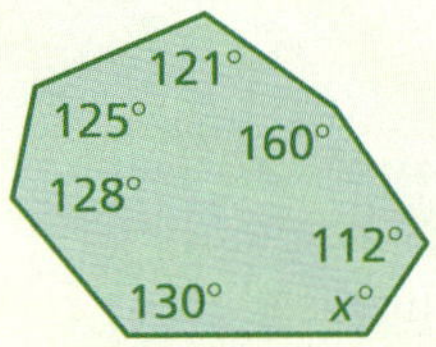

124

On Your Own

3. 105

4. 75

5. 35

Extra Example 3

What if the cloud system in Example 3 is in the approximate shape of a regular polygon with 12 sides? Find the measure of each interior angle of the polygon.
150°

English Language Learners

Vocabulary

Preview the *Key Vocabulary* in this chapter. Understanding geometry depends on understanding the terminology used. Have students write key vocabulary words in their notebooks. Include definitions and examples to help distinguish between words (e.g., convex polygon and concave polygon).

Laurie's Notes

Example 2

- **Connection:** This example integrates equation solving with finding a missing angle.
- ? "How many sides does the polygon have?" 7
- ? "How do you find the sum of the measures of all of the interior angles of a 7-gon?" Solve $(7 - 2)180 = 900$.
- Once the sum is known, write and solve the equation as shown. Caution students to be careful with their arithmetic.

On Your Own

- **Make Sense of Problems and Persevere in Solving Them:** Students should check with their neighbors to make sure they are setting up the equation correctly. Each problem has two parts: determining the sum of all of the interior angle measures and then writing the equation to solve for the missing angle.
- In Question 4, remind students that the symbol for a right angle means the angle measures 90°.
- In Question 5, two angles are missing, each with a measure of $2x°$. The sum of the interior angle measures of this pentagon is 540°, so $2x + 145 + 145 + 2x + 110 = 540$. The steps are to combine like terms, isolate the variable, and solve.
- **Common Error:** Students will solve for the variable correctly, but then forget to substitute this value back into the variable expression to solve for the angle measure. In Question 5, students were only asked to solve for *x*. If they had been asked to find the measure of the angle, there would be one last step. In this case, $x = 35$ and the two missing angle measures are each 70°.

Example 3

- Review the definition of a regular polygon. Point out to students that squares and equilateral triangles are examples of regular polygons.
- **Make Sense of Problems and Persevere in Solving Them:** A regular hexagon has 6 congruent angles. If the angle measures of a hexagon sum to 720° and the 6 angles are congruent, it should make sense to students why they divide 720 by 6.
- Look back to the honeycomb you drew at the beginning of the lesson. There are three 120° angles about one point.
- You can show a video of the cloud system from the website *jpl.nasa.gov*.

EXAMPLE 2 Finding an Interior Angle Measure of a Polygon

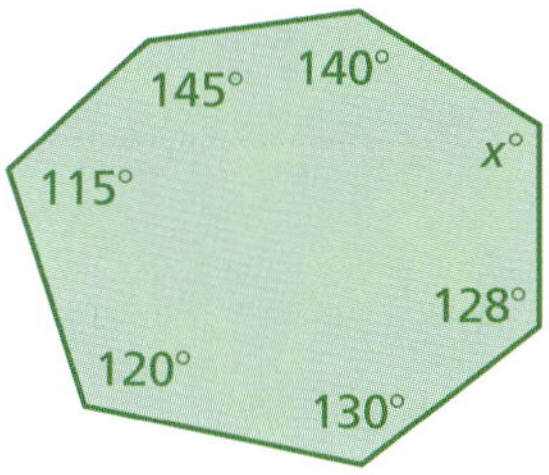

Find the value of x.

Step 1: The polygon has 7 sides. Find the sum of the interior angle measures.

$S = (n - 2) \cdot 180°$ Write the formula.

$= (7 - 2) \cdot 180°$ Substitute 7 for n.

$= 900°$ Simplify. The sum of the interior angle measures is 900°.

Step 2: Write and solve an equation.

$$140 + 145 + 115 + 120 + 130 + 128 + x = 900$$

$$778 + x = 900$$

$$x = 122$$

The value of x is 122.

On Your Own

Now You're Ready
Exercises 12–14

Find the value of x.

3.

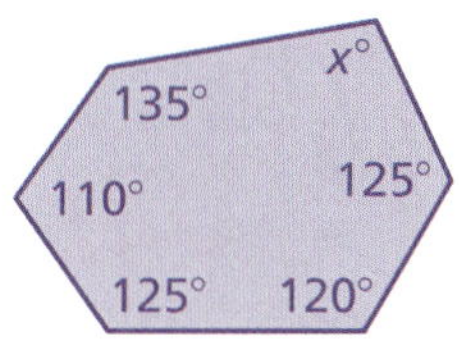

4.

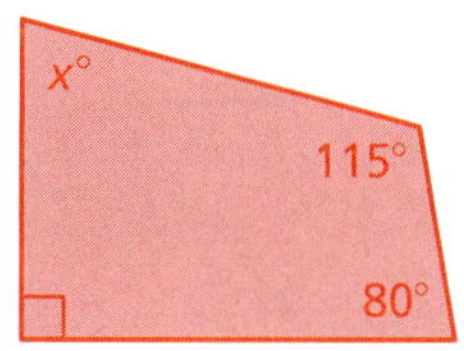

5.

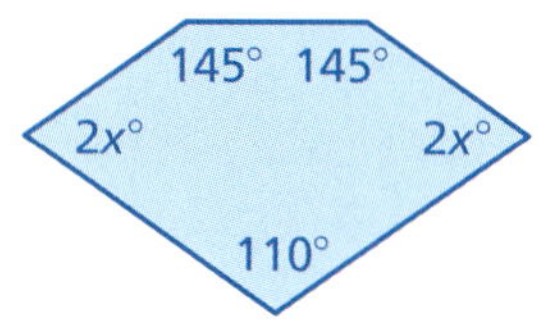

In a **regular polygon**, all the sides are congruent, and all the interior angles are congruent.

EXAMPLE 3 Real-Life Application

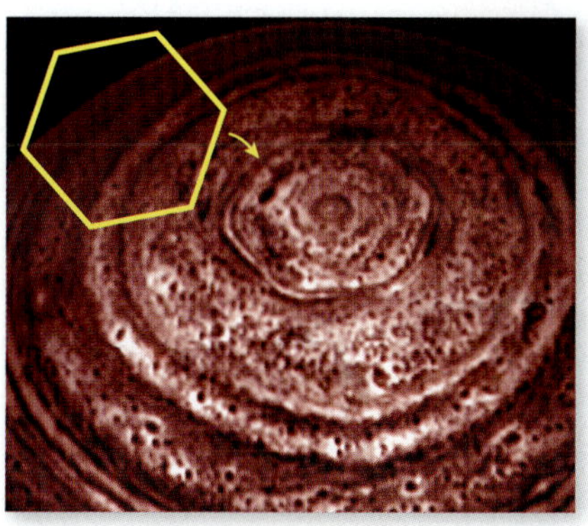

The hexagon is about 15,000 miles across. Approximately four Earths could fit inside it.

A cloud system discovered on Saturn is in the approximate shape of a regular hexagon. Find the measure of each interior angle of the hexagon.

Step 1: A hexagon has 6 sides. Find the sum of the interior angle measures.

$S = (n - 2) \cdot 180°$ Write the formula.

$= (6 - 2) \cdot 180°$ Substitute 6 for n.

$= 720°$ Simplify. The sum of the interior angle measures is 720°.

Step 2: Divide the sum by the number of interior angles, 6.

$720° \div 6 = 120°$

The measure of each interior angle is 120°.

On Your Own

Now You're Ready
Exercises 16–18

Find the measure of each interior angle of the regular polygon.

6. octagon **7.** decagon **8.** 18-gon

Key Idea

Exterior Angle Measures of a Polygon

Words The sum of the measures of the exterior angles of a convex polygon is 360°.

Algebra $w + x + y + z = 360$

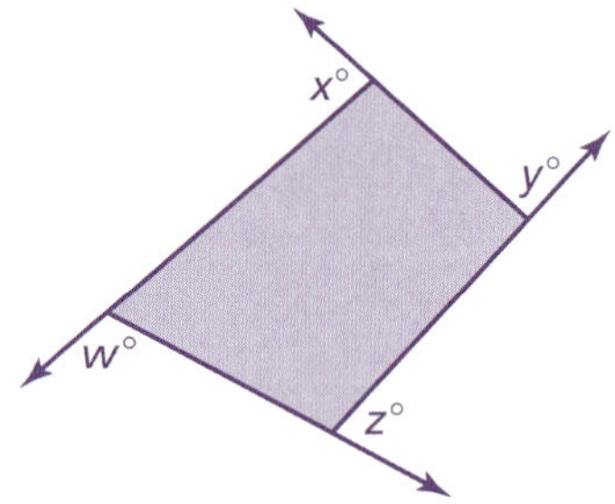

EXAMPLE 4 Finding Exterior Angle Measures

Find the measures of the exterior angles of each polygon.

a.

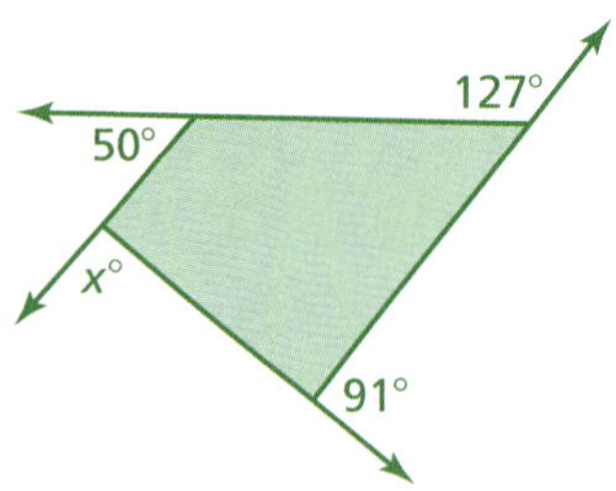

Write and solve an equation for x.

$$x + 50 + 127 + 91 = 360$$
$$x + 268 = 360$$
$$x = 92$$

So, the measures of the exterior angles are 92°, 50°, 127°, and 91°.

b.

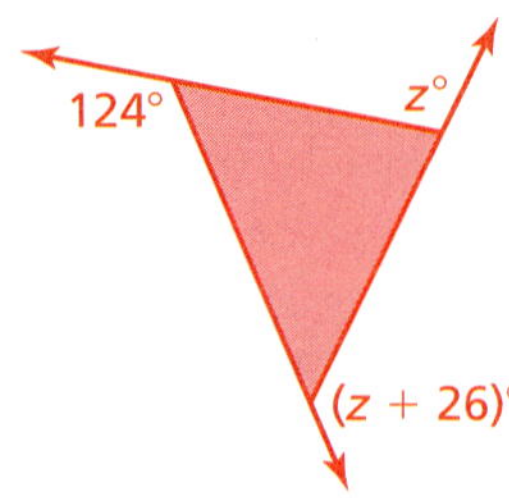

Write and solve an equation for z.

$$124 + z + (z + 26) = 360$$
$$2z + 150 = 360$$
$$z = 105$$

So, the measures of the exterior angles are 124°, 105°, and $(105 + 26)° = 131°$.

On Your Own

Now You're Ready
Exercises 22–28

9. Find the measures of the exterior angles of the polygon.

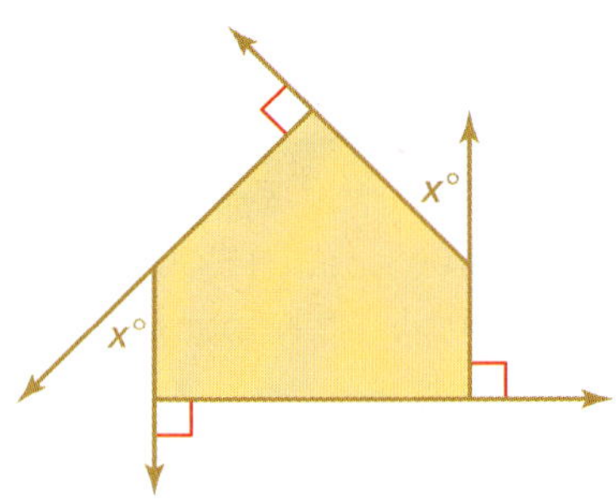

Laurie's Notes

On Your Own

- **Think-Pair-Share:** Students should read each question independently and then work in pairs to answer the questions. When they have answered the questions, the pair should compare their answers with another group and discuss any discrepancies.

Key Idea

- Write the Key Idea. Draw a quadrilateral that does not have special attributes, and extend the sides so that an exterior angle is formed, one at each vertex.
- Remind students that although there are two exterior angles at each vertex, this formula considers only one exterior angle at each vertex. This is customary.
- Stress that the formula is true for *any* polygon, not just the quadrilateral.

Example 4

- Draw the example in part (a) on the board.
- ? "How can you find the measure of the missing exterior angle?" Set the sum of the four angle measures equal to 360°, and solve the equation for x.
- Draw the example in part (b) on the board.
- ? "How can we find the measures of the missing exterior angles?" Set the sum of the three angle measures equal to 360°, and solve the equation for z.
- The last step in this problem is to remember to substitute the value of z into the expression $z + 26$ to find the measure of last exterior angle.

On Your Own

- Students should recognize that three of the exterior angles are right angles and the other two angles are congruent.

Closure

- A pentagon has two interior right angles and the other three interior angles are all congruent. What is the measure of one of the missing angles? 120°

On Your Own

6. 135°

7. 144°

8. 160°

Extra Example 4

Find the measures of the exterior angles of each polygon.

a.

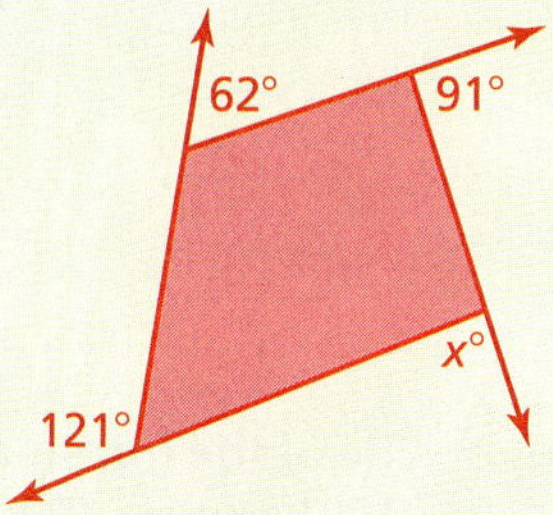

121°, 62°, 91°, and 86°

b.

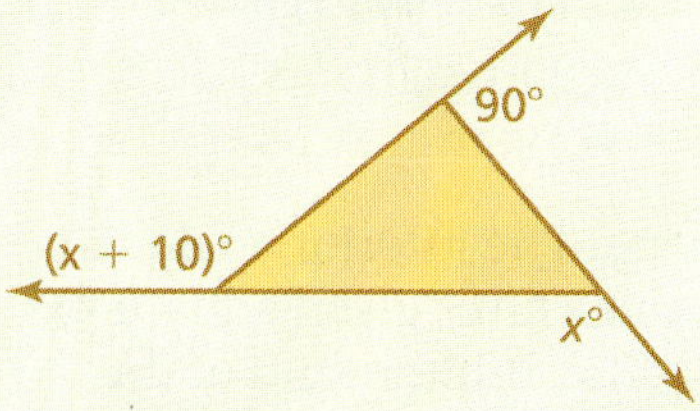

140°, 90°, and 130°

On Your Own

9. 90°, 45°, 90°, 45°, 90°

Vocabulary and Concept Check

1. *Sample answer:*

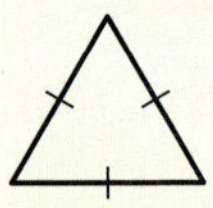

2. The second figure does not belong because it is not made up entirely of line segments.
3. What is the measure of an interior angle of a regular pentagon?; 108°; 540°

Practice and Problem Solving

4. 360°
5. 1260°
6. 900°
7. 360°
8. 1080°
9. 1260°
10. The right side of the formula is $(n - 2) \cdot 180°$, not $n \cdot 180°$.
$S = (n - 2) \cdot 180°$
$= (13 - 2) \cdot 180°$
$= 11 \cdot 180°$
$= 1980°$
11. no; The interior angle measures given add up to 535°, but the sum of the interior angle measures of a pentagon is 540°.

Assignment Guide and Homework Check

Level	Assignment	Homework Check
Accelerated	1–6, 10–18 even, 19, 20–26 even, 28–38	14, 20, 24, 28, 30

Common Errors

- **Exercises 4–6** Students may struggle dividing the polygon into triangles. Encourage them to trace the polygon in pen in their notebooks and then to draw triangles with a pencil so that they can erase lines if necessary.
- **Exercises 7–9** Students may forget to subtract 2 from the number of sides when using the formula to find the sum of the interior angle measures. Remind them of the formula and encourage them to write the formula before substituting the number of sides.
- **Exercise 11** Students may say that because the sum of the interior angle measures is close to the value found when using the formula, a pentagon can have these angle measures. Remind them that the sum of the interior angle measures must be *exactly* the same as the sum found with the formula for the polygon to be drawn with the given angles.

12.3 Record and Practice Journal

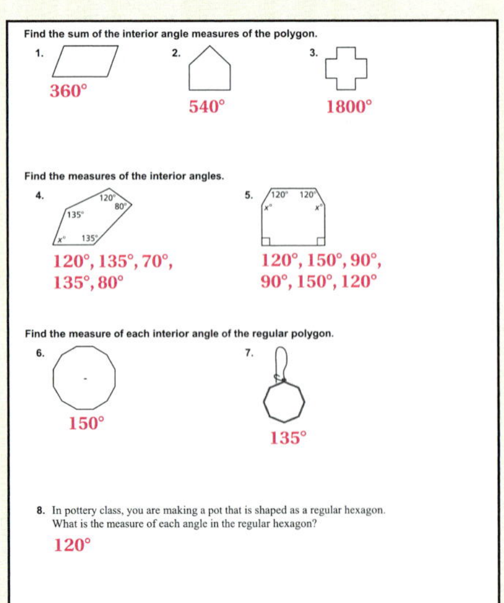

Find the sum of the interior angle measures of the polygon.

1. 360°
2. 540°
3. 1800°

Find the measures of the interior angles.

4. 120°, 135°, 70°, 135°, 80°
5. 120°, 150°, 90°, 90°, 150°, 120°

Find the measure of each interior angle of the regular polygon.

6. 150°
7. 135°

8. In pottery class, you are making a pot that is shaped as a regular hexagon. What is the measure of each angle in the regular hexagon?
120°

12.3 Exercises

Check It Out
Help with Homework
BigIdeasMath.com

Vocabulary and Concept Check

1. **VOCABULARY** Draw a regular polygon that has three sides.

2. **WHICH ONE DOESN'T BELONG?** Which figure does *not* belong with the other three? Explain your reasoning.

3. **DIFFERENT WORDS, SAME QUESTION** Which is different? Find "both" answers.

What is the measure of an interior angle of a regular pentagon?	What is the sum of the interior angle measures of a convex pentagon?
What is the sum of the interior angle measures of a regular pentagon?	What is the sum of the interior angle measures of a concave pentagon?

Practice and Problem Solving

Use triangles to find the sum of the interior angle measures of the polygon.

4. 

5. 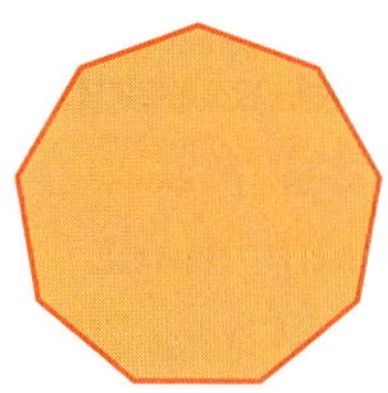

6. 

Find the sum of the interior angle measures of the polygon.

7. 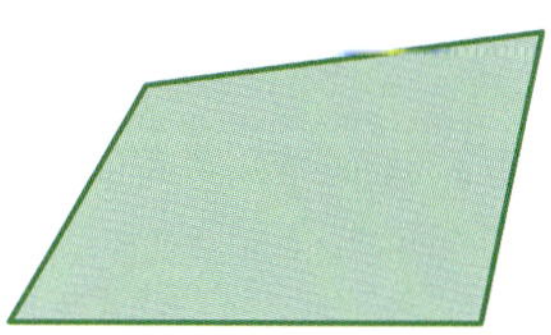

8. 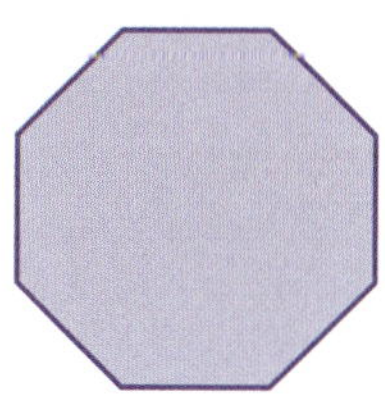

9. 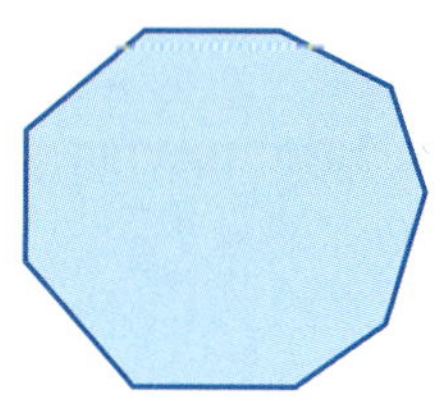

10. **ERROR ANALYSIS** Describe and correct the error in finding the sum of the interior angle measures of a 13-gon.

$$\begin{aligned} S &= n \cdot 180^\circ \\ &= 13 \cdot 180^\circ \\ &= 2340^\circ \end{aligned}$$

11. **NUMBER SENSE** Can a pentagon have interior angles that measure 120°, 105°, 65°, 150°, and 95°? Explain.

Find the measures of the interior angles.

2 **12.**

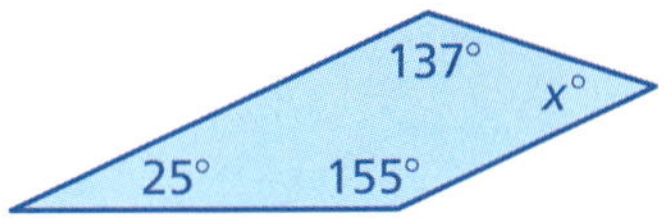

13.

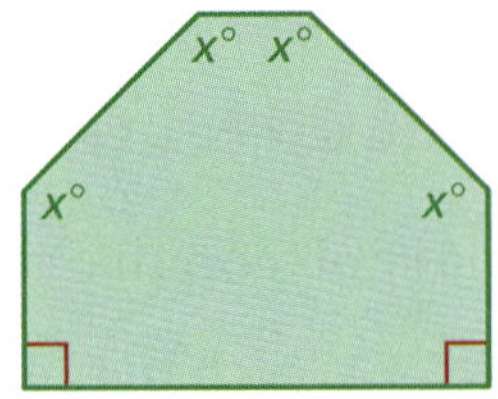

14.

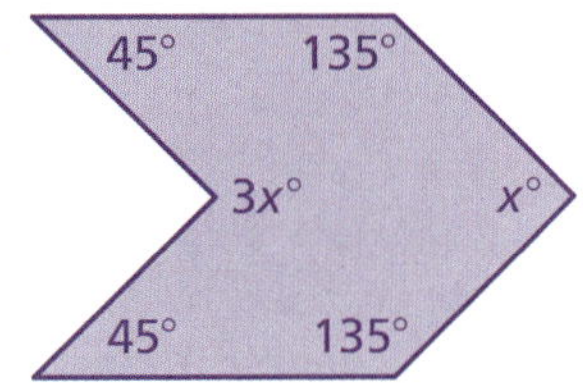

15. REASONING The sum of the interior angle measures in a regular polygon is 1260°. What is the measure of one of the interior angles of the polygon?

Find the measure of each interior angle of the regular polygon.

3 **16.**

17.

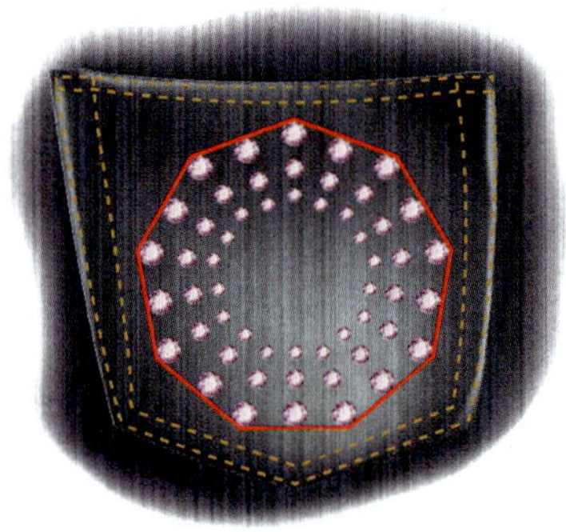

18.

19. ERROR ANALYSIS Describe and correct the error in finding the measure of each interior angle of a regular 20-gon.

✗

$S = (n - 2) \cdot 180°$

$= (20 - 2) \cdot 180°$

$= 18 \cdot 180°$

$= 3240°$

$3240° \div 18 = 180$

The measure of each interior angle is 180°.

20. FIRE HYDRANT A fire hydrant bolt is in the shape of a regular pentagon.

a. What is the measure of each interior angle?

b. Why are fire hydrants made this way?

21. PROBLEM SOLVING The interior angles of a regular polygon each measure 165°. How many sides does the polygon have?

Find the measures of the exterior angles of the polygon.

4 **22.**

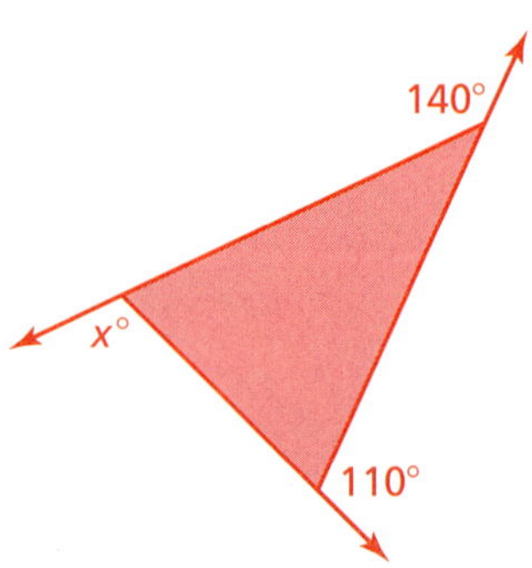

23.

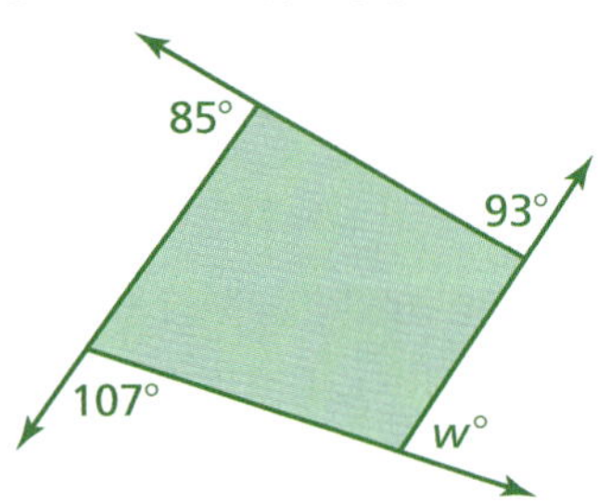

24.

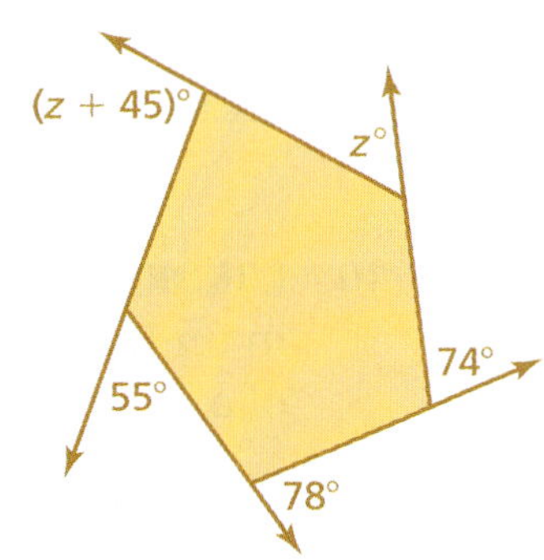

25. REASONING What is the measure of an exterior angle of a regular hexagon? Explain.

Common Errors

- **Exercises 12–14** Students may forget to include one or more of the given angles when writing an equation for the missing angles. For example, in Exercise 13, students may write $4x = 720$. Remind them to include all of the angles. Encourage them to write the equation and then count the number of terms to make sure that there are the same number of terms as angles before simplifying.
- **Exercises 16–18** Students may find the sum of the interior angle measures of the regular polygon, but forget to divide by the number of angles to answer the question. Remind them that they are finding the measure of *one* angle. Because all the angles are congruent (by the definition of a regular polygon), they can divide the sum of the interior angle measures by the number of angles.
- **Exercises 22–24, 26–28** Students may solve for the variable, but forget to find the measures of the angles. Remind them to read the directions carefully and to answer the question.

English Language Learners

Vocabulary

Many English mathematical names are influenced by Latin and Greek names for numbers.

English	Latin	Greek
one	unus	heis
two	duo	duo
three	tres	tria
four	quattuor	tettara
five	quinque	pente
six	sex	hex
seven	septe	hepta
eight	octo	okto
nine	nove	ennea
ten	dece	deka

Spanish speakers will recognize the similarity of numbers in Spanish to the numbers in Latin.

Practice and Problem Solving

12. 25°, 137°, 43°, 155°

13. 90°, 135°, 135°, 135°, 135°, 90°

14. 45°, 135°, 90°, 135°, 45°, 270°

15. 140° **16.** 60°

17. 140° **18.** 150°

19. The sum of the interior angle measures should have been divided by the number of angles, 20. $3240° \div 20 = 162°$; The measure of each interior angle is 162°.

20. a. 108°

b. *Sample answer:* to deter people from tampering with fire hydrants, because most wrenches are hexagonal.

21. 24 sides

22. 110°, 110°, 140°

23. 75°, 93°, 85°, 107°

24. 54°, 74°, 78°, 55°, 99°

25. See Additional Answers.

Practice and Problem Solving

26. 90°, 45°, 45°, 90°, 45°, 45°

27. 120°, 120°, 120°

28. 125°, 125°, 55°, 55°

29. interior: 135°; exterior: 45°

30. See *Taking Math Deeper*.

31. 120°

32. **a.** 11 sides

b. 147°

33. See Additional Answers.

Fair Game Review

34. 9 **35.** 2

36. 3 **37.** 6

38. D

Mini-Assessment

Find the sum of the interior angle measures of the polygon.

1.

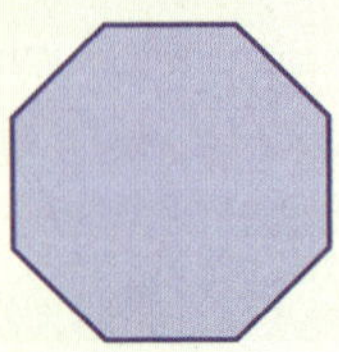

1080°

2.

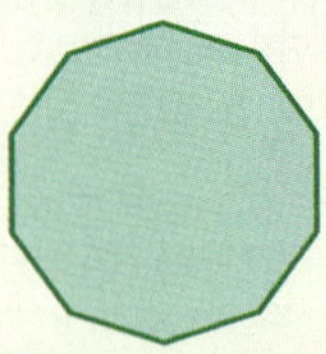

1440°

3. Find the measure of each interior angle of a 16-gon. 157.5°

4. Find the measures of the exterior angles of the polygon.

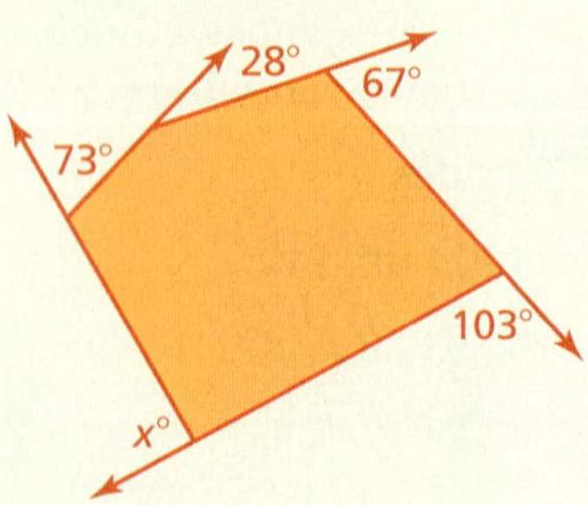

89°, 73°, 28°, 67°, and 103°

Taking Math Deeper

Exercise 30

Instead of trying to draw the angles one at a time, begin by drawing a different shape and manipulating it.

1 The desired pentagon has two right angles. So, you could start with a rectangle.

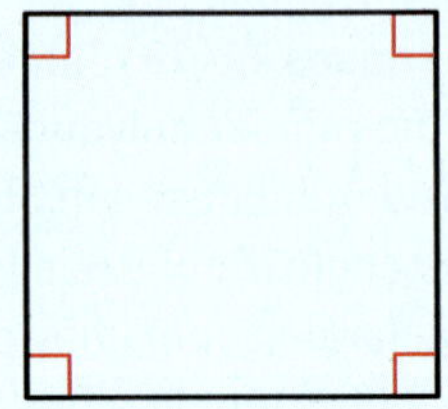

2 Erase one of the sides.

This leaves you with three sides and two right angles. You need two more sides and two 45° angles, so draw two sides that form 45° interior angles that are long enough to meet and create a fifth vertex.

3 Check the remaining angle.

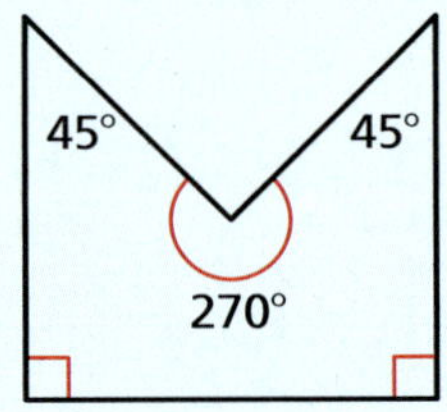

Using a protractor, you can see that the remaining interior angle measure is 270°.

Notice that the result is a concave pentagon.

Project

Tell students to use the method above to draw a pentagon that has two right interior angles, two 15° interior angels, and one 330° angle. Ask them if they can start with *any* rectangle. If not, ask them to describe the dimensions of rectangles that would not work.

Reteaching and Enrichment Strategies

If students need help. . .	If students got it. . .
Resources by Chapter • Practice A and Practice B • Puzzle Time Record and Practice Journal Practice Differentiating the Lesson Lesson Tutorials Skills Review Handbook	Resources by Chapter • Enrichment and Extension • Technology Connection Start the next section

Find the measures of the exterior angles of the polygon.

26. 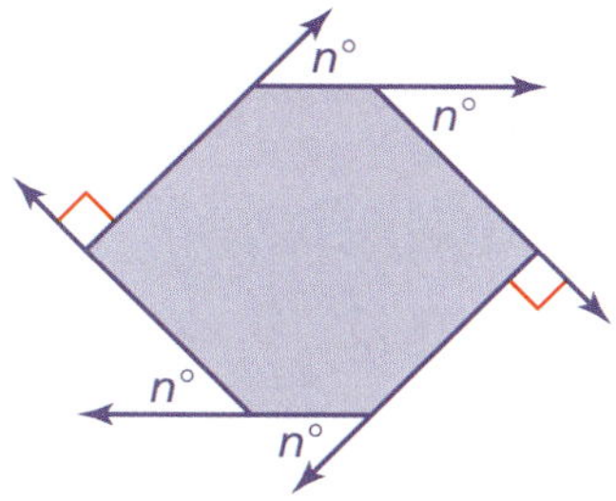

27. 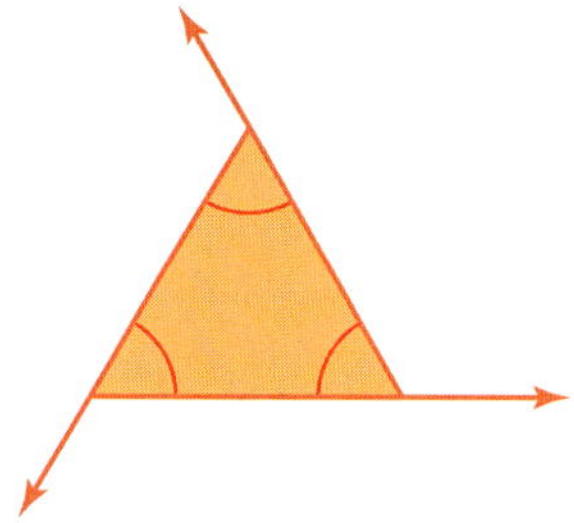

28. 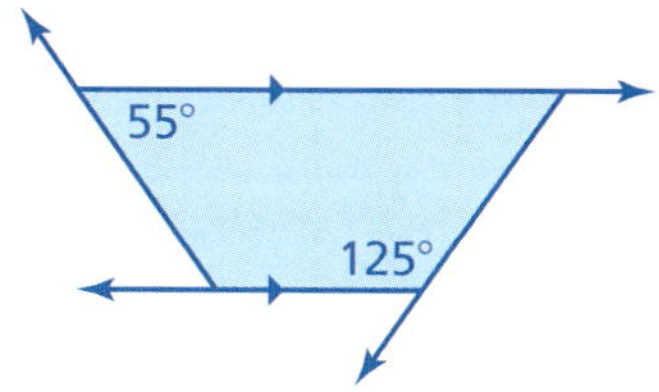

29. STAINED GLASS The center of the stained glass window is in the shape of a regular polygon. What is the measure of each interior angle of the polygon? What is the measure of each exterior angle?

30. PENTAGON Draw a pentagon that has two right interior angles, two 45° interior angles, and one 270° interior angle.

31. GAZEBO The floor of a gazebo is in the shape of a heptagon. Four of the interior angles measure 135°. The other interior angles have equal measures. Find their measures.

32. MONEY The border of a Susan B. Anthony dollar is in the shape of a regular polygon.

a. How many sides does the polygon have?

b. What is the measure of each interior angle of the border? Round your answer to the nearest degree.

33. Geometry When tiles can be used to cover a floor with no empty spaces, the collection of tiles is called a *tessellation*.

a. Create a tessellation using equilateral triangles.

b. Find two more regular polygons that form tessellations.

c. Create a tessellation that uses two different regular polygons.

d. Use what you know about interior and exterior angles to explain why the polygons in part (c) form a tessellation.

Fair Game Review What you learned in previous grades & lessons

Solve the proportion. *(Section 5.4)*

34. $\frac{x}{12} = \frac{3}{4}$ **35.** $\frac{14}{21} = \frac{x}{3}$ **36.** $\frac{9}{x} = \frac{6}{2}$ **37.** $\frac{10}{4} = \frac{15}{x}$

38. MULTIPLE CHOICE The ratio of tulips to daisies is 3 : 5. Which of the following could be the total number of tulips and daisies? *(Skills Review Handbook)*

Ⓐ 6 Ⓑ 10 Ⓒ 15 Ⓓ 16

12.4 Using Similar Triangles

Essential Question How can you use angles to tell whether triangles are similar?

1 ACTIVITY: Constructing Similar Triangles

Work with a partner.

- **Use a straightedge to draw a line segment that is 4 centimeters long.**
- **Then use the line segment and a protractor to draw a triangle that has a 60° and a 40° angle, as shown. Label the triangle *ABC*.**

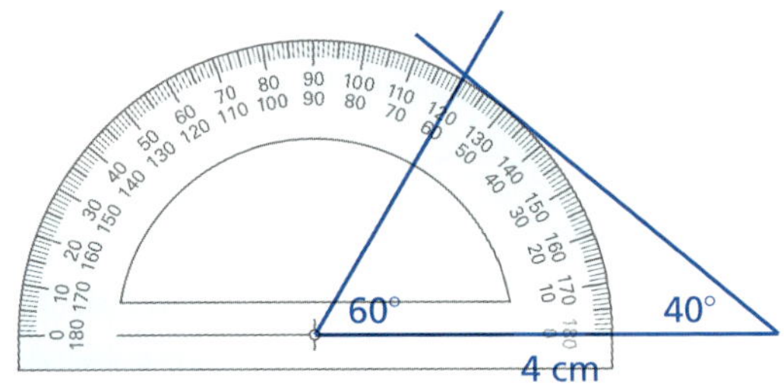

a. Explain how to draw a larger triangle that has the same two angle measures. Label the triangle *JKL*.

b. Explain how to draw a smaller triangle that has the same two angle measures. Label the triangle *PQR*.

c. Are all of the triangles similar? Explain.

2 ACTIVITY: Using Technology to Explore Triangles

Work with a partner. Use geometry software to draw the triangle below.

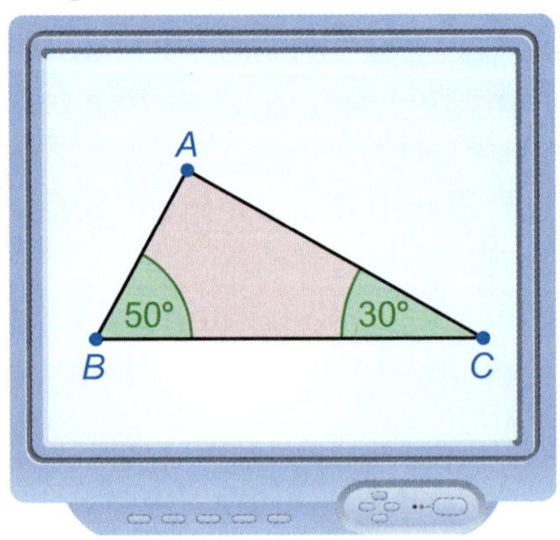

a. Dilate the triangle by the following scale factors.

$2 \qquad \frac{1}{2} \qquad \frac{1}{4} \qquad 2.5$

b. Measure the third angle in each triangle. What do you notice?

c. **REASONING** You have two triangles. Two angles in the first triangle are congruent to two angles in the second triangle. Can you conclude that the triangles are similar? Explain.

Geometry

In this lesson, you will

- understand the concept of similar triangles.
- identify similar triangles.
- use indirect measurement to find missing measures.

Laurie's Notes

Introduction

Applying Mathematical Practices

- **Use Appropriate Tools Strategically:** Similar triangles can be investigated using protractor and ruler or dynamic geometry software. Mathematically proficient students consider the available tools when solving a mathematics problem.

Motivate

- Each student will need a protractor and a ruler for today's activities.
- Ask students to work with their partners to construct a triangle with side lengths of 4 inches, 5 inches and 6 inches.
- Without a compass, they will need to work together, using both rulers to locate the third vertex.
- Have students measure the angles in the triangle they constructed.

- Have students hold up their constructions and look at the work of others. Discuss the results; namely, that all of the triangles are congruent.

Activity Notes

Activity 1

- **Use Appropriate Tools Strategically:** The goal of Activities 1 and 2 is the same. The approach is different depending upon the tools selected.
- After students have finished drawing the first triangle, stop to compare results. The orientations may be different, however all of the triangles should be congruent.
- ? "Do you know the measure of the third angle of your triangle without measuring? Explain." yes; 80° because the three angle measures need to sum to 180° and the first two angle measures already total 100°.
- Circulate as students work on parts (a) and (b). They should not have difficulty drawing the triangles.
- ? "Are all of the triangles similar? Explain." same shape, just a different size
- ? "How are all the triangles alike?" same angle measures, 40°, 60°, and 80°
- Discuss the definition of similar triangles. It is unlikely that any students checked to see that corresponding sides were in the same ratio. This is done easily using dynamic software. Otherwise, take the time for students to measure the sides, and allow for a bit of human error in measuring. Check to see that corresponding sides are proportional.

Activity 2

- The length of side *BC* is not specified.
- You may need to demonstrate how the dilate function of the software is used to complete part (a).
- Students can use the measure function to find the measure of angle *A* in each triangle. Students can also measure sides to see that corresponding sides are proportional.

What Your Students Will Learn

- Understand the concept of similar triangles.
- Identify similar triangles by using congruent angles.
- Use indirect measurement and proportions to find missing measures.

Previous Learning

Students should know the definition of similar triangles.

Technology for the Teacher

Dynamic Classroom

Lesson Plans
Complete Materials List

12.4 Record and Practice Journal

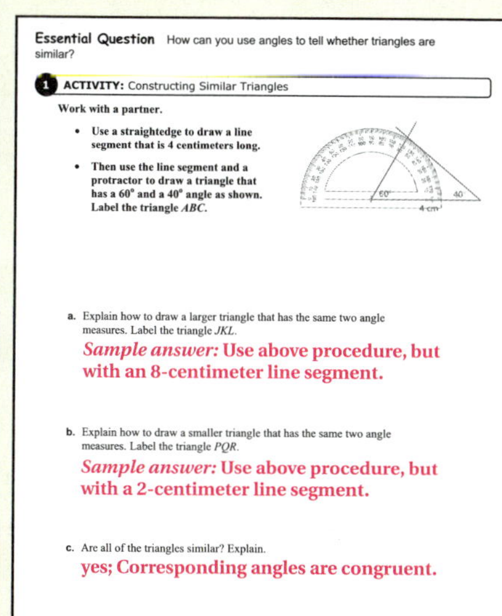

Essential Question How can you use angles to tell whether triangles are similar?

1 ACTIVITY: Constructing Similar Triangles

Work with a partner.

- Use a straightedge to draw a line segment that is 4 centimeters long.
- Then use the line segment and a protractor to draw a triangle that has a 60° and a 40° angle as shown. Label the triangle *ABC*.

a. Explain how to draw a larger triangle that has the same two angle measures. Label the triangle *JKL*.
Sample answer: Use above procedure, but with an 8-centimeter line segment.

b. Explain how to draw a smaller triangle that has the same two angle measures. Label the triangle *PQR*.
Sample answer: Use above procedure, but with a 2-centimeter line segment.

c. Are all of the triangles similar? Explain.
yes; Corresponding angles are congruent.

Differentiated Instruction

Kinesthetic

When setting up a proportion, have students write each of the three known values and the one unknown value with their units on index cards. On a fifth index card, have the students write an equal sign. Students should then place the cards on their desks to set up the proportion. Discuss the different ways to set up a proportion.

12.4 Record and Practice Journal

2 ACTIVITY: Using Technology to Explore Triangles

Work with a partner. Use geometry software to draw the triangle shown.

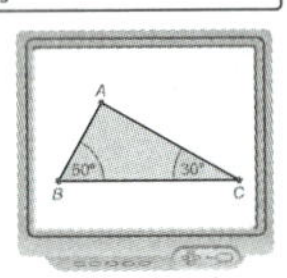

a. Dilate the triangle by the following scale factors.

2 $\quad \frac{1}{2} \quad \frac{1}{4} \quad$ 2.5

See Additional Answers.

b. Measure the third angle in each triangle. What do you notice?

It is 100° for each triangle.

c. **REASONING** When two angles in one triangle are congruent to two angles in another triangle, can you conclude that the triangles are similar? Explain.

yes

3 ACTIVITY: Indirect Measurement

Work with a partner.

a. Use the fact that two rays from the Sun are parallel to explain why $\triangle ABC$ and $\triangle DEF$ are similar.

Rays make same angle with ground, and triangles are right triangles, so other pair of corresponding angles are congruent.

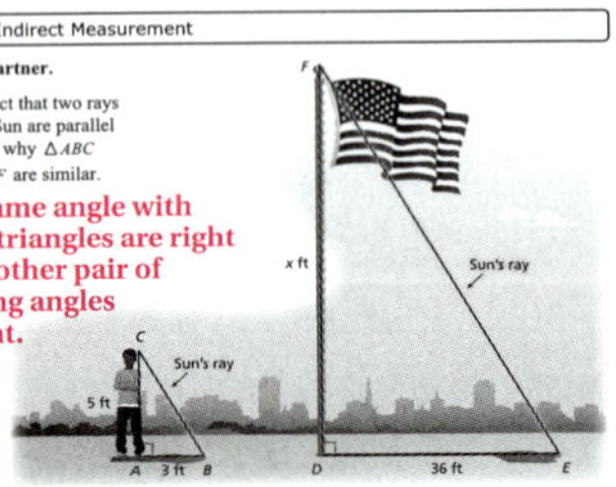

b. Explain how to use similar triangles to find the height of the flagpole.

Solve $\frac{x}{5} = \frac{36}{3}$.

What Is Your Answer?

4. **IN YOUR OWN WORDS** How can you use angles to tell whether triangles are similar?

When two angles of one triangle are congruent to two angles of another, the third angles are congruent and the triangles are similar.

5. **PROJECT** Work with a partner or in a small group.

a. Explain why the process in Activity 3 is called "indirect" measurement.

You are not measuring the flagpole directly.

b. **CHOOSE TOOLS** Use indirect measurement to measure the height of something outside your school (a tree, a building, a flagpole). Before going outside, decide what materials you need to take with you.

Check students' work.

c. **MODELING** Draw a diagram of the indirect measurement process you used. In the diagram, label the lengths that you actually measured and also the lengths that you calculated.

Check students' work.

6. **PRECISION** Look back at Exercise 17 in Section 11.5. Explain how you can show that the two triangles are similar.

See Additional Answers.

Laurie's Notes

Activity 3

- Students will need to use the results of the first two activities. When two triangles have two congruent angles, the triangles are similar.
- ? "Is your shadow shorter at noon or 5 P.M.? Explain." Noon; The sun is overhead, not at a lower position in the sky.
- ? "Do adjacent objects of different heights cast the same length shadow? Explain." no; Taller objects cast longer shadows.
- The triangles are similar because they both have a right angle and the parallel rays of the sun are at the same angle to the ground.

What Is Your Answer?

- Question 4 summarizes the results of today's activities.
- If you assign Question 5, it will take planning and discussions before students go outside.

Closure

- **Exit Ticket:** Are all 40°-60°-80° triangles congruent? Similar? Explain. no; For triangles to be congruent, corresponding sides must be congruent. yes; For similar triangles, only the corresponding angles need to be congruent.

3 ACTIVITY: Indirect Measurement

Math Practice

Make Sense of Quantities

What do you know about the sides of the triangles when the triangles are similar?

Work with a partner.

a. Use the fact that two rays from the Sun are parallel to explain why $\triangle ABC$ and $\triangle DEF$ are similar.

b. Explain how to use similar triangles to find the height of the flagpole.

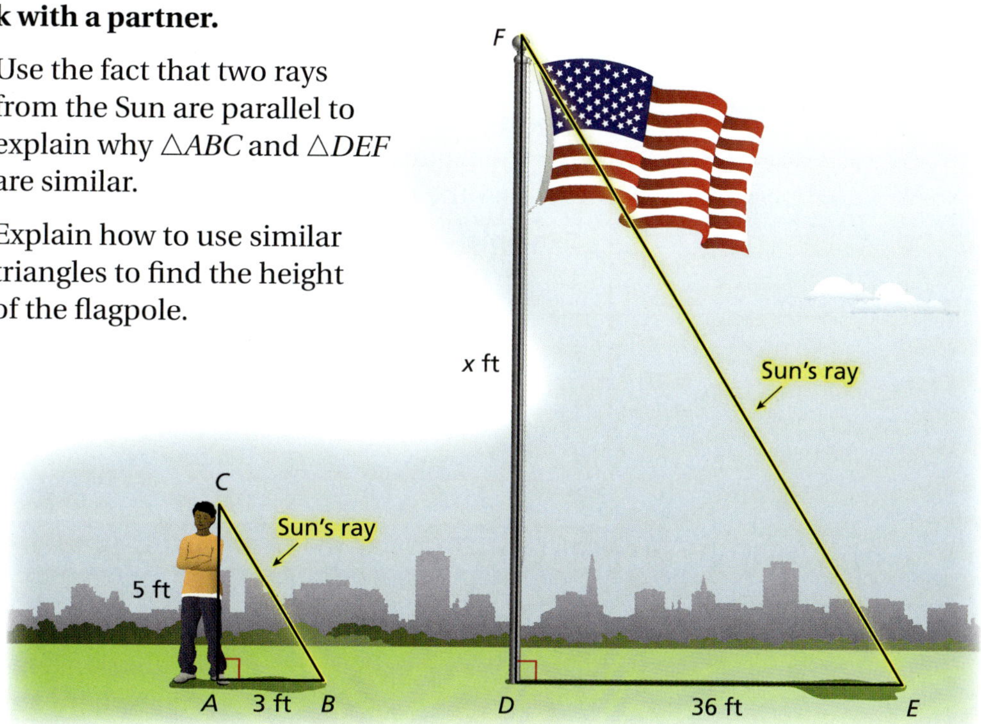

What Is Your Answer?

4. **IN YOUR OWN WORDS** How can you use angles to tell whether triangles are similar?

5. **PROJECT** Work with a partner or in a small group.

   a. Explain why the process in Activity 3 is called "indirect" measurement.

   b. **CHOOSE TOOLS** Use indirect measurement to measure the height of something outside your school (a tree, a building, a flagpole). Before going outside, decide what materials you need to take with you.

   c. **MODELING** Draw a diagram of the indirect measurement process you used. In the diagram, label the lengths that you actually measured and also the lengths that you calculated.

6. **PRECISION** Look back at Exercise 17 in Section 11.5. Explain how you can show that the two triangles are similar.

Practice Use what you learned about similar triangles to complete Exercises 4 and 5 on page 554.

Check It Out
Lesson Tutorials
BigIdeasMath.com

Key Vocabulary
indirect measurement, *p. 553*

Key Idea

Angles of Similar Triangles

Words When two angles in one triangle are congruent to two angles in another triangle, the third angles are also congruent and the triangles are similar.

Example

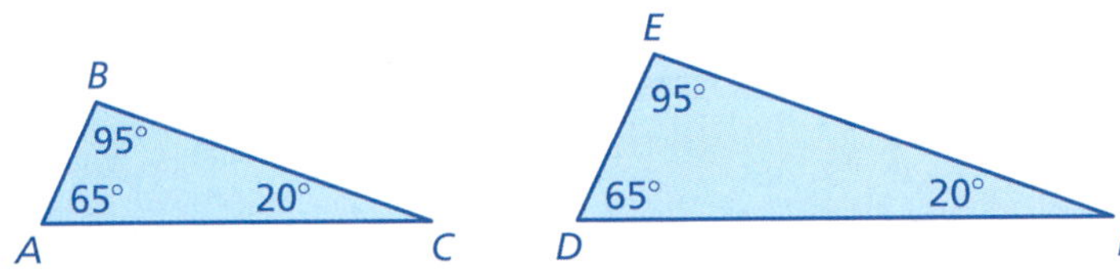

Triangle ABC is similar to Triangle DEF: $\triangle ABC \sim \triangle DEF$.

EXAMPLE 1 Identifying Similar Triangles

Tell whether the triangles are similar. Explain.

a.

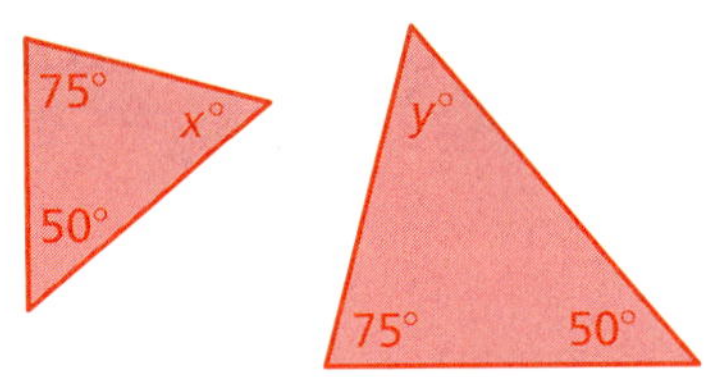

The triangles have two pairs of congruent angles.

So, the third angles are congruent, and the triangles are similar.

b.

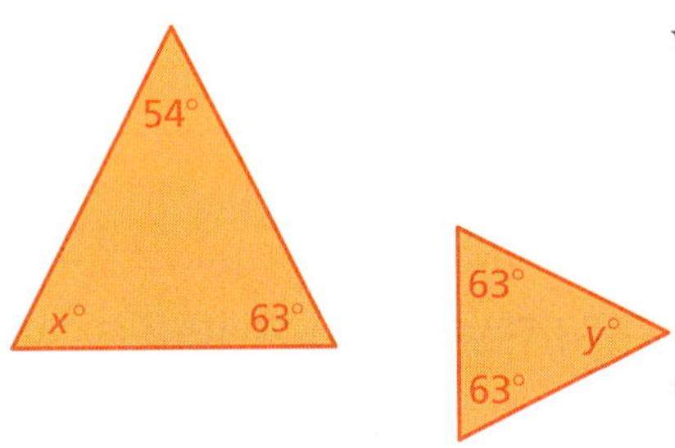

Write and solve an equation to find x.

$$x + 54 + 63 = 180$$

$$x + 117 = 180$$

$$x = 63$$

The triangles have two pairs of congruent angles.

So, the third angles are congruent, and the triangles are similar.

c.

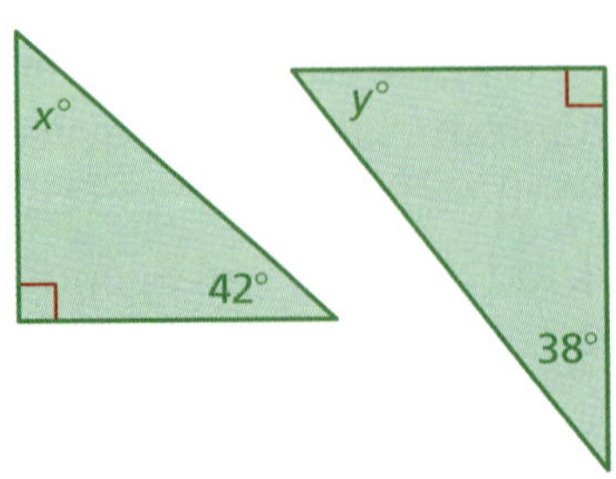

Write and solve an equation to find x.

$$x + 90 + 42 = 180$$

$$x + 132 = 180$$

$$x = 48$$

The triangles do not have two pairs of congruent angles.

So, the triangles are not similar.

Multi-Language Glossary at BigIdeasMath.com

Laurie's Notes

Introduction

Connect

- **Yesterday:** Students explored special properties of similar triangles.
- **Today:** Students will use similar triangles to solve real-life problems.

Motivate

- Have pairs of students put a visual barrier (i.e., a notebook) between them. One student draws a triangle using a straightedge. This student now gives directions to the second student who will draw a triangle based on the information given. The only information the first student may give is angle measure! In fact, after the second angle measure is given, the second student should know the measure of the third angle.
- The triangles should be similar.

? "What do you notice about the triangles?" similar

? "How do you know they are similar?" Listen for same shape, different size; students may also mention yesterday's activities.

Lesson Notes

Key Idea

- Write the informal definition (same shape, not necessarily the same size), and draw examples of similar triangles.

? "What is the formal definition of similar triangles?" Similar triangles have corresponding sides that are proportional and corresponding angles that are congruent.

- Write the Key Idea.
- Make sure students understand that there are two parts stated in the Key Idea. When you have two angles in one triangle congruent to two angles in another triangle, then:
  1) the third angles are congruent.
  2) the two triangles are similar.

Example 1

- Draw the two triangles and label the given information. Ask students to solve for the missing angle measure of each triangle.

? "Are the triangles in part (a) similar? Explain." yes; The triangles have two pairs of congruent angles.

? "Are the triangles in part (b) similar? Explain." yes; First, solve for the missing angles. The triangles have two pairs of congruent angles, so they are similar.

? "Do you need to solve for *y* in part (b)?" no; Once you solve for *x*, you know that the triangles have two pairs of congruent angles.

? "What type of triangles are in part (c)?" Both are right triangles.

? "Are the triangles similar?" no; Solving for the missing angles, the triangles only have one pair of congruent angles. So the triangles are not similar.

Goal Today's lesson is using similar triangles to solve problems.

Technology for the Teacher

Dynamic Classroom

Lesson Tutorials
Lesson Plans
Answer Presentation Tool

Extra Example 1

Tell whether the triangles are similar. Explain.

a.

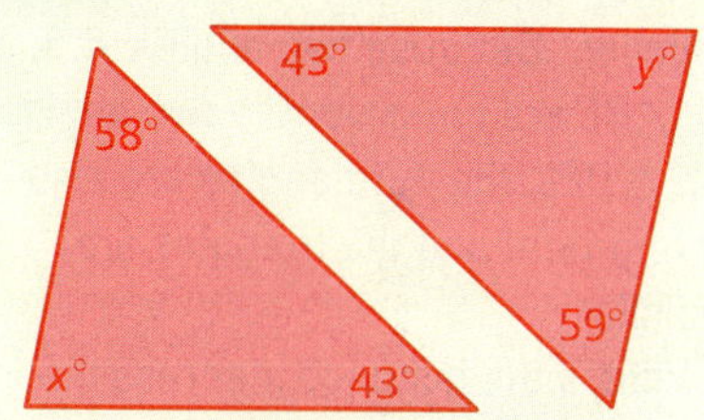

no; The triangles do not have the same angle measures.

b.

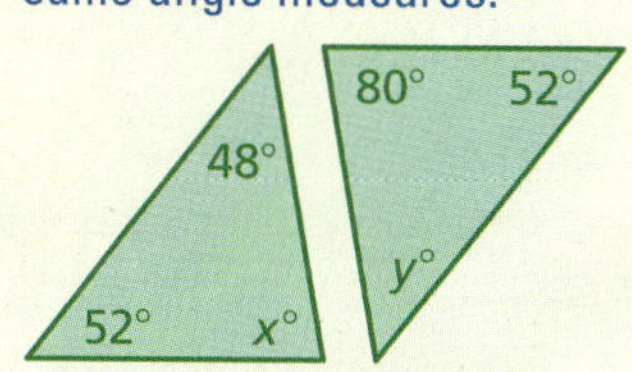

yes; The triangles have the same angle measures, 52°, 48°, and 80°.

c.

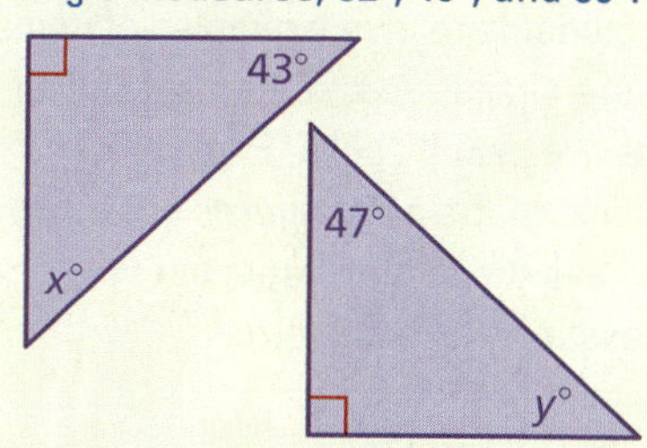

yes; The triangles have the same angle measures, 47°, 90°, and 43°.

On Your Own

1. no; The triangles do not have the same angle measures.
2. yes; The triangles have the same angle measures, 90°, 66°, and 24°.

Extra Example 2

You plan to cross a river and want to know how far it is to the other side. You take measurements on your side of the river and make the drawing shown.

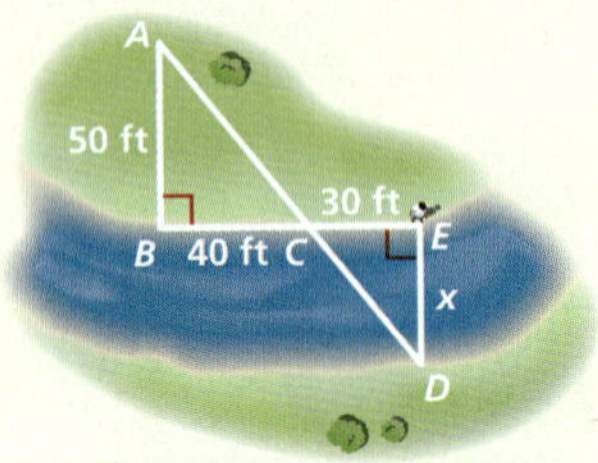

a. Explain why $\triangle ABC$ and $\triangle DEC$ are similar. Because two angles in $\triangle ABC$ are congruent to two angles in $\triangle DEC$, the third angles are also congruent and the triangles are similar.

b. What is the distance x across the river? 37.5 ft

On Your Own

3. 44 ft

English Language Learners

Build on Past Knowledge

Ask students to give examples of items that are similar. Ask students if similar items are exactly alike. Explain to students that *similar figures* are figures that have the same shape, but not necessarily the same size.

Laurie's Notes

On Your Own

- **Think-Pair-Share:** Students should read each question independently and then work in pairs to answer the questions. When they have answered the questions, the pair should compare their answers with another group and discuss any discrepancies.

Example 2

- Indirect measurement is used when you want to know the measurement of some length (or angle) and you cannot measure the object directly.
- Ask a volunteer to read the problem. Make a rough sketch of the diagram.

? "What do you know about the angles in either triangle?" $\angle B$ and $\angle E$ are right angles. The vertical angles are congruent (mark the diagram to show the congruent angles).

? "What do you know about the third angle in each triangle?" They are congruent.

- **Attend to Precision:** Because the triangles are similar, the corresponding sides will have the same ratio. Setting up the ratios is challenging for students. Talk about the sides in terms of being the shorter leg of the right triangle and the longer leg of the right triangle.
- Use the Multiplication Property of Equality or the Cross Products Property to solve. Check the reasonableness of the answer.

On Your Own

- **Think-Pair-Share:** Students should read the question independently and then work in pairs to answer the question. When they have answered the question, the pair should compare their answer with another group and discuss any discrepancies.

Closure

- **Exit Ticket:** Are the two triangles similar? Explain.

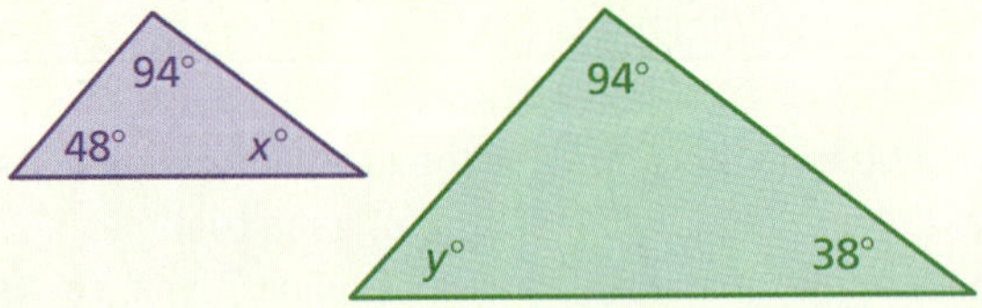

The triangles have the same angle measures, 94°, 48°, and 38°. So, the triangles are similar.

On Your Own

Now You're Ready
Exercises 6–9

Tell whether the triangles are similar. Explain.

1.

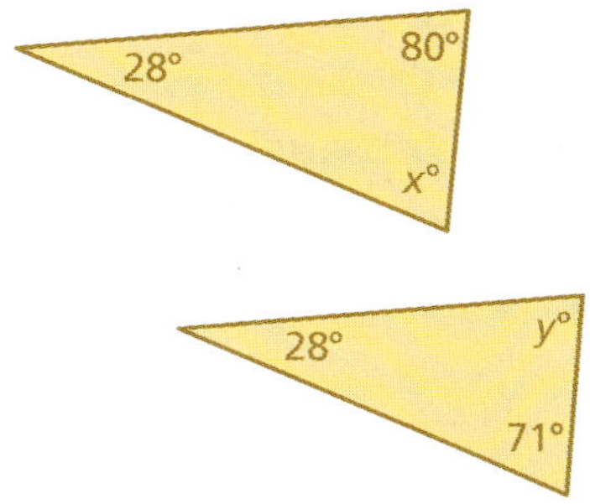

2.

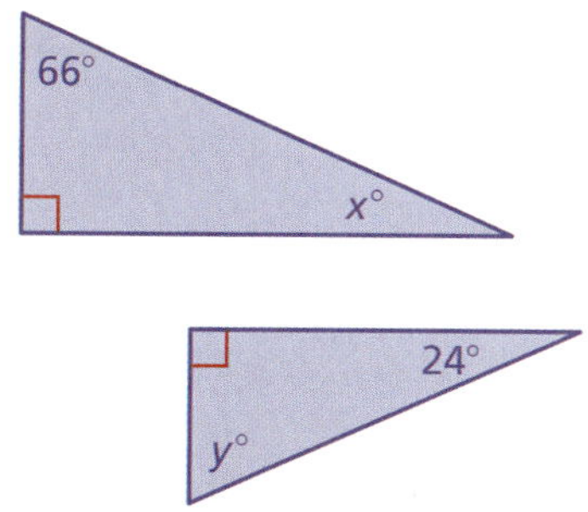

Indirect measurement uses similar figures to find a missing measure when it is difficult to find directly.

EXAMPLE 2 Using Indirect Measurement

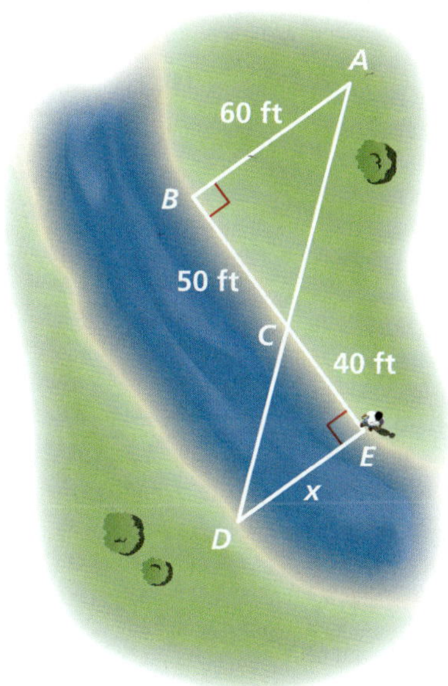

You plan to cross a river and want to know how far it is to the other side. You take measurements on your side of the river and make the drawing shown. (a) Explain why $\triangle ABC$ and $\triangle DEC$ are similar. (b) What is the distance x across the river?

a. $\angle B$ and $\angle E$ are right angles, so they are congruent. $\angle ACB$ and $\angle DCE$ are vertical angles, so they are congruent.

Because two angles in $\triangle ABC$ are congruent to two angles in $\triangle DEC$, the third angles are also congruent and the triangles are similar.

b. The ratios of the corresponding side lengths in similar triangles are equal. Write and solve a proportion to find x.

$$\frac{x}{60} = \frac{40}{50} \qquad \text{Write a proportion.}$$

$$60 \cdot \frac{x}{60} = 60 \cdot \frac{40}{50} \qquad \text{Multiplication Property of Equality}$$

$$x = 48 \qquad \text{Simplify.}$$

So, the distance across the river is 48 feet.

On Your Own

Now You're Ready
Exercise 13

3. **WHAT IF?** The distance from vertex A to vertex B is 55 feet. What is the distance across the river?

12.4 Exercises

Check It Out
Help with Homework
BigIdeasMath.com

Vocabulary and Concept Check

1. **REASONING** How can you use similar triangles to find a missing measurement?

2. **WHICH ONE DOESN'T BELONG?** Which triangle does *not* belong with the other three? Explain your reasoning.

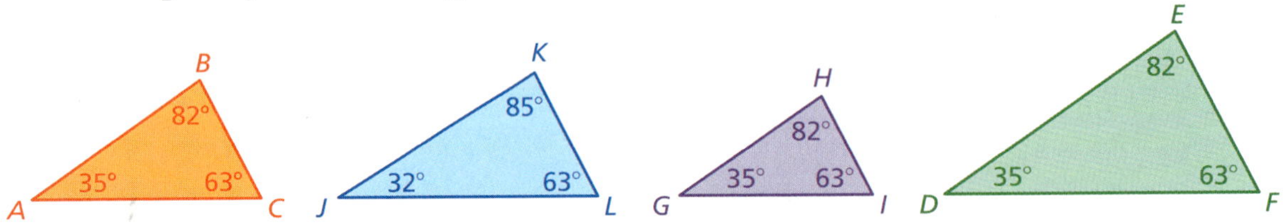

3. **WRITING** Two triangles have two pairs of congruent angles. In your own words, explain why you do not need to find the measures of the third pair of angles to determine that they are congruent.

Practice and Problem Solving

Make a triangle that is larger or smaller than the one given and has the same angle measures. Find the ratios of the corresponding side lengths.

4.

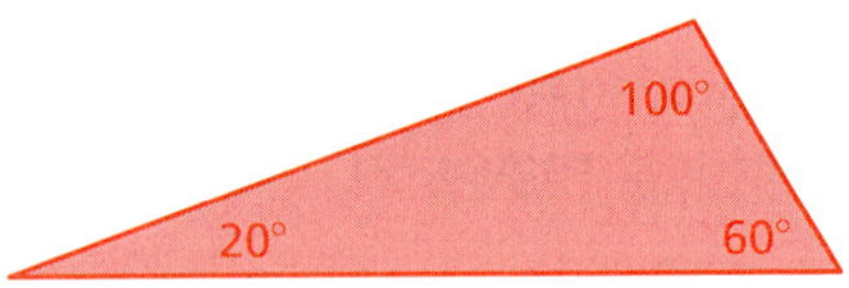

5.

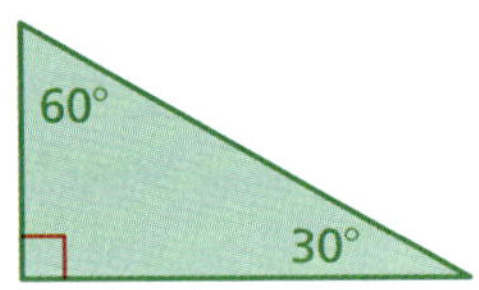

Tell whether the triangles are similar. Explain.

1 6.

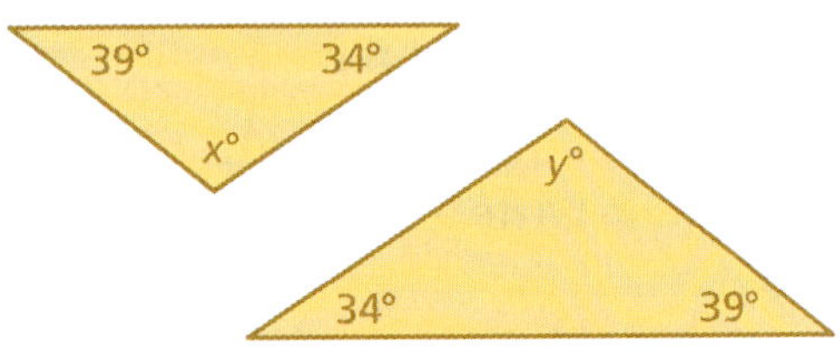

7.

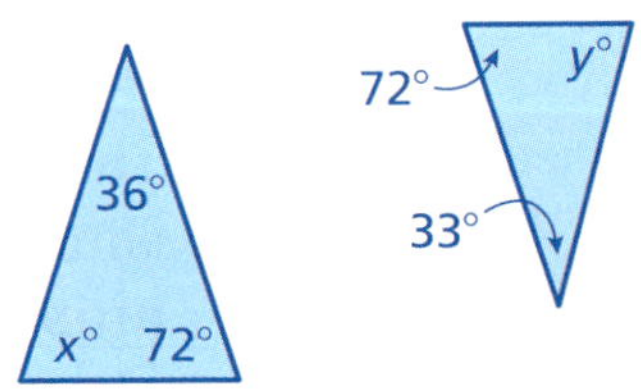

8.

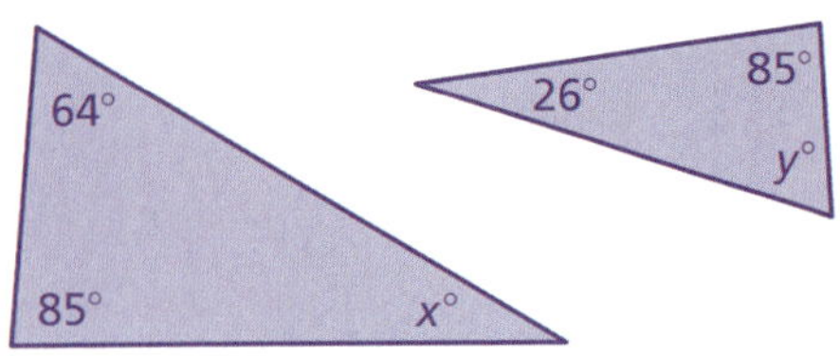

9.

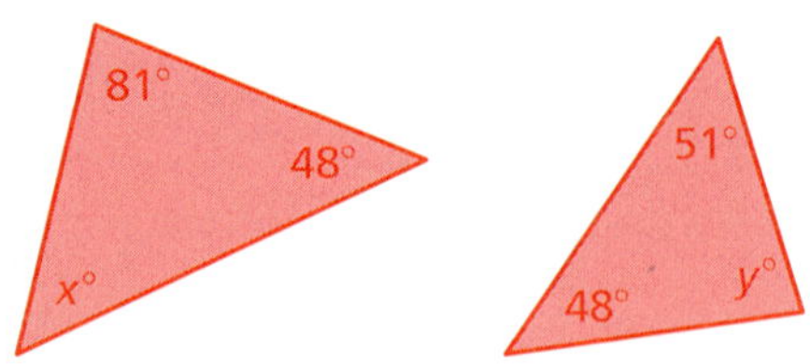

10. **RULERS** Which of the rulers are similar in shape? Explain.

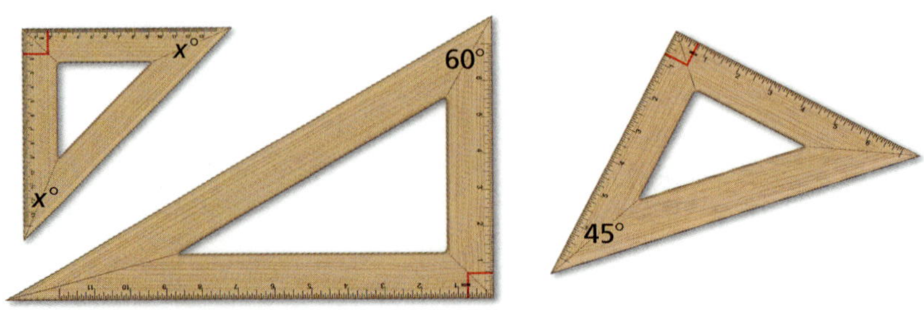

Assignment Guide and Homework Check

Level	Assignment	Homework Check
Accelerated	1–22	8, 12, 14, 15, 17

Common Errors

- **Exercises 6–9** Students may find the missing angle measure for one of the triangles and then make a decision about the similarity of the triangles. While it is possible to use this method, encourage them to find the missing angles of both triangles to verify that they are correct.

12.4 Record and Practice Journal

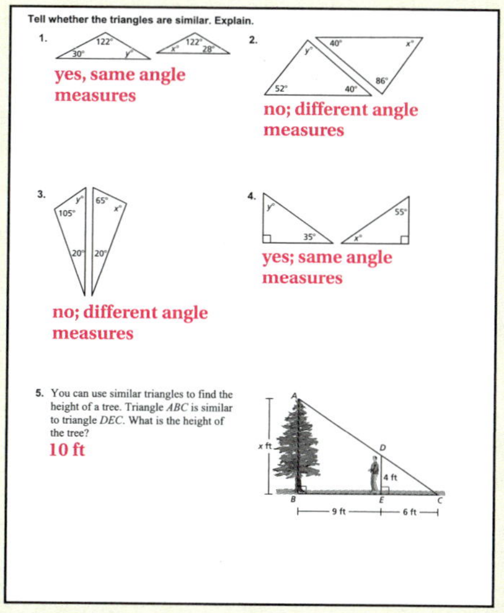
Tell whether the triangles are similar. Explain.

1. yes, same angle measures

2. no; different angle measures

3. no; different angle measures

4. yes; same angle measures

5. You can use similar triangles to find the height of a tree. Triangle *ABC* is similar to triangle *DEC*. What is the height of the tree?
10 ft

Vocabulary and Concept Check

1. Write a proportion that uses the missing measurement because the ratios of corresponding side lengths are equal.

2. $\triangle JKL$ because the other three triangles are similar.

3. *Sample answer:* Two of the angles are congruent, so they have the same sum. When you subtract this from 180°, you will get the same third angle.

Practice and Problem Solving

4–5. Student should draw a triangle with the same angle measures as the ones given in the textbook. If the student's triangle is larger than the one given, then the ratio of the corresponding side lengths, $\frac{\text{student's triangle length}}{\text{book's triangle length}}$, should be greater than 1. If the student's triangle is smaller than the one given, then the ratio of the corresponding side lengths, $\frac{\text{student's triangle length}}{\text{book's triangle length}}$, should be less than 1.

6. yes; The triangles have two pairs of congruent angles.

7. no; The triangles do not have two pairs of congruent angles.

8. no; The triangles do not have the same angle measures.

9. yes; The triangles have the same angle measures, 81°, 51°, and 48°.

10. the leftmost and rightmost; They both are right triangles with 45° angles.

Practice and Problem Solving

11. yes; The triangles have two pairs of congruent angles.
12. no; The triangles do not have two pairs of congruent angles.
13. See *Taking Math Deeper*.
14–18. See Additional Answers.

Fair Game Review

19. $y = 5x + 3$
20. $y = -\frac{2}{3}x + 2$
21. $y = 8x - 4$
22. B

Mini-Assessment

Tell whether the triangles are similar. Explain.

1.

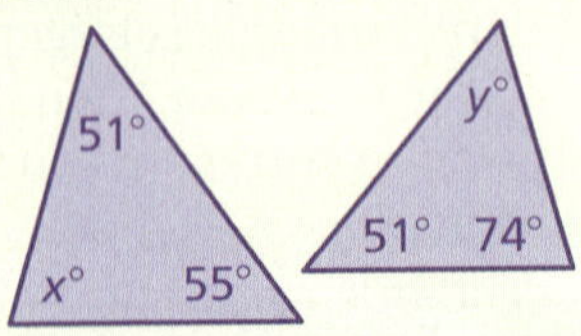

yes; The triangles have the same angle measures, 51°, 55°, and 74°.

2.

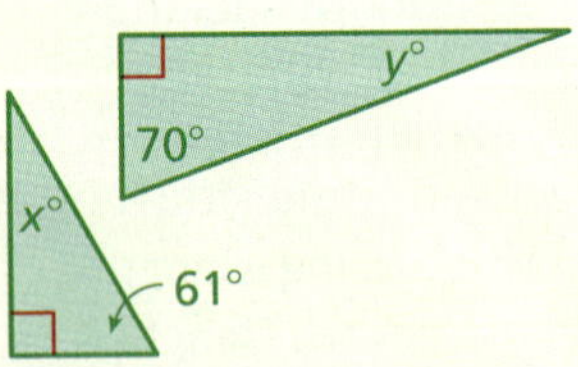

no; The triangles do not have the same angle measures.

3.

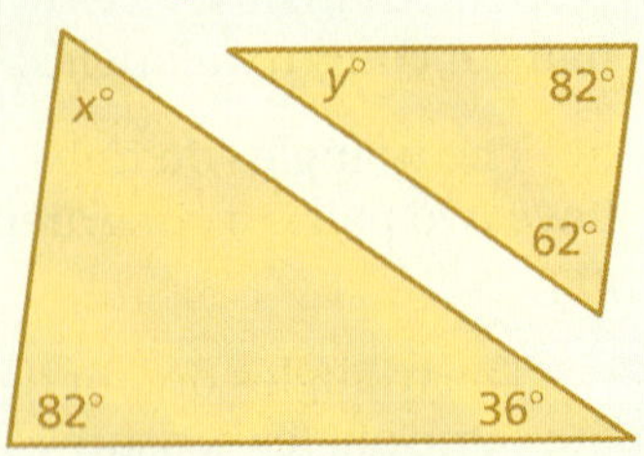

yes; The triangles have the same angle measures, 82°, 62°, and 36°.

4. A person that is 5 feet tall casts a 3-foot-long shadow. A nearby telephone pole casts a 12-foot-long shadow. What is the height h of the telephone pole? 20 ft

Taking Math Deeper

Exercise 13

In Chapter 11, students learned that two figures are similar when one can be obtained from the other by a sequence of translations, reflections, rotations, and dilations. So, one way to solve this problem is to draw the situation in a coordinate plane.

1 Explain why the triangles are similar.

Because the triangles have a pair of vertical angles and a pair of right angles, the triangles have the same interior angle measures. So, the triangles are similar.

2 Draw the larger triangle in a coordinate plane. Let the vertex of the vertical angles be located at (0, 0).

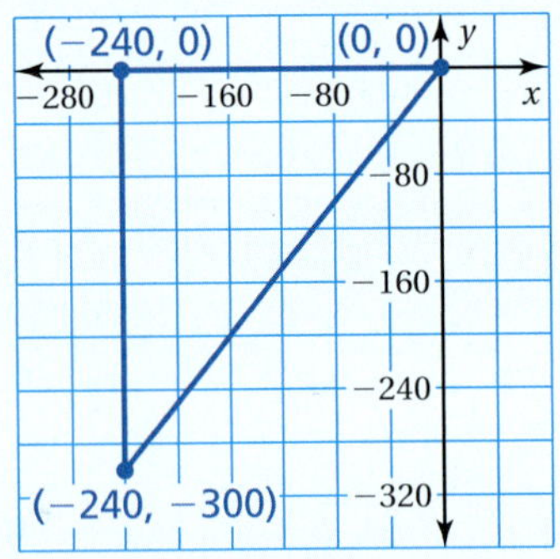

Rotate the triangle 180° about the origin. Then dilate the image using a scale factor of $\frac{1}{3}$ to obtain the smaller triangle.

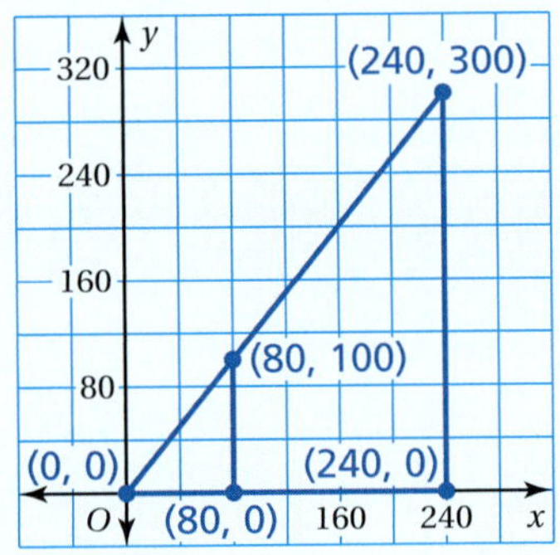

3 Answer the question.

By obtaining the smaller triangle through transformations, you can see that the treasure is located at (80, 100).

So, you take 100 steps from the pyramids to the treasure.

Project

Have students research the Pythagorean Theorem, and then use it to find the number of steps you would have by taking the "straight-line" approach.

Reteaching and Enrichment Strategies

If students need help. . .	If students got it. . .
Resources by Chapter • Practice A and Practice B • Puzzle Time Record and Practice Journal Practice Differentiating the Lesson Lesson Tutorials Skills Review Handbook	Resources by Chapter • Enrichment and Extension • Technology Connection Start the next section

Tell whether the triangles are similar. Explain.

11.

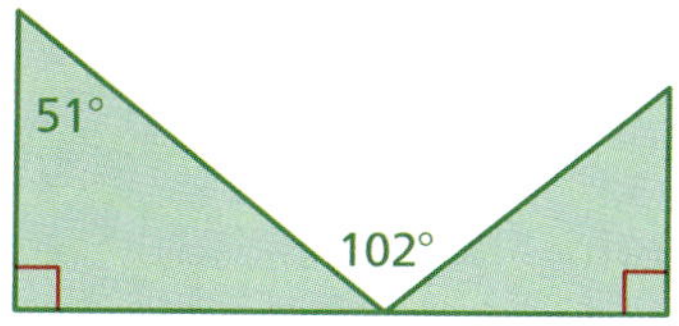

12.

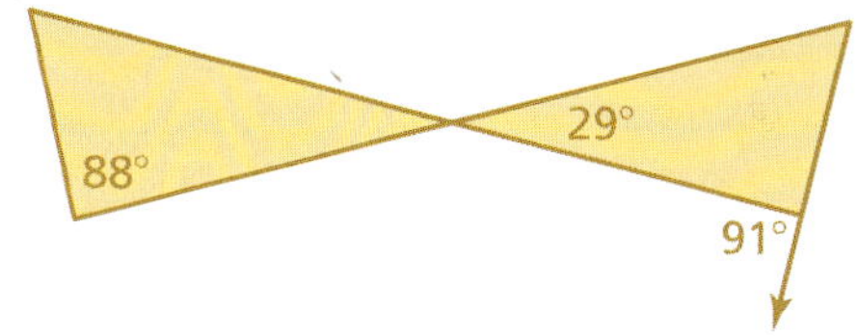

13. TREASURE The map shows the number of steps you must take to get to the treasure. However, the map is old, and the last dimension is unreadable. Explain why the triangles are similar. How many steps do you take from the pyramids to the treasure?

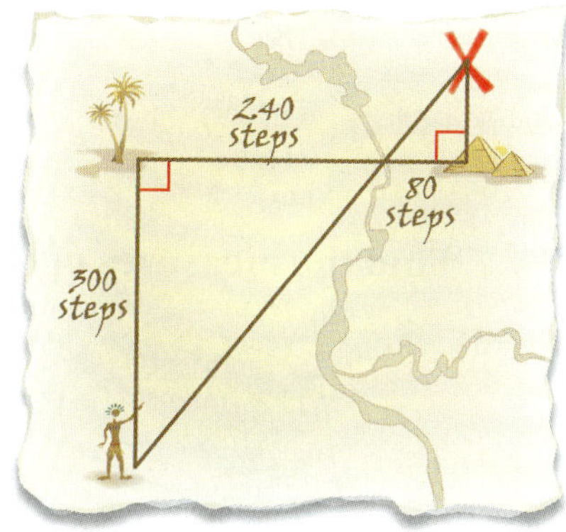

14. CRITICAL THINKING The side lengths of a triangle are increased by 50% to make a similar triangle. Does the area increase by 50% as well? Explain.

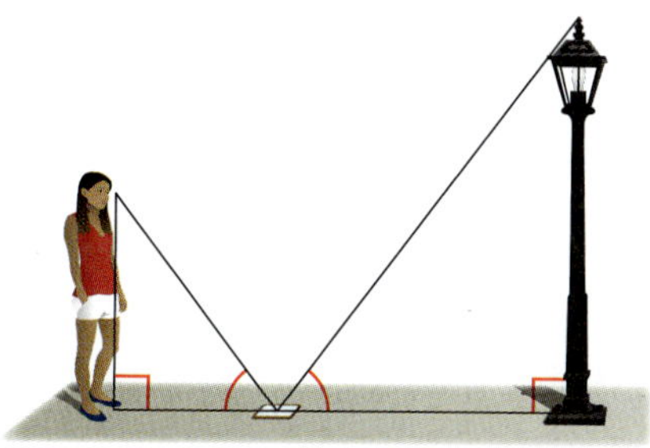

15. PINE TREE A person who is 6 feet tall casts a 3-foot-long shadow. A nearby pine tree casts a 15-foot-long shadow. What is the height h of the pine tree?

16. OPEN-ENDED You place a mirror on the ground 6 feet from the lamppost. You move back 3 feet and see the top of the lamppost in the mirror. What is the height of the lamppost?

17. REASONING In each of two right triangles, one angle measure is two times another angle measure. Are the triangles similar? Explain your reasoning.

18. Geometry In the diagram, segments BG, CF, and DE are parallel. The length of segment BD is 6.32 feet, and the length of segment DE is 6 feet. Name all pairs of similar triangles in the diagram. Then find the lengths of segments BG and CF.

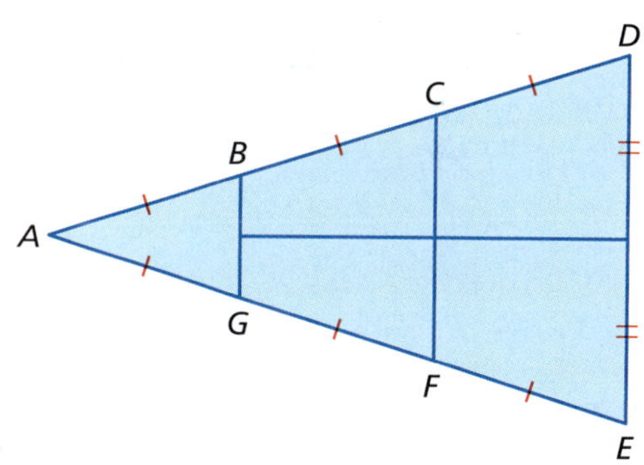

Fair Game Review What you learned in previous grades & lessons

Solve the equation for *y*. *(Topic 3)*

19. $y - 5x = 3$

20. $4x + 6y = 12$

21. $2x - \frac{1}{4}y = 1$

22. MULTIPLE CHOICE What is the value of x? *(Section 12.2)*

Ⓐ 17 Ⓑ 62

Ⓒ 118 Ⓓ 152

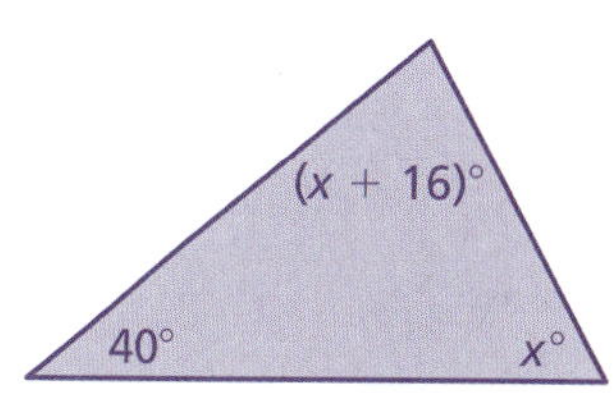

12.3–12.4 Quiz

Check It Out
Progress Check
BigIdeasMath.com

Find the sum of the interior angle measures of the polygon. *(Section 12.3)*

1. 

2. 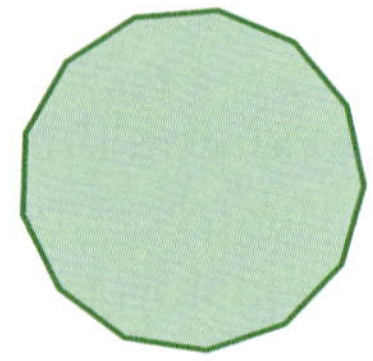

Find the measures of the interior angles of the polygon. *(Section 12.3)*

3. 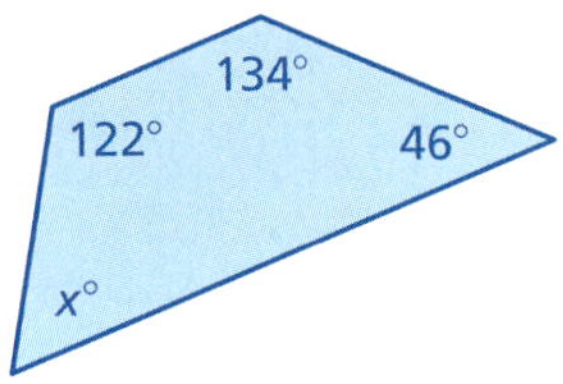

4. 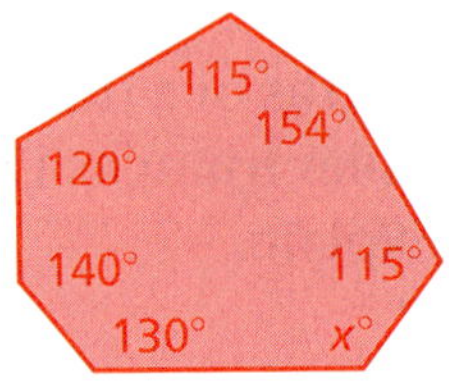

5. 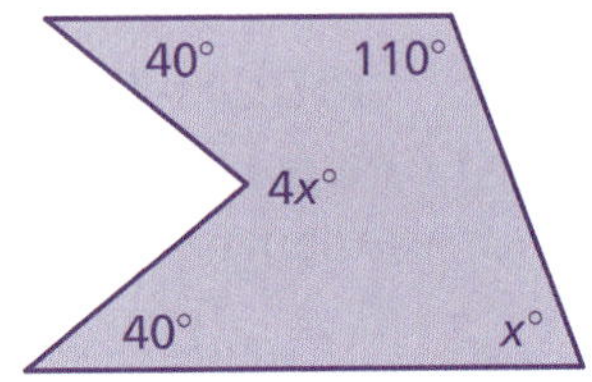

Find the measures of the exterior angles of the polygon. *(Section 12.3)*

6. 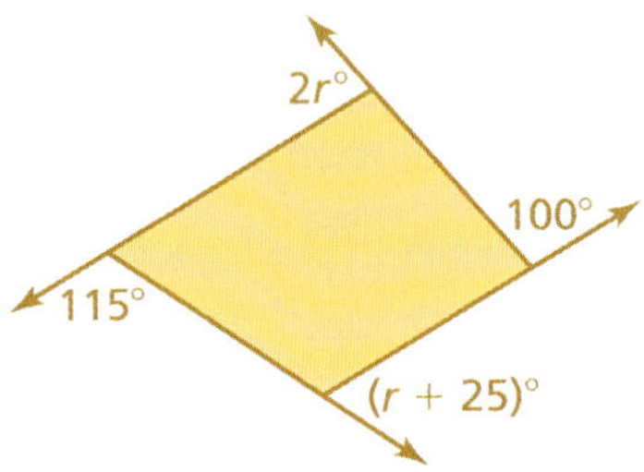

7. 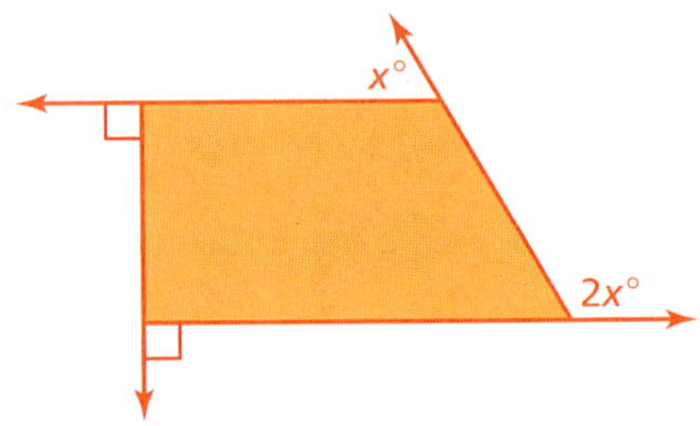

Tell whether the triangles are similar. Explain. *(Section 12.4)*

8. 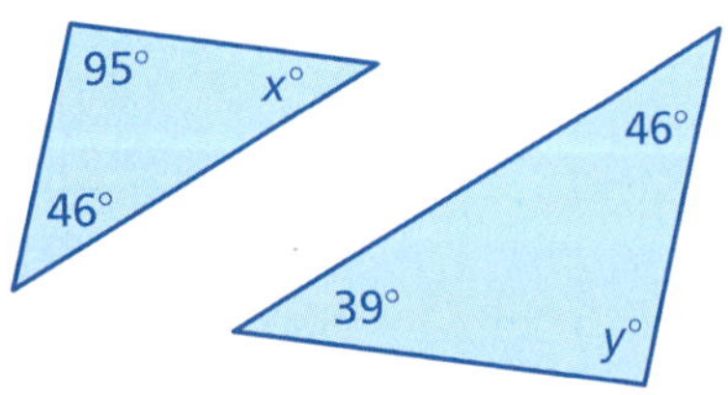

9. 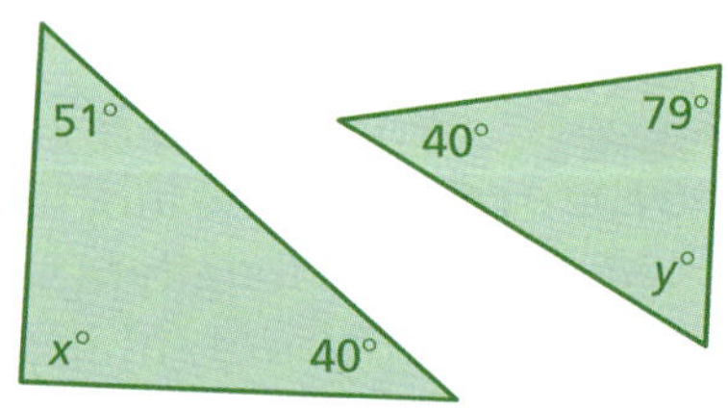

10. REASONING The sum of the interior angle measures of a polygon is 4140°. How many sides does the polygon have? *(Section 12.3)*

11. SWAMP You are trying to find the distance ℓ across a patch of swamp water. *(Section 12.4)*

a. Explain why $\triangle VWX$ and $\triangle YZX$ are similar.

b. What is the distance across the patch of swamp water?

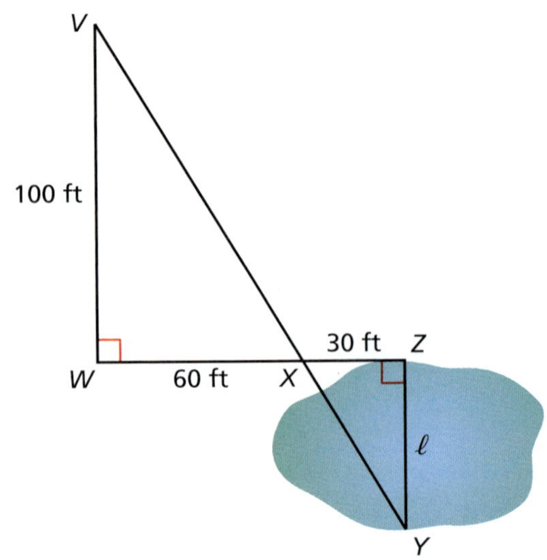

Alternative Assessment Options

Math Chat	Student Reflective Focus Question
Structured Interview	Writing Prompt

Math Chat

- Put students in pairs to complete and discuss the exercises from the quiz.
- The discussion should include terms such as sums and measures of interior angles, measures of exterior angles, and similar triangles.
- The teacher should walk around the classroom listening to the pairs and ask questions to ensure understanding.

Study Help Sample Answers

Remind students to complete Graphic Organizers for the rest of the chapter.

3. *Interior angles of a polygon*

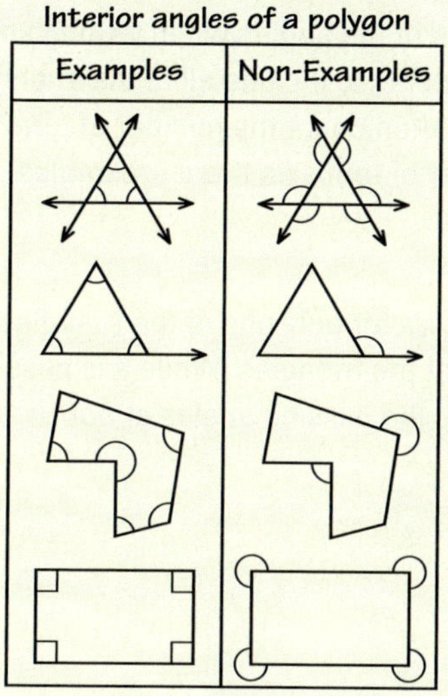

4–6. Available at *BigIdeasMath.com.*

Reteaching and Enrichment Strategies

If students need help. . .	If students got it. . .
Resources by Chapter • Practice A and Practice B • Puzzle Time Lesson Tutorials *BigIdeasMath.com*	Resources by Chapter • Enrichment and Extension • Technology Connection Game Closet at *BigIdeasMath.com* Start the Chapter Review

Answers

1. 1440°
2. 1980°
3. 58°; 122°; 134°; 46°
4. 126°; 130°; 140°; 120°; 115°; 154°; 115°
5. 280°; 40°; 110°; 70°; 40°
6. 80°; 100°; 65°; 115°
7. 60°; 120°; 90°; 90°
8. yes; The triangles have the same angle measures, 95°, 46°, and 39°.
9. no; The triangles do not have the same angle measures.
10. 25 sides
11. **a.** Angles W and Z are right angles, so they are congruent. Angles WXV and ZXY are vertical angles, so they are congruent. Because two angles in $\triangle VWX$ are congruent to two angles in $\triangle YZX$, the third angles are also congruent and the triangles are similar.

    b. 50 ft

Technology for the Teacher

Online Assessment
Assessment Book
ExamView® Assessment Suite

For the Teacher
Additional Review Options

- *BigIdeasMath.com*
- Online Assessment
- Game Closet at *BigIdeasMath.com*
- Vocabulary Help
- Resources by Chapter

Answers

1. 140°; $\angle 8$ and the given angle are alternate exterior angles.
2. 140°; $\angle 8$ and $\angle 5$ are vertical angles.
3. 40°; $\angle 8$ and $\angle 7$ are supplementary.
4. 40°; $\angle 2$ and the given angle are supplementary.

Review of Common Errors

Exercises 1–4

- Students may not understand alternate interior and exterior angles and think that an exterior angle is congruent to the alternate interior angle. Use corresponding angles to show that this is not necessarily true.

Exercise 8

- Students may forget to combine like terms when solving for *t*. Remind them that because there are two variables on the same side of the equal sign, they should start by combining like terms.
- Students may solve for the variable, but forget to find the measures of the angles. Remind them to read the directions carefully and to answer the question.

Exercises 9–11

- Students may forget to include one or more of the angles when writing an equation to find the value of *x*. Remind students to include all of the angles. Encourage them to write the equation and then count the number of terms to make sure that there is the same number of terms as there are angles before solving.

Exercises 14 and 15

- Students may find the missing angle measure for only one of the triangles and then make a decision about the similarity of the triangles. While it is possible to use this method, encourage them to find the missing angles of *both* triangles to verify that they are correct.

12 Chapter Review

Check It Out
Vocabulary Help
BigIdeasMath.com

Review Key Vocabulary

transversal, *p. 528*
interior angles, *p. 529*
exterior angles, *p. 529*
interior angles of a polygon, *p. 536*
exterior angles of a polygon, *p. 536*
convex polygon, *p. 543*
concave polygon, *p. 543*
regular polygon, *p. 545*
indirect measurement, *p. 553*

Review Examples and Exercises

12.1 Parallel Lines and Transversals *(pp. 526–533)*

Use the figure to find the measure of $\angle 6$.

$\angle 2$ and the 55° angle are supplementary.
So, the measure of $\angle 2$ is $180° - 55° = 125°$.

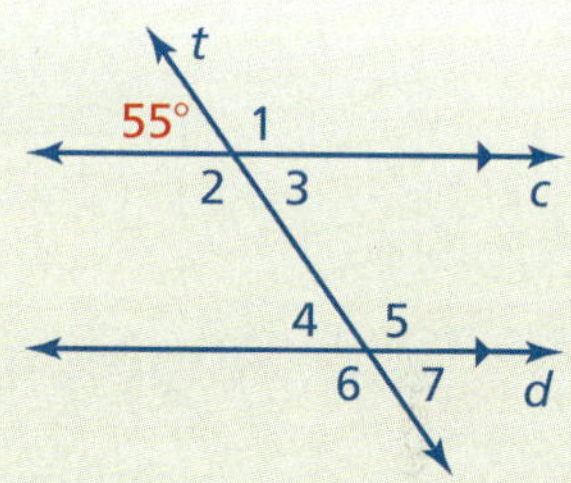

$\angle 2$ and $\angle 6$ are corresponding angles.
They are congruent.

∴ So, the measure of $\angle 6$ is 125°.

Exercises

Use the figure to find the measure of the angle. Explain your reasoning.

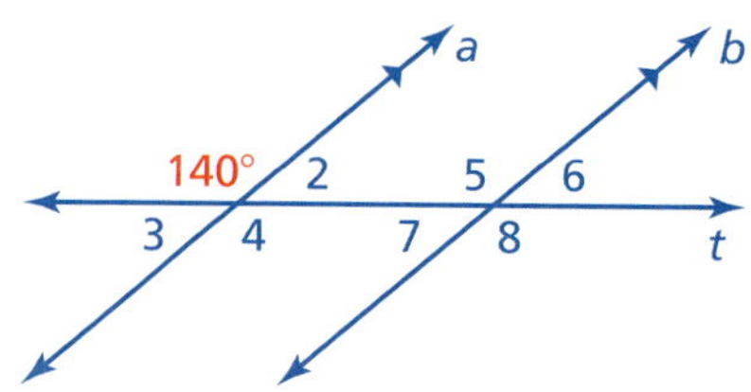

1. $\angle 8$
2. $\angle 5$
3. $\angle 7$
4. $\angle 2$

12.2 Angles of Triangles *(pp. 534–539)*

a. Find the value of x.

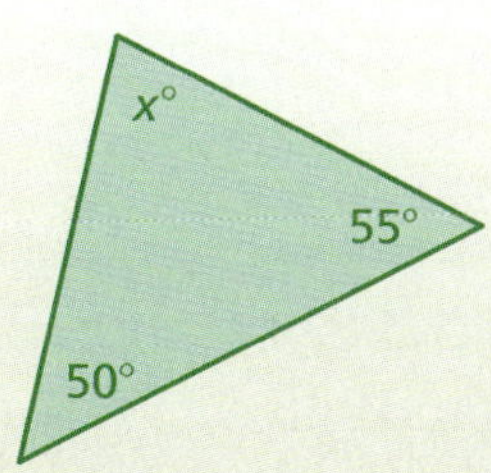

$$x + 50 + 55 = 180$$
$$x + 105 = 180$$
$$x = 75$$

∴ The value of x is 75.

b. Find the measure of the exterior angle.

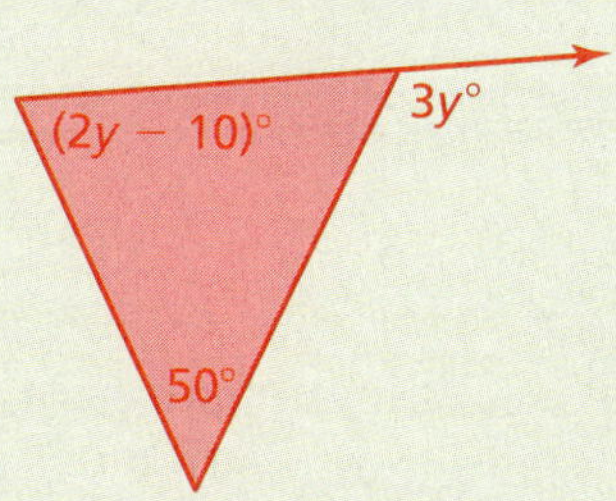

$$3y = (2y - 10) + 50$$
$$3y = 2y + 40$$
$$y = 40$$

∴ So, the measure of the exterior angle is $3(40)° = 120°$.

Exercises

Find the measures of the interior angles.

5.

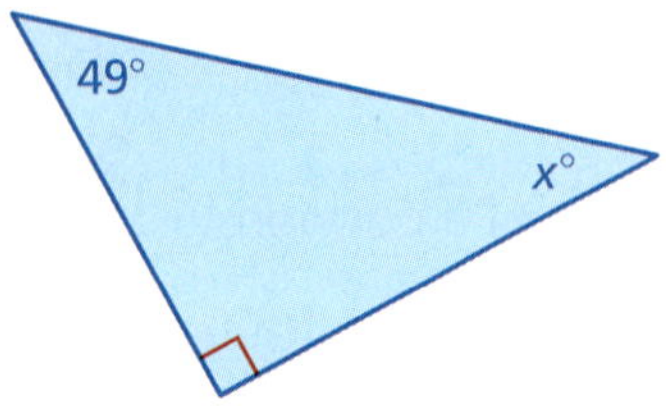

6.

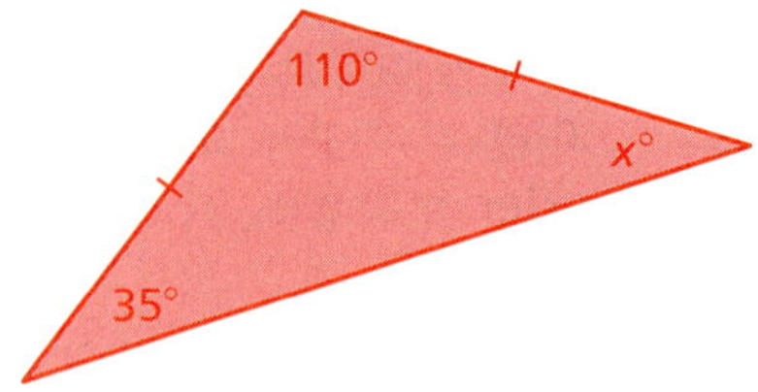

Find the measure of the exterior angle.

7.

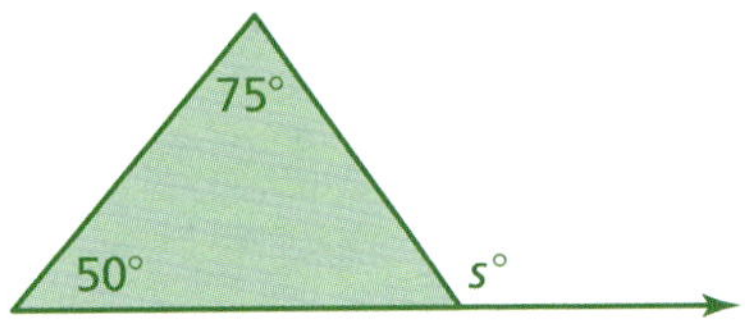

8.

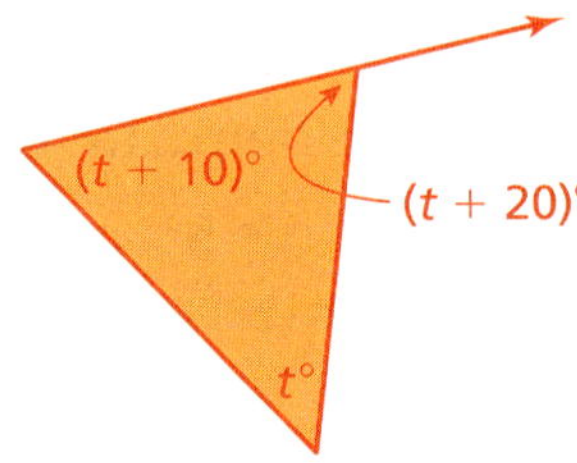

12.3 Angles of Polygons *(pp. 542–549)*

a. Find the value of x.

Step 1: The polygon has 6 sides. Find the sum of the interior angle measures.

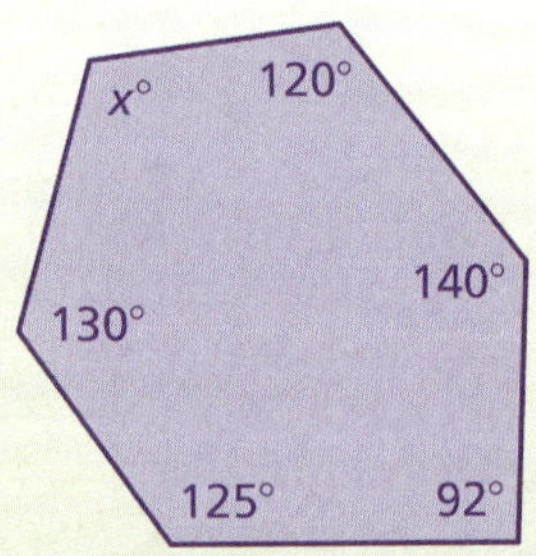

$S = (n - 2) \cdot 180°$ — Write the formula.

$= (6 - 2) \cdot 180°$ — Substitute 6 for n.

$= 720$ — Simplify. The sum of the interior angle measures is 720°.

Step 2: Write and solve an equation.

$$130 + 125 + 92 + 140 + 120 + x = 720$$
$$607 + x = 720$$
$$x = 113$$

The value of x is 113.

b. Find the measures of the exterior angles of the polygon.

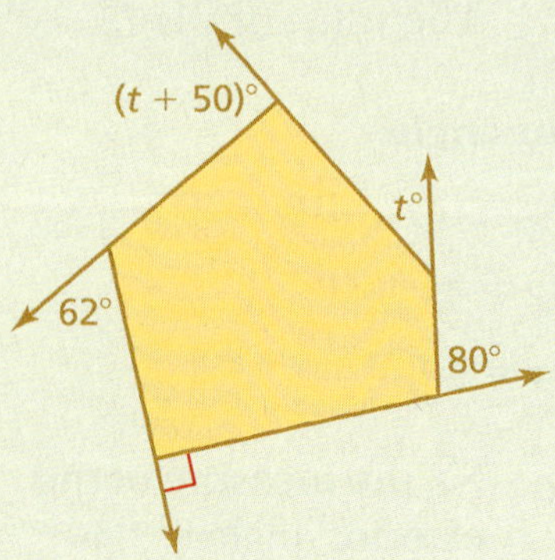

Write and solve an equation for t.

$$t + 80 + 90 + 62 + (t + 50) = 360$$
$$2t + 282 = 360$$
$$2t = 78$$
$$t = 39$$

So, the measures of the exterior angles are 39°, 80°, 90°, 62°, and $(39 + 50)° = 89°$.

Review Game

Finding Angle Measures

Materials per Group

- deck of playing cards
- paper
- pencil
- stopwatch

Directions

Divide the class into equally sized groups. A group member lays down two cards next to each other, and below this pair lays down another two cards next to each other. Then they multiply the values of the cards in each pair. (Count kings, queens, jacks, and aces as 10.) These are used to represent the measures of two angles of a triangle. The group member then finds the angle measure of the third angle. Other members time the one working and make sure the computed angle is correct. Each group member takes a turn going through the deck as fast as he or she can. If there is a combination that is impossible to use, they must identify this and move on.

Who wins?

The fastest member in a group after 2 rounds competes against the fastest members in the other groups. The winner is the fastest student.

For the Student Additional Practice

- Lesson Tutorials
- Multi-Language Glossary
- Self-Grading Progress Check
- *BigIdeasMath.com*
  Dynamic Student Edition
  Student Resources

Answers

5. 41°, 49°, 90°

6. 35°, 35°, 110°

7. 125°

8. 110°

9. 77°, 60°, 128°, 95°

10. 110°, 135°, 125°, 135°, 105°, 150°, 140°

11. 125°, 100°, 120°, 60°, 250°, 65°

12. 135°, 100°, 125°

13. 60°, 60°, 60°, 60°, 60°, 60°

14. yes; The triangles have the same angle measures, 90°, 68°, and 22°.

15. yes; The triangles have two pairs of congruent angles.

My Thoughts on the Chapter

What worked. . .

Teacher Tip

Not allowed to write in your teaching edition? Use sticky notes to record your thoughts.

What did not work. . .

What I would do differently. . .

Exercises

Find the measures of the interior angles of the polygon.

9.

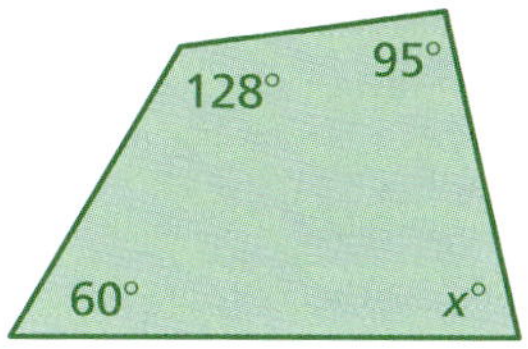

10.

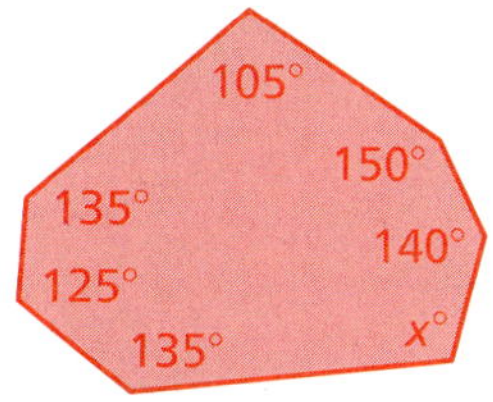

11.

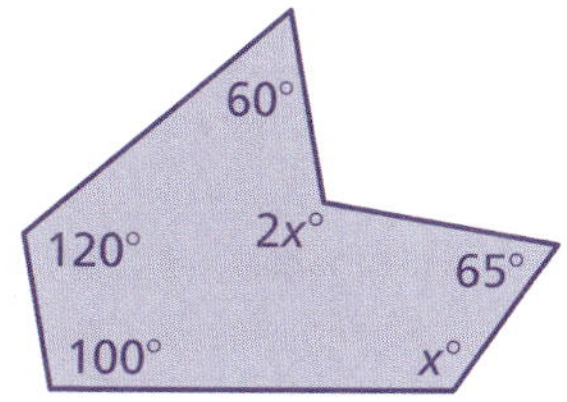

Find the measures of the exterior angles of the polygon.

12.

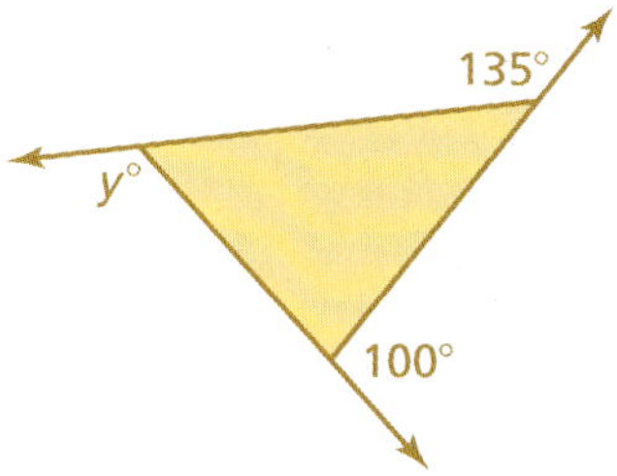

13.

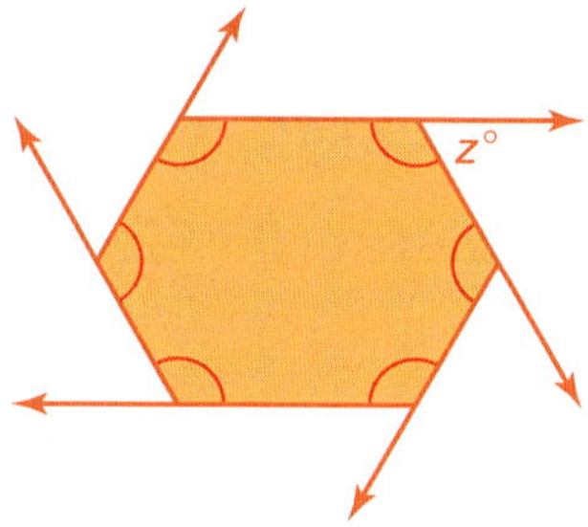

12.4 Using Similar Triangles *(pp. 550–555)*

Tell whether the triangles are similar. Explain.

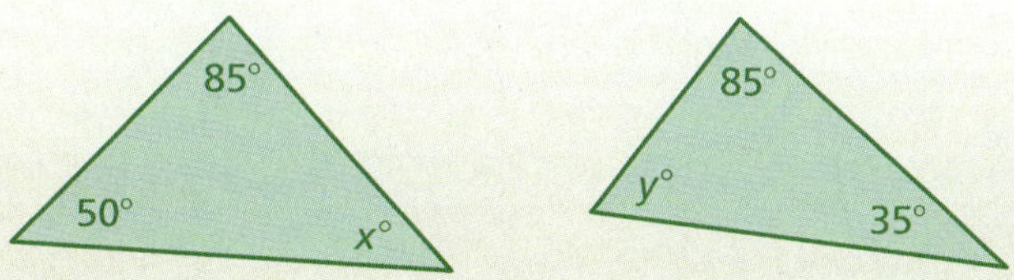

Write and solve an equation to find x.

$$50 + 85 + x = 180$$

$$135 + x = 180$$

$$x = 45$$

∴ The triangles do not have two pairs of congruent angles. So, the triangles are not similar.

Exercises

Tell whether the triangles are similar. Explain.

14.

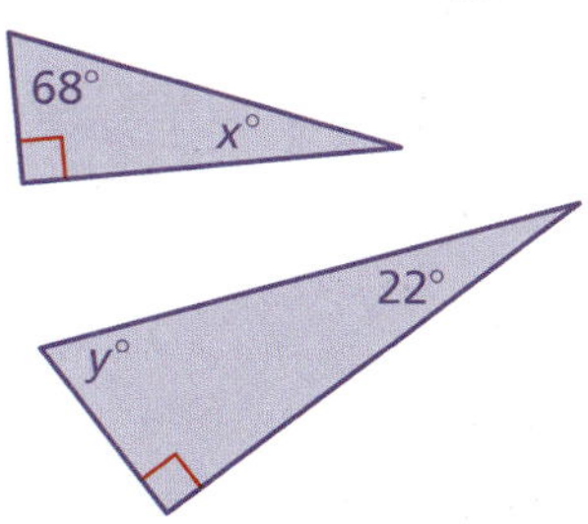

15.

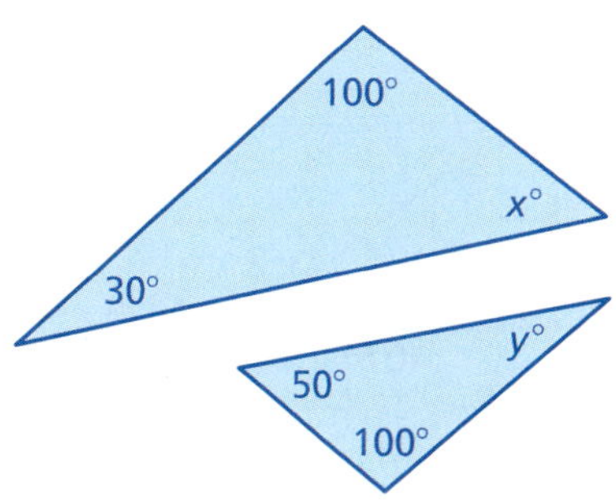

12 Chapter Test

Check It Out
Test Practice
BigIdeasMath.com

Use the figure to find the measure of the angle. Explain your reasoning.

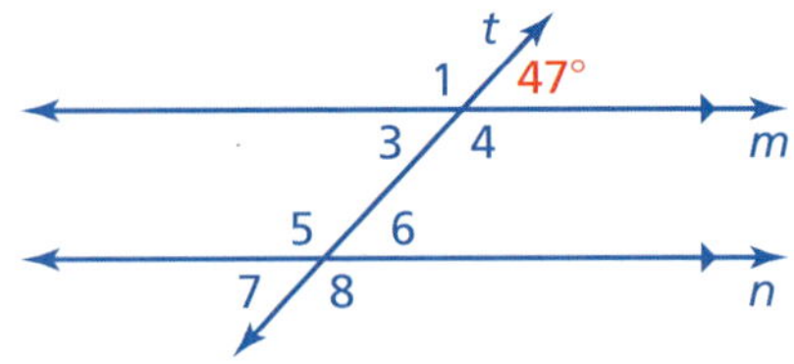

1. $\angle 1$ **2.** $\angle 8$

3. $\angle 4$ **4.** $\angle 5$

Find the measures of the interior angles.

5.

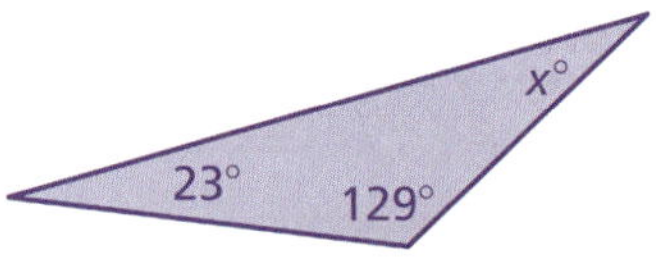

6.

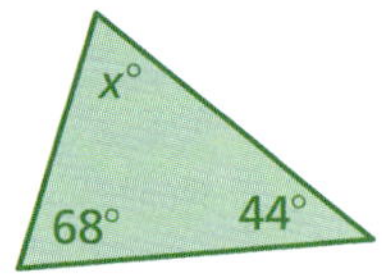

7.

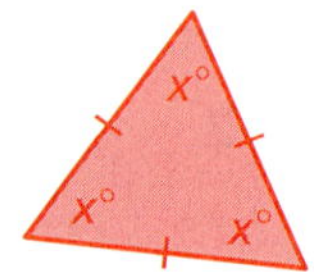

Find the measure of the exterior angle.

8.

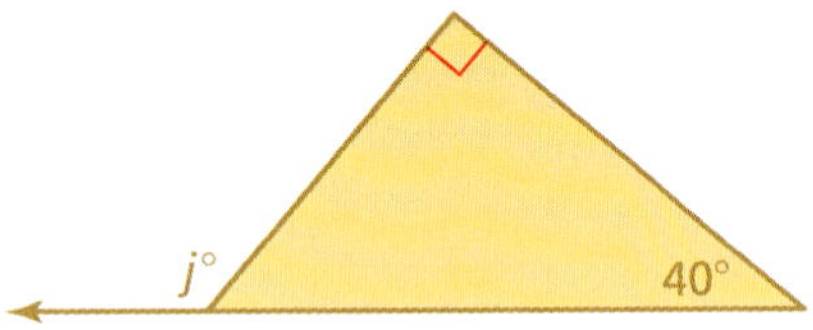

9.

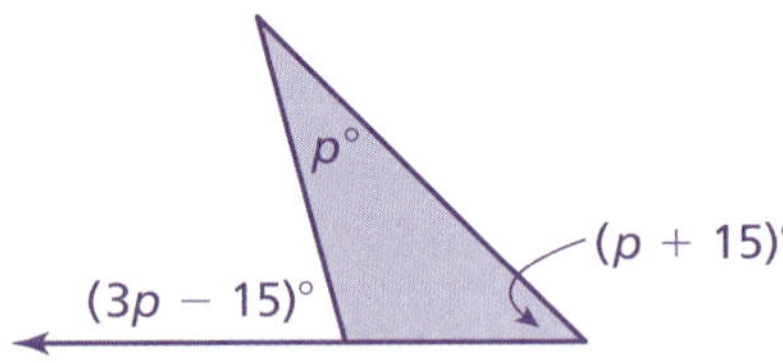

10. Find the measures of the interior angles of the polygon.

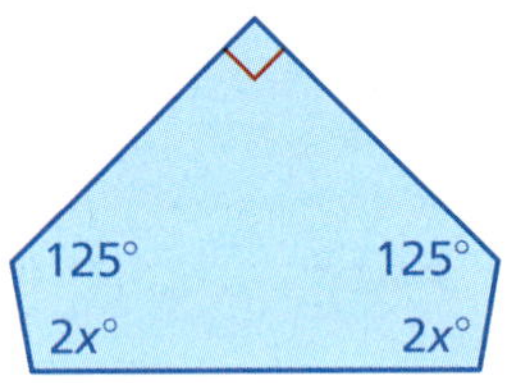

11. Find the measures of the exterior angles of the polygon.

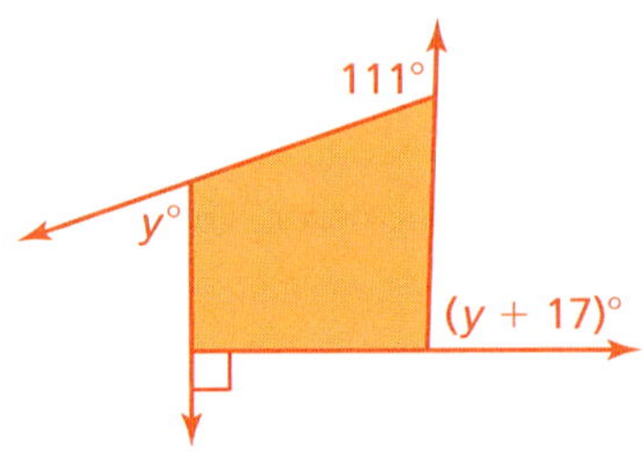

Tell whether the triangles are similar. Explain.

12.

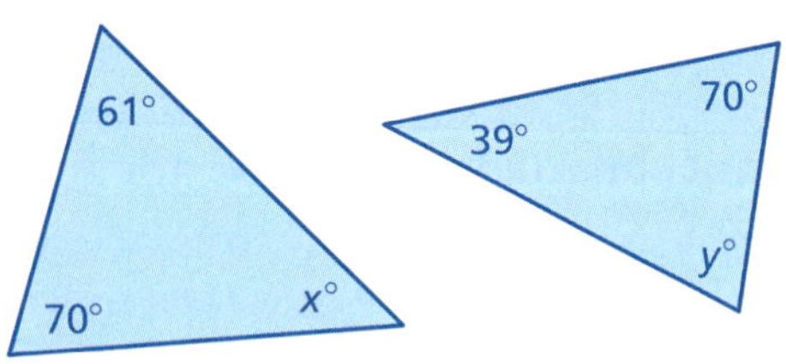

13.

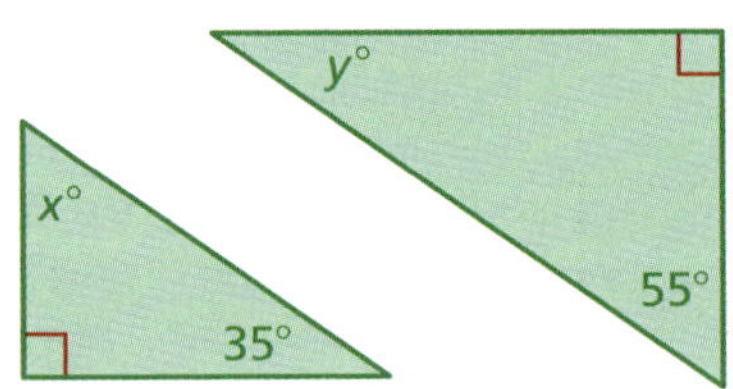

14. WRITING Describe two ways you can find the measure of $\angle 5$.

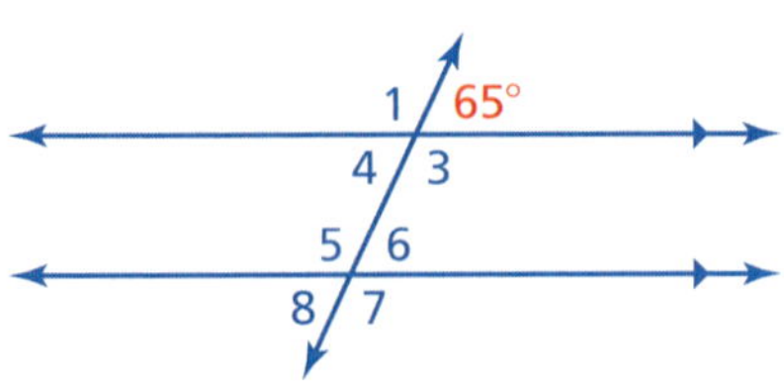

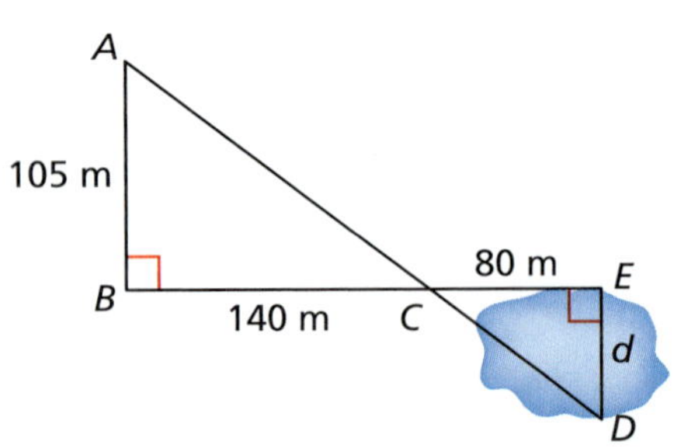

15. POND Use the given measurements to find the distance d across the pond.

Test Item References

Chapter Test Questions	Section to Review
1–4, 14	12.1
5–9	12.2
10, 11	12.3
12, 13, 15	12.4

Test-Taking Strategies

Remind students to quickly look over the entire test before they start so that they can budget their time. Students should jot down the formula for the sum of interior angles of a polygon on the back of the test before they begin. Students need to use the **Stop** and **Think** strategy before answering questions.

Common Errors

- **Exercises 9–11** Students may forget to combine like terms when solving for *x*. Remind them that because there are two variables on the same side of the equal sign, they should start by combining like terms.
- **Exercises 9–11** Students may solve for the variable, but forget to find the measures of the angles. Remind them to read the directions carefully and to answer the question.
- **Exercise 10** Students may forget to include one or more of the angles when writing an equation to find the value of *x*. Remind them to include all the angles. Encourage students to write the equation and then count the number of terms to make sure that there is the same number of terms as there are angles before solving.
- **Exercises 12 and 13** Students may find only one missing angle measure and then make a decision about the similarity of the triangles. While it is possible to use this method, encourage them to find *both* missing angle measures to verify their answers.

Reteaching and Enrichment Strategies

If students need help...	If students got it...
Resources by Chapter • Practice A and Practice B • Puzzle Time Record and Practice Journal Practice Differentiating the Lesson Lesson Tutorials *BigIdeasMath.com* Skills Review Handbook	Resources by Chapter • Enrichment and Extension • Technology Connection Game Closet at *BigIdeasMath.com* Start Cumulative Assessment

Answers

1. 133°; $\angle 1$ and the given angle are supplementary.
2. 133°; $\angle 8$ and $\angle 1$ are alternate exterior angles.
3. 133°; $\angle 1$ and $\angle 4$ are vertical angles.
4. 133°; $\angle 4$ and $\angle 5$ are alternate interior angles.
5. 28°, 129°, 23°
6. 68°, 68°, 44°
7. 60°, 60°, 60°
8. 130°
9. 75°
10. 90°, 125°, 100°, 100°, 125°
11. 71°, 111°, 88°, 90°
12. no; The triangles do not have the same angle measures.
13. yes; The two triangles have two pairs of congruent angles.
14. *Sample answer:*

    1) The given angle and $\angle 3$ are supplementary, so $\angle 3 = 115°$; $\angle 3$ and $\angle 5$ are alternate interior angles, so $\angle 3 = \angle 5 = 115°$.

    2) The given angle and $\angle 8$ are alternate exterior angles, so $\angle 8 = 65°$; $\angle 5$ and $\angle 8$ are supplementary, so $\angle 5 = 115°$.
15. 60 m

Technology for the Teacher

Online Assessment
Assessment Book
ExamView® Assessment Suite

Test-Taking Strategies

Available at *BigIdeasMath.com*

After Answering Easy Questions, Relax
Answer Easy Questions First
Estimate the Answer
Read All Choices before Answering
Read Question before Answering
Solve Directly or Eliminate Choices
Solve Problem before Looking at Choices
Use Intelligent Guessing
Work Backwards

About this Strategy

When taking a multiple choice test, be sure to read each question carefully and thoroughly. Sometimes it is easier to solve the problem and then look for the answer among the choices.

Answers

1. 147°
2. B
3. I
4. C

Technology for the Teacher

Performance Tasks
Online Assessment
Assessment Book
ExamView® Assessment Suite

Item Analysis

1. **Gridded Response:** Correct answer: 147°

   Common Error: The student might divide 180 by 11.

2. **A.** The student adds 11 and 1.6 together before dividing.

   B. Correct answer

   C. The student divides first and then subtracts.

   D. The student subtracts 1.6 instead of dividing.

3. **F.** The student subtracts 20 from the right side instead of adding 20.

   G. The student did not distribute the 5 to both terms of $(x - 4)$.

   H. The student adds $3x$ to the left side instead of subtracting $3x$.

   I. Correct answer

4. **A.** The student does not correctly match corresponding side lengths, instead using the proportion $\frac{PQ}{QR} = \frac{US}{ST}$.

   B. The student does not correctly match corresponding side lengths, instead using the proportion $\frac{PQ}{QR} = \frac{TU}{ST}$.

   C. Correct answer

   D. The student does not correctly match corresponding side lengths, instead using the proportion $\frac{PQ}{QR} = \frac{ST}{US}$.

12 Cumulative Assessment

Test-Taking Strategy

Solve Problem Before Looking at Choices

"Solve the problem before looking at the choices. You know 180 − 2(70) = 40. So, the answer is C."

1. The border of a Canadian one-dollar coin is shaped like an 11-sided regular polygon. The shape was chosen to help visually impaired people identify the coin. How many degrees are in each angle along the border? Round your answer to the nearest degree.

2. A public utility charges its residential customers for natural gas based on the number of therms used each month. The formula below shows how the monthly cost C in dollars is related to the number t of therms used.

$$C = 11 + 1.6t$$

Solve this formula for t.

A. $t = \dfrac{C}{12.6}$

B. $t = \dfrac{C - 11}{1.6}$

C. $t = \dfrac{C}{1.6} - 11$

D. $t = C - 12.6$

3. What is the value of x?

$$5(x - 4) = 3x$$

F. -10

G. 2

H. $2\frac{1}{2}$

I. 10

4. In the figures below, $\triangle PQR$ and $\triangle STU$ are similar.

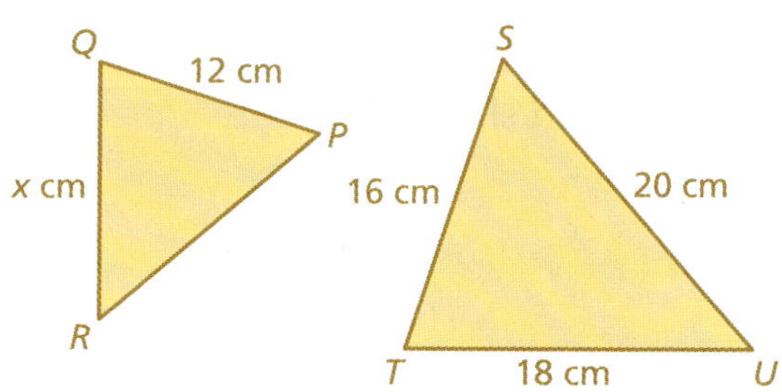

What is the value of x?

A. 9.6

B. $10\frac{2}{3}$

C. 13.5

D. 15

5. What is the value of x?

6. Olga was solving an equation in the box shown.

$$-\frac{2}{5}(10x - 15) = -30$$
$$10x - 15 = -30\left(-\frac{2}{5}\right)$$
$$10x - 15 = 12$$
$$10x - 15 + 15 = 12 + 15$$
$$10x = 27$$
$$\frac{10x}{10} = \frac{27}{10}$$
$$x = \frac{27}{10}$$

What should Olga do to correct the error that she made?

F. Multiply both sides by $-\frac{5}{2}$ instead of $-\frac{2}{5}$.

G. Multiply both sides by $\frac{2}{5}$ instead of $-\frac{2}{5}$.

H. Distribute $-\frac{2}{5}$ to get $-4x - 6$.

I. Add 15 to -30.

Item Analysis (continued)

5. **Gridded Response:** Correct answer: 55

   Common Error: The student thinks the angles are congruent.

6. **F.** Correct answer

   G. The student thinks that multiplying by $\frac{2}{5}$ is the inverse operation of multiplying by $-\frac{2}{5}$.

   H. The student does not distribute the negative sign to the second term.

   I. The student makes an order of operations error by not first distributing the multiplication.

Answers

5. 55

6. F

Answers

7. B

8. *Part A* $S = (n - 2) \cdot 180$

Part B 80°

Part C
The sum of the angle measures of a triangle is 180 degrees. Because the pentagon can be divided into three triangles, the sum of the angle measures of a pentagon is

$180 + 180 + 180 = 540$ or

$(n - 2) \cdot 180 =$

$(5 - 2) \cdot 180 =$

$3 \cdot 180 = 540.$

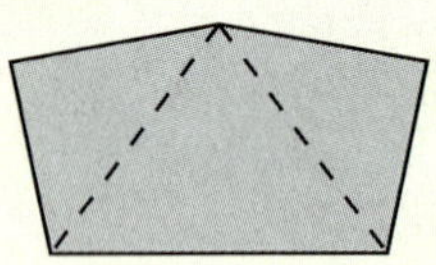

Item Analysis (continued)

7. **A.** The student reflects the figure in the *x*-axis.

B. Correct answer

C. The student translates the figure 7 units right.

D. The student translates the figure 6 units up.

8. **4 points** The student demonstrates a thorough understanding of writing and applying the angle sum formula for polygons, as well as how it relates to the fact that there are 180 degrees in a triangle. The student's work in Part B shows step-by-step how the fourth angle measures 80 degrees. The student's explanation in part C makes the algebraic-geometric connection clear.

3 points The student demonstrates an essential but less than thorough understanding. In particular, Parts A and B should be completed fully and clearly, but Part C may lack full explanation of the algebraic-geometric connection.

2 points The student's work and explanations demonstrate a lack of essential understanding. The formula in Part A should be properly stated, but Part B may show an error in application. Part C lacks any explanation.

1 point The student demonstrates limited understanding. The student's response is incomplete and exhibits many flaws, including, but not limited to, the inability to state the proper formula in Part A.

0 points The student provides no response, a completely incorrect or incomprehensible response, or a response that demonstrates insufficient understanding of writing, applying, and understanding the angle sum formula for polygons.

7. In the coordinate plane below, $\triangle XYZ$ is plotted and its vertices are labeled.

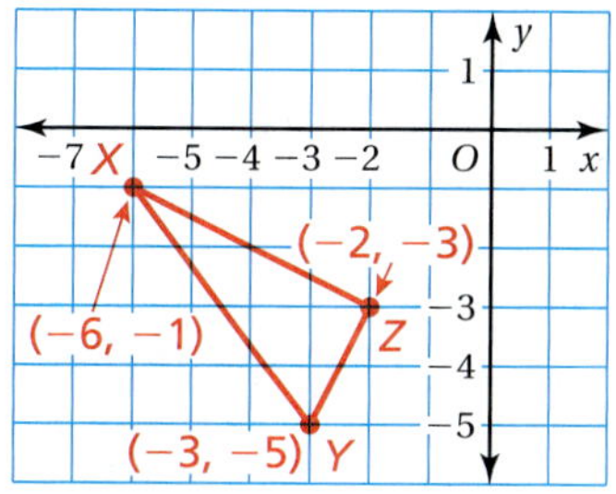

Which of the following shows $\triangle X'Y'Z'$, the image of $\triangle XYZ$ after it is reflected in the y-axis?

A.

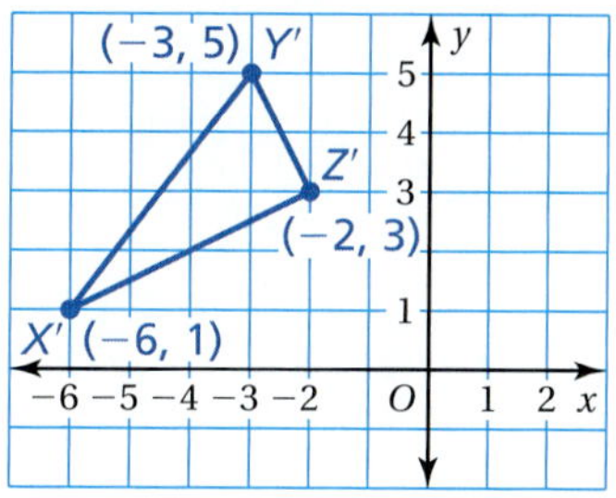

C.

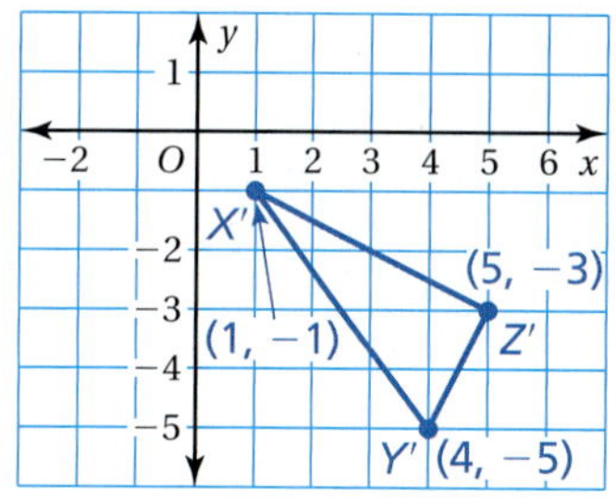

B.

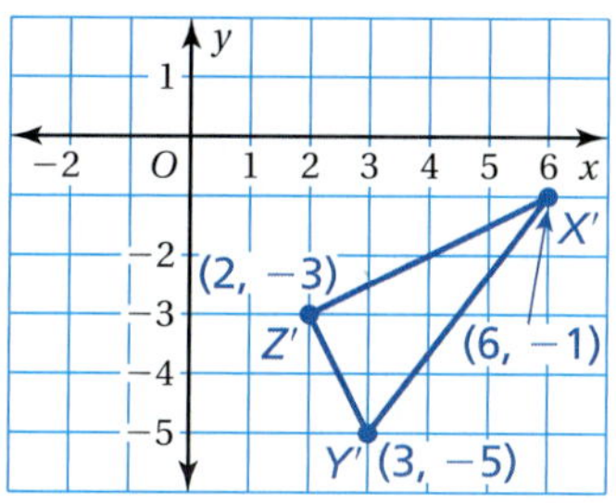

D.

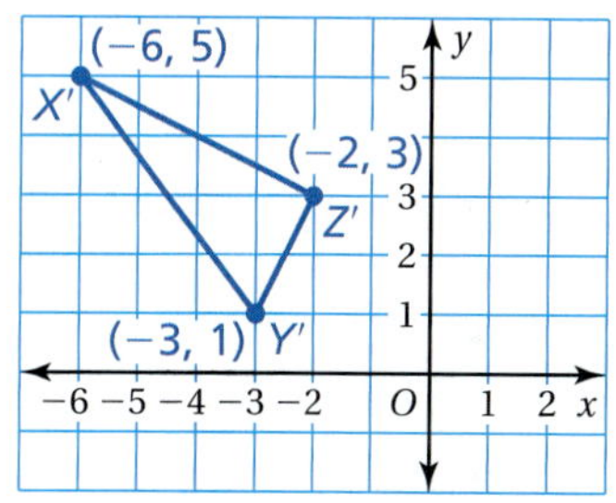

8. The sum S of the interior angle measures of a polygon with n sides can be found by using a formula.

Think Solve Explain

Part A Write the formula.

Part B A quadrilateral has angles measuring 100°, 90°, and 90°. Find the measure of its fourth angle. Show your work and explain your reasoning.

Part C The sum of the measures of the angles of the pentagon shown is 540°. Divide the pentagon into triangles to show why this must be true. Show your work and explain your reasoning.

13 Graphing and Writing Linear Equations

"Okay Descartes, stand on the *y*-axis and try to intercept the pass when I throw."

"Here's an easy example of a line with a slope of 1."

"You eat one mouse treat the first day. Two treats the second day. And so on. Get it?"

What Your Students Have Learned

- Make tables of equivalent ratios and plot the pairs of values in a coordinate plane.
- Write an equation in two variables and analyze the relationship between the independent and dependent variables using graphs and tables.
- Identify the constant of proportionality (unit rate) in tables, graphs, equations, diagrams, and verbal descriptions.
- Represent proportional relationships with equations.

What Your Students Will Learn

- Understand that lines represent solutions of linear equations.
- Use similar triangles to explain why the slope is the same between any two points on a line.
- Identify parallel and perpendicular lines.
- Graph proportional relationships, interpreting the unit rate as the slope.
- Compare proportional relationships represented in different ways.
- Write and graph linear equations in slope-intercept, standard, and point-slope form.
- Derive $y = mx$ and $y = mx + b$.

Pacing Guide for Chapter 13

Chapter Opener Accelerated	1 Day
Section 1 Accelerated	1 Day
Section 2 Accelerated	2 Days
Section 3 Accelerated	1 Day
Study Help / Quiz Accelerated	1 Day
Section 4 Accelerated	1 Day
Section 5 Accelerated	1 Day
Section 6 Accelerated	1 Day
Section 7 Accelerated	1 Day
Chapter Review/ Chapter Tests Accelerated	2 Days
Total Chapter 13 Accelerated	12 Days
Year-to-Date Accelerated	124 Days

Technology for the Teacher

BigIdeasMath.com
Chapter at a Glance
Complete Materials List
Parent Letters: English and Spanish

What Your Students Have Learned

- Evaluate expressions using substitution and the order of operations.
- Identify points and write ordered pairs in all four quadrants of the coordinate plane.

Additional Topics for Review

- Order of Operations
- Exponents
- Similar Triangles
- Parallel and Perpendicular Lines
- Proportional Relationships
- Slope
- Unit Rate

Try It Yourself

1. −12 **2.** −23

3. 15 **4.** $4\frac{3}{4}$

5. (0, 4) **6.** (4, 2)

7. Point *R* **8.** Point *N*

Record and Practice Journal Fair Game Review

1. 5 **2.** 16

3. −5 **4.** $-38\frac{1}{2}$

5. 108 **6.** 65

7. $-3\frac{7}{19}$ **8.** 262

9. \$50.00 **10.** (−5, 0)

11. (3, −5) **12.** Point *F*

13. Point *G*

14. Point *B*, Point *H*

15. Point *C*, Point *E*

16–20. See Additional Answers.

Math Background Notes

Vocabulary Review

- Evaluate
- Expression
- Order of Operations
- Substitute
- Coordinates

Evaluating Expressions Using Order of Operations

- Students should know how to substitute values into algebraic expressions and evaluate the results using order of operations.
- **Teaching Tip:** Sometimes color coding substitutions can help students to evaluate expressions. Each time you want to substitute a number in place of a variable, you must substitute your lead pencil for a colored pencil.
- Remind students that after they substitute values in for the variables, they must use the correct order of operations to continue evaluating the expression.
- **Common Error:** Encourage students to use a set of parentheses whenever they do a substitution. This will help students distinguish between subtraction and multiplication.

Plotting Points

- Students should know how to plot points in all four quadrants.
- **Common Error:** Students may write the coordinates backwards. Remind them that coordinates are written in alphabetical order with the *x* move (horizontal) written before the *y* move (vertical).
- **Common Error:** Students may also have difficulty with the negative numbers associated with plotting outside Quadrant I. Remind them that the negatives are directional. A negative *x*-coordinate communicates a move to the left of the origin and a negative *y*-coordinate communicates a move downward from the origin.

Reteaching and Enrichment Strategies

If students need help...	If students got it...
Record and Practice Journal • Fair Game Review Skills Review Handbook Lesson Tutorials	Game Closet at *BigIdeasMath.com* Start the next section

What You Learned Before

"I estimate that we are on a slope of about −0.625. What do you think?"

Evaluating Expressions Using Order of Operations

Example 1 **Evaluate $2xy + 3(x + y)$ when $x = 4$ and $y = 7$.**

$2xy + 3(x + y) = 2(4)(7) + 3(4 + 7)$	Substitute 4 for x and 7 for y.
$= 8(7) + 3(4 + 7)$	Use order of operations.
$= 56 + 3(11)$	Simplify.
$= 56 + 33$	Multiply.
$= 89$	Add.

Try It Yourself

Evaluate the expression when $a = \frac{1}{4}$ and $b = 6$.

1. $-8ab$

2. $16a^2 - 4b$

3. $\frac{5b}{32a^2}$

4. $12a + (b - a - 4)$

Plotting Points

Example 2 **Write the ordered pair that corresponds to point U.**

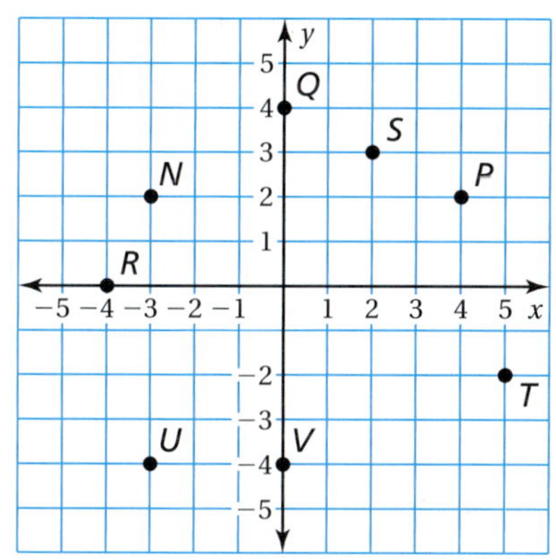

Point U is 3 units to the left of the origin and 4 units down. So, the x-coordinate is -3, and the y-coordinate is -4.

∴ The ordered pair $(-3, -4)$ corresponds to point U.

Example 3 **Which point is located at $(5, -2)$?**

Start at the origin. Move 5 units right and 2 units down.

∴ Point T is located at $(5, -2)$.

Try It Yourself

Use the graph to answer the question.

5. Write the ordered pair that corresponds to point Q.

6. Write the ordered pair that corresponds to point P.

7. Which point is located at $(-4, 0)$?

8. Which point is located in Quadrant II?

13.1 Graphing Linear Equations

Essential Question How can you recognize a linear equation? How can you draw its graph?

1 ACTIVITY: Graphing a Linear Equation

Work with a partner.

a. Use the equation $y = \frac{1}{2}x + 1$ to complete the table. (Choose any two x-values and find the y-values.)

	Solution Points	
x		
$y = \frac{1}{2}x + 1$		

b. Write the two ordered pairs given by the table. These are called *solution points* of the equation.

c. **PRECISION** Plot the two solution points. Draw a line *exactly* through the two points.

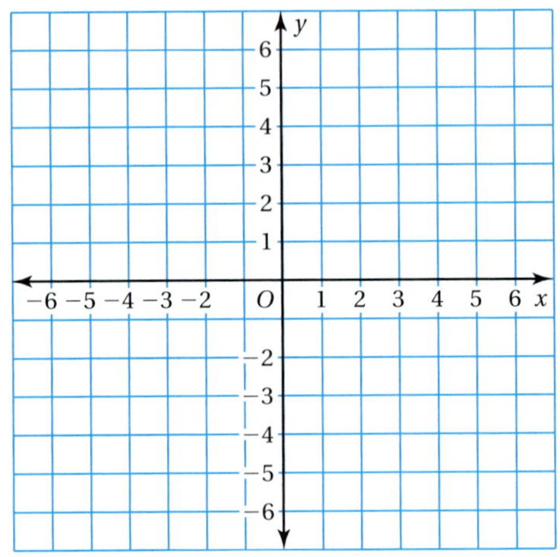

d. Find a different point on the line. Check that this point is a solution point of the equation $y = \frac{1}{2}x + 1$.

e. **LOGIC** Do you think it is true that *any* point on the line is a solution point of the equation $y = \frac{1}{2}x + 1$? Explain.

f. Choose five additional x-values for the table. (Choose positive and negative x-values.) Plot the five corresponding solution points. Does each point lie on the line?

	Solution Points				
x					
$y = \frac{1}{2}x + 1$					

g. **LOGIC** Do you think it is true that *any* solution point of the equation $y = \frac{1}{2}x + 1$ is a point on the line? Explain.

h. Why do you think $y = ax + b$ is called a *linear equation*?

Graphing Equations
In this lesson, you will
- understand that lines represent solutions of linear equations.
- graph linear equations.

Laurie's Notes

Introduction

Applying Mathematical Practices

- **Model with Mathematics:** The goal is for students to use a table of values and a graph to model a linear equation. Using multiple representations of linear equations deepens students' understanding and supports learning.
- Throughout this chapter, you may encounter applications that show a graph of discrete data with a line through the points. At this point in the text, we think it is acceptable for students to draw a line through discrete data points to help them solve an exercise. They will determine whether non-integer values or negative values are valid in various contexts. However, the terms discrete and continuous are not used at this time.

Motivate

- Play a game of coordinate BINGO.
- Distribute small coordinate grids to students. They should plot ten ordered pairs, where the x- and y-coordinates are integers between -4 and 4.
- Generate a random ordered pair in the grid. Write the integers from -4 to 4 on slips of paper and place them in a bag. Draw and replace two slips of paper to generate the ordered pair, then write it on the board.
- Each time you record a new ordered pair, the students check to see if it is one of their 10 ordered pairs. If it is, they put an X over the point on their grids. The goal is to be the first person with three Xs. A student who calls "BINGO" reads the three ordered pairs for you to check against the master list.

? "Are there ordered pairs that are not on lattice points, meaning the x- or y-coordinate is not an integer? Explain." yes; It's possible for the ordered pair to be $\left(3.5, \frac{1}{2}\right)$. Plot whatever examples students give.

- Remind students that the ordered pairs are always (x, y), where x is the horizontal direction and y is the vertical direction.

Activity Notes

Activity 1

- Some students will recognize right away that if they substitute an even number for x, the y-coordinate will not be a fraction. It is likely that students will only try positive x-values. Encourage them to try negative values for x.
- In part (d), suggest that students consider only those ordered pairs that appear to be lattice points.
- **Construct Viable Arguments:** Listen and discuss student responses to the generalizations in parts (e) and (g).
- **Big Idea:** The goal of this activity is for students to recognize and understand two related, but different, ideas. 1) *All* solution points of a linear equation lie on the same line. 2) *All* points on the line are solution points of the equation.

What Your Students Will Learn

- Understand that lines represent solutions of linear equations.
- Graph linear equations using a table of values.

Previous Learning

Students should know how to plot ordered pairs from a table.

Technology for the Teacher

Dynamic Classroom

Lesson Plans
Complete Materials List

13.1 Record and Practice Journal

Essential Question How can you recognize a linear equation? How can you draw its graph?

1 ACTIVITY: Graphing a Linear Equation

Work with a partner.

Sample answer:

a. Use the equation $y = \frac{1}{2}x + 1$ to complete the table. (Choose any two x-values and find the y-values).

	Solution Points	
x	-2	2
$y = \frac{1}{2}x + 1$	0	2

b. Write the two ordered pairs given by the table. These are called *solution points* of the equation.
$(-2, 0), (2, 2)$

c. **PRECISION** Plot the two solution points. Draw a line *exactly* through the two points.

d. Find a different point on the line. Check that this point is a solution point of the equation $y = \frac{1}{2}x + 1$.
$(0, 1)$

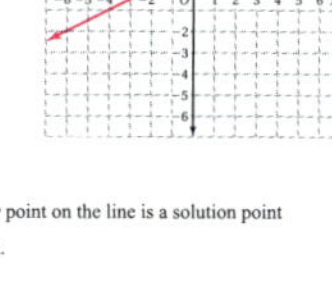

e. **LOGIC** Do you think it is true that *any* point on the line is a solution point of the equation $y = \frac{1}{2}x + 1$? Explain.
yes

Differentiated Instruction

Kinesthetic

For students that are kinesthetic learners and have difficulty in plotting points in the coordinate plane, suggest they use a finger for tracing. Have each student place a finger at the origin and trace left or right along the x-axis to the first coordinate, then trace up or down to the second coordinate. Students should also practice writing the ordered pair of a plotted point. Guide students with questions such as, "Should you move left or right? How far? Should you move up or down? How far?"

13.1 Record and Practice Journal

f. Choose five additional x-values for the table. (Choose positive and negative x-values.) Plot the five corresponding solution points on the previous page. Does each point lie on the line? *Sample answer:*

	Solution Points				
x	−6	−4	1	4	6
$y = \frac{1}{2}x + 1$	−2	−1	$1\frac{1}{2}$	3	4

g. **LOGIC** Do you think it is true that *any* solution point of the equation $y = \frac{1}{2}x + 1$ is a point on the line? Explain.

yes

h. Why do you think $y = ax + b$ is called a *linear equation*?

The graph is a line.

2 ACTIVITY: Using a Graphing Calculator

Use a graphing calculator to graph $y = 2x + 5$.

a. Enter the equation $y = 2x + 5$ into your calculator.

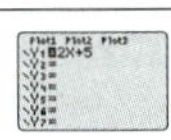

b. Check the settings of the *viewing window*. The boundaries of the graph are set by the minimum and maximum x- and y-values. The numbers of units between the tick marks are set by the x- and y-scales.

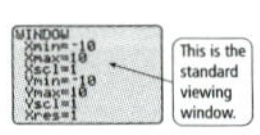

c. Graph $y = 2x + 5$ on your calculator.

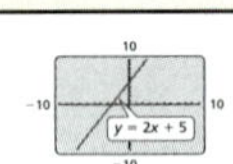

d. Change the settings of the viewing window to match those shown. Compare the two graphs.

Easier to see x- and y-intercepts on the second graph.

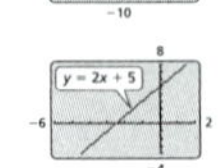

What Is Your Answer?

3. **IN YOUR OWN WORDS** How can you recognize a linear equation? How can you draw its graph? Write an equation that is linear. Write an equation that is *not* linear.

The form $y = ax + b$; The graph is a line.

4. Use a graphing calculator to graph $y = 5x - 12$ in the standard viewing window.

a. Can you tell where the line crosses the x-axis? Can you tell where the line crosses the y-axis?

yes; no

b. How can you adjust the viewing window so that you can determine where the line crosses the x- and y-axes?

Choose a lower minimum y-value.

5. **CHOOSE TOOLS** You want to graph $y = 2.5x - 3.8$. Would you graph it by hand or by using a graphing calculator? Why?

graphing calculator

Laurie's Notes

Activity 2

- This may be a student's first experience with using a graphing calculator to graph a linear equation. Explain that the calculator can graph equations that are entered in the equation editor.
- Explain how to set the *standard viewing window*, or *standard viewing rectangle*.
- Because the viewing screen is a rectangle, one unit in the x-direction appears longer than one unit in the y-direction. When graphing by hand, you generally are using a square grid. It is important to point out this distinction to students because they will eventually graph $y = x$, which is a 45° line. It will not appear this way in the standard viewing rectangle.
- ? After the graph of $y = 2x + 5$ appears, ask, "Can you name a solution of this equation from looking at the graph?" Students may name solutions such as (0, 5) or (1, 7).
- ? After the viewing rectangle changes, ask, "Did the solutions of the equation change?" The graph appears less steep; however, it is the same graph in a new view. The solutions have not changed.

What Is Your Answer?

- Discuss students' responses to the first question.
- Have students share the viewing window they selected for Question 4b.
- **Use Appropriate Tools Strategically:** This year students will be asked to graph many linear equations. In certain contexts they will need an accurate graph, while in other contexts a rough sketch will be sufficient. Students should use appropriate tools (paper and pencil versus technology) strategically.

Closure

- Find three ordered pairs that are solutions of the equation $y = 2x - 3$. Draw the graph. *Sample answer:* (−1, −5), (0, −3), and (1, −1)

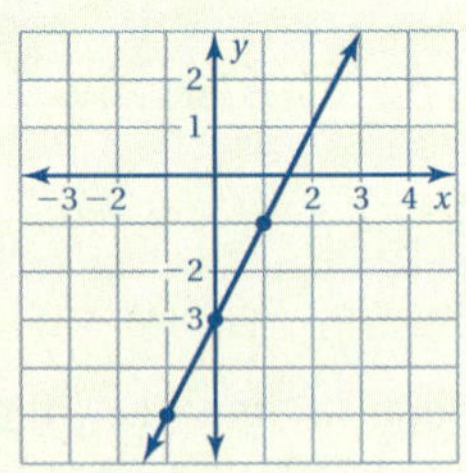

2 ACTIVITY: Using a Graphing Calculator

Use a graphing calculator to graph $y = 2x + 5$.

a. Enter the equation $y = 2x + 5$ into your calculator.

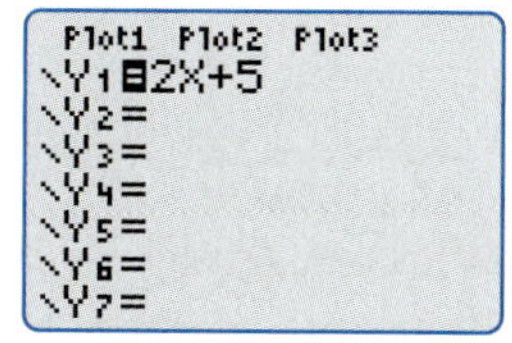

> **Math Practice**
>
> **Recognize Usefulness of Tools**
>
> What are some advantages and disadvantages of using a graphing calculator to graph a linear equation?

b. Check the settings of the *viewing window*. The boundaries of the graph are set by the minimum and the maximum x- and y-values. The numbers of units between the tick marks are set by the x- and y-scales.

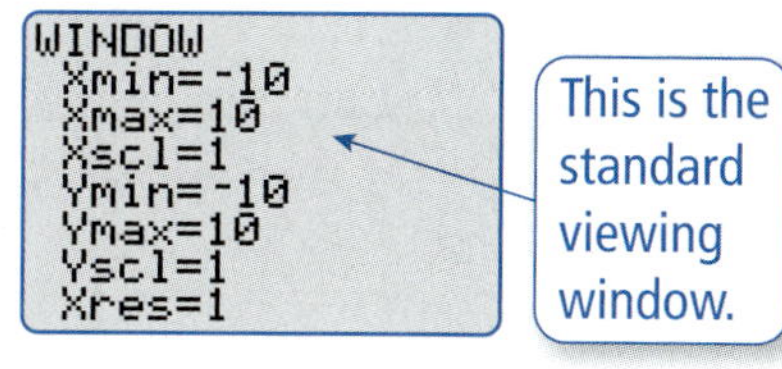

c. Graph $y = 2x + 5$ on your calculator.

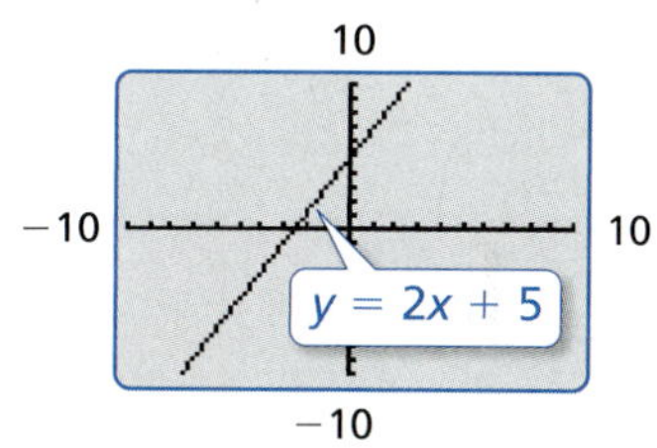

d. Change the settings of the viewing window to match those shown.

Compare the two graphs.

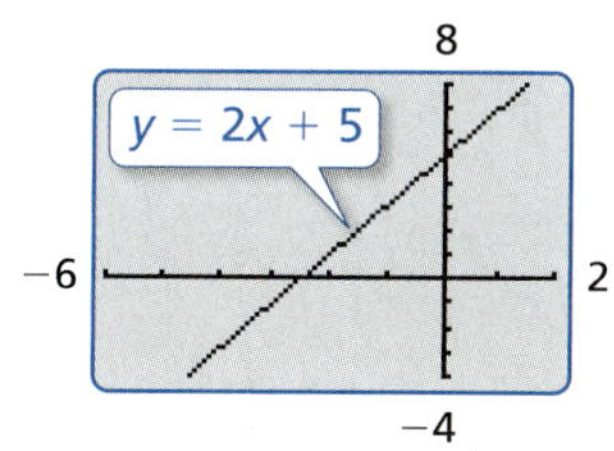

What Is Your Answer?

3. **IN YOUR OWN WORDS** How can you recognize a linear equation? How can you draw its graph? Write an equation that is linear. Write an equation that is *not* linear.

4. Use a graphing calculator to graph $y = 5x - 12$ in the standard viewing window.
   a. Can you tell where the line crosses the x-axis? Can you tell where the line crosses the y-axis?
   b. How can you adjust the viewing window so that you can determine where the line crosses the x- and y-axes?

5. **CHOOSE TOOLS** You want to graph $y = 2.5x - 3.8$. Would you graph it by hand or by using a graphing calculator? Why?

Practice Use what you learned about graphing linear equations to complete Exercises 3 and 4 on page 570.

13.1 Lesson

Check It Out
Lesson Tutorials
BigIdeasMath.com

Key Vocabulary
linear equation, *p. 568*
solution of a linear equation, *p. 568*

Remember
An ordered pair (x, y) is used to locate a point in a coordinate plane.

Key Idea

Linear Equations

A **linear equation** is an equation whose graph is a line. The points on the line are **solutions** of the equation.

You can use a graph to show the solutions of a linear equation. The graph below represents the equation $y = x + 1$.

x	y	(x, y)
-1	0	$(-1, 0)$
0	1	$(0, 1)$
2	3	$(2, 3)$

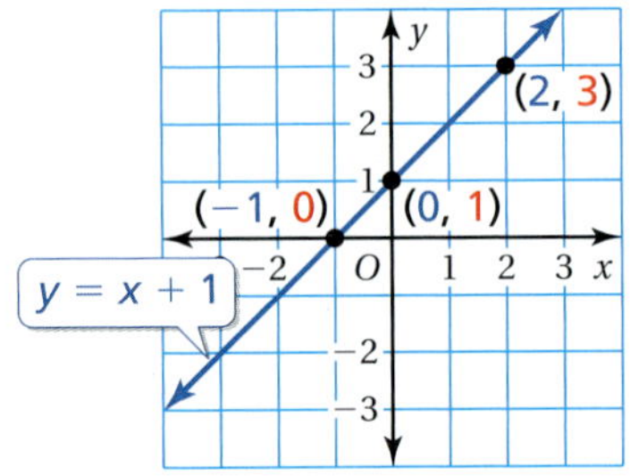

EXAMPLE 1 Graphing a Linear Equation

Graph $y = -2x + 1$.

Step 1: Make a table of values.

x	$y = -2x + 1$	y	(x, y)
-1	$y = -2(-1) + 1$	3	$(-1, 3)$
0	$y = -2(0) + 1$	1	$(0, 1)$
2	$y = -2(2) + 1$	-3	$(2, -3)$

Step 2: Plot the ordered pairs.

Step 3: Draw a line through the points.

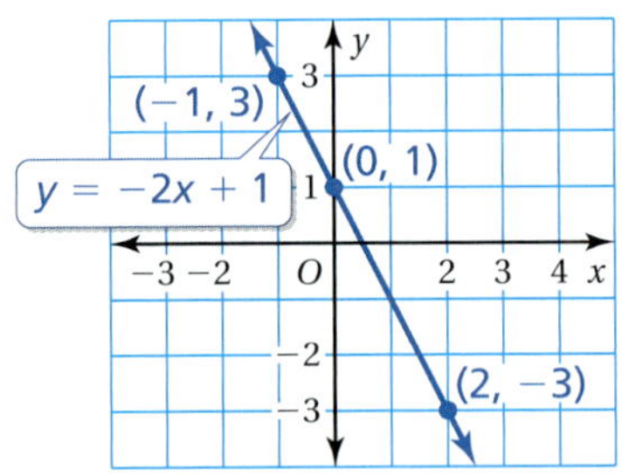

Check

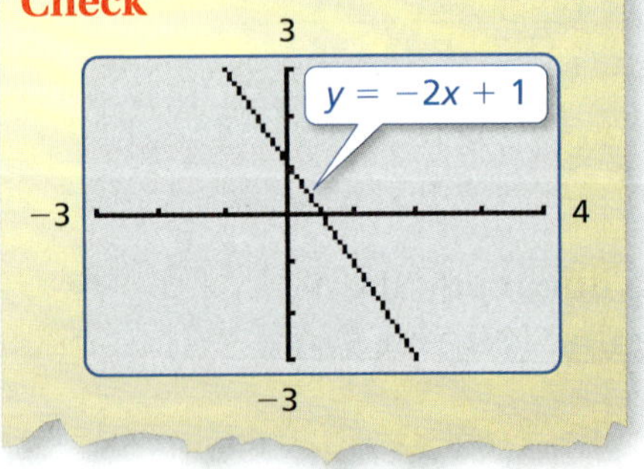

Key Idea

Graphing Horizontal and Vertical Lines

The graph of $y = b$ is a horizontal line passing through $(0, b)$.

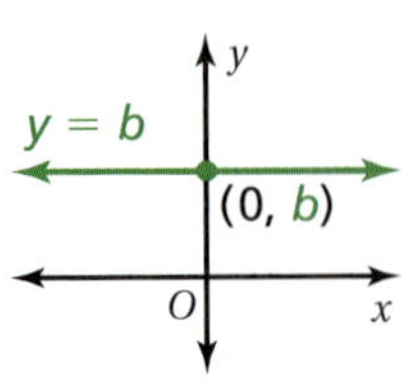

The graph of $x = a$ is a vertical line passing through $(a, 0)$.

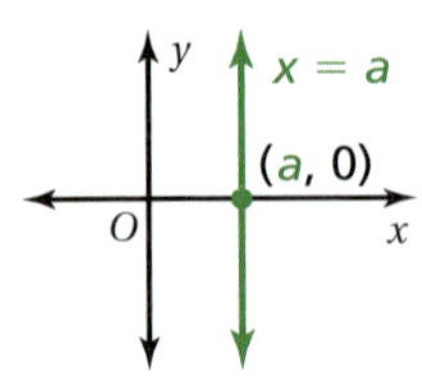

Multi-Language Glossary at BigIdeasMath.com

Laurie's Notes

Introduction

Connect

- **Yesterday:** Students explored the graphs of linear equations.
- **Today:** Students will graph linear equations using a table of values.

Motivate

- Discuss a fact about wind speeds related to Example 3. During a wild April storm in 1934, a wind gust of 231 miles per hour (372 kilometers per hour) pushed across the summit of Mt. Washington in New Hampshire.

Lesson Notes

Key Idea

- Define *linear equation* and *solutions* of the equation.
- Note the use of color in the table. The equation used is a simple equation that helps students focus on the representation of the solutions as ordered pairs. The *y*-coordinate is always 1 greater than the *x*-coordinate, just as the equation states.

Example 1

? As a quick review, ask a volunteer to review the rules for integer multiplication. If the factors have the same sign, the product is positive. If the factors have different signs, the product is negative.

- Write the 4-column table. Take the time to show how the *x*-coordinate is being substituted in the second column. The number in blue is the only quantity that varies (variable); the other quantities are always the same (constant). Values from the first and third columns form the ordered pair.

? "From the graph, can you estimate the solution when $x = \frac{1}{2}$? Verify your answer by solving the equation when $x = \frac{1}{2}$." yes; $\left(\frac{1}{2}, 0\right)$

- **Use Appropriate Tools Strategically:** Students can check their graphs on a graphing calculator.

Key Idea

- Students are sometimes confused by the equations $x = a$ and $y = b$. Explain to students that a and b can equal any number.
- **Teaching Tip:** Another way to discuss the equation $y = b$ is to say that "y always equals a certain number, while x can equal anything." For example, if $y = -4$, the table of values may look like this:

x	−1	0	1	2
y	−4	−4	−4	−4

- **Teaching Tip:** Another way to discuss the equation $x = a$ is to say that "x always equals a certain number, while y can equal anything." For example, if $x = -2$, the table of values may look like this:

x	−2	−2	−2	−2
y	−1	0	1	2

Goal Today's lesson is graphing **linear equations**.

Technology for the Teacher

Dynamic Classroom

Lesson Tutorials
Lesson Plans
Answer Presentation Tool

Extra Example 1

Graph $y = \frac{1}{2}x - 3$.

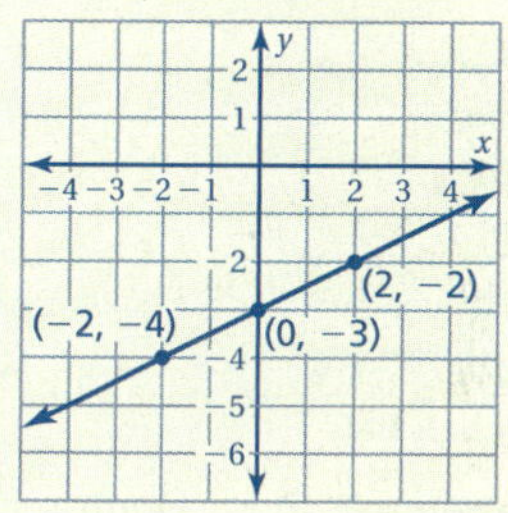

Extra Example 2

a. Graph $y = 4$.

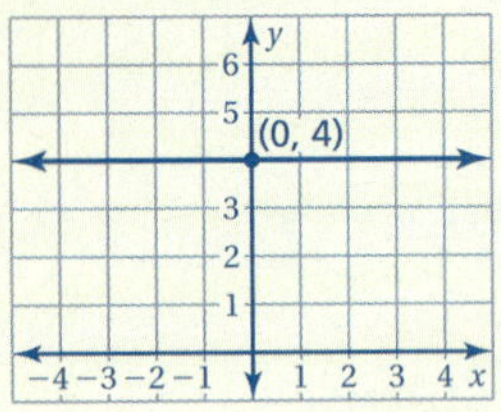

b. Graph $x = -1$.

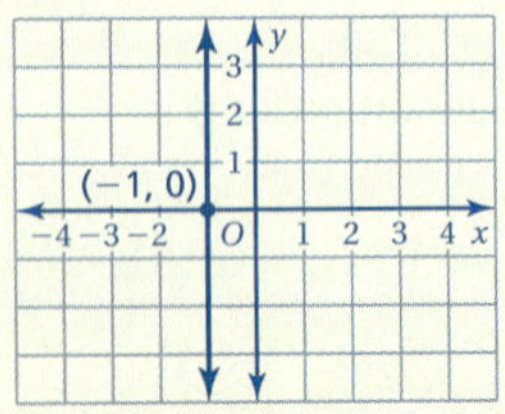

On Your Own

1–4. See Additional Answers.

Extra Example 3

The cost y (in dollars) for making friendship bracelets is $y = 0.5x + 2$, where x is the number of bracelets.

a. Graph the equation.

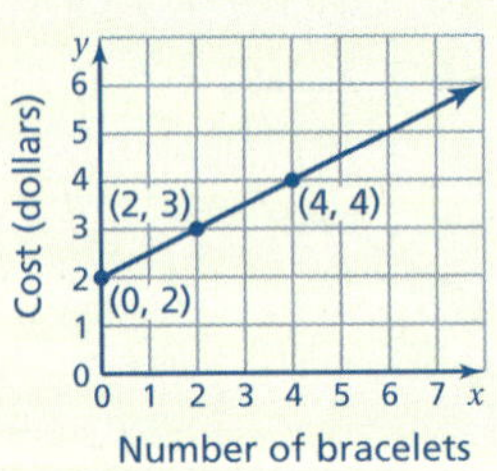

b. How many bracelets can be made for $10? 16

On Your Own

5. 8 hours after it enters the Gulf of Mexico

English Language Learners

Vocabulary

Make sure students understand that the graph of a *linear* equation is a *line*. Only two points are needed to graph a line, but if one of the points is incorrect the wrong line will be graphed. Plotting three points for a line in the coordinate plane and making sure that the points form a line provide students with a check when graphing.

Laurie's Notes

Example 2

? "What are other points on the line $y = -3$?" *Sample answer:* (5, −3), or anything of the form (x, −3)

? "What are other points on the line $x = 2$?" *Sample answer:* (2, −3), or anything of the form (2, y)

On Your Own

- Ask volunteers to share their graphs at the board.
- Students may ask how to graph $x = -4$ on their calculators. To create a graph using a graphing calculator, the equation must begin with "$y =$." So, vertical lines cannot be graphed on a calculator.

Example 3

? "What does x represent in the problem? What does y represent?" x = number of hours after the storm enters the Gulf of Mexico; y = wind speed

- Work through the problem using the 4-column table to generate solutions of the equation.
- Note that the y-coordinate is much greater than the x-coordinate. For this reason, a broken vertical axis is used. Students should *not* scale the y-axis beginning at 0.

? "Why are only non-negative numbers substituted for x?" Because x equals the number of hours after the storm enters the Gulf of Mexico, you do not know if the equation makes sense for x-values before that.

- Note that the ordered pairs are all located in Quadrant I because x is non-negative. Even though this restriction was not stated explicitly, you know from reading the description of x that it needs to be non-negative.
- In part (b), help students read the graph. Starting with a y-value of 74 on the y-axis, trace horizontally until you reach the graph of the line, and then trace straight down (vertically) to the x-axis. The x-coordinate is 4.

On Your Own

- **Neighbor Check:** Have students work independently and then have their neighbors check their work. Have students discuss any discrepancies.

Closure

- Explain how you know if an equation is linear. *Sample answer:* The graph of the equation is a line.

EXAMPLE 2 Graphing a Horizontal Line and a Vertical Line

a. Graph $y = -3$.

The graph of $y = -3$ is a horizontal line passing through $(0, -3)$. Draw a horizontal line through this point.

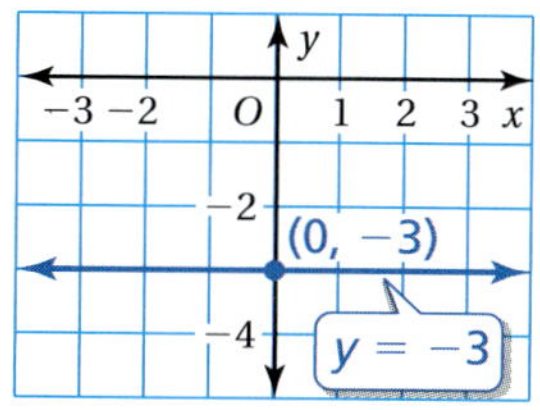

b. Graph $x = 2$.

The graph of $x = 2$ is a vertical line passing through $(2, 0)$. Draw a vertical line through this point.

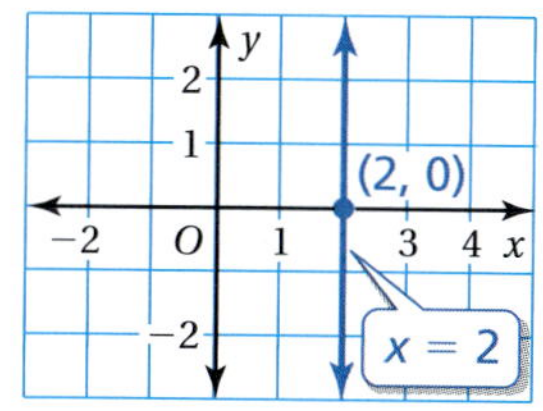

On Your Own

Now You're Ready
Exercises 5–16

Graph the linear equation. Use a graphing calculator to check your graph, if possible.

1. $y = 3x$
2. $y = -\frac{1}{2}x + 2$
3. $x = -4$
4. $y = -1.5$

EXAMPLE 3 Real-Life Application

The wind speed y (in miles per hour) of a tropical storm is $y = 2x + 66$, where x is the number of hours after the storm enters the Gulf of Mexico.

a. Graph the equation.

b. When does the storm become a hurricane?

A tropical storm becomes a hurricane when wind speeds are at least 74 miles per hour.

a. Make a table of values.

x	$y = 2x + 66$	y	(x, y)
0	$y = 2(0) + 66$	66	(0, 66)
1	$y = 2(1) + 66$	68	(1, 68)
2	$y = 2(2) + 66$	70	(2, 70)
3	$y = 2(3) + 66$	72	(3, 72)

Plot the ordered pairs and draw a line through the points.

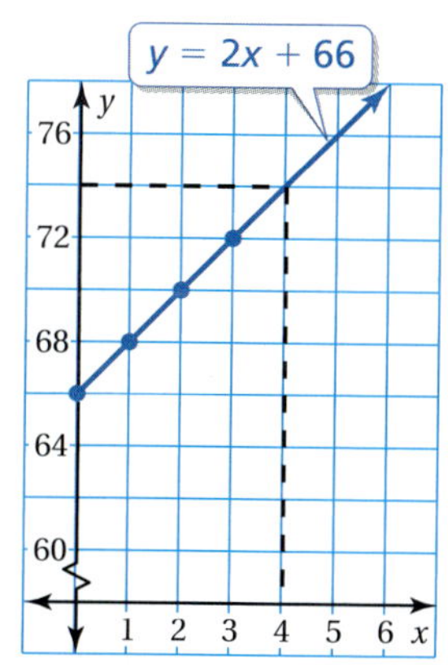

b. From the graph, you can see that $y = 74$ when $x = 4$. So, the storm becomes a hurricane 4 hours after it enters the Gulf of Mexico.

On Your Own

5. **WHAT IF?** The wind speed of the storm is $y = 1.5x + 62$. When does the storm become a hurricane?

13.1 Exercises

Check It Out
Help with Homework
BigIdeasMath.com

Vocabulary and Concept Check

1. **VOCABULARY** What type of graph represents the solutions of the equation $y = 2x + 4$?

2. **WHICH ONE DOESN'T BELONG?** Which equation does *not* belong with the other three? Explain your reasoning.

$y = 0.5x - 0.2$ $4x + 3 = y$ $y = x^2 + 6$ $\frac{3}{4}x + \frac{1}{3} = y$

Practice and Problem Solving

PRECISION Copy and complete the table. Plot the two solution points and draw a line *exactly* through the two points. Find a different solution point on the line.

3.

x		
$y = 3x - 1$		

4.

x		
$y = \frac{1}{3}x + 2$		

Graph the linear equation. Use a graphing calculator to check your graph, if possible.

1 2 5. $y = -5x$ 6. $y = \frac{1}{4}x$ 7. $y = 5$ 8. $x = -6$

9. $y = x - 3$ 10. $y = -7x - 1$ 11. $y = -\frac{x}{3} + 4$ 12. $y = \frac{3}{4}x - \frac{1}{2}$

13. $y = -\frac{2}{3}$ 14. $y = 6.75$ 15. $x = -0.5$ 16. $x = \frac{1}{4}$

17. **ERROR ANALYSIS** Describe and correct the error in graphing the equation.

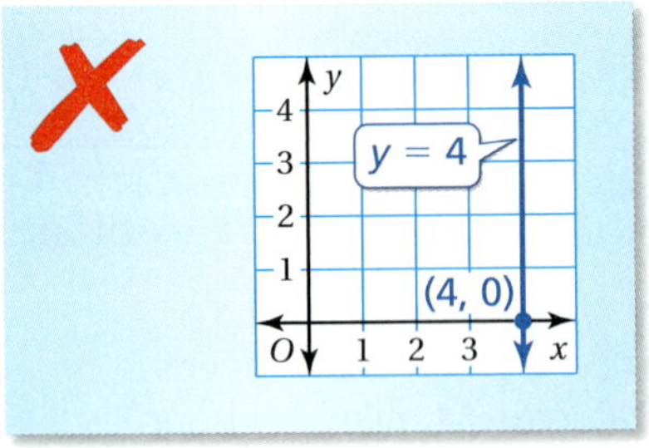

18. **MESSAGING** You sign up for an unlimited text-messaging plan for your cell phone. The equation $y = 20$ represents the cost y (in dollars) for sending x text messages. Graph the equation. What does the graph tell you?

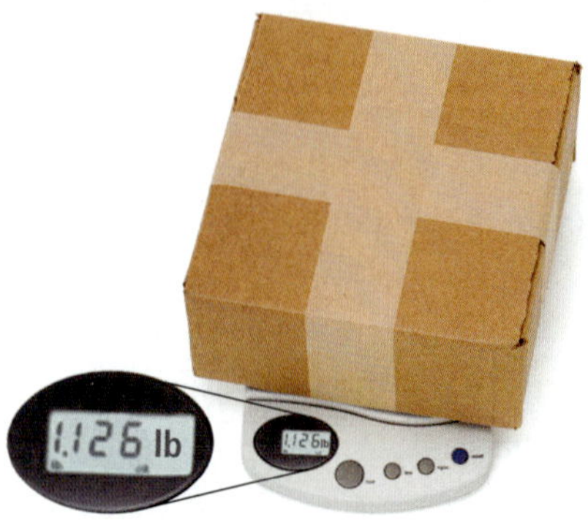

19. **MAIL** The equation $y = 2x + 3$ represents the cost y (in dollars) of mailing a package that weighs x pounds.

   a. Graph the equation.

   b. Use the graph to estimate how much it costs to mail the package.

   c. Use the equation to find exactly how much it costs to mail the package.

Assignment Guide and Homework Check

Level	Assignment	Homework Check
Accelerated	1–4, 6–16 even, 17, 18–26 even, 27–32	12, 18, 22, 24, 26

Common Errors

- **Exercises 5–16** Students may make calculation errors when solving for ordered pairs. If they only find two ordered pairs for the graph, they may not recognize their mistakes. Encourage them to find at least three ordered pairs when drawing a graph.
- **Exercises 7, 13, and 14** Students may draw vertical lines through points on the *x*-axis. Remind them that the graph of the equation is a horizontal line. Ask them to identify the *y*-coordinate for several *x*-coordinates. For example, what is the *y*-coordinate for $x = 5$? $x = 6$? $x = -4$? Students should answer with the same *y*-coordinate each time.
- **Exercises 8, 15, and 16** Students may draw horizontal lines through points on the *y*-axis. Remind them that the graph of the equation is a vertical line. Ask them to identify the *x*-coordinate for several *y*-coordinates. For example, what is the *x*-coordinate for $y = 3$? $y = -1$? $y = 0$? Students should answer with the same *x*-coordinate each time.
- **Exercises 20–23** Students may make mistakes in solving for *y*, such as using the same operation instead of the opposite operation.

13.1 Record and Practice Journal

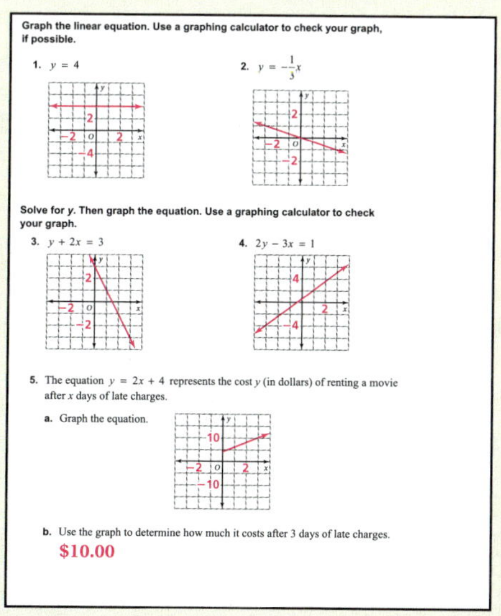
Graph the linear equation. Use a graphing calculator to check your graph, if possible.

1. $y = 4$
2. $y = -\frac{1}{3}x$

Solve for *y*. Then graph the equation. Use a graphing calculator to check your graph.

3. $y + 2x = 3$
4. $2y - 3x = 1$

5. The equation $y = 2x + 4$ represents the cost *y* (in dollars) of renting a movie after *x* days of late charges.

a. Graph the equation.

b. Use the graph to determine how much it costs after 3 days of late charges.
$10.00

Vocabulary and Concept Check

1. a line
2. $y = x^2 + 6$ does not belong because it is not a linear equation.

Practice and Problem Solving

3. *Sample answer:*

x	0	1
$y = 3x - 1$	-1	2

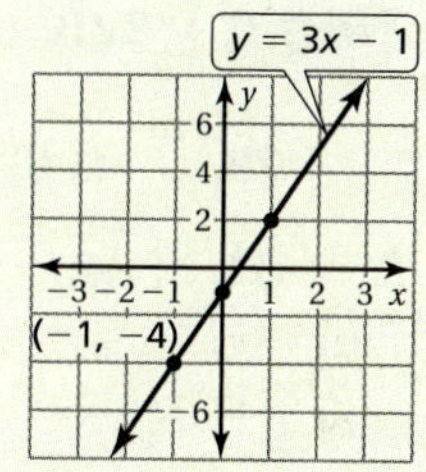

4. *Sample answer:*

x	0	3
$y = \frac{1}{3}x + 2$	2	3

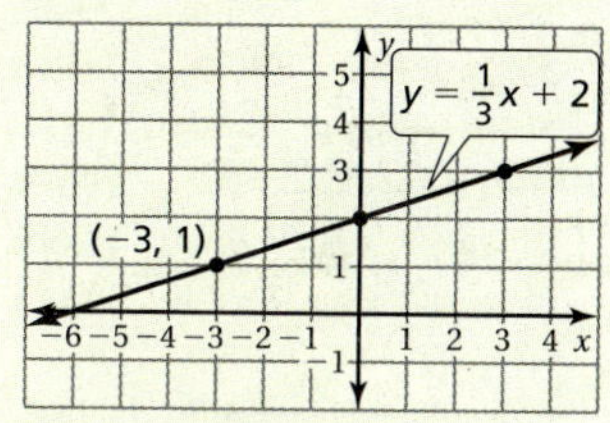

5.
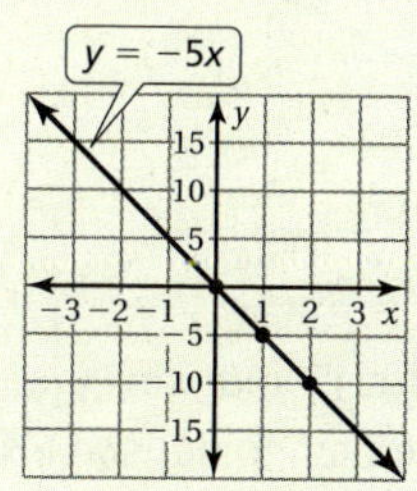

6.
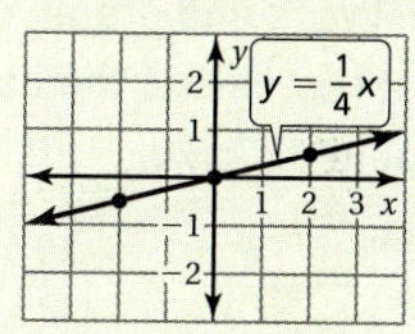

7–19. See Additional Answers.

Practice and Problem Solving

20. $y = 3x + 1$

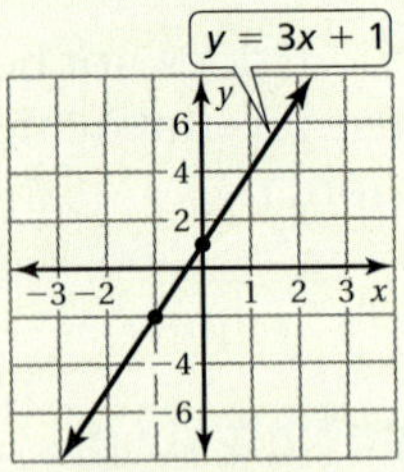

21–26. See Additional Answers.

27. See *Taking Math Deeper*.

Fair Game Review

28. (5, 3) **29.** (−6, 6)

30. (2, −2) **31.** (−4, −3)

32. B

Mini-Assessment

1. Graph $y = \frac{1}{2}x - 2$.

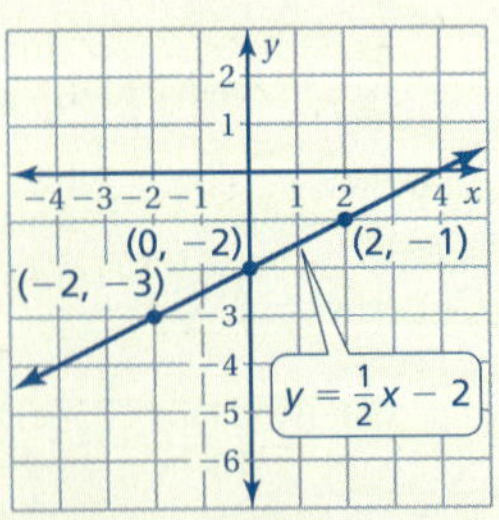

2. You have \$100 in your savings account and plan to deposit \$20 each month. Write and graph a linear equation that represents the balance in your account. $y = 20x + 100$

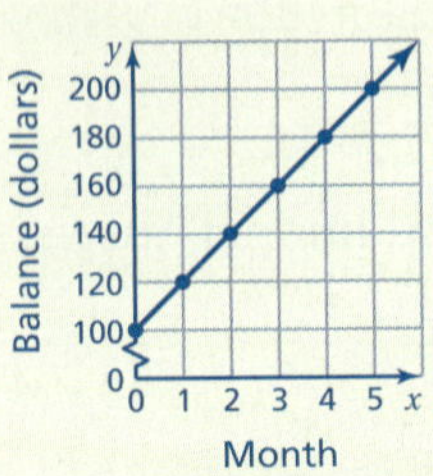

Taking Math Deeper

Exercise 27

Some of the information for this exercise is given in the photo and some is given in the text. It is a good idea to start by listing all of the given information.

1 List the given information.

- The camera can store 250 pictures.
- 1 second of video = 2 pictures.
- Video time used = 90 seconds.
- Let y = number of pictures.
- Let x = number of seconds of video.

2 **a.** Write and graph an equation for x and y.

$$y + 2x = 250$$
$$y = -2x + 250$$

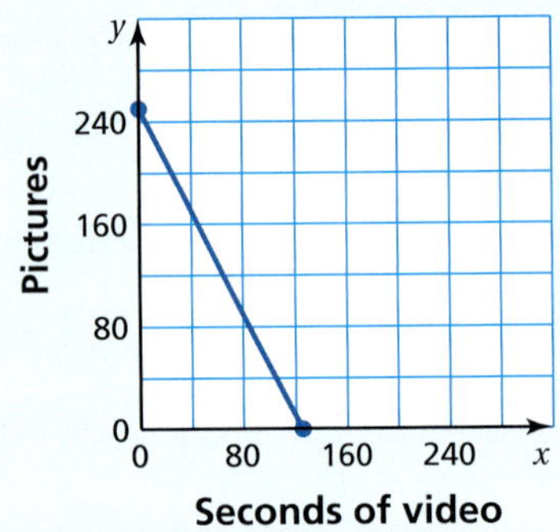

3 **b.** Answer the question.

When $x = 90$, the value of y is as follows.

$$y = -2x + 250$$
$$= -2(90) + 250$$
$$= 70$$

Your camera can store 70 pictures in addition to the 90-second video.

Project

Research digital cameras. Find the number of pictures that can be stored on five different cameras. Compare the prices of the cameras. Which do you consider to be the better buy? Why?

Reteaching and Enrichment Strategies

If students need help...	If students got it...
Resources by Chapter • Practice A and Practice B • Puzzle Time Record and Practice Journal Practice Differentiating the Lesson Lesson Tutorials Skills Review Handbook	Resources by Chapter • Enrichment and Extension • Technology Connection Start the next section

Solve for y. Then graph the equation. Use a graphing calculator to check your graph.

20. $y - 3x = 1$ **21.** $5x + 2y = 4$

22. $-\frac{1}{3}y + 4x = 3$ **23.** $x + 0.5y = 1.5$

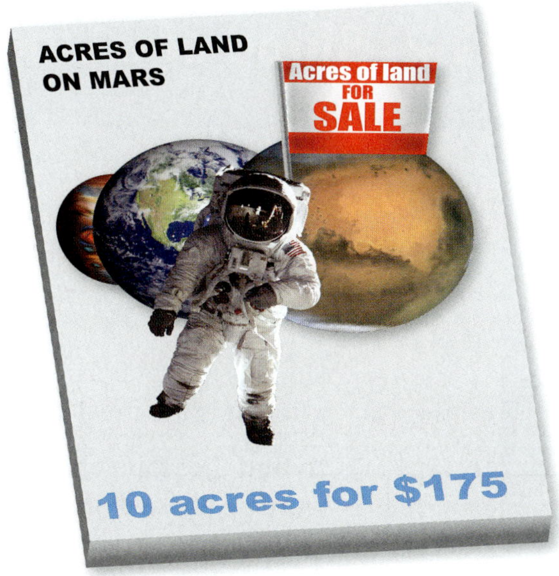

24. SAVINGS You have \$100 in your savings account and plan to deposit \$12.50 each month.

a. Graph a linear equation that represents the balance in your account.

b. How many months will it take you to save enough money to buy 10 acres of land on Mars?

25. GEOMETRY The sum S of the interior angle measures of a polygon with n sides is $S = (n - 2) \cdot 180°$.

a. Plot four points (n, S) that satisfy the equation. Is the equation a linear equation? Explain your reasoning.

b. Does the value $n = 3.5$ make sense in the context of the problem? Explain your reasoning.

26. SEA LEVEL Along the U.S. Atlantic coast, the sea level is rising about 2 millimeters per year. How many millimeters has sea level risen since you were born? How do you know? Use a linear equation and a graph to justify your answer.

27. Problem Solving One second of video on your digital camera uses the same amount of memory as two pictures. Your camera can store 250 pictures.

a. Write and graph a linear equation that represents the number y of pictures your camera can store when you take x seconds of video.

b. How many pictures can your camera store in addition to the video shown?

Fair Game Review What you learned in previous grades & lessons

Write the ordered pair corresponding to the point. *(Skills Review Handbook)*

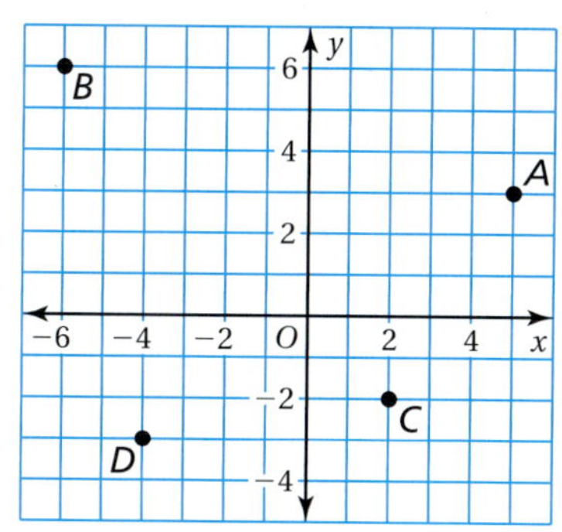

28. point A **29.** point B

30. point C **31.** point D

32. MULTIPLE CHOICE A debate team has 15 female members. The ratio of females to males is 3 : 2. How many males are on the debate team? *(Skills Review Handbook)*

Ⓐ 6 Ⓑ 10 Ⓒ 22 Ⓓ 25

13.2 Slope of a Line

Essential Question How can you use the slope of a line to describe the line?

Slope is the rate of change between any two points on a line. It is the measure of the *steepness* of the line.

To find the slope of a line, find the ratio of the change in y (vertical change) to the change in x (horizontal change).

$$\text{slope} = \frac{\text{change in } y}{\text{change in } x}$$

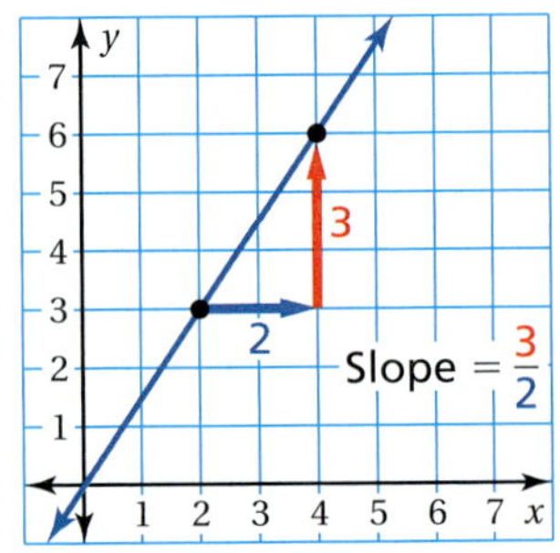

1 ACTIVITY: Finding the Slope of a Line

Work with a partner. Find the slope of each line using two methods.

Method 1: Use the two black points. •

Method 2: Use the two pink points. •

Do you get the same slope using each method? Why do you think this happens?

a.

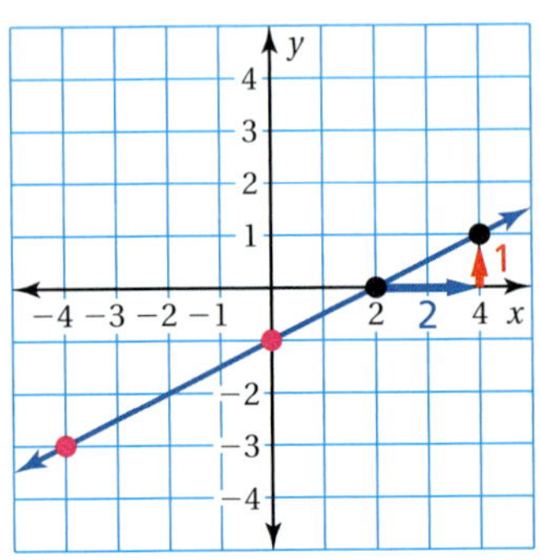

b.

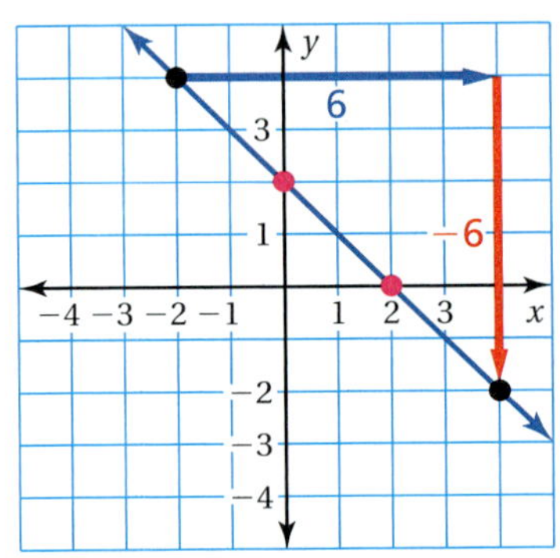

c.

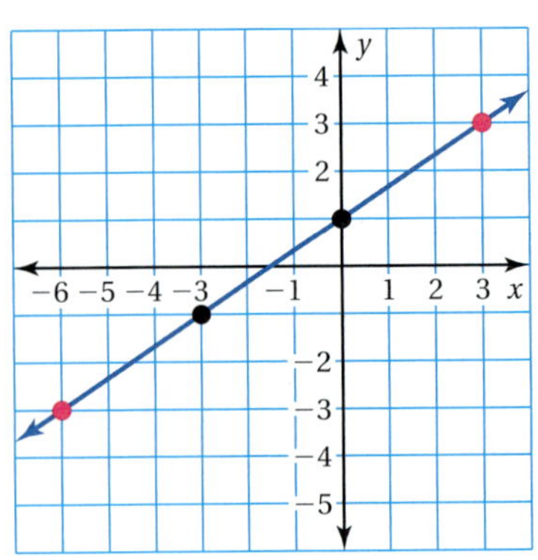

d.

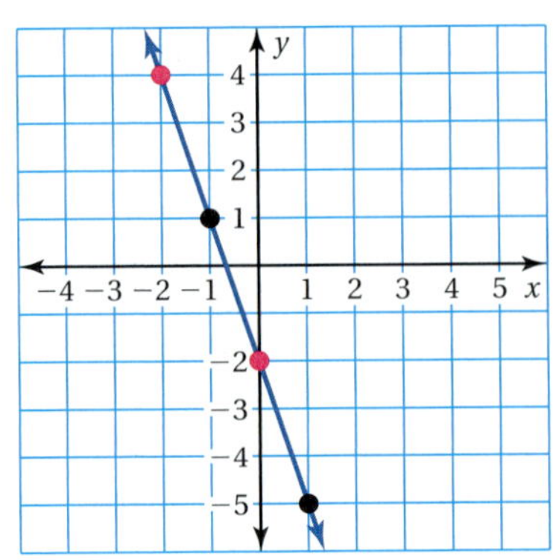

Graphing Equations
In this lesson, you will
- find slopes of lines by using two points.
- find slopes of lines from tables.

Laurie's Notes

Introduction

Applying Mathematical Practices

- **Make Sense of Problems:** The goal is for students to use different pairs of points on a line to find the line's slope. Drawing the arrow diagrams will help students to visualize the *slope triangle.* Students will recognize the triangles are similar and proportions can be formed.

Motivate

? "How many of you have been on a roller coaster?"

- Discuss with students what makes one roller coaster more thrilling than another. Students will usually describe how quickly the coaster drops or the steepness of the hill. This is similar to the *change in y* of a line when finding the slope.

Activity Notes

Discuss

? "Does anyone remember what is meant by slope of a line?" At least one student should recall that it measures the steepness of a line.

- Write the definition for slope. Sketch the graph shown to demonstrate what is meant by change in *y* (red vertical arrow) and change in *x* (blue horizontal arrow).
- Remind students that slope is always the change in *y* in the numerator and the change in *x* in the denominator. This can be confusing for students because graphs are read from left to right, and we have a tendency to move in the *x*-direction first. For this reason, students want to write the change in *x* in the numerator.

? "Can the change in *x* be negative? Explain." yes; Moving to the left horizontally is negative.

? "Can the change in *y* be negative? Explain." yes; Moving down vertically is negative.

Activity 1

- Encourage students to draw the change arrows for each pair of points. Label the change in *x* (or *y*) next to the arrow.
- **Big Idea:** The slope of a line is always the same regardless of what two ordered pairs are selected.
- **Common Error:** Students may forget to make the change negative when moving downward in the *y*-direction.
- **Reason Abstractly and Quantitatively** and **Critique the Reasoning of Others:** Students are asked to reason why the slopes are the same. They will form their own conjectures and listen to reasons offered by other students. Students will use geometry to explain why this is true in Activity 2.

What Your Students Will Learn

- Describe and find slopes of lines using two points on the line.
- Find slopes of horizontal and vertical lines.
- Find slopes of lines from tables.

Previous Learning

Students should know the relationship between corresponding sides of similar triangles.

Technology for the Teacher

Dynamic Classroom

Lesson Plans
Complete Materials List

13.2 Record and Practice Journal

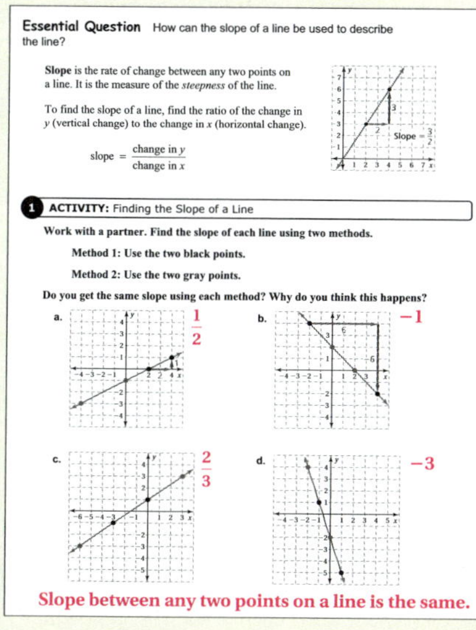

Essential Question How can the slope of a line be used to describe the line?

Slope is the rate of change between any two points on a line. It is the measure of the *steepness* of the line.

To find the slope of a line, find the ratio of the change in *y* (vertical change) to the change in *x* (horizontal change).

$$\text{slope} = \frac{\text{change in } y}{\text{change in } x}$$

Slope = $\frac{3}{2}$

1 ACTIVITY: Finding the Slope of a Line

Work with a partner. Find the slope of each line using two methods.

Method 1: Use the two black points.

Method 2: Use the two gray points.

Do you get the same slope using each method? Why do you think this happens?

a. $\frac{1}{2}$

b. -1

c. $\frac{2}{3}$

d. -3

Slope between any two points on a line is the same.

Differentiated Instruction

Kinesthetic

Help students develop number sense about slope. Have them draw lines in the coordinate plane through the following pairs of points.

(0, 0) and (3, 5) (0, 0) and (3, 4)
(0, 0) and (3, 3) (0, 0) and (3, 2)
(0, 0) and (3, 1)

Next have students find the slope of each line. Point out that the line passing through (3, 3) has a slope of 1. The lines with y-coordinates greater than 3 have a slope greater than 1. The lines with y-coordinates less than 3 have a slope less than 1. For positive slopes, the steeper lines will have a greater slope.

13.2 Record and Practice Journal

2 ACTIVITY: Using Similar Triangles

Work with a partner. Use the figure shown.

a. ΔABC is a right triangle formed by drawing a horizontal line segment from point A and a vertical line segment from point B. Use this method to draw another right triangle, ΔDEF.

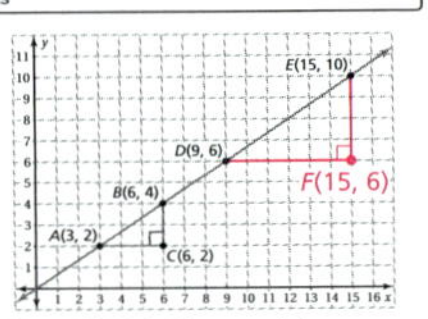

b. What can you conclude about ΔABC and ΔDEF? Justify your conclusion.
They are similar.

c. For each triangle, find the ratio of the length of the vertical side to the length of the horizontal side. What do these ratios represent?
$\frac{2}{3}$; the slope of the line between points A and B and between points D and E

d. What can you conclude about the slope between any two points on the line?
***Sample answer:* It is constant.**

3 ACTIVITY: Drawing Lines with Given Slopes

Work with a partner.

a. Draw two lines with slope $\frac{3}{4}$. One line passes through (−4, 1), and the other line passes through (4, 0). What do you notice about the two lines?
The two lines are parallel.

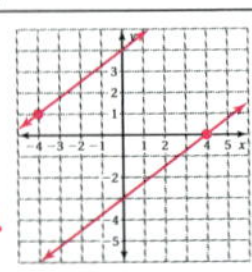

b. Draw two lines with slope $-\frac{4}{3}$. One line passes through (2, 1), and the other line passes through (−1, −1). What do you notice about the two lines?
The two lines are parallel.

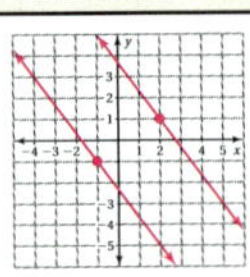

c. **CONJECTURE** Make a conjecture about two different nonvertical lines in the same plane that have the same slope.
They are parallel.

d. Graph one line from part (a) and one line from part (b) in the same coordinate plane. Describe the angle formed by the two lines. What do you notice about the product of the slopes of the two lines?
The two lines form a right angle. The product is −1.

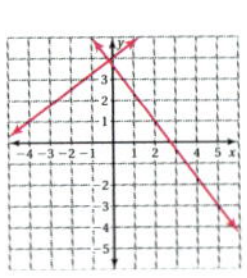

e. **REPEATED REASONING** Repeat part (d) for the two lines you did *not* choose. Based on your results, make a conjecture about two lines in the same plane whose slopes have a product of −1.
They are perpendicular.

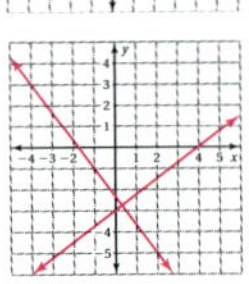

What Is Your Answer?

4. **IN YOUR OWN WORDS** How can you use the slope of a line to describe the line?
tells you whether the line rises or falls from left to right and how steep the line is

Laurie's Notes

Activity 2

- In part (a), check to see that students have located point F correctly.
- In part (b), students should use what they know about parallel lines and transversals to conclude that the triangles are similar.
- Explain that it is common to refer to $\triangle ABC$ and $\triangle DEF$ as *slope triangles.* They are formed by the line and the change in x and change in y arrows.
- Choose additional pairs of points on the line and repeat the example. Help students realize that they can pick any two points on the line and construct a right triangle that is similar to the ones shown. Then lead them to the conclusion that the ratio of the side lengths, which gives the slope, is the same regardless of which two points you choose.

Activity 3

- In addition to being able to determine the slope of a line, you want students to be able to draw a line that has a particular slope.

? "What does it mean for a line to have a slope of $\frac{3}{4}$?" For every 3 units of change in the y-direction, there is a change of 4 units in the x-direction.

? "What does it mean for a line to have a slope of $-\frac{4}{3}$?" For every −4 units of change in the y-direction, there is a change of 3 units in the x-direction. This is the same as 4 units in the y-direction and −3 units in the x-direction.

? "What do you notice about the lines drawn in part (a) and the lines drawn in part (b)?" Students should observe that the lines in part (a) are parallel and the lines in part (b) are parallel. They may also notice that positive slopes rise (from left to right) and negative slopes fall (from left to right).

- In part (c), students should realize that two different lines with the same slope are parallel.
- **Attend to Precision:** In parts (d) and (e), if students have graphed carefully they should observe that the lines form a right angle. They may describe the lines as being perpendicular. The product of the slopes is –1 and they may refer to the slopes as negative (or opposite) reciprocals.
- **Connection:** This activity provides the discovery for Extension 13.2.

What Is Your Answer?

- Discuss students' responses. Listen for the words steepness, rising, falling, and so on.

Closure

- Plot the point (0, 3). Draw the line through this point that has a slope of $\frac{1}{3}$.

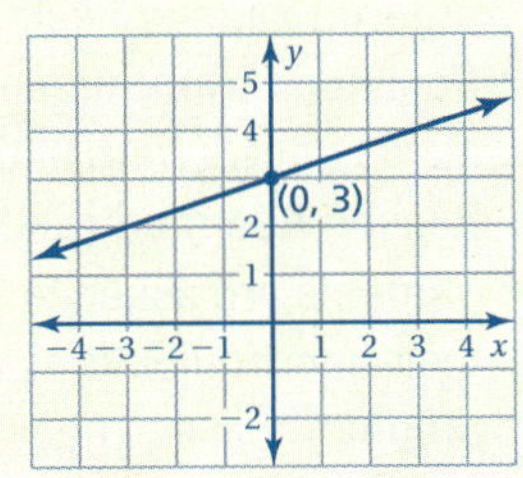

2 ACTIVITY: Using Similar Triangles

Work with a partner. Use the figure shown.

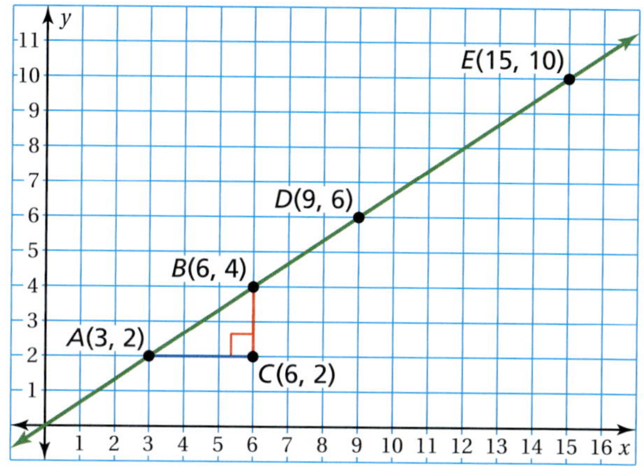

a. $\triangle ABC$ is a right triangle formed by drawing a horizontal line segment from point A and a vertical line segment from point B. Use this method to draw another right triangle, $\triangle DEF$.

b. What can you conclude about $\triangle ABC$ and $\triangle DEF$? Justify your conclusion.

c. For each triangle, find the ratio of the length of the vertical side to the length of the horizontal side. What do these ratios represent?

d. What can you conclude about the slope between any two points on the line?

3 ACTIVITY: Drawing Lines with Given Slopes

Work with a partner.

a. Draw two lines with slope $\frac{3}{4}$. One line passes through $(-4, 1)$, and the other line passes through $(4, 0)$. What do you notice about the two lines?

b. Draw two lines with slope $-\frac{4}{3}$. One line passes through $(2, 1)$, and the other line passes through $(-1, -1)$. What do you notice about the two lines?

c. **CONJECTURE** Make a conjecture about two different nonvertical lines in the same plane that have the same slope.

d. Graph one line from part (a) and one line from part (b) in the same coordinate plane. Describe the angle formed by the two lines. What do you notice about the product of the slopes of the two lines?

e. **REPEATED REASONING** Repeat part (d) for the two lines you did *not* choose. Based on your results, make a conjecture about two lines in the same plane whose slopes have a product of -1.

Math Practice

Interpret a Solution

What does the slope tell you about the graph of the line? Explain.

What Is Your Answer?

4. **IN YOUR OWN WORDS** How can you use the slope of a line to describe the line?

Practice Use what you learned about the slope of a line to complete Exercises 4–6 on page 577.

13.2 Lesson

Check It Out
Lesson Tutorials
BigIdeasMath.com

Key Vocabulary
slope, *p. 574*
rise, *p. 574*
run, *p. 574*

Key Idea

Slope

The **slope** m of a line is a ratio of the change in y (the **rise**) to the change in x (the **run**) between any two points, (x_1, y_1) and (x_2, y_2), on the line.

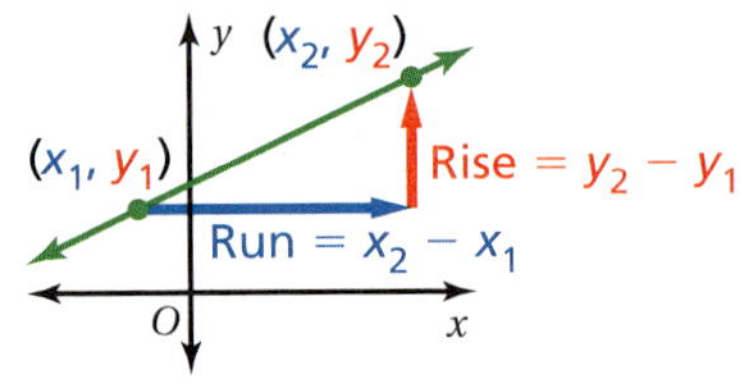

$$m = \frac{\text{rise}}{\text{run}} = \frac{\text{change in } y}{\text{change in } x} = \frac{y_2 - y_1}{x_2 - x_1}$$

Positive Slope

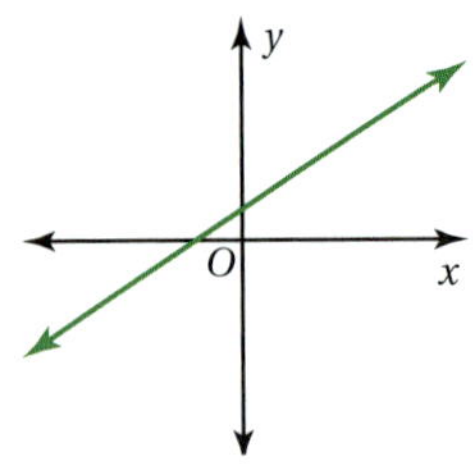

The line rises from left to right.

Negative Slope

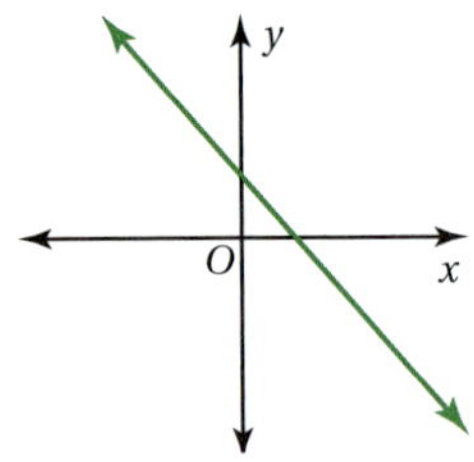

The line falls from left to right.

Reading

In the slope formula, x_1 is read as "x sub one," and y_2 is read as "y sub two." The numbers 1 and 2 in x_1 and y_2 are called *subscripts*.

EXAMPLE 1 Finding the Slope of a Line

Describe the slope of the line. Then find the slope.

a.

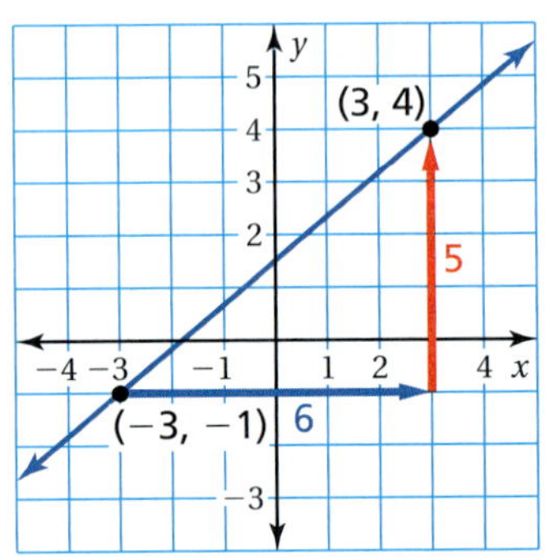

The line rises from left to right. So, the slope is positive. Let $(x_1, y_1) = (-3, -1)$ and $(x_2, y_2) = (3, 4)$.

$$m = \frac{y_2 - y_1}{x_2 - x_1}$$

$$= \frac{4 - (-1)}{3 - (-3)}$$

$$= \frac{5}{6}$$

b.

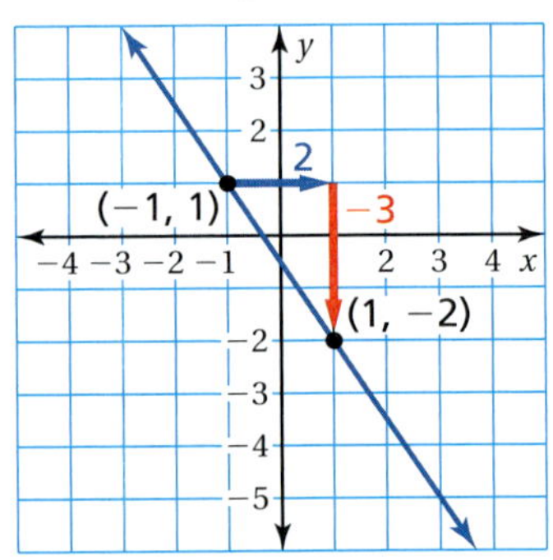

The line falls from left to right. So, the slope is negative. Let $(x_1, y_1) = (-1, 1)$ and $(x_2, y_2) = (1, -2)$.

$$m = \frac{y_2 - y_1}{x_2 - x_1}$$

$$= \frac{-2 - 1}{1 - (-1)}$$

$$= \frac{-3}{2}, \text{ or } -\frac{3}{2}$$

Study Tip

When finding slope, you can label either point as (x_1, y_1) and the other point as (x_2, y_2).

 Multi-Language Glossary at BigIdeasMath.com

Laurie's Notes

Introduction

Connect

- **Yesterday:** Students explored slopes of lines.
- **Today:** Students will find slopes of lines from graphs and tables.

Motivate

- Have students plot four points: $A(5, 0)$, $B(0, 5)$, $C(-5, 0)$, and $D(0, -5)$. Connect the points to form the quadrilateral *ABCD*.

? "What type of quadrilateral is *ABCD*?" Without proof, students should say square.

? "What is the slope of each side, meaning the slopes of the lines through *AB*, *BC*, *CD*, and *DA*?" Slopes of *AB* and *CD* are both -1. Slopes of *BC* and *DA* are both 1.

Lesson Notes

Key Idea

- Write the Key Idea. Define slope of a line.
- Tell students that it is traditional to use *m* to represent slope. They will also see this in future mathematics classes.
- Note the use of color in the definition and on the graph. The *change in y* and the *vertical change arrow* are both red. The *change in x* and the *horizontal change arrow* are both blue.
- Discuss the difference in positive and negative slopes, a concept students explored yesterday.
- Remind students that graphs are read from left to right.
- Explain to students that you can also subtract coordinates to find the rise and run in addition to finding rise and run graphically.

Example 1

- **Make Sense of Problems:** Drawing the arrow diagrams will help students visualize the *slope triangle*.
- Students often ask if they can move in the *y*-direction first, followed by the *x*-direction. The answer is yes. Demonstrate this on either graph.
  - In part (a), start at $(-3, -1)$ and move up 5 units in the *y*-direction and then to the right 6 units in the *x*-direction. You will end at $(3, 4)$.
  - In part (b), start at $(-1, 1)$ and move down 3 units in the *y*-direction and then to the right 2 units in the *x*-direction. You will end at $(1, -2)$.
- Discuss the Study Tip. You can move in either direction first, and the labeling of the ordered pairs is arbitrary. Either point can be (x_1, y_1).

Goal Today's lesson is finding the **slope** of a line.

Technology for the Teacher

Dynamic Classroom

Lesson Tutorials
Lesson Plans
Answer Presentation Tool

Extra Example 1

Describe the slope of the line. Then find the slope.

a.

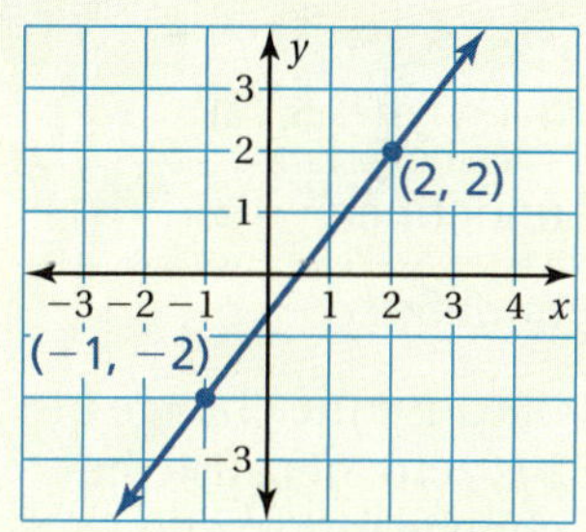

rises from left to right, so it is positive; $\frac{4}{3}$

b.

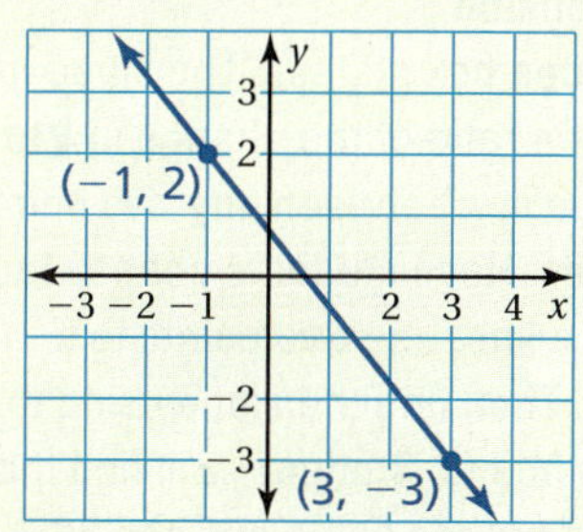

falls from left to right, so it is negative; $-\frac{5}{4}$

On Your Own

1. $-\frac{1}{5}$ **2.** $\frac{1}{3}$

3. $\frac{5}{2}$

Extra Example 2

Find the slope of the line.

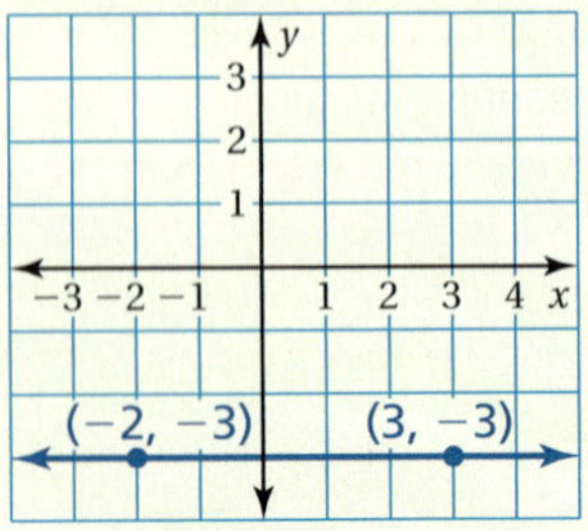

0

Extra Example 3

Find the slope of the line.

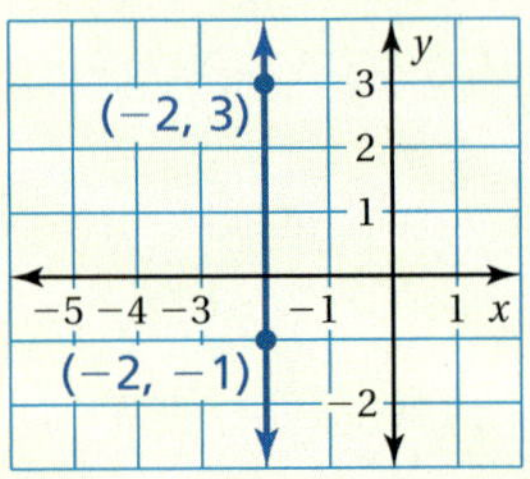

undefined

On Your Own

4. 0 **5.** 0

6. undefined

7. undefined

8. because the change in y is zero; because the change in x is zero

English Language Learners

Comprehension

The Key Idea box states, "The slope m of a line is a ratio of the change in y to the change in x between any two points on the line." Have students choose four points on a line. Use two points to find the slope. Then find the slope using the other two points. Students will find that the slopes are the same and should understand that the slope of the line is the same for the entire infinite length of the line.

Laurie's Notes

On Your Own

- **Neighbor Check:** Have students work independently and then have their neighbors check their work. Have students discuss any discrepancies.

Example 2

? "How does a slope of $\frac{1}{2}$ compare to a slope of $\frac{1}{5}$? Describe the lines." A slope of $\frac{1}{2}$ runs 2 units for every 1 unit it rises. A slope of $\frac{1}{5}$ runs 5 units for each 1 unit it rises. A slope of $\frac{1}{5}$ is not as steep.

? "What would a slope of $\frac{1}{10}$ look like?" A slope of $\frac{1}{10}$ is less steep than a slope of $\frac{1}{5}$, so it is almost flat.

? "How steep do you think a horizontal line is?" Listen for students to describe a horizontal line as having no rise. In this example, they will see it has a slope of 0.

- This progression of questions is to help students visualize that as slopes of lines get less steep, the lines become horizontal.
- Work through the example. There is no change in y. So, the change in y is 0.

Example 3

- Ask a series of questions similar to Example 2.

? "How does a slope of $\frac{9}{2}$ compare to a slope of $\frac{3}{2}$? Describe the lines." A slope of $\frac{9}{2}$ rises 9 units for every 2 units it runs. A slope of $\frac{3}{2}$ rises 3 units for every 2 units it runs. A slope of $\frac{9}{2}$ is steeper.

? "What would a slope of 10 look like?" A slope of 10 is steeper than a slope of $\frac{9}{2}$, so it is almost vertical.

? "How steep do you think a vertical line is?" Listen for students to describe a vertical line as having a slope of infinity.

- Work through the example. There is no change in x. So, the change in x is 0. Since division by zero is undefined, the slope of the line is undefined.

On Your Own

? Listen to students' explanations of Question 8. If necessary, you might ask,

- "What is true about every y-coordinate for points on a horizontal line?" y-values are all the same
- "When you compute the rise, what will you get?" 0
- "What is true about every x-coordinate for points on a vertical line?" x-values are all the same
- "When you compute the run, what will you get?" 0

- Students should recall that 0 can be divided by a non-zero number and the quotient is 0. Division by 0 is undefined.

On Your Own

Now You're Ready
Exercises 7–9

Find the slope of the line.

1.

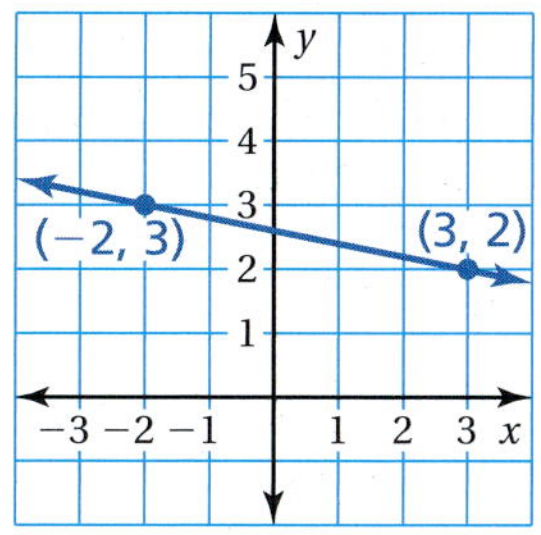

2.

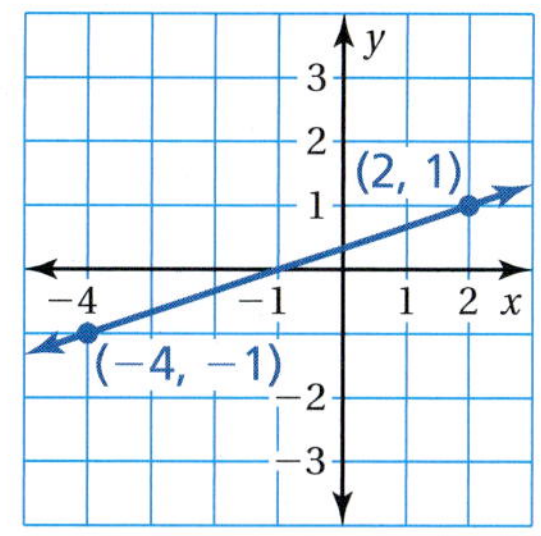

3.

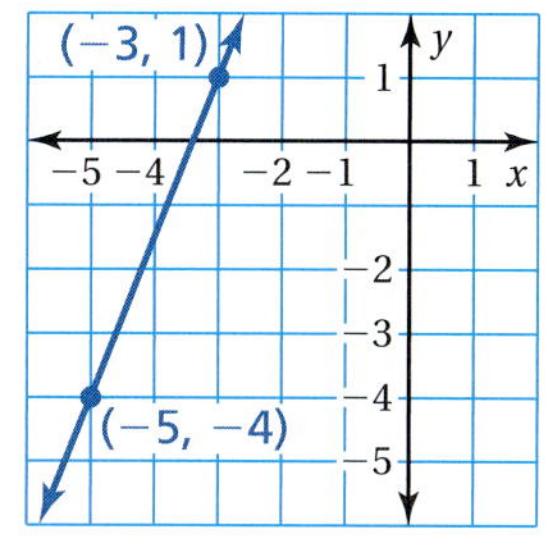

EXAMPLE 2 Finding the Slope of a Horizontal Line

Find the slope of the line.

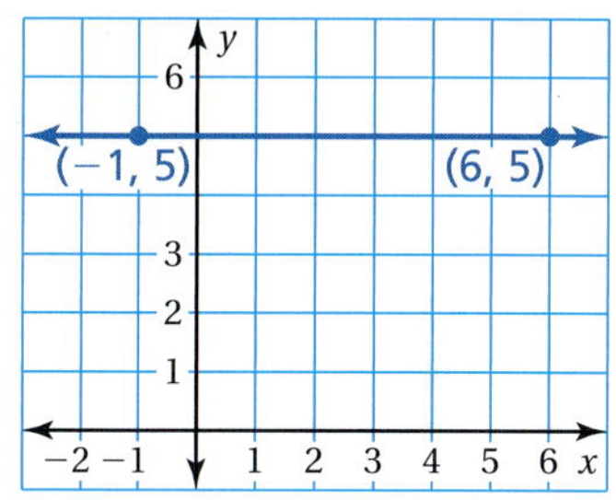

$$m = \frac{y_2 - y_1}{x_2 - x_1}$$

$$= \frac{5 - 5}{6 - (-1)}$$

$$= \frac{0}{7}, \text{ or } 0$$

∴ The slope is 0.

EXAMPLE 3 Finding the Slope of a Vertical Line

Find the slope of the line.

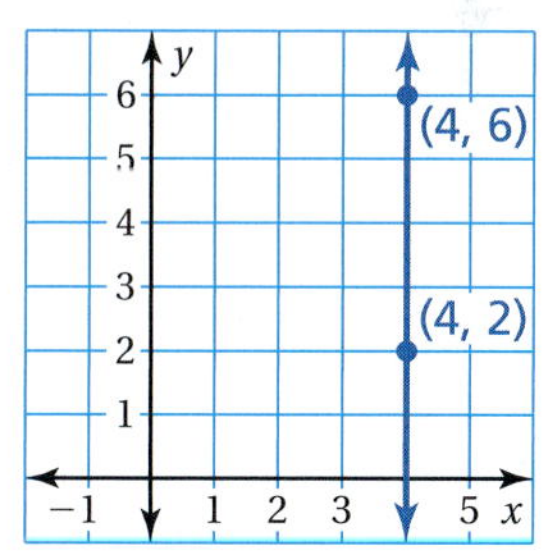

$$m = \frac{y_2 - y_1}{x_2 - x_1}$$

$$= \frac{6 - 2}{4 - 4}$$

$$= \frac{4}{0} \quad ✗$$

∴ Because division by zero is undefined, the slope of the line is undefined.

Study Tip

The slope of every horizontal line is 0. The slope of every vertical line is undefined.

On Your Own

Now You're Ready
Exercises 13–15

Find the slope of the line through the given points.

4. (1, −2), (7, −2)

5. (−2, 4), (3, 4)

6. (−3, −3), (−3, −5)

7. (0, 8), (0, 0)

8. How do you know that the slope of every horizontal line is 0? How do you know that the slope of every vertical line is undefined?

EXAMPLE 4 Finding Slope from a Table

The points in the table lie on a line. How can you find the slope of the line from the table? What is the slope?

x	1	4	7	10
y	8	6	4	2

Choose any two points from the table and use the slope formula.

Use the points $(x_1, y_1) = (1, 8)$ and $(x_2, y_2) = (4, 6)$.

$$m = \frac{y_2 - y_1}{x_2 - x_1}$$

$$= \frac{6 - 8}{4 - 1}$$

$$= \frac{-2}{3}, \text{ or } -\frac{2}{3}$$

The slope is $-\frac{2}{3}$.

Check

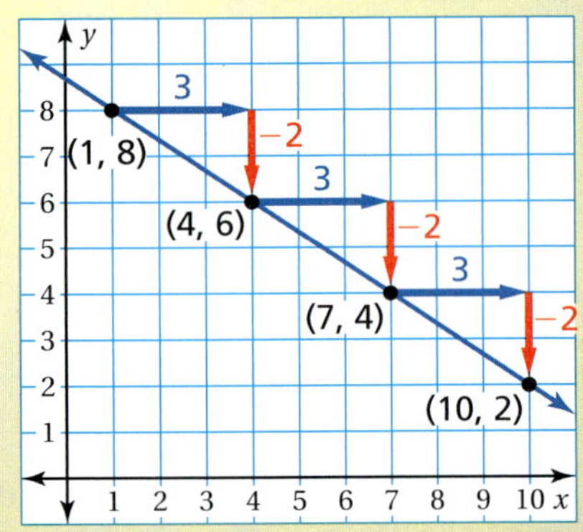

On Your Own

Now You're Ready Exercises 21–24

The points in the table lie on a line. Find the slope of the line.

9.

x	1	3	5	7
y	2	5	8	11

10.

x	−3	−2	−1	0
y	6	4	2	0

Summary

Slope

Positive Slope — The line rises from left to right.

Negative Slope — The line falls from left to right.

Slope of 0 — The line is horizontal.

Undefined Slope — The line is vertical.

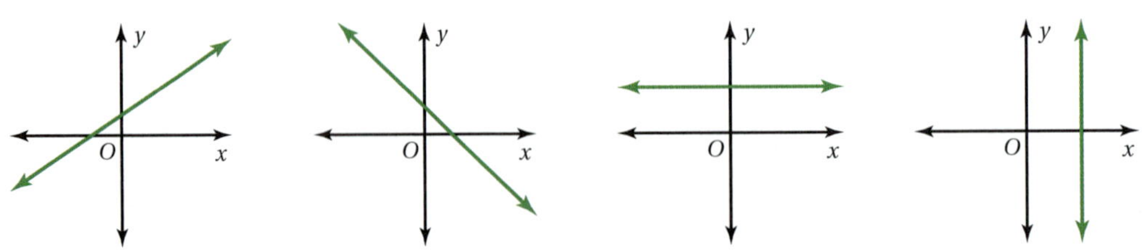

Laurie's Notes

Example 4

? "What do you notice about the x-values and the y-values?" The x-values are increasing by 3 and the y-values are decreasing by 2.

- **Connection:** Show the changes in x and y from each column to the next (i.e., $+3$, $+3$, $+3$ on the top and -2, -2, -2 on the bottom). Ask students to use the table and graph to determine if x and y are in a proportional relationship (no). Ask again in Question 10 (yes). This review of proportionality will help prepare for Section 13.3.
- Compute the slope between any two points in the table.
- Using (1, 8) and (4, 6): $m = \frac{y_2 - y_1}{x_2 - x_1} = \frac{6 - 8}{4 - 1} = -\frac{2}{3}$
- Using (1, 8) and (7, 4): $m = \frac{y_2 - y_1}{x_2 - x_1} = \frac{4 - 8}{7 - 1} = -\frac{4}{6} = -\frac{2}{3}$
- Remind students about the activity, where they found that the slope is the same regardless of which two points are selected. The slope triangles that are formed are similar.

? "The line has a negative slope. What do you notice about the line?" The line falls from left to right.

On Your Own

- **Neighbor Check:** Have students work independently and then have their neighbors check their work. Have students discuss any discrepancies.
- **Connection:** In Question 9, students may recognize from the table that both x and y are increasing and the slope is positive. In Question 10, as x increases, the y-values are decreasing and the slope is negative.
- Students should observe by inspection the change in x and the change in y. This is *reading* the table.

Summary

- Students have computed the slopes of many lines today. Discuss the Summary by referring to previous examples.

Closure

- The points in the table lie on a line. How can you find the slope of the line from the table? What is the slope?

x	-1	0	1	2
y	-3	-1	1	3

Choose any two points from the table and use the slope formula. $m = 2$

Extra Example 4

The points in the table lie on a line. How can you find the slope of the line from the table? What is the slope?

x	-2	-1	0	1
y	-8	-5	-2	1

Choose any two points from the table and use the slope formula; 3

On Your Own

9. $\frac{3}{2}$ **10.** -2

Vocabulary and Concept Check

1. **a.** B and C

   b. A

   c. no; None of the lines are vertical.

2. *Sample answer:* When constructing a wheelchair ramp, you need to know the slope.

3. The line is horizontal.

Practice and Problem Solving

4.

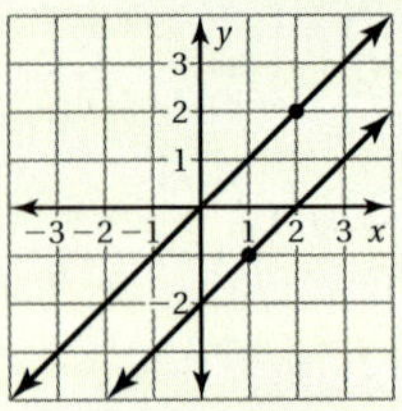

The lines are parallel.

5.

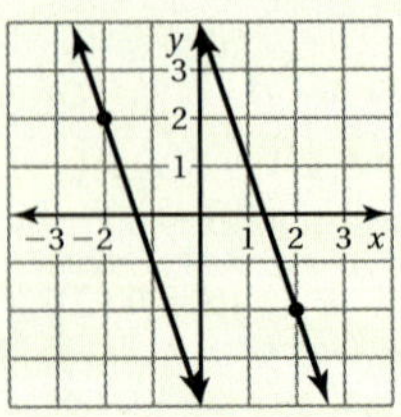

The lines are parallel.

6.

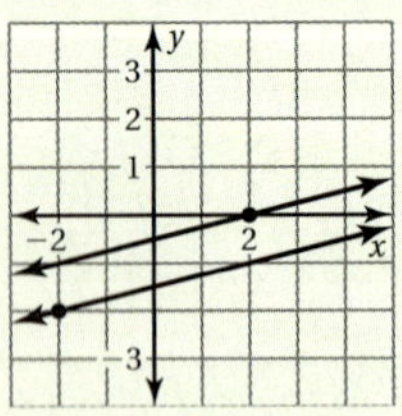

The lines are parallel.

7. $\frac{3}{4}$ **8.** $-\frac{5}{4}$

9. $-\frac{3}{5}$ **10.** $\frac{1}{6}$

11. 0

12. undefined

Assignment Guide and Homework Check

Level	Assignment	Homework Check
Accelerated	1–6, 12, 18, 19, 20–30 even, 31–40	18, 24, 26, 34, 35

Common Errors

- **Exercises 7–12** Students may forget negatives or include them when they are not needed. Remind them that if the line rises from left to right the slope is positive, and if the line falls from left to right the slope is negative.
- **Exercises 7–18** Students may find the reciprocal of the slope because they mix up rise and run. Remind them that the change in y is the numerator and the change in x is the denominator.

13.2 Record and Practice Journal

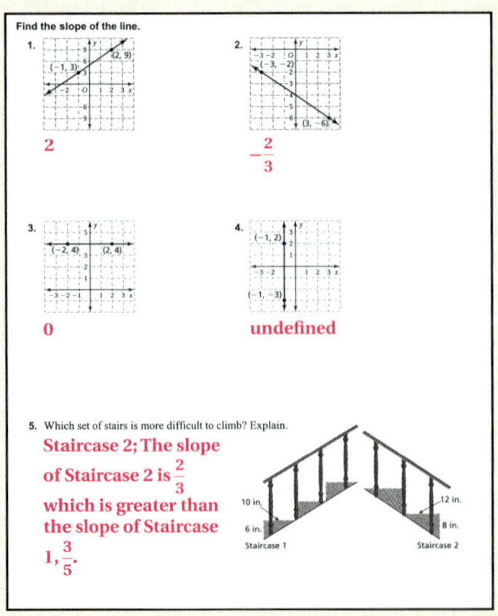

Find the slope of the line.

1. 2

2. $-\frac{2}{3}$

3. 0

4. undefined

5. Which set of stairs is more difficult to climb? Explain.

Staircase 2; The slope of Staircase 2 is $\frac{2}{3}$ which is greater than the slope of Staircase 1, $\frac{3}{5}$.

13.2 Exercises

Check It Out
Help with Homework
BigIdeasMath.com

Vocabulary and Concept Check

1. **CRITICAL THINKING** Refer to the graph.

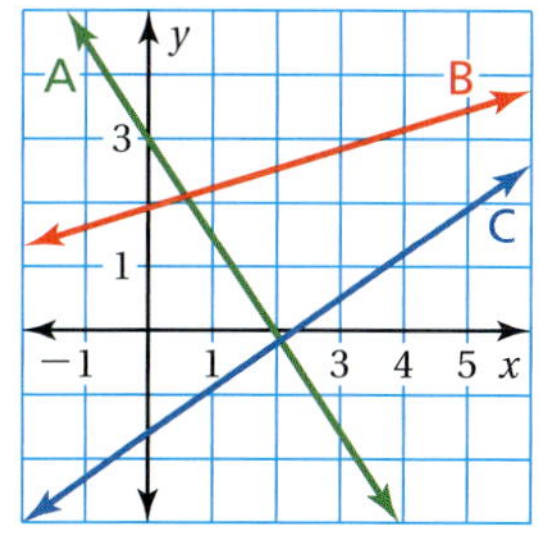

   a. Which lines have positive slopes?
   b. Which line has the steepest slope?
   c. Do any lines have an undefined slope? Explain.

2. **OPEN-ENDED** Describe a real-life situation in which you need to know the slope.

3. **REASONING** The slope of a line is 0. What do you know about the line?

Practice and Problem Solving

Draw a line through each point using the given slope. What do you notice about the two lines?

4. slope = 1

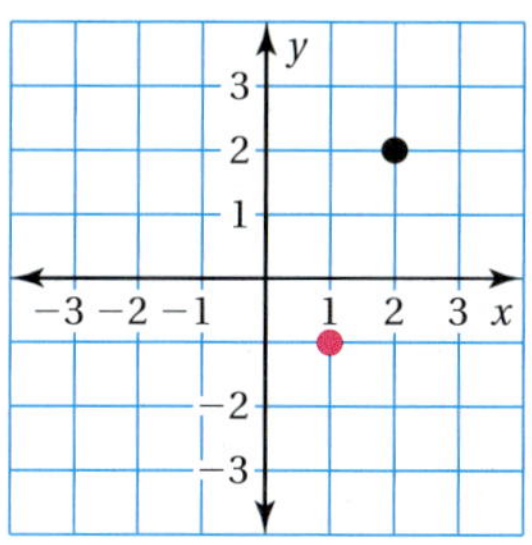

5. slope = −3

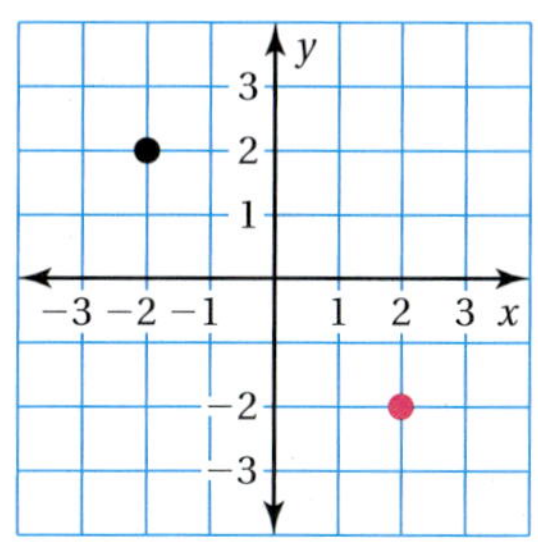

6. slope = $\frac{1}{4}$

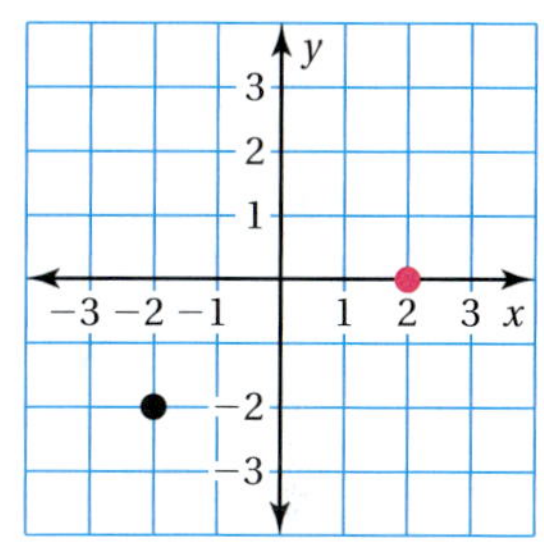

Find the slope of the line.

7.

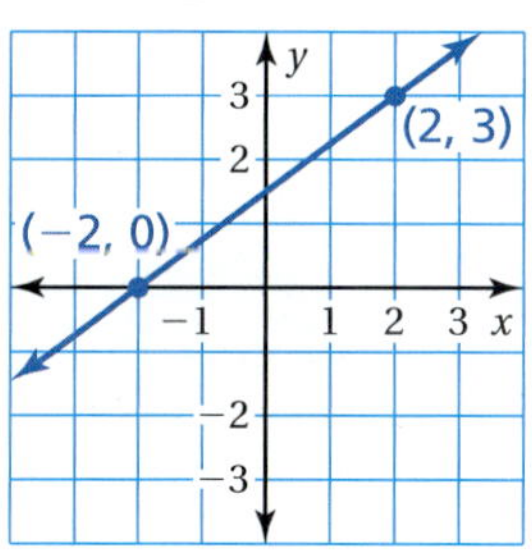

8.

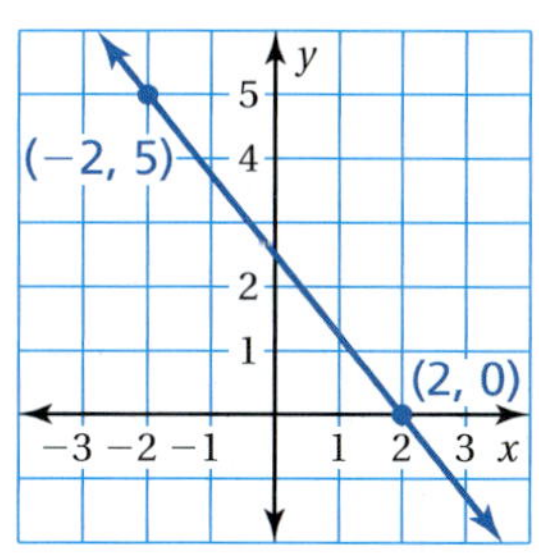

9.

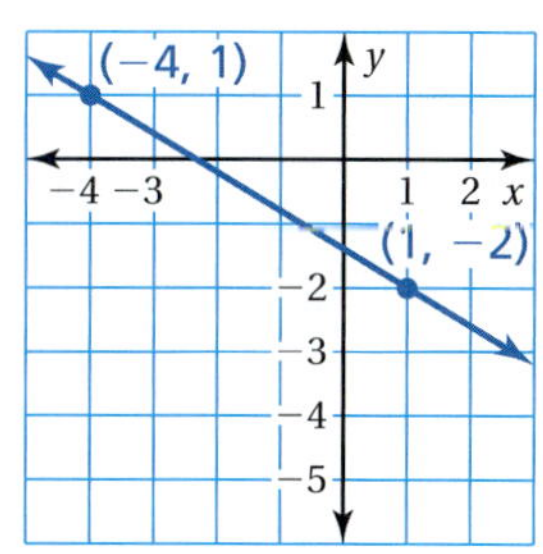

10.

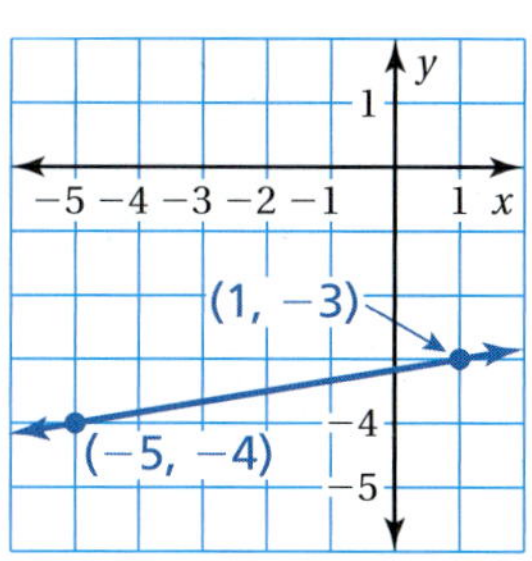

11.

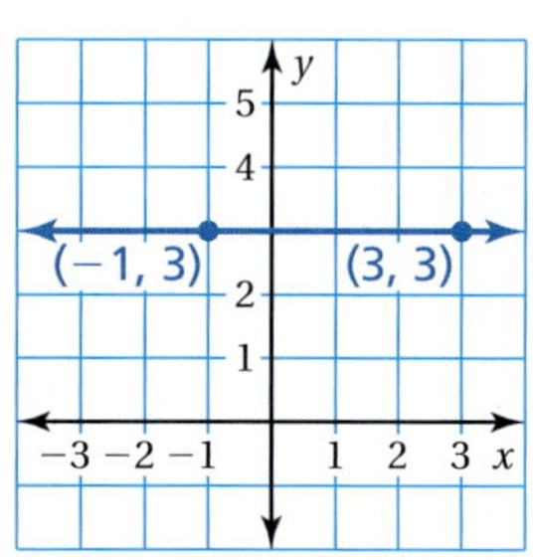

12.

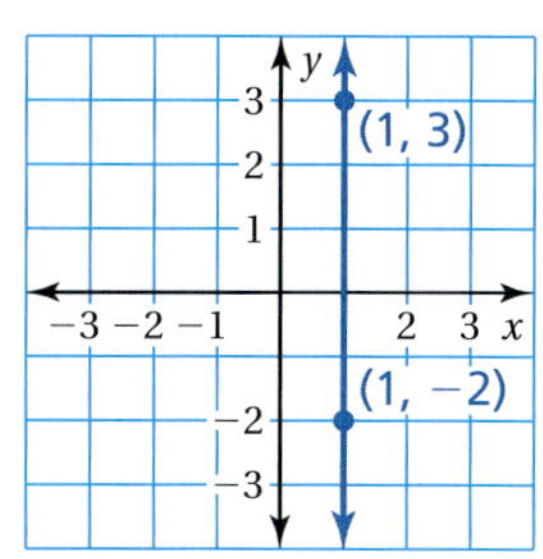

Find the slope of the line through the given points.

2 3 **13.** $(4, -1), (-2, -1)$ **14.** $(5, -3), (5, 8)$ **15.** $(-7, 0), (-7, -6)$

16. $(-3, 1), (-1, 5)$ **17.** $(10, 4), (4, 15)$ **18.** $(-3, 6), (2, 6)$

19. ERROR ANALYSIS Describe and correct the error in finding the slope of the line.

$$m = \frac{3-1}{4-2} = \frac{2}{2} = 1$$

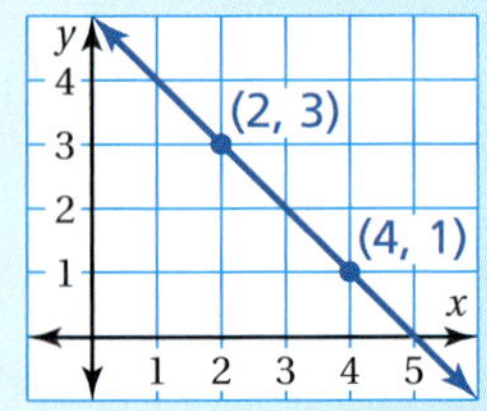

20. CRITICAL THINKING Is it more difficult to walk up the ramp or the hill? Explain.

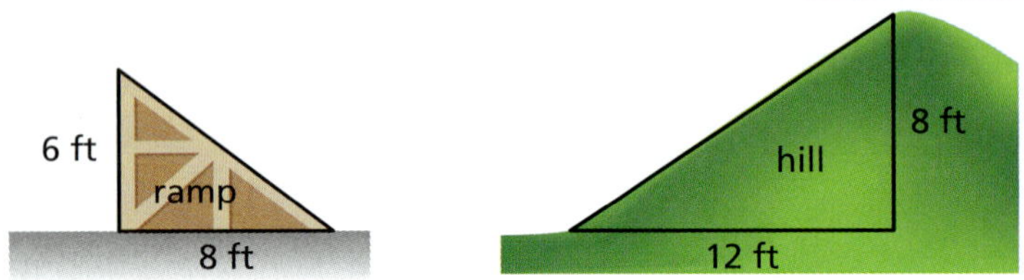

The points in the table lie on a line. Find the slope of the line.

4 **21.**

x	1	3	5	7
y	2	10	18	26

22.

x	-3	2	7	12
y	0	2	4	6

23.

x	-6	-2	2	6
y	8	5	2	-1

24.

x	-8	-2	4	10
y	8	1	-6	-13

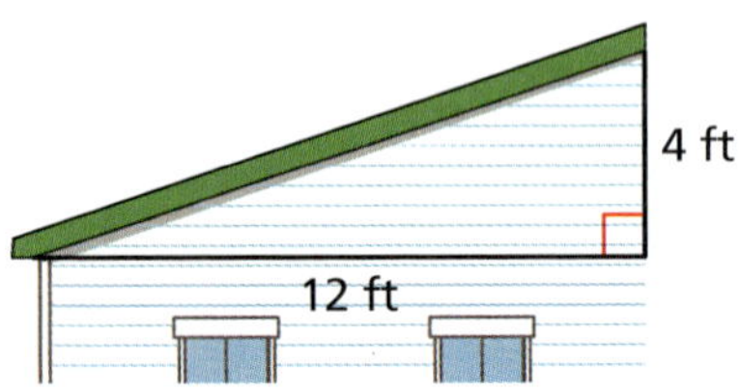

25. PITCH Carpenters refer to the slope of a roof as the *pitch* of the roof. Find the pitch of the roof.

26. PROJECT The guidelines for a wheelchair ramp suggest that the ratio of the rise to the run be no greater than 1 : 12.

a. CHOOSE TOOLS Find a wheelchair ramp in your school or neighborhood. Measure its slope. Does the ramp follow the guidelines?

b. Design a wheelchair ramp that provides access to a building with a front door that is 2.5 feet above the sidewalk. Illustrate your design.

Use an equation to find the value of *k* so that the line that passes through the given points has the given slope.

27. $(1, 3), (5, k); m = 2$ **28.** $(-2, k), (2, 0); m = -1$

29. $(-4, k), (6, -7); m = -\frac{1}{5}$ **30.** $(4, -4), (k, -1); m = \frac{3}{4}$

Common Errors

- **Exercise 20** Students may get confused because one of the slopes is negative and the other is positive. Tell them to think of the absolute values of the slopes when comparing. Encourage them to graph the slopes on a number line to check their answers.
- **Exercises 21–24** Students may find the change in *x* over the change in *y*. Remind them that slope is the change in *y* over the change in *x*.

Practice and Problem Solving

13. 0

14. undefined

15. undefined

16. 2

17. $-\frac{11}{6}$

18. 0

19. The denominator should be $2-4$.
$m=-1$

20. The ramp because its slope is steeper.

21. 4

22. $\frac{2}{5}$

23. $-\frac{3}{4}$

24. $-\frac{7}{6}$

25. $\frac{1}{3}$

26. See Additional Answers.

27. $k=11$

28. $k=4$

29. $k=-5$

30. $k=8$

Differentiated Instruction

Auditory

Discuss how the rate of change in a rate problem is related to slope. For example, the cost to travel on a turnpike (cost per mile) can be expressed as $\frac{\text{cost (in dollars)}}{\text{miles driven}}$, where the cost is the change in *y*-values and the miles driven is the change in *x*-values.

Practice and Problem Solving

31. **a.** $\frac{3}{40}$

b. The cost increases by $3 for every 40 miles you drive, or the cost increases by $0.075 for every mile you drive.

32. The boat ramp, because it has a 16.67% grade.

33. yes; The slopes are the same between the points.

34. $2750 per month

35. When you switch the coordinates, the differences in the numerator and denominator are the opposite of the numbers when using the slope formula. You still get the same slope.

36. See *Taking Math Deeper.*

Fair Game Review

37. $b = 25$ **38.** $n = 56$

39. $x = 7.5$ **40.** B

Mini-Assessment

Find the slope of the line.

1.

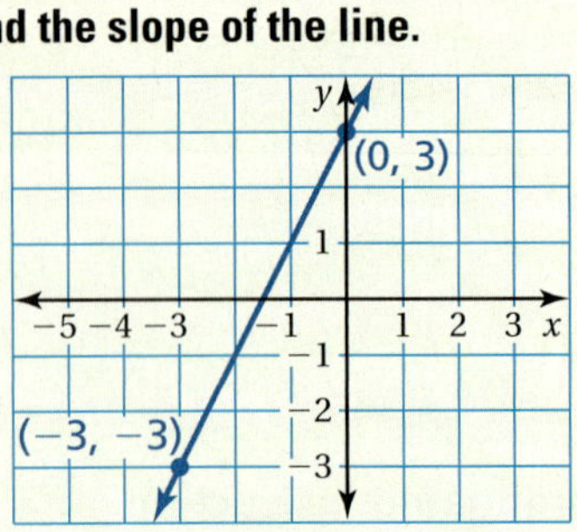

$m = 2$

2.

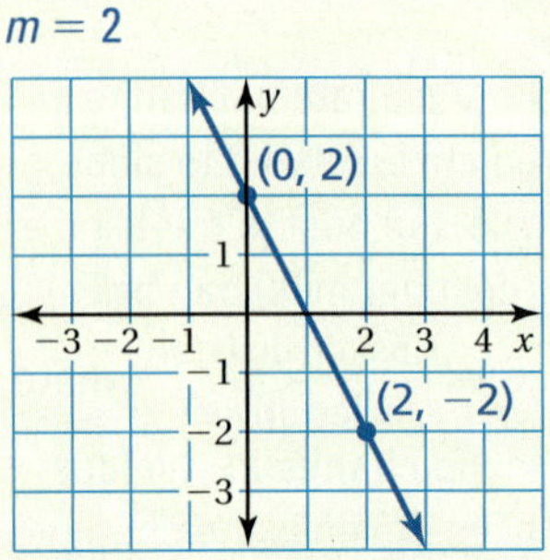

$m = -2$

Taking Math Deeper

Exercise 36

This exercise is a nice example of the power of a diagram. Instead of using the drawing of the slide, encourage students to draw the slide in a coordinate plane. Once that is done, the question is easier to answer.

1 Draw a diagram.

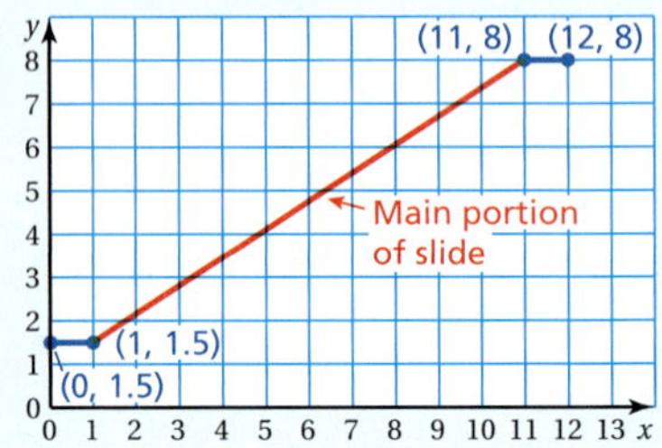

2 **a.** Find the slope of the slide.

$$m = \frac{y_2 - y_1}{x_2 - x_1} = \frac{8 - 1.5}{11 - 1} = \frac{6.5}{10} = 0.65$$

3 Compare the slopes.

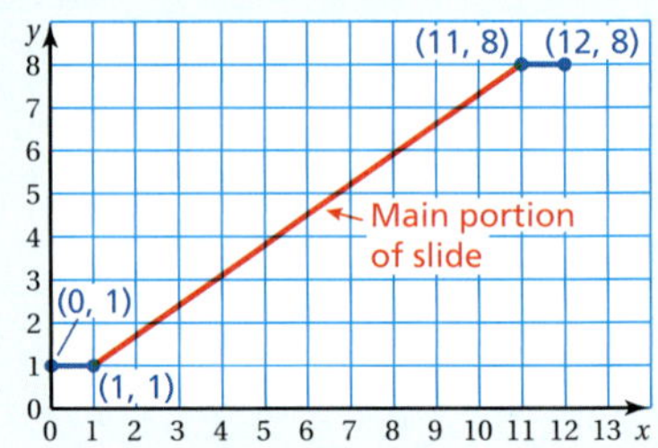

b. $$m = \frac{y_2 - y_1}{x_2 - x_1} = \frac{8 - 1}{11 - 1} = \frac{7}{10} = 0.7$$

Because 0.7 > 0.65, the slope increased and the slide is steeper.

Project

Many water parks and amusement parks have water slides. Find the height of a slide and calculate the slope of the main part of the slide.

Reteaching and Enrichment Strategies

If students need help...	If students got it...
Resources by Chapter • Practice A and Practice B • Puzzle Time Record and Practice Journal Practice Differentiating the Lesson Lesson Tutorials Skills Review Handbook	Resources by Chapter • Enrichment and Extension • Technology Connection Start the next section

31. **TURNPIKE TRAVEL** The graph shows the cost of traveling by car on a turnpike.

   a. Find the slope of the line.

   b. Explain the meaning of the slope as a rate of change.

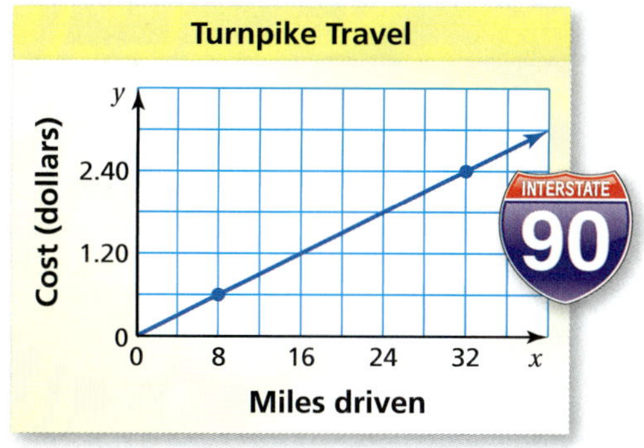

32. **BOAT RAMP** Which is steeper: the boat ramp or a road with a 12% grade? Explain. (*Note:* Road grade is the vertical increase divided by the horizontal distance.)

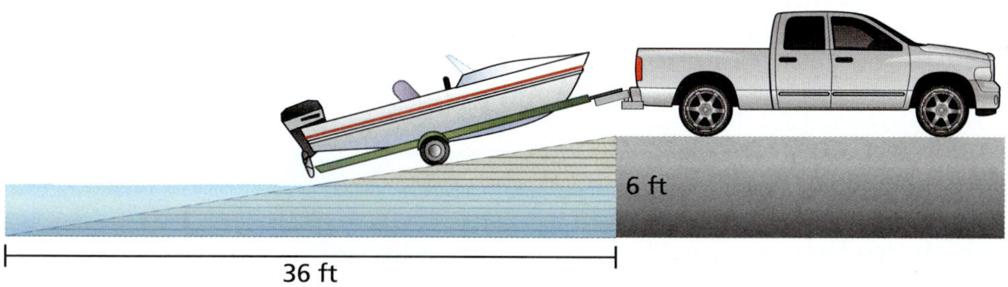

33. **REASONING** Do the points $A(-2, -1)$, $B(1, 5)$, and $C(4, 11)$ lie on the same line? Without using a graph, how do you know?

34. **BUSINESS** A small business earns a profit of \$6500 in January and \$17,500 in May. What is the rate of change in profit for this time period?

35. **STRUCTURE** Choose two points in the coordinate plane. Use the slope formula to find the slope of the line that passes through the two points. Then find the slope using the formula $\frac{y_1 - y_2}{x_1 - x_2}$. Explain why your results are the same.

36. Critical Thinking The top and the bottom of the slide are level with the ground, which has a slope of 0.

   a. What is the slope of the main portion of the slide?

   b. How does the slope change when the bottom of the slide is only 12 inches above the ground? Is the slide steeper? Explain.

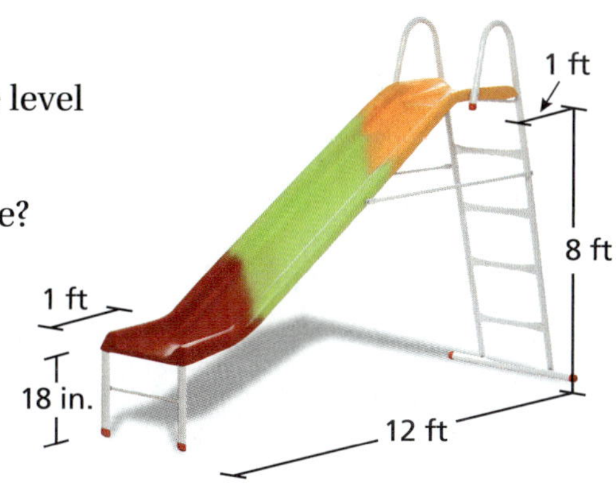

Fair Game Review What you learned in previous grades & lessons

Solve the proportion. *(Section 5.4)*

37. $\frac{b}{30} = \frac{5}{6}$

38. $\frac{7}{4} = \frac{n}{32}$

39. $\frac{3}{8} = \frac{x}{20}$

40. **MULTIPLE CHOICE** What is the prime factorization of 84? *(Skills Review Handbook)*

   (A) $2 \times 3 \times 7$ (B) $2^2 \times 3 \times 7$ (C) $2 \times 3^2 \times 7$ (D) $2^2 \times 21$

Extension 13.2 Slopes of Parallel and Perpendicular Lines

Check It Out
Lesson Tutorials
BigIdeasMath.com

Key Idea

Graphing Equations
In this extension, you will
- identify parallel and perpendicular lines.

Parallel Lines and Slopes

Lines in the same plane that do not intersect are parallel lines. Nonvertical parallel lines have the same slope.

All vertical lines are parallel.

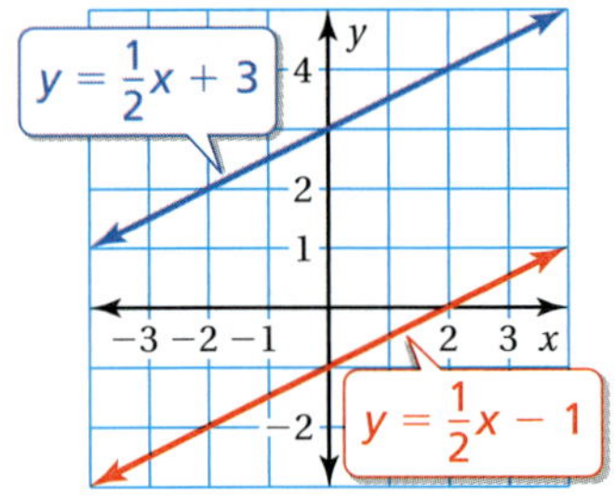

EXAMPLE 1 Identifying Parallel Lines

Which two lines are parallel? How do you know?

Find the slope of each line.

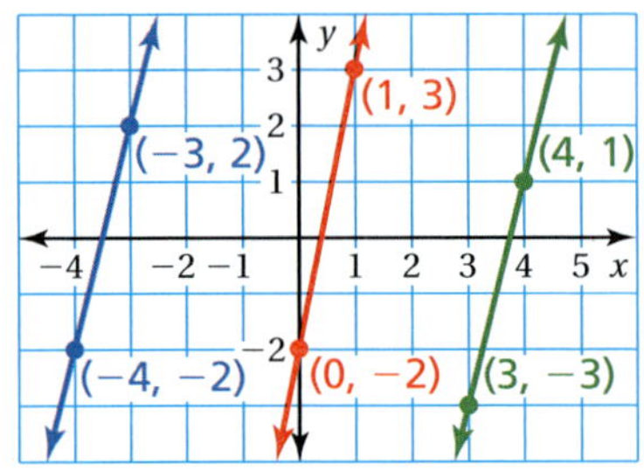

Blue Line

$$m = \frac{y_2 - y_1}{x_2 - x_1} = \frac{-2 - 2}{-4 - (-3)} = \frac{-4}{-1}, \text{ or } 4$$

Red Line

$$m = \frac{y_2 - y_1}{x_2 - x_1} = \frac{-2 - 3}{0 - 1} = \frac{-5}{-1}, \text{ or } 5$$

Green Line

$$m = \frac{y_2 - y_1}{x_2 - x_1} = \frac{-3 - 1}{3 - 4} = \frac{-4}{-1}, \text{ or } 4$$

The slopes of the blue and green lines are 4. The slope of the red line is 5.

∴ The blue and green lines have the same slope, so they are parallel.

Practice

Which lines are parallel? How do you know?

1.

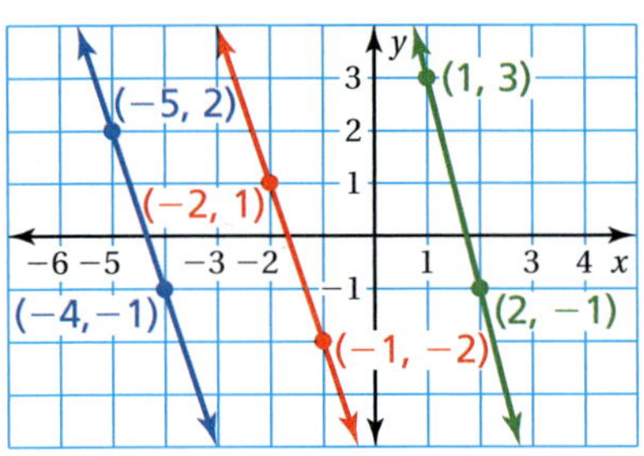

2.

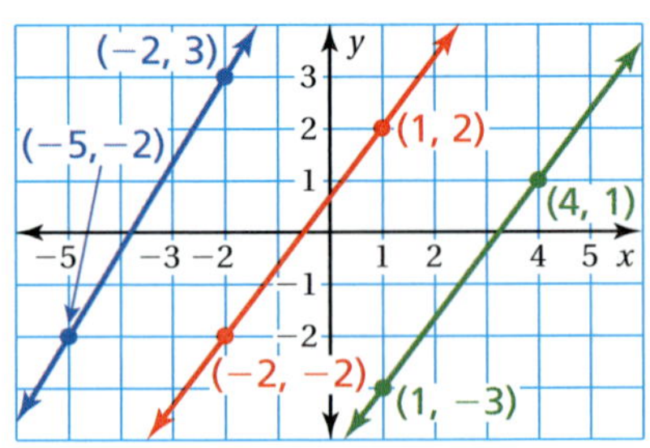

Are the given lines parallel? Explain your reasoning.

3. $y = -5, y = 3$

4. $y = 0, x = 0$

5. $x = -4, x = 1$

6. **GEOMETRY** The vertices of a quadrilateral are $A(-5, 3)$, $B(2, 2)$, $C(4, -3)$, and $D(-2, -2)$. How can you use slope to determine whether the quadrilateral is a parallelogram? Is it a parallelogram? Justify your answer.

Laurie's Notes

Introduction

Connect

- **Yesterday:** Students found slopes of lines.
- **Today:** Students will use slope to determine if lines are parallel or perpendicular.

Motivate

- The words *parallel* and *perpendicular* can be confusing for students and difficult to remember.
- The word parallel comes from para-, "beside," and allelois, "each other," so parallel lines are lines that are beside each other. For example, parallel bars in gymnastics have bars that are beside each other.
- The word perpendicular comes from perpendicularis, meaning "vertical, as a plumb line." A plumb line is a cord with a weight attached that is used to determine perpendicularity.
- You can refer back to the Motivate in Section 13.2 and ask about the angles of the quadrilateral and compare the slopes of the lines that form the quadrilateral.

Lesson Notes

Discuss

- The concepts presented in this extension were explored by students in Activity 3. Students graphed lines with a given slope through different ordered pairs.
- Refer to conjectures made regarding parallel and perpendicular lines.

Key Idea

- Write the Key Idea on the board.
- Model what a slope of $\frac{1}{2}$ means. Start at a point on the line and run 2 units for each unit you rise. Repeat for each line.

Example 1

? "How do you compute the slope for each line?" Students should use rise over run language, along with the formal definition, $\frac{y_2 - y_1}{x_2 - x_1}$.

- Work through the example. Students may look quickly and believe that all of the lines are parallel. They should compute the slope of each line to prove which lines are parallel.

Practice

- **Construct Viable Arguments and Critique the Reasoning of Others:** In Exercise 6, students are asked to make a conjecture and then justify their answers. Students need opportunities to construct viable arguments and share their thinking with other students.

What Your Students Will Learn

- Identify parallel and perpendicular lines using slope.

Goal Today's lesson is using slope to determine whether lines are parallel or perpendicular.

Technology for the Teacher

Dynamic Classroom

Lesson Tutorials
Lesson Plans
Answer Presentation Tool

Extra Example 1

Which lines are parallel? How do you know?

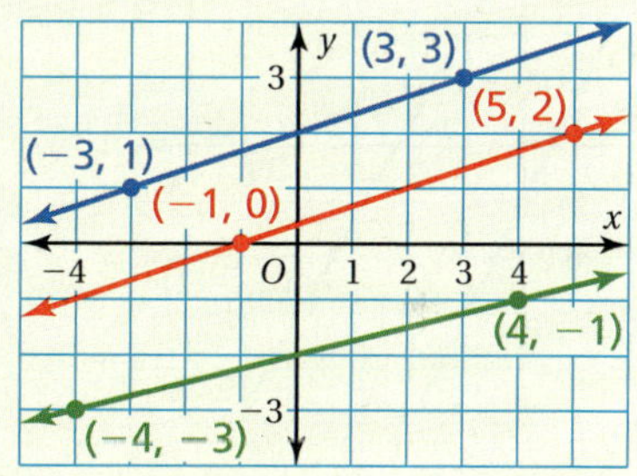

blue and red; They both have a slope of $\frac{1}{3}$.

Practice

1. blue and red; They both have a slope of -3.
2. red and green; They both have a slope of $\frac{4}{3}$.

3–6. See Additional Answers.

Record and Practice Journal Extension 13.2 Practice

1–10. See Additional Answers.

Extra Example 2

Which lines are perpendicular? How do you know?

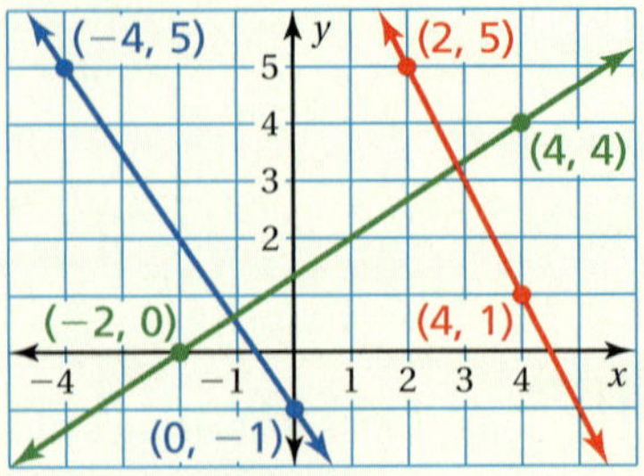

See Additional Answers.

Practice

7–12. See Additional Answers.

Mini-Assessment

1. Which two lines are parallel? How do you know?

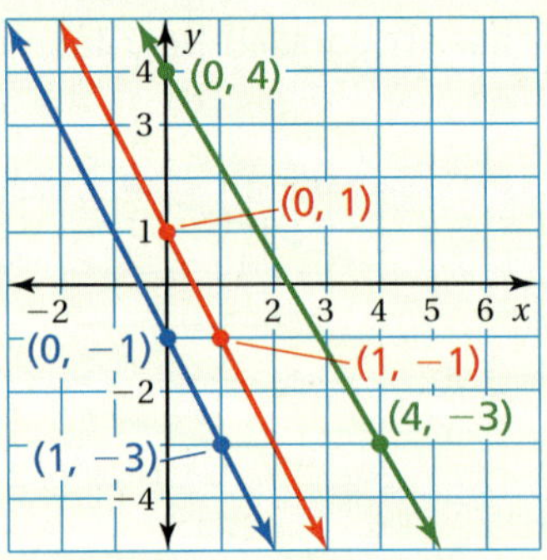

red and blue; Each has a slope of −2.

2. Which two lines are perpendicular? How do you know?

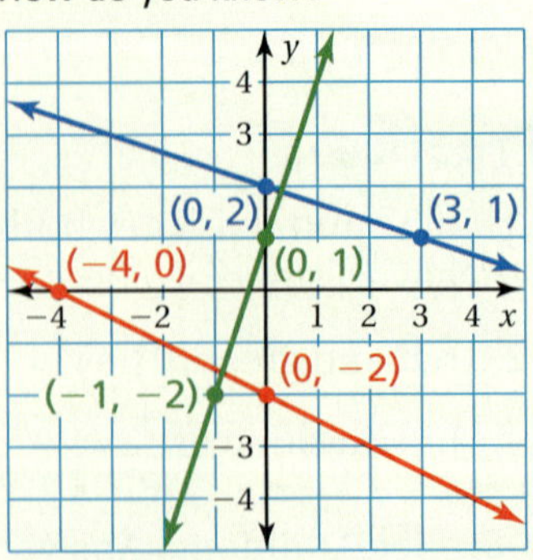

blue and green; The blue line has a slope of 3. The green line has a slope of $-\frac{1}{3}$. The product of their slopes is $3 \cdot -\frac{1}{3} = -1$.

Laurie's Notes

Key Idea

- Write the Key Idea on the board.
- **Teaching Tip:** Use the corner of a piece of paper, placed at the point of intersection, to provide a visual model of what perpendicular means.

Example 2

- Work through the example as shown.
- Students may also refer to the slopes of perpendicular lines as being opposite reciprocals.

Practice

- Because vertical lines have undefined slope, students cannot multiply slopes to determine if the lines are perpendicular. In Exercises 9 and 11, students should give reasoning that states horizontal and vertical lines are perpendicular. In Exercise 10, students should give reasoning that states that the lines are not perpendicular because they are both vertical lines.
- **Construct Viable Arguments and Critique the Reasoning of Others:** In Exercise 12, students are asked to make conjectures and then justify their answers. Students need opportunities to construct viable arguments and share their thinking with other students.

Closure

- **Exit Ticket:** Which lines are parallel? Which lines are perpendicular? How do you know?

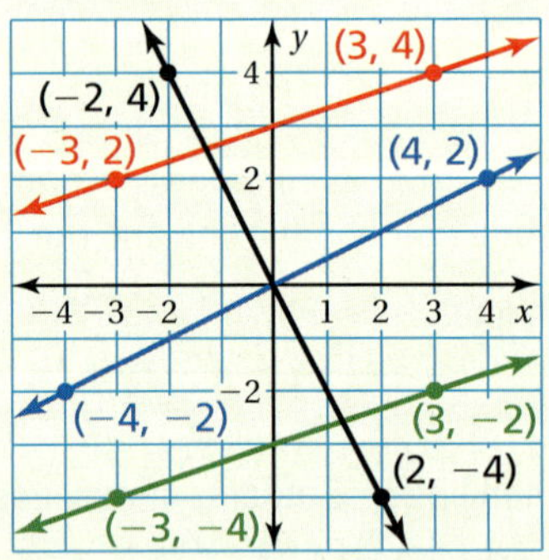

The red and green lines are parallel. Each has a slope of $\frac{1}{3}$. The black and blue lines are perpendicular. The product of their slopes is −1.

Key Idea

Perpendicular Lines and Slope

Lines in the same plane that intersect at right angles are perpendicular lines. Two nonvertical lines are perpendicular when the product of their slopes is -1.

Vertical lines are perpendicular to horizontal lines.

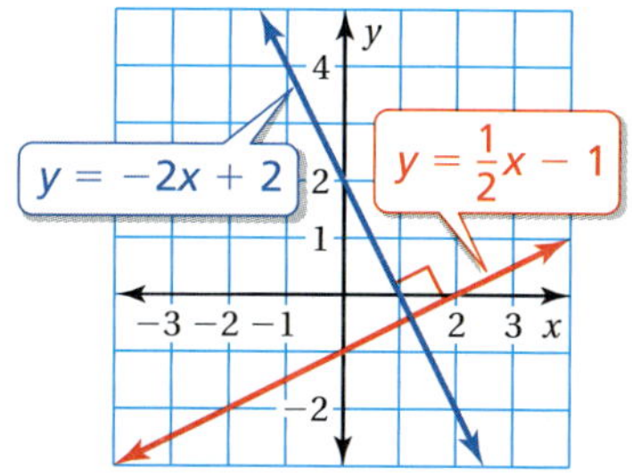

EXAMPLE 2 Identifying Perpendicular Lines

Which two lines are perpendicular? How do you know?

Find the slope of each line.

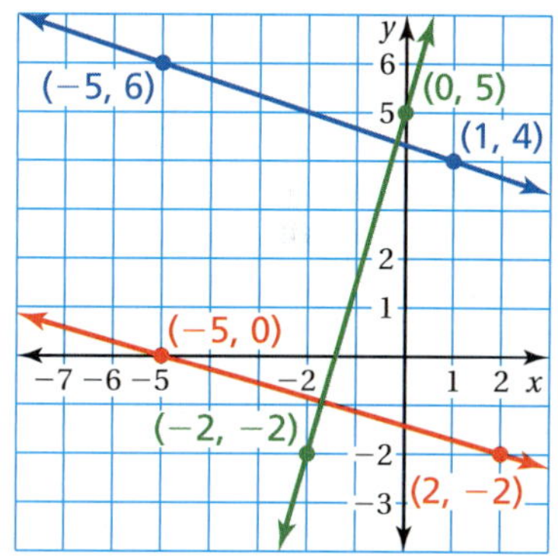

Blue Line

$$m = \frac{y_2 - y_1}{x_2 - x_1} = \frac{4 - 6}{1 - (-5)} = \frac{-2}{6}, \text{ or } -\frac{1}{3}$$

Red Line

$$m = \frac{y_2 - y_1}{x_2 - x_1} = \frac{-2 - 0}{2 - (-5)} = -\frac{2}{7}$$

Green Line

$$m = \frac{y_2 - y_1}{x_2 - x_1} = \frac{5 - (-2)}{0 - (-2)} = \frac{7}{2}$$

The slope of the red line is $-\frac{2}{7}$. The slope of the green line is $\frac{7}{2}$.

Because $-\frac{2}{7} \cdot \frac{7}{2} = -1$, the red and green lines are perpendicular.

Practice

Which lines are perpendicular? How do you know?

7.

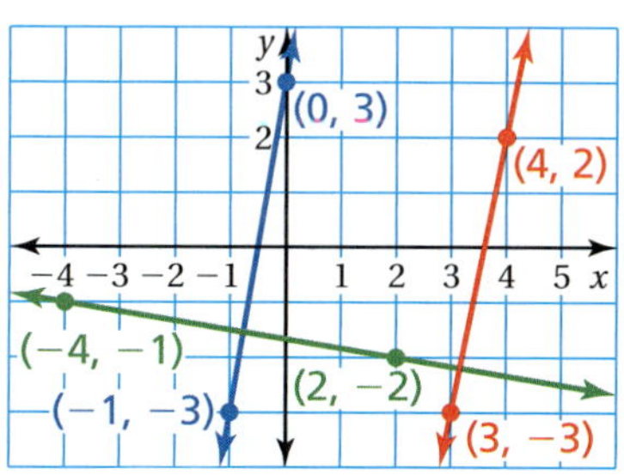

8.

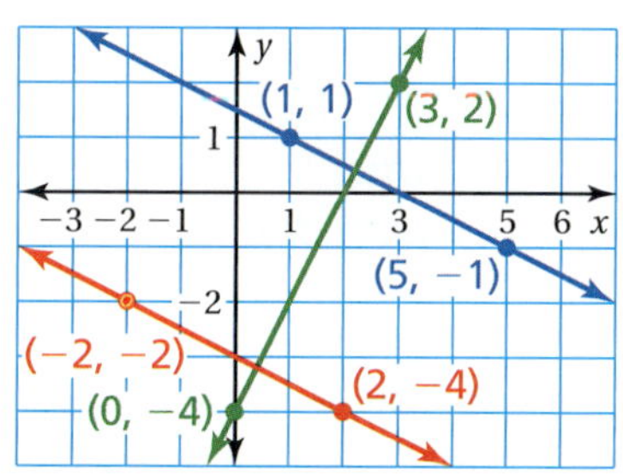

Are the given lines perpendicular? Explain your reasoning.

9. $x = -2, y = 8$ **10.** $x = -8, x = 7$ **11.** $y = 0, x = 0$

12. **GEOMETRY** The vertices of a parallelogram are $J(-5, 0)$, $K(1, 4)$, $L(3, 1)$, and $M(-3, -3)$. How can you use slope to determine whether the parallelogram is a rectangle? Is it a rectangle? Justify your answer.

13.3 Graphing Proportional Relationships

Essential Question How can you describe the graph of the equation $y = mx$?

1 ACTIVITY: Identifying Proportional Relationships

Work with a partner. Tell whether x and y are in a proportional relationship. Explain your reasoning.

a. **Money**

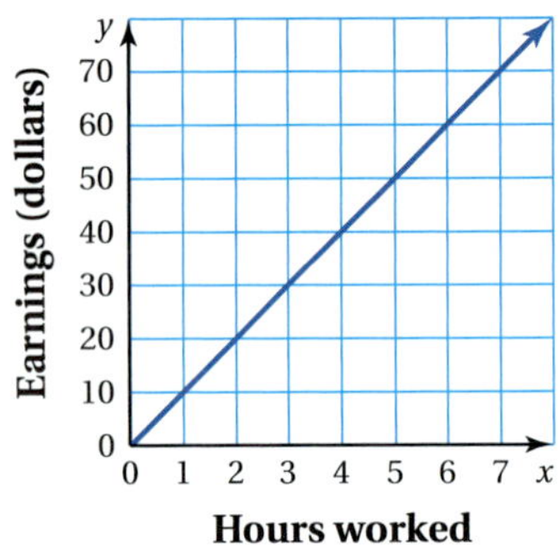

b. **Helicopter**

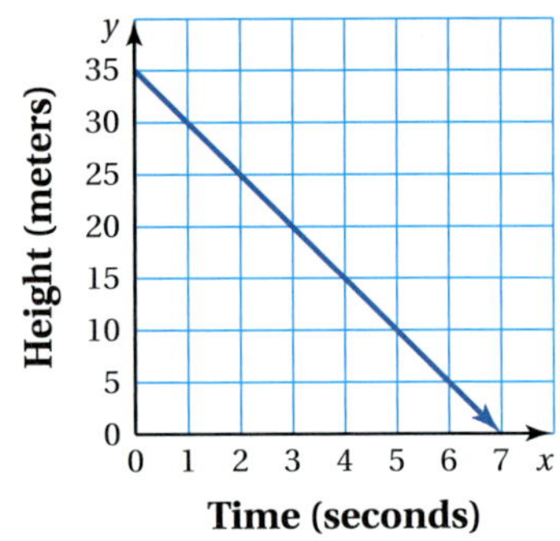

c. **Tickets**

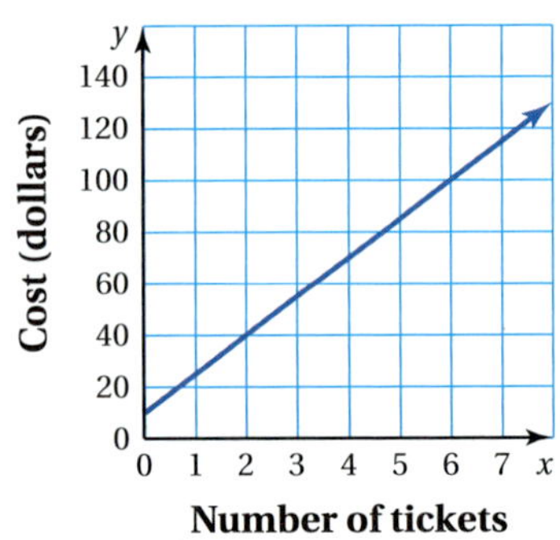

d. **Pizzas**

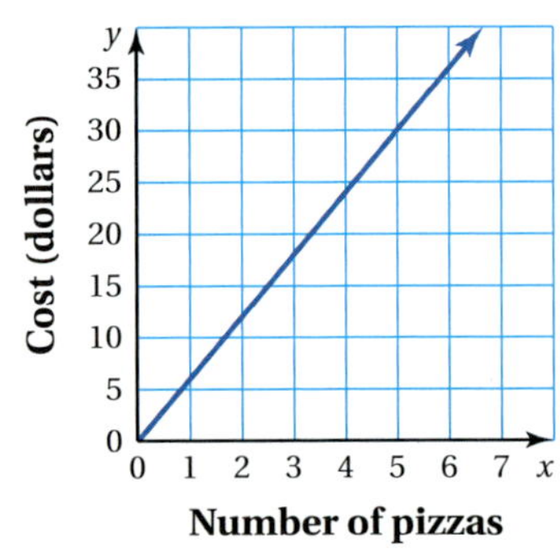

e.

Laps, x	1	2	3	4
Time (seconds), y	90	200	325	480

f.

Cups of Sugar, x	$\frac{1}{2}$	1	$1\frac{1}{2}$	2
Cups of Flour, y	1	2	3	4

Graphing Equations
In this lesson, you will
- write and graph proportional relationships.

2 ACTIVITY: Analyzing Proportional Relationships

Work with a partner. Use only the proportional relationships in Activity 1 to do the following.

- **Find the slope of the line.**
- **Find the value of y for the ordered pair $(1, y)$.**

What do you notice? What does the value of y represent?

Laurie's Notes

Introduction

Applying Mathematical Practices

- **Reason Abstractly and Quantitatively:** Mathematically proficient students make sense of the quantities and their relationships in problem situations. To develop this proficiency, students must be asked to interpret the meaning of a slope as a unit rate.

Motivate

- As a warm-up and to connect to prior content, tell students that x and y are in a proportional relationship. Have students fill in the ratio table.

Minutes, x	1	2	4	6
Gallons, y	8.5			

- **Reason Abstractly and Quantitatively** and **Construct Viable Arguments:** Ask volunteers to justify their procedures and explain why their procedures show a proportional relationship.

Activity Notes

Discuss

- You may wish to do a quick review before students start the activity.
- ? "What is a proportion?" an equation stating that two ratios are equivalent
- ? "How can you tell if two ratios form a proportion?" There are different methods. Students may mention the Cross Products Property.
- ? "Does anyone remember what a direct variation equation is?" an equation that can be written in the form $y = kx$; its graph passes through (0, 0)

Activity 1

- Expect students to solve parts (a)–(d) by observing whether the graph passes through the origin.
- **Reason Abstractly and Quantitatively:** Alternative reasoning in part (a) could be that the person earns $10 in 1 hour, $20 in 2 hours, $30 in 3 hours, and so on. This is $10 every hour, which describes a proportional relationship.
- Some students may want to graph the ordered pairs in parts (e) and (f). Consider asking them how they can solve without graphing (check for equivalent ratios).

Activity 2

- The goal of this activity is for students to connect the unit rate, the slope m, and the point $(1, y)$ on the graph of a proportional relationship.
- Students may correctly find the slope and say that the y-coordinate in $(1, y)$ is the slope, but they may not make the connection to the unit rate. One way to help them see the connection is to tell them to interpret the ordered pair. The ordered pair $(1, y)$ in part (a) means (1 hour, $10 earned).

What Your Students Will Learn

- Write, graph, and compare proportional relationships.
- Write and use direct variation equations.

Previous Learning

Students have determined whether two quantities are proportional. Students have also graphed linear equations and found slopes of lines.

Technology for the Teacher

Dynamic Classroom

Lesson Plans
Complete Materials List

13.3 Record and Practice Journal

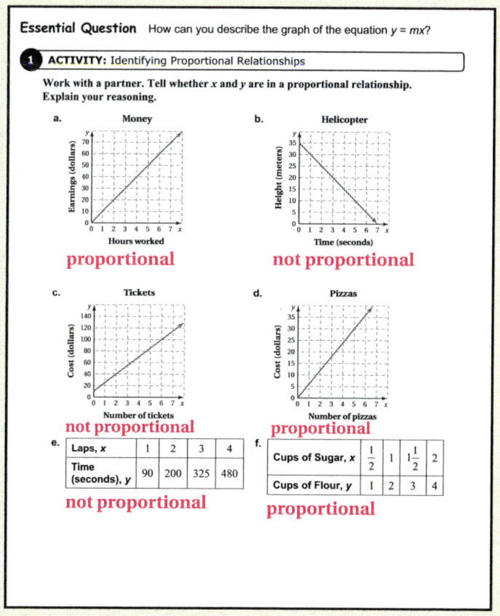

Essential Question How can you describe the graph of the equation $y = mx$?

1 ACTIVITY: Identifying Proportional Relationships

Work with a partner. Tell whether x and y are in a proportional relationship. Explain your reasoning.

a. Money — proportional

b. Helicopter — not proportional

c. Tickets — not proportional

d. Pizzas — proportional

e.

Laps, x	1	2	3	4
Time (seconds), y	90	200	325	480

not proportional

f.

Cups of Sugar, x	$\frac{1}{2}$	1	$1\frac{1}{2}$	2
Cups of Flour, y	1	2	3	4

proportional

English Language Learners

Word Problems

Most word problems follow a standard format that allows English learners to recognize key words that are integral to writing a mathematical statement of the problem. Most numbers given in a word problem are used. Analyzing the units in the mathematical statement and determining the units of the solution give students confidence that they are on the right path to solving the problem.

13.3 Record and Practice Journal

2 ACTIVITY: Analyzing Proportional Relationships

Work with a partner. Use only the proportional relationships in Activity 1 to do the following.

- **Find the slope of the line.**
- **Find the value of y for the ordered pair $(1, y)$.**

What do you notice? What does the value of y represent?

See Additional Answers.

3 ACTIVITY: Deriving an Equation

Work with a partner. Let (x, y) represent any point on the graph of a proportional relationship.

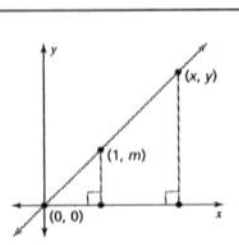

a. Explain why the two triangles are similar.

Angles are congruent.

b. Because the triangles are similar, the corresponding side lengths are proportional. Use the vertical and horizontal side lengths to complete the steps below.

$\dfrac{y - 0}{x - 0} = \dfrac{m}{1}$ Ratios of side lengths

$\dfrac{y}{x} = m$ Simplify.

$y = m \cdot x$ Multiplication Property of Equality

What does the final equation represent?

general equation for a proportional relationship

c. Use your result in part (b) to write an equation that represents each proportional relationship in Activity 1.

(a) $y = 10x$
(d) $y = 6x$
(f) $y = 2x$

What Is Your Answer?

4. **IN YOUR OWN WORDS** How can you describe the graph of the equation $y = mx$? How does the value of m affect the graph of the equation?

a line with slope m that passes through the origin; steepness of the line

5. Give a real-life example of two quantities that are in a proportional relationship. Write an equation that represents the relationship and sketch its graph.

Sample answer: amount of gasoline purchased (gallons) and the amount spent (dollars); $y = 3.5x$

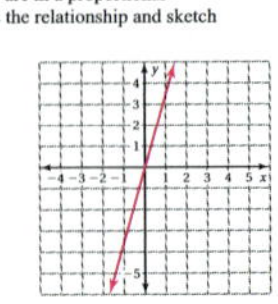

Laurie's Notes

Activity 3

- This activity uses the result of Activity 2. If you know the direct variation equation representing a proportional relationship, then you can find the slope by solving the equation for y when $x = 1$. Therefore, you can label the ordered pair $(1, m)$.
- **Reason Abstractly and Quantitatively:** In part (a), students need to use what they learned in Section 12.4, that two triangles are similar when they have two pairs of congruent angles.
- The triangles in the diagram share an angle (at the origin) and both have a right angle. So, the triangles have two pairs of congruent angles and are similar.

? "When two triangles are similar, what do you know about their side lengths?" Corresponding side lengths are proportional.

- It may not be obvious to students that the vertical side of the larger triangle has a length of y units and the horizontal side of the larger triangle has a length of x units. It is less obvious to students because the ordered pair is represented by variables rather than numbers.
- Ask students to share their responses for part (c).

What Is Your Answer?

- **Think-Pair-Share:** Students should read each question independently and then work in pairs to answer the questions. When they have answered the questions, the pair should compare their answers with another group and discuss any discrepancies.

Closure

- **Exit Ticket:** Refer back to the ratio table in the Motivate with minutes and gallons.
  a. Write an equation that represents this situation.
  b. What is the value of x when y is 85 gallons? Explain.

3 ACTIVITY: Deriving an Equation

Work with a partner. Let (x, y) represent any point on the graph of a proportional relationship.

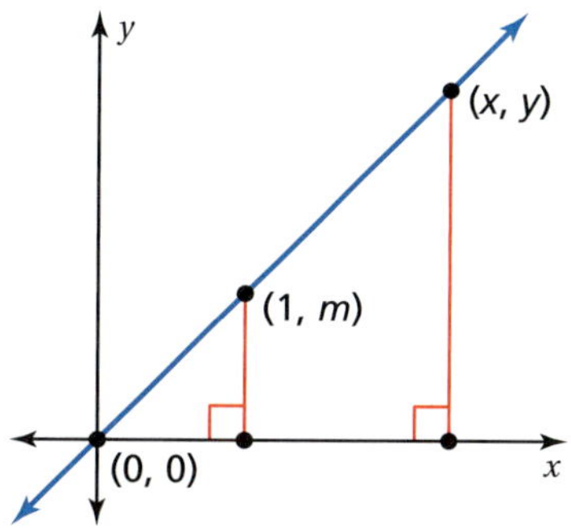

a. Explain why the two triangles are similar.

b. Because the triangles are similar, the corresponding side lengths are proportional. Use the vertical and horizontal side lengths to complete the steps below.

$\frac{\square}{\square} = \frac{m}{1}$ Ratios of side lengths

$\frac{\square}{\square} = m$ Simplify.

$\square = m \cdot \square$ Multiplication Property of Equality

What does the final equation represent?

c. Use your result in part (b) to write an equation that represents each proportional relationship in Activity 1.

Math Practice

View as Components

What part of the graph can you use to find the side lengths?

What Is Your Answer?

4. **IN YOUR OWN WORDS** How can you describe the graph of the equation $y = mx$? How does the value of m affect the graph of the equation?

5. Give a real-life example of two quantities that are in a proportional relationship. Write an equation that represents the relationship and sketch its graph.

Practice Use what you learned about proportional relationships to complete Exercises 3–6 on page 586.

13.3 Lesson

Check It Out
Lesson Tutorials
BigIdeasMath.com

Key Idea

Study Tip

In the direct variation equation $y = mx$, m represents the constant of proportionality, the slope, and the unit rate.

Direct Variation

Words When two quantities x and y are proportional, the relationship can be represented by the direct variation equation $y = mx$, where m is the constant of proportionality.

Graph The graph of $y = mx$ is a line with a slope of m that passes through the origin.

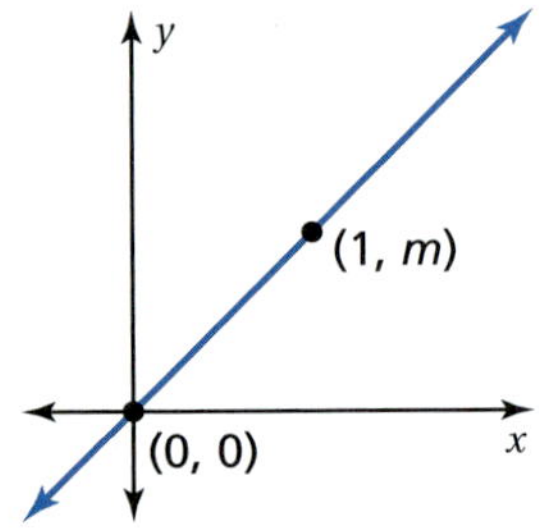

EXAMPLE 1 Graphing a Proportional Relationship

The cost y (in dollars) for x gigabytes of data on an Internet plan is represented by $y = 10x$. Graph the equation and interpret the slope.

The equation shows that the slope m is 10. So, the graph passes through (0, 0) and (1, 10).

Plot the points and draw a line through the points. Because negative values of x do not make sense in this context, graph in the first quadrant only.

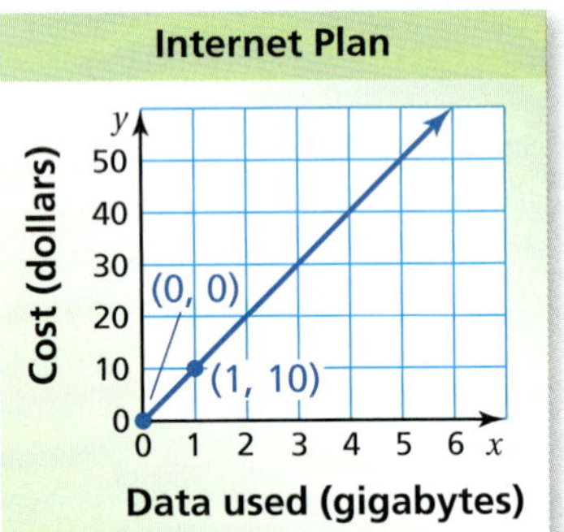

∴ The slope indicates that the unit cost is \$10 per gigabyte.

EXAMPLE 2 Writing and Using a Direct Variation Equation

The weight y of an object on Titan, one of Saturn's moons, is proportional to the weight x of the object on Earth. An object that weighs 105 pounds on Earth would weigh 15 pounds on Titan.

a. Write an equation that represents the situation.

Use the point (105, 15) to find the slope of the line.

$y = mx$	Direct variation equation
$15 = m(105)$	Substitute 15 for y and 105 for x.
$\frac{1}{7} = m$	Simplify.

∴ So, an equation that represents the situation is $y = \frac{1}{7}x$.

Study Tip

In Example 2, the slope indicates that the weight of an object on Titan is one-seventh its weight on Earth.

b. How much would a chunk of ice that weighs 3.5 pounds on Titan weigh on Earth?

$3.5 = \frac{1}{7}x$	Substitute 3.5 for y.
$24.5 = x$	Multiply each side by 7.

∴ So, the chunk of ice would weigh 24.5 pounds on Earth.

Laurie's Notes

Introduction

Connect

- **Yesterday:** Students explored slopes of proportional relationships.
- **Today:** Students will write and graph proportional relationships.

Motivate

- Ask if anyone has heard of Saturn's largest known moon, Titan. One of today's examples will be about Titan.
- Share some of the interesting information below. You can find more information about Titan at *www.nasa.gov*.
  - Titan is larger than our moon and the planet Mercury.
  - Titan has clouds and a thick atmosphere, like some planets.
  - It may rain "gasoline-like" liquids on Titan.

Lesson Notes

Key Idea

- Write the Key Idea. Draw the graph and discuss the two labeled points.
- Make sure students understand that a direct variation equation is a special kind of linear equation, one whose graph passes through the origin.
- Discuss the Study Tip to help students connect prior concepts. In the equation $y = mx$, you can think of the coefficient m as the unit rate, constant of proportionality, and slope.

Example 1

? "Do x and y show direct variation?" Yes, the equation is of the form $y = mx$.

? "In the context of the problem, can x be negative?" no

- There are many points that can be graphed. Use (0, 0) and (1, m).
- **Attend to Precision:** Ask students to interpret the slope of the line. Students should precisely refer to the *unit* cost of \$10 per gigabyte.
- **Extension:** Many of these plans are tiered such that once you exceed a whole number of gigabytes, the cost jumps \$10 at once. Share this with students and ask if this is also a proportional relationship. Discuss the differences.

Example 2

- Read the problem. The first sentence explains how the ordered pairs will be written (weight on Earth, weight on Titan).

? "What two ordered pairs do we know?" (0, 0) and (105, 15)

- **Reason Abstractly and Quantitatively:** In part (a), students should recognize that they know an ordered pair that satisfies the equation $y = mx$. Substitute the x- and y-values and solve for m.
- In part (b), you need to find x when y is 3.5. Substitute and solve.
- **Connection:** Have students solve part (b) using a ratio table.

Goal Today's lesson is graphing proportional relationships.

Technology for the Teacher

Dynamic Classroom

Lesson Tutorials
Lesson Plans
Answer Presentation Tool

Differentiated Instruction

Visual

Encourage students to write down notes when solving word problems or to underline relevant information and cross out irrelevant information. Allow students to do this on handouts and tests.

Extra Example 1

The cost y (in dollars) to rent x video games is represented by $y = 4x$. Graph the equation and interpret the slope.

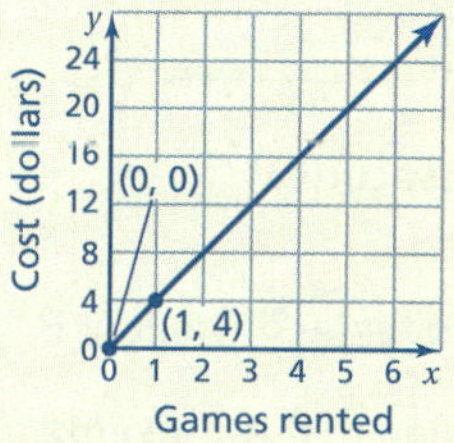

The slope indicates that the unit cost is \$4 per video game.

Extra Example 2

The daily wage y (in dollars) of a factory worker is proportional to the number of parts x assembled in a day. A worker who assembles 250 parts in a day earns \$75.

a. Write an equation that represents the situation. $y = \frac{3}{10}x$

b. How much does a worker earn who assembles 300 parts in a day? \$90

On Your Own

1. 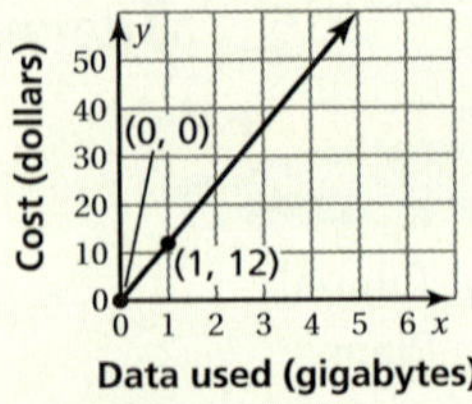

The slope indicates that unit cost is \$12 per gigabyte.

2. 500 kg

Extra Example 3

At a track event, the distance y (in meters) traveled by Student A in x seconds is represented by the equation $y = 7x$. The graph shows the distance traveled by Student B.

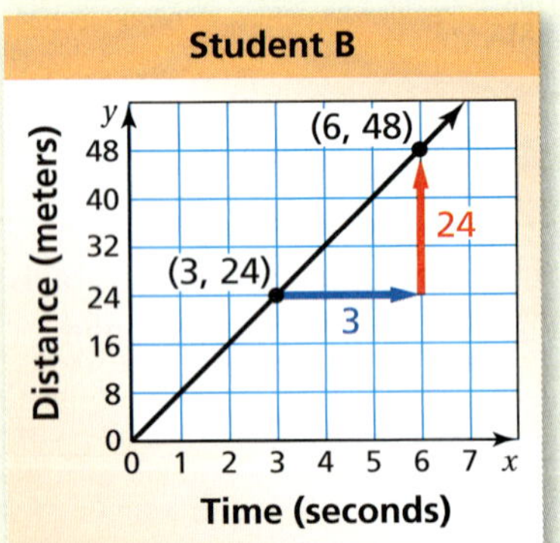

a. Which student is faster? Student B

b. Graph the equation that represents Student A in the same coordinate plane as Student B. Compare the steepness of the graphs. What does this mean in the context of the problem? See Additional Answers.

On Your Own

3. The T-bar ski lift speed of 2.25 meters per second is faster than the two-person lift, but slower than the four-person lift.

Laurie's Notes

On Your Own

- **Neighbor Check:** Have students work independently and then have their neighbors check their work. Have students discuss any discrepancies.

Example 3

- Discuss why these relationships are proportional. For instance, the two-person lift starts at 0, travels 2 meters in 1 second, 4 meters in 2 seconds, 6 meters in 3 seconds, and so on. $\frac{2\text{ m}}{1\text{ sec}} = \frac{4\text{ m}}{2\text{ sec}} = \frac{6\text{ m}}{3\text{ sec}}$.
- **Attend to Precision:** Make sure that students recognize the unit labels for each axis. The x-axis represents time (in seconds) and the y-axis represents distance (in meters).
- **Connection:** The question asks which ski lift is faster. Students are looking for the rate, or the speed of each lift. The rate is the slope of the line. For the two-person lift, the slope can be found using the ordered pairs in the graph. For the four-person lift, the slope is given in the equation.
- In part (b), point out to students that the graphs do not represent the steepness of the lifts, but rather the distance traveled (y-axis) over a period of time (x-axis).

? "If a vertical line is drawn through the graph in part (b) at $x = 4$, then it will intersect the two lines. What do these points of intersection mean in the context of the problem?" The y-value is the distance traveled by each lift in 4 seconds. The four-person lift travels farther in 4 seconds.

On Your Own

? "Is this a ratio table? Explain." yes; The ratios are all equivalent with a unit rate of 2.25 meters per second.

Closure

- **Writing Prompt:** "If a relationship is proportional, then . . ." the graph of the relationship goes through the origin and can be represented by an equation of the form $y = mx$, where m is the slope of the line.

On Your Own

Now You're Ready
Exercises 7–8

1. **WHAT IF?** In Example 1, the cost is represented by $y = 12x$. Graph the equation and interpret the slope.

2. In Example 2, how much would a spacecraft that weighs 3500 kilograms on Earth weigh on Titan?

EXAMPLE 3 Comparing Proportional Relationships

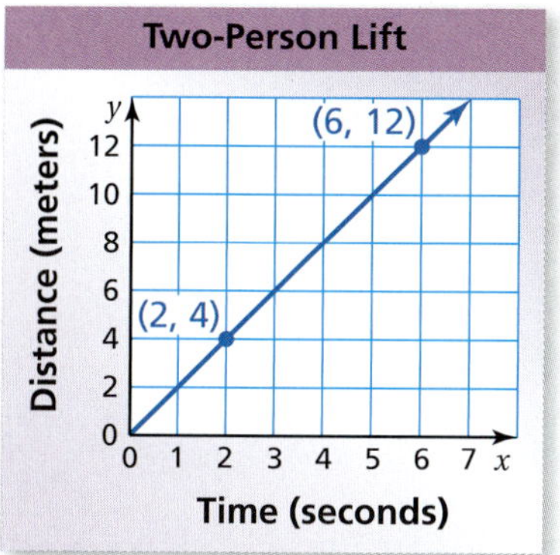

The distance y (in meters) that a four-person ski lift travels in x seconds is represented by the equation $y = 2.5x$. The graph shows the distance that a two-person ski lift travels.

a. Which ski lift is faster?

Interpret each slope as a unit rate.

Four-Person Lift	***Two-Person Lift***
$y = 2.5x$ The slope is 2.5.	$\text{slope} = \dfrac{\text{change in } y}{\text{change in } x}$ $= \dfrac{8}{4} = 2$
The four-person lift travels 2.5 meters per second.	The two-person lift travels 2 meters per second.

So, the four-person lift is faster than the two-person lift.

b. Graph the equation that represents the four-person lift in the same coordinate plane as the two-person lift. Compare the steepness of the graphs. What does this mean in the context of the problem?

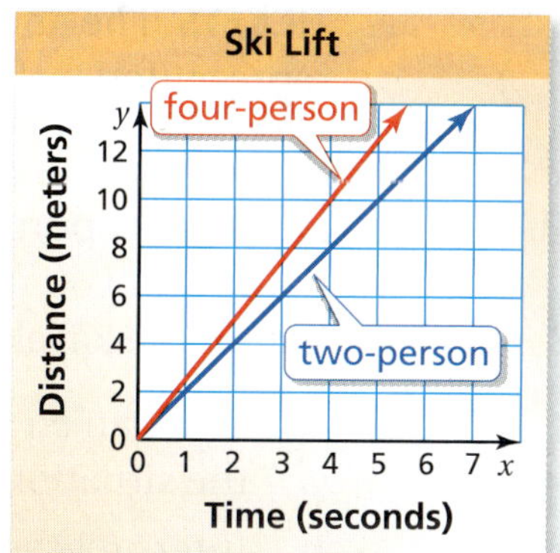

The graph that represents the four-person lift is steeper than the graph that represents the two-person lift. So, the four-person lift is faster.

On Your Own

Now You're Ready
Exercise 9

3. The table shows the distance y (in meters) that a T-bar ski lift travels in x seconds. Compare its speed to the ski lifts in Example 3.

x (seconds)	1	2	3	4
y (meters)	$2\frac{1}{4}$	$4\frac{1}{2}$	$6\frac{3}{4}$	9

13.3 Exercises

Check It Out
Help with Homework
BigIdeasMath.com

Vocabulary and Concept Check

1. **VOCABULARY** What point is on the graph of every direct variation equation?

2. **REASONING** Does the equation $y = 2x + 3$ represent a proportional relationship? Explain.

Practice and Problem Solving

Tell whether x and y are in a proportional relationship. Explain your reasoning. If so, write an equation that represents the relationship.

3. 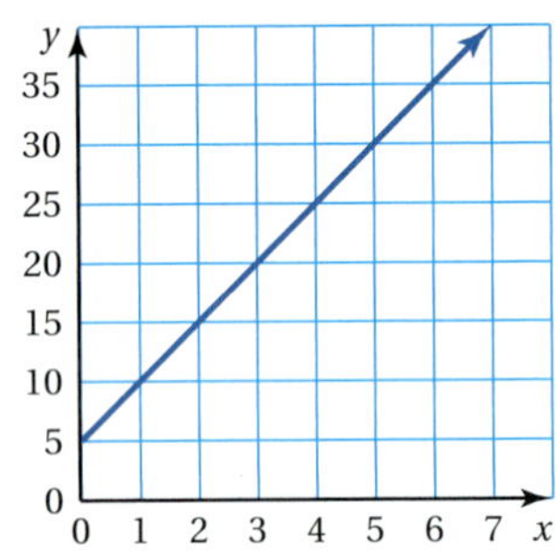

4. 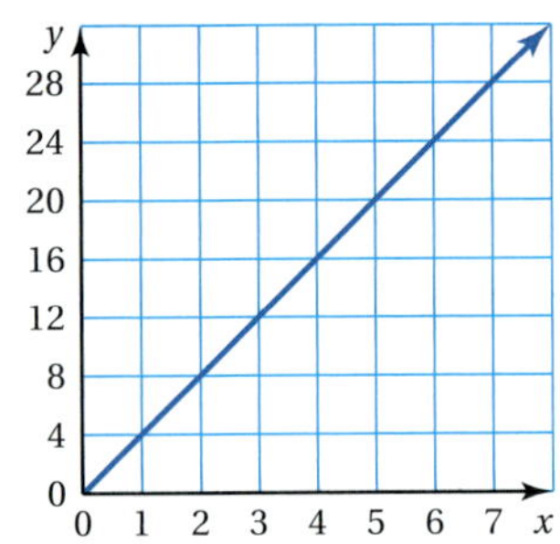

5.

x	3	6	9	12
y	1	2	3	4

6.

x	2	5	8	10
y	4	8	13	23

1 7. **TICKETS** The amount y (in dollars) that you raise by selling x fundraiser tickets is represented by the equation $y = 5x$. Graph the equation and interpret the slope.

2 8. **KAYAK** The cost y (in dollars) to rent a kayak is proportional to the number x of hours that you rent the kayak. It costs \$27 to rent the kayak for 3 hours.

   a. Write an equation that represents the situation.

   b. Interpret the slope.

   c. How much does it cost to rent the kayak for 5 hours?

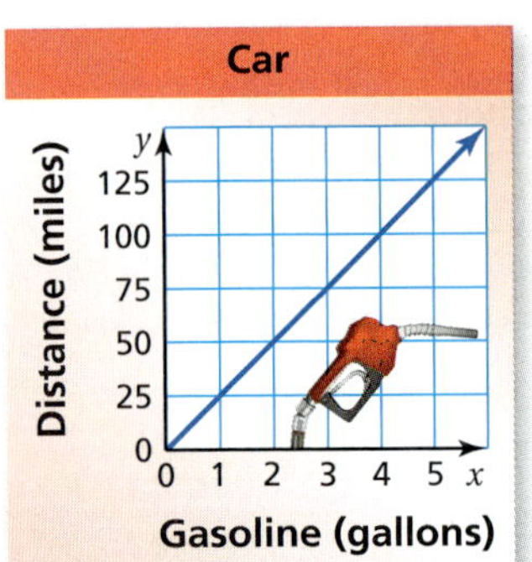

3 9. **MILEAGE** The distance y (in miles) that a truck travels on x gallons of gasoline is represented by the equation $y = 18x$. The graph shows the distance that a car travels.

   a. Which vehicle gets better gas mileage? Explain how you found your answer.

   b. How much farther can the vehicle you chose in part (a) travel than the other vehicle on 8 gallons of gasoline?

Assignment Guide and Homework Check

Level	Assignment	Homework Check
Accelerated	1–17	8, 10, 11, 12

Common Errors

- **Exercises 3 and 4** Students may think that the slope of each line is 1. Remind them to pay attention to the scales on the axes.
- **Exercise 8** Students may switch the *x*- and *y*-values, substituting incorrectly in the equation $y = mx$ when finding the slope. Remind them to make sure they are substituting the correct value for each variable.

13.3 Record and Practice Journal

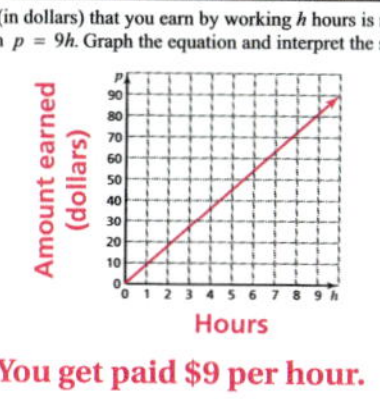

1. The amount p (in dollars) that you earn by working h hours is represented by the equation $p = 9h$. Graph the equation and interpret the slope.

You get paid $9 per hour.

2. The cost c (in dollars) to rent a bicycle is proportional to the number h of hours that you rent the bicycle. It costs $20 to rent the bicycle for 4 hours.

   a. Write an equation that represents the situation.
   $C = 5h$

   b. Interpret the slope.
   It cost $5 an hour to rent a bicycle.

   c. How much does it cost to rent the bicycle for 6 hours?
   $30

Vocabulary and Concept Check

1. (0, 0)

2. no; *Sample answer:* The graph of the equation does not pass through the origin.

Practice and Problem Solving

3. no; *Sample answer:* The graph of the equation does not pass through the origin.

4. yes; $y = 4x$; *Sample answer:* The graph is a line through the origin.

5. yes; $y = \frac{1}{3}x$; *Sample answer:* The rate of change in the table is constant.

6. no; *Sample answer:* The rate of change in the table is not constant.

7.

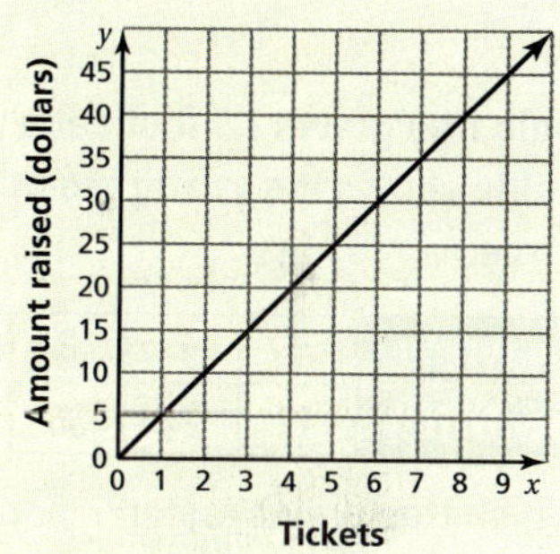

Each ticket costs $5.

8. **a.** $y = 9x$

b. It costs $9 per hour to rent a kayak.

c. $45

9. **a.** the car; *Sample answer:* The equation for the car is $y = 25x$. Because 25 is greater than 18, the car gets better gas mileage.

b. 56 miles

Practice and Problem Solving

10. See Additional Answers.

11. See *Taking Math Deeper*.

12. **a.** yes; The line passes through the origin.

b. $y = -3.5x$; The temperature decreases by 3.5°F for each 1000-foot increase in altitude.

c. 54.75°F

13. See Additional Answers.

Fair Game Review

14–16. See Additional Answers.

17. B

Mini-Assessment

A maple tree grows 1.5 feet each year. The table shows the yearly growth for a pine tree.

Time (yr)	1	2	3	4
Growth (in.)	12	24	36	48

1. Which tree grows faster? maple
2. Write and graph equations that represent the growth rates of each tree. Compare the steepness of the graphs. What does this mean in the context of the problem?

   Maple tree: $y = 1.5x$; Pine tree: $y = x$

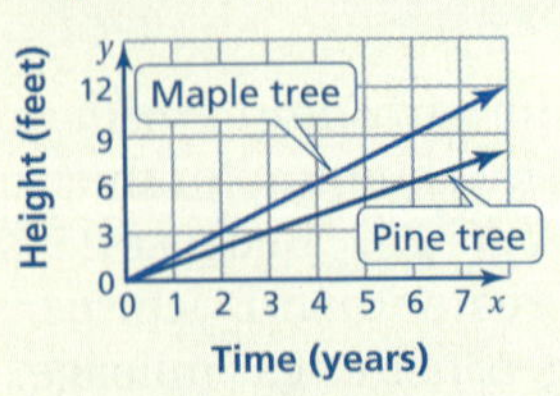

The graph that represents the maple tree is steeper than the graph that represents the pine tree. So, the maple tree grows faster than the pine tree.

Taking Math Deeper

Exercise 11

This exercise is leading students to the realization that for any point (x, y) on the graph of a proportional relationship, the ratio y to x is constant. There are several ways that students could come to this conclusion.

1 Use the equation of a proportional relationship.

The quantities x and y are in a proportional relationship, so $y = mx$ for some constant m. Solve for m.

$y = mx$ Write the equation.

$\frac{y}{x} = m$ Divide each side by x.

This tells you that the ratio of y to x is equal to the slope of the line, m, which is constant.

2 Use the slope formula. Use the point (0, 0) because the graph of any proportional relationship passes through the origin.

$m = \frac{y_2 - y_1}{x_2 - x_1}$ Slope formula

$= \frac{y - 0}{x - 0}$ Substitute.

$= \frac{y}{x}$ Simplify.

The result is the same, showing the ratio of y to x is constant.

3 Make sure students understand that this means they can use any *single* point on the graph of a proportional relationship to find the slope, unit rate, or constant of proportionality.

Project

Have students research *inverse variation*. Does it represent a linear relationship? Ask students to compare and contrast direct variation and inverse variation.

Reteaching and Enrichment Strategies

If students need help...	If students got it...
Resources by Chapter • Practice A and Practice B • Puzzle Time Record and Practice Journal Practice Differentiating the Lesson Lesson Tutorials Skills Review Handbook	Resources by Chapter • Enrichment and Extension • Technology Connection Start the next section

10. **BIOLOGY** Toenails grow about 13 millimeters per year. The table shows fingernail growth.

Weeks	1	2	3	4
Fingernail Growth (millimeters)	0.7	1.4	2.1	2.8

a. Do fingernails or toenails grow faster? Explain.

b. In the same coordinate plane, graph equations that represent the growth rates of toenails and fingernails. Compare the steepness of the graphs. What does this mean in the context of the problem?

11. **REASONING** The quantities x and y are in a proportional relationship. What do you know about the ratio of y to x for any point (x, y) on the line?

12. **PROBLEM SOLVING** The graph relates the temperature change y (in degrees Fahrenheit) to the altitude change x (in thousands of feet).

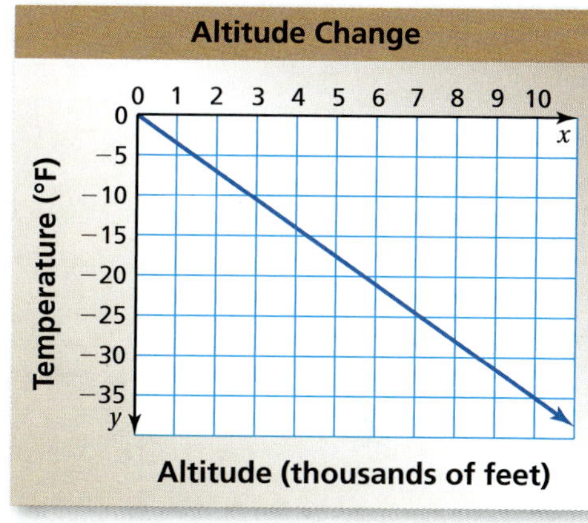

a. Is the relationship proportional? Explain.

b. Write an equation of the line. Interpret the slope.

c. You are at the bottom of a mountain where the temperature is 74°F. The top of the mountain is 5500 feet above you. What is the temperature at the top of the mountain?

13. **Critical Thinking** Consider the distance equation $d = rt$, where d is the distance (in feet), r is the rate (in feet per second), and t is the time (in seconds).

a. You run 6 feet per second. Are distance and time proportional? Explain. Graph the equation.

b. You run for 50 seconds. Are distance and rate proportional? Explain. Graph the equation.

c. You run 300 feet. Are rate and time proportional? Explain. Graph the equation.

d. One of these situations represents *inverse variation*. Which one is it? Why do you think it is called inverse variation?

Fair Game Review What you learned in previous grades & lessons

Graph the linear equation. *(Section 13.1)*

14. $y = -\frac{1}{2}x$

15. $y = 3x - \frac{3}{4}$

16. $y = -\frac{x}{3} - \frac{3}{2}$

17. **MULTIPLE CHOICE** What is the value of x? *(Section 12.3)*

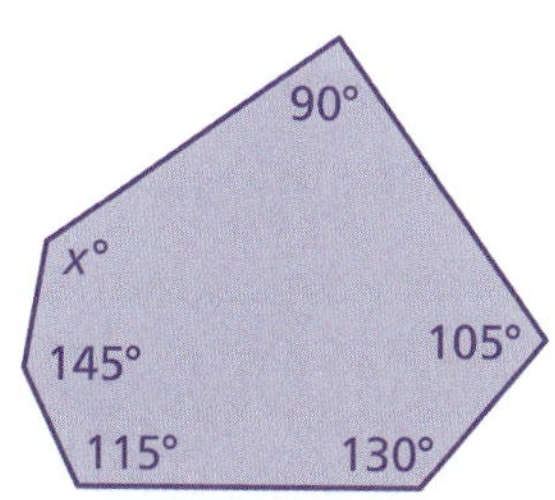

Ⓐ 110 Ⓑ 135

Ⓒ 315 Ⓓ 522

13 Study Help

Check It Out
Graphic Organizer
BigIdeasMath.com

You can use a **process diagram** to show the steps involved in a procedure. Here is an example of a process diagram for graphing a linear equation.

Graphing a linear equation	Example: Graph $y = 2x - 1$.
↓ Make a table of values.	↓ (table below)
↓ Plot the ordered pairs.	↓ 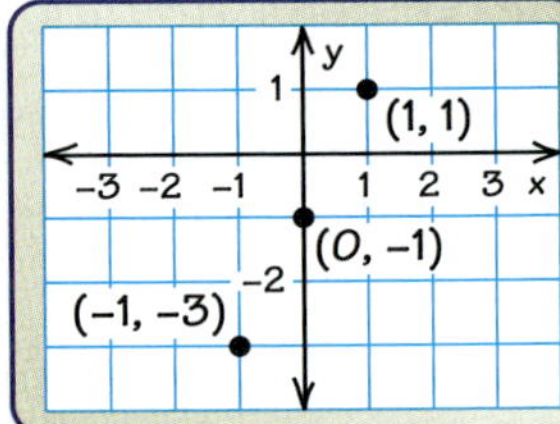
↓ Draw a line through the points.	↓ 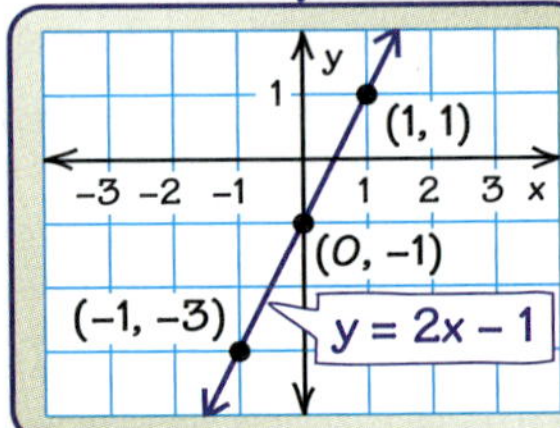

x	−1	0	1
y	−3	−1	1

On Your Own

Make process diagrams with examples to help you study these topics.

1. finding the slope of a line
2. graphing a proportional relationship

After you complete this chapter, make process diagrams for the following topics.

3. graphing a linear equation using
   a. slope and y-intercept
   b. x- and y-intercepts
4. writing equations in slope-intercept form
5. writing equations in point-slope form

"Here is a process diagram with suggestions for what to do if a hyena knocks on your door."

Sample Answers

1.

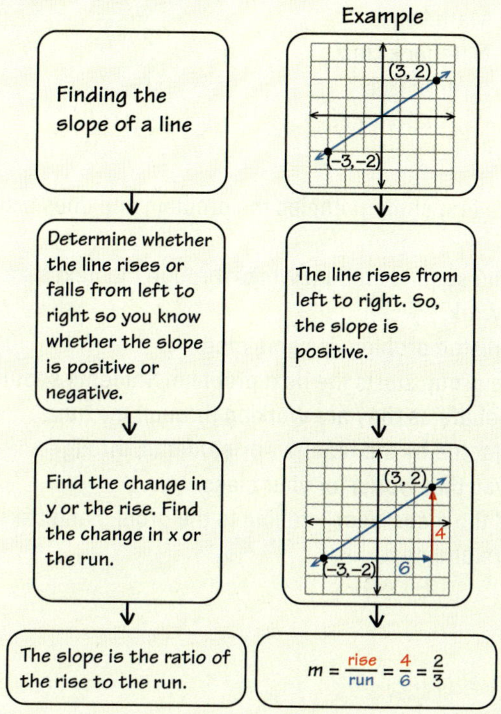

2.

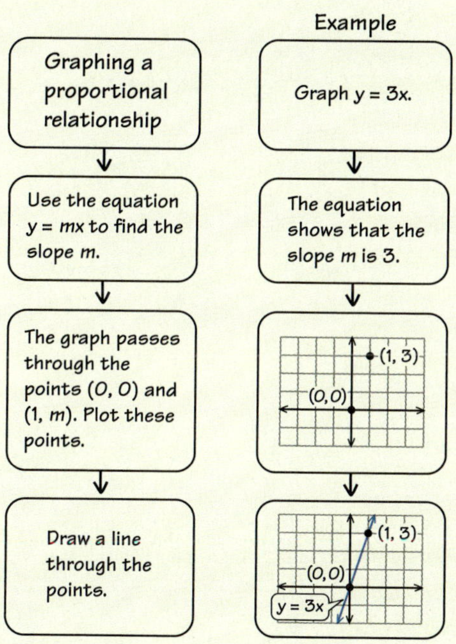

List of Organizers

Available at *BigIdeasMath.com*

Comparison Chart
Concept Circle
Definition (Idea) and Example Chart
Example and Non-Example Chart
Formula Triangle
Four Square
Information Frame
Information Wheel
Notetaking Organizer
Process Diagram
Summary Triangle
Word Magnet
Y Chart

About this Organizer

A **Process Diagram** can be used to show the steps involved in a procedure. Process diagrams are particularly useful for illustrating procedures with two or more steps, and they can have one or more branches. As shown, students' process diagrams can have two parallel parts, in which the procedure is stepped out in one part and an example illustrating each step is shown in the other part. Or, the diagram can be made up of just one part, with example(s) included in the last "bubble" to illustrate the steps that precede it.

Technology for the Teacher

Editable Graphic Organizer

Answers

1–4. See Additional Answers.

5. $-\frac{1}{2}$

6. 2

7. undefined

8. 0

9. parallel slope: $-\frac{1}{2}$

perpendicular slope: 2

10. no; yes; The line $x = 1$ is vertical. The line $y = -1$ is horizontal. A vertical line is perpendicular to a horizontal line.

11.

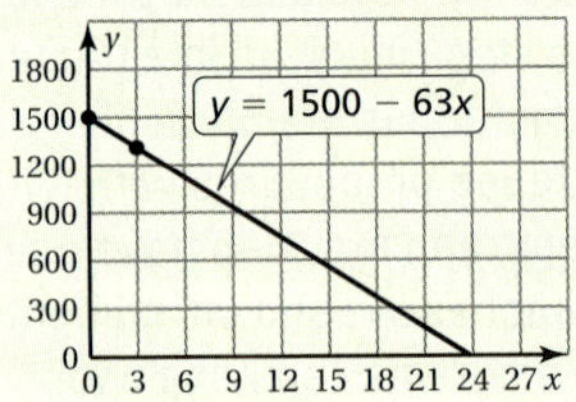

12.

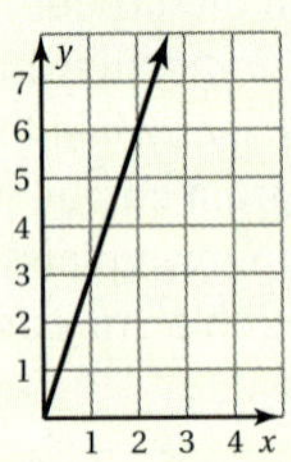

You take 3 hours of cello lessons per week.

13. **a.** $y = 7.50x$

b. The slope indicates that the unit cost is $7.50 per guest.

c. $75

Technology for the Teacher

Online Assessment
Assessment Book
ExamView® Assessment Suite

Alternative Quiz Ideas

100% Quiz
Error Notebook
Group Quiz
Homework Quiz
Math Log
Notebook Quiz
Partner Quiz
Pass the Paper

Pass the Paper

- Work in groups of four. The first student copies the problem and does a step, explaining his or her work.
- The paper is passed and the second student works through the next step, also explaining his or her work.
- This process continues until the problem is completed.
- The second member of the group starts the next problem. Students should be allowed to question and debate as they are working through the quiz.
- Student groups can be selected by the teacher, by students, through a random process, or any way that works for your class.
- The teacher walks around the classroom listening to the groups and asks questions to ensure understanding.

Reteaching and Enrichment Strategies

If students need help. . .	If students got it. . .
Resources by Chapter • Practice A and Practice B • Puzzle Time Lesson Tutorials *BigIdeasMath.com*	Resources by Chapter • Enrichment and Extension • Technology Connection Game Closet at *BigIdeasMath.com* Start the next section

13.1–13.3 Quiz

Check It Out
Progress Check
BigIdeasMath.com

Graph the linear equation. *(Section 13.1)*

1. $y = -x + 8$

2. $y = \frac{x}{3} - 4$

3. $x = -1$

4. $y = 3.5$

Find the slope of the line. *(Section 13.2)*

5.

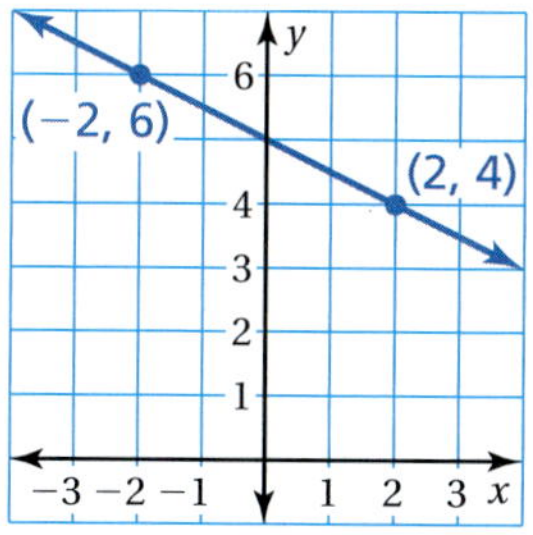

6.

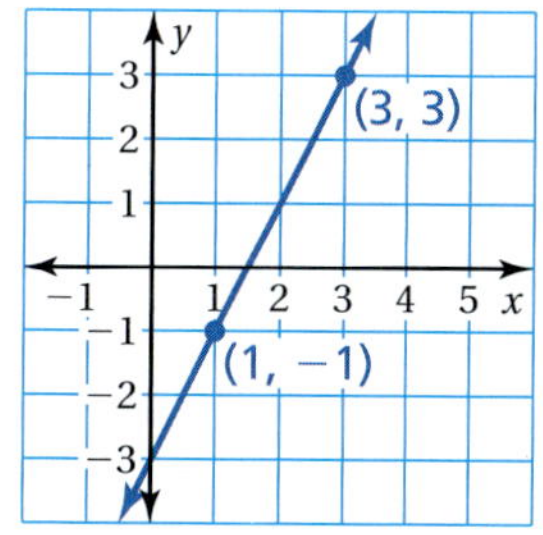

7.

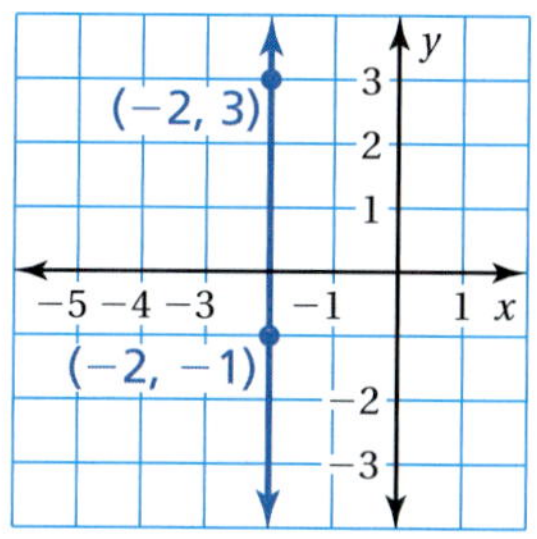

8.

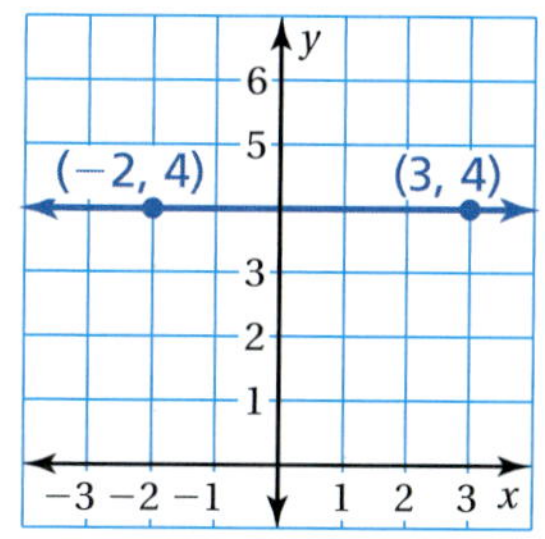

9. What is the slope of a line that is parallel to the line in Exercise 5? What is the slope of a line that is perpendicular to the line in Exercise 5? *(Section 13.2)*

10. Are the lines $y = -1$ and $x = 1$ parallel? Are they perpendicular? Justify your answer. *(Section 13.2)*

11. BANKING A bank charges $3 each time you use an out-of-network ATM. At the beginning of the month, you have $1500 in your bank account. You withdraw $60 from your bank account each time you use an out-of-network ATM. Graph a linear equation that represents the balance in your account after you use an out-of-network ATM x times. *(Section 13.1)*

12. MUSIC The number y of hours of cello lessons that you take after x weeks is represented by the equation $y = 3x$. Graph the equation and interpret the slope. *(Section 13.3)*

13. DINNER PARTY The cost y (in dollars) to provide food for guests at a dinner party is proportional to the number x of guests attending the party. It costs $30 to provide food for 4 guests. *(Section 13.3)*

a. Write an equation that represents the situation.

b. Interpret the slope.

c. How much does it cost to provide food for 10 guests?

13.4 Graphing Linear Equations in Slope-Intercept Form

Essential Question How can you describe the graph of the equation $y = mx + b$?

1 ACTIVITY: Analyzing Graphs of Lines

Work with a partner.

- **Graph each equation.**
- **Find the slope of each line.**
- **Find the point where each line crosses the *y*-axis.**
- **Complete the table.**

Equation	Slope of Graph	Point of Intersection with *y*-axis
a. $y = -\frac{1}{2}x + 1$		
b. $y = -x + 2$		
c. $y = -x - 2$		
d. $y = \frac{1}{2}x + 1$		
e. $y = x + 2$		
f. $y = x - 2$		
g. $y = \frac{1}{2}x - 1$		
h. $y = -\frac{1}{2}x - 1$		
i. $y = 3x + 2$		
j. $y = 3x - 2$		

Graphing Equations
In this lesson, you will
- find slopes and *y*-intercepts of graphs of linear equations.
- graph linear equations written in slope-intercept form.

k. Do you notice any relationship between the slope of the graph and its equation? between the point of intersection with the *y*-axis and its equation? Compare the results with those of other students in your class.

Laurie's Notes

Introduction

Applying Mathematical Practices

- **Construct Viable Arguments** and **Look for and Express Regularity in Repeated Reasoning:** The goal is for students to discover that when equations are written in slope-intercept form, the coefficient of the x-term is the slope and the constant is the y-intercept. Students graph the lines using a table of values, a good review of this skill.

Motivate

- **Preparation:** Make three demonstration cards on 8.5"x 11" paper. The x-axis is labeled "time" and the y-axis is labeled "distance from home."
- Sample cards A, B, and C are shown.

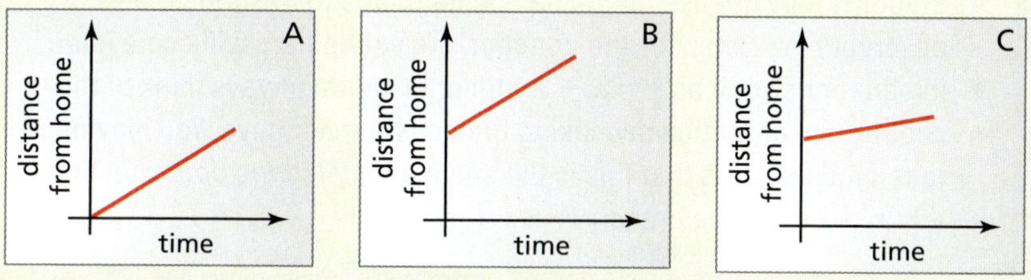

- Ask 3 students to hold the cards for the class to see.
- ? "Consider how the axes are labeled. What does the slope of the line represent?" $\frac{\text{distance}}{\text{time}} = \text{rate}$
- ? "What story does each card tell? How are the stories similar and different?" A: you begin at home; B: you travel at the same rate, but you start away from home; C: you start away from home, but you travel at a slower rate
- **Management Tip:** If you plan to use the demonstration cards again next year, laminate them.

Activity Notes

Activity 1

- When making a table of values, remind students to think about what values of x they should substitute. When the coefficient of x is a fraction, it is wise to select x-values that are multiples of the denominator. This may help eliminate fractional values.
- ? "How many points do you need in order to graph the equation?" Minimum is 2. Plot 3 to be safe.
- Remind students to solve the equation when $x = 0$. This will ensure that they find the point where the graph crosses the y-axis.
- Students should begin to observe patterns as they complete the table.
- Encourage students to draw the directed arrows in order to help them find the slope of the line.

What Your Students Will Learn

- Identify slopes and x- and y-intercepts of graphs of linear equations.
- Graph linear equations written in slope-intercept form.

Previous Learning

Students should know how to graph linear equations and find slopes of lines.

Technology for the Teacher

Dynamic Classroom

Lesson Plans
Complete Materials List

13.4 Record and Practice Journal

Essential Question How can you describe the graph of the equation $y = mx + b$?

1 ACTIVITY: Analyzing Graphs of Lines

Work with a partner. See Additional Answers.

- Graph each equation.
- Find the slope of each line.
- Find the point where each line crosses the y-axis.
- Complete the table.

Equation	Slope of Graph	Point of Intersection with y-axis
a. $y = -\frac{1}{2}x + 1$	$-\frac{1}{2}$	(0, 1)
b. $y = -x + 2$	-1	(0, 2)
c. $y = -x - 2$	-1	(0, −2)
d. $y = \frac{1}{2}x + 1$	$\frac{1}{2}$	(0, 1)
e. $y = x + 2$	1	(0, 2)
f. $y = x - 2$	1	(0, −2)
g. $y = \frac{1}{2}x - 1$	$\frac{1}{2}$	(0, −1)
h. $y = -\frac{1}{2}x - 1$	$-\frac{1}{2}$	(0, −1)
i. $y = 3x + 2$	3	(0, 2)

English Language Learners

Build on Past Knowledge

Remind students from their study of rational numbers that the slope -2 can be written as the fraction $\frac{-2}{1}$. By writing the integer as a fraction, students can see that the slope has a run of 1 and a rise of -2. This will help in graphing linear equations.

13.4 Record and Practice Journal

Equation	Slope of Graph	Point of Intersection with y-axis
j. $y = 3x - 2$	3	$(0, -2)$

k. Do you notice any relationship between the slope of the graph and its equation? Between the point of intersection with the y-axis and its equation? Compare the results with those of other students in your class.
Check students' work.

2 ACTIVITY: Deriving an Equation

Work with a partner.

a. Look at the graph of each equation in Activity 1. Do any of the graphs represent a proportional relationship? Explain. no

b. For a nonproportional linear relationship, the graph crosses the y-axis at some point $(0, b)$, where b does not equal 0.

Let (x, y) represent any other point on the graph. You can use the formula for slope to write the equation for a nonproportional linear relationship.

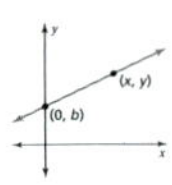

Use the graph to complete the steps.

$\frac{y_2 - y_1}{x_2 - x_1} = m$	Slope formula
$\frac{y - \boxed{b}}{x - \boxed{0}} = m$	Substitute values.
$\frac{\boxed{y - b}}{\boxed{x}} = m$	Simplify.
$\frac{\boxed{y - b}}{\boxed{x}} \cdot \boxed{x} = m \cdot \boxed{x}$	Multiplication Property of Equality
$y - \boxed{b} = m \cdot \boxed{x}$	Simplify.
$y = m\boxed{x} + \boxed{b}$	Addition Property of Equality

c. What do m and b represent in the equation?
m represents the slope; b represents the y-coordinate of the point of intersection with the y-axis.

What Is Your Answer?

3. IN YOUR OWN WORDS How can you describe the graph of the equation $y = mx + b$?
a line with slope m that crosses the y-axis at $(0, b)$

a. How does the value of m affect the graph of the equation?
steepness of line and whether it rises or falls from left to right

b. How does the value of b affect the graph of the equation?
where the graph crosses the y-axis

c. Check your answers to parts (a) and (b) with three equations that are not in Activity 1.
Check students' work.

4. LOGIC Why do you think $y = mx + b$ is called the *slope-intercept form* of the equation of a line? Use drawings or diagrams to support your answer.
m is the slope and b is the y-intercept.

Laurie's Notes

Activity 1 (continued)

- Have students put a few graphs on the board to help facilitate discussion.

? In part (k), ask a series of summary questions.
  - "Compare certain pairs of graphs such as (e) and (f) or (i) and (j). What do you observe?" They have the same steepness (slope) and the number at the end of the equation is the y-coordinate of where the graph crosses the y-axis.
  - "Where does the equation $y = x + 7$ cross the y-axis?" at $(0, 7)$
  - "Compare certain groups of graphs such as (b), (e), and (i) or (c), (f), and (j). What do you observe?" They cross the y-axis at the same point, but the slopes are different; the coefficient of x is the slope of the line.
  - "What is the slope of the equation $y = 7x + 2$?" slope $= 7$

Words of Wisdom

- Students may not use mathematical language to describe their observations. Listen for the concept, the vocabulary will come later.
- In equations such as $y = x - 2$, students do not always think of the subtraction operation as making the constant negative. You may need to remind students that this is the same as *adding the opposite*. So, $y = x - 2$ is equivalent to $y = x + (-2)$.

Activity 2

- The goal of this activity is for students to derive the equation of a nonproportional linear relationship.
- **Make Sense of Problems:** Students are led through the steps necessary to derive the equation of the line. The ordered pairs $(0, b)$ and (x, y) are substituted into the slope formula.
- **Reason Abstractly and Quantitatively:** Students must reason abstractly to solve this equation for y. Working with the expression $(y - b)$ can feel very different to students than working with $(y - 4)$.

? "Assuming that there are no restrictions on x, does every nonvertical line pass through the y-axis?" yes

What Is Your Answer?

- These answers should follow immediately from discussing student observations.

Closure

- Refer back to the demonstration cards A, B, and C. Have students describe how the equations would be similar and how they would be different. *Sample answer:* A and B would have the same coefficient of x, but different constants. B and C would have different coefficients of x, but the same constant.

2 ACTIVITY: Deriving an Equation

Work with a partner.

a. Look at the graph of each equation in Activity 1. Do any of the graphs represent a proportional relationship? Explain.

b. For a nonproportional linear relationship, the graph crosses the y-axis at some point $(0, b)$, where b does not equal 0. Let (x, y) represent any other point on the graph. You can use the formula for slope to write the equation for a nonproportional linear relationship.

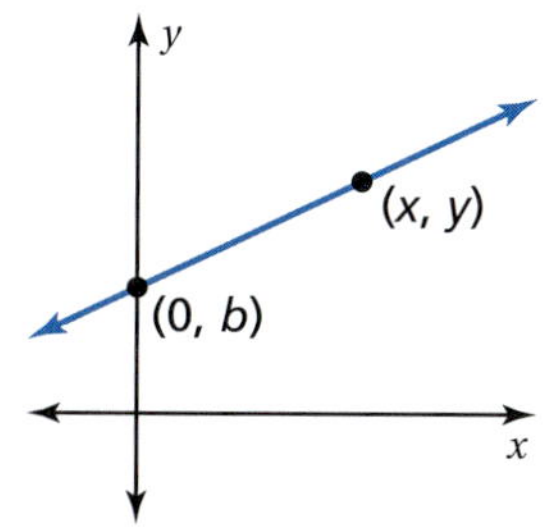

Use the graph to complete the steps.

$$\frac{y_2 - y_1}{x_2 - x_1} = m$$ Slope formula

$$\frac{y - \square}{x - \square} = m$$ Substitute values.

$$\frac{\square}{\square} = m$$ Simplify.

$$\frac{\square}{\square} \bullet \square = m \bullet \square$$ Multiplication Property of Equality

$$y - \square = m \bullet \square$$ Simplify.

$$y = m\square + \square$$ Addition Property of Equality

Math Practice

Use Prior Results

How can you use the results of Activity 1 to help support your answer?

c. What do m and b represent in the equation?

What Is Your Answer?

3. IN YOUR OWN WORDS How can you describe the graph of the equation $y = mx + b$?

a. How does the value of m affect the graph of the equation?

b. How does the value of b affect the graph of the equation?

c. Check your answers to parts (a) and (b) with three equations that are not in Activity 1.

4. LOGIC Why do you think $y = mx + b$ is called the *slope-intercept form* of the equation of a line? Use drawings or diagrams to support your answer.

Practice Use what you learned about graphing linear equations in slope-intercept form to complete Exercises 4–6 on page 594.

Check It Out
Lesson Tutorials
BigIdeasMath.com

Key Vocabulary

x-intercept, *p. 592*
y-intercept, *p. 592*
slope-intercept form, *p. 592*

Key Ideas

Intercepts

The **x-intercept** of a line is the x-coordinate of the point where the line crosses the x-axis. It occurs when $y = 0$.

The **y-intercept** of a line is the y-coordinate of the point where the line crosses the y-axis. It occurs when $x = 0$.

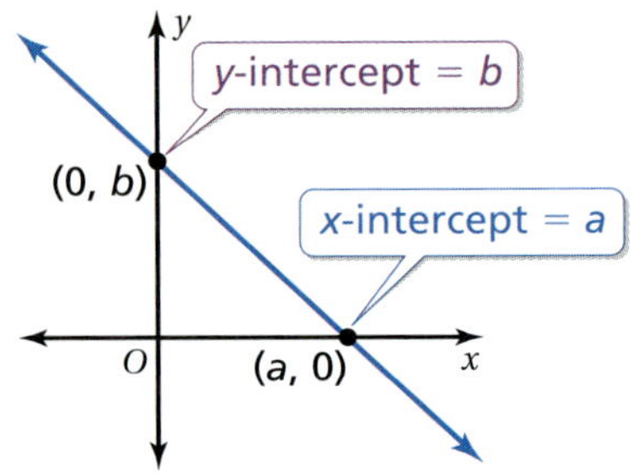

Slope-Intercept Form

Words A linear equation written in the form $y = mx + b$ is in **slope-intercept form**. The slope of the line is m, and the y-intercept of the line is b.

Algebra

$$y = mx + b$$

slope (m) — y-intercept (b)

Study Tip

Linear equations can, but do not always, pass through the origin. So, proportional relationships are a special type of linear equation in which $b = 0$.

EXAMPLE 1 Identifying Slopes and y-Intercepts

Find the slope and the y-intercept of the graph of each linear equation.

a. $y = -4x - 2$

$y = -4x + (-2)$ Write in slope-intercept form.

The slope is -4, and the y-intercept is -2.

b. $y - 5 = \frac{3}{2}x$

$y = \frac{3}{2}x + 5$ Add 5 to each side.

The slope is $\frac{3}{2}$, and the y-intercept is 5.

On Your Own

Now You're Ready
Exercises 7–15

Find the slope and the y-intercept of the graph of the linear equation.

1. $y = 3x - 7$

2. $y - 1 = -\frac{2}{3}x$

Multi-Language Glossary at BigIdeasMath.com

Laurie's Notes

Introduction

Connect

- **Yesterday:** Students explored the connection between the equation of a line and its graph.
- **Today:** Students will graph equations in slope-intercept form.

Motivate

- Share the following taxi information. All trips start at a convention center.

Destination	Distance	Taxi Fare
Football stadium	18.7 mi	\$39 approx.
Airport	12 mi	\$32 flat fee
Shopping district	9.5 mi	\$20 approx.

? "How do you think taxi fares are determined?" Answers will vary; listen for distance, number of passengers, tolls.

- Discuss why some locations, often involving airports, have flat fees associated with them.

Lesson Notes

Key Ideas

- Write the Key Ideas on the board. Draw the graph and discuss the vocabulary of this lesson: *x*-intercept, *y*-intercept, and slope-intercept form.
- Explain to students that the equation must be written with y in terms of x. This means that the equation must be solved for y.
- Discuss the Study Tip that connects to earlier lessons in this chapter.

Example 1

? "What is a linear equation?" an equation whose graph is a line

- Write part (a). This is written in the form $y = mx + b$, enabling students to quickly identify the slope and *y*-intercept.
- Write part (b).

? "Is $y - 5 = \frac{3}{2}x$ in slope-intercept form?" no "Can you rewrite it so that it is?" yes; Add 5 to each side of the equation.

- Make sure students understand that you can use the Commutative Property of Addition to write $y = b + mx$ as $y = mx + b$.

On Your Own

- **Think-Pair-Share:** Students should read each question independently and then work in pairs to answer the questions. When they have answered the questions, the pair should compare their answers with another group and discuss any discrepancies.

Goal Today's lesson is graphing the equation of a line written in **slope-intercept form**.

Technology for the Teacher

Dynamic Classroom

Lesson Tutorials
Lesson Plans
Answer Presentation Tool

Extra Example 1

Find the slope and *y*-intercept of the graph of each linear equation.

a. $y = \frac{3}{4}x - 5$

slope: $\frac{3}{4}$; *y*-intercept: -5

b. $y + \frac{1}{2} = -6x$

slope: -6; *y*-intercept: $-\frac{1}{2}$

On Your Own

1. slope: 3; *y*-intercept: -7
2. slope: $-\frac{2}{3}$; *y*-intercept: 1

Extra Example 2

Graph $y = -\frac{2}{3}x - 2$. Identify the x-intercept.

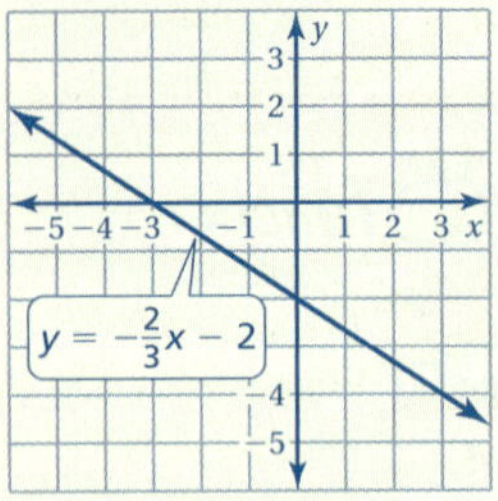

−3

Extra Example 3

The cost y (in dollars) for making friendship bracelets is $y = 0.5x + 2$, where x is the number of bracelets.

a. Graph the equation.

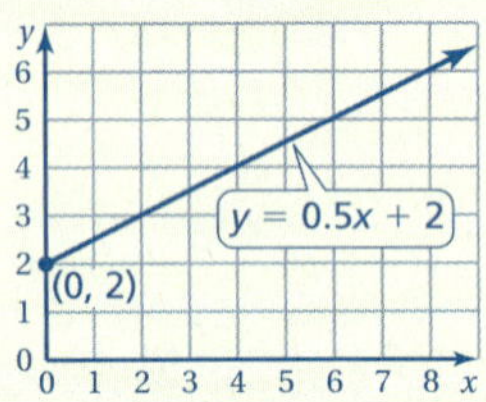

b. Interpret the slope and y-intercept.
The slope is 0.5. So, the cost per bracelet is \$0.50. The y-intercept is 2. So, there is an initial cost of \$2 to make the bracelets.

On Your Own

3–5. See Additional Answers.

Differentiated Instruction

Kinesthetic

When graphing a linear equation using the slope-intercept form, students must apply the slope correctly after plotting the point for the y-intercept. Have students plot (0, 3) in the coordinate plane. Then graph the lines $y = 4x + 3$, $y = \frac{1}{4}x + 3$, $y = -4x + 3$, and $y = -\frac{1}{4}x + 3$ in the same coordinate plane using (0, 3) as the starting point. Make sure students identify the correct rise and run for each line.

Laurie's Notes

Example 2

- ? "How can knowing the slope and the y-intercept help you graph a line?" Listen for student understanding of what slope and y-intercept mean.
- Remind students that a slope of −3 can be interpreted as $\frac{-3}{1} = \frac{3}{-1}$. Starting at the y-intercept, you can move to the right 1 unit and down 3 units, or to the left 1 unit and up 3 units. In both cases, you land on a point which satisfies the equation.
- ? **FYI:** "In this problem, you used the slope to plot a point that coincidentally landed on the x-axis. How would you find the x-intercept without using a graph?" Set $y = 0$ and solve for x.

Example 3

- Write the equation $y = 2.5x + 2$ on the board.
- ? "What is the slope and what does it mean in the context of this problem?" 2.5; It costs \$2.50 for each mile you travel in the taxi. "What is the y-intercept and what does it mean in the context of this problem?" 2; The initial fee is \$2 when you sit down in the taxi.
- **Reason Abstractly and Quantitatively:** Mathematically proficient students make sense of the quantities and their relationships in problem situations. To develop this proficiency, students must be asked to interpret the meaning of the symbols.
- **Attend to Precision:** Suggest to students that because the slope is 2.5, any ratio equivalent to 2.5 can also be used, such as $\frac{2.5}{1} = \frac{5}{2}$. Using whole numbers instead of decimals improves the accuracy of graphing.
- Explain that the graph of this equation will only be in Quadrant I because it does not make sense to have a negative number of miles or a negative cost.
- ? "What is the cost for a 2-mile taxi ride? a 10-mile taxi ride?" \$7; \$27

On Your Own

- Students are asked to check their answers using a graphing calculator. It is helpful to build proficiency with using the graphing calculator so that the calculator becomes a useful tool in problem solving.

Closure

- **Exit Ticket:** Graph $y - 4 = 2x$ and identify the slope and y-intercept.

slope = 2; y-intercept = 4

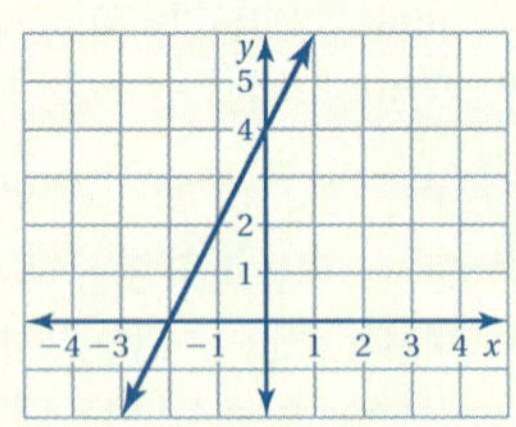

EXAMPLE 2 Graphing a Linear Equation in Slope-Intercept Form

Graph $y = -3x + 3$. Identify the x-intercept.

Step 1: Find the slope and the y-intercept.

$$y = -3x + 3$$

slope: -3; y-intercept: 3

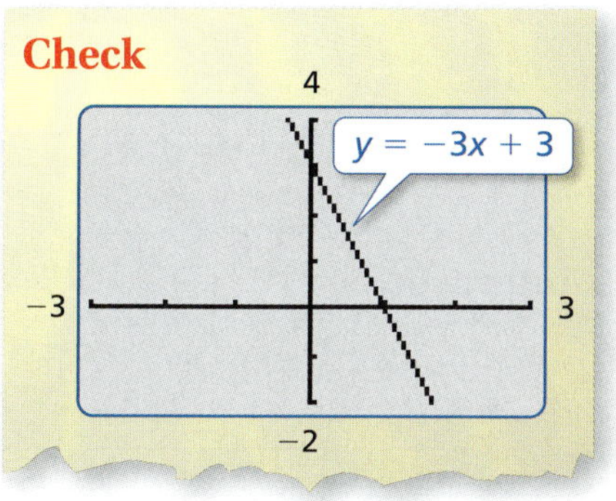

Step 2: The y-intercept is 3. So, plot $(0, 3)$.

Step 3: Use the slope to find another point and draw the line.

$$m = \frac{\text{rise}}{\text{run}} = \frac{-3}{1}$$

Plot the point that is 1 unit right and 3 units down from $(0, 3)$. Draw a line through the two points.

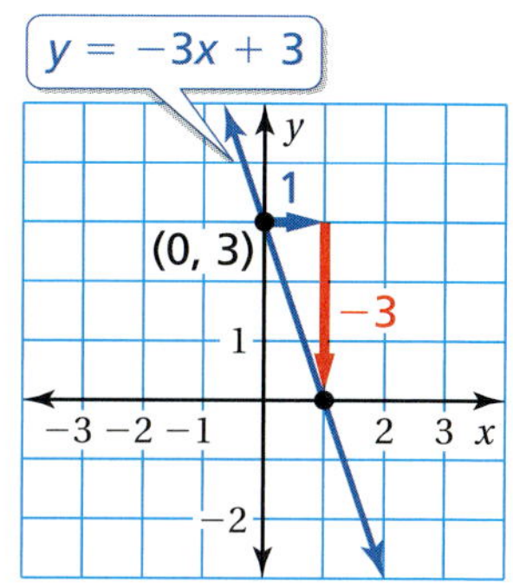

∴ The line crosses the x-axis at $(1, 0)$. So, the x-intercept is 1.

EXAMPLE 3 Real-Life Application

The cost y (in dollars) of taking a taxi x miles is $y = 2.5x + 2$. (a) Graph the equation. (b) Interpret the y-intercept and the slope.

a. The slope of the line is $2.5 = \frac{5}{2}$. Use the slope and the y-intercept to graph the equation.

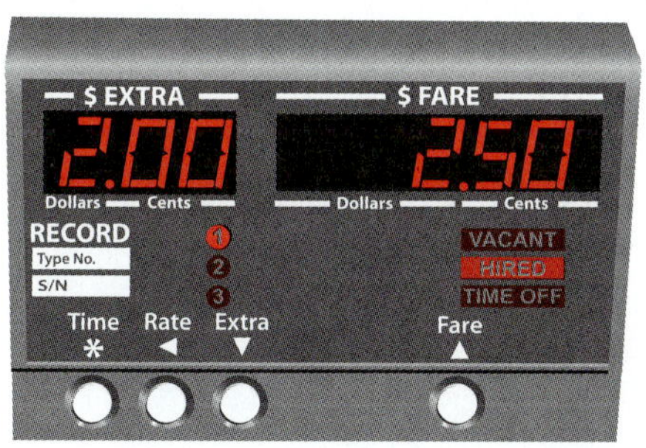

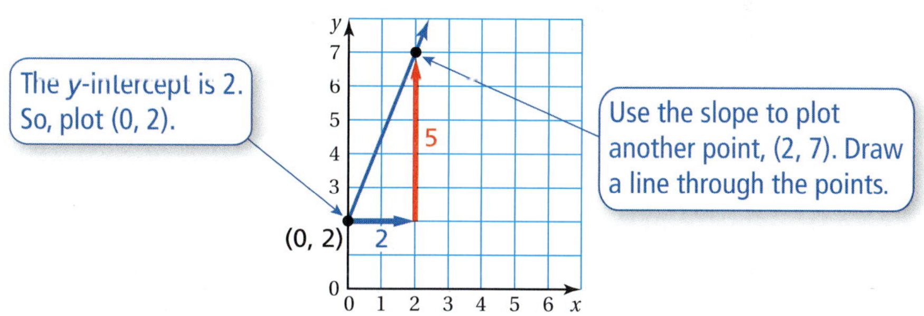

b. The slope is 2.5. So, the cost per mile is \$2.50. The y-intercept is 2. So, there is an initial fee of \$2 to take the taxi.

On Your Own

Now You're Ready
Exercises 18–23

Graph the linear equation. Identify the x-intercept. Use a graphing calculator to check your answer.

3. $y = x - 4$

4. $y = -\frac{1}{2}x + 1$

5. In Example 3, the cost y (in dollars) of taking a different taxi x miles is $y = 2x + 1.5$. Interpret the y-intercept and the slope.

13.4 Exercises

Check It Out
Help with Homework
BigIdeasMath.com

Vocabulary and Concept Check

1. **VOCABULARY** How can you find the x-intercept of the graph of $2x + 3y = 6$?
2. **CRITICAL THINKING** Is the equation $y = 3x$ in slope-intercept form? Explain.
3. **OPEN-ENDED** Describe a real-life situation that you can model with a linear equation. Write the equation. Interpret the y-intercept and the slope.

Practice and Problem Solving

Match the equation with its graph. Identify the slope and the y-intercept.

4. $y = 2x + 1$
5. $y = \frac{1}{3}x - 2$
6. $y = -\frac{2}{3}x + 1$

A.
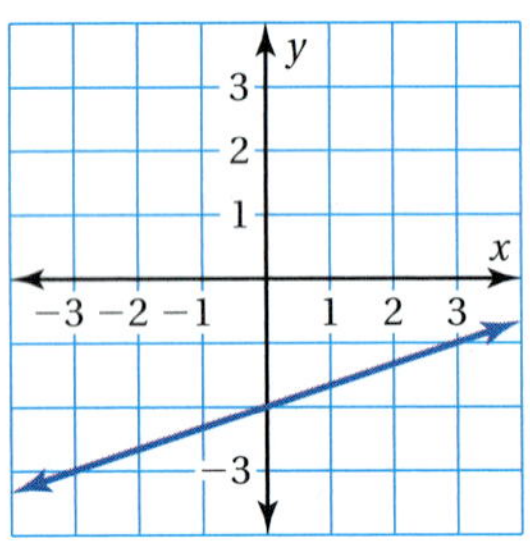

B.
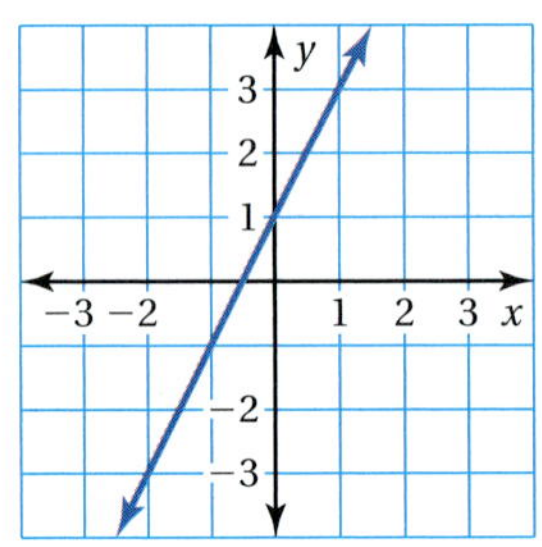

C.
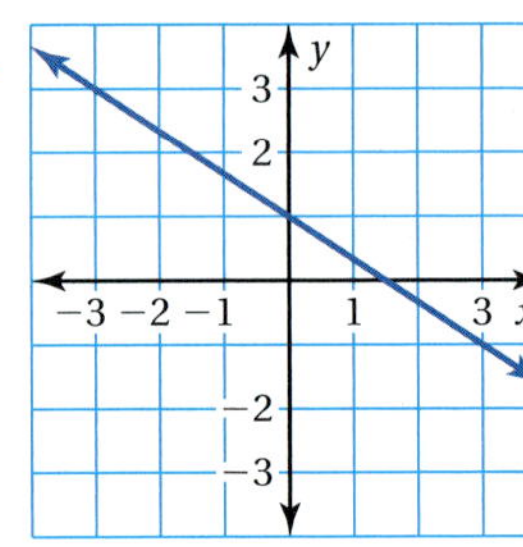

Find the slope and the y-intercept of the graph of the linear equation.

1 7. $y = 4x - 5$
8. $y = -7x + 12$
9. $y = -\frac{4}{5}x - 2$
10. $y = 2.25x + 3$
11. $y + 1 = \frac{4}{3}x$
12. $y - 6 = \frac{3}{8}x$
13. $y - 3.5 = -2x$
14. $y = -5 - \frac{1}{2}x$
15. $y = 11 + 1.5x$

16. **ERROR ANALYSIS** Describe and correct the error in finding the slope and the y-intercept of the graph of the linear equation.

> ✗ $y = 4x - 3$
> The slope is 4, and the y-intercept is 3.

17. **SKYDIVING** A skydiver parachutes to the ground. The height y (in feet) of the skydiver after x seconds is $y = -10x + 3000$.
    a. Graph the equation.
    b. Interpret the x-intercept and the slope.

Assignment Guide and Homework Check

Level	Assignment	Homework Check
Accelerated	1–6, 8–16 even, 18–31	12, 14, 22, 24, 25

Common Errors

- **Exercises 7–15** Students may forget to include negatives with the slope and/or *y*-intercept. Remind them to look at the sign in front of the slope and the *y*-intercept. Also remind students that slope-intercept form is $y = mx + b$. This means that if the linear equation has "minus *b*," then the *y*-intercept is negative.
- **Exercises 11–13** Students may identify the opposite *y*-intercept because they forget to solve for *y*. Remind them that slope-intercept form has *y* by itself, so they must solve for *y* before identifying the slope and *y*-intercept.
- **Exercises 18–23** Students may use the reciprocal of the slope when graphing and may find an incorrect *x*-intercept. Remind them that slope is *rise* over *run*, so the numerator represents vertical change, not horizontal.

13.4 Record and Practice Journal

Find the slope and *y*-intercept of the graph of the linear equation.

1. $y = -3x + 9$ slope: -3 *y*-intercept: 9
2. $y = 4 - \frac{2}{5}x$ slope: $-\frac{2}{5}$ *y*-intercept: 4
3. $6 + y = 8x$ slope: 8 *y*-intercept: -6

Graph the linear equation. Identify the *x*-intercept. Use a graphing calculator to check your answer.

4. $y = \frac{2}{3}x + 6$ *x*-intercept: -9

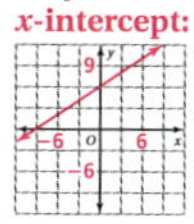

5. $y - 10 = -5x$ *x*-intercept: 2

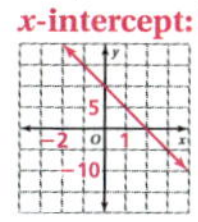

6. The equation $y = -90x + 1440$ represents the time (in minutes) left after *x* games of a tournament.

   a. Graph the equation.

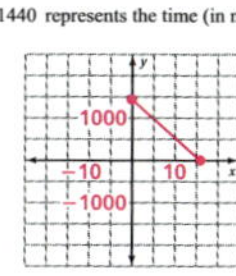

   b. Interpret the *x*-intercept and slope.
   The slope is -90. So, the length of each game is 90 minutes. The *x*-intercept is 16. So, there are 16 games in the tournament.

Vocabulary and Concept Check

1. Find the *x*-coordinate of the point where the graph crosses the *x*-axis.
2. yes; The slope is 3 and the *y*-intercept is 0.
3. *Sample answer:* The amount of gasoline *y* (in gallons) left in your tank after you travel *x* miles is $y = -\frac{1}{20}x + 20$. The slope of $-\frac{1}{20}$ means the car uses 1 gallon of gas for every 20 miles driven. The *y*-intercept of 20 means there is originally 20 gallons of gas in the tank.

Practice and Problem Solving

4. B; slope: 2; *y*-intercept: 1
5. A; slope: $\frac{1}{3}$; *y*-intercept: -2
6. C; slope: $-\frac{2}{3}$; *y*-intercept: 1
7. slope: 4; *y*-intercept: -5
8. slope: -7; *y*-intercept: 12
9. slope: $-\frac{4}{5}$; *y*-intercept: -2
10. slope: 2.25; *y*-intercept: 3
11. slope: $\frac{4}{3}$; *y*-intercept: -1
12. slope: $\frac{3}{8}$; *y*-intercept: 6
13. slope: -2; *y*-intercept: 3.5
14. slope: $-\frac{1}{2}$; *y*-intercept: -5
15. slope: 1.5; *y*-intercept: 11
16. The *y*-intercept should be -3.
    $y = 4x - 3$
    The slope is 4 and the *y*-intercept is -3.
17. See Additional Answers.

Practice and Problem Solving

18.

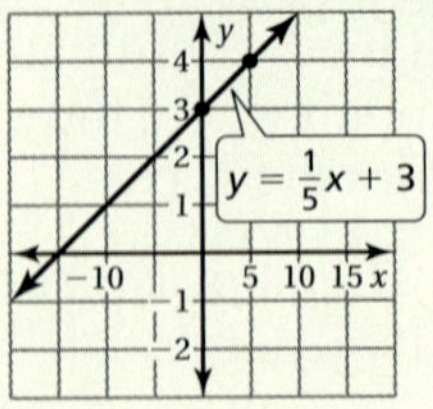

x-intercept: -15

19–25. See Additional Answers.

26. See *Taking Math Deeper*.

Fair Game Review

27. $y = 2x + 3$

28. $y = -\frac{4}{5}x + \frac{13}{5}$

29. $y = \frac{2}{3}x - 2$

30. $y = -\frac{7}{4}x + 2$

31. B

Mini-Assessment

Find the slope and y-intercept of the graph of the equation. Then graph the equation.

1. $y = -5x + 3$

slope $= -5$, y-intercept $= 3$

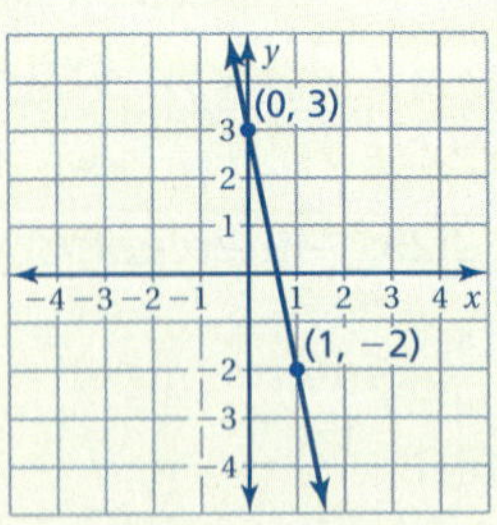

2. $y - 4 = \frac{1}{2}x$

slope $= \frac{1}{2}$, y-intercept $= 4$

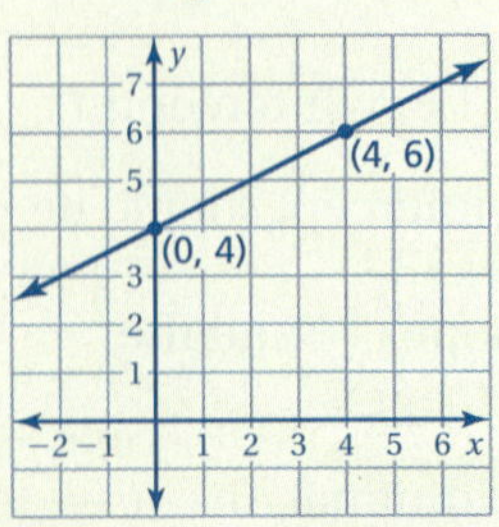

Taking Math Deeper

Exercise 26

This is a classic business problem. The business has monthly costs. The question is how many clicks are needed to cover the costs and start making a profit.

1 Organize the given information.

- The site has 5 banner ads.
- Monthly income is \$0.005 per click.
- It costs \$120 per month to run the site.
- Let y be the monthly income (in dollars).
- Let x be the number of clicks per month.

2 **a.** Write an equation for the income.

$y = 0.005x$

3 **b.** Graph the equation.

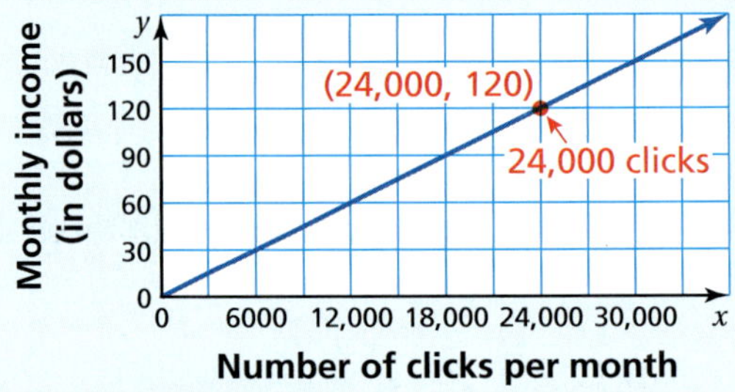

When the ads start to get 24,000 clicks a month, the income will be \$120 per month. Each banner ad needs to average $\frac{24{,}000}{5} = 4800$ clicks. Any additional clicks per month will start earning a profit.

Project

Use the Internet or the school library to research methods for determining the number of clicks on a website.

Reteaching and Enrichment Strategies

If students need help. . .	If students got it. . .
Resources by Chapter • Practice A and Practice B • Puzzle Time Record and Practice Journal Practice Differentiating the Lesson Lesson Tutorials Skills Review Handbook	Resources by Chapter • Enrichment and Extension • Technology Connection Start the next section

Graph the linear equation. Identify the x-intercept. Use a graphing calculator to check your answer.

2 **18.** $y = \frac{1}{5}x + 3$ **19.** $y = 6x - 7$ **20.** $y = -\frac{8}{3}x + 9$

21. $y = -1.4x - 1$ **22.** $y + 9 = -3x$ **23.** $y = 4 - \frac{3}{5}x$

24. APPLES You go to a harvest festival and pick apples.

a. Which equation represents the cost (in dollars) of going to the festival and picking x pounds of apples? Explain.

$y = 5x + 0.75$ $y = 0.75x + 5$

b. Graph the equation you chose in part (a).

25. REASONING Without graphing, identify the equations of the lines that are (a) parallel and (b) perpendicular. Explain your reasoning.

$y = 2x + 4$	$y = -\frac{1}{3}x - 1$	$y = -3x - 2$	$y = \frac{1}{2}x + 1$
$y = 3x + 3$	$y = -\frac{1}{2}x + 2$	$y = -3x + 5$	$y = 2x - 3$

26. Critical Thinking Six friends create a website. The website earns money by selling banner ads. The site has 5 banner ads. It costs $120 a month to operate the website.

a. A banner ad earns $0.005 per click. Write a linear equation that represents the monthly income y (in dollars) for x clicks.

b. Graph the equation in part (a). On the graph, label the number of clicks needed for the friends to start making a profit.

Fair Game Review What you learned in previous grades & lessons

Solve the equation for y. *(Topic 3)*

27. $y - 2x = 3$ **28.** $4x + 5y = 13$ **29.** $2x - 3y = 6$ **30.** $7x + 4y = 8$

31. MULTIPLE CHOICE Which point is a solution of the equation $3x - 8y = 11$? *(Section 13.1)*

Ⓐ $(1, 1)$ Ⓑ $(1, -1)$ Ⓒ $(-1, 1)$ Ⓓ $(-1, -1)$

13.5 Graphing Linear Equations in Standard Form

Essential Question How can you describe the graph of the equation $ax + by = c$?

1 ACTIVITY: Using a Table to Plot Points

Work with a partner. You sold a total of $16 worth of tickets to a school concert. You lost track of how many of each type of ticket you sold.

$\frac{\square}{\text{adult}}$ • Number of adult tickets + $\frac{\square}{\text{student}}$ • Number of student tickets = ☐

a. Let x represent the number of adult tickets.
Let y represent the number of student tickets.
Write an equation that relates x and y.

b. Copy and complete the table showing the different combinations of tickets you might have sold.

Number of Adult Tickets, x					
Number of Student Tickets, y					

c. Plot the points from the table. Describe the pattern formed by the points.

d. If you remember how many adult tickets you sold, can you determine how many student tickets you sold? Explain your reasoning.

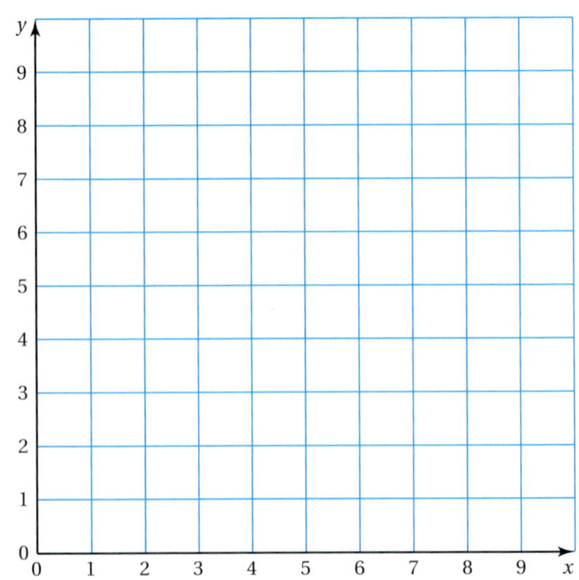

Graphing Equations
In this lesson, you will
- graph linear equations written in standard form.

Laurie's Notes

Introduction

Applying Mathematical Practices

- **Look for and Make Use of Structure:** In this lesson students will graph a linear equation in a new form. Mathematically proficient students discern a pattern or structure. Recognizing the equivalence of equations written in different forms requires that students be able to manipulate equations.

Motivate

- **Preparation:** Make a set of equation cards on strips of paper. The equations are all the same when simplified and need to be written large enough to be read by students sitting at the back of the classroom.
- Here is a sample set of equations: $y = 2x + 1$, $-2x + y = 1$, $2x - y = -1$, $4x - 2y = -2$
- Ask 4 students to stand at the front of the room and hold the cards so only they can see the equations.
- As you state an ordered pair, each student holding a card determines whether it is a solution to the equation on the card and if it is, raises his or her hand. If not, they do nothing. State several ordered pairs, four that are solutions and two that are not. Plot all of the points that you state. The four ordered pairs that are solutions will lie on a line.

? "How many lines can pass through any two points?" one "How many lines pass through the four solutions points?" Students will say 4; now is the time to discuss the idea of one line written in different forms.

- Have the students holding the cards reveal the equations to the class and read them aloud. Write each of the equations on the board.
- Explain to students that equations can be written in different forms. Today they will explore a new form of a linear equation.

Activity Notes

Activity 1

- Read the problem aloud. Note that a verbal model is shown for the equation $4x + 2y = 16$.
- Discuss what the variables x and y represent.

? "Could you have sold 5 adult tickets? Explain." No; 5 adult tickets would be $20, which is too much.

- Students may say that they do not know how to figure out x and y. Students may not realize that there is more than one solution. Remind students that *Guess, Check, and Revise* would be an appropriate strategy to use.
- Discuss part (c). The points lie on a line.
- Discuss part (d). Students may not recognize that in knowing x, they can substitute and solve for y. This is not an obvious step for students.

? "Can $x = 1.5$? Explain." No, you cannot sell 1.5 tickets.

? "What are the different numbers of adult tickets that are possible to sell?" 0, 1, 2, 3, 4

- **Note:** This is an example of a discrete domain; there are only 5 possible values for the variable x. Discrete domains are not taught at this time.

What Your Students Will Learn

- Graph linear equations written in standard form by using the slope and y-intercept or by finding the x- and y-intercepts.

Previous Learning

Students should know how to graph linear equations in slope-intercept form.

Technology for the Teacher

Dynamic Classroom

Lesson Plans
Complete Materials List

13.5 Record and Practice Journal

Essential Question How can you describe the graph of the equation $ax + by = c$?

1 ACTIVITY: Using a Table to Plot Points

Work with a partner. You sold a total of $16 worth of tickets to a school concert. You lost track of how many of each type of ticket you sold.

$4 adult • Number of adult tickets + $2 student • Number of student tickets = $16

a. Let x represent the number of adult tickets. Let y represent the number of student tickets. Write an equation that relates x and y. $4x + 2y = 16$

b. Complete the table showing the different combinations of tickets you might have sold.

Number of Adult Tickets, x	0	1	2	3	4
Number of Student Tickets, y	8	6	4	2	0

c. Plot the points from the table. Describe the pattern formed by the points. form a line

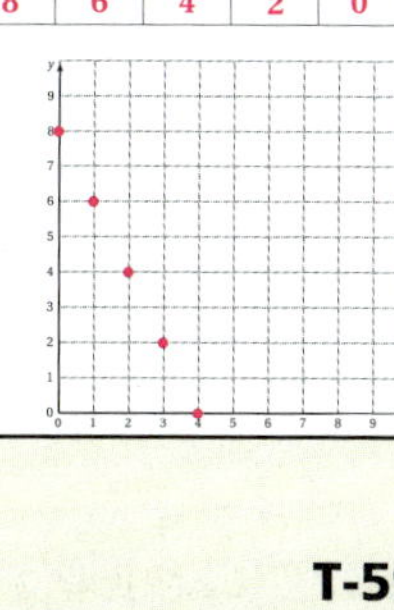

d. If you remember how many adult tickets you sold, can you determine how many student tickets you sold? Explain your reasoning. yes

Differentiated Instruction

Visual

Have students create charts in their notebooks of the equation forms and how to graph them.

Slope-intercept form $y = mx + b$	• Plot $(0, b)$. • Use the slope m to plot a second point. • Draw a line through the two points.
Horizontal line $y = c$	• Draw a horizontal line through $(0, c)$.
Vertical line $x = c$	• Draw a vertical line through $(c, 0)$.
Standard form $ax + by = c$	• Find the y-intercept. • Find the x-intercept. • Plot the associated points. Draw a line through the two points.

13.5 Record and Practice Journal

2 ACTIVITY: Rewriting an Equation

Work with a partner. You sold a total of $16 worth of cheese. You forgot how many pounds of each type of cheese you sold.

$\frac{\$4}{\text{lb}}$ • Pounds of swiss + $\frac{\$2}{\text{lb}}$ • Pound of cheddar = $16

a. Let x represent the number of pounds of swiss cheese. Let y represent the number of pounds of cheddar cheese. Write an equation that relates x and y.

$4x + 2y = 16$

b. Rewrite the equation in slope-intercept form. Then graph the equation.

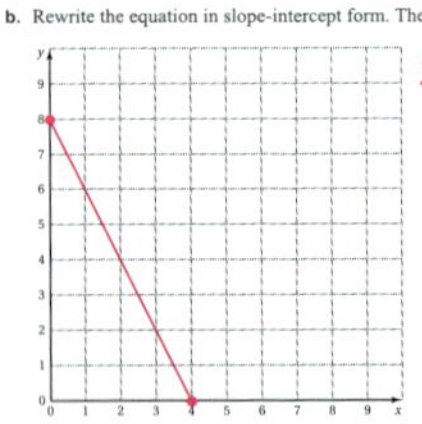

$y = -2x + 8$

c. You sold 2 pounds of cheddar cheese. How many pounds of swiss cheese did you sell?

3 lb

d. Does the value $x = 2.5$ make sense in the context of the problem? Explain.

yes

What Is Your Answer?

3. **IN YOUR OWN WORDS** How can you describe the graph of the equation $ax + by = c$?

Sample answer: a line with a slope of $-\frac{a}{b}$ and y-intercept of $\frac{c}{b}$

4. Activities 1 and 2 show two different methods for graphing $ax + by = c$. Describe the two methods. Which method do you prefer? Explain.

Check students' work.

5. Write a real-life problem that is similar to those shown in Activities 1 and 2.

Check students' work.

6. Why do you think it might be easier to graph $x + y = 10$ without rewriting it in slope-intercept form and then graphing?

You can see that when $x = 0$, $y = 10$, and when $y = 0$, $x = 10$. You can graph the equation through its x- and y-intercepts.

Laurie's Notes

Activity 2

- Read the problem aloud. Note that a verbal model is shown for the equation $4x + 2y = 16$.
- Discuss what the variables x and y represent.
- ? "Could you have sold 5 pounds of swiss cheese? Explain." No; 5 pounds of swiss cheese would be $20, which is too much.
- Give time for students to work with their partners. While this may be the same equation as Activity 1, the approach is different. Students are asked to write the equation in slope-intercept form. After the equation is in slope-intercept form, students can substitute a value for x and find y. This is generally not the case for equations written in standard form.
- Students may solve part (c) in several different ways. They may try to "read" the solution from their graphs or they may substitute $y = 2$ into either of the equations in parts (a) and (b).
- ? "Can $x = 2.5$? Explain." yes; You can buy a portion of a pound.
- **Note:** This is an example of a continuous domain; all numbers $0 \le x \le 4$ are possible. Continuous domains are not taught at this time.
- Students might observe that both examples have graphs in the first quadrant. This is common for real-life examples.

What Is Your Answer?

- **Question 3:** Students may guess that the graph is linear from Activity 1. However, some students may not be secure with this knowledge yet.
- Question 6 asks students to think about the process of graphing a line. Students consider the structure of the equation $x + y = 10$. When sharing their answers, listen for students to translate the equation as "the sum of two number is 10."

Closure

- Refer back to the equation cards. Rewrite the last three equations in slope-intercept form. $y = 2x + 1$

2 ACTIVITY: Rewriting an Equation

Work with a partner. You sold a total of $16 worth of cheese. You forgot how many pounds of each type of cheese you sold.

$\dfrac{\square}{\text{pound}} \cdot$ Pounds of swiss $+ \dfrac{\square}{\text{pound}} \cdot$ Pounds of cheddar $= \square$

Math Practice

Understand Quantities

What do the equation and the graph represent? How can you use this information to solve the problem?

a. Let x represent the number of pounds of swiss cheese.

Let y represent the number of pounds of cheddar cheese.

Write an equation that relates x and y.

b. Rewrite the equation in slope-intercept form. Then graph the equation.

c. You sold 2 pounds of cheddar cheese. How many pounds of swiss cheese did you sell?

d. Does the value $x = 2.5$ make sense in the context of the problem? Explain.

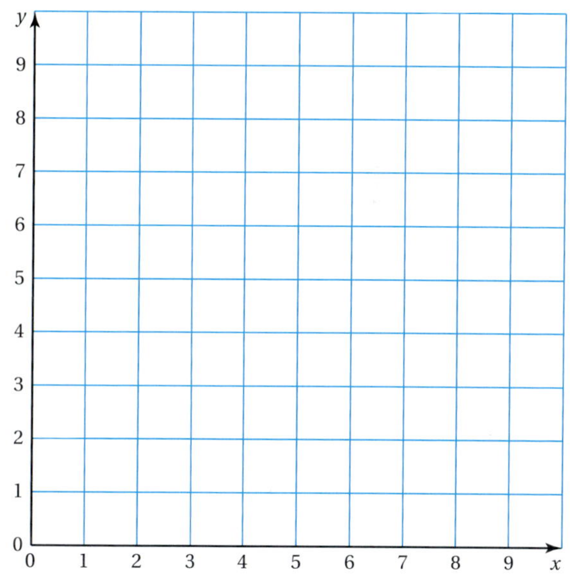

What Is Your Answer?

3. **IN YOUR OWN WORDS** How can you describe the graph of the equation $ax + by = c$?

4. Activities 1 and 2 show two different methods for graphing $ax + by = c$. Describe the two methods. Which method do you prefer? Explain.

5. Write a real-life problem that is similar to those shown in Activities 1 and 2.

6. Why do you think it might be easier to graph $x + y = 10$ without rewriting it in slope-intercept form and then graphing?

Practice Use what you learned about graphing linear equations in standard form to complete Exercises 3 and 4 on page 600.

13.5 Lesson

Check It Out
Lesson Tutorials
BigIdeasMath.com

Key Vocabulary
standard form, *p. 598*

Study Tip

Any linear equation can be written in standard form.

Key Idea

Standard Form of a Linear Equation

The **standard form** of a linear equation is

$$ax + by = c$$

where a and b are not both zero.

EXAMPLE 1 Graphing a Linear Equation in Standard Form

Graph $-2x + 3y = -6$.

Step 1: Write the equation in slope-intercept form.

$-2x + 3y = -6$	Write the equation.
$3y = 2x - 6$	Add $2x$ to each side.
$y = \frac{2}{3}x - 2$	Divide each side by 3.

Step 2: Use the slope and the y-intercept to graph the equation.

$$y = \frac{2}{3}x + (-2)$$

slope: $\frac{2}{3}$; y-intercept: -2

Check

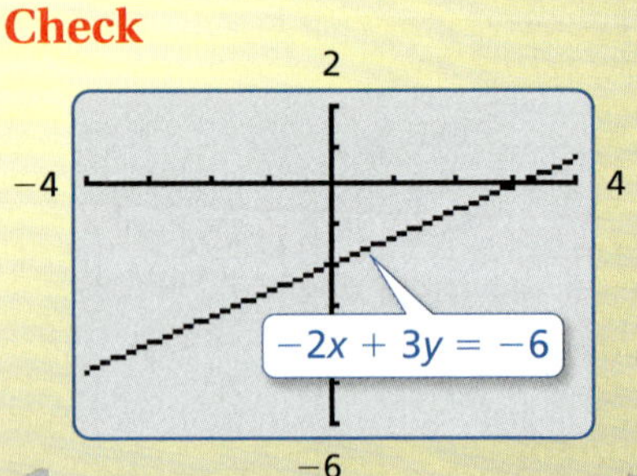

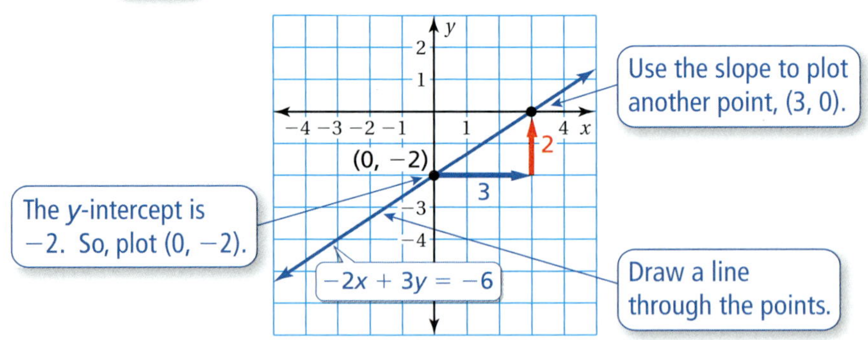

On Your Own

Now You're Ready
Exercises 5–10

Graph the linear equation. Use a graphing calculator to check your graph.

1. $x + y = -2$

2. $-\frac{1}{2}x + 2y = 6$

3. $-\frac{2}{3}x + y = 0$

4. $2x + y = 5$

Multi-Language Glossary at BigIdeasMath.com

Laurie's Notes

Introduction

Connect

- **Yesterday:** Students explored the graph of an equation written in standard form.
- **Today:** Students will graph equations written in standard form.

Motivate

? "How many pairs of numbers can you think of that have a sum of 5?" Encourage students to write their numbers on paper as ordered pairs. Example: (2, 3)

? "Did any of you include numbers that are not whole numbers?" Check to see if anyone had negative numbers or rational numbers.

- Ask one student to name an x-coordinate and another student to provide the y-coordinate. Plot the ordered pairs in a coordinate plane.

? "What do you think the equation of this line would be?" $x + y = 5$

Lesson Notes

Key Idea

- Define the standard form of a linear equation.
- Students may ask why both a and b cannot be zero. Explain that if $a = 0$ and $b = 0$, you would not have the equation of a line.
- **Teaching Tip:** Students are often confused when the standard form is written with parameters a, b, and c. Students see 5 variables. Show examples of equations written in standard form and identify a, b, and c.
- Ask students why they think $ax + by = c$ is called *standard* form. Students might suggest that the variables are on the left and a constant on the right.

Example 1

- Have students identify a, b, and c. $a = -2$, $b = 3$, and $c = -6$

? "How do you solve for y?" Add $2x$ to each side, then divide both sides by 3.

- Explain that the reason for rewriting the equation in slope-intercept form is so that the slope and the y-intercept can be used to graph the equation.
- **Common Error:** Students only divide one of the two terms on the right side of the equation by 3. Relate this to fraction operations. You are separating the expression into two terms and then simplifying.

? "Now that the equation is in slope-intercept form, explain how to graph the equation." Plot the ordered pair for the y-intercept. To plot another point, start at $(0, -2)$ and move to the right 3 units and up 2 units. Note that you can also move 3 units to the left and down 2 units. Connect these points with a line.

- Substitute the additional ordered pairs into the original equation to verify that they are solutions of the equation.

On Your Own

- In Questions 2 and 3, the fractional coefficients may present a problem.
- Remind students that equations must be solved for y in order to enter them in the equation editor of the graphing calculator.

Goal Today's lesson is graphing a linear equation written in **standard form**.

Technology for the Teacher

Dynamic Classroom

Lesson Tutorials
Lesson Plans
Answer Presentation Tool

Extra Example 1

Graph $3x - 2y = 2$.

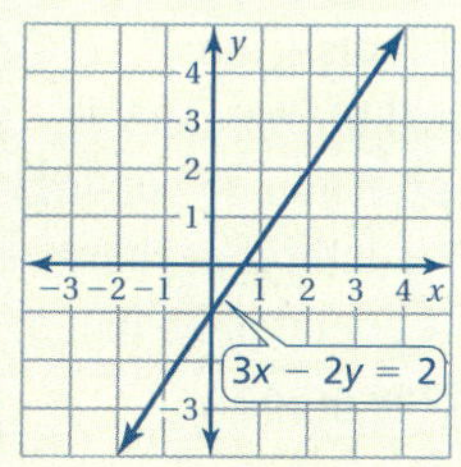

On Your Own

1.
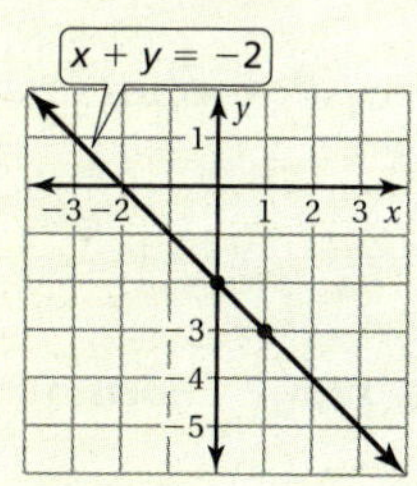

2.
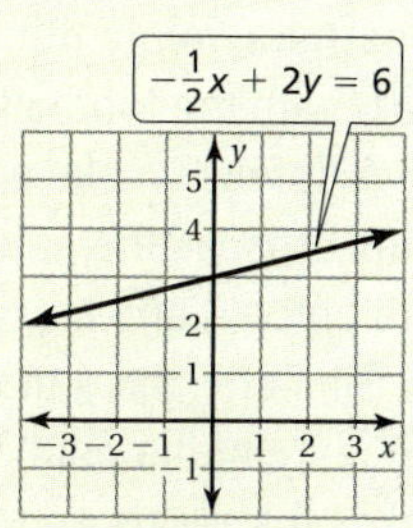

3–4. See Additional Answers.

Extra Example 2

Graph $5x - y = -5$ using intercepts.

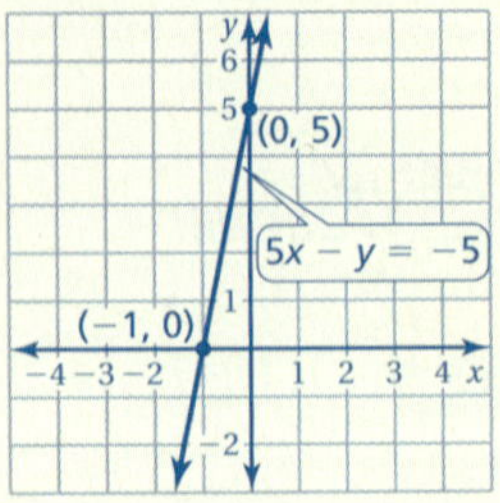

Extra Example 3

You have $2.40 to spend on grapes and bananas.

a. Graph the equation $1.2x + 0.6y = 2.4$, where x is the number of pounds of grapes and y is the number of pounds of bananas.

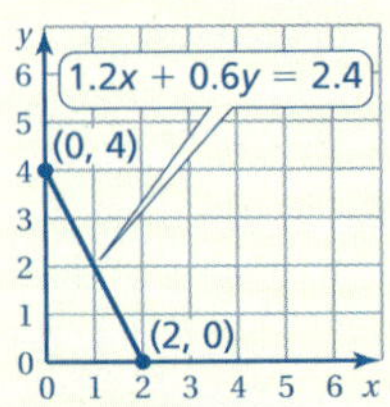

b. Interpret the intercepts. The x-intercept shows that you can buy 2 pounds of grapes, if you do not buy any bananas. The y-intercept shows that you can buy 4 pound of bananas, if you do not buy any grapes.

On Your Own

5–7. See Additional Answers.

English Language Learners

Vocabulary

For English learners, relate the word *intercept* with the football term *interception.* A defensive player on a football team crosses the path of the football to catch it and make an interception. Similarly, the y-intercept is the y-coordinate of the point where the line crosses the y-axis and the x-intercept is the x-coordinate of the point where the line crosses the x-axis.

Laurie's Notes

Example 2

- Start with a simple equation in standard form, such as $x + y = 4$. In this example, $a = 1$, $b = 1$, and $c = 4$. Explain to students that this could be solved for y by subtracting x from each side of the equation. Instead, you want to leave the equation as it was written.
- ? "Another way to think of this equation is *the sum of two numbers is 4.* Can you name some ordered pairs that would satisfy the equation?" Students should give many, including (0, 4) and (4, 0).
- Explain to students that sometimes an equation in standard form is graphed by using the two intercepts, instead of rewriting the equation in slope-intercept form.
- Write the equation shown: $x + 3y = -3$.
- ? "To find the x-intercept, what is the value of y? To find the y-intercept, what is the value of x?" 0; 0
- Finish the problem as shown.
- **Big Idea:** When the equation is in standard form, you can plot the points for the two intercepts and then draw the line through them.

Example 3

- Read the problem. Write the equation $1.5x + 0.6y = 6$ on the board.
- ? "What are the intercepts for this equation?" The x-intercept is 4 and the y-intercept is 10.
- Interpreting the intercepts in part (b) is an important step, particularly for real-life applications.
- Explain to students that negative values of x and y are not included in the graph because it does not make sense to have negative pounds of apples and bananas.
- ? "What is the cost of 2 pounds of apples and 5 pounds of bananas?" $6
- ? **Make Sense of Problems:** "What can you buy for $6?" *Sample answer:* 4 pounds of apples, or 10 pounds of bananas, or some other combination that is a solution. You are helping students make sense of the problem by asking them to interpret the symbolic representation.

On Your Own

- Students should work in pairs.

Closure

- **Writing Prompt:** To graph the equation $2x + y = 4$ … *Sample answer:* Find and plot the points for the x- and y-intercepts, then draw a line through these two points.

EXAMPLE 2 Graphing a Linear Equation in Standard Form

Graph $x + 3y = -3$ using intercepts.

Step 1: To find the x-intercept, substitute 0 for y.

$$x + 3y = -3$$
$$x + 3(0) = -3$$
$$x = -3$$

To find the y-intercept, substitute 0 for x.

$$x + 3y = -3$$
$$0 + 3y = -3$$
$$y = -1$$

Step 2: Graph the equation.

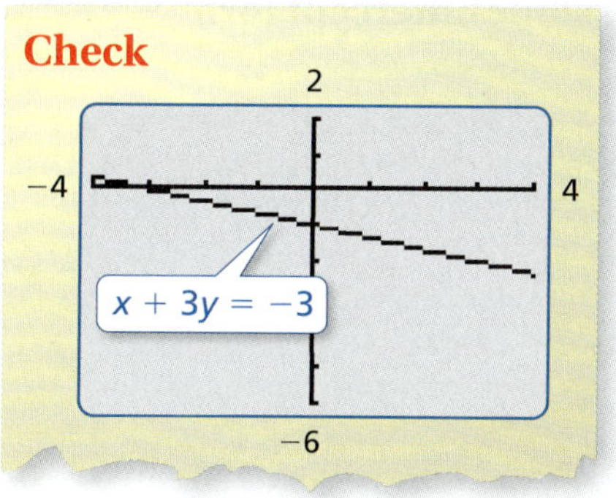

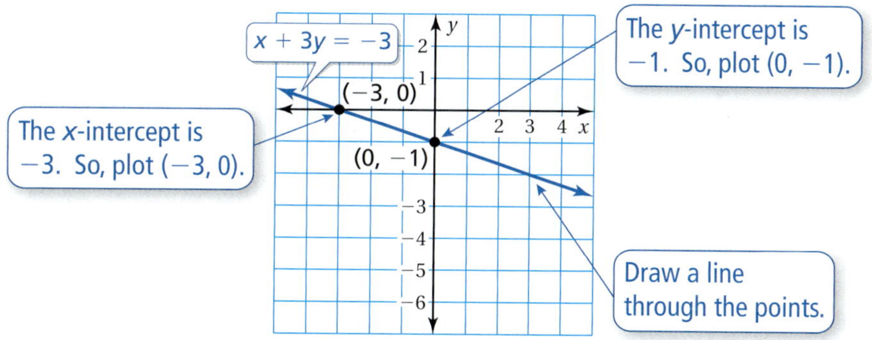

EXAMPLE 3 Real-Life Application

You have \$6 to spend on apples and bananas. (a) Graph the equation $1.5x + 0.6y = 6$, where x is the number of pounds of apples and y is the number of pounds of bananas. (b) Interpret the intercepts.

a. Find the intercepts and graph the equation.

***x*-intercept**	***y*-intercept**
$1.5x + 0.6y = 6$	$1.5x + 0.6y = 6$
$1.5x + 0.6(0) = 6$	$1.5(0) + 0.6y = 6$
$x = 4$	$y = 10$

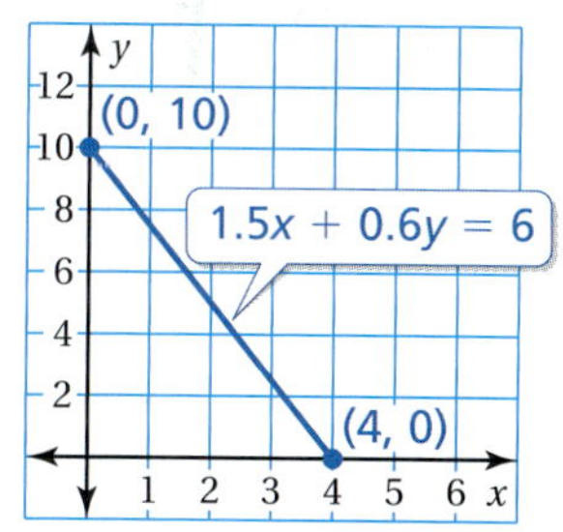

b. The x-intercept shows that you can buy 4 pounds of apples when you do not buy any bananas. The y-intercept shows that you can buy 10 pounds of bananas when you do not buy any apples.

On Your Own

Now You're Ready
Exercises 16–18

Graph the linear equation using intercepts. Use a graphing calculator to check your graph.

5. $2x - y = 8$

6. $x + 3y = 6$

7. **WHAT IF?** In Example 3, you buy y pounds of oranges instead of bananas. Oranges cost \$1.20 per pound. Graph the equation $1.5x + 1.2y = 6$. Interpret the intercepts.

13.5 Exercises

Check It Out
Help with Homework
BigIdeasMath.com

Vocabulary and Concept Check

1. **VOCABULARY** Is the equation $y = -2x + 5$ in standard form? Explain.
2. **WRITING** Describe two ways to graph the equation $4x + 2y = 6$.

Practice and Problem Solving

Define two variables for the verbal model. Write an equation in slope-intercept form that relates the variables. Graph the equation.

3. $\frac{\$2.00}{\text{pound}}$ • Pounds of peaches + $\frac{\$1.50}{\text{pound}}$ • Pounds of apples = \$15

4. $\frac{16 \text{ miles}}{\text{hour}}$ • Hours biked + $\frac{2 \text{ miles}}{\text{hour}}$ • Hours walked = 32 miles

Write the linear equation in slope-intercept form.

1 5. $2x + y = 17$ 6. $5x - y = \frac{1}{4}$ 7. $-\frac{1}{2}x + y = 10$

Graph the linear equation. Use a graphing calculator to check your graph.

8. $-18x + 9y = 72$ 9. $16x - 4y = 2$ 10. $\frac{1}{4}x + \frac{3}{4}y = 1$

Match the equation with its graph.

11. $15x - 12y = 60$ 12. $5x + 4y = 20$ 13. $10x + 8y = -40$

A.
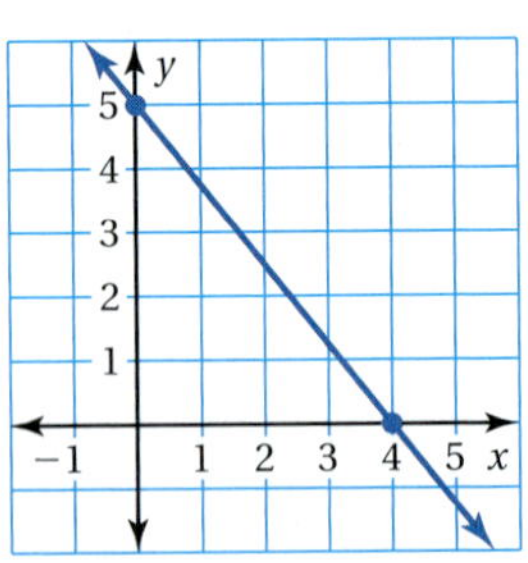

B.
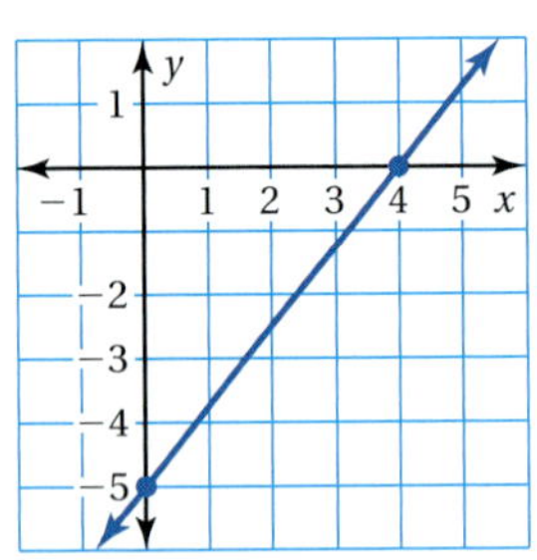

C.
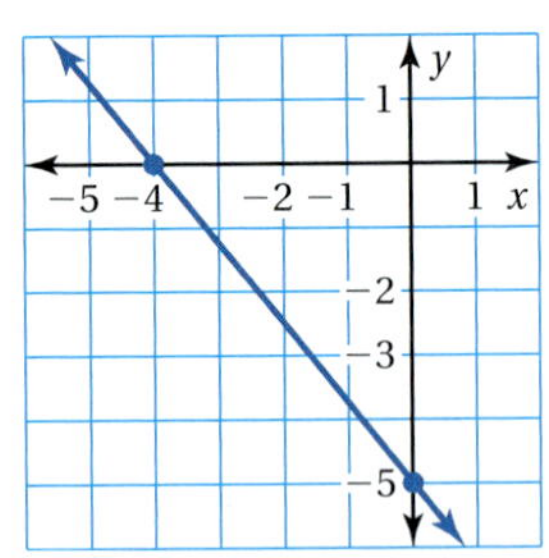

14. **ERROR ANALYSIS** Describe and correct the error in finding the x-intercept.

$$-2x + 3y = 12$$
$$-2(0) + 3y = 12$$
$$3y = 12$$
$$y = 4$$

15. **BRACELET** A charm bracelet costs \$65, plus \$25 for each charm. The equation $-25x + y = 65$ represents the cost y of the bracelet, where x is the number of charms.
   a. Graph the equation.
   b. How much does the bracelet shown cost?

Assignment Guide and Homework Check

Level	Assignment	Homework Check
Accelerated	1–4, 6–14 even, 16–26	10, 18, 19, 20, 22

Common Errors

- **Exercises 5–10** Students may use the same operation instead of the opposite operation when rewriting the equation in slope-intercept form.
- **Exercises 11–13, 16–18** Students may mix up the *x*- and *y*-intercepts. Remind them that the *x*-intercept is the *x*-coordinate of where the line crosses the *x*-axis and the *y*-intercept is the *y*-coordinate of where the line crosses the *y*-axis.

13.5 Record and Practice Journal

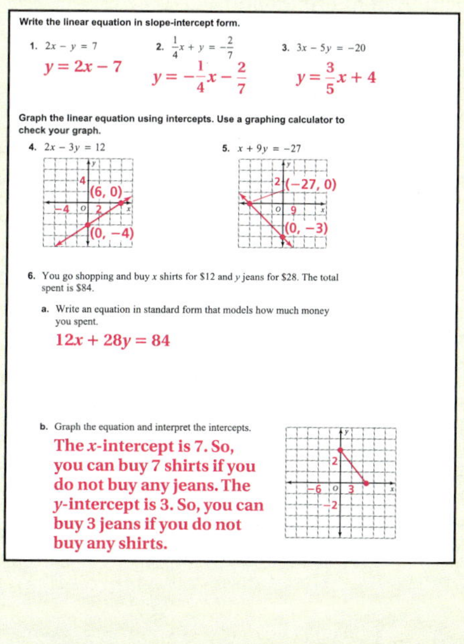
Write the linear equation in slope-intercept form.

1. $2x - y = 7$ $\quad y = 2x - 7$
2. $\frac{1}{4}x + y = -\frac{2}{7}$ $\quad y = -\frac{1}{4}x - \frac{2}{7}$
3. $3x - 5y = -20$ $\quad y = \frac{3}{5}x + 4$

Graph the linear equation using intercepts. Use a graphing calculator to check your graph.

4. $2x - 3y = 12$ $\quad (6, 0)$, $(0, -4)$
5. $x + 9y = -27$ $\quad (-27, 0)$, $(0, -3)$

6. You go shopping and buy x shirts for \$12 and y jeans for \$28. The total spent is \$84.

 a. Write an equation in standard form that models how much money you spent.

 $12x + 28y = 84$

 b. Graph the equation and interpret the intercepts.

 The x-intercept is 7. So, you can buy 7 shirts if you do not buy any jeans. The y-intercept is 3. So, you can buy 3 jeans if you do not buy any shirts.

Vocabulary and Concept Check

1. no; The equation is in slope-intercept form.
2. *Sample answer:*
 1) Write the equation in slope-intercept form and use the slope and *y*-intercept to graph the equation.
 2) Find the *x*- and *y*-intercepts, plot the points representing the intercepts, and draw a line through the points.

Practice and Problem Solving

3. $x =$ pounds of peaches
 $y =$ pounds of apples
 $y = -\frac{4}{3}x + 10$

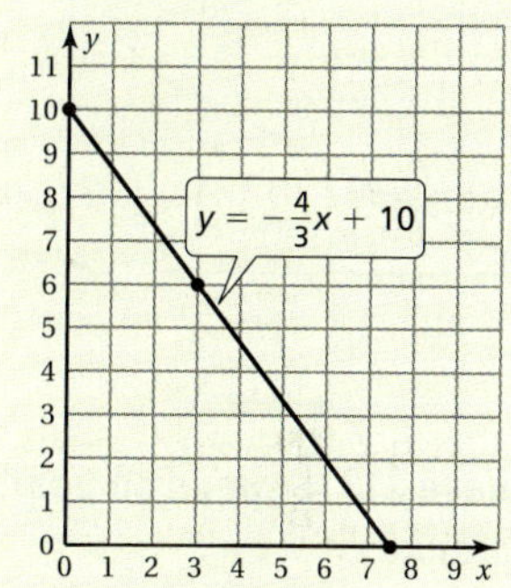

4. $x =$ hours biked
 $y =$ hours walked
 $y = -8x + 16$

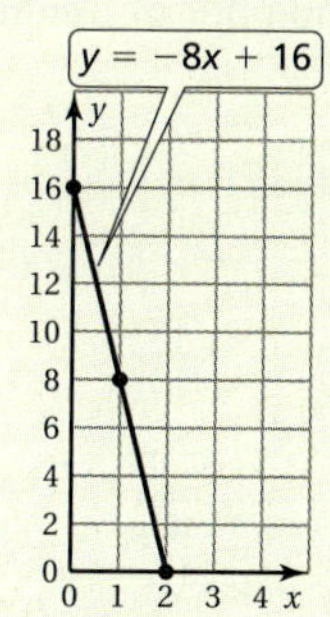

5. $y = -2x + 17$
6. $y = 5x - \frac{1}{4}$
7. $y = \frac{1}{2}x + 10$

8–15. See Additional Answers.

Practice and Problem Solving

16–19. See Additional Answers.

20. See *Taking Math Deeper*.

21–23. See Additional Answers.

Fair Game Review

24. 4 **25.** $\frac{1}{2}$

26. D

Mini-Assessment

1. Graph $-2x + 4y = 16$ using intercepts.

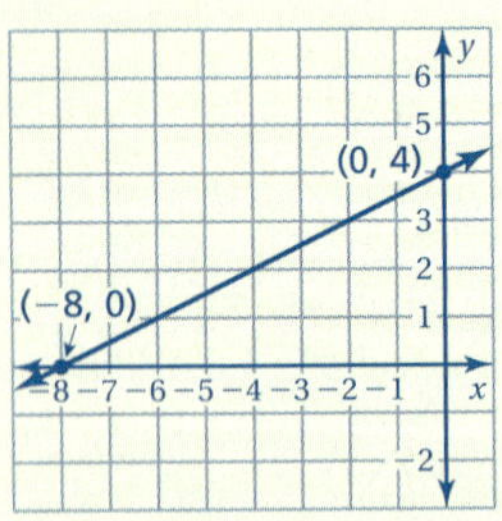

2. **You have \$12 to spend on pears and oranges.**
   a. Graph the equation $1.2x + 0.8y = 12$, where x is the number of pounds of pears and y is the number of pounds of oranges.

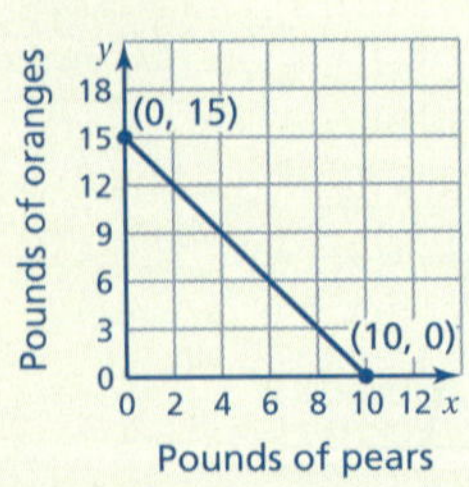

   b. Interpret the intercepts.
   The x-intercept shows that you can buy 10 pounds of pears if you do not buy any oranges. The y-intercept shows that you can buy 15 pounds of oranges if you do not buy any pears.

Taking Math Deeper

Exercise 20

As with many real-life problems, it helps to start by summarizing the given information.

1. Summarize the given information.
   - Let x = days for renting boat.
   - Let y = days for renting scuba gear.
   - Cost of boat = \$250 per day.
   - Cost of scuba gear = \$50 per day.
   - Total spent = \$1000.

2. **a.** Write an equation.

$$250x + 50y = 1000$$

3. **b.** Graph the equation and interpret the intercepts.

$$y = -5x + 20$$

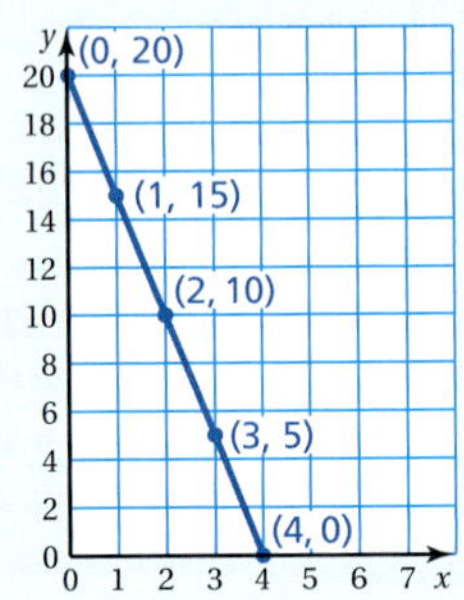

If $x = 0$, then the group rented only the scuba gear for 20 days.
If $y = 0$, then the group rented only the boat for 4 days.

Project

To go on a professional scuba diving tour, you need to be a certified diver. Use the school library or the Internet to research the requirements to become certified in scuba diving.

Reteaching and Enrichment Strategies

If students need help. . .	If students got it. . .
Resources by Chapter • Practice A and Practice B • Puzzle Time Record and Practice Journal Practice Differentiating the Lesson Lesson Tutorials Skills Review Handbook	Resources by Chapter • Enrichment and Extension • Technology Connection Start the next section

Graph the linear equation using intercepts. Use a graphing calculator to check your graph.

2 **16.** $3x - 4y = -12$ **17.** $2x + y = 8$ **18.** $\frac{1}{3}x - \frac{1}{6}y = -\frac{2}{3}$

19. SHOPPING The amount of money you spend on x CDs and y DVDs is given by the equation $14x + 18y = 126$. Find the intercepts and graph the equation.

20. SCUBA Five friends go scuba diving. They rent a boat for x days and scuba gear for y days. The total spent is $1000.

a. Write an equation in standard form that represents the situation.

b. Graph the equation and interpret the intercepts.

21. MODELING You work at a restaurant as a host and a server. You earn $9.45 for each hour you work as a host and $7.65 for each hour you work as a server.

a. Write an equation in standard form that models your earnings.

b. Graph the equation.

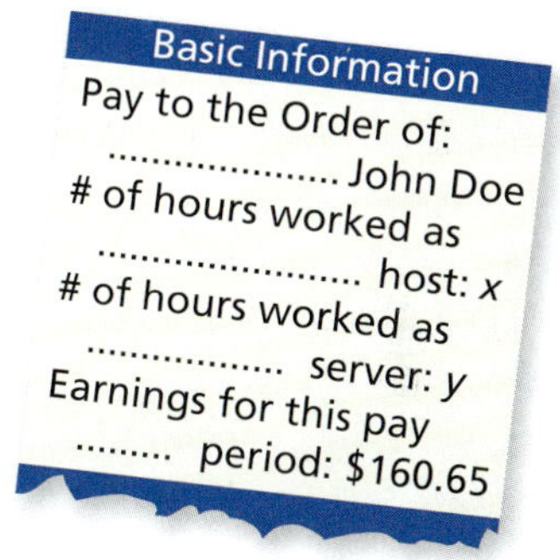

22. LOGIC Does the graph of every linear equation have an x-intercept? Explain your reasoning. Include an example.

23. **Critical Thinking** For a house call, a veterinarian charges $70, plus $40 an hour.

a. Write an equation that represents the total fee y (in dollars) the veterinarian charges for a visit lasting x hours.

b. Find the x-intercept. Does this value make sense in this context? Explain your reasoning.

c. Graph the equation.

Fair Game Review What you learned in previous grades & lessons

The points in the table lie on a line. Find the slope of the line. *(Section 13.2)*

24.

x	−2	−1	0	1
y	−10	−6	−2	2

25.

x	2	4	6	8
y	2	3	4	5

26. MULTIPLE CHOICE Which value of x makes the equation $4x - 12 = 3x - 9$ true? *(Topic 2)*

(A) −1 (B) 0 (C) 1 (D) 3

13.6 Writing Equations in Slope-Intercept Form

Essential Question How can you write an equation of a line when you are given the slope and the *y*-intercept of the line?

1 ACTIVITY: Writing Equations of Lines

Work with a partner.

- **Find the slope of each line.**
- **Find the *y*-intercept of each line.**
- **Write an equation for each line.**
- **What do the three lines have in common?**

a.

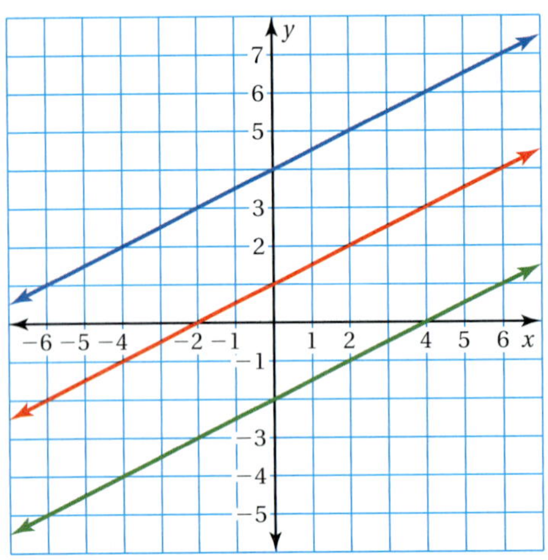

b.

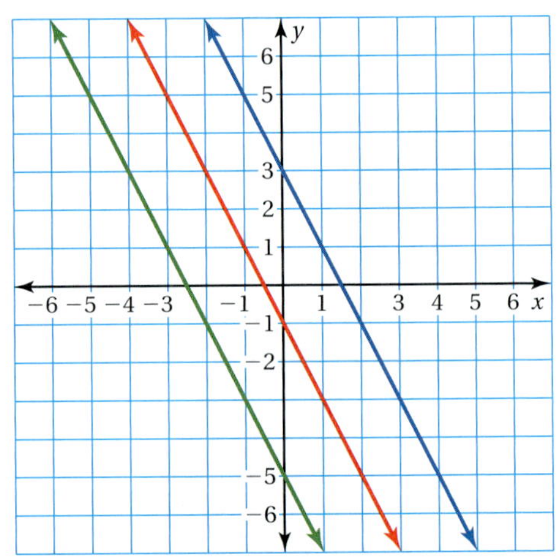

c.

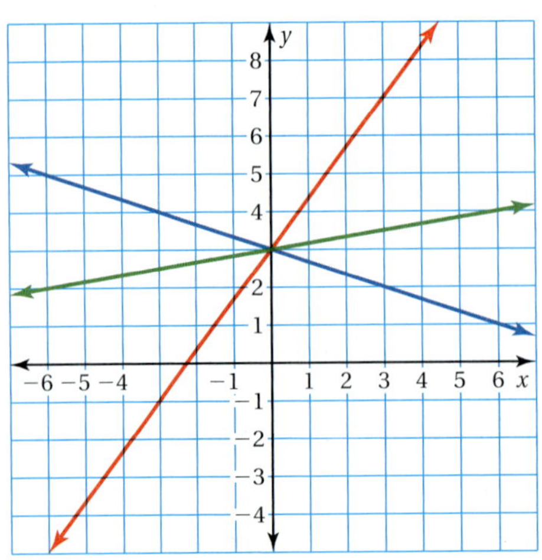

d.

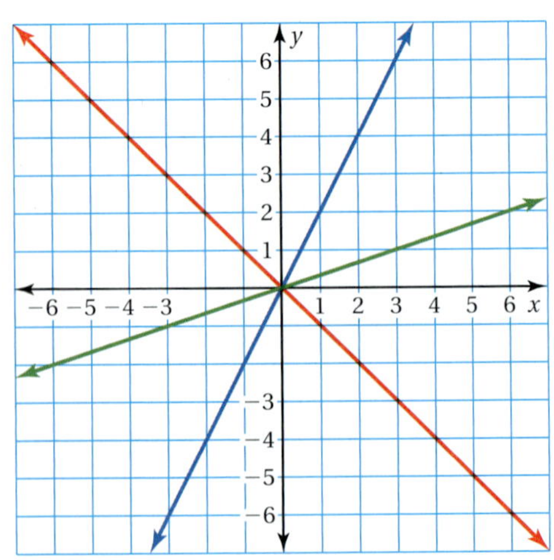

Writing Equations
In this lesson, you will
- write equations of lines in slope-intercept form.

Laurie's Notes

Introduction

Applying Mathematical Practices

- **Make Sense of Problems and Persevere in Solving Them** and **Model with Mathematics:** The goal of this lesson is for students to write equations of lines by first determining the slope and *y*-intercept from a graph. The visual model helps students make sense of and solve the problem. The graphical representation (model) helps students identify the important features of the line.

Motivate

- If there is sufficient space in your classroom, hallway, or school foyer, make coordinate axes using masking tape. Use a marker to scale each axis with integers -5 through 5.
- Take turns having two students be the *rope anchors* who then will make a line on the coordinate axes while other students observe.
- Here are a series of directions you can give and some follow-up questions. Remind students that slope is rise over run and that the equation of a line in slope-intercept form is $y = mx + b$.
  - ? Make the line $y = x$. "What is the slope?" 1 "What is the *y*-intercept?" 0
  - ? Keep the same slope, but make the *y*-intercept 2. "What is the equation of this line?" $y = x + 2$
  - ? Use the *y*-intercept 2, but make the slope steeper. "What is the slope of this line?" Answers will vary.
  - ? Keep the same *y*-intercept, but make the slope $\frac{1}{2}$. "What is the equation?" $y = \frac{1}{2}x + 2$
- **Management Tip:** This activity can also be done by drawing the axes on the board and having the students hold the rope against the board.

Activity Notes

Activity 1

- ? "How do you determine the slope of a line drawn in a coordinate plane?" Use two points that you are sure are on the graph and find the rise and run between the points.
- ? "Does it matter whether you move left-to-right or right-to-left when you are finding the rise and run? Explain." no; Either way the slope will be the same.
- Students may have difficulty writing the equation in slope-intercept form. They think it should be harder to do!
- **FYI:** When the *y*-intercept is negative, students may leave their equation as $y = 3x + (-4)$ instead of $y = 3x - 4$. Remind students that it is more common to represent the equation as $y = 3x - 4$.
- **Teaching Tip:** If you have a student who is color blind, refer to the lines by a number or letter scheme (1, 2, 3 or A, B, C).
- Ask students to share what they found in common for each trio of lines.

What Your Students Will Learn

- Write equations of lines in slope-intercept form using the slope and *y*-intercept of given graphs.

Previous Learning

Students should know how to find the slope of a line. Students should know about parallel lines.

Technology for the Teacher

Dynamic Classroom

Lesson Plans
Complete Materials List

13.6 Record and Practice Journal

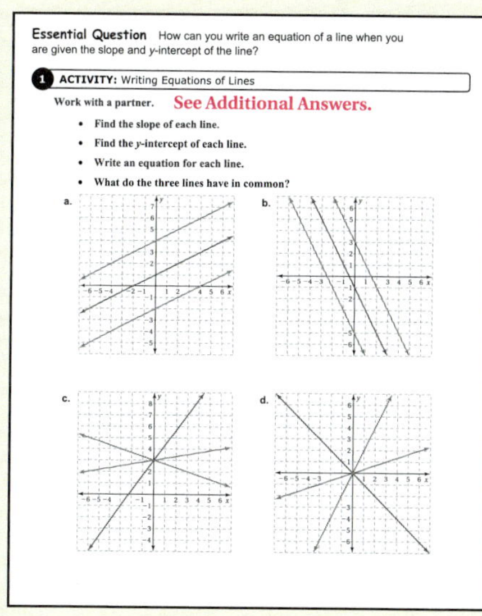

Essential Question How can you write an equation of a line when you are given the slope and *y*-intercept of the line?

1 ACTIVITY: Writing Equations of Lines

Work with a partner. See Additional Answers.

- Find the slope of each line.
- Find the *y*-intercept of each line.
- Write an equation for each line.
- What do the three lines have in common?

Differentiated Instruction

Visual

To avoid mistakes when substituting the variables, have students color code the slope and *y*-intercept of an equation.

slope: 3 $\quad$ *y*-intercept: 4

$y = mx + b$

$y = 3x + 4$

13.6 Record and Practice Journal

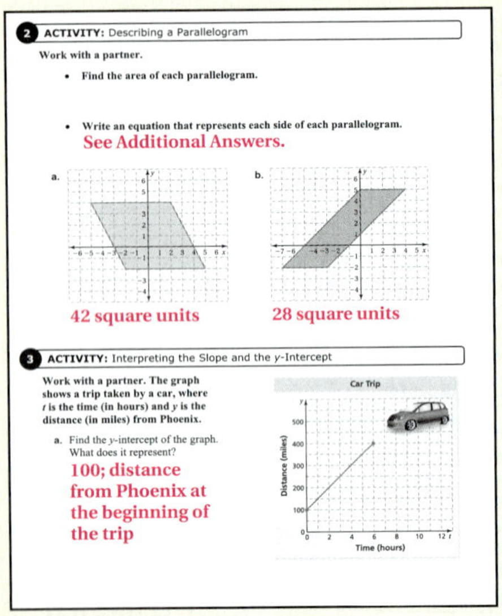
2 ACTIVITY: Describing a Parallelogram

Work with a partner.

- Find the area of each parallelogram.
- Write an equation that represents each side of each parallelogram.

See Additional Answers.

a. 42 square units $\quad$ b. 28 square units

3 ACTIVITY: Interpreting the Slope and the *y*-Intercept

Work with a partner. The graph shows a trip taken by a car, where *t* is the time (in hours) and *y* is the distance (in miles) from Phoenix.

a. Find the *y*-intercept of the graph. What does it represent?

100; distance from Phoenix at the beginning of the trip

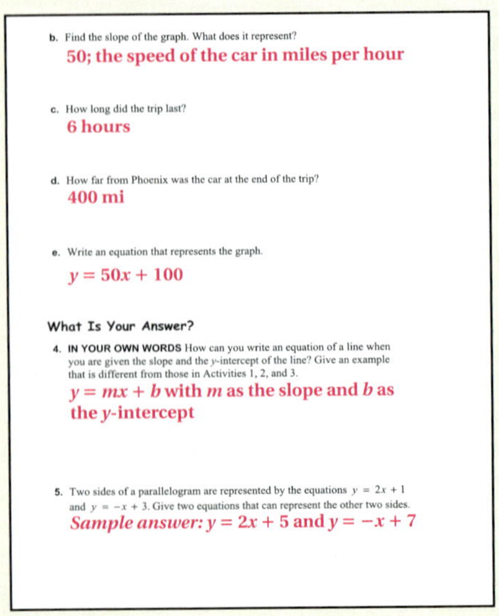
b. Find the slope of the graph. What does it represent?

50; the speed of the car in miles per hour

c. How long did the trip last?

6 hours

d. How far from Phoenix was the car at the end of the trip?

400 mi

e. Write an equation that represents the graph.

$y = 50x + 100$

What Is Your Answer?

4. IN YOUR OWN WORDS How can you write an equation of a line when you are given the slope and the *y*-intercept of the line? Give an example that is different from those in Activities 1, 2, and 3.

$y = mx + b$ with m as the slope and b as the *y*-intercept

5. Two sides of a parallelogram are represented by the equations $y = 2x + 1$ and $y = -x + 3$. Give two equations that can represent the other two sides.

Sample answer: $y = 2x + 5$ and $y = -x + 7$

Laurie's Notes

Activity 2

- ? "How do you find the area of a parallelogram?" area = base × height
- ? "Are the base and height the sides of the parallelogram?" They could be if it's a rectangle. Otherwise, height is the perpendicular distance between the two bases.
- **Make Sense of Problems and Persevere in Solving Them:** This is a good example of where students have the necessary skills but they will need to make sense of the problem and then persevere in working through the problem.
- To find the base and height, students simply count the units on the diagram. Note that the height for the parallelogram in part (b) is outside the parallelogram.
- **Common Error:** The slope of the horizontal sides is zero. Students may say that you cannot find the slope *for a flat line.*
- ? "What is the equation of a horizontal line?" $y = b$
- The challenge in this activity is writing the equations for the diagonal sides of the figure in part (a). Suggest that by extending the sides using the slope, the students should be able to determine the *y*-intercept.
- This activity reviews positive, negative, and zero slope. Area of a parallelogram is also reviewed.

Activity 3

- The graph in this problem represents a real-life context. Mathematically proficient students are able to interpret the mathematical results in the context of the situation or problem.
- If students have difficulty getting started with this activity, remind them to read the labels on the axes and interpret the *y*-intercept. The car was 100 miles from Phoenix at the beginning of the trip.
- Discuss answers to each part of the problem as a class.
- ? **Extension:** Draw the segment from (6, 400) to (12, 0) and explain that this represents the return trip. Ask the following questions.
  - "What is the slope of this line segment? What does the slope mean in the context of the problem?" slope ≈ −67; returning at a rate of about 67 mi/h
  - "What does the point (12, 0) mean in the context of the problem?" You have arrived in Phoenix and drove 12 hours.
  - "What would the graph look like if the car had stopped for 1 hour?" horizontal segment of length 1 unit

What Is Your Answer?

- Discuss answers to Question 5 as a class.

Closure

- **Exit Ticket:** What is the slope and *y*-intercept of the equation $y = 2x + 4$? slope = 2, *y*-intercept = 4 Write an equation of a line with a slope of 3 and a *y*-intercept of 1. $y = 3x + 1$

2 ACTIVITY: Describing a Parallelogram

Math Practice

Analyze Givens

What do you need to know to write an equation?

Work with a partner.

- **Find the area of each parallelogram.**
- **Write an equation that represents each side of each parallelogram.**

a.

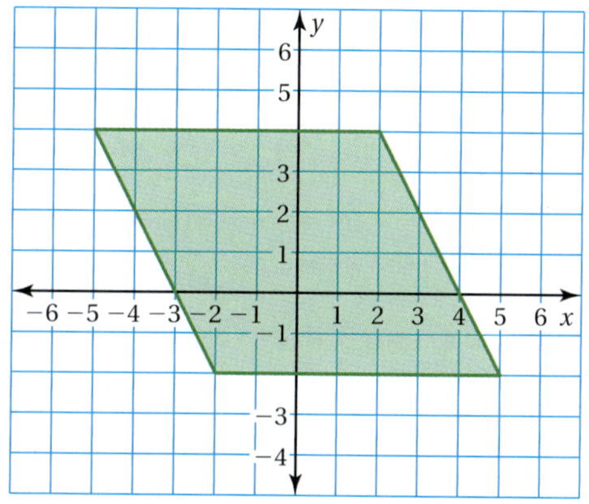

b.

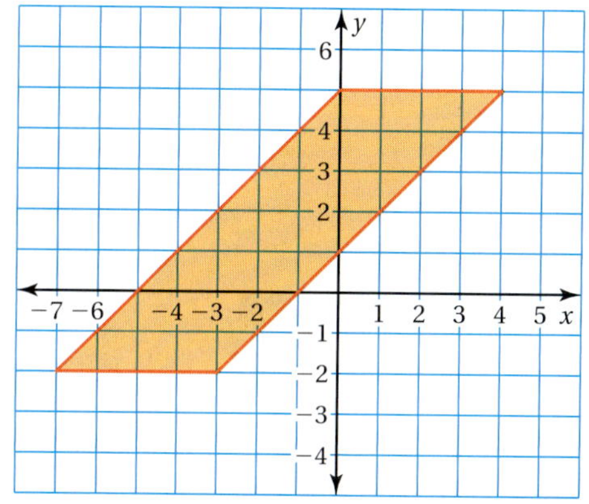

3 ACTIVITY: Interpreting the Slope and the *y*-Intercept

Work with a partner. The graph shows a trip taken by a car, where *t* is the time (in hours) and *y* is the distance (in miles) from Phoenix.

a. Find the *y*-intercept of the graph. What does it represent?

b. Find the slope of the graph. What does it represent?

c. How long did the trip last?

d. How far from Phoenix was the car at the end of the trip?

e. Write an equation that represents the graph.

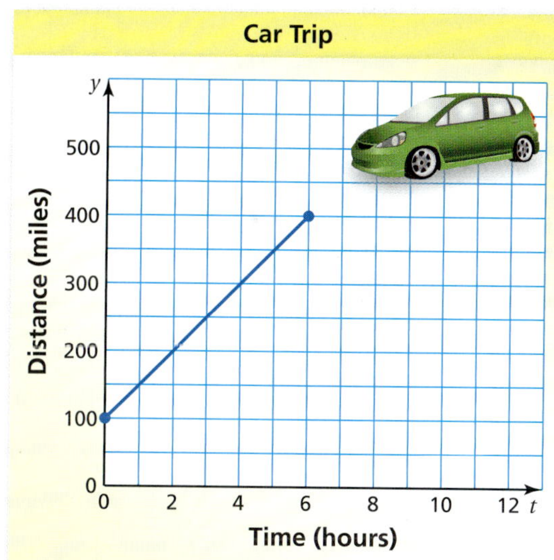

What Is Your Answer?

4. **IN YOUR OWN WORDS** How can you write an equation of a line when you are given the slope and the *y*-intercept of the line? Give an example that is different from those in Activities 1, 2, and 3.

5. Two sides of a parallelogram are represented by the equations $y = 2x + 1$ and $y = -x + 3$. Give two equations that can represent the other two sides.

Practice Use what you learned about writing equations in slope-intercept form to complete Exercises 3 and 4 on page 606.

13.6 Lesson

Check It Out
Lesson Tutorials
BigIdeasMath.com

EXAMPLE 1 Writing Equations in Slope-Intercept Form

Write an equation of the line in slope-intercept form.

a.

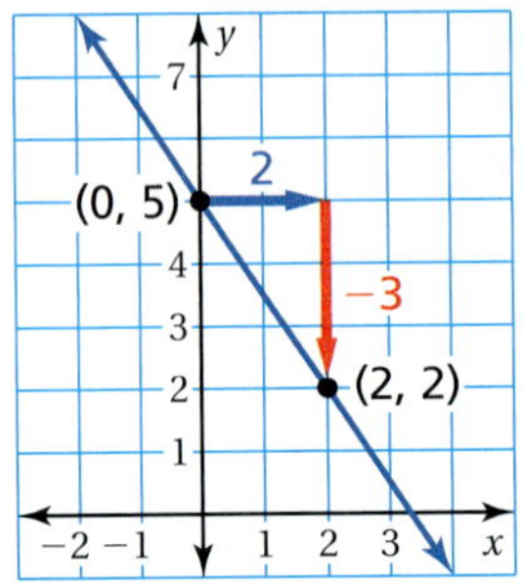

Find the slope and the y-intercept.

$$m = \frac{y_2 - y_1}{x_2 - x_1}$$

$$= \frac{2 - 5}{2 - 0}$$

$$= \frac{-3}{2}, \text{ or } -\frac{3}{2}$$

Because the line crosses the y-axis at (0, 5), the y-intercept is 5.

slope y-intercept

So, the equation is $y = -\frac{3}{2}x + 5$.

Study Tip

After writing an equation, check that the given points are solutions of the equation.

b.

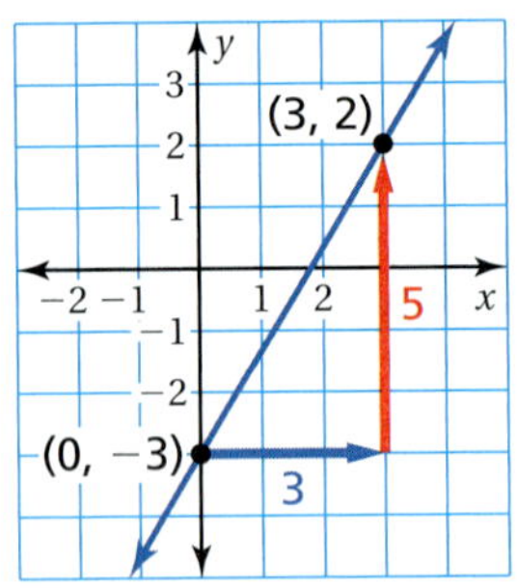

Find the slope and the y-intercept.

$$m = \frac{y_2 - y_1}{x_2 - x_1}$$

$$= \frac{-3 - 2}{0 - 3}$$

$$= \frac{-5}{-3}, \text{ or } \frac{5}{3}$$

Because the line crosses the y-axis at (0, −3), the y-intercept is −3.

slope y-intercept

So, the equation is $y = \frac{5}{3}x + (-3)$, or $y = \frac{5}{3}x - 3$.

On Your Own

Now You're Ready
Exercises 5–10

Write an equation of the line in slope-intercept form.

1.

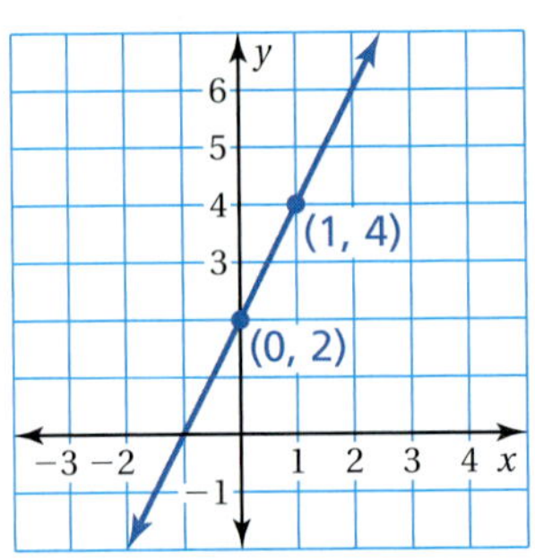

2.

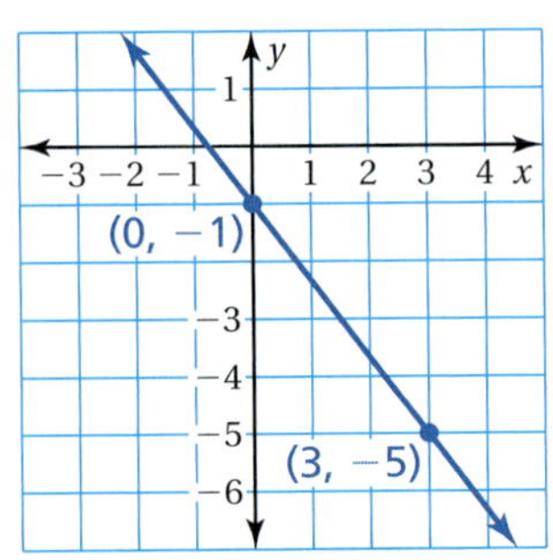

Laurie's Notes

Introduction

Connect

- **Yesterday:** Students developed an understanding of how to write an equation of a line using its slope and y-intercept.
- **Today:** Students will write an equation of a line using the slope and y-intercept.

Motivate

- **Story Time:** Tell students that as a child you loved to dig tunnels in the sand. Ask if any of them like to dig tunnels or if they have traveled through tunnels. Hold a paper towel tube or other similar model to pique student interest. Share some facts about tunnels.
  - The world's longest overland tunnel is a 21-mile-long rail link under the Alps in Switzerland. The tunnel took eight years to build and cost \$3.5 billion. It reduces the time trains need to cross between Germany and Italy from 3.5 hours to just under 2 hours.
  - The world's longest underwater tunnel is the Seikan Tunnel in Japan. It is about 33.5 miles long and runs under the Tsugaru Strait. It opened in 1988 and took 17 years to construct.
  - The Channel Tunnel (Chunnel) connects England and France. It is 31 miles long and travels under the English Channel.

Lesson Notes

Discuss

- **Make Sense of Problems** and **Model with Mathematics:** In this lesson students will make quick visual inspection of linear graphs to approximate the slope and the y-intercept. This approximation is a helpful check when the slope and y-intercept are computed.

Example 1

- Write the slope-intercept form of an equation, $y = mx + b$. Review with students that the coefficient of the x-term is the slope, and the constant b is the y-intercept. Also, review how to compute slope.
- ? "What do you know about the slope of the line in part (a) by inspection? Explain." Slope is negative because the graph falls from left to right.
- ? "In part (a), what are the coordinates of the point where the line crosses the y-axis?" (0, 5)
- Use the slope and the y-intercept to write the equation.
- Work through part (b). Remind students that you want the more simplified equation $y = \frac{5}{3}x - 3$ instead of $y = \frac{5}{3}x + (-3)$. Stress that while both forms are correct, the simplified version is preferred.

On Your Own

- Before students begin these two problems, they should do a visual inspection. They should make a note of the sign of the slope and y-intercept. It is very easy to have the wrong sign(s) when the equation is written.
- Encourage students to draw the slope triangle and label the horizontal and vertical lengths.

Goal Today's lesson is writing an equation of a line in slope-intercept form.

Technology for the Teacher

Dynamic Classroom

Lesson Tutorials
Lesson Plans
Answer Presentation Tool

Extra Example 1

Write an equation of the line in slope-intercept form.

a.

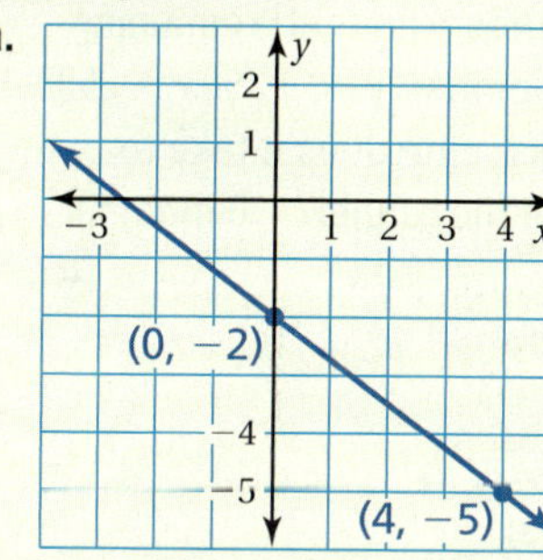

$y = -\frac{3}{4}x - 2$

b.

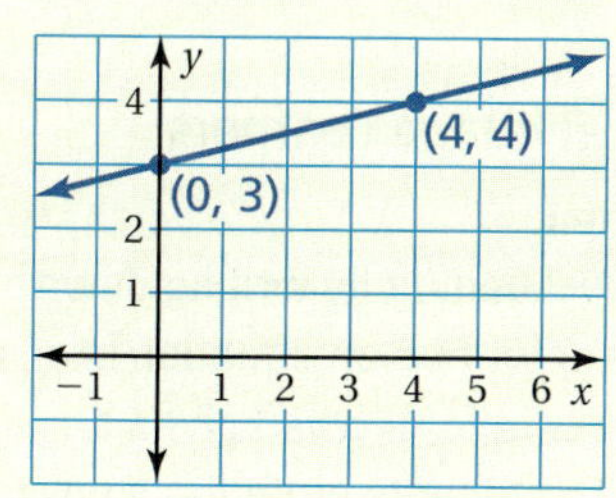

$y = \frac{1}{4}x + 3$

On Your Own

1. $y = 2x + 2$
2. $y = -\frac{4}{3}x - 1$

Laurie's Notes

Extra Example 2

Write an equation of the line that passes through the points $(0, -1)$ and $(4, -1)$.
$y = -1$

Extra Example 3

Repeat Example 3 with the points (0, 3500) and (5, 1750).

a. Write an equation that represents the distance y (in feet) remaining after x months. $y = -350x + 3500$

b. How much time does it take to complete the tunnel? 10 months

On Your Own

3. $y = 5$

4. $8\frac{3}{4}$ mo

English Language Learners

Organization

Students will benefit by writing down the steps for writing an equation in slope-intercept form when given a graph. Have students write the steps in their notebooks. A poster with the steps could be posted in the classroom.

Step 1: Write the slope-intercept form of an equation.

Step 2: Determine the slope of the line.

Step 3: Determine the y-intercept of the line.

Step 4: Write the equation in slope-intercept form.

Example 2

- Make a quick sketch of the graph to reference as you work the problem.
- When finding the slope, students are unsure of how to simplify $\frac{0}{3}$. This is a good time to review the difference between $\frac{0}{3}$ and $\frac{3}{0}$.
- **Teaching Tip:** To explain why $\frac{3}{0}$ is undefined, first write the problem $8 \div 4 = 2$ on the board. Then rewrite it as $4\overline{)8}$. To check, multiply the quotient (2) times the divisor (4) and you get the dividend (8). In other words, 2 multiplied by 4 is 8. Do the same thing with $\frac{3}{0}$. Rewrite it using long division, $0\overline{)3}$. What do you multiply 0 by to get 3? There is no quotient, so you say $\frac{3}{0}$ is undefined. You cannot divide by 0.
- **Look for and Make Use of Structure:** Students don't always recognize that $y = -4$ is a linear equation written in slope-intercept form. It helps to write the extra step of $y = (0)x + (-4)$ so students can see that the slope is 0. Students should recognize that $y = -4$ and $y = (0)x + (-4)$ are equivalent.

Example 3

- Ask a volunteer to read the problem. Discuss information that can be *read* from the graph.
- ? "By visual inspection, what do you know about the sign of the slope and the y-intercept in this problem?" The slope is negative. The y-intercept is positive.
- ? "What does a slope of -500 mean in the context of this problem?" A slope of -500 means that for each additional month of work, the distance left to complete is 500 feet less.
- The x-intercept for this graph is 7.
- Note that the graph is in Quadrant I. In the context of this problem, it doesn't make sense for time or distance to be negative.

On Your Own

- **Attend to Precision:** For Question 3, encourage students to sketch a graph of the line through the two points to give them a clue as to how to begin. The visual model is an approximation that can be used to check their final answer. This technique will help students start Question 4.

Closure

- **Writing Prompt:** For a line that has been graphed in a coordinate plane, you can write the equation by … finding the slope and y-intercept

EXAMPLE 2 Writing an Equation

Which equation is shown in the graph?

Ⓐ $y = -4$ Ⓑ $y = -3$

Ⓒ $y = 0$ Ⓓ $y = -3x$

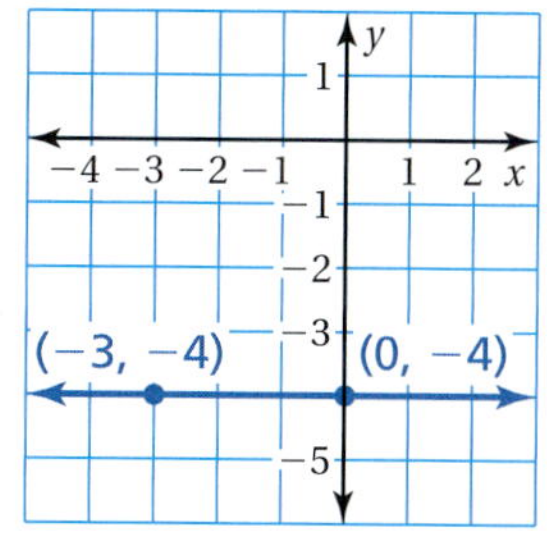

Remember

The graph of $y = a$ is a horizontal line that passes through $(0, a)$.

Find the slope and the y-intercept.

The line is horizontal, so the change in y is 0.

$$m = \frac{\text{change in } y}{\text{change in } x} = \frac{0}{3} = 0$$

Because the line crosses the y-axis at $(0, -4)$, the y-intercept is -4.

∴ So, the equation is $y = 0x + (-4)$, or $y = -4$. The correct answer is Ⓐ.

EXAMPLE 3 Real-Life Application

The graph shows the distance remaining to complete a tunnel. (a) Write an equation that represents the distance y (in feet) remaining after x months. (b) How much time does it take to complete the tunnel?

Engineers used tunnel boring machines like the ones shown above to dig an extension of the Metro Gold Line in Los Angeles. The new tunnels are 1.7 miles long and 21 feet wide.

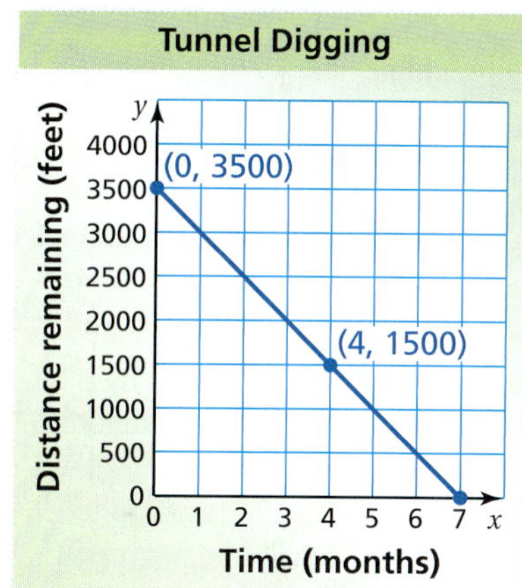

a. Find the slope and the y-intercept.

$$m = \frac{\text{change in } y}{\text{change in } x} = \frac{-2000}{4} = -500$$

Because the line crosses the y-axis at $(0, 3500)$, the y-intercept is 3500.

∴ So, the equation is $y = -500x + 3500$.

b. The tunnel is complete when the distance remaining is 0 feet. So, find the value of x when $y = 0$.

$y = -500x + 3500$	Write the equation.
$0 = -500x + 3500$	Substitute 0 for y.
$-3500 = -500x$	Subtract 3500 from each side.
$7 = x$	Divide each side by -500.

∴ It takes 7 months to complete the tunnel.

On Your Own

Now You're Ready
Exercises 13–15

3. Write an equation of the line that passes through $(0, 5)$ and $(4, 5)$.

4. **WHAT IF?** In Example 3, the points are $(0, 3500)$ and $(5, 1500)$. How long does it take to complete the tunnel?

13.6 Exercises

Check It Out
Help with Homework
BigIdeasMath.com

Vocabulary and Concept Check

1. **PRECISION** Explain how to find the slope of a line given the intercepts of the line.

2. **WRITING** Explain how to write an equation of a line using its graph.

Practice and Problem Solving

Write an equation that represents each side of the figure.

3.

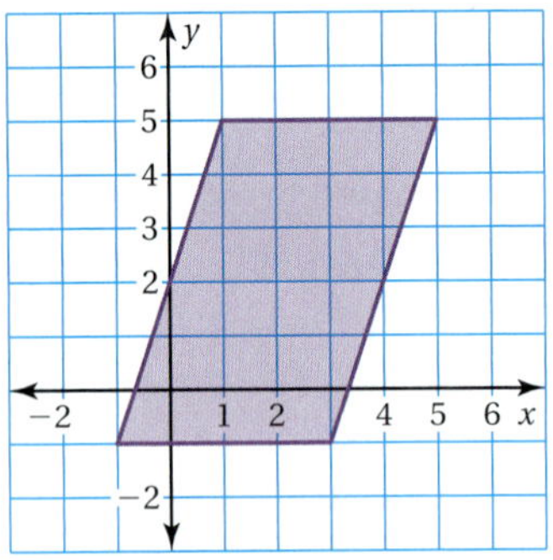

4.

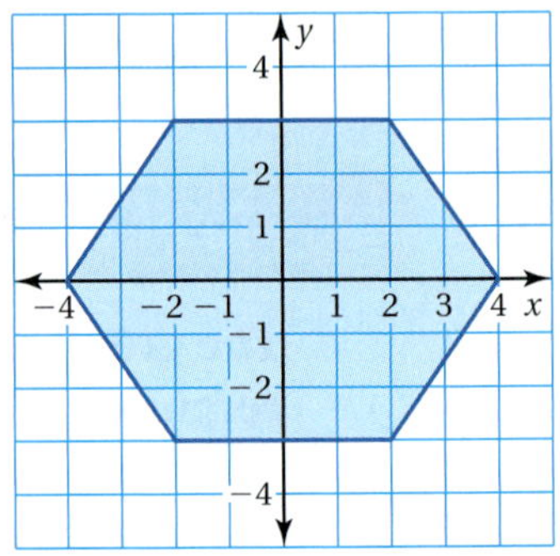

Write an equation of the line in slope-intercept form.

1 5.

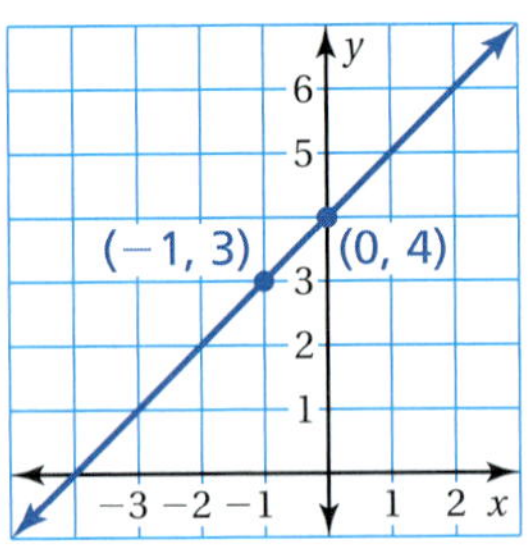

6.

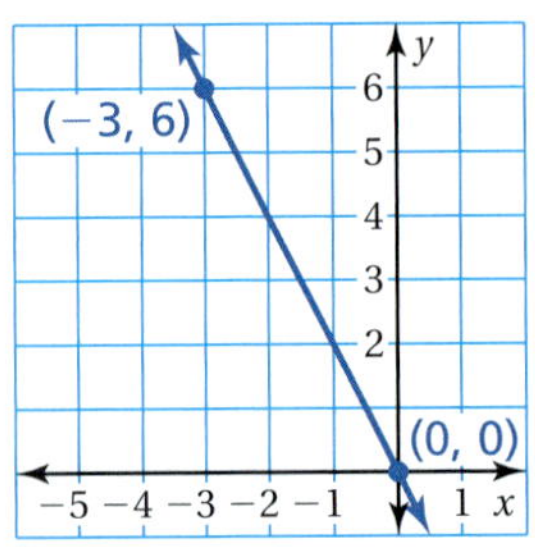

7.

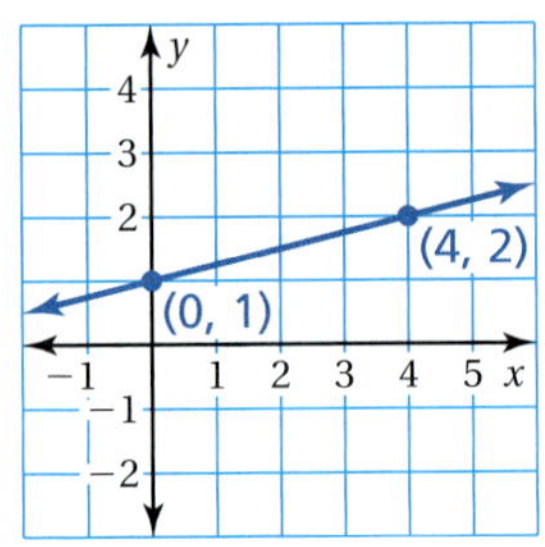

8.

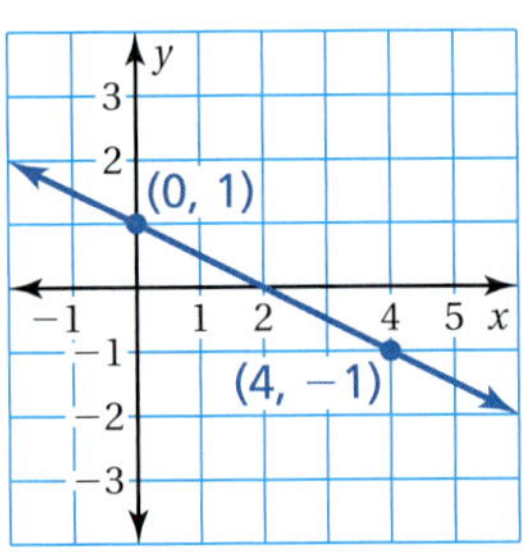

9.

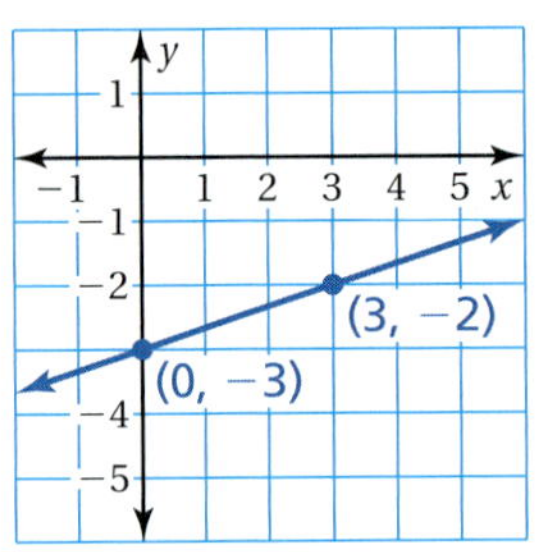

10.

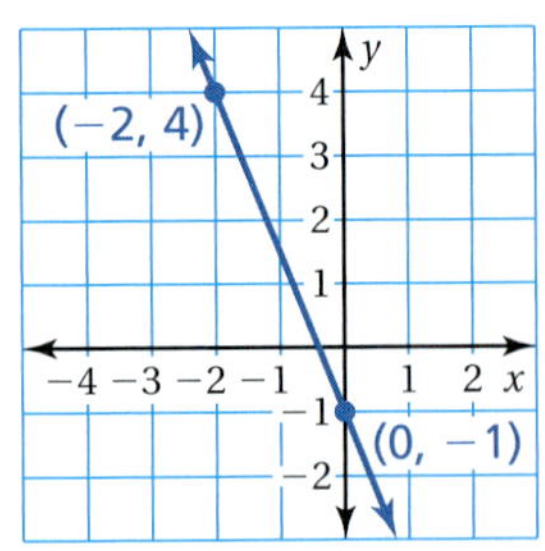

11. **ERROR ANALYSIS** Describe and correct the error in writing an equation of the line.

✗ $y = \frac{1}{2}x + 4$

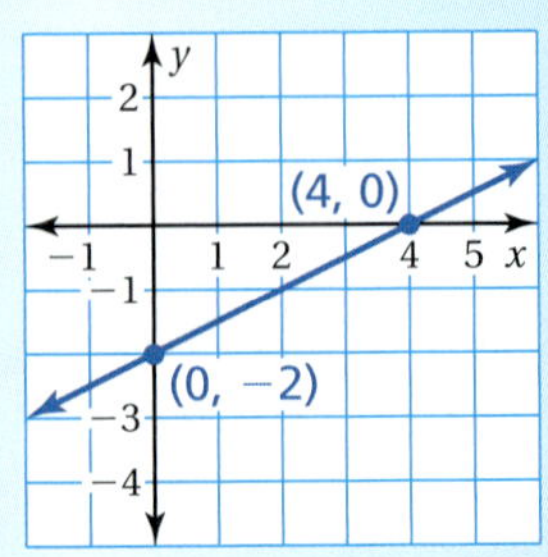

12. **BOA** A boa constrictor is 18 inches long at birth and grows 8 inches per year. Write an equation that represents the length y (in feet) of a boa constrictor that is x years old.

Assignment Guide and Homework Check

Level	Assignment	Homework Check
Accelerated	1–4, 6–10 even, 11–24	8, 12, 14, 17, 19

Common Errors

- **Exercises 5–10** Students may write the reciprocal of the slope or forget a negative sign. Remind them of the definition of slope. Ask students to predict the sign of the slope based on the rise or fall of the line.
- **Exercises 13–15** Students may write the wrong equation when the slope is zero. For example, instead of $y = 5$, students may write $x = 5$. Ask them what is the rise of the graph (zero) and write this in slope-intercept form with the y-intercept as well, such as $y = 0x + 5$. Then ask students what happens when a variable (or any number) is multiplied by zero. Rewrite the equation as $y = 5$.

13.6 Record and Practice Journal

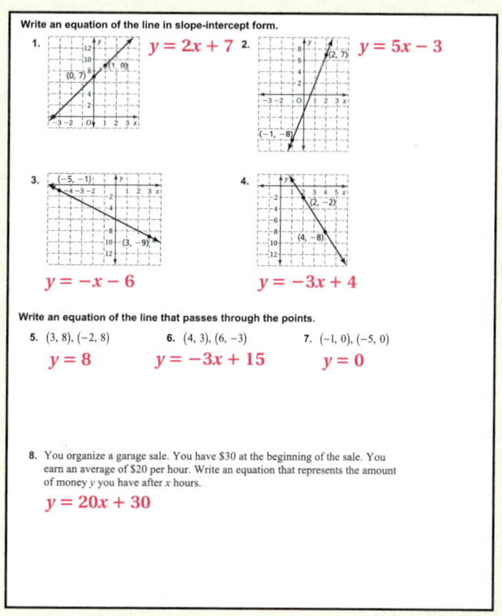

Write an equation of the line in slope-intercept form.

1. $y = 2x + 7$
2. $y = 5x - 3$
3. $y = -x - 6$
4. $y = -3x + 4$

Write an equation of the line that passes through the points.

5. (3, 8), (−2, 8) $y = 8$
6. (4, 3), (6, −3) $y = -3x + 15$
7. (−1, 0), (−5, 0) $y = 0$

8. You organize a garage sale. You have \$30 at the beginning of the sale. You earn an average of \$20 per hour. Write an equation that represents the amount of money y you have after x hours.
$y = 20x + 30$

Vocabulary and Concept Check

1. *Sample answer:* Find the ratio of the rise to the run between the intercepts.
2. *Sample answer:* Find the slope of the line between any two points. Then find the y-intercept. The equation of the line is $y = mx + b$, where m is the slope and b is the y-intercept.

Practice and Problem Solving

3. $y = 3x + 2$; $y = 3x - 10$; $y = 5$; $y = -1$
4. $y = \frac{3}{2}x + 6$; $y = 3$; $y = -\frac{3}{2}x + 6$; $y = \frac{3}{2}x - 6$; $y = -3$; $y = -\frac{3}{2}x - 6$
5. $y = x + 4$
6. $y = -2x$
7. $y = \frac{1}{4}x + 1$
8. $y = -\frac{1}{2}x + 1$
9. $y = \frac{1}{3}x - 3$
10. $y = -\frac{5}{2}x - 1$
11. The x-intercept was used instead of the y-intercept. $y = \frac{1}{2}x - 2$
12. $y = \frac{2}{3}x + \frac{3}{2}$

Practice and Problem Solving

13–17. See Additional Answers.

18. $y = -140x + 500$

19. See *Taking Math Deeper*.

Fair Game Review

20–23.

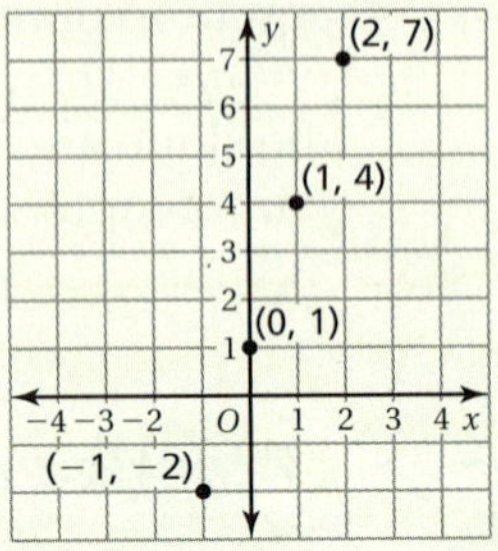

24. C

Mini-Assessment

Write an equation of the line in slope-intercept form.

1.

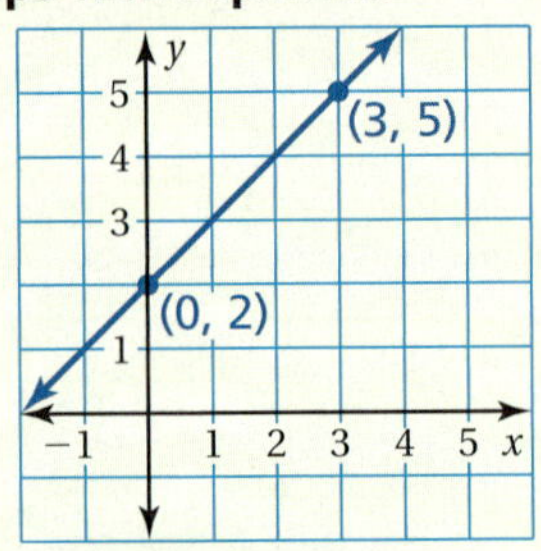

$y = x + 2$

2.

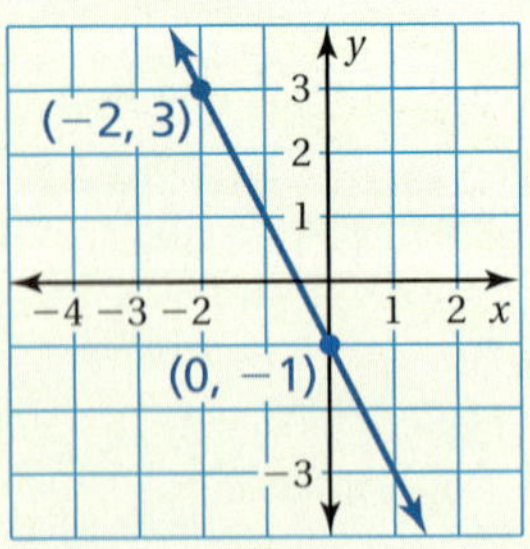

$y = -2x - 1$

3.

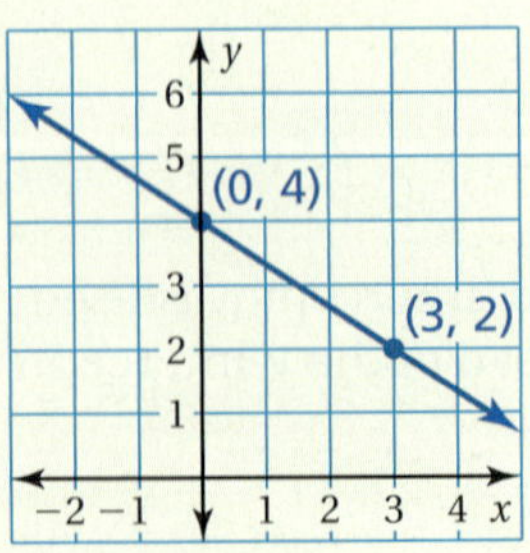

$y = -\frac{2}{3}x + 4$

Taking Math Deeper

Exercise 19

This is a nice real-life problem using estimation. For this problem, remember that you are not looking for exact solutions. You want to know *about* how much the trees grow each year so that you can predict their approximate heights.

1 Estimate the heights in the photograph.

a. Height of 10-year-old tree: about 18 ft
Height of 8-year-old tree: about 14 ft

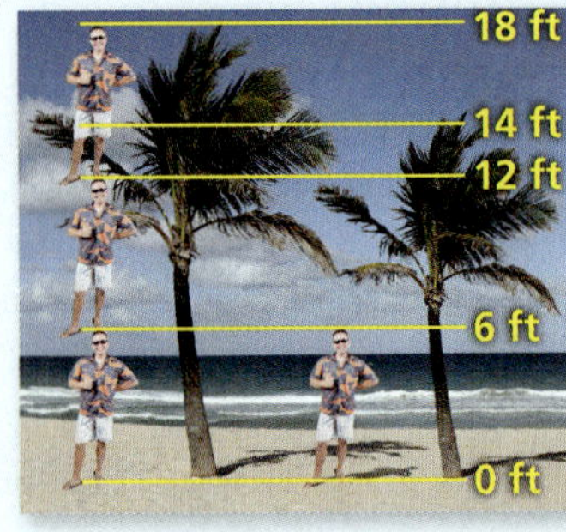

2 **b.** Plot the points that represent the two trees.

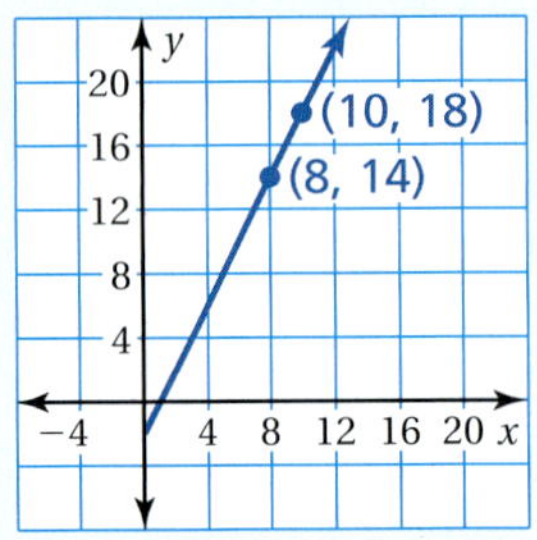

3 **c.** The trees are growing at a rate of about 2 feet per year. Because this would put the height of a 0-year-old tree at −2, it is better to adjust the rate of growth to be about 1.8 feet per year.

d. A possible equation for the growth rate is $y = 1.8x$.

Project

Research information about the palm tree. Pick any kind of palm tree that interests you. How old is the longest living palm tree?

Reteaching and Enrichment Strategies

If students need help...	If students got it...
Resources by Chapter • Practice A and Practice B • Puzzle Time Record and Practice Journal Practice Differentiating the Lesson Lesson Tutorials Skills Review Handbook	Resources by Chapter • Enrichment and Extension • Technology Connection Start the next section

Write an equation of the line that passes through the points.

2 **13.** (2, 5), (0, 5) **14.** (−3, 0), (0, 0) **15.** (0, −2), (4, −2)

16. WALKATHON One of your friends gives you \$10 for a charity walkathon. Another friend gives you an amount per mile. After 5 miles, you have raised \$13.50 total. Write an equation that represents the amount y of money you have raised after x miles.

17. BRAKING TIME During each second of braking, an automobile slows by about 10 miles per hour.

a. Plot the points (0, 60) and (6, 0). What do the points represent?

b. Draw a line through the points. What does the line represent?

c. Write an equation of the line.

18. PAPER You have 500 sheets of notebook paper. After 1 week, you have 72% of the sheets left. You use the same number of sheets each week. Write an equation that represents the number y of pages remaining after x weeks.

19. Critical Thinking The palm tree on the left is 10 years old. The palm tree on the right is 8 years old. The trees grow at the same rate.

a. Estimate the height y (in feet) of each tree.

b. Plot the two points (x, y), where x is the age of each tree and y is the height of each tree.

c. What is the rate of growth of the trees?

d. Write an equation that represents the height of a palm tree in terms of its age.

Fair Game Review What you learned in previous grades & lessons

Plot the ordered pair in a coordinate plane. *(Skills Review Handbook)*

20. (1, 4) **21.** (−1, −2) **22.** (0, 1) **23.** (2, 7)

24. MULTIPLE CHOICE Which of the following statements is true? *(Section 13.4)*

(A) The x-intercept is 5.

(B) The x-intercept is −2.

(C) The y-intercept is 5.

(D) The y-intercept is −2.

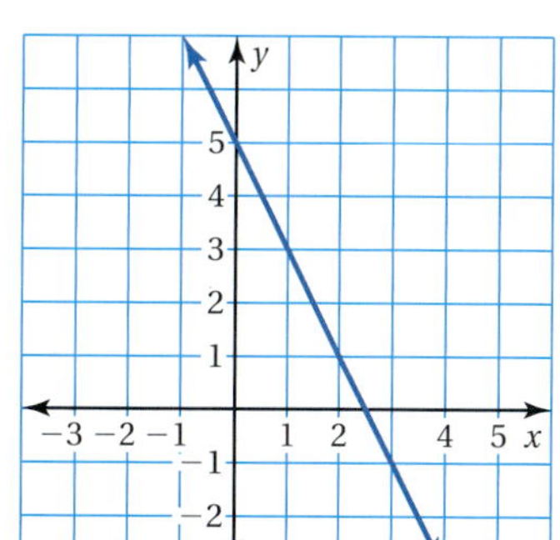

13.7 Writing Equations in Point-Slope Form

Essential Question How can you write an equation of a line when you are given the slope and a point on the line?

1 ACTIVITY: Writing Equations of Lines

Work with a partner.

- **Sketch the line that has the given slope and passes through the given point.**
- **Find the *y*-intercept of the line.**
- **Write an equation of the line.**

a. $m = -2$

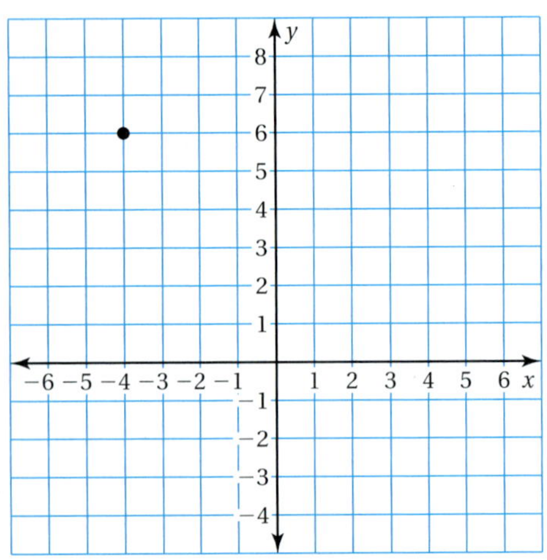

b. $m = \frac{1}{3}$

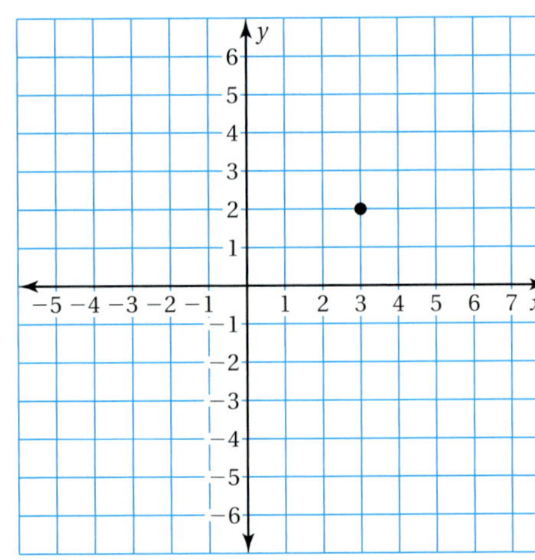

c. $m = -\frac{2}{3}$

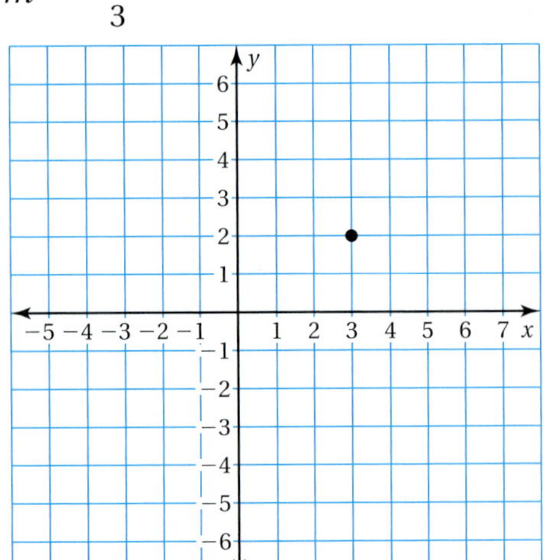

d. $m = \frac{5}{2}$

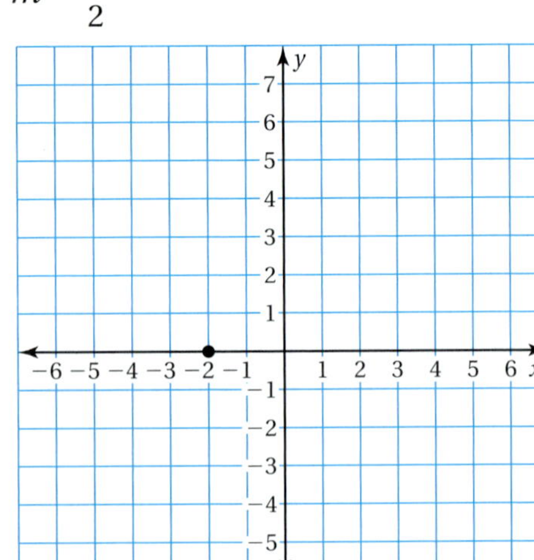

Writing Equations

In this lesson, you will

- write equations of lines using a slope and a point.
- write equations of lines using two points.

Laurie's Notes

Introduction

Applying Mathematical Practices

- **Make Sense of Problems and Persevere in Solving Them:** The goal is for students to write equations of lines given a slope and a point. The slope may be stated explicitly or determined from a contextual setting. Students use the slope to graph the line and work backwards to find the *y*-intercept. There are different approaches students may use as they make sense of the problem.
- **Construct Viable Arguments and Critique the Reasoning of Others:** Take time for discussions and explanations so that students' reasoning is revealed.

Motivate

- Hold a piece of ribbon and a pair of scissors in your hands. Snip off a one-foot piece of ribbon. Repeat once or twice more.

? "Do you know how long my ribbon was when I first started?" no

- Your question should prompt students to ask two obvious questions: "How much are you cutting off each time?" and "How many times have you made a cut?" How much you cut off is the slope (-1). How many times you cut the ribbon helps students work backwards to find the length before any cuts were made, which is the *y*-intercept.

Activity Notes

Activity 1

? "What does it mean for a line to have a slope of -2? A slope of $\frac{1}{3}$?"

For every unit it runs, it falls 2 units. For every 3 units it runs, it rises 1.

- Students may also answer the last question by saying, "Over 1, down 2," and "Over 3, up 1." These geometric answers are fine. Students will need this level of understanding to locate additional points on a line in order to find the *y*-intercept.
- You cannot sketch the line immediately. You must first find additional points on the line. Students should start at the given point and use the slope to find additional points on the line. One of the points will give the *y*-intercept.
- For part (b), it might be helpful to think of the slope of $\frac{1}{3}$ as $\frac{-1}{-3}$. So, start at the point given and move left 3 units and then down 1 unit.
- **Common Error:** Students may interchange the rise and run. Have students look back at their graphs, to see if the slope looks correct to them.
- **Teaching Tip:** Encourage students to lightly trace the rise and run direction arrows with their pencils as they locate additional points.

? "What made it possible to write the equation of the line?" The slope was given and by using the slope, it was possible to find the *y*-intercept. Then substitute into the formula $y = mx + b$.

- **Make Sense of Problems and Persevere in Solving Them:** Asking the last question about slope and having students state their understanding helps them make sense of the problem.

What Your Students Will Learn

- Write equations of lines in point-slope form using the slope and a point on the line or two points on the line.

Previous Learning

Students should know how to plot ordered pairs and apply the definition of slope.

Technology for the Teacher

Dynamic Classroom

Lesson Plans
Complete Materials List

13.7 Record and Practice Journal

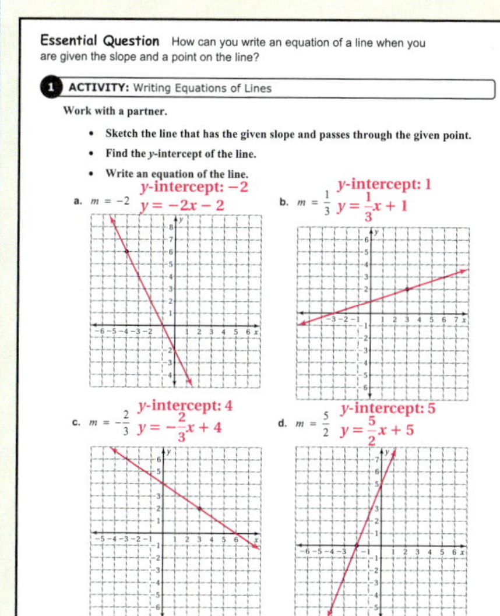

Differentiated Instruction

Kinesthetic

Write a list of linear equations on the board or overhead. Have students copy the equations onto index cards. On the back of each card students are to write the slope and y-intercept of the line. After the cards are completed, students can work in pairs to check each other's work. Finally, students can quiz each other with the flash cards they made.

13.7 Record and Practice Journal

2 ACTIVITY: Deriving an Equation

Work with a partner.

Sample answer:

a. Draw a nonvertical line that passes through the point (x_1, y_1).

b. Plot another point on your line. Label this point as (x, y). This point represents any other point on the line.

c. Label the rise and run of the line through the points (x_1, y_1) and (x, y).

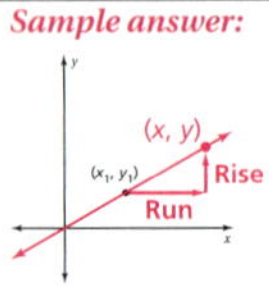

d. The rise can be written as $y - y_1$. The run can be written as $x - x_1$. Explain why this is true.

The rise is the change in y, or difference in the y-coordinates. The run is the change in x, or difference in the x-coordinates.

e. Write an equation for the slope m of the line using the expressions from part (d).

$$m = \frac{y - y_1}{x - x_1}$$

f. Multiply each side of the equation by the expression in the denominator. Write your result. What does this result represent?

$y - y_1 = m(x - x_1)$; the equation of a line with slope m that passes through the point (x_1, y_1)

3 ACTIVITY: Writing an Equation

Work with a partner.

For 4 months, you saved \$25 a month. You now have \$175 in your savings account.

- Draw a graph that shows the balance in your account after t months.
- Use your result from Activity 2 to write an equation that represents the balance A after t months.

$A - 175 = 25(t - 4)$, or $A = 25t + 75$

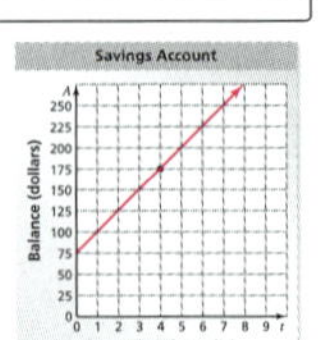

What Is Your Answer?

4. Redo Activity 1 using the equation you found in Activity 2. Compare the results. What do you notice?

same results; The formula can be used to write equations in slope-intercept form.

5. Why do you think $y - y_1 = m(x - x_1)$ is called the *point-slope form* of the equation of a line? Why do you think this is important?

You use the given point and slope to write an equation of a line.

6. **IN YOUR OWN WORDS** How can you write an equation of a line when you are given the slope and a point on the line? Give an example that is different from those in Activity 1.

Use the point-slope form and substitute.

Laurie's Notes

Activity 2

- **Big Idea:** This activity derives the point-slope form of the equation of a line. Students should see the relationship between the slope formula and the point-slope form.
- **Make Sense of Problems and Persevere in Solving Them:** The steps of the derivation are provided. Encourage students to read carefully, discuss with their partners, and think about the process. Do not jump in too quickly to rescue students!
- Allow sufficient time before having a class discussion of this activity. Ask volunteers to share their work on the board. Sharing their process aloud helps students become more confident in their reasoning.

Activity 3

- If students do not understand what the slope is, suggest they work backwards and make a table of values.

Month, t	0	1	2	3	4
Balance in Account, A	\$75	\$100	\$125	\$150	\$175

- Students can use the table to draw the graph.
- ? Ask a few questions to guide students' understanding:
  - "What is the slope for this problem? What point is given?" slope = 25; given point is (4, 175)
  - "Do you have enough information to write an equation?" yes; $A - 175 = 25(t - 4)$, or $A = 25t + 75$
  - "Explain why the slope is positive." You are putting money in the bank. Your account is growing.
- **Make Sense of Problems and Persevere in Solving Them:** Take time for students to transform the equation into slope-intercept form. Students should interpret the slope and y-intercept in the context of the problem.

What Is Your Answer?

- **Neighbor Check:** Have students work independently and then have their neighbors check their work. Have students discuss any discrepancies.
- **Look for and Make Use of Structure:** In Question 4, the equations in point-slope form and slope-intercept form are equivalent, but structurally they look different. Take time for students to appreciate what information is known about the line from the form in which the equation is written.

Closure

- Refer back to the ribbon and scissors. If the ribbon is now 7 feet after making 4 equal cuts of 1-foot length, write an equation that gives the length of the ribbon R after n cuts. $R = 11 - n$ or $R = -n + 11$

2 ACTIVITY: Deriving an Equation

Work with a partner.

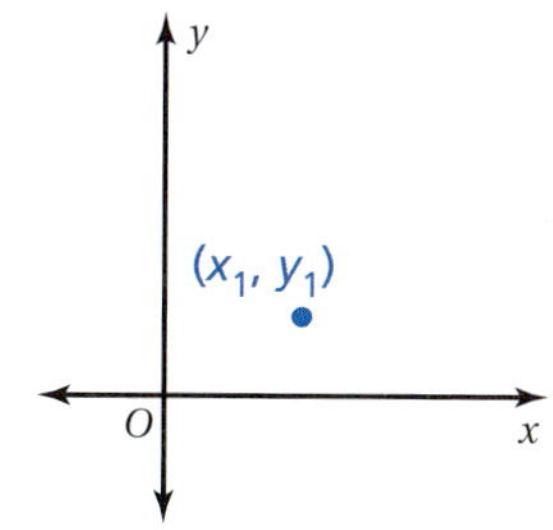

a. Draw a nonvertical line that passes through the point (x_1, y_1).

b. Plot another point on your line. Label this point as (x, y). This point represents any other point on the line.

c. Label the rise and the run of the line through the points (x_1, y_1) and (x, y).

d. The rise can be written as $y - y_1$. The run can be written as $x - x_1$. Explain why this is true.

e. Write an equation for the slope m of the line using the expressions from part (d).

f. Multiply each side of the equation by the expression in the denominator. Write your result. What does this result represent?

Math Practice

Construct Arguments

How does a graph help you derive an equation?

3 ACTIVITY: Writing an Equation

Work with a partner.

For 4 months, you saved $25 a month. You now have $175 in your savings account.

- Draw a graph that shows the balance in your account after t months.
- Use your result from Activity 2 to write an equation that represents the balance A after t months.

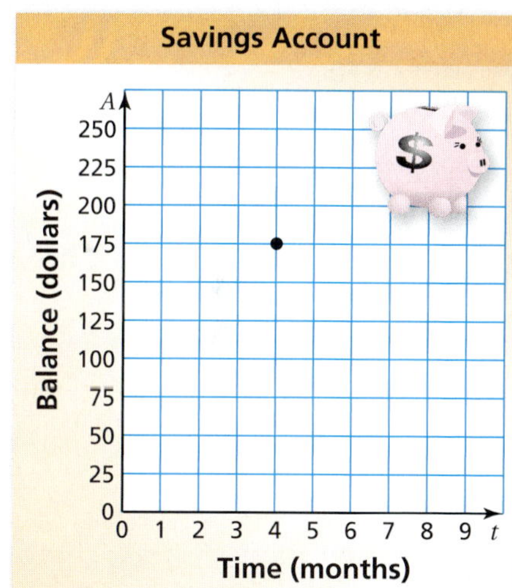

What Is Your Answer?

4. Redo Activity 1 using the equation you found in Activity 2. Compare the results. What do you notice?

5. Why do you think $y - y_1 = m(x - x_1)$ is called the *point-slope form* of the equation of a line? Why do you think it is important?

6. **IN YOUR OWN WORDS** How can you write an equation of a line when you are given the slope and a point on the line? Give an example that is different from those in Activity 1.

Practice Use what you learned about writing equations using a slope and a point to complete Exercises 3–5 on page 612.

13.7 Lesson

Check It Out
Lesson Tutorials
BigIdeasMath.com

Key Vocabulary
point-slope form, *p. 610*

Key Idea

Point-Slope Form

Words A linear equation written in the form $y - y_1 = m(x - x_1)$ is in **point-slope form**. The line passes through the point (x_1, y_1), and the slope of the line is m.

Algebra

slope

$$y - y_1 = m(x - x_1)$$

passes through (x_1, y_1)

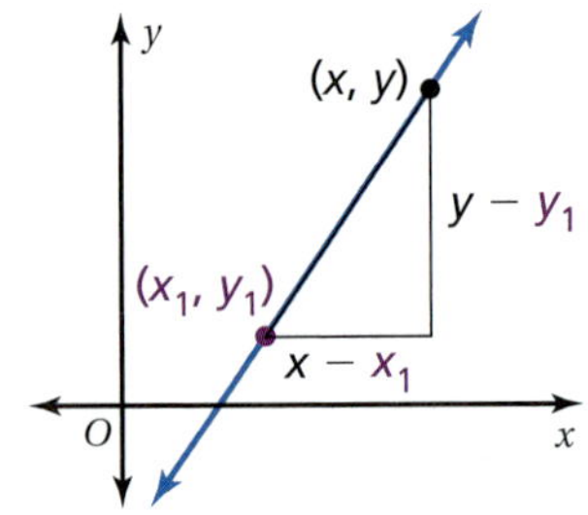

EXAMPLE 1 Writing an Equation Using a Slope and a Point

Write in point-slope form an equation of the line that passes through the point (−6, 1) with slope $\frac{2}{3}$.

$y - y_1 = m(x - x_1)$ — Write the point-slope form.

$y - 1 = \frac{2}{3}[x - (-6)]$ — Substitute $\frac{2}{3}$ for m, -6 for x_1, and 1 for y_1.

$y - 1 = \frac{2}{3}(x + 6)$ — Simplify.

So, the equation is $y - 1 = \frac{2}{3}(x + 6)$.

Check Check that (−6, 1) is a solution of the equation.

$y - 1 = \frac{2}{3}(x + 6)$ — Write the equation.

$1 - 1 \stackrel{?}{=} \frac{2}{3}(-6 + 6)$ — Substitute.

$0 = 0$ ✓ — Simplify.

On Your Own

Now You're Ready
Exercises 6–11

Write in point-slope form an equation of the line that passes through the given point and has the given slope.

1. $(1, 2);\ m = -4$ **2.** $(7, 0);\ m = 1$ **3.** $(-8, -5);\ m = -\frac{3}{4}$

Laurie's Notes

Introduction

Connect

- **Yesterday:** Students developed an intuitive understanding of how to write an equation of a line given the slope and a point.
- **Today:** Students will write an equation of a line given the slope and a point.

Motivate

- ? "Have you seen an airplane come in for a landing either in real life, on television, or in movies?" Most will answer yes.
- ? "Can you describe in words or with a picture what it looks like?" Listen for a smooth approach, meaning a constant rate of descent.
- ? "A plane begins its descent. After 10 seconds, the height of the plane is 1000 feet. If the plane descends 200 feet per second, how long before it lands?" 5 sec
- Make a sketch of this scenario and ask if it is possible to write an equation that models the height h of the airplane, t seconds after it begins its descent.

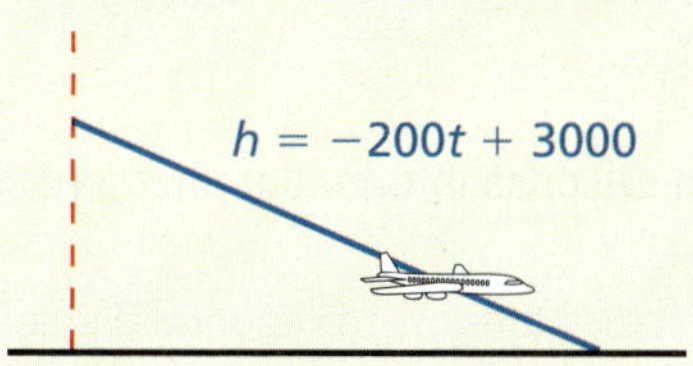

Lesson Notes

Key Idea

- ? Draw a coordinate plane and graph a point. "How many lines go through this point with a slope of $\frac{1}{2}$?" only one line
- Explain that the *point-slope form* of the equation of a line is equivalent to the slope-intercept form and is the equation of a unique line.
- Write the Key Idea on the board. Use of color is helpful.
- **Teaching Tip:** On a side board, write the formula for slope as $\frac{y - y_1}{x - x_1} = m$ so students are reminded of how point-slope form was derived.

Discuss

- **Make Sense of Problems:** Although students derived point-slope form in the activity, they will have lingering questions about the use of subscripts. They might ask why the first point was not labeled (x, y) and the second point (x_1, y_1). The labels are arbitrary, the line could be sloping downward, and the points could be located in any quadrant.

Example 1

- Write the point-slope form of a linear equation.
- ? "What is the slope of the line?" $\frac{2}{3}$ "What point do we know the line passes through?" $(-6, 1)$
- Substitute the known information. Remind students that they are subtracting a negative, so they have $x + 6$ inside the parentheses.
- ? "How can we check if our equation is reasonable?" Students might suggest a quick sketch or rewriting the equation in slope-intercept form to see if the y-intercept makes sense.

Goal Today's lesson is writing an equation of a line given a slope and a point or two points.

Technology for the Teacher

Dynamic Classroom

Lesson Tutorials
Lesson Plans
Answer Presentation Tool

Extra Example 1

Write in point-slope form an equation of the line that passes through the given point and has the given slope.

a. $(2, 2)$; $m = \frac{5}{2}$ $y - 2 = \frac{5}{2}(x - 2)$

b. $(3, -6)$; $m = -\frac{4}{3}$ $y + 6 = -\frac{4}{3}(x - 3)$

On Your Own

1. $y - 2 = -4(x - 1)$
2. $y - 0 = 1(x - 7)$
3. $y + 5 = -\frac{3}{4}(x + 8)$

$y = -\frac{3}{4}x - 11$

$3x + 4y = -44$

Extra Example 2

Write in slope-intercept form an equation of the line that passes through the points $(-3, 0)$ and $(6, 3)$. $y = \frac{1}{3}x + 1$

Extra Example 3

You are pulling down your kite at a rate of 2 feet per second. After 3 seconds, your kite is 54 feet above you.

a. Write and graph an equation that represents the height y (in feet) of the kite above you after x seconds. $y = -2x + 60$

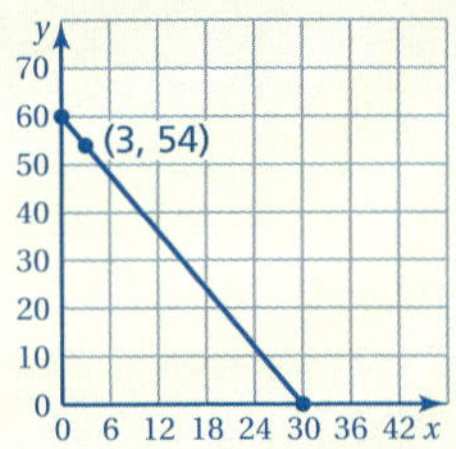

b. At what height was the kite flying before you began pulling it down? 60 ft

On Your Own

4. $y = -x - 1$
5. $y = 4x + 15$
6. $y = \frac{1}{2}x + 10$
7. $y = -10x + 55$

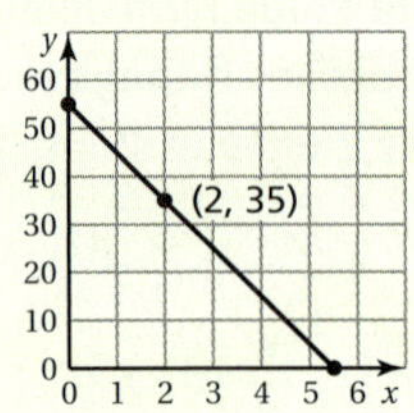

English Language Learners

Visual

Encourage English learners to plot the given point in the coordinate plane and then use the slope. The graph will give them a visual reference they can use when writing the equation.

Laurie's Notes

On Your Own

- **Neighbor Check:** Have students work independently and then have their neighbors check their work. Have students discuss any discrepancies.

Example 2

- Plot both points. Draw the line through the two points.
- ? "Is the slope positive or negative?" negative
- ? "How can you find the slope exactly?" Use the slope formula.
- ? "Can you estimate the y-intercept?" Listen for a positive number greater than 4.
- Continue to work through the problem as shown.
- ? "Do you think we would get the same equation if we had used $(5, -2)$ instead of $(2, 4)$?" yes; Students may be unsure.
- **Make Sense of Problems:** Work the problem again using $(5, -2)$ as shown in the Study Tip. Mathematically proficient students can make sense of why either point will result in the same equation.

Example 3

- Ask a volunteer to read the problem. Discuss information that can be *read* from the illustration.
- ? "Have any of you parasailed?" Wait for students to respond. Explain that when parasailing, you want a smooth descent, like an airplane.
- ? "What is the slope for this problem? How did you know?" Slope is -10. The arrow pointing down means the slope is negative.
- ? "Do we know a point that satisfies the equation?" yes, (2, 25)
- Write the point-slope form of a linear equation. Substitute the known information.
- **Extension:** Have students determine when you reach the boat, meaning $y = 0$.

On Your Own

- Discuss student solutions. Check that signs of numbers are correct for Questions 4–6.

Closure

- **Exit Ticket:** Write an equation of the line with a slope of 2 that passes through the point $(-1, 4)$ in point-slope form and slope-intercept form. $y - 4 = 2(x + 1)$; $y = 2x + 6$

EXAMPLE 2 Writing an Equation Using Two Points

Study Tip

You can use either of the given points to write the equation of the line.

Use $m = -2$ and $(5, -2)$.

$y - (-2) = -2(x - 5)$

$y + 2 = -2x + 10$

$y = -2x + 8$ ✓

Write in slope-intercept form an equation of the line that passes through the points (2, 4) and (5, −2).

Find the slope: $m = \frac{y_2 - y_1}{x_2 - x_1} = \frac{-2 - 4}{5 - 2} = \frac{-6}{3} = -2$

Then use the slope $m = -2$ and the point (2, 4) to write an equation of the line.

$y - y_1 = m(x - x_1)$	Write the point-slope form.
$y - 4 = -2(x - 2)$	Substitute −2 for m, 2 for x_1, and 4 for y_1.
$y - 4 = -2x + 4$	Distributive Property
$y = -2x + 8$	Write in slope-intercept form.

EXAMPLE 3 Real-Life Application

You finish parasailing and are being pulled back to the boat. After 2 seconds, you are 25 feet above the boat. (a) Write and graph an equation that represents your height *y* (in feet) above the boat after *x* seconds. (b) At what height were you parasailing?

a. You are being pulled down at the rate of 10 feet per second. So, the slope is −10. You are 25 feet above the boat after 2 seconds. So, the line passes through (2, 25). Use the point-slope form.

$y - 25 = -10(x - 2)$	Substitute for m, x_1, and y_1.
$y - 25 = -10x + 20$	Distributive Property
$y = -10x + 45$	Write in slope-intercept form.

∴ So, the equation is $y = -10x + 45$.

b. You start descending when $x = 0$. The y-intercept is 45. So, you were parasailing at a height of 45 feet.

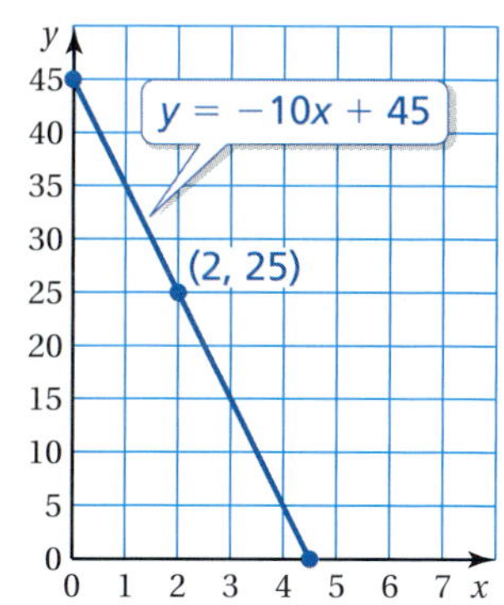

On Your Own

Now You're Ready
Exercises 12–17

Write in slope-intercept form an equation of the line that passes through the given points.

4. (−2, 1), (3, −4) **5.** (−5, −5), (−3, 3) **6.** (−8, 6), (−2, 9)

7. **WHAT IF?** In Example 3, you are 35 feet above the boat after 2 seconds. Write and graph an equation that represents your height *y* (in feet) above the boat after *x* seconds.

13.7 Exercises

Check It Out
Help with Homework
BigIdeasMath.com

Vocabulary and Concept Check

1. **VOCABULARY** From the equation $y - 3 = -2(x + 1)$, identify the slope and a point on the line.

2. **WRITING** Describe how to write an equation of a line using (a) its slope and a point on the line and (b) two points on the line.

Practice and Problem Solving

Use the point-slope form to write an equation of the line with the given slope that passes through the given point.

3. $m = \frac{1}{2}$

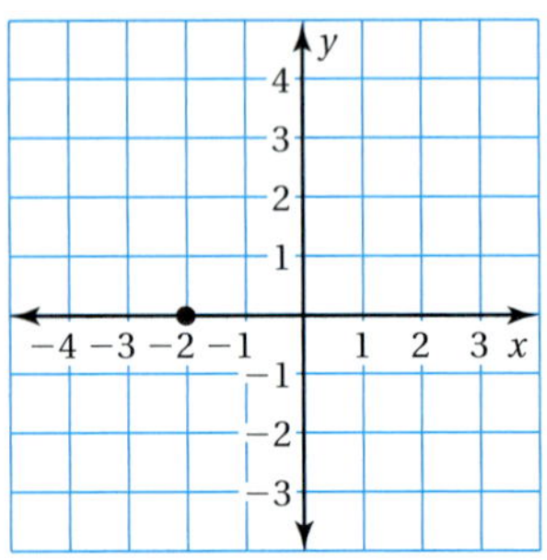

4. $m = -\frac{3}{4}$

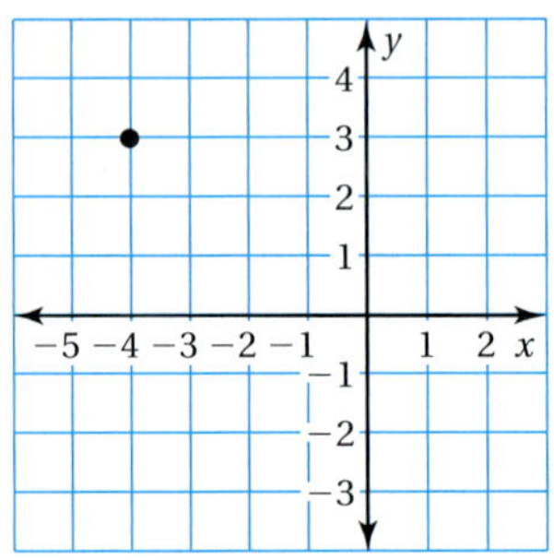

5. $m = -3$

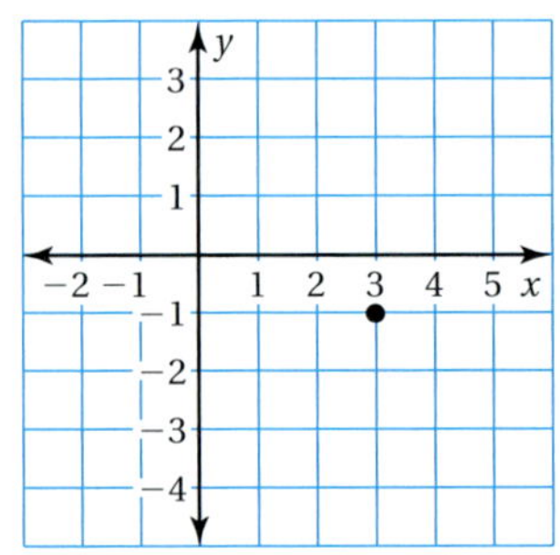

Write in point-slope form an equation of the line that passes through the given point and has the given slope.

1 6. $(3, 0);\ m = -\frac{2}{3}$

7. $(4, 8);\ m = \frac{3}{4}$

8. $(1, -3);\ m = 4$

9. $(7, -5);\ m = -\frac{1}{7}$

10. $(3, 3);\ m = \frac{5}{3}$

11. $(-1, -4);\ m = -2$

Write in slope-intercept form an equation of the line that passes through the given points.

2 12. $(-1, -1), (1, 5)$

13. $(2, 4), (3, 6)$

14. $(-2, 3), (2, 7)$

15. $(4, 1), (8, 2)$

16. $(-9, 5), (-3, 3)$

17. $(1, 2), (-2, -1)$

18. **CHEMISTRY** At 0°C, the volume of a gas is 22 liters. For each degree the temperature T (in degrees Celsius) increases, the volume V (in liters) of the gas increases by $\frac{2}{25}$. Write an equation that represents the volume of the gas in terms of the temperature.

Assignment Guide and Homework Check

Level	Assignment	Homework Check
Accelerated	1–5, 6–18 even, 19–27	16, 18, 19, 21, 22

Common Errors

- **Exercises 6–17** Students may forget to include negatives with the slope and coordinates, or they may apply them incorrectly. Remind them that when the coordinates are negative, they will be subtracting a negative after substituting in point-slope form, which results in adding a positive.
- **Exercises 12–17** Students may use the reciprocal of the slope when writing the equation. Remind them that slope is the change in *y* over the change in *x*.
- **Exercise 18** Students might have trouble knowing which variable can be compared with *x* and *y* and may write the given point backwards. Review what the words "in terms of" mean when writing an equation. In this problem, *V* could be replaced by *y* and *T* could be replaced by *x*. Remind students to check their equations by substituting the given point and checking that it is a solution of the equation.

13.7 Record and Practice Journal

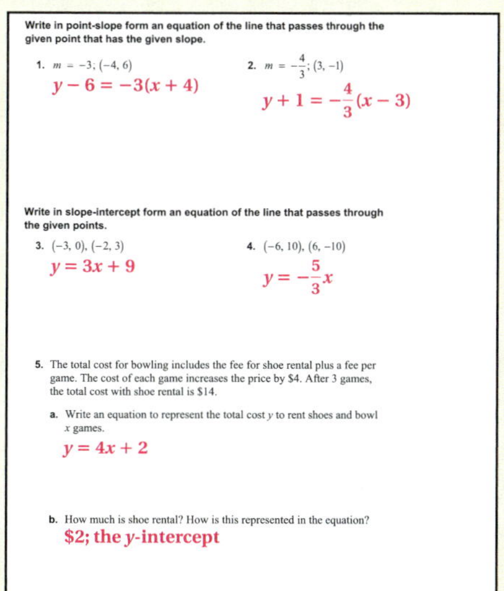
Write in point-slope form an equation of the line that passes through the given point that has the given slope.

1. $m = -3$; $(-4, 6)$
   $y - 6 = -3(x + 4)$
2. $m = -\frac{4}{3}$; $(3, -1)$
   $y + 1 = -\frac{4}{3}(x - 3)$

Write in slope-intercept form an equation of the line that passes through the given points.

3. $(-3, 0)$, $(-2, 3)$
   $y = 3x + 9$
4. $(-6, 10)$, $(6, -10)$
   $y = -\frac{5}{3}x$

5. The total cost for bowling includes the fee for shoe rental plus a fee per game. The cost of each game increases the price by \$4. After 3 games, the total cost with shoe rental is \$14.
   a. Write an equation to represent the total cost y to rent shoes and bowl x games.
   $y = 4x + 2$
   b. How much is shoe rental? How is this represented in the equation?
   \$2; the y-intercept

Vocabulary and Concept Check

1. $m = -2$; $(-1, 3)$
2. a. Write the point-slope form. Substitute the slope for m and the point for (x_1, y_1). Simplify and check your work.
   b. First use the two points to find the slope. Then write the point-slope form. Substitute the slope for m and one of the points for (x_1, y_1). Simplify and check your work.

Practice and Problem Solving

3. $y - 0 = \frac{1}{2}(x + 2)$
4. $y - 3 = -\frac{3}{4}(x + 4)$
5. $y + 1 = -3(x - 3)$
6. $y - 0 = -\frac{2}{3}(x - 3)$
7. $y - 8 = \frac{3}{4}(x - 4)$
8. $y + 3 = 4(x - 1)$
9. $y + 5 = -\frac{1}{7}(x - 7)$
10. $y - 3 = \frac{5}{3}(x - 3)$
11. $y + 4 = -2(x + 1)$
12. $y = 3x + 2$
13. $y = 2x$
14. $y = x + 5$
15. $y = \frac{1}{4}x$
16. $y = -\frac{1}{3}x + 2$
17. $y = x + 1$
18. $V = \frac{2}{25}T + 22$

Practice and Problem Solving

19. **a.** $V = -4000x + 30{,}000$

b. $30,000

20. **a.** $y = 4x - 30$

b. $y = -\frac{1}{4}x + 4$

21. See *Taking Math Deeper.*

22. **a.** $y = -2x + 68$

b. 68 ounces

c. after 34 seconds

23. **a.** $y = 14x - 108.5$

b. 4 meters

Fair Game Review

24–26. See Additional Answers.

27. D

Mini-Assessment

Write in point-slope form an equation of the line that passes through the given point and has the given slope.

1. $(1, 4)$; $m = 3$ $y - 4 = 3(x - 1)$
2. $(-2, 1)$; $m = -2$ $y - 1 = -2(x + 2)$
3. $(3, 5)$; $m = 1$ $y - 5 = 1(x - 3)$
4. $(2, -1)$; $m = \frac{1}{2}$ $y + 1 = \frac{1}{2}(x - 2)$
5. **You rent a floor sander for $24 per day. You pay $82 for 3 days.**
   - **a.** Write an equation that represents your total cost y (in dollars) after x days. $y = 24x + 10$
   - **b.** Interpret the y-intercept. The y-intercept is 10. This means you paid an initial fee of $10 to rent the sander.

Taking Math Deeper

Exercise 21

The challenge in this biology problem is to interpret the given information as a rate of change (or slope) and as an ordered pair.

1 Translate the given information into math.

T = temperature (°F)
x = chirps per minute
Rate of change = 0.25 degree per chirp

2 Write an equation.

Given point: $(x, T) = (40, 50)$

With a slope of 0.25, you can determine that the T-intercept of the line is 40. So, the equation is

a. $T = 0.25x + 40.$

3 Use the equation.
If $x = 100$ chirps per minute, then

$$T = 0.25(100) + 40$$
b. $$= 65°\text{F}.$$

If $T = 96$, then you can find the number of chirps per minute as follows.

$$96 = 0.25x + 40$$
$$56 = 0.25x$$
$$224 = x$$

c. So, you would expect the cricket to make 224 chirps in one minute.

This relationship between temperature and cricket chirps was first published by Amos Dolbear in 1897 in an article called *The Cricket as a Thermometer.*

Project

Research other plants or animals that predict the temperature or weather.

Reteaching and Enrichment Strategies

If students need help. . .	If students got it. . .
Resources by Chapter • Practice A and Practice B • Puzzle Time Record and Practice Journal Practice Differentiating the Lesson Lesson Tutorials Skills Review Handbook	Resources by Chapter • Enrichment and Extension • Technology Connection Start the next section

19. **CARS** After it is purchased, the value of a new car decreases \$4000 each year. After 3 years, the car is worth \$18,000.

   a. Write an equation that represents the value V (in dollars) of the car x years after it is purchased.

   b. What was the original value of the car?

20. **REASONING** Write an equation of a line that passes through the point (8, 2) that is (a) parallel and (b) perpendicular to the graph of the equation $y = 4x - 3$.

21. **CRICKETS** According to Dolbear's law, you can predict the temperature T (in degrees Fahrenheit) by counting the number x of chirps made by a snowy tree cricket in 1 minute. For each rise in temperature of 0.25°F, the cricket makes an additional chirp each minute.

   a. A cricket chirps 40 times in 1 minute when the temperature is 50°F. Write an equation that represents the temperature in terms of the number of chirps in 1 minute.

   b. You count 100 chirps in 1 minute. What is the temperature?

   c. The temperature is 96°F. How many chirps would you expect the cricket to make?

Leaning Tower of Pisa

22. **WATERING CAN** You water the plants in your classroom at a constant rate. After 5 seconds, your watering can contains 58 ounces of water. Fifteen seconds later, the can contains 28 ounces of water.

   a. Write an equation that represents the amount y (in ounces) of water in the can after x seconds.

   b. How much water was in the can when you started watering the plants?

   c. When is the watering can empty?

23. Problem Solving The Leaning Tower of Pisa in Italy was built between 1173 and 1350.

   a. Write an equation for the yellow line.

   b. The tower is 56 meters tall. How far off center is the top of the tower?

Fair Game Review What you learned in previous grades & lessons

Graph the linear equation. *(Section 13.4)*

24. $y = 4x$ 25. $y = -2x + 1$ 26. $y = 3x - 5$

27. **MULTIPLE CHOICE** What is the x-intercept of the equation $3x + 5y = 30$? *(Section 13.5)*

   Ⓐ −10 Ⓑ −6 Ⓒ 6 Ⓓ 10

13.4–13.7 Quiz

Check It Out
Progress Check
BigIdeasMath.com

Find the slope and the y-intercept of the graph of the linear equation. *(Section 13.4)*

1. $y = \frac{1}{4}x - 8$

2. $y = -x + 3$

Find the x- and y-intercepts of the graph of the equation. *(Section 13.5)*

3. $3x - 2y = 12$

4. $x + 5y = 15$

Write an equation of the line in slope-intercept form. *(Section 13.6)*

5.

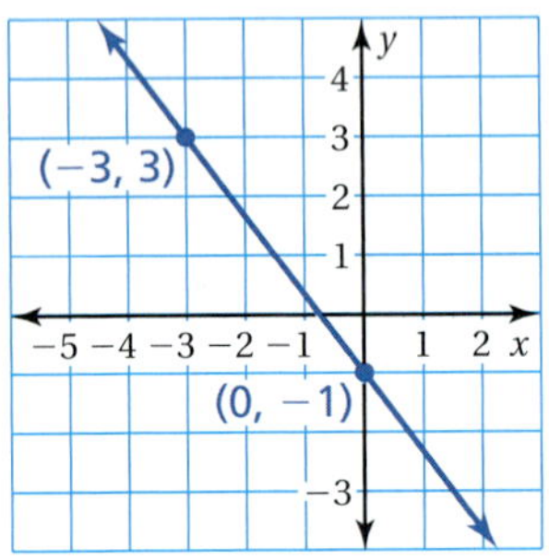

6.

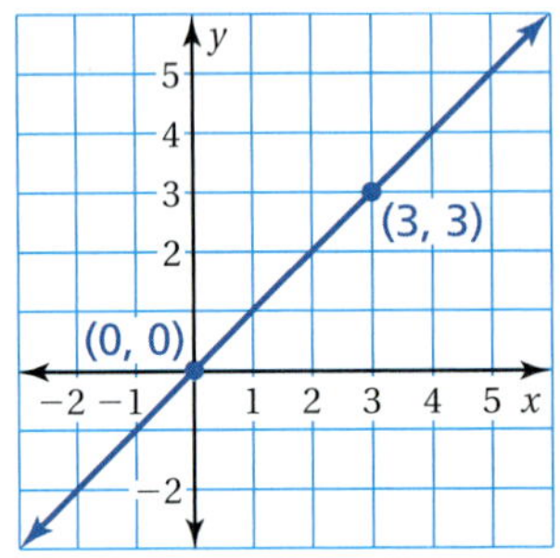

7.

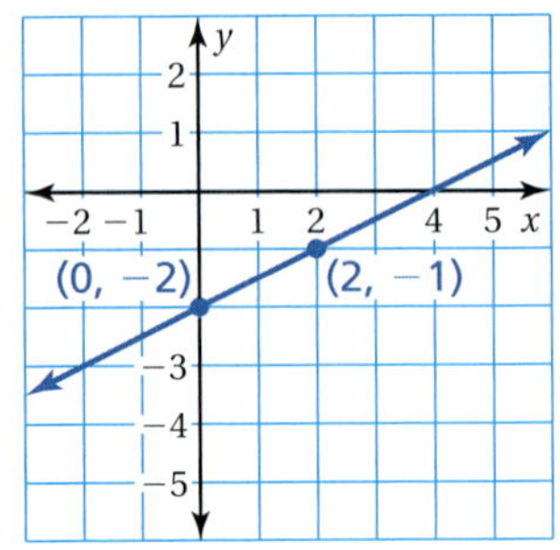

Write in point-slope form an equation of the line that passes through the given point and has the given slope. *(Section 13.7)*

8. $(1, 3);\ m = 2$

9. $(-3, -2);\ m = \frac{1}{3}$

10. $(-1, 4);\ m = -1$

11. $(8, -5);\ m = -\frac{1}{8}$

Write in slope-intercept form an equation of the line that passes through the given points. *(Section 13.7)*

12. $\left(0, -\frac{2}{3}\right) \left(-3, -\frac{2}{3}\right)$

13. $(4, 0), (0, 4)$

14. **STATE FAIR** The cost y (in dollars) of one person buying admission to a fair and going on x rides is $y = x + 12$. *(Section 13.4)*

a. Graph the equation.

b. Interpret the y-intercept and the slope.

15. **PAINTING** You used \$90 worth of paint for a school float. *(Section 13.5)*

a. Graph the equation $18x + 15y = 90$, where x is the number of gallons of blue paint and y is the number of gallons of white paint.

b. Interpret the intercepts.

16. **CONSTRUCTION** A construction crew is extending a highway sound barrier that is 13 miles long. The crew builds $\frac{1}{2}$ of a mile per week. Write an equation that represents the length y (in miles) of the barrier after x weeks. *(Section 13.6)*

Alternative Assessment Options

Math Chat	**Student Reflective Focus Question**
Structured Interview	Writing Prompt

Student Reflective Focus Question

Ask students to summarize the similarities and differences of proportional relationships and nonproportional linear relationships. Be sure that they include examples. Select students at random to present their summaries to the class.

Study Help Sample Answers

Remind students to complete Graphic Organizers for the rest of the chapter.

3a.

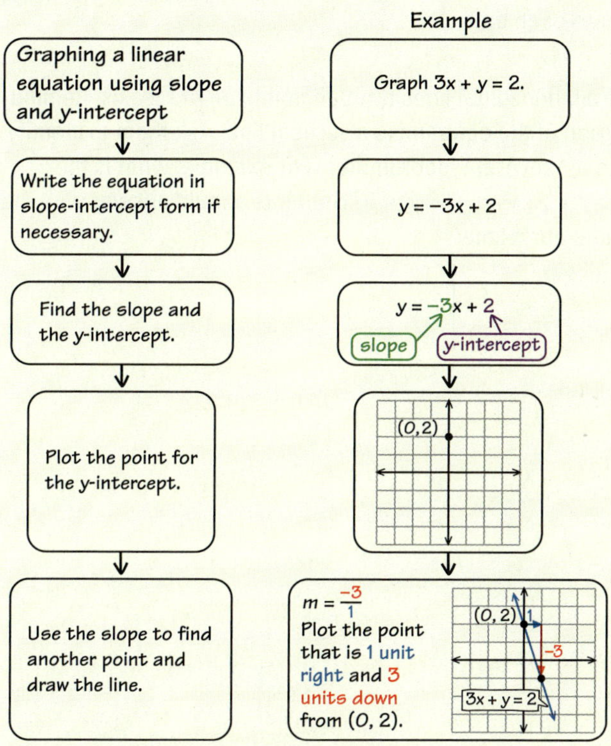

3b, 4–5. Available at *BigIdeasMath.com.*

Reteaching and Enrichment Strategies

If students need help. . .	If students got it. . .
Resources by Chapter • Practice A and Practice B • Puzzle Time Lesson Tutorials *BigIdeasMath.com*	Resources by Chapter • Enrichment and Extension • Technology Connection Game Closet at *BigIdeasMath.com* Start the Chapter Review

Answers

1. slope: $\frac{1}{4}$
y-intercept: -8

2. slope: -1
y-intercept: 3

3. x-intercept: 4
y-intercept: -6

4. x-intercept: 15
y-intercept: 3

5. $y = -\frac{4}{3}x - 1$

6. $y = x$

7. $y = \frac{1}{2}x - 2$

8. $y - 3 = 2(x - 1)$

9. $y + 2 = \frac{1}{3}(x + 3)$

10. $y - 4 = -1(x + 1)$

11. $y + 5 = -\frac{1}{8}(x - 8)$

12. $y = -\frac{2}{3}$

13. $y = -x + 4$

14. a.

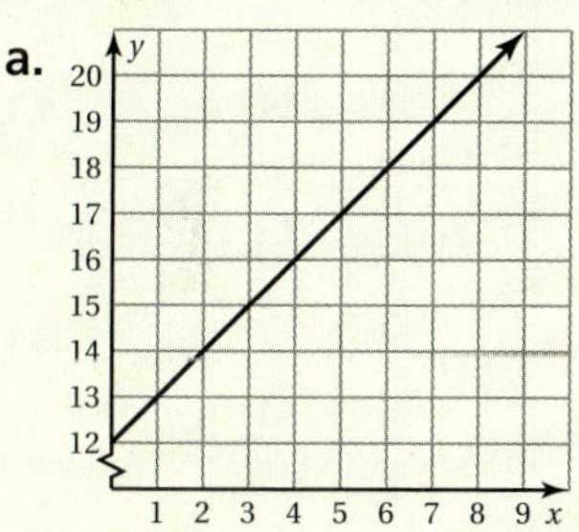

b. The y-intercept represents the admission price of \$12 and the slope represents the unit cost of \$1 per ride.

15. See Additional Answers.

16. $y = \frac{1}{2}x + 13$

Technology for the Teacher

Online Assessment
Assessment Book
ExamView® Assessment Suite

For the Teacher
Additional Review Options

- *BigIdeasMath.com*
- Online Assessment
- Game Closet at *BigIdeasMath.com*
- Vocabulary Help
- Resources by Chapter

Answers

1. 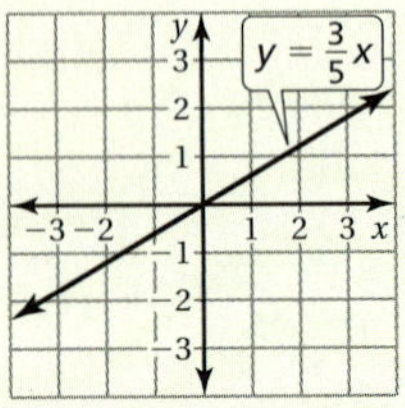

2. 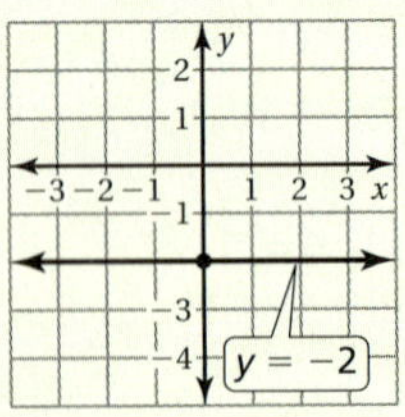

3. 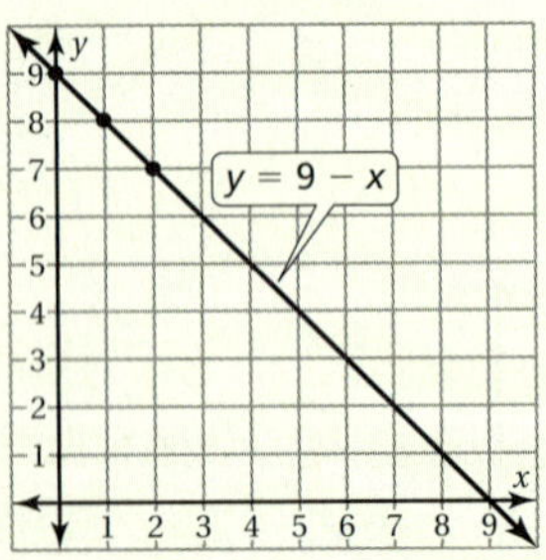

4. 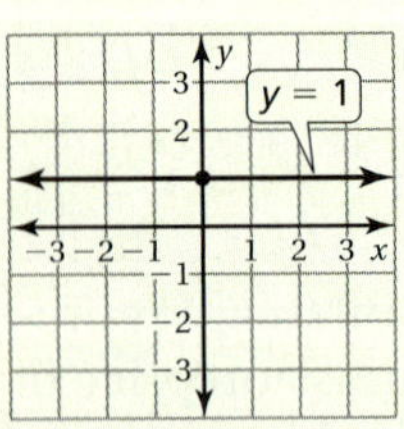

5. 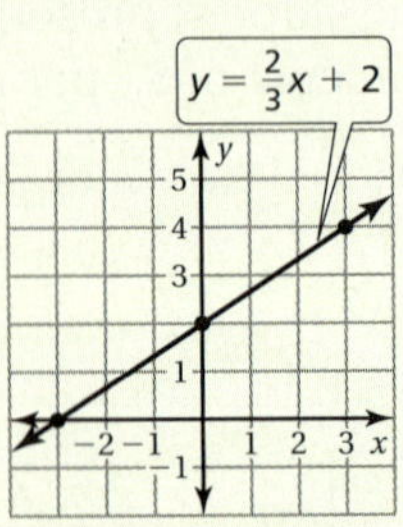

6. 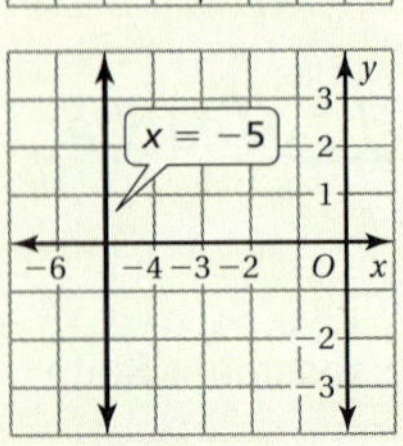

Review of Common Errors

Exercises 1–6

- Students may make calculation errors when solving for ordered pairs. If they only find two ordered pairs for the graph, they may not recognize their mistakes. Encourage them to find at least three ordered pairs when drawing a graph.

Exercises 2 and 4

- Students may draw vertical lines through points on the x-axis. Remind them that the graph of the equation is a horizontal line. Ask them to identify the y-coordinate for several x-coordinates. For example, what is the y-coordinate for $x = 3$? $x = 8$? $x = -5$? Students should answer with the same y-coordinate each time.

Exercise 6

- Students may draw horizontal lines through points on the y-axis. Remind them that the graph of the equation is a vertical line. Ask them to identify the x-coordinate for several y-coordinates. For example, what is the x-coordinate for $y = 2$? $y = -6$? $y = 0$? Students should answer with the same x-coordinate each time.

13 Chapter Review

Check It Out Vocabulary Help BigIdeasMath.com

Review Key Vocabulary

linear equation *p. 568*
solution of a linear equation, *p. 568*
slope, *p. 574*
rise, *p. 574*
run, *p. 574*
x-intercept, *p. 592*
y-intercept, *p. 592*
slope-intercept form, *p. 592*
standard form, *p. 598*
point-slope form, *p. 610*

Review Examples and Exercises

13.1 Graphing Linear Equations *(pp. 566–571)*

Graph $y = 3x - 1$.

Step 1: Make a table of values.

x	$y = 3x - 1$	y	(x, y)
-2	$y = 3(-2) - 1$	-7	$(-2, -7)$
-1	$y = 3(-1) - 1$	-4	$(-1, -4)$
0	$y = 3(0) - 1$	-1	$(0, -1)$
1	$y = 3(1) - 1$	2	$(1, 2)$

Step 2: Plot the ordered pairs.

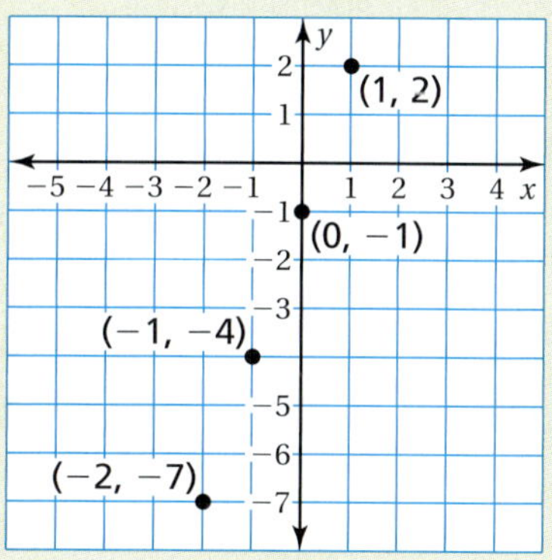

Step 3: Draw a line through the points.

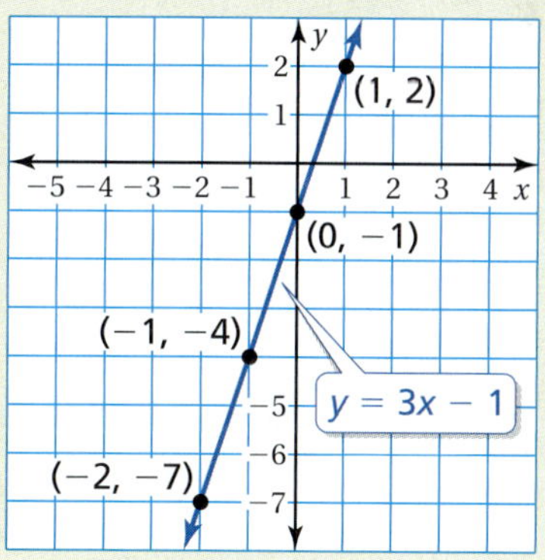

Exercises

Graph the linear equation.

1. $y = \frac{3}{5}x$

2. $y = -2$

3. $y = 9 - x$

4. $y = 1$

5. $y = \frac{2}{3}x + 2$

6. $x = -5$

13.2 Slope of a Line (pp. 572–581)

Find the slope of each line in the graph.

Red Line: $m = \frac{y_2 - y_1}{x_2 - x_1} = \frac{5 - (-3)}{2 - 2} = \frac{8}{0}$

∴ The slope of the red line is undefined.

Blue Line: $m = \frac{y_2 - y_1}{x_2 - x_1} = \frac{-1 - 2}{4 - (-3)} = \frac{-3}{7}$, or $-\frac{3}{7}$

Green Line: $m = \frac{y_2 - y_1}{x_2 - x_1} = \frac{4 - 4}{5 - 0} = \frac{0}{5}$, or 0

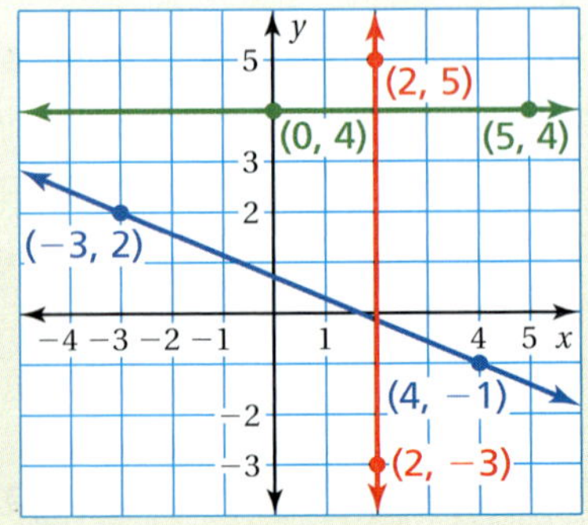

Exercises

The points in the table lie on a line. Find the slope of the line.

7.

x	0	1	2	3
y	−1	0	1	2

8.

x	−2	0	2	4
y	3	4	5	6

9. Are the lines $x = 2$ and $y = 4$ parallel? Are they perpendicular? Explain.

13.3 Graphing Proportional Relationships (pp. 582–587)

The cost y (in dollars) for x tickets to a movie is represented by the equation $y = 7x$. Graph the equation and interpret the slope.

The equation shows that the slope m is 7. So, the graph passes through (0, 0) and (1, 7).

Plot the points and draw a line through the points. Because negative values of x do not make sense in this context, graph in the first quadrant only.

∴ The slope indicates that the unit cost is $7 per ticket.

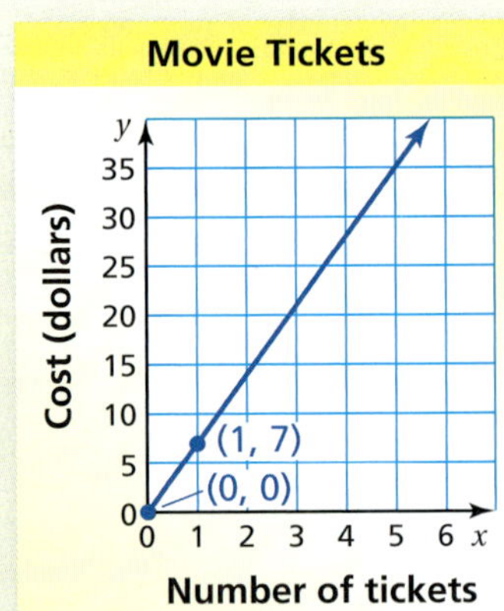

Exercises

10. **RUNNING** The number y of miles you run after x weeks is represented by the equation $y = 8x$. Graph the equation and interpret the slope.

11. **STUDYING** The number y of hours that you study after x days is represented by the equation $y = 1.5x$. Graph the equation and interpret the slope.

Review of Common Errors (continued)

Exercises 7 and 8

- Students may find the reciprocal of the slope instead of the slope. Remind them that slope is change in *y* over change in *x*.

Answers

7. 1 **8.** $\frac{1}{2}$

9. no; yes; The line $x = 2$ is vertical. The line $y = 4$ is horizontal. A vertical line is perpendicular to a horizontal line.

10.

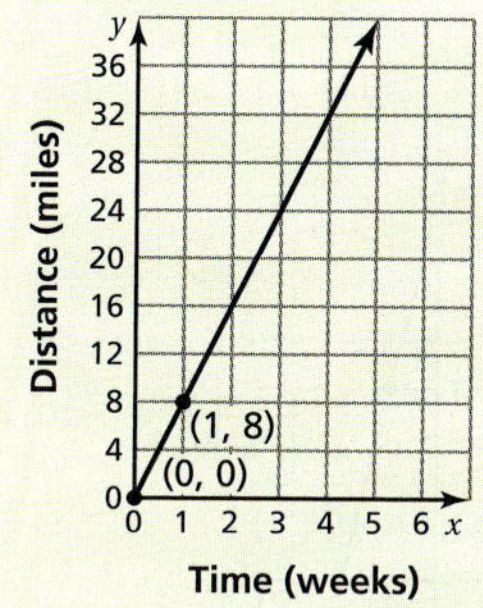

You run 8 miles per week.

11.

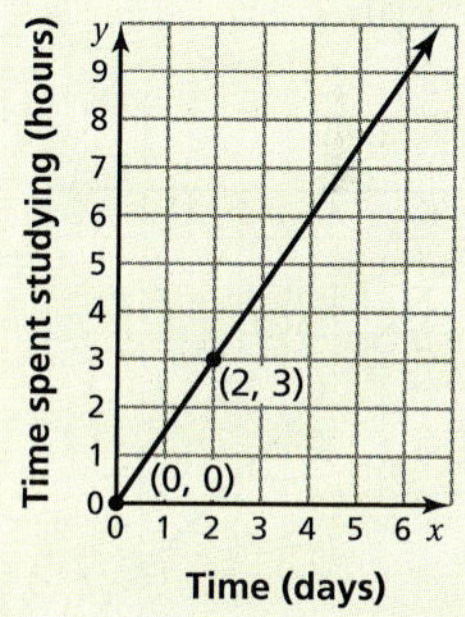

You study 1.5 hours per day.

Answers

12.

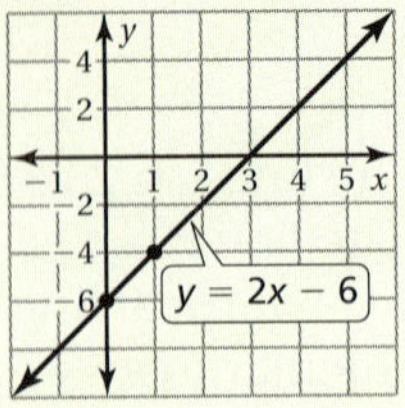

x-intercept: 3

13.

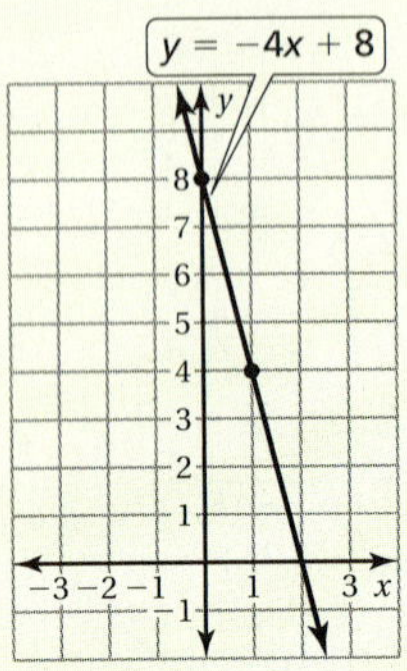

x-intercept: 2

14.

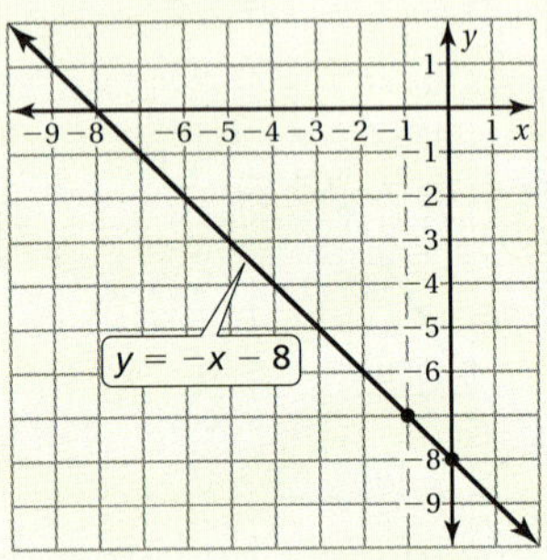

x-intercept: -8

15.

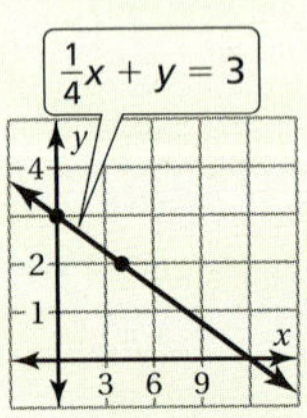

16.

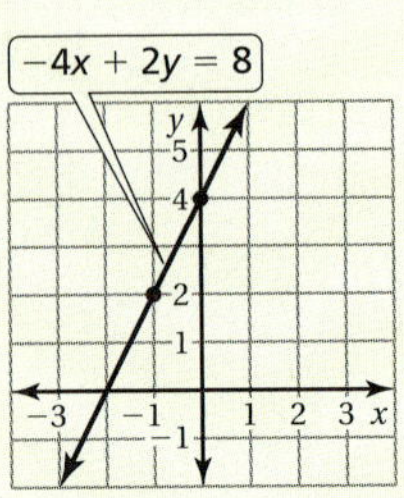

17.

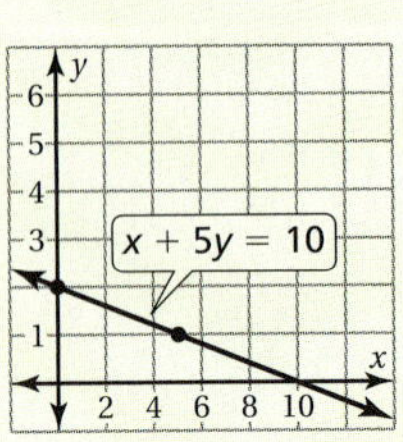

Review of Common Errors (continued)

Exercises 12–14

- Students may forget to include negatives with the slope and/or y-intercept. Remind them to look at the sign in front of the slope and the y-intercept. Also remind students that slope-intercept form is $y = mx + b$. This means that if the linear equation has "minus b," then the y-intercept is negative.
- Students may use the reciprocal of the slope when graphing and may find an incorrect x-intercept. Remind them that slope is *rise* over *run*, so the numerator represents vertical change, not horizontal.

Exercises 15–19

- Students may use the same operation instead of the inverse operation when rewriting the equation in slope-intercept form. Remind them of the steps to rewrite an equation.
- Students may mix up the x- and y-intercepts. Remind them that the x-intercept is the x-coordinate of where the line crosses the x-axis and the y-intercept is the y-coordinate of where the line crosses the y-axis.

Exercises 20–23

- Students may write the reciprocal of the slope or forget a negative sign. Remind them of the definition of slope. Ask them to predict the sign of the slope based on the rise or fall of the line.

Exercises 24 and 25

- Students may write the wrong equation when the slope is zero. For example, instead of $y = 5$, students may write $x = 5$. Ask them what is the rise of the graph (zero) and write this in slope-intercept form with the y-intercept as well, such as $y = 0x + 5$. Then ask students what happens when a variable (or any number) is multiplied by zero. Rewrite the equation as $y = 5$.

Exercises 26 and 27

- Students may use the reciprocal of the slope when writing the equation. Remind them that slope is the change in y over the change in x.

13.4 Graphing Linear Equations in Slope-Intercept Form *(pp. 590–595)*

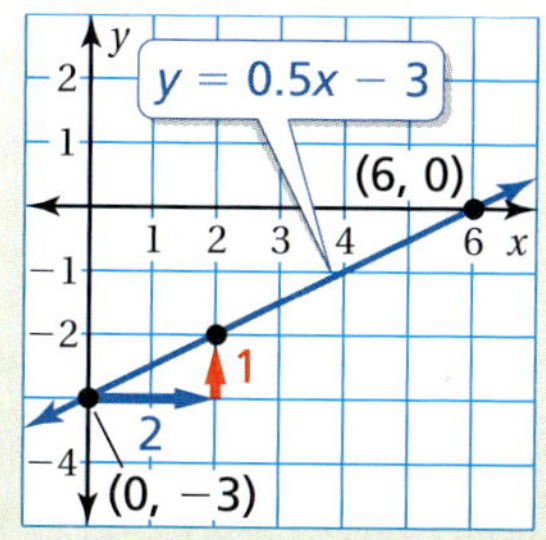

Graph $y = 0.5x - 3$. Identify the x-intercept.

Step 1: Find the slope and the y-intercept.

$$y = 0.5x + (-3)$$

slope → 0.5; y-intercept → −3

Step 2: The y-intercept is −3. So, plot (0, −3).

Step 3: Use the slope to find another point and draw the line.

$$m = \frac{\text{rise}}{\text{run}} = \frac{1}{2}$$

Plot the point that is 2 units right and 1 unit up from (0, –3). Draw a line through the two points.

∴ The line crosses the x-axis at (6, 0). So, the x-intercept is 6.

Exercises

Graph the linear equation. Identify the x-intercept. Use a graphing calculator to check your answer.

12. $y = 2x - 6$ **13.** $y = -4x + 8$ **14.** $y = -x - 8$

13.5 Graphing Linear Equations in Standard Form *(pp. 596–601)*

Graph $8x + 4y = 16$.

Step 1: Write the equation in slope-intercept form.

$8x + 4y = 16$	Write the equation.
$4y = -8x + 16$	Subtract $8x$ from each side.
$y = -2x + 4$	Divide each side by 4.

Step 2: Use the slope and the y-intercept to graph the equation.

$$y = -2x + 4$$

slope → −2; y-intercept → 4

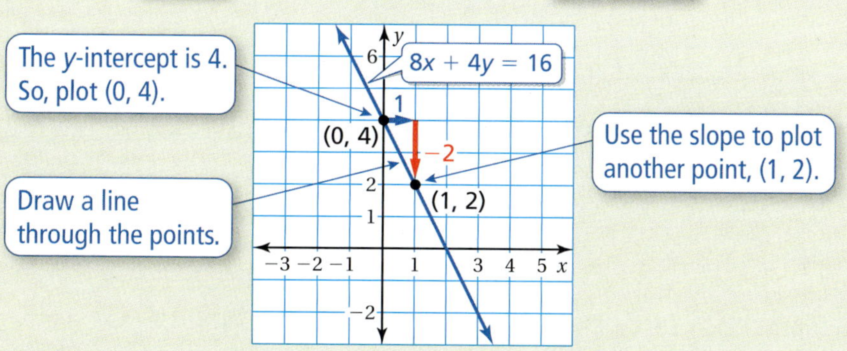

Exercises

Graph the linear equation.

15. $\frac{1}{4}x + y = 3$

16. $-4x + 2y = 8$

17. $x + 5y = 10$

18. $-\frac{1}{2}x + \frac{1}{8}y = \frac{3}{4}$

19. A dog kennel charges \$30 per night to board your dog and \$6 for each hour of playtime. The amount of money you spend is given by $30x + 6y = 180$, where x is the number of nights and y is the number of hours of playtime. Graph the equation and interpret the intercepts.

13.6 Writing Equations in Slope-Intercept Form *(pp. 602–607)*

Write an equation of the line in slope-intercept form.

a.

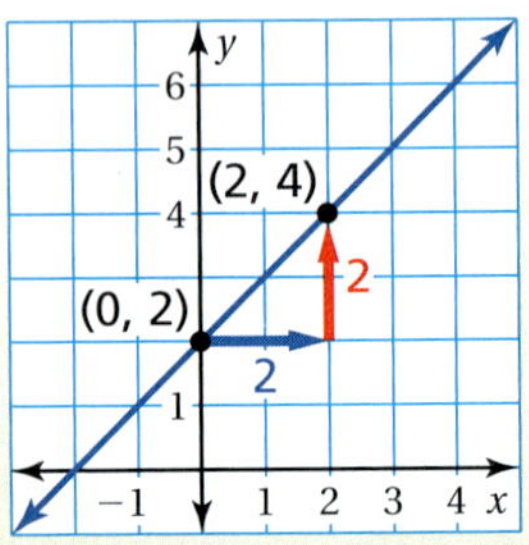

Find the slope and the y-intercept.

$$m = \frac{y_2 - y_1}{x_2 - x_1} = \frac{4 - 2}{2 - 0} = \frac{2}{2}, \text{ or } 1$$

Because the line crosses the y-axis at (0, 2), the y-intercept is 2.

slope — y-intercept

So, the equation is $y = 1x + 2$, or $y = x + 2$.

b.

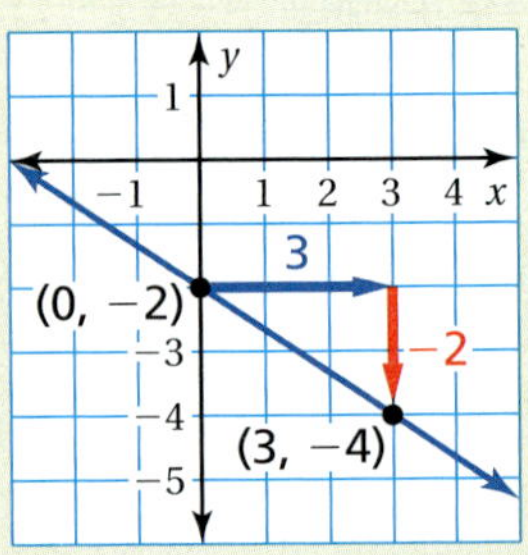

Find the slope and the y-intercept.

$$m = \frac{y_2 - y_1}{x_2 - x_1} = \frac{-4 - (-2)}{3 - 0} = \frac{-2}{3}, \text{ or } -\frac{2}{3}$$

Because the line crosses the y-axis at (0, −2), the y-intercept is −2.

slope — y-intercept

So, the equation is $y = -\frac{2}{3}x + (-2)$, or $y = -\frac{2}{3}x - 2$.

Review Game

Graphing Linear Equations

Materials per Group:

- map of the United States
- pencil
- straightedge

Directions:

On a map of the United States, students will place a coordinate plane with the origin located at Wichita, Kansas. The x-axis will go from -1700 miles to 1700 miles and the y-axis will go from -625 miles to 625 miles. These are roughly the dimensions of the United States.

The teacher will write equations and cities, in jumbled order, on the board. Students will work in groups and graph the equations to determine which line goes through which city.

Examples:

Dallas	$y = \frac{156}{50}x - 156$
Denver	$y = \frac{312}{625}x + 312$
Orlando	$y = \frac{100}{200}x + 100$
Chicago	$y = \frac{280}{600}x$
Las Vegas	$y = \frac{625}{1275}x - 625$

Who Wins?

The first group to correctly graph the lines and match the cities wins.

For the Student
Additional Practice

- Lesson Tutorials
- Multi-Language Glossary
- Self-Grading Progress Check
- *BigIdeasMath.com*
  Dynamic Student Edition
  Student Resources

Answers

18. 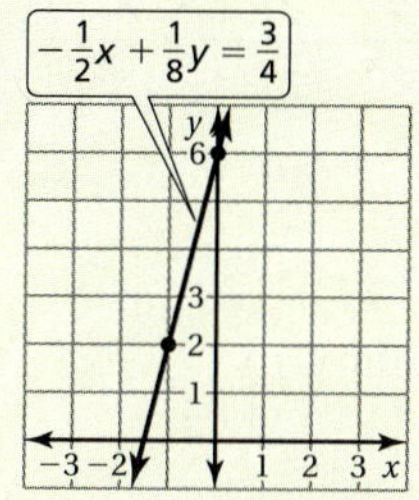

19. 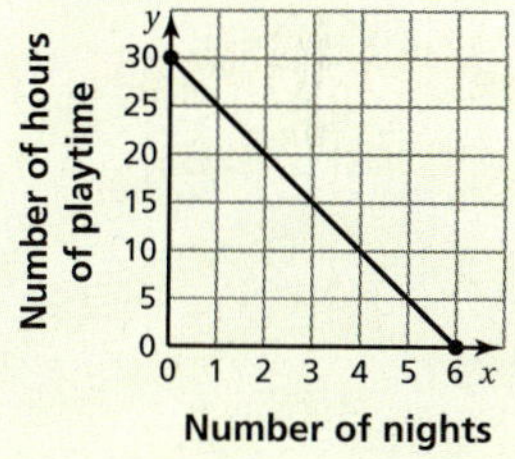

The x-intercept shows that you can board your dog for 6 nights when there are no hours of playtime. The y-intercept shows that you can have 30 hours of playtime for your dog when you do not leave your dog at the kennel for any nights.

20. $y = x - 2$

21. $y = -\frac{1}{2}x + 4$

22. $y = -2x + 1$

23. $y = 2x - 3$

24. $y = 8$

25. $y = -5$

26. $y - 4 = 3(x - 4)$

27. $y = -\frac{1}{2}x$

My Thoughts on the Chapter

Teacher Tip

Not allowed to write in your teaching edition? Use sticky notes to record your thoughts.

What worked. . .

What did not work. . .

What I would do differently. . .

Exercises

Write an equation of the line in slope-intercept form.

20.

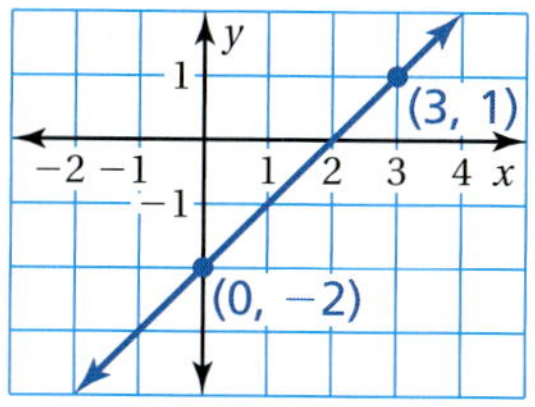

21.

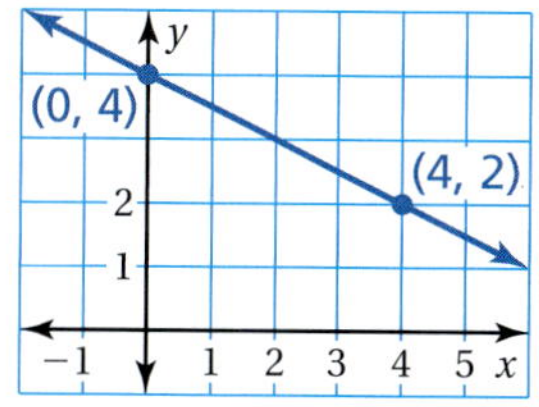

22.

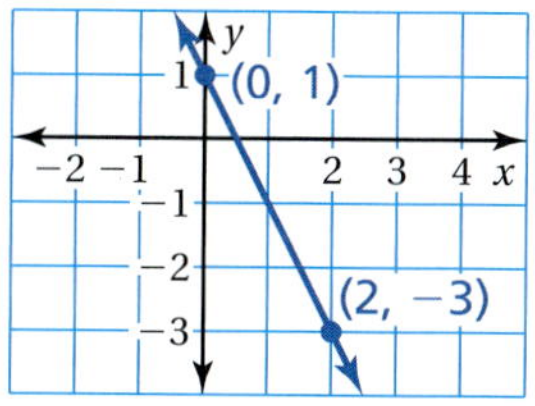

23.

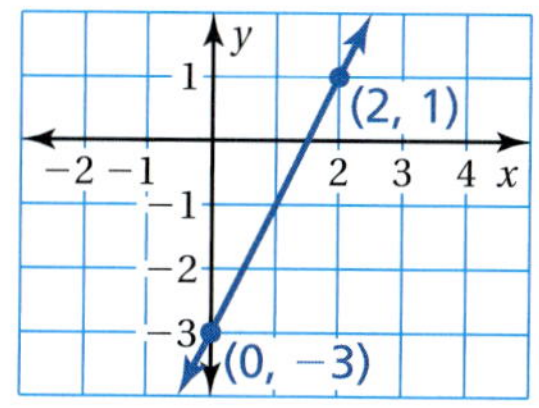

24. Write an equation of the line that passes through (0, 8) and (6, 8).

25. Write an equation of the line that passes through (0, −5) and (−5, −5).

13.7 Writing Equations in Point-Slope Form *(pp. 608–613)*

Write in slope-intercept form an equation of the line that passes through the points (2, 1) and (3, 5).

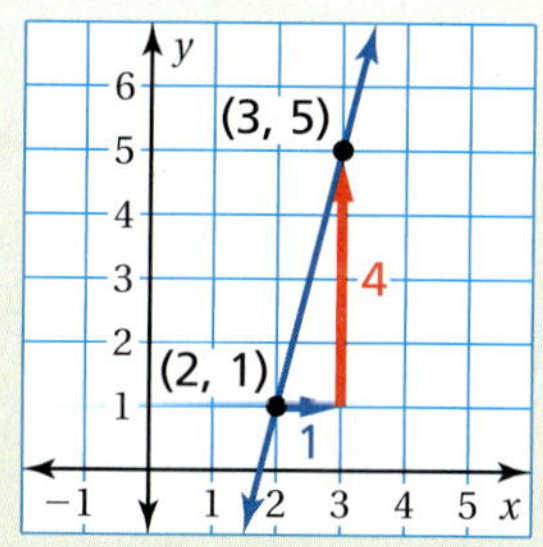

Find the slope.

$$m = \frac{y_2 - y_1}{x_2 - x_1} = \frac{5 - 1}{3 - 2} = \frac{4}{1}, \text{ or } 4$$

Then use the slope and onc of the given points to write an equation of the line.

Use $m = 4$ and (2, 1).

$y - y_1 = m(x - x_1)$	Write the point-slope form.
$y - 1 = 4(x - 2)$	Substitute 4 for m, 2 for x_1, and 1 for y_1.
$y - 1 = 4x - 8$	Distributive Property
$y = 4x - 7$	Write in slope-intercept form.

So, the equation is $y = 4x - 7$.

Exercises

26. Write in point-slope form an equation of the line that passes through the point (4, 4) with slope 3.

27. Write in slope-intercept form an equation of the line that passes through the points (−4, 2) and (6, −3).

13 Chapter Test

Check It Out
Test Practice
BigIdeasMath.com

Find the slope and the *y*-intercept of the graph of the linear equation.

1. $y = 6x - 5$

2. $y = 20x + 15$

3. $y = -5x - 16$

4. $y - 1 = 3x + 8.4$

5. $y + 4.3 = 0.1x$

6. $-\frac{1}{2}x + 2y = 7$

Graph the linear equation.

7. $y = 2x + 4$

8. $y = -\frac{1}{2}x - 5$

9. $-3x + 6y = 12$

10. Which lines are parallel? Which lines are perpendicular? Explain.

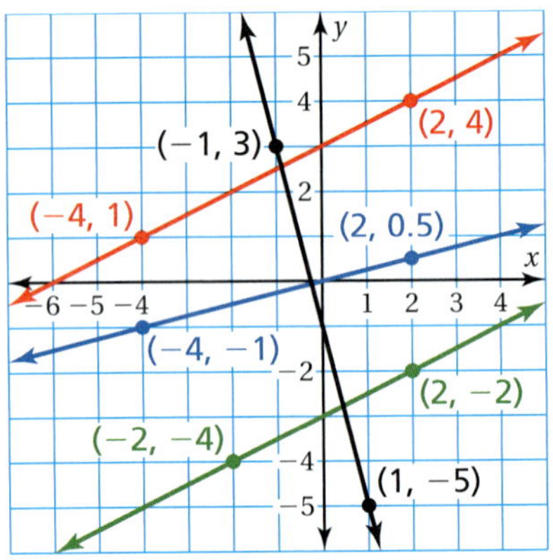

11. The points in the table lie on a line. Find the slope of the line.

x	*y*
−1	−4
0	−1
1	2
2	5

Write an equation of the line in slope-intercept form.

12.

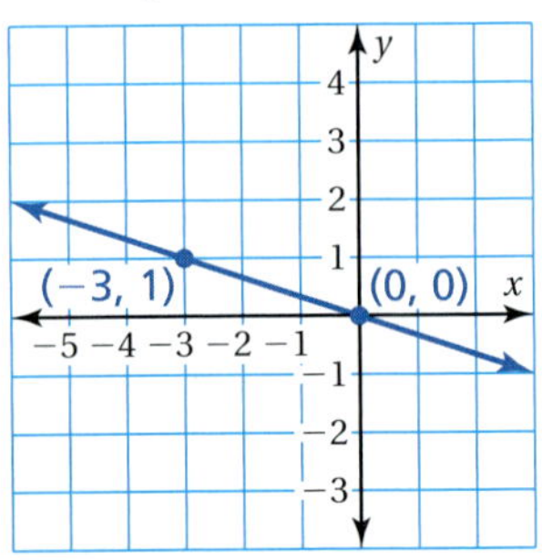

13.

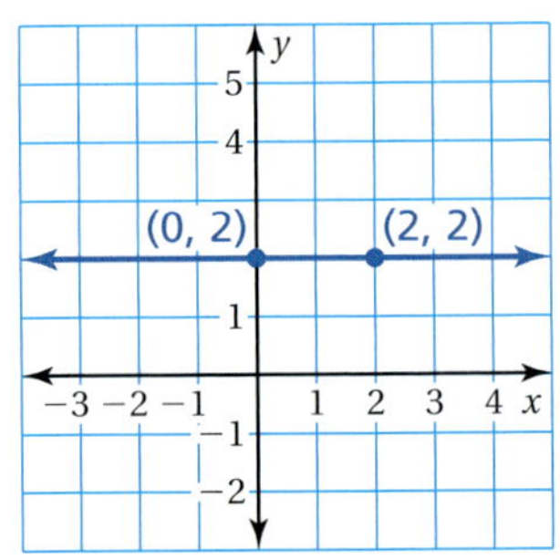

Write in slope-intercept form an equation of the line that passes through the given points.

14. $(-1, 5), (3, -3)$

15. $(-4, 1), (4, 3)$

16. $(-2, 5), (-1, 1)$

17. VOCABULARY The number y of new vocabulary words that you learn after x weeks is represented by the equation $y = 15x$.

a. Graph the equation and interpret the slope.

b. How many new vocabulary words do you learn after 5 weeks?

c. How many more vocabulary words do you learn after 6 weeks than after 4 weeks?

Test Item References

Chapter Test Questions	Section to Review
7, 8	13.1
10, 11	13.2
17	13.3
1–6	13.4
9	13.5
12, 13	13.6
14–16	13.7

Test-Taking Strategies

Remind students to quickly look over the entire test before they start so that they can budget their time. Students should jot down the formulas for slope-intercept form and point-slope form on the back of their test before they begin. Teach students to use the Stop and Think strategy before answering. **Stop** and carefully read the question, and **Think** about what the answer should look like.

Common Errors

- **Exercises 1–6** Students may use the reciprocal of the slope when graphing and may find an incorrect x-intercept. Remind them that slope is *rise* over *run*, so the numerator represents vertical change, not horizontal.
- **Exercises 7–9** Students may make calculation errors when solving for ordered pairs. If they only find two ordered pairs for the graph, they may not recognize their mistakes. Encourage them to find at least three ordered pairs when drawing a graph.
- **Exercise 12** Students may write the reciprocal of the slope or forget a negative sign. Ask them to predict the sign of the slope based on the rise or fall of the line.
- **Exercise 14–16** Students may use the reciprocal of the slope when writing the equation. Remind them that slope is the change in y over the change in x.

Reteaching and Enrichment Strategies

If students need help. . .	If students got it. . .
Resources by Chapter • Practice A and Practice B • Puzzle Time Record and Practice Journal Practice Differentiating the Lesson Lesson Tutorials *BigIdeasMath.com* Skills Review Handbook	Resources by Chapter • Enrichment and Extension • Technology Connection Game Closet at *BigIdeasMath.com* Start Cumulative Assessment

Answers

1. slope: 6; y-intercept: -5
2. slope: 20; y-intercept: 15
3. slope: -5; y-intercept: -16
4. slope: 3; y-intercept: 9.4
5. slope: 0.1; y-intercept: -4.3
6. slope: $\frac{1}{4}$; y-intercept: $\frac{7}{2}$

7–9. See Additional Answers.

10. The red and green lines are parallel. They both have a slope of $\frac{1}{2}$. The black and blue lines are perpendicular. The product of their slopes is -1.
11. 3
12. $y = -\frac{1}{3}x$
13. $y = 2$
14. $y = -2x + 3$
15. $y = \frac{1}{4}x + 2$
16. $y = -4x - 3$
17. a.

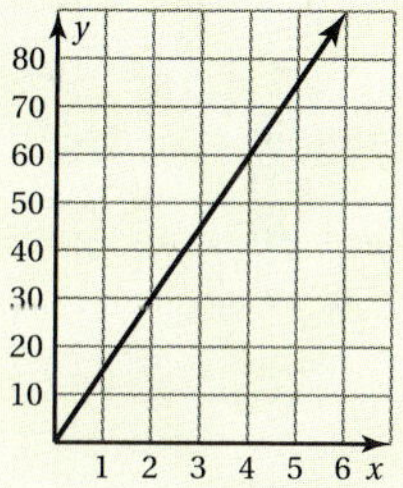

You learn 15 new vocabulary words per week.

b. 75 new vocabulary words

c. 30 more words

Technology for the Teacher

Online Assessment
Assessment Book
ExamView® Assessment Suite

Test-Taking Strategies

Available at *BigIdeasMath.com*

After Answering Easy Questions, Relax
Answer Easy Questions First
Estimate the Answer
Read All Choices before Answering
Read Question before Answering
Solve Directly or Eliminate Choices
Solve Problem before Looking at Choices
Use Intelligent Guessing
Work Backwards

About this Strategy

When taking a multiple choice test, be sure to read each question carefully and thoroughly. After reading the question, estimate the answer before trying to solve.

Answers

1. A
2. H
3. D

Technology for the Teacher

Performance Tasks
Online Assessment
Assessment Book
ExamView® Assessment Suite

Item Analysis

1. **A.** Correct answer

 B. The student reads the slope correctly, but uses the wrong point to identify the *y*-intercept.

 C. The student reads the *y*-intercept correctly, but miscalculates the slope.

 D. The student finds the slope and *y*-intercept incorrectly.

2. **F.** The student interchanges correct values for *x* and *y*.

 G. The student makes two errors: interchanging *x* and *y*, and assigning a negative sign incorrectly.

 H. Correct answer

 I. The student makes a mistake with a correct solution (4, 2) and assigns a negative value to 2, forgetting that there is already a minus sign in the equation.

3. **A.** The student mistakes slope for meaning that a line passes through (0, 0).

 B. The student mistakes a vertical line for zero slope.

 C. The student mistakes slope for meaning that a line passes through (0, 0).

 D. Correct answer

13 Cumulative Assessment

1. Which equation matches the line shown in the graph?

A. $y = 2x - 2$

B. $y = 2x + 1$

C. $y = x - 2$

D. $y = x + 1$

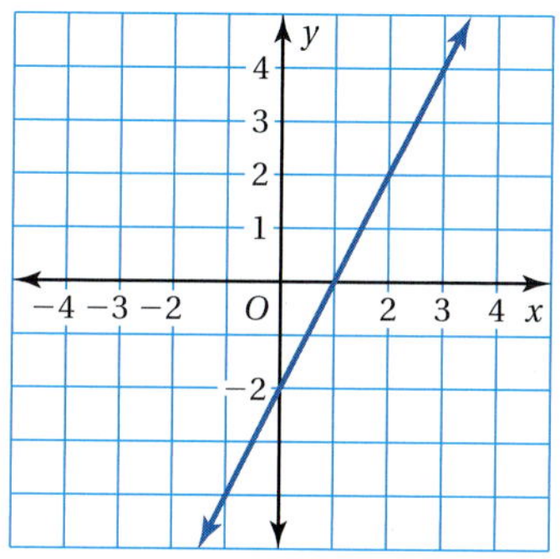

Test-Taking Strategy
Estimate the Answer

"Using estimation, you can see that there are about 300 hairs. So, it has to be C."

2. The equation $6x - 5y = 14$ is written in standard form. Which point lies on the graph of this equation?

F. $(-4, -1)$

G. $(-2, 4)$

H. $(-1, -4)$

I. $(4, -2)$

3. Which line has a slope of 0?

A.

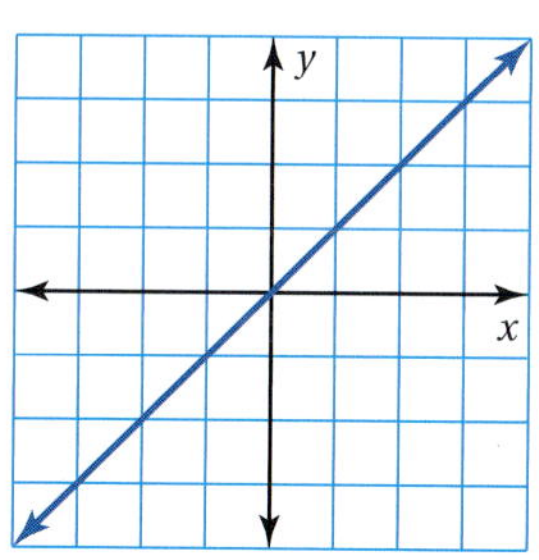

C.

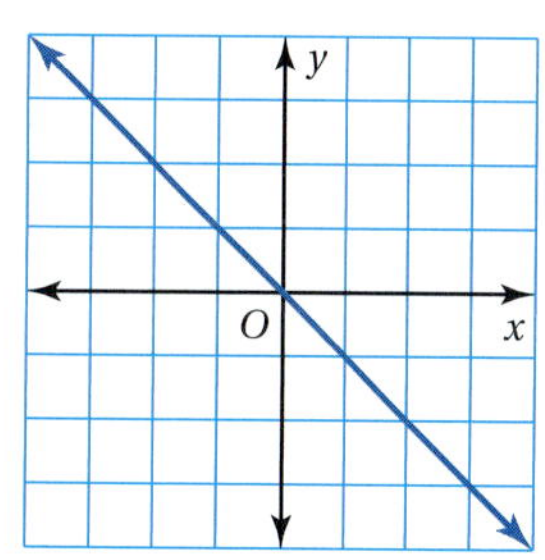

B.

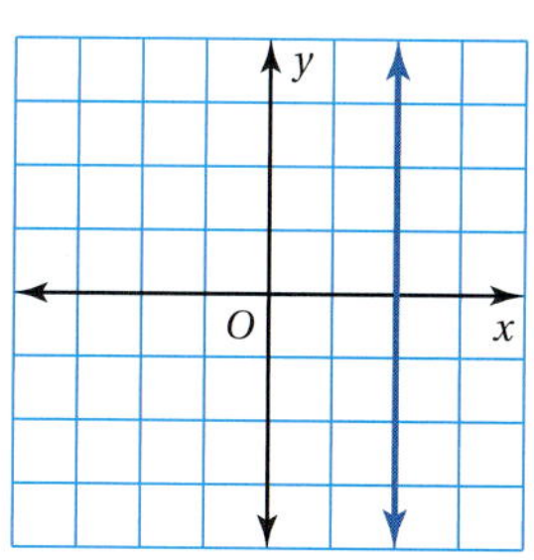

D.

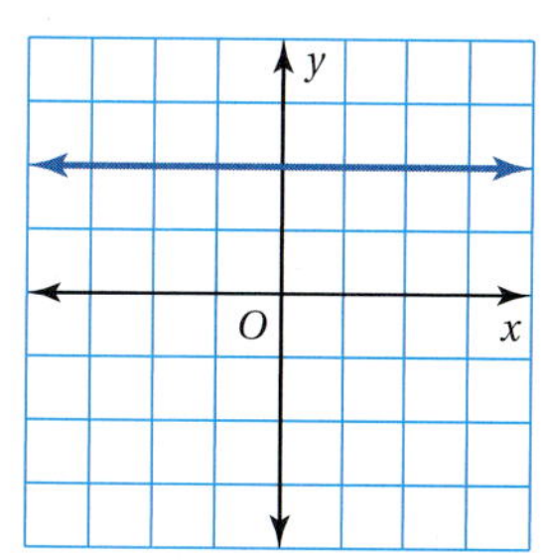

4. Which of the following is the equation of a line perpendicular to the line shown in the graph?

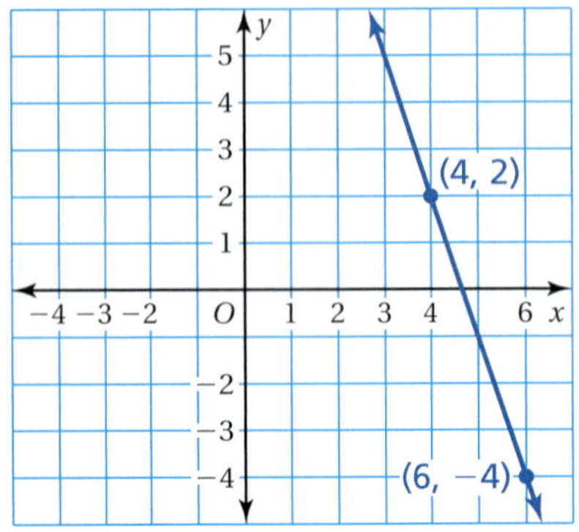

F. $y = 3x - 10$

G. $y = \frac{1}{3}x + 12$

H. $y = -3x + 5$

I. $y = -\frac{1}{3}x - 18$

5. What is the slope of the line that passes through the points $(2, -2)$ and $(8, 1)$?

6. A cell phone plan costs \$10 per month plus \$0.10 for each minute used. Last month, you spent \$18.50 using this plan. This can be modeled by the equation below, where m represents the number of minutes used.

$$0.1m + 10 = 18.5$$

How many minutes did you use last month?

A. 8.4 min

B. 85 min

C. 185 min

D. 285 min

7. Think Solve Explain It costs \$40 to rent a car for one day. In addition, the rental agency charges you for each mile driven, as shown in the graph.

Part A Determine the slope of the line joining the points on the graph.

Part B Explain what the slope represents.

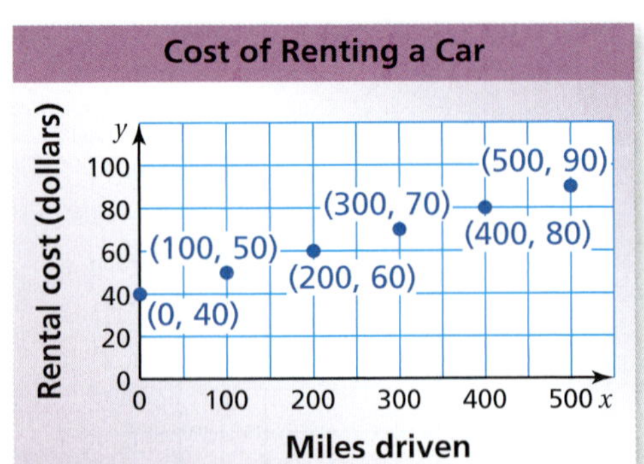

Item Analysis (continued)

4. **F.** The student thinks perpendicular lines have slopes that are opposites.

   G. Correct answer

   H. The student thinks perpendicular lines have the same slope.

   I. The student thinks perpendicular lines have slopes that are reciprocals of each other.

5. **Gridded Response:** Correct answer: 0.5, or $\frac{1}{2}$

   Common Error: The student performs subtraction incorrectly for the *y*-terms, yielding an answer of $\frac{1}{6}$ or $-\frac{1}{6}$.

6. **A.** The student correctly subtracts 10 from both sides, but then subtracts 0.1 instead of dividing.

   B. Correct answer

   C. The student ignores 10 and simply divides by 0.1.

   D. The student adds 10 to both sides, and then divides by 0.1.

7. **2 points** The student demonstrates a thorough understanding of the slope of a line and what it represents, explains the work fully, and calculates the slope accurately. The slope of the line is $\frac{50-40}{100-0} = \frac{10}{100} = \frac{1}{10} = 0.10$. The slope represents the rental cost per mile driven, \$0.10 per mile.

   1 point The student's work and explanations demonstrate a lack of essential understanding. The formula for the slope of a line is misstated, or the student incorrectly states what the slope of the line represents.

   0 points The student provides no response, a completely incorrect or incomprehensible response, or a response that demonstrates insufficient understanding of the slope of a line and what it represents.

Answers

4. G

5. $\frac{1}{2}$

6. B

7. *Part A* 0.10

Part B \$0.10 per mile

Answers

8. 6

9. F

10. B

11. H

Item Analysis (continued)

8. Gridded Response: Correct answer: 6

Common Error: The student correctly subtracts 7 from each side, but incorrectly adds $4x$ to $2x$ instead of subtracting.

9. F. Correct answer

G. The student reflects the figure in the y-axis.

H. The student rotates the figure 180°.

I. The student rotates the figure 90° counterclockwise.

10. A. The student divides M by 3, but fails to divide $(K + 7)$ by 3.

B. Correct answer

C. The student divides K by 3, but fails to divide 7 by 3.

D. The student subtracts 7 instead of adding it to both sides.

11. F. The student incorrectly sets up the proportion as $\frac{30}{100} = \frac{d}{12}$ or $\frac{100}{30} = \frac{12}{d}$.

G. The student thinks the corresponding side lengths are the same length.

H. Correct answer

I. The student incorrectly sets up the proportion as $\frac{d}{100} = \frac{30}{12}$ or $\frac{100}{d} = \frac{12}{30}$.

8. What value of x makes the equation below true?

$$7 + 2x = 4x - 5$$

9. Trapezoid *KLMN* is graphed in the coordinate plane shown.

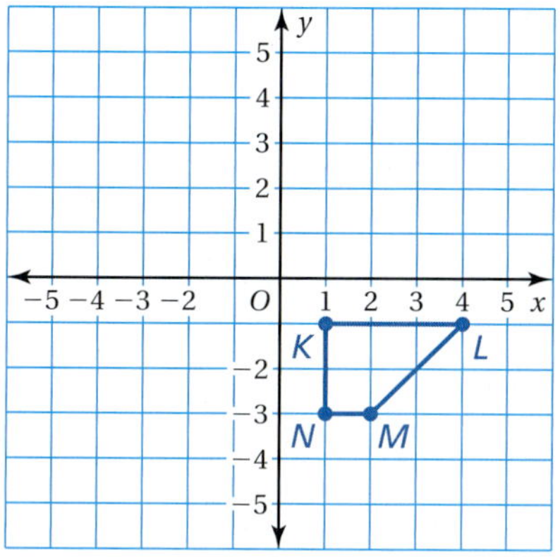

Rotate Trapezoid *KLMN* 90° clockwise about the origin. What are the coordinates of point M', the image of point M after the rotation?

F. $(-3, -2)$

G. $(-2, -3)$

H. $(-2, 3)$

I. $(3, 2)$

10. Solve the formula $K = 3M - 7$ for M.

A. $M = K + 7$

B. $M - \dfrac{K+7}{3}$

C. $M = \dfrac{K}{3} + 7$

D. $M = \dfrac{K-7}{3}$

11. What is the distance d across the canyon?

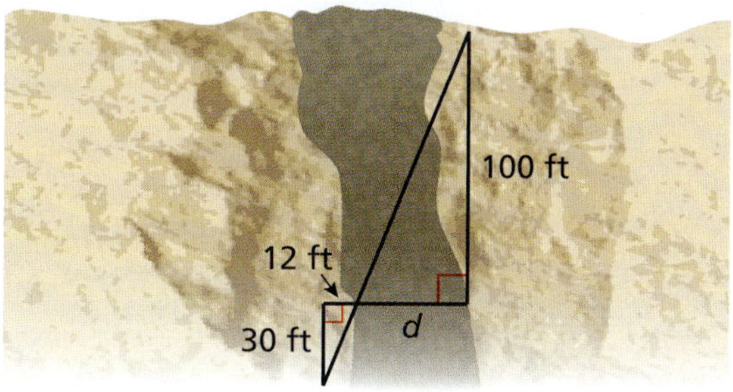

F. 3.6 ft

G. 12 ft

H. 40 ft

I. 250 ft

14 Real Numbers and the Pythagorean Theorem

What Your Students Have Learned

- Fluently divide whole numbers.
- Evaluate expressions with whole-number exponents.
- Divide fractions.
- Multiply and divide by powers of 10.
- Fluently add, subtract, multiply, and divide decimals.
- Find distances between points with the same *x*- or *y*-coordinate.
- Convert rational numbers to decimals using long division.
- Add, subtract, multiply, and divide rational numbers.
- Understand that every quotient of integers (non-zero divisor) is a rational number.

What Your Students Will Learn

- Understand that every rational number has a decimal expansion that terminates or repeats.
- Understand that numbers that are not rational are irrational.
- Compare irrational numbers using rational exponents.
- Evaluate square roots and cube roots, including those resulting from solving equations.
- Explain a proof of the Pythagorean Theorem and its converse.
- Use the Pythagorean Theorem to find missing measures of right triangles and distances between points in the coordinate plane.
- Approximate values of expressions involving square roots and irrational numbers.
- Write repeating decimals as fractions.

Pacing Guide for Chapter 14

Chapter Opener Accelerated	1 Day
Section 1 Accelerated	1 Day
Section 2 Accelerated	1 Day
Section 3 Accelerated	1 Day
Section 4 Accelerated	2 Days
Section 5 Accelerated	1 Day
Chapter Review/ Chapter Tests Accelerated	2 Days
Total Chapter 14 Accelerated	9 Days
Year-to-Date Accelerated	133 Days

Technology for the Teacher

BigIdeasMath.com
Chapter at a Glance
Complete Materials List
Parent Letters: English and Spanish

What Your Students Have Learned

- Compare decimals by converting them to fractions with a common denominator.
- Use the order of operations to evaluate expressions with whole number exponents.

Additional Topics for Review

- Number line
- Converting decimals to fractions
- Order of operations
- Exponents
- Compare and order decimals and fractions

Try It Yourself

1. =
2. <
3. <
4. *Sample answer:* $-0.009, -0.001, 0.01$
5. *Sample answer:* $-1.75, -1.74, 1.74$
6. *Sample answer:* $-0.75, 0.74, 0.75$
7. -3
8. 181
9. 99

Record and Practice Journal Fair Game Review

1. <
2. >
3. =
4. >

5–8. Sample answers are given.

5. $-5.2, -5.3, -6.5$
6. $2.56, 2.3, -3.2$
7. $-3.18, -3.1, -2.05$
8. $0.05, 0.3, 1.55$
9. 12.49; 12.495; 12.55; 12.60; 12.63
10. 167
11. 3

12–16. See Additional Answers.

Math Background Notes

Vocabulary Review

- Greater Than
- Less Than
- Order of Operations

Comparing Decimals

- Students should know how to compare decimals.
- **Teaching Tip:** Some students will have difficulty determining which decimal is greater simply by looking. Encourage these students to convert the decimals to fractions with a common denominator and compare the numerators.
- **Common Error:** Some students will have difficulty with Example 3 because there is not one "right" answer. Remind them that any number that makes the comparison true is a correct answer. Encourage creativity and remind students their answers will not always match the teacher's answers!

Using Order of Operations

- Students should know how to use the order of operations.
- You may want to review the correct order of operations with students. Many students probably learned the pneumonic device *Please Excuse My Dear Aunt Sally.* Ask a volunteer to explain why this phrase is helpful.
- You may want to review exponents with students. Remind students that the exponent tells you how many times the base is a factor. Exponents express repeated multiplication.

Reteaching and Enrichment Strategies

If students need help. . .	If students got it. . .
Record and Practice Journal • Fair Game Review Skills Review Handbook Lesson Tutorials	Game Closet at *BigIdeasMath.com* Start the next section

What You Learned Before

"I just remember that the sum of the squares of the shorter sides is equal to the square of the longest side."

Comparing Decimals

Complete the number sentence with <, >, or =.

Example 1 **1.1** $\square$ **1.01**

Because $\frac{110}{100}$ is greater than $\frac{101}{100}$, 1.1 is greater than 1.01.

So, $1.1 > 1.01$.

Example 2 **−0.3** $\square$ **−0.003**

Because $-\frac{300}{1000}$ is less than $-\frac{3}{1000}$, −0.3 is less than −0.003.

So, $-0.3 < -0.003$.

Example 3 **Find three decimals that make the number sentence $-5.12 > \square$ true.**

Any decimal less than −5.12 will make the sentence true.

Sample answer: −10.1, −9.05, −8.25

Try It Yourself

Complete the number sentence with <, >, or =.

1. $2.10 \ \square \ 2.1$

2. $-4.5 \ \square \ -4.25$

3. $\pi \ \square \ 3.2$

Find three decimals that make the number sentence true.

4. $-0.01 \leq \square$

5. $1.75 > \square$

6. $0.75 \geq \square$

Using Order of Operations

Example 4 **Evaluate $8^2 \div (32 \div 2) - 2(3 - 5)$.**

First:	Parentheses	$8^2 \div (32 \div 2) - 2(3 - 5) = 8^2 \div 16 - 2(-2)$
Second:	Exponents	$= 64 \div 16 - 2(-2)$
Third:	Multiplication and Division (from left to right)	$= 4 + 4$
Fourth:	Addition and Subtraction (from left to right)	$= 8$

Try It Yourself

Evaluate the expression.

7. $15\left(\frac{12}{3}\right) - 7^2 - 2 \cdot 7$

8. $3^2 \cdot 4 \div 18 + 30 \cdot 6 - 1$

9. $-1 + \left(\frac{4}{2}(6 - 1)\right)^2$

14.1 Finding Square Roots

Essential Question How can you find the dimensions of a square or a circle when you are given its area?

When you multiply a number by itself, you square the number.

Symbol for squaring is the exponent 2. → $4^2 = 4 \cdot 4$

$= 16$ 4 squared is 16.

To "undo" this, take the *square root* of the number.

Symbol for square root is a *radical sign*, $\sqrt{\ }$. → $\sqrt{16} = \sqrt{4^2} = 4$ The square root of 16 is 4.

1 ACTIVITY: Finding Square Roots

Work with a partner. Use a square root symbol to write the side length of the square. Then find the square root. Check your answer by multiplying.

a. Sample: $s = \sqrt{121} = 11$ ft

Area = 121 ft^2

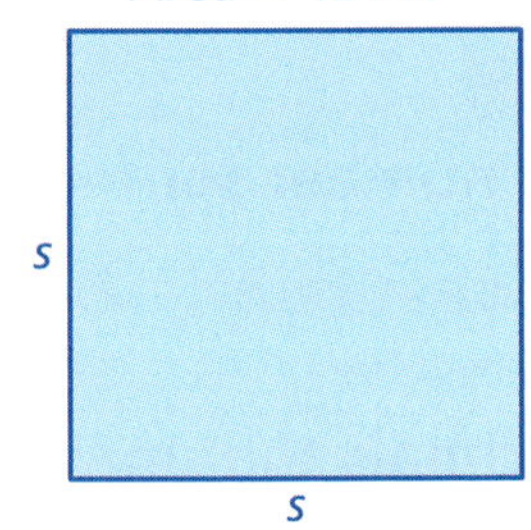

Check

$$\begin{array}{r} 11 \\ \times\ 11 \\ \hline 11 \\ 110 \\ \hline 121 \end{array} \checkmark$$

The side length of the square is 11 feet.

b. Area = 81 yd^2

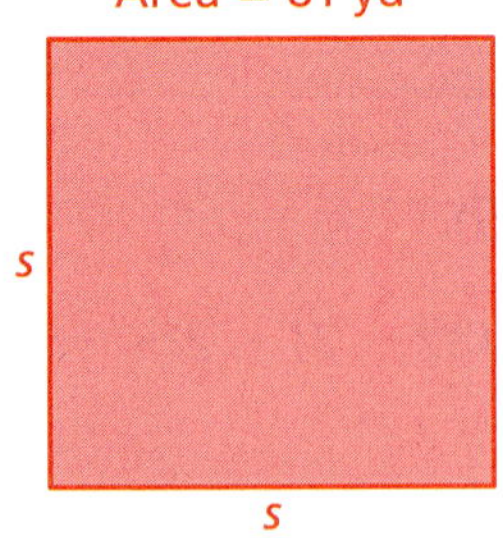

c. Area = 324 cm^2

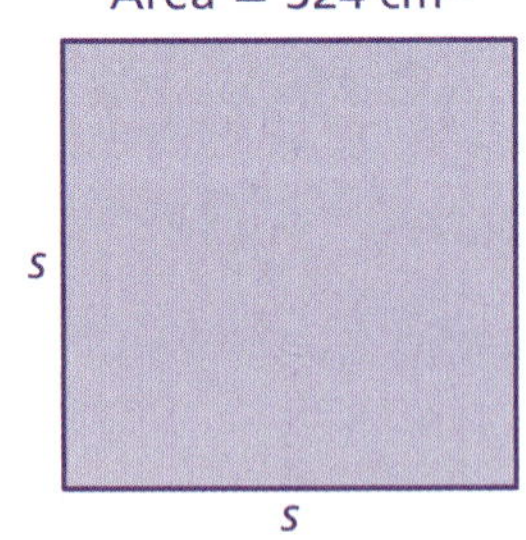

d. Area = 361 mi^2

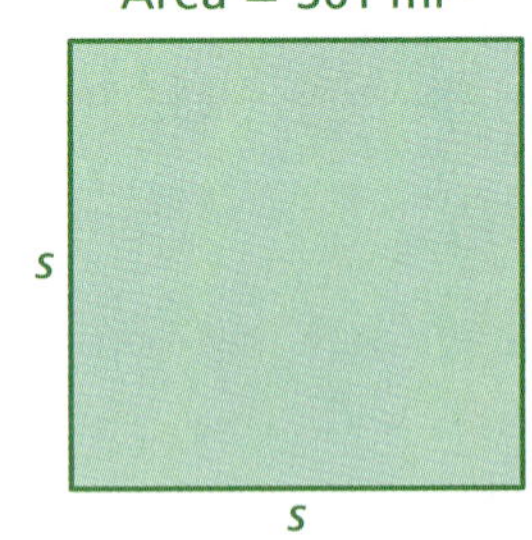

e. Area = 225 mi^2

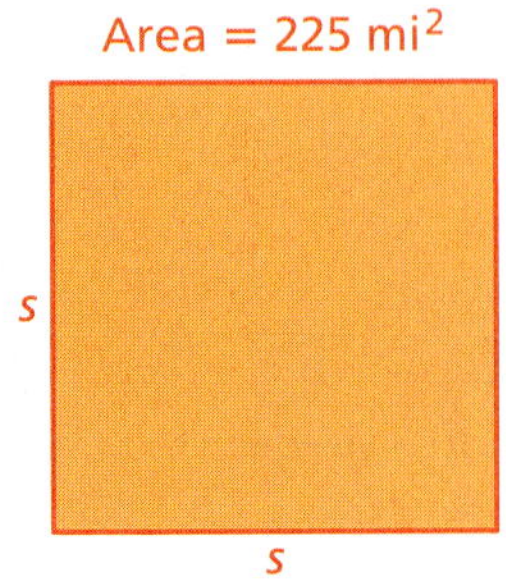

f. Area = 2.89 in.^{2}

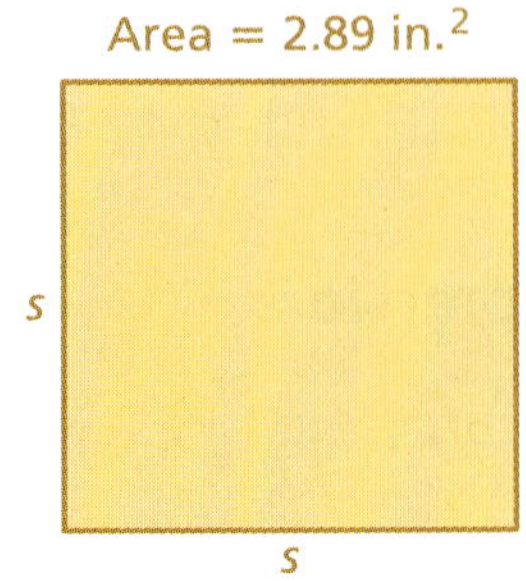

g. Area = $\frac{4}{9}$ ft^2

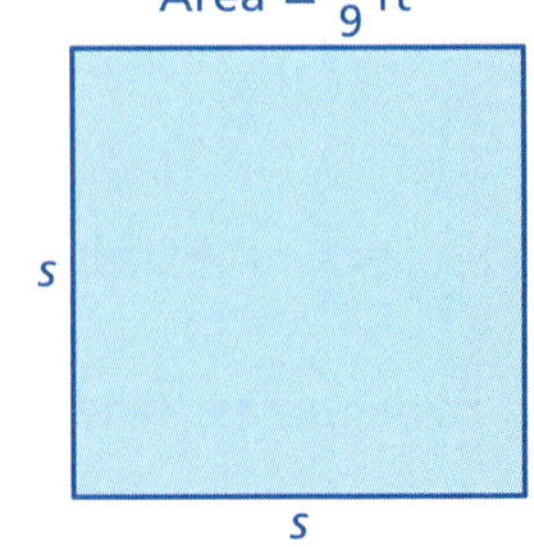

Square Roots

In this lesson, you will

- find square roots of perfect squares.
- evaluate expressions involving square roots.
- use square roots to solve equations.

Laurie's Notes

Introduction

Applying Mathematical Practices

- **Attend to Precision:** In these activities, students will need to multiply fractions and decimals and recall perfect squares. When graphing, accuracy will be necessary in order to recognize that the plotted points are not linear.

Motivate

- **Preparation:** Make two (or more) pendulums of different lengths.
- Swing the two pendulums back and forth a few times while telling a story.
- ? "Does it take the same amount of time for the pendulum to go back and forth for each length?" Answers will vary.
- **Extension:** Have a student time you as you swing the pendulum through 10 periods. Divide the total time by 10 to find the time of one period. Repeat for a different length pendulum.

Activity Notes

Activity 1

- **Preparation:** Cut a number of squares from paper. Calculate the areas of several and write the areas on the figures. Make one of the squares 6 cm by 6 cm.
- ? "Can you find the area of any of these squares? Explain." Yes, multiply length by width.
- ? "There are no dimensions marked on the square, so how do you find the area?" Measure the side lengths, then multiply to find the area.
- Hold up the 6-cm-by-6-cm square and show students that it has Area = 36 cm^2 recorded on it.
- ? "If you know the area, how can you find the dimensions without measuring the side lengths?" Ideas will vary.
- **Common Error:** Students will say to divide by 4 (perimeter) or divide by 2. Using the 6-cm-by-6-cm square, remind students that the side lengths are the same and they are multiplied together to get 36. So, the side lengths must be 6 because 6(6) = 36.
- Students should now be ready to begin the activity.
- When students have finished, discuss the answers and the strategies they used.

What Your Students Will Learn

- Find square roots of perfect squares.
- Evaluate expressions involving square roots and square roots of rational numbers.
- Use square roots to solve algebraic equations.

Previous Learning

Students have found squares of numbers. They have also found areas of squares and circles.

Technology for the Teacher

Dynamic Classroom

Lesson Plans
Complete Materials List

14.1 Record and Practice Journal

Essential Question How can you find the dimensions of a square or a circle when you are given its area?

When you multiply a number by itself, you square the number.

Symbol for squaring is the exponent 2. → $4^2 = 4 \bullet 4 = 16$ 4 squared is 16.

To "undo" this, take the *square root* of the number.

Symbol for square root is a *radical sign*, $\sqrt{\ }$. → $\sqrt{16} = \sqrt{4^2} = 4$ The square root of 16 is 4.

1 ACTIVITY: Finding Square Roots

Work with a partner. Use a square root symbol to write the side length of the square. Then find the square root. Check your answer by multiplying.

a. **Sample:** $s = \sqrt{121} =$ **Check:**

Area = 121 ft^2

The length of each side of the square is 11 ft.

b. Area = 81 yd^2 $\sqrt{81} = 9$ yd

c. Area = 324 cm^2 $\sqrt{324} = 18$ cm

d. Area = 361 mi^2 $\sqrt{361} = 19$ mi

English Language Learners

Build on Past Knowledge

Remind students of inverse operations. Addition and subtraction are inverse operations, as are multiplication and division. Taking the square root of a number is the inverse of squaring a number and squaring a number is the inverse of taking the square root of a number.

14.1 Record and Practice Journal

e. Area = 225 mi² f. Area = 2.89 in.² g. Area = $\frac{4}{9}$ ft²

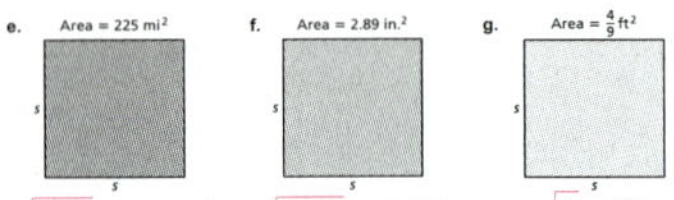

$\sqrt{225} = 15$ mi $\sqrt{2.89} = 1.7$ in. $\sqrt{\frac{4}{9}} = \frac{2}{3}$ ft

2 ACTIVITY: Using Square Roots

Work with a partner. Find the radius of each circle.

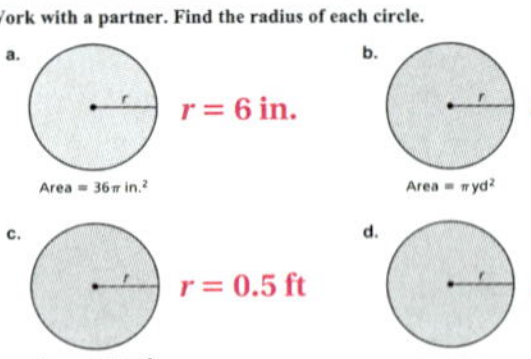

a. $r = 6$ in. Area = 36π in.²

b. $r = 1$ yd Area = π yd²

c. $r = 0.5$ ft Area = 0.25π ft²

d. $r = \frac{3}{4}$ m Area = $\frac{9}{16}$π m²

3 ACTIVITY: The Period of a Pendulum

Work with a partner.

The period of a pendulum is the time (in seconds) it takes the pendulum to swing back *and* forth.

The period T is represented by $T = 1.1\sqrt{L}$, where L is the length of the pendulum (in feet).

Complete the table. Then graph the ordered pairs on the next page. Is the equation linear?

not linear

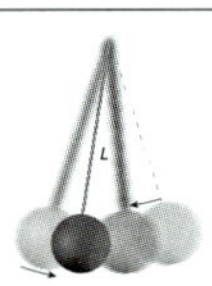

L	1.00	1.96	3.24	4.00	4.84	6.25	7.29	7.84	9.00
T	1.1	1.54	1.98	2.2	2.42	2.75	2.97	3.08	3.3

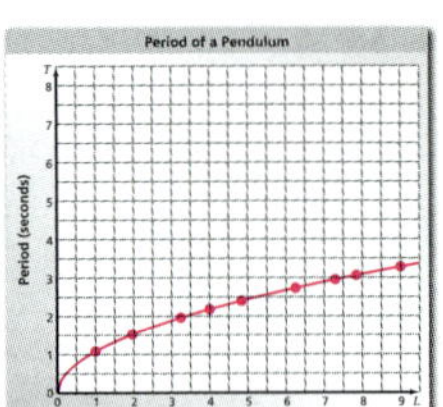

What Is Your Answer?

4. **IN YOUR OWN WORDS** How can you find the dimensions of a square or circle when you are given its area? Give an example of each. How can you check your answers?

For a square, find a number whose square is the area.
For a circle, find a number whose square times π is the area.

Laurie's Notes

Activity 2

? **Make Sense of Problems:** "How do you find the area of a circle?"
$A = \pi r^2$

? "If you know the area of a circle, can you solve for the radius?" yes

- Do not give away too much at this point. Let students think through the problems. The first problem involves an obvious perfect square.
- **Common Error:** For part (c), students will often answer 0.05, forgetting how decimals are multiplied.

Activity 3

- Define and model the period of a pendulum, if you did not do the opening Motivate.
- Write the formula, $T = 1.1\sqrt{L}$. Explain that if the length is 9 feet, you can find out the time of one period by evaluating the equation for $L = 9$. So, $T = 1.1\sqrt{9} = 1.1(3) = 3.3$ seconds.
- For each of the values in the table, remind students to think about the whole numbers 100, 196, 324, 400, and so on, to help find $\sqrt{1.00}$, $\sqrt{1.96}$, $\sqrt{3.24}$, $\sqrt{4.00}$, and so on.
- **Attend to Precision:** When plotting the ordered pairs, students will need to be precise to see the curvature in the graph.
- **Connection:** The equation they are graphing is $y = 1.1\sqrt{x}$. This is not a linear equation, so the graph is not a line. Find the slope between two different pairs of points on the graph. For (1, 1.1) and (4, 2.2), the slope is $\frac{1.1}{3} = 0.3\overline{6}$. For (4, 2.2) and (9, 3.3), the slope is $\frac{1.1}{5} = 0.22$. The slopes are not the same, so the graph cannot be a line.

What Is Your Answer?

- **Neighbor Check:** Have students work independently and then have their neighbors check their work. Have students discuss any discrepancies.

Closure

- **Matching Activity:** Match each square root with the correct answer.

1. $\sqrt{1600}$ D		**A.** 12
2. $\sqrt{400}$ B		**B.** 20
3. $\sqrt{144}$ A		**C.** 6
4. $\sqrt{36}$ C		**D.** 40

2 ACTIVITY: Using Square Roots

Work with a partner. Find the radius of each circle.

a.

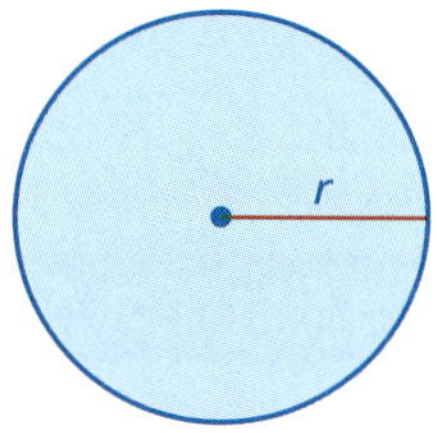

Area $= 36\pi$ in.^{2}

b.

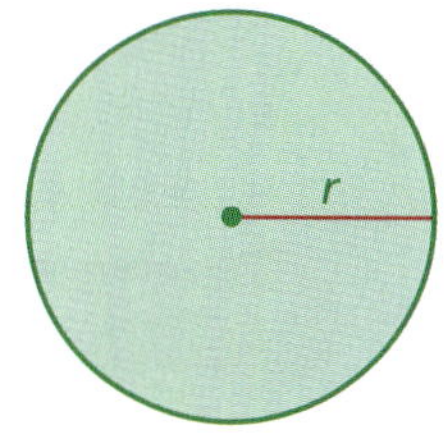

Area $= \pi$ yd^2

c.

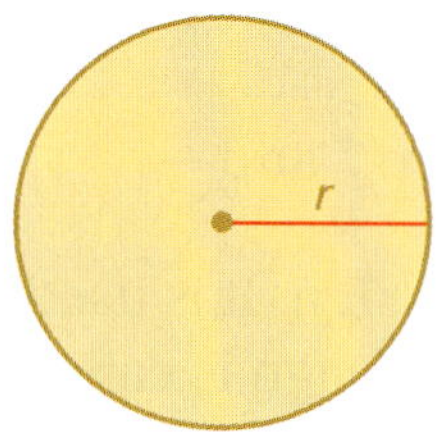

Area $= 0.25\pi$ ft^2

d.

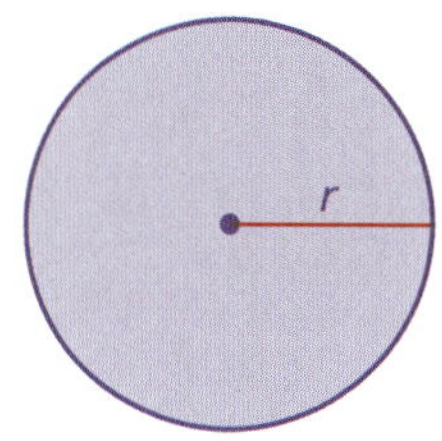

Area $= \frac{9}{16}\pi$ m^2

3 ACTIVITY: The Period of a Pendulum

Math Practice

Calculate Accurately

How can you use the graph to help you determine whether you calculated the values of T correctly?

Work with a partner.

The period of a pendulum is the time (in seconds) it takes the pendulum to swing back *and* forth.

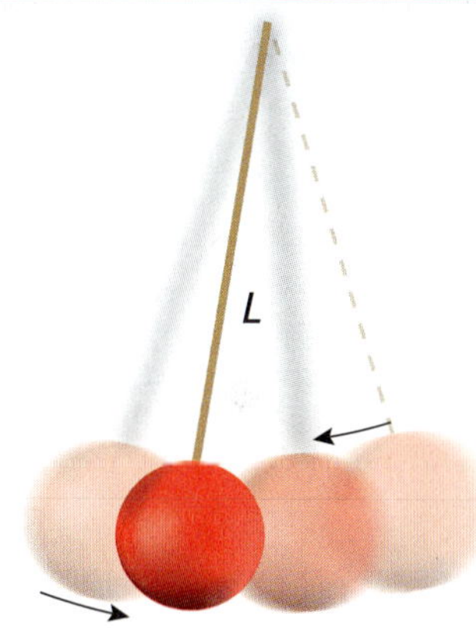

The period T is represented by $T = 1.1\sqrt{L}$, where L is the length of the pendulum (in feet).

Copy and complete the table. Then graph the ordered pairs. Is the equation linear?

L	1.00	1.96	3.24	4.00	4.84	6.25	7.29	7.84	9.00
T									

What Is Your Answer?

4. **IN YOUR OWN WORDS** How can you find the dimensions of a square or a circle when you are given its area? Give an example of each. How can you check your answers?

Practice Use what you learned about finding square roots to complete Exercises 4–6 on page 630.

14.1 Lesson

Check It Out
Lesson Tutorials
BigIdeasMath.com

Key Vocabulary
square root, *p. 628*
perfect square, *p. 628*
radical sign, *p. 628*
radicand, *p. 628*

A **square root** of a number is a number that, when multiplied by itself, equals the given number. Every positive number has a positive *and* a negative square root. A **perfect square** is a number with integers as its square roots.

EXAMPLE 1 Finding Square Roots of a Perfect Square

Find the two square roots of 49.

$7 \cdot 7 = 49$ and $(-7) \cdot (-7) = 49$

So, the square roots of 49 are 7 and -7.

Study Tip
Zero has one square root, which is 0.

The symbol $\sqrt{\ }$ is called a **radical sign**. It is used to represent a square root. The number under the radical sign is called the **radicand**.

Positive Square Root, $\sqrt{\ }$	Negative Square Root, $-\sqrt{\ }$	Both Square Roots, $\pm\sqrt{\ }$
$\sqrt{16} = 4$	$-\sqrt{16} = -4$	$\pm\sqrt{16} = \pm 4$

EXAMPLE 2 Finding Square Roots

Find the square root(s).

a. $\sqrt{25}$

Because $5^2 = 25$, $\sqrt{25} = \sqrt{5^2} = 5$.

$\sqrt{25}$ represents the *positive* square root.

b. $-\sqrt{\frac{9}{16}}$

Because $\left(\frac{3}{4}\right)^2 = \frac{9}{16}$, $-\sqrt{\frac{9}{16}} = -\sqrt{\left(\frac{3}{4}\right)^2} = -\frac{3}{4}$.

$-\sqrt{\frac{9}{16}}$ represents the *negative* square root.

c. $\pm\sqrt{2.25}$

$\pm\sqrt{2.25}$ represents both the *positive* and the *negative* square roots.

Because $1.5^2 = 2.25$, $\pm\sqrt{2.25} = \pm\sqrt{1.5^2} = 1.5$ and -1.5.

On Your Own

Now You're Ready
Exercises 7–18

Find the two square roots of the number.

1. 36 **2.** 100 **3.** 121

Find the square root(s).

4. $-\sqrt{1}$ **5.** $\pm\sqrt{\frac{4}{25}}$ **6.** $\sqrt{12.25}$

Laurie's Notes

Introduction

Connect

- **Yesterday:** Students explored square roots.
- **Today:** Students will find square roots and evaluate expressions involving square roots.

Motivate

- Play the game *Keep it Going!*
- Give the students the first 3 to 4 numbers in a sequence and have them *Keep it Going.* If students are sitting in a row, each person in the row says the next number in the pattern. Keep the pattern going until it becomes too difficult to continue. For example, use the sequence 4, 400, 40,000, … (4,000,000, 400,000,000…)

Lesson Notes

Discuss

- Write and discuss the definitions of square root of a number and perfect square. Mention the *Study Tip*.
- Students are often confused when you say "every positive number has a positive and a negative square root." Use Example 1 to explain.

Example 1

- Note that the direction line is written in words without the square root symbol. The notation is introduced after this example.
- ? "What is the product of two positives? two negatives?" Both are positive.

Discuss

- **Attend to Precision:** It is important for students to use correct language and be familiar with the symbols $\sqrt{\ }$, $-\sqrt{\ }$, and $\pm\sqrt{\ }$.
- The square root symbol is called a **radical sign** and the number under the radical sign is the **radicand**.
- Write and discuss the three examples in the table. Explain to students that the symbol $\pm$ is read as *plus or minus*.

Example 2

- Remind students to pay attention to the signs that may precede the radical sign.
- ? "How do you multiply fractions?" Write the product of the numerators over the product of the denominators.
- ? "What fraction is multiplied by itself to get $\frac{9}{16}$?" $\frac{3}{4}$

On Your Own

- **Think-Pair-Share:** Students should read each question independently and then work in pairs to answer the questions. When they have answered the questions, the pair should compare their answers with another group and discuss any discrepancies.

Goal Today's lesson is finding **square roots**.

Technology for the Teacher

Dynamic Classroom

Lesson Tutorials
Lesson Plans
Answer Presentation Tool

Extra Example 1

Find the two square roots of 64. 8 and −8

Extra Example 2

Find the square root(s).

a. $-\sqrt{81}$ −9

b. $\pm\sqrt{\frac{9}{64}}$ $\frac{3}{8}$ and $-\frac{3}{8}$

c. $\sqrt{4.84}$ 2.2

On Your Own

1. 6 and −6
2. 10 and −10
3. 11 and −11
4. −1
5. $\pm\frac{2}{5}$
6. 3.5

Extra Example 3

Evaluate each expression.

a. $2\sqrt{144} - 30$ -6

b. $\sqrt{\frac{36}{4}} + \frac{1}{6}$ $3\frac{1}{6}$

c. $49 - \left(\sqrt{49}\right)^2$ 0

Extra Example 4

What is the radius of the circle? Use 3.14 for π. about 4 in.

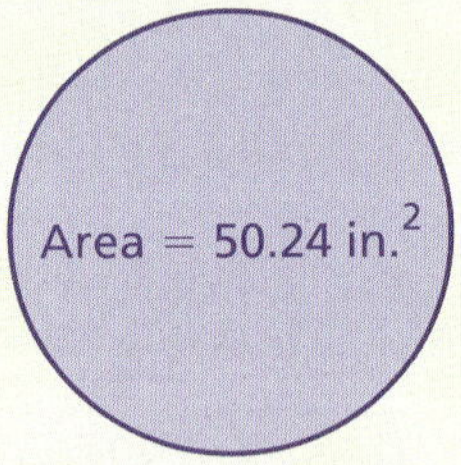

On Your Own

7. -3
8. 4.4
9. 11
10. $3.14r^2 = 2826$; about 30 ft

Differentiated Instruction

Auditory

Ask students to use mental math to answer the following verbal questions.

- "What is the sum of the square root of 9 and 3?" 6
- "What is the difference of 12 and the square root of 144?" 0
- "What is twice the square root of 16?" 8
- "What is one-fourth of the square root of 64?" 2

Laurie's Notes

Example 3

- **Teaching Tip:** In these examples, remind students that square roots are numbers, so you can evaluate numerical expressions that include square roots. Students see a *symbol*, think *variable*, and suddenly they forget things like the order of operations. In part (a), some students think of this as $5x + 7$ and will not know what to do.
- Write the expression in part (a).
- ? "Which operations are involved in this problem?" taking a square root, multiplication, and addition
- Work through parts (a) and (b) as shown.
- Write the expression in part (c). Discuss that squaring and taking the square root are inverse operations, as explained prior to the example.
- ? "What is $\sqrt{81}$?" 9 "What is 9^2?" 81
- Say, "So, taking the square root of 81 and then squaring the answer results in 81." Finish evaluating the expression as shown.
- ? "Can you name inverse operations?" addition and subtraction, multiplication and division, squaring and taking the square root

Example 4

- ? "How do you find the area of a circle?" $A = \pi r^2$
- The numbers involved may be overwhelming to students. Reassure them that this is an equation with one variable and that they know how to solve equations!
- Use a calculator or long division to divide 45,216 by 3.14.
- When you get to the step $14{,}400 = r^2$, remind students that whatever you do to one side of an equation, you must do to the other side. So, to get r by itself, you need to undo the squaring. So, take the square root of each side.
- Discuss with students why a negative square root does not make sense in this context. You cannot have a negative radius, so you can "ignore" the negative square root.

On Your Own

- Ask volunteers to share their work at the board for each of the problems.
- Question 8 looks more difficult because of the fraction. You may want to point out that $\frac{28}{7} = 4$.

Closure

- Write 3 numbers of which you know how to take the square root.
- Write 3 numbers of which you do not know how to take the square root.

Squaring a positive number and finding a square root are inverse operations. You can use this relationship to evaluate expressions and solve equations involving squares.

EXAMPLE 3 Evaluating Expressions Involving Square Roots

Evaluate each expression.

a. $5\sqrt{36} + 7 = 5(6) + 7$ — Evaluate the square root.

$= 30 + 7$ — Multiply.

$= 37$ — Add.

b. $\frac{1}{4} + \sqrt{\frac{18}{2}} = \frac{1}{4} + \sqrt{9}$ — Simplify.

$= \frac{1}{4} + 3$ — Evaluate the square root.

$= 3\frac{1}{4}$ — Add.

c. $\left(\sqrt{81}\right)^2 - 5 = 81 - 5$ — Evaluate the power using inverse operations.

$= 76$ — Subtract.

EXAMPLE 4 Real-Life Application

The area of a crop circle is 45,216 square feet. What is the radius of the crop circle? Use 3.14 for π.

$A = \pi r^2$ — Write the formula for the area of a circle.

$45{,}216 \approx 3.14r^2$ — Substitute 45,216 for A and 3.14 for π.

$14{,}400 = r^2$ — Divide each side by 3.14.

$\sqrt{14{,}400} = \sqrt{r^2}$ — Take positive square root of each side.

$120 = r$ — Simplify.

The radius of the crop circle is about 120 feet.

On Your Own

Now You're Ready
Exercises 20–27

Evaluate the expression.

7. $12 - 3\sqrt{25}$

8. $\sqrt{\frac{28}{7}} + 2.4$

9. $15 - \left(\sqrt{4}\right)^2$

10. The area of a circle is 2826 square feet. Write and solve an equation to find the radius of the circle. Use 3.14 for π.

14.1 Exercises

Check It Out
Help with Homework
BigIdeasMath.com

Vocabulary and Concept Check

1. **VOCABULARY** Is 26 a perfect square? Explain.
2. **REASONING** Can the square of an integer be a negative number? Explain.
3. **NUMBER SENSE** Does $\sqrt{256}$ represent the positive square root of 256, the negative square root of 256, or both? Explain.

Practice and Problem Solving

Find the dimensions of the square or circle. Check your answer.

4. Area = 441 cm^2

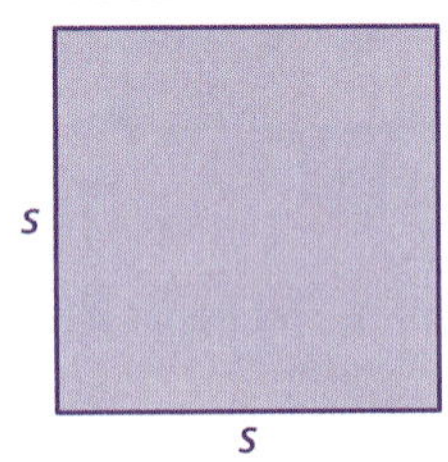

5. Area = 1.69 km^2

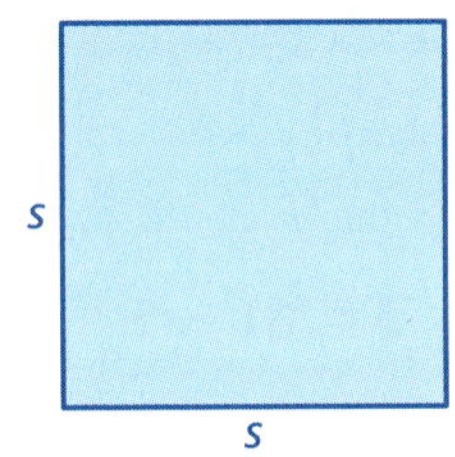

6. Area = $64\pi \text{ in.}^2$

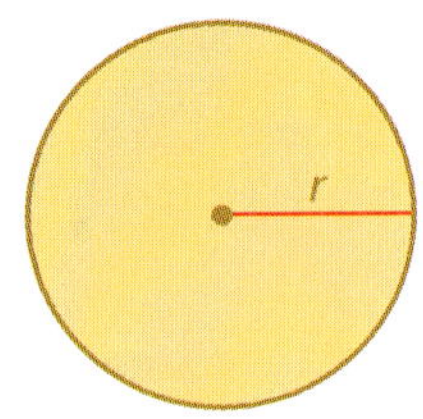

Find the two square roots of the number.

1 7. 9 8. 64 9. 4 10. 144

Find the square root(s).

2 11. $\sqrt{625}$ 12. $\pm\sqrt{196}$ 13. $\pm\sqrt{\frac{1}{961}}$ 14. $-\sqrt{\frac{9}{100}}$

15. $\pm\sqrt{4.84}$ 16. $\sqrt{7.29}$ 17. $-\sqrt{361}$ 18. $-\sqrt{2.25}$

19. **ERROR ANALYSIS** Describe and correct the error in finding the square roots.

✗ $\pm\sqrt{\frac{1}{4}} = \frac{1}{2}$

Evaluate the expression.

3 20. $(\sqrt{9})^2 + 5$ 21. $28 - (\sqrt{144})^2$ 22. $3\sqrt{16} - 5$ 23. $10 - 4\sqrt{\frac{1}{16}}$

24. $\sqrt{6.76} + 5.4$ 25. $8\sqrt{8.41} + 1.8$ 26. $2\left(\sqrt{\frac{80}{5}} - 5\right)$ 27. $4\left(\sqrt{\frac{147}{3}} + 3\right)$

28. **NOTEPAD** The area of the base of a square notepad is 2.25 square inches. What is the length of one side of the base of the notepad?

29. **CRITICAL THINKING** There are two square roots of 25. Why is there only one answer for the radius of the button?

$A = 25\pi \text{ mm}^2$

Assignment Guide and Homework Check

Level	Assignment	Homework Check
Accelerated	1–6, 10, 18, 19, 20–28 even, 30–42	18, 24, 32, 34, 37

Common Errors

- **Exercises 7–10** Students may only find the positive square root of the number given. Remind them that a square root can be positive or negative, and the question is asking for both answers.
- **Exercises 11–18** Students may divide the number by two instead of finding a number that, when multiplied by itself, gives the radicand. Remind them that taking the square root of a number is the inverse of squaring a number.
- **Exercises 20–27** Students may not follow the order of operations when evaluating the expression. Remind them of the order of operations. Because taking a square root is the inverse of squaring, it is evaluated before multiplication and division.

14.1 Record and Practice Journal

Find the two square roots of the number.

1. 16 — **4 and −4**
2. 100 — **10 and −10**
3. 196 — **14 and −14**

Find the square root(s).

4. $\sqrt{169}$ — **13**
5. $\sqrt{\frac{4}{225}}$ — $\frac{2}{15}$
6. $-\sqrt{12.25}$ — **−3.5**

Evaluate the expression.

7. $2\sqrt{36} + 9$ — **21**
8. $8 - 11\sqrt{\frac{25}{121}}$ — **3**
9. $3\left(\sqrt{\frac{125}{5}} - 8\right)$ — **−9**

10. A trampoline has an area of 49π square feet. What is the diameter of the trampoline?
14 ft

Vocabulary and Concept Check

1. no; There is no integer whose square is 26.
2. no; A positive number times a positive number is a positive number, and a negative number times a negative number is a positive number.
3. $\sqrt{256}$ represents the positive square root because there is not a − or a ± in front.

Practice and Problem Solving

4. $s = 21$ cm
5. $s = 1.3$ km
6. $r = 8$ in.
7. 3 and −3
8. 8 and −8
9. 2 and −2
10. 12 and −12
11. 25
12. ±14
13. $\frac{1}{31}$ and $-\frac{1}{31}$
14. $-\frac{3}{10}$
15. 2.2 and −2.2
16. 2.7
17. −19
18. −1.5
19. The positive and negative square roots should have been given.
$\pm\sqrt{\frac{1}{4}} = \frac{1}{2}$ and $-\frac{1}{2}$
20. 14
21. −116
22. 7
23. 9
24. 8
25. 25
26. −2
27. 40
28. 1.5 in.
29. because a negative radius does not make sense

Practice and Problem Solving

30. >

31. =

32. <

33. 9 ft

34. yes; *Sample answer:* Consider the perfect squares, a^2 and b^2. Their product can be written as $a^2b^2 = a \cdot a \cdot b \cdot b = (a \cdot b) \cdot (a \cdot b) = (a \cdot b)^2$.

35. 8 m/sec

36. See *Taking Math Deeper.*

37. 2.5 ft

38. 8 cm

Fair Game Review

39. $y = 3x - 2$

40. $y = -2x + 5$

41. $y = \frac{3}{5}x + 1$

42. B

Mini-Assessment

Find the square root(s).

1. $\sqrt{169}$ 13
2. $\sqrt{225}$ 15
3. $\pm\sqrt{4.41}$ 2.1 and -2.1
4. $-\sqrt{\frac{16}{25}}$ $-\frac{4}{5}$
5. $\sqrt{\frac{512}{2}}$ 16

Taking Math Deeper

Exercise 36

In this problem, students are given the area of the smaller watch face and are asked to find the radius of the larger watch face.

1 Summarize the given information and find the radius of the smaller watch face.

Area $= 4\pi$ cm^2
Ratio of areas is 16 to 25.

2 Answer the question. The two watch faces are similar, so the ratio of their areas is equal to the square of the ratio of their radii.

$$\frac{\text{Area of small}}{\text{Area of large}} = \left(\frac{\text{Radius of small}}{\text{Radius of large}}\right)^2$$

$$\frac{16}{25} = \left(\frac{\text{Radius of small}}{\text{Radius of large}}\right)^2$$

$$\sqrt{\frac{16}{25}} = \frac{\text{Radius of small}}{\text{Radius of large}}$$

$$\frac{4}{5} = \frac{\text{Radius of small}}{\text{Radius of large}}$$

a. The ratio of the radius of the smaller watch face to the radius of the larger watch face is $\frac{4}{5}$.

3 **b.** Write and solve a proportion to find R.

$$\frac{4}{5} = \frac{r}{R}$$

$$\frac{4}{5} = \frac{2}{R}$$

$$R = \frac{10}{4}, \text{ or } \frac{5}{2}$$

The radius of the larger watch face is $\frac{5}{2}$, or 2.5 centimeters.

Reteaching and Enrichment Strategies

If students need help. . .	If students got it. . .
Resources by Chapter • Practice A and Practice B • Puzzle Time Record and Practice Journal Practice Differentiating the Lesson Lesson Tutorials Skills Review Handbook	Resources by Chapter • Enrichment and Extension • Technology Connection Start the next section

Copy and complete the statement with <, >, or =.

30. $\sqrt{81}$ $\square$ 8 **31.** 0.5 $\square$ $\sqrt{0.25}$ **32.** $\frac{3}{2}$ $\square$ $\sqrt{\frac{25}{4}}$

33. SAILBOAT The area of a sail is $40\frac{1}{2}$ square feet. The base and the height of the sail are equal. What is the height of the sail (in feet)?

34. REASONING Is the product of two perfect squares always a perfect square? Explain your reasoning.

35. ENERGY The kinetic energy K (in joules) of a falling apple is represented by $K = \frac{v^2}{2}$, where v is the speed of the apple (in meters per second). How fast is the apple traveling when the kinetic energy is 32 joules?

36. PRECISION The areas of the two watch faces have a ratio of 16 : 25.

a. What is the ratio of the radius of the smaller watch face to the radius of the larger watch face?

b. What is the radius of the larger watch face?

37. WINDOW The cost C (in dollars) of making a square window with a side length of n inches is represented by $C = \frac{n^2}{5} + 175$. A window costs \$355. What is the length (in feet) of the window?

38. Geometry The area of the triangle is represented by the formula $A = \sqrt{s(s-21)(s-17)(s-10)}$, where s is equal to half the perimeter. What is the height of the triangle?

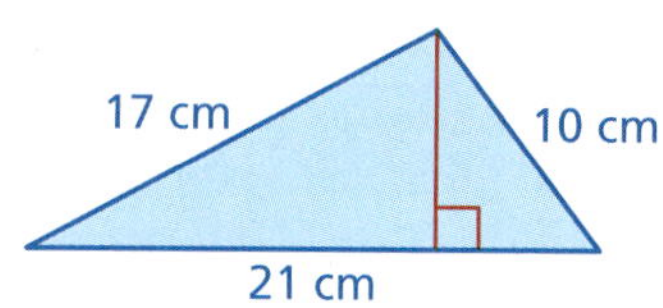

Fair Game Review What you learned in previous grades & lessons

Write in slope-intercept form an equation of the line that passes through the given points. *(Section 13.7)*

39. (2, 4), (5, 13) **40.** (−1, 7), (3, −1) **41.** (−5, −2), (5, 4)

42. MULTIPLE CHOICE What is the value of x? *(Section 12.2)*

Ⓐ 41 Ⓑ 44

Ⓒ 88 Ⓓ 134

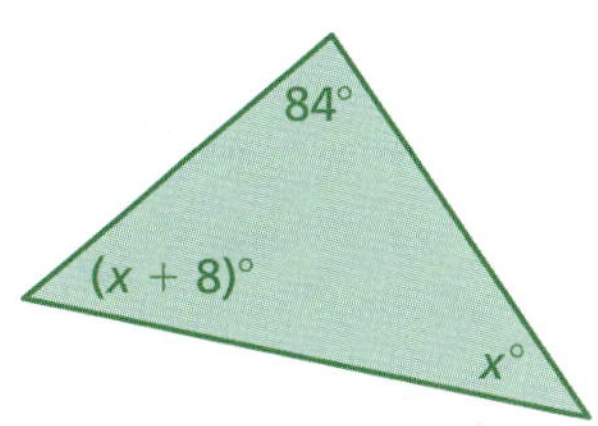

14.2 Finding Cube Roots

Essential Question How is the cube root of a number different from the square root of a number?

When you multiply a number by itself twice, you cube the number.

Symbol for cubing is the exponent 3. → $4^3 = 4 \cdot 4 \cdot 4$

$= 64$ 4 cubed is 64.

To "undo" this, take the *cube root* of the number.

Symbol for cube root is $\sqrt[3]{\ }$. → $\sqrt[3]{64} = \sqrt[3]{4^3} = 4$ The cube root of 64 is 4.

1 ACTIVITY: Finding Cube Roots

Work with a partner. Use a cube root symbol to write the edge length of the cube. Then find the cube root. Check your answer by multiplying.

a. Sample:

$s = \sqrt[3]{343} = \sqrt[3]{7^3} = 7$ inches

Volume = 343 in.^{3}

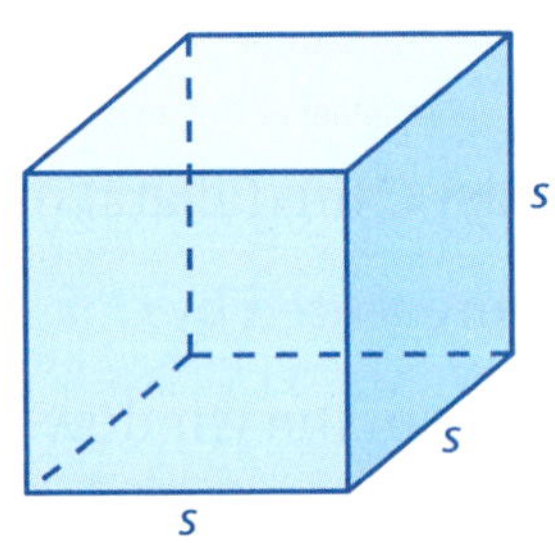

Check

$7 \cdot 7 \cdot 7 = 49 \cdot 7$

$= 343$ ✓

The edge length of the cube is 7 inches.

b. Volume = 27 ft^3

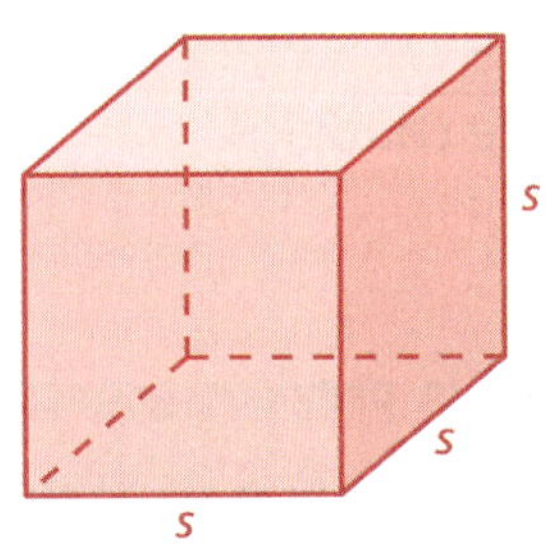

c. Volume = 125 m^3

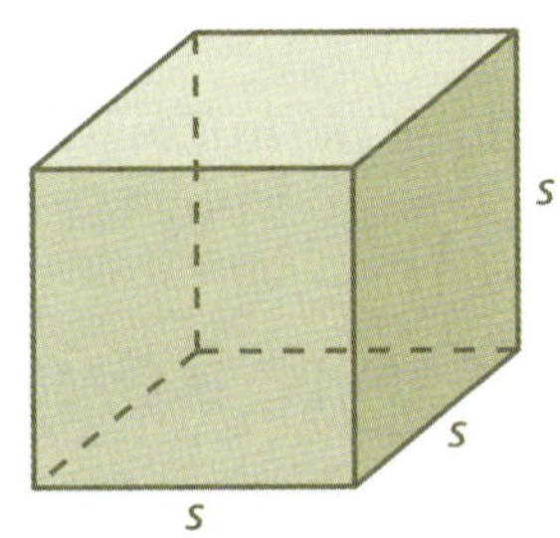

d. Volume = 0.001 cm^3

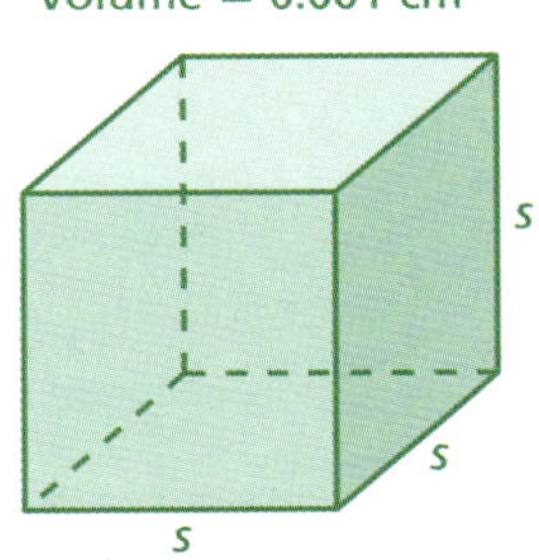

e. Volume = $\frac{1}{8}$ yd^3

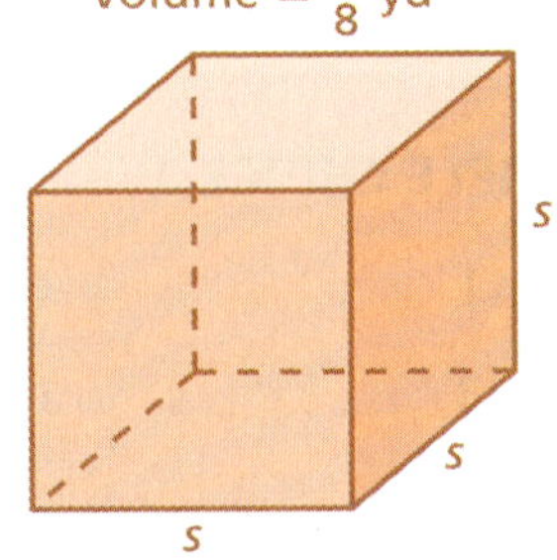

Cube Roots

In this lesson, you will

- find cube roots of perfect cubes.
- evaluate expressions involving cube roots.
- use cube roots to solve equations.

Laurie's Notes

Introduction

Applying Mathematical Practices

- **Reason Abstractly and Quantitatively:** Students have found square roots of a number. When finding a cube root, they must think about multiplying a number by itself twice.

Motivate

- Use 8 cubes to build a $2 \times 2 \times 2$ cube. Inch-cubes or larger are best for visibility in the class. Display the large cube by holding it up on a rigid surface so that all students are able to see it.

? "Could someone describe this shape?" It is a cube.

? "Can you describe any of the numerical attributes of the cube?" The dimensions are $2 \times 2 \times 2$ and the volume is 8.

? "If the edge length of each smaller cube is 1 inch, what are the units for the volume of the large cube?" cubic inches

- Summarize by saying, "So, $8 = 2 \times 2 \times 2 = 2^3$. Cubing the edge length gives the volume."
- Hold another cube that is not made of smaller cubes, such as a cube of sticky notes.

? "If you knew the volume of this cube, do you think you could find the dimensions—without measuring?" Give students time to think about this question and then introduce the activities.

Activity Notes

Discuss

- Review the meaning of the notation in the expression 4^3. Introduce the cube root symbol, $\sqrt[3]{\ }$, and how it is read.

Activity 1

- **Make Sense of Problems and Persevere in Solving Them:** Following the sample, students should be able to do the rest of the activity. Before you jump in with answers, remind students to think about the Motivate problem and the relationship between edge length and volume.
- Students may think that a calculator is necessary to solve these problems, but they may not find the cube root key (if the calculator has one). If a calculator helps them explore the problem, let them use it.
- **Attend to Precision:** Remind students that answers need correct labels (units).
- **Reason Abstractly and Quantitatively:** When students have finished, ask volunteers to describe their thinking behind the solutions. They should describe thinking about what number could be multiplied by itself twice to get the volume.
- Students will often say that they need a number that, when multiplied by itself 3 times, will be the volume. Clarify that the number (edge length) is used as a factor 3 times, but it is multiplied by itself only 2 times.

What Your Students Will Learn

- Find cube roots of perfect cubes.
- Evaluate expressions involving cube roots.
- Use cube roots to solve algebraic equations.

Previous Learning

Students have found cubes of numbers. They have also found volumes of cubes.

Technology for the Teacher

Dynamic Classroom

Lesson Plans
Complete Materials List

14.2 Record and Practice Journal

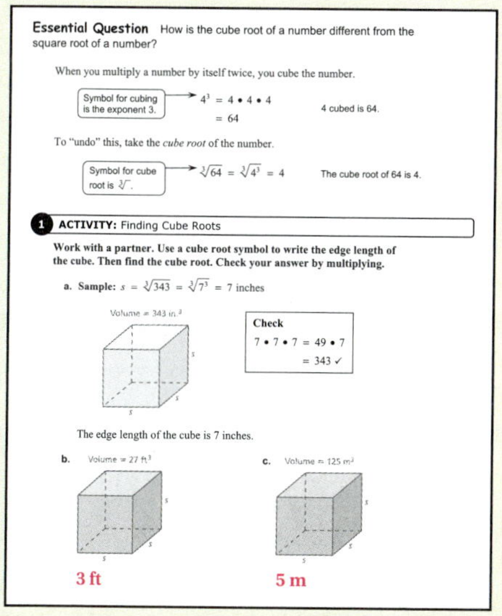

Essential Question How is the cube root of a number different from the square root of a number?

When you multiply a number by itself twice, you cube the number.

Symbol for cubing is the exponent 3. → $4^3 = 4 \cdot 4 \cdot 4 = 64$ 4 cubed is 64.

To "undo" this, take the *cube root* of the number.

Symbol for cube root is $\sqrt[3]{\ }$. → $\sqrt[3]{64} = \sqrt[3]{4^3} = 4$ The cube root of 64 is 4.

1 ACTIVITY: Finding Cube Roots

Work with a partner. Use a cube root symbol to write the edge length of the cube. Then find the cube root. Check your answer by multiplying.

a. **Sample:** $s = \sqrt[3]{343} = \sqrt[3]{7^3} = 7$ inches

Check
$7 \cdot 7 \cdot 7 = 49 \cdot 7$
$= 343$ ✓

The edge length of the cube is 7 inches.

b. Volume = 27 ft³ **3 ft**

c. Volume = 125 m³ **5 m**

English Language Learners

Vocabulary

For English learners, note the different meanings of the word *radical*. In everyday language, it can be used to describe a considerable departure from the usual or traditional view. In slang, *radical* is something that is excellent or cool. In mathematics, it is a symbol for roots—square roots, cube roots, and so on.

14.2 Record and Practice Journal

d. Volume = 0.001 cm³

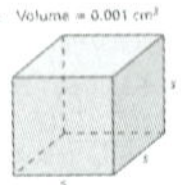

0.1 cm

e. Volume = $\frac{1}{8}$ yd³

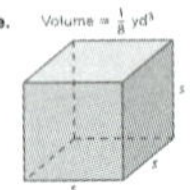

$\frac{1}{2}$ yd

2 ACTIVITY: Use Prime Factorizations to Find Cube Roots

Work with a partner. Write the prime factorization of each number. Then use the prime factorization to find the cube root of the number.

a. 216

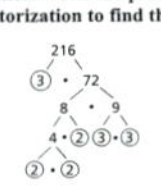

$216 = 3 \cdot 2 \cdot 3 \cdot 3 \cdot 2 \cdot 2$ Prime factorization

$= (3 \cdot \boxed{2}) \cdot (3 \cdot \boxed{2}) \cdot (3 \cdot \boxed{2})$ Commutative Property of Multiplication

$= \boxed{6} \cdot \boxed{6} \cdot \boxed{6}$ Simplify.

The cube root of 216 is 6.

b. 1000

$1000 = 5 \cdot 2 \cdot 2 \cdot 5 \cdot 2 \cdot 5$; 10

c. 3375

$3375 = 3 \cdot 5 \cdot 3 \cdot 3 \cdot 5 \cdot 5$; 15

d. **STRUCTURE** Does this procedure work for every number? Explain why or why not.

no

What Is Your Answer?

3. Complete each statement using *positive* or *negative*.
   a. A positive number times a positive number is a positive number.
   b. A negative number times a negative number is a positive number.
   c. A positive number multiplied by itself twice is a positive number.
   d. A negative number multiplied by itself twice is a negative number.
4. **REASONING** Can a negative number have a cube root? Give an example to support your explanation.

   yes
5. **IN YOUR OWN WORDS** How is the cube root of a number different from the square root of a number?

   Check students' work.
6. Give an example of a number whose square root and cube root are equal.

   Sample answers: 0 or 1
7. A cube has a volume of 13,824 cubic meters. Use a calculator to find the edge length.

   24 m

Laurie's Notes

Activity 2

- **Connection:** This activity shows an interesting application of prime factorizations.

? "What is a prime number?" a number with exactly two factors, 1 and itself

? "What does it mean to write the prime factorization of a number?" Write the number as a product of only prime factors.

- Work through the factor tree in part (a), which will remind students of one method for finding the prime factorization of a number. The factor tree results in three 3s and three 2s.
- Make sure students understand why they rearrange the factors using the Commutative Property of Multiplication. You want to be able to write 216 as a number times itself twice. So, you want to group the prime factors into three identical groups.
- Work through the rest of part (a), showing three groups of $3 \cdot 2$, or $6 \cdot 6 \cdot 6$, meaning that 216 is a *perfect cube* whose cube root is 6. Perfect cube is not defined until the lesson, but students should understand what it means from their work with perfect squares.
- Students may quickly recognize 1000 as 10^3 without making a factor tree, and that is fine. It is unlikely that they will recognize 3375 as 15^3 without a bit of factoring!
- Part (d) may not be as obvious as you might imagine. Not all students will immediately understand that the procedure doesn't work for all numbers. Even if they do, they may not be able to explain correctly that not all numbers are *perfect cubes*.
- **Extension:** Ask students to describe how they could use a similar procedure to find square roots of large perfect squares.

What Is Your Answer?

- Questions 3–5 are all related. Take time to discuss students' responses and understanding of these problems.
- **Reason Abstractly and Quantitatively** and **Use Appropriate Tools Strategically:** Question 7 may be approached as a *Guess, Check, and Revise* problem or a prime factorization problem, versus students using the cube root key on their calculators.

Closure

- Find the edge length of the cube. $\frac{1}{3}$ in.

Volume = $\frac{1}{27}$ in.³

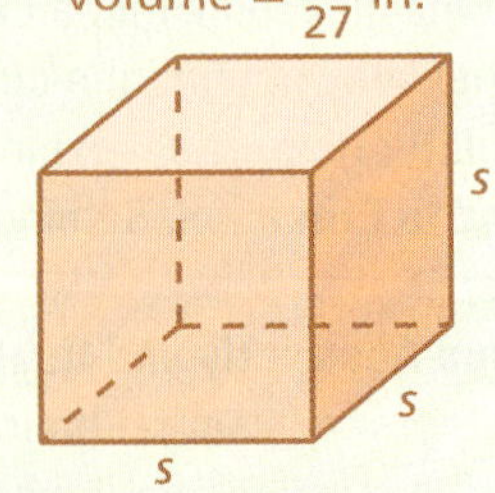

2 ACTIVITY: Using Prime Factorizations to Find Cube Roots

Math Practice

View as Components

When writing the prime factorizations in Activity 2, how many times do you expect to see each factor? Why?

Work with a partner. Write the prime factorization of each number. Then use the prime factorization to find the cube root of the number.

a. 216

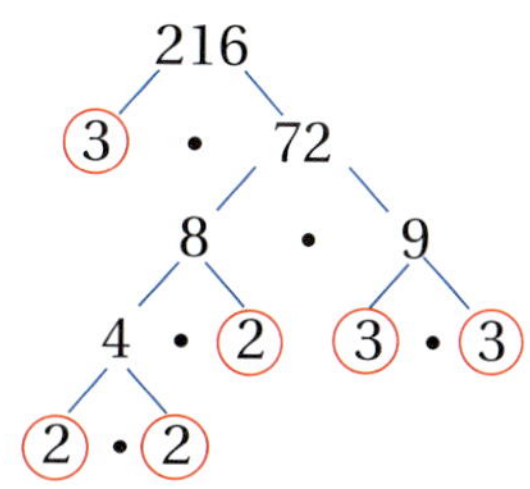

$216 = 3 \cdot 2 \cdot 3 \cdot 3 \cdot 2 \cdot 2$ Prime factorization

$= (3 \cdot \_\_\_) \cdot (3 \cdot \_\_\_) \cdot (3 \cdot \_\_\_)$ Commutative Property of Multiplication

$= \_\_\_ \cdot \_\_\_ \cdot \_\_\_$ Simplify.

The cube root of 216 is ___.

b. 1000

c. 3375

d. **STRUCTURE** Does this procedure work for every number? Explain why or why not.

What Is Your Answer?

3. Complete each statement using *positive* or *negative*.

   a. A positive number times a positive number is a ________ number.

   b. A negative number times a negative number is a ________ number.

   c. A positive number multiplied by itself twice is a ________ number.

   d. A negative number multiplied by itself twice is a ________ number.

4. **REASONING** Can a negative number have a cube root? Give an example to support your explanation.

5. **IN YOUR OWN WORDS** How is the cube root of a number different from the square root of a number?

6. Give an example of a number whose square root and cube root are equal.

7. A cube has a volume of 13,824 cubic meters. Use a calculator to find the edge length.

Practice Use what you learned about cube roots to complete Exercises 3–5 on page 636.

14.2 Lesson

Check It Out
Lesson Tutorials
BigIdeasMath.com

Key Vocabulary
cube root, *p. 634*
perfect cube, *p. 634*

A **cube root** of a number is a number that, when multiplied by itself, and then multiplied by itself again, equals the given number. A **perfect cube** is a number that can be written as the cube of an integer. The symbol $\sqrt[3]{}$ is used to represent a cube root.

EXAMPLE 1 Finding Cube Roots

Find each cube root.

a. $\sqrt[3]{8}$

Because $2^3 = 8$, $\sqrt[3]{8} = \sqrt[3]{2^3} = 2$.

b. $\sqrt[3]{-27}$

Because $(-3)^3 = -27$, $\sqrt[3]{-27} = \sqrt[3]{(-3)^3} = -3$.

c. $\sqrt[3]{\dfrac{1}{64}}$

Because $\left(\dfrac{1}{4}\right)^3 = \dfrac{1}{64}$, $\sqrt[3]{\dfrac{1}{64}} = \sqrt[3]{\left(\dfrac{1}{4}\right)^3} = \dfrac{1}{4}$.

Cubing a number and finding a cube root are inverse operations. You can use this relationship to evaluate expressions and solve equations involving cubes.

EXAMPLE 2 Evaluating Expressions Involving Cube Roots

Evaluate each expression.

a. $2\sqrt[3]{-216} - 3 = 2(-6) - 3$ — Evaluate the cube root.

$= -12 - 3$ — Multiply.

$= -15$ — Subtract.

b. $\left(\sqrt[3]{125}\right)^3 + 21 = 125 + 21$ — Evaluate the power using inverse operations.

$= 146$ — Add.

On Your Own

Now You're Ready
Exercises 6–17

Find the cube root.

1. $\sqrt[3]{1}$
2. $\sqrt[3]{-343}$
3. $\sqrt[3]{-\dfrac{27}{1000}}$

Evaluate the expression.

4. $18 - 4\sqrt[3]{8}$
5. $\left(\sqrt[3]{-64}\right)^3 + 43$
6. $5\sqrt[3]{512} - 19$

Laurie's Notes

Introduction

Connect

- **Yesterday:** Students explored cube roots.
- **Today:** Students will find cube roots and evaluate expressions involving cube roots.

Motivate

- Write the following sequences on the board:
  a. 1, 4, 9, 16, 25, . . . **b.** 1, 8, 27, 64, 125, . . .

? "Describe each sequence in words. What number comes next in each sequence?" Part (a) is a sequence of perfect squares and 36 is next; Part (b) is a sequence of perfect cubes and 216 is next.

- Explain to students that today they are going to be looking at perfect cubes and their cube roots.

Lesson Notes

Example 1

- Define the vocabulary: cube root and perfect cube.
- **FYI:** Every real number has three cube roots, one real and two imaginary. Students only need to be concerned with the real cube root in this lesson.

? Write part (a) and ask, "What number times itself twice equals 8?" 2

? Write part (b) and ask, "What number times itself twice equals -27?" -3

? Write part (c) and ask, "What number times itself twice equals $\frac{1}{64}$?" $\frac{1}{4}$

Example 2

- The cube root of a perfect cube is an integer and can be evaluated as part of an expression.

? Write part (a) and ask, "What is the cube root of -216?" -6

- Replace $\sqrt[3]{-216}$ with -6 and continue to evaluate the expression.
- Emphasize that cubing a number and finding a cube root are inverse operations, as stated before the example.
- Work through part (b), which helps students make the connection between cubing and taking the cube root.

? **Reason Abstractly and Quantitatively** and **Construct Viable Arguments:** "In general, do you think $\left(\sqrt[3]{n}\right)^3 = n$? Explain." Listen for an understanding of how taking the cube root of a number and then cubing the result gives the original number.

On Your Own

- Students are sometimes intimidated by the notation. Assure them that a calculator is not needed. The radicands are from the list of perfect cubes: 1, 8, 27, 64 125, ...
- Ask volunteers to share their work for Questions 4–6 at the board.

Goal Today's lesson is finding **cube roots**.

Technology for the Teacher

Dynamic Classroom

Lesson Tutorials
Lesson Plans
Answer Presentation Tool

Extra Example 1

Find each cube root.

a. $\sqrt[3]{512}$ 8

b. $\sqrt[3]{-729}$ -9

c. $\sqrt[3]{-\frac{125}{343}}$ $-\frac{5}{7}$

Extra Example 2

Evaluate each expression.

a. $3\sqrt[3]{125} - 8$ 7

b. $\left(\sqrt[3]{27}\right)^3 - 4$ 23

On Your Own

1. 1 **2.** -7

3. $-\frac{3}{10}$ **4.** 10

5. -21 **6.** 21

Extra Example 3

Evaluate $\frac{w}{30} - \sqrt[3]{\frac{w}{5}}$ when $w = 1080$. 30

On Your Own

7. 72 **8.** -3

Extra Example 4

Find the surface area of a cube with a volume of 27 cubic feet. 54 ft^2

On Your Own

9. 384 cm^2

Differentiated Instruction

Visual

Students may have a difficult time understanding that the square root or a cube root of a number between 0 and 1 is greater than the number. They may think that $\sqrt{\frac{1}{4}} = \frac{1}{16}$, when actually $\sqrt{\frac{1}{16}} = \frac{1}{4}$. Help students to understand this concept by using a 10-by-10 grid to represent the number 1. A square of $\frac{49}{100}$ square units has a side length of $\frac{7}{10}$. Because $\sqrt{\frac{49}{100}} = \frac{7}{10}$, $\frac{7}{10} > \frac{49}{100}$ (0.7 > 0.49).

Laurie's Notes

Example 3

- Write the problem.
- ? "Is 192 a perfect cube?" no "Is $\frac{192}{3}$ a perfect cube?" Yes because it equals 64, which is a perfect cube.
- Substitute 192 for *x* and evaluate the expression as shown.
- ? "In general, do you think any value of *x* could be used in this problem? Explain." The correct answer is yes, but students may answer that you can only use values of *x* in which $\frac{1}{3}$ of *x* is a perfect cube. They are unaware that a cube root can be an *irrational number*, which is okay at this stage. They will learn about irrational numbers in Section 7.4.

On Your Own

- **Think-Pair-Share:** Students should read each question independently and then work in pairs to answer the questions. When they have answered the questions, the pair should compare their answers with another group and discuss any discrepancies.

Example 4

- **Teaching Tip:** Some tissue boxes are close to the shape of a cube. Use a tissue box or an actual cube as a prop. Pose the question: If you know the volume of the cube, can you find the surface area?
- ? "How do you find the volume of a cube?" Cube the edge length; $V = s^3$
- ? "How do you find the surface area of a cube?" Find the area of one face and multiply by 6; $S = 6s^2$
- The first step is to find the edge length of the cube. Students should be thinking, "What number multiplied by itself twice equals 125?"
- Continue to work through the problem as shown.

On Your Own

- **Neighbor Check:** Have students work independently and then have their neighbors check their work. Have students discuss any discrepancies.

Closure

- Explain the difference between $\sqrt{64}$ and $\sqrt[3]{64}$. $\sqrt{64}$ is a number that when multiplied by itself is equal to 64, and $\sqrt[3]{64}$ is a number that when multiplied by itself twice is equal to 64.

EXAMPLE 3 Evaluating an Algebraic Expression

Evaluate $\frac{x}{4} + \sqrt[3]{\frac{x}{3}}$ when $x = 192$.

$$\frac{x}{4} + \sqrt[3]{\frac{x}{3}} = \frac{192}{4} + \sqrt[3]{\frac{192}{3}}$$ Substitute 192 for *x*.

$$= 48 + \sqrt[3]{64}$$ Simplify.

$$= 48 + 4$$ Evaluate the cube root.

$$= 52$$ Add.

On Your Own

Now You're Ready
Exercises 18–20

Evaluate the expression for the given value of the variable.

7. $\sqrt[3]{8y} + y,\ y = 64$

8. $2b - \sqrt[3]{9b},\ b = -3$

EXAMPLE 4 Real-Life Application

Find the surface area of the baseball display case.

The baseball display case is in the shape of a cube. Use the formula for the volume of a cube to find the edge length *s*.

Volume = 125 in.^{3}

$$V = s^3$$ Write formula for volume.

$$125 = s^3$$ Substitute 125 for *V*.

$$\sqrt[3]{125} = \sqrt[3]{s^3}$$ Take the cube root of each side.

$$5 = s$$ Simplify.

Remember

The volume V of a cube with edge length s is given by $V = s^3$. The surface area S is given by $S - 6s^2$.

The edge length is 5 inches. Use a formula to find the surface area of the cube.

$$S = 6s^2$$ Write formula for surface area.

$$= 6(5)^2$$ Substitute 5 for *s*.

$$= 150$$ Simplify.

So, the surface area of the baseball display case is 150 square inches.

On Your Own

9. The volume of a music box that is shaped like a cube is 512 cubic centimeters. Find the surface area of the music box.

14.2 Exercises

Check It Out
Help with Homework
BigIdeasMath.com

Vocabulary and Concept Check

1. **VOCABULARY** Is 25 a perfect cube? Explain.
2. **REASONING** Can the cube of an integer be a negative number? Explain.

Practice and Problem Solving

Find the edge length of the cube.

3. Volume = 125,000 in.^{3}

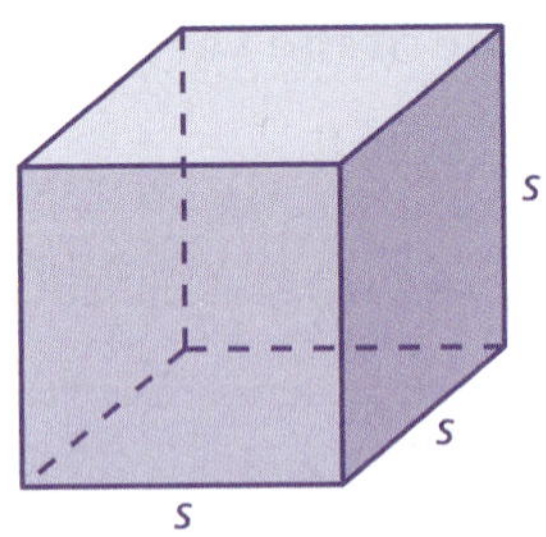

4. Volume = $\frac{1}{27}$ ft^3

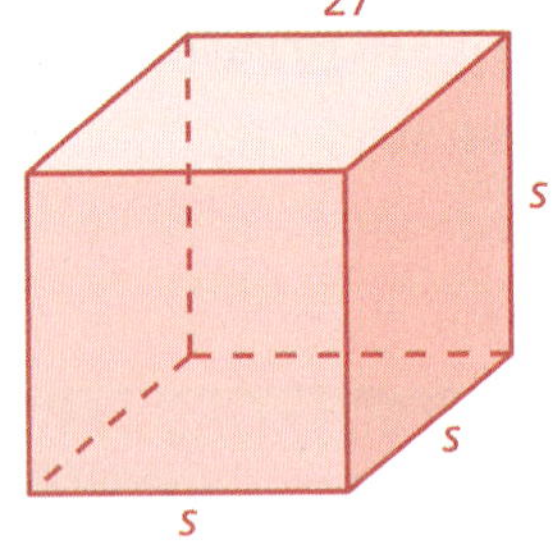

5. Volume = 0.064 m^3

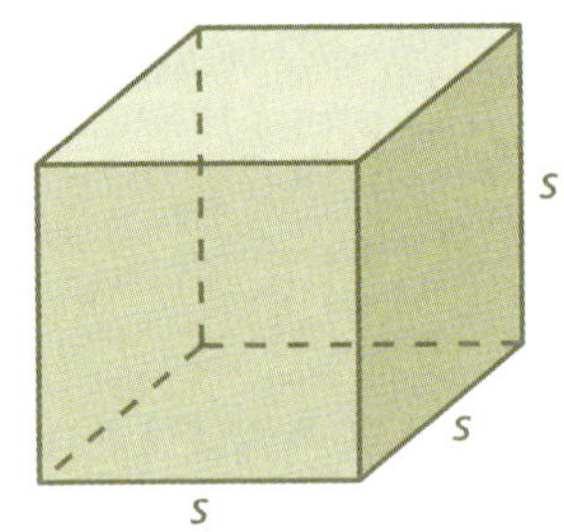

Find the cube root.

1 6. $\sqrt[3]{729}$

7. $\sqrt[3]{-125}$

8. $\sqrt[3]{-1000}$

9. $\sqrt[3]{1728}$

10. $\sqrt[3]{-\frac{1}{512}}$

11. $\sqrt[3]{\frac{343}{64}}$

Evaluate the expression.

2 12. $18 - \left(\sqrt[3]{27}\right)^3$

13. $\left(\sqrt[3]{-\frac{1}{8}}\right)^3 + 3\frac{3}{4}$

14. $5\sqrt[3]{729} - 24$

15. $\frac{1}{4} - 2\sqrt[3]{-\frac{1}{216}}$

16. $54 + \sqrt[3]{-4096}$

17. $4\sqrt[3]{8000} - 6$

Evaluate the expression for the given value of the variable.

3 18. $\sqrt[3]{\frac{n}{4}} + \frac{n}{10}$, $n = 500$

19. $\sqrt[3]{6w} - w$, $w = 288$

20. $2d + \sqrt[3]{-45d}$, $d = 75$

21. **STORAGE CUBE** The volume of a plastic storage cube is 27,000 cubic centimeters. What is the edge length of the storage cube?

22. **ICE SCULPTURE** The volume of a cube of ice for an ice sculpture is 64,000 cubic inches.
    a. What is the edge length of the cube of ice?
    b. What is the surface area of the cube of ice?

Assignment Guide and Homework Check

Level	Assignment	Homework Check
Accelerated	1–5, 10–24 even, 26–38	10, 16, 24, 26, 28

Common Errors

- **Exercises 6–11** Students may disregard negatives in the radicand. Remind them to pay attention to the sign of the radicand and to check the sign of their answers.
- **Exercises 12–17** Students may not follow the order of operations when evaluating the expression. Remind them of the order of operations. Because taking a cube root is the inverse of cubing, it is evaluated before multiplication and division.

14.2 Record and Practice Journal

Find the cube root.

1. $\sqrt[3]{27}$
   3
2. $\sqrt[3]{8}$
   2
3. $\sqrt[3]{-64}$
   -4
4. $\sqrt[3]{-\frac{125}{216}}$
   $-\frac{5}{6}$

Evaluate the expression.

5. $10 - \left(\sqrt[3]{12}\right)^3$
   -2
6. $2\sqrt[3]{512} + 10$
   26
7. The volume of a cube is 1000 cubic inches. What is the edge length of the cube?
   10 in.

Vocabulary and Concept Check

1. no; There is no integer that equals 25 when cubed.
2. yes; When the integer is negative, the cube of the negative integer is negative. For example, $(-4)^3 = (-4) \cdot (-4) \cdot (-4) = -64$.

Practice and Problem Solving

3. 50 in.
4. $\frac{1}{3}$ ft
5. 0.4 m
6. 9
7. -5
8. -10
9. 12
10. $-\frac{1}{8}$
11. $\frac{7}{4}$
12. -9
13. $3\frac{5}{8}$
14. 21
15. $\frac{7}{12}$
16. 38
17. 74
18. 55
19. -276
20. 135
21. 30 cm
22. a. 40 in.
    b. 9600 $in.^2$

Practice and Problem Solving

23. >

24. >

25. <

26. 183 miles per hour

27. $-1, 0, 1$

28. **a.** not true; *Sample answer:* $\sqrt[3]{-8} = -2$

b. not true; *Sample answer:* 64 has only a positive cube root.

29. The side length of the square base is 18 inches and the height of the pyramid is 9 inches.

30. See *Taking Math Deeper.*

31. $x = 3$

32. $x = \frac{3}{2}$

33. $x = 4$

Fair Game Review

34. 25 **35.** 289

36. 144 **37.** 49

38. C

Mini-Assessment

1. Find $\sqrt[3]{-512}$. -8

Evaluate the expression.

2. $3 + 4\sqrt[3]{27}$ 15
3. $\sqrt[3]{-216} - 16$ -22
4. $\left(\sqrt[3]{-8}\right)^3 + 15$ 7
5. Evaluate $\left(\sqrt[3]{5b}\right) - \frac{3b}{5}$ when $b = 25$. -10

Taking Math Deeper

Exercise 30

One way to solve this problem is to write and solve a proportion. You can use what you know about factors to avoid calculations with large numbers.

1 Write a proportion.

$$\frac{125}{x} = \frac{x^2}{125}$$

Using the Cross Products Property, you obtain $x^3 = 15{,}625$. Rather than work with large numbers, rewrite the quantities in the original proportion as products of factors before solving.

$$\frac{125}{x} = \frac{x^2}{125} \Rightarrow \frac{5 \cdot 5 \cdot 5}{x} = \frac{x \cdot x}{5 \cdot 5 \cdot 5}$$

2 Solve the proportion. Multiply each side by the least common denominator, $5 \cdot 5 \cdot 5 \cdot x$.

$$(5 \cdot 5 \cdot 5 \cdot \cancel{x}) \cdot \frac{5 \cdot 5 \cdot 5}{\cancel{x}} = (\cancel{5} \cdot \cancel{5} \cdot \cancel{5} \cdot x) \cdot \frac{x \cdot x}{\cancel{5} \cdot \cancel{5} \cdot \cancel{5}}$$

$$5 \cdot 5 \cdot 5 \cdot 5 \cdot 5 \cdot 5 = x \cdot x \cdot x$$

$$(5 \cdot 5) \cdot (5 \cdot 5) \cdot (5 \cdot 5) = x \cdot x \cdot x$$

$$25 \cdot 25 \cdot 25 = x \cdot x \cdot x$$

$$25^3 = x^3$$

$$\sqrt[3]{25^3} = \sqrt[3]{x^3}$$

$$25 = x$$

So, x is 25.

3 Check the solution.

$$\frac{125}{x} = \frac{x^2}{125}$$

$$\frac{125}{25} \stackrel{?}{=} \frac{25^2}{125}$$

$$5 \stackrel{?}{=} \frac{625}{125}$$

$$5 = 5 \checkmark$$

Reteaching and Enrichment Strategies

If students need help. . .	If students got it. . .
Resources by Chapter • Practice A and Practice B • Puzzle Time Record and Practice Journal Practice Differentiating the Lesson Lesson Tutorials Skills Review Handbook	Resources by Chapter • Enrichment and Extension • Technology Connection Start the next section

Copy and complete the statement with <, >, or =.

23. $-\frac{1}{4}$ ☐ $\sqrt[3]{-\frac{8}{125}}$

24. $\sqrt[3]{0.001}$ ☐ 0.01

25. $\sqrt[3]{64}$ ☐ $\sqrt{64}$

26. DRAG RACE The estimated velocity v (in miles per hour) of a car at the end of a drag race is $v = 234\sqrt[3]{\frac{p}{w}}$, where p is the horsepower of the car and w is the weight (in pounds) of the car. A car has a horsepower of 1311 and weighs 2744 pounds. Find the velocity of the car at the end of a drag race. Round your answer to the nearest whole number.

27. NUMBER SENSE There are three numbers that are their own cube roots. What are the numbers?

28. LOGIC Each statement below is true for square roots. Determine whether the statement is also true for cube roots. Explain your reasoning and give an example to support your explanation.

a. You cannot find the square root of a negative number.

b. Every positive number has a positive square root and a negative square root.

29. GEOMETRY The pyramid has a volume of 972 cubic inches. What are the dimensions of the pyramid?

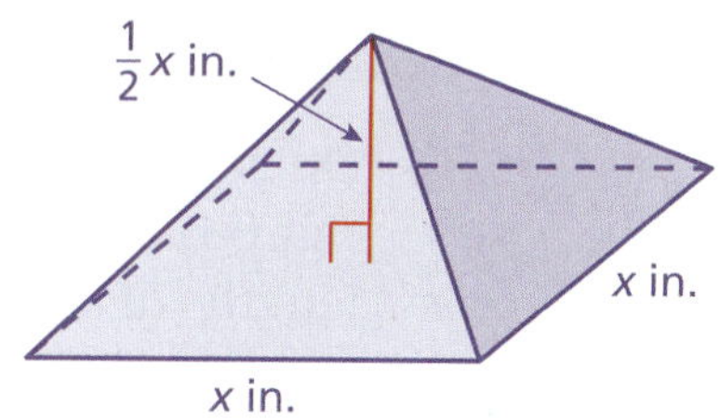

30. RATIOS The ratio $125 : x$ is equivalent to the ratio $x^2 : 125$. What is the value of x?

Critical Thinking **Solve the equation.**

31. $(3x + 4)^3 = 2197$

32. $(8x^3 - 9)^3 = 5832$

33. $\left((5x - 16)^3 - 4\right)^3 = 216{,}000$

Fair Game Review What you learned in previous grades & lessons

Evaluate the expression. *(Skills Review Handbook)*

34. $3^2 + 4^2$

35. $8^2 + 15^2$

36. $13^2 - 5^2$

37. $25^2 - 24^2$

38. MULTIPLE CHOICE Which linear equation is shown by the table? *(Section 13.6)*

x	0	1	2	3
y	1	4	7	10

Ⓐ $y = \frac{1}{3}x + 1$ Ⓑ $y = 4x$ Ⓒ $y = 3x + 1$ Ⓓ $y = \frac{1}{4}x$

14.3 The Pythagorean Theorem

Essential Question How are the lengths of the sides of a right triangle related?

Pythagoras was a Greek mathematician and philosopher who discovered one of the most famous rules in mathematics. In mathematics, a rule is called a **theorem**. So, the rule that Pythagoras discovered is called the Pythagorean Theorem.

Pythagoras
(c. 570–c. 490 B.C.)

1 ACTIVITY: Discovering the Pythagorean Theorem

Work with a partner.

a. On grid paper, draw any right triangle. Label the lengths of the two shorter sides a and b.

b. Label the length of the longest side c.

c. Draw squares along each of the three sides. Label the areas of the three squares a^2, b^2, and c^2.

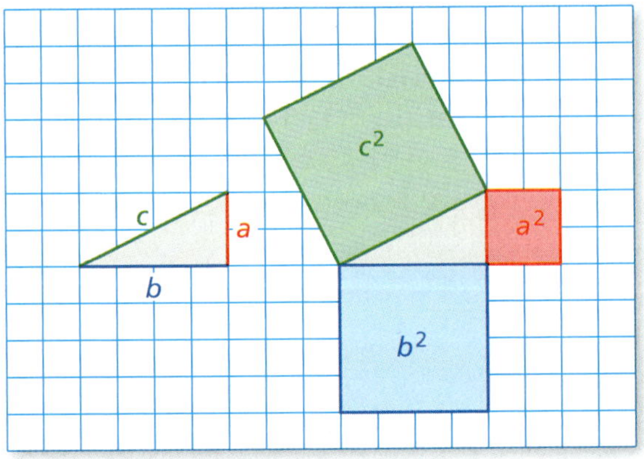

d. Cut out the three squares. Make eight copies of the right triangle and cut them out. Arrange the figures to form two identical larger squares.

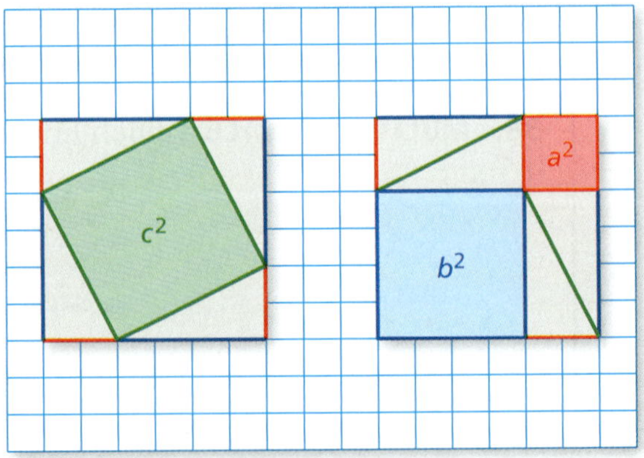

e. MODELING The Pythagorean Theorem describes the relationship among a^2, b^2, and c^2. Use your result from part (d) to write an equation that describes this relationship.

Pythagorean Theorem
In this lesson, you will
- provide geometric proof of the Pythagorean Theorem.
- use the Pythagorean Theorem to find missing side lengths of right triangles.
- solve real-life problems.

Laurie's Notes

Introduction

Applying Mathematical Practices

- **Model with Mathematics:** In Activity 1, students will analyze the relationship between two models they create. This represents an "informal proof" of the Pythagorean Theorem.

Motivate

- Share information about Pythagoras, who was born in Greece in 569 B.C.
  - He is known as the *Father of Numbers*.
  - He traveled extensively in Egypt, learning math, astronomy, and music.
  - Pythagoras urged the citizens of Cretona to follow his religious, political, and philosophical goals.
  - His followers were known as Pythagoreans. They observed a rule of silence called *echemythia.* One had to remain silent for *five years* before he could contribute to the group. Breaking this silence was punishable by death!

Activity Notes

Activity 1

- **Suggestions:** Use centimeter grid paper for ease of manipulating the cut pieces. Suggest to students that they draw the original triangle in the upper left of the grid paper, and then make a working copy of the triangle towards the middle of the paper. This gives enough room for the squares to be drawn on each side of the triangle.
- Vertices of the triangle need to be on lattice points. You do not want every student in the room to use the same triangle. Suggest other lengths for the shorter sides (3 and 4, 3 and 6, 2 and 4, 2 and 3, and so on).
- **Model:** Drawing the square on the longest side of the triangle is the challenging step. Model one technique for accomplishing the task using a right triangle with shorter side lengths of 2 units and 5 units.
  - Notice that the longest side has a slope of "right 5 units, up 2 units."
  - Place your pencil on the upper right endpoint and rotate the paper 90° clockwise. Move your pencil right 5 units and up 2 units. Mark a point.
  - Repeat rotating and moving "right 5 units, up 2 units" until you get back to the longest side of the triangle.
  - Use a straightedge to connect the four points (two that you marked and two on the endpoints of the longest side) to form the square.
- Before students cut anything, check that they have 3 squares of the correct size.
- **Big Idea:** The two large squares in part (d) have equal area. Referring to areas, if $c^2 + (4 \text{ triangles}) = a^2 + b^2 + (4 \text{ triangles})$, then $c^2 = a^2 + b^2$ by subtracting the 4 triangles from each side of the equation.
- The work in this activity constitutes an "informal proof" of the Pythagorean Theorem. There are many proofs of this theorem, and this version is generally understood by middle school students.

What Your Students Will Learn

- Provide geometric proof of the Pythagorean Theorem using grid paper.
- Use the Pythagorean Theorem in two and three dimensions to find the missing side lengths or the hypotenuse of right triangles.

Previous Learning

Students should know how to evaluate algebraic expressions for given values of the variables.

Technology for the Teacher

Dynamic Classroom

Lesson Plans
Complete Materials List

14.3 Record and Practice Journal

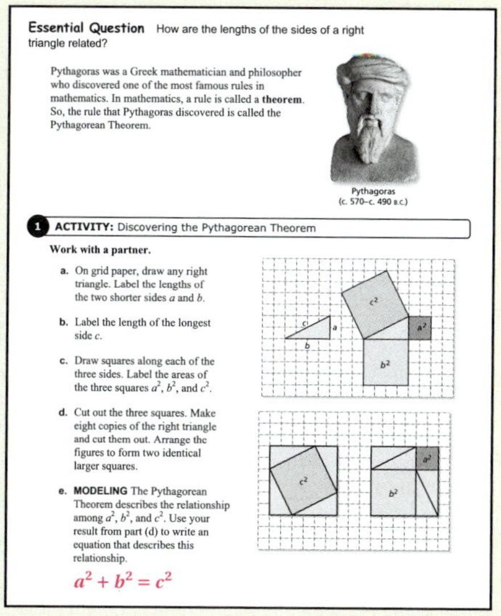

Essential Question How are the lengths of the sides of a right triangle related?

Pythagoras was a Greek mathematician and philosopher who discovered one of the most famous rules in mathematics. In mathematics, a rule is called a **theorem**. So, the rule that Pythagoras discovered is called the Pythagorean Theorem.

Pythagoras (c. 570–c. 490 B.C.)

1 ACTIVITY: Discovering the Pythagorean Theorem

Work with a partner.

a. On grid paper, draw any right triangle. Label the lengths of the two shorter sides a and b.

b. Label the length of the longest side c.

c. Draw squares along each of the three sides. Label the areas of the three squares a^2, b^2, and c^2.

d. Cut out the three squares. Make eight copies of the right triangle and cut them out. Arrange the figures to form two identical larger squares.

e. **MODELING** The Pythagorean Theorem describes the relationship among a^2, b^2, and c^2. Use your result from part (d) to write an equation that describes this relationship.

$a^2 + b^2 = c^2$

English Language Learners

Vocabulary

Help English learners understand the meanings of the words that make up a definition. Provide students with statements containing blanks and a list of the words used to fill in the blanks.

- In any right __, the __ is the side __ the right __.
  Word list: angle, hypotenuse, opposite, triangle
  triangle, hypotenuse, opposite, angle
- In any right __, the __ are the __ sides and the __ is always the __ side.
  Word list: hypotenuse, legs, longest, shorter, triangle
  triangle, legs, shorter, hypotenuse, longest

14.3 Record and Practice Journal

2 ACTIVITY: Using the Pythagorean Theorem in Two Dimensions

Work with a partner. Use a ruler to measure the longest side of each right triangle. Verify the result of Activity 1 for each right triangle.

a. 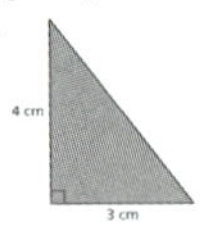

5 cm

b. 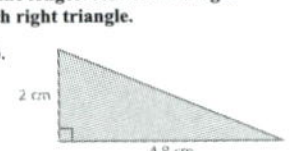

5.2 cm

c. 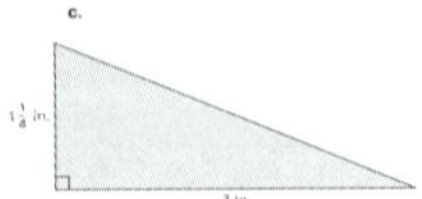

$3\frac{1}{4}$ in.

d. 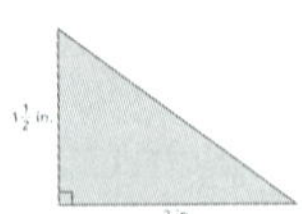

$2\frac{1}{2}$ in.

3 ACTIVITY: Using the Pythagorean Theorem in Three Dimensions

Work with a partner. A guy wire attached 24 feet above ground level on a telephone pole provides support for the pole.

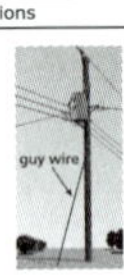

a. **PROBLEM SOLVING** Describe a procedure that you could use to find the length of the guy wire without directly measuring the wire.

Check students' work.

b. Find the length of the wire when it meets the ground 10 feet from the base of the pole.

26 ft

What Is Your Answer?

4. **IN YOUR OWN WORDS** How are the lengths of the sides of a right triangle related? Give an example using whole numbers.

Sum of the squares of the lengths of the legs is equal to the square of the length of the hypotenuse.

Laurie's Notes

Activity 2

- Begin by asking students to summarize what they learned from Activity 1.
- Distribute rulers with both metric and customary measures.
- **Attend to Precision:** Caution students to pay attention to the units and to measure accurately. You may choose to have them begin by checking the accuracy of the side lengths already given.
- You may want students to use calculators to square some of the side lengths.
- When students have finished, ask them to share their measures for each of the longest sides *and* their calculations. Students should not simply measure. You want to ensure that they have performed the calculations to verify the results of Activity 1.

Activity 3

- Explain what a guy wire is if students are not familiar with the term.
- **Construct Viable Arguments and Critique the Reasoning of Others:** Students are asked to explain a method for finding the length of a guy wire without measuring directly. They may incorrectly assume that they can't measure anything. The length of one of the shorter sides of the right triangle is known (24 feet), and the other shorter side can be measured directly (distance from bottom of guy wire to the base of the pole).
- You may want to discuss the results of part (a), and then have students work through part (b) with a partner.

What Is Your Answer?

- **Neighbor Check:** Have students work independently and then have their neighbors check their work. Have students discuss any discrepancies.

Closure

- **Exit Ticket:** If you drew a right triangle with shorter side lengths of 4 and 6 on grid paper, what would be the area of the square drawn on the longest side of the triangle? 52 square units

2 ACTIVITY: Using the Pythagorean Theorem in Two Dimensions

Work with a partner. Use a ruler to measure the longest side of each right triangle. Verify the result of Activity 1 for each right triangle.

a.

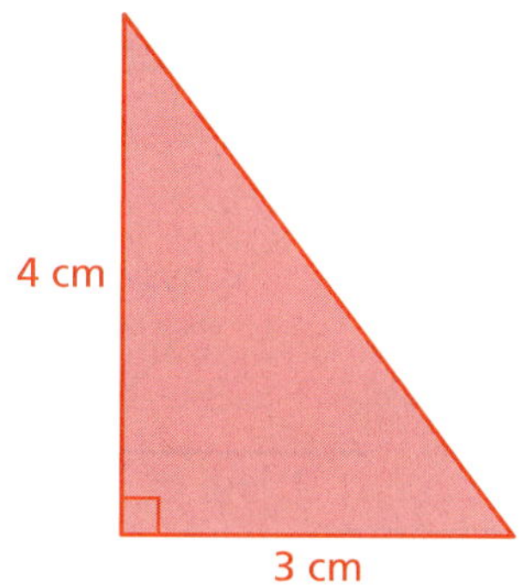

b.

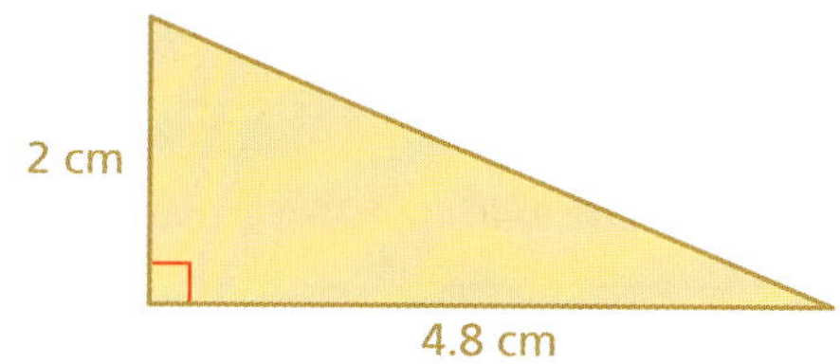

c.

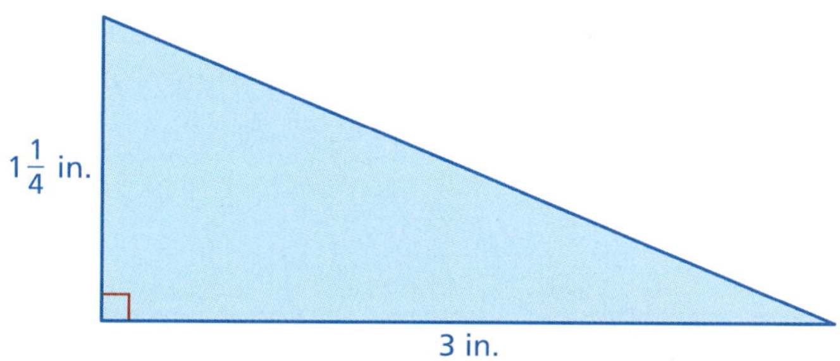

d.

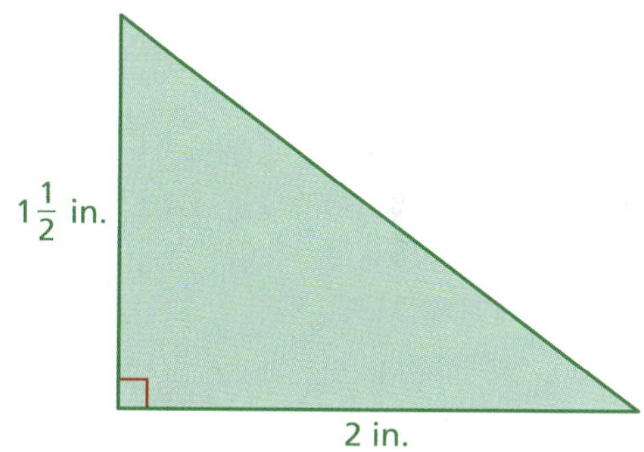

3 ACTIVITY: Using the Pythagorean Theorem in Three Dimensions

Math Practice

Use Definitions

How can you use what you know about the Pythagorean Theorem to describe the procedure for finding the length of the guy wire?

Work with a partner. A guy wire attached 24 feet above ground level on a telephone pole provides support for thc pole.

a. **PROBLEM SOLVING** Describe a procedure that you could use to find the length of the guy wire without directly measuring the wire.

b. Find the length of the wire when it meets the ground 10 feet from the base of the pole.

What Is Your Answer?

4. **IN YOUR OWN WORDS** How are the lengths of the sides of a right triangle related? Give an example using whole numbers.

Practice Use what you learned about the Pythagorean Theorem to complete Exercises 3 and 4 on page 642.

14.3 Lesson

Check It Out
Lesson Tutorials
BigIdeasMath.com

Key Vocabulary
theorem, *p. 638*
legs, *p. 640*
hypotenuse, *p. 640*
Pythagorean Theorem, *p. 640*

Key Ideas

Sides of a Right Triangle

The sides of a right triangle have special names.

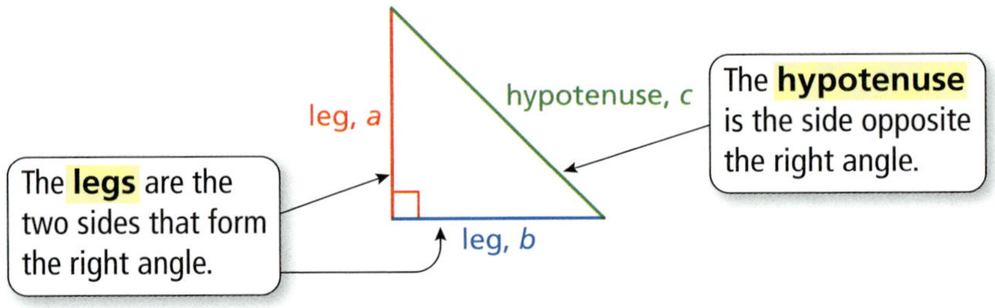

The Pythagorean Theorem

Words In any right triangle, the sum of the squares of the lengths of the legs is equal to the square of the length of the hypotenuse.

Algebra $a^2 + b^2 = c^2$

Study Tip
In a right triangle, the legs are the shorter sides and the hypotenuse is always the longest side.

EXAMPLE 1 Finding the Length of a Hypotenuse

Find the length of the hypotenuse of the triangle.

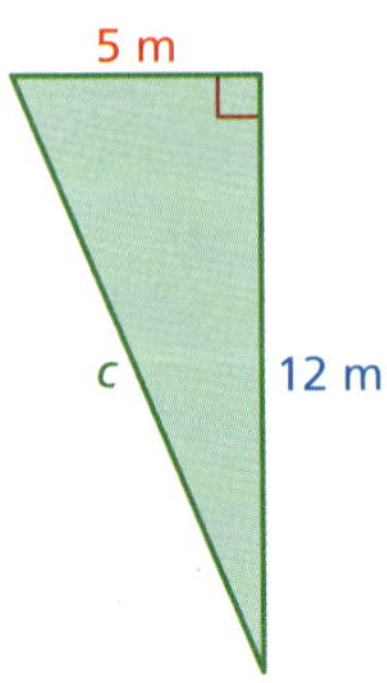

$a^2 + b^2 = c^2$	Write the Pythagorean Theorem.
$5^2 + 12^2 = c^2$	Substitute 5 for a and 12 for b.
$25 + 144 = c^2$	Evaluate powers.
$169 = c^2$	Add.
$\sqrt{169} = \sqrt{c^2}$	Take positive square root of each side.
$13 = c$	Simplify.

The length of the hypotenuse is 13 meters.

On Your Own

Now You're Ready
Exercises 3 and 4

Find the length of the hypotenuse of the triangle.

1.

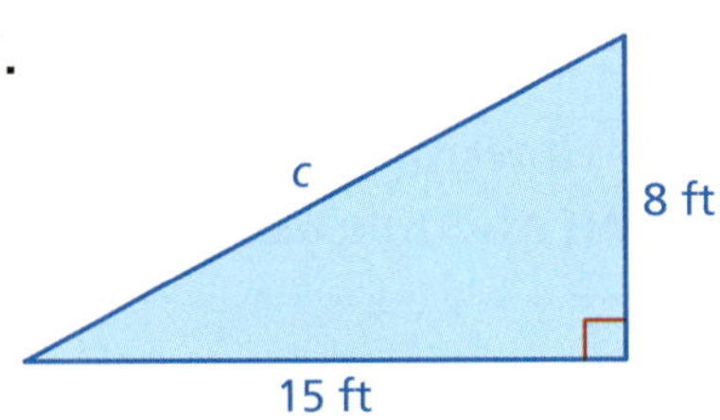

2.

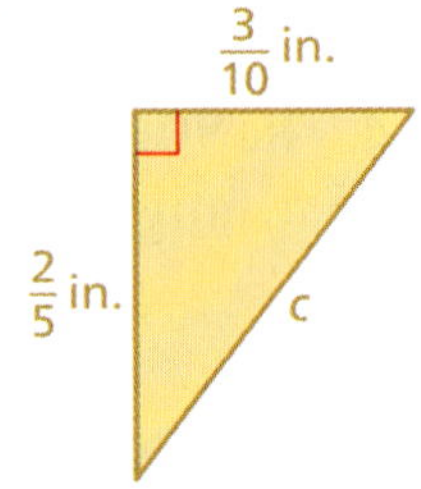

Multi-Language Glossary at BigIdeasMath.com

Laurie's Notes

Introduction

Connect

- **Yesterday:** Students investigated a visual proof of the Pythagorean Theorem.
- **Today:** Students will use the Pythagorean Theorem to find missing side lengths of right triangles.

Motivate

- **Preparation:** Cut coffee stirrers (or carefully break spaghetti) so that triangles with the following side lengths can be made: 2-3-4; 3-4-5; 4-5-6.

? "What are consecutive numbers?" numbers in sequential order

- With student aid, use the coffee stirrers to make three triangles: 2-3-4; 3-4-5; and 4-5-6 on a document camera or overhead projector. If arranged carefully, all 3 will fit on the screen.
- Ask students to make observations about the 3 triangles. Students may mention that all triangles are scalene; one triangle appears to be acute, one right, one obtuse.
- They should observe that the change in the side lengths seems to have made a big change in the angle measures.

Lesson Notes

Key Ideas

- Draw a right triangle and label the *legs* and the *hypotenuse*. The **hypotenuse** is always opposite the right angle and is the longest side of a right triangle.
- Try not to have all right triangles in the same orientation.
- Write the Pythagorean Theorem.
- **Common Error:** Students often forget that the Pythagorean Theorem is a relationship that is *only* true for right triangles.

Example 1

- Draw and label the triangle. Review the symbol used to show that an angle is a right angle.

? "What information is known for this triangle?" The legs are 5 m and 12 m.

- Substitute and solve as shown. Explain that you disregard the negative square root because the length is positive.

On Your Own

- Give time for students to work the problems. Knowing their perfect squares is helpful.
- **Reason Abstractly and Quantitatively:** In Question 2, if students recognize that the decimal equivalents of the given fractions are 0.3 and 0.4, finding the hypotenuse may be quick for them.

Goal Today's lesson is using the **Pythagorean Theorem** to solve for side lengths of right triangles.

Technology for the Teacher

Dynamic Classroom

Lesson Tutorials
Lesson Plans
Answer Presentation Tool

Extra Example 1

Find the length of the hypotenuse of the triangle. 5 in.

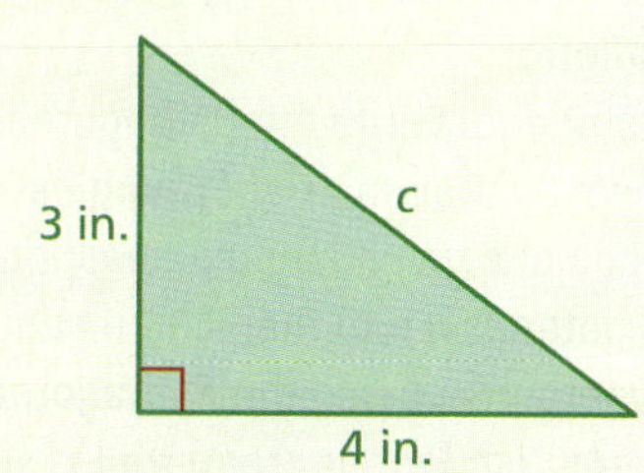

On Your Own

1. 17 ft
2. $\frac{1}{2}$ in.

Extra Example 2

Find the missing length of the triangle.
24 ft

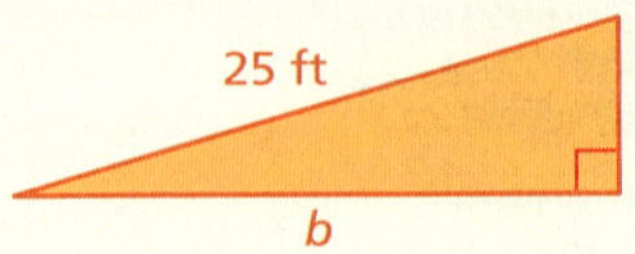

Extra Example 3

You and your cousin are planning to go to an amusement park. You live 36 miles south of the amusement park and 15 miles west of your cousin. How far away from the amusement park does your cousin live? 39 miles

On Your Own

3. 30 yd **4.** 4 m

5. 100 yd

Differentiated Instruction

Kinesthetic

Have students verify the Pythagorean Theorem by drawing right triangles with legs of a given length, measuring the hypotenuse, and then calculating the hypotenuse using the Pythagorean Theorem. Use Pythagorean triples so that students work only with whole numbers.

Leg Lengths	*Hypotenuse Length*
3, 4	5
6, 8	10
5, 12	13
8, 15	17

Laurie's Notes

Example 2

? "What information is known for this triangle?" One leg is 2.1 centimeters and the hypotenuse is 2.9 centimeters.

- Substitute and solve as shown.
- **Common Error:** Students need to be careful with decimal multiplication. It is very common for students to multiply the decimal by 2 instead of multiplying the decimal by itself.
- **FYI:** The triangle is similar to a 20-21-29 right triangle.

Example 3

- Ask a student to read the example.

? "Given the compass directions stated, what is a reasonable way to represent this information?" coordinate plane

- **Model with Mathematics:** Explain that east is the positive x-direction and north is the positive y-direction. Draw the situation in a coordinate plane.

? "Is there enough information to use the Pythagorean Theorem? Explain." yes; The legs of the triangle can be found and then used to solve for the hypotenuse.

- **FYI:** This example previews the *distance formula,* which will be presented in Section 14.5.

On Your Own

- **Think-Pair-Share:** Students should read each question independently and then work in pairs to answer the questions. When they have answered the questions, the pair should compare their answers with another group and discuss any discrepancies.

Closure

- **Exit Ticket:** Solve for the missing side length. $x = 15$ cm, $y = 0.5$ m

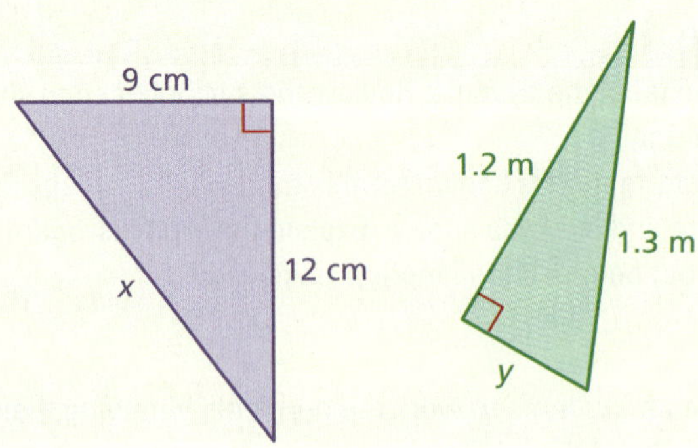

EXAMPLE 2 Finding the Length of a Leg

Find the missing length of the triangle.

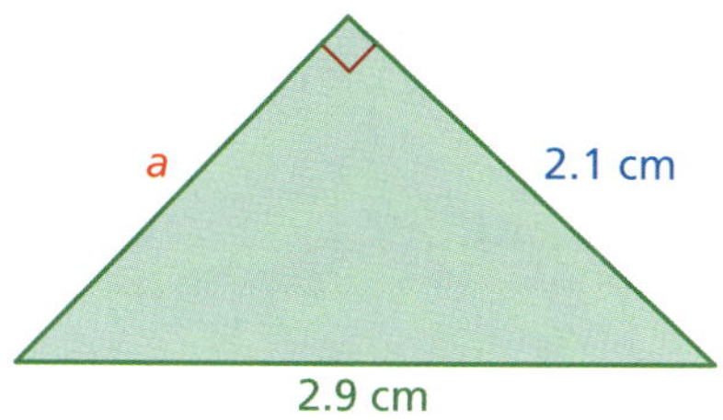

$a^2 + b^2 = c^2$	Write the Pythagorean Theorem.
$a^2 + 2.1^2 = 2.9^2$	Substitute 2.1 for *b* and 2.9 for *c*.
$a^2 + 4.41 = 8.41$	Evaluate powers.
$a^2 = 4$	Subtract 4.41 from each side.
$a = 2$	Take positive square root of each side.

∴ The missing length is 2 centimeters.

EXAMPLE 3 Real-Life Application

You are playing capture the flag. You are 50 yards north and 20 yards east of your team's base. The other team's base is 80 yards north and 60 yards east of your base. How far are you from the other team's base?

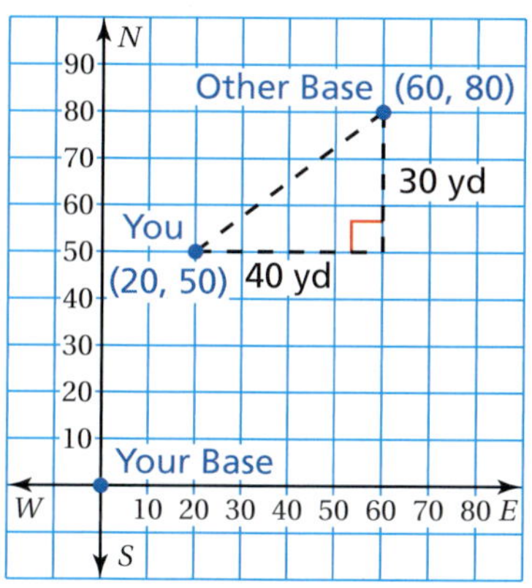

Step 1: Draw the situation in a coordinate plane. Let the origin represent your team's base. From the descriptions, you are at (20, 50) and the other team's base is at (60, 80).

Step 2: Draw a right triangle with a hypotenuse that represents the distance between you and the other team's base. The lengths of the legs are 30 yards and 40 yards.

Step 3: Use the Pythagorean Theorem to find the length of the hypotenuse.

$a^2 + b^2 = c^2$	Write the Pythagorean Theorem.
$30^2 + 40^2 = c^2$	Substitute 30 for *a* and 40 for *b*.
$900 + 1600 = c^2$	Evaluate powers.
$2500 = c^2$	Add.
$50 = c$	Take positive square root of each side.

∴ So, you are 50 yards from the other team's base.

On Your Own

Now You're Ready
Exercises 5–8

Find the missing length of the triangle.

3.

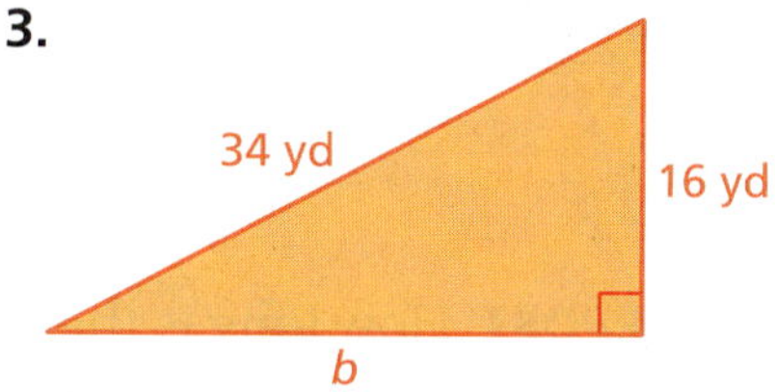

4.

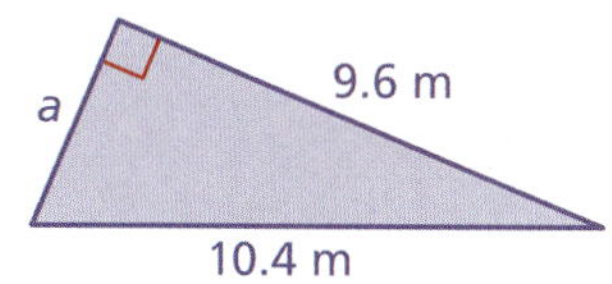

5. In Example 3, what is the distance between the bases?

14.3 Exercises

Check It Out
Help with Homework
BigIdeasMath.com

Vocabulary and Concept Check

1. **VOCABULARY** In a right triangle, how can you tell which sides are the legs and which side is the hypotenuse?

2. **DIFFERENT WORDS, SAME QUESTION** Which is different? Find "both" answers.

Which side is the hypotenuse?

Which side is the longest?

Which side is a leg?

Which side is opposite the right angle?

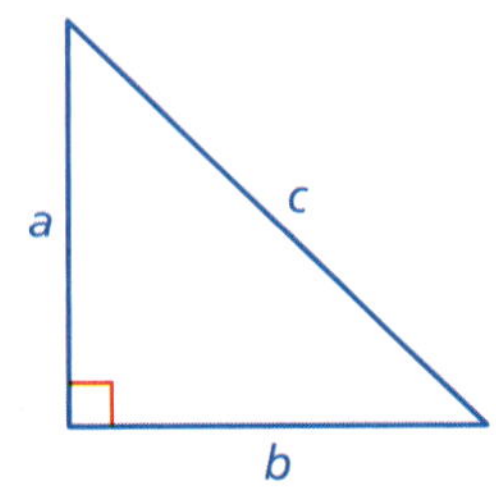

Practice and Problem Solving

Find the missing length of the triangle.

1 2 3.

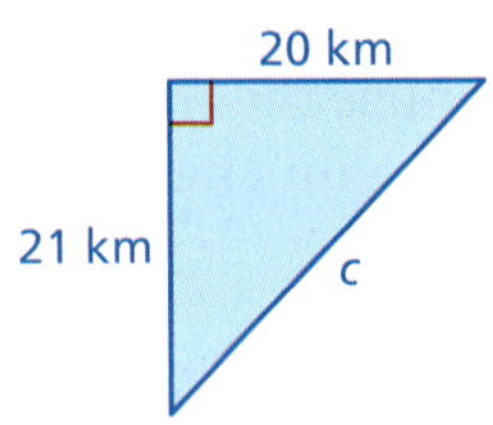

4.

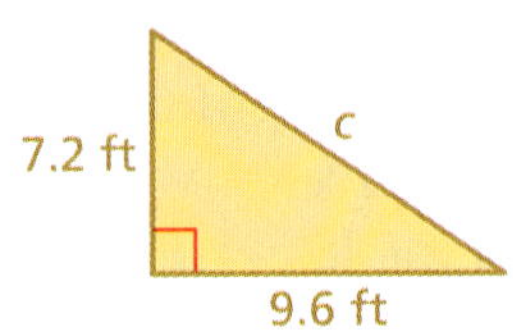

5.

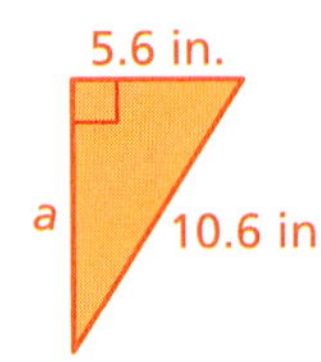

6.

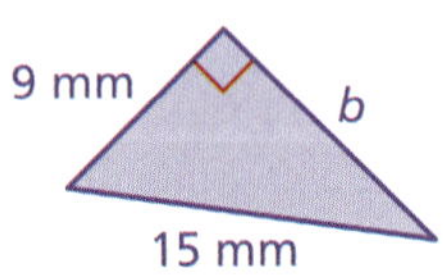

7.

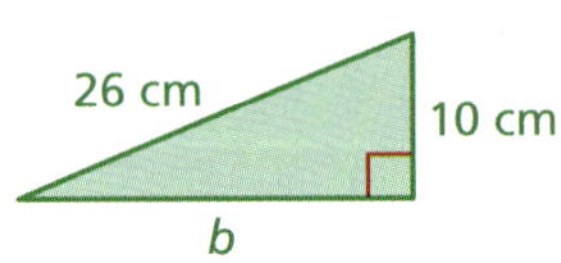

8.

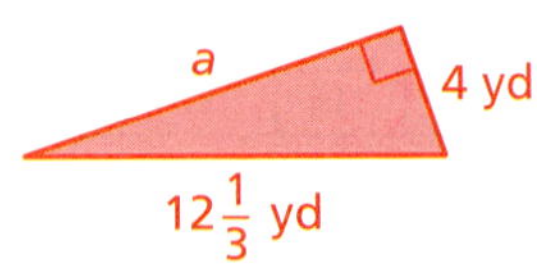

9. **ERROR ANALYSIS** Describe and correct the error in finding the missing length of the triangle.

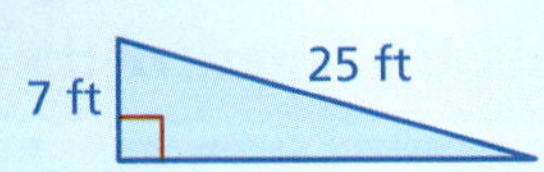

$$a^2 + b^2 = c^2$$
$$7^2 + 25^2 = c^2$$
$$674 = c^2$$
$$\sqrt{674} = c$$

10. **TREE SUPPORT** How long is the wire that supports the tree?

Assignment Guide and Homework Check

Level	Assignment	Homework Check
Accelerated	1–4, 6, 8–23	8, 10, 12, 14, 18

For Your Information

- **Exercise 17** There is more than one correct drawing for this exercise. Encourage students to start at the origin and move along an axis to begin.

Common Errors

- **Exercises 3–8** Students may substitute the given lengths in the wrong part of the formula. For example, if they are finding one of the legs, they may write $5^2 + 13^2 = c^2$ instead of $5^2 + b^2 = 13^2$. Remind them that the side opposite the right angle is the hypotenuse *c*.
- **Exercises 3–8** Students may multiply each side length by two instead of squaring the side length. Remind them of the definition of exponents.
- **Exercises 11 and 12** Students may think that there is not enough information to find the value of *x*. Tell them that it is possible to find *x*; however, they may have to make an extra calculation before writing an equation for *x*.

14.3 Record and Practice Journal

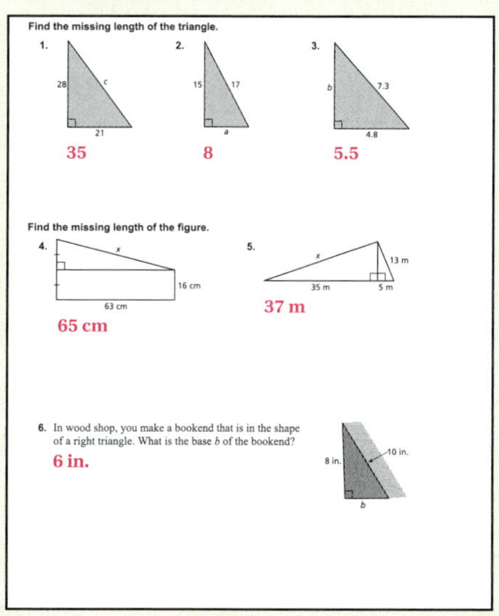
Find the missing length of the triangle.

1. 35
2. 8
3. 5.5

Find the missing length of the figure.

4. 65 cm
5. 37 m

6. In wood shop, you make a bookend that is in the shape of a right triangle. What is the base *b* of the bookend? 6 in.

Vocabulary and Concept Check

1. The hypotenuse is the longest side and the legs are the other two sides.
2. Which side is a leg?; *a* or *b*; *c*

Practice and Problem Solving

3. 29 km
4. 12 ft
5. 9 in.
6. 12 mm
7. 24 cm
8. $11\frac{2}{3}$ yd
9. The length of the hypotenuse was substituted for the wrong variable.
$$a^2 + b^2 = c^2$$
$$7^2 + b^2 = 25^2$$
$$49 + b^2 = 625$$
$$b^2 = 576$$
$$b = 24$$
10. 6.5 ft

Practice and Problem Solving

11. 16 cm

12. 37 mm

13. See *Taking Math Deeper*.

14. yes; The distance from the player's mouth to the referee's ear is 25 feet.

15. *Sample answer:* length = 20 ft, width = 48 ft, height = 10 ft; $BC = 52$ ft, $AB = \sqrt{2804}$ ft

16. 7

17. See Additional Answers.

18. yes; The rod is 25 inches long and the diagonal from a top corner to the opposite bottom corner is 26 inches long.

Fair Game Review

19. 6 and −6 **20.** −11

21. 13 **22.** −15

23. C

Mini-Assessment

Find the missing length of the triangle.

1. 50 ft

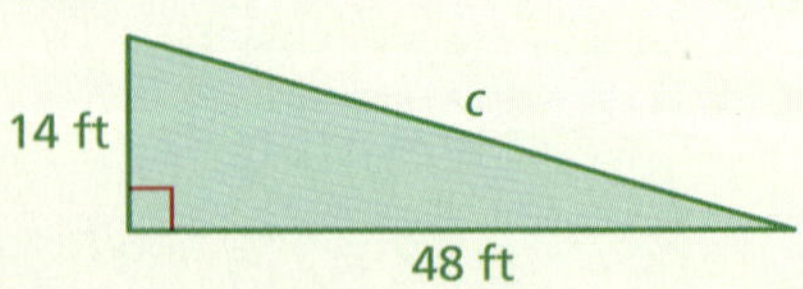

2. 24 mm

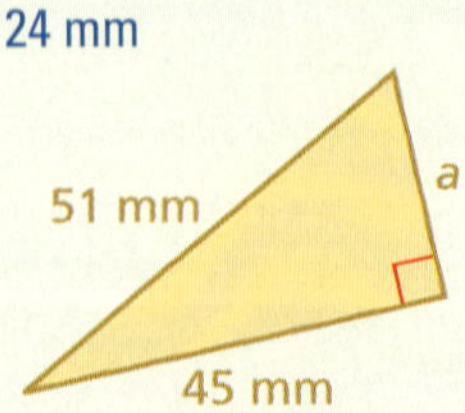

3. 12 in.

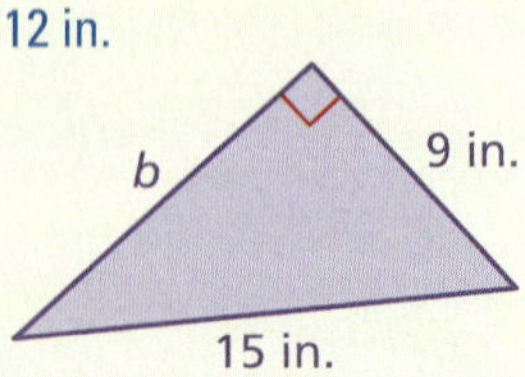

Taking Math Deeper

Exercise 13

The challenging part of this problem is realizing that the hypotenuse of the right triangle is given as 181 yards below the diagram.

1 Draw and label a diagram with the given information.

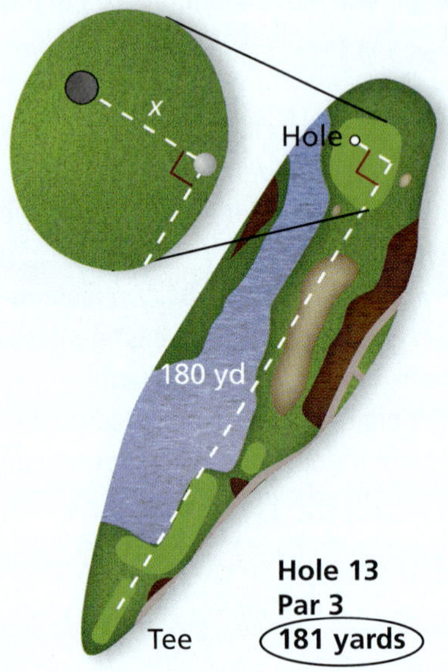

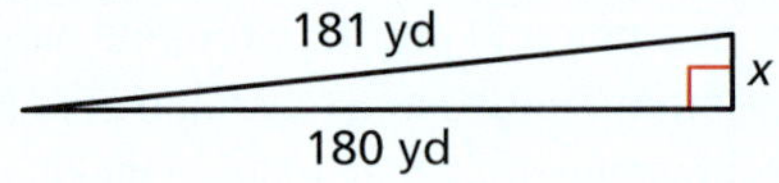

2 Use the Pythagorean Theorem.

$$180^2 + x^2 = 181^2$$
$$32{,}400 + x^2 = 32{,}761$$
$$x^2 = 361$$
$$x = 19$$

3 Answer the question.

The ball is 19 yards from the hole. Using the relationship of 3 feet = 1 yard, $19 \text{ yd} \times \frac{3 \text{ ft}}{1 \text{ yd}} = 57$ ft. So, the ball is 57 feet from the hole.

Reteaching and Enrichment Strategies

If students need help...	If students got it...
Resources by Chapter • Practice A and Practice B • Puzzle Time Record and Practice Journal Practice Differentiating the Lesson Lesson Tutorials Skills Review Handbook	Resources by Chapter • Enrichment and Extension • Technology Connection Start the next section

Find the missing length of the figure.

11.

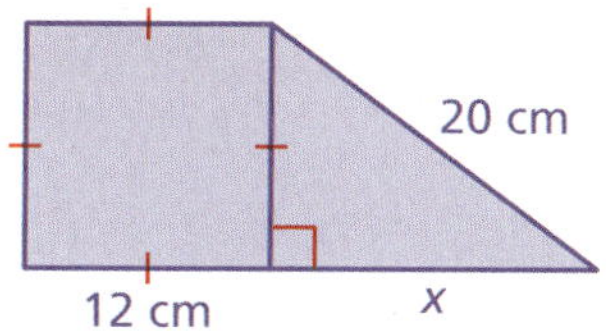

12.

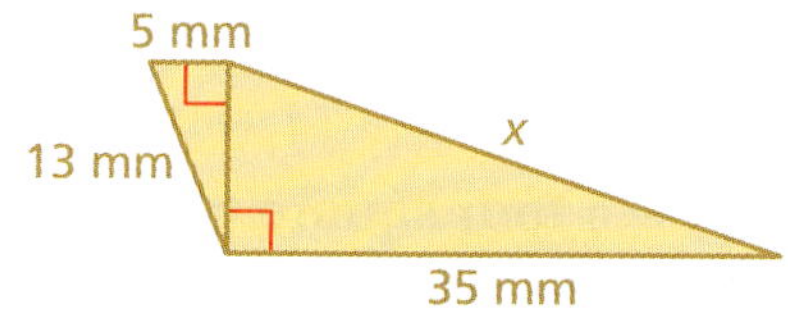

13. **GOLF** The figure shows the location of a golf ball after a tee shot. How many feet from the hole is the ball?

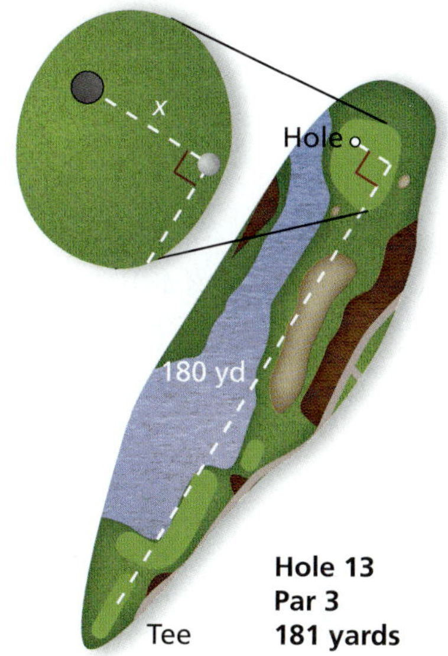

14. **TENNIS** A tennis player asks the referee a question. The sound of the player's voice travels only 30 feet. Can the referee hear the question? Explain.

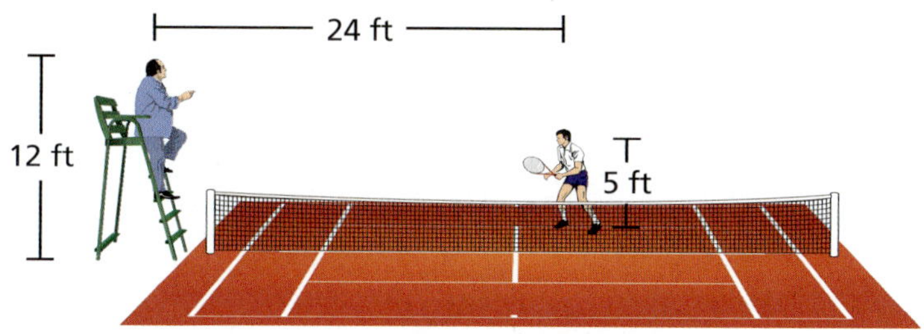

15. **PROJECT** Measure the length, width, and height of a rectangular room. Use the Pythagorean Theorem to find length BC and length AB.

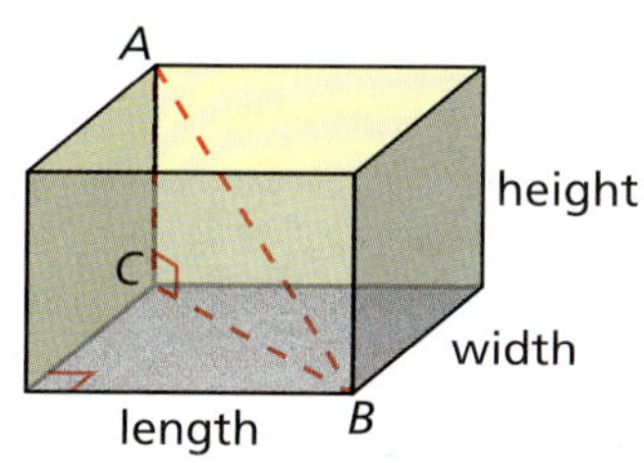

16. **ALGEBRA** The legs of a right triangle have lengths of 28 meters and 21 meters. The hypotenuse has a length of $5x$ meters. What is the value of x?

17. **SNOWBALLS** You and a friend stand back-to-back. You run 20 feet forward, then 15 feet to your right. At the same time, your friend runs 16 feet forward, then 12 feet to her right. She stops and hits you with a snowball.

   a. Draw the situation in a coordinate plane.

   b. How far does your friend throw the snowball?

18. **Precision** A box has a length of 6 inches, a width of 8 inches, and a height of 24 inches. Can a cylindrical rod with a length of 63.5 centimeters fit in the box? Explain your reasoning.

Fair Game Review What you learned in previous grades & lessons

Find the square root(s). *(Section 14.1)*

19. $\pm\sqrt{36}$

20. $-\sqrt{121}$

21. $\sqrt{169}$

22. $-\sqrt{225}$

23. **MULTIPLE CHOICE** Which equation represents a proportional relationship? *(Section 13.3)*

Ⓐ $y = x + 1$ Ⓑ $y = x - 1$ Ⓒ $y = 0.5x$ Ⓓ $y = 5 - 0.5x$

14 Study Help

Check It Out
Graphic Organizer
BigIdeasMath.com

You can use a **four square** to organize information about a topic. Each of the four squares can be a category, such as *definition, vocabulary, example, non-example, words, algebra, table, numbers, visual, graph,* or *equation*. Here is an example of a four square for the Pythagorean Theorem.

Pythagorean Theorem

Definition

In any right triangle, the sum of the squares of the lengths of the legs is equal to the square of the length of the hypotenuse.

Visual

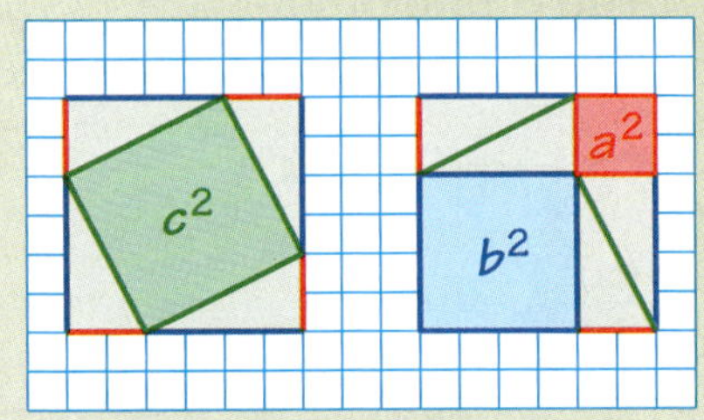

Algebra

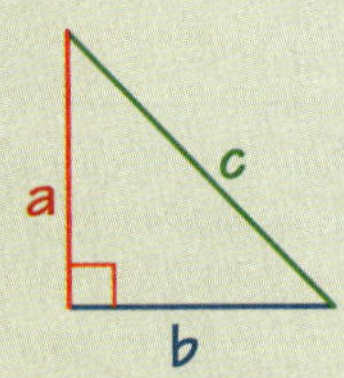

$a^2 + b^2 = c^2$

Example

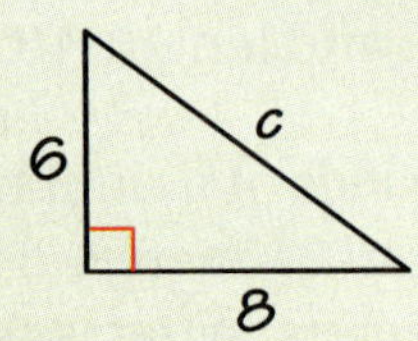

$$6^2 + 8^2 = c^2$$
$$36 + 64 = c^2$$
$$100 = c^2$$
$$10 = c$$

On Your Own

Make four squares to help you study these topics.

1. square roots
2. cube roots

After you complete this chapter, make four squares for the following topics.

3. irrational numbers
4. real numbers
5. converse of the Pythagorean Theorem
6. distance formula

"I'm taking a survey for my four square. How many fleas do you have?"

Sample Answers

1.

Definition	Symbol
A *square root* of a number is a number that, when multiplied by itself, equals the given number. Every positive number has a positive and a negative square root.	The symbol $\sqrt{\ }$ is called a *radical sign*. It is used to represent a square root. The number under the radical sign is called the *radicand*.
Square roots	
Example	**Example**
The two square roots of 100 are 10 and −10 because $10^2 = 100$ and $(-10)^2 = 100$. Positive square root: $\sqrt{100} = 10$ Negative square root: $-\sqrt{100} = -10$ Both square roots: $\pm\sqrt{100} = \pm 10$	Find $\sqrt{\frac{4}{9}}$. Because $\left(\frac{2}{3}\right)^2 = \frac{4}{9}$, $\sqrt{\frac{4}{9}} = \sqrt{\left(\frac{2}{3}\right)^2} = \frac{2}{3}$.

2.

Definition	Symbol
A *cube root* of a number is a number that, when multiplied by itself, and then multiplied by itself again, equals the given number.	The symbol $\sqrt[3]{\ }$ is used to represent a cube root.
Cube roots	
Algebra	**Examples**
If $a^3 = b$, then $a = \sqrt[3]{b}$, and a is a cube root of b.	a. Because $4^3 = 64$, $\sqrt[3]{64} = \sqrt[3]{4^3} = 4$. b. Because $(-5)^3 = -125$, $\sqrt[3]{-125} = \sqrt[3]{(-5)^3} = -5$. c. Because $\left(\frac{2}{3}\right)^3 = \frac{8}{27}$, $\sqrt[3]{\frac{8}{27}} = \sqrt[3]{\left(\frac{2}{3}\right)^3} = \frac{2}{3}$.

List of Organizers

Available at *BigIdeasMath.com*

Comparison Chart
Concept Circle
Definition (Idea) and Example Chart
Example and Non-Example Chart
Formula Triangle
Four Square
Information Frame
Information Wheel
Notetaking Organizer
Process Diagram
Summary Triangle
Word Magnet
Y Chart

About this Organizer

A **Four Square** can be used to organize information about a topic. Students write the topic in the "bubble" in the middle of the four square. Then students write concepts related to the topic in the four squares surrounding the bubble. Any concept related to the topic can be used. Encourage students to include concepts that will help them learn the topic. Students can place their four squares on note cards to use as a quick study reference.

Technology for the Teacher

Editable Graphic Organizer

Answers

1. -2
2. $\frac{4}{5}$
3. 2.5 and -2.5
4. 4
5. -6
6. $-\frac{7}{10}$
7. 26
8. -6
9. $5\frac{1}{4}$
10. 34
11. 30
12. -23
13. 41 ft
14. 28 in.
15. 6.3 cm
16. $\frac{1}{2}$ yd
17. $3.14r^2 = 314$; about 20 feet
18. 18 in.
19. 53 in.

Technology for the Teacher

Online Assessment
Assessment Book
ExamView® Assessment Suite

Alternative Quiz Ideas

100% Quiz	**Math Log**
Error Notebook	Notebook Quiz
Group Quiz	Partner Quiz
Homework Quiz	Pass the Paper

Math Log

Ask students to keep a math log for the chapter. Have them include diagrams, definitions, and examples. Everything should be clearly labeled. It might be helpful if they put the information in a chart. Students can add to the log as they are introduced to new topics.

Reteaching and Enrichment Strategies

If students need help...	If students got it...
Resources by Chapter • Practice A and Practice B • Puzzle Time Lesson Tutorials *BigIdeasMath.com*	Resources by Chapter • Enrichment and Extension • Technology Connection Game Closet at *BigIdeasMath.com* Start the next section

14.1–14.3 Quiz

Check It Out
Progress Check
BigIdeasMath.com

Find the square root(s). *(Section 14.1)*

1. $-\sqrt{4}$

2. $\sqrt{\frac{16}{25}}$

3. $\pm\sqrt{6.25}$

Find the cube root. *(Section 14.2)*

4. $\sqrt[3]{64}$

5. $\sqrt[3]{-216}$

6. $\sqrt[3]{-\frac{343}{1000}}$

Evaluate the expression. *(Section 14.1 and Section 14.2)*

7. $3\sqrt{49} + 5$

8. $10 - 4\sqrt{16}$

9. $\frac{1}{4} + \sqrt{\frac{100}{4}}$

10. $\left(\sqrt[3]{-27}\right)^3 + 61$

11. $15 + 3\sqrt[3]{125}$

12. $2\sqrt[3]{-729} - 5$

Find the missing length of the triangle. *(Section 14.3)*

13.

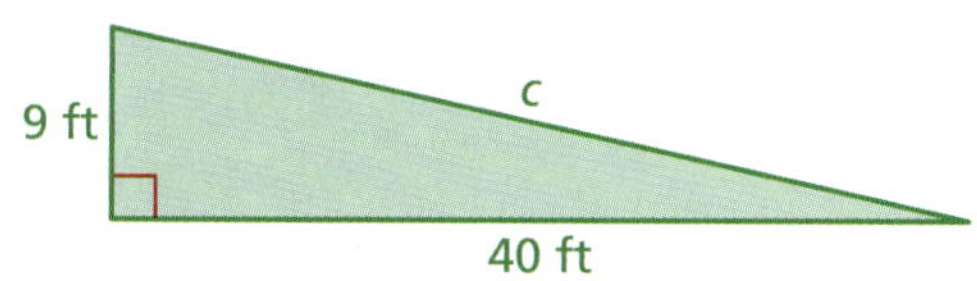

14.

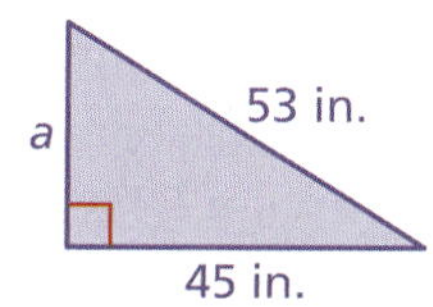

15.

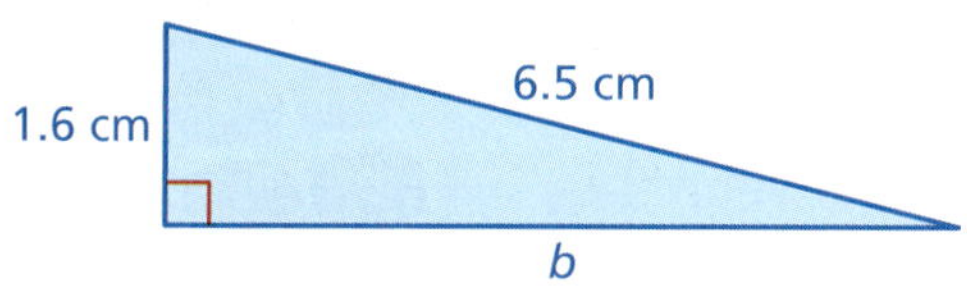

16.

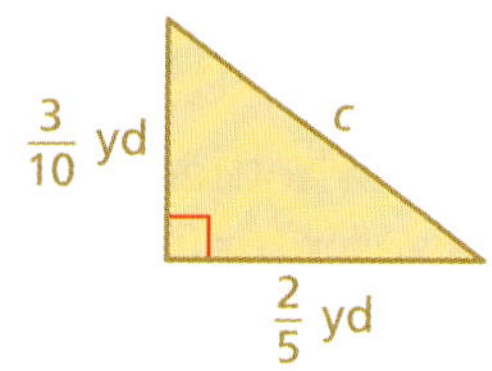

17. POOL The area of a circular pool cover is 314 square feet. Write and solve an equation to find the diameter of the pool cover. Use 3.14 for π. *(Section 14.1)*

18. PACKAGE A cube-shaped package has a volume of 5832 cubic inches. What is the edge length of the package? *(Section 14.2)*

19. FABRIC You are cutting a rectangular piece of fabric in half along the diagonal. The fabric measures 28 inches wide and $1\frac{1}{4}$ yards long. What is the length (in inches) of the diagonal? *(Section 14.3)*

14.4 Approximating Square Roots

Essential Question How can you find decimal approximations of square roots that are not rational?

1 ACTIVITY: Approximating Square Roots

Work with a partner. Archimedes was a Greek mathematician, physicist, engineer, inventor, and astronomer. He tried to find a rational number whose square is 3. Two that he tried were $\frac{265}{153}$ and $\frac{1351}{780}$.

a. Are either of these numbers equal to $\sqrt{3}$? Explain.

b. Use a calculator to approximate $\sqrt{3}$. Write the number on a piece of paper. Enter it into the calculator and square it. Then subtract 3. Do you get 0? What does this mean?

c. The value of $\sqrt{3}$ is between which two integers?

d. Tell whether the value of $\sqrt{3}$ is between the given numbers. Explain your reasoning.

1.7 and 1.8 | 1.72 and 1.73 | 1.731 and 1.732

2 ACTIVITY: Approximating Square Roots Geometrically

Work with a partner. Refer to the square on the number line below.

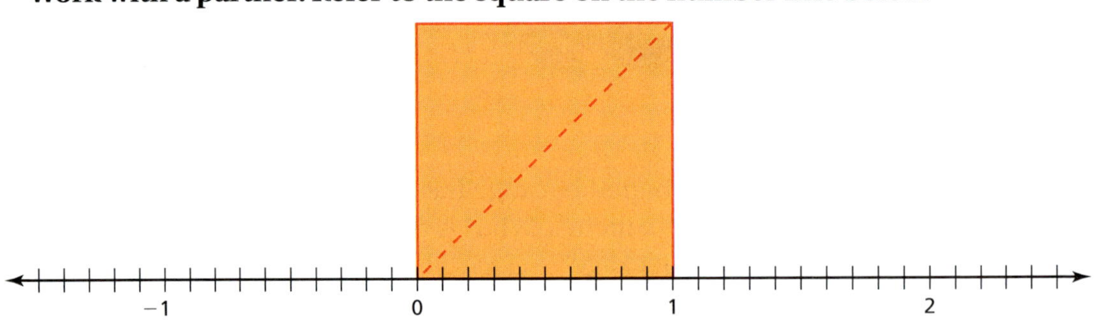

Square Roots
In this lesson, you will
- define irrational numbers.
- approximate square roots.
- approximate values of expressions involving irrational numbers.

a. What is the length of the diagonal of the square?

b. Copy the square and its diagonal onto a piece of transparent paper. Rotate it about zero on the number line so that the diagonal aligns with the number line. Use the number line to estimate the length of the diagonal.

c. **STRUCTURE** How do you think your answers in parts (a) and (b) are related?

Laurie's Notes

Introduction

Applying Mathematical Practices

- **Make Sense of Problems:** To help students make sense of square roots they will use calculators and construction tools to estimate square roots.

Motivate

- Make a large Venn diagram based on student characteristics.
- The diagram can be made on the floor with yarn. Have students write their names on index cards.
- Use the diagram shown for students to place themselves. Sample labels for the groups: A = girls in our class, B = boys in our class, C = wears glasses/contacts, D = brown hair, E = taller than 5′ 4″, F = wearing a short sleeved T-shirt

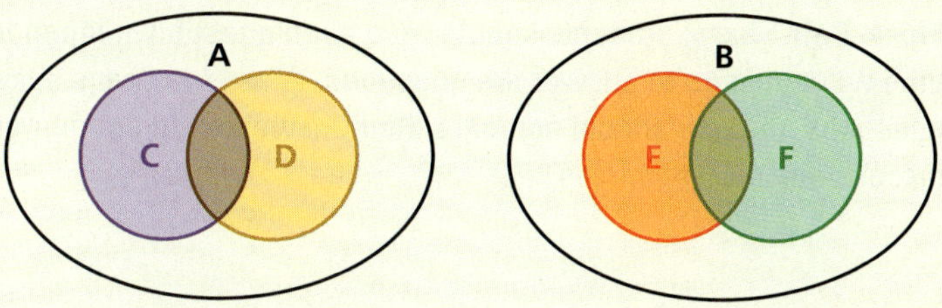

- Discuss what it means to be in certain sets and not in other sets.

Activity Notes

Activity 1

- **FYI:** Archimedes wanted a number x, such that $x^2 = 3$. If $x^2 = 3$, then $x = \pm\sqrt{3}$. Archimedes wanted to find a rational number that equals $\sqrt{3}$.
- **Part (a):** To square a fraction using a calculator, you can (1) write the fraction as a decimal by dividing, then square the decimal, or (2) write the fraction inside parentheses and use the exponent key.
- **Note:** In part (b), if you get anything other than 0, it means you did not have the exact value for $\sqrt{3}$, because $\sqrt{3}$ is irrational.
- **Make Sense of Problems and Persevere in Solving Them:** Part (d) checks understanding of place value as well as how much students persevered in approximating $\sqrt{3}$. For a given pair of numbers that $\sqrt{3}$ is not between, have students adjust the numbers so that it is between them.

Activity 2

- **Teaching Tip:** Students can use transparencies or tracing paper.
- The goal is for students to recognize that *irrational numbers* (defined in this lesson) can have a length associated with them. They are numbers that you can approximate. When students approximate the length of the diagonal, it should be close to 1.4 (approximation of $\sqrt{2}$).

What Your Students Will Learn

- Define irrational numbers and classify real numbers.
- Approximate square roots using squares and number lines.
- Compare real numbers using a number line.
- Approximate values of expressions involving rational numbers.

Previous Learning

Students should know how to find square roots of perfect squares.

Technology for the Teacher

Dynamic Classroom

Lesson Plans
Complete Materials List

14.4 Record and Practice Journal

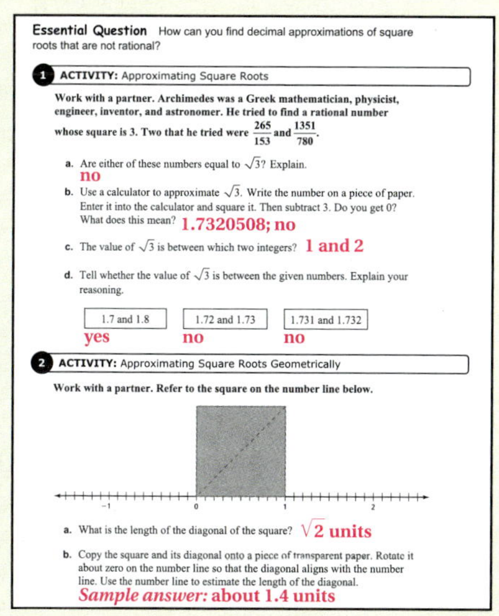

Essential Question How can you find decimal approximations of square roots that are not rational?

1 ACTIVITY: Approximating Square Roots

Work with a partner. Archimedes was a Greek mathematician, physicist, engineer, inventor, and astronomer. He tried to find a rational number whose square is 3. Two that he tried were $\frac{265}{153}$ and $\frac{1351}{780}$.

a. Are either of these numbers equal to $\sqrt{3}$? Explain.
no

b. Use a calculator to approximate $\sqrt{3}$. Write the number on a piece of paper. Enter it into the calculator and square it. Then subtract 3. Do you get 0? What does this mean? 1.7320508; no

c. The value of $\sqrt{3}$ is between which two integers? 1 and 2

d. Tell whether the value of $\sqrt{3}$ is between the given numbers. Explain your reasoning.

1.7 and 1.8	1.72 and 1.73	1.731 and 1.732
yes	no	no

2 ACTIVITY: Approximating Square Roots Geometrically

Work with a partner. Refer to the square on the number line below.

a. What is the length of the diagonal of the square? $\sqrt{2}$ units

b. Copy the square and its diagonal onto a piece of transparent paper. Rotate it about zero on the number line so that the diagonal aligns with the number line. Use the number line to estimate the length of the diagonal.
Sample answer: about 1.4 units

Differentiated Instruction

Visual

On the board, create a large Venn diagram. Have students place numbers they are familiar with, such as natural numbers, whole numbers, integers, and rational numbers in the correct spaces on the Venn diagram. Reinforce that a number such as 5 is a natural number, as well as a whole number, an integer, and a rational number.

14.4 Record and Practice Journal

c. STRUCTURE How do you think your answers in parts (a) and (b) are related?
Sample answer: The length of 1.4 units is an approximation of $\sqrt{2}$ units.

3 ACTIVITY: Approximating Square Roots Geometrically

Work with a partner.

a. Use grid paper and the given scale to draw a horizontal line segment 1 unit in length. Draw your segment near the bottom of the grid. Label this segment *AC*.

b. Draw a vertical line segment 2 units in length. Draw your segment near the left edge of the grid. Label this segment *DC*.

c. Set the point of a compass on *A*. Set the compass to 2 units. Swing the compass to intersect segment *DC*. Label this intersection as *B*.

d. Use the Pythagorean Theorem to find the length of segment *BC*.

$$(BC)^2 + (AC)^2 = (AB)^2$$
$$(BC)^2 + 1^2 = 2^2$$
$$(BC)^2 + 1 = 4$$
$$(BC)^2 = 3$$
$$BC = \sqrt{3}$$

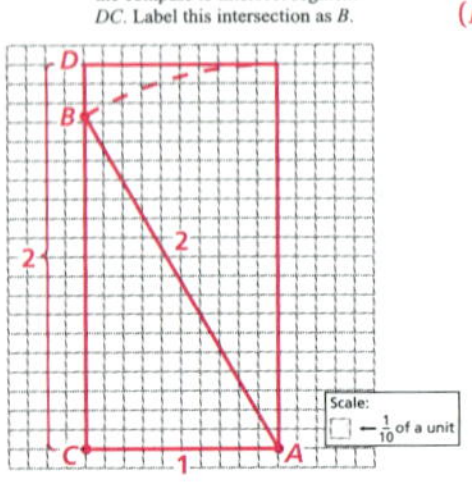

e. Use the grid paper to approximate $\sqrt{3}$ to the nearest tenth.
about 1.7

4. Compare your approximation in Activity 3 with your results from Activity 1. Approximations are about the same.

What Is Your Answer?

5. Repeat Activity 3 for a triangle in which segment *AC* is 2 units and segment *BA* is 3 units. Use the Pythagorean Theorem to find the length of segment *BC*. Use the grid paper to approximate $\sqrt{5}$ to the nearest tenth.

$$(BC)^2 + (AC)^2 = (AB)^2$$
$$(BC)^2 + 2^2 = 3^2$$
$$(BC)^2 + 4 = 9$$
$$(BC)^2 = 5$$
$$BC = \sqrt{5}$$

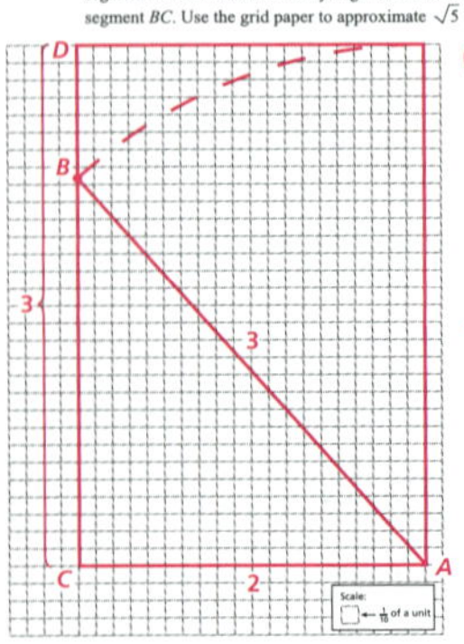

$\sqrt{5} \approx 2.2$

6. IN YOUR OWN WORDS How can you find decimal approximations of square roots that are not rational?
Use geometry.

Laurie's Notes

Activity 3

- **Use Appropriate Tools Strategically:** In Activity 1, students looked at $\sqrt{3}$ as a number. In this activity, students will look at $\sqrt{3}$ geometrically, as a length of a line segment.
- Students should be able to follow the written directions to construct segment *BC*.

? "Why does swinging the compass make *AB* equal 2 units?" *AB* is a radius of a circle that you know equals 2.

? "What type of triangle is *ABC*?" right

? "What information do you know about the triangle?" The hypotenuse equals 2 and one leg equals 1.

- Ask a pair of students to show how they used the Pythagorean Theorem to find the length of segment *BC*.

? "Is $\sqrt{3}$ greater than or less than 1.5?" greater than

What Is Your Answer?

- **Think-Pair-Share:** Students should read each question independently and then work in pairs to answer the questions. When they have answered the questions, the pair should compare their answers with another group and discuss any discrepancies.

Closure

- **Exit Ticket:** Describe how you would approximate $\sqrt{5}$.
Listen for students to describe a procedure similar to that used in Activity 3, except with segments 2 units and 3 units in length.

3 ACTIVITY: Approximating Square Roots Geometrically

Math Practice

Recognize Usefulness of Tools

Why is the Pythagorean Theorem a useful tool when approximating a square root?

Work with a partner.

a. Use grid paper and the given scale to draw a horizontal line segment 1 unit in length. Label this segment *AC*.

b. Draw a vertical line segment 2 units in length. Label this segment *DC*.

c. Set the point of a compass on *A*. Set the compass to 2 units. Swing the compass to intersect segment *DC*. Label this intersection as *B*.

d. Use the Pythagorean Theorem to find the length of segment *BC*.

e. Use the grid paper to approximate $\sqrt{3}$ to the nearest tenth.

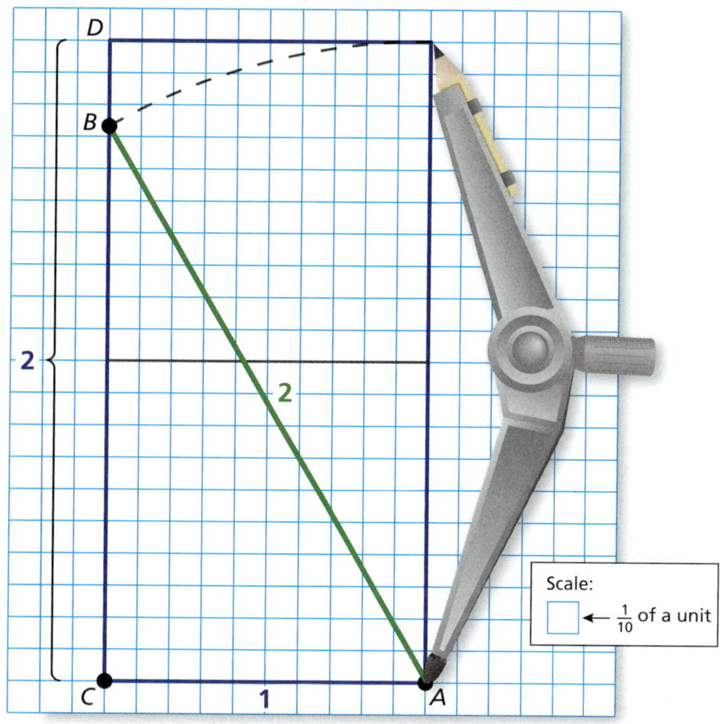

What Is Your Answer?

4. Compare your approximation in Activity 3 with your results from Activity 1.

5. Repeat Activity 3 for a triangle in which segment *AC* is 2 units and segment *BA* is 3 units. Use the Pythagorean Theorem to find the length of segment *BC*. Use the grid paper to approximate $\sqrt{5}$ to the nearest tenth.

6. **IN YOUR OWN WORDS** How can you find decimal approximations of square roots that are not rational?

Practice Use what you learned about approximating square roots to complete Exercises 5–8 on page 651.

14.4 Lesson

Check It Out
Lesson Tutorials
BigIdeasMath.com

Key Vocabulary
irrational number, *p. 648*
real numbers, *p. 648*

A rational number is a number that can be written as the ratio of two integers. An **irrational number** cannot be written as the ratio of two integers.

- The square root of any whole number that is not a perfect square is irrational. The cube root of any integer that is not a perfect cube is irrational.
- The decimal form of an irrational number neither terminates nor repeats.

Key Idea

Real Numbers

Rational numbers and irrational numbers together form the set of **real numbers**.

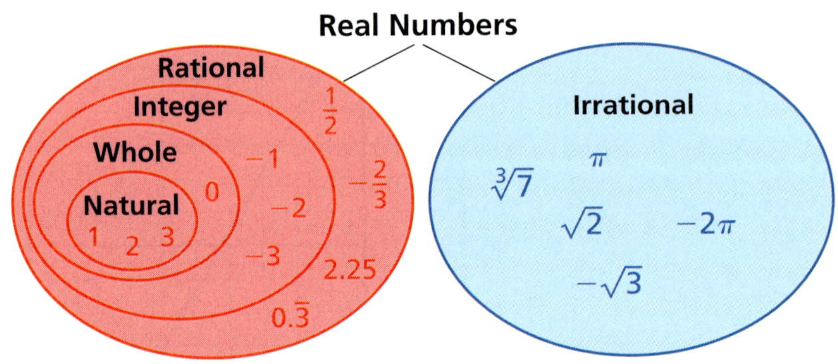

Remember
The decimal form of a rational number either terminates or repeats.

EXAMPLE 1 Classifying Real Numbers

Classify each real number.

	Number	Subset(s)	Reasoning
a.	$\sqrt{12}$	Irrational	12 is not a perfect square.
b.	$-0.\overline{25}$	Rational	$-0.\overline{25}$ is a repeating decimal.
c.	$-\sqrt{9}$	Integer, Rational	$-\sqrt{9}$ is equal to -3.
d.	$\frac{72}{4}$	Natural, Whole, Integer, Rational	$\frac{72}{4}$ is equal to 18.
e.	π	Irrational	The decimal form of π neither terminates nor repeats.

Study Tip
When classifying a real number, list all the subsets in which the number belongs.

On Your Own

Now You're Ready
Exercises 9–16

Classify the real number.

1. $0.121221222\ldots$ **2.** $-\sqrt{196}$ **3.** $\sqrt[3]{2}$

Multi-Language Glossary at BigIdeasMath.com

Laurie's Notes

Introduction

Connect

- **Yesterday:** Students investigated $\sqrt{3}$ numerically and geometrically.
- **Today:** Students will approximate square roots.

Motivate

- Discuss applications of a periscope and share the following information.
  - A periscope is an optical device for conducting observations from a concealed or protected position.
  - Simple periscopes consist of reflecting mirrors and/or prisms at opposite ends of a tube container. The reflecting surfaces are parallel to each other and at a 45° angle to the axis of the tube.
  - The Navy attributes the invention of the periscope (1902) to Simon Lake and the perfection of the periscope to Sir Howard Grubb.

Lesson Notes

Key Idea

- Explain the definitions of *rational* and *irrational* numbers. Use the Venn diagram to give several examples of each.
- Write the Key Idea. It is important for students to understand that any real number is either rational or irrational. The two sets do not intersect.
- Explain that a *subset* is a set in which every element is contained within a larger set. The sets of rational numbers and irrational numbers are subsets of the set of real numbers. The sets of natural numbers, whole numbers, and integers are subsets of the set of rational numbers.
- Natural numbers are also called counting numbers because they are used to count objects. Whole numbers include the natural numbers and 0. Integers include the natural numbers, 0, and the opposites of the natural numbers.
- Mention that all square roots of perfect squares are rational.

? "Can you think of a repeating decimal and its fractional equivalent?"

Sample answer: $0.\overline{3} = \frac{1}{3}$

? "Can you think of a terminating decimal and its fractional equivalent?"

Sample answer: $0.5 = \frac{1}{2}$

Example 1

- Students will gain a better understanding of how to classify real numbers in this example.
- Discuss the Study Tip with students. Point out in part (c), for instance, that because $-\sqrt{9} = -3 = \frac{-3}{1}$, $-\sqrt{9}$ is an integer *and* a rational number.

On Your Own

- **Think-Pair-Share:** Students should read each question independently and then work in pairs to answer the questions. When they have answered the questions, the pair should compare their answers with another group and discuss any discrepancies.

Goal Today's lesson is approximating square roots.

Technology for the Teacher

Dynamic Classroom

Lesson Tutorials
Lesson Plans
Answer Presentation Tool

Extra Example 1

Classify each real number.

a. $\sqrt{15}$ irrational

b. 0.35 rational

On Your Own

1. irrational
2. integer, rational
3. irrational

Laurie's Notes

Extra Example 2

Estimate $\sqrt{23}$ to the nearest (a) integer and (b) tenth.

a. 5

b. 4.8

On Your Own

4.	**a.** 3	**b.** 2.8
5.	**a.** -4	**b.** -3.6
6.	**a.** -5	**b.** -4.9
7.	**a.** 10	**b.** 10.5

Extra Example 3

Which is greater, $\sqrt{0.49}$ or 0.71? 0.71

English Language Learners

Vocabulary

Point out to students that the prefix *ir-* means *not.* An *irrational* number is a number that is not rational. Here are other common prefixes that also mean *not.*

dis-	disadvantage, disagree
il-	illiterate, illogical
im-	impolite, improper
in-	independent, indirect
ir-	irrational, irregular
un-	unfair, unfriendly

Example 2

- It is important for students to make an estimate before using a calculator. Use reasoning first!
- ? "What are the first 10 perfect squares?" 1, 4, 9, 16, 25, 36, 49, 64, 81, 100
- ? "What type of number do you get when you take the square root of any of these perfect squares?" integer
- ? **Construct Viable Arguments and Critique the Reasoning of Others:** "Between what two whole numbers is $\sqrt{71}$, and how do you know?" 8 and 9 because $\sqrt{64} = 8$ and $\sqrt{81} = 9$, so $\sqrt{71}$ has to be a number between 8 and 9.
- ? "Is $\sqrt{71}$ closer to 8 or 9? Why?" It is closer to 8 because 71 is closer to 64 than to 81.
- You may wish to allow students to calculate squares of decimals using a calculator.
- You could explore more about square roots using a calculator approximation. For example, $\sqrt{71} \approx 8.4261498$. So, you can rationalize that $\sqrt{71}$ is between 8 and 9, between 8.4 and 8.5, between 8.42 and 8.43, etc., by truncating the decimal.

On Your Own

- Ask volunteers to share their thinking about each problem.

Example 3

- **Model with Mathematics:** A number line is used as a visual model.
- Students will ask where to place $\sqrt{5}$. Knowing that $\sqrt{5}$ is between $\sqrt{4}$ and $\sqrt{9}$ does not tell you where to graph it on the number line. Explain that you know it has to be closer to $\sqrt{4}$ than $\sqrt{9}$ because 5 is closer to 4 than to 9.
- The fraction $2\frac{2}{3}$ is greater than $2\frac{1}{2}$, so it is closer to $\sqrt{9}$.

EXAMPLE 2 Approximating a Square Root

Estimate $\sqrt{71}$ to the nearest (a) integer and (b) tenth.

a. Make a table of numbers whose squares are close to 71.

Number	7	8	9	10
Square of Number	49	64	81	100

The table shows that 71 is between the perfect squares 64 and 81. Because 71 is closer to 64 than to 81, $\sqrt{71}$ is closer to 8 than to 9.

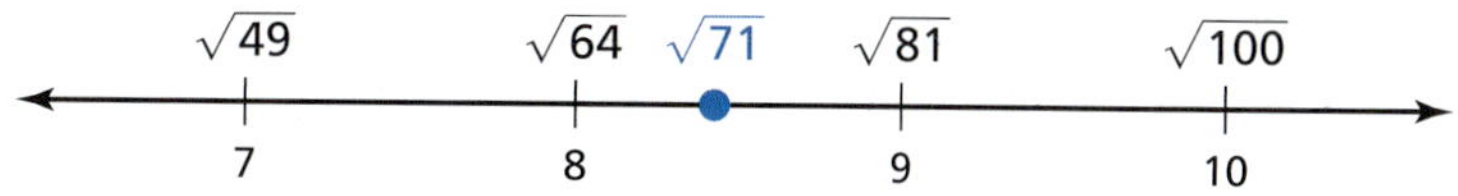

So, $\sqrt{71} \approx 8$.

b. Make a table of numbers between 8 and 9 whose squares are close to 71.

Number	8.3	8.4	8.5	8.6
Square of Number	68.89	70.56	72.25	73.96

Because 71 is closer to 70.56 than to 72.25, $\sqrt{71}$ is closer to 8.4 than to 8.5.

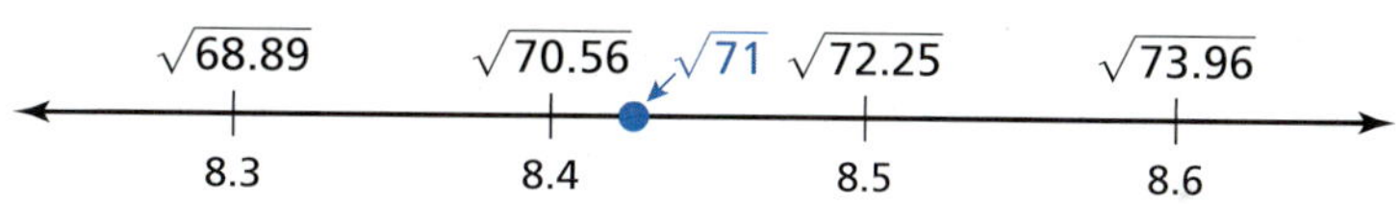

So, $\sqrt{71} \approx 8.4$.

Study Tip

You can continue the process shown in Example 2 to approximate square roots using more decimal places.

On Your Own

Now You're Ready
Exercises 20–25

Estimate the square root to the nearest (a) integer and (b) tenth.

4. $\sqrt{8}$ **5.** $-\sqrt{13}$ **6.** $-\sqrt{24}$ **7.** $\sqrt{110}$

EXAMPLE 3 Comparing Real Numbers

Which is greater, $\sqrt{5}$ or $2\frac{2}{3}$?

Estimate $\sqrt{5}$ to the nearest integer. Then graph the numbers on a number line.

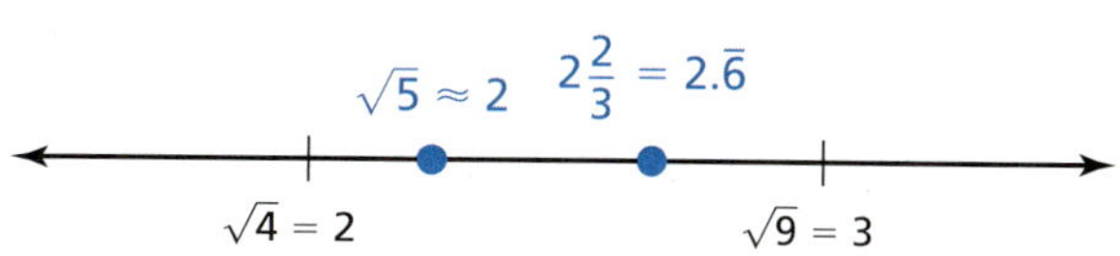

$2\frac{2}{3}$ is to the right of $\sqrt{5}$. So, $2\frac{2}{3}$ is greater.

EXAMPLE 4 Approximating the Value of an Expression

The radius of a circle with area A is approximately $\sqrt{\frac{A}{3}}$. The area of a circular mouse pad is 51 square inches. Estimate its radius to the nearest integer.

$$\sqrt{\frac{A}{3}} = \sqrt{\frac{51}{3}} \qquad \text{Substitute 51 for } A.$$

$$= \sqrt{17} \qquad \text{Divide.}$$

The nearest perfect square less than 17 is 16. The nearest perfect square greater than 17 is 25.

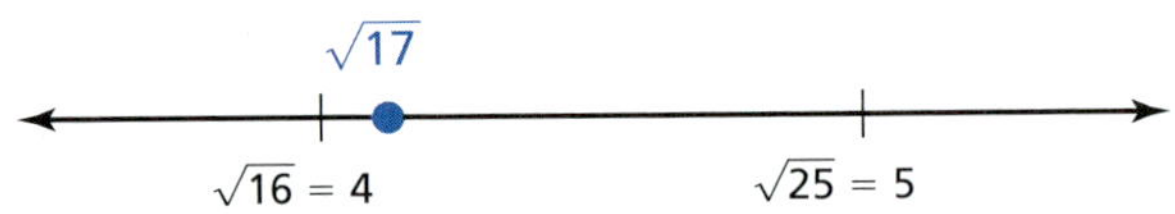

Because 17 is closer to 16 than to 25, $\sqrt{17}$ is closer to 4 than to 5.

∴ So, the radius is about 4 inches.

EXAMPLE 5 Real-Life Application

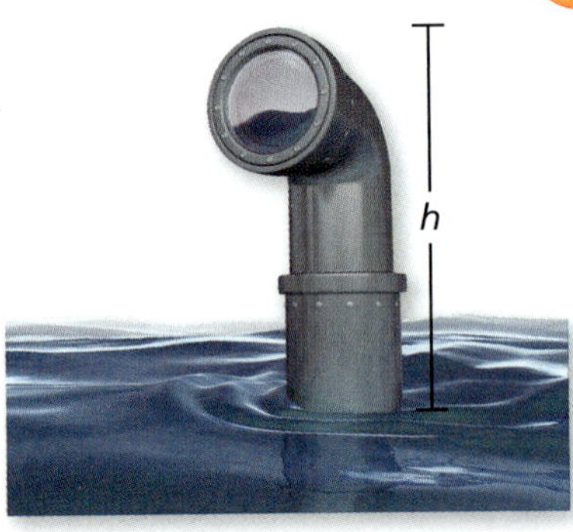

The distance (in nautical miles) you can see with a periscope is $1.17\sqrt{h}$, where h is the height of the periscope above the water. Can you see twice as far with a periscope that is 6 feet above the water than with a periscope that is 3 feet above the water? Explain.

Use a calculator to find the distances.

3 Feet Above Water		*6 Feet Above Water*
$1.17\sqrt{h} = 1.17\sqrt{3}$	Substitute for h.	$1.17\sqrt{h} = 1.17\sqrt{6}$
≈ 2.03	Use a calculator.	≈ 2.87

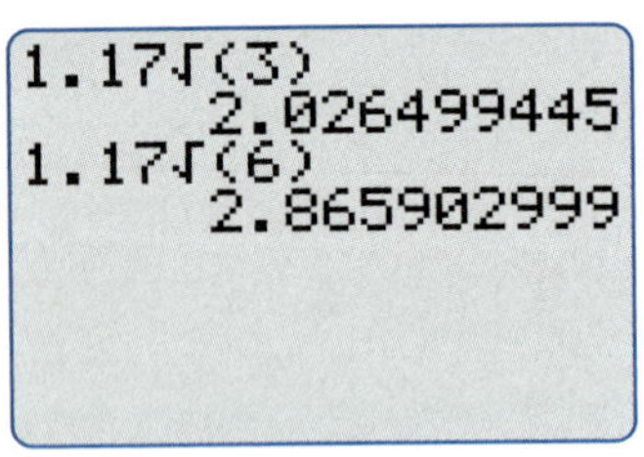

You can see $\frac{2.87}{2.03} \approx 1.41$ times farther with the periscope that is 6 feet above the water than with the periscope that is 3 feet above the water.

∴ No, you cannot see twice as far with the periscope that is 6 feet above the water.

On Your Own

Now You're Ready
Exercises 26–31

Which number is greater? Explain.

8. $4\frac{1}{5}, \sqrt{23}$ **9.** $\sqrt{10}, -\sqrt{5}$ **10.** $-\sqrt{2}, -2$

11. The area of a circular mouse pad is 64 square inches. Estimate its radius to the nearest integer.

12. In Example 5, you use a periscope that is 10 feet above the water. Can you see farther than 4 nautical miles? Explain.

Laurie's Notes

Example 4

- A simple diagram of a circle helps students focus on what is being asked.

? "What is the formula for the area of a circle?" $A = \pi r^2$

- Review with students how to solve this formula for *r*.

$A = \pi r^2$ Write the area formula.

$\frac{A}{\pi} = r^2$ Divide each side by π.

$\sqrt{\frac{A}{\pi}} = r$ Take positive square root of each side.

Because π is close to 3, when approximating the radius, you can replace π with 3. This is the formula presented in this example.

? "What information is known in this problem?" You have a circle with an area of 51 square inches.

? "What are you trying to find?" an estimate for the radius

- Radicands that are fractions can be intimidating to students.
- Draw the number line and work the problem as shown.

Example 5

- Ask a student to read through the problem.
- You want to compare the distances for a periscope at two different heights above water, so the equation is used twice.

? "Do you think you can see twice as far at 6 feet than at 3 feet?" Most will say yes.

- Write the expression $1.17\sqrt{h}$ on the board and evaluate it for each height, as shown.

On Your Own

- Ask volunteers to share their work at the board.
- **Question 9:** A positive number is always greater than a negative number.
- **Question 10:** A number line is helpful for this question.
- **Question 11:** The quotient $\frac{64}{3}$ is not a whole number, but the question is still completed in the same fashion.

Closure

- Order the numbers from least to greatest: $\sqrt{38}$, $\sqrt{\frac{100}{3}}$, $6.\overline{5}$ $\sqrt{\frac{100}{3}}$, $\sqrt{38}$, $6.\overline{5}$

Extra Example 4

In Example 4, estimate the radius of a circular mouse pad with an area of 45 square inches. Round your answer to the nearest integer. about 4 in.

Extra Example 5

In Example 5, a periscope is 8 feet above the water. Can you see farther than 3 nautical miles? Explain. yes; You can see about 3.3 nautical miles.

On Your Own

8. $\sqrt{23}$; $\sqrt{23}$ is to the right of $4\frac{1}{5}$.
9. $\sqrt{10}$; $\sqrt{10}$ is positive and $-\sqrt{5}$ is negative.
10. $-\sqrt{2}$; $-\sqrt{2}$ is to the right of -2.
11. about 5 in.
12. no; You can only see about 3.7 nautical miles.

Vocabulary and Concept Check

1. A rational number can be written as the ratio of two integers. An irrational number cannot be written as the ratio of two integers.
2. 32 is between the perfect squares 25 and 36, but is closer to 36, so $\sqrt{32} \approx 6$.
3. all rational and irrational numbers; *Sample answer:* $-2, \frac{1}{8}, \sqrt{7}$
4. $\sqrt{8}$; $\sqrt{8}$ is irrational and the other three numbers are rational.

Practice and Problem Solving

5. yes
6. no
7. no
8. yes
9. whole, integer, rational
10. natural, whole, integer, rational
11. irrational
12. integer, rational
13. rational
14. natural, whole, integer, rational
15. irrational
16. irrational
17. 144 is a perfect square. So, $\sqrt{144}$ is rational.
18. no; 52 is not a perfect square.
19. a. If the last digit is 0, it is a whole number. Otherwise, it is a natural number.
    b. irrational number
    c. irrational number

20–25. See Additional Answers.

Assignment Guide and Homework Check

Level	Assignment	Homework Check
Accelerated	1–8, 17, 24–50 even, 51–54	24, 30, 34, 38, 50

Common Errors

- **Exercises 9–16** Students may not classify the real number in as many ways as possible. For instance, they may classify $\frac{52}{13}$ as rational only, because it is written as a fraction. Remind students that real numbers can have more than one classification. Point out that they should simplify the number, if possible, before classifying it.
- **Exercises 12 and 13** Students may think that all negative numbers are irrational. Remind them of the integers and that negative numbers can be rational or irrational.
- **Exercises 12, 15, and 16** Students may think that all square roots and cube roots are irrational. Remind them that square roots of perfect squares are rational and that cube roots of perfect cubes are also rational.
- **Exercises 20–25** Students may struggle with knowing what integer is closest to the given number. To help make comparisons, encourage them to write the first 10 perfect squares. If the number under the radical is greater than 100, then students should use *Guess, Check, and Revise* to find two integers on either side of the number. When determining which integer is closer to the rational number, encourage students to use a number line.

14.4 Record and Practice Journal

Classify the real number.

1. $\sqrt{14}$ — irrational
2. $-\frac{3}{7}$ — rational
3. $\frac{153}{3}$ — natural, whole, integer, rational

Estimate the square root to the nearest (a) integer and (b) tenth.

4. $\sqrt{8}$ — a. 3 b. 2.8
5. $\sqrt{60}$ — a. 8 b. 7.7
6. $-\sqrt{\frac{172}{25}}$ — a. -3 b. -2.6

Which number is greater? Explain.

7. $\sqrt{88}$, 12 — 12
8. $-\sqrt{18}$, -6 — $-\sqrt{18}$
9. 14.5, $\sqrt{220}$ — $\sqrt{220}$

10. The velocity in meters per second of a ball that is dropped from a window at a height of 10.5 meters is represented by the equation $v = \sqrt{2(9.8)(10.5)}$. Estimate the velocity of the ball. Round your answer to the nearest tenth.
14.3 m/sec

14.4 Exercises

Check It Out
Help with Homework
BigIdeasMath.com

Vocabulary and Concept Check

1. **VOCABULARY** How are rational numbers and irrational numbers different?
2. **WRITING** Describe a method of approximating $\sqrt{32}$.
3. **VOCABULARY** What are real numbers? Give three examples.
4. **WHICH ONE DOESN'T BELONG?** Which number does *not* belong with the other three? Explain your reasoning.

   $-\frac{11}{12}$ 25.075 $\sqrt{8}$ $-3.\overline{3}$

Practice and Problem Solving

Tell whether the rational number is a reasonable approximation of the square root.

5. $\frac{559}{250}$, $\sqrt{5}$
6. $\frac{3021}{250}$, $\sqrt{11}$
7. $\frac{678}{250}$, $\sqrt{28}$
8. $\frac{1677}{250}$, $\sqrt{45}$

Classify the real number.

1 9. 0
10. $\sqrt[3]{343}$
11. $\frac{\pi}{6}$
12. $-\sqrt{81}$
13. -1.125
14. $\frac{52}{13}$
15. $\sqrt[3]{-49}$
16. $\sqrt{15}$

17. **ERROR ANALYSIS** Describe and correct the error in classifying the number.

   ✗ $\sqrt{144}$ is irrational.

18. **SCRAPBOOKING** You cut a picture into a right triangle for your scrapbook. The lengths of the legs of the triangle are 4 inches and 6 inches. Is the length of the hypotenuse a rational number? Explain.

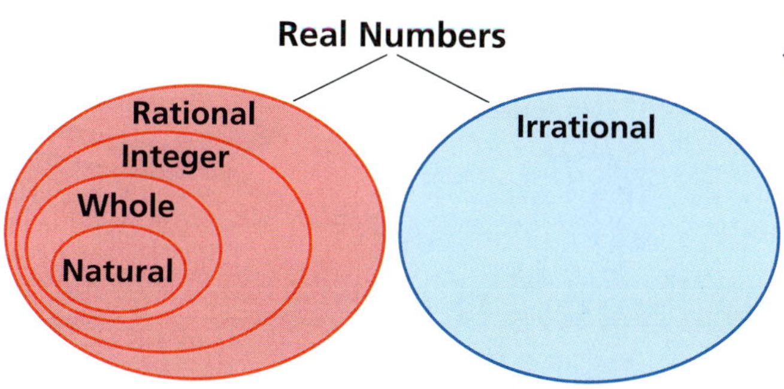

19. **VENN DIAGRAM** Place each number in the correct area of the Venn Diagram.
    a. the last digit of your phone number
    b. the square root of any prime number
    c. the ratio of the circumference of a circle to its diameter

Estimate the square root to the nearest (a) integer and (b) tenth.

2 20. $\sqrt{46}$
21. $\sqrt{685}$
22. $-\sqrt{61}$
23. $-\sqrt{105}$
24. $\sqrt{\frac{27}{4}}$
25. $-\sqrt{\frac{335}{2}}$

Which number is greater? Explain.

26. $\sqrt{20}$, 10 **27.** $\sqrt{15}$, -3.5 **28.** $\sqrt{133}$, $10\frac{3}{4}$

29. $\frac{2}{3}$, $\sqrt{\frac{16}{81}}$ **30.** $-\sqrt{0.25}$, -0.25 **31.** $-\sqrt{182}$, $-\sqrt{192}$

Use the graphing calculator screen to determine whether the statement is *true* or *false*.

```
√(10)
          3.16227766
√(14)
         3.741657387
```

32. To the nearest tenth, $\sqrt{10} = 3.1$.

33. The value of $\sqrt{14}$ is between 3.74 and 3.75.

34. $\sqrt{10}$ lies between 3.1 and 3.16 on a number line.

35. FOUR SQUARE The area of a four square court is 66 square feet. Estimate the side length s to the nearest tenth of a foot.

36. CHECKERS A checkers board is 8 squares long and 8 squares wide. The area of each square is 14 square centimeters. Estimate the perimeter of the checkers board to the nearest tenth of a centimeter.

Approximate the length of the diagonal of the square or rectangle to the nearest tenth.

37.

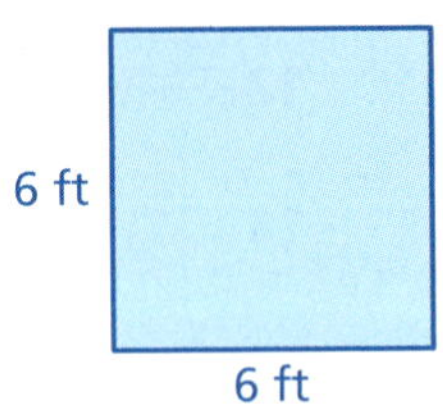

38.

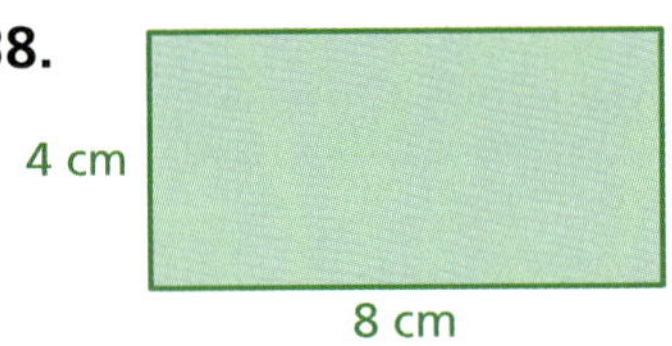

39.

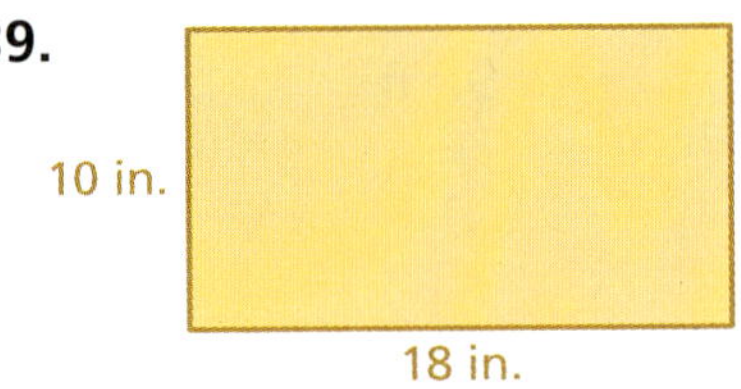

40. WRITING Explain how to continue the method in Example 2 to estimate $\sqrt{71}$ to the nearest hundredth.

41. REPEATED REASONING Describe a method that you can use to estimate a cube root to the nearest tenth. Use your method to estimate $\sqrt[3]{14}$ to the nearest tenth.

42. RADIO SIGNAL The maximum distance (in nautical miles) that a radio transmitter signal can be sent is represented by the expression $1.23\sqrt{h}$, where h is the height (in feet) above the transmitter.

Estimate the maximum distance x (in nautical miles) between the plane that is receiving the signal and the transmitter. Round your answer to the nearest tenth.

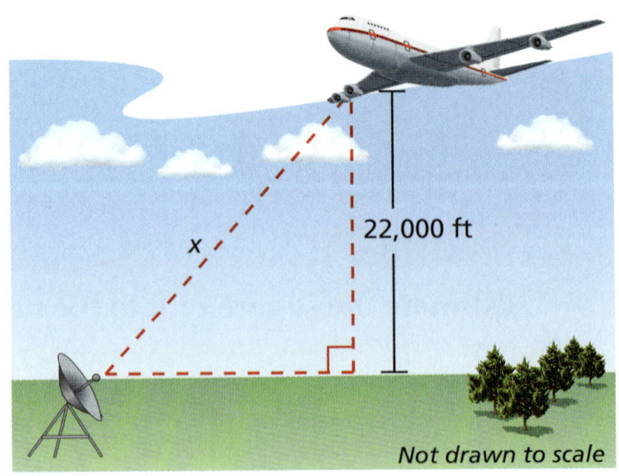

Common Errors

- **Exercises 26–31** Students may guess which is greater just by looking at the numbers. Encourage them to use a number line to compare the numbers. Also remind them to simplify and/or estimate the numbers so that they are easier to compare.
- **Exercises 44–46** Students may struggle estimating the square roots because of the decimals. Remind them of the method they used in Exercises 20–25. Tell them to use the list they wrote, but with the decimal points moved two places to the left. These new perfect squares will help to estimate the square roots.

English Language Learners

Illustrate

Draw a number line on the board. Have students locate $\sqrt{1}$, $\sqrt{4}$, and $\sqrt{9}$ on the number line. Now have them approximate the location of $\sqrt{5}$. Visualizing the $\sqrt{5}$ between 2 and 3 helps students form an approximation. Continue doing this with other square roots.

Practice and Problem Solving

26. 10; 10 is to the right of $\sqrt{20}$.

27. $\sqrt{15}$; $\sqrt{15}$ is positive and -3.5 is negative.

28. $\sqrt{133}$; $\sqrt{133}$ is to the right of $10\frac{3}{4}$.

29. $\frac{2}{3}$; $\frac{2}{3}$ is to the right of $\sqrt{\frac{16}{81}}$.

30. -0.25; -0.25 is to the right of $-\sqrt{0.25}$.

31. $-\sqrt{182}$; $-\sqrt{182}$ is to the right of $-\sqrt{192}$.

32. false **33.** true

34. false **35.** 8.1 ft

36. 118.4 cm **37.** 8.5 ft

38. 8.9 cm **39.** 20.6 in.

40. Create a table of numbers between 8.4 and 8.5 whose squares are close to 71, and then determine which square is closest to 71.

41. Create a table of integers whose cubes are close to the radicand. Determine which two integers the cube root is between. Then create another table of numbers between those two integers whose cubes are close to the radicand. Determine which cube is closest to the radicand; 2.4

42. 182.4 nautical miles

Practice and Problem Solving

43. *Sample answer:* $a = 82$, $b = 97$

44. 0.6 **45.** 1.1

46. 1.2 **47.** 30.1 m/sec

48. yes; $\left(\frac{1}{2}\right)^2 = \frac{1}{4}$, so $\sqrt{\frac{1}{4}} = \frac{1}{2}$.
no; $\left(\frac{\sqrt{3}}{4}\right)^2 = \frac{3}{16}$, and $\sqrt{3}$ is irrational.

49. See *Taking Math Deeper*.

50. Sample answers are given.

- **a.** always; The product of two fractions is a fraction. $\frac{2}{3} \cdot \frac{3}{4} = \frac{1}{2}$
- **b.** sometimes; $\pi \cdot 0 = 0$ is rational, but $2 \cdot \sqrt{3}$ is irrational.
- **c.** sometimes; $\sqrt{2} \cdot \pi$ is irrational, but $\pi \cdot \frac{1}{\pi}$ is rational.

Fair Game Review

51. 40 m **52.** 24 in.

53. 9 cm **54.** D

Mini-Assessment

Estimate to the nearest integer and the nearest tenth.

1. $\sqrt{65}$ 8; 8.1
2. $\sqrt{99}$ 10; 9.9
3. $\sqrt{\frac{15}{2}}$ 3; 2.7

Which number is greater?

4. $2\frac{11}{12}$, $\sqrt{8}$ $2\frac{11}{12}$
5. $\frac{4}{5}$, $\sqrt{\frac{49}{64}}$ $\sqrt{\frac{49}{64}}$

Taking Math Deeper

Exercise 49

This is a nice science problem. Students learn from this problem that objects do not fall at a linear rate. Their speed increases with each second they are falling.

1 Understand the problem.

A water balloon is dropped from a height of 14 meters. How long does it take the balloon to fall to the ground?

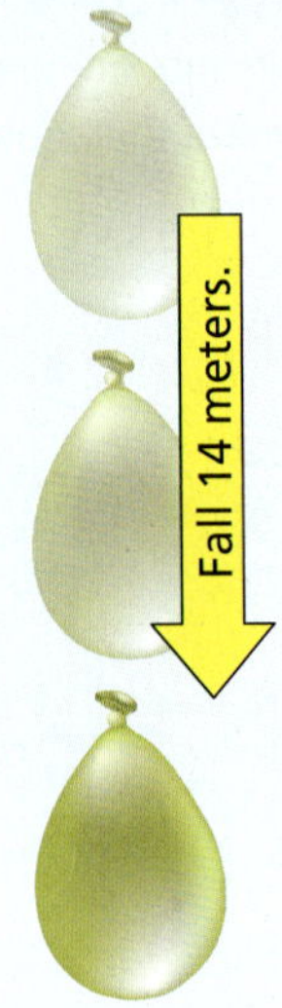

2 Use the given formula.

$$t = \sqrt{\frac{d}{4.9}} = \sqrt{\frac{14}{4.9}} \approx \sqrt{2.86} \approx 1.7$$

3 Answer the question.

The water balloon will hit the ground in about 1.7 seconds.

Project

Use a stop watch, a metric tape measure, and several basketballs. Measure the distance from the top of the bleachers at your school. Drop a ball and record the time it takes to fall to the ground. Use the formula in the problem to calculate the time it should take. Compare.

Reteaching and Enrichment Strategies

If students need help. . .	If students got it. . .
Resources by Chapter • Practice A and Practice B • Puzzle Time Record and Practice Journal Practice Differentiating the Lesson Lesson Tutorials Skills Review Handbook	Resources by Chapter • Enrichment and Extension • Technology Connection Start the next section

43. OPEN-ENDED Find two numbers a and b that satisfy the diagram.

Estimate the square root to the nearest tenth.

44. $\sqrt{0.39}$ **45.** $\sqrt{1.19}$ **46.** $\sqrt{1.52}$

47. ROLLER COASTER The speed s (in meters per second) of a roller-coaster car is approximated by the equation $s = 3\sqrt{6r}$, where r is the radius of the loop. Estimate the speed of a car going around the loop. Round your answer to the nearest tenth.

48. STRUCTURE Is $\sqrt{\frac{1}{4}}$ a rational number? Is $\sqrt{\frac{3}{16}}$ a rational number? Explain.

49. WATER BALLOON The time t (in seconds) it takes a water balloon to fall d meters is represented by the equation $t = \sqrt{\frac{d}{4.9}}$. Estimate the time it takes the balloon to fall to the ground from a window that is 14 meters above the ground. Round your answer to the nearest tenth.

50. Number Sense Determine if the statement is *sometimes, always,* or *never* true. Explain your reasoning and give an example of each.

a. A rational number multiplied by a rational number is rational.

b. A rational number multiplied by an irrational number is rational.

c. An irrational number multiplied by an irrational number is rational.

Fair Game Review What you learned in previous grades & lessons

Find the missing length of the triangle. *(Section 14.3)*

51.

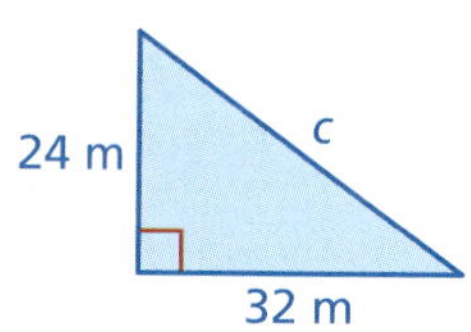

52.

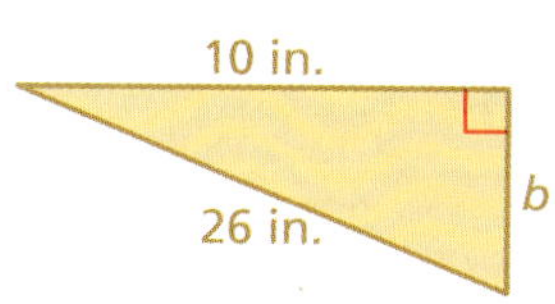

53.

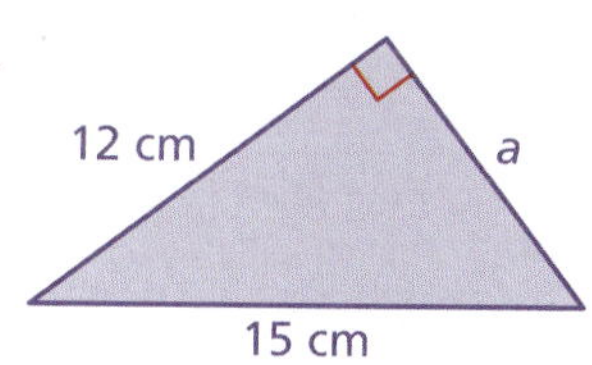

54. MULTIPLE CHOICE What is the ratio (red to blue) of the corresponding side lengths of the similar triangles? *(Section 11.5)*

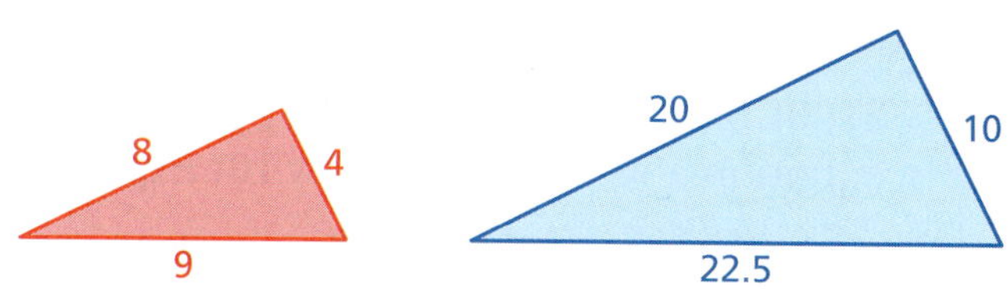

Ⓐ 1:3 Ⓑ 5:2

Ⓒ 3:4 Ⓓ 2:5

Extension 14.4 Repeating Decimals

Check It Out Lesson Tutorials BigIdeasMath.com

You have written terminating decimals as fractions. Because repeating decimals are rational numbers, you can also write repeating decimals as fractions.

Rational Numbers

In this extension, you will

- write a repeating decimal as a fraction.

Key Idea

Writing a Repeating Decimal as a Fraction

Let a variable x equal the repeating decimal d.

Step 1: Write the equation $x = d$.

Step 2: Multiply each side of the equation by 10^n to form a new equation, where n is the number of repeating digits.

Step 3: Subtract the original equation from the new equation.

Step 4: Solve for x.

EXAMPLE 1 Writing a Repeating Decimal as a Fraction (1 Digit Repeats)

Write $0.\overline{4}$ as a fraction in simplest form.

Let $x = 0.\overline{4}$.

$x = 0.\overline{4}$	Step 1: Write the equation.
$10 \cdot x = 10 \cdot 0.\overline{4}$	Step 2: There is 1 repeating digit, so multiply each side by $10^1 = 10$.
$10x = 4.\overline{4}$	Simplify.
$-(x = 0.\overline{4})$	Step 3: Subtract the original equation.
$9x = 4$	Simplify.
$x = \frac{4}{9}$	Step 4: Solve for x.

So, $0.\overline{4} = \frac{4}{9}$.

Check

$$\begin{array}{r} 0.44\ldots \\ 9\overline{)4.00} \\ \underline{36} \\ 40 \\ \underline{36} \\ 40 \end{array}$$ ✓

Practice

Write the decimal as a fraction or a mixed number.

1. $0.\overline{1}$
2. $-0.\overline{5}$
3. $-1.\overline{2}$
4. $5.\overline{8}$

5. **STRUCTURE** In Example 1, why can you subtract the original equation from the new equation after multiplying by 10? Explain why these two steps are performed.

6. **REPEATED REASONING** Compare the repeating decimals and their equivalent fractions in Exercises 1–4. Describe the pattern. Use the pattern to explain how to write a repeating decimal as a fraction when only the tenths digit repeats.

Laurie's Notes

Introduction

Connect

- **Yesterday:** Students approximated irrational numbers.
- **Today:** Students will write repeating decimals as fractions.

Motivate

- Ask students to use a calculator to help write the fractions $\frac{1}{11}, \frac{2}{11}, \frac{3}{11}, \ldots, \frac{10}{11}$ as decimals. To save time, have students work in groups on different fractions instead of trying to write decimal equivalents for all of them.
- Record the results on the board.
- ? "What patterns do you observe?" All of the decimals have two repeating digits, and the sum of those two digits is equal to 9. Also, the first of the two repeating digits is one less than the numerator of the fraction.
- Explain that today they are going to learn how to do the reverse of this process, begin with a repeating decimal, and write it as a fraction.

Lesson Notes

Key Idea

- It may be difficult for students to understand all of these steps until they actually try a problem. Once a problem has been completed, the steps may make more sense. Refer to these steps as you work through the examples.

Example 1

- Write the example. Some students may immediately tell you the correct fraction. When asked how they know, they may say to divide 4 by 9 to get $0.\overline{4}$. But knowing the answer is not the same as showing that $0.\overline{4} = \frac{4}{9}$.
- ? "If you let $x = 0.\overline{4}$, then what is $10x$?" $4.\overline{4}$; Because multiplying a decimal by 10 moves the decimal point one place value to the right.
- Say, "We now have two equations: $x = 0.\overline{4}$ and $10x = 4.\overline{4}$. We are going to subtract one equation from the other."
- **Look for and Make Use of Structure:** When subtracting the original equation, align the repeating decimal portion of each equation as shown.
- Explain that subtracting $0.\overline{4}$ from $4.\overline{4}$ gives a difference of 4.

Practice

- For Exercises 2 and 3, students can ignore the negative sign. The negative sign can be "put back in" after converting the decimal.
- For Exercises 3 and 4, students can "separate" the number into two parts: the integer and the repeating decimal. Then convert the repeating decimal to a fraction, and combine it back with the integer. Alternatively, they can leave the whole number and still begin by multiplying by 10.

What Your Students Will Learn

- Write decimals with one or two digit repeats as fractions.

Goal Today's lesson is writing repeating decimals as fractions.

Technology for the Teacher

Dynamic Classroom

Lesson Tutorials
Lesson Plans
Answer Presentation Tool

Extra Example 1

Write $1.\overline{7}$ as a fraction in simplest form.

$1\frac{7}{9}$

Practice

1. $\frac{1}{9}$
2. $-\frac{5}{9}$
3. $-1\frac{2}{9}$
4. $5\frac{8}{9}$
5. Because the solution does not change when adding/subtracting two equivalent equations; Multiply by 10 so that when you subtract the original equation, the repeating part is removed.
6. Write the digit that repeats in the numerator and use 9 in the denominator

Record and Practice Journal Extension 14.4 Practice

1–11. See Additional Answers.

Extra Example 2

Write $0.4\overline{6}$ as a fraction in simplest form.

$\frac{7}{15}$

Extra Example 3

Write $5.\overline{12}$ as a mixed number. $5\frac{4}{33}$

Practice

7. $-\frac{13}{30}$
8. $2\frac{1}{15}$
9. $\frac{3}{11}$
10. $-4\frac{50}{99}$
11. Pattern: Digits that repeat are in the numerator and 99 is in the denominator; Use 9 as the integer part, 4 as the numerator, and 99 as the denominator of the fractional part.

Mini-Assessment

1. Write $0.\overline{2}$ as a fraction in simplest form. $\frac{2}{9}$
2. Write $0.7\overline{5}$ as a fraction in simplest form. $\frac{34}{45}$
3. Write $-3.\overline{81}$ as a mixed number. $-3\frac{9}{11}$

Laurie's Notes

Example 2

- In many cases, students may be unsure what to do when the repeating digit is not in the tenths place or what to do when there is more than one repeating digit. Examples 2 and 3 address these concerns.
- Write the example.
- ? "What portion of the decimal repeats?" only the digit 3
- ? "If you let $x = -0.2\overline{3}$, then what is $10x$?" $-2.\overline{3}$
- Write the two equations. Some students find it helpful when I write $-2.\overline{3}$ as $-2.3\overline{3}$ so that the bar is not over the tenths place.
- To help better understand Step 3, encourage students to write more digits for the repeating decimals to convince themselves that they will be left with a terminating decimal after subtracting.
- Subtract the equations and solve for x.
- Use a calculator to check the answer or check by dividing 7 by 30.

Example 3

- ? "What is 1.25 written as a fraction in simplest form?" $1\frac{1}{4}$
- ? "What does this fraction tell you?" The answer is approximately $1\frac{1}{4}$
- Let $x = 1.\overline{25}$. Because there are two repeating digits, multiply by 100.
- ? "If $x = 1.\overline{25}$ then what is $100x$?" $125.\overline{25}$
- Subtract the equations and solve for x.
- Use a calculator to check the answer.
- **Extension:** If time permits, you can explore with students what happens when you use the given steps for a repeating decimal that is not written in its "most abbreviated" form. For example, a repeating decimal written as $0.\overline{44}$ instead of $0.\overline{4}$. These steps still work, but the student must simplify a more difficult fraction.
- To challenge students, ask them why multiplying by 10^n works in Step 2. Tell them to focus on the portion that repeats and how the decimal point moves. Encourage them to try some examples with more repeating digits. They may see that it allows for the repeating portion to be "aligned" when subtracting in Step 3. So, it gets eliminated and produces a terminating decimal.

Practice

- Give students sufficient time to work through the exercises. Ask for volunteers to explain their work at the board.

Closure

- Explain the steps to write $2.\overline{35}$ as a mixed number.

EXAMPLE 2 Writing a Repeating Decimal as a Fraction (1 Digit Repeats)

Write $-0.2\overline{3}$ as a fraction in simplest form.

Let $x = -0.2\overline{3}$.

$x = -0.2\overline{3}$	Step 1: Write the equation.
$10 \cdot x = 10 \cdot (-0.2\overline{3})$	Step 2: There is 1 repeating digit, so multiply each side by $10^1 = 10$.
$10x = -2.\overline{3}$	Simplify.
$-(x = -0.2\overline{3})$	Step 3: Subtract the original equation.
$9x = -2.1$	Simplify.
$x = \frac{-2.1}{9}$	Step 4: Solve for *x*.

So, $-0.2\overline{3} = \frac{-2.1}{9} = -\frac{21}{90} = -\frac{7}{30}$.

Check

```
-7/30
  -.2333333333
```

EXAMPLE 3 Writing a Repeating Decimal as a Fraction (2 Digits Repeat)

Write $1.\overline{25}$ as a mixed number.

Let $x = 1.\overline{25}$.

$x = 1.\overline{25}$	Step 1: Write the equation.
$100 \cdot x = 100 \cdot 1.\overline{25}$	Step 2: There are 2 repeating digits, so multiply each side by $10^2 = 100$.
$100x = 125.\overline{25}$	Simplify.
$-(x = 1.\overline{25})$	Step 3: Subtract the original equation.
$99x = 124$	Simplify.
$x = \frac{124}{99}$	Step 4: Solve for *x*.

So, $1.\overline{25} = \frac{124}{99} = 1\frac{25}{99}$.

Check

```
124/99
   1.252525253
```

Practice

Write the decimal as a fraction or a mixed number.

7. $-0.4\overline{3}$ **8.** $2.0\overline{6}$ **9.** $0.\overline{27}$ **10.** $-4.\overline{50}$

11. **REPEATED REASONING** Find a pattern in the fractional representations of repeating decimals in which only the tenths and hundredths digits repeat. Use the pattern to explain how to write $9.\overline{04}$ as a mixed number.

14.5 Using the Pythagorean Theorem

Essential Question In what other ways can you use the Pythagorean Theorem?

The *converse* of a statement switches the hypothesis and the conclusion.

Statement: If p, then q.	Converse of the statement: If q, then p.

1 ACTIVITY: Analyzing Converses of Statements

Work with a partner. Write the converse of the true statement. Determine whether the converse is *true* or *false*. If it is true, justify your reasoning. If it is false, give a counterexample.

a. If $a = b$, then $a^2 = b^2$.

b. If $a = b$, then $a^3 = b^3$.

c. If one figure is a translation of another figure, then the figures are congruent.

d. If two triangles are similar, then the triangles have the same angle measures.

Is the converse of a true statement always true? always false? Explain.

2 ACTIVITY: The Converse of the Pythagorean Theorem

Work with a partner. The converse of the Pythagorean Theorem states: "If the equation $a^2 + b^2 = c^2$ is true for the side lengths of a triangle, then the triangle is a right triangle."

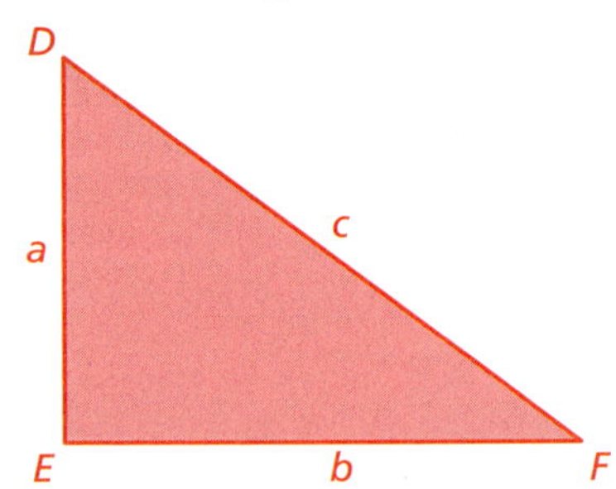

a. Do you think the converse of the Pythagorean Theorem is *true* or *false*? How could you use deductive reasoning to support your answer?

b. Consider $\triangle DEF$ with side lengths a, b, and c, such that $a^2 + b^2 = c^2$. Also consider $\triangle JKL$ with leg lengths a and b, where $\angle K = 90°$.

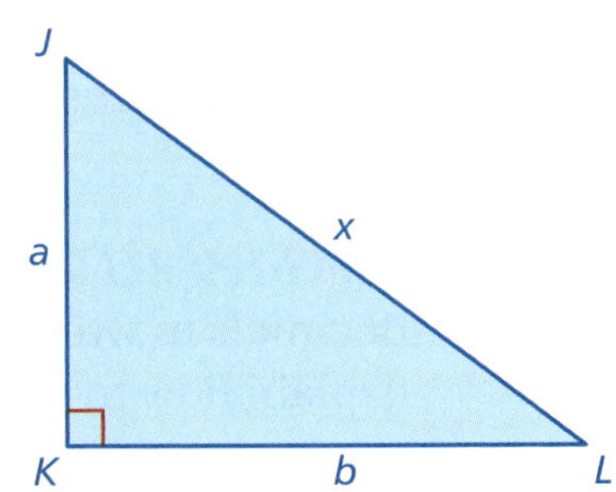

- What does the Pythagorean Theorem tell you about $\triangle JKL$?
- What does this tell you about c and x?
- What does this tell you about $\triangle DEF$ and $\triangle JKL$?
- What does this tell you about $\angle E$?
- What can you conclude?

Pythagorean Theorem

In this lesson, you will

- use the converse of the Pythagorean Theorem to identify right triangles.
- use the Pythagorean Theorem to find distances in a coordinate plane.
- solve real-life problems.

Laurie's Notes

Introduction

Applying Mathematical Practices

- **Construct Viable Arguments and Critique the Reasoning of Others:** Students will develop a "proof" of the converse of the Pythagorean Theorem, and they will derive the distance formula. It is important that students be able to explain the steps in their work and compare it to the reasoning of their classmates.

Motivate

- Explain what the converse of a statement is.
- **Example:** If I live in Moab, then I live in Utah.
  The converse is: If I live in Utah, then I live in Moab.
  The original statement is true, but the converse is false.
- Ask students to write four true if-then statements, two with true converses and two with false converses. Have students share some examples.

Activity Notes

Activity 1

- **Construct Viable Arguments and Critique the Reasoning of Others:** For each true statement, students construct a valid explanation versus simply saying the statement is true. Classmates are expected to listen and critique the explanation offered.
- Review the meaning of the word *counterexample*.
- It is important for students to understand that a statement might be false even if they cannot think of a counterexample.
- **Big Idea:** Even when a conditional statement is true, its converse does not have to be true. Students should keep this in mind for the next activity.

Activity 2

- Students often say that the Pythagorean Theorem is simply $a^2 + b^2 = c^2$. Tell them that it is actually a conditional statement. If a and b are the lengths of the legs and c is the length of the hypotenuse of a right triangle, then $a^2 + b^2 = c^2$.
- ❓ "What is the converse of the Pythagorean Theorem?" If $a^2 + b^2 = c^2$, then a triangle with side lengths a, b, and c is a right triangle.
- ❓ "Do you think the converse of the Pythagorean Theorem is true?" Students may simply guess at this point.
- **Construct Viable Arguments and Critique the Reasoning of Others:** In this activity, students use deductive reasoning to show that the converse is true. They may need guidance in linking their reasoning.
- **FYI:** Strategies for proofs are taught in later courses, so the framework of a proof is provided.
- In the second bullet, listen for $c^2 = x^2$ and then $c = x$. Students found this was not always true in Activity 1 part (a), but point out that it is true here because c and x must be positive.
- Students learned in Chapter 7 that you can only construct one triangle given three side lengths. Review this concept for the third bullet.

What Your Students Will Learn

- Use the converse of the Pythagorean Theorem to identify right triangles.
- Use the Pythagorean Theorem to derive the Distance Formula.
- Find distances between two points in a coordinate plane.

Previous Learning

Students should know how to use the Pythagorean Theorem.

Technology for the Teacher

Dynamic Classroom

Lesson Plans
Complete Materials List

14.5 Record and Practice Journal

Essential Question In what other ways can you use the Pythagorean Theorem?

The *converse* of a statement switches the hypothesis and the conclusion.

Statement: If p, then q.	Converse of the statement: If q, then p.

1 ACTIVITY: Analyzing Converses of Statements

Work with a partner. Write the converse of the true statement. Determine whether the converse is *true* or *false*. If it is true, justify your reasoning. If it is false, give a counterexample.

a. If $a = b$, then $a^2 = b^2$.
Converse: If $a^2 = b^2$, then $a = b$. false

b. If $a = b$, then $a^3 = b^3$.
Converse: If $a^3 = b^3$, then $a = b$. true

c. If one figure is a translation of another figure, then the figures are congruent.
Converse: If two figures are congruent, then one is a translation of the other figure. false

d. If two triangles are similar, then the triangles have the same angle measures.
Converse: If two triangles have the same angle measures, then the triangles are similar. true

Is the converse of a true statement always true? always false? Explain.
neither

English Language Learners

Comprehension

English language learners may struggle with the concept of the converse of a statement. Some may think that the converse is always true. Give an example where the converse of a statement is not true.

Statement: *If a figure is a square, then it has four right angles.*

Converse of the Statement: *If a figure has four right angles, then the figure is a square.*

The converse of the statement is not always true. The figure could be a rectangle.

14.5 Record and Practice Journal

2 ACTIVITY: The Converse of the Pythagorean Theorem

Work with a partner. The converse of the Pythagorean Theorem states: "If the equation $a^2 + b^2 = c^2$ is true for the side lengths of a triangle, then the triangle is a right triangle."

a. Do you think the converse of the Pythagorean Theorem is *true* or *false*? How could you use deductive reasoning to support your answer?

true

b. Consider $\triangle DEF$ with side lengths a, b, and c, such that $a^2 + b^2 = c^2$. Also consider $\triangle JKL$ with leg lengths a and b, where $\angle K = 90°$.

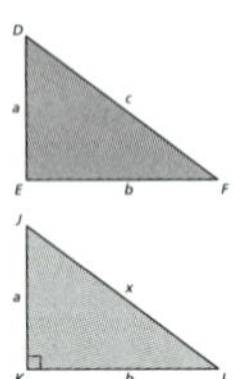

- What does the Pythagorean Theorem tell you about $\triangle JKL$?

  $a^2 + b^2 = x^2$

- What does this tell you about c and x?

  $c = x$

- What does this tell you about $\triangle DEF$ and $\triangle JKL$?

  $\triangle DEF \cong \triangle JKL$

- What does this tell you about $\angle E$?

  $\angle E = 90°$

- What can you conclude?

  $\triangle DEF$ is a right triangle.

3 ACTIVITY: Developing the Distance Formula

Work with a partner. Follow the steps below to write a formula that you can use to find the distance between and two points in a coordinate plane.

Step 1: Choose two points in the coordinate plane that do not lie on the same horizontal or vertical line. Label the points (x_1, y_1) and (x_2, y_2).

Sample answer:

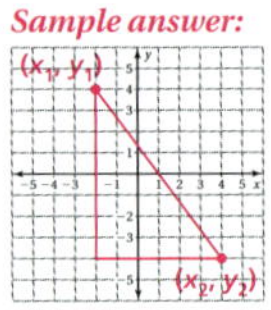

Step 2: Draw a line segment connecting the points. This will be the hypotenuse of a right triangle.

Step 3: Draw horizontal and vertical line segments from the points to form the legs of the right triangle.

Step 4: Use the x-coordinates to write an expression for the length of the horizontal leg.

$|x_2 - x_1|$

Step 5: Use the y-coordinates to write an expression for the length of the vertical leg.

$|y_2 - y_1|$

Step 6: Substitute the expressions for the lengths of the legs into the Pythagorean Theorem.

$(x_2 - x_1)^2 + (y_2 - y_1)^2 = c^2$

Step 7: Solve the equation in Step 6 for the hypotenuse c.

$c = \sqrt{(x_2 - x_1)^2 + (y_2 - y_1)^2}$

What does the length of the hypotenuse tell you about the two points?

It is the distance between the points.

What Is Your Answer?

4. **IN YOUR OWN WORDS** In what other ways can you use the Pythagorean Theorem?

   Show a triangle is a right triangle; Find the distance between two points in a plane.

5. What kind of real-life problems do you think the converse of the Pythagorean Theorem can help you solve?

   Check students' work.

Laurie's Notes

Activity 3

- Visually, it will be helpful for students to select lattice points.
- In Step 3, point out that there are two distinct ways they can draw the legs. They can draw them either way because the two possible triangles are congruent.
- **Look for and Make Use of Structure:** In Steps 4 and 5, students may ask about the order in which the subtraction is performed. Tell them that in Step 6 these expressions are squared. So, the order in which the subtraction is performed does not matter.
- Give students adequate time to read carefully and work through the steps on their own. Resist the temptation to jump in and solve it for them.

What Is Your Answer?

- **Think-Pair-Share:** Students should read each question independently and then work in pairs to answer the questions. When they have answered the questions, the pair should compare their answers with another group and discuss any discrepancies.

Closure

- **Writing Prompt:** The Pythagorean Theorem can be used to . . . find missing side lengths of right triangles, find the distances between points in a coordinate plane, determine whether given side lengths form a right triangle, etc.

3 ACTIVITY: Developing the Distance Formula

Work with a partner. Follow the steps below to write a formula that you can use to find the distance between any two points in a coordinate plane.

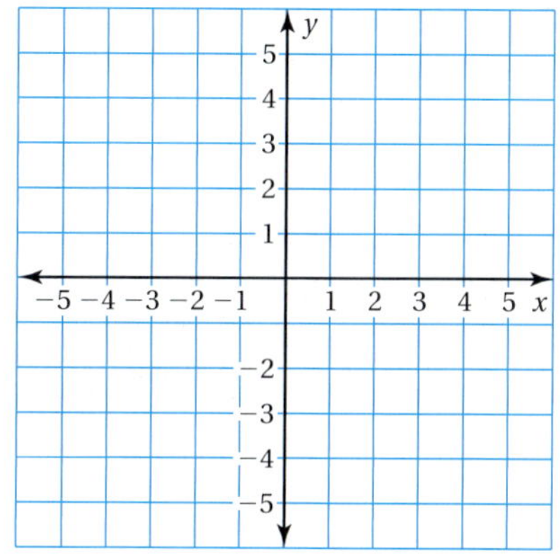

> **Math Practice**
>
> **Communicate Precisely**
>
> What steps can you take to make sure that you have written the distance formula accurately?

Step 1: Choose two points in the coordinate plane that do not lie on the same horizontal or vertical line. Label the points (x_1, y_1) and (x_2, y_2).

Step 2: Draw a line segment connecting the points. This will be the hypotenuse of a right triangle.

Step 3: Draw horizontal and vertical line segments from the points to form the legs of the right triangle.

Step 4: Use the x-coordinates to write an expression for the length of the horizontal leg.

Step 5: Use the y-coordinates to write an expression for the length of the vertical leg.

Step 6: Substitute the expressions for the lengths of the legs into the Pythagorean Theorem.

Step 7: Solve the equation in Step 6 for the hypotenuse c.

What does the length of the hypotenuse tell you about the two points?

What Is Your Answer?

4. **IN YOUR OWN WORDS** In what other ways can you use the Pythagorean Theorem?

5. What kind of real-life problems do you think the converse of the Pythagorean Theorem can help you solve?

Practice Use what you learned about the converse of a true statement to complete Exercises 3 and 4 on page 660.

14.5 Lesson

Check It Out
Lesson Tutorials
BigIdeasMath.com

Key Vocabulary
distance formula, *p. 658*

Key Ideas

Converse of the Pythagorean Theorem

If the equation $a^2 + b^2 = c^2$ is true for the side lengths of a triangle, then the triangle is a right triangle.

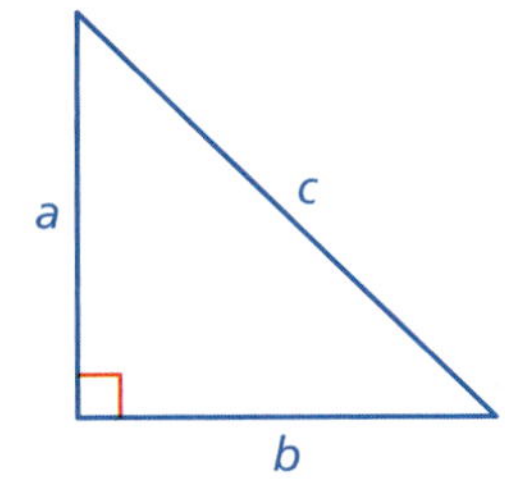

EXAMPLE 1 Identifying a Right Triangle

Tell whether each triangle is a right triangle.

a.

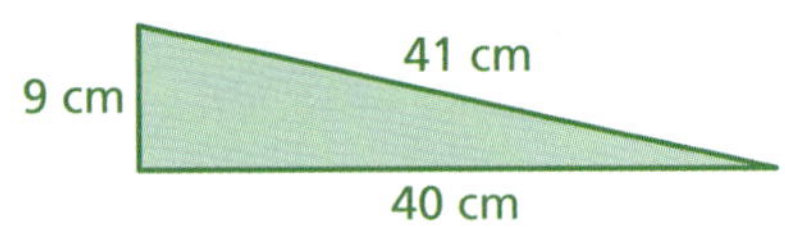

$$a^2 + b^2 = c^2$$
$$9^2 + 40^2 \stackrel{?}{=} 41^2$$
$$81 + 1600 \stackrel{?}{=} 1681$$
$$1681 = 1681 \checkmark$$

∴ It *is* a right triangle.

b.

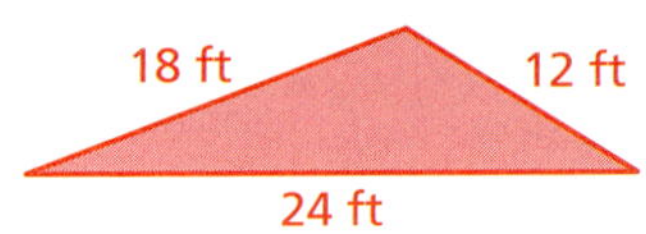

$$a^2 + b^2 = c^2$$
$$12^2 + 18^2 \stackrel{?}{=} 24^2$$
$$144 + 324 \stackrel{?}{=} 576$$
$$468 \neq 576 \quad ✗$$

∴ It is *not* a right triangle.

Study Tip
A *Pythagorean triple* is a set of three positive integers *a*, *b*, and *c*, where $a^2 + b^2 = c^2$.

Common Error
When using the converse of the Pythagorean Theorem, always substitute the length of the longest side for *c*.

On Your Own

Tell whether the triangle with the given side lengths is a right triangle.

1. 28 in., 21 in., 20 in.

2. 1.25 mm, 1 mm, 0.75 mm

Now You're Ready
Exercises 5–10

On page 657, you used the Pythagorean Theorem to develop the *distance formula*. You can use the **distance formula** to find the distance between any two points in a coordinate plane.

Key Idea

Distance Formula

The distance d between any two points (x_1, y_1) and (x_2, y_2) is given by the formula

$$d = \sqrt{(x_2 - x_1)^2 + (y_2 - y_1)^2}.$$

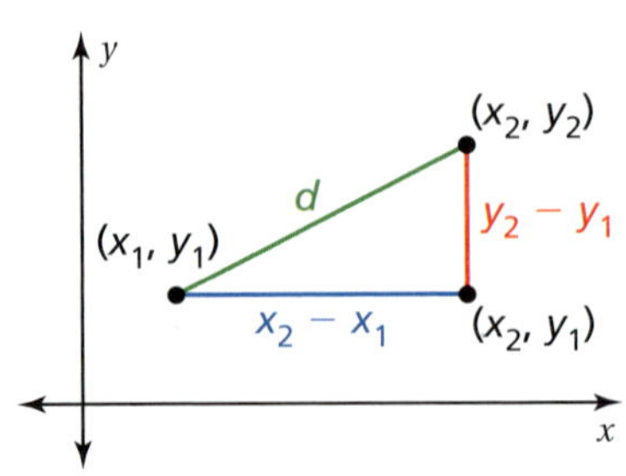

Multi-Language Glossary at BigIdeasMath.com

Laurie's Notes

Introduction

Connect

- **Yesterday:** Students proved the converse of the Pythagorean Theorem and developed the distance formula.
- **Today:** Students will use the converse of the Pythagorean Theorem and the distance formula.

Motivate

- Write the following numbers on the board: 3-4-5, 5-12-13, and 8-15-17. Tell students that these are called *Pythagorean triples.* They are positive integers that satisfy the Pythagorean Theorem.
- Multiples of these examples also satisfy the Pythagorean Theorem. Have students try to name a few more Pythagorean triples.

Lesson Notes

Key Idea

- Write the Key Idea stating the converse of the Pythagorean Theorem. It is one way of determining whether given side lengths form a right triangle.

Example 1

- If the three numbers satisfy the theorem, then it is a right triangle. Explain that you are not using eyesight to decide if it is a right triangle.
- Substitute the side lengths for each triangle. Remind students that the longest side is substituted for c.
- Because 9, 40, and 41 satisfy the theorem, they form a Pythagorean triple and the triangle is a right triangle.
- Because 12, 18, and 24 do not satisfy the theorem, it is not a right triangle.
- **Extension:** The 12, 18, 24 triangle is similar to a triangle with side lengths 2, 3, and 4 (scale factor is 6). The 2, 3, 4 triangle is not a right triangle either.

On Your Own

- **Common Error:** Students may not substitute the longest side for c. This is particularly true because the measures are listed longest to shortest.

Key Idea

- Write the Key Idea that states the distance formula, derived from the Pythagorean Theorem.
- Note the use of colors in the diagram. The same colors can be used when writing the formula.
- **Look for and Make Use of Structure:** Discuss with students that the order in which the subtraction is performed is not important because the difference is squared.

Goal Today's lesson is using the converse of the Pythagorean Theorem and the **distance formula**.

Technology for the Teacher

Dynamic Classroom

Lesson Tutorials
Lesson Plans
Answer Presentation Tool

Extra Example 1

Tell whether each triangle is a right triangle.

a.

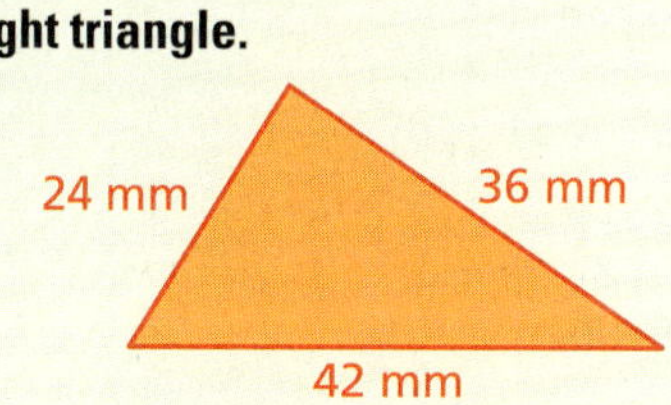

not a right triangle

b.

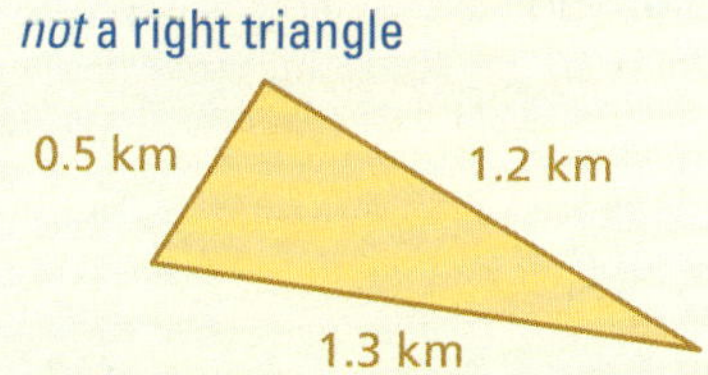

right triangle

On Your Own

1. no
2. yes

Laurie's Notes

Extra Example 2

Find the distance between (1, −5) and (7, 4). $\sqrt{117}$

Example 2

- Although it is not necessary, you can plot the two points as a visual aid.
- The choice of which point is (x_1, y_1) and which point is (x_2, y_2) is arbitrary. If time permits, do the problem both ways to show that the result is the same.
- Caution students to be careful with the subtraction. It is easy to make a careless calculation mistake.
- There are no units associated with the answer. In a real-life example, students would need to label the units in their answers.

Extra Example 3

In Example 3, your friend starts at (40, 40). Did your friend make a 90° turn? no

Example 3

- Ask a student to read the example.
- **Model with Mathematics:** Sketch the coordinate plane shown with the ordered pairs identified.

? "How do you determine if the receiver ran the play as designed?" Check whether the triangle is a right triangle.

? "How do you determine if the triangle is a right triangle?" Use the converse of the Pythagorean Theorem.

- Use the distance formula to find the length of each side of the triangle.
- Use the converse of the Pythagorean Theorem to show that it is a right triangle.
- Note that students could also use what they know about slopes of perpendicular lines to solve this problem.

On Your Own

3. $\sqrt{41}$
4. $\sqrt{85}$
5. 5
6. no

On Your Own

- There are multiple steps required in each question. You may wish to divide the class into groups and assign a different question to each group.

Closure

- **Exit Ticket:** Find the distance between (−4, 6) and (3, −2). $\sqrt{113}$

Differentiated Instruction

Inclusion

Encourage students to learn the Pythagorean Theorem using the language of the triangle, $\text{leg}^2 + \text{leg}^2 = \text{hypotenuse}^2$. Have them label each side as a leg or hypotenuse before substituting the numbers into the equation.

EXAMPLE 2 Finding the Distance Between Two Points

Find the distance between (1, 5) and (−4, −2).

Let $(x_1, y_1) = (1, 5)$ and $(x_2, y_2) = (-4, -2)$.

$$d = \sqrt{(x_2 - x_1)^2 + (y_2 - y_1)^2} \quad \text{Write the distance formula.}$$

$$= \sqrt{(-4 - 1)^2 + (-2 - 5)^2} \quad \text{Substitute.}$$

$$= \sqrt{(-5)^2 + (-7)^2} \quad \text{Simplify.}$$

$$= \sqrt{25 + 49} \quad \text{Evaluate powers.}$$

$$= \sqrt{74} \quad \text{Add.}$$

EXAMPLE 3 Real-Life Application

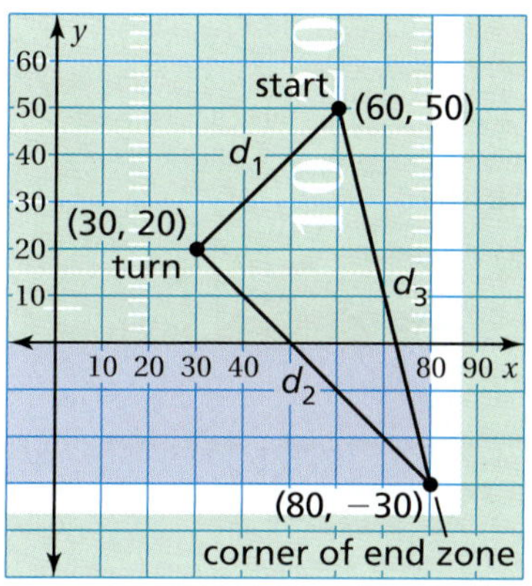

You design a football play in which a player runs down the field, makes a 90° turn, and runs to the corner of the end zone. Your friend runs the play as shown. Did your friend make a 90° turn? Each unit of the grid represents 10 feet.

Use the distance formula to find the lengths of the three sides.

$$d_1 = \sqrt{(60 - 30)^2 + (50 - 20)^2} = \sqrt{30^2 + 30^2} = \sqrt{1800} \text{ feet}$$

$$d_2 = \sqrt{(80 - 30)^2 + (-30 - 20)^2} = \sqrt{50^2 + (-50)^2} = \sqrt{5000} \text{ feet}$$

$$d_3 = \sqrt{(80 - 60)^2 + (-30 - 50)^2} = \sqrt{20^2 + (-80)^2} = \sqrt{6800} \text{ feet}$$

Use the converse of the Pythagorean Theorem to determine if the side lengths form a right triangle.

$$(\sqrt{1800})^2 + (\sqrt{5000})^2 \stackrel{?}{=} (\sqrt{6800})^2$$

$$1800 + 5000 \stackrel{?}{=} 6800$$

$$6800 = 6800 \checkmark$$

The sides form a right triangle.

∴ So, your friend made a 90° turn.

On Your Own

Now You're Ready
Exercises 11–16

Find the distance between the two points.

3. (0, 0), (4, 5) **4.** (7, −3), (9, 6) **5.** (−2, −3), (−5, 1)

6. **WHAT IF?** In Example 3, your friend made the turn at (20, 10). Did your friend make a 90° turn?

14.5 Exercises

Check It Out
Help with Homework
BigIdeasMath.com

Vocabulary and Concept Check

1. **WRITING** Describe two ways to find the distance between two points in a coordinate plane.

2. **WHICH ONE DOESN'T BELONG?** Which set of numbers does *not* belong with the other three? Explain your reasoning.

3, 6, 8	6, 8, 10	5, 12, 13	7, 24, 25

Practice and Problem Solving

Write the converse of the true statement. Determine whether the converse is *true* or *false*. If it is true, justify your reasoning. If it is false, give a counterexample.

3. If a is an odd number, then a^2 is odd.

4. If $ABCD$ is a square, then $ABCD$ is a parallelogram.

Tell whether the triangle with the given side lengths is a right triangle.

(1) 5.
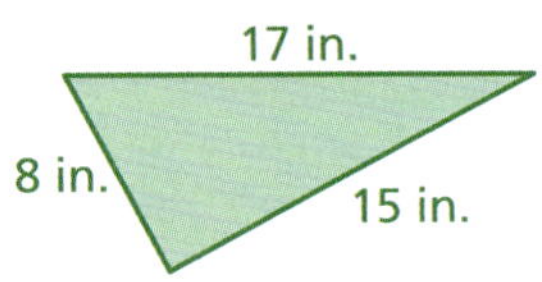

6.
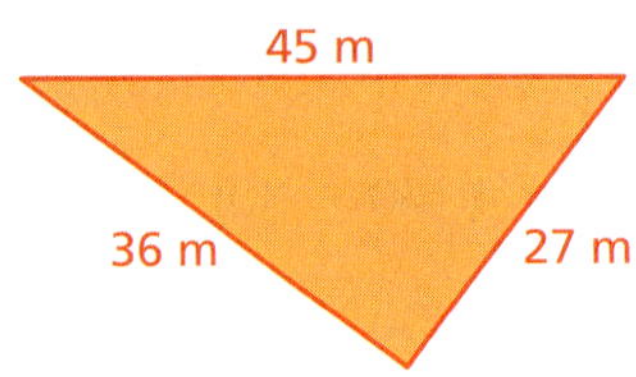

7.
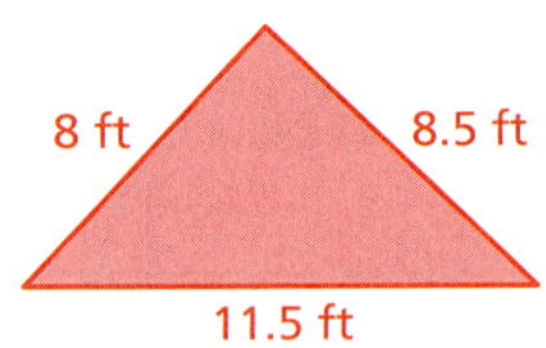

8. 14 mm, 19 mm, 23 mm

9. $\frac{9}{10}$ mi, $1\frac{1}{5}$ mi, $1\frac{1}{2}$ mi

10. 1.4 m, 4.8 m, 5 m

Find the distance between the two points.

(2) 11. (1, 2), (7, 6)

12. (4, −5), (−1, 7)

13. (2, 4), (7, 2)

14. (−1, −3), (1, 3)

15. (−6, −7), (0, 0)

16. (12, 5), (−12, −2)

17. **ERROR ANALYSIS** Describe and correct the error in finding the distance between the points (−3, −2) and (7, 4).

$$d = \sqrt{[7-(-3)]^2 - [4-(-2)]^2}$$
$$= \sqrt{100 - 36}$$
$$= \sqrt{64} = 8$$

18. **CONSTRUCTION** A post and beam frame for a shed is shown in the diagram. Does the brace form a right triangle with the post and beam? Explain.

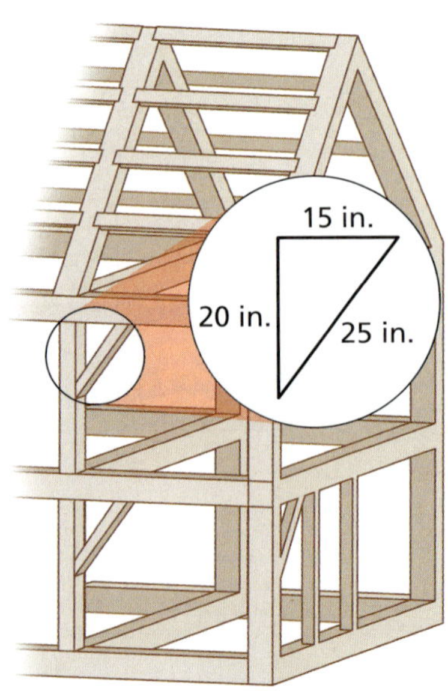

Assignment Guide and Homework Check

Level	Assignment	Homework Check
Accelerated	1–4, 8–16 even, 17, 18, 20, 22–30	10, 16, 20, 25, 26

Common Errors

- **Exercises 5–10** Students may substitute the wrong value for c in the Pythagorean Theorem. Remind them that c will be the longest side, so they should substitute the greatest value for c.
- **Exercises 11–16** Students may mismatch the x-values and y-values when using the distance formula. This will result in students subtracting an x from a y, or vice versa. Encourage students to pair the numbers properly.
- **Exercises 11–16** Students may get careless when squaring negative numbers. Remind them that everything inside the parentheses is squared, including the minus sign, and that the square of a negative is a positive.
- **Exercise 25** Students may pick Plane A because it appears to be closer. Remind students that the drawing is not to scale. Tell them to calculate the distances before answering the question.

14.5 Record and Practice Journal

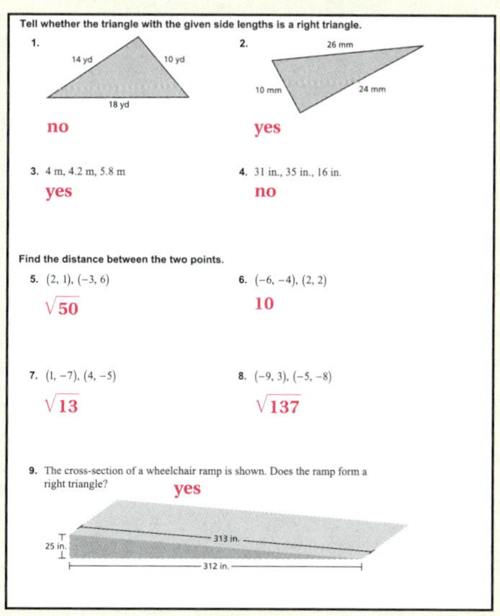

Tell whether the triangle with the given side lengths is a right triangle.

1. (triangle: 14 yd, 10 yd, 18 yd) **no**
2. (triangle: 26 mm, 10 mm, 24 mm) **yes**
3. 4 m, 4.2 m, 5.8 m **yes**
4. 31 in., 35 in., 16 in. **no**

Find the distance between the two points.

5. (2, 1), (−3, 6) $\sqrt{50}$
6. (−6, −4), (2, 2) **10**
7. (1, −7), (4, −5) $\sqrt{13}$
8. (−9, 3), (−5, −8) $\sqrt{137}$
9. The cross-section of a wheelchair ramp is shown. Does the ramp form a right triangle? **yes**

Vocabulary and Concept Check

1. the Pythagorean Theorem and the distance formula
2. 3, 6, 8; It is the only set that is not a Pythagorean triple.

Practice and Problem Solving

3. If a^2 is odd, then a is an odd number; true when a is an integer; A product of two integers is odd only when each integer is odd.
4. If $ABCD$ is a parallelogram, then $ABCD$ is a square; false; counterexample: any parallelogram that does not have right angles
5. yes
6. yes
7. no
8. no
9. yes
10. yes
11. $2\sqrt{13}$
12. 13
13. $\sqrt{29}$
14. $2\sqrt{10}$
15. $\sqrt{85}$
16. 25
17. The squared quantities under the radical should be added not subtracted; $2\sqrt{34}$
18. yes; The side lengths satisfy the converse of the Pythagorean Theorem.

Practice and Problem Solving

19. yes **20.** no

21. yes

22. yes; Use the distance formula to find the lengths of the three sides. Use the converse of the Pythagorean Theorem to show they form a right triangle.

23. no; The measures of the side lengths are $\sqrt{5000}$, $\sqrt{3700}$, and $\sqrt{8500}$ and $(\sqrt{5000})^2 + (\sqrt{3700})^2 \neq (\sqrt{8500})^2$.

24. yes; $\sqrt{58}$; Because you square the differences $(x_2 - x_1)$ and $(y_2 - y_1)$, it does not matter if the differences are positive or negative. The squares of opposite numbers are equivalent.

25. See *Taking Math Deeper*.

26. See Additional Answers.

Fair Game Review

27. mean: 13; median: 12.5; mode: 12

28. mean: 21; median: 21; no mode

29. mean: 58; median: 59; mode: 59

30. B

Mini-Assessment

Tell whether the triangle with the given side lengths is a right triangle.

1. 32 m, 56 m, 64 m no

2. 1.8 mi, 8 mi, 8.2 mi yes

Find the distance between the two points.

3. $(-3, -1), (6, 2)$ $\sqrt{90}$

4. $(2, 10), (5, -4)$ $\sqrt{205}$

Taking Math Deeper

Exercise 25

At first this seems like a simple question. Students may think that Plane A is 5 kilometers from the base of the tower, and Plane B is 7 kilometers from the base of the tower. So, Plane A seems closer. However, on second glance, you see that Plane A is much higher than Plane B. So, to see which is closer, you need to compute the diagonal distance of each.

1 Find the distance for Plane A.

$$x^2 \approx 5^2 + 6.1^2$$
$$x^2 = 62.21$$
$$x \approx 7.89 \text{ km}$$

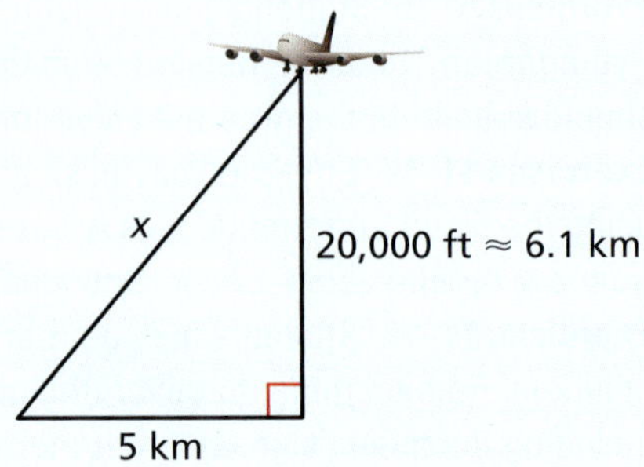

2 Find the distance for Plane B.

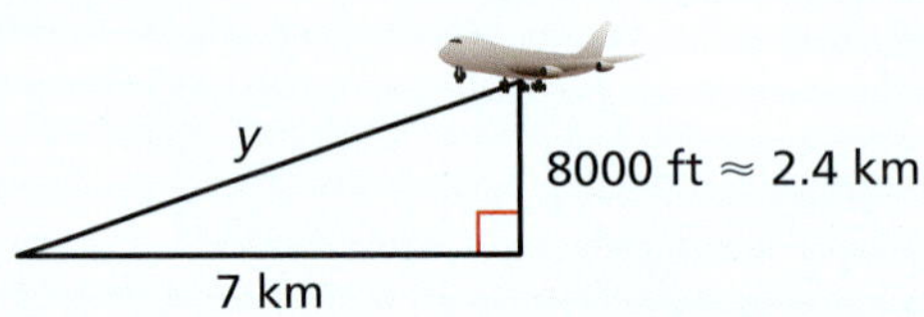

$$y^2 \approx 7^2 + 2.4^2$$
$$y^2 = 54.76$$
$$y = 7.4 \text{ km}$$

3 Answer the question.

Plane B is slightly closer to the base of the tower.

Reteaching and Enrichment Strategies

If students need help...	If students got it...
Resources by Chapter • Practice A and Practice B • Puzzle Time Record and Practice Journal Practice Differentiating the Lesson Lesson Tutorials Skills Review Handbook	Resources by Chapter • Enrichment and Extension • Technology Connection Start the next section

Tell whether a triangle with the given side lengths is a right triangle.

19. $\sqrt{63}$, 9, 12

20. 4, $\sqrt{15}$, 6

21. $\sqrt{18}$, $\sqrt{24}$, $\sqrt{42}$

22. REASONING Plot the points (−1, 3), (4, −2), and (1, −5) in a coordinate plane. Are the points the vertices of a right triangle? Explain.

23. GEOCACHING You spend the day looking for hidden containers in a wooded area using a Global Positioning System (GPS). You park your car on the side of the road, and then locate Container 1 and Container 2 before going back to the car. Does your path form a right triangle? Explain. Each unit of the grid represents 10 yards.

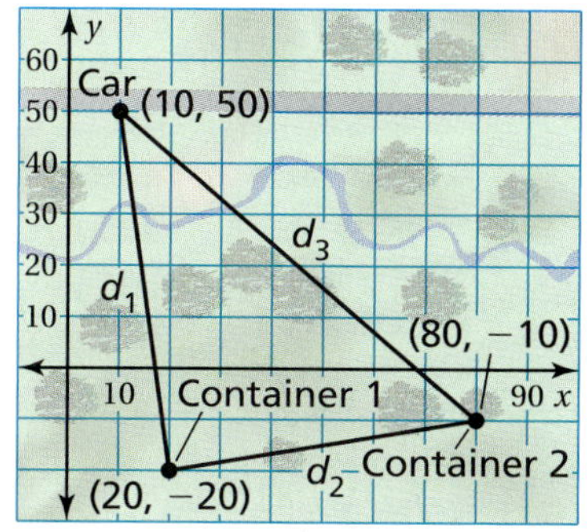

24. REASONING Your teacher wants the class to find the distance between the two points (2, 4) and (9, 7). You use (2, 4) for (x_1, y_1), and your friend uses (9, 7) for (x_1, y_1). Do you and your friend obtain the same result? Justify your answer.

25. AIRPORT Which plane is closer to the base of the airport tower? Explain.

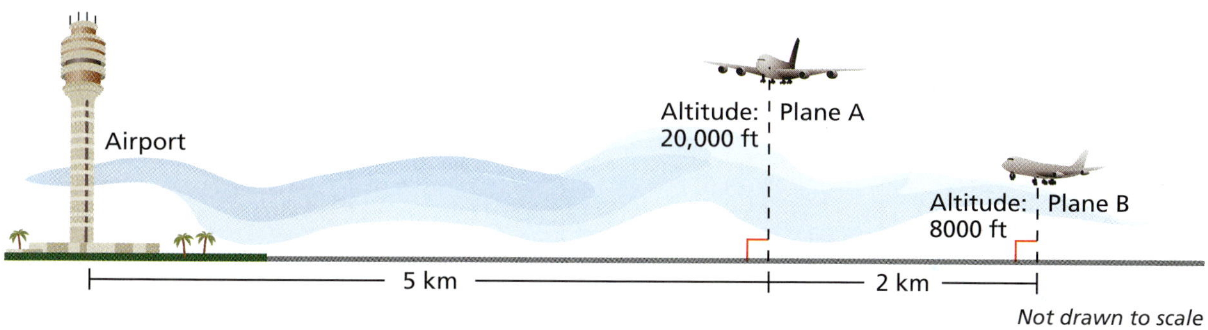

26. Structure Consider the two points (x_1, y_1) and (x_2, y_2) in the coordinate plane. How can you find the point (x_m, y_m) located in the middle of the two given points? Justify your answer using the distance formula.

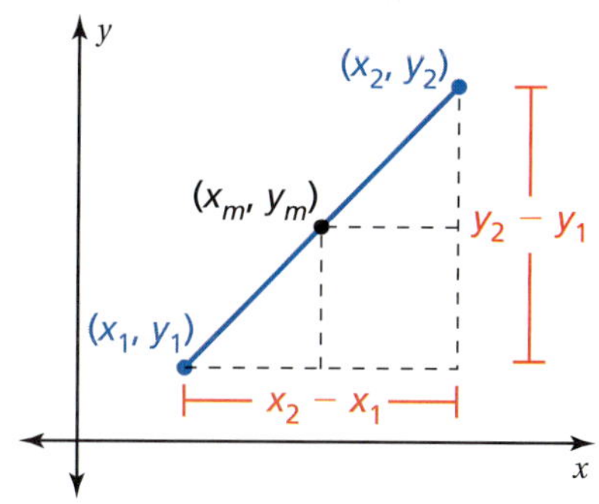

Fair Game Review What you learned in previous grades & lessons

Find the mean, median, and mode of the data. *(Skills Review Handbook)*

27. 12, 9, 17, 15, 12, 13

28. 21, 32, 16, 27, 22, 19, 10

29. 67, 59, 34, 71, 59

30. MULTIPLE CHOICE What is the sum of the interior angle measures of an octagon? *(Section 12.3)*

Ⓐ 720° Ⓑ 1080° Ⓒ 1440° Ⓓ 1800°

14.4–14.5 Quiz

Check It Out
Progress Check
BigIdeasMath.com

Classify the real number. *(Section 14.4)*

1. $-\sqrt{225}$

2. $-1\frac{1}{9}$

3. $\sqrt{41}$

4. $\sqrt{17}$

Estimate the square root to the nearest (a) integer and (b) tenth. *(Section 14.4)*

5. $\sqrt{38}$

6. $-\sqrt{99}$

7. $\sqrt{172}$

8. $\sqrt{115}$

Which number is greater? Explain. *(Section 14.4)*

9. $\sqrt{11}, 3\frac{3}{5}$

10. $\sqrt{1.44}, 1.1\overline{8}$

Write the decimal as a fraction or a mixed number. *(Section 14.4)*

11. $0.\overline{7}$

12. $-1.\overline{63}$

Tell whether the triangle with the given side lengths is a right triangle. *(Section 14.5)*

13.

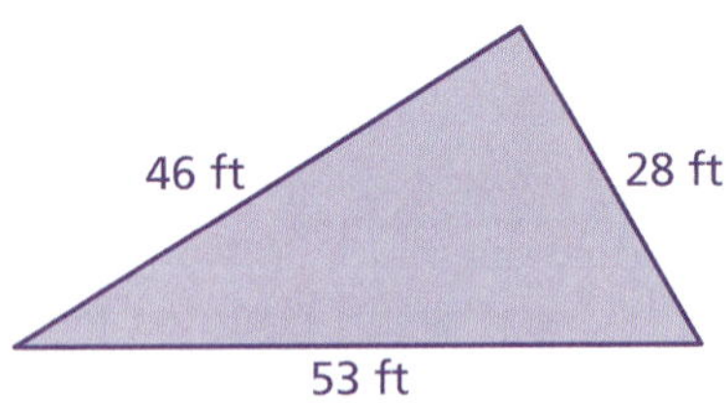

14.

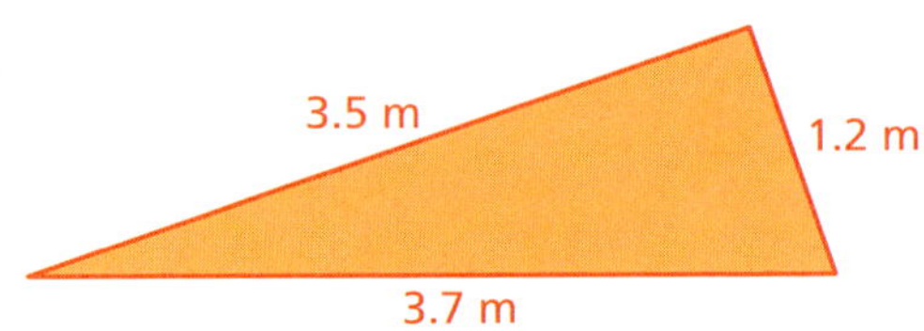

Find the distance between the two points. *(Section 14.5)*

15. $(-3, -1), (-1, -5)$

16. $(-4, 2), (5, 1)$

17. $(1, -2), (4, -5)$

18. $(-1, 1), (7, 4)$

19. $(-6, 5), (-4, -6)$

20. $(-1, 4), (1, 3)$

Use the figure to answer Exercises 21–24. Round your answer to the nearest tenth. *(Section 14.5)*

21. How far is the cabin from the peak?

22. How far is the fire tower from the lake?

23. How far is the lake from the peak?

24. You are standing at $(-5, -6)$. How far are you from the lake?

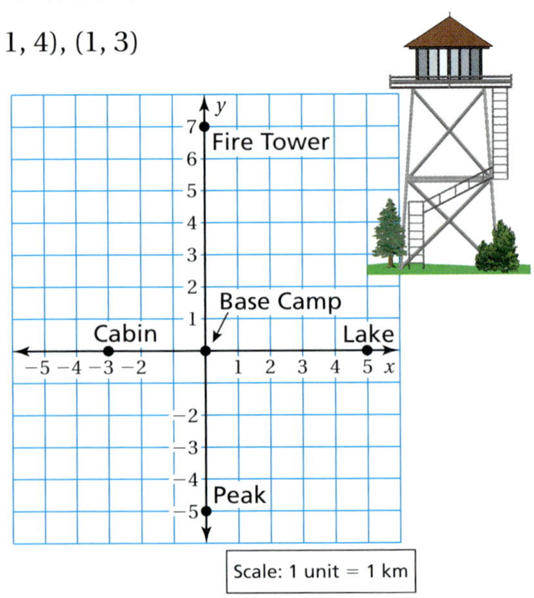

Alternative Assessment Options

Math Chat | **Student Reflective Focus Question**
Structured Interview | Writing Prompt

Student Reflective Focus Question

Ask students to summarize approximating square roots and using the Pythagorean Theorem. Be sure that they include examples. Select students at random to present their summaries to the class.

Study Help Sample Answers

Remind students to complete Graphic Organizers for the rest of the chapter.

3.

Irrational numbers

Definition

An *irrational number* cannot be written as the ratio of two integers.

Words

- The square root of any whole number that is not a perfect square is irrational. The cube root of any integer that is not a perfect cube is irrational.
- The decimal form of an irrational number neither terminates nor repeats.

Examples

Number	Reasoning
0.010010001...	The decimal neither terminates nor repeats.
π	The decimal form of π neither terminates nor repeats.
$\sqrt{18}$	18 is not a perfect square.
$\sqrt[3]{-9}$	−9 is not a perfect cube.

Non-Examples

Number	Reasoning
−17	−17 can be written as a ratio of two integers: $\frac{-17}{1}$.
$0.\overline{13}$	$0.\overline{13}$ is a repeating decimal.
$\sqrt[3]{-27}$	$\sqrt[3]{-27}$ is equal to −3 because −27 is a perfect cube.
$\sqrt{144}$	$\sqrt{144}$ is equal to 12 because 144 is a perfect square.

4–6. Available at *BigIdeasMath.com.*

Reteaching and Enrichment Strategies

If students need help. . .	If students got it. . .
Resources by Chapter • Practice A and Practice B • Puzzle Time Lesson Tutorials *BigIdeasMath.com*	Resources by Chapter • Enrichment and Extension • Technology Connection Game Closet at *BigIdeasMath.com* Start the Chapter Review

Answers

1. integer, rational
2. rational
3. irrational
4. irrational
5. **a.** 6
   b. 6.2
6. **a.** −10
   b. −9.9
7. **a.** 13
   b. 13.1
8. **a.** 11
   b. 10.7
9. $3\frac{3}{5}$; $3\frac{3}{5}$ is to the right of $\sqrt{11}$.
10. $\sqrt{1.44}$; $\sqrt{1.44}$ is to the right of $1.1\overline{8}$.

11. $\frac{7}{9}$ **12.** $-1\frac{7}{11}$

13. no **14.** yes

15. $2\sqrt{5}$ **16.** $\sqrt{82}$

17. $3\sqrt{2}$ **18.** $\sqrt{73}$

19. $5\sqrt{5}$ **20.** $\sqrt{5}$

21. 5.8 km **22.** 8.6 km

23. 7.1 km **24.** 11.7 km

Technology for the Teacher

Online Assessment
Assessment Book
ExamView® Assessment Suite

For the Teacher
Additional Review Options

- *BigIdeasMath.com*
- Online Assessment
- Game Closet at *BigIdeasMath.com*
- Vocabulary Help
- Resources by Chapter

Answers

1. 1
2. $-\frac{3}{5}$
3. 1.3 and -1.3
4. -9
5. $3\frac{2}{3}$
6. -30
7. 9
8. $\frac{4}{7}$
9. $-\frac{2}{3}$
10. -13
11. 17
12. -42

Review of Common Errors

Exercises 1–3

- Students may divide the number by two instead of finding a number that, when multiplied by itself, gives the radicand. Remind them that taking the square root of a number is the inverse of squaring a number.

Exercises 4–6

- Remind students of the order of operations. Because taking a square root is the inverse of squaring, it is evaluated before multiplication and division.

Exercise 9

- Students may disregard the negative in the radicand. Remind them to pay attention to the sign of the radicand and to check the sign of the answer.

Exercises 10–12

- Remind students of the order of operations. Because taking a cube root is the inverse of cubing, it is evaluated before multiplication and division.

Exercises 13, 14, 24, 25

- Students may substitute the given lengths in the wrong part of the formula. Remind them that the side opposite the right angle is the hypotenuse c.

Exercises 15–17

- Students may not classify the real number in as many ways as possible. Remind students that real numbers can have more than one classification.

Exercises 18–20

- Encourage students to write the first 10 perfect squares at the top of their papers as a reminder and a reference.

Exercises 26 and 27

- Students may mismatch the x-values and y-values when using the distance formula. This will result in students subtracting an x from a y, or vice versa. Encourage students to pair the numbers properly.
- Students may get careless when squaring negative numbers. Remind them that everything inside the parentheses is squared, including the minus sign, and that the square of a negative is a positive.

14 Chapter Review

Check It Out
Vocabulary Help
BigIdeasMath.com

Review Key Vocabulary

square root, *p. 628*
perfect square, *p. 628*
radical sign, *p. 628*
radicand, *p. 628*
cube root, *p. 634*
perfect cube, *p. 634*
theorem, *p. 638*
legs, *p. 640*
hypotenuse, *p. 640*
Pythagorean Theorem, *p. 640*
irrational number, *p. 648*
real numbers, *p. 648*
distance formula, *p. 658*

Review Examples and Exercises

14.1 Finding Square Roots (*pp. 626–631*)

Find $-\sqrt{36}$.

$-\sqrt{36}$ represents the *negative* square root.

Because $6^2 = 36$, $-\sqrt{36} = -\sqrt{6^2} = -6$.

Exercises

Find the square root(s).

1. $\sqrt{1}$

2. $-\sqrt{\frac{9}{25}}$

3. $\pm\sqrt{1.69}$

Evaluate the expression.

4. $15 - 4\sqrt{36}$

5. $\sqrt{\frac{54}{6}} + \frac{2}{3}$

6. $10(\sqrt{81} - 12)$

14.2 Finding Cube Roots (*pp. 632–637*)

Find $\sqrt[3]{\frac{125}{216}}$.

Because $\left(\frac{5}{6}\right)^3 = \frac{125}{216}$, $\sqrt[3]{\frac{125}{216}} = \sqrt[3]{\left(\frac{5}{6}\right)^3} = \frac{5}{6}$.

Exercises

Find the cube root.

7. $\sqrt[3]{729}$

8. $\sqrt[3]{\frac{64}{343}}$

9. $\sqrt[3]{-\frac{8}{27}}$

Evaluate the expression.

10. $\sqrt[3]{27} - 16$

11. $25 + 2\sqrt[3]{-64}$

12. $3\sqrt[3]{-125} - 27$

14.3 The Pythagorean Theorem (pp. 638–643)

Find the length of the hypotenuse of the triangle.

$a^2 + b^2 = c^2$	Write the Pythagorean Theorem.
$7^2 + 24^2 = c^2$	Substitute.
$49 + 576 = c^2$	Evaluate powers.
$625 = c^2$	Add.
$\sqrt{625} = \sqrt{c^2}$	Take positive square root of each side.
$25 = c$	Simplify.

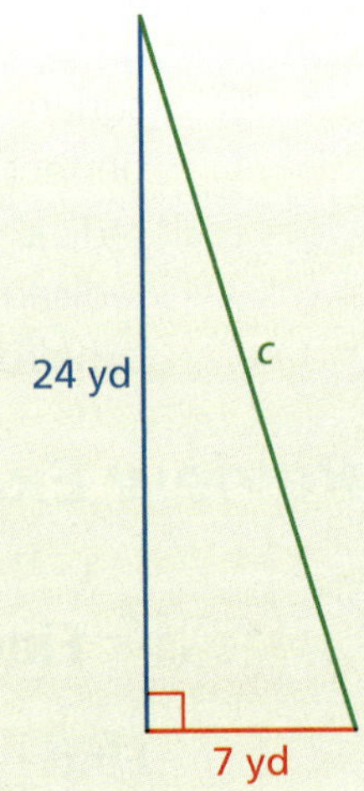

The length of the hypotenuse is 25 yards.

Exercises

Find the missing length of the triangle.

13. 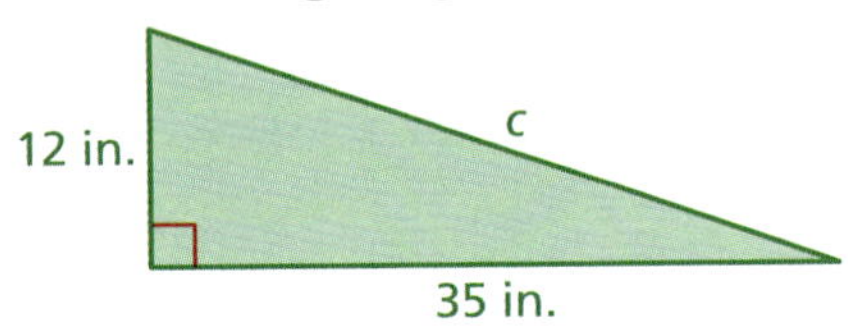

14. 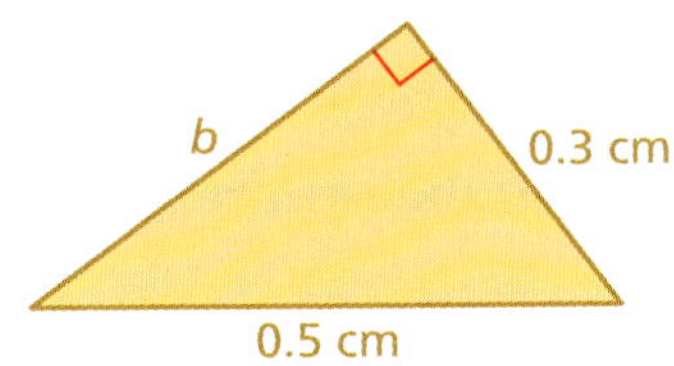

14.4 Approximating Square Roots (pp. 646–655)

a. Classify $\sqrt{19}$.

The number $\sqrt{19}$ is irrational because 19 is not a perfect square.

b. Estimate $\sqrt{34}$ to the nearest integer.

Make a table of numbers whose squares are close to the radicand, 34.

Number	4	5	6	7
Square of Number	16	25	36	49

The table shows that 34 is between the perfect squares 25 and 36. Because 34 is closer to 36 than to 25, $\sqrt{34}$ is closer to 6 than to 5.

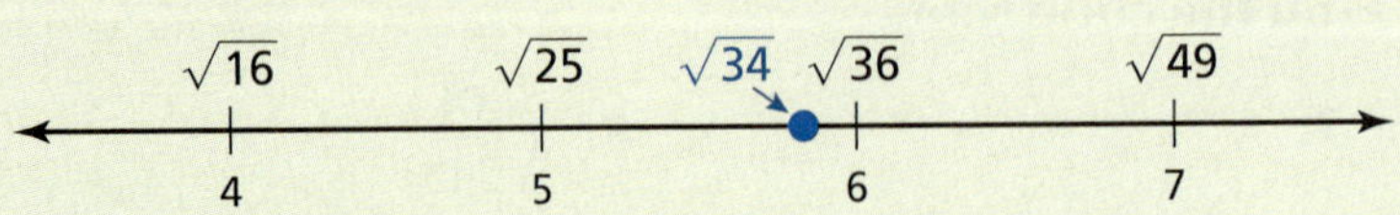

So, $\sqrt{34} \approx 6$.

Review Game

Significant Square Roots

Materials per Group

- piece of paper
- pencil

Directions

Divide the class into groups of 3 or 4.

Each group is to come up with 5 significant numbers and compute the exact square root of each number.

Examples of significant numbers:

School address: 1764 Knowledge Road; $\sqrt{1764} = 42$

Year of presidential election: 1936—Franklin D. Roosevelt elected to his second term; $\sqrt{1936} = 44$

Age for driver's license: 16 years old; $\sqrt{16} = 4$

Who wins?

The first group to come up with five significant numbers and correct square roots wins.

For the Student
Additional Practice

- Lesson Tutorials
- Multi-Language Glossary
- Self-Grading Progress Check
- *BigIdeasMath.com*
  Dynamic Student Edition
  Student Resources

Answers

13. 37 in.

14. 0.4 cm

15. rational

16. irrational

17. natural, whole, integer, rational

18. **a.** 4

b. 3.7

19. **a.** 9

b. 9.5

20. **a.** 13

b. 13.2

21. $\frac{8}{9}$

22. $\frac{4}{11}$

23. $-1\frac{2}{3}$

24. yes

25. no

26. $5\sqrt{5}$

27. $\sqrt{113}$

My Thoughts on the Chapter

Teacher Tip

Not allowed to write in your teaching edition? Use sticky notes to record your thoughts.

What worked. . .

What did not work. . .

What I would do differently. . .

Test Item References

Chapter Test Questions	Section to Review
1–3, 7, 8	14.1
4–6, 9, 10	14.2
11, 21	14.3
12–17	14.4
18–20	14.5

Test-Taking Strategies

Remind students to quickly look over the entire test before they start so that they can budget their time. Students should estimate and check their answers for reasonableness as they work through the test. Teach the students to use the Stop and Think strategy before answering. **Stop** and carefully read the question, and **Think** about what the answer should look like.

Common Errors

- **Exercises 1–3** Remind students that a square root can be positive or negative.
- **Exercises 7 and 8** Remind students of the order of operations. Because taking a square root is the inverse of squaring, it is evaluated before multiplication and division.
- **Exercises 11, 18, 21** Students may substitute the given lengths in the wrong part of the formula. Remind them that the side opposite the right angle is the hypotenuse *c*.
- **Exercises 12 and 13** Students may not classify the real number in as many ways as possible. Remind students that real numbers can have more than one classification.

Reteaching and Enrichment Strategies

If students need help. . .	If students got it. . .
Resources by Chapter • Practice A and Practice B • Puzzle Time Record and Practice Journal Practice Differentiating the Lesson Lesson Tutorials *BigIdeasMath.com* Skills Review Handbook	Resources by Chapter • Enrichment and Extension • Technology Connection Game Closet at *BigIdeasMath.com* Start Cumulative Assessment

Answers

1. -40
2. $\frac{5}{7}$
3. $\frac{10}{3}$ and $-\frac{10}{3}$
4. -3
5. $\frac{2}{5}$
6. $-2\frac{1}{4}$
7. 44
8. $6\frac{1}{2}$
9. -50
10. 54
11. 10 in.
12. irrational
13. integer, rational
14. **a.** 8
    b. 7.6
15. **a.** 9
    b. 9.1
16. $-\frac{1}{3}$
17. $1\frac{8}{33}$
18. yes
19. 10
20. $2\sqrt{13}$
21. 66 ft

Technology for the Teacher

Online Assessment
Assessment Book
ExamView® Assessment Suite

Test-Taking Strategies

Available at *BigIdeasMath.com*

After Answering Easy Questions, Relax
Answer Easy Questions First
Estimate the Answer
Read All Choices before Answering
Read Question before Answering
Solve Directly or Eliminate Choices
Solve Problem before Looking at Choices
Use Intelligent Guessing
Work Backwards

About this Strategy

When taking a multiple choice test, be sure to read each question carefully and thoroughly. When taking a timed test, it is often best to skim the test and answer the easy questions first. Be careful that you record your answer in the correct position on the answer sheet.

Answers

1. D
2. H
3. B
4. F

Technology for the Teacher

Performance Tasks
Online Assessment
Assessment Book
ExamView® Assessment Suite

Item Analysis

1. **A.** The student adds 1.1 and 4.
   B. The student multiplies 1.1 by 4.
   C. The student adds 1.1 and 2.
   D. Correct answer

2. **F.** The student thinks that the parallelograms are similar because the corresponding angles are congruent. But, the corresponding sides are not proportional.
   G. The student thinks that the parallelograms are similar because the corresponding sides are proportional. But, the corresponding angles are not congruent.
   H. Correct answer
   I. The student thinks that the parallelograms are similar because the corresponding sides are proportional. But, the corresponding angles are not congruent.

3. **A.** The student does not realize that all graphs of proportional relationships contain the point (0, 0).
   B. Correct answer
   C. The student does not realize that (2, 6) is a multiple of (1, 3).
   D. The student does not realize that (6, 18) is a multiple of (1, 3).

4. **F.** Correct answer
   G. The student has the correct slope but the wrong y-intercept, perhaps confusing the x- and y-intercepts.
   H. The student has the correct y-intercept but the wrong slope.
   I. The student has the wrong slope and the wrong y-intercept, perhaps confusing the x- and y-intercepts.

14 Cumulative Assessment

Test-Taking Strategy
Answer Easy Questions First

"Scan the test and **answer the easy questions first**. You know the square root of 4 is 2."

1. The period T of a pendulum is the time, in seconds, it takes the pendulum to swing back and forth. The period can be found using the formula $T = 1.1\sqrt{L}$, where L is the length, in feet, of the pendulum. A pendulum has a length of 4 feet. Find its period.

A. 5.1 sec **C.** 3.1 sec

B. 4.4 sec **D.** 2.2 sec

2. Which parallelogram is a dilation of parallelogram $JKLM$? (Figures not drawn to scale.)

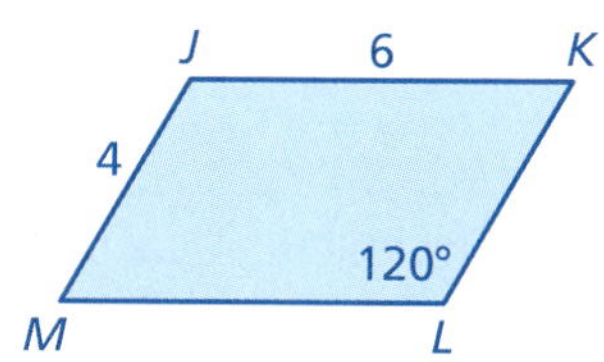

F.

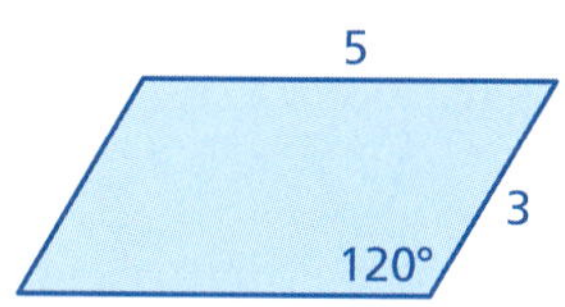

H.

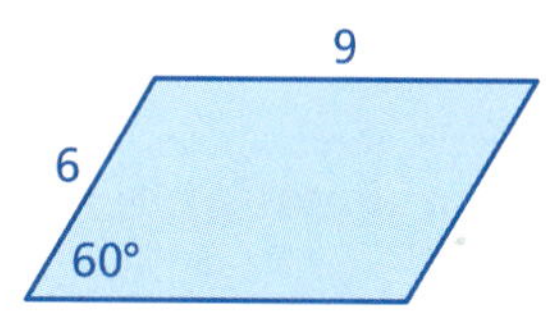

G.

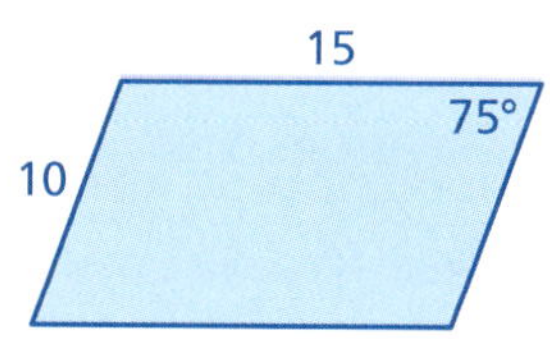

I.

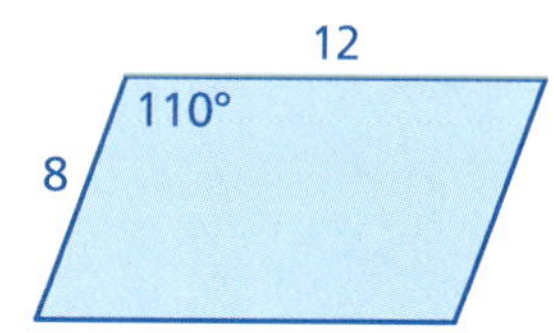

3. The point (1, 3) is on the graph of a proportional relationship. Which point is *not* on the graph?

A. (0, 0) **C.** (2, 6)

B. (2, 4) **D.** (6, 18)

4. Which linear equation matches the line shown in the graph?

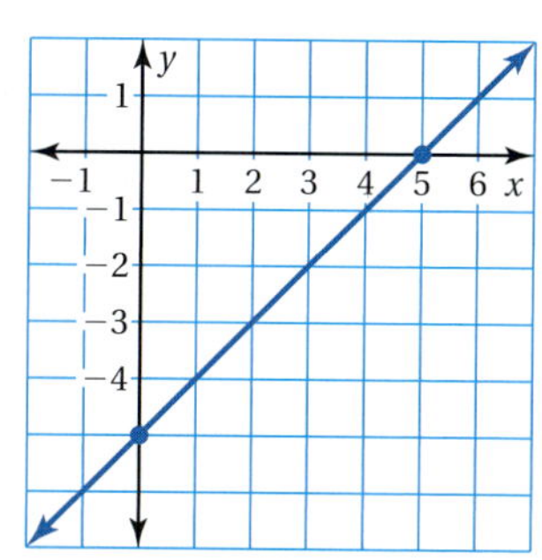

F. $y = x - 5$ **H.** $y = -x - 5$

G. $y = x + 5$ **I.** $y = -x + 5$

5. A football field is 40 yards wide and 120 yards long. Find the distance between opposite corners of the football field. Show your work and explain your reasoning.

Think Solve Explain

6. A computer consultant charges \$50 plus \$40 for each hour she works. The consultant charged \$650 for one job. This can be represented by the equation below, where h represents the number of hours worked.

$$40h + 50 = 650$$

How many hours did the consultant work?

7. You can use the formula below to find the sum S of the interior angle measures of a polygon with n sides. Solve the formula for n.

$$S = 180(n - 2)$$

A. $n = 180(S - 2)$

B. $n = \frac{S}{180} + 2$

C. $n = \frac{S}{180} - 2$

D. $n = \frac{S}{180} + \frac{1}{90}$

8. Which linear equation relates y to x?

x	1	2	3	4	5
y	4	2	0	−2	−4

F. $y = 2x + 2$

G. $y = 4x$

H. $y = -2x + 2$

I. $y = -2x + 6$

9. An airplane flies from City 1 at (0, 0) to City 2 at (33, 56) and then to City 3 at (23, 32). What is the total number of miles it flies? Each unit of the coordinate grid represents 1 mile.

10. What is the missing length of the right triangle shown?

A. 16 cm

B. 18 cm

C. 24 cm

D. $\sqrt{674}$ cm

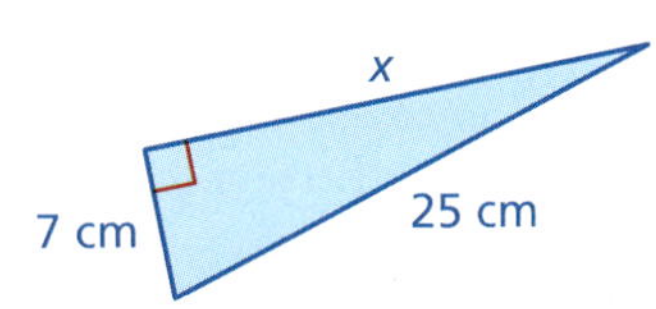

Item Analysis (continued)

5. **2 points** The student demonstrates a thorough understanding of how to apply the Pythagorean Theorem to the problem, explains the work fully, and calculates the distance accurately. The distance between opposite corners is $\sqrt{16,000} \approx 126.5$ yards.

   1 point The student's work and explanations demonstrate a lack of essential understanding. The Pythagorean Theorem is misstated or, if stated correctly, is applied incorrectly to the problem.

   0 points The student provides no response, a completely incorrect or incomprehensible response, or a response that demonstrates insufficient understanding of the Pythagorean Theorem.

6. **Gridded Response:** Correct answer: 15 hours

   Common Error: The student adds 50 to the right hand side of the equation instead of subtracting, yielding an answer of 17.5.

7. **A.** The student interchanges the roles of *n* and *S*.

   B. Correct answer

   C. The student subtracts 2 from *S* instead of adding it.

   D. The student adds 2 first, then divides through by 180.

8. **F.** The student has the wrong sign for the slope and makes a mistake finding the *y*-intercept.

   G. The student bases the equation on the first column only.

   H. The student finds the correct slope, but makes a mistake finding the *y*-intercept.

   I. Correct answer

9. **Gridded Response:** Correct answer: 91

   Common Error: The student only finds the distance between City 1 and City 2, yielding an answer of 65.

10. **A.** The student takes the average of the two sides.

    B. The student subtracts the shorter from the longer side.

    C. Correct answer

    D. The student treats the missing leg as the hypotenuse.

Answers

5. about 126.5 yards
6. 15 hours
7. B
8. I
9. 91 miles
10. C

Answers

11. H

12. D

13. F

Item Analysis (continued)

11. **F.** The student uses a scale factor of 2 instead of $\frac{1}{2}$.

G. The student uses a scale factor of $\frac{1}{2}$ instead of 2.

H. Correct answer

I. The student uses a scale factor of 2 instead of $\frac{1}{2}$.

12. **A.** The student fails to see that these are supplementary, not congruent angles.

B. The student thinks these are alternate interior (or alternate exterior) angles, and believes they are congruent.

C. The student thinks these are corresponding angles, and believes they are congruent.

D. Correct answer

13. **F.** Correct answer

G. The student picks slope of 2 instead of -2.

H. The student picks y-intercept of 2 instead of -2.

I. The student picks y-intercept of 2 instead of -2 and slope of 2 instead of -2.

11. Which sequence of transformations shows that the triangles in the coordinate plane are similar?

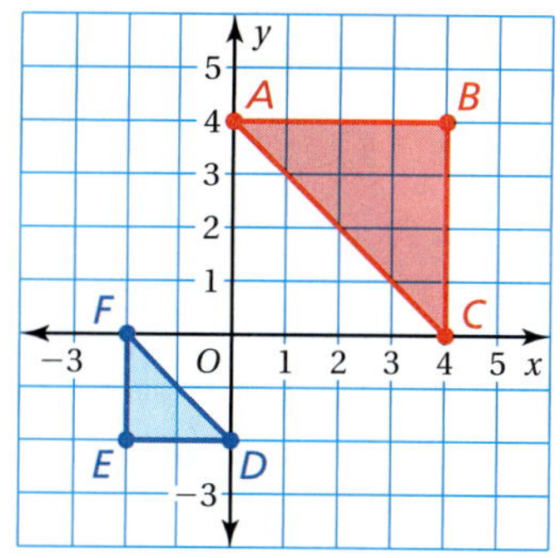

F. Dilate $\triangle ABC$ using a scale factor of 2 and then rotate 180° about the origin.

G. Dilate $\triangle DEF$ using a scale factor of $\frac{1}{2}$ and then rotate 180° about the origin.

H. Reflect $\triangle DEF$ in both axes and then dilate using a scale factor of 2.

I. Rotate $\triangle ABC$ 180° about the origin and then dilate using a scale factor of 2.

12. In the diagram, lines ℓ and m are parallel. Which angle has the same measure as $\angle 1$?

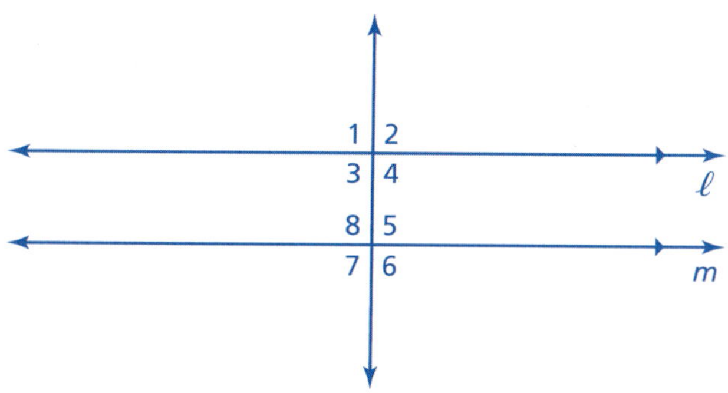

A. $\angle 2$

B. $\angle 5$

C. $\angle 7$

D. $\angle 8$

13. Which graph represents the linear equation $y = -2x - 2$?

F.

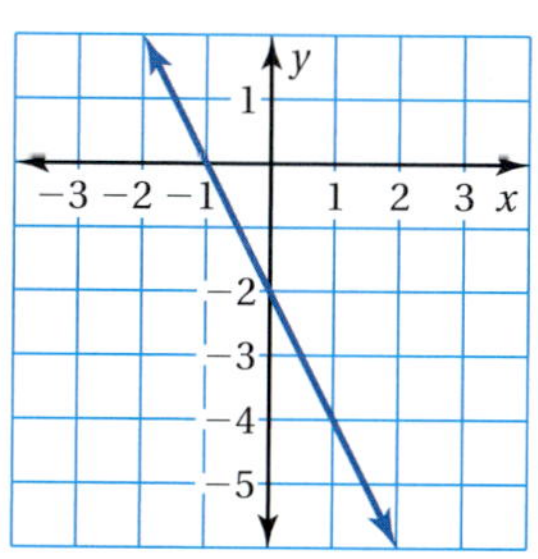

H.

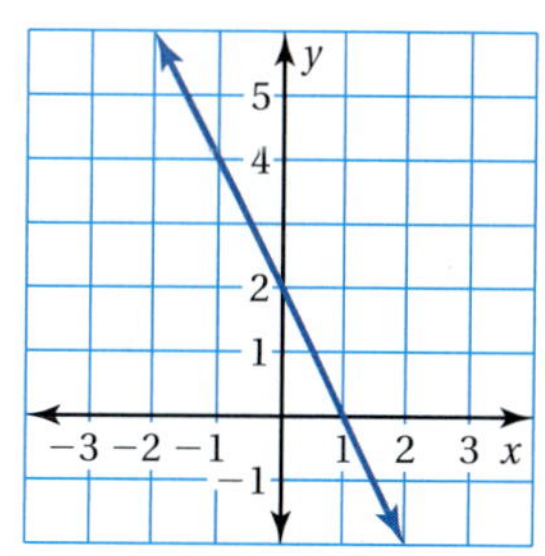

G.

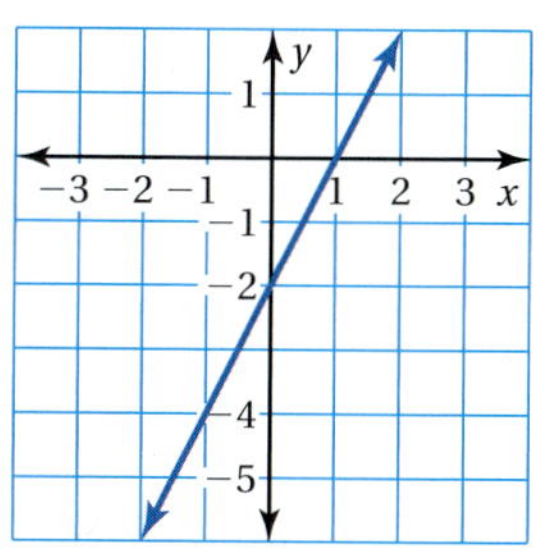

I.

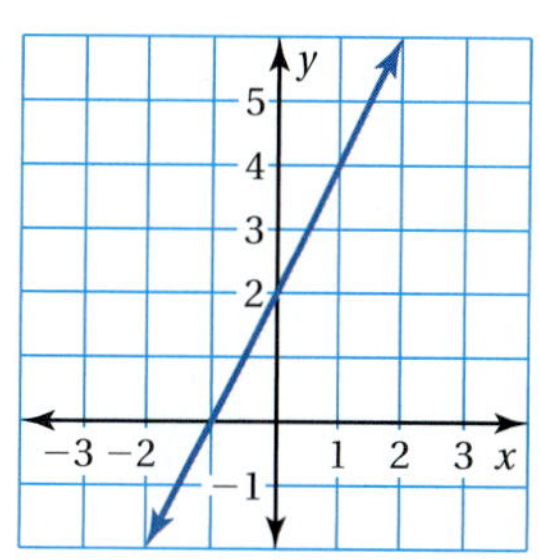

15 Volume and Similar Solids

15.1 Volumes of Cylinders

15.2 Volumes of Cones

15.3 Volumes of Spheres

15.4 Surface Areas and Volumes of Similar Solids

What Your Students Have Learned

- Find volumes of right rectangular prisms using formulas.
- Find areas of triangles, special quadrilaterals, and polygons.
- Use nets made up of rectangles and triangles to find surface areas.
- Find areas and circumferences of circles.
- Solve problems involving area, volume, and surface area of objects composed of triangles, quadrilaterals, cubes, and right prisms.
- Represent proportional relationships with equations.
- Decide whether two quantities are proportional.
- Use scale drawings to compute actual lengths and areas.

What Your Students Will Learn

- Know and apply the formulas for the volumes of cones, cylinders, and spheres.
- Describe a sequence that exhibits similarity between two figures.
- Understand the relationship between surface areas or volumes of similar solids.

Pacing Guide for Chapter 15

Chapter Opener Accelerated	1 Day
Section 1 Accelerated	1 Day
Section 2 Accelerated	1 Day
Section 3 Accelerated	1 Day
Section 4 Accelerated	2 Days
Chapter Review/ Chapter Tests Accelerated	2 Days
Total Chapter 15 Accelerated	8 Days
Year-to-Date Accelerated	141 Days

Technology for the Teacher

BigIdeasMath.com
Chapter at a Glance
Complete Materials List
Parent Letters: English and Spanish

What Your Students Have Learned

- Find the areas of composite figures using the area formulas of known shapes that make up the figures.
- Find the areas of circles using a formula.

Additional Topics for Review

- Square Roots and Cube Roots
- Volumes of Prisms and Pyramids
- Cross Sections of Three-Dimensional Figures
- Writing and Solving Proportions
- Perimeters and Areas of Similar Figures
- Surface Areas of Prisms, Pyramids, and Cylinders

Try It Yourself

1. about 145.12 m^2
2. 86 cm^2
3. about 78.5 ft^2
4. about 530.66 $in.^2$
5. about 38.465 cm^2

Record and Practice Journal Fair Game Review

1. 51 m^2
2. about 146.93 m^2
3. 74 $in.^2$
4. 171 $in.^2$
5. 81 ft^2
6. 88 $in.^2$
7. $444
8. about 314 $in.^2$
9. about 113.04 m^2
10. about 452.16 cm^2
11. about 153.86 ft^2
12. about 490.625 yd^2
13. about 706.5 mm^2
14. about 502.4 cm^2

Math Background Notes

Vocabulary Review

- Area
- Composite figures
- Pi
- Radius
- Diameter

Finding the Area of a Composite Figure

- Students should be able to compute areas of composite figures.
- Remind students to identify the basic figures contained in the composite figure before they consider the area.
- Remind students that to find the area of a composite figure, all they need do is sum the areas of the basic figures.
- **Teaching Tip:** Sometimes students find it helpful to "break up" the composite figure. For instance, rather than working with the composite figure in Example 1, have students draw the triangle separately from the square and mark the dimensions on each figure. Ask students to find the area of each figure and then sum these quantities to determine the area of the composite figure.
- **Common Error:** Students will often think that the problem does not provide enough information to be solved. In Example 1, some students may think they have not been given the base of the triangle. Try to help students see that the basic shapes contained in the figure are just as important as how the shapes fit together. Because the base of the triangle stretches the same length as the top of the square, the base must measure 10 inches.

Finding the Areas of Circles

- Students should be able to compute areas of circles.
- You may wish to review the concept of pi with students. Pi is the ratio of a circle's circumference (perimeter) to its diameter. This ratio is constant regardless of the size of the circle. As a result of its frequent appearance in mathematics, the symbol π is used to represent the ratio. Students should be familiar with using $\frac{22}{7}$ or 3.14 as approximate values of pi.
- **Common Error:** You may want to review the relationship between a circle's diameter and radius before completing Example 3. Students will often substitute a circle's diameter rather than its radius into the formula.

Reteaching and Enrichment Strategies

If students need help...	If students got it...
Record and Practice Journal • Fair Game Review Skills Review Handbook Lesson Tutorials	Game Closet at *BigIdeasMath.com* Start the next section

What You Learned Before

"I just figured out how to find your volume. We'll immerse you in a barrel of water and measure the water that overflows."

Finding the Area of a Composite Figure

Example 1 Find the area of the figure.

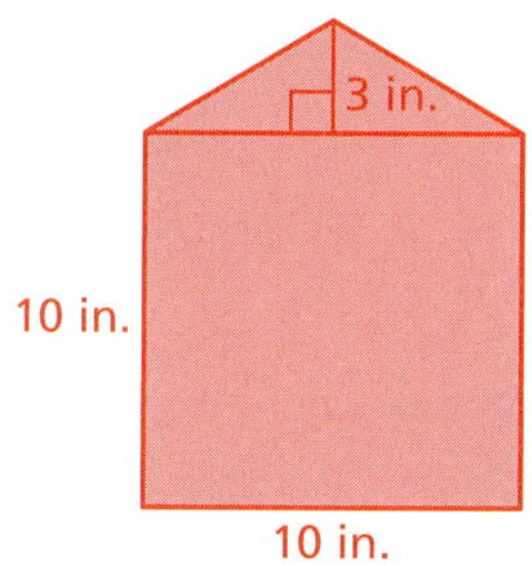

Area = Area of square + Area of triangle

$$A = s^2 + \frac{1}{2}bh$$

$$= 10^2 + \left(\frac{1}{2} \cdot 10 \cdot 3\right)$$

$$= 100 + 15$$

$$= 115 \text{ in.}^2$$

Try It Yourself

Find the area of the figure.

1.

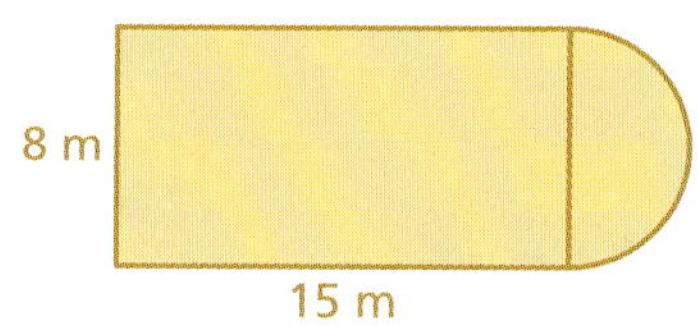

2.

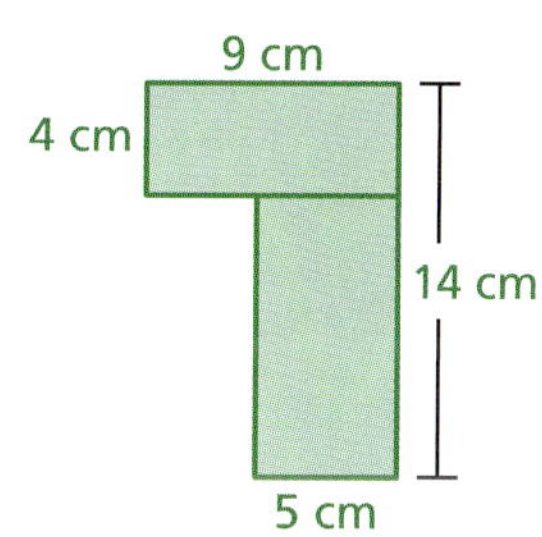

Finding the Areas of Circles

Example 2 Find the area of the circle.

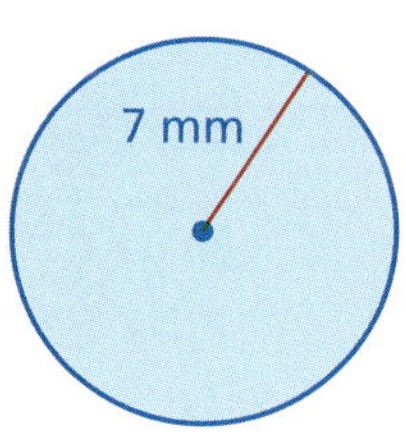

$$A = \pi r^2$$

$$\approx \frac{22}{7} \cdot 7^2$$

$$= \frac{22}{7} \cdot 49$$

$$= 154 \text{ mm}^2$$

Example 3 Find the area of the circle.

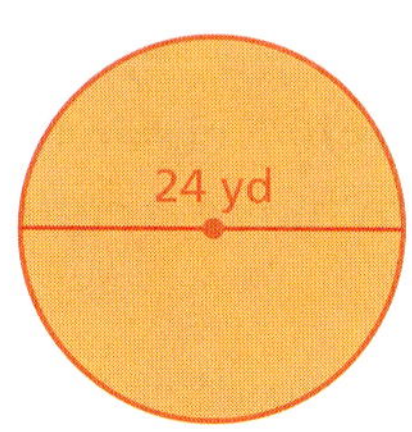

$$A = \pi r^2$$

$$\approx 3.14 \cdot 12^2$$

$$= 3.14 \cdot 144$$

$$= 452.16 \text{ yd}^2$$

Try It Yourself

Find the area of the circle.

3.

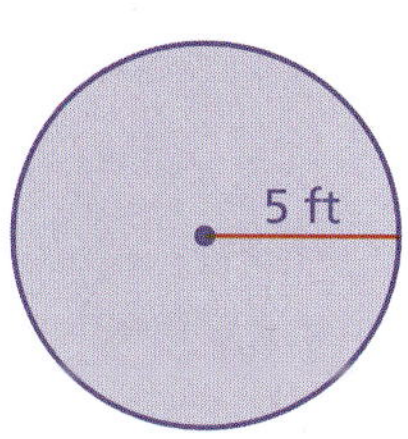

4.

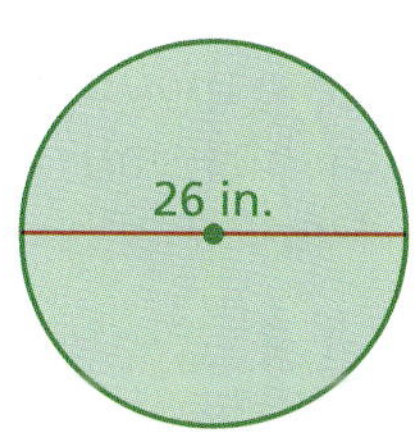

5.

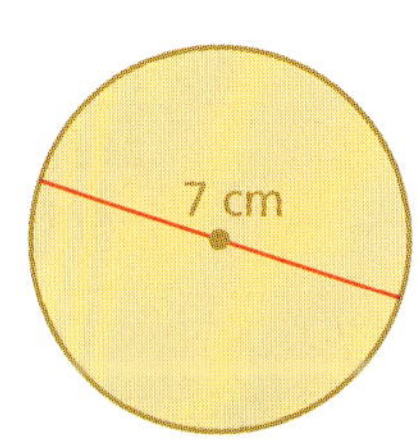

15.1 Volumes of Cylinders

Essential Question How can you find the volume of a cylinder?

1 ACTIVITY: Finding a Formula Experimentally

Work with a partner.

a. Find the area of the face of a coin.

b. Find the volume of a stack of a dozen coins.

c. Write a formula for the volume of a cylinder.

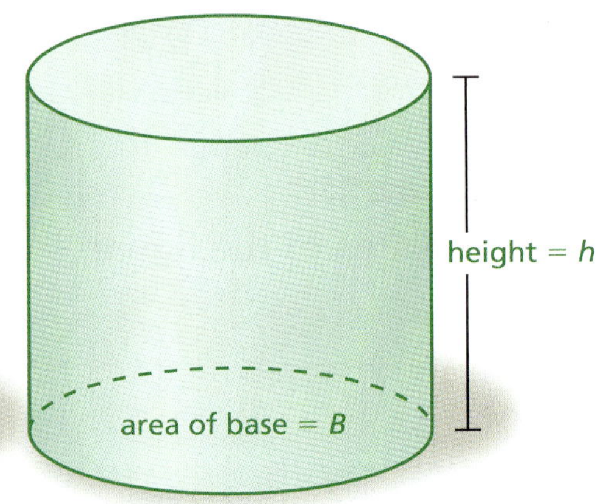

2 ACTIVITY: Making a Business Plan

Work with a partner. You are planning to make and sell three different sizes of cylindrical candles. You buy 1 cubic foot of candle wax for \$20 to make 8 candles of each size.

a. Design the candles. What are the dimensions of each size of candle?

b. You want to make a profit of \$100. Decide on a price for each size of candle.

c. Did you set the prices so that they are proportional to the volume of each size of candle? Why or why not?

Geometry

In this lesson, you will

- find the volumes of cylinders.
- find the heights of cylinders given the volumes.
- solve real-life problems.

Laurie's Notes

Introduction

Applying Mathematical Practices

- **Model with Mathematics:** In these activities, students will model and work with volumes of cylinders. In Activity 1, they will use a layering approach to write a formula for the volume of a cylinder. You want students to make a connection to the formula they learned for volume of a prism, where they found the area of the base and multiplied by the height.

Motivate

- Use round crackers, wafer candies, coins, or circular metal washers to model layers of very thin cylinders being stacked to make thicker cylinders.
- Display the models for students to see.
- ? "What do these items have in common?" *Sample answer:* They are cylinders made up of thin layers.
- Explain that today they will explore the volume of a cylinder using a similar approach.

Activity Notes

Activity 1

- ? "How will you find the area of the base?" Area of a circle $= \pi r^2$; Students will likely measure the diameter to the nearest tenth of a centimeter.
- Discuss with students how they found the volume of the stack of 12 coins.
- **Big Idea:** Area is measured in square units. Volume is measured in cubic units.

Activity 2

- **Make Sense of Problems and Persevere in Solving Them:** This is an open-ended activity that you can adjust to the skill level of your students.
- You can make the following assumptions: all of the candles have the same base with different heights: 2 inches, 3 inches, and 5 inches. Because there are 8 of each, the total height is 80 inches.
- ? There are 12^3 or 1728 cubic inches of wax available. "What does the radius of the candle need to be to use up the wax?"
- **Reason Abstractly and Quantitatively:** At this stage, students need to think through how they can find the volume of a cylinder. They need to find the area of the base (πr^2) and multiply by 80, the total height of the 24 candles. With a calculator, they will find that a radius of about 2.6 inches uses 1728 cubic inches of wax.
- Pricing the candles involves proportions. The 3-inch tall candle should cost 50% more than the 2-inch tall candle. The 5-inch tall candle should cost 150% more than the 2-inch tall candle.

What Your Students Will Learn

- Find the volumes of cylinders using a formula.
- Find the heights of cylinders given the volumes and the radii or diameters.

Previous Learning

Students should know that cylinders are composed of 2 circular bases and a rectangle.

Technology for the Teacher

Dynamic Classroom

Lesson Plans
Complete Materials List

15.1 Record and Practice Journal

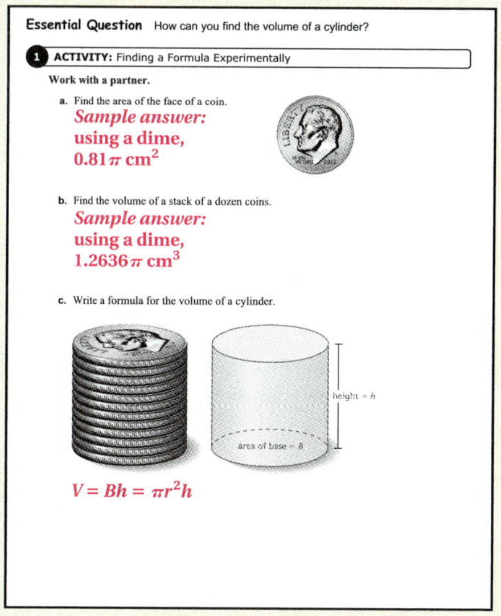

Essential Question How can you find the volume of a cylinder?

1 ACTIVITY: Finding a Formula Experimentally

Work with a partner.

a. Find the area of the face of a coin. *Sample answer:* using a dime, 0.81π cm^2

b. Find the volume of a stack of a dozen coins. *Sample answer:* using a dime, 1.2636π cm^3

c. Write a formula for the volume of a cylinder.

$V = Bh = \pi r^2 h$

Laurie's Notes

Activity 2 (continued)

- **Construct Viable Arguments and Critique the Reasoning of Others:** It is important for students to present their results to classmates. How did they make decisions about size and pricing? What assumptions did they make? What calculations did they perform and did they perform them accurately?

Activity 3

- Perhaps students have found volume by displacement in a science class.
- **Note:** This way of measuring volume is similar to the story of Archimedes who yelled "Eureka" when getting into a bathtub of water. He realized that he had displaced a volume of water equal to the volume of his body.
- If possible, have graduated cylinders and a large stone available to model this problem.

Activity 4

- Encourage students to take a guess even if their reasoning is no more than "it looks like it would hold a lot more."

? "What is the area of each base (leave answers in terms of π)?" 9π and 4π

? "So, which is greater, 4 layers of 9π or 9 layers of 4π?" They are the same.

What Is Your Answer?

- **Think-Pair-Share:** Students should read each question independently and then work in pairs to answer the questions. When they have answered the questions, the pair should compare their answers with another group and discuss any discrepancies.

Closure

- Refer to one of the cylinders used to motivate the activity and ask students to describe how they would find the volume.

Differentiated Instruction

Money

Have students bring in coins of different denominations from their country of origin. Repeat Activity 1 using the measurements of the coins.

15.1 Record and Practice Journal

2 ACTIVITY: Making a Business Plan

Work with a partner. You are planning to make and sell three different sizes of cylindrical candles. You buy 1 cubic foot of candle wax for $20 to make 8 candles of each size. Check students' work.

a. Design the candles. What are the dimensions of each size of candle?

b. You want to make a profit of $100. Decide on a price for each size of candle.

c. Did you set the prices so that they are proportional to the volume of each size of candle? Why or why not?

3 ACTIVITY: Science Experiment

Work with a partner. Use the diagram to describe how you can find the volume of a small object.

Pour water into a beaker until it flows out the side tube. Place an empty cylinder at the end of the side tube. Gently lower the object into the beaker. The volume of the object is equal to the amount of water that flows into the cylinder.

4 ACTIVITY: Comparing Cylinders

Work with a partner.

a. Just by looking at the two cylinders, which one do you think has the greater volume? Explain your reasoning.

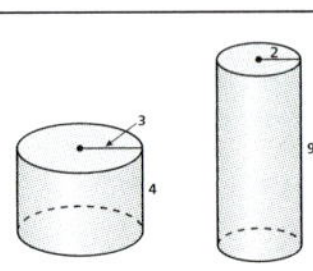

Check students' work.

b. Find the volume of each cylinder. Was your prediction in part (a) correct? Explain your reasoning.

Each cylinder has a volume of 36π cubic units.

What Is Your Answer?

5. **IN YOUR OWN WORDS** How can you find the volume of a cylinder?

Find the area of the base and multiply by the height.

6. Compare your formula for the volume of a cylinder with the formula for the volume of a prism. How are they the same?

Both formulas are Bh, where B is the area of the base and h is the height.

3 ACTIVITY: Science Experiment

Work with a partner. Use the diagram to describe how you can find the volume of a small object.

4 ACTIVITY: Comparing Cylinders

Math Practice

Consider Similar Problems

How can you use the results of Activity 1 to find the volumes of the cylinders?

Work with a partner.

a. Just by looking at the two cylinders, which one do you think has the greater volume? Explain your reasoning.

b. Find the volume of each cylinder. Was your prediction in part (a) correct? Explain your reasoning.

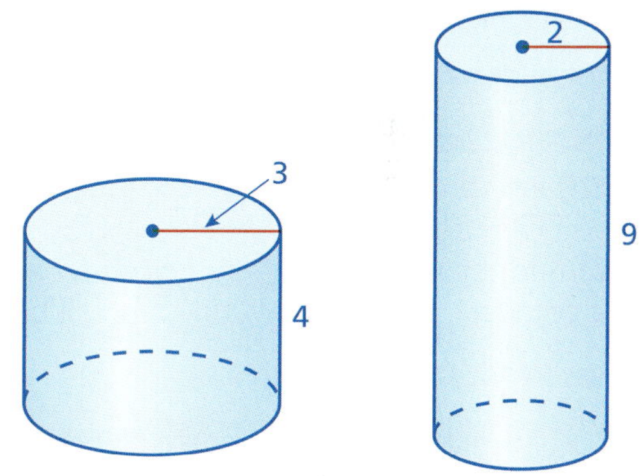

What Is Your Answer?

5. **IN YOUR OWN WORDS** How can you find the volume of a cylinder?

6. Compare your formula for the volume of a cylinder with the formula for the volume of a prism. How are they the same?

"Here's how I remember how to find the volume of any prism or cylinder." "Base times tall, will fill 'em all."

Practice Use what you learned about the volumes of cylinders to complete Exercises 3–5 on page 676.

15.1 Lesson

Check It Out
Lesson Tutorials
BigIdeasMath.com

Key Idea

Volume of a Cylinder

Words The volume V of a cylinder is the product of the area of the base and the height of the cylinder.

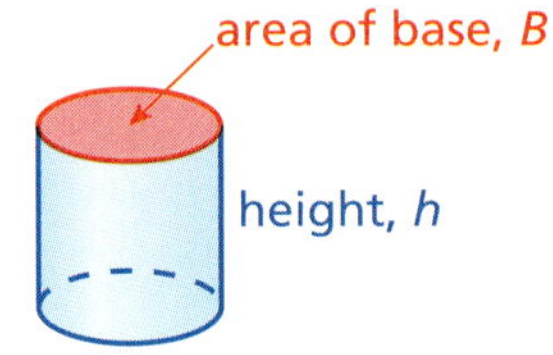

Algebra $V = Bh$

Area of base → B; Height of cylinder → h

EXAMPLE 1 Finding the Volume of a Cylinder

Find the volume of the cylinder. Round your answer to the nearest tenth.

$V = Bh$	Write formula for volume.
$= \pi(3)^2(6)$	Substitute.
$= 54\pi \approx 169.6$	Use a calculator.

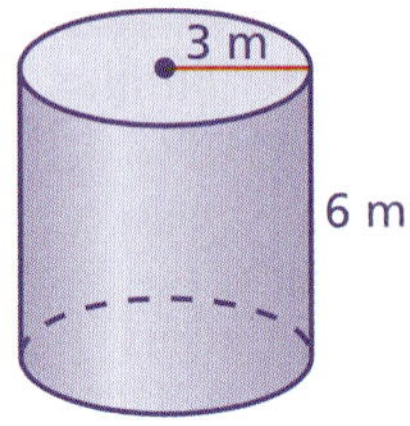

The volume is about 169.6 cubic meters.

Study Tip

Because $B = \pi r^2$, you can use $V = \pi r^2 h$ to find the volume of a cylinder.

EXAMPLE 2 Finding the Height of a Cylinder

Find the height of the cylinder. Round your answer to the nearest whole number.

The diameter is 10 inches. So, the radius is 5 inches.

$V = Bh$	Write formula for volume.
$314 = \pi(5)^2(h)$	Substitute.
$314 = 25\pi h$	Simplify.
$4 \approx h$	Divide each side by 25π.

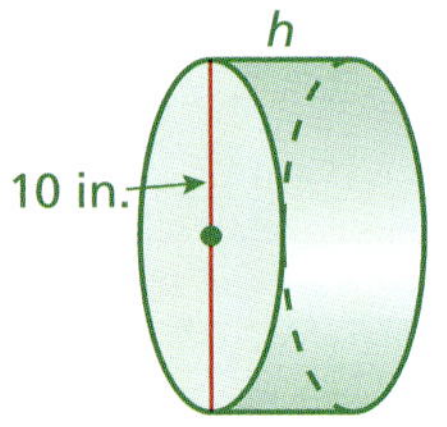

The height is about 4 inches.

On Your Own

Now You're Ready
Exercises 3–11 and 13–15

Find the volume V or height h of the cylinder. Round your answer to the nearest tenth.

1. 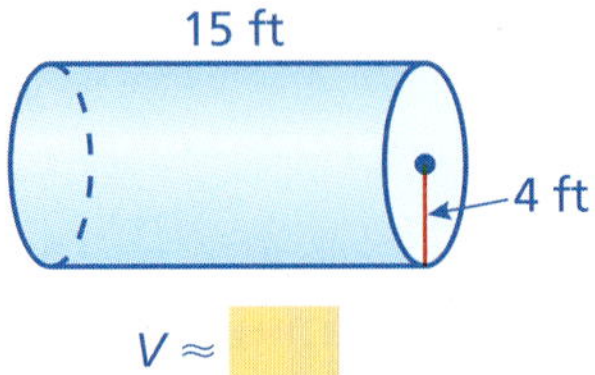

2. 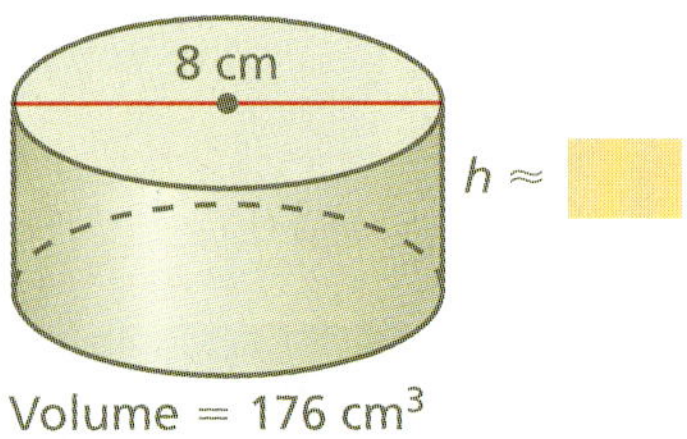

Laurie's Notes

Introduction

Connect

- **Yesterday:** Students discovered how to find the volume of a cylinder by considering the layers that make up a cylinder.
- **Today:** Students will work with a formula for the volume of a cylinder.

Motivate

- Hold two different-sized cans for the class to see. I often use cans of whole tomatoes—not only are the dimensions what I want but the contents are the same, so you can do a little cost analysis at the end of class.
- ? "How do the volumes of these two cans compare?" The purpose here is to get students thinking about dimensions, not to do computations.

Lesson Notes

Key Idea

- ? "How are cylinders and rectangular prisms alike?" Each has two congruent bases and a lateral portion.
- ? "How are cylinders and rectangular prisms different?" Cylinders have circular bases, but rectangular prisms have rectangular bases.
- Write the formula in words.
- Before writing the formula in symbols, ask how to find the area of the base.
- Note the use of color to identify the base in the formula and the diagram.

Example 1

- Model good problem solving by writing the formula first.
- Notice that the values of the variables are substituted, simplified, and left in terms of π. The last step is to use the π key on a calculator. If the calculator doesn't have a π key, use 3.14. Note that if 3.14 is used, the answers may not agree exactly with the answers shown in this text.
- ? Write "$\approx$" on the board. "What does this symbol mean?" approximately equal to "Why do you use this symbol here?" π is an irrational number.
- **Extension:** Discuss how big this cylinder is. Its diameter and height (6 meters) are wider than my classroom and more than twice its height.
- **Big Idea:** Volume $= Bh$ is the general formula for both prisms and cylinders. The base of a cylinder is a circle, so the general formula can be rewritten as the specific formula $V = \pi r^2 h$.

Example 2

- **Look for and Make Use of Structure:** Students may be unsure of how to divide 314 by 25π. One way is to find the product 25π, and then divide 314 by the product. Another way is to divide 314 by π, and then divide that answer by 25.

On Your Own

- **Neighbor Check:** Have students work independently and then have their neighbors check their work. Have students discuss any discrepancies.

Goal Today's lesson is finding the volumes of cylinders.

Technology for the Teacher

Dynamic Classroom

Lesson Tutorials
Lesson Plans
Answer Presentation Tool

Extra Example 1

Find the volume of a cylinder with a radius of 6 feet and a height of 3 feet. Round your answer to the nearest tenth. $108\pi \approx 339.3 \text{ ft}^3$

Extra Example 2

Find the height of a cylinder with a diameter of 4 yards and a volume of 88 cubic yards. Round your answer to the nearest whole number. $\frac{22}{\pi} \approx 7$ yd

On Your Own

1. $240\pi \approx 754.0 \text{ ft}^3$
2. $\frac{11}{\pi} \approx 3.5 \text{ cm}$

Laurie's Notes

Example 3

- "What percent of the salsa is missing and what percent remains?" 60%; 40%
- Work through the problem.
- **Extension:** Find the original volume without using the volume formula. *Hint:* You could use the percent equation or set up a proportion.

Example 4

- It is helpful to have 3 rulers to model what a cubic foot looks like. Hold the 3 rulers so they form 3 edges of a cube that meet at a vertex.
- "About how many gallons do you think would fill a cubic foot?" There will be a range of answers.
- Work through the problem, finding the volume of the watercooler in cubic feet.
- The second part of the problem involves dimensional analysis, a technique used earlier in the text.
- **Attend to Precision:** Students should be comfortable with the term "conversion factor."
- Estimate first. If 1 $ft^3 \approx 7.5$ gal, how many gallons would 1.3352 cubic feet be? An estimate of 10 gallons is reasonable and that is choice B.

On Your Own

- Give students sufficient time to do their work before asking volunteers to share their work at the board.

Closure

- Hold the two cans used to motivate the lesson and ask students to find the volume of each. If the contents are the same (or pretend that they are the same), how should the prices compare?

Extra Example 3

A jelly jar has a radius of 3 centimeters and a height of 8 centimeters. The jelly remaining in the jar has a height of 3 centimeters. How much jelly is missing from the jar? $45\pi \approx 141.4$ cm^3

Extra Example 4

About how many gallons of water does the watercooler bottle in Example 4 contain if the bottle is 1.25 feet tall? about 7.4 gal

On Your Own

3. $125\pi \approx 392.7$ cm^3

4. about 233,263 gal

English Language Learners

Vocabulary

Discuss the meanings of the words *volume* and *cubic units*. Have students add these words to their notebooks.

EXAMPLE 3 Real-Life Application

How much salsa is missing from the jar?

The empty space in the jar is a cylinder with a height of $10 - 4 = 6$ centimeters and a radius of 5 centimeters.

$V = Bh$ Write formula for volume.

$= \pi(5)^2(6)$ Substitute.

$= 150\pi \approx 471$ Use a calculator.

So, about 471 cubic centimeters of salsa are missing from the jar.

EXAMPLE 4 Real-Life Application

About how many gallons of water does the watercooler bottle contain? ($1\ \text{ft}^3 \approx 7.5$ gal)

Ⓐ 5.3 gallons Ⓑ 10 gallons Ⓒ 17 gallons Ⓓ 40 gallons

Find the volume of the cylinder. The diameter is 1 foot. So, the radius is 0.5 foot.

$V = Bh$ Write formula for volume.

$= \pi(0.5)^2(1.7)$ Substitute.

$= 0.425\pi \approx 1.3352$ Use a calculator.

So, the bottle contains about 1.3352 cubic feet of water. To find the number of gallons it contains, multiply by the conversion factor $\frac{7.5\ \text{gal}}{1\ \text{ft}^3}$.

$$1.3352\ \cancel{\text{ft}^3} \times \frac{7.5\ \text{gal}}{1\ \cancel{\text{ft}^3}} \approx 10\ \text{gal}$$

The watercooler bottle contains about 10 gallons of water. So, the correct answer is Ⓑ.

On Your Own

Now You're Ready
Exercise 12

3. **WHAT IF?** In Example 3, the height of the salsa in the jar is 5 centimeters. How much salsa is missing from the jar?

4. A cylindrical water tower has a diameter of 15 meters and a height of 5 meters. About how many gallons of water can the tower contain? ($1\ \text{m}^3 \approx 264$ gal)

15.1 Exercises

Check It Out
Help with Homework
BigIdeasMath.com

Vocabulary and Concept Check

1. **DIFFERENT WORDS, SAME QUESTION** Which is different? Find "both" answers.

How much does it take to fill the cylinder?

What is the capacity of the cylinder?

How much does it take to cover the cylinder?

How much does the cylinder contain?

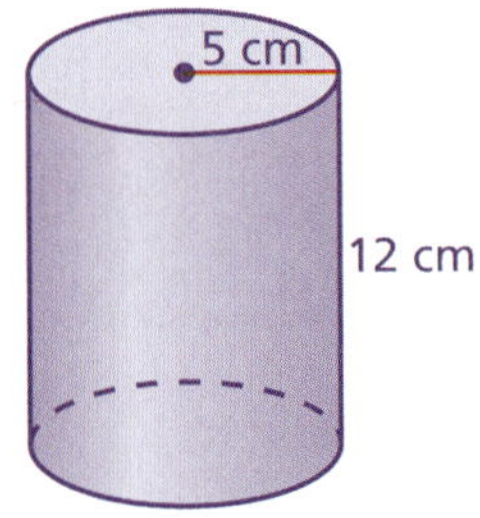

2. **REASONING** Without calculating, which of the solids has the greater volume? Explain.

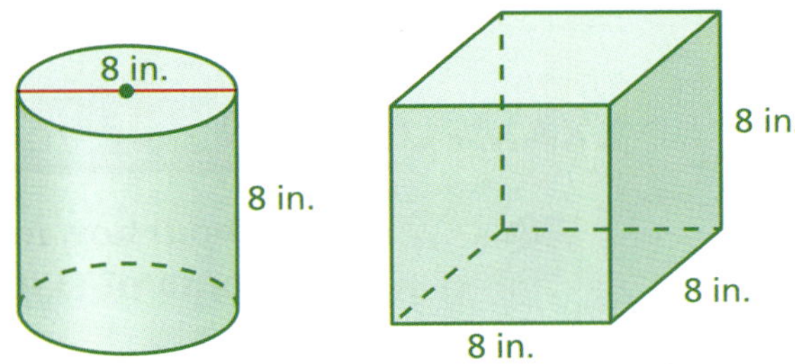

Practice and Problem Solving

Find the volume of the cylinder. Round your answer to the nearest tenth.

1 3.

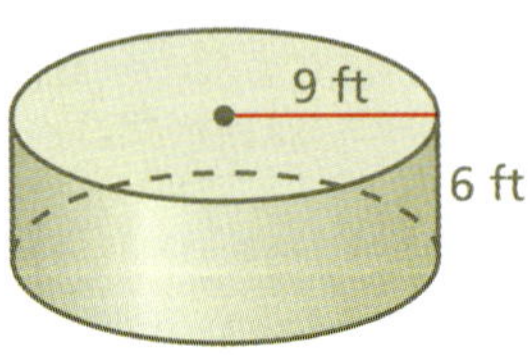

4.

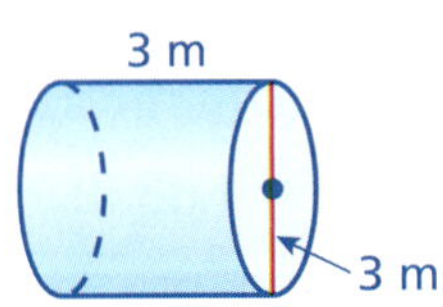

5.

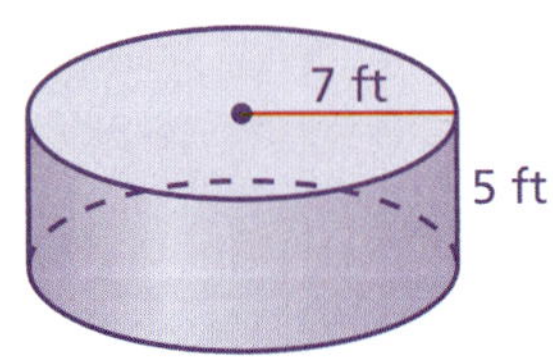

6.

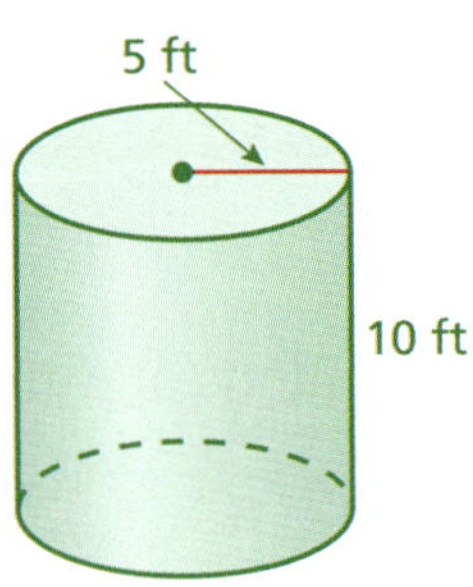

7.

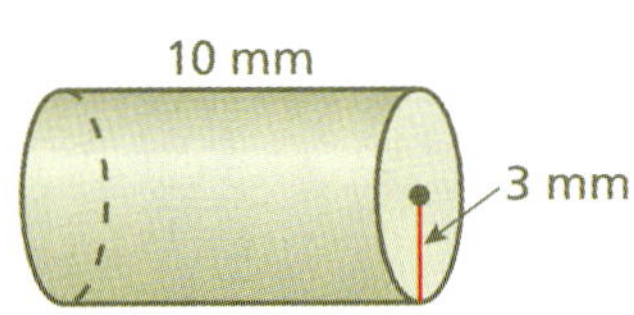

8.

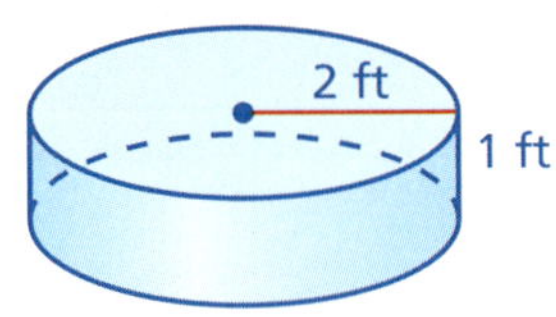

9.

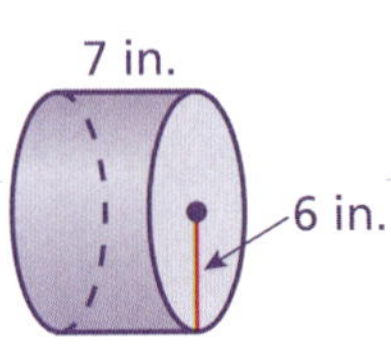

10.

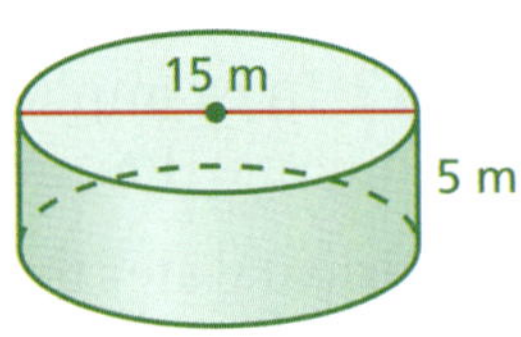

11.

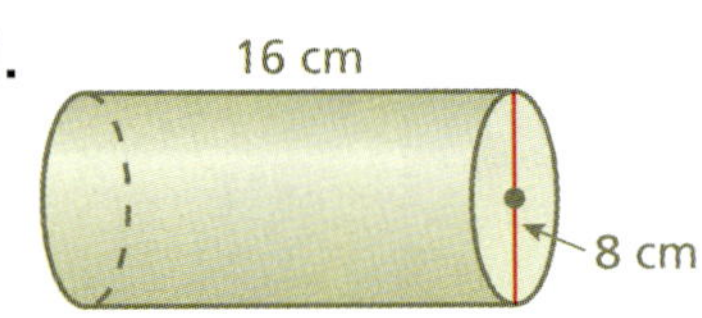

4 12. **SWIMMING POOL** A cylindrical swimming pool has a diameter of 16 feet and a height of 4 feet. About how many gallons of water can the pool contain? Round your answer to the nearest whole number. ($1 \text{ ft}^3 \approx 7.5$ gal)

Assignment Guide and Homework Check

Level	Assignment	Homework Check
Accelerated	1–5, 9–11, 14–24	2, 10, 14, 16

Common Errors

- **Exercises 3–11** Students may forget to square the radius when finding the area of the base. Remind them of the formula for the area of a circle.
- **Exercises 4, 10–14** Students may use the diameter in the formula for the area of a circle that calls for the radius. Encourage them to write the dimensions that they are given before attempting to find the volume. For example, in Exercise 4 a student would write: diameter = 3 m, height = 3 m.
- **Exercise 12** Students may find the volume of the pool, but forget to find how many gallons of water that the pool contains. Encourage them to write the information that they know about the problem and also what they are trying to find. This should help them answer each part of the question.

15.1 Record and Practice Journal

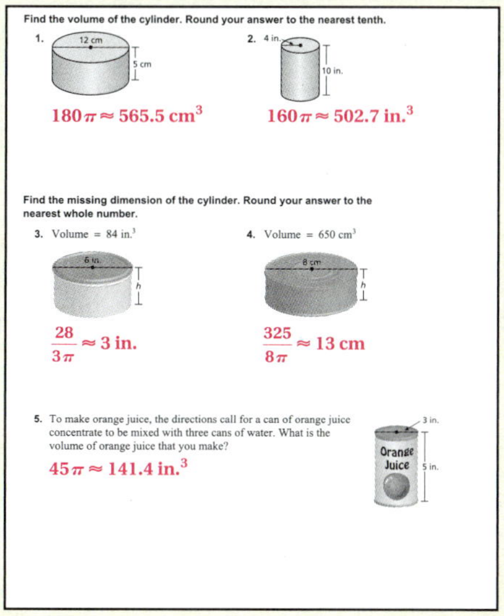
Find the volume of the cylinder. Round your answer to the nearest tenth.

1. $180\pi \approx 565.5 \text{ cm}^3$

2. $160\pi \approx 502.7 \text{ in.}^3$

Find the missing dimension of the cylinder. Round your answer to the nearest whole number.

3. Volume = 84 in.^{3}

$\frac{28}{3\pi} \approx 3$ in.

4. Volume = 650 cm^3

$\frac{325}{8\pi} \approx 13$ cm

5. To make orange juice, the directions call for a can of orange juice concentrate to be mixed with three cans of water. What is the volume of orange juice that you make?

$45\pi \approx 141.4 \text{ in.}^3$

Vocabulary and Concept Check

1. How much does it take to cover the cylinder?; $170\pi \approx 534.1 \text{ cm}^2$; $300\pi \approx 942.5 \text{ cm}^3$
2. The cube has a greater volume because the cylinder could fit inside the cube and there is still room in the corners of the cube that are not in the cylinder.

Practice and Problem Solving

3. $486\pi \approx 1526.8 \text{ ft}^3$
4. $\frac{27}{4}\pi \approx 21.2 \text{ m}^3$
5. $245\pi \approx 769.7 \text{ ft}^3$
6. $250\pi \approx 785.4 \text{ ft}^3$
7. $90\pi \approx 282.7 \text{ mm}^3$
8. $4\pi \approx 12.6 \text{ ft}^3$
9. $252\pi \approx 791.7 \text{ in.}^3$
10. $\frac{1125}{4}\pi \approx 883.6 \text{ m}^3$
11. $256\pi \approx 804.2 \text{ cm}^3$
12. about 6032 gal

Practice and Problem Solving

13. $\frac{125}{8\pi} \approx 5$ ft

14. $\frac{625}{16} \approx 39$ in.

15. $\sqrt{\frac{150{,}000}{19\pi}} \approx 50$ cm

16. The volume is $\frac{1}{4}$ of the original volume. Because the diameter is halved, the radius is also halved.

    So, $V = \pi\left(\frac{r}{2}\right)^2 h = \frac{1}{4}\pi r^2 h.$

17. See *Taking Math Deeper*.

18. about 4712 lb

19. $8325 - 729\pi \approx 6035$ m^3

20. a. $384\pi \approx 1206.37$ in.^{3}

    b. about 14.22 in.

    c. about 19 min

Fair Game Review

21. yes 22. yes

23. no 24. C

Mini-Assessment

Find the volume of the cylinder. Round your answer to the nearest tenth.

1.

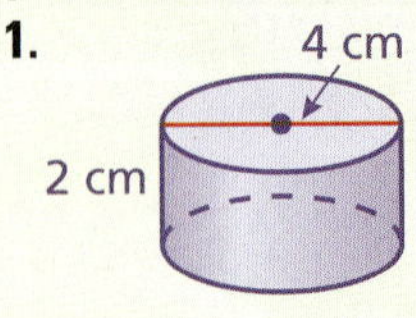

$8\pi \approx 25.1$ cm^3

2.

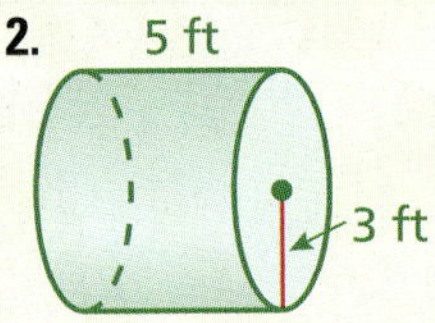

$45\pi \approx 141.4$ ft^3

3. Find the volume of the can of beans. Round your answer to the nearest whole number.

$\frac{81}{8}\pi \approx 32$ in.^{3}

Taking Math Deeper

Exercise 17

When driving through farm country, you can sometimes see hay and straw fields with large round bales. You seldom see the smaller *square bales*, but they are still used. The reason you don't see them is that they are usually moved to a storage shed as soon as they are baled.

1. Find the volume of the round bale.

$$V = \pi r^2 h$$
$$= \pi \cdot 2^2 \cdot 5$$
$$\approx 62.8 \text{ ft}^3$$

2. Find the volume of the square bale.

$$V = \ell wh$$
$$= 2 \cdot 2 \cdot 4$$
$$= 16 \text{ ft}^3$$

3. Find the number of square bales in a round bale.

$$\frac{62.8 \text{ ft}^3}{16 \text{ ft}^3} = 3.925$$

There are about 4 square bales in a round bale.

Both square and round bales vary in size. Suppose the square bale is only 18 in. by 18 in. by 3 ft. Its volume would be 6.75 cubic feet and there would be about 9.3 square bales in a round bale.

Project

In Germany, some farmers use large rolls of hay to make Mr. and Mrs. Hay people—like snow people. Draw a poster of two hay people. Assuming the rolls are the size of those in Exercise 17, how much hay would be needed to make your *people*?

Reteaching and Enrichment Strategies

If students need help...	If students got it...
Resources by Chapter • Practice A and Practice B • Puzzle Time Record and Practice Journal Practice Differentiating the Lesson Lesson Tutorials Skills Review Handbook	Resources by Chapter • Enrichment and Extension • Technology Connection Start the next section

Find the missing dimension of the cylinder. Round your answer to the nearest whole number.

2 **13.** Volume = 250 ft^3

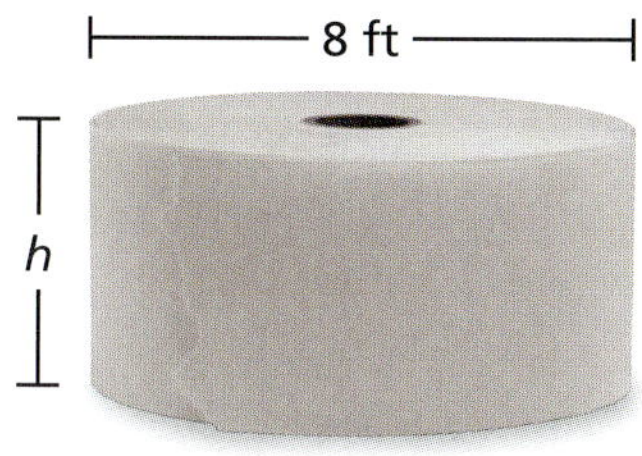

14. Volume = $10{,}000\pi$ in.^3

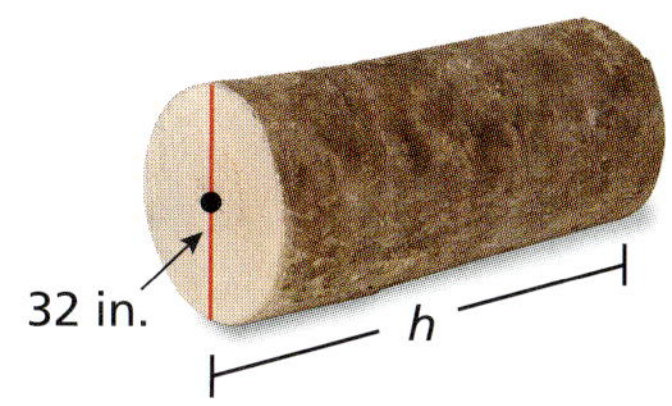

15. Volume = 600,000 cm^3

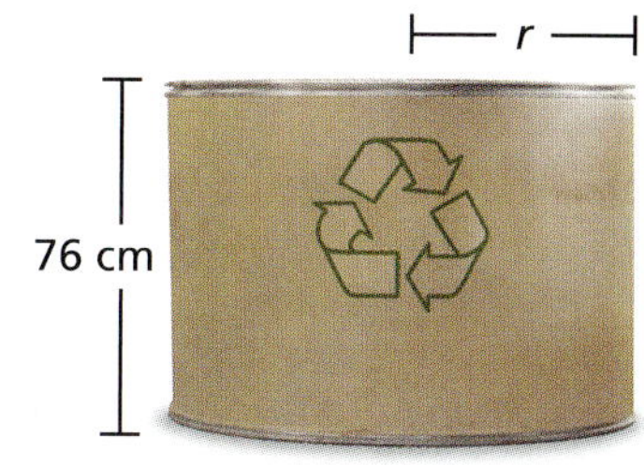

16. CRITICAL THINKING How does the volume of a cylinder change when its diameter is halved? Explain.

Round hay bale

17. MODELING A traditional "square" bale of hay is actually in the shape of a rectangular prism. Its dimensions are 2 feet by 2 feet by 4 feet. How many square bales contain the same amount of hay as one large "round" bale?

18. ROAD ROLLER A tank on a road roller is filled with water to make the roller heavy. The tank is a cylinder that has a height of 6 feet and a radius of 2 feet. One cubic foot of water weighs 62.5 pounds. Find the weight of the water in the tank.

19. VOLUME A cylinder has a surface area of 1850 square meters and a radius of 9 meters. Estimate the volume of the cylinder to the nearest whole number.

20. Problem Solving Water flows at 2 feet per second through a pipe with a diameter of 8 inches. A cylindrical tank with a diameter of 15 feet and a height of 6 feet collects the water.

a. What is the volume, in cubic inches, of water flowing out of the pipe every second?

b. What is the height, in inches, of the water in the tank after 5 minutes?

c. How many minutes will it take to fill 75% of the tank?

Fair Game Review What you learned in previous grades & lessons

Tell whether the triangle with the given side lengths is a right triangle. *(Section 14.5)*

21. 20 m, 21 m, 29 m

22. 1 in., 2.4 in., 2.6 in.

23. 5.6 ft, 8 ft, 10.6 ft

24. MULTIPLE CHOICE What is the area of a circle with a diameter of 10 meters? *(Section 8.3)*

Ⓐ $10\pi\ \text{m}^2$ Ⓑ $20\pi\ \text{m}^2$ Ⓒ $25\pi\ \text{m}^2$ Ⓓ $100\pi\ \text{m}^2$

15.2 Volumes of Cones

Essential Question How can you find the volume of a cone?

You already know how the volume of a pyramid relates to the volume of a prism. In this activity, you will discover how the volume of a cone relates to the volume of a cylinder.

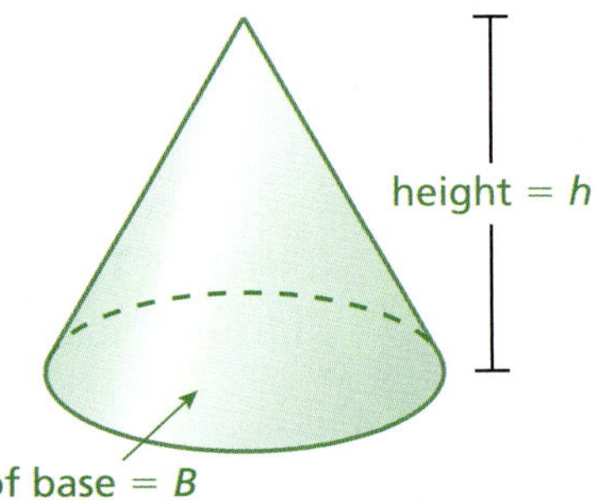

1 ACTIVITY: Finding a Formula Experimentally

Work with a partner. Use a paper cup that is shaped like a cone.

- Estimate the height of the cup.
- Trace the top of the cup on a piece of paper. Find the diameter of the circle.
- Use these measurements to draw a net for a cylinder with the same base and height as the paper cup.
- Cut out the net. Then fold and tape it to form an open cylinder.
- Fill the paper cup with rice. Then pour the rice into the cylinder. Repeat this until the cylinder is full. How many cones does it take to fill the cylinder?
- Use your result to write a formula for the volume of a cone.

2 ACTIVITY: Summarizing Volume Formulas

Geometry

In this lesson, you will

- find the volumes of cones.
- find the heights of cones given the volumes.
- solve real-life problems.

Work with a partner. You can remember the volume formulas for prisms, cylinders, pyramids, and cones with just two concepts.

Volumes of Prisms and Cylinders

Volume = Area of base × [ ]

Volumes of Pyramids and Cones

Volume = [ ] Volume of prism or cylinder with same base and height

Make a list of all the formulas you need to remember to find the area of a base. Talk about strategies for remembering these formulas.

Laurie's Notes

Introduction

Applying Mathematical Practices

- **Model with Mathematics:** In the first activity, students will explore the relationship between the volume of a cone and the volume of a cylinder. Modeling such relationships is a powerful teaching tool.
- **Big Idea:** It is not a big stretch for students to accept that the volume relationship between the cone and cylinder is the same as the volume relationship between the pyramid and prism.

  Volume of a Cylinder = (Area of Base)(Height)

  Volume of a Cone $= \frac{1}{3}$(Area of Base)(Height)

Motivate

- Give pairs of students two minutes to look around the classroom and make a list of all the geometric solids they see that are prisms, cylinders, pyramids, or cones.
- Make a column on the board for each type of solid. Ask one pair of students to list an item in each column. Continue to have pairs of students add to the lists, but only items that are not in the lists already.
- Was every group able to list 4 new items? If your classroom is like most, there are fewer pyramids and cones than prisms and cylinders.

Activity Notes

Activity 1

- **Management Tip:** Instead of having the whole class perform this activity, you may want to have a pair of students demonstrate it. If you have a cone and a cylinder with the same base and height from a commercially-available (fillable) geometric solids kit, then you could use these instead of having students make them. Or, you could show a video demonstration, such as the one at http://www.youtube.com/watch?v=QnVr_x7c79w.
- **Model with Mathematics:** Regardless, you want students to *see* the relationship and not just hear what it is.
- If necessary, have materials ready for students: heavy-weight paper, tape, scissors, ruler, and uncooked rice.
- **Common Error:** Students may measure the slant height of the cone. Point out that the height of a cone is shown in the diagram.
- The net students make for the cylinder will most likely have two bases. If so, then they can just tear off one of the bases to get the open cylinder.
- Have students guess the relationship before they see it. If they recall the relationship between the volume of a pyramid and the volume of a prism with the same base and height, then they may guess correctly.
- ? "How many times does it take to fill the cylinder with the contents of the cone?" three
- ? "How does the volume of the cone compare with the volume of the cylinder?" The volume of the cone is $\frac{1}{3}$ the volume of the cylinder.

What Your Students Will Learn

- Find the volumes of cones using a formula.
- Find the heights of cones given the volumes and the radii or diameters.

Previous Learning

Students should know how to find the surface area of a cone.

Technology for the Teacher

Dynamic Classroom

Lesson Plans
Complete Materials List

15.2 Record and Practice Journal

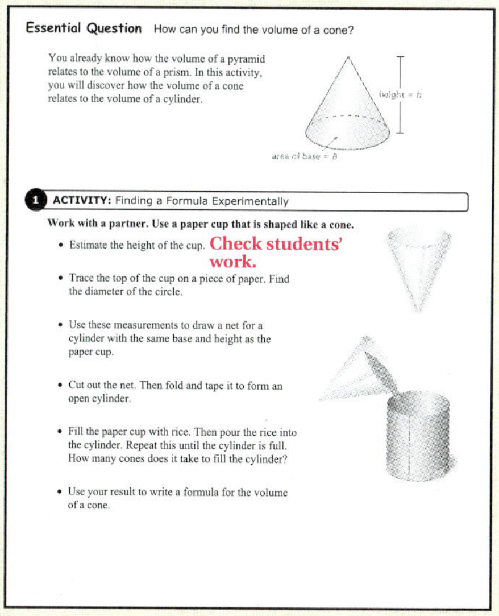

Essential Question How can you find the volume of a cone?

You already know how the volume of a pyramid relates to the volume of a prism. In this activity, you will discover how the volume of a cone relates to the volume of a cylinder.

height = h

area of base = B

1 ACTIVITY: Finding a Formula Experimentally

Work with a partner. Use a paper cup that is shaped like a cone.

- Estimate the height of the cup. Check students' work.
- Trace the top of the cup on a piece of paper. Find the diameter of the circle.
- Use these measurements to draw a net for a cylinder with the same base and height as the paper cup.
- Cut out the net. Then fold and tape it to form an open cylinder.
- Fill the paper cup with rice. Then pour the rice into the cylinder. Repeat this until the cylinder is full. How many cones does it take to fill the cylinder?
- Use your result to write a formula for the volume of a cone.

English Language Learners

Visual

Show students models of a pyramid and a prism. Then show them models of a cone and a cylinder. Have students note that the relationship between the volume of a cone and the volume of a cylinder with the same base and height is the same as the relationship between the volume of a pyramid and the volume of a prism with the same base and height.

15.2 Record and Practice Journal

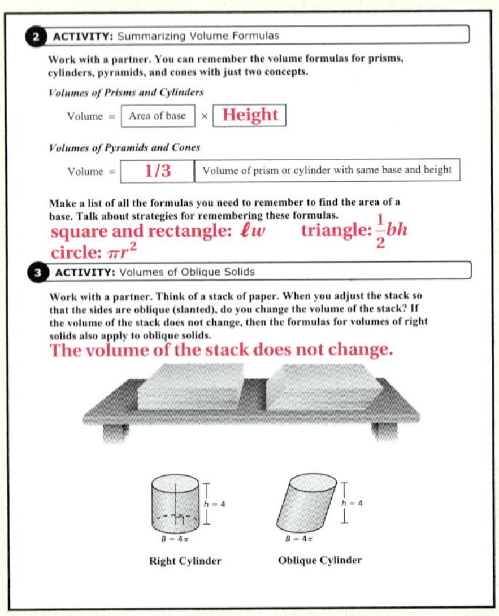

2 ACTIVITY: Summarizing Volume Formulas

Work with a partner. You can remember the volume formulas for prisms, cylinders, pyramids, and cones with just two concepts.

Volumes of Prisms and Cylinders

Volume = Area of base × **Height**

Volumes of Pyramids and Cones

Volume = **1/3** Volume of prism or cylinder with same base and height

Make a list of all the formulas you need to remember to find the area of a base. Talk about strategies for remembering these formulas.

square and rectangle: ℓw triangle: $\frac{1}{2}bh$
circle: πr^2

3 ACTIVITY: Volumes of Oblique Solids

Work with a partner. Think of a stack of paper. When you adjust the stack so that the sides are oblique (slanted), do you change the volume of the stack? If the volume of the stack does not change, then the formulas for volumes of right solids also apply to oblique solids.

The volume of the stack does not change.

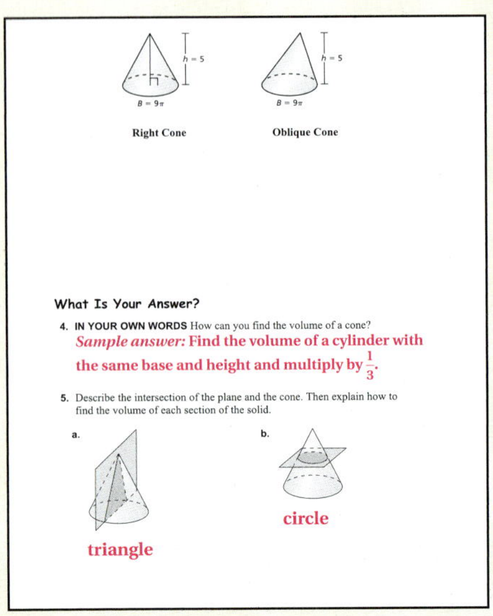

What Is Your Answer?

4. **IN YOUR OWN WORDS** How can you find the volume of a cone?
   Sample answer: **Find the volume of a cylinder with the same base and height and multiply by $\frac{1}{3}$.**

5. Describe the intersection of the plane and the cone. Then explain how to find the volume of each section of the solid.

   a. **triangle**

   b. **circle**

Laurie's Notes

Activity 2

- **Reason Abstractly and Quantitatively:** You do not want students to think that it is necessary to memorize a lot of formulas. Instead, students need to consider the structure of the shape. Prisms and cylinders have the same structure (two congruent bases and a lateral portion) and pyramids and cones have the same structure (one base and a lateral portion that contains a vertex). There is one general formula for each pair of solids. Moreover, the two general formulas have a 1 : 3 relationship.
- **Teaching Tip:** Make a poster of each of the two volume formulas for your classroom, or ask a student to make them.

? "How many general volume formulas are there?" two

? "In each formula, you need to find the area of a base. What types of bases have you studied?" Most were squares, rectangles, triangles, or circles.

- Have pairs of students share their lists and strategies. Collect information at the board.

Activity 3

- It is helpful to have a stack of paper or a deck of cards to model this activity.
- Give students time to discuss their thinking and then have them share their thoughts with the rest of the class.
- **Common Misconception:** Students may believe that the height decreases as the stack of paper is slanted to one side. The thickness doesn't change, so the height remains the same. The volume remains the same because no sheets are removed.

? "What is changing in this problem?" Some students may recognize that the surface area is changing because more area is being exposed.

What Is Your Answer?

- Have students work in pairs.

Closure

- Sketch a cube with edge length 2 centimeters, and a cylinder with height and diameter each 2 centimeters. Compare the volumes of the cube and cylinder.
- Sketch a cylinder and a cone, each with height and diameter of 3 centimeters. Compare the volumes of the cylinder and cone.

3 ACTIVITY: Volumes of Oblique Solids

Work with a partner. Think of a stack of paper. When you adjust the stack so that the sides are oblique (slanted), do you change the volume of the stack? If the volume of the stack does not change, then the formulas for volumes of right solids also apply to oblique solids.

Math Practice

Use Equations

What equation would you use to find the volume of the oblique solid? Explain.

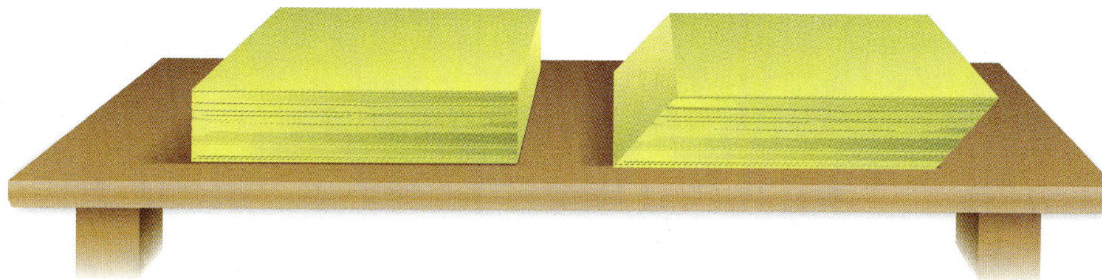

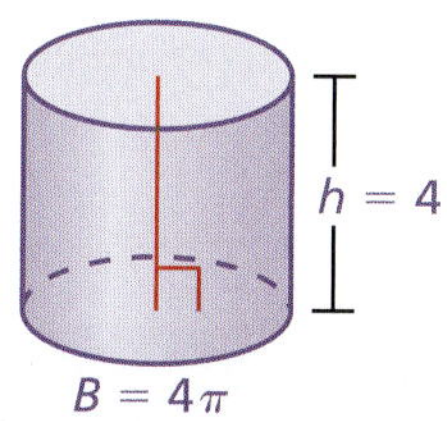

Right cylinder

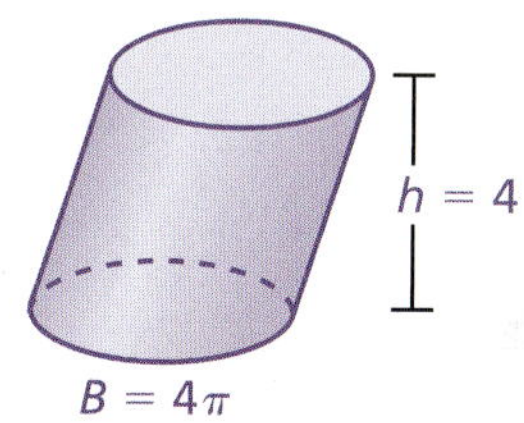

Oblique cylinder

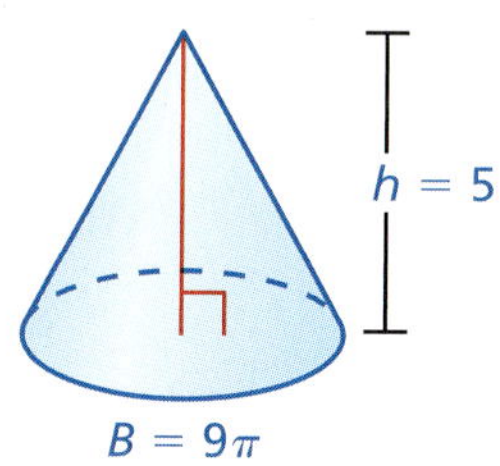

Right cone

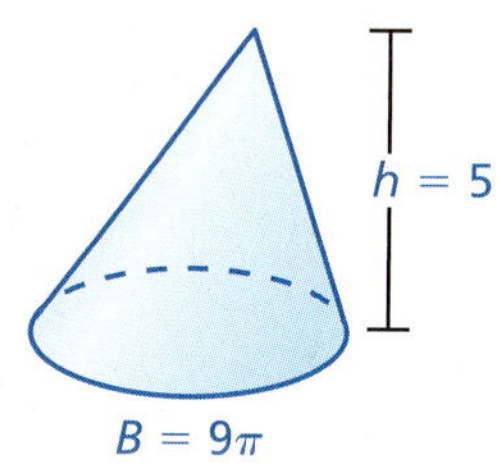

Oblique cone

What Is Your Answer?

4. **IN YOUR OWN WORDS** How can you find the volume of a cone?
5. Describe the intersection of the plane and the cone. Then explain how to find the volume of each section of the solid.

a.

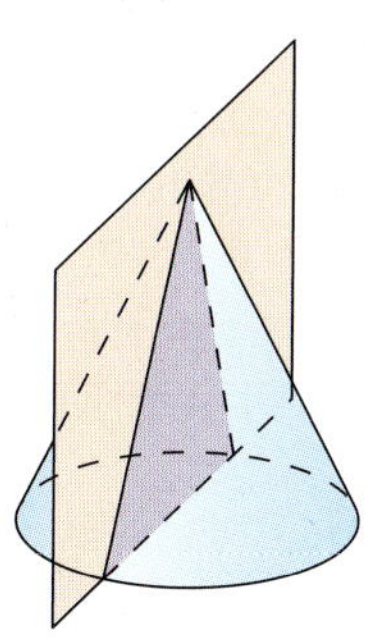

b.

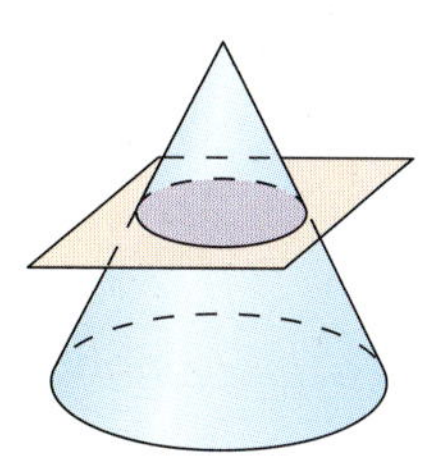

Practice Use what you learned about the volumes of cones to complete Exercises 4–6 on page 682.

15.2 Lesson

Check It Out
Lesson Tutorials
BigIdeasMath.com

Key Idea

Volume of a Cone

Words The volume V of a cone is one-third the product of the area of the base and the height of the cone.

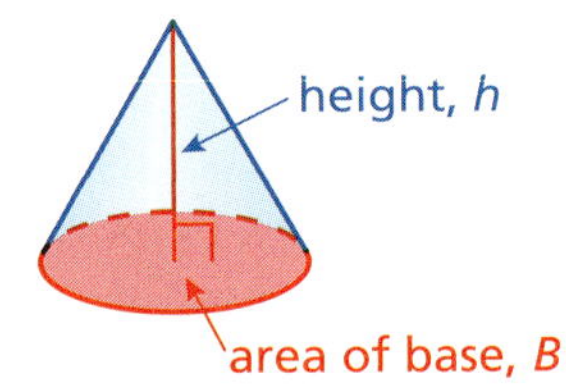

Algebra $V = \frac{1}{3}Bh$

Area of base (points to B)

Height of cone (points to h)

Study Tip

The *height* of a cone is the perpendicular distance from the base to the vertex.

EXAMPLE 1 Finding the Volume of a Cone

Find the volume of the cone. Round your answer to the nearest tenth.

The diameter is 4 meters. So, the radius is 2 meters.

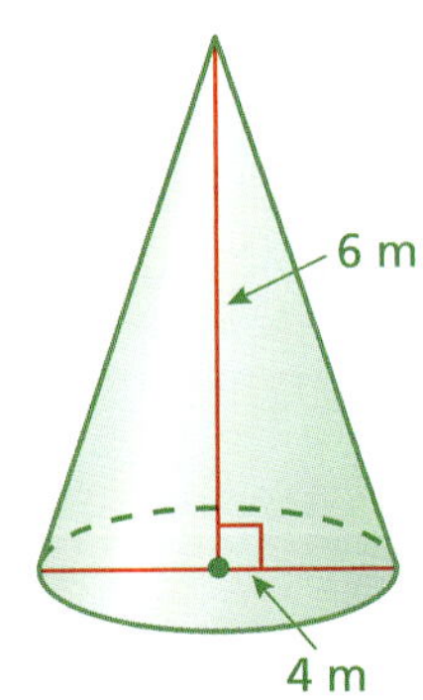

$V = \frac{1}{3}Bh$	Write formula for volume.
$= \frac{1}{3}\pi(2)^2(6)$	Substitute.
$= 8\pi \approx 25.1$	Use a calculator.

∴ The volume is about 25.1 cubic meters.

Study Tip

Because $B = \pi r^2$, you can use $V = \frac{1}{3}\pi r^2 h$ to find the volume of a cone.

EXAMPLE 2 Finding the Height of a Cone

Find the height of the cone. Round your answer to the nearest tenth.

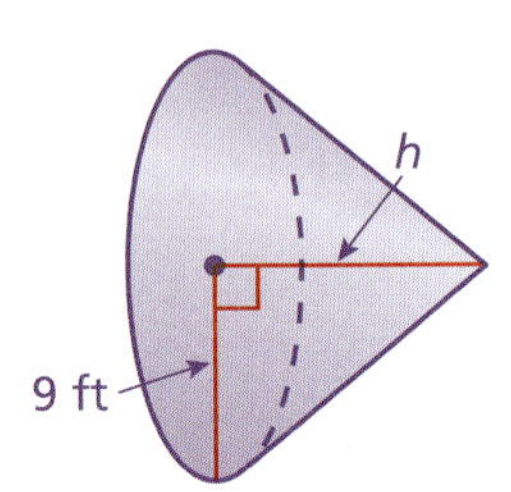

Volume = 956 ft^3

$V = \frac{1}{3}Bh$	Write formula for volume.
$956 = \frac{1}{3}\pi(9)^2(h)$	Substitute.
$956 = 27\pi h$	Simplify.
$11.3 \approx h$	Divide each side by 27π.

∴ The height is about 11.3 feet.

Laurie's Notes

Introduction

Connect

- **Yesterday:** Students developed a strategy to summarize volume and surface area formulas.
- **Today:** Students will work with the formula for the volume of a cone.

Motivate

- Bring an ice cream cone and ice cream scoop to class.
- **FYI:** An ice cream scoop with a radius of 1 inch will make a round (spherical) scoop of ice cream that is a little more than 4 cubic inches. Share this information with students. You may also want to mention that students will study volumes of spheres in the next section.

? "If I place a scoop of ice cream (with a 1 inch radius) on this cone, and the ice cream melts (because I received a phone call), will the ice cream overflow the cone?" This question is posed only to get students thinking about the volume of a cone. The volume of the cone will be found at the end of the lesson.

Lesson Notes

Key Idea

- Write the Key Idea.
- Write the formula in words. Draw the cone with the dimensions labeled.
- Write the symbolic formula.

? "What shape is the base?" circle "How do you find its area A?" $A = \pi r^2$

Example 1

- Notice that the work is done in terms of π. It is not until the last step that you use the π key on a calculator.
- **Representation:** Encourage students to use the parentheses to represent multiplication. Using the $\times$ symbol would make the expression confusing.
- **Common Misconception:** Remind students that π is a number and because multiplication is commutative and associative, this expression could be rewritten as $\frac{1}{3}(6)(2)^2\pi$, making the computation less confusing.

? "What is being squared in this expression?" only the 2

Example 2

- This example requires students to solve an equation for a variable.
- Work through the problem, annotating the steps as shown in the book.

? "How does $\frac{1}{3}\pi(9)^2h$ equal $27\pi h$?" Only the 9 is being squared, which is 81. One-third of 81 is 27. The order of the factors doesn't matter.

- **Look for and Make Use of Structure:** Students may have difficulty with the last step, dividing by 27π. It can be done in two steps—divide by 27 then divide by π. Or, divide 956 by the product 27π, which is about 84.82.

Goal Today's lesson is finding the volume of a cone.

Technology for the Teacher

Dynamic Classroom

Lesson Tutorials
Lesson Plans
Answer Presentation Tool

Extra Example 1

Find the volume of a cone with a diameter of 6 feet and a height of 3 feet. Round your answer to the nearest tenth. $9\pi \approx 28.3 \text{ ft}^3$

Extra Example 2

Find the height of a cone with a radius of 6 yards and a volume of 75 cubic yards. Round your answer to the nearest whole number. $\frac{25}{4\pi} \approx 2$ yd

On Your Own

1. $180\pi \approx 565.5 \text{ cm}^3$
2. $\frac{96}{\pi} \approx 30.6 \text{ yd}$

Extra Example 3

In Example 3, the height of the sand is 36 millimeters and the radius is 15 millimeters. The sand falls at a rate of 150 cubic millimeters per second. How much time do you have to answer the question? about 57 sec

On Your Own

3. about 42 sec
4. about 6 sec

Differentiated Instruction

Organization

Some students might benefit from first finding the area of the base B of the cone. Then they can substitute this value into the formula, $V = \frac{1}{3}Bh$.

Laurie's Notes

On Your Own

- Ask volunteers to share their work at the board.

Example 3

- If you have a timer of this type, use it as a model.
- Ask a volunteer to read the problem. Ask for ideas as to how the problem can be solved.

? "How long is 30 millimeters?" 30 millimeters is equal to 3 centimeters, which is a little more than 1 inch. This helps students form a visual image of the actual size of the sand timer.

- **Teaching Tip**: Again, explain that $\frac{1}{3}(24)$ is a whole number. Then multiply $8(10)^2 = 800$.
- Be sure to use units in labeling answers. Dimensional analysis shows that the answer will have units of seconds.
- **Extension:** I have a sand timer in my classroom. Students calculate the volume, measure the amount of time it takes to fall to the bottom, and use this information to calculate the rate at which the sand is falling.

On Your Own

- **Extension:** Question 4 is a preview of an upcoming lesson. The height and radius have each been decreased by a factor of 2. (They are $\frac{1}{2}$ the original dimensions). What happens to the volume? It is decreased by a factor of 8, or 2^3.

Closure

- **Exit Ticket:** Have students find the volume of the ice cream cone used to motivate the lesson.

On Your Own

Now You're Ready
Exercises 4–12 and 15–17

Find the volume V or height h of the cone. Round your answer to the nearest tenth.

1. 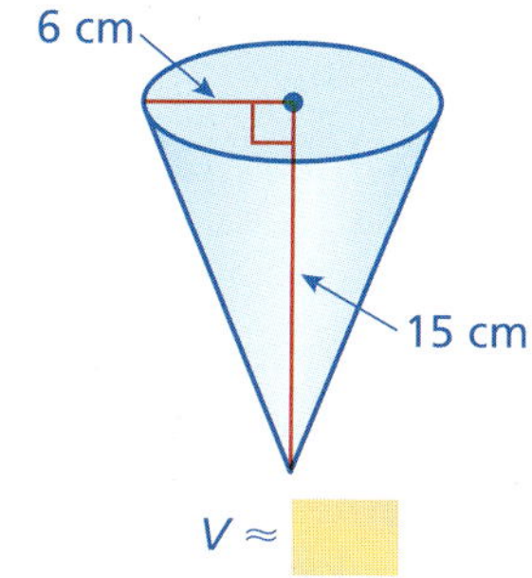

2. 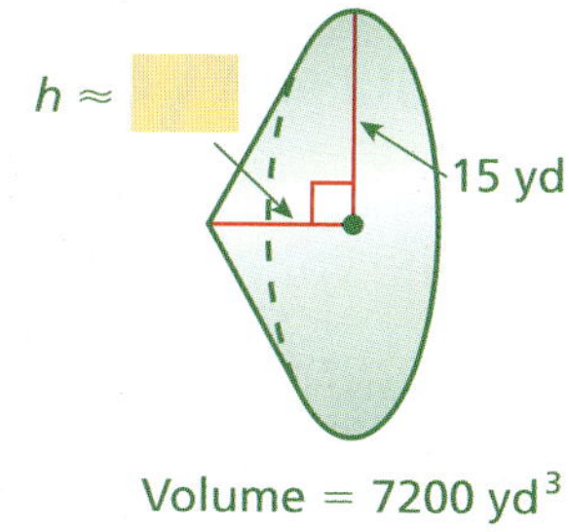

EXAMPLE 3 Real-Life Application

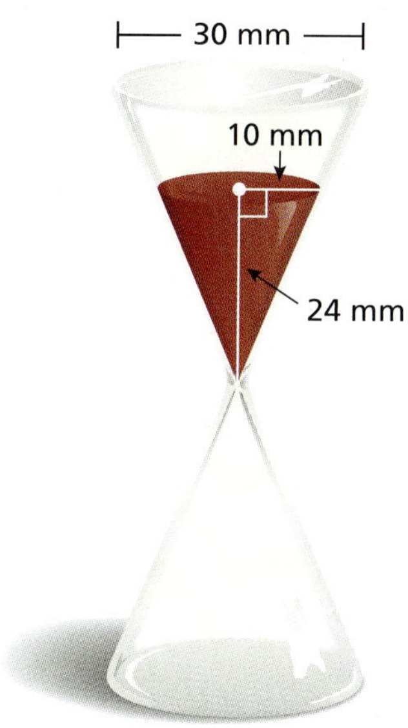

You must answer a trivia question before the sand in the timer falls to the bottom. The sand falls at a rate of 50 cubic millimeters per second. How much time do you have to answer the question?

Use the formula for the volume of a cone to find the volume of the sand in the timer.

$V = \frac{1}{3}Bh$ Write formula for volume.

$= \frac{1}{3}\pi(10)^2(24)$ Substitute.

$= 800\pi \approx 2513$ Use a calculator.

The volume of the sand is about 2513 cubic millimeters. To find the amount of time you have to answer the question, multiply the volume by the rate at which the sand falls.

$$2513 \cancel{\text{mm}^3} \times \frac{1 \text{ sec}}{50 \cancel{\text{mm}^3}} = 50.26 \text{ sec}$$

So, you have about 50 seconds to answer the question.

On Your Own

3. **WHAT IF?** The sand falls at a rate of 60 cubic millimeters per second. How much time do you have to answer the question?

4. **WHAT IF?** The height of the sand in the timer is 12 millimeters, and the radius is 5 millimeters. How much time do you have to answer the question?

15.2 Exercises

Check It Out
Help with Homework
BigIdeasMath.com

Vocabulary and Concept Check

1. **VOCABULARY** Describe the height of a cone.
2. **WRITING** Compare and contrast the formulas for the volume of a pyramid and the volume of a cone.
3. **REASONING** You know the volume of a cylinder. How can you find the volume of a cone with the same base and height?

Practice and Problem Solving

Find the volume of the cone. Round your answer to the nearest tenth.

1 **4.**

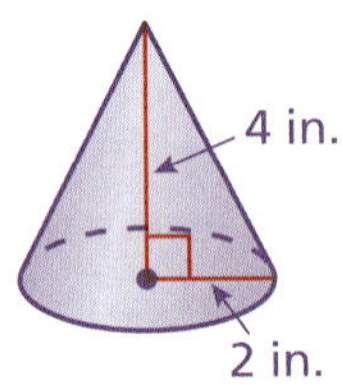

5.

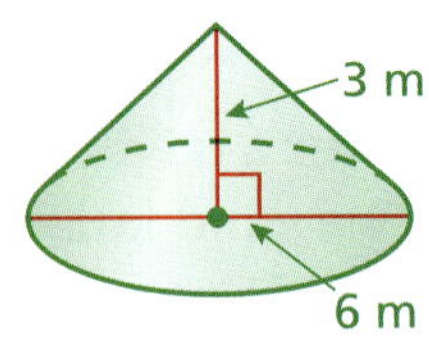

6.

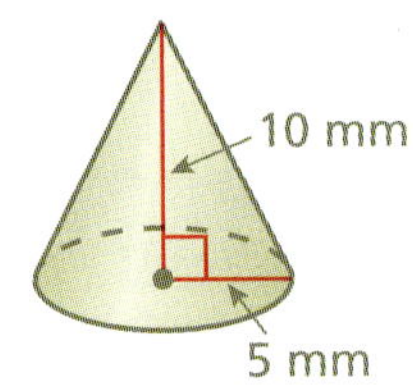

7.

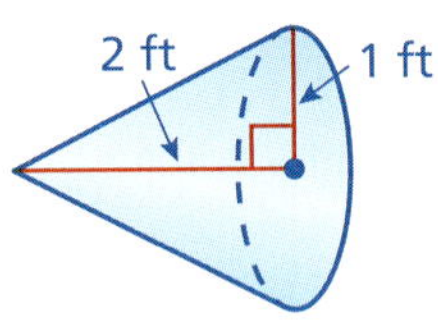

8.

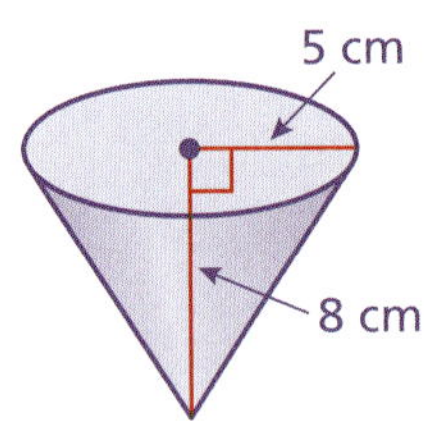

9.

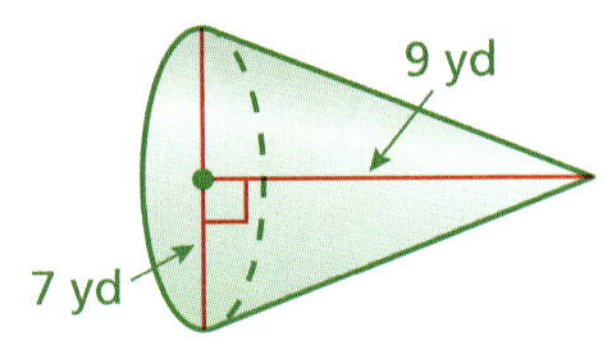

10.

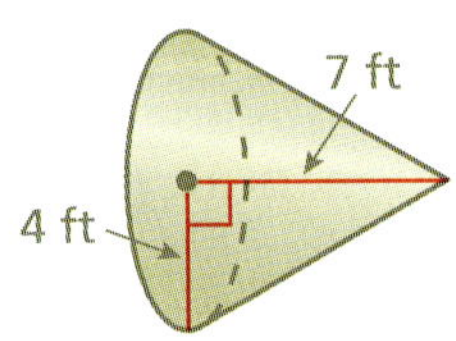

11.

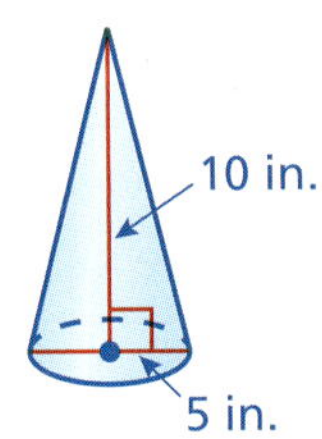

12.

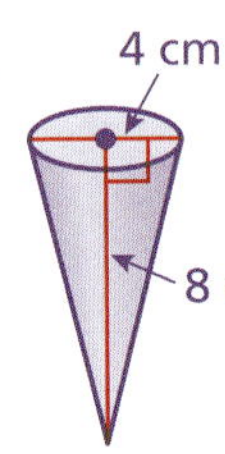

13. **ERROR ANALYSIS** Describe and correct the error in finding the volume of the cone.

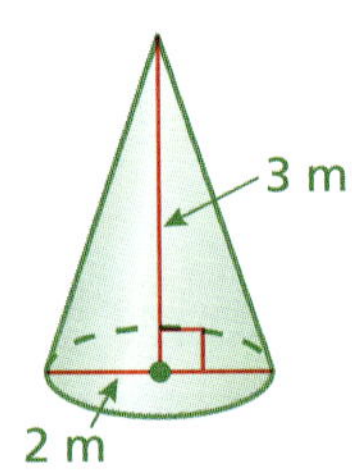

✗
$$V = \frac{1}{3}Bh$$
$$= \frac{1}{3}(\pi)(2)^2(3)$$
$$= 4\pi \text{ m}^3$$

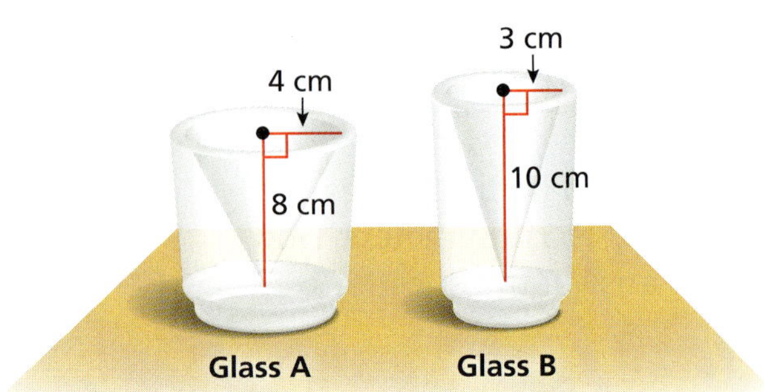

14. **GLASS** The inside of each glass is shaped like a cone. Which glass can hold more liquid? How much more?

Assignment Guide and Homework Check

Level	Assignment	Homework Check
Accelerated	1–6, 13, 16–25	2, 16, 18, 21

For Your Information

- **Exercise 15** Because the volume is given in terms of π, students should not substitute for π.

Common Errors

- **Exercises 4–12** Students may write linear or square units for volume rather than cubic units. Remind them that part of writing a correct answer is including the correct units.
- **Exercises 4–12** When finding the area of the base, students may not square the radius, or they may use the diameter when the formula calls for the radius. Remind them of the formula that they learned for the area of a circle.
- **Exercises 15 and 16** Students may try to use the Distributive Property before solving for h. For example, in Exercise 16, a student may incorrectly write $225 = \frac{1}{3}\pi(5^2) \cdot \frac{1}{3}h$. Remind them that factors are multiplied.
- **Exercise 17** The solution of this exercise has many parts that can be confusing to students. It may be helpful to go over it together in class.

15.2 Record and Practice Journal

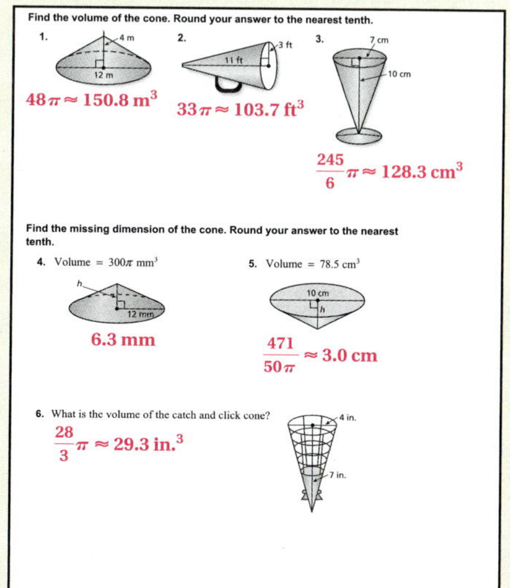
Find the volume of the cone. Round your answer to the nearest tenth.

1. $48\pi \approx 150.8 \text{ m}^3$
2. $33\pi \approx 103.7 \text{ ft}^3$
3. $\frac{245}{6}\pi \approx 128.3 \text{ cm}^3$

Find the missing dimension of the cone. Round your answer to the nearest tenth.

4. Volume = $300\pi \text{ mm}^3$ — 6.3 mm
5. Volume = 78.5 cm^3 — $\frac{471}{50\pi} \approx 3.0 \text{ cm}$
6. What is the volume of the catch and click cone? $\frac{28}{3}\pi \approx 29.3 \text{ in.}^3$

Vocabulary and Concept Check

1. The height of a cone is the perpendicular distance from the base to the vertex.
2. Both formulas are $\frac{1}{3}Bh$, but the base of a cone is always a circle.
3. Divide by 3.

Practice and Problem Solving

4. $\frac{16\pi}{3} \approx 16.8 \text{ in.}^3$
5. $9\pi \approx 28.3 \text{ m}^3$
6. $\frac{250\pi}{3} \approx 261.8 \text{ mm}^3$
7. $\frac{2\pi}{3} \approx 2.1 \text{ ft}^3$
8. $\frac{200\pi}{3} \approx 209.4 \text{ cm}^3$
9. $\frac{147\pi}{4} \approx 115.5 \text{ yd}^3$
10. $\frac{112\pi}{3} \approx 117.3 \text{ ft}^3$
11. $\frac{125\pi}{6} \approx 65.4 \text{ in.}^3$
12. $\frac{32\pi}{3} \approx 33.5 \text{ cm}^3$
13. The diameter was used instead of the radius;

    $V = \frac{1}{3}(\pi)(1)^2(3) = \pi \text{ m}^3$
14. Glass A; $\frac{38\pi}{3} \approx 39.8 \text{ cm}^3$

Practice and Problem Solving

15. 1.5 ft

16. $\frac{27}{\pi} \approx 8.6$ cm

17. $2\sqrt{\frac{10.8}{4.2\pi}} \approx 1.8$ in.

18. 60π m^3

19. 24.1 min

20. See *Taking Math Deeper.*

21. $3y$

22. about 98 seconds

Fair Game Review

23. $A'(-1, 1), B'(-3, 4), C'(-1, 4)$

24. $E'(4, -1), F'(3, -3), G'(2, -3), H'(1, -1)$

25. D

Mini-Assessment

Find the volume of the cone. Round your answer to the nearest tenth.

1.

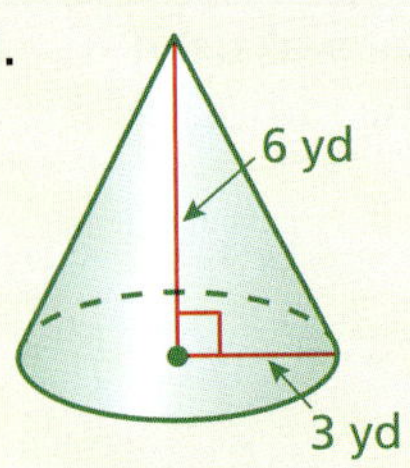

$18\pi \approx 56.5$ yd^3

2.

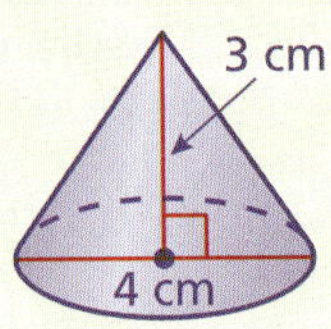

$4\pi \approx 12.6$ cm^3

3. The volume of the ice cream cone is 4.71 cubic inches. Find the height of the cone.

$\frac{14.13}{\pi} \approx 4.5$ in.

Taking Math Deeper

Exercise 20

This is a great type of problem to help students understand the importance of *planning ahead.* Also, in planning, remind students that you can't plan *exactly* how many cups will be used, nor can you plan how full each cup will be. So, the answers to the questions are just "ball park" figures.

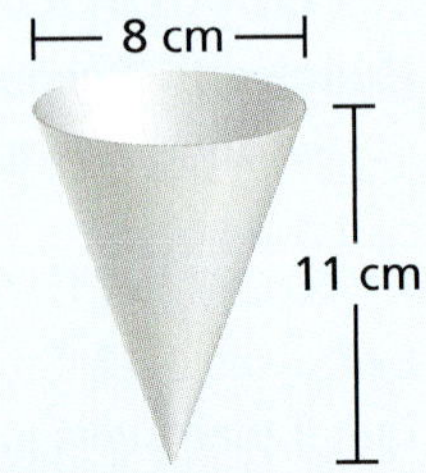

1 How many paper cups will you need?

$$\text{Volume of Cup} = \frac{1}{3}\pi(4)^2(11) \approx 184.3 \text{ cm}^3$$

$$\text{Amount of Lemonade} = (10 \text{ gal})\left(3785 \frac{\text{cm}^3}{\text{gal}}\right) = 37{,}850 \text{ cm}^3$$

$$\text{Number of Cups} \approx \frac{37{,}850}{184.3} \approx 205.4$$

a. You need about 206 cups.

Think Outside the Box

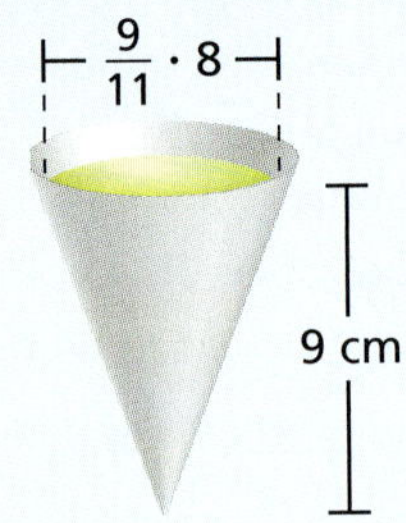

2 How many packs of 50 cups?

b. You should order 5 packs of 50 cups. This will give you 250 cups.

Suppose each cup is not filled to the brim, but only to a height of 9 centimeters. This could mean each cup has a volume of 101 cubic centimeters, which would imply that you would use about 375 cups. . . so 5 packs would *not* be enough.

3 How many cups are left over if you sell only 80% of the lemonade? 80% of 37,850 = 30,280 cm^3

$$\text{Number of Cups} \approx \frac{30{,}280}{184.3} \approx 164.3$$

c. You would have about 250 − 165 = 85 cups left over.

Project

You open a lemonade stand. The lemonade costs you $5.00 per gallon and cups are $6.00 per 50 cups. Create an advertisement including the price of your lemonade. How did you determine the price to charge customers?

Reteaching and Enrichment Strategies

If students need help. . .	If students got it. . .
Resources by Chapter • Practice A and Practice B • Puzzle Time Record and Practice Journal Practice Differentiating the Lesson Lesson Tutorials Skills Review Handbook	Resources by Chapter • Enrichment and Extension • Technology Connection Start the next section

Find the missing dimension of the cone. Round your answer to the nearest tenth.

2 **15.** Volume $= \frac{1}{18}\pi$ ft^3

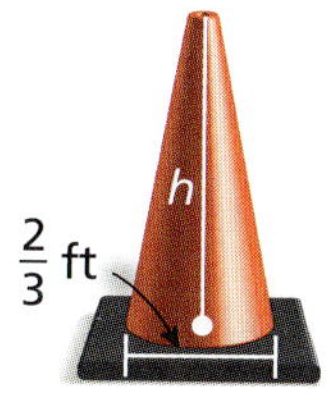

16. Volume $= 225$ cm^3

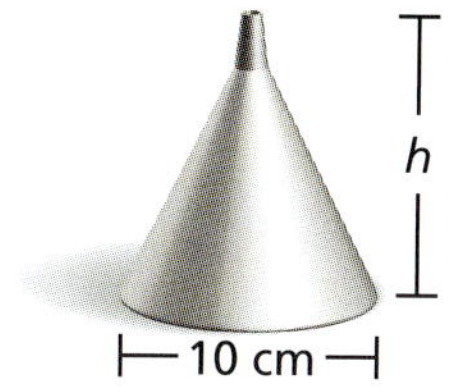

17. Volume $= 3.6$ in.^{3}

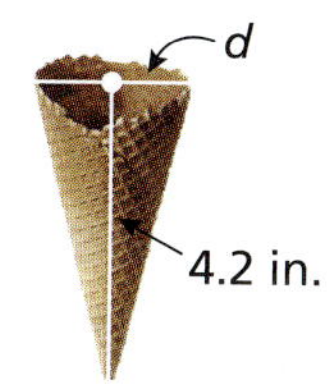

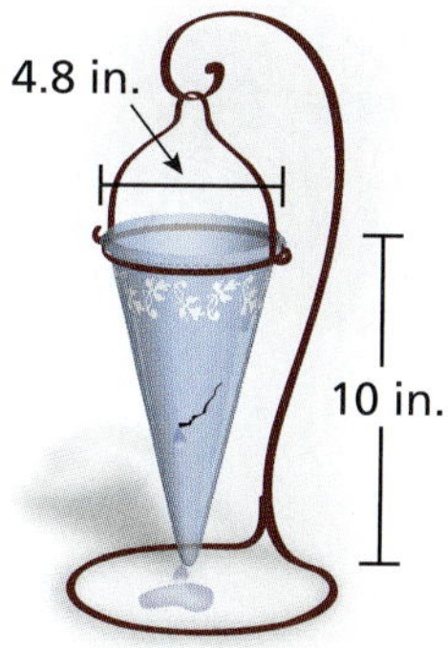

18. REASONING The volume of a cone is 20π cubic meters. What is the volume of a cylinder with the same base and height?

19. VASE Water leaks from a crack in a vase at a rate of 0.5 cubic inch per minute. How long does it take for 20% of the water to leak from a full vase?

20. LEMONADE STAND You have 10 gallons of lemonade to sell. (1 gal $\approx$ 3785 cm^3)

a. Each customer uses one paper cup. How many paper cups will you need?

b. The cups are sold in packages of 50. How many packages should you buy?

c. How many cups will be left over if you sell 80% of the lemonade?

21. STRUCTURE The cylinder and the cone have the same volume. What is the height of the cone?

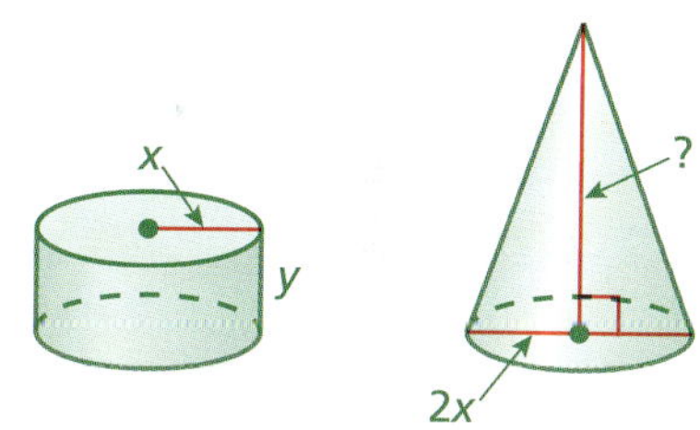

22. Critical Thinking In Example 3, you use a different timer with the same dimensions. The sand in this timer has a height of 30 millimeters. How much time do you have to answer the question?

Fair Game Review What you learned in previous grades & lessons

The vertices of a figure are given. Rotate the figure as described. Find the coordinates of the image. *(Section 11.4)*

23. $A(-1, 1)$, $B(2, 3)$, $C(2, 1)$
90° counterclockwise about vertex A

24. $E(-4, 1)$, $F(-3, 3)$, $G(-2, 3)$, $H(-1, 1)$
180° about the origin

25. MULTIPLE CHOICE $\triangle ABC$ is similar to $\triangle XYZ$. How many times greater is the area of $\triangle XYZ$ than the area of $\triangle ABC$? *(Section 11.6)*

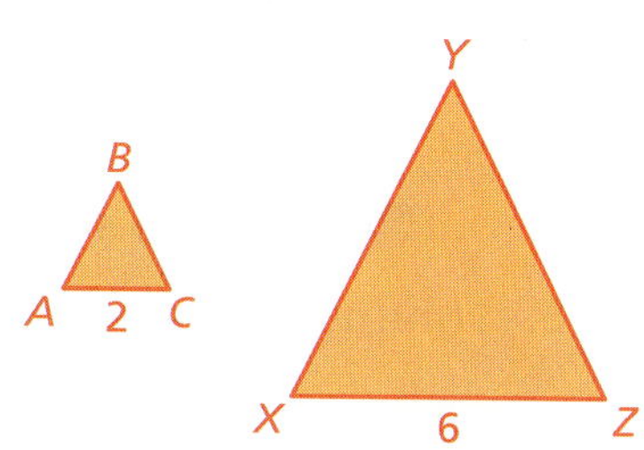

Ⓐ $\frac{1}{9}$ Ⓑ $\frac{1}{3}$

Ⓒ 3 Ⓓ 9

15 Study Help

Check It Out
Graphic Organizer
BigIdeasMath.com

You can use a **formula triangle** to arrange variables and operations of a formula. Here is an example of a formula triangle for the volume of a cylinder.

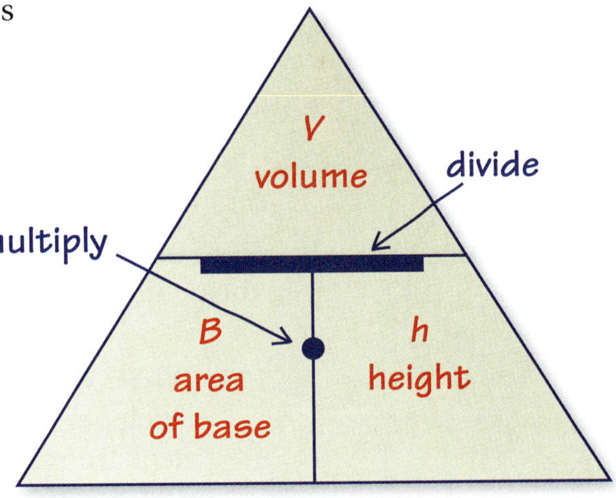

To find an unknown variable, use the other variables and the operation between them. For example, to find the area B of the base, cover up the B. Then you can see that you divide the volume V by the height h.

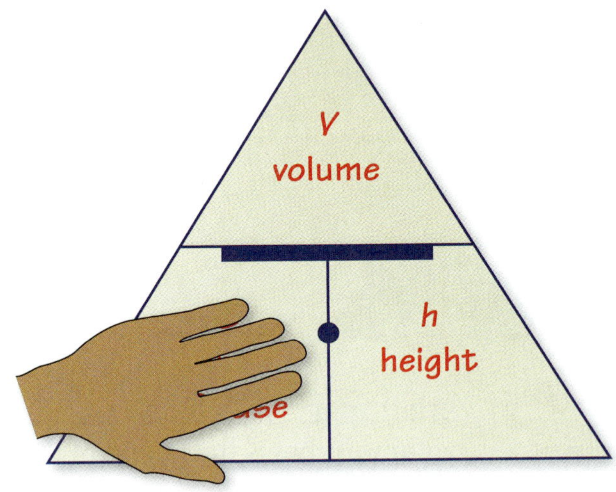

On Your Own

Make a formula triangle to help you study this topic. (*Hint:* Your formula triangle may have a different form than what is shown in the example.)

1. volume of a cone

After you complete this chapter, make formula triangles for the following topics.

2. volume of a sphere
3. volume of a composite solid
4. surface areas of similar solids
5. volumes of similar solids

"See how a formula triangle works? Cover any variable and you get its formula."

Sample Answer

1. Volume of a cone

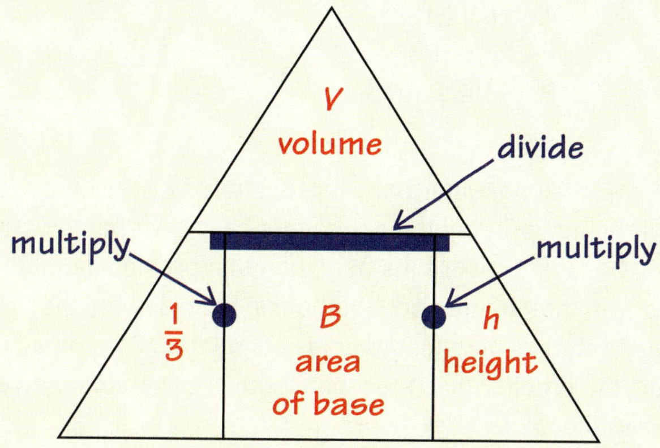

List of Organizers

Available at *BigIdeasMath.com*

Comparison Chart
Concept Circle
Example and Non-Example Chart
Formula Triangle
Four Square
Idea (Definition) and Examples Chart
Information Frame
Information Wheel
Notetaking Organizer
Process Diagram
Summary Triangle
Word Magnet
Y Chart

About this Organizer

A **Formula Triangle** can be used to arrange variables and operations of a formula. Students divide a triangle into the same number of parts as there are variables and factors in a formula. Then students write the variables and factors in the parts of the triangle and place either a multiplication or a division symbol, as appropriate, between the parts. This type of organizer can help students learn the formulas as well as see how the variables in the formulas are related. Students can place their formula triangles on note cards to use as a quick study reference.

Technology for the Teacher

Editable Graphic Organizer

Answers

1. $14\pi \approx 44.0 \text{ yd}^3$
2. $36\pi \approx 113.1 \text{ ft}^3$
3. $50\pi \approx 157.1 \text{ cm}^3$
4. $132\pi \approx 414.7 \text{ in.}^3$
5. $\frac{340}{9\pi} \approx 12.0 \text{ ft}$
6. $\sqrt{\frac{2814}{4.7\pi}} \approx 13.8 \text{ cm}$
7. $\frac{28.26}{\pi} \approx 9 \text{ cm}$
8. The volume is 27 times greater.
9. about 42.45 in.^3
10. 13.5 in.

Technology for the Teacher

Online Assessment
Assessment Book
ExamView® Assessment Suite

Alternative Quiz Ideas

100% Quiz
Error Notebook
Group Quiz
Homework Quiz
Math Log
Notebook Quiz
Partner Quiz
Pass the Paper

Homework Quiz

A homework notebook provides an opportunity for teachers to check that students are doing their homework regularly. Students keep their homework in notebooks. They should be told to record the page number, problem number, and copy the problem exactly in their homework notebooks. Each day the teacher walks around and visually checks that homework is completed. Periodically, without advance notice, the teacher tells the students to put everything away except their homework notebooks.

Questions are from students' homework.

1. What are the answers to Exercises 13–15 on page 677?
2. What are the answers to Exercises 7–9 on page 682?
3. What are the answers to Exercises 3–5 on page 690?
4. What are the answers to Exercises 10–12 on page 698?

Reteaching and Enrichment Strategies

If students need help. . .	If students got it. . .
Resources by Chapter • Practice A and Practice B • Puzzle Time Lesson Tutorials *BigIdeasMath.com*	Resources by Chapter • Enrichment and Extension • Technology Connection Game Closet at *BigIdeasMath.com* Start the next section

15.1–15.2 Quiz

Check It Out
Progress Check
BigIdeasMath.com

Find the volume of the solid. Round your answer to the nearest tenth. *(Section 15.1 and Section 15.2)*

1.

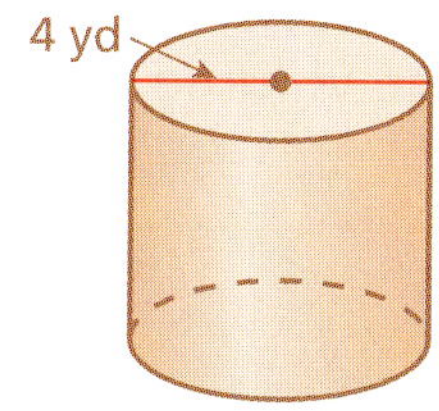

2.

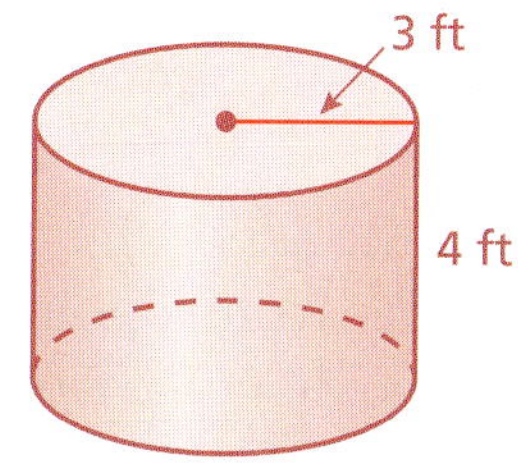

3.

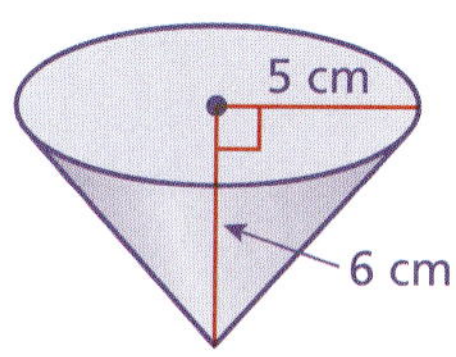

4.

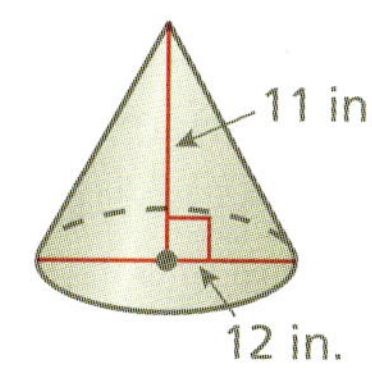

Find the missing dimension of the solid. Round your answer to the nearest tenth. *(Section 15.1 and Section 15.2)*

5.

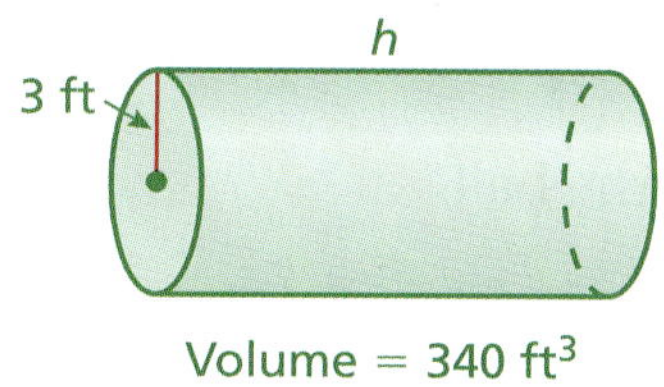

6.

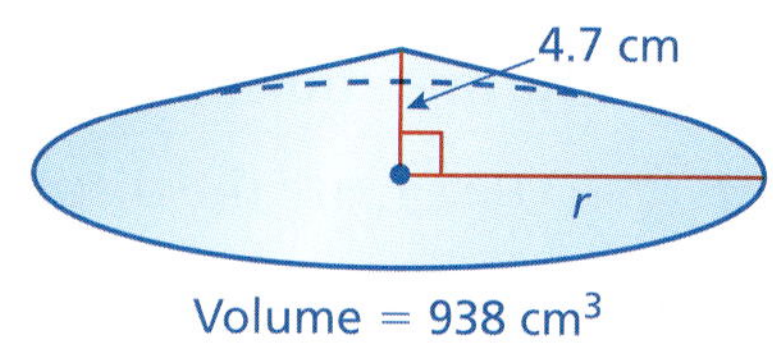

7. **PAPER CONE** The paper cone can hold 84.78 cubic centimeters of water. What is the height of the cone? *(Section 15.2)*

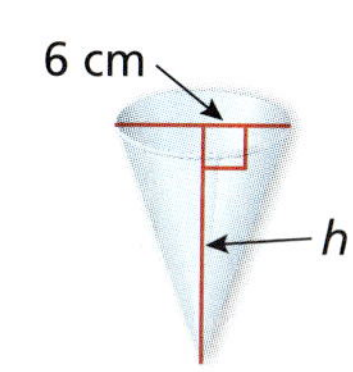

8. **GEOMETRY** Triple both dimensions of the cylinder. How many times greater is the volume of the new cylinder than the volume of the original cylinder? *(Section 15.1)*

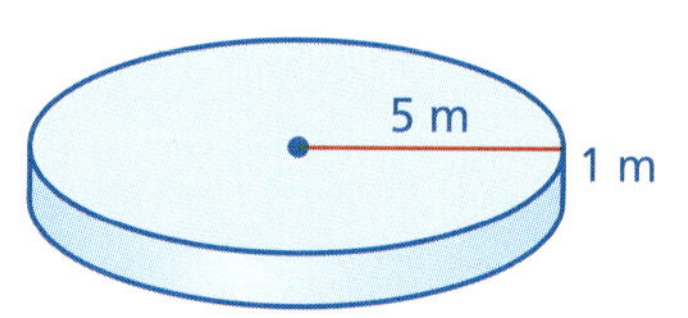

9. **SAND ART** There are 42.39 cubic inches of blue sand and 28.26 cubic inches of red sand in the cylindrical container. How many cubic inches of white sand are in the container? *(Section 15.1)*

10. **JUICE CAN** You are buying two cylindrical cans of juice. Each can holds the same amount of juice. What is the height of Can B? *(Section 15.1)*

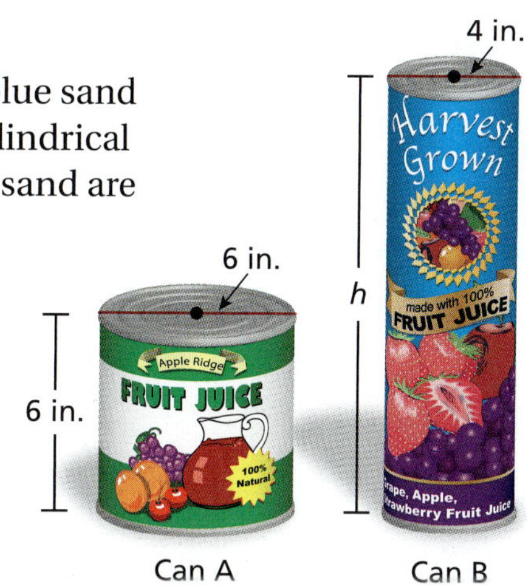

15.3 Volumes of Spheres

Essential Question How can you find the volume of a sphere?

A **sphere** is the set of all points in space that are the same distance from a point called the *center*. The *radius* r is the distance from the center to any point on the sphere.

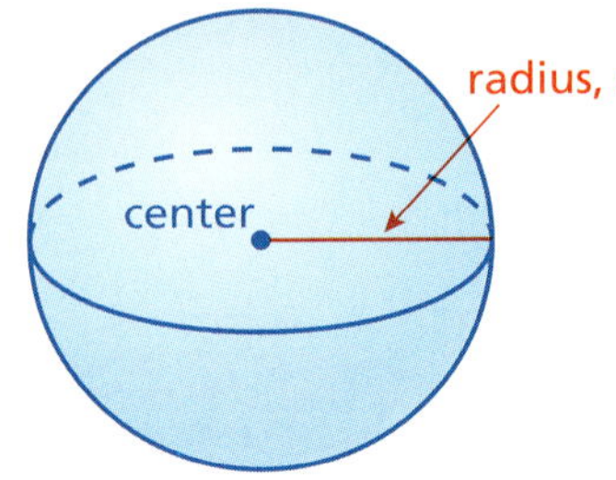

A sphere is different from the other solids you have studied so far because it does not have a base. To discover the volume of a sphere, you can use an activity similar to the one in the previous section.

1 ACTIVITY: Exploring the Volume of a Sphere

Work with a partner. Use a plastic ball similar to the one shown.

- Estimate the diameter and the radius of the ball.

- Use these measurements to draw a net for a cylinder with a diameter and a height equal to the diameter of the ball. How is the height h of the cylinder related to the radius r of the ball? Explain.

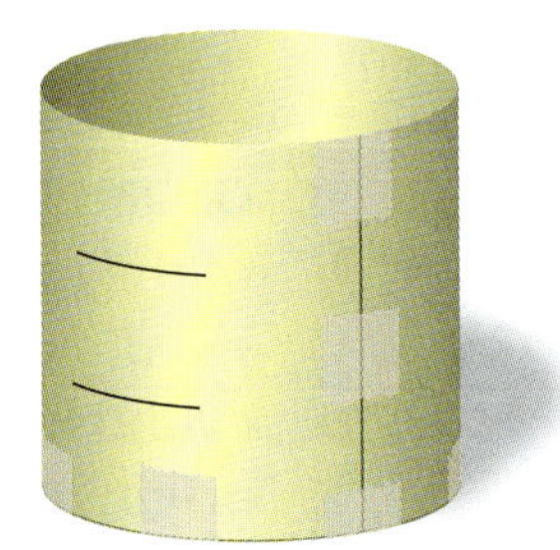

- Cut out the net. Then fold and tape it to form an open cylinder. Make two marks on the cylinder that divide it into thirds, as shown.

- Cover the ball with aluminum foil or tape. Leave one hole open. Fill the ball with rice. Then pour the rice into the cylinder. What fraction of the cylinder is filled with rice?

Geometry

In this lesson, you will

- find the volumes of spheres.
- find the radii of spheres given the volumes.
- solve real-life problems.

Laurie's Notes

Introduction

Applying Mathematical Practices

- **Model with Mathematics:** In the first activity, students will explore the relationship between the volume of a sphere and volume of a cylinder. Modeling such relationships is a powerful teaching tool.

Motivate

- Bring a collection of spherical objects, such as rubber balls, to class. The objects should be of different sizes.
- ? "What is the geometric name for these solids?" sphere
- ? "What linear dimension or dimensions does a sphere have?" radius
- Discuss with students the fact that spheres have only one linear dimension. Other solids they studied have two or three linear dimensions.
- Hold several of the objects and ask students which has the greatest volume and which has the least volume. Ask them to explain their reasoning. Because a sphere has only one linear dimension, the object with the greatest radius has the greatest volume and the object with the least radius has the least volume.

Activity Notes

Activity 1

- **Management Tip:** Instead of having the whole class perform this activity, you may want to have a pair of students demonstrate it. If you have a sphere and a cylinder from a commercially-available (fillable) geometric solids kit, then you could use these instead of having students make the cylinder. The sphere and the cylinder must have equal diameters and the height of the cylinder must also equal the diameter. Alternatively, you could show a video demonstration, such as the one at http://www.youtube.com/watch?v=aLyQddyY8ik.
- **Model with Mathematics:** Regardless, you want students to see the relationship and not just hear what it is.
- If necessary, have materials ready for students: heavy-weight paper, tape, scissors, ruler, and uncooked rice.
- The net students make for the cylinder will most likely have two bases. If so, then they can just tear off one of the bases to get the open cylinder.
- Be sure the cylinder(s) are marked by thirds so students can clearly see the relationship between the volume of the cylinder and the volume of the sphere.
- Students can use a piece of paper and shape it like a funnel to fill the ball with rice.
- Have students guess the relationship before they see it. Because the sphere fits inside the cylinder, students at least know that the cylinder has a greater volume.
- ? "How much of the cylinder is filled by the contents of the sphere?" $\frac{2}{3}$

What Your Students Will Learn

- Find the volumes of spheres using a formula.
- Find the radii of spheres given the volumes.
- Find the volume of composite solids using volume formulas of known solids that make up the composite solid.

Previous Learning

Students should know how to find volumes of cylinders and cones.

Technology for the Teacher

Dynamic Classroom

Lesson Plans
Complete Materials List

15.3 Record and Practice Journal

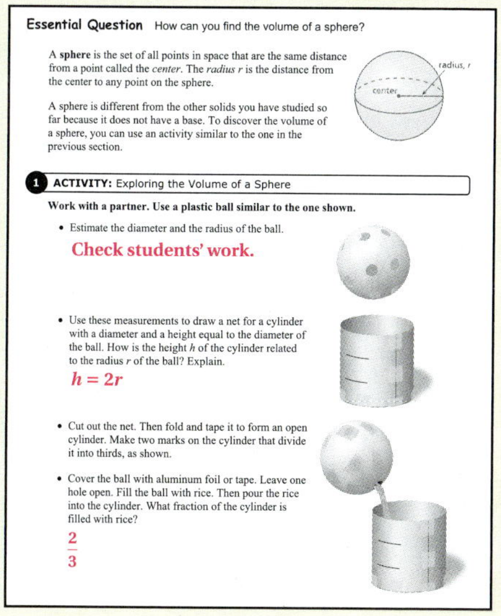

Essential Question How can you find the volume of a sphere?

A **sphere** is the set of all points in space that are the same distance from a point called the *center*. The *radius* r is the distance from the center to any point on the sphere.

A sphere is different from the other solids you have studied so far because it does not have a base. To discover the volume of a sphere, you can use an activity similar to the one in the previous section.

1 ACTIVITY: Exploring the Volume of a Sphere

Work with a partner. Use a plastic ball similar to the one shown.

- Estimate the diameter and the radius of the ball.

  Check students' work.

- Use these measurements to draw a net for a cylinder with a diameter and a height equal to the diameter of the ball. How is the height h of the cylinder related to the radius r of the ball? Explain.

  $h = 2r$

- Cut out the net. Then fold and tape it to form an open cylinder. Make two marks on the cylinder that divide it into thirds, as shown.
- Cover the ball with aluminum foil or tape. Leave one hole open. Fill the ball with rice. Then pour the rice into the cylinder. What fraction of the cylinder is filled with rice?

  $\frac{2}{3}$

English Language Learners

Vocabulary

English learners may be familiar with the word *net* in everyday context. In finance, the net profit describes the bottom line of a financial transaction. In fishing, a net is a collection of knotted strings used to catch fish. In a mathematical context, such as in Activity 1, *net* or *geometric net* is used to mean the two-dimensional representation of a solid object.

15.3 Record and Practice Journal

2 ACTIVITY: Deriving the Formula for the Volume of a Sphere

Work with a partner. Use the results from Activity 1 and the formula for the volume of a cylinder to complete the steps.

$V = \pi r^2 h$ — Write formula for volume of a cylinder.

$= \frac{2}{3}\pi r^2 h$ — Multiply by $\frac{2}{3}$ because the volume of a sphere is $\frac{2}{3}$ of the volume of the cylinder.

$= \frac{2}{3}\pi r^2 (2r)$ — Substitute $2r$ for h.

$= \frac{4}{3}\pi r^3$ — Simplify.

3 ACTIVITY: Deriving the Formula for the Volume of a Sphere

Work with a partner. Imagine filling the inside of a sphere with n small pyramids. The vertex of each pyramid is at the center of the sphere and the height of each pyramid is approximately equal to r, as shown. Complete the steps. (The surface area of a sphere is equal to $4\pi r^2$.)

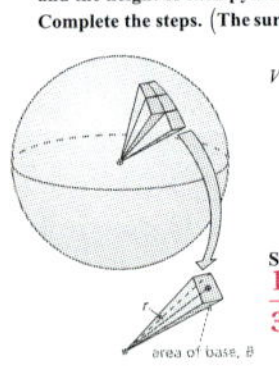

$V = \frac{1}{3}Bh$ — Write formula for volume of a pyramid.

$= n\frac{1}{3}B\,r$ — Multiply by the number of small pyramids n and substitute r for h.

$= \frac{1}{3}(4\pi r^2)\,r$ — $4\pi r^2 \approx n \cdot B$.

Show how this result is equal to the result in Activity 2.

$$\frac{1}{3}(4\pi r^2)r = \frac{1}{3}(4 \cdot \pi \cdot r \cdot r)r = \left(\frac{1}{3} \cdot 4\right)(\pi \cdot r \cdot r \cdot r) = \frac{4}{3}\pi r^3$$

What Is Your Answer?

4. **IN YOUR OWN WORDS** How can you find the volume of a sphere?
Sample answer: Use the formula $V = \frac{4}{3}\pi r^3$.

5. Describe the intersection of the plane and the sphere. Then explain how to find the volume of each section of the solid.
circle

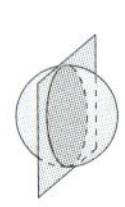

Laurie's Notes

Activity 2

- This activity uses the results from Activity 1, so it is important for students to complete that activity before moving on to this activity.
- ? "What is the formula for the volume V of a cylinder?" $V = Bh = 2\pi r^2 h$, where B is the area of the base, h is the height, and r is the radius of the cylinder.
- **Model with Mathematics:** Students will use the relationship between the volume of a sphere and the volume of a cylinder, and the formula for the volume of a cylinder, to derive the formula for the volume of a sphere.
- **Make Sense of Problems and Persevere in Solving Them:** Resist the urge to tell students how to fill in the blanks. Students should talk with their partners and use available resources to help them fill in the blanks.
- Students may struggle with substituting $2r$ for the height of the cylinder and then simplifying.
- After students perform the simplification in the last step, they will have the formula for the volume of a sphere.

Activity 3

- This activity, which is optional, takes students through an alternate derivation of the formula for the volume of a sphere.
- The symbolic manipulation should not be beyond the ability of your students. Students may, however, have difficulty visualizing a sphere composed of small pyramids.
- The curvature of the sphere is ignored because it is assumed that the base areas of the pyramids are small.
- Students are given the formula for the surface area of a sphere. Tell them that they will learn more about the surface area of a sphere in later courses.
- Animations of the approach shown in this activity can be viewed online, such as the one at http://www.youtube.com/watch?v=xuPl_8o_j7k. You may wish to view the animation yourself or show it to the class.

What Is Your Answer?

- Students should check their answers with their neighbors.
- Problem 5 suggests the idea of a sphere composed of two hemispheres. Point out that a hemisphere is half a sphere. Students will need to find the volume of a hemisphere in problems such as Example 3.

Closure

- **Exit Ticket:** The radius of a sphere is 2 centimeters. Find the volume of a cube that the sphere fits snugly within and find the volume of the sphere.
64 cm^3; $\frac{32}{3}\pi \approx 33.5\text{ cm}^3$

2 ACTIVITY: Deriving the Formula for the Volume of a Sphere

Work with a partner. Use the results from Activity 1 and the formula for the volume of a cylinder to complete the steps.

$V = \pi r^2 h$	Write formula for volume of a cylinder.
$= \dfrac{\square}{\square}\pi r^2 h$	Multiply by $\dfrac{\square}{\square}$ because the volume of a sphere is $\dfrac{\square}{\square}$ of the volume of the cylinder.
$= \dfrac{\square}{\square}\pi r^2 \square$	Substitute $\square$ for h.
$= \dfrac{\square}{\square}\pi \square$	Simplify.

Math Practice

Analyze Relationships

What is the relationship between the volume of a sphere and the volume of a cylinder? How does this help you derive a formula for the volume of a sphere?

3 ACTIVITY: Deriving the Formula for the Volume of a Sphere

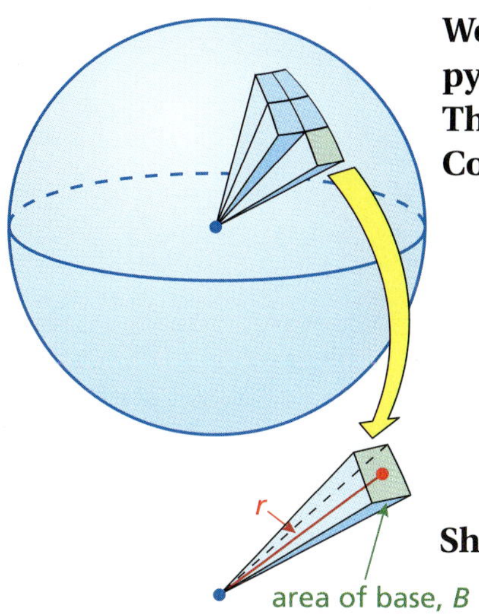

Work with a partner. Imagine filling the inside of a sphere with n small pyramids. The vertex of each pyramid is at the center of the sphere. The height of each pyramid is approximately equal to r, as shown. Complete the steps. (The surface area of a sphere is equal to $4\pi r^2$.)

$V = \dfrac{1}{3}Bh$	Write formula for volume of a pyramid.
$= n\dfrac{1}{3}B\square$	Multiply by the number of small pyramids n and substitute $\square$ for h.
$= \dfrac{1}{3}(4\pi r^2)\square$	$4\pi r^2 \approx n \cdot \square$

Show how this result is equal to the result in Activity 2.

What Is Your Answer?

4. **IN YOUR OWN WORDS** How can you find the volume of a sphere?

5. Describe the intersection of the plane and the sphere. Then explain how to find the volume of each section of the solid.

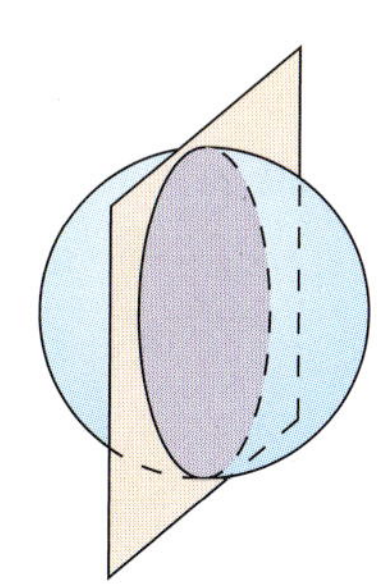

Practice Use what you learned about the volumes of spheres to complete Exercises 3–5 on page 690.

15.3 Lesson

Check It Out
Lesson Tutorials
BigIdeasMath.com

Key Vocabulary

sphere, *p. 686*

hemisphere, *p. 689*

Key Idea

Volume of a Sphere

Words The volume V of a sphere is the product of $\frac{4}{3}\pi$ and the cube of the radius of the sphere.

Algebra $V = \frac{4}{3}\pi r^3$

Cube of radius of sphere

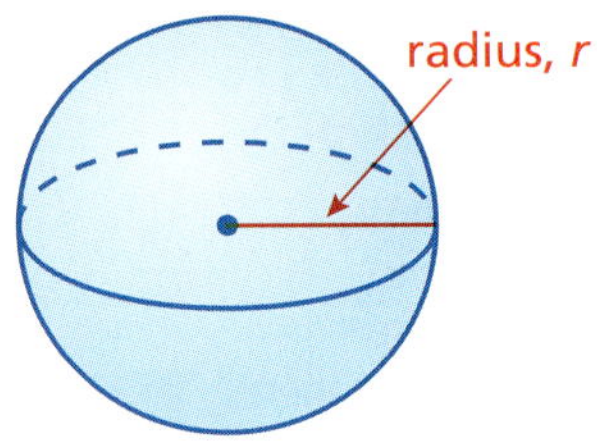

EXAMPLE 1 Finding the Volume of a Sphere

Find the volume of the sphere. Round your answer to the nearest tenth.

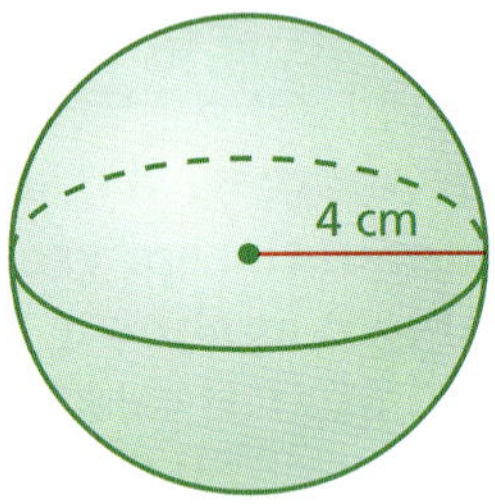

$V = \frac{4}{3}\pi r^3$ Write formula for volume.

$= \frac{4}{3}\pi(4)^3$ Substitute 4 for r.

$= \frac{256}{3}\pi$ Simplify.

≈ 268.1 Use a calculator.

The volume is about 268.1 cubic centimeters.

EXAMPLE 2 Finding the Radius of a Sphere

Find the radius of the sphere.

Volume = 288π in.^{3}

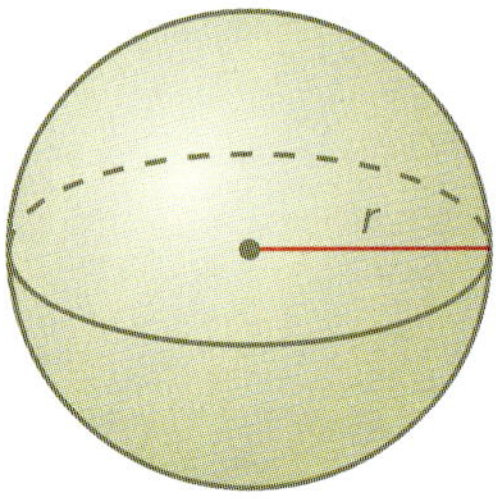

$V = \frac{4}{3}\pi r^3$ Write formula.

$288\pi = \frac{4}{3}\pi r^3$ Substitute.

$288\pi = \frac{4\pi}{3}r^3$ Multiply.

$\frac{3}{4\pi} \cdot 288\pi = \frac{3}{4\pi} \cdot \frac{4\pi}{3}r^3$ Multiplication Property of Equality

$216 = r^3$ Simplify.

$6 = r$ Take the cube root of each side.

The radius is 6 inches.

Multi-Language Glossary at BigIdeasMath.com

Laurie's Notes

Introduction

Connect

- **Yesterday:** Students derived the formula for the volume of a sphere.
- **Today:** Students will find volumes of spheres.

Motivate

- In some places you can buy ground meat in a cylindrical casing. The casing with one pound of ground meat has a diameter of 3 inches and a height of 6 inches.
- Tell a story about making meatballs that have a 1-inch diameter using the ground meat from the cylindrical casing.
- ? "How can I figure out the number of meatballs I can make?" Find the volume of the ground meat in the cylindrical casing and divide it by the volume of a meatball.
- Explain that in today's lesson they will find the volumes of spheres.

Lesson Notes

Key Idea

- Have physical objects, such as a ball or a globe, available to reference.
- Draw the sphere and write the formula in words.
- Students may derive this formula again in a high school geometry course.

Example 1

- Notice that the work is done in terms of π. It is not until the last step that you use the π key on a calculator.
- ? "What information do you know?" The radius of the sphere is 4 centimeters.
- **Representation:** Encourage students to use the parentheses to represent multiplication. Using the $\times$ symbol would make the expression confusing.
- **Common Misconception:** Remind students that π is a number and because multiplication is commutative and associative, this expression could be rewritten as $\frac{4}{3}(4)^3\pi$, making the computation less confusing.
- ? "What is being cubed in this expression?" only the 4
- **Extension:** Ask students how big they think a cubic centimeter is. To help students visualize, tell them that a cubic centimeter is about the size of a sugar cube.

Example 2

- This example requires students to solve an equation for a variable.
- Work through the problem, annotating the steps as shown in the book.

Goal Today's lesson is finding volumes of **spheres**.

Technology for the Teacher

Dynamic Classroom

Lesson Tutorials
Lesson Plans
Answer Presentation Tool

Extra Example 1

Find the volume of a sphere with a radius of 11 meters. Round your answer to the nearest tenth.

$\frac{5324}{3}\pi \approx 5575.3 \text{ m}^3$

Extra Example 2

Find the radius of a sphere with a volume of 2304π cubic centimeters.

12 cm

On Your Own

1. $\frac{2048}{3}\pi \approx 2144.7 \text{ ft}^3$
2. 3 m

Extra Example 3

In Example 3, the radius of the silo is 9 feet and the overall height is 48 feet. What is the volume of the silo? Round your answer to the nearest thousand. $3645\pi \approx 11{,}000 \text{ ft}^3$

On Your Own

3. $\frac{80}{3}\pi \approx 83.8 \text{ in.}^3$
4. $96\pi \approx 301.6 \text{ m}^3$

Differentiated Instruction

Visual

Students can check the volume of a sphere for reasonableness by finding the volume of a cube with side length $2r$, as suggested in the Closure on page T-687. An example is shown below.

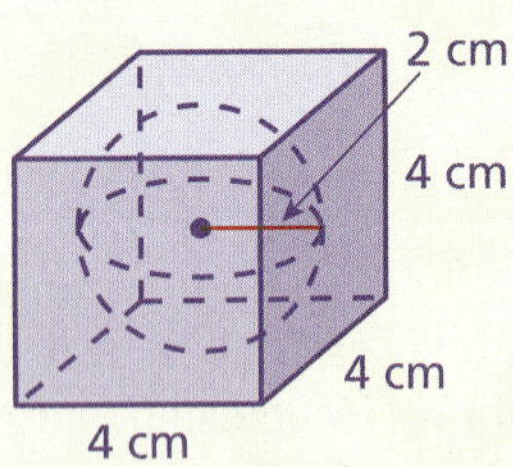

This will give an overestimate of the volume of the sphere. Also, it will help students remember to use the radius, not the diameter, in the given formula for volume of a sphere.

Laurie's Notes

Example 2 (continued)

- **Look for and Make Use of Structure:** Discuss why students can write $\frac{4}{3}\pi$ as $\frac{4\pi}{3}$. You want them to see the connection to fraction multiplication: $\frac{4}{3} \cdot 7 = \frac{4 \cdot 7}{3}$, so likewise $\frac{4}{3} \cdot \pi = \frac{4 \cdot \pi}{3}$.
- **Look for and Make Use of Structure:** Students may have difficulty solving for r in this example. As shown, you can accomplish this by multiplying each side of the equation by the reciprocal of the variable term's coefficient and then taking the cube root of each side. Provide a quick review of cube roots if necessary.

On Your Own

- Ask volunteers to share their work at the board.

Example 3

- Students have previously found areas of composite figures. Students generally have little difficulty understanding how to find volumes of composite solids.
- Ask a student to read the problem. Sketch the solid as the student is reading.

? "What information do you know?" the height and radius of the silo

- Discuss the Study Tip.

? "How can you find the volume of the silo?" Find the volume of the cylinder and the volume of the hemisphere and add.

- Use the good problem solving technique of writing the formulas first and then substituting the values of the variables.
- Work through the problem as shown. Notice that the solution is kept in terms of π until the last step.

On Your Own

- **Think-Pair-Share:** Students should read each question independently and then work in pairs to answer the questions. When they have answered the questions, the pair should compare their answers with another group and discuss any discrepancies.

Closure

- Have students answer the following question:
  You have an ice cream scoop with a 2-inch diameter. You have an ice cream cone with a 2-inch diameter and a height of 5 inches. If you place one scoop of ice cream on the cone and let the ice cream melt, will it spill over the cone? Explain. no; The volume of the cone is greater than the volume of the ice cream.

On Your Own

Now You're Ready
Exercises 3–11

Find the volume V or radius r of the sphere. Round your answer to the nearest tenth, if necessary.

1.

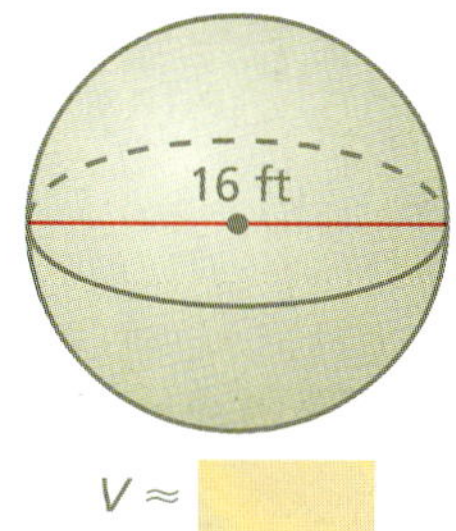

2.

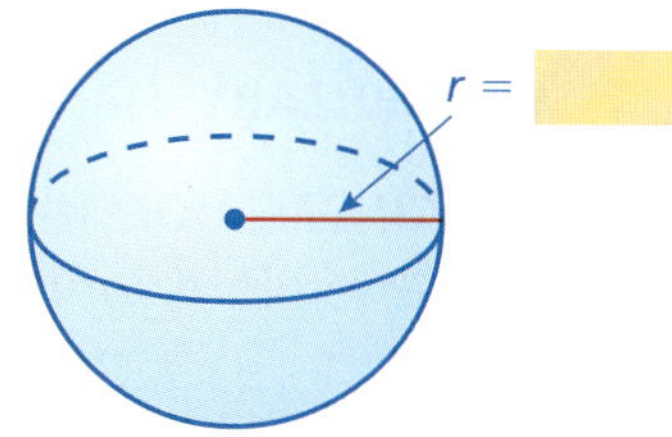

Volume $= 36\pi$ m^3

EXAMPLE 3 Finding the Volume of a Composite Solid

A hemisphere is one-half of a sphere. The top of the silo is a hemisphere with a radius of 12 feet. What is the volume of the silo? Round your answer to the nearest thousand.

The silo is made up of a cylinder and a hemisphere. Find the volume of each solid.

Cylinder

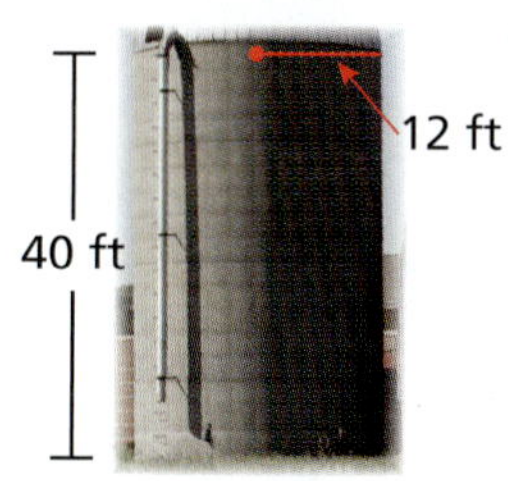

$$V = Bh$$
$$= \pi(12)^2(40)$$
$$= 5760\pi$$

Hemisphere

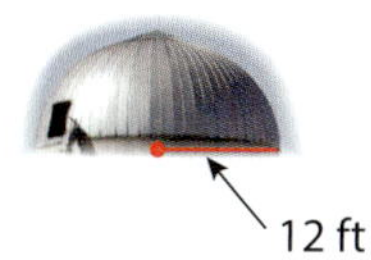

$$V = \frac{1}{2} \cdot \frac{4}{3}\pi r^3$$
$$= \frac{1}{2} \cdot \frac{4}{3}\pi(12)^3$$
$$= 1152\pi$$

So, the volume is $5760\pi + 1152\pi = 6912\pi \approx 22{,}000$ cubic feet.

Study Tip

In Example 3, the height of the cylindrical part of the silo is the difference of the silo height and the radius of the hemisphere.

$52 - 12 = 40$ ft

On Your Own

Now You're Ready
Exercises 14–16

Find the volume of the composite solid. Round your answer to the nearest tenth.

3.

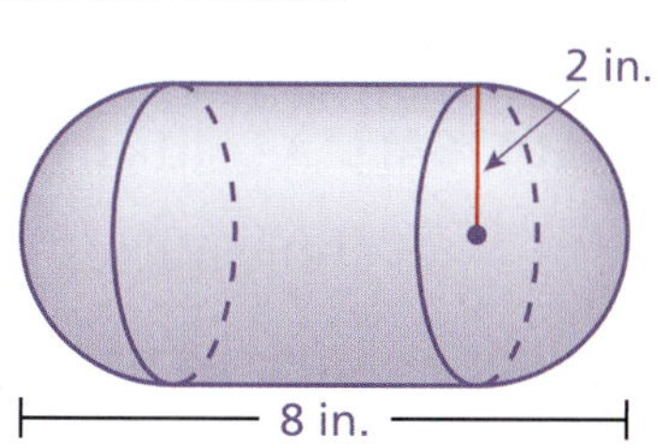

4.

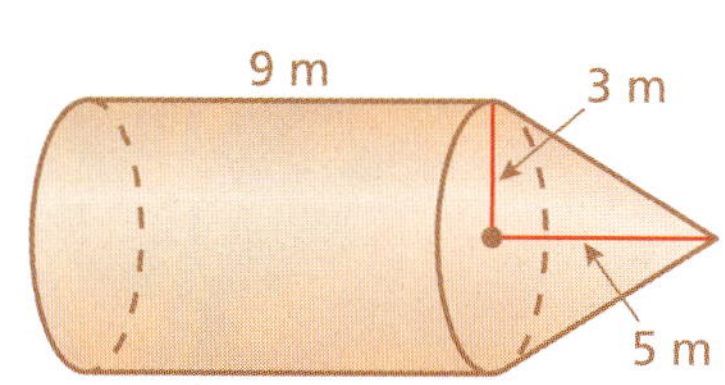

15.3 Exercises

Check It Out
Help with Homework
BigIdeasMath.com

Vocabulary and Concept Check

1. **VOCABULARY** How is a sphere different from a hemisphere?

2. **WHICH ONE DOESN'T BELONG?** Which figure does *not* belong with the other three? Explain your reasoning.

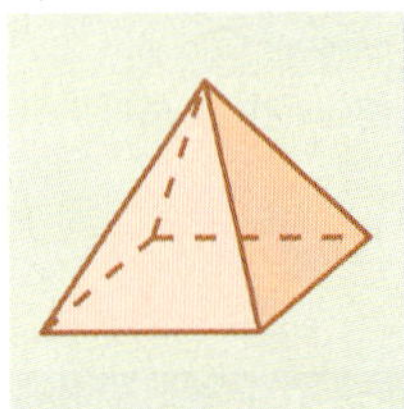
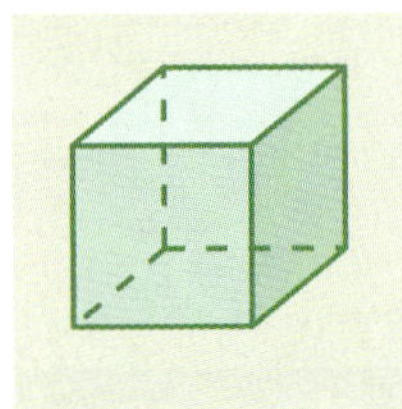
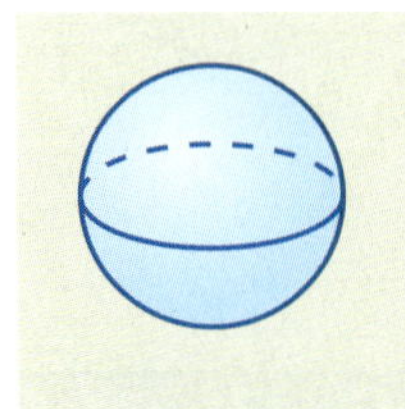
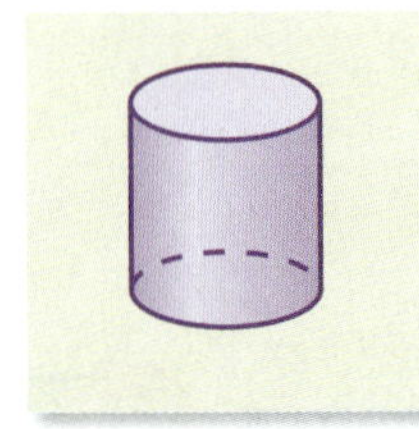

Practice and Problem Solving

Find the volume of the sphere. Round your answer to the nearest tenth.

1 **3.**
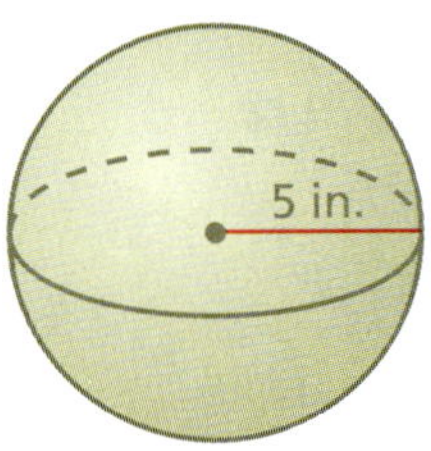

4.
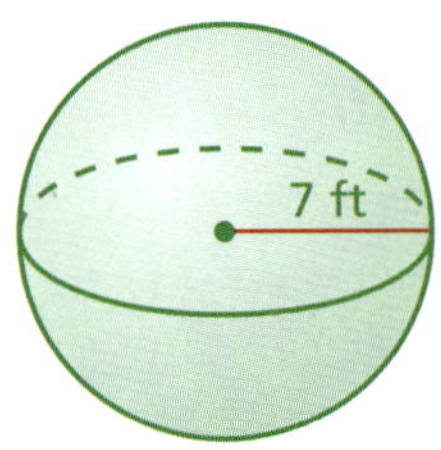

5.
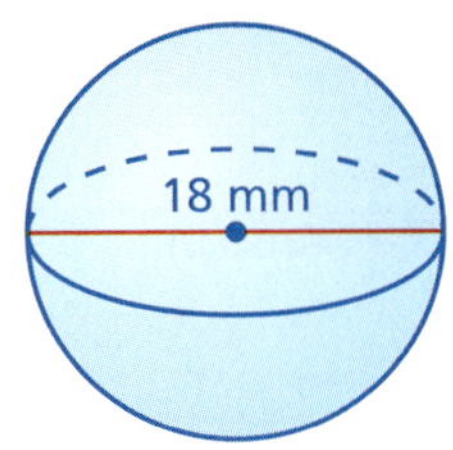

6.
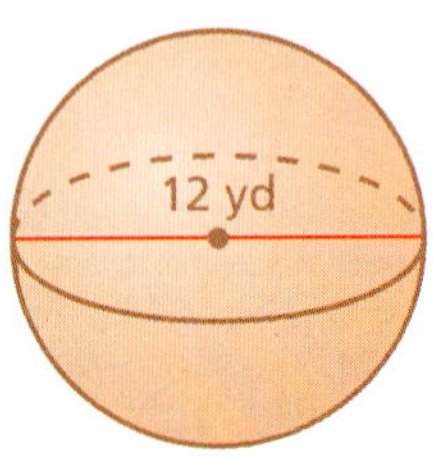

7.
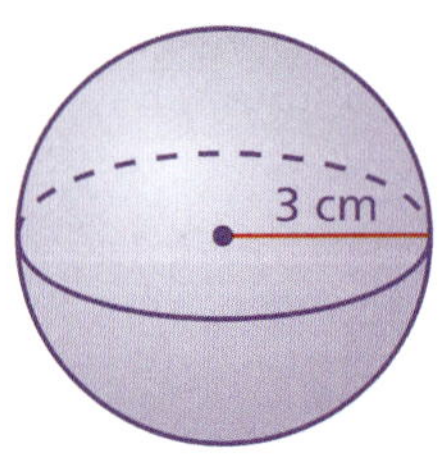

8.
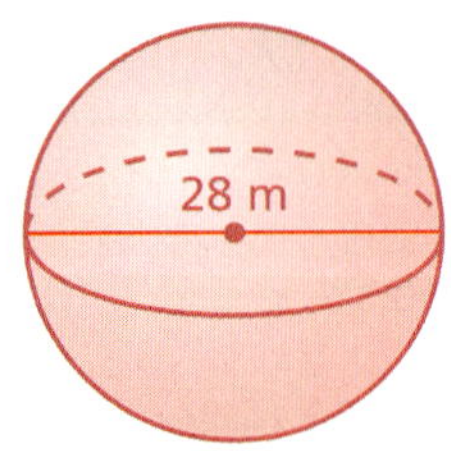

Find the radius of the sphere with the given volume.

2 **9.** Volume $= 972\pi$ mm^3 **10.** Volume $= 4.5\pi$ cm^3 **11.** Volume $= 121.5\pi$ ft^3

12. GLOBE The globe of the Moon has a radius of 10 inches. Find the volume of the globe. Round your answer to the nearest whole number.

13. SOFTBALL A softball has a volume of $\frac{125}{6}\pi$ cubic inches. Find the radius of the softball.

Assignment Guide and Homework Check

Level	Assignment	Homework Check
Accelerated	1–23	8, 10, 13, 15, 19

Common Errors

- **Exercises 3–8 and 12** When finding the volume of a sphere, students may forget to multiply by $\frac{4}{3}$ or cube the radius, they may use diameter when the formula calls for radius, or they may write the incorrect units. Remind them of the given formula for volume of a sphere and that part of writing a correct answer is including the correct units.
- **Exercises 9–11 and 13** Students may not complete the solution; they may solve for the *cube* of the radius instead of the radius. Point out that they need to find a cube root, which they learned in Section 14.2, and provide a quick review if necessary.
- **Exercises 14–16** Students may think that there is not enough information to solve the problem. It may help to have them "break up" the composite solid into two parts, whose volumes they know how to find, and mark the dimensions on each part.
- **Exercise 16** Students may add instead of subtract the volume of the hemisphere. Point out that in contrast to the solid in Example 3, this solid is made up of a cylinder with a hemisphere *removed*, not added.

15.3 Record and Practice Journal

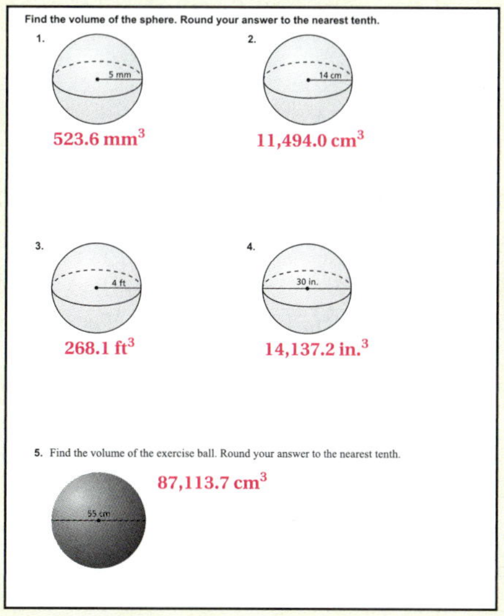
Find the volume of the sphere. Round your answer to the nearest tenth.

1. 5 mm — 523.6 mm^3
2. 14 cm — $11{,}494.0 \text{ cm}^3$
3. 4 ft — 268.1 ft^3
4. 30 in. — $14{,}137.2 \text{ in.}^3$
5. Find the volume of the exercise ball. Round your answer to the nearest tenth. 55 cm — $87{,}113.7 \text{ cm}^3$

Vocabulary and Concept Check

1. A hemisphere is one-half of a sphere.
2. sphere; It does not have a base.

Practice and Problem Solving

3. $\frac{500\pi}{3} \approx 523.6 \text{ in.}^3$
4. $\frac{1372\pi}{3} \approx 1436.8 \text{ ft}^3$
5. $972\pi \approx 3053.6 \text{ mm}^3$
6. $288\pi \approx 904.8 \text{ yd}^3$
7. $36\pi \approx 113.1 \text{ cm}^3$
8. $\frac{10{,}976\pi}{3} \approx 11{,}494.0 \text{ m}^3$
9. 9 mm
10. 1.5 cm
11. 4.5 ft
12. $\frac{4000\pi}{3} \approx 4189 \text{ in.}^3$
13. 2.5 in.
14. $512 + \frac{128\pi}{3} \approx 646.0 \text{ cm}^3$

Practice and Problem Solving

15. $256\pi + 128\pi = 384\pi \approx 1206.4 \text{ ft}^3$

16. $99\pi - 18\pi = 81\pi \approx 254.5 \text{ in.}^3$

17. $r = \frac{3}{4}h$

18. $162\pi - 108\pi = 54\pi \approx 170 \text{ cm}^3$

19. 5400 in.^2; $27{,}000 \text{ in.}^3$

20. See *Taking Math Deeper*.

Fair Game Review

21. enlargement; 2

22. reduction; $\frac{1}{3}$

23. A

Mini-Assessment

Find the volume of the sphere. Round your answer to the nearest tenth.

1.

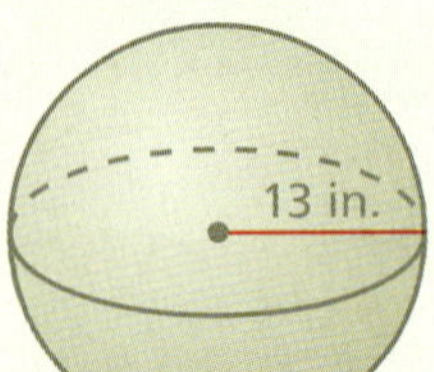

$\frac{8788\pi}{3} \approx 9202.8 \text{ in.}^3$

2.

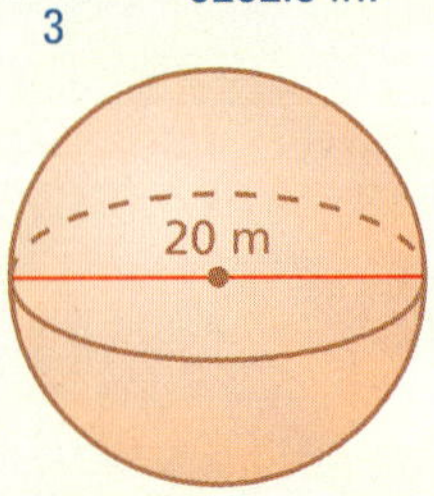

$\frac{4000\pi}{3} \approx 4188.8 \text{ m}^3$

3. Find the radius of a sphere with a volume of 972π cubic millimeters.
9 mm

4. In Example 3, the diameter of the silo is 18 meters and the overall height is 62 meters. What is the volume of the silo? Round your answer to the nearest thousand.
$4779\pi \approx 15{,}000 \text{ m}^3$

Taking Math Deeper

Exercise 20

This exercise is a good review of rewriting literal equations. Start by writing the volume formula for each solid.

1 Draw the solids and write the corresponding volume formulas.

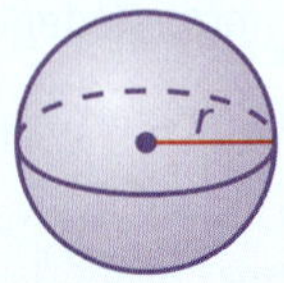

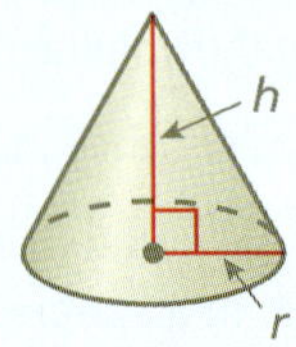

$V = \frac{4}{3}\pi r^3$ $\quad$ $V = \frac{1}{3}\pi r^2 h$

2 Write and solve an equation.

Volume of sphere = 4 • Volume of cone

$$\frac{4}{3}\pi r^3 = 4 \cdot \frac{1}{3}\pi r^2 h$$

$$\frac{4}{3}\pi r^3 = \frac{4}{3}\pi r^2 h$$

$$r^3 = r^2 h$$

$$r \cdot r \cdot r = r \cdot r \cdot h$$

$$r = h$$

So, the volume of a sphere with radius r is four times the volume of a cone with radius r when the height of the cone is equal to the radius.

3 Check your solution by choosing a value for r. Let $r = 3$, so $h = 3$.

Sphere: $V = \frac{4}{3}\pi(3)^3 = 36\pi$ $\quad$ **Cone:** $V = \frac{1}{3}\pi(3)^2(3) = 9\pi$

Because $36\pi = 4 \cdot 9\pi$, the solution checks.

Reteaching and Enrichment Strategies

If students need help...	If students got it...
Resources by Chapter • Practice A and Practice B • Puzzle Time Record and Practice Journal Practice Differentiating the Lesson Lesson Tutorials Skills Review Handbook	Resources by Chapter • Enrichment and Extension • Technology Connection Start the next section

Find the volume of the composite solid. Round your answer to the nearest tenth.

3 **14.**

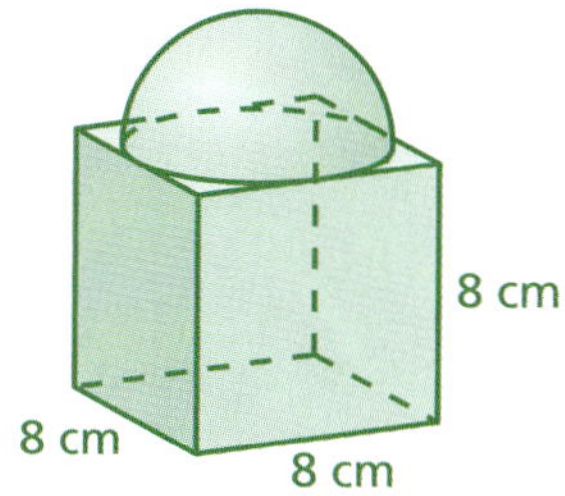

15.

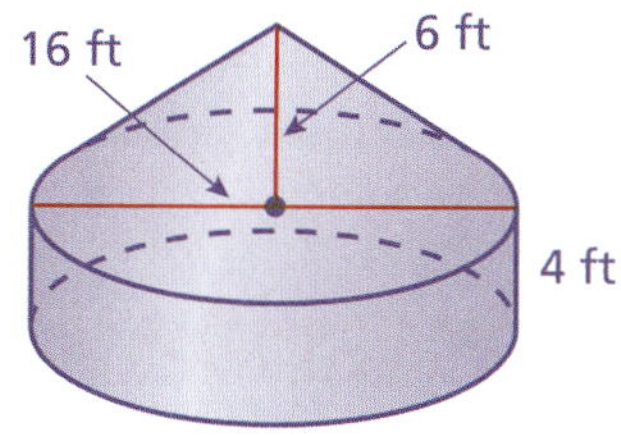

16.

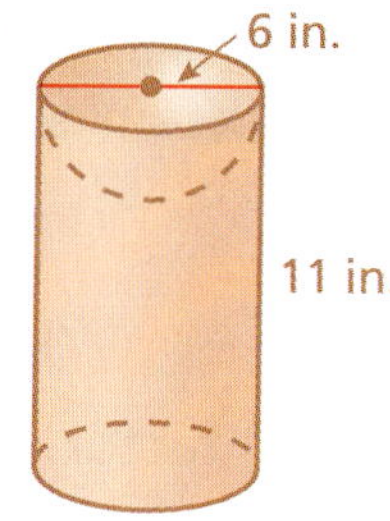

17. **REASONING** A sphere and a right cylinder have the same radius and volume. Find the radius r in terms of the height h of the cylinder.

18. **PACKAGING** A cylindrical container of three rubber balls has a height of 18 centimeters and a diameter of 6 centimeters. Each ball in the container has a radius of 3 centimeters. Find the amount of space in the container that is not occupied by rubber balls. Round your answer to the nearest whole number.

Volume = 4500π in.^{3}

19. **BASKETBALL** The basketball shown is packaged in a box that is in the shape of a cube. The edge length of the box is equal to the diameter of the basketball. What is the surface area and the volume of the box?

20. **Logic** Your friend says that the volume of a sphere with radius r is four times the volume of a cone with radius r. When is this true? Justify your answer.

Fair Game Review What you learned in previous grades & lessons

The blue figure is a dilation of the red figure. Identify the type of dilation and find the scale factor. *(Section 11.7)*

21.

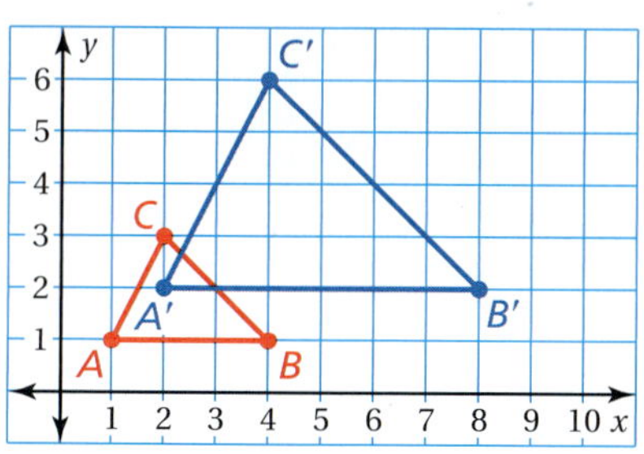

22.

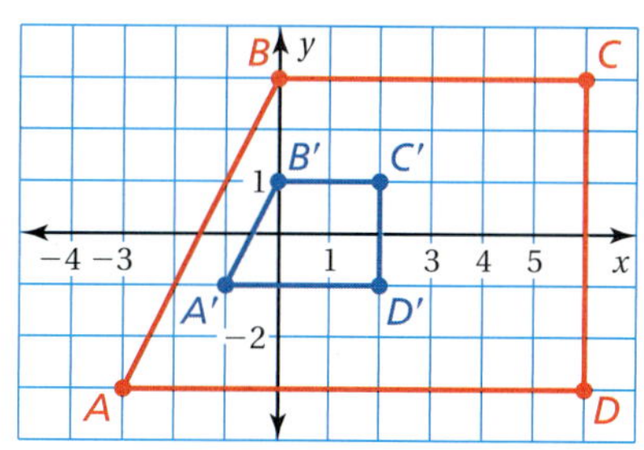

23. **MULTIPLE CHOICE** A person who is 5 feet tall casts a 6-foot-long shadow. A nearby flagpole casts a 30-foot-long shadow. What is the height of the flagpole? *(Section 12.4)*

Ⓐ 25 ft Ⓑ 29 ft Ⓒ 36 ft Ⓓ 40 ft

15.4 Surface Areas and Volumes of Similar Solids

Essential Question When the dimensions of a solid increase by a factor of k, how does the surface area change? How does the volume change?

1 ACTIVITY: Comparing Surface Areas and Volumes

Work with a partner. Copy and complete the table. Describe the pattern. Are the dimensions proportional? Explain your reasoning.

a.

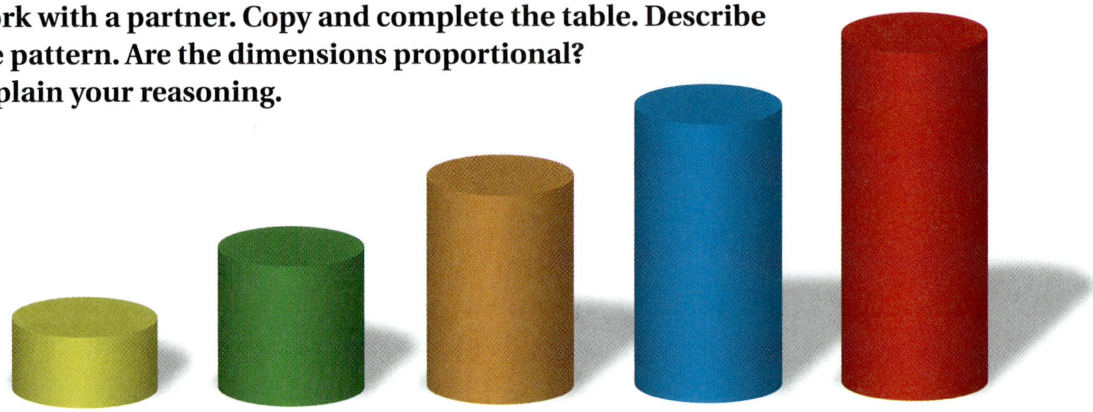

Radius	1	1	1	1	1
Height	1	2	3	4	5
Surface Area					
Volume					

b.

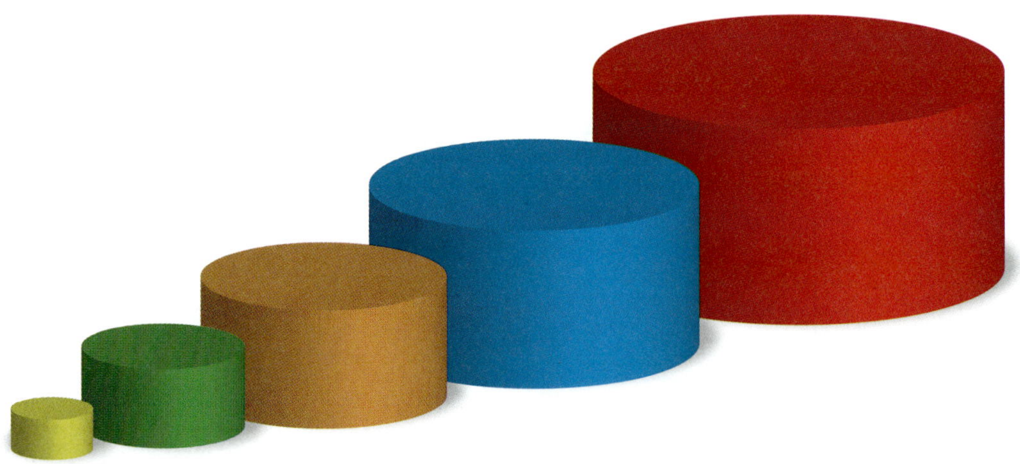

Radius	1	2	3	4	5
Height	1	2	3	4	5
Surface Area					
Volume					

Geometry

In this lesson, you will

- identify similar solids.
- use properties of similar solids to find missing measures.
- understand the relationship between surface areas of similar solids.
- understand the relationship between volumes of similar solids.
- solve real-life problems.

Laurie's Notes

Introduction

Applying Mathematical Practices

- **Look for and Express Regularity in Repeated Reasoning:** Students may recognize and apply patterns in repeated computations they are performing, which will make these activities go quickly.

Motivate

- **Story Time:** Retell a portion of the story of *Goldilocks and the Three Bears.* Focus on the three sizes of porridge bowls, chairs, and beds.
- Share with students that Papa Bear's mattress was twice as long, twice as wide, and twice as high as Baby Bear's mattress. So, are there twice as many feathers in Papa Bear's feather bed mattress?
- This question will be answered at the end of the class.
- If you sense it is necessary, provide a review of how to find surface areas of cylinders, and surface areas and volumes of pyramids, before starting the activities.

Activity Notes

Activity 1

- Tell students to leave their answers in terms of π.
- ? To help students see the pattern in the first table, ask the following questions.
  - "Describe the changes in the dimensions." radius same, height increases by 1
  - "How does the surface area change?" increases by 2π
  - "How does the volume change?" increases by π
  - "Compare each figure's height to the original (first) figure's height. Do the same for surface areas and volumes. What do you notice?" The volumes are multiplied by the same number as the heights.
- ? To help students see the pattern in the second table, ask the following questions.
  - "Describe the changes in the dimensions." radius and height each increase by 1
  - "Compare each figure's height to the original figure's height. Do the same for radii, surface areas, and volumes. What do you notice?" The heights and radii are multiplied by the same number. The surface areas are multiplied by the square of this number, and the volumes are multiplied by the cube of this number.
- ? "Are the dimensions proportional in part (a)? Explain." No, only the height increases, not the radius.
- ? "Are the dimensions proportional in part (b)? Explain." Yes, both the radius and height increase by the same factor.

What Your Students Will Learn

- Identify similar solids using proportions.
- Use properties of similar solids to find missing measures.
- Understand the relationship between surface areas of similar solids.
- Understand the relationship between volumes of similar solids.

Previous Learning

Students should be familiar with similar figures, surface area formulas, and volume formulas.

Technology for the Teacher

Dynamic Classroom

Lesson Plans

Complete Materials List

15.4 Record and Practice Journal

Essential Question When the dimensions of a solid increase by a factor of k, how does the surface area change? How does the volume change?

1 ACTIVITY: Comparing Surface Areas and Volumes

Work with a partner. Complete the table. Describe the pattern. Are the dimensions proportional? Explain your reasoning.

a.

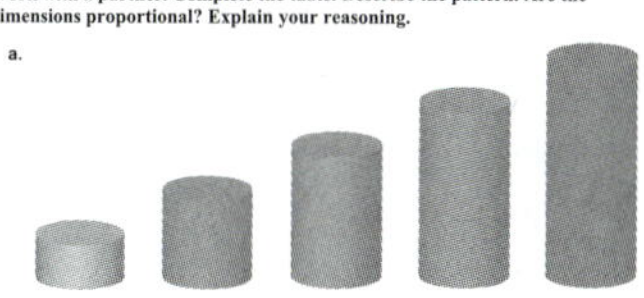

Radius	1	1	1	1	1
Height	1	2	3	4	5
Surface Area	4π	6π	8π	10π	12π
Volume	π	2π	3π	4π	5π

The surface area increases by 2π and the volume increases by π; not proportional

Differentiated Instruction

In the examples, check to be sure that students are correctly identifying corresponding sides. Remind them that they have to identify corresponding linear measures to write proportions before solving them.

15.4 Record and Practice Journal

b.

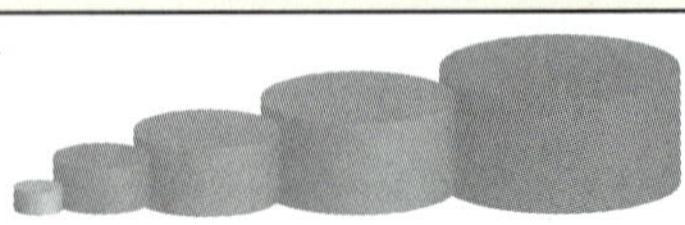

Radius	1	2	3	4	5
Height	1	2	3	4	5
Surface Area	4π	16π	36π	64π	100π
Volume	π	8π	27π	64π	125π

proportional

2 ACTIVITY: Comparing Surface Areas and Volumes

Work with a partner. Complete the table. Describe the pattern. Are the dimensions proportional? Explain.

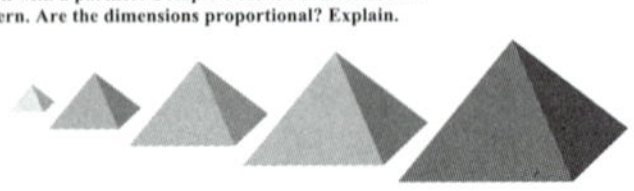

Base Side	6	12	18	24	30
Height	4	8	12	16	20
Slant Height	5	10	15	20	25
Surface Area	96	384	864	1536	2400
Volume	48	384	1296	3072	6000

proportional

What Is Your Answer?

3. **IN YOUR OWN WORDS** When the dimensions of a solid increase by a factor of k, how does the surface area change?

   The surface area increases by a factor of k^2.

4. **IN YOUR OWN WORDS** When the dimensions of a solid increase by a factor of k, how does the volume change?

   The volume increases by a factor of k^3.

5. **REPEATED REASONING** All the dimensions of a prism increase by a factor of 5.

   a. How many times greater is the surface area? Explain.

   5 | 10 | 25 | 125

   25; because $5^2 = 25$

   b. How many times greater is the volume? Explain.

   5 | 10 | 25 | 125

   125; because $5^3 = 125$

Laurie's Notes

Activity 2

- You want students to see a pattern. It may be helpful for students to use a calculator.
- **Look for and Express Regularity in Repeated Reasoning:** Ask students to describe patterns they see. Remind them to think about factors (multiplication) versus addition. The first pyramid should be referred to as the original pyramid. Describe any patterns in terms of the original pyramid.

? "Are the dimensions proportional? Explain." yes; The three dimensions are all changing by factors of 2, 3, 4, and 5 times the original dimensions.

- To help students see the factor by which the surface areas and volumes are multiplied, they should divide the new surface area (or volume) by the original surface area (or volume).

  Example for Blue Pyramid

Base Side	$24 \div 6 = 4$	Multiplied by a scale factor of 4
Height	$16 \div 4 = 4$	Multiplied by a scale factor of 4
Slant Height	$20 \div 5 = 4$	Multiplied by a scale factor of 4
Surface Area	$1536 \div 96 = 16$	Multiplied by a scale factor of 4^2
Volume	$3072 \div 48 = 64$	Multiplied by a scale factor of 4^3

- **Big Idea:** When the dimensions of a solid are all multiplied by a scale factor of k, the surface area is multiplied by a scale factor of k^2, and the volume is multiplied by a scale factor of k^3.

What Is Your Answer?

- **Think-Pair-Share:** Students should read each question independently and then work in pairs to answer the questions. When they have answered the questions, the pair should compare their answers with another group and discuss any discrepancies.

Closure

- Refer to Papa Bear's feather bed mattress. If the dimensions are all double Baby Bear's mattress, how many times more feathers are there? 8 times more feathers

2 ACTIVITY: Comparing Surface Areas and Volumes

Work with a partner. Copy and complete the table. Describe the pattern. Are the dimensions proportional? Explain.

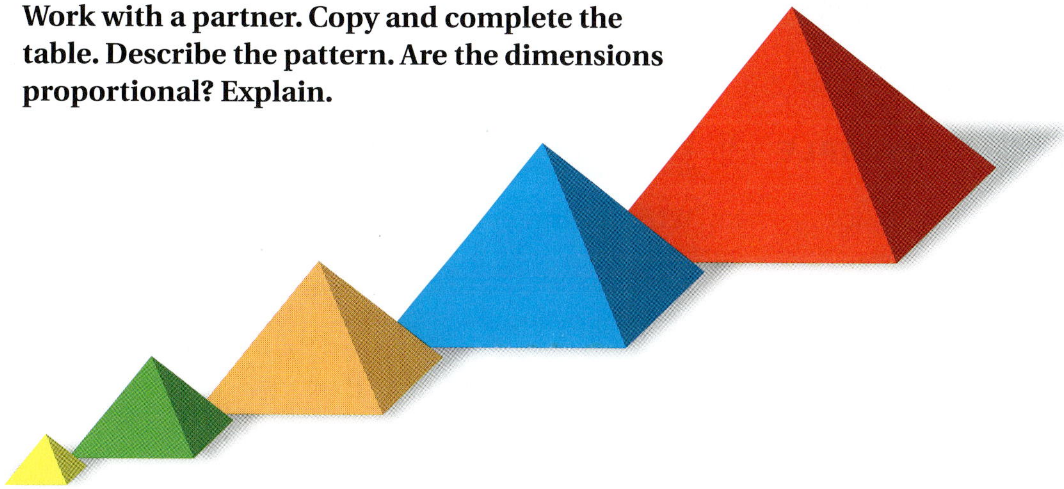

Math Practice

Repeat Calculations

Which calculations are repeated? How does this help you describe the pattern?

Base Side	6	12	18	24	30
Height	4	8	12	16	20
Slant Height	5	10	15	20	25
Surface Area					
Volume					

What Is Your Answer?

3. **IN YOUR OWN WORDS** When the dimensions of a solid increase by a factor of k, how does the surface area change?
4. **IN YOUR OWN WORDS** When the dimensions of a solid increase by a factor of k, how does the volume change?
5. **REPEATED REASONING** All the dimensions of a prism increase by a factor of 5.

   a. How many times greater is the surface area? Explain.

   5   10   25   125

   b. How many times greater is the volume? Explain.

   5   10   25   125

Practice Use what you learned about surface areas and volumes of similar solids to complete Exercise 3 on page 697.

15.4 Lesson

Check It Out
Lesson Tutorials
BigIdeasMath.com

Key Vocabulary
similar solids, *p. 694*

Similar solids are solids that have the same shape and proportional corresponding dimensions.

EXAMPLE 1 Identifying Similar Solids

Cylinder B

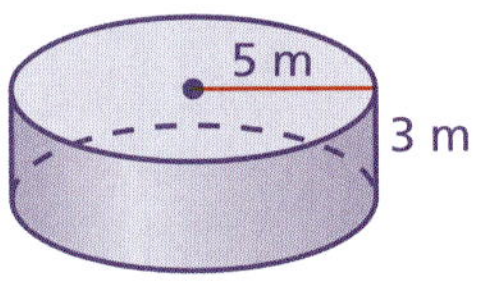

Cylinder C

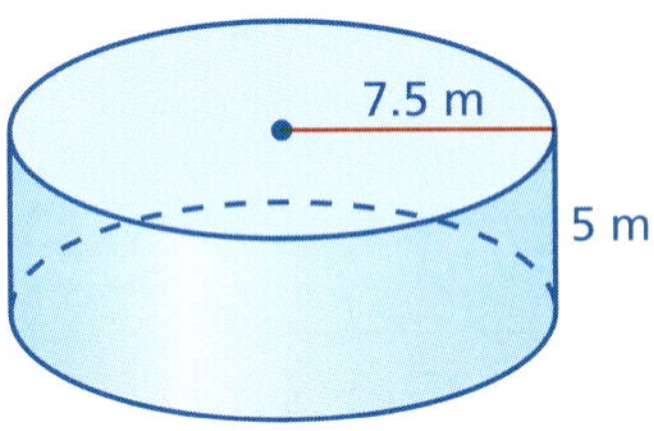

Which cylinder is similar to Cylinder A?

Cylinder A

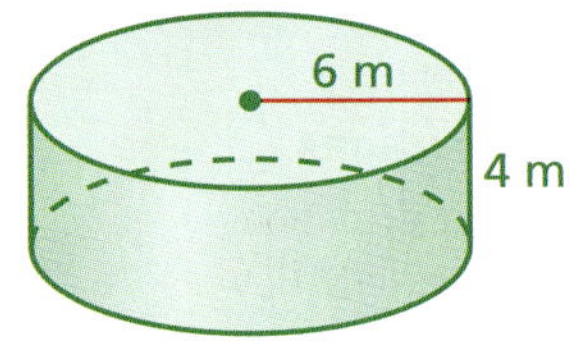

Check to see if corresponding dimensions are proportional.

Cylinder A and Cylinder B

$$\frac{\text{Height of A}}{\text{Height of B}} = \frac{4}{3} \qquad \frac{\text{Radius of A}}{\text{Radius of B}} = \frac{6}{5}$$

Not proportional

Cylinder A and Cylinder C

$$\frac{\text{Height of A}}{\text{Height of C}} = \frac{4}{5} \qquad \frac{\text{Radius of A}}{\text{Radius of C}} = \frac{6}{7.5} = \frac{4}{5}$$

Proportional

So, Cylinder C is similar to Cylinder A.

EXAMPLE 2 Finding Missing Measures in Similar Solids

Cone X

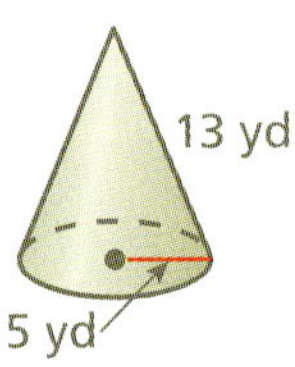

Cone Y

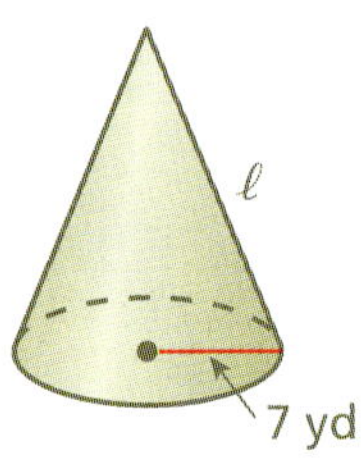

The cones are similar. Find the missing slant height ℓ.

$$\frac{\text{Radius of X}}{\text{Radius of Y}} = \frac{\text{Slant height of X}}{\text{Slant height of Y}}$$

$$\frac{5}{7} = \frac{13}{\ell} \qquad \text{Substitute.}$$

$$5\ell = 91 \qquad \text{Cross Products Property}$$

$$\ell = 18.2 \qquad \text{Divide each side by 5.}$$

The slant height is 18.2 yards.

On Your Own

Now You're Ready
Exercises 4–9

1. Cylinder D has a radius of 7.5 meters and a height of 4.5 meters. Which cylinder in Example 1 is similar to Cylinder D?
2. The prisms at the right are similar. Find the missing width and length.

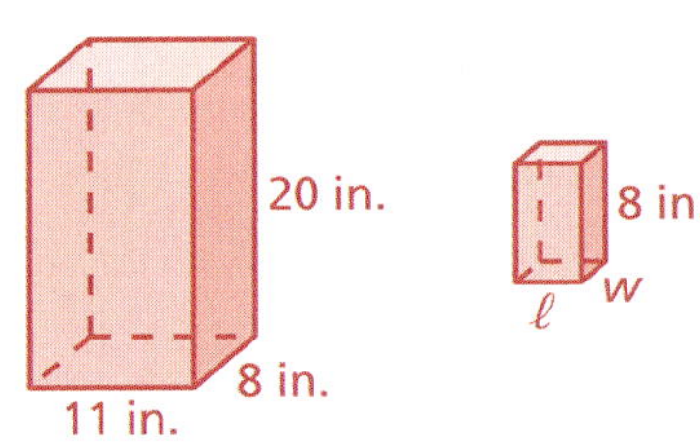

Multi-Language Glossary at BigIdeasMath.com

Laurie's Notes

Introduction

Connect

- **Yesterday:** Students explored what happens to the surface areas and volumes of solids when the dimensions are multiplied by a factor of *k*.
- **Today:** Students will use properties of similar solids to solve problems.
- **FYI:** This is a long lesson and it may take more than one day to cover. Take your time and present the concepts well.

Motivate

- **Movie Time:** Hold an object that is miniature in size (model car, doll house item, statue, and so on). Tell students that this is a prop from a movie set. The movie plot is about giants. In order to make people look large, all of the props have been shrunk proportionally.
- Spend some time talking about movie making. Creating props larger than normal will make people appear smaller than normal, and vice versa.

Lesson Notes

Example 1

- Note that the definition simply states that the corresponding linear measures must be proportional. This means that similar solids are proportional in size.

? "What is a proportion?" an equation of two equal ratios

? "How do you know if two ratios are equal?" Students might say by eyesight; by simple arithmetic, like $\frac{1}{2} = \frac{2}{4}$; that the ratios simplify to the same ratio. Students should recall the Cross Products Property.

- Work through the example.
- Be sure to write the words and the numbers. Use language such as "the ratio of the height of A to the height of B is 4 to 3."

? "How do you know $\frac{6}{7.5} = \frac{4}{5}$?" Answers may vary depending upon students' number sense. By the Cross Products Property $6 \times 5 = 7.5 \times 4$.

Example 2

? "By the definition of similar solids, what can you determine about two similar cones?" Corresponding linear measures are proportional.

- Set up the proportion and solve for the missing slant height.

On Your Own

- **Think-Pair-Share:** Students should read each question independently and then work in pairs to answer the questions. When they have answered the questions, the pair should compare their answers with another group and discuss any discrepancies.
- Ask volunteers to put their work on the board.

Goal Today's lesson is finding the surface areas and volumes of **similar solids**.

Technology for the Teacher

Dynamic Classroom

Lesson Tutorials
Lesson Plans
Answer Presentation Tool

Extra Example 1

Which prism is similar to Prism A?

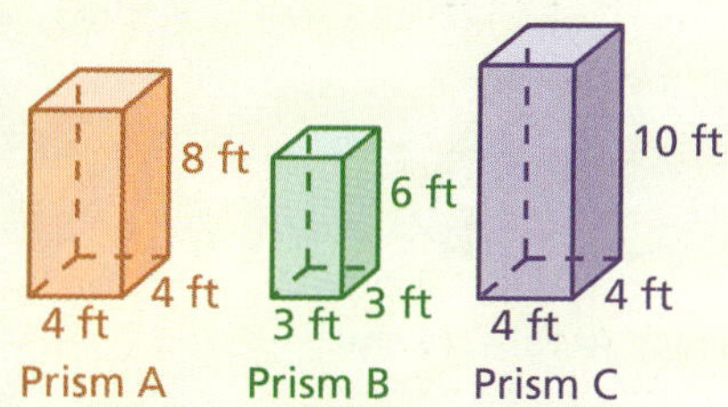

Prism B

Extra Example 2

The square pyramids are similar. Find the length of the base of Pyramid E.

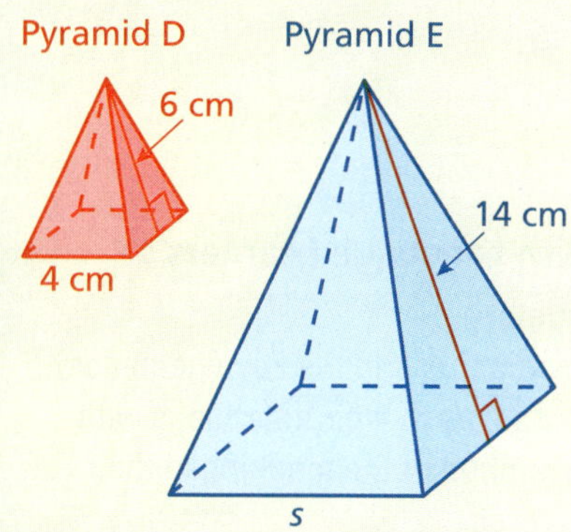

$9\frac{1}{3}$ cm

On Your Own

1. Cylinder B
2. $w = 3.2$ in.
   $\ell = 4.4$ in.

Extra Example 3

The cones are similar. What is the surface area of Cone G? Round your answer to the nearest tenth.

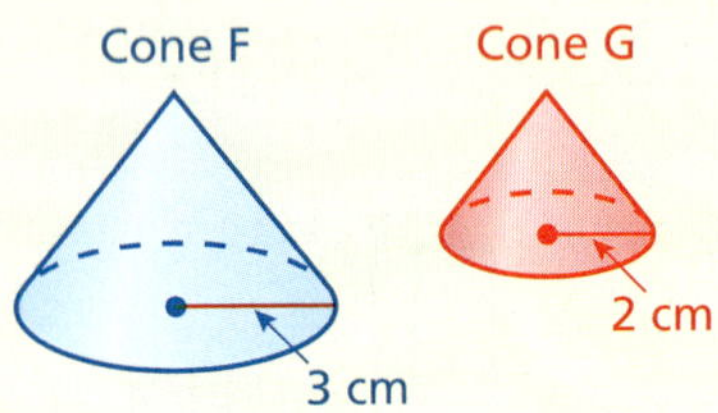

33.3 cm^2

On Your Own

3. 237.5 m^2

4. 171.9 cm^2

English Language Learners

Vocabulary

Have students add the key vocabulary *similar solids* to their notebooks with a description of the meaning in their own words.

Laurie's Notes

Key Ideas

- Write the Key Ideas.
- **Example:** When the linear dimensions of B are double A, the dimensions are in the ratio of $\frac{1}{2}$, and the surface areas are in the ratio of $\left(\frac{1}{2}\right)^2$ or $\frac{1}{4}$.
- Refer to yesterday's activity to confirm that this relationship was found in Activity 1, part (b) and Activity 2.
- Students may not know how to find surface areas of cones or spheres. However, they do not need this skill to determine whether two cones or two spheres are similar. The relationship given in the Key Idea applies to all similar solids.

Example 3

? "Do you have enough information to solve this problem? Explain." yes; The heights are in the ratio of $\frac{6}{10}$, so the surface areas are in the ratio of $\left(\frac{6}{10}\right)^2$.

- Set up the problem and solve.
- **FYI:** Notice that the problem is solved using the Multiplication Property of Equality. It could also be solved using the Cross Products Property.
- **Connection:** The ratio:

$$\frac{\text{dimension of A}}{\text{dimension of B}}$$

is the scale factor. The square of the scale factor is used to find the unknown surface area.

On Your Own

- Students should first identify the ratio of the corresponding linear measurements. Question 3: $\frac{5}{8}$; Question 4: $\frac{5}{4}$
- Ask volunteers to share their work at the board.

Key Ideas

Linear Measures

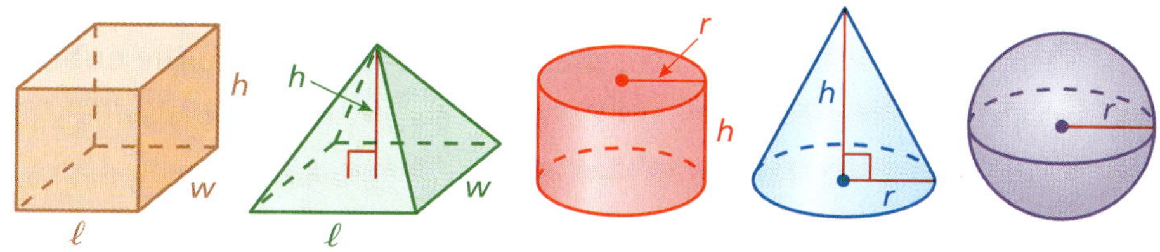

Surface Areas of Similar Solids

When two solids are similar, the ratio of their surface areas is equal to the square of the ratio of their corresponding linear measures.

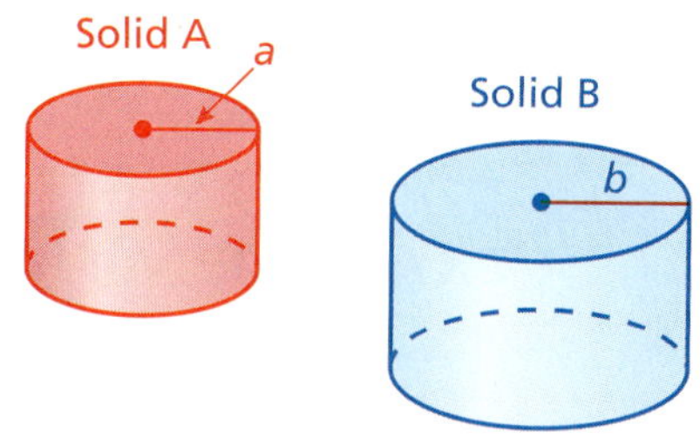

$$\frac{\text{Surface Area of A}}{\text{Surface Area of B}} = \left(\frac{a}{b}\right)^2$$

EXAMPLE 3 Finding Surface Area

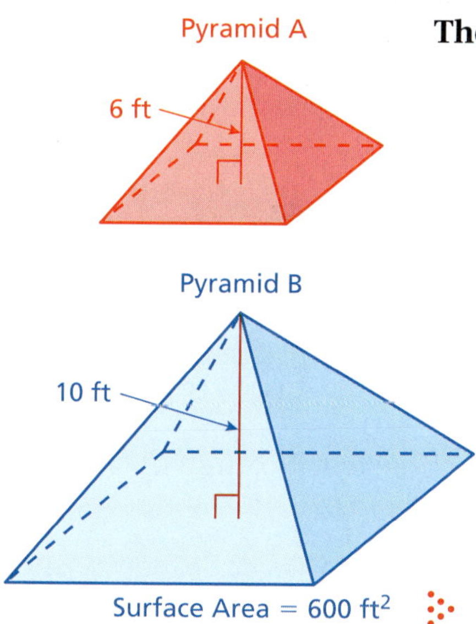

The pyramids are similar. What is the surface area of Pyramid A?

$$\frac{\text{Surface Area of A}}{\text{Surface Area of B}} = \left(\frac{\text{Height of A}}{\text{Height of B}}\right)^2$$

$$\frac{S}{600} = \left(\frac{6}{10}\right)^2 \qquad \text{Substitute.}$$

$$\frac{S}{600} = \frac{36}{100} \qquad \text{Evaluate.}$$

$$\frac{S}{600} \cdot 600 = \frac{36}{100} \cdot 600 \qquad \text{Multiplication Property of Equality}$$

$$S = 216 \qquad \text{Simplify.}$$

∴ The surface area of Pyramid A is 216 square feet.

On Your Own

The solids are similar. Find the surface area of the red solid. Round your answer to the nearest tenth.

3.

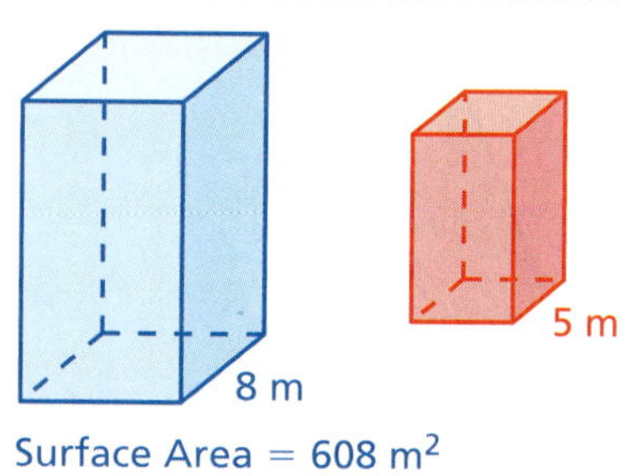

4.

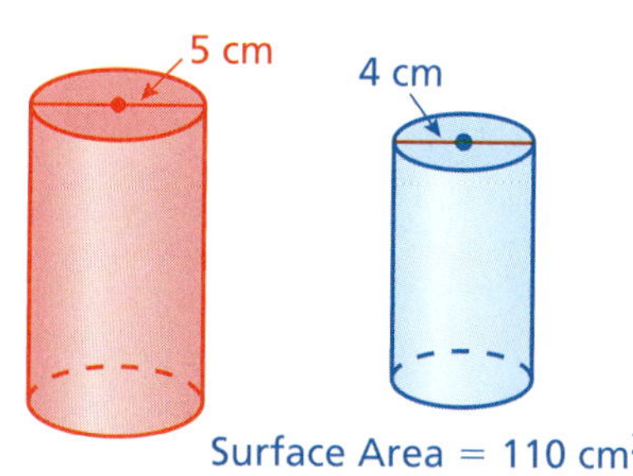

Key Idea

Volumes of Similar Solids

When two solids are similar, the ratio of their volumes is equal to the cube of the ratio of their corresponding linear measures.

$$\frac{\text{Volume of A}}{\text{Volume of B}} = \left(\frac{a}{b}\right)^3$$

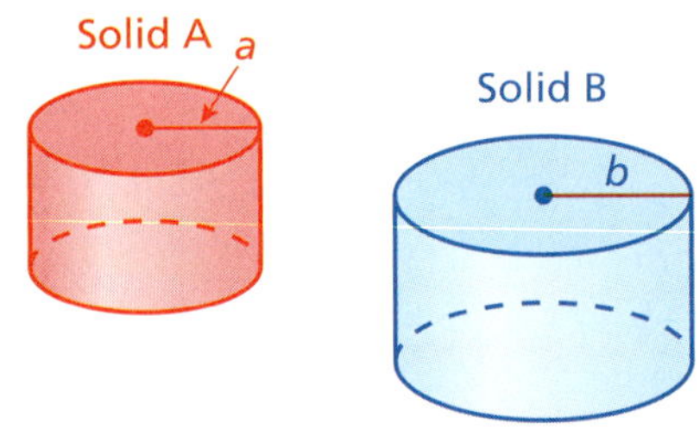

EXAMPLE 4 Finding Volume

Original Tank

Volume = 2000 ft^3

The dimensions of the touch tank at an aquarium are doubled. What is the volume of the new touch tank?

Ⓐ 150 ft^3 Ⓑ 4000 ft^3

Ⓒ 8000 ft^3 Ⓓ 16,000 ft^3

The dimensions are doubled, so the ratio of the dimensions of the original tank to the dimensions of the new tank is 1 : 2.

$$\frac{\text{Original volume}}{\text{New volume}} = \left(\frac{\text{Original dimension}}{\text{New dimension}}\right)^3$$

$$\frac{2000}{V} = \left(\frac{1}{2}\right)^3 \quad \text{Substitute.}$$

$$\frac{2000}{V} = \frac{1}{8} \quad \text{Evaluate.}$$

$$16{,}000 = V \quad \text{Cross Products Property}$$

∴ The volume of the new tank is 16,000 cubic feet. So, the correct answer is Ⓓ.

Study Tip

When the dimensions of a solid are multiplied by k, the surface area is multiplied by k^2 and the volume is multiplied by k^3.

On Your Own

Now You're Ready
Exercises 10–13

The solids are similar. Find the volume of the red solid. Round your answer to the nearest tenth.

5.

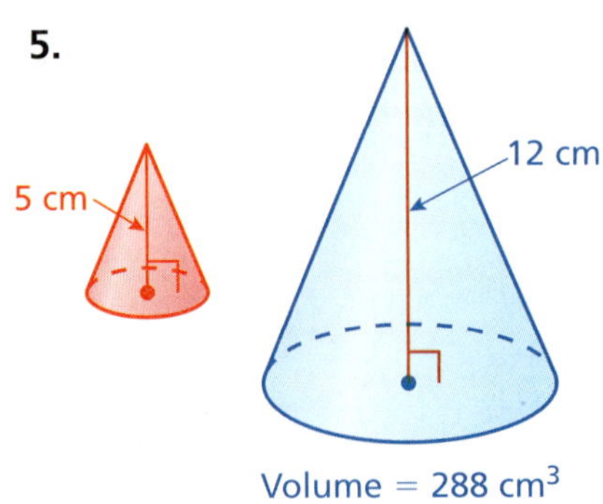

6.

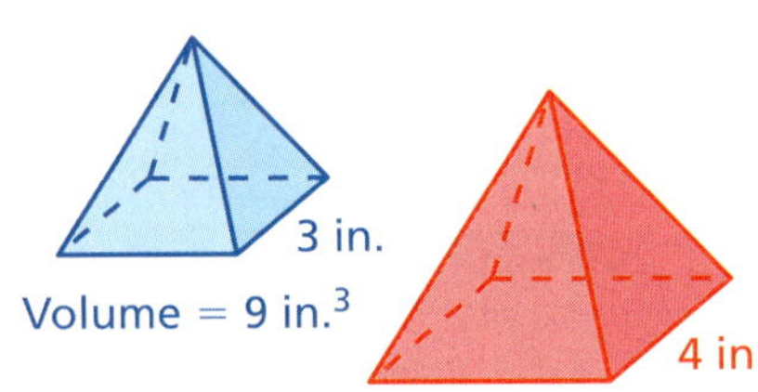

Laurie's Notes

Key Idea

- Write the Key Idea.
- **Example:** When the linear dimensions of B are double A, the dimensions are in the ratio of $\frac{1}{2}$, and the volumes are in the ratio of $\left(\frac{1}{2}\right)^3$ or $\frac{1}{8}$.
- Refer to the activity to confirm that this relationship was found in Activity 1, part (b) and Activity 2.

Example 4

- Write the problem in words to help students recognize how the numbers are being substituted.
- **Common Misconception:** Many students think that when you double the dimensions, the surface area and volume also double. This Big Idea takes time for students to fully understand.
- **Connection:** The ratio:

  $$\frac{\text{original dimension}}{\text{new dimension}}$$

  is the scale factor. The cube of the scale factor is used to find the new volume.

On Your Own

- Students should first identify the ratio of the corresponding linear measurements. Question 5: $\frac{5}{12}$; Question 6: $\frac{4}{3}$
- Ask volunteers to share their work at the board.

Closure

- Use one of the miniature items used to motivate the lesson and ask a question related to surface area or volume. Some miniature items have a scale printed on the item.

Extra Example 4

The cylinders are similar. Find the volume of Cylinder J. Round your answer to the nearest tenth.

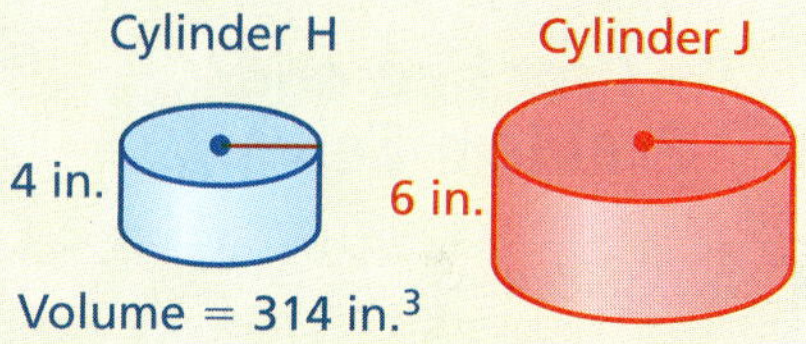

1059.8 in.^{3}

On Your Own

5. 20.8 cm^3

6. 21.3 in.^{3}

Vocabulary and Concept Check

1. Similar solids are solids of the same type that have proportional corresponding linear measures.
2. *Sample answer:*

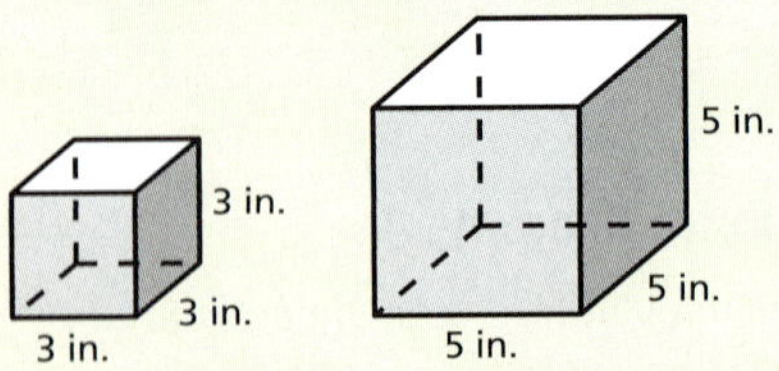

Practice and Problem Solving

3. a. $\frac{9}{4}$; because $\left(\frac{3}{2}\right)^2 = \frac{9}{4}$

   b. $\frac{27}{8}$; because $\left(\frac{3}{2}\right)^3 = \frac{27}{8}$
4. yes
5. no
6. yes
7. no
8. 25 in.
9. $b = 18$ m
   $c = 19.5$ m
   $h = 9$ m

Assignment Guide and Homework Check

Level	Assignment	Homework Check
Accelerated	1–3, 7–22	7, 9, 13, 15, 19

Common Errors

- **Exercises 4–6** Students may compare only two pairs of corresponding dimensions instead of all three. The bases may be similar, but the heights may not be proportional. Remind them to check all of the corresponding dimensions when determining whether two solids are similar. Ask them how many ratios they need to write for each type of solid.
- **Exercises 8 and 9** Students may write the proportion incorrectly. For example, in Exercise 8, they may write $\frac{10}{4} = \frac{10}{d}$. Remind them to write the proportion correctly and to check their work to make sure it makes sense.

15.4 Record and Practice Journal

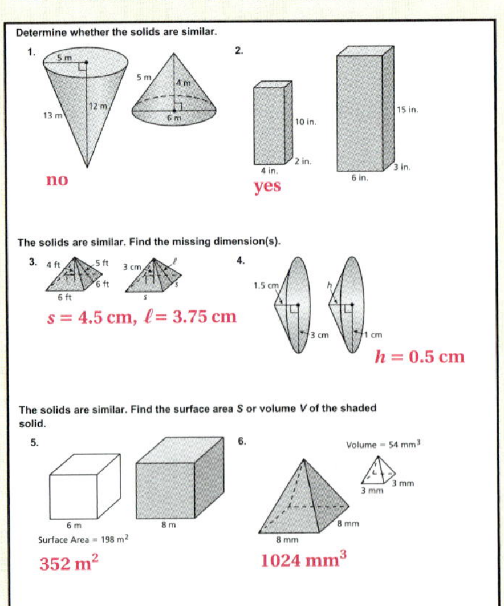
Determine whether the solids are similar.

1. no
2. yes

The solids are similar. Find the missing dimension(s).

3. $s = 4.5$ cm, $\ell = 3.75$ cm
4. $h = 0.5$ cm

The solids are similar. Find the surface area S or volume V of the shaded solid.

5. Surface Area = 198 m² — 352 m²
6. Volume = 54 mm³ — 1024 mm³

15.4 Exercises

Check It Out
Help with Homework
BigIdeasMath.com

Vocabulary and Concept Check

1. **VOCABULARY** What are similar solids?
2. **OPEN-ENDED** Draw two similar solids and label their corresponding linear measures.

Practice and Problem Solving

3. **NUMBER SENSE** All the dimensions of a cube increase by a factor of $\frac{3}{2}$.
   a. How many times greater is the surface area? Explain.
   b. How many times greater is the volume? Explain.

Determine whether the solids are similar.

1 4.

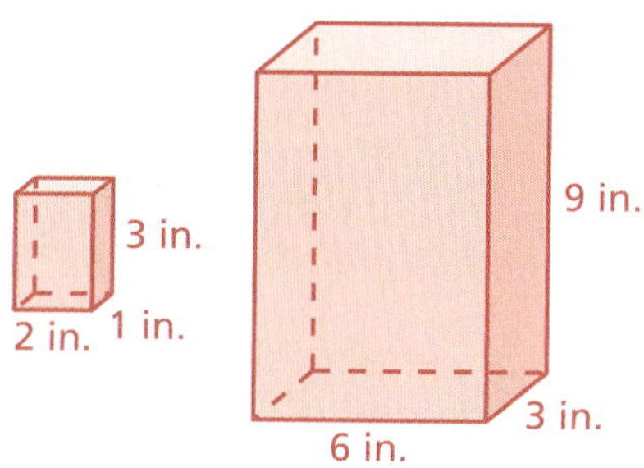

5.

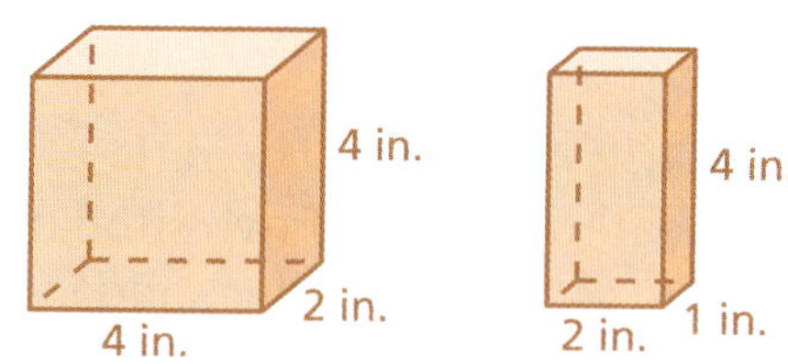

6.

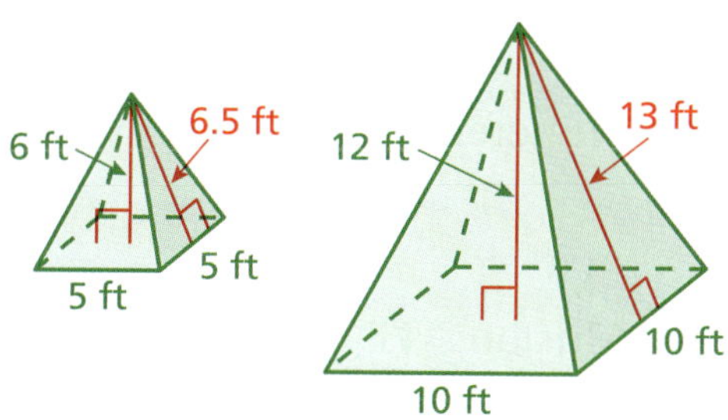

7.

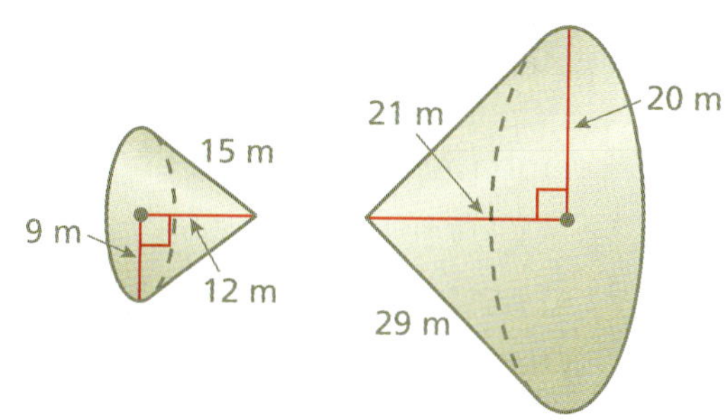

The solids are similar. Find the missing dimension(s).

2 8.

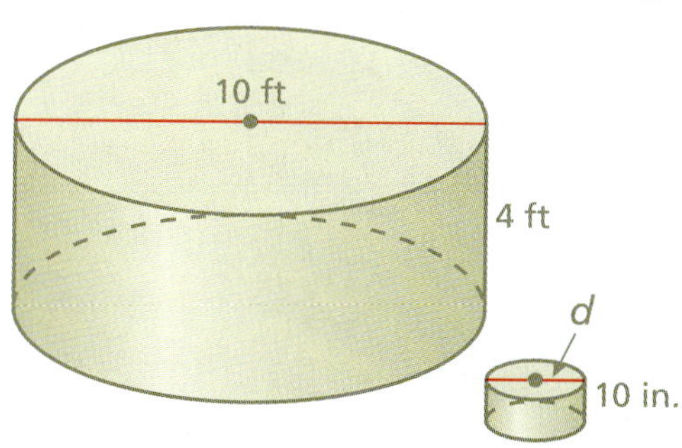

9.

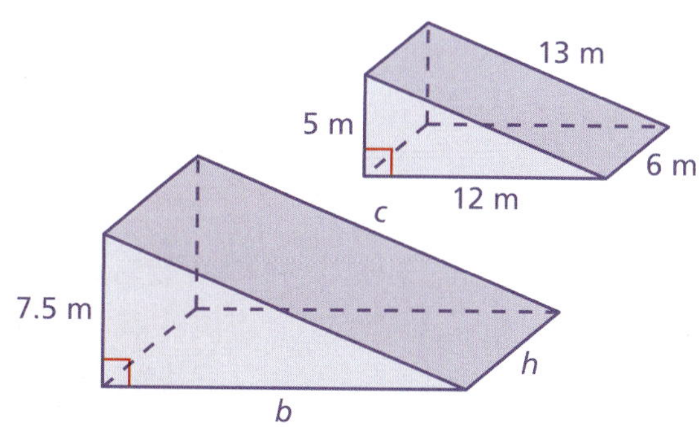

The solids are similar. Find the surface area S or volume V of the red solid. Round your answer to the nearest tenth.

3 4 **10.**

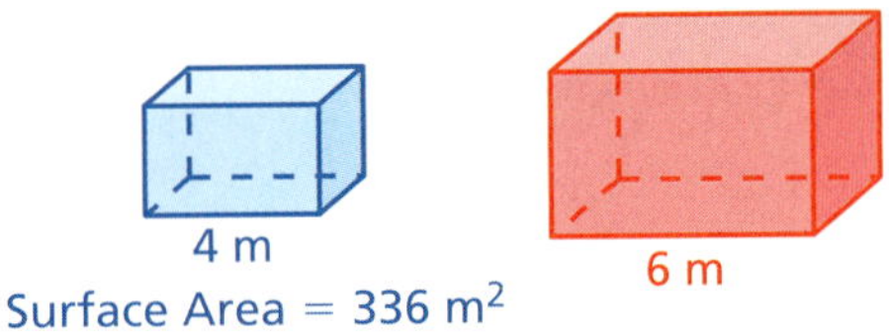

11.

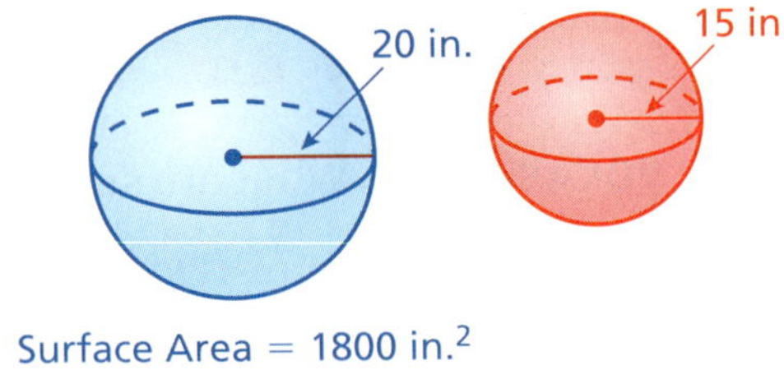

12.

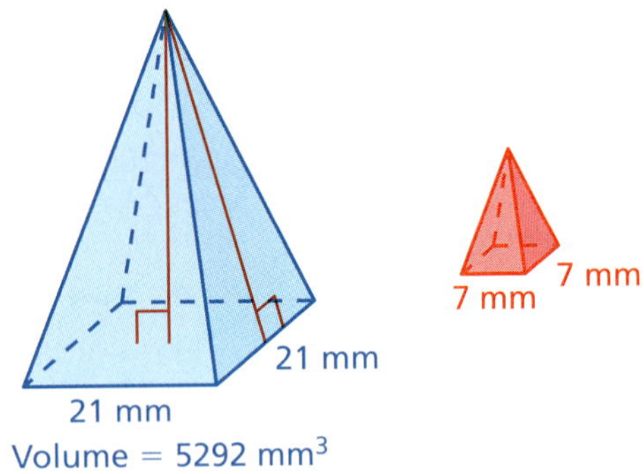

13.

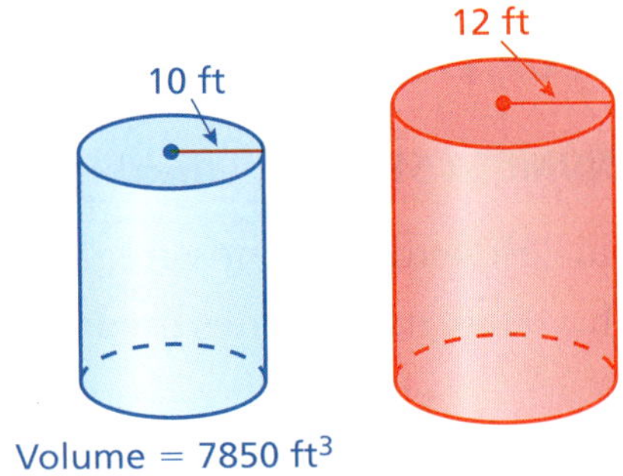

14. ERROR ANALYSIS The ratio of the corresponding linear measures of two similar solids is 3 : 5. The volume of the smaller solid is 108 cubic inches. Describe and correct the error in finding the volume of the larger solid.

✗

$$\frac{108}{V} = \left(\frac{3}{5}\right)^2$$

$$\frac{108}{V} = \frac{9}{25}$$

$$300 = V$$

The volume of the larger solid is 300 cubic inches.

15. MIXED FRUIT The ratio of the corresponding linear measures of two similar cans of fruit is 4 to 7. The smaller can has a surface area of 220 square centimeters. Find the surface area of the larger can.

16. CLASSIC MUSTANG The volume of a 1968 Ford Mustang GT engine is 390 cubic inches. Which scale model of the Mustang has the greater engine volume, a 1 : 18 scale model or a 1 : 24 scale model? How much greater is it?

Common Errors

- **Exercises 10–13** Students may forget to square or cube the ratio of the corresponding linear measures when finding the surface area or volume of the red solid. Remind them of the Key Ideas in this section.
- **Exercises 10–13** Students may cube the ratio of corresponding linear measures when finding surface area or square the ratio of corresponding linear measures when finding volume. Remind them that the ratio of corresponding linear measures is squared for surface area and cubed for volume.
- **Exercise 15** Students may write the ratio of the surface areas incorrectly in the proportion. When they look at the ratio of corresponding linear measures as a fraction, ask whether the numerator or denominator corresponds to the smaller figure. This should help them write the ratio of the surface areas correctly.

Practice and Problem Solving

10. 756 m^2

11. 1012.5 in.^2

12. 196 mm^3

13. $13{,}564.8 \text{ ft}^3$

14. The ratio of the volumes of two similar solids is equal to the cube of the ratio of their corresponding linear measures.

$$\frac{108}{V} = \left(\frac{3}{5}\right)^3$$

$$\frac{108}{V} = \frac{27}{125}$$

$$V = 500 \text{ in.}^3$$

15. 673.75 cm^2

16. 1 : 18 scale model; about 0.04 in.^3

English Language Learners

Pair Activity

As a confidence booster for English learners, have students write their own math problems. Place students in pairs, one English learner and one English speaker, and ask each pair to write a problem involving surface areas or volumes of similar solids. On a separate piece of paper, students solve their own problem. Next, students exchange their problem with another pair of students. Each pair then solves the new problem. Finally, the four students discuss the problems and the solution methods.

Practice and Problem Solving

17. See Additional Answers.

18. See *Taking Math Deeper.*

19. a. yes; Because all circles are similar, the slant height and the circumference of the base of the cones are proportional.

b. no; because the ratio of the volumes of similar solids is equal to the cube of the ratio of their corresponding linear measures

Fair Game Review

20–22. See Additional Answers.

Mini-Assessment

The solids are similar. Find the surface area *S* of the red solid. Round your answer to the nearest tenth.

1. 288 m^2

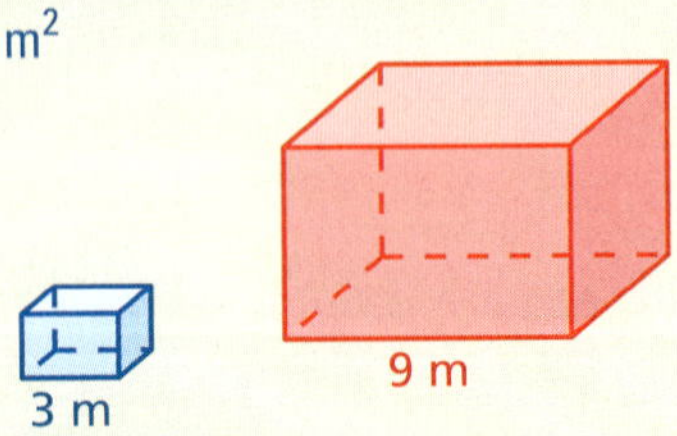

Surface Area = 32 m^2

2. 37.8 $in.^2$

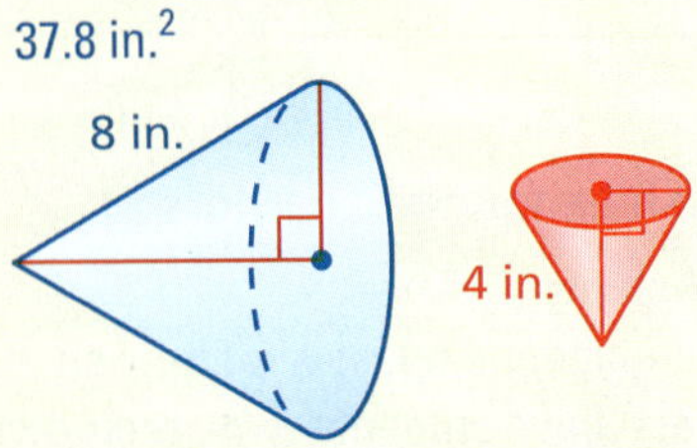

Surface Area = 151 $in.^2$

3. The cylinders are similar. Find the volume *V* of the red cylinder. 600 $in.^3$

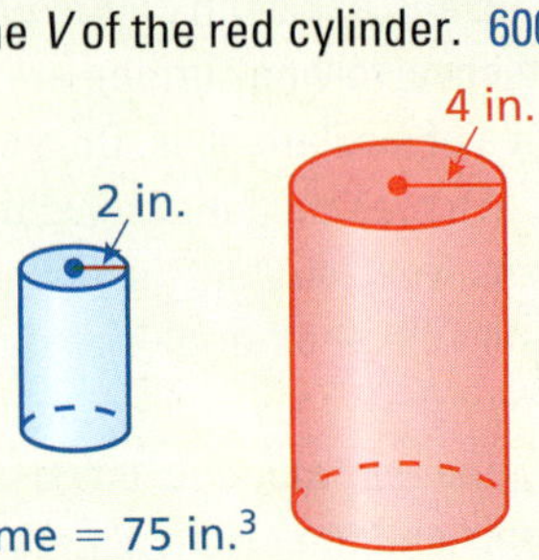

Volume = 75 $in.^3$

Taking Math Deeper

Exercise 18

This problem is a straightforward application of the two main concepts of the lesson. That is, with similar solids, the ratio of their surface areas is equal to the square of the ratio of their corresponding linear measures. Also, the ratio of their volumes is equal to the cube of the ratio of their corresponding linear measures. Even so, students have trouble with this problem because they do not see that they can let the surface area and volume of the shortest doll be *S* and *V*.

1 Make a table. Include the height of each doll.

Height	1	2	3	4	5	6	7
Surface Area	S	$4S$	$9S$	$16S$	$25S$	$36S$	$49S$
Volume	V	$8V$	$27V$	$64V$	$125V$	$216V$	$343V$

2 Compare the surface areas of the dolls.

3 Compare the volumes of the dolls.

Matryoshka dolls, or Russian nested dolls, are also called stacking dolls. A set of matryoshkas consists of a wooden figure which can be pulled apart to reveal another figure of the same sort inside. It has, in turn, another figure inside, and so on. The number of nested figures is usually five or more.

Reteaching and Enrichment Strategies

If students need help. . .	If students got it. . .
Resources by Chapter • Practice A and Practice B • Puzzle Time Record and Practice Journal Practice Differentiating the Lesson Lesson Tutorials Skills Review Handbook	Resources by Chapter • Enrichment and Extension • Technology Connection Start the next section

17. **MARBLE STATUE** You have a small marble statue of Wolfgang Mozart. It is 10 inches tall and weighs 16 pounds. The original statue is 7 feet tall.

   a. Estimate the weight of the original statue. Explain your reasoning.

   b. If the original statue were 20 feet tall, how much would it weigh?

Wolfgang Mozart

18. **REPEATED REASONING** The largest doll is 7 inches tall. Each of the other dolls is 1 inch shorter than the next larger doll. Make a table that compares the surface areas and the volumes of the seven dolls.

19. **Precision** You and a friend make paper cones to collect beach glass. You cut out the largest possible three-fourths circle from each piece of paper.

   a. Are the cones similar? Explain your reasoning.

   b. Your friend says that because your sheet of paper is twice as large, your cone will hold exactly twice the volume of beach glass. Is this true? Explain your reasoning.

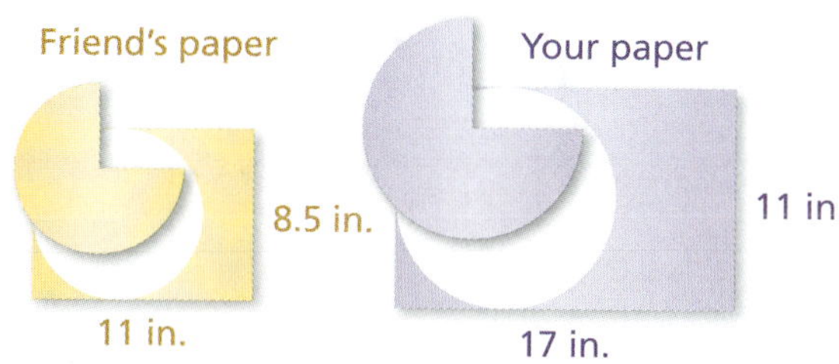

Fair Game Review What you learned in previous grades & lessons

Draw the figure and its reflection in the x-axis. Identify the coordinates of the image. *(Section 11.3)*

20. $A(1, 1)$, $B(3, 4)$, $C(4, 2)$

21. $J(-3, 0)$, $K(-4, 3)$, $L(-1, 4)$

22. **MULTIPLE CHOICE** Which pair of lines have the same slope, but different y-intercepts? *(Section 13.4 and Section 13.5)*

   Ⓐ $y = 4x + 1$, $y = -4x + 1$

   Ⓑ $y = 2x - 7$, $y = 2x + 7$

   Ⓒ $3x + y = 1$, $6x + 2y = 2$

   Ⓓ $5x + y = 3$, $x + 5y = 15$

15.3 – 15.4 Quiz

Check It Out
Progress Check
BigIdeasMath.com

Find the volume of the sphere. Round your answer to the nearest tenth. *(Section 15.3)*

1.

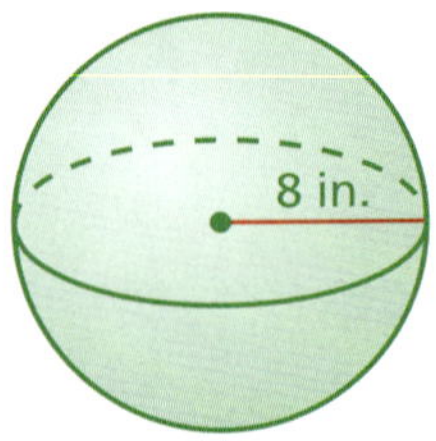

2.

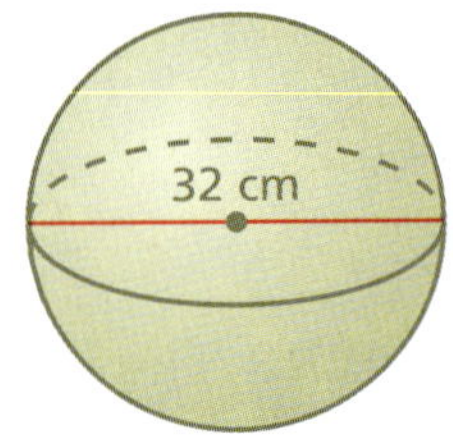

Find the radius of the sphere with the given volume. *(Section 15.3)*

3. Volume $= 4500\pi$ yd^3

4. Volume $= \frac{32}{3}\pi$ ft^3

5. Find the volume of the composite solid. Round your answer to the nearest tenth. *(Section 15.3)*

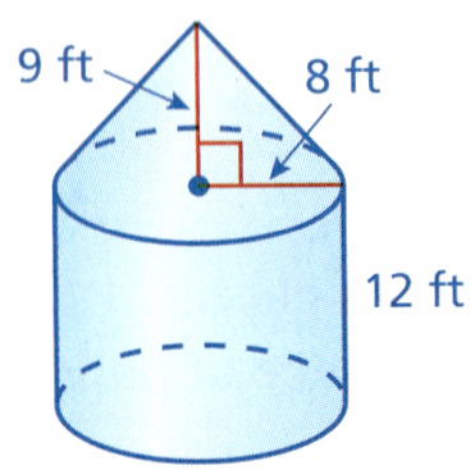

6. Determine whether the solids are similar. *(Section 15.4)*

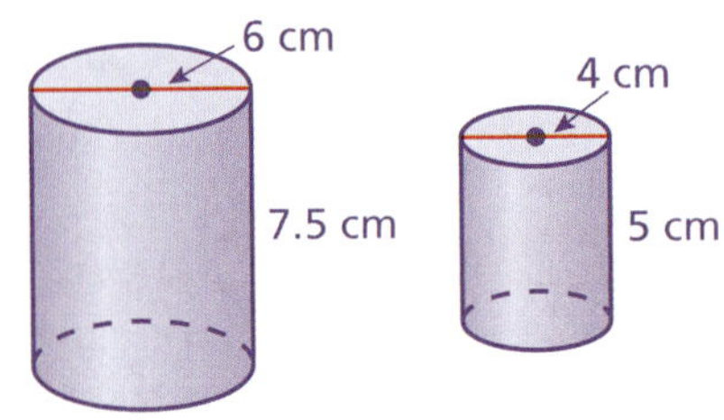

7. The prisms are similar. Find the missing width and height. *(Section 15.4)*

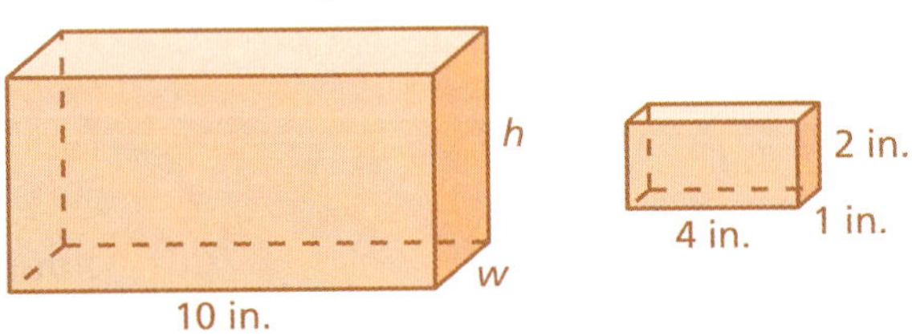

8. The solids are similar. Find the surface area of the red solid. *(Section 15.4)*

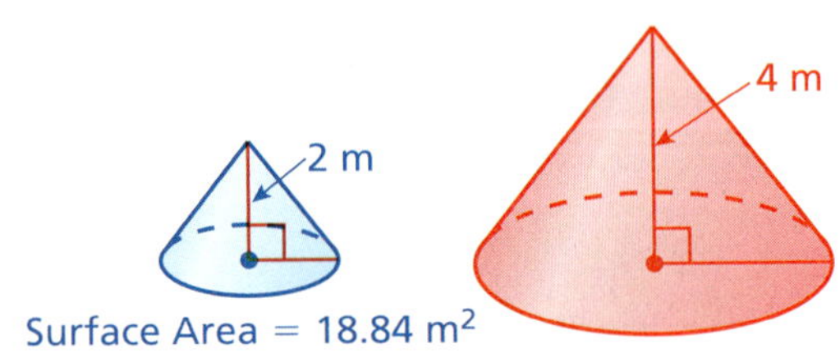

9. **HAMSTER** A hamster toy is in the shape of a sphere. What is the volume of the toy? Round your answer to the nearest whole number. *(Section 15.3)*

10. **JEWELRY BOXES** The ratio of the corresponding linear measures of two similar jewelry boxes is 2 to 3. The larger box has a volume of 162 cubic inches. Find the volume of the smaller jewelry box. *(Section 15.4)*

11. **ARCADE** You win a token after playing an arcade game. What is the volume of the gold ring? Round your answer to the nearest tenth. *(Section 15.3)*

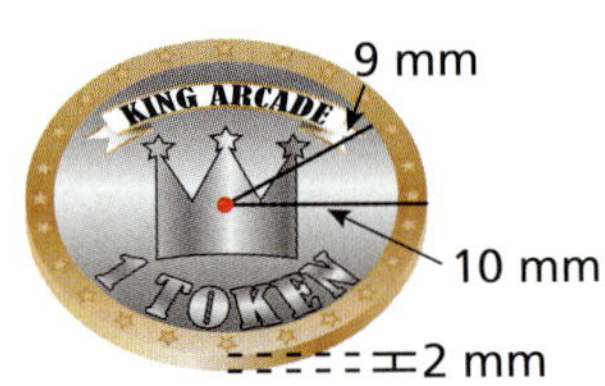

Alternative Assessment Options

Math Chat	Student Reflective Focus Question
Structured Interview	Writing Prompt

Math Chat

Ask students to use their own words to summarize what they know about finding volumes of spheres and finding surface areas and volumes of similar solids. Be sure that they include examples. Select students at random to present their summaries to the class.

Study Help Sample Answers

Remind students to complete Graphic Organizers for the rest of the chapter.

2. Volume of a sphere

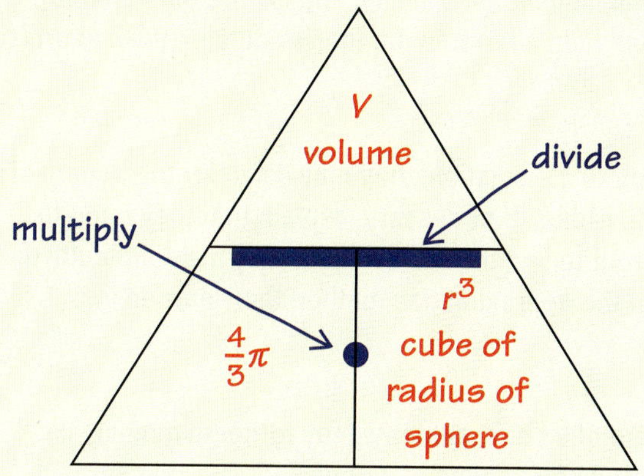

3. Volume of a composite solid

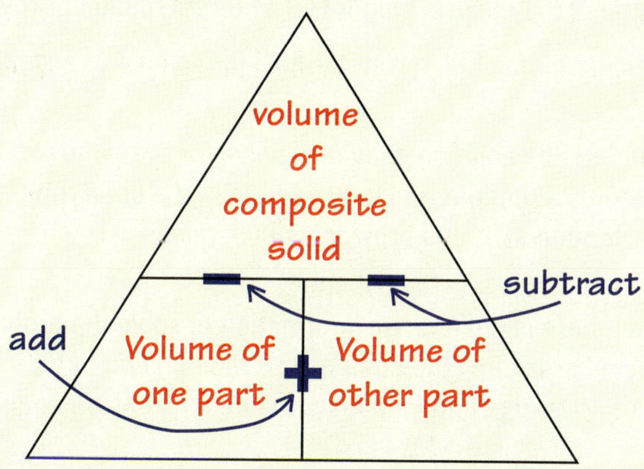

4–5. Available at *BigIdeasMath.com.*

Reteaching and Enrichment Strategies

If students need help. . .	If students got it. . .
Resources by Chapter • Practice A and Practice B • Puzzle Time Lesson Tutorials *BigIdeasMath.com*	Resources by Chapter • Enrichment and Extension • Technology Connection Game Closet at *BigIdeasMath.com* Start the Chapter Review

Answers

1. $\frac{2048\pi}{3} \approx 2144.7 \text{ in.}^3$
2. $\frac{16{,}384\pi}{3} \approx 17{,}157.3 \text{ cm}^3$
3. 15 yd
4. 2 ft
5. $768\pi + 192\pi = 960\pi$

   $\approx 3015.9 \text{ ft}^3$
6. yes
7. $w = 2.5$ in.
   $h = 5$ in.
8. 75.36 m^2
9. $\frac{32\pi}{3} \approx 34 \text{ cm}^3$
10. 48 in.^3
11. $38\pi \approx 119.4 \text{ mm}^3$

Technology for the Teacher

Online Assessment
Assessment Book
ExamView® Assessment Suite

For the Teacher
Additional Review Options

- *BigIdeasMath.com*
- Online Assessment
- Game Closet at *BigIdeasMath.com*
- Vocabulary Help
- Resources by Chapter

Answers

1. $\frac{1575\pi}{4} \approx 1237.0 \text{ ft}^3$
2. $40\pi \approx 125.7 \text{ in.}^3$
3. $108\pi \approx 339.3 \text{ yd}^3$
4. $1458\pi \approx 4580.4 \text{ in.}^3$
5. $\frac{25}{2.25\pi} \approx 4 \text{ in.}$
6. $\sqrt{\frac{7599}{20\pi}} \approx 11 \text{ m}$

Review of Common Errors

Exercises 1–4, 7, 8, 10, and 12–15

- Students may write linear or square units for volume rather than cubic units. Remind them that part of writing a correct answer is including the correct units.

Exercises 1–9, 12, and 14

- When finding the area of the circular base of a cylinder or a cone, students may not square the radius, or they may use the diameter when the formula calls for the radius. Remind them of the formula that they learned for the area of a circle.

Exercises 5, 6, 9, and 11

- Students may be confused about how to find the indicated dimension. Remind them to write the formula, substitute, simplify, and isolate the variable. Some students may need help with isolating the variable in these exercises, particularly in Exercises 6 and 11.

Exercise 6

- Students may not complete the solution; they may solve for the *square* of the radius instead of the radius. If necessary, explain that they need to approximate a square root to find the radius and if they have difficulty doing so, then remind them of the approximation method they learned in Section 14.4.

Exercises 7, 8, and 12

- When finding the volume of a cone, students may forget to multiply by $\frac{1}{3}$. Remind them of the formula for volume of a cone.

Exercise 10

- When finding the volume of the sphere, students may forget to multiply by $\frac{4}{3}$ or cube the radius. Remind them of the given formula for volume of a sphere.

Exercise 11

- Students may not complete the solution; they may solve for the *cube* of the radius instead of the radius. Point out that they need to find a cube root, which they learned in Section 14.2, and provide a quick review if necessary.

Exercises 12–14

- Students may think that there is not enough information to solve the problem. It may help to have them "break up" the composite solid into two parts, whose volumes they know how to find, and mark the dimensions on each part.

Exercise 14

- Students may not realize that the top part of the solid is a hemisphere. Point this out to students.

Exercises 15 and 16

- Students may raise the ratio of the linear measures to the wrong exponent, or forget to square or cube the ratio altogether. Discuss why squaring or cubing the ratio makes sense.

15 Chapter Review

Check It Out
Vocabulary Help
BigIdeasMath.com

Review Key Vocabulary

sphere, *p. 686* hemisphere, *p. 689* similar solids, *p. 694*

Review Examples and Exercises

15.1 Volumes of Cylinders *(pp. 672–677)*

Find the volume of the cylinder. Round your answer to the nearest tenth.

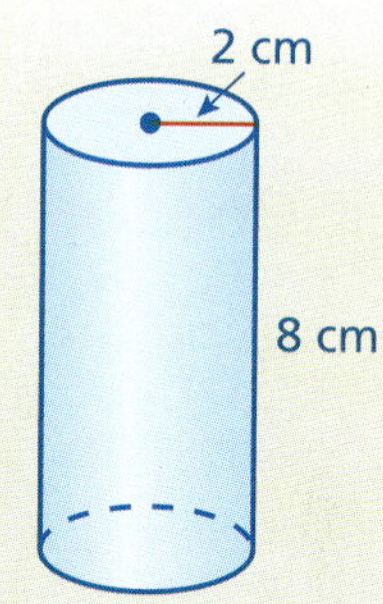

$V = Bh$ Write formula for volume.

$= \pi(2)^2(8)$ Substitute.

$= 32\pi \approx 100.5$ Use a calculator.

The volume is about 100.5 cubic centimeters.

Exercises

Find the volume of the cylinder. Round your answer to the nearest tenth.

1. 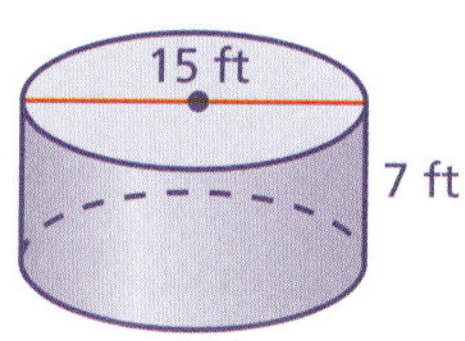

2. 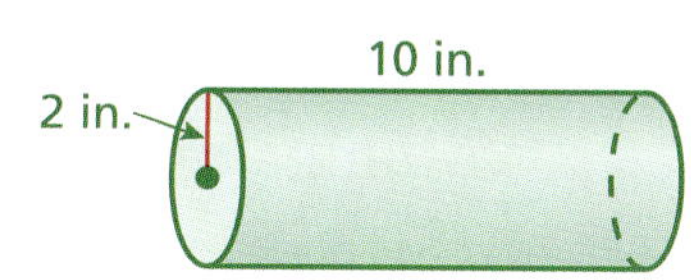

3. 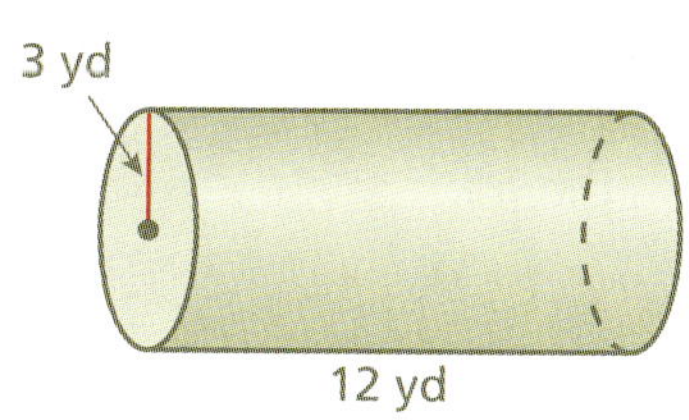

4. 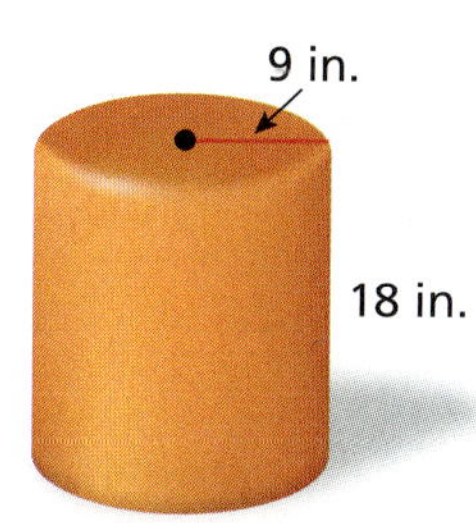

Find the missing dimension of the cylinder. Round your answer to the nearest whole number.

5. Volume = 25 in.^{3}

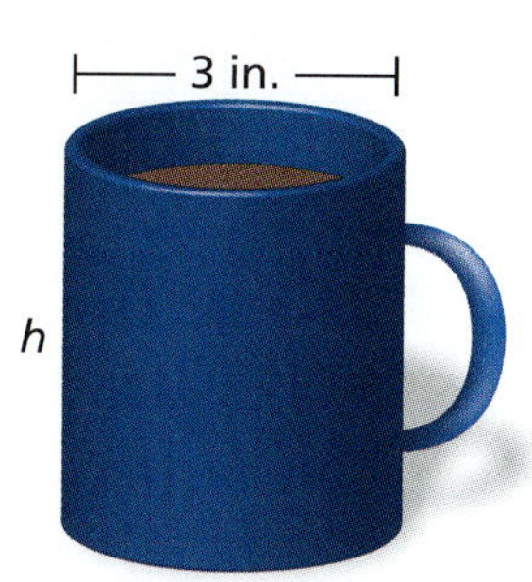

6. Volume = 7599 m^3

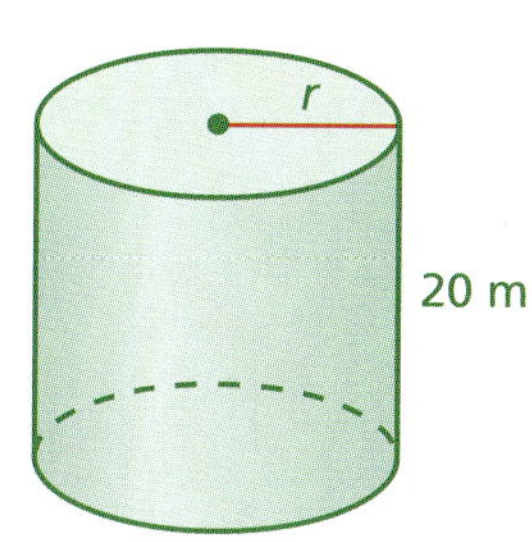

15.2 Volumes of Cones (pp. 678–683)

Find the height of the cone. Round your answer to the nearest tenth.

$V = \frac{1}{3}Bh$	Write formula for volume.
$900 = \frac{1}{3}\pi(6)^2(h)$	Substitute.
$900 = 12\pi h$	Simplify.
$23.9 \approx h$	Divide each side by 12π.

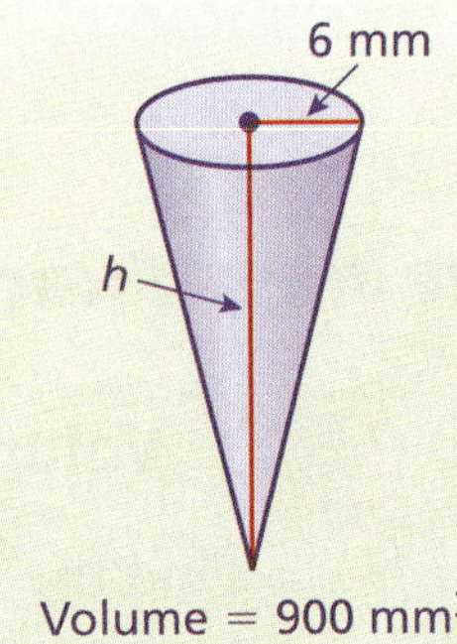

The height is about 23.9 millimeters.

Exercises

Find the volume V or height h of the cone. Round your answer to the nearest tenth.

7.

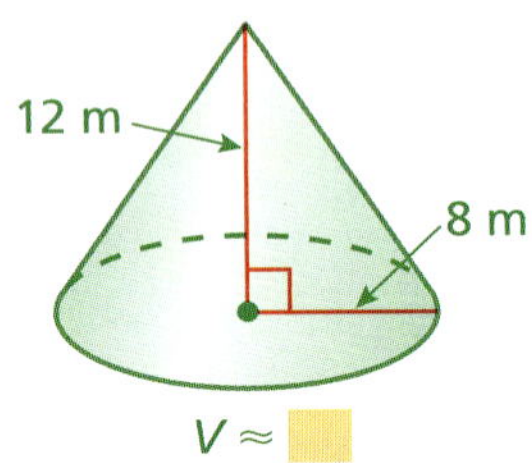

8.

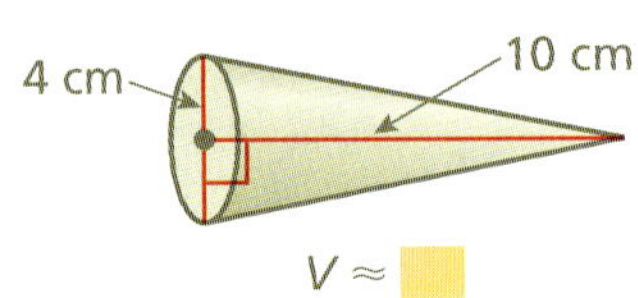

9.

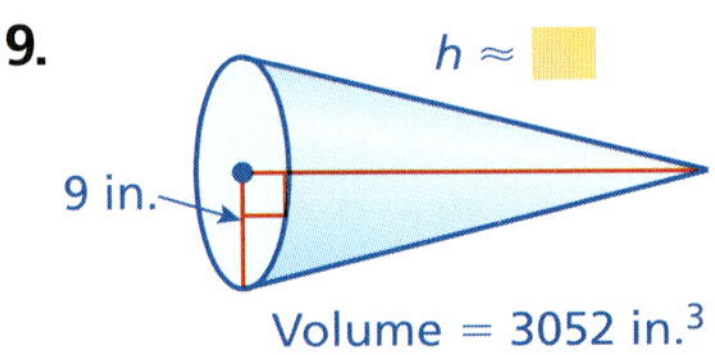

15.3 Volumes of Spheres (pp. 686–691)

a. Find the volume of the sphere. Round your answer to the nearest tenth.

$V = \frac{4}{3}\pi r^3$	Write formula for volume.
$= \frac{4}{3}\pi(11)^3$	Substitute 11 for r.
$= \frac{5324}{3}\pi$	Simplify.
≈ 5575.3	Use a calculator.

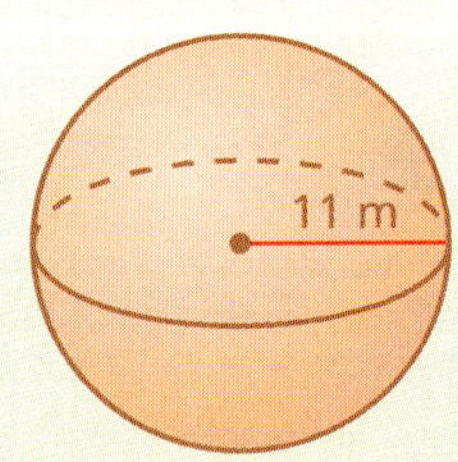

The volume is about 5575.3 cubic meters.

b. Find the volume of the composite solid. Round your answer to the nearest tenth.

Square Prism	*Cylinder*
$V = Bh$	$V = Bh$
$= (12)(12)(9)$	$= \pi(5)^2(9)$
$= 1296$	$= 225\pi \approx 706.9$

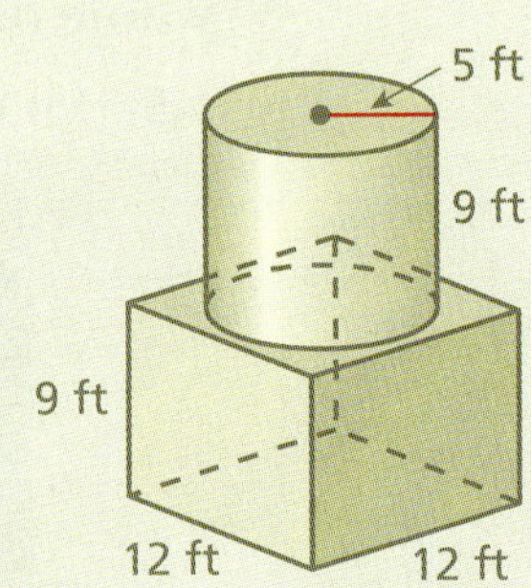

So, the volume is about $1296 + 706.9 = 2002.9$ cubic feet.

Review Game

Volume

Materials

- a variety of containers of different shapes and sizes
- liquid measuring devices
- water

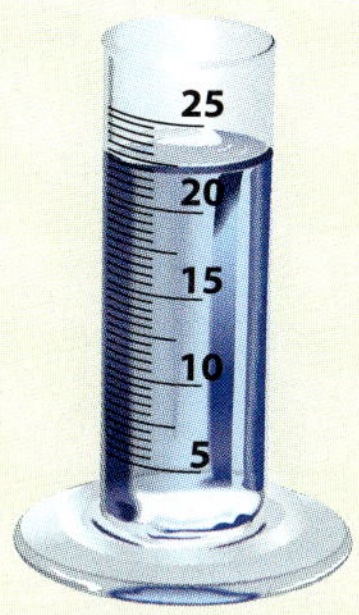

Directions

At the start of the chapter, have students bring in containers of different shapes and sizes. The containers should be shaped the same as the solids being studied in the chapter, they should be able to hold water, and they should be small enough so that they do not require a lot of water to fill. Collect containers until you have a sufficient variety of shapes and enough for the number of groups you want to have.

Work in groups. Give each group a container. Each group calculates the volume of the container and passes it to another group. Continue until all groups have calculated the volumes of the containers. When calculations are completed, each group measures the volume of a container using water. Measured volumes are shared with the class and compared to the calculated volumes.

Who Wins?

The group whose calculated volume is closest to the correct measured volume receives 1 point. The group with the most points wins.

For the Student
Additional Practice

- Lesson Tutorials
- Multi-Language Glossary
- Self-Grading Progress Check
- *BigIdeasMath.com*
  Dynamic Student Edition
  Student Resources

Answers

7. $256\pi \approx 804.2 \text{ m}^3$

8. $\frac{40}{3}\pi \approx 41.9 \text{ cm}^3$

9. $\frac{3052}{27\pi} \approx 36.0 \text{ in.}$

10. $2304\pi \approx 7238.2 \text{ ft}^3$

11. 21 in.

12. $360\pi \approx 1131.0 \text{ m}^3$

13. 132 ft^3

14. $16\pi + \frac{16\pi}{3} = \frac{64\pi}{3} \approx 67.0 \text{ cm}^3$

15. 576 m^3

16. 86.6 yd^2

My Thoughts on the Chapter

What worked. . .

Teacher Tip

Not allowed to write in your teaching edition? Use sticky notes to record your thoughts.

What did not work. . .

What I would do differently. . .

Exercises

Find the volume V or radius r of the sphere. Round your answer to the nearest tenth, if necessary.

10.

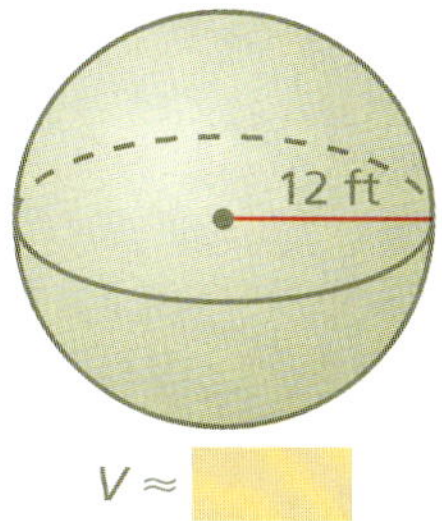

11.

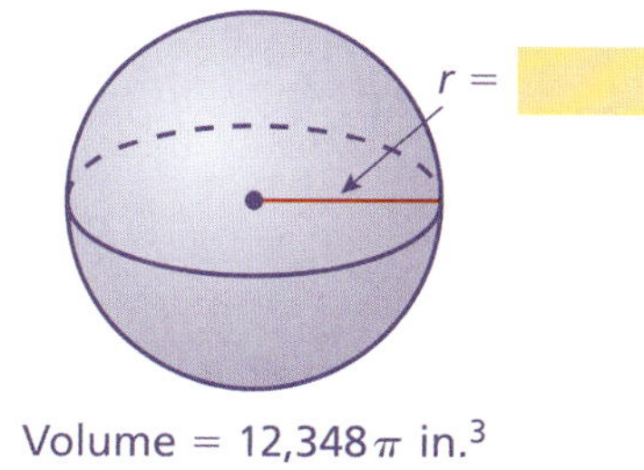

Find the volume of the composite solid. Round your answer to the nearest tenth.

12.

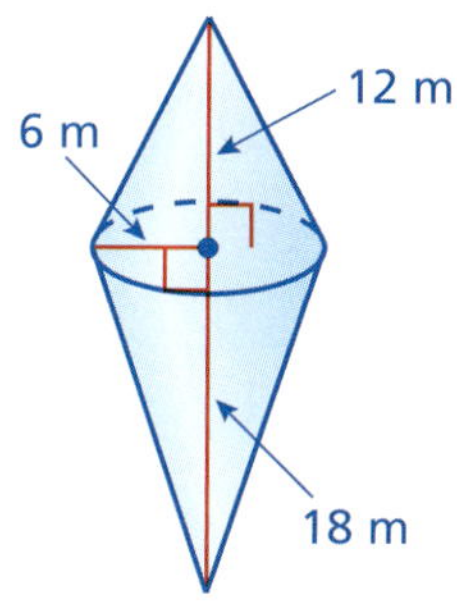

13.

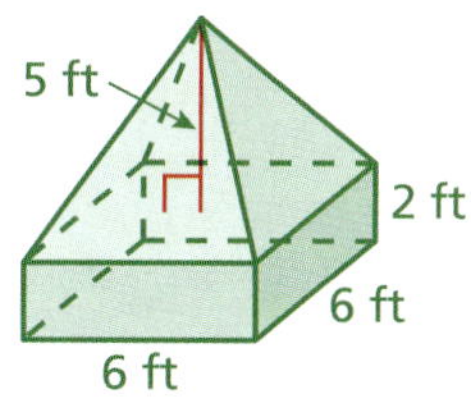

14.

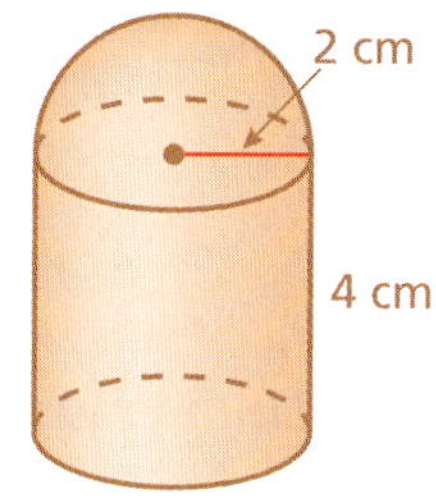

15.4 Surface Areas and Volumes of Similar Solids *(pp. 692–699)*

The cones are similar. What is the volume of the red cone? Round your answer to the nearest tenth.

$\dfrac{\text{Volume of A}}{\text{Volume of B}} = \left(\dfrac{\text{Height of A}}{\text{Height of B}}\right)^3$

$\dfrac{V}{157} = \left(\dfrac{4}{6}\right)^3$ Substitute.

$\dfrac{V}{157} = \dfrac{64}{216}$ Evaluate.

$V \approx 46.5$ Solve for V.

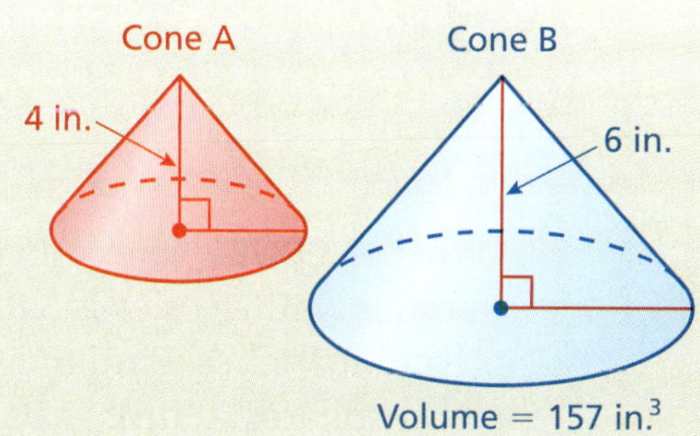

The volume is about 46.5 cubic inches.

Exercises

The solids are similar. Find the surface area S or volume V of the red solid. Round your answer to the nearest tenth.

15.

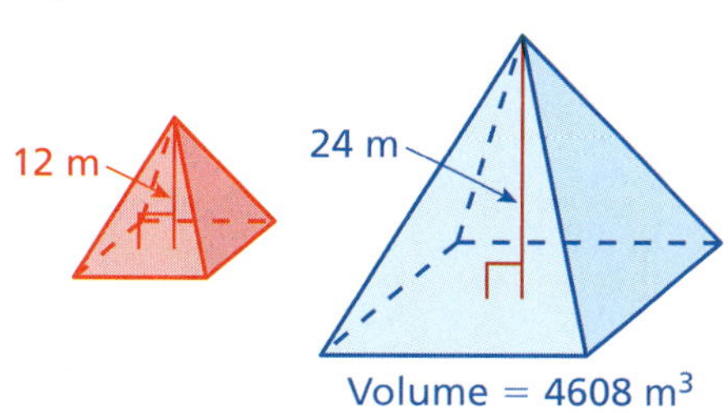

16.

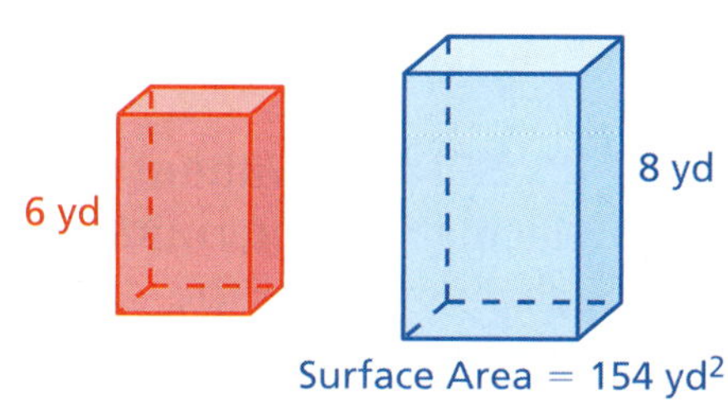

15 Chapter Test

Check It Out
Test Practice
BigIdeasMath.com

Find the volume of the solid. Round your answer to the nearest tenth.

1.

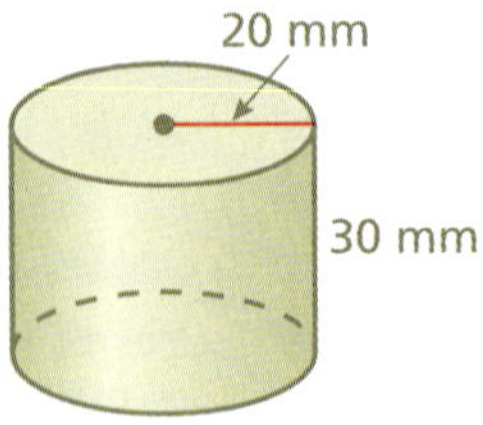

2.

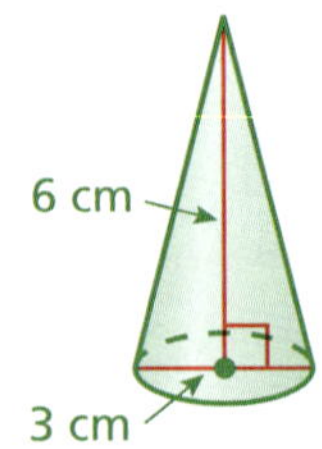

3.

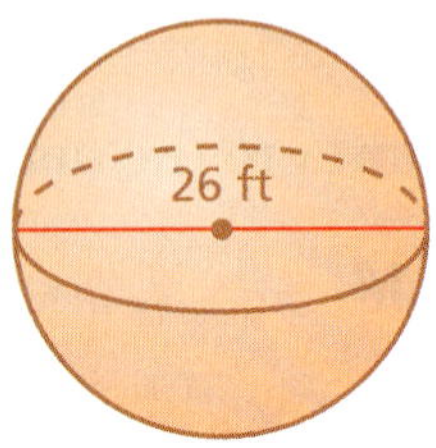

4.

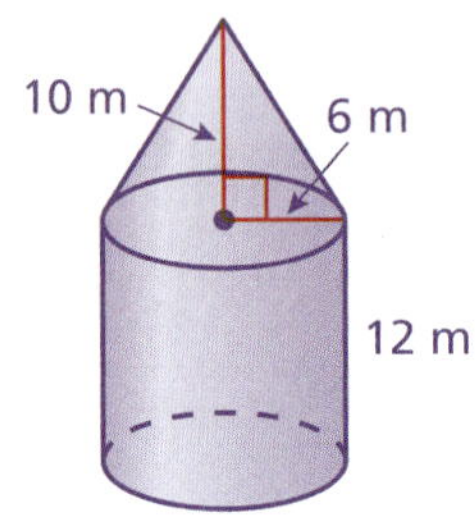

5. The pyramids are similar.

a. Find the missing dimension.

b. Find the surface area of the red pyramid.

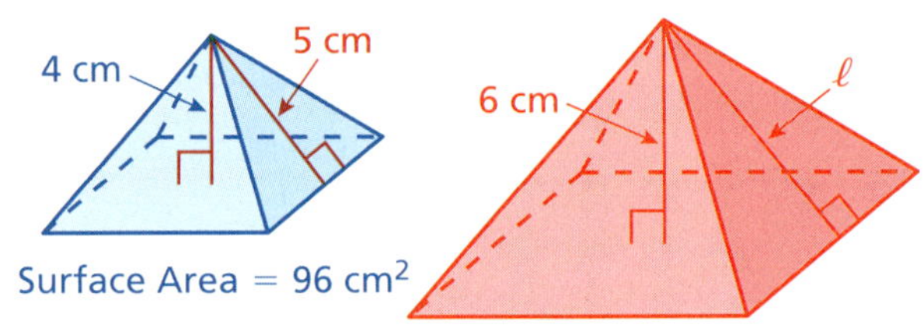

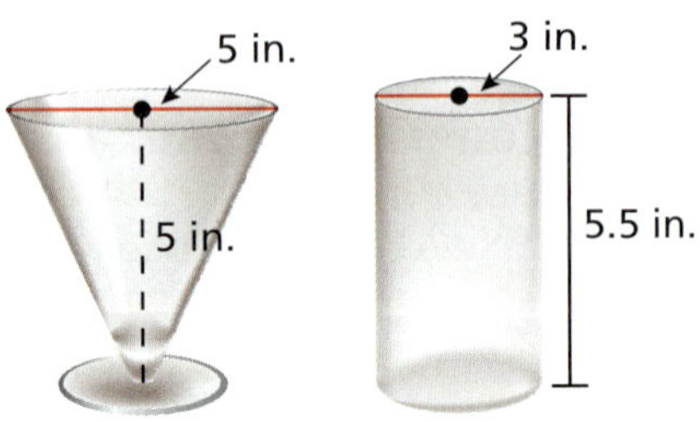

6. SMOOTHIES You are making smoothies. You will use either the cone-shaped glass or the cylindrical glass. Which glass holds more? About how much more?

7. WAFFLE CONES The ratio of the corresponding linear measures of two similar waffle cones is 3 to 4. The smaller cone has a volume of about 18 cubic inches. Find the volume of the larger cone. Round your answer to the nearest tenth.

8. OPEN-ENDED Draw two different composite solids that have the same volume but different surface areas. Explain your reasoning.

9. MILK Glass A has a diameter of 3.5 inches and a height of 4 inches. Glass B has a radius of 1.5 inches and a height of 5 inches. Which glass can hold more milk?

10. REASONING Without calculating, determine which solid has the greater volume. Explain your reasoning.

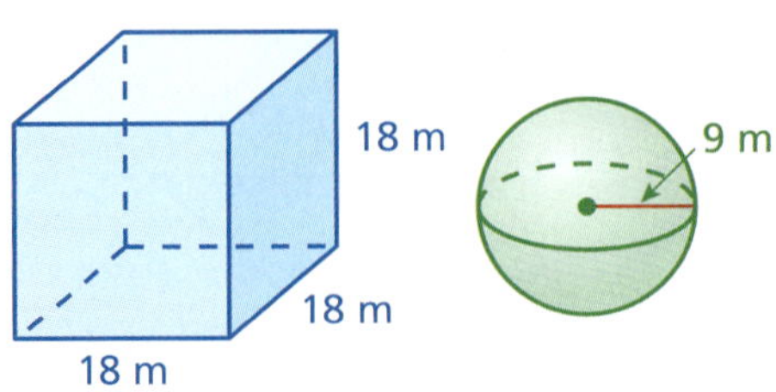

Test Item References

Chapter Test Questions	Section to Review
1, 6, 9	15.1
2, 6	15.2
3, 4, 8, 10	15.3
5, 7	15.4

Test-Taking Strategies

Remind students to quickly look over the entire test before they start so that they can budget their time. This test is very visual and requires that students remember many terms. It might be helpful for them to jot down some of the terms on the back of the test before they start. Students should make sketches and diagrams to help them.

Common Errors

- **Exercises 1–4** Students may write linear or square units for volume rather than cubic units. Remind them that part of writing a correct answer is including the correct units.
- **Exercises 1, 2, 4, 6, and 9** When finding the area of the circular base of a cylinder or a cone, students may not square the radius, or they may use the diameter when the formula calls for the radius. Remind them of the formula that they learned for the area of a circle.
- **Exercises 2, 4, and 6** When finding the volume of a cone, students may forget to multiply by $\frac{1}{3}$. Remind them of the formula for volume of a cone.
- **Exercise 3** When finding the volume of a sphere, students may forget to multiply by $\frac{4}{3}$ or cube the radius, they may use diameter when the formula calls for radius, or they may write the incorrect units. Remind them of the given formula for volume of a sphere and that part of writing a correct answer is including the correct units.
- **Exercises 5 and 7** Students may raise the ratio of the linear measures to the wrong exponent or forget to square or cube the ratio altogether. Discuss why squaring or cubing the ratio makes sense.

Reteaching and Enrichment Strategies

If students need help. . .	If students got it. . .
Resources by Chapter • Practice A and Practice B • Puzzle Time Record and Practice Journal Practice Differentiating the Lesson Lesson Tutorials *BigIdeasMath.com* Skills Review Handbook	Resources by Chapter • Enrichment and Extension • Technology Connection Game Closet at *BigIdeasMath.com* Start Cumulative Assessment

Answers

1. $12{,}000\pi \approx 37{,}699.1\text{ mm}^3$
2. $4.5\pi \approx 14.1\text{ cm}^3$
3. $\frac{8788\pi}{3} \approx 9202.8\text{ ft}^3$
4. $552\pi \approx 1734.2\text{ m}^3$
5. **a.** $\ell = 7.5$ cm

   b. 216 cm^2
6. cylindrical glass; about 6.2 in.^3
7. 42.7 in.^3
8. *Sample answer:*

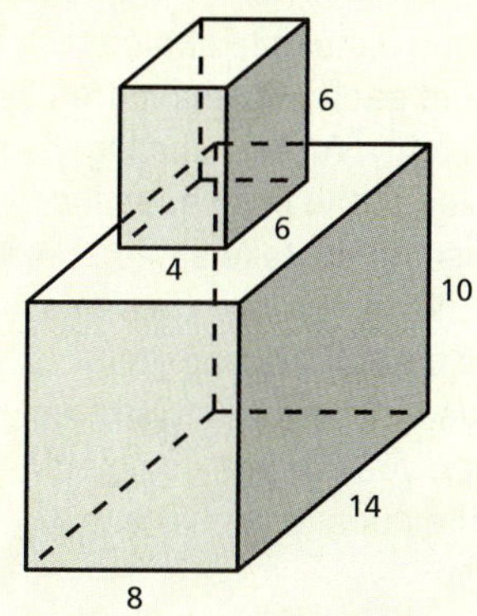

$V = 1264$ cubic units
$S = 784$ square units

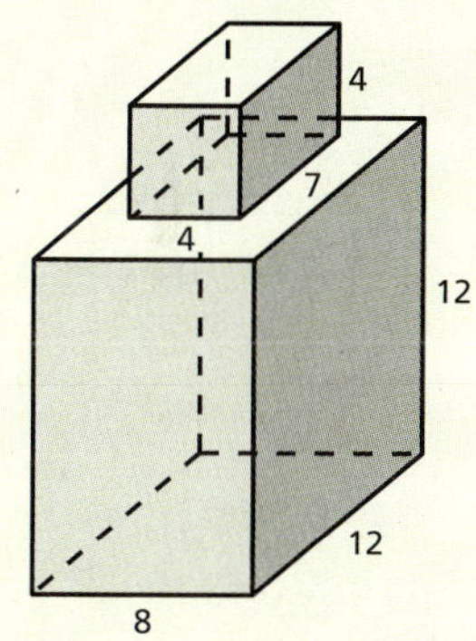

$V = 1264$ cubic units
$S = 760$ square units

9. Glass A
10. cube; The sphere could fit inside the cube and there would still be extra space outside the sphere but inside the cube.

Technology for the Teacher

Online Assessment
Assessment Book
ExamView® Assessment Suite

Test-Taking Strategies
Available at *BigIdeasMath.com*

After Answering Easy Questions, Relax

Answer Easy Questions First
Estimate the Answer
Read All Choices before Answering
Read Question before Answering
Solve Directly or Eliminate Choices
Solve Problem before Looking at Choices
Use Intelligent Guessing
Work Backwards

About this Strategy

When taking a multiple choice test, be sure to read each question carefully and thoroughly. After skimming the test and answering the easy questions, stop for a few seconds, take a deep breath, and relax. Work through the remaining questions carefully, using your knowledge and test-taking strategies. Remember, you already completed many of the questions on the test!

Answers

1. D
2. H
3. C

Technology for the Teacher

Performance Tasks
Online Assessment
Assessment Book
ExamView® Assessment Suite

Item Analysis

1. **A.** The student multiplies both sides by 3, distributes the 3 correctly, but does not distribute the 9 correctly. Alternatively, the student starts by distributing the 3 on the right side correctly. But when multiplying both sides by 3, the student does not distribute it across the expression $3w - 4$ correctly.

   B. The student starts by distributing the 3 on the right side incorrectly.

   C. The student multiplies both sides by 3, but fails to distribute the 3 correctly on the right side.

   D. Correct answer

2. **F.** The student incorrectly uses half the radius.

   G. The student incorrectly uses half the radius and also uses the formula for a right circular cylinder, neglecting to multiply by $\frac{1}{3}$.

   H. Correct answer

   I. The student uses the formula for a right circular cylinder, neglecting to multiply by $\frac{1}{3}$.

3. **A.** The student makes an order of operations error by not first distributing the multiplication.

   B. The student does not distribute the negative sign to the second term.

   C. Correct answer

   D. The student thinks that multiplying by $\frac{3}{2}$ is the inverse operation of multiplying by $-\frac{3}{2}$.

15 Cumulative Assessment

Test-Taking Strategy
After Answering Easy Questions, Relax

"After answering the easy questions, relax and try the harder ones. For this, $\pi r^2 h = 2\pi$. So, it's A."

1. What value of w makes the equation below true?

$$\frac{w}{3} = 3(w - 1) - 1$$

A. $\frac{1}{2}$

B. $\frac{3}{4}$

C. $\frac{5}{4}$

D. $\frac{3}{2}$

2. A right circular cone and its dimensions are shown below.

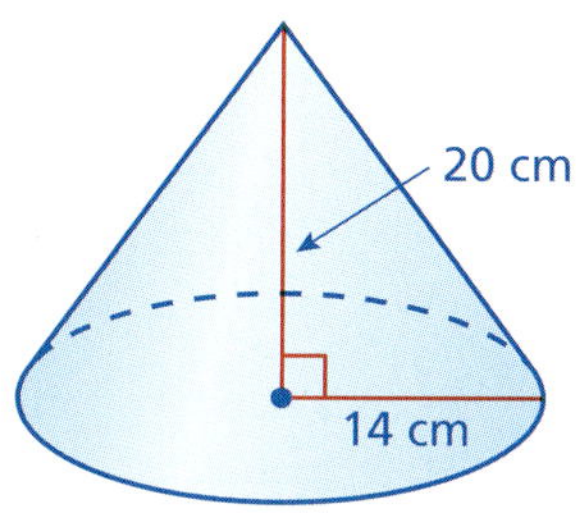

What is the volume of the right circular cone? $\left(\text{Use } \frac{22}{7} \text{ for } \pi.\right)$

F. $1{,}026\frac{2}{3}$ cm³

G. 3,080 cm³

H. $4{,}106\frac{2}{3}$ cm³

I. 12,320 cm³

3. Patricia solved the equation in the box shown.

$$-\frac{3}{2}(8x - 10) = -20$$
$$8x - 10 = -20\left(-\frac{3}{2}\right)$$
$$8x - 10 = 30$$
$$8x - 10 + 10 = 30 + 10$$
$$8x = 40$$
$$\frac{8x}{8} = \frac{40}{8}$$
$$x = 5$$

What should Patricia do to correct the error that she made?

A. Add 10 to -20.

B. Distribute $-\frac{3}{2}$ to get $-12x - 15$.

C. Multiply both sides by $-\frac{2}{3}$ instead of $-\frac{3}{2}$.

D. Multiply both sides by $\frac{3}{2}$ instead of $-\frac{3}{2}$.

4. On the grid below, Rectangle *EFGH* is plotted and its vertices are labeled.

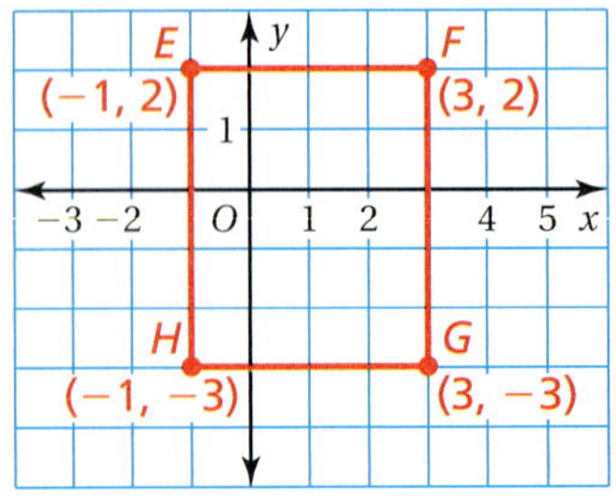

Which of the following shows Rectangle $E'F'G'H'$, the image of Rectangle *EFGH* after it is reflected in the x-axis?

F.

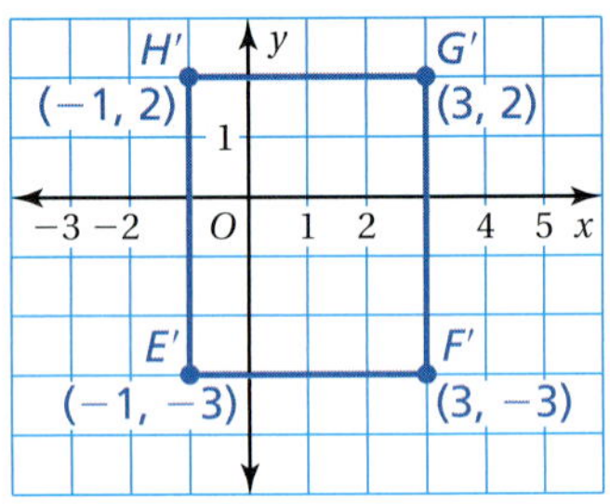

H.

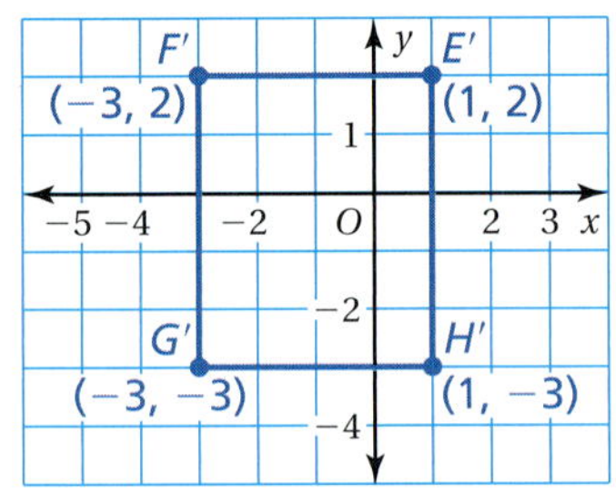

G.

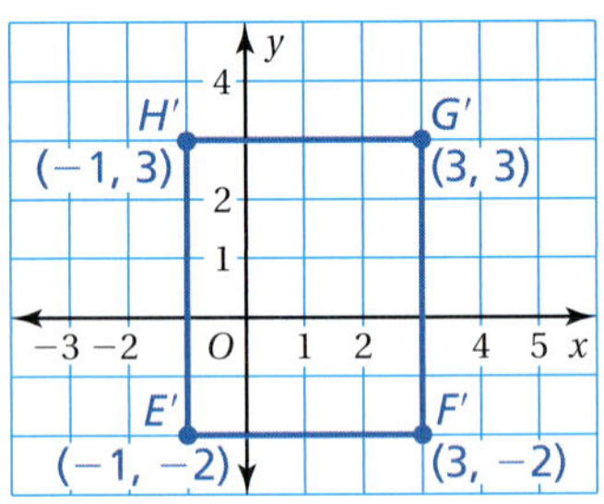

I.

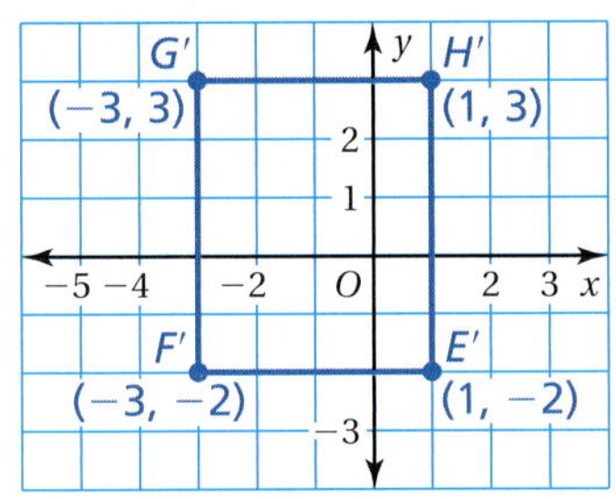

5. What is the measure of the exterior angle in the triangle below?

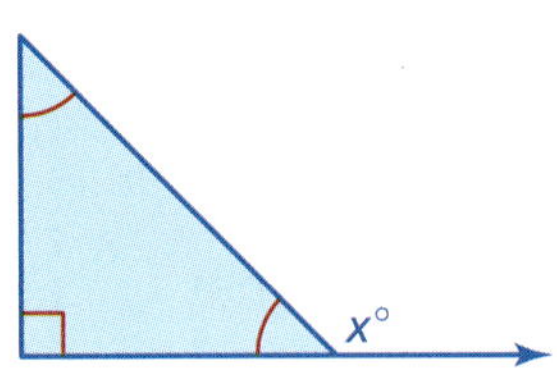

A. 45°

B. 90°

C. 135°

D. 145°

6. The temperature fell from 54 degrees Fahrenheit to 36 degrees Fahrenheit over a 6-hour period. The temperature fell by the same number of degrees each hour. How many degrees Fahrenheit did the temperature fall each hour?

Item Analysis (continued)

4. **F.** The student chooses the same rectangle with the vertices labeled with different points.
   G. Correct answer
   H. The student reflects the rectangle in the *y*-axis.
   I. The student rotates the rectangle 180°.

5. **A.** The student thinks that the exterior angle is congruent to one of the nonadjacent interior angles.
   B. The student thinks that the exterior angle is congruent to one of the nonadjacent interior angles.
   C. Correct answer
   D. The student makes a calculation error.

6. **Gridded Response:** Correct answer: 3°F

   Common Error: The student subtracts 36 from 54, but fails to divide the result by 6.

Answers

4. G

5. C

6. 3°F

Answers

7. I
8. 9 in.^{3}
9. B
10. 113.04 in.^{3}

Item Analysis (continued)

7. **F.** The student performs the first step correctly, but then subtracts P rather than dividing.

   G. The student performs the first step correctly, but then divides A by P instead of dividing the entire expression by P.

   H. The student performs the first step correctly, but then divides P by P instead of dividing the entire expression by P.

   I. Correct answer

8. **Gridded Response:** Correct answer: 9 in.^{3}

   Common Error: The student divides the volume by r instead of r^2, getting an answer of 108 cubic inches.

9. **A.** The student finds the slope incorrectly.

   B. Correct answer

   C. The student finds the slope incorrectly.

   D. The student finds the slope incorrectly.

10. **2 points** The student demonstrates a thorough understanding of finding the volume of a composite solid made up of a right circular cylinder and a right circular cone. The student correctly finds a volume of $36\pi \approx 113.04$ cubic inches. The student provides clear and complete work and explanations.

    1 point The student demonstrates a partial understanding of finding the volume of a composite solid made up of a right circular cylinder and a right circular cone. The student provides some correct work and explanation toward finding the volume.

    0 points The student demonstrates insufficient understanding of finding the volume of a composite solid made up of a right circular cylinder and a right circular cone. The student is unable to make any meaningful progress toward finding the volume.

7. Solve the formula below for I.

$$A = P + PI$$

F. $I = A - 2P$

G. $I = \dfrac{A}{P} - P$

H. $I = A - \dfrac{P}{P}$

I. $I = \dfrac{A - P}{P}$

8. A right circular cylinder has a volume of 1296 cubic inches. If you divide the radius of the cylinder by 12, what would be the volume, in cubic inches, of the smaller cylinder?

9. Which line has a slope of -2?

A.

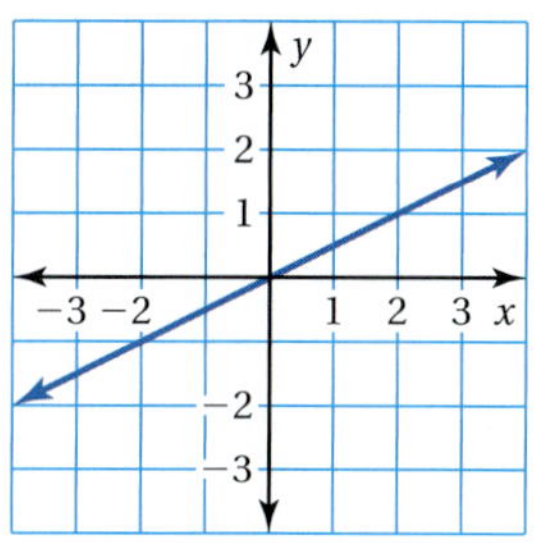

C.

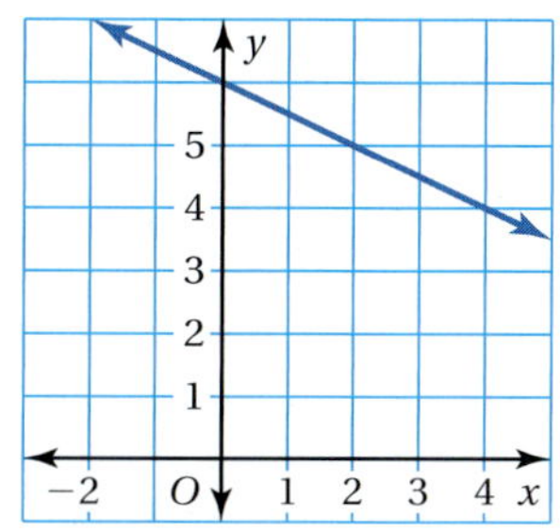

B.

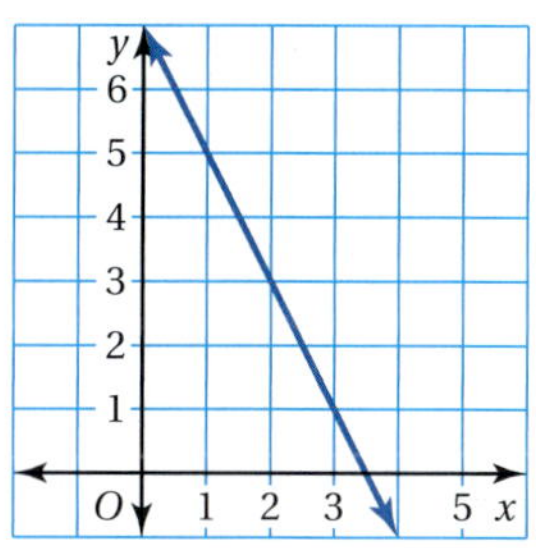

D.

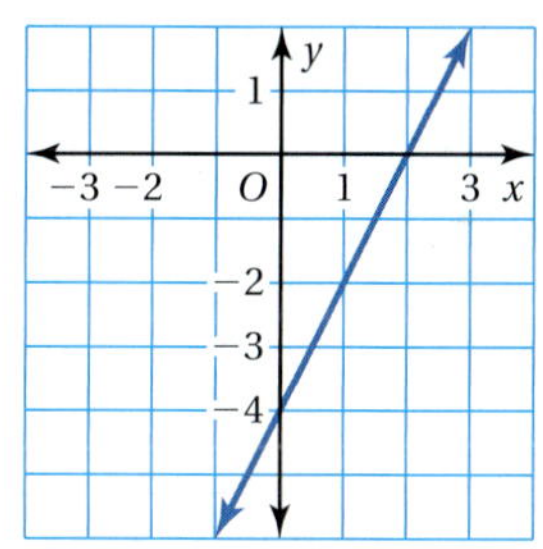

10. The figure below is a diagram for making a tin lantern.

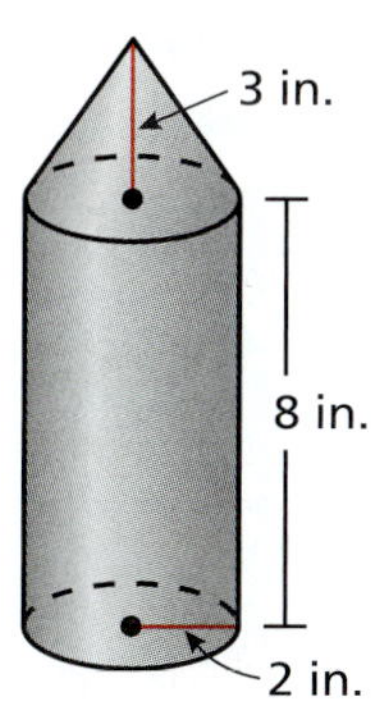

The figure consists of a right circular cylinder without its top base and a right circular cone without its base. What is the volume, in cubic inches, of the entire lantern? Show your work and explain your reasoning. (Use 3.14 for π.)

16 Exponents and Scientific Notation

What Your Students Have Learned

- Write and evaluate numerical expressions involving whole-number exponents.
- Perform arithmetic operations, including those involving whole-number exponents, using the order of operations.
- Apply the properties of operations to generate equivalent expressions.
- Solve problems involving operations with rational numbers.
- Understand that rewriting expressions in different forms can show how quantities are related.

What Your Students Will Learn

- Use the properties of integer exponents to generate equivalent expressions.
- Use scientific notation to estimate very large or very small quantities.
- Perform operations with numbers expressed in scientific notation and other forms.
- Interpret scientific notation that has been generated by technology.

Pacing Guide for Chapter 16

Chapter Opener Accelerated	1 Day
Section 1 Accelerated	1 Day
Section 2 Accelerated	1 Day
Section 3 Accelerated	1 Day
Section 4 Accelerated	1 Day
Study Help / Quiz Accelerated	1 Day
Section 5 Accelerated	1 Day
Section 6 Accelerated	1 Day
Section 7 Accelerated	1 Day
Chapter Review/ Chapter Tests Accelerated	2 Days
Total Chapter 16 Accelerated	11 Days
Year-to-Date Accelerated	152 Days

Technology for the Teacher

BigIdeasMath.com
Chapter at a Glance
Complete Materials List
Parent Letters: English and Spanish

What Your Students Have Learned

- Use the order of operations to evaluate expressions.
- Multiply and divide decimals.

Additional Topics for Review

- Factors
- Simplifying Expressions
- Multiplicative Inverse Property
- Place Value
- Reciprocals
- Converting Measures
- Distributive Property
- Commutative and Associative Properties of Multiplication

Try It Yourself

1. 13 **2.** 10

3. 48 **4.** 0.35

5. 0.84 **6.** 30.229

7. 0.1788 **8.** 60

9. 13.9 **10.** 24

11. 1800

Record and Practice Journal Fair Game Review

1. 14 **2.** 3

3. 394 **4.** 86

5. 76 **6.** 16

7. **a.** 386

b. $4(2) + 2(5^2) + 3^2(6^2) + 2^2 = 386$

8. 2.352 **9.** 0.1014

10. 6.0048 **11.** 9

12. 1.5 **13.** 2700

14. $6.93

Math Background Notes

Vocabulary Review

- Order of Operations
- Expression
- Evaluating an Expression
- Exponent
- Decimal
- Dividend
- Product
- Quotient

Using Order of Operations

- Students should know the order of operations, but you may want to review it.
- Many students probably learned the mnemonic device *Please Excuse My Dear Aunt Sally*. Ask a volunteer to explain why this phrase is helpful.
- **Common Error:** Students may misinterpret the exponent in Example 1. Watch for students who incorrectly compute $6^2 = 6 \cdot 2 = 12$.

Multiplying and Dividing Decimals

- Students should be able to multiply and divide decimals.
- Remind students to multiply as they would with whole numbers. After the multiplication is complete, remind them to count the total number of digits in both factors that appear to the right of the decimal point, and then put that many digits to the right of the decimal point in the answer.
- **Common Error:** In a horizontal division problem, some students have difficulty determining which number is the dividend and which number is the divisor. Encourage students to read the problem aloud as they rewrite it. Each time a student says the words "divided by," he or she should be trapping the first number inside the division box.
- **Common Error:** Some students may try to clear the decimal point from the dividend instead of from the divisor. Remind students that having a decimal inside the division box as part of the dividend is fine. Also remind students that the divisor and dividend must both be multiplied by the same power of 10, effectively multiplying by 1, so they do not change the problem.

Reteaching and Enrichment Strategies

If students need help. . .	If students got it. . .
Record and Practice Journal • Fair Game Review Skills Review Handbook Lesson Tutorials	Game Closet at *BigIdeasMath.com* Start the next section

What You Learned Before

"It's called the Power of Negative One, Descartes!"

Using Order of Operations

Example 1 Evaluate $6^2 \div 4 - 2(9 - 5)$.

First:	Parentheses	$6^2 \div 4 - 2(9 - 5) = 6^2 \div 4 - 2 \cdot 4$
Second:	Exponents	$= 36 \div 4 - 2 \cdot 4$
Third:	Multiplication and Division (from left to right)	$= 9 - 8$
Fourth:	Addition and Subtraction (from left to right)	$= 1$

Try It Yourself

Evaluate the expression.

1. $15\left(\frac{8}{4}\right) + 2^2 - 3 \cdot 7$ **2.** $5^2 \cdot 2 \div 10 + 3 \cdot 2 - 1$ **3.** $3^2 - 1 + 2(4(3 + 2))$

Multiplying and Dividing Decimals

Example 2 Find 2.1 • 0.35.

$$\begin{array}{r} 2.1 \\ \times\ 0.35 \\ \hline 105 \\ 63 \\ \hline 0.735 \end{array}$$

2.1 ← 1 decimal place
0.35 ← + 2 decimal places
0.735 ← 3 decimal places

Example 3 Find 1.08 ÷ 0.9.

$0.9\overline{)1.08}$ Multiply each number by 10.

$$\begin{array}{r} 1.2 \\ 9\overline{)10.8} \\ -\ 9 \\ \hline 1\,8 \\ -\ 1\,8 \\ \hline 0 \end{array}$$

Place the decimal point above the decimal point in the dividend 10.8.

Try It Yourself

Find the product or quotient.

4. $1.75 \cdot 0.2$ **5.** $1.4 \cdot 0.6$ **6.** $\begin{array}{r} 7.03 \\ \times\ 4.3 \\ \hline \end{array}$ **7.** $\begin{array}{r} 0.894 \\ \times\ 0.2 \\ \hline \end{array}$

8. $5.40 \div 0.09$ **9.** $4.17 \div 0.3$ **10.** $0.15\overline{)3.6}$ **11.** $0.004\overline{)7.2}$

16.1 Exponents

Essential Question How can you use exponents to write numbers?

The expression 3^5 is called a *power*. The *base* is 3. The *exponent* is 5.

1 ACTIVITY: Using Exponent Notation

Work with a partner.

a. Copy and complete the table.

Power	Repeated Multiplication Form	Value
$(-3)^1$	-3	-3
$(-3)^2$	$(-3) \cdot (-3)$	9
$(-3)^3$		
$(-3)^4$		
$(-3)^5$		
$(-3)^6$		
$(-3)^7$		

b. **REPEATED REASONING** Describe what is meant by the expression $(-3)^n$. How can you find the value of $(-3)^n$?

2 ACTIVITY: Using Exponent Notation

Exponents

In this lesson, you will

- write expressions using integer exponents.
- evaluate expressions involving integer exponents.

Work with a partner.

a. The cube at the right has $3 in each of its small cubes. Write a power that represents the total amount of money in the large cube.

b. Evaluate the power to find the total amount of money in the large cube.

Laurie's Notes

Introduction

Applying Mathematical Practices

- **Look for and Express Regularity in Repeated Reasoning:** In Activity 1, students will raise a negative number to whole number exponents and observe patterns that emerge.

Motivate

? "How big is a cubic millimeter?" A grain of salt is a reasonable estimate.

- Use a metric ruler and your fingers to show what a millimeter is. A cubic millimeter is 1 mm × 1 mm × 1 mm.

? "How big is a cubic meter?" about the size of a baby's play pen

- Use 3 meter sticks to demonstrate 1 m × 1 m × 1 m.

? "How many cubic millimeters are in a cubic meter?" Students may or may not have an answer.

- Give students time to think. Someone might ask how many millimeters there are in a meter. The prefix milli- means $\frac{1}{1000}$.
- The volume of a cubic meter in terms of cubic millimeters is $1000 \times 1000 \times 1000 = 1{,}000{,}000{,}000 = 1$ billion mm^3.

? "Can 1 billion be expressed using exponents?" 1000^3 or 10^9

Activity Notes

Activity 1

- **Attend to Precision:** Review the vocabulary associated with exponents.
- Have students work with their partners to complete the table. Students should recognize that a calculator is not necessary. They only need to multiply their previous product by −3.
- When students have finished, discuss the problem.

? "What do you notice about the values in the third column?" Values alternate—negative, positive, negative…

- **Look for and Express Regularity in Repeated Reasoning:** Students might also mention: all values are odd and divisible by 3; last digits repeat in a cluster of 4: 3, 9, 7, 1, 3, 9, 7, …; there are two 1-digit numbers, two 2-digit numbers, two 3-digit numbers, and predict two 4-digit numbers
- **Part (b):** Listen for students to describe the exponent n in $(-3)^n$ as how many times −3 is multiplied by itself. Try to have students say that the exponent tells the number of times the base is used as a factor.

? "How can you find the value of $(-3)^n$?" Multiply −3 by itself n times.

Activity 2

- Although not all of the cubes are visible, students generally know that the cube contains 3^3 or 27 smaller cubes. So, at \$3 per small cube, $3 \times 3^3 = 3^4$.
- The expression is 3^4 and the answer is \$81.

What Your Students Will Learn

- Rewrite expressions with repeated factors using integer exponents.
- Evaluate expressions involving integer exponents using repeated multiplication or the order of operations.

Previous Learning

Students should know how to raise a number to an exponent.

Technology for the Teacher

Dynamic Classroom

Lesson Plans
Complete Materials List

16.1 Record and Practice Journal

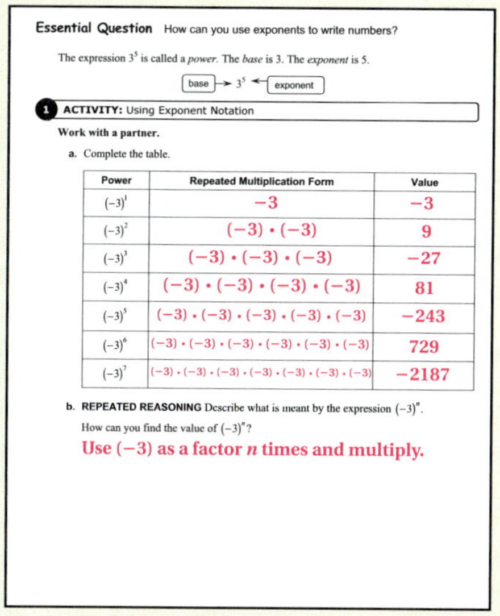

Essential Question How can you use exponents to write numbers?

The expression 3^5 is called a *power*. The *base* is 3. The *exponent* is 5.

base → 3^5 ← exponent

1 ACTIVITY: Using Exponent Notation

Work with a partner.

a. Complete the table.

Power	Repeated Multiplication Form	Value
$(-3)^1$	−3	−3
$(-3)^2$	(−3) • (−3)	9
$(-3)^3$	(−3) • (−3) • (−3)	−27
$(-3)^4$	(−3) • (−3) • (−3) • (−3)	81
$(-3)^5$	(−3) • (−3) • (−3) • (−3) • (−3)	−243
$(-3)^6$	(−3) • (−3) • (−3) • (−3) • (−3) • (−3)	729
$(-3)^7$	(−3) • (−3) • (−3) • (−3) • (−3) • (−3) • (−3)	−2187

b. **REPEATED REASONING** Describe what is meant by the expression $(-3)^n$. How can you find the value of $(-3)^n$?

Use (−3) as a factor n times and multiply.

Differentiated Instruction

Visual

Have students create pyramids of factors -10, -5, -2, 2, 5, and 10. Ask them to write the exponential form and evaluate each row. Ask, "What is the product of 3 factors of -2?" and "What is the product of 4 factors of 5?" -8; 625

$$-2$$
$$(-2)\times(-2)$$
$$(-2)\times(-2)\times(-2)$$

$$5$$
$$5\times 5$$
$$5\times 5\times 5$$
$$5\times 5\times 5\times 5$$

16.1 Record and Practice Journal

2 ACTIVITY: Using Exponent Notation

Work with a partner.

a. The cube at the right has \$3 in each of its small cubes. Write a power that represents the total amount of money in the large cube.

3^4

b. Evaluate the power to find the total amount of money in the large cube.

\$81

3 ACTIVITY: Writing Powers as Whole Numbers

Work with a partner. Write each distance as a whole number. Which numbers do you know how to write in words? For instance, in words, 10^3 is equal to *one thousand*.

a. 10^{26} meters: diameter of observable universe

b. 10^{21} meters: diameter of Milky Way galaxy

a. 100,000,000,000,000,000,000,000,000 m

b. 1,000,000,000,000,000,000,000 m

c. 10^{16} meters: diameter of solar system

10,000,000,000,000,000 m

d. 10^7 meters: diameter of Earth

10,000,000 m

e. 10^6 meters: length of Lake Erie shoreline

1,000,000 m

f. 10^5 meters: width of Lake Erie

100,000 m

4 ACTIVITY: Writing a Power

Work with a partner. Write the number of kits, cats, sacks, and wives as a power.

As I was going to St. Ives
I met a man with seven wives
Each wife had seven sacks
Each sack had seven cats
Each cat had seven kits
Kits, cats, sacks, wives
How many were going to St. Ives?

Nursery Rhyme, 1730

wives: 7^1, sacks: 7^2, cats: 7^3, kits: 7^4

What Is Your Answer?

5. **IN YOUR OWN WORDS** How can you use exponents to write numbers? Give some examples of how exponents are used in real life.

represents repeated multiplication of the same factor

Laurie's Notes

Activity 3

- The distances in this activity represent magnitudes, powers of 10, not exact distances.
- **FYI:** There is a classic video made by two designers in the late 1970s called *Powers of Ten,* which is easily available on the Internet.

? "Which numbers do you know the names for in parts (a)–(f)?" part (d): ten million; part (e): one million; part (f): one hundred thousand

- Share vocabulary that might be of interest:

million $= 10^6$	billion $= 10^9$	trillion $= 10^{12}$
quadrillion $= 10^{15}$	quintillion $= 10^{18}$	hexillion $= 10^{21}$
heptillion $= 10^{24}$	octillion $= 10^{27}$	nonillion $= 10^{30}$
decillion $= 10^{33}$	unodecillion $= 10^{36}$	duodecillion $= 10^{39}$

- Some of the above prefixes are the same ones used in naming polygons, so they should look familiar.

Activity 4

- This is a classic rhyme. In the original rhyme, the answer is one because the man and his wives were coming *from* St. Ives.
- For this problem, suggest to students that they draw a picture or diagram to help them solve the problem.
- **Summary:** wives $= 7^1$, sacks $= 7^2$, cats $= 7^3$, kits $= 7^4$

What Is Your Answer?

- **Neighbor Check:** Have students work independently and then have their neighbors check their work. Have students discuss any discrepancies.

Closure

- Compare the following powers using >, <, or =.

1. 2^{10} _____ 10^2 > **2.** 10^3 _____ 3^{10} <

3 ACTIVITY: Writing Powers as Whole Numbers

Work with a partner. Write each distance as a whole number. Which numbers do you know how to write in words? For instance, in words, 10^3 is equal to *one thousand*.

a. 10^{26} meters: diameter of observable universe

b. 10^{21} meters: diameter of Milky Way galaxy

c. 10^{16} meters: diameter of solar system

d. 10^{7} meters: diameter of Earth

e. 10^{6} meters: length of Lake Erie shoreline

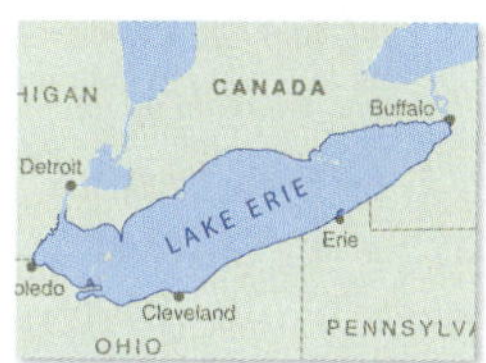

f. 10^{5} meters: width of Lake Erie

4 ACTIVITY: Writing a Power

Math Practice

Analyze Givens

What information is given in the poem? What are you trying to find?

Work with a partner. Write the numbers of kits, cats, sacks, and wives as powers.

As I was going to St. Ives
I met a man with seven wives
Each wife had seven sacks
Each sack had seven cats
Each cat had seven kits
Kits, cats, sacks, wives
How many were going to St. Ives?

Nursery Rhyme, 1730

What Is Your Answer?

5. **IN YOUR OWN WORDS** How can you use exponents to write numbers? Give some examples of how exponents are used in real life.

Practice Use what you learned about exponents to complete Exercises 3–5 on page 714.

16.1 Lesson

Check It Out
Lesson Tutorials
BigIdeasMath.com

Key Vocabulary
power, *p. 712*
base, *p. 712*
exponent, *p. 712*

A **power** is a product of repeated factors. The **base** of a power is the common factor. The **exponent** of a power indicates the number of times the base is used as a factor.

$$\left(\frac{1}{2}\right)^5 = \frac{1}{2} \cdot \frac{1}{2} \cdot \frac{1}{2} \cdot \frac{1}{2} \cdot \frac{1}{2}$$

base: $\frac{1}{2}$; exponent: 5; power: $\left(\frac{1}{2}\right)^5$

$\frac{1}{2}$ is used as a factor 5 times.

EXAMPLE 1 Writing Expressions Using Exponents

Study Tip
Use parentheses to write powers with negative bases.

Write each product using exponents.

a. $(-7) \cdot (-7) \cdot (-7)$

Because -7 is used as a factor 3 times, its exponent is 3.

So, $(-7) \cdot (-7) \cdot (-7) = (-7)^3$.

b. $\pi \cdot \pi \cdot r \cdot r \cdot r$

Because π is used as a factor 2 times, its exponent is 2. Because r is used as a factor 3 times, its exponent is 3.

So, $\pi \cdot \pi \cdot r \cdot r \cdot r = \pi^2 r^3$.

On Your Own

Now You're Ready
Exercises 3–10

Write the product using exponents.

1. $\frac{1}{4} \cdot \frac{1}{4} \cdot \frac{1}{4} \cdot \frac{1}{4} \cdot \frac{1}{4}$

2. $0.3 \cdot 0.3 \cdot 0.3 \cdot 0.3 \cdot x \cdot x$

EXAMPLE 2 Evaluating Expressions

Evaluate each expression.

a. $(-2)^4$

The base is -2.

$(-2)^4 = (-2) \cdot (-2) \cdot (-2) \cdot (-2)$ Write as repeated multiplication.

$= 16$ Simplify.

b. -2^4

The base is 2.

$-2^4 = -(2 \cdot 2 \cdot 2 \cdot 2)$ Write as repeated multiplication.

$= -16$ Simplify.

Multi-Language Glossary at BigIdeasMath.com

Laurie's Notes

Introduction

Connect

- **Yesterday:** Students explored writing numbers with exponents.
- **Today:** Students will write expressions involving exponents and evaluate powers.

Motivate

- Because U.S. currency has coins and bills which are powers of 10, find out what your students know about the people on the coins and bills. You can create a matching activity or have a few questions and answers ready for the class.

Lesson Notes

Example 1

- Write the definitions of power, base, and exponent. Note the use of *factor* instead of *multiplying the base by itself.*
- Write the example shown. When this power is evaluated, the answer is $\frac{1}{32}$. You say, "$\frac{1}{2}$ to the fifth power is $\frac{1}{32}$."
- Exponents are used to rewrite an expression involving repeated factors.

? "Is it necessary to write the multiplication dot between the factors in part (a)?" no

- **Common Error:** Parentheses *must* be used when you write a power with a negative base. This is a common error that students will make. For example: $(-2)^2 = (-2)(-2) = 4$
  $-2^2 = -(2)(2) = -4$
  Without the parentheses, the number being squared is 2, and then you multiply the product by -1. With the parentheses, the number being squared is -2, and the product is 4. The underlying property is the order of operations. Exponents are performed before multiplication.
- **Part (b):** Variables and constants are expressed using exponents in a similar fashion.

On Your Own

- **Attend to Precision:** In Question 1, $\frac{1}{4}$ needs to be written with parentheses so that both the numerator and the denominator are raised to an exponent. Without parentheses, it could be read as $\frac{1^5}{4}$.

Example 2

? "What is the base in each problem? What will be used as the factor in each problem?" Bases are -2 and 2. Factors are -2 and 2.

- **Attend to Precision:** This example addresses the need to write the base within parentheses when the base is negative.

Goal Today's lesson is writing and evaluating **powers**.

Technology for the Teacher

Dynamic Classroom

Lesson Tutorials
Lesson Plans
Answer Presentation Tool

Extra Example 1

Write each product using exponents.

a. $4 \cdot 4 \cdot 4 \cdot 4$ 4^4

b. $(-2) \cdot (-2) \cdot x \cdot x \cdot x$ $(-2)^2x^3$

On Your Own

1. $\left(\frac{1}{4}\right)^5$
2. $(0.3)^4x^2$

Extra Example 2

Evaluate each expression.

a. $(-3)^3$ -27

b. -3^3 -27

Extra Example 3

Evaluate each expression.

a. $2 - 4 \cdot 5^2$ -98

b. $5 + 6^2 \div 4$ 14

On Your Own

3. -625 **4.** $-\frac{1}{216}$

5. 1 **6.** -7

Extra Example 4

In Example 4, the diameter of the inner sphere is 2.2 meters. What is the volume of the inflated space? about 8.56 m^3

On Your Own

7. about 11.08 m^3

English Language Learners

Labels

Have students label and practice reading statements of powers.

base → 2 (base); exponent → 3 (exponent)

$2^3 = 2 \times 2 \times 2 = 8$

power: 2^3; factor: each 2

"Two to the third equals two times two times two."

"The base is 2, the exponent is 3, and 2 is written as a factor 3 times."

Laurie's Notes

Example 3

- You may wish to review the order of operations or wait to see how students evaluate the expressions.
- **Common Error:** Students may evaluate the problem left to right, performing the addition first. They need to be reminded of the order of operations.

? "How do you start to evaluate this expression?" powers first

- Continue to evaluate the problem as shown.
- In part (b), there are two powers to evaluate. After that is done, division is performed before subtraction.

On Your Own

- Encourage students to write out the steps in their solutions. Discourage them from performing multiple steps in their heads.

Example 4

- **FYI:** Ask if any of your students have heard of sphering. You can find information about this sport online at *zorb.com.*
- You want to find the volume of the inflated space.
- Write the formula for the volume V of a sphere with radius r: $V = \frac{4}{3}\pi r^3$. From the photo, the diameter of each sphere is known. The radius is half the diameter. Substitute the radius of each sphere and simplify.
- Let students use the context of the problem to decide how to round their answers.
- **Extension:** The inner sphere has a volume of a little more than 4 cubic meters. Find a region of your classroom that is approximately 4 cubic meters.

On Your Own

- Before students do any calculations, ask them whether the answer will be less than or greater than the answer to Example 4. greater than

Closure

- **Exit Ticket:** Evaluate.

  1. $4^2 - 8(2) + 3^3$ 27 **2.** $\left(-\frac{2}{3}\right)^3 + \left|5^2 - 2 \cdot 15\right|$ $\frac{127}{27} \approx 4.7$

EXAMPLE 3 Using Order of Operations

Evaluate each expression.

a. $3 + 2 \cdot 3^4 = 3 + 2 \cdot 81$ — Evaluate the power.

$= 3 + 162$ — Multiply.

$= 165$ — Add.

b. $3^3 - 8^2 \div 2 = 27 - 64 \div 2$ — Evaluate the powers.

$= 27 - 32$ — Divide.

$= -5$ — Subtract.

On Your Own

Now You're Ready
Exercises 11–16 and 21–26

Evaluate the expression.

3. -5^4 **4.** $\left(-\frac{1}{6}\right)^3$ **5.** $\left|-3^3 \div 27\right|$ **6.** $9 - 2^5 \cdot 0.5$

EXAMPLE 4 Real-Life Application

In sphering, a person is secured inside a small, hollow sphere that is surrounded by a larger sphere. The space between the spheres is inflated with air. What is the volume of the inflated space?

You can find the radius of each sphere by dividing each diameter given in the diagram by 2.

Outer Sphere		***Inner Sphere***
$V = \frac{4}{3}\pi r^3$	Write formula.	$V = \frac{4}{3}\pi r^3$
$= \frac{4}{3}\pi\left(\frac{3}{2}\right)^3$	Substitute.	$= \frac{4}{3}\pi(1)^3$
$= \frac{4}{3}\pi\left(\frac{27}{8}\right)$	Evaluate the power.	$= \frac{4}{3}\pi(1)$
$= \frac{9}{2}\pi$	Multiply.	$= \frac{4}{3}\pi$

So, the volume of the inflated space is $\frac{9}{2}\pi - \frac{4}{3}\pi = \frac{19}{6}\pi$, or about 10 cubic meters.

On Your Own

7. **WHAT IF?** The diameter of the inner sphere is 1.8 meters. What is the volume of the inflated space?

16.1 Exercises

Check It Out
Help with Homework
BigIdeasMath.com

Vocabulary and Concept Check

1. **NUMBER SENSE** Describe the difference between -3^4 and $(-3)^4$.

2. **WHICH ONE DOESN'T BELONG?** Which one does *not* belong with the other three? Explain your reasoning.

5^3 The exponent is 3.	5^3 The power is 5.	5^3 The base is 5.	5^3 Five is used as a factor 3 times.

Practice and Problem Solving

Write the product using exponents.

1 3. $3 \cdot 3 \cdot 3 \cdot 3$

4. $(-6) \cdot (-6)$

5. $\left(-\frac{1}{2}\right) \cdot \left(-\frac{1}{2}\right) \cdot \left(-\frac{1}{2}\right)$

6. $\frac{1}{3} \cdot \frac{1}{3} \cdot \frac{1}{3}$

7. $\pi \cdot \pi \cdot \pi \cdot x \cdot x \cdot x \cdot x$

8. $(-4) \cdot (-4) \cdot (-4) \cdot y \cdot y$

9. $6.4 \cdot 6.4 \cdot 6.4 \cdot 6.4 \cdot b \cdot b \cdot b$

10. $(-t) \cdot (-t) \cdot (-t) \cdot (-t) \cdot (-t)$

Evaluate the expression.

2 11. 5^2

12. -11^3

13. $(-1)^6$

14. $\left(\frac{1}{2}\right)^6$

15. $\left(-\frac{1}{12}\right)^2$

16. $-\left(\frac{1}{9}\right)^3$

17. **ERROR ANALYSIS** Describe and correct the error in evaluating the expression.

✗ $-6^2 = (-6) \cdot (-6) = 36$

18. **PRIME FACTORIZATION** Write the prime factorization of 675 using exponents.

19. **STRUCTURE** Write $-\left(\frac{1}{4} \cdot \frac{1}{4} \cdot \frac{1}{4} \cdot \frac{1}{4}\right)$ using exponents.

20. **RUSSIAN DOLLS** The largest doll is 12 inches tall. The height of each of the other dolls is $\frac{7}{10}$ the height of the next larger doll. Write an expression involving a power for the height of the smallest doll. What is the height of the smallest doll?

Assignment Guide and Homework Check

Level	Assignment	Homework Check
Accelerated	1–5, 6–16 even, 17, 18–26 even, 27–33	8, 16, 18, 26, 28

Common Errors

- **Exercises 3–10** Students may count the wrong number of factors. Remind them to check their work.
- **Exercises 11–16** Students may have the wrong sign in their answers. Remind them that when the negative sign is inside the parentheses, it is part of the base. When the negative sign is outside the parentheses, it is not part of the base.
- **Exercises 21–26** Students may not remember the definition of absolute value or the correct order of operations. Review these topics with students.

16.1 Record and Practice Journal

Write the product using exponents.

1. $4 \cdot 4 \cdot 4 \cdot 4 \cdot 4$
   4^5
2. $\left(-\frac{1}{8}\right) \cdot \left(-\frac{1}{8}\right) \cdot \left(-\frac{1}{8}\right)$
   $\left(-\frac{1}{8}\right)^3$
3. $5 \cdot 5 \cdot (-x) \cdot (-x) \cdot (-x) \cdot (-x)$
   $5^2(-x)^4$
4. $9 \cdot 9 \cdot y \cdot y \cdot y \cdot y \cdot y \cdot y$
   9^2y^6

Evaluate the expression.

5. 10^3
   1000
6. $(-7)^4$
   2401
7. $-\left(\frac{1}{6}\right)^5$
   $-\frac{1}{7776}$
8. $3 + 6 \cdot (-5)^2$
   153
9. $\left|-\frac{1}{3}(1^{10} + 9 - 2^3)\right|$
   $\frac{2}{3}$
10. A foam toy is 2 inches wide. It doubles in size for every minute it is in water. Write an expression for the width of the toy after 5 minutes. What is the width after 5 minutes?
    2^6; 64 inches

Vocabulary and Concept Check

1. -3^4 is the negative of 3^4, so the base is 3, the exponent is 4, and its value is -81. $(-3)^4$ has a base of -3, an exponent of 4, and a value of 81.
2. 5^3, The power is 5; The power is 5^3. Five is the base.

Practice and Problem Solving

3. 3^4
4. $(-6)^2$
5. $\left(-\frac{1}{2}\right)^3$
6. $\left(\frac{1}{3}\right)^3$
7. π^3x^4
8. $(-4)^3y^2$
9. $(6.4)^4b^3$
10. $(-t)^5$
11. 25
12. -1331
13. 1
14. $\frac{1}{64}$
15. $\frac{1}{144}$
16. $-\frac{1}{729}$
17. The negative sign is not part of the base; $-6^2 = -(6 \cdot 6) = -36$.
18. $3^3 \cdot 5^2$
19. $-\left(\frac{1}{4}\right)^4$
20. $12 \cdot \left(\frac{7}{10}\right)^3$; 4.116 in.

Practice and Problem Solving

21. 29

22. 65

23. 5

24. 5

25. 66

26. 2

27. See Additional Answers.

28. **a.** about 99.95 g

b. 99.95%

29. See *Taking Math Deeper*.

Fair Game Review

30. Commutative Property of Multiplication

31. Associative Property of Multiplication

32. Identity Property of Multiplication

33. B

Mini-Assessment

Evaluate the expression.

1. 7^2 49
2. -3^4 -81
3. $(-3)^4$ 81
4. $4 + 6 \cdot (-2)^3$ -44
5. $\frac{3}{4}\left(2^5 - 6 \div \left(\frac{1}{2}\right)^2\right)$ 6

Taking Math Deeper

Exercise 29

This exercise is based on the 12-note chromatic scale used in western music. Asian music uses a 9-note scale. Other cultures use 8-note and 10-note scales.

1 Use a calculator or spreadsheet to calculate the frequency of each note.

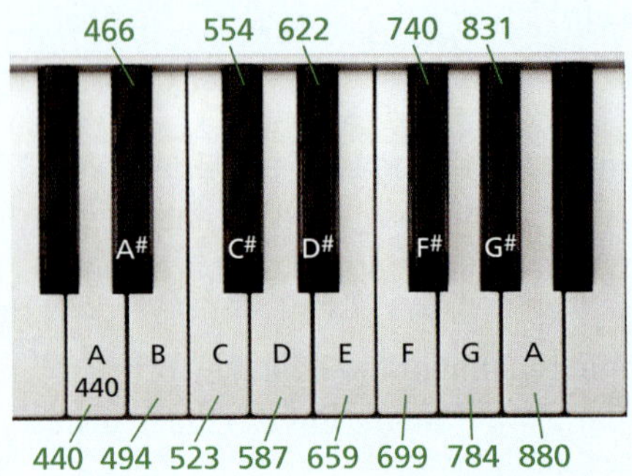

In the formula, the number 1.0595 is the 12th root of 2. A more accurate representation of this number is 1.0594630944.

2 **a.** It takes 12 notes to travel from A-440 to A-880. That is why the western scale is called a 12-note scale. The term *octave,* which is based on the number 8, refers only to the white keys on the piano.

3 **b.** The frequency of the A above A-440 is about 880 vibrations per second. Piano tuners use A-440 as the basic key to tune all the other notes. The A below A-440 has a frequency of 220 vibrations per second.

c. Because $(1.0594630944)^{12} \approx 2$, it follows that the A above A-440 has twice the frequency of A. For a 12-note increase, the frequency approximately doubles.

Note: The scale given by this formula is called the *equal temperament scale*. Notes whose frequencies have many common divisors "harmonize" more.

Project

Use the school library or the Internet to research instruments used in other countries, such as China and Japan. Many have different scales. Compare the scales of the instruments you find to that of the piano.

Reteaching and Enrichment Strategies

If students need help. . .	If students got it. . .
Resources by Chapter • Practice A and Practice B • Puzzle Time Record and Practice Journal Practice Differentiating the Lesson Lesson Tutorials Skills Review Handbook	Resources by Chapter • Enrichment and Extension • Technology Connection Start the next section

Evaluate the expression.

21. $5 + 3 \cdot 2^3$

22. $2 + 7 \cdot (-3)^2$

23. $(13^2 - 12^2) \div 5$

24. $\frac{1}{2}(4^3 - 6 \cdot 3^2)$

25. $\left| \frac{1}{2}(7 + 5^3) \right|$

26. $\left| \left(-\frac{1}{2}\right)^3 \div \left(\frac{1}{4}\right)^2 \right|$

27. MONEY You have a part-time job. One day your boss offers to pay you either $2^h - 1$ or 2^{h-1} dollars for each hour h you work that day. Copy and complete the table. Which option should you choose? Explain.

h	1	2	3	4	5
$2^h - 1$					
2^{h-1}					

28. CARBON-14 DATING Scientists use carbon-14 dating to determine the age of a sample of organic material.

a. The amount C (in grams) of a 100-gram sample of carbon-14 remaining after t years is represented by the equation $C = 100(0.99988)^t$. Use a calculator to find the amount of carbon-14 remaining after 4 years.

b. What percent of the carbon-14 remains after 4 years?

29. **Critical Thinking** The frequency (in vibrations per second) of a note on a piano is represented by the equation $F = 440(1.0595)^n$, where n is the number of notes above A-440. Each black or white key represents one note.

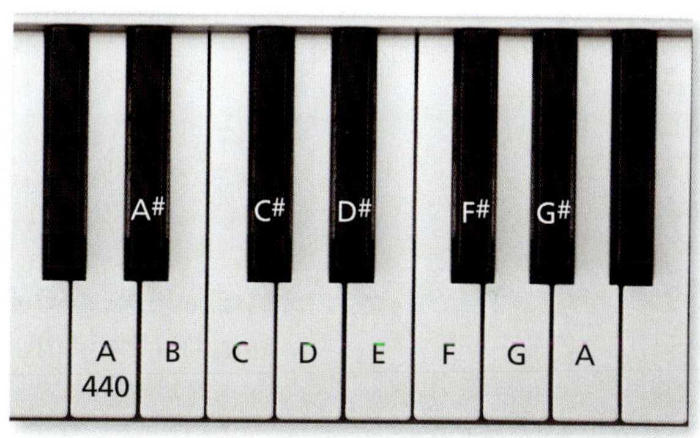

a. How many notes do you take to travel from A-440 to A?

b. What is the frequency of A?

c. Describe the relationship between the number of notes between A-440 and A and the increase in frequency.

Fair Game Review What you learned in previous grades & lessons

Tell which property is illustrated by the statement. *(Skills Review Handbook)*

30. $8 \cdot x = x \cdot 8$

31. $(2 \cdot 10)x = 2(10 \cdot x)$

32. $3(x \cdot 1) = 3x$

33. MULTIPLE CHOICE The polygons are similar. What is the value of x? *(Section 11.5)*

Ⓐ 15 Ⓑ 16

Ⓒ 17 Ⓓ 36

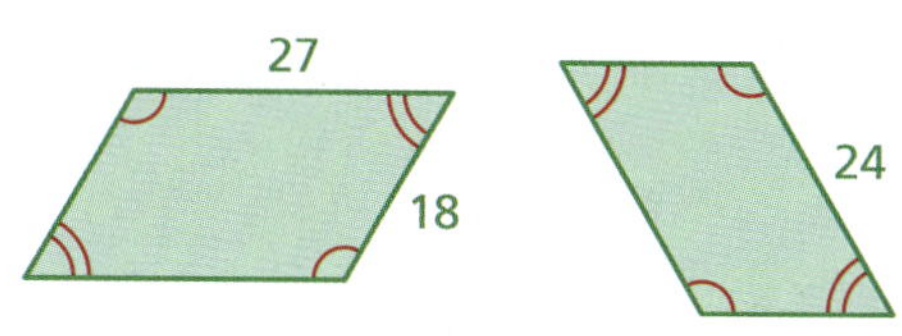

16.2 Product of Powers Property

Essential Question How can you use inductive reasoning to observe patterns and write general rules involving properties of exponents?

1 ACTIVITY: Finding Products of Powers

Work with a partner.

a. Copy and complete the table.

Product	Repeated Multiplication Form	Power
$2^2 \cdot 2^4$		
$(-3)^2 \cdot (-3)^4$		
$7^3 \cdot 7^2$		
$5.1^1 \cdot 5.1^6$		
$(-4)^2 \cdot (-4)^2$		
$10^3 \cdot 10^5$		
$\left(\frac{1}{2}\right)^5 \cdot \left(\frac{1}{2}\right)^5$		

b. **INDUCTIVE REASONING** Describe the pattern in the table. Then write a *general rule* for multiplying two powers that have the same base.

$$a^m \cdot a^n = a^{\square}$$

c. Use your rule to simplify the products in the first column of the table above. Does your rule give the results in the third column?

d. Most calculators have *exponent* keys that you can use to evaluate powers. Use a calculator with an exponent key to evaluate the products in part (a).

Exponents

In this lesson, you will

- multiply powers with the same base.
- find a power of a power.
- find a power of a product.

2 ACTIVITY: Writing a Rule for Powers of Powers

Work with a partner. Write the expression as a single power. Then write a *general rule* for finding a power of a power.

a. $(3^2)^3 = (3 \cdot 3)(3 \cdot 3)(3 \cdot 3) = \square^{\square}$

b. $(2^2)^4 =$ ____ **c.** $(7^3)^2 =$ ____

d. $(y^3)^3 =$ ____ **e.** $(x^4)^2 =$ ____

Laurie's Notes

Introduction

Applying Mathematical Practices

- **Look for and Express Regularity in Repeated Reasoning:** In the first three activities, students will find products of powers, powers of powers, and powers of products, and observe patterns that emerge.

Motivate

- **Story Time:** Tell students that the superintendent has agreed to put you on a special salary schedule for one month. On day 1 you will receive 1¢, on day 2 you will receive 2¢, day 3 is 4¢, and so on, with your salary doubling every school day for the month. There are 23 school days this month. Should you take the new salary?
- Give time for students to start the tabulation. Let them use a calculator for speed. The table below shows the daily pay.

1	$2 = 2^1$	$4 = 2^2$	$8 = 2^3$
$16 = 2^4$	$32 = 2^5$	$64 = 2^6$	$128 = 2^7$
$256 = 2^8$	$512 = 2^9$	$1024 = 2^{10}$	$2048 = 2^{11}$
$4096 = 2^{12}$	$8192 = 2^{13}$	$16,384 = 2^{14}$	$32,768 = 2^{15}$
$65,536 = 2^{16}$	$131,072 = 2^{17}$	$262,144 = 2^{18}$	$524,288 = 2^{19}$
$1,048,576 = 2^{20}$	$2,097,152 = 2^{21}$	$4,194,304 = 2^{22}$	

- If the superintendent is looking for additional math teachers, they will be lined up at the door.
- In this penny doubling problem, each day you are paid a power of 2. Your salary is actually the *sum* of all of these amounts.

Activity Notes

Activity 1

- Have students work with their partners to complete the table. It may help students if you point out that in the middle column, *repeated multiplication form* means the expanded form of each power in the product.
- ? "Why is the first column labeled **Product**?" Two powers are being multiplied.
- ? "What do you notice about the number of factors in the middle column and the exponent used to write the power?" same number
- **Part (b):** Students will recognize that the exponents are added together, but it may not be obvious to them how to write this fact using variables.
- **Use Appropriate Tools Strategically: Part (d):** Have students evaluate the products in the first column and the powers in the last column to confirm their answers and to confirm that they are using their calculators correctly.
- **Big Idea:** Write the summary statement: $a^m \cdot a^n = a^{m+n}$. Stress that the bases must be the same. That is why a is the base for both powers. This rule says nothing about how to simplify a product such as $3^3 \cdot 4^2$.

What Your Students Will Learn

- Multiply powers with the same base using the Product of Powers Property.
- Find powers of powers using the Power of a Power Property.
- Find powers of products using the Power of a Product Property.

Previous Learning

Students should know how to raise a number to an exponent.

Technology for the Teacher

Dynamic Classroom

Lesson Plans
Complete Materials List

16.2 Record and Practice Journal

Essential Question How can you use inductive reasoning to observe patterns and write general rules involving properties of exponents?

1 ACTIVITY: Finding Products of Powers

Work with a partner.

a. Complete the table.

Product	Repeated Multiplication Form	Power
$2^2 \cdot 2^4$	$2 \cdot 2 \cdot 2 \cdot 2 \cdot 2 \cdot 2$	2^6
$(-3)^2 \cdot (-3)^4$	$(-3) \cdot (-3) \cdot (-3) \cdot (-3) \cdot (-3) \cdot (-3)$	$(-3)^6$
$7^3 \cdot 7^2$	$7 \cdot 7 \cdot 7 \cdot 7 \cdot 7$	7^5
$5.1^1 \cdot 5.1^6$	$(5.1) \cdot (5.1) \cdot (5.1) \cdot (5.1) \cdot (5.1) \cdot (5.1) \cdot (5.1)$	$(5.1)^7$
$(-4)^2 \cdot (-4)^2$	$(-4) \cdot (-4) \cdot (-4) \cdot (-4)$	$(-4)^4$
$10^3 \cdot 10^5$	$10 \cdot 10 \cdot 10 \cdot 10 \cdot 10 \cdot 10 \cdot 10 \cdot 10$	10^8
$\left(\frac{1}{2}\right)^5 \cdot \left(\frac{1}{2}\right)^5$	$\frac{1}{2} \cdot \frac{1}{2} \cdot \frac{1}{2} \cdot \frac{1}{2} \cdot \frac{1}{2} \cdot \frac{1}{2} \cdot \frac{1}{2} \cdot \frac{1}{2} \cdot \frac{1}{2} \cdot \frac{1}{2}$	$\left(\frac{1}{2}\right)^{10}$

b. **INDUCTIVE REASONING** Describe the pattern in the table. Then write a *general rule* for multiplying two powers that have the same base.

Number of factors = sum of exponents

$a^m \cdot a^n = a^{m+n}$

c. Use your rule to simplify the products in the first column of the table above. Does your rule give the results in the third column?
yes

d. Most calculators have *exponent* keys that are used to evaluate powers. Use a calculator with an exponent key to evaluate the products in part (a).
Check students' work.

English Language Learners

Pair Activity

Have students work in pairs to simplify exponential expressions. Each student simplifies a different expression. When both students are done, they take turns explaining the solution while the other person follows along.

16.2 Record and Practice Journal

2 ACTIVITY: Writing a Rule for Powers of Powers

Work with a partner. Write the expression as a single power. Then write a *general rule* for finding a power of a power.

a. $(3^2)^3 = 3^6$

b. $(2^2)^4 = 2^8$

c. $(7^3)^2 = 7^6$

d. $(y^3)^3 = y^9$

e. $(x^4)^2 = x^8$

$$(a^m)^n = a^{mn}$$

3 ACTIVITY: Writing a Rule for Powers of Products

Work with a partner. Write the expression as the product of two powers. Then write a *general rule* for finding a power of a product.

a. $(2 \cdot 3)^3 = (2^3)(3^3)$

b. $(2 \cdot 5)^2 = (2^2)(5^2)$

c. $(5 \cdot 4)^3 = (5^3)(4^3)$

d. $(6a)^4 = (6^4)(a^4)$

e. $(3x)^2 = (3^2)(x^2)$

$$(ab)^m = a^m b^m$$

4 ACTIVITY: The Penny Puzzle

Work with a partner.

- The rows y and columns x of a chess board are numbered as shown.
- Each position on the chess board has a stack of pennies. (Only the first row is shown.)
- The number of pennies in each stack is $2^x \cdot 2^y$.

a. How many pennies are in the stack in location (3, 5)?
256

b. Which locations have 32 pennies in their stacks?
(1, 4), (2, 3), (3, 2), (4, 1)

c. How much money (in dollars) is in the location with the tallest stack?
$655.36

d. A penny is about 0.06 inch thick. About how tall (in inches) is the tallest stack?
3932.16 in.

What Is Your Answer?

5. **IN YOUR OWN WORDS** How can you use inductive reasoning to observe patterns and write general rules involving properties of exponents?
Check students' work.

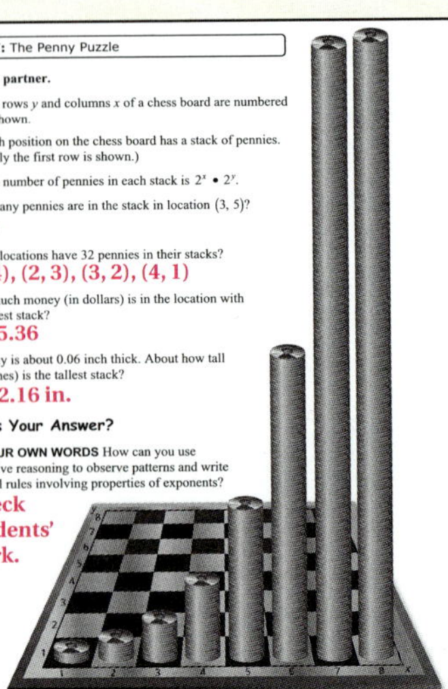

Laurie's Notes

Activity 2

- Students often confuse this property with the Product of Powers Property (Activity 1). Make sure students can explain how $4^3 \cdot 4^2$ is different from $(4^3)^2$. Expanding the expressions helps to demonstrate this.
- When students have difficulty recognizing a pattern, have them expand the power in stages: $(2^2)^4 = 2^2 \cdot 2^2 \cdot 2^2 \cdot 2^2 = (2 \cdot 2)(2 \cdot 2)(2 \cdot 2)(2 \cdot 2) = 2^8$.
- **Attend to Precision:** Expect correct vocabulary, such as "When you raise a power to an exponent, multiply the exponents and keep the same base."

Activity 3

- "Products of powers" and "powers of products" sound very similar to students. Ask a volunteer to describe the difference.
- In an expression such as $(6a)^4$, students are often unsure how to simplify $6a \cdot 6a \cdot 6a \cdot 6a$. The Commutative and Associative Properties allow you to rewrite this as $6 \cdot 6 \cdot 6 \cdot 6 \cdot a \cdot a \cdot a \cdot a$.
- **Attend to Precision:** Expect correct vocabulary, such as "when you raise a product to an exponent, raise each base to the exponent and multiply the two powers."

Activity 4

- Take time to discuss the notation. In position (1, 1), the number of pennies is $2^1 \cdot 2^1 = 4$. In position (2, 1), the number of pennies is $2^2 \cdot 2^1 = 8$. Answer any questions about notation or how to find the number of pennies on any square.
- There are many patterns and interesting extensions to this problem that may surface as they explore the questions presented.
- **Part (a):** There are $2^3 \cdot 2^5 = 2^8 = 256$ pennies in location (3, 5).
- **Part (b):** Because $32 = 2^5$, the exponents need to sum to 5. The locations include (1, 4), (4, 1), (2, 3), and (3, 2).
- **Part (c):** The most money will be in the location where x and y have the greatest sum. This will occur at (8, 8), where the value is $2^8 \cdot 2^8 = 2^{16} = 65{,}536 = \$655.36$.
- **Part (d):** Multiply the number of pennies by the thickness, $65{,}536 \times 0.06 = 3932.16$ inches.

What Is Your Answer?

- **Neighbor Check:** Have students work independently and then have their neighbors check their work. Have students discuss any discrepancies.

Closure

- Refer back to the penny doubling problem from the beginning of the lesson. On what days was your salary more than $1000? days 18–23

3 ACTIVITY: Writing a Rule for Powers of Products

Work with a partner. Write the expression as the product of two powers. Then write a *general rule* for finding a power of a product.

a. $(2 \cdot 3)^3 = (2 \cdot 3)(2 \cdot 3)(2 \cdot 3) = \square^{\square} \cdot \square^{\square}$

b. $(2 \cdot 5)^2 = \square$

c. $(5 \cdot 4)^3 = \square$

d. $(6a)^4 = \square$

e. $(3x)^2 = \square$

4 ACTIVITY: The Penny Puzzle

Work with a partner.

- The rows y and columns x of a chessboard are numbered as shown.
- Each position on the chessboard has a stack of pennies. (Only the first row is shown.)
- The number of pennies in each stack is $2^x \cdot 2^y$.

a. How many pennies are in the stack in location (3, 5)?

b. Which locations have 32 pennies in their stacks?

c. How much money (in dollars) is in the location with the tallest stack?

d. A penny is about 0.06 inch thick. About how tall (in inches) is the tallest stack?

Math Practice

Look for Patterns

What patterns do you notice? How does this help you determine which stack is the tallest?

What Is Your Answer?

5. **IN YOUR OWN WORDS** How can you use inductive reasoning to observe patterns and write general rules involving properties of exponents?

Practice Use what you learned about properties of exponents to complete Exercises 3–5 on page 720.

16.2 Lesson

Check It Out
Lesson Tutorials
BigIdeasMath.com

Key Ideas

Product of Powers Property

Words To multiply powers with the same base, add their exponents.

Numbers $4^2 \cdot 4^3 = 4^{2+3} = 4^5$ **Algebra** $a^m \cdot a^n = a^{m+n}$

Power of a Power Property

Words To find a power of a power, multiply the exponents.

Numbers $(4^6)^3 = 4^{6 \cdot 3} = 4^{18}$ **Algebra** $(a^m)^n = a^{mn}$

Power of a Product Property

Words To find a power of a product, find the power of each factor and multiply.

Numbers $(3 \cdot 2)^5 = 3^5 \cdot 2^5$ **Algebra** $(ab)^m = a^m b^m$

EXAMPLE 1 Multiplying Powers with the Same Base

a. $2^4 \cdot 2^5 = 2^{4+5}$ Product of Powers Property

$= 2^9$ Simplify.

Study Tip

When a number is written without an exponent, its exponent is 1.

b. $-5 \cdot (-5)^6 = (-5)^1 \cdot (-5)^6$ Rewrite -5 as $(-5)^1$.

$= (-5)^{1+6}$ Product of Powers Property

$= (-5)^7$ Simplify.

c. $x^3 \cdot x^7 = x^{3+7}$ Product of Powers Property

$= x^{10}$ Simplify.

EXAMPLE 2 Finding a Power of a Power

a. $(3^4)^3 = 3^{4 \cdot 3}$ Power of a Power Property

$= 3^{12}$ Simplify.

b. $(w^5)^4 = w^{5 \cdot 4}$ Power of a Power Property

$= w^{20}$ Simplify.

Laurie's Notes

Introduction

Connect

- **Yesterday:** Students explored exponents.
- **Today:** Students will use the Product of Powers Property, the Power of a Power Property, and the Power of a Product Property to simplify expressions.

Motivate

- More money talk! The $10,000 bill, which is no longer in circulation, would be much easier to carry than the same amount in pennies.

? Ask a few questions about money.

- "How many pennies equal $10,000?" $100 \times 10{,}000 = 1{,}000{,}000$ or 10^6
- "How many dimes equal $10,000?" $10 \times 10{,}000 = 100{,}000$ or 10^5
- "How many $10 bills equal $10,000?" $\frac{1}{10} \times 10{,}000 = 1000$ or 10^3
- "How many $100 bills equal $10,000?" $\frac{1}{100} \times 10{,}000 = 100$ or 10^2

Lesson Notes

Key Ideas

- Write the Key Ideas. Explain the Words, Numbers, and Algebra.
- Explain that this and the following section mirror each other. However, this section contains more properties because the additional ones follow directly from the Product of Powers Property. Essentially, the properties in this section are the "multiplication properties" of exponents and the Quotient of Powers Property (in the next section) is the "division property" of exponents.

Example 1

- **Part (a)**: Write and simplify the expression. The base is 2 for each power, so add the exponents.

? "In part (b), what is the base for each power? To what exponent is each base raised?" -5; 1 and 6

- **Common Error:** When the exponent is 1, it is not written. When it is not written, students will sometimes forget to add 1 in their answer.
- **Part (c):** The properties apply to variables as well as numbers.

Example 2

? "In the expression $(3^4)^3$, what does the exponent of 3 tell you to do?" Use 3^4 as a factor three times.

- Use the Power of a Power Property and multiply the exponents.

Goal Today's lesson is using the Product of Powers Property, the Power of a Power Property, and the Power of a Product Property to simplify expressions.

Technology for the Teacher

Dynamic Classroom

Lesson Tutorials
Lesson Plans
Answer Presentation Tool

Extra Example 1

Simplify each expression. Write your answer as a power.

a. $6^2 \cdot 6^7$ 6^9

b. $-2 \cdot (-2)^3$ $(-2)^4$

c. $x^2 \cdot x^5$ x^7

Extra Example 2

Simplify each expression. Write your answer as a power.

a. $(5^2)^3$ 5^6

b. $(y^4)^6$ y^{24}

Laurie's Notes

Extra Example 3

Simplify each expression.

a. $(4x)^2$ $16x^2$

b. $(wz)^3$ w^3z^3

On Your Own

1. 6^6 **2.** $\left(-\frac{1}{2}\right)^9$

3. z^{13} **4.** 4^{12}

5. y^8 **6.** $(-4)^6$

7. $625y^4$ **8.** a^5b^5

9. $0.25m^2n^2$

Extra Example 4

In Example 4, the total storage space of a computer is 32 gigabytes. How many bytes of total storage space does the computer have? 2^{35} bytes

On Your Own

10. 2^{34} bytes

Differentiated Instruction

Visual

Remind students that the Product of Powers Property can only be applied to powers having the same base. Have students highlight each unique base with a different color. Then add the exponents.

$$2^4 \cdot 2^5 - 3^2 \cdot 3^2 = 2^{4+5} - 3^{2+2}$$
$$= 2^9 - 3^4$$
$$= 512 - 81$$
$$= 431$$

Example 3

- Be careful and deliberate with language when simplifying these expressions.
- ? "In part (a), what does the exponent of 3 tell you to do in the expression $(2x)^3$?" Use $2x$ as a factor three times.
- **Look for and Make Use of Structure:** To verify the solution of part (a), you could write the factor $2x$ three times. Properties of Multiplication (Associative and Commutative) allow you to reorder the terms. You can identify six factors: three 2's and three x's. Use exponents to write the factors. Finally, 2^3 is rewritten as 8 and the final answer is $8x^3$.
- **Part (b):** Follow the same procedure to verify this solution, by writing $3xy$ as a factor twice. It is very common for students to write $x \cdot x = 2x$. Do not assume that students will see this error.

On Your Own

- Encourage students to write out the steps in their solutions.
- **Common Error:** In Question 9, $(0.5)^2 \neq 1$; $(0.5)^2 = 0.25$.
- Have volunteers write their solutions on the board.

Example 4

- Writing the verbal model is necessary in this problem because the terms gigabytes and bytes may not be familiar to all. The first sentence is a conversion fact: 1 GB $= 2^{30}$ bytes. There are 64 GB of total storage. Students may naturally think 64×2^{30} to solve the problem.
- Rewrite 64 as a power with a base of 2, $64 = 2^6$, and solve the problem.

On Your Own

- Students may respond with $\frac{1}{4}$ the total storage space. In fact, $\frac{1}{4}$ of $2^{36} = 2^{34}$. However, this is not an obvious step. Students should model the problem after Example 4.

Closure

- **Exit Ticket:** Simplify. $5^3 \cdot 5^4$ 5^7 $(-3x)^3$ $-27x^3$

EXAMPLE 3 Finding a Power of a Product

a. $(2x)^3 = 2^3 \cdot x^3$ — Power of a Product Property

$= 8x^3$ — Simplify.

b. $(3xy)^2 = 3^2 \cdot x^2 \cdot y^2$ — Power of a Product Property

$= 9x^2y^2$ — Simplify.

On Your Own

Now You're Ready Exercises 3–14 and 17–22

Simplify the expression.

1. $6^2 \cdot 6^4$	**2.** $\left(-\frac{1}{2}\right)^3 \cdot \left(-\frac{1}{2}\right)^6$	**3.** $z \cdot z^{12}$
4. $(4^4)^3$	**5.** $\left(y^2\right)^4$	**6.** $\left((-4)^3\right)^2$
7. $(5y)^4$	**8.** $(ab)^5$	**9.** $(0.5mn)^2$

EXAMPLE 4 Simplifying an Expression

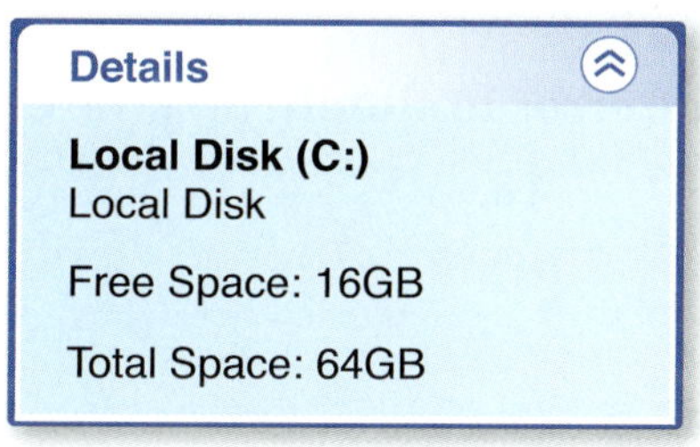

A gigabyte (GB) of computer storage space is 2^{30} bytes. The details of a computer are shown. How many bytes of total storage space does the computer have?

Ⓐ 2^{34} Ⓑ 2^{36} Ⓒ 2^{180} Ⓓ 128^{30}

The computer has 64 gigabytes of total storage space. Notice that you can write 64 as a power, 2^6. Use a model to solve the problem.

Total number of bytes = Number of bytes in a gigabyte • Number of gigabytes

$= 2^{30} \cdot 2^6$ — Substitute.

$= 2^{30+6}$ — Product of Powers Property

$= 2^{36}$ — Simplify.

The computer has 2^{36} bytes of total storage space. The correct answer is Ⓑ.

On Your Own

10. How many bytes of free storage space does the computer have?

16.2 Exercises

Check It Out
Help with Homework
BigIdeasMath.com

Vocabulary and Concept Check

1. **REASONING** When should you use the Product of Powers Property?

2. **CRITICAL THINKING** Can you use the Product of Powers Property to multiply $5^2 \cdot 6^4$? Explain.

Practice and Problem Solving

Simplify the expression. Write your answer as a power.

1 2 **3.** $3^2 \cdot 3^2$ **4.** $8^{10} \cdot 8^4$ **5.** $(-4)^5 \cdot (-4)^7$

6. $a^3 \cdot a^3$ **7.** $h^6 \cdot h$ **8.** $\left(\frac{2}{3}\right)^2 \cdot \left(\frac{2}{3}\right)^6$

9. $\left(-\frac{5}{7}\right)^8 \cdot \left(-\frac{5}{7}\right)^9$ **10.** $(-2.9) \cdot (-2.9)^7$ **11.** $(5^4)^3$

12. $(b^{12})^3$ **13.** $(3.8^3)^4$ **14.** $\left(\left(-\frac{3}{4}\right)^5\right)^2$

ERROR ANALYSIS **Describe and correct the error in simplifying the expression.**

15. ✗ $5^2 \cdot 5^9 = (5 \cdot 5)^{2+9} = 25^{11}$

16. ✗ $(r^6)^4 = r^{6+4} = r^{10}$

Simplify the expression.

3 **17.** $(6g)^3$ **18.** $(-3v)^5$ **19.** $\left(\frac{1}{5}k\right)^2$

20. $(1.2m)^4$ **21.** $(rt)^{12}$ **22.** $\left(-\frac{3}{4}p\right)^3$

23. **PRECISION** Is $3^2 + 3^3$ equal to 3^5? Explain.

24. **ARTIFACT** A display case for the artifact is in the shape of a cube. Each side of the display case is three times longer than the width of the artifact.

a. Write an expression for the volume of the case. Write your answer as a power.

b. Simplify the expression.

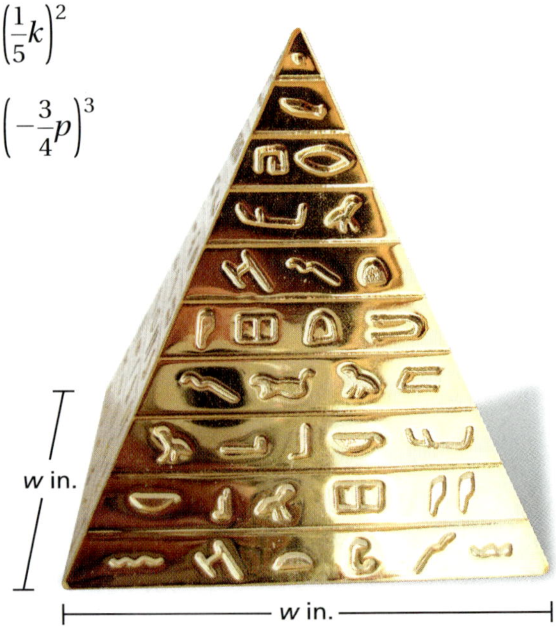

Assignment Guide and Homework Check

Level	Assignment	Homework Check
Accelerated	1–5, 6–24 even, 25–37	14, 22, 26, 28, 30

Common Errors

- **Exercises 3–10** Students may multiply the bases. Remind them that the base stays the same and only the exponent changes when using the Product of Powers Property.
- **Exercises 3–14** Students may confuse the Product of Powers Property and the Power of a Power Property. Explain to them why it makes sense to add exponents when using the Product of Powers Property and multiply exponents when using the Power of a Power Property.
- **Exercises 17–22** Students may forget to find the power of the constant factor and write, for example, $(6g)^3 = 6g^3$. Remind them of the Power of a Product Property.

16.2 Record and Practice Journal

Simplify the expression. Write your answer as a power.

1. $(-6)^5 \bullet (-6)^4$ — $(-6)^9$
2. $x^1 \bullet x^9$ — x^{10}
3. $\left(\frac{4}{5}\right)^3 \bullet \left(\frac{4}{5}\right)^{12}$ — $\left(\frac{4}{5}\right)^{15}$
4. $(-1.5)^{11} \bullet (-1.5)^{11}$ — $(-1.5)^{22}$
5. $(y^{10})^{20}$ — y^{200}
6. $\left(\left(-\frac{2}{9}\right)^8\right)^7$ — $\left(-\frac{2}{9}\right)^{56}$

Simplify the expression.

7. $(2a)^6$ — $64a^6$
8. $(-4b)^4$ — $256b^4$
9. $\left(-\frac{9}{10}p\right)^2$ — $\frac{81}{100}p^2$
10. $(xy)^{15}$ — $x^{15}y^{15}$
11. $10^5 \bullet 10^3 - (10^1)^8$ — 0
12. $7^2(7^4 \bullet 7^4)$ — 282,475,249
13. The surface area of the Sun is about $4 \times 3.141 \times (7 \times 10^5)^2$ square kilometers. Simplify the expression.
    6,156,360,000,000 square kilometers

Vocabulary and Concept Check

1. when multiplying powers with the same base
2. no; The bases are not the same.

Practice and Problem Solving

3. 3^4
4. 8^{14}
5. $(-4)^{12}$
6. a^6
7. h^7
8. $\left(\frac{2}{3}\right)^8$
9. $\left(-\frac{5}{7}\right)^{17}$
10. $(-2.9)^8$
11. 5^{12}
12. b^{36}
13. 3.8^{12}
14. $\left(-\frac{3}{4}\right)^{10}$
15. The bases should not be multiplied.
    $5^2 \bullet 5^9 = 5^{2+9}$
    $= 5^{11}$
16. The exponents should not be added. Write the expression as repeated multiplication.
    $(r^6)^4 = r^6 \bullet r^6 \bullet r^6 \bullet r^6$
    $= r^{6+6+6+6}$
    $= r^{24}$
17. $216g^3$
18. $-243v^5$
19. $\frac{1}{25}k^2$
20. $2.0736m^4$
21. $r^{12}t^{12}$
22. $-\frac{27}{64}p^3$
23. no; $3^2 + 3^3 = 9 + 27 = 36$ and $3^5 = 243$
24. a. $(3w)^3$
    b. $27w^3$

Practice and Problem Solving

25. 496

26. x^4

27. 78,125

28. 3^9 ft

29. **a.** $16\pi \approx 50.27$ in.^{3}

b. $192\pi \approx 603.19$ in.^{3} Squaring each of the dimensions causes the volume to be 12 times larger.

30. $V = \frac{3}{4}b^2h$

31. See *Taking Math Deeper.*

32. **a.** 3

b. 4

Fair Game Review

33. 4 **34.** 25

35. 3 **36.** 6

37. B

Mini-Assessment

Simplify the expression. Write your answer as a power.

1. $b^2 \cdot b^6$ b^8

2. $(-2)^3 \cdot (-2)^2$ $(-2)^5$

3. $(c^8)^3$ c^{24}

Simplify the expression.

4. $(-5w)^4$ $625w^4$

5. $(st)^{11}$ $s^{11}t^{11}$

Taking Math Deeper

Exercise 31

This exercise gives students some practice in representing large numbers as powers.

1. Summarize the given information.

Mail delivered each second: $2^8 \cdot 5^2 = 6400$
Seconds in 6 days: $2^8 \cdot 3^4 \cdot 5^2 = 518{,}400$
How many pieces of mail in 6 days?

2. Multiply to find the number of pieces of mail delivered in 6 days.

$$(2^8 \cdot 5^2)(2^8 \cdot 3^4 \cdot 5^2) = 2^8 \cdot 5^2 \cdot 2^8 \cdot 3^4 \cdot 5^2$$
$$= 2^{16} \cdot 3^4 \cdot 5^4$$

3. Write the number in normal decimal form.

If you expand this number, you find that the U.S. postal service delivers about 3 billion pieces of mail each week (6 days not counting Sunday). This is an average of 10 pieces of mail per week for each person in the United States!

Project

Research the price of a postage stamp. How many times has it changed? How often has it changed? What has been the range in the cost over the last one hundred years?

Reteaching and Enrichment Strategies

If students need help. . .	If students got it. . .
Resources by Chapter • Practice A and Practice B • Puzzle Time Record and Practice Journal Practice Differentiating the Lesson Lesson Tutorials Skills Review Handbook	Resources by Chapter • Enrichment and Extension • Technology Connection Start the next section

Simplify the expression.

25. $2^4 \cdot 2^5 - (2^2)^2$

26. $16\left(\frac{1}{2}x\right)^4$

27. $5^2(5^3 \cdot 5^2)$

28. **CLOUDS** The lowest altitude of an altocumulus cloud is about 3^8 feet. The highest altitude of an altocumulus cloud is about 3 times the lowest altitude. What is the highest altitude of an altocumulus cloud? Write your answer as a power.

29. **PYTHON EGG** The volume V of a python egg is given by the formula $V = \frac{4}{3}\pi abc$. For the python eggs shown, $a = 2$ inches, $b = 2$ inches, and $c = 3$ inches.

a. Find the volume of a python egg.

b. Square the dimensions of the python egg. Then evaluate the formula. How does this volume compare to your answer in part (a)?

30. **PYRAMID** A square pyramid has a height h and a base with side length b. The side lengths of the base increase by 50%. Write a formula for the volume of the new pyramid in terms of b and h.

31. **MAIL** The United States Postal Service delivers about $2^8 \cdot 5^2$ pieces of mail each second. There are $2^8 \cdot 3^4 \cdot 5^2$ seconds in 6 days. How many pieces of mail does the United States Postal Service deliver in 6 days? Write your answer as an expression involving powers.

32. Critical Thinking Find the value of x in the equation without evaluating the power.

a. $2^5 \cdot 2^x = 256$

b. $\left(\frac{1}{3}\right)^2 \cdot \left(\frac{1}{3}\right)^x = \frac{1}{729}$

Fair Game Review What you learned in previous grades & lessons

Simplify. *(Skills Review Handbook)*

33. $\frac{4 \cdot 4}{4}$

34. $\frac{5 \cdot 5 \cdot 5}{5}$

35. $\frac{2 \cdot 3}{2}$

36. $\frac{8 \cdot 6 \cdot 6}{6 \cdot 8}$

37. **MULTIPLE CHOICE** What is the measure of each interior angle of the regular polygon? *(Section 12.3)*

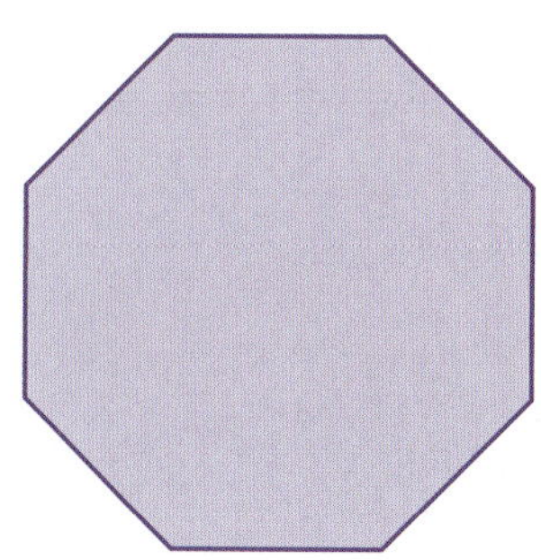

Ⓐ 45°   Ⓑ 135°

Ⓒ 1080°   Ⓓ 1440°

16.3 Quotient of Powers Property

Essential Question How can you divide two powers that have the same base?

1 ACTIVITY: Finding Quotients of Powers

Work with a partner.

a. Copy and complete the table.

Quotient	Repeated Multiplication Form	Power
$\frac{2^4}{2^2}$		
$\frac{(-4)^5}{(-4)^2}$		
$\frac{7^7}{7^3}$		
$\frac{8.5^9}{8.5^6}$		
$\frac{10^8}{10^5}$		
$\frac{3^{12}}{3^4}$		
$\frac{(-5)^7}{(-5)^5}$		
$\frac{11^4}{11^1}$		

Exponents

In this lesson, you will

- divide powers with the same base.
- simplify expressions involving the quotient of powers.

b. **INDUCTIVE REASONING** Describe the pattern in the table. Then write a rule for dividing two powers that have the same base.

$$\frac{a^m}{a^n} = a^{\square}$$

c. Use your rule to simplify the quotients in the first column of the table above. Does your rule give the results in the third column?

Laurie's Notes

Introduction

Applying Mathematical Practices

- **Look for and Express Regularity in Repeated Reasoning:** In Activity 1, students will find quotients of powers and observe the pattern that emerges.
- Remember to use correct vocabulary in this lesson. The numbers are not *canceling*. The factors that are common in the numerator and the denominator are being divided out, similar to simplifying fractions. The fraction $\frac{2}{4} = \frac{1}{2}$ because there is a common factor of 2 in both the numerator and denominator that divide out. This same concept of dividing out common factors is why the Quotient of Powers Property works.

Motivate

- Tell students that you spent last evening working on a very long problem and you want them to give it a try. Write the problem on the board.

$$\frac{1}{2} \cdot \frac{2}{3} \cdot \frac{3}{4} \cdot \frac{4}{5} \cdot \frac{5}{6} \cdot \frac{6}{7} \cdot \frac{7}{8} \cdot \frac{8}{9} \cdot \frac{9}{10}$$

- It is likely that at least one of your students will recognize the answer immediately after you finish writing the problem. Act surprised and ask for their strategy...because you spent a long time on the problem.
- You want all students to recognize that the common factors in the numerator divide out with common factors in the denominator, leaving only $\frac{1}{10}$ as the final answer.

Activity Notes

Activity 1

- Have students work with their partners to complete the table. It may help students if you point out that in the middle column, *repeated multiplication form* means the expanded form of each power in the quotient.
- Notice that integers and decimals are used as bases.
- ? "Why is the first column labeled **Quotient**?" Two powers are being divided.
- ? "What do you notice about the number of factors in the numerator and denominator of the middle column, and the exponent used to write the power?" When you subtract the number of factors in the denominator from the number of factors in the numerator, it equals the exponent in the power.
- Students may need help in writing the summary statement: $\frac{a^m}{a^n} = a^{m-n}$. Stress that the bases must be the same in order to use this property.

What Your Students Will Learn

- Divide powers with the same base using the Quotient of Powers Property.
- Simplify expressions involving the quotient of powers.

Previous Learning

Students should know how to simplify fractions by dividing out common factors.

Technology for the Teacher

Dynamic Classroom

Lesson Plans
Complete Materials List

16.3 Record and Practice Journal

Essential Question How can you divide two powers that have the same base?

1 ACTIVITY: Finding Quotients of Powers

Work with a partner.

a. Complete the table.

Quotient	Repeated Multiplication Form	Power
$\frac{2^4}{2^2}$	Check students' work.	2^2
$\frac{(-4)^5}{(-4)^2}$		$(-4)^3$
$\frac{7^7}{7^3}$		7^4
$\frac{8.5^9}{8.5^6}$		8.5^3
$\frac{10^8}{10^5}$		10^3
$\frac{3^{12}}{3^4}$		3^8
$\frac{(-5)^7}{(-5)^5}$		$(-5)^2$
$\frac{11^4}{11^1}$		11^3

b. INDUCTIVE REASONING Describe the pattern in the table. Then write a rule for dividing two powers that have the same base.

$\frac{a^m}{a^n} = a^{m-n}$ the difference of the exponents (numerator to denominator) in the quotient column

Differentiated Instruction

Kinesthetic

Use algebra tiles or slips of paper to help students understand the Quotient of Powers Property. Have students model the quotient $\frac{x^4}{x^2}$.

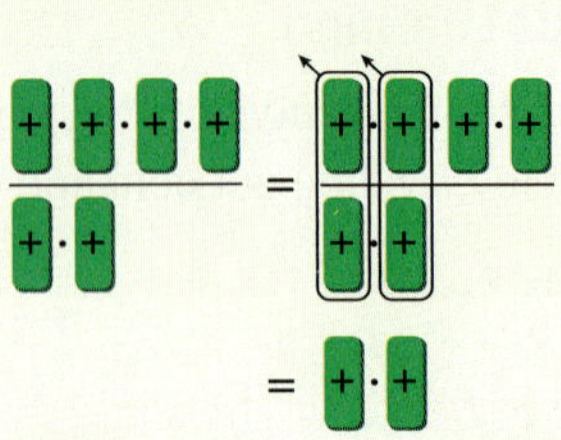

So, $\frac{x^4}{x^2} = x^2$.

16.3 Record and Practice Journal

c. Use your rule to simplify the quotients in the first column of the table on the previous page. Does your rule give the results in the third column?

yes

2 ACTIVITY: Comparing Volumes

Work with a partner.

How many of the smaller cubes will fit inside the larger cube? Record your results in the table on the next page. Describe the pattern in the table.

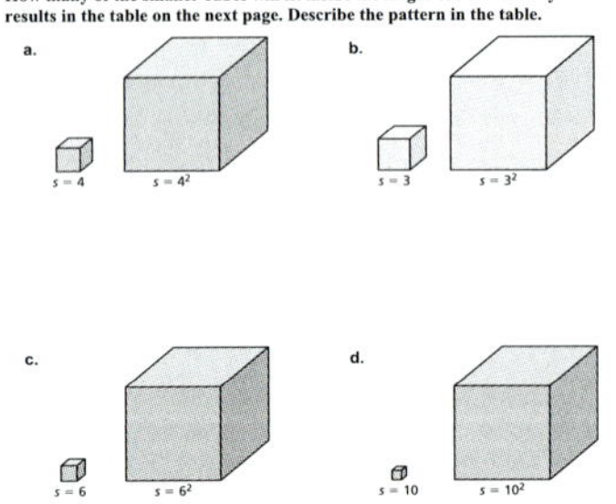

	Volume of Smaller Cube	Volume of Larger Cube	$\frac{\text{Larger Volume}}{\text{Smaller Volume}}$	Answer
a.	4^3	$(4^2)^3 = 4^6$	$\frac{4^6}{4^3}$	4^3
b.	3^3	$(3^2)^3 = 3^6$	$\frac{3^6}{3^3}$	3^3
c.	6^3	$(6^2)^3 = 6^6$	$\frac{6^6}{6^3}$	6^3
d.	10^3	$(10^2)^3 = 10^6$	$\frac{10^6}{10^3}$	10^3

What Is Your Answer?

3. **IN YOUR OWN WORDS** How can you divide two powers that have the same base? Give two examples of your rule.

Subtract their exponents.

Laurie's Notes

Activity 2

- **Model with Mathematics:** If you have small wooden or plastic cubes available, model one of these problems or a similar problem to start.
- Point out to students that $s = 4$ means the edge (or side) length is 4. In part (a), the side length of the larger cube is 4^2 or 16, which is 4 times as long as the smaller cube.
- When completing the table, it is necessary for students to simplify a power raised to an exponent. Recall, $(4^2)^3 = 4^{2 \cdot 3} = 4^6$.
- Students will work with their partners to complete the table.

? "How do you find the volume of the smaller cube?" Cube the side length; s^3.

? "How do you find the volume of the larger cube?" Cube the side length; s^3; The side length for the larger cube, however, is expressed as a power.

- When finding the ratio of the volumes, students will need to divide out the common factors.

? "What do you notice about the volume of the smaller cube and the answer?" The answer is always the same as the volume of the smaller cube.

- **Extension:** Discuss how the relationship between volumes of similar solids is connected to this activity.

What Is Your Answer?

- **Think-Pair-Share:** Students should read the question independently and then work in pairs to answer the question. When they have answered the questions, the pair should compare their answers with another group and discuss any discrepancies.

Closure

- Simplify.

1. $\frac{2^2}{2} \cdot \frac{2^3}{2^2} \cdot \frac{2^4}{2^3}$ 2^3 **2.** $\frac{(-3)^7}{(-3)^4}$ $(-3)^3$

2 ACTIVITY: Comparing Volumes

Math Practice

Repeat Calculations

What calculations are repeated in the table?

Work with a partner.

How many of the smaller cubes will fit inside the larger cube? Record your results in the table. Describe the pattern in the table.

a.

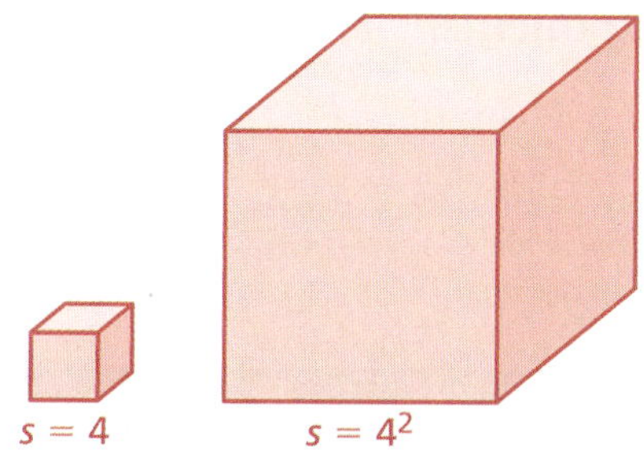

b.

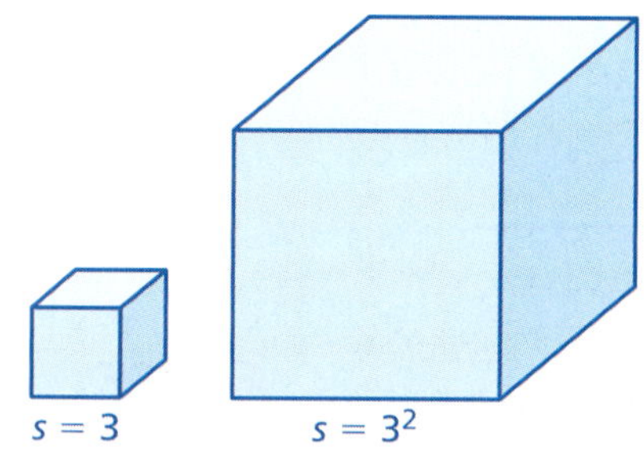

c.

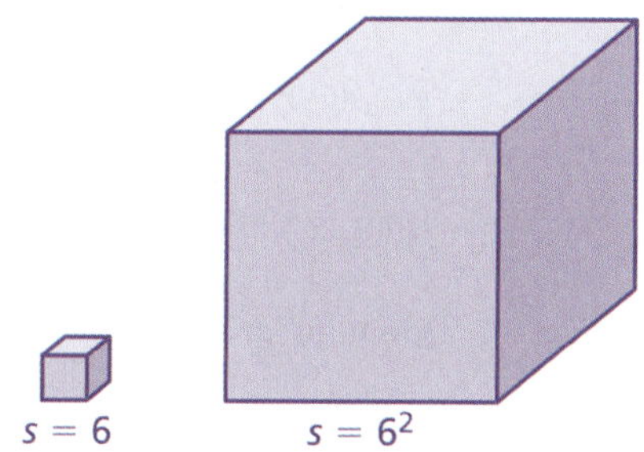

d.

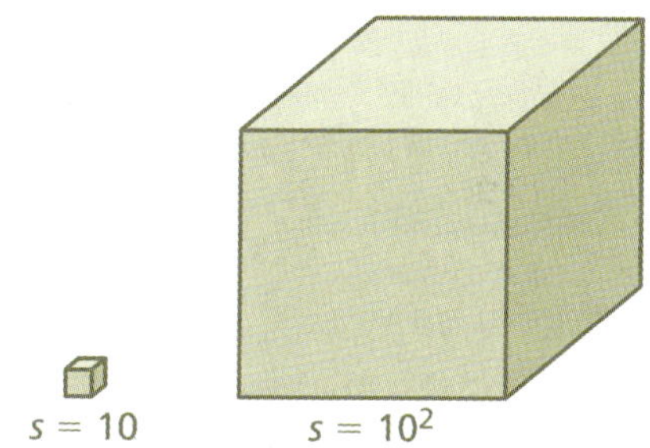

	Volume of Smaller Cube	Volume of Larger Cube	$\frac{\text{Larger Volume}}{\text{Smaller Volume}}$	Answer
a.				
b.				
c.				
d.				

What Is Your Answer?

3. **IN YOUR OWN WORDS** How can you divide two powers that have the same base? Give two examples of your rule.

Practice Use what you learned about dividing powers with the same base to complete Exercises 3–6 on page 726.

16.3 Lesson

Check It Out
Lesson Tutorials
BigIdeasMath.com

Key Idea

Quotient of Powers Property

Words To divide powers with the same base, subtract their exponents.

Numbers $\frac{4^5}{4^2} = 4^{5-2} = 4^3$ **Algebra** $\frac{a^m}{a^n} = a^{m-n}$, where $a \neq 0$

EXAMPLE 1 Dividing Powers with the Same Base

a. $\frac{2^6}{2^4} = 2^{6-4}$ Quotient of Powers Property

$= 2^2$ Simplify.

b. $\frac{(-7)^9}{(-7)^3} = (-7)^{9-3}$ Quotient of Powers Property

$= (-7)^6$ Simplify.

c. $\frac{h^7}{h^6} = h^{7-6}$ Quotient of Powers Property

$= h^1 = h$ Simplify.

Common Error

When dividing powers, do not divide the bases.

$\frac{2^6}{2^4} = 2^2$, not 1^2.

On Your Own

Now You're Ready
Exercises 7–14

Simplify the expression. Write your answer as a power.

1. $\frac{9^7}{9^4}$ **2.** $\frac{4.2^6}{4.2^5}$ **3.** $\frac{(-8)^8}{(-8)^4}$ **4.** $\frac{x^8}{x^3}$

EXAMPLE 2 Simplifying an Expression

Simplify $\frac{3^4 \cdot 3^2}{3^3}$. Write your answer as a power.

The numerator is a product of powers. Add the exponents in the numerator.

$\frac{3^4 \cdot 3^2}{3^3} = \frac{3^{4+2}}{3^3}$ Product of Powers Property

$= \frac{3^6}{3^3}$ Simplify.

$= 3^{6-3}$ Quotient of Powers Property

$= 3^3$ Simplify.

Laurie's Notes

Introduction

Connect

- **Yesterday:** Students explored exponents.
- **Today:** Students will use the Quotient of Powers Property to simplify expressions.

Motivate

- **Preparation:** Find the area of your classroom in square feet. Select two smaller regions of your room that make logical sense given the shape of your room. My classroom is shown. I found the area of the entire room (A + B); the area of B; and the area of C.

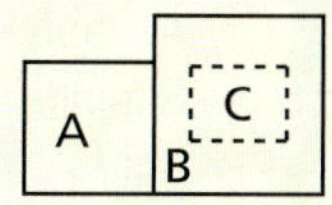

- First, have students stand around the room so that they are an arm's length away from everyone else.
- Then, ask them to stand only in region B (which includes C).
- Finally, ask them to move into region C. It should be very tight.
- Ask them to describe the three regions and how they felt about personal space. Then discuss population density and compute it for each region.
- Explain that one of the examples in today's lesson involves finding the population density of Tennessee.

Lesson Notes

Key Idea

- Write the Key Idea. Discuss the Words, Numbers, and Algebra.

Example 1

? **Attend to Precision: Part (a):** Write and simplify the expression. The base is 2 for each power. Ask the following questions to help develop correct vocabulary.

- "How many factors of 2 are in the numerator?" 6
- "How many factors of 2 are in the denominator?" 4
- "How many factors of 2 are common in *both* the numerator and denominator?" 4
- "How many factors of 2 remain after you divide out the common factors?" 2

- Repeat similar questions for parts (b) and (c).

On Your Own

- Ask students to explain their answers.

Example 2

? This example combines two properties. Ask the following questions.

- "How many factors of 3 are in the numerator?" $4 + 2 = 6$
- "How many factors of 3 are in the denominator?" 3
- "How many factors of 3 are common in *both* the numerator and denominator?" 3
- **Look for and Make Use of Structure:** "How many factors of 3 remain after you divide out the common factors?" 3

Goal Today's lesson is using the Quotient of Powers Property to simplify expressions.

Technology for the Teacher

Dynamic Classroom

Lesson Tutorials
Lesson Plans
Answer Presentation Tool

Extra Example 1

Simplify each expression. Write your answer as a power.

a. $\frac{4^5}{4^2}$ 4^3

b. $\frac{(-2)^{10}}{(-2)^3}$ $(-2)^7$

c. $\frac{p^7}{p^6}$ p

On Your Own

1. 9^3 **2.** 4.2

3. $(-8)^4$ **4.** x^5

Extra Example 2

Simplify $\frac{5^6 \cdot 5^2}{5^4}$. Write your answer as a power. 5^4

Laurie's Notes

Extra Example 3

Simplify $\frac{z^6}{z^2} \cdot \frac{z^8}{z^5}$. Write your answer as a power. z^7

On Your Own

5. 2^7 **6.** d^5

7. 5^8

Extra Example 4

The projected population of Hawaii in 2030 is about $5.59 \cdot 2^{18}$. The land area of Hawaii is about 2^{14} square kilometers. Predict the average number of people per square kilometer in 2030. about 89 people per km^2

On Your Own

8. 36 people per km^2

English Language Learners

Organization

Have students organize the *Key Ideas* of this chapter in their notebooks. This will provide them with easy access to the material and concepts of the chapter.

Example 3

- This example also combines two properties.
- Work through the problem as shown.
- Discuss the approach with students. Each quotient was simplified first and then the product of the two expressions was found.

? "Will the answer be the same if the product of the two expressions is found and then the quotient is simplified? Explain." yes; It is similar to multiplying two fractions and then simplifying the answer.

- Simplify the expression using the alternate approach in the Study Tip.

$$\frac{a^{10}}{a^6} \cdot \frac{a^7}{a^4} = \frac{a^{10} \cdot a^7}{a^6 \cdot a^4} = \frac{a^{10+7}}{a^{6+4}} = \frac{a^{17}}{a^{10}} = a^{17-10} = a^7$$

On Your Own

- There is more than one way to simplify these expressions. Remind students to think about the number of factors as they work the problems.
- **Question 6:** Students may forget that $d = d^1$.
- Have volunteers write their solutions on the board.

Example 4

- This problem is about population density, the number of people per square unit. In this case, it is the projected number of people in Tennessee per square mile in 2030.
- When working through this problem, notice that the factor 5 in the numerator does not have the same base as the two powers.

? **Look for and Make Use of Structure:** "Why can you move 5 out of the numerator and write it as a whole number times the quotient of $(5.9)^8$ and $(5.9)^6$?" definition of multiplying fractions

- Simplify the quotient and multiply by 5.
- Use local landmarks to help students visualize the size of a square mile.

On Your Own

- **Neighbor Check:** Have students work independently and then have their neighbors check their work. Have students discuss any discrepancies.

Closure

- Explain how the Quotient of Powers Property is related to simplifying fractions. You divide out the common factors.

EXAMPLE 3 Simplifying an Expression

Simplify $\frac{a^{10}}{a^6} \cdot \frac{a^7}{a^4}$. Write your answer as a power.

$$\frac{a^{10}}{a^6} \cdot \frac{a^7}{a^4} = a^{10-6} \cdot a^{7-4} \quad \text{Quotient of Powers Property}$$

$$= a^4 \cdot a^3 \quad \text{Simplify.}$$

$$= a^{4+3} \quad \text{Product of Powers Property}$$

$$= a^7 \quad \text{Simplify.}$$

Study Tip

You can also simplify the expression in Example 3 as follows.

$$\frac{a^{10}}{a^6} \cdot \frac{a^7}{a^4} = \frac{a^{10} \cdot a^7}{a^6 \cdot a^4}$$

$$= \frac{a^{17}}{a^{10}}$$

$$= a^{17-10}$$

$$= a^7$$

On Your Own

Now You're Ready
Exercises 16–21

Simplify the expression. Write your answer as a power.

5. $\frac{2^{15}}{2^3 \cdot 2^5}$

6. $\frac{d^5}{d} \cdot \frac{d^9}{d^8}$

7. $\frac{5^9}{5^4} \cdot \frac{5^5}{5^2}$

EXAMPLE 4 Real-Life Application

The projected population of Tennessee in 2030 is about $5 \cdot 5.9^8$. Predict the average number of people per square mile in 2030.

Use a model to solve the problem.

$$\text{People per square mile} = \frac{\text{Population in 2030}}{\text{Land area}}$$

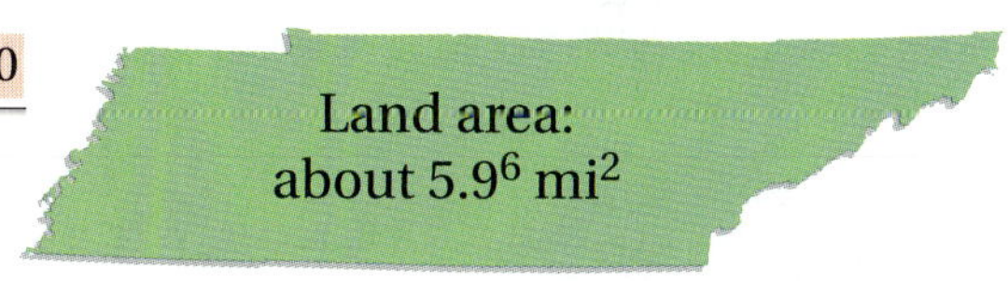

$$= \frac{5 \cdot 5.9^8}{5.9^6} \quad \text{Substitute.}$$

$$= 5 \cdot \frac{5.9^8}{5.9^6} \quad \text{Rewrite.}$$

$$= 5 \cdot 5.9^2 \quad \text{Quotient of Powers Property}$$

$$= 174.05 \quad \text{Evaluate.}$$

So, there will be about 174 people per square mile in Tennessee in 2030.

On Your Own

Now You're Ready
Exercises 23–28

8. The projected population of Alabama in 2030 is about $2.25 \cdot 2^{21}$. The land area of Alabama is about 2^{17} square kilometers. Predict the average number of people per square kilometer in 2030.

16.3 Exercises

Check It Out
Help with Homework
BigIdeasMath.com

Vocabulary and Concept Check

1. **WRITING** Describe in your own words how to divide powers.

2. **WHICH ONE DOESN'T BELONG?** Which quotient does *not* belong with the other three? Explain your reasoning.

$\frac{(-10)^7}{(-10)^2}$	$\frac{6^3}{6^2}$	$\frac{(-4)^8}{(-3)^4}$	$\frac{5^6}{5^3}$

Practice and Problem Solving

Simplify the expression. Write your answer as a power.

3. $\frac{6^{10}}{6^4}$
4. $\frac{8^9}{8^7}$
5. $\frac{(-3)^4}{(-3)^1}$
6. $\frac{4.5^5}{4.5^3}$

1 7. $\frac{5^9}{5^3}$
8. $\frac{64^4}{64^3}$
9. $\frac{(-17)^5}{(-17)^2}$
10. $\frac{(-7.9)^{10}}{(-7.9)^4}$

11. $\frac{(-6.4)^8}{(-6.4)^6}$
12. $\frac{\pi^{11}}{\pi^7}$
13. $\frac{b^{24}}{b^{11}}$
14. $\frac{n^{18}}{n^7}$

15. **ERROR ANALYSIS** Describe and correct the error in simplifying the quotient.

✗ $\frac{6^{15}}{6^5} = 6^{\frac{15}{5}}$
$= 6^3$

Simplify the expression. Write your answer as a power.

2 3 16. $\frac{7^5 \cdot 7^3}{7^2}$
17. $\frac{2^{19} \cdot 2^5}{2^{12} \cdot 2^3}$
18. $\frac{(-8.3)^8}{(-8.3)^7} \cdot \frac{(-8.3)^4}{(-8.3)^3}$

19. $\frac{\pi^{30}}{\pi^{18} \cdot \pi^4}$
20. $\frac{c^{22}}{c^8 \cdot c^9}$
21. $\frac{k^{13}}{k^5} \cdot \frac{k^{17}}{k^{11}}$

22. **SOUND INTENSITY** The sound intensity of a normal conversation is 10^6 times greater than the quietest noise a person can hear. The sound intensity of a jet at takeoff is 10^{14} times greater than the quietest noise a person can hear. How many times more intense is the sound of a jet at takeoff than the sound of a normal conversation?

Assignment Guide and Homework Check

Level	Assignment	Homework Check
Accelerated	1–6, 12, 14, 15, 16–28 even, 29–37	14, 20, 26, 30, 31

Common Errors

- **Exercises 3–14, 16–21** Students may divide the exponents when they should be subtracting them. Remind them that the Quotient of Powers Property states that the exponents are subtracted.
- **Exercises 3–14, 16–21** Students may multiply and/or divide the bases when simplifying the expression. Remind them that the base does not change when they use the Quotient of Powers or Product of Powers Property.
- **Exercises 23–28** Students may try to combine unlike bases when simplifying. Remind them that the Quotient of Powers and Product of Powers Properties only apply to powers with the same base.

16.3 Record and Practice Journal

Simplify the expression. Write your answer as a power.

1. $\frac{7^6}{7^5}$ **7**
2. $\frac{(-21)^{15}}{(-21)^9}$ $(-21)^6$
3. $\frac{(3.9)^{20}}{(3.9)^{10}}$ $(3.9)^{10}$
4. $\frac{t^7}{t^3}$ t^4
5. $\frac{8^7 \bullet 8^4}{8^9}$ 8^2
6. $\frac{(-1.1)^{13} \bullet (-1.1)^{12}}{(-1.1)^{10} \bullet (-1.1)^1}$ $(-1.1)^{14}$

Simplify the expression.

7. $\frac{k \bullet 3^9}{3^5}$ $81k$
8. $\frac{x^4 \bullet y^{10} \bullet 2^{11}}{y^8 \bullet 2^7}$ $16x^4y^2$
9. The radius of a basketball is about 3.6 times greater than the radius of a tennis ball. How many times greater is the volume of a basketball than the volume of a tennis ball? $\left(\text{Note: The volume of a sphere is } V = \frac{4}{3}\pi r^3.\right)$
   46.656

Vocabulary and Concept Check

1. To divide powers means to divide out the common factors of the numerator and denominator. To divide powers with the same base, write the power with the common base and an exponent found by subtracting the exponent in the denominator from the exponent in the numerator.
2. $\frac{(-4)^8}{(-3)^4}$; The other quotients have powers with the same base.

Practice and Problem Solving

3. 6^6
4. 8^2
5. $(-3)^3$
6. 4.5^2
7. 5^6
8. 64
9. $(-17)^3$
10. $(-7.9)^6$
11. $(-6.4)^2$
12. π^4
13. b^{13}
14. n^{11}
15. You should subtract the exponents instead of dividing them.
$$\frac{6^{15}}{6^5} = 6^{15-5}$$
$$= 6^{10}$$
16. 7^6
17. 2^9
18. $(-8.3)^2$
19. π^8
20. c^5
21. k^{14}
22. 10^8 times

Practice and Problem Solving

23. $64x$

24. $6w$

25. $125a^3b^2$

26. $125cd^2$

27. x^7y^6

28. m^9n

29. See *Taking Math Deeper*.

30. **a.** *Sample answer:* $m = 5$, $n = 3$

b. infinitely many solutions; Any two numbers that satisfy the equation $m - n = 2$ are solutions.

31. 10^{13} galaxies

32. 10; The difference in the exponents needs to be 9. To find x, solve the equation $3x - (2x + 1) = 9$.

Fair Game Review

33. -9 **34.** -8

35. 61 **36.** -4

37. B

Mini-Assessment

Simplify the expression. Write your answer as a power.

1. $\dfrac{(-4)^3}{(-4)^1}$ $(-4)^2$
2. $\dfrac{9.7^7}{9.7^3}$ 9.7^4
3. $\dfrac{5^4 \cdot 5^2}{5^3}$ 5^3
4. $\dfrac{m^{10}}{m^5 \cdot m^2}$ m^3
5. $\dfrac{y^{17}}{y^{10}} \cdot \dfrac{y^6}{y^3}$ y^{10}

Taking Math Deeper

Exercise 29

This is an interesting problem. The memory in the different styles of MP3 players increases exponentially, but the price increases linearly.

1 Compare Player D with Player B.

a. $\dfrac{2^4}{2^2} = 2^2 = 4$ times more memory

2 Compare the memory with the price.

If you plot the five points representing the memory and the prices, you get the following graph.

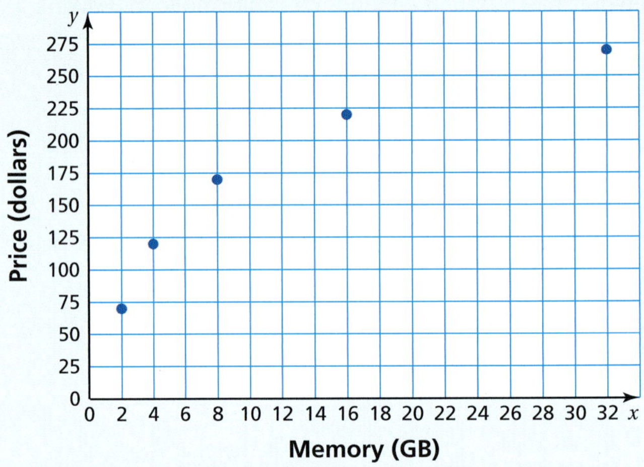

3 Answer the question.

b. This graph does not show a constant rate of change. So, memory and price do not show a linear relationship. However, the differences in price between consecutive sizes reflect a constant rate of change.

Project

What changes in technology have occurred over the past 50 years? What do you predict will change over the next 50 years?

Reteaching and Enrichment Strategies

If students need help. . .	If students got it. . .
Resources by Chapter • Practice A and Practice B • Puzzle Time Record and Practice Journal Practice Differentiating the Lesson Lesson Tutorials Skills Review Handbook	Resources by Chapter • Enrichment and Extension • Technology Connection Start the next section

Simplify the expression.

4 23. $\dfrac{x \cdot 4^8}{4^5}$

24. $\dfrac{6^3 \cdot w}{6^2}$

25. $\dfrac{a^3 \cdot b^4 \cdot 5^4}{b^2 \cdot 5}$

26. $\dfrac{5^{12} \cdot c^{10} \cdot d^2}{5^9 \cdot c^9}$

27. $\dfrac{x^{15}y^9}{x^8y^3}$

28. $\dfrac{m^{10}n^7}{m^1n^6}$

MP3 Player	Memory (GB)	Price
A	2^1	\$70
B	2^2	\$120
C	2^3	\$170
D	2^4	\$220
E	2^5	\$270

29. **MEMORY** The memory capacities and prices of five MP3 players are shown in the table.

a. How many times more memory does MP3 Player D have than MP3 Player B?

b. Do memory and price show a linear relationship? Explain.

30. **CRITICAL THINKING** Consider the equation $\dfrac{9^m}{9^n} = 9^2$.

a. Find two numbers m and n that satisfy the equation.

b. Describe the number of solutions that satisfy the equation. Explain your reasoning.

Milky Way galaxy
$10 \cdot 10^{10}$ stars

31. **STARS** There are about 10^{24} stars in the universe. Each galaxy has approximately the same number of stars as the Milky Way galaxy. About how many galaxies are in the universe?

32. **Number Sense** Find the value of x that makes $\dfrac{8^{3x}}{8^{2x+1}} = 8^9$ true. Explain how you found your answer.

Fair Game Review What you learned in previous grades & lessons

Subtract. *(Section 1.3)*

33. $-4 - 5$

34. $-23 - (-15)$

35. $33 - (-28)$

36. $18 - 22$

37. **MULTIPLE CHOICE** What is the value of x? *(Section 7.1 and Section 7.2)*

Ⓐ 20 Ⓑ 30

Ⓒ 45 Ⓓ 60

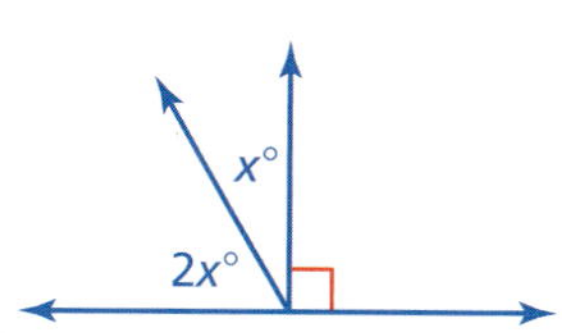

16.4 Zero and Negative Exponents

Essential Question How can you evaluate a nonzero number with an exponent of zero? How can you evaluate a nonzero number with a negative integer exponent?

1 ACTIVITY: Using the Quotient of Powers Property

Work with a partner.

a. Copy and complete the table.

Quotient	Quotient of Powers Property	Power
$\frac{5^3}{5^3}$		
$\frac{6^2}{6^2}$		
$\frac{(-3)^4}{(-3)^4}$		
$\frac{(-4)^5}{(-4)^5}$		

b. **REPEATED REASONING** Evaluate each expression in the first column of the table. What do you notice?

c. How can you use these results to define a^0 where $a \neq 0$?

2 ACTIVITY: Using the Product of Powers Property

Work with a partner.

a. Copy and complete the table.

Product	Product of Powers Property	Power
$3^0 \cdot 3^4$		
$8^2 \cdot 8^0$		
$(-2)^3 \cdot (-2)^0$		
$\left(-\frac{1}{3}\right)^0 \cdot \left(-\frac{1}{3}\right)^5$		

b. Do these results support your definition in Activity 1(c)?

Exponents

In this lesson, you will

- evaluate expressions involving numbers with zero as an exponent.
- evaluate expressions involving negative integer exponents.

Laurie's Notes

Introduction

Applying Mathematical Practices

- **Look for and Express Regularity in Repeated Reasoning:** In the first three activities, students will use previously learned properties to define zero and negative exponents.

Motivate

- Writing a number in expanded form should be familiar to students.
- ? "How do you write 234 in expanded form?" $200 + 30 + 4$
- ? "How do you write the expanded form using powers of 10?" $2 \times 10^2 + 3 \times 10 + 4 \times 1$
- ? "Do you think it is possible to write 234.56 in expanded form using powers of 10?" Answers will vary.
- Explain that in today's activities, students will explore zero and negative exponents, including zero and negative powers of 10, which can be used to write a decimal in expanded notation.

Activity Notes

Activity 1

- Have students work with their partners to complete the table. To fill in the middle column, students should use the Quotient of Powers Property to rewrite the quotients in the first column.
- As students work through the activity, they should discover that all of the powers in the last column of the table have an exponent of zero.
- **Part (b):** Students should notice that each quotient in the first column of the table is equivalent to 1.
- **Part (c):** Students should summarize their findings by writing $a^0 = 1$.
- Discuss the fact that some of the language used in earlier sections does not apply here. For instance, when referring to 5^0, you do not say, "5 used as a factor 0 times."

Activity 2

- This activity presents another way of showing that $a^0 = 1$ where $a \neq 0$.
- Have students work with their partners to complete the table. To fill in the middle column, students should use the Product of Powers Property to rewrite the products in the first column.
- **Construct Viable Arguments and Critique the Reasoning of Others: Part (b):** Students should explain their findings, particularly that the results support their definition in Activity 1(c). They should explain that each power in the last column of the table has the same exponent as the corresponding nonzero exponent in the first column.

What Your Students Will Learn

- Evaluate expressions involving nonzero numbers with zero as an exponent.
- Evaluate expressions involving negative integer exponents.

Previous Learning

Students should know how to apply the properties of exponents.

Technology for the Teacher

Dynamic Classroom

Lesson Plans
Complete Materials List

16.4 Record and Practice Journal

Essential Question How can you evaluate a nonzero number with an exponent of zero? How can you evaluate a nonzero number with a negative integer exponent?

1 ACTIVITY: Using the Quotient of Powers Property

Work with a partner.

a. Complete the table.

Quotient	Quotient of Powers Property	Power
$\frac{5^3}{5^3}$	5^{3-3}	5^0
$\frac{6^2}{6^2}$	6^{2-2}	6^0
$\frac{(-3)^4}{(-3)^4}$	$(-3)^{4-4}$	$(-3)^0$
$\frac{(-4)^5}{(-4)^5}$	$(-4)^{5-5}$	$(-4)^0$

b. **REPEATED REASONING** Evaluate each expression in the first column of the table. What do you notice?

They are all equal to 1.

c. How can you use these results to define a^0 where $a \neq 0$?

$a^0 = 1$, where $a \neq 0$

Differentiated Instruction

Visual

Help students to understand zero and negative exponents using methods already known to them.

Evaluating and then simplifying:

$$\frac{3^2}{3^3} = \frac{9}{27} = \frac{9 \div 9}{27 \div 9} = \frac{1}{3}$$

Dividing out common factors:

$$\frac{3^2}{3^3} = \frac{\cancel{3}^1 \cdot \cancel{3}^1}{\cancel{3}^1 \cdot \cancel{3}^1 \cdot 3} = \frac{1}{3}$$

Quotient of Powers Property:

$$\frac{3^2}{3^3} = 3^{2-3} = 3^{-1} = \frac{1}{3}$$

16.4 Record and Practice Journal

2 ACTIVITY: Using the Product of Powers Property

Work with a partner.

a. Complete the table.

Product	Product of Powers Property	Power
$3^0 \cdot 3^4$	3^{0+4}	3^4
$8^2 \cdot 8^0$	8^{2+0}	8^2
$(-2)^3 \cdot (-2)^0$	$(-2)^{3+0}$	$(-2)^3$
$\left(-\frac{1}{3}\right)^0 \cdot \left(-\frac{1}{3}\right)^5$	$\left(-\frac{1}{3}\right)^{0+5}$	$\left(-\frac{1}{3}\right)^5$

b. Do these results support your definition in Activity 1(c)?

yes

3 ACTIVITY: Using the Product of Powers Property

Work with a partner.

a. Complete the table.

Product	Product of Powers Property	Power
$5^{-3} \cdot 5^3$	5^{-3+3}	5^0
$6^2 \cdot 6^{-2}$	$6^{2+(-2)}$	6^0
$(-3)^4 \cdot (-3)^{-4}$	$(-3)^{4+(-4)}$	$(-3)^0$
$(-4)^{-5} \cdot (-4)^5$	$(-4)^{-5+5}$	$(-4)^0$

b. According to your results from Activities 1 and 2, the products in the first column are equal to what value?

1

c. **REASONING** How does the Multiplicative Inverse Property help you to rewrite the numbers with negative exponents?

The product of a number and its reciprocal is 1.

d. **STRUCTURE** Use these results to define a^{-n} where $a \neq 0$ and n is an integer.

a^{-n} is equal to the reciprocal of a^n, or $\frac{1}{a^n}$.

4 ACTIVITY: Using a Place Value Chart

Work with a partner. Use the place value chart that shows the number 3452.867.

ones: 10^0
tenths: 10^{-1}
hundredths: 10^{-2}
thousandths: 10^{-3}

Place Value Chart

thousands	hundreds	tens	ones	and	tenths	hundredths	thousandths
10^3	10^2	10^1	$10^{\square}$		$10^{\square}$	$10^{\square}$	$10^{\square}$
3	4	5	2	.	8	6	7

a. **REPEATED REASONING** What pattern do you see in the exponents? Continue the pattern to find the other exponents.

decrease by 1

b. **STRUCTURE** Show how to write the expanded form of 3452.867.

See Additional Answers.

What Is Your Answer?

5. **IN YOUR OWN WORDS** How can you evaluate a nonzero number with an exponent of zero? How can you evaluate a nonzero number with a negative integer exponent?

$a^0 = 1; a^{-n} = \frac{1}{a^n}$

Laurie's Notes

Activity 3

- As in Activity 2, this activity uses the Product of Powers Property, but this time to help students come up with the definition of a negative exponent.
- Have students work with their partners to complete the table. To fill in the middle column, students should use the Product of Powers Property to rewrite the products in the first column.
- **Part (c):** The Multiplicative Inverse Property states that the product of a number and its reciprocal is 1. Make sure students see the connection between negative exponents and reciprocals.
- **Construct Viable Arguments and Critique the Reasoning of Others: Part (d):** Ask several students to explain their definitions.
- Discuss the fact that some of the language used in earlier sections does not apply here. For instance, when referring to 5^{-3}, you do not say, "5 used as a factor -3 times."

Activity 4

- Students should be familiar with place value charts.
- Students reviewed writing a whole number in expanded form at the beginning of class.
- By observing and continuing the pattern in the exponents in the given place value chart, students should be able to write the expanded form of a decimal.

What Is Your Answer?

- Have students share their understandings of zero and negative exponents.

Closure

- **Exit Ticket:** Write 234.56 in expanded notation using powers of 10.
  $2 \times 10^2 + 3 \times 10^1 + 4 \times 10^0 + 5 \times 10^{-1} + 6 \times 10^{-2}$

3 ACTIVITY: Using the Product of Powers Property

Work with a partner.

a. Copy and complete the table.

Product	Product of Powers Property	Power
$5^{-3} \cdot 5^{3}$		
$6^{2} \cdot 6^{-2}$		
$(-3)^{4} \cdot (-3)^{-4}$		
$(-4)^{-5} \cdot (-4)^{5}$		

b. According to your results from Activities 1 and 2, the products in the first column are equal to what value?

c. **REASONING** How does the Multiplicative Inverse Property help you rewrite the numbers with negative exponents?

d. **STRUCTURE** Use these results to define a^{-n} where $a \neq 0$ and n is an integer.

4 ACTIVITY: Using a Place Value Chart

Math Practice

Use Operations

What operations are used when writing the expanded form?

Work with a partner. Use the place value chart that shows the number 3452.867.

Place Value Chart

thousands	hundreds	tens	ones	and	tenths	hundredths	thousandths
10^{3}	10^{2}	10^{1}	$10^{\square}$		$10^{\square}$	$10^{\square}$	$10^{\square}$
3	4	5	2	.	8	6	7

a. **REPEATED REASONING** What pattern do you see in the exponents? Continue the pattern to find the other exponents.

b. **STRUCTURE** Show how to write the expanded form of 3452.867.

What Is Your Answer?

5. **IN YOUR OWN WORDS** How can you evaluate a nonzero number with an exponent of zero? How can you evaluate a nonzero number with a negative integer exponent?

Practice Use what you learned about zero and negative exponents to complete Exercises 5–8 on page 732.

16.4 Lesson

Check It Out
Lesson Tutorials
BigIdeasMath.com

Key Ideas

Zero Exponents

Words For any nonzero number a, $a^0 = 1$. The power 0^0 is *undefined*.

Numbers $4^0 = 1$ **Algebra** $a^0 = 1$, where $a \neq 0$

Negative Exponents

Words For any integer n and any nonzero number a, a^{-n} is the reciprocal of a^n.

Numbers $4^{-2} = \frac{1}{4^2}$ **Algebra** $a^{-n} = \frac{1}{a^n}$, where $a \neq 0$

EXAMPLE 1 Evaluating Expressions

a. $3^{-4} = \frac{1}{3^4}$ Definition of negative exponent

$= \frac{1}{81}$ Evaluate power.

b. $(-8.5)^{-4} \cdot (-8.5)^4 = (-8.5)^{-4+4}$ Product of Powers Property

$= (-8.5)^0$ Simplify.

$= 1$ Definition of zero exponent

c. $\frac{2^6}{2^8} = 2^{6-8}$ Quotient of Powers Property

$= 2^{-2}$ Simplify.

$= \frac{1}{2^2}$ Definition of negative exponent

$= \frac{1}{4}$ Evaluate power.

On Your Own

Now You're Ready
Exercises 5–16

Evaluate the expression.

1. 4^{-2}
2. $(-2)^{-5}$
3. $6^{-8} \cdot 6^8$
4. $\frac{(-3)^5}{(-3)^6}$
5. $\frac{1}{5^7} \cdot \frac{1}{5^{-4}}$
6. $\frac{4^5 \cdot 4^{-3}}{4^2}$

Laurie's Notes

Introduction

Connect

- **Yesterday:** Students explored zero and negative exponents.
- **Today:** Students will use the definitions of zero and negative exponents to evaluate and simplify expressions.

Motivate

? "Did you know that a faucet dripping 30 times per minute wastes 54 gallons of water per month?"

? "Can you visualize 54 gallons? Is that more than a standard kitchen sink? Is it more than a bathtub? To what does it compare?" Most students have difficulty estimating capacity. If there is a 50-gallon waste barrel in your school's cafeteria, you could compare it to that.

- Explain that one of the examples in today's lesson involves finding the amount of water that leaks from a faucet in a given period of time.

Lesson Notes

Key Ideas

- The definition of zero exponents is easily understood by most students. Writing a negative exponent as a unit fraction with a positive exponent in the denominator takes time for some students to understand. They need to see multiple examples of simplifying fractions using the Quotient of Powers Property and of dividing out common factors, where the exponent in the denominator is greater than the exponent in the numerator.
- Example: $\frac{5^2}{5^3} = 5^{2-3} = 5^{-1}$ $\qquad \frac{5^2}{5^3} = \frac{\cancel{5} \cdot \cancel{5}}{\cancel{5} \cdot \cancel{5} \cdot 5} = \frac{1}{5}$

  The Quotient of Powers Property is used on the left. Dividing out common factors is used on the right. Because both starting expressions are the same, the results must be equivalent. So, $5^{-1} = \frac{1}{5^1}$.
- Note that 0^0 is undefined.

Example 1

- **Part (a):** This is a direct application of the definition of negative exponents.
- **Part (b):** The bases are the same, so the exponents are added.
- **Part (c):** The Quotient of Powers Property is used first, resulting in a negative exponent. Use the definition of negative exponent and simplify.

On Your Own

- Question 5 can be done by thinking about simple fractions and how they are multiplied. The product of these two fractions is $\frac{1}{5^7 \cdot 5^{-4}} = \frac{1}{5^3}$.
- **Reason Abstractly and Quantitatively** and **Look for and Make Use of Structure:** Several of these expressions can be evaluated in more than one way. Ask students to share their methods.

Goal Today's lesson is evaluating and simplifying expressions with negative and zero exponents.

Technology for the Teacher

Dynamic Classroom

Lesson Tutorials
Lesson Plans
Answer Presentation Tool

Extra Example 1

Evaluate each expression.

a. 4^{-3} $\frac{1}{64}$

b. $(-3.7)^{-2} \cdot (-3.7)^2$ 1

c. $\frac{3^6}{3^9}$ $\frac{1}{27}$

On Your Own

1. $\frac{1}{16}$
2. $-\frac{1}{32}$
3. 1
4. $-\frac{1}{3}$
5. $\frac{1}{125}$
6. 1

Extra Example 2

Simplify. If possible, write the expression using only positive exponents.

a. $-2x^0$ -2

b. $\frac{4b^{-4}}{b^7}$ $\frac{4}{b^{11}}$

On Your Own

7. $\frac{8}{x^2}$

8. $\frac{1}{b^{10}}$

9. $\frac{1}{15z^3}$

Extra Example 3

In Example 3, the faucet leaks water at a rate of 4^{-6} liter per second. How many liters of water leak from the faucet in 1 hour? about 0.88 L

On Your Own

10. 1.152 L

English Language Learners

Vocabulary

Remind English learners that when they see a negative exponent, they should think *reciprocal*. Review the meaning of the word reciprocal. Students often think that because the exponent is negative, the expression is negative. Remind them that a number of the form x^n cannot be negative unless the base is negative.

Laurie's Notes

Example 2

- **Attend to Precision:** It is assumed that the variables are nonzero in expressions such as those given in this example, so that the rules for zero and negative exponents can be used. Discuss this with students.
- **Common Error:** In part (a), some students see the zero exponent and immediately think the answer is 1. Remind students that only the variable is being raised to the 0 exponent; -5 is not.
- **Common Error:** In part (b), the constant 9 is not being raised to an exponent, only the variables are. Students need to distinguish this. In the step where the expression has been simplified to $9y^{-8}$ ask, "What is being raised to the -8?" In other words, what is the base for the exponent?
- **Make Sense of Problems:** Work through the steps slowly. It takes time for students to make sense of all that is going on in each problem. Because there is often more than one approach to simplifying the expression, it can confuse students. Instead of seeing it as a way to show that the properties are all connected, students see it as a way of trying to confuse them.

On Your Own

- Have students share their work at the board, *and* explain aloud what they did. Students need to hear the words and see the work. It also helps students become better communicators when they have the opportunity to practice their skills.

Example 3

- Ask a student to read the problem. The information known in this problem is the rate of leaking (in drops per second) and the amount of liters in a drop (from the illustration).

? "What are you asked to find in this problem?" amount faucet leaks in 1 hour

- First, use unit analysis to convert 1 hour to 3600 seconds. Then, because the faucet leaks 50^{-2} liter every second and there are 3600 seconds in an hour, multiply to find how many liters leak in 1 hour.

On Your Own

- Before students begin, ask them which faucet leaks more, the faucet that leaks 50^{-2} liter per second or the faucet that leaks 5^{-5} liter per second.

Closure

- **Exit Ticket:** Simplify.

1. $4x^{-3}$ $\frac{4}{x^3}$ **2.** $\frac{6^3}{6^5}$ $\frac{1}{36}$ **3.** $\frac{4n^0}{n^2}$ $\frac{4}{n^2}$

EXAMPLE 2 Simplifying Expressions

a. $-5x^0 = -5(1)$ — Definition of zero exponent

$= -5$ — Multiply.

b. $\frac{9y^{-3}}{y^5} = 9y^{-3-5}$ — Quotient of Powers Property

$= 9y^{-8}$ — Simplify.

$= \frac{9}{y^8}$ — Definition of negative exponent

On Your Own

Now You're Ready
Exercises 20–27

Simplify. Write the expression using only positive exponents.

7. $8x^{-2}$ **8.** $b^0 \cdot b^{-10}$ **9.** $\frac{z^6}{15z^9}$

EXAMPLE 3 Real-Life Application

A drop of water leaks from a faucet every second. How many liters of water leak from the faucet in 1 hour?

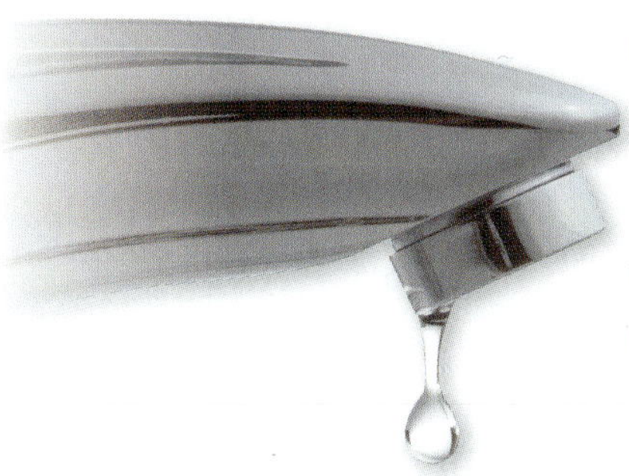

Drop of water: 50^{-2} liter

Convert 1 hour to seconds.

$$1\ \cancel{h} \times \frac{60\ \cancel{min}}{1\ \cancel{h}} \times \frac{60\ sec}{1\ \cancel{min}} = 3600\ sec$$

Water leaks from the faucet at a rate of 50^{-2} liter per second. Multiply the time by the rate.

$3600\ sec \cdot 50^{-2}\frac{L}{sec} = 3600 \cdot \frac{1}{50^2}$ — Definition of negative exponent

$= 3600 \cdot \frac{1}{2500}$ — Evaluate power.

$= \frac{3600}{2500}$ — Multiply.

$= 1\frac{11}{25} = 1.44\ L$ — Simplify.

So, 1.44 liters of water leak from the faucet in 1 hour.

On Your Own

10. **WHAT IF?** The faucet leaks water at a rate of 5^{-5} liter per second. How many liters of water leak from the faucet in 1 hour?

16.4 Exercises

Check It Out
Help with Homework
BigIdeasMath.com

Vocabulary and Concept Check

1. **VOCABULARY** If a is a nonzero number, does the value of a^0 depend on the value of a? Explain.
2. **WRITING** Explain how to evaluate 10^{-3}.
3. **NUMBER SENSE** Without evaluating, order 5^0, 5^4, and 5^{-5} from least to greatest.
4. **DIFFERENT WORDS, SAME QUESTION** Which is different? Find "both" answers.

Rewrite $\frac{1}{3 \cdot 3 \cdot 3}$ using a negative exponent.	Write 3 to the negative third.
Write $\frac{1}{3}$ cubed as a power.	Write $(-3) \cdot (-3) \cdot (-3)$ as a power.

Practice and Problem Solving

Evaluate the expression.

1 5. $\frac{8^7}{8^7}$ 6. $5^0 \cdot 5^3$ 7. $(-2)^{-8} \cdot (-2)^8$ 8. $9^4 \cdot 9^{-4}$

9. 6^{-2} 10. 158^0 11. $\frac{4^3}{4^5}$ 12. $\frac{-3}{(-3)^2}$

13. $4 \cdot 2^{-4} + 5$ 14. $3^{-3} \cdot 3^{-2}$ 15. $\frac{1}{5^{-3}} \cdot \frac{1}{5^6}$ 16. $\frac{(1.5)^2}{(1.5)^{-2} \cdot (1.5)^4}$

17. **ERROR ANALYSIS** Describe and correct the error in evaluating the expression.

$$(4)^{-3} = (-4)(-4)(-4)$$
$$= -64$$

18. **SAND** The mass of a grain of sand is about 10^{-3} gram. About how many grains of sand are in the bag of sand?

19. **CRITICAL THINKING** How can you write the number 1 as 2 to a power? 10 to a power?

Simplify. Write the expression using only positive exponents.

2 20. $6y^{-4}$ 21. $8^{-2} \cdot a^7$ 22. $\frac{9c^3}{c^{-4}}$ 23. $\frac{5b^{-2}}{b^{-3}}$

24. $\frac{8x^3}{2x^9}$ 25. $3d^{-4} \cdot 4d^4$ 26. $m^{-2} \cdot n^3$ 27. $\frac{3^{-2} \cdot k^0 \cdot w^0}{w^{-6}}$

Assignment Guide and Homework Check

Level	Assignment	Homework Check
Accelerated	1–8, 14, 16, 17, 18–28 even, 29–40	16, 27, 32, 34

Common Errors

- **Exercises 5–16** Students may think that a power with a zero exponent is equal to zero. Remind them of the definition of zero exponents.
- **Exercises 5–16** Students may think that a negative exponent makes the power negative. Remind them of the definition of negative exponents.
- **Exercises 5–16** Students may forget to complete the solution; they may simplify the expression, but leave the expression with exponents. Point out that they need to evaluate any powers to complete the solution.
- **Exercises 20–27** In an expression such as $6y^{-4}$, students may think that both the constant and the variable have the exponent -4. Make sure students understand that in such an expression, the base for the exponent is y and not $6y$.

16.4 Record and Practice Journal

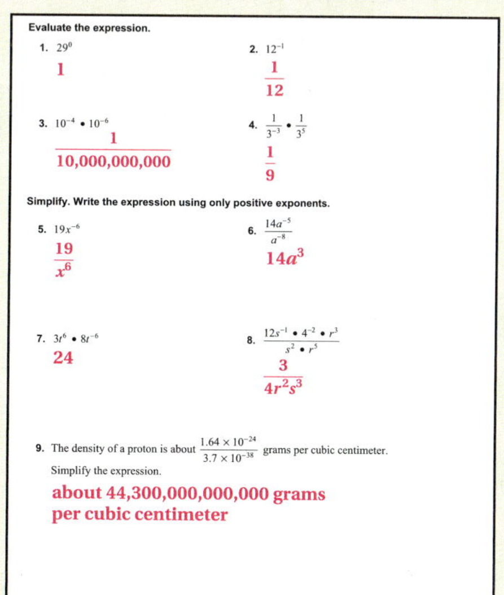

Evaluate the expression.

1. 29^0 **1**
2. 12^{-1} $\frac{1}{12}$
3. $10^{-4} \cdot 10^{-6}$ $\frac{1}{10{,}000{,}000{,}000}$
4. $\frac{1}{3^{-3}} \cdot \frac{1}{3^5}$ $\frac{1}{9}$

Simplify. Write the expression using only positive exponents.

5. $19x^{-6}$ $\frac{19}{x^6}$
6. $\frac{14a^{-5}}{a^{-8}}$ $14a^3$
7. $3t^6 \cdot 8t^{-6}$ **24**
8. $\frac{12s^{-1} \cdot 4^{-2} \cdot r^3}{s^2 \cdot r^5}$ $\frac{3}{4r^2s^3}$
9. The density of a proton is about $\frac{1.64 \times 10^{-24}}{3.7 \times 10^{-38}}$ grams per cubic centimeter. Simplify the expression.
   about 44,300,000,000,000 grams per cubic centimeter

Vocabulary and Concept Check

1. no; Any nonzero base raised to a zero exponent is always 1.
2. Use the definition of negative exponents to rewrite it as $\frac{1}{10^3}$. Then evaluate the power to get $\frac{1}{1000}$.
3. $5^{-5}, 5^0, 5^4$
4. Write $(-3) \cdot (-3) \cdot (-3)$ as a power.; $(-3)^3$; 3^{-3}

Practice and Problem Solving

5. 1
6. 125
7. 1
8. 1
9. $\frac{1}{36}$
10. 1
11. $\frac{1}{16}$
12. $-\frac{1}{3}$
13. $5\frac{1}{4}$
14. $\frac{1}{243}$
15. $\frac{1}{125}$
16. 1
17. The negative sign goes with the exponent, not the base. $(4)^{-3} = \frac{1}{4^3} = \frac{1}{64}$
18. 10,000,000 grains of sand
19. 2^0; 10^0
20. $\frac{6}{y^4}$
21. $\frac{a^7}{64}$
22. $9c^7$
23. $5b$
24. $\frac{4}{x^6}$
25. 12
26. $\frac{n^3}{m^2}$
27. $\frac{w^6}{9}$

Practice and Problem Solving

28. *Sample answer:* 2^{-4}; 4^{-2}

29. 100 mm

30. 10,000 micrometers

31. 1,000,000 nanometers

32. 1,000,000 micrometers

33. **a.** 10^{-9} m

b. equal to

34. See *Taking Math Deeper*.

35. Write the power as 1 divided by the power and use a negative exponent. Justifications will vary.

36. If $a = 0$, then $0^n = 0$. Because you can not divide by 0, the expression $\frac{1}{0}$ is undefined.

Fair Game Review

37. 10^9	**38.** 10^3
39. 10^4	**40.** C

Mini-Assessment

Evaluate the expression.

1. 5^{-3} $\frac{1}{125}$

2. 9^0 1

3. $\frac{3^6}{3^{10}}$ $\frac{1}{81}$

4. $\frac{1}{2^{-2}} \cdot \frac{1}{2^5}$ $\frac{1}{8}$

5. $\frac{(2.3)^6}{(2.3)^{-3} \cdot (2.3)^8}$ 2.3

Taking Math Deeper

Exercise 34

To solve this problem, students need to notice that the blood sample shown has a volume of 500 milliliters. This is a good problem to help students with *unit analysis.*

1 **a.** How many white blood cells are in the donation?

$$500 \cancel{\text{mL}} \cdot \frac{1 \cancel{\text{mm}^3}}{10^{-3} \cancel{\text{mL}}} \cdot \frac{10^4 \text{ white blood cells}}{1 \cancel{\text{mm}^3}}$$
$$= 500 \cdot \frac{10^4 \text{ white blood cells}}{10^{-3}}$$
$$= 500 \cdot 10^7 \text{ white blood cells}$$
$$= 5{,}000{,}000{,}000 \text{ white blood cells}$$
$$= 5 \text{ billion white blood cells}$$

2 **b.** How many red blood cells are in the donation?

$$500 \cancel{\text{mL}} \cdot \frac{1 \cancel{\text{mm}^3}}{10^{-3} \cancel{\text{mL}}} \cdot \frac{5 \cdot 10^6 \text{ red blood cells}}{1 \cancel{\text{mm}^3}}$$
$$= 2500 \cdot \frac{10^6 \text{ red blood cells}}{10^{-3}}$$
$$= 2500 \cdot 10^9 \text{ red blood cells}$$
$$= 2{,}500{,}000{,}000{,}000 \text{ red blood cells}$$
$$= 2.5 \text{ trillion red blood cells}$$

3 **c.** The ratio of red blood cells to white blood cells is

$$\frac{2{,}500{,}000{,}000{,}000 \text{ red blood cells}}{5{,}000{,}000{,}000 \text{ white blood cells}} = \frac{500}{1}.$$

Red blood cells are responsible for picking up carbon dioxide from our blood and for transporting oxygen. White blood cells are responsible for fighting foreign organisms that enter the body.

Reteaching and Enrichment Strategies

If students need help. . .	If students got it. . .
Resources by Chapter • Practice A and Practice B • Puzzle Time Record and Practice Journal Practice Differentiating the Lesson Lesson Tutorials Skills Review Handbook	Resources by Chapter • Enrichment and Extension • Technology Connection Start the next section

28. **OPEN-ENDED** Write two different powers with negative exponents that have the same value.

METRIC UNITS **In Exercises 29–32, use the table.**

29. How many millimeters are in a decimeter?
30. How many micrometers are in a centimeter?
31. How many nanometers are in a millimeter?
32. How many micrometers are in a meter?

Unit of Length	Length (meter)
Decimeter	10^{-1}
Centimeter	10^{-2}
Millimeter	10^{-3}
Micrometer	10^{-6}
Nanometer	10^{-9}

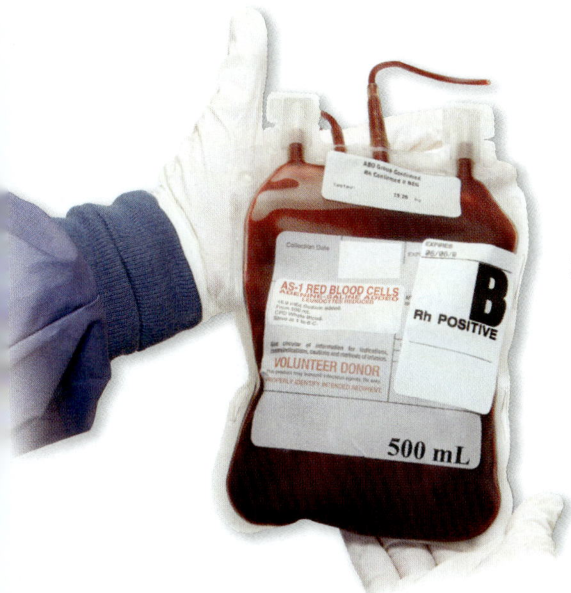

33. **BACTERIA** A species of bacteria is 10 micrometers long. A virus is 10,000 times smaller than the bacteria.
    a. Using the table above, find the length of the virus in meters.
    b. Is the answer to part (a) *less than, greater than,* or *equal to* one nanometer?

34. **BLOOD DONATION** Every 2 seconds, someone in the United States needs blood. A sample blood donation is shown. ($1\text{ mm}^3 = 10^{-3}\text{ mL}$)
    a. One cubic millimeter of blood contains about 10^4 white blood cells. How many white blood cells are in the donation? Write your answer in words.
    b. One cubic millimeter of blood contains about 5×10^6 red blood cells. How many red blood cells are in the donation? Write your answer in words.
    c. Compare your answers for parts (a) and (b).

35. **PRECISION** Describe how to rewrite a power with a positive exponent so that the exponent is in the denominator. Use the definition of negative exponents to justify your reasoning.

36. **Reasoning** The rule for negative exponents states that $a^{-n} = \frac{1}{a^n}$. Explain why this rule does not apply when $a = 0$.

Fair Game Review What you learned in previous grades & lessons

Simplify the expression. Write your answer as a power. *(Section 16.2 and Section 16.3)*

37. $10^3 \cdot 10^6$

38. $10^2 \cdot 10$

39. $\frac{10^8}{10^4}$

40. **MULTIPLE CHOICE** What is the volume of a cylinder with a radius of 4 inches and a height of 5 inches? *(Section 15.1)*

    (A) $16\pi\text{ in.}^3$ (B) $72\pi\text{ in.}^3$

    (C) $80\pi\text{ in.}^3$ (D) $100\pi\text{ in.}^3$

16 Study Help

Check It Out
Graphic Organizer
BigIdeasMath.com

You can use an **information wheel** to organize information about a topic. Here is an example of an information wheel for exponents.

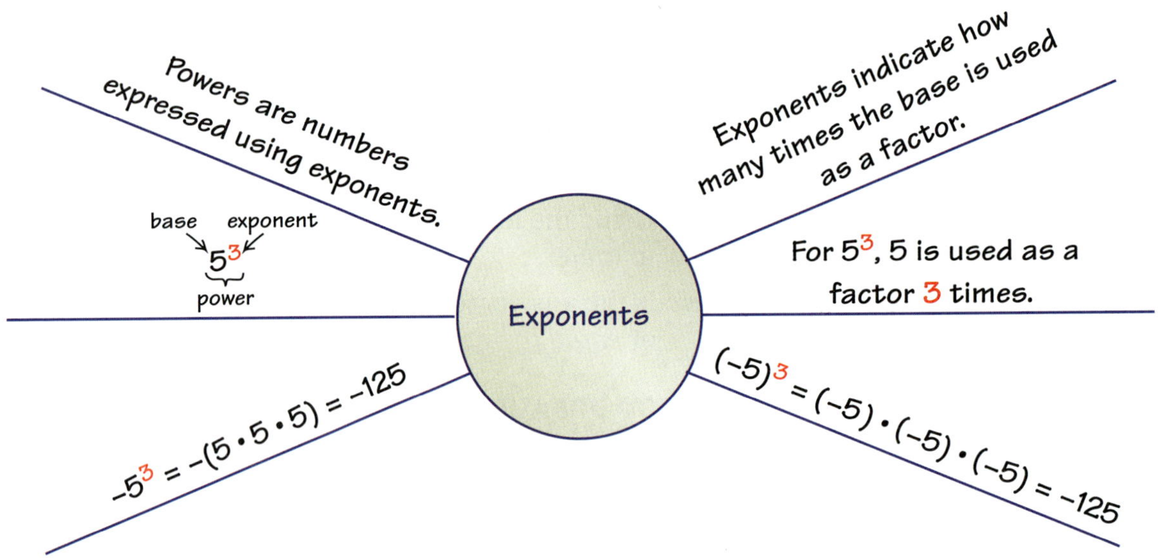

On Your Own

Make information wheels to help you study these topics.

1. Product of Powers Property
2. Quotient of Powers Property
3. zero and negative exponents

After you complete this chapter, make information wheels for the following topics.

4. writing numbers in scientific notation
5. writing numbers in standard form
6. adding and subtracting numbers in scientific notation
7. multiplying and dividing numbers in scientific notation
8. Choose three other topics you studied earlier in this course. Make an information wheel for each topic to summarize what you know about them.

"I decided to color code the different flavors in my information wheel."

Sample Answers

1. 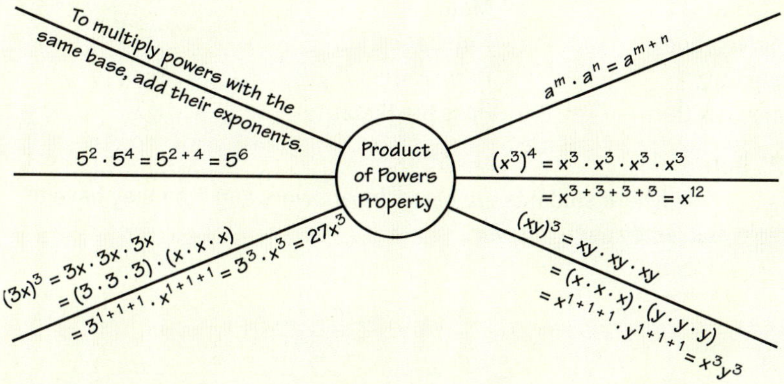

2. 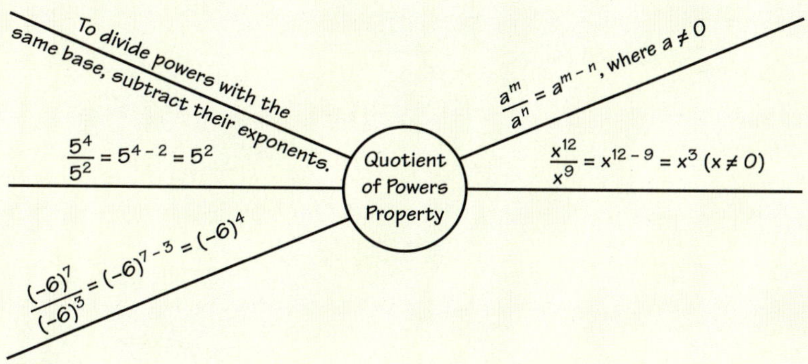

3. 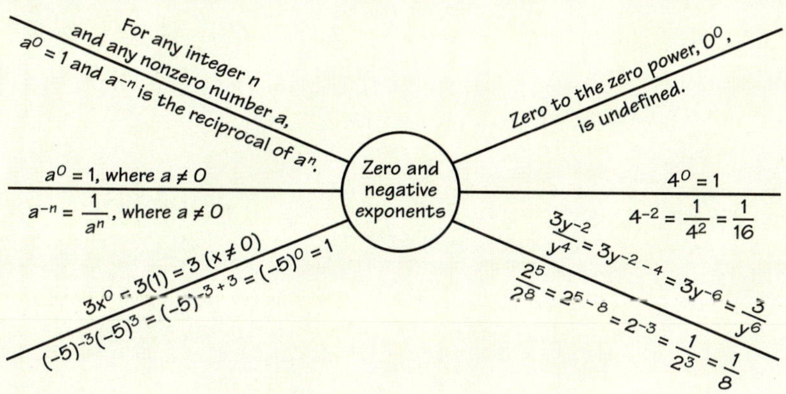

List of Organizers

Available at *BigIdeasMath.com*

Comparison Chart
Concept Circle
Definition (Idea) and Example Chart
Example and Non-Example Chart
Formula Triangle
Four Square
Information Frame
Information Wheel
Notetaking Organizer
Process Diagram
Summary Triangle
Word Magnet
Y Chart

About this Organizer

An **Information Wheel** can be used to organize information about a concept. Students write the concept in the middle of the "wheel." Then students write information related to the concept on the "spokes" of the wheel. Related information can include, but is not limited to: vocabulary words or terms, definitions, formulas, procedures, examples, and visuals. This type of organizer serves as a good summary tool because any information related to a concept can be included.

Technology for the Teacher

Editable Graphic Organizer

Answers

1. $(-5)^4$
2. 7^2m^3
3. 625
4. 64
5. 1
6. $\frac{1}{125}$
7. 3^9
8. a^{15}
9. $81c^4$
10. $\frac{4}{49}p^2$
11. 8^3
12. 6^8
13. π^3
14. t^{10}
15. $\frac{8}{d^6}$
16. $\frac{3}{x^2}$
17. a 10^{-3} m

    b. 1 millimeter; The length is less than 1 meter and a millimeter is smaller than a meter.
18. 10^5 times

Technology for the Teacher

Online Assessment
Assessment Book
ExamView® Assessment Suite

Alternative Quiz Ideas

100% Quiz	Math Log
Error Notebook	Notebook Quiz
Group Quiz	Partner Quiz
Homework Quiz	Pass the Paper

100% Quiz

This is a quiz where students are given the answers and then they have to explain and justify each answer.

Reteaching and Enrichment Strategies

If students need help. . .	If students got it. . .
Resources by Chapter • Practice A and Practice B • Puzzle Time Lesson Tutorials *BigIdeasMath.com*	Resources by Chapter • Enrichment and Extension • Technology Connection Game Closet at *BigIdeasMath.com* Start the next section

16.1–16.4 Quiz

Write the product using exponents. *(Section 16.1)*

1. $(-5) \cdot (-5) \cdot (-5) \cdot (-5)$

2. $7 \cdot 7 \cdot m \cdot m \cdot m$

Evaluate the expression. *(Section 16.1 and Section 16.4)*

3. 5^4

4. $(-2)^6$

5. $(-4.8)^{-9} \cdot (-4.8)^9$

6. $\dfrac{5^4}{5^7}$

Simplify the expression. Write your answer as a power. *(Section 16.2)*

7. $3^8 \cdot 3$

8. $(a^5)^3$

Simplify the expression. *(Section 16.2)*

9. $(3c)^4$

10. $\left(-\dfrac{2}{7}p\right)^2$

Simplify the expression. Write your answer as a power. *(Section 16.3)*

11. $\dfrac{8^7}{8^4}$

12. $\dfrac{6^3 \cdot 6^7}{6^2}$

13. $\dfrac{\pi^{15}}{\pi^3 \cdot \pi^9}$

14. $\dfrac{t^{13}}{t^5} \cdot \dfrac{t^8}{t^6}$

Simplify. Write the expression using only positive exponents. *(Section 16.4)*

15. $8d^{-6}$

16. $\dfrac{12x^5}{4x^7}$

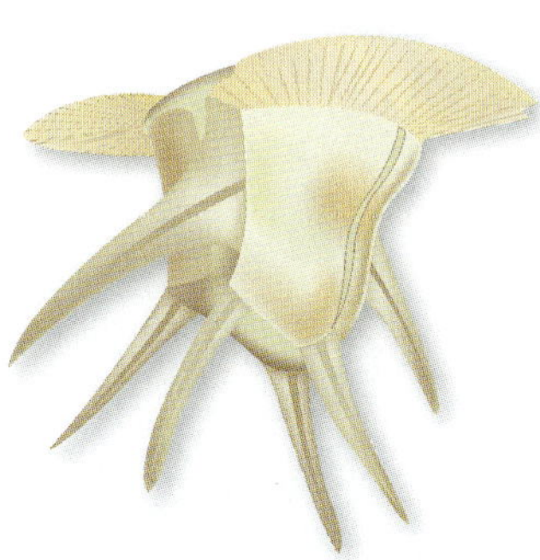

17. ORGANISM A one-celled, aquatic organism called a dinoflagellate is 1000 micrometers long. *(Section 16.4)*

a. One micrometer is 10^{-6} meter. What is the length of the dinoflagellate in meters?

b. Is the length of the dinoflagellate equal to 1 millimeter or 1 kilometer? Explain.

18. EARTHQUAKES An earthquake of magnitude 3.0 is 10^2 times stronger than an earthquake of magnitude 1.0. An earthquake of magnitude 8.0 is 10^7 times stronger than an earthquake of magnitude 1.0. How many times stronger is an earthquake of magnitude 8.0 than an earthquake of magnitude 3.0? *(Section 16.3)*

16.5 Reading Scientific Notation

Essential Question How can you read numbers that are written in scientific notation?

1 ACTIVITY: Very Large Numbers

Work with a partner.

- Use a calculator. Experiment with multiplying large numbers until your calculator displays an answer that is *not* in standard form.
- When the calculator at the right was used to multiply 2 billion by 3 billion, it listed the result as

  6.0E+18.
- Multiply 2 billion by 3 billion by hand. Use the result to explain what 6.0E+18 means.
- Check your explanation by calculating the products of other large numbers.
- Why didn't the calculator show the answer in standard form?
- Experiment to find the maximum number of digits your calculator displays. For instance, if you multiply 1000 by 1000 and your calculator shows 1,000,000, then it can display seven digits.

2 ACTIVITY: Very Small Numbers

Work with a partner.

- Use a calculator. Experiment with multiplying very small numbers until your calculator displays an answer that is *not* in standard form.
- When the calculator at the right was used to multiply 2 billionths by 3 billionths, it listed the result as

  6.0E–18.
- Multiply 2 billionths by 3 billionths by hand. Use the result to explain what 6.0E–18 means.
- Check your explanation by calculating the products of other very small numbers.

Scientific Notation

In this lesson, you will
- identify numbers written in scientific notation.
- write numbers in standard form.
- compare numbers in scientific notation.

Laurie's Notes

Introduction

Applying Mathematical Practices

- **Use Appropriate Tools Strategically:** In the first two activities, students will use calculators to multiply very large and very small numbers. Students will then explain the meanings of the resulting calculator displays.

Motivate

? "Have you had your millionth heartbeat?" Answers will vary.

- Assume that your heart beats once per second, and it has since you were born. Convert 1 million seconds to days.

$$10^6 \text{ sec} \cdot \frac{1 \text{ min}}{60 \text{ sec}} \cdot \frac{1 \text{ h}}{60 \text{ min}} \cdot \frac{1 \text{ day}}{24 \text{ h}} \approx 11.57 \text{ days}$$

- Clearly, all of your students have had their millionth heart beat. But have they had their billionth? Because there are 1000 million in 1 billion,

$$11.57 \text{ days} \times 1000 = 11{,}570 \text{ days and}$$

$$11{,}570 \text{ days} \cdot \frac{1 \text{ year}}{365 \text{ days}} \approx 31.7 \text{ years.}$$

Activity Notes

Activity 1

? "What does standard form mean in the context of this activity?" Numbers are written using digits; example: 123.

? "What does expanded form mean?" Numbers are written showing the value of each digit; example: $123 = 1 \times 100 + 2 \times 10 + 3 \times 1$.

- **Use Appropriate Tools Strategically:** This activity gives students time to explore how scientific notation is displayed on their calculators.
- Students should be able to explain how they determined the number of digits the calculator displays.

? "When the display on a calculator reads 4.5 E+8, what does this mean?" The E+8 means that the decimal point should be moved 8 places to the right.

- Write on the board: 4.5 E+8 = 450,000,000.

Activity 2

- This activity is the same as Activity 1, except with very small numbers.
- Review place values less than 1.
- **Summary:** Check to see that everyone was successful in getting both large and small numbers to display. Has everyone figured out what notations such as E+4 and E−6 mean on the calculator? Do students know how many digits their calculators display?

? "When the display on a calculator reads 6.2 E−6, what does this mean?" The E−6 means that the decimal point should be moved 6 places to the left.

- Write on the board: 6.2 E−6 = 0.0000062.

What Your Students Will Learn

- Identify numbers written in scientific notation.
- Convert numbers in scientific notation to standard form.
- Compare numbers in scientific notation.

Previous Learning

Students should know the base 10 place value system.

Technology for the Teacher

Dynamic Classroom

Lesson Plans

Complete Materials List

16.5 Record and Practice Journal

Essential Question How can you read numbers that are written in scientific notation?

1 ACTIVITY: Very Large Numbers

Work with a partner.

- Use a calculator. Experiment with multiplying large numbers until your calculator displays an answer that is *not* in standard form.
- When the calculator at the right was used to multiply 2 billion by 3 billion, it listed the result as

  6.0E+18.

- Multiply 2 billion by 3 billion by hand. Use the result to explain what 6.0E+18 means.

  6 followed by 18 zeros

- Check your explanation using products of other large numbers.
- Why didn't the calculator show the answer in standard form?

  not enough display places for all the zeros

- Experiment to find the maximum number of digits your calculator displays. For instance, if you multiply 1000 by 1000 and your calculator shows 1,000,000, then it can display 7 digits.

  Check students' work.

Differentiated Instruction

Kinesthetic

Have students use grid paper when converting numbers from scientific notation to standard form and from standard form to scientific notation. Write the number with one digit in each square. Place the decimal point on the line segment between the squares. Students may find it easier to count the number of squares than the number of digits.

16.5 Record and Practice Journal

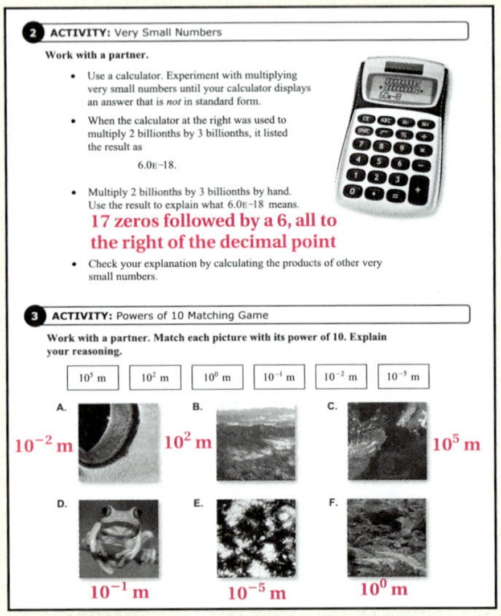

2 ACTIVITY: Very Small Numbers

Work with a partner.

- Use a calculator. Experiment with multiplying very small numbers until your calculator displays an answer that is *not* in standard form.
- When the calculator at the right was used to multiply 2 billionths by 3 billionths, it listed the result as

  6.0E−18.

- Multiply 2 billionths by 3 billionths by hand. Use the result to explain what 6.0E−18 means.

  17 zeros followed by a 6, all to the right of the decimal point

- Check your explanation by calculating the products of other very small numbers.

3 ACTIVITY: Powers of 10 Matching Game

Work with a partner. Match each picture with its power of 10. Explain your reasoning.

10^5 m	10^2 m	10^0 m	10^{-1} m	10^{-2} m	10^{-5} m

A. 10^{-2} m B. 10^2 m C. 10^5 m

D. 10^{-1} m E. 10^{-5} m F. 10^0 m

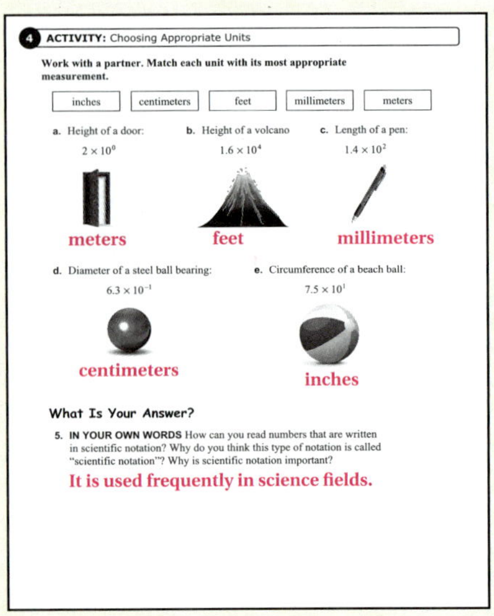

4 ACTIVITY: Choosing Appropriate Units

Work with a partner. Match each unit with its most appropriate measurement.

inches	centimeters	feet	millimeters	meters

a. Height of a door: 2×10^0 meters

b. Height of a volcano 1.6×10^4 feet

c. Length of a pen: 1.4×10^2 millimeters

d. Diameter of a steel ball bearing: 6.3×10^{-1} centimeters

e. Circumference of a beach ball: 7.5×10^1 inches

What Is Your Answer?

5. **IN YOUR OWN WORDS** How can you read numbers that are written in scientific notation? Why do you think this type of notation is called "scientific notation"? Why is scientific notation important?

   It is used frequently in science fields.

Laurie's Notes

Activity 3

- Most students will enjoy trying to figure out what is in each picture.
- Make a transparency of the six photos or display them under a document camera to facilitate discussion.

? "What clues did you use to order the photos?" open ended

Activity 4

- Most students will enjoy this activity.
- Students need to think about the standard form of each measurement *and* which unit makes sense.
- **Reason Abstractly and Quantitatively:** Students may successfully identify the standard form of each measurement but they may have difficulty reasoning about the appropriate unit. Ask questions such as, "Would a reasonable height of a door be 2 inches, 2 centimeters, 2 feet, 2 millimeters, or 2 meters?
- Ask a volunteer to share his or her answers.

What Is Your Answer?

- **Think-Pair-Share:** Students should read the question independently and then work in pairs to answer the questions. When they have answered the questions, the pair should compare their answer with another group and discuss any discrepancies.

Closure

- What examples have you read or heard about that involve very large or very small numbers?

3 ACTIVITY: Powers of 10 Matching Game

Math Practice

Analyze Relationships

How are the pictures related? How can you order the pictures to find the correct power of 10?

Work with a partner. Match each picture with its power of 10. Explain your reasoning.

10^5 m $\quad$ 10^2 m $\quad$ 10^0 m $\quad$ 10^{-1} m $\quad$ 10^{-2} m $\quad$ 10^{-5} m

A. 

B. 

C. 

D. 

E. 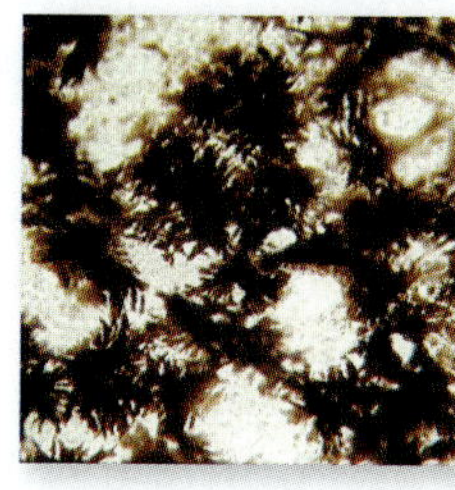

F. 

4 ACTIVITY: Choosing Appropriate Units

Work with a partner. Match each unit with its most appropriate measurement.

inches $\quad$ centimeters $\quad$ feet $\quad$ millimeters $\quad$ meters

A. Height of a door: 2×10^0

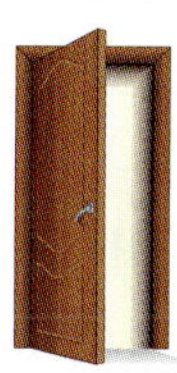

B. Height of a volcano: 1.6×10^4

C. Length of a pen: 1.4×10^2

D. Diameter of a steel ball bearing: 6.3×10^{-1}

E. Circumference of a beach ball: 7.5×10^1

What Is Your Answer?

5. **IN YOUR OWN WORDS** How can you read numbers that are written in scientific notation? Why do you think this type of notation is called *scientific notation*? Why is scientific notation important?

Practice Use what you learned about reading scientific notation to complete Exercises 3–5 on page 740.

16.5 Lesson

Check It Out
Lesson Tutorials
BigIdeasMath.com

Key Vocabulary
scientific notation, *p. 738*

Study Tip
Scientific notation is used to write very small and very large numbers.

Key Idea

Scientific Notation

A number is written in **scientific notation** when it is represented as the product of a factor and a power of 10. The factor must be greater than or equal to 1 and less than 10.

The factor is greater than or equal to 1 and less than 10. → 8.3×10^{-7} ← The power of 10 has an integer exponent.

EXAMPLE 1 Identifying Numbers Written in Scientific Notation

Tell whether the number is written in scientific notation. Explain.

a. 5.9×10^{-6}

The factor is greater than or equal to 1 and less than 10. The power of 10 has an integer exponent. So, the number is written in scientific notation.

b. 0.9×10^{8}

The factor is less than 1. So, the number is not written in scientific notation.

Key Idea

Writing Numbers in Standard Form

The absolute value of the exponent indicates how many places to move the decimal point.

- If the exponent is negative, move the decimal point to the left.
- If the exponent is positive, move the decimal point to the right.

EXAMPLE 2 Writing Numbers in Standard Form

a. Write 3.22×10^{-4} in standard form.

$3.22 \times 10^{-4} = 0.000322$ (4) Move decimal point $|-4| = 4$ places to the left.

b. Write 7.9×10^{5} in standard form.

$7.9 \times 10^{5} = 790{,}000$ (5) Move decimal point $|5| = 5$ places to the right.

Laurie's Notes

Introduction

Connect

- **Yesterday:** Students explored very large and very small numbers written in scientific notation.
- **Today:** Students will read numbers in scientific notation and write them in standard form.

Motivate

- Share some information about the Florida Keys.
- The Florida Keys are made up of approximately 1700 islands, or keys, that stretch about 126 miles from the mainland to the last key, Key West. Most of the islands are uninhabited, but the populated keys are connected by a highway that crosses 42 bridges.
- ? "Do you know how big a square foot is?" Answers will vary.
- The Florida Keys are approximately 3.83×10^9 square feet.
- ? "Is the area of the Keys more than or less than a billion square feet?" more than; Students will probably need to write 3.83×10^9 square feet in standard form to answer.

Lesson Notes

Key Idea

- Write the Key Idea. There are two parts to the definition; the factor is a number n, with $1 \le n < 10$, and it is multiplied by a power of 10 with an integer exponent.

Example 1

- Work through each part of the example as shown.

Key Idea

- Write the Key Idea. From the activities yesterday, students should find it reasonable that the exponent of 10 is connected to place value. If the exponent is positive, the number will be larger, so the decimal point moves to the right. Conversely, if the exponent is negative, the number will be smaller, so the decimal point moves to the left.
- ? Have students fill in the blanks.
  - "A power of 10 with a positive exponent is _____ 1." greater than or equal to
  - "A power of 10 with a negative exponent is _____ 1." less than

Example 2

- In part (a), 3.22 is the factor and -4 is the exponent. The number in standard form will be less than 3.22, so the decimal point moves to the left 4 places.
- In part (b), 7.9 is the factor and 5 is the exponent. The number in standard form will be greater than 7.9, so the decimal point moves to the right 5 places.

Goal Today's lesson is reading numbers in **scientific notation** and writing them in standard form.

Technology for the Teacher

Dynamic Classroom

Lesson Tutorials
Lesson Plans
Answer Presentation Tool

Extra Example 1

Tell whether the number is written in scientific notation. Explain.

a. 2.5×10^{-9} yes; The factor is greater than or equal to 1 and less than 10. The power of 10 has an integer exponent.

b. 0.5×10^{6} no; The factor is less than 1.

Extra Example 2

a. Write 2.75×10^{-3} in standard form. 0.00275

b. Write 6.38×10^{7} in standard form. 63,800,000

On Your Own

1. no; The factor is greater than 10.
2. 60,000,000
3. 0.000099 4. 12,850

Extra Example 3

In Example 3, the density of an ear of corn is 7.21×10^2 kilograms per cubic meter. What happens when an ear of corn is placed in water? The ear of corn is less dense than water, so it will float.

Extra Example 4

In Example 4, a dog has 50 female fleas. How much blood do the fleas consume per day? 0.7 milliliter of blood per day

On Your Own

5. It will sink.
6. 1.05 mL

English Language Learners

Word Problems

Have students work in groups of 3 or 4, including both English learners and English speakers. Provide them with poster board and markers. Assign each group a problem-solving exercise with scientific notation. Each group is to solve their problem showing all of the steps and using scientific notation. English learners will benefit by having the opportunity to restate the problem and gain a deeper understanding of the concept.

Laurie's Notes

On Your Own

? **Extension:** "How could you write the number in Question 1 in scientific notation?" 1.2×10^5

- **Think-Pair-Share:** Students should read each question independently and then work in pairs to answer the questions. When they have answered the questions, the pair should compare their answers with another group and discuss any discrepancies.

Example 3

- Ask a student to read the problem. Discuss the density of a substance if they have encountered it in science class, otherwise give an explanation of what density means. Density equals mass divided by volume.
- Work through the example as shown.
- **Look for and Make Use of Structure:** Show students an alternate way to solve Example 3. You can compare numbers without writing them in standard form. Rewrite the numbers so that they all have the same exponent, then compare the factor of each number. For example, 1.0×10^3, 1.84×10^3, and 0.641×10^3.

Example 4

- Before students get too concerned about fleas, remind them that 1.4×10^{-5} liter is a small amount!
- Students should be comfortable multiplying by powers of 10.

? "Can you multiply 10^{-5} by 100 first, and then multiply by 1.4? Explain." yes; Multiplication is commutative.

On Your Own

- **Neighbor Check:** Have students work independently and then have their neighbors check their work. Have students discuss any discrepancies.

Closure

- **Exit Ticket:** Write the number in standard form.
  1. 1.56×10^7 15,600,000
  2. 6.3×10^{-5} 0.000063

On Your Own

Now You're Ready
Exercises 6–23

1. Is 12×10^4 written in scientific notation? Explain.

Write the number in standard form.

2. 6×10^7
3. 9.9×10^{-5}
4. 1.285×10^4

EXAMPLE 3 Comparing Numbers in Scientific Notation

An object with a lesser density than water will float. An object with a greater density than water will sink. Use each given density (in kilograms per cubic meter) to explain what happens when you place a brick and an apple in water.

Water: 1.0×10^3 **Brick:** 1.84×10^3 **Apple:** 6.41×10^2

You can compare the densities by writing each in standard form.

Water	Brick	Apple
$1.0 \times 10^3 = 1000$	$1.84 \times 10^3 = 1840$	$6.41 \times 10^2 = 641$

∴ The apple is less dense than water, so it will float. The brick is denser than water, so it will sink.

EXAMPLE 4 Real-Life Application

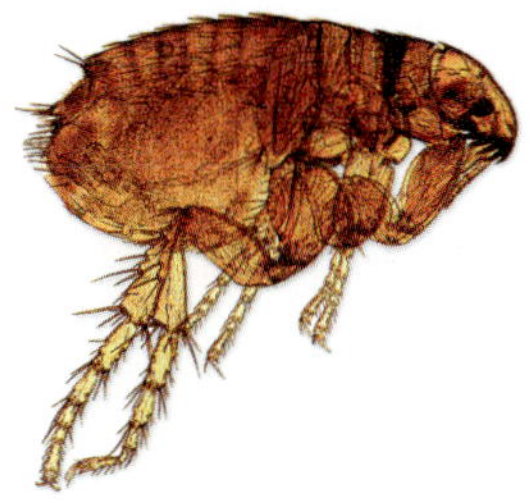

A female flea consumes about 1.4×10^{-5} liter of blood per day.

A dog has 100 female fleas. How much blood do the fleas consume per day?

$1.4 \times 10^{-5} \cdot 100 = 0.000014 \cdot 100$ Write in standard form.

$= 0.0014$ Multiply.

∴ The fleas consume about 0.0014 liter, or 1.4 milliliters of blood per day.

On Your Own

Now You're Ready
Exercise 27

5. **WHAT IF?** In Example 3, the density of lead is 1.14×10^4 kilograms per cubic meter. What happens when you place lead in water?
6. **WHAT IF?** In Example 4, a dog has 75 female fleas. How much blood do the fleas consume per day?

16.5 Exercises

Check It Out
Help with Homework
BigIdeasMath.com

Vocabulary and Concept Check

1. **WRITING** Describe the difference between scientific notation and standard form.

2. **WHICH ONE DOESN'T BELONG?** Which number does *not* belong with the other three? Explain.

2.8×10^{15} $\quad$ 4.3×10^{-30} $\quad$ 1.05×10^{28} $\quad$ 10×9.2^{-13}

Practice and Problem Solving

Write the number shown on the calculator display in standard form.

3. 5.6E12

4. 2.1E-10

5. 8.73E16

Tell whether the number is written in scientific notation. Explain.

1

6. 1.8×10^{9}

7. 3.45×10^{14}

8. 0.26×10^{-25}

9. 10.5×10^{12}

10. 46×10^{-17}

11. 5×10^{-19}

12. 7.814×10^{-36}

13. 0.999×10^{42}

14. 6.022×10^{23}

Write the number in standard form.

2

15. 7×10^{7}

16. 8×10^{-3}

17. 5×10^{2}

18. 2.7×10^{-4}

19. 4.4×10^{-5}

20. 2.1×10^{3}

21. 1.66×10^{9}

22. 3.85×10^{-8}

23. 9.725×10^{6}

24. **ERROR ANALYSIS** Describe and correct the error in writing the number in standard form.

✗ $4.1 \times 10^{-6} = 4{,}100{,}000$

2.7×10^{8} platelets per milliliter

25. **PLATELETS** Platelets are cell-like particles in the blood that help form blood clots.

    a. How many platelets are in 3 milliliters of blood? Write your answer in standard form.

    b. An adult human body contains about 5 liters of blood. How many platelets are in an adult human body?

Assignment Guide and Homework Check

Level	Assignment	Homework Check
Accelerated	1–5, 12–26 even, 27–36	12, 22, 27, 30, 31

Common Errors

- **Exercises 6–14** Students may think that all of the numbers are written in scientific notation because all of the exponents are integers. Remind them that the factor must be greater than or equal to 1 and less than 10 in order for the number to be in scientific notation.
- **Exercises 15–23** Students may move the decimal point in the wrong direction. Remind them that when the exponent is negative they move the decimal point to the left, and when it is positive they move the decimal point to the right.
- **Exercise 27** Students may order the surface temperatures by the factor without considering the power of 10. Encourage them to write the numbers in standard form, or rewrite the numbers so that they all have the same exponent, before comparing the numbers.

16.5 Record and Practice Journal

Tell whether the number is written in scientific notation. Explain.

1. 14×10^8 no; The factor is greater than 10.
2. 2.6×10^{12} yes; The factor is greater than 1 and less than 10.
3. 4.79×10^{-8} yes; The factor is greater than 1 and less than 10.
4. 3.99×10^{16} yes; The factor is greater than 1 and less than 10.
5. 0.15×10^{22} no; The factor is less than 1.
6. 6×10^3 yes; The factor is greater than 1 and less than 10.

Write the number in standard form.

7. 4×10^9 4,000,000,000
8. 2×10^{-5} 0.00002
9. 3.7×10^6 3,700,000
10. 4.12×10^{-3} 0.00412
11. 7.62×10^{10} 76,200,000,000
12. 9.908×10^{-12} 0.000000000009908
13. Light travels at 3×10^8 meters per second.
    a. Write the speed of light in standard form. 300,000,000
    b. How far has light traveled after 5 seconds? 1,500,000,000 meters

Vocabulary and Concept Check

1. Scientific notation uses a factor greater than or equal to 1 but less than 10 multiplied by a power of 10. A number in standard form is written out with all the zeros and place values included.
2. 10×9.2^{-13}; All of the other numbers are written in scientific notation.

Practice and Problem Solving

3. 5,600,000,000,000
4. 0.00000000021
5. 87,300,000,000,000,000
6. yes; The factor is greater than or equal to 1 and less than 10. The power of 10 has an integer exponent.
7. yes; The factor is greater than or equal to 1 and less than 10. The power of 10 has an integer exponent.
8. no; The factor is less than 1.
9. no; The factor is greater than 10.
10. no; The factor is greater than 10.
11. yes; The factor is greater than or equal to 1 and less than 10. The power of 10 has an integer exponent.
12. yes; The factor is greater than or equal to 1 and less than 10. The power of 10 has an integer exponent.
13. no; The factor is less than 1.
14. yes; The factor is greater than or equal to 1 and less than 10. The power of 10 has an integer exponent.

Practice and Problem Solving

15. 70,000,000 **16.** 0.008

17. 500 **18.** 0.00027

19. 0.000044 **20.** 2100

21. 1,660,000,000

22. 0.0000000385

23. 9,725,000

24. The negative exponent means the decimal point will move left, not right, when the number is written in standard form. $4.1 \times 10^{-6} = 0.0000041$

25. **a.** 810,000,000 platelets

b. 1,350,000,000,000 platelets

26. 100 zeros

27. **a.** Bellatrix

b. Betelgeuse

28. The value of the number is 10 times greater.

29. 1555.2 km^2

30. 5×10^{12} km^2

31. 35,000,000 km^3

32. See *Taking Math Deeper.*

Fair Game Review

33. 4^5

34. 3^3y^3

35. $(-2)^3$

36. B

Mini-Assessment

Write the number in standard form.

1. 5×10^{-4} 0.0005
2. 2.5×10^{-3} 0.0025
3. 1.66×10^3 1660
4. 3.89×10^{-5} 0.0000389
5. 4.576×10^8 457,600,000

Taking Math Deeper

Exercise 32

This is an interesting problem in physics. It is about the speed of light in different media. The problem gives students practice in unit analysis, and it also points out that to compare speeds, you need to compare apples to apples, not apples to oranges.

1 Make a table and convert measures.

Medium	Speed	Speed (m per sec)
Air	$\frac{6.7 \times 10^8 \text{ mi}}{\text{h}}$	$\frac{3.0 \times 10^8 \text{ m}}{\text{sec}}$
Glass	$\frac{6.6 \times 10^8 \text{ ft}}{\text{sec}}$	$\frac{2.0 \times 10^8 \text{ m}}{\text{sec}}$
Ice	$\frac{2.3 \times 10^5 \text{ km}}{\text{sec}}$	$\frac{2.3 \times 10^8 \text{ m}}{\text{sec}}$
Vacuum	$\frac{3.0 \times 10^8 \text{ m}}{\text{sec}}$	$\frac{3.0 \times 10^8 \text{ m}}{\text{sec}}$
Water	$\frac{2.3 \times 10^{10} \text{ cm}}{\text{sec}}$	$\frac{2.3 \times 10^8 \text{ m}}{\text{sec}}$

2 **a.** For the significant digits given, the speed of light is the same in air or in a vacuum. Light is fastest in these two media.

b. Of the five media listed, light travels the slowest in glass.

3 If students take a course in physics, they will learn that light is slowed down in transparent media such as air, water, ice, and glass. The ratio by which it is slowed is called the *refractive index* of the medium and is always greater than one. This was discovered by Jean Foucault in 1850. The refractive index of air is 1.0003, which means that light travels slightly slower in air than in a vacuum.

When people talk about "the speed of light" in a general context, they usually mean "the speed of light in a vacuum." This quantity is also refered to as c. It is famous from Einstein's equation $E = mc^2$.

Reteaching and Enrichment Strategies

If students need help. . .	If students got it. . .
Resources by Chapter • Practice A and Practice B • Puzzle Time Record and Practice Journal Practice Differentiating the Lesson Lesson Tutorials Skills Review Handbook	Resources by Chapter • Enrichment and Extension • Technology Connection Start the next section

26. **REASONING** A googol is 1.0×10^{100}. How many zeros are in a googol?

3 27. **STARS** The table shows the surface temperatures of five stars.

a. Which star has the highest surface temperature?

b. Which star has the lowest surface temperature?

Star	Betelgeuse	Bellatrix	Sun	Aldebaran	Rigel
Surface Temperature (°F)	6.2×10^3	3.8×10^4	1.1×10^4	7.2×10^3	2.2×10^4

28. **NUMBER SENSE** Describe how the value of a number written in scientific notation changes when you increase the exponent by 1.

29. **CORAL REEF** The area of the Florida Keys National Marine Sanctuary is about 9.6×10^3 square kilometers. The area of the Florida Reef Tract is about 16.2% of the area of the sanctuary. What is the area of the Florida Reef Tract in square kilometers?

30. **REASONING** A gigameter is 1.0×10^6 kilometers. How many square kilometers are in 5 square gigameters?

31. **WATER** There are about 1.4×10^9 cubic kilometers of water on Earth. About 2.5% of the water is fresh water. How much fresh water is on Earth?

32. **Critical Thinking** The table shows the speed of light through five media.

a. In which medium does light travel the fastest?

b. In which medium does light travel the slowest?

Medium	Speed
Air	6.7×10^8 mi/h
Glass	6.6×10^8 ft/sec
Ice	2.3×10^5 km/sec
Vacuum	3.0×10^8 m/sec
Water	2.3×10^{10} cm/sec

Fair Game Review What you learned in previous grades & lessons

Write the product using exponents. *(Section 16.1)*

33. $4 \cdot 4 \cdot 4 \cdot 4 \cdot 4$

34. $3 \cdot 3 \cdot 3 \cdot y \cdot y \cdot y$

35. $(-2) \cdot (-2) \cdot (-2)$

36. **MULTIPLE CHOICE** What is the length of the hypotenuse of the right triangle? *(Section 14.3)*

Ⓐ $\sqrt{18}$ in. Ⓑ $\sqrt{41}$ in.

Ⓒ 18 in. Ⓓ 41 in.

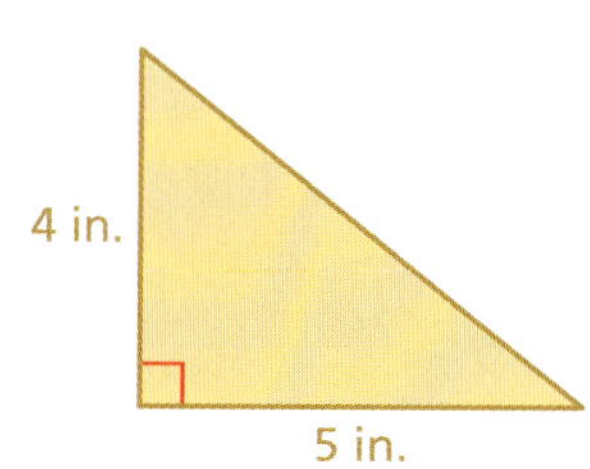

16.6 Writing Scientific Notation

Essential Question How can you write a number in scientific notation?

1 ACTIVITY: Finding pH Levels

Work with a partner. In chemistry, pH is a measure of the activity of dissolved hydrogen ions (H^+). Liquids with low pH values are called *acids*. Liquids with high pH values are called *bases*.

Find the pH of each liquid. Is the liquid a base, neutral, or an acid?

a. Lime juice:
$[H^+] = 0.01$

b. Egg:
$[H^+] = 0.00000001$

c. Distilled water:
$[H^+] = 0.0000001$

d. Ammonia water:
$[H^+] = 0.00000000001$

e. Tomato juice:
$[H^+] = 0.0001$

f. Hydrochloric acid:
$[H^+] = 1$

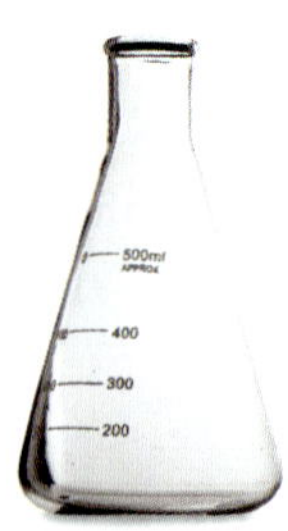

pH	$[H^+]$	
14	1×10^{-14}	Bases
13	1×10^{-13}	
12	1×10^{-12}	
11	1×10^{-11}	
10	1×10^{-10}	
9	1×10^{-9}	
8	1×10^{-8}	
7	1×10^{-7}	Neutral
6	1×10^{-6}	Acids
5	1×10^{-5}	
4	1×10^{-4}	
3	1×10^{-3}	
2	1×10^{-2}	
1	1×10^{-1}	
0	1×10^{0}	

Scientific Notation

In this lesson, you will

- write large and small numbers in scientific notation.
- perform operations with numbers written in scientific notation.

Laurie's Notes

Introduction

Applying Mathematical Practices

- **Make Sense of Problems** and **Use Appropriate Tools Strategically:** Visual models are used to help students make sense of numbers written in scientific notation.

Motivate

- **Preparation:** If possible, borrow a pH meter or a few strips of litmus paper from the science department. If these items are not available, move on to Activity 1.
- Without explanation, have 3 containers (coffee cups or paper cups) containing different liquids at the front of the room.
- Use the litmus paper or pH meter to test the liquids. Students should guess that you are doing a pH test, which leads into Activity 1.

Words of Wisdom

- **Safety:** For reasons of safety you should not consume, nor allow the students to consume, any of the liquids.

Activity Notes

Activity 1

? "How many of you have heard of pH or studied pH in science?"

? "What does pH level refer to? Can anyone explain?" Listen for the measure of concentration of dissolved hydrogen ions; liquids with low pH are called acids; liquids with high pH are called bases.

- **Use Appropriate Tools Strategically:** Have students refer to the pH chart in the book. The pH level is a number from 0 to 14, the opposite of the exponent (0 to -14) measuring the concentration of the dissolved hydrogen ions. At the middle of the chart, a pH of 7 is called neutral.
- **FYI:** Pure water is neutral with a pH value of 7. Low on the scale are acids, which have a sour taste like lemons. High on the scale are bases, which have a bitter taste like soap.
- Notice that all of the pH values are given in standard form. Students will need to think about how these numbers would be written in scientific notation.

? "How did you compare the numbers in standard form with the scale in scientific notation?" Listen for idea of the number of place values away from 1.

What Your Students Will Learn

- Write large and small numbers in scientific notation by moving the decimal point.
- Perform operations with numbers written in scientific notation.

Previous Learning

Students should be familiar with the metric system and powers of ten.

Technology for the Teacher

Dynamic Classroom

Lesson Plans
Complete Materials List

16.6 Record and Practice Journal

Essential Question How can you write a number in scientific notation?

1 ACTIVITY: Finding pH Levels

Work with a partner. In chemistry, pH is a measure of the activity of dissolved hydrogen ions (H^+). Liquids with low pH values are called *acids*. Liquids with high pH values are called *bases*.

Find the pH of each liquid. Is the liquid a base, neutral, or an acid?

a. Lime juice: $[H^+] = 0.01$
2; acid

b. Egg: $[H^+] = 0.00000001$
8; base

c. Distilled water: $[H^+] = 0.0000001$
7; neutral

d. Ammonia water: $[H^+] = 0.00000000001$
11; base

e. Tomato juice: $[H^+] = 0.0001$
4; acid

f. Hydrochloric acid: $[H^+] = 1$
0; acid

pH	$[H^+]$	
14	1×10^{-14}	Bases
13	1×10^{-13}	
12	1×10^{-12}	
11	1×10^{-11}	
10	1×10^{-10}	
9	1×10^{-9}	
8	1×10^{-8}	
7	1×10^{-7}	Neutral
6	1×10^{-6}	Acids
5	1×10^{-5}	
4	1×10^{-4}	
3	1×10^{-3}	
2	1×10^{-2}	
1	1×10^{-1}	
0	1×10^{0}	

English Language Learners

Culture

Have each student use the Internet or library resources to find the land area of his or her native country, the state where he or she currently lives, and two other states in the U.S. that have approximately the same land area as the native country. Each student can organize the information in a table giving the land area, the land area rounded to the nearest thousand square miles, and the land area in scientific notation. Create a classroom display of the tables.

16.6 Record and Practice Journal

2 ACTIVITY: Writing Scientific Notation

Work with a partner. Match each planet with its distance from the Sun. Then write each distance in scientific notation. Do you think it is easier to match the distances when they are written in standard form or in scientific notation? Explain.

a. 1,800,000,000 miles
Uranus; 1.8×10^9 mi

b. 67,000,000 miles
Venus; 6.7×10^7 mi

c. 890,000,000 miles
Saturn; 8.9×10^8 mi

d. 93,000,000 miles
Earth; 9.3×10^7 mi

e. 140,000,000 miles
Mars; 1.4×10^8 mi

f. 2,800,000,000 miles
Neptune; 2.8×10^9 mi

g. 480,000,000 miles
Jupiter; 4.8×10^8 mi

h. 36,000,000 miles
Mercury; 3.6×10^7 mi

Scientific notation; You can compare the exponents.

3 ACTIVITY: Making a Scale Drawing

Work with a partner. The illustration in Activity 2 is not drawn to scale. Use the instructions below to make a scale drawing of the distances in our solar system.

- **Cut a sheet of paper into three strips of equal width. Tape the strips together.**
- **Draw a long number line. Label the number line in hundreds of millions of miles.**
- **Locate each planet's position on the number line.**

Check students' work.

What Is Your Answer?

4. **IN YOUR OWN WORDS** How can you write a number in scientific notation?

See Additional Answers.

Laurie's Notes

Activity 2

- This activity demonstrates one way scientific notation is helpful. The distance of each planet from the sun is a very large number (in magnitude).
- Students should be able to get started right away without much explanation.

? "How do you match the planets with the distances?" Listen for ordering distances least to greatest and looking at the illustration.

? "How do you write the distances from the sun in scientific notation?" Move the decimal point to the left until you have a number that is at least 1 and less than 10. Count the number of places you moved the decimal point. This number becomes the exponent of the power of 10.

- **Model with Mathematics:** Because of the context, and the visual illustration, students should not have difficulty with this activity. They understand how to write the numbers in scientific notation.

Activity 3

- To help facilitate this activity, you may want to prepare strips of paper in advance.
- Even though the scale is given (hundreds of millions of miles), just writing multiples of this scale is a challenge for students. You may want to make a model on the front board.
- If students have studied the solar system, they should have some sense about how far away Neptune is.
- **Make Sense of Problems** and **Reason Abstractly and Quantitatively:** To make a number line model of the distances to scale, you can write the distances in scientific notation. Each time the exponent increases by 1, the distance increases by a factor of 10.

What Is Your Answer?

- **Neighbor Check:** Have students work independently and then have their neighbors check their work. Have students discuss any discrepancies.

Closure

- **Writing Prompt:** Why is it useful to write very large or very small numbers in scientific notation?

2 ACTIVITY: Writing Scientific Notation

Work with a partner. Match each planet with its distance from the Sun. Then write each distance in scientific notation. Do you think it is easier to match the distances when they are written in standard form or in scientific notation? Explain.

a. 1,800,000,000 miles

b. 67,000,000 miles

c. 890,000,000 miles

d. 93,000,000 miles

e. 140,000,000 miles

f. 2,800,000,000 miles

g. 480,000,000 miles

h. 36,000,000 miles

3 ACTIVITY: Making a Scale Drawing

Math Practice

Calculate Accurately

How can you verify that you have accurately written each distance in scientific notation?

Work with a partner. The illustration in Activity 2 is not drawn to scale. Use the instructions below to make a scale drawing of the distances in our solar system.

- **Cut a sheet of paper into three strips of equal width. Tape the strips together to make one long piece.**
- **Draw a long number line. Label the number line in hundreds of millions of miles.**
- **Locate each planet's position on the number line.**

What Is Your Answer?

4. **IN YOUR OWN WORDS** How can you write a number in scientific notation?

Practice Use what you learned about writing scientific notation to complete Exercises 3–5 on page 746.

16.6 Lesson

Check It Out
Lesson Tutorials
BigIdeasMath.com

Key Idea

Writing Numbers in Scientific Notation

Step 1: Move the decimal point so it is located to the right of the leading nonzero digit.

Step 2: Count the number of places you moved the decimal point. This indicates the exponent of the power of 10, as shown below.

Number Greater Than or Equal to 10

Use a positive exponent when you move the decimal point to the left.

$$8600 = 8.6 \times 10^3$$

3

Number Between 0 and 1

Use a negative exponent when you move the decimal point to the right.

$$0.0024 = 2.4 \times 10^{-3}$$

3

Study Tip

When you write a number greater than or equal to 1 and less than 10 in scientific notation, use zero as the exponent.

$$6 = 6 \times 10^0$$

EXAMPLE 1 Writing Large Numbers in Scientific Notation

Google purchased YouTube for $1,650,000,000. Write this number in scientific notation.

Move the decimal point 9 places to the left.

$$1{,}650{,}000{,}000 = 1.65 \times 10^9$$

9

The number is greater than 10. So, the exponent is positive.

EXAMPLE 2 Writing Small Numbers in Scientific Notation

The 2004 Indonesian earthquake slowed the rotation of Earth, making the length of a day 0.00000268 second shorter. Write this number in scientific notation.

Move the decimal point 6 places to the right.

$$0.00000268 = 2.68 \times 10^{-6}$$

6

The number is between 0 and 1. So, the exponent is negative.

On Your Own

Now You're Ready
Exercises 3–11

Write the number in scientific notation.

1. 50,000 **2.** 25,000,000 **3.** 683

4. 0.005 **5.** 0.00000033 **6.** 0.000506

Laurie's Notes

Introduction

Connect

- **Yesterday:** Students explored very large and very small numbers written in standard form.
- **Today:** Students will convert numbers in standard form to scientific notation.

Motivate

- Share with students some information about the U.S. economy.
- Numbers related to the U.S. economy are often so large and so common, many become numb to their magnitude. The place values, millions, billions, and trillions, differ by only a letter or two, yet their magnitudes are significantly different.
- The reason to bring these numbers up is that after you move beyond a million, many calculators will not display the number in standard form.

Lesson Notes

Key Idea

- Review with students the definition of scientific notation.
- Write the Key Idea. Relate the steps to the activities from the investigation.

? "Why is the decimal point moved to the right of the first nonzero digit?" Need the factor n to be in the interval $1 \le n < 10$.

Example 1

- **Teaching Tip:** Have students underline the first nonzero digit and the digit to its right. In scientific notation, the decimal point is placed between these two digits.

? "How do you read the number?" one billion six hundred fifty million dollars

- **FYI:** Drawing the movement of the decimal point under the numbers helps students keep track of their counting.

? **Construct Viable Arguments:** "How do you know if the exponent for the power of 10 will be positive or negative?" If the standard form of the number is greater than or equal to 10, positive exponent; if the standard form of the number is between 0 and 1, negative exponent.

Example 2

- Discuss the context of this number.
- Note that reading the number takes time, and you have to count place values. The number is 268 hundred millionths.

? "Why is the exponent negative?" Original number is between 0 and 1.

On Your Own

- **Think-Pair-Share:** Students should read each question independently and then work in pairs to answer the questions. When they have answered the questions, the pair should compare their answers with another group and discuss any discrepancies.

Goal Today's lesson is writing numbers in scientific notation.

Technology for the Teacher

Dynamic Classroom

Lesson Tutorials
Lesson Plans
Answer Presentation Tool

Extra Example 1

Write 2,450,000 in scientific notation.
2.45×10^6

Extra Example 2

Write 0.0000045 in scientific notation.
4.5×10^{-6}

On Your Own

1. 5×10^4
2. 2.5×10^7
3. 6.83×10^2
4. 5×10^{-3}
5. 3.3×10^{-7}
6. 5.06×10^{-4}

Laurie's Notes

Example 3

- Encourage students to read the numbers when they are in standard form: 10 million; 8 million 780 thousand.
- The subtraction is performed on the numbers in standard form, with the result written in scientific notation.
- **Look for and Make Use of Structure:** Ask students if it is possible to subtract numbers in scientific notation. The answer is no, unless the powers of 10 are the same. Students will learn more about this in the next section.

Example 4

- Students have compared numbers written in scientific notation by writing each number in standard form. Another method is presented in this example. Students first compare the powers of ten, followed by comparing the decimal numbers.
- **Common Error:** Note that in order to use this method, *all* the numbers must be written in scientific notation.
- **Extension:** Create a proportionally correct timeline of the geologic periods. This can be done in a hallway or within the classroom. There are many resources online for this project.

On Your Own

- **Think-Pair-Share:** Students should read each question independently and then work in pairs to answer the questions. When they have answered the questions, the pair should compare their answers with another group and discuss any discrepancies.

Closure

- The land area of Virginia is about 39,500 square miles. The land area of Alaska is about 570,000 square miles. The United States land area is about 3,500,000 square miles. Write each of these in scientific notation. 3.95×10^4, 5.7×10^5, 3.5×10^6

Extra Example 3

In Example 3, an album has sold 750,000 copies. How many more copies does it need to sell to receive the award? Write your answer in scientific notation. 9.25×10^6 copies

Extra Example 4

Order 6.8×10^4, 2.04×10^5, and 5.65×10^4 from least to greatest. 5.65×10^4, 6.8×10^4, 2.04×10^5

On Your Own

7. 9.045×10^6
8. Mesozoic era

Differentiated Instruction

Kinesthetic

Students may incorrectly count the number of zeros in a number and use that as the exponent in scientific notation. Have students write the number and place an arrow where the decimal point will be placed in the factor. Then have students count the number of places the decimal point moves.

$54{,}000 = 5.4 \times 10^4$ (arrow after the 5)

$0.00000675 = 6.75 \times 10^{-6}$ (arrow after the 6)

EXAMPLE 3 Using Scientific Notation

An album receives an award when it sells 10,000,000 copies.

An album has sold 8,780,000 copies. How many more copies does it need to sell to receive the award?

Ⓐ 1.22×10^{-7} Ⓑ 1.22×10^{-6}

Ⓒ 1.22×10^{6} Ⓓ 1.22×10^{7}

Use a model to solve the problem.

Remaining sales needed for award = Sales required for award − Current sales total

$= 10{,}000{,}000 - 8{,}780{,}000$

$= 1{,}220{,}000$

$= 1.22 \times 10^{6}$

The album must sell 1.22×10^{6} more copies to receive the award. So, the correct answer is Ⓒ.

EXAMPLE 4 Real-Life Application

The table shows when the last three geologic eras began. Order the eras from earliest to most recent.

Era	Began
Paleozoic	5.42×10^{8} years ago
Cenozoic	6.55×10^{7} years ago
Mesozoic	2.51×10^{8} years ago

Step 1: Compare the powers of 10.

Because $10^{7} < 10^{8}$,

$6.55 \times 10^{7} < 5.42 \times 10^{8}$ and

$6.55 \times 10^{7} < 2.51 \times 10^{8}$.

Step 2: Compare the factors when the powers of 10 are the same.

Because $2.51 < 5.42$,

$2.51 \times 10^{8} < 5.42 \times 10^{8}$.

From greatest to least, the order is 5.42×10^{8}, 2.51×10^{8}, and 6.55×10^{7}.

So, the eras in order from earliest to most recent are the Paleozoic era, Mesozoic era, and Cenozoic era.

Common Error

To use the method in Example 4, the numbers must be written in scientific notation.

On Your Own

Now You're Ready
Exercises 14–19

7. **WHAT IF?** In Example 3, an album has sold 955,000 copies. How many more copies does it need to sell to receive the award? Write your answer in scientific notation.

8. The *Tyrannosaurus rex* lived 7.0×10^{7} years ago. Consider the eras given in Example 4. During which era did the *Tyrannosaurus rex* live?

16.6 Exercises

Check It Out
Help with Homework
BigIdeasMath.com

Vocabulary and Concept Check

1. **REASONING** How do you know whether a number written in standard form will have a positive or a negative exponent when written in scientific notation?
2. **WRITING** When is it appropriate to use scientific notation instead of standard form?

Practice and Problem Solving

Write the number in scientific notation.

1 2 3. 0.0021 | 4. 5,430,000 | 5. 321,000,000

6. 0.00000625 | 7. 0.00004 | 8. 10,700,000

9. 45,600,000,000 | 10. 0.000000000009256 | 11. 840,000

ERROR ANALYSIS **Describe and correct the error in writing the number in scientific notation.**

12.

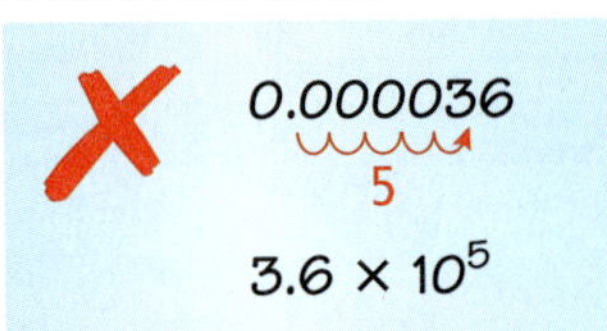

13.

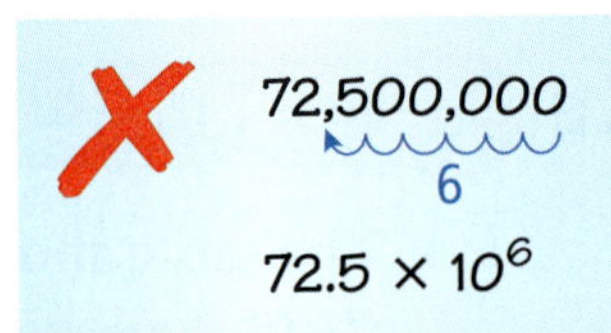

Order the numbers from least to greatest.

4 14. 1.2×10^8, 1.19×10^8, 1.12×10^8

15. 6.8×10^{-5}, 6.09×10^{-5}, 6.78×10^{-5}

16. 5.76×10^{12}, 9.66×10^{11}, 5.7×10^{10}

17. 4.8×10^{-6}, 4.8×10^{-5}, 4.8×10^{-8}

18. 9.9×10^{-15}, 1.01×10^{-14}, 7.6×10^{-15}

19. 5.78×10^{23}, 6.88×10^{-23}, 5.82×10^{23}

20. **HAIR** What is the diameter of a human hair written in scientific notation?

Diameter: 0.000099 meter

21. **EARTH** What is the circumference of Earth written in scientific notation?

Circumference at the equator: about 40,100,000 meters

22. **CHOOSING UNITS** In Exercise 21, name a unit of measurement that would be more appropriate for the circumference. Explain.

Assignment Guide and Homework Check

Level	Assignment	Homework Check
Accelerated	1–5, 6–26 even, 27–34	10, 18, 24, 27, 29

Common Errors

- **Exercises 3–11** Students may write an exponent with the opposite sign of what is correct. Remind them that large numbers have a positive exponent in scientific notation and that small numbers have a negative exponent in scientific notation.
- **Exercises 14–19** Students may order the numbers without taking into account the power of 10. Encourage them to order the numbers by powers first and then by decimal factors.

16.6 Record and Practice Journal

Write the number in scientific notation.

1. 4,200,000
   4.2×10^6
2. 0.038
   3.8×10^{-2}
3. 600,000
   6×10^5
4. 0.0000808
   8.08×10^{-5}
5. 0.0007
   7×10^{-4}
6. 29,010,000,000
   2.901×10^{10}

Order the numbers from least to greatest.

7. $6.4 \times 10^8, 5.3 \times 10^9, 2.3 \times 10^8$
   2.3×10^8
   6.4×10^8
   5.3×10^9
8. $9.1 \times 10^{-3}, 9.6 \times 10^{-3}, 9.02 \times 10^{-3}$
   9.02×10^{-3}
   9.1×10^{-3}
   9.6×10^{-3}
9. $7.3 \times 10^7, 5.6 \times 10^{10}, 3.7 \times 10^9$
   7.3×10^7
   3.7×10^9
   5.6×10^{10}
10. $1.4 \times 10^{-5}, 2.01 \times 10^{-15}, 6.3 \times 10^{-2}$
    2.01×10^{-15}
    1.4×10^{-5}
    6.3×10^{-2}
11. A patient has 0.0000075 gram of iron in 1 liter of blood. The normal level is between 6×10^{-7} gram and 1.6×10^{-5} gram. Is the patient's iron level normal? Write the patient's amount of iron in scientific notation.
    yes; 7.5×10^{-6}

Vocabulary and Concept Check

1. If the number is greater than or equal to 10, the exponent will be positive. If the number is less than 1 and greater than 0, the exponent will be negative.
2. It is appropriate to use scientific notation instead of standard form when a number is very large or very small.

Practice and Problem Solving

3. 2.1×10^{-3}
4. 5.43×10^6
5. 3.21×10^8
6. 6.25×10^{-6}
7. 4×10^{-5}
8. 1.07×10^7
9. 4.56×10^{10}
10. 9.256×10^{-12}
11. 8.4×10^5
12. The decimal point moved 5 places to the right, so the exponent should be negative. 3.6×10^{-5}
13. 72.5 is not less than 10. The decimal point needs to move one more place to the left. 7.25×10^7
14. $1.12 \times 10^8, 1.19 \times 10^8, 1.2 \times 10^8$
15. $6.09 \times 10^{-5}, 6.78 \times 10^{-5}, 6.8 \times 10^{-5}$
16. $5.7 \times 10^{10}, 9.66 \times 10^{11}, 5.76 \times 10^{12}$
17. $4.8 \times 10^{-8}, 4.8 \times 10^{-6}, 4.8 \times 10^{-5}$
18. $7.6 \times 10^{-15}, 9.9 \times 10^{-15}, 1.01 \times 10^{-14}$
19. $6.88 \times 10^{-23}, 5.78 \times 10^{23}, 5.82 \times 10^{23}$
20. 9.9×10^{-5} m
21. 4.01×10^7 m

Practice and Problem Solving

22. *Sample answer:* kilometers or miles; They are both larger units of length, so the number would be smaller.

23. $680,\ 6.8 \times 10^3,\ \dfrac{68{,}500}{10}$

24. $0.02,\ \dfrac{5}{241},\ 2.1 \times 10^{-2}$

25. $6.25 \times 10^{-3},\ 6.3\%,\ 0.625,\ 6\frac{1}{4}$

26. $305\%,\ 3.3 \times 10^2,\ 3033.4,\ \dfrac{10{,}000}{3}$

27. 1.99×10^9 watts

28. *Sample answer:* Enter 1.174E10 − 9.75E9.

29. carat; Because 1 carat $= 1.2 \times 10^{23}$ atomic mass units and 1 milligram $= 6.02 \times 10^{20}$ atomic mass units, and $1.2 \times 10^{23} > 6.02 \times 10^{20}$.

30. See *Taking Math Deeper*.

Fair Game Review

31. natural, whole, integer, rational

32. integer, rational

33. irrational

34. D

Mini-Assessment

Write the number in scientific notation.

1. 0.00035 3.5×10^{-4}
2. 0.0000000000567 5.67×10^{-11}
3. 25,500,000 2.55×10^7

Order the numbers from least to greatest.

4. $3 \times 10^4,\ 6.1 \times 10^3,\ 1.6 \times 10^4$
   $6.1 \times 10^3,\ 1.6 \times 10^4,\ 3 \times 10^4$
5. $5.8 \times 10^{-6},\ 2.8 \times 10^{-7},\ 5.9 \times 10^{-6}$
   $2.8 \times 10^{-7},\ 5.8 \times 10^{-6},\ 5.9 \times 10^{-6}$

Taking Math Deeper

Exercise 30

This is an interesting question that deals with the different eras in Earth's geological history.

1 Order the eras from oldest to youngest. Write each beginning date in standard form.

Era	Began	Standard Form
Paleozoic era	5.42×10^8 years ago	542,000,000 years ago
Mesozoic era	2.51×10^8 years ago	251,000,000 years ago
Cenozoic era	6.55×10^7 years ago	65,500,000 years ago

2 Find the length of each era.

Paleozoic era = 542,000,000 − 251,000,000 = 291,000,000 yr
Mesozoic era = 251,000,000 − 65,500,000 = 185,500,000 yr
Cenozoic era = 65,500,000 − 0 = 65,500,000 yr

In scientific notation, these lengths from least to greatest are:

a. Cenozoic era = 6.55×10^7 yr
Mesozoic era = 1.855×10^8 yr
Paleozoic era = 2.91×10^8 yr

3 **b.** You can make a time line that is either horizontal or vertical. Here is an example of a vertical time line.

c. For these three eras, the older the era, the longer it is. This is also true of the next three eras.
Neoproterozoic:
began 1.0×10^9 years ago
Mesoproterozoic:
began 1.6×10^9 years ago
Paleoproterozoic:
began 2.5×10^9 years ago

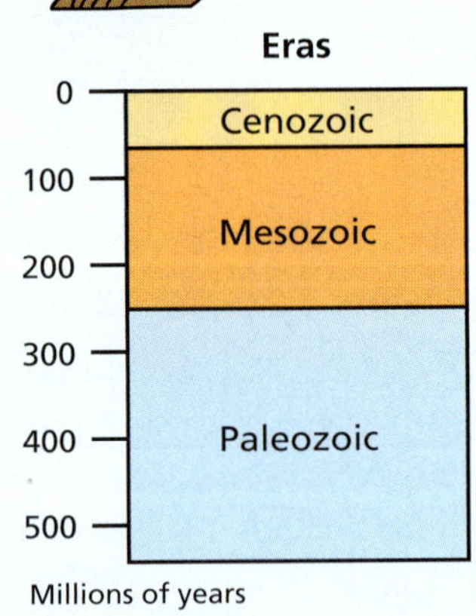

Reteaching and Enrichment Strategies

If students need help. . .	If students got it. . .
Resources by Chapter • Practice A and Practice B • Puzzle Time Record and Practice Journal Practice Differentiating the Lesson Lesson Tutorials Skills Review Handbook	Resources by Chapter • Enrichment and Extension • Technology Connection Start the next section

Order the numbers from least to greatest.

23. $\frac{68{,}500}{10}$, 680, 6.8×10^3

24. $\frac{5}{241}$, 0.02, 2.1×10^{-2}

25. 6.3%, 6.25×10^{-3}, $6\frac{1}{4}$, 0.625

26. 3033.4, 305%, $\frac{10{,}000}{3}$, 3.3×10^2

27. SPACE SHUTTLE The total power of a space shuttle during launch is the sum of the power from its solid rocket boosters and the power from its main engines. The power from the solid rocket boosters is 9,750,000,000 watts. What is the power from the main engines?

Total power = 1.174×10^{10} watts

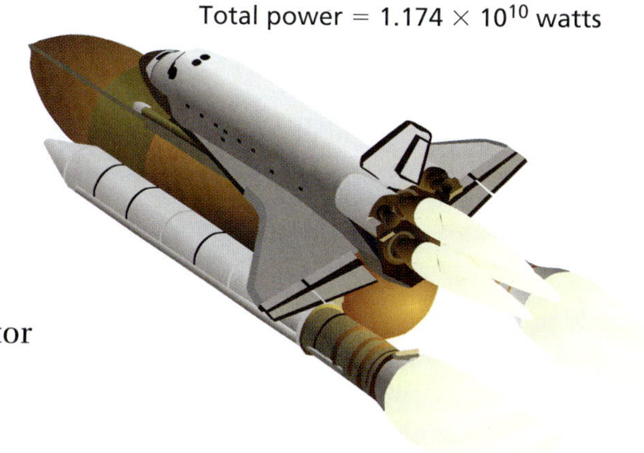

28. CHOOSE TOOLS Explain how to use a calculator to verify your answer to Exercise 27.

Equivalent to 1 Atomic Mass Unit
8.3×10^{-24} carat
1.66×10^{-21} milligram

29. ATOMIC MASS The mass of an atom or molecule is measured in atomic mass units. Which is greater, a *carat* or a *milligram*? Explain.

30. Reasoning In Example 4, the Paleozoic era ended when the Mesozoic era began. The Mesozoic era ended when the Cenozoic era began. The Cenozoic era is the current era.

a. Write the lengths of the three eras in scientific notation. Order the lengths from least to greatest.

b. Make a time line to show when the three eras occurred and how long each era lasted.

c. What do you notice about the lengths of the three eras? Use the Internet to determine whether your observation is true for *all* the geologic eras. Explain your results.

Fair Game Review What you learned in previous grades & lessons

Classify the real number. *(Section 14.4)*

31. 15

32. $\sqrt[3]{-8}$

33. $\sqrt{73}$

34. What is the surface area of the prism? *(Section 9.1)*

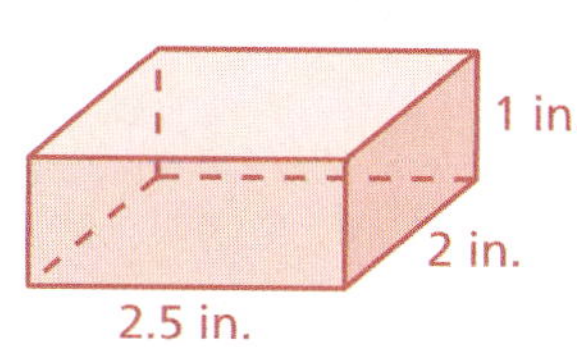

Ⓐ 5 in.^{2}

Ⓑ 5.5 in.^{2}

Ⓒ 10 in.^{2}

Ⓓ 19 in.^{2}

16.7 Operations in Scientific Notation

Essential Question How can you perform operations with numbers written in scientific notation?

1 ACTIVITY: Adding Numbers in Scientific Notation

Work with a partner. Consider the numbers 2.4×10^3 and 7.1×10^3.

a. Explain how to use order of operations to find the sum of these numbers. Then find the sum.

$$2.4 \times 10^3 + 7.1 \times 10^3$$

b. The factor ______ is common to both numbers. How can you use the Distributive Property to rewrite the sum $(2.4 \times 10^3) + (7.1 \times 10^3)$?

$(2.4 \times 10^3) + (7.1 \times 10^3) =$ ______ Distributive Property

c. Use order of operations to evaluate the expression you wrote in part (b). Compare the result with your answer in part (a).

d. **STRUCTURE** Write a rule you can use to add numbers written in scientific notation where the powers of 10 are the same. Then test your rule using the sums below.

- $(4.9 \times 10^5) + (1.8 \times 10^5) =$ ______
- $(3.85 \times 10^4) + (5.72 \times 10^4) =$ ______

2 ACTIVITY: Adding Numbers in Scientific Notation

Work with a partner. Consider the numbers 2.4×10^3 and 7.1×10^4.

a. Explain how to use order of operations to find the sum of these numbers. Then find the sum.

$$2.4 \times 10^3 + 7.1 \times 10^4$$

b. How is this pair of numbers different from the pairs of numbers in Activity 1?

c. Explain why you cannot immediately use the rule you wrote in Activity 1(d) to find this sum.

d. **STRUCTURE** How can you rewrite one of the numbers so that you can use the rule you wrote in Activity 1(d)? Rewrite one of the numbers. Then find the sum using your rule and compare the result with your answer in part (a).

e. **REASONING** Do these procedures work when subtracting numbers written in scientific notation? Justify your answer by evaluating the differences below.

- $(8.2 \times 10^5) - (4.6 \times 10^5) =$ ______
- $(5.88 \times 10^5) - (1.5 \times 10^4) =$ ______

Scientific Notation

In this lesson, you will

- add, subtract, multiply, and divide numbers written in scientific notation.

Laurie's Notes

Introduction

Applying Mathematical Practices

- **Construct Viable Arguments:** Mathematically proficient students are able to give explanations for how a computation is performed. They reference definitions and properties in establishing the validity of their argument.

Motivate

- Write the following problem on the board and ask student to evaluate it.
  $40 \times 10 + 8 \times 10 + 0.5 \times 10 + 0.07 \times 10$
- ? "How did you evaluate this expression?" It is likely that students found the sum of the four products.
- ? "Is there another method that could be used?" Students might mention factoring out the 10.
- **Look for and Make Use of Structure:** Underline the 10s and ask students how the Distributive Property could be used to evaluate the expression. Then write: $10(40 + 8 + 0.5 + 0.07)$.
- Tell students to use mental math to find the product: $10(48.57) = 485.7$.

Activity Notes

Activity 1

- After working through the Motivate, students should have a strategy for working through this activity. What students might not recognize is that 10^3 is a common factor, just like 10 was a common factor in the Motivate.
- ? "What is 10^3 in standard form?" 1000
- **Construct Viable Arguments:** In part (d), listen for students explaining the rule for adding numbers in scientific notation.

Activity 2

- The written problem looks very similar to the problem in Activity 1. What students will note when they examine it closely is that the powers of 10 are not the same.
- **Make Sense of Problems and Persevere in Solving Them:** Resist the urge to jump in and tell students how to proceed. Reassure them that they have the knowledge and skills to work through the problem.
- ? "Why can't you immediately use the rule you wrote in Activity 1?" There is no common factor because the powers of 10 are different.
- ? "What number or numbers did you rewrite in part (d)?" rewrite 2.4×10^3 as 0.24×10^4, or 7.1×10^4 as 71×10^3
- **Big Idea:** Students need to pay extra attention to the powers of 10 when adding and subtracting numbers in scientific notation.

What Your Students Will Learn

- Add, subtract, multiply, and divide numbers written in scientific notation.

Previous Learning

Students have written numbers in scientific notation.

Technology for the Teacher

Dynamic Classroom

Lesson Plans
Complete Materials List

16.7 Record and Practice Journal

Essential Question How can you perform operations with numbers written in scientific notation?

1 ACTIVITY: Adding Numbers in Scientific Notation

Work with a partner. Consider the numbers 2.4×10^3 and 7.1×10^3.

a. Explain how to use order of operations to find the sum of these numbers. Then find the sum.

$2.4 \times 10^3 + 7.1 \times 10^3$ **9500**

b. The factor 10^3 is common to both numbers. How can you use the Distributive Property to rewrite the sum $(2.4 \times 10^3) + (7.1 \times 10^3)$?

$(2.4 \times 10^3) + (7.1 \times 10^3) =$ $(2.4 + 7.1) \times 10^3$ Distributive Property

c. Use order of operations to evaluate the expression you wrote in part (b). Compare the result with your answer in part (a).

9500; same

d. STRUCTURE Write a rule you can use to add numbers written in scientific notation where the powers of 10 are the same. Then test your rule using the sums below. **Add the factors.**

- $(4.9 \times 10^5) + (1.8 \times 10^5) = (4.9 + 1.8) \times 10^5 = 6.7 \times 10^5$
- $(3.85 \times 10^4) + (5.72 \times 10^4) = (3.85 + 5.72) \times 10^4 = 9.57 \times 10^4$

2 ACTIVITY: Adding Numbers in Scientific Notation

Work with a partner. Consider the numbers 2.4×10^3 and 7.1×10^4.

a. Explain how to use order of operations to find the sum of these numbers. Then find the sum.

$2.4 \times 10^3 + 7.1 \times 10^4$ **73,400**

English Language Learners

Group Activity

Have students work in groups that include both English learners and English speakers. Assign each group a scientific notation application problem. Have each group solve their problem showing all of the steps. English learners will benefit by having the opportunity to restate the problem and gain a deeper understanding of the concept.

16.7 Record and Practice Journal

b. How is this pair of numbers different from the pair of numbers in Activity 1?

The powers of 10 are not the same.

c. Explain why you cannot immediately use the rule you wrote in Activity 1(d) to find this sum.

There is no common factor to factor out.

d. **STRUCTURE** How can you rewrite one of the numbers so that you can use the rule you wrote in Activity 1(d)? Rewrite one of the numbers. Then find the sum using your rule and compare the result with your answer in part (a).

0.24×10^4; 73,400; same

e. **REASONING** Does this procedure work when subtracting numbers written in scientific notation? Justify your answer by evaluating the differences below. **yes**

- $(8.2 \times 10^5) - (4.6 \times 10^5) = (8.2 - 4.6) \times 10^5 = 3.6 \times 10^5$
- $(5.88 \times 10^5) - (1.5 \times 10^4) = (5.88 \times 10^5) - (0.15 \times 10^5)$ $= (5.88 - 0.15) \times 10^5 = 5.73 \times 10^5$

3 ACTIVITY: Multiplying Numbers in Scientific Notation

Work with a partner. Match each step with the correct description.

Step		Description	
$(2.4 \times 10^3) \times (7.1 \times 10^3)$		Original expression	
1. $= 2.4 \times 7.1 \times 10^3 \times 10^3$	**D**	A.	Write in standard form.
2. $= (2.4 \times 7.1) \times (10^3 \times 10^3)$	**F**	B.	Product of Powers Property
3. $= 17.04 \times 10^6$	**E**	C.	Write in scientific notation.
4. $= 1.704 \times 10^1 \times 10^6$	**C**	D.	Commutative Property of Multiplication
5. $= 1.704 \times 10^7$	**B**	E.	Simplify.
6. $= 17{,}040{,}000$	**A**	F.	Associative Property of Multiplication

Does this procedure work when the numbers have different powers of 10? Justify your answer by using this procedure to evaluate the products below. **yes**

- $(1.9 \times 10^2) \times (2.3 \times 10^5) = (1.9 \times 2.3) \times (10^2 \times 10^5)$ $= 4.37 \times 10^7 = 43{,}700{,}000$
- $(8.4 \times 10^6) \times (5.7 \times 10^{-4}) = (8.4 \times 5.7) \times (10^6 \times 10^{-4})$ $= 47.88 \times 10^2$

4 ACTIVITY: Using Scientific Notation to Estimate

Work with a partner. A person normally breathes about 6 liters of air per minute. The life expectancy of a person in the United States at birth is about 80 years. Use scientific notation to estimate the total amount of air a person born in the United States breathes over a lifetime.

***Sample answer:* 2.7×10^8 liters of air**

What Is Your Answer?

5. **IN YOUR OWN WORDS** How can you perform operations with numbers written in scientific notation?

See Additional Answers.

6. Use a calculator to evaluate the expression. Write your answer in scientific notation and in standard form.

a. $(1.5 \times 10^4) + (6.3 \times 10^4)$

7.8×10^4; 78,000

b. $(7.2 \times 10^5) - (2.2 \times 10^3)$

7.178×10^5; 717,800

c. $(4.1 \times 10^{-3}) \times (4.3 \times 10^{-3})$

1.763×10^{-5}; 0.00001763

d. $(4.75 \times 10^{-6}) \times (1.34 \times 10^7)$

6.365×10^1; 63.65

Laurie's Notes

Activity 3

- Give students time to match the descriptions and steps with their partners. Remind students that the question they need to ask themselves is, "What was done in moving from one step to the next? What justifies, or explains each step?"
- This can be a challenging exercise for students. They can read through the steps and make sense of them, but they're unsure of what justifies each step.
- When you discuss this as a class, you could put the steps and descriptions on strips of paper. See if students can order the steps *and* the descriptions.

Activity 4

- This activity is similar to the heartbeat question students answered in the Activity Motivate in Section 16.5.
- **Use Appropriate Tools Strategically:** You may wish to give students access to calculators for this activity.
- **Teaching Tip:** Provide transparencies or blank paper so that students can display their answers on the overhead or at the document camera.
- Ask volunteers to share their work. Look for evidence of determining the total number of minutes in 80 years, or the number of liters of air breathed per year.

What Is Your Answer?

- **Use Appropriate Tools Strategically:** Question 6 gives students extra practice with entering numbers written in scientific notation into a calculator.

Closure

- Evaluate:
  1. $6.4 \times 10^4 + 1.3 \times 10^4$ 7.7×10^4
  2. $(1.2 \times 10^3) \times (4.3 \times 10^3)$ 5.16×10^6

3 ACTIVITY: Multiplying Numbers in Scientific Notation

Math Practice

Justify Conclusions

Which step of the procedure would be affected if the powers of 10 were different? Explain.

Work with a partner. Match each step with the correct description.

Step	Description
$(2.4 \times 10^3) \times (7.1 \times 10^3)$	Original expression
1. $= 2.4 \times 7.1 \times 10^3 \times 10^3$	**A.** Write in standard form.
2. $= (2.4 \times 7.1) \times (10^3 \times 10^3)$	**B.** Product of Powers Property
3. $= 17.04 \times 10^6$	**C.** Write in scientific notation.
4. $= 1.704 \times 10^1 \times 10^6$	**D.** Commutative Property of Multiplication
5. $= 1.704 \times 10^7$	**E.** Simplify.
6. $= 17{,}040{,}000$	**F.** Associative Property of Multiplication

Does this procedure work when the numbers have different powers of 10? Justify your answer by using this procedure to evaluate the products below.

- $(1.9 \times 10^2) \times (2.3 \times 10^5) =$
- $(8.4 \times 10^6) \times (5.7 \times 10^{-4}) =$

4 ACTIVITY: Using Scientific Notation to Estimate

Work with a partner. A person normally breathes about 6 liters of air per minute. The life expectancy of a person in the United States at birth is about 80 years. Use scientific notation to estimate the total amount of air a person born in the United States breathes over a lifetime.

What Is Your Answer?

5. **IN YOUR OWN WORDS** How can you perform operations with numbers written in scientific notation?

6. Use a calculator to evaluate the expression. Write your answer in scientific notation and in standard form.

a. $(1.5 \times 10^4) + (6.3 \times 10^4)$ **b.** $(7.2 \times 10^5) - (2.2 \times 10^3)$

c. $(4.1 \times 10^{-3}) \times (4.3 \times 10^{-3})$ **d.** $(4.75 \times 10^{-6}) \times (1.34 \times 10^7)$

Practice Use what you learned about evaluating expressions involving scientific notation to complete Exercises 3–6 on page 752.

16.7 Lesson

Check It Out
Lesson Tutorials
BigIdeasMath.com

To add or subtract numbers written in scientific notation with the same power of 10, add or subtract the factors. When the numbers have different powers of 10, first rewrite the numbers so they have the same power of 10.

EXAMPLE 1 Adding and Subtracting Numbers in Scientific Notation

Find the sum or difference. Write your answer in scientific notation.

a. $(4.6 \times 10^3) + (8.72 \times 10^3)$

$= (4.6 + 8.72) \times 10^3$ — Distributive Property

$= 13.32 \times 10^3$ — Add.

$= (1.332 \times 10^1) \times 10^3$ — Write 13.32 in scientific notation.

$= 1.332 \times 10^4$ — Product of Powers Property

b. $(3.5 \times 10^{-2}) - (6.6 \times 10^{-3})$

Rewrite 6.6×10^{-3} so that it has the same power of 10 as 3.5×10^{-2}.

$6.6 \times 10^{-3} = 6.6 \times 10^{-1} \times 10^{-2}$ — Rewrite 10^{-3} as $10^{-1} \times 10^{-2}$.

$= 0.66 \times 10^{-2}$ — Rewrite 6.6×10^{-1} as 0.66.

Subtract the factors.

$(3.5 \times 10^{-2}) - (0.66 \times 10^{-2})$

$= (3.5 - 0.66) \times 10^{-2}$ — Distributive Property

$= 2.84 \times 10^{-2}$ — Subtract.

Study Tip

In Example 1(b), you will get the same answer when you start by rewriting 3.5×10^{-2} as 35×10^{-3}.

On Your Own

Now You're Ready
Exercises 7–14

Find the sum or difference. Write your answer in scientific notation.

1. $(8.2 \times 10^2) + (3.41 \times 10^{-1})$
2. $(7.8 \times 10^{-5}) - (4.5 \times 10^{-5})$

To multiply or divide numbers written in scientific notation, multiply or divide the factors and powers of 10 separately.

EXAMPLE 2 Multiplying Numbers in Scientific Notation

Find $(3 \times 10^{-5}) \times (5 \times 10^{-2})$. Write your answer in scientific notation.

$(3 \times 10^{-5}) \times (5 \times 10^{-2})$

$= 3 \times 5 \times 10^{-5} \times 10^{-2}$ — Commutative Property of Multiplication

$= (3 \times 5) \times (10^{-5} \times 10^{-2})$ — Associative Property of Multiplication

$= 15 \times 10^{-7}$ — Simplify.

$= 1.5 \times 10^1 \times 10^{-7}$ — Write 15 in scientific notation.

$= 1.5 \times 10^{-6}$ — Product of Powers Property

Study Tip

You can check your answer using standard form.

(3×10^{-5})
$\times (5 \times 10^{-2})$
$= 0.00003 \times 0.05$
$= 0.0000015$
$= 1.5 \times 10^{-6}$

Laurie's Notes

Introduction

Connect

- **Yesterday:** Students explored how to use properties to perform operations with numbers written in scientific notation.
- **Today:** Students will perform operations with numbers written in scientific notation.

Motivate

- Discuss the meaning of national debt. Compare the national debt of the United States *x* number of years ago with the national debt today.

Year	2000	2002	2004	2006	2008	2010	2012
Debt (in trillions)	5.7	6.2	7.4	8.5	10.0	13.6	16.1

You can also discuss each student's share of the U.S. national debt. Ask students how they might calculate their own share.

? "How do you write the 2000 and 2012 national debts in scientific notation?" 5.7×10^{12}; 1.61×10^{13}

- Pose questions about the national debt, such as the difference in the national debt between 2000 and 2012, and how many times greater the national debt is in 2012 compared to 2000. Explain that this type of arithmetic is the focus of this lesson.

Lesson Notes

Example 1

- **Note:** In part (a), students do not always recognize the Distributive Property when it is used to pull out a common factor, such as 10^3.

? "Why not leave the answer as 13.32×10^3?" It is not in scientific notation.

- In part (b), rewriting 6.6×10^{-3} can be confusing to students. Tell them this step is similar to rewriting 24×6 as $12 \times 2 \times 6$.
- Show students that this expression can also be simplified by rewriting 3.5×10^{-2} as 35×10^{-3}.

? "Why can't the numbers just be subtracted?" The factors are multiplied by different powers of 10, so you cannot use the Distributive Property.

- **Alternative Method:** In part (b) write the numbers in standard form, subtract, and then write the answer in scientific notation.

On Your Own

- **Neighbor Check:** Have students work independently and then have their neighbors check their work. Have students discuss any discrepancies.

Example 2

- Even though you cannot add or subtract numbers in scientific notation (unless they have the same power of 10), you can multiply them.

? "How would you multiply the numbers?" Most students will immediately suggest multiplying the factors, and then multiplying the powers of 10.

- **Construct Viable Arguments:** Make sure that students realize that the Commutative and Associative Properties allow this to happen. The Product of Powers Property is used to multiply the powers of 10.

Goal Today's lesson is adding, subtracting, multiplying, and dividing numbers in scientific notation.

Technology for the Teacher

Dynamic Classroom

Lesson Tutorials
Lesson Plans
Answer Presentation Tool

Extra Example 1

Find the sum or difference. Write your answer in scientific notation.

a. $(2.1 \times 10^{-4}) + (9.74 \times 10^{-4})$
1.184×10^{-3}

b. $(4.7 \times 10^{5}) - (7.2 \times 10^{3})$
4.628×10^{5}

On Your Own

1. 8.20341×10^{2}

2. 3.3×10^{-5}

Extra Example 2

Find $(2 \times 10^{-4}) \times (6 \times 10^{-3})$. Write your answer in scientific notation.
1.2×10^{-6}

Extra Example 3

Find $\frac{5.3 \times 10^8}{4 \times 10^{-3}}$. Write your answer in scientific notation. 1.325×10^{11}

On Your Own

3. 4.8×10^{-4}
4. 2.1×10^{8}
5. 2×10^{12}
6. 2×10^{-6}

Extra Example 4

The diameter of the Moon is about 3.48×10^3 kilometers. Using the information in Example 4, how many times greater is the diameter of the Sun than the diameter of the Moon? about 402 times greater

On Your Own

7. 693,600 km

Differentiated Instruction

Kinesthetic

When writing a number in scientific notation, have students underline the first nonzero digit and the digit to its right. The decimal point will be placed between these two digits to create the factor.

$\underline{27},000 = 2.7 \times 10^4$

$0.00000\underline{48} = 4.8 \times 10^{-6}$

Laurie's Notes

Example 3

- Before this example, work through a simple, but related problem such as: $\frac{2}{3} \cdot \frac{9}{10} = \frac{\cancel{2}^1 \cdot \cancel{9}^3}{{}_1\cancel{3} \cdot {}_5\cancel{10}} = \frac{3}{5}$. Point out how the common factors divide out.
- Write the example and relate it to the problem above.
- ? "Why not leave the answer as 0.25×10^{-15}?" It is not in scientific notation.

On Your Own

- **Common Error:** Students may use the Quotient of Powers Property incorrectly when simplifying the fraction with the powers of 10.

Example 4

- ? "Does anyone know the approximate diameter of Earth? the Sun?" Answers will vary.
- While students may not know the diameters, they should know that the Sun's diameter is much greater than Earth's diameter. This example will determine how many times greater.
- Explain to students that the answer is written in standard form to make the comparison more meaningful. It is easier to understand that the Sun's diameter is about 109 times greater than Earth's diameter, instead of about 1.09×10^2 times greater.

On Your Own

- In Question 7, have students think about which form they should convert to first. Have them do it both ways. Ask them which they prefer and why.

Closure

- **Exit Ticket:** Add or divide. Write your answer in scientific notation.
  a. $(3.5 \times 10^4) + (7.6 \times 10^4)$ 1.11×10^5
  b. $\frac{8.4 \times 10^3}{4.2 \times 10^{-2}}$ 2×10^5

EXAMPLE 3 Dividing Numbers in Scientific Notation

Find $\frac{1.5 \times 10^{-8}}{6 \times 10^{7}}$. Write your answer in scientific notation.

$$\frac{1.5 \times 10^{-8}}{6 \times 10^{7}} = \frac{1.5}{6} \times \frac{10^{-8}}{10^{7}}$$ Rewrite as a product of fractions.

$$= 0.25 \times \frac{10^{-8}}{10^{7}}$$ Divide 1.5 by 6.

$$= 0.25 \times 10^{-15}$$ Quotient of Powers Property

$$= 2.5 \times 10^{-1} \times 10^{-15}$$ Write 0.25 in scientific notation.

$$= 2.5 \times 10^{-16}$$ Product of Powers Property

On Your Own

Now You're Ready
Exercises 16–23

Find the product or quotient. Write your answer in scientific notation.

3. $6 \times (8 \times 10^{-5})$

4. $(7 \times 10^{2}) \times (3 \times 10^{5})$

5. $(9.2 \times 10^{12}) \div 4.6$

6. $(1.5 \times 10^{-3}) \div (7.5 \times 10^{2})$

EXAMPLE 4 Real-Life Application

Diameter = 1,400,000 km

How many times greater is the diameter of the Sun than the diameter of Earth?

Diameter = 1.28×10^4 km

Write the diameter of the Sun in scientific notation.

$$\underbrace{1{,}400{,}000}_{6} = 1.4 \times 10^{6}$$

Divide the diameter of the Sun by the diameter of Earth.

$$\frac{1.4 \times 10^{6}}{1.28 \times 10^{4}} = \frac{1.4}{1.28} \times \frac{10^{6}}{10^{4}}$$ Rewrite as a product of fractions.

$$= 1.09375 \times \frac{10^{6}}{10^{4}}$$ Divide 1.4 by 1.28.

$$= 1.09375 \times 10^{2}$$ Quotient of Powers Property

$$= 109.375$$ Write in standard form.

The diameter of the Sun is about 109 times greater than the diameter of Earth.

On Your Own

7. How many more kilometers is the radius of the Sun than the radius of Earth? Write your answer in standard form.

16.7 Exercises

Check It Out
Help with Homework
BigIdeasMath.com

Vocabulary and Concept Check

1. **WRITING** Describe how to subtract two numbers written in scientific notation with the same power of 10.

2. **NUMBER SENSE** You are multiplying two numbers written in scientific notation with different powers of 10. Do you have to rewrite the numbers so they have the same power of 10 before multiplying? Explain.

Practice and Problem Solving

Evaluate the expression using two different methods. Write your answer in scientific notation.

3. $(2.74 \times 10^7) + (5.6 \times 10^7)$
4. $(8.3 \times 10^6) + (3.4 \times 10^5)$
5. $(5.1 \times 10^5) \times (9.7 \times 10^5)$
6. $(4.5 \times 10^4) \times (6.2 \times 10^3)$

Find the sum or difference. Write your answer in scientific notation.

(1) 7. $(2 \times 10^5) + (3.8 \times 10^5)$
8. $(6.33 \times 10^{-9}) - (4.5 \times 10^{-9})$
9. $(9.2 \times 10^8) - (4 \times 10^8)$
10. $(7.2 \times 10^{-6}) + (5.44 \times 10^{-6})$
11. $(7.8 \times 10^7) - (2.45 \times 10^6)$
12. $(5 \times 10^{-5}) + (2.46 \times 10^{-3})$
13. $(9.7 \times 10^6) + (6.7 \times 10^5)$
14. $(2.4 \times 10^{-1}) - (5.5 \times 10^{-2})$

15. **ERROR ANALYSIS** Describe and correct the error in finding the sum of the numbers.

✗ $(2.5 \times 10^9) + (5.3 \times 10^8) = (2.5 + 5.3) \times (10^9 \times 10^8)$
$= 7.8 \times 10^{17}$

Find the product or quotient. Write your answer in scientific notation.

(2) (3) 16. $5 \times (7 \times 10^7)$
17. $(5.8 \times 10^{-6}) \div (2 \times 10^{-3})$
18. $(1.2 \times 10^{-5}) \div 4$
19. $(5 \times 10^{-7}) \times (3 \times 10^6)$
20. $(3.6 \times 10^7) \div (7.2 \times 10^7)$
21. $(7.2 \times 10^{-1}) \times (4 \times 10^{-7})$
22. $(6.5 \times 10^8) \times (1.4 \times 10^{-5})$
23. $(2.8 \times 10^4) \div (2.5 \times 10^6)$

24. **MONEY** How many times greater is the thickness of a dime than the thickness of a dollar bill?

Thickness = 0.135 cm

Thickness = 1.0922×10^{-2} cm

Assignment Guide and Homework Check

Level	Assignment	Homework Check
Accelerated	1–6, 8–14 even, 15, 16–26 even, 28–35	14, 20, 22, 26, 28

For Your Information

- **Exercise 30** Remind students that population density is the average number of people for some amount of land area. To find the population density, you divide the population by the land area.

Common Errors

- **Exercises 11–14** Students may incorrectly rewrite a number when adding or subtracting numbers with different powers of 10.
- **Exercises 16–23** Students may multiply the factors and leave the number greater than 10. Remind them that the factor in scientific notation must be at least one and less than 10. If the number is less than 10 and greater than or equal to 1, the exponent will be 0.
- **Exercises 16–23** Students may use the Quotient of Powers Property incorrectly when simplifying fractions containing powers of 10.

16.7 Record and Practice Journal

Find the sum or difference. Write your answer in scientific notation.

1. $(2 \times 10^4) + (7.2 \times 10^4)$
   9.2×10^4
2. $(3.2 \times 10^{-2}) + (9.4 \times 10^{-2})$
   1.26×10^{-1}
3. $(6.7 \times 10^5) - (4.3 \times 10^5)$
   2.4×10^5
4. $(8.9 \times 10^{-3}) - (1.9 \times 10^{-3})$
   7×10^{-3}

Find the product or quotient. Write your answer in scientific notation.

5. $(6 \times 10^8) \times (4 \times 10^6)$
   2.4×10^{15}
6. $(9 \times 10^{-3}) \times (9 \times 10^{-3})$
   8.1×10^{-5}
7. $(8 \times 10^3) \div (2 \times 10^2)$
   4.0×10^1
8. $(2.34 \times 10^5) \div (7.8 \times 10^5)$
   3×10^{-1}
9. How many times greater is the radius of a basketball than the radius of a marble?

Radius = 1.143×10^1 cm   Radius = 5×10^{-1} cm

about 23 times greater

Vocabulary and Concept Check

1. Use the Distributive Property to group the factors together. Then subtract the factors and write it with the power of 10. The number may need to be rewritten so that it is still in scientific notation.
2. no; You can use the Commutative and Associative Properties of Multiplication to group the factors and the powers of 10. Then, you multiply the factors and multiply the powers of 10.

Practice and Problem Solving

3. 8.34×10^7
4. 8.64×10^6
5. 4.947×10^{11}
6. 2.79×10^8
7. 5.8×10^5
8. 1.83×10^{-9}
9. 5.2×10^8
10. 1.264×10^{-5}
11. 7.555×10^7
12. 2.51×10^{-3}
13. 1.037×10^7
14. 1.85×10^{-1}
15. You have to rewrite the numbers so they have the same power of 10 before adding; 3.03×10^9
16. 3.5×10^8
17. 2.9×10^{-3}
18. 3×10^{-6}
19. 1.5×10^0
20. 5×10^{-1}
21. 2.88×10^{-7}
22. 9.1×10^3
23. 1.12×10^{-2}
24. about 12 times greater

Practice and Problem Solving

25. 4.006×10^9

26. 2.9×10^{-2}

27. 1.962×10^8 cm

28. 2.65×10^6 L; 1.85×10^8 L; Justifications will vary.

29. See *Taking Math Deeper.*

30. *Answer should include, but is not limited to:* Make sure calculations using scientific notation are done correctly.

31. 3×10^8 m/sec

Fair Game Review

32. -9 **33.** $\frac{1}{8}$

34. $-\frac{5}{7}$ **35.** C

Mini-Assessment

Evaluate the expression. Write your answer in scientific notation.

1. $(3.4 \times 10^6) + (8.1 \times 10^6)$ 1.15×10^7
2. $(4.3 \times 10^{-3}) + (7.8 \times 10^{-4})$ 5.08×10^{-3}
3. $(5.6 \times 10^{-8}) - (1.9 \times 10^{-8})$ 3.7×10^{-8}
4. $(1.7 \times 10^2) \times (4.3 \times 10^4)$ 7.31×10^6
5. $(6.2 \times 10^5) \div (2 \times 10^{-4})$ 3.1×10^9
6. The mass of Earth is about 6.58×10^{21} tons. The mass of Mars is about 7.08×10^{20} tons. How much greater is the mass of Earth than the mass of Mars? about 5.872×10^{21} tons

Taking Math Deeper

Exercise 29

You are probably familiar with the look of a DVD, but do you know what the surface of a DVD looks like? This problem gives students some idea of what this digital storage device actually looks like.

1 Summarize the given information.

Width of each ridge = 0.000032 cm
Width of each valley = 0.000074 cm
Diameter of center portion = 4.26 cm

2 Find the diameter of the DVD.

$$\begin{aligned}\text{ridges} + \text{valleys} &= 73{,}000(0.000032) + 73{,}000(0.000074)\\ &= 73{,}000(0.000032 + 0.000074)\\ &= 73{,}000(0.000106)\\ &= 7.738 \text{ cm}\end{aligned}$$

OR using scientific notation

$$\begin{aligned}\text{ridges} + \text{valleys} &= (7.3 \times 10^4)(3.2 \times 10^{-5}) + (7.3 \times 10^4)(7.4 \times 10^{-5})\\ &= (7.3 \times 10^4)(3.2 \times 10^{-5} + 7.4 \times 10^{-5})\\ &= (7.3 \times 10^4)(10.6 \times 10^{-5})\\ &= 77.38 \times 10^{-1}\\ &= 7.738 \text{ cm}\end{aligned}$$

$$\begin{aligned}\text{diameter} &= \text{ridges} + \text{valleys} + \text{center portion}\\ &= 7.738 + 4.26\\ &= 11.998 \text{ cm}\\ &\approx 12 \text{ cm}\end{aligned}$$

3 Here's a fun fact.

The microscopic dimensions of the bumps make the spiral track on a DVD extremely long. If you could lift the data track off a single layer of a DVD, and stretch it out into a straight line, it would be almost 7.5 miles long!

Project

Write a report on the invention of the DVD and the DVD player.

Reteaching and Enrichment Strategies

If students need help. . .	If students got it. . .
Resources by Chapter • Practice A and Practice B • Puzzle Time Record and Practice Journal Practice Differentiating the Lesson Lesson Tutorials Skills Review Handbook	Resources by Chapter • Enrichment and Extension • Technology Connection Start the next section

Evaluate the expression. Write your answer in scientific notation.

25. $5{,}200{,}000 \times (8.3 \times 10^2) - (3.1 \times 10^8)$

26. $(9 \times 10^{-3}) + (2.4 \times 10^{-5}) \div 0.0012$

27. **GEOMETRY** Find the perimeter of the rectangle.

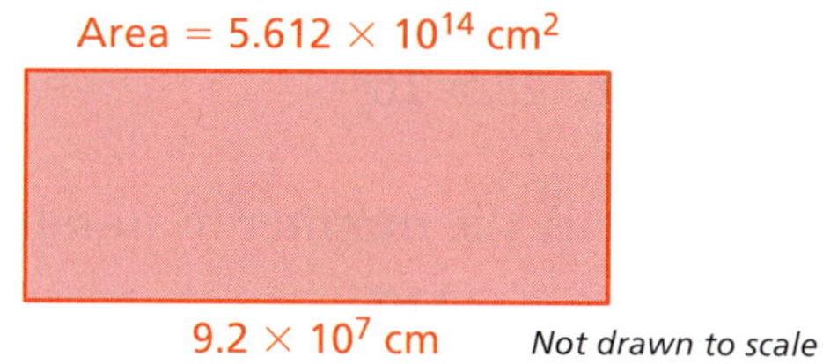

28. **BLOOD SUPPLY** A human heart pumps about 7×10^{-2} liter of blood per heartbeat. The average human heart beats about 72 times per minute. How many liters of blood does a heart pump in 1 year? in 70 years? Write your answers in scientific notation. Then use estimation to justify your answers.

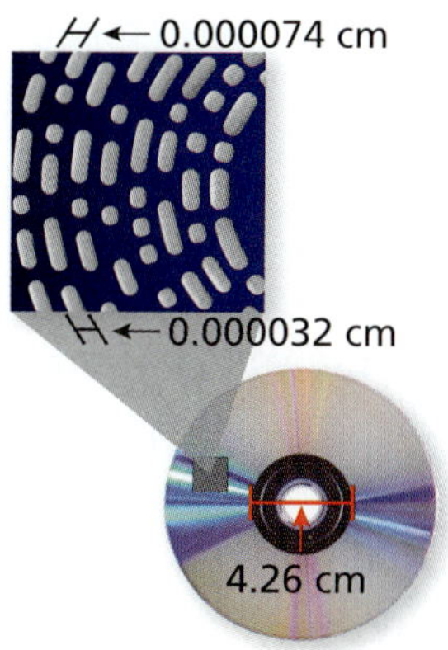

29. **DVDS** On a DVD, information is stored on bumps that spiral around the disk. There are 73,000 ridges (with bumps) and 73,000 valleys (without bumps) across the diameter of the DVD. What is the diameter of the DVD in centimeters?

30. **PROJECT** Use the Internet or some other reference to find the populations and areas (in square miles) of India, China, Argentina, the United States, and Egypt. Round each population to the nearest million and each area to the nearest thousand square miles.

a. Write each population and area in scientific notation.

b. Use your answers to part (a) to find and order the population densities (people per square mile) of each country from least to greatest.

31. Critical Thinking Albert Einstein's most famous equation is $E = mc^2$, where E is the energy of an object (in joules), m is the mass of an object (in kilograms), and c is the speed of light (in meters per second). A hydrogen atom has 15.066×10^{-11} joule of energy and a mass of 1.674×10^{-27} kilogram. What is the speed of light? Write your answer in scientific notation.

Fair Game Review What you learned in previous grades & lessons

Find the cube root. *(Section 14.2)*

32. $\sqrt[3]{-729}$

33. $\sqrt[3]{\frac{1}{512}}$

34. $\sqrt[3]{-\frac{125}{343}}$

35. **MULTIPLE CHOICE** What is the volume of the cone? *(Section 15.2)*

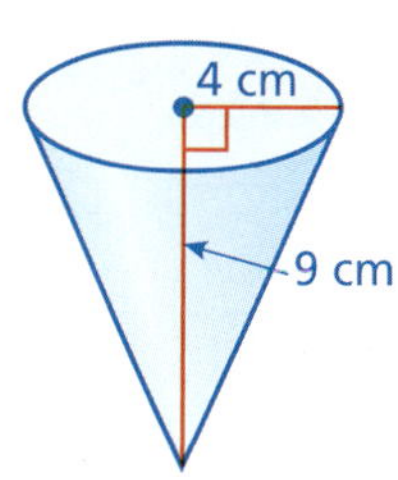

Ⓐ 16π cm^3

Ⓑ 108π cm^3

Ⓒ 48π cm^3

Ⓓ 144π cm^3

16.5 – 16.7 Quiz

Check It Out
Progress Check
BigIdeasMath.com

Tell whether the number is written in scientific notation. Explain. *(Section 16.5)*

1. 23×10^9
2. 0.6×10^{-7}

Write the number in standard form. *(Section 16.5)*

3. 8×10^6
4. 1.6×10^{-2}

Write the number in scientific notation. *(Section 16.6)*

5. 0.00524
6. 892,000,000

Evaluate the expression. Write your answer in scientific notation. *(Section 16.7)*

7. $(7.26 \times 10^4) + (3.4 \times 10^4)$
8. $(2.8 \times 10^{-5}) - (1.6 \times 10^{-6})$
9. $(2.4 \times 10^4) \times (3.8 \times 10^{-6})$
10. $(5.2 \times 10^{-3}) \div (1.3 \times 10^{-12})$

11. **PLANETS** The table shows the equatorial radii of the eight planets in our solar system. *(Section 16.5)*

    a. Which planet has the second-smallest equatorial radius?

    b. Which planet has the second-largest equatorial radius?

Planet	Equatorial Radius (km)
Mercury	2.44×10^3
Venus	6.05×10^3
Earth	6.38×10^3
Mars	3.4×10^3
Jupiter	7.15×10^4
Saturn	6.03×10^4
Uranus	2.56×10^4
Neptune	2.48×10^4

12. **OORT CLOUD** The Oort cloud is a spherical cloud that surrounds our solar system. It is about 2×10^5 astronomical units from the Sun. An astronomical unit is about 1.5×10^8 kilometers. How far is the Oort cloud from the Sun in kilometers? *(Section 16.6)*

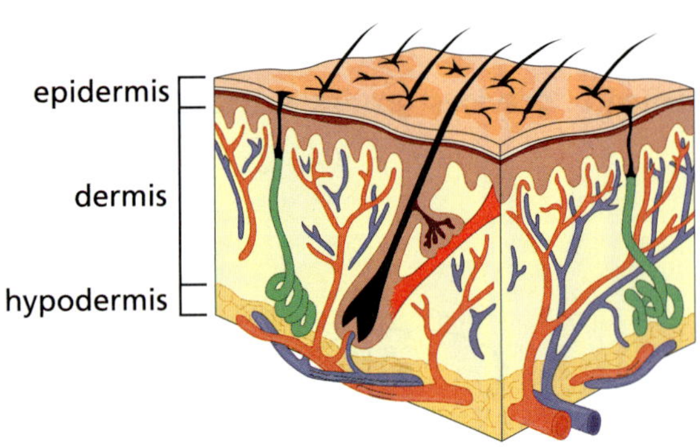

13. **EPIDERMIS** The outer layer of skin is called the *epidermis*. On the palm of your hand, the epidermis is 0.0015 meter thick. Write this number in scientific notation. *(Section 16.6)*

14. **ORBITS** It takes the Sun about 2.3×10^8 years to orbit the center of the Milky Way. It takes Pluto about 2.5×10^2 years to orbit the Sun. How many times does Pluto orbit the Sun while the Sun completes one orbit around the Milky Way? Write your answer in standard form. *(Section 16.7)*

Alternative Assessment Options

Math Chat	Student Reflective Focus Question
Structured Interview	Writing Prompt

Math Chat

- Have individual students work problems from the quiz on the board. The student explains the process used and justifies each step. Students in the class ask questions of the student presenting.
- The teacher probes the thought process of the student presenting, but does not teach or ask leading questions.

Study Help Sample Answers

Remind students to complete Graphic Organizers for the rest of the chapter.

4.

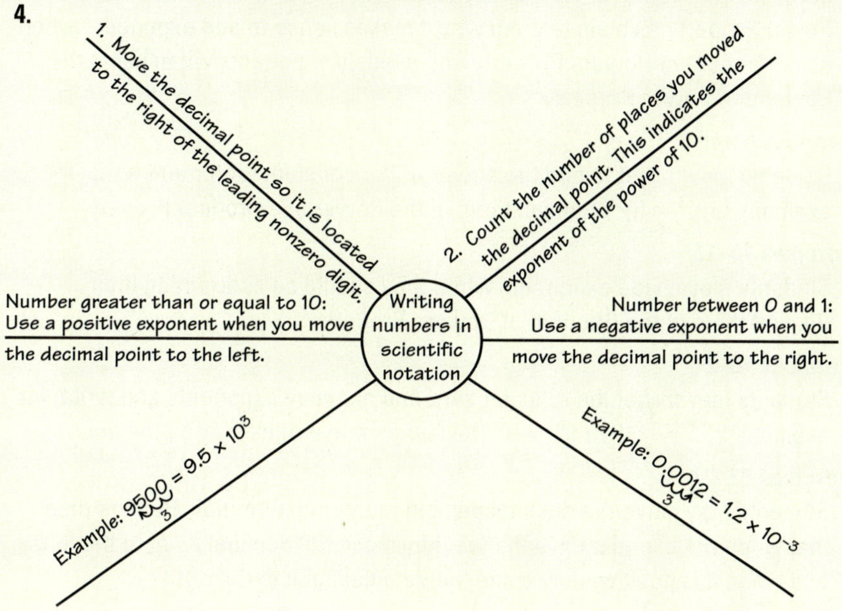

5–8. Available at *BigIdeasMath.com.*

Answers

1. no; The factor is greater than 10.
2. no; The factor is less than 1.
3. 8,000,000
4. 0.016
5. 5.24×10^{-3}
6. 8.92×10^{8}
7. 1.066×10^{5}
8. 2.64×10^{-5}
9. 9.12×10^{-2}
10. 4×10^{9}
11. **a.** Mars

    b. Saturn
12. 3×10^{13} km
13. 1.5×10^{-3} m
14. 920,000

Reteaching and Enrichment Strategies

If students need help. . .	If students got it. . .
Resources by Chapter • Practice A and Practice B • Puzzle Time Lesson Tutorials *BigIdeasMath.com*	Resources by Chapter • Enrichment and Extension • Technology Connection Game Closet at *BigIdeasMath.com* Start the Chapter Review

Technology for the Teacher

Online Assessment
Assessment Book
ExamView® Assessment Suite

For the Teacher
Additional Review Options

- *BigIdeasMath.com*
- Online Assessment
- Game Closet at *BigIdeasMath.com*
- Vocabulary Help
- Resources by Chapter

Answers

1. $(-9)^5$
2. 2^3n^2
3. 216
4. $-\frac{1}{16}$
5. 100
6. p^7
7. n^{22}
8. $125y^3$
9. $16k^4$

Review of Common Errors

Exercises 1 and 2

- Students may count the wrong number of factors. Remind them to check their work.

Exercises 3–5

- Students may treat the exponent as a factor. For example, some may think that 6^3 means 6×3. Remind them of the definition of an exponent.
- Students may have the wrong sign in their answers. Remind them that when a negative sign is outside the parentheses, it is not part of the base.
- Students may not use the correct order of operations when they evaluate the expression in Exercise 5. Review the order of operations with students.

Exercises 6 and 7

- Students may confuse the Product of Powers Property and the Power of a Power Property. Explain to them why it makes sense to add exponents when using Product of Powers Property and multiply exponents when using the Power of a Power Property.

Exercises 8 and 9

- Students may forget to find the power of the constant factor and write, for example, $(5y)^3 = 5y^3$. Remind them of the Power of a Product Property.

Exercises 10–15

- Students may divide exponents when they should be subtracting them. Remind them of the Quotient of Powers Property.

Exercises 16–21

- Students may forget the rules for zero and negative exponents and write, for example, $2^{-4} = -16$ or $95^0 = 0$. Reviewing these rules may be helpful.

Exercises 22–27

- Students may move the decimal point in the wrong direction. Remind them that when the exponent is negative they move the decimal point to the left, and when it is positive they move the decimal point to the right.

Exercises 28–30

- Students may write an exponent with the incorrect sign. Remind them that large numbers have a positive exponent in scientific notation and small numbers have a negative exponent in scientific notation.

Exercises 31–34

- Students may leave the decimal factor greater than 10 and/or combine exponents incorrectly, coming up with answers such as 10.1×10^{17} for Exercise 31, 4.1×10^0 for Exercise 32, 37.73×10^{-40} for Exercise 33, and $2 \times 10^{5/9}$ for Exercise 34. Remind them what scientific notation means and point out that the directions call for the answer in scientific notation. Also, if necessary, provide students with a quick review of the properties they learned in this chapter, as well as how to correctly perform operations in scientific notation. In particular, point out that in Exercise 31, students should rewrite one of the addends so that the powers of 10 are the same before adding.

16 Chapter Review

Check It Out
Vocabulary Help
BigIdeasMath.com

Review Key Vocabulary

power, *p. 712*
base, *p. 712*
exponent, *p. 712*
scientific notation, *p. 738*

Review Examples and Exercises

16.1 Exponents (pp. 710–715)

Write $(-4) \cdot (-4) \cdot (-4) \cdot y \cdot y$ using exponents.

Because -4 is used as a factor 3 times, its exponent is 3. Because y is used as a factor 2 times, its exponent is 2.

So, $(-4) \cdot (-4) \cdot (-4) \cdot y \cdot y = (-4)^3 y^2$.

Exercises

Write the product using exponents.

1. $(-9) \cdot (-9) \cdot (-9) \cdot (-9) \cdot (-9)$
2. $2 \cdot 2 \cdot 2 \cdot n \cdot n$

Evaluate the expression.

3. 6^3
4. $-\left(\frac{1}{2}\right)^4$
5. $\left|\frac{1}{2}(16 - 6^3)\right|$

16.2 Product of Powers Property (pp. 716–721)

a. $\left(-\frac{1}{8}\right)^7 \cdot \left(-\frac{1}{8}\right)^4 = \left(-\frac{1}{8}\right)^{7+4}$ Product of Powers Property

$= \left(-\frac{1}{8}\right)^{11}$ Simplify.

b. $(2.5^7)^2 = 2.5^{7 \cdot 2}$ Power of a Power Property

$= 2.5^{14}$ Simplify.

c. $(3m)^2 = 3^2 \cdot m^2$ Power of a Product Property

$= 9m^2$ Simplify.

Exercises

Simplify the expression.

6. $p^5 \cdot p^2$
7. $(n^{11})^2$
8. $(5y)^3$
9. $(-2k)^4$

16.3 Quotient of Powers Property (pp. 722–727)

a. $\dfrac{(-4)^9}{(-4)^6} = (-4)^{9-6}$ — Quotient of Powers Property

$= (-4)^3$ — Simplify.

b. $\dfrac{x^4}{x^3} = x^{4-3}$ — Quotient of Powers Property

$= x^1$

$= x$ — Simplify.

Exercises

Simplify the expression. Write your answer as a power.

10. $\dfrac{8^8}{8^3}$ **11.** $\dfrac{5^2 \cdot 5^9}{5}$ **12.** $\dfrac{w^8}{w^7} \cdot \dfrac{w^5}{w^2}$

Simplify the expression.

13. $\dfrac{2^2 \cdot 2^5}{2^3}$ **14.** $\dfrac{(6c)^3}{c}$ **15.** $\dfrac{m^8}{m^6} \cdot \dfrac{m^{10}}{m^9}$

16.4 Zero and Negative Exponents (pp. 728–733)

a. $10^{-3} = \dfrac{1}{10^3}$ — Definition of negative exponent

$= \dfrac{1}{1000}$ — Evaluate power.

b. $(-0.5)^{-5} \cdot (-0.5)^5 = (-0.5)^{-5+5}$ — Product of Powers Property

$= (-0.5)^0$ — Simplify.

$= 1$ — Definition of zero exponent

Exercises

Evaluate the expression.

16. 2^{-4} **17.** 95^0 **18.** $\dfrac{8^2}{8^4}$

19. $(-12)^{-7} \cdot (-12)^7$ **20.** $\dfrac{1}{7^9} \cdot \dfrac{1}{7^{-6}}$ **21.** $\dfrac{9^4 \cdot 9^{-2}}{9^2}$

Review Game

Comparing Values in Scientific Notation

Materials per Group

- pencil
- paper
- computer with Internet access

Directions

- The game can be completed in one to two class periods, but you may want to give the students one or two days to complete the necessary research.
- Each student comes up with three different values written in scientific notation. (Example: the length of an ant in meters, the weight of a person in ounces, the volume of a car's gas tank in cups)
- One value should be length, one value should be weight, and one value should be volume.
- Divide the class into an even number of groups.
- Randomly call on two groups and have them complete the following:
  - Each group writes the length (in scientific notation) of one of their items on the board.
  - The members of each group work together to write each number in standard form and determine which value is the least.
  - They write their answer on a piece of paper and submit it to the teacher.
  - One point is awarded to each group that answers correctly.
  - Note: Be sure students are aware that when they are comparing these values, the units of the values must be considered.
- Repeat this process for the remaining length, weight, and volume values.

Who wins?

After all groups have compared their values, the group(s) with the most points wins.

For the Student Additional Practice

- Lesson Tutorials
- Multi-Language Glossary
- Self-Grading Progress Check
- *BigIdeasMath.com*
  Dynamic Student Edition
  Student Resources

Answers

10. 8^5

11. 5^{10}

12. w^4

13. 16

14. $216c^2$

15. m^3

16. $\frac{1}{16}$

17. 1

18. $\frac{1}{64}$

19. 1

20. $\frac{1}{343}$

21. 1

22. 20,000,000

23. 0.034

24. 0.0000000015

25. 59,000,000,000

26. 0.0048

27. 625,000

28. 3.6×10^{-4}

29. 8×10^5

30. 7.92×10^7

31. 6.32×10^9

32. 4.1×10^{-4}

33. 3.773×10^4

34. 2×10^{-4}

My Thoughts on the Chapter

Teacher Tip

Not allowed to write in your teaching edition? Use sticky notes to record your thoughts.

What worked. . .

What did not work. . .

What I would do differently. . .

16.5 Reading Scientific Notation *(pp. 736–741)*

Write (a) 5.9×10^4 and (b) 7.31×10^{-6} in standard notation.

a. $5.9 \times 10^4 = 59{,}000$ (4)

Move decimal point $|4| = 4$ places to the right.

b. $7.31 \times 10^{-6} = 0.00000731$ (6)

Move decimal point $|-6| = 6$ places to the left.

Exercises

Write the number in standard form.

22. 2×10^7

23. 3.4×10^{-2}

24. 1.5×10^{-9}

25. 5.9×10^{10}

26. 4.8×10^{-3}

27. 6.25×10^5

16.6 Writing Scientific Notation *(pp. 742–747)*

Write (a) 309,000,000 and (b) 0.00056 in scientific notation.

a. $309{,}000{,}000 = 3.09 \times 10^8$ (8)

The number is greater than 10. So, the exponent is positive.

b. $0.00056 = 5.6 \times 10^{-4}$ (4)

The number is between 0 and 1. So, the exponent is negative.

Exercises

Write the number in scientific notation.

28. 0.00036

29. 800,000

30. 79,200,000

16.7 Operations in Scientific Notation *(pp. 748–753)*

Find $(2.6 \times 10^5) + (3.1 \times 10^5)$.

$$(2.6 \times 10^5) + (3.1 \times 10^5) = (2.6 + 3.1) \times 10^5 \quad \text{Distributive Property}$$
$$= 5.7 \times 10^5 \quad \text{Add.}$$

Exercises

Evaluate the expression. Write your answer in scientific notation.

31. $(4.2 \times 10^8) + (5.9 \times 10^9)$

32. $(5.9 \times 10^{-4}) - (1.8 \times 10^{-4})$

33. $(7.7 \times 10^8) \times (4.9 \times 10^{-5})$

34. $(3.6 \times 10^5) \div (1.8 \times 10^9)$

16 Chapter Test

Check It Out
Test Practice
BigIdeasMath.com

Write the product using exponents.

1. $(-15) \cdot (-15) \cdot (-15)$

2. $\left(\frac{1}{12}\right) \cdot \left(\frac{1}{12}\right) \cdot \left(\frac{1}{12}\right) \cdot \left(\frac{1}{12}\right) \cdot \left(\frac{1}{12}\right)$

Evaluate the expression.

3. -2^3

4. $10 + 3^3 \div 9$

Simplify the expression. Write your answer as a power.

5. $9^{10} \cdot 9$

6. $(6^6)^5$

7. $\frac{10^4}{10^0}$

8. $\frac{(-3.5)^{13}}{(-3.5)^9}$

Evaluate the expression.

9. $5^{-2} \cdot 5^2$

10. $\frac{-8}{(-8)^3}$

Write the number in standard form.

11. 3×10^7

12. 9.05×10^{-3}

Evaluate the expression. Write your answer in scientific notation.

13. $(7.8 \times 10^7) + (9.9 \times 10^7)$

14. $(6.4 \times 10^5) - (5.4 \times 10^4)$

15. $(3.1 \times 10^6) \times (2.7 \times 10^{-2})$

16. $(9.6 \times 10^7) \div (1.2 \times 10^{-4})$

17. **CRITICAL THINKING** Is $(xy^2)^3$ the same as $(xy^3)^2$? Explain.

18. **RICE** A grain of rice weighs about 3^3 milligrams. About how many grains of rice are in one scoop?

19. **TASTE BUDS** There are about 10,000 taste buds on a human tongue. Write this number in scientific notation.

One scoop of rice weighs about 3^9 milligrams.

20. **LEAD** From 1978 to 2008, the amount of lead allowed in the air in the United States was 1.5×10^{-6} gram per cubic meter. In 2008, the amount allowed was reduced by 90%. What is the new amount of lead allowed in the air?

Test Item References

Chapter Test Questions	Section to Review
1–4	16.1
5–7, 17	16.2
8, 18	16.3
9, 10	16.4
11, 12, 20	16.5
19	16.6
13–16	16.7

Test-Taking Strategies

Remind students to quickly look over the entire test before they start so that they can budget their time. Have students use the **Stop** and **Think** strategy before they answer each question.

Common Errors

- **Exercises 3, 8, and 10** Students may have the wrong sign in their answers. Remind them that when the negative sign is inside parentheses, it is part of the base. Point out that in Exercise 3, the negative sign is not part of the base.
- **Exercise 4** Students may not use the correct order of operations when they evaluate the expression. Review the order of operations with students.
- **Exercises 5, 6, and 9** Students may confuse the Product of Powers Property and the Power of a Power Property. Explain to them why it makes sense to add exponents when using the Product of Powers Property and multiply exponents when using the Power of a Power Property.
- **Exercises 8 and 10** Students may divide the exponents instead of subtracting them. Remind them of the Quotient of Powers Property.
- **Exercises 11 and 12** Students may move the decimal point in the wrong direction. Remind them that when the exponent is negative they move the decimal point to the left, and when the exponent is positive they move the decimal point to the right.

Reteaching and Enrichment Strategies

If students need help. . .	If students got it. . .
Resources by Chapter • Practice A and Practice B • Puzzle Time Record and Practice Journal Practice Differentiating the Lesson Lesson Tutorials *BigIdeasMath.com* Skills Review Handbook	Resources by Chapter • Enrichment and Extension • Technology Connection Game Closet at *BigIdeasMath.com* Start Cumulative Assessment

Answers

1. $(-15)^3$
2. $\left(\frac{1}{12}\right)^5$
3. -8
4. 13
5. 9^{11}
6. 6^{30}
7. 10^4
8. $(-3.5)^4$
9. 1
10. $\frac{1}{64}$
11. 30,000,000
12. 0.00905
13. 1.77×10^8
14. 5.86×10^5
15. 8.37×10^4
16. 8×10^{11}
17. no; $(xy^2)^3 = (xy^2) \cdot (xy^2) \cdot (xy^2) = x \cdot x \cdot x \cdot y^2 \cdot y^2 \cdot y^2 = x^3y^6$
    $(xy^3)^2 = (xy^3) \cdot (xy^3) = x \cdot x \cdot y^3 \cdot y^3 = x^2y^6$
18. 3^6 or 729 grains
19. 1×10^4
20. 1.5×10^{-7} gram per cubic meter

Technology for the Teacher

Online Assessment
Assessment Book
ExamView® Assessment Suite

Test-Taking Strategies

Available at *BigIdeasMath.com*

After Answering Easy Questions, Relax
Answer Easy Questions First
Estimate the Answer
Read All Choices before Answering
Read Question before Answering
Solve Directly or Eliminate Choices
Solve Problem before Looking at Choices
Use Intelligent Guessing
Work Backwards

About this Strategy

When taking a multiple choice test, be sure to read each question carefully and thoroughly. Sometimes you don't know the answer. So . . . guess intelligently! Look at the choices and choose the ones that are possible answers.

Answers

1. C
2. I
3. D
4. 40

Technology for the Teacher

Performance Tasks
Online Assessment
Assessment Book
ExamView® Assessment Suite

Item Analysis

1. **A.** The student assumes there should be 7 zeros.
   B. The student miscounts when adding zeros.
   C. Correct answer
   D. The student miscounts when adding zeros.
2. **F.** The student is finding the sum of the angle measures of a quadrilateral.
   G. The student is finding the sum of the measures of two acute angles of a right triangle.
   H. The student finds the measure of the wrong angle.
   I. Correct answer
3. **A.** The student multiplies exponents instead of adding them.
   B. The student multiplies the bases and adds the exponents.
   C. The student multiplies all of the numbers in the expression.
   D. Correct answer
4. **Gridded Response:** Correct answer: 40

   Common Error: The student applies the ratio of the corresponding sides to the corresponding angles, getting an answer of 80.

16 Cumulative Assessment

Test-Taking Strategy
Use Intelligent Guessing

"It can't be 40 or 5000 because they aren't divisible by 3. So, you can intelligently guess between 3000 and 3072."

1. Mercury's distance from the Sun is approximately 5.79×10^7 kilometers. What is this distance in standard form?

   A. 5,790,000,000 km **C.** 57,900,000 km

   B. 579,000,000 km **D.** 5,790,000 km

2. The steps Jim took to answer the question are shown below. What should Jim change to correctly answer the question?

   How many degrees are in the largest angle in the triangle below?

   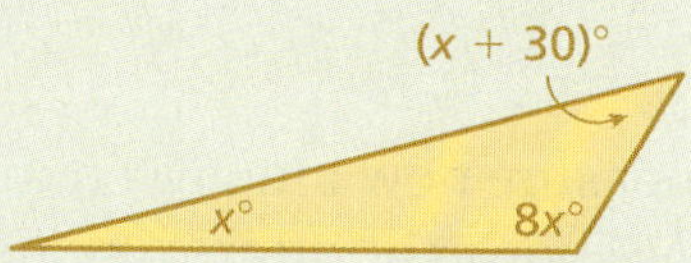

   $$x + 8x + x + 30 = 180$$
   $$10x = 150$$
   $$x = 15$$

   F. The left side of the equation should equal 360° instead of 180°.

   G. The sum of the acute angles should equal 90°.

   H. Evaluate the smallest angle when $x = 15$.

   I. Evaluate the largest angle when $x = 15$.

3. Which expression is equivalent to the expression below?

   $$2^4 2^3$$

   A. 2^{12} **C.** 48

   B. 4^7 **D.** 128

4. In the figure below, $\triangle ABC$ is a dilation of $\triangle DEF$.

   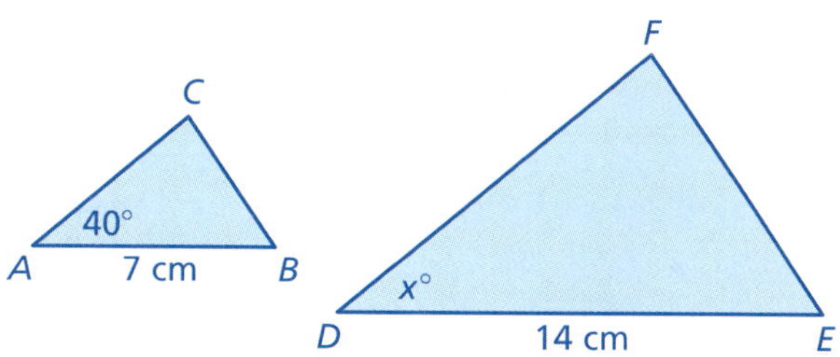

   What is the value of x?

5. A bank account pays interest so that the amount in the account doubles every 10 years. The account started with $5,000 in 1940. Which expression represents the amount (in dollars) in the account n decades later?

F. $2^n \cdot 5000$

G. $5000(n + 1)$

H. 5000^n

I. $2^n + 5000$

6. The formula for the volume V of a pyramid is $V = \frac{1}{3}Bh$. Solve the formula for the height h.

A. $h = \frac{1}{3}VB$

B. $h = \frac{3V}{B}$

C. $h = \frac{V}{3B}$

D. $h = V - \frac{1}{3}B$

7. The gross domestic product (GDP) is a way to measure how much a country produces economically in a year. The table below shows the approximate population and GDP for the United States.

Think Solve Explain

United States 2012	
Population	312 million (312,000,000)
GDP	15.1 trillion dollars ($15,100,000,000,000)

Part A Find the GDP per person for the United States. Show your work and explain your reasoning.

Part B Write the population and the GDP using scientific notation.

Part C Find the GDP per person for the United States using your answers from Part B. Write your answer in scientific notation. Show your work and explain your reasoning.

8. What is the equation of the line shown in the graph?

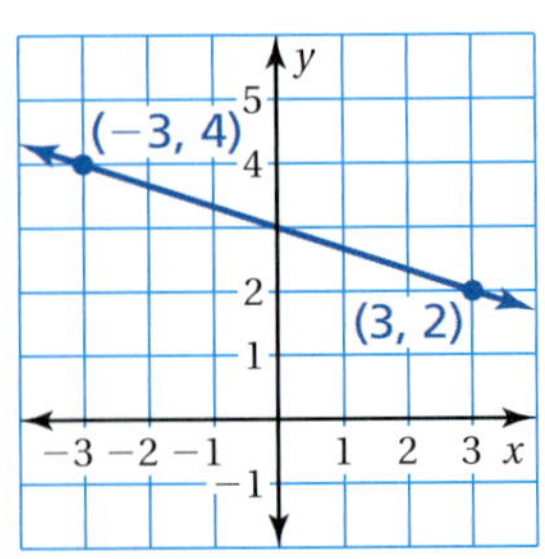

F. $y = -\frac{1}{3}x + 3$

G. $y = \frac{1}{3}x + 1$

H. $y = -3x + 3$

I. $y = 3x - \frac{1}{3}$

Item Analysis (continued)

5. **F.** Correct answer

   G. The student multiplies 5000 by one more than the number of decades.

   H. The student uses 5000 as the base.

   I. The student adds 2^n and 5000 instead of multiplying.

6. **A.** The student does not understand inverse operations.

   B. Correct answer

   C. The student does not understand inverse operations.

   D. The student subtracts $\frac{1}{3}B$ instead of dividing by $\frac{1}{3}B$.

7. **4 points** The student demonstrates a thorough understanding of how to work arithmetically with large numbers both in standard form and in scientific notation. In Part A, an answer of \$48,397.44 is obtained. In Part B, the data are written as 3.12×10^8 and 1.51×10^{13}. In Part C, the student works out the quotient $\frac{1.51 \times 10^{13}}{3.12 \times 10^8}$ step-by-step.

   3 points The student demonstrates an essential but less than thorough understanding. In particular, Parts A and B should be completed correctly, but the steps taken in Part C may show gaps or be incomplete.

   2 points The student demonstrates an understanding of how to write the data using scientific notation, but is otherwise limited in understanding how to approach the problem arithmetically.

   1 point The student demonstrates limited understanding of working with large numbers arithmetically. The student's response is incomplete and exhibits many flaws.

   0 points The student provides no response, a completely incorrect or incomprehensible response, or a response that demonstrates insufficient understanding of how to work with large numbers.

8. **F.** Correct answer

   G. The student miscalculates slope as positive, and then chooses the equation that has (3, 2) as a solution.

   H The student reverses the roles of x and y in finding slope, and then uses the correct intercept, interpolated from the graph.

   I. The student reverses slope and y-intercept in the equation.

Answers

5. F

6. B

7. *Part A* about \$48,397.44

   Part B 3.12×10^8; 1.51×10^{13} dollars

   Part C about 4.84×10^4 dollars

8. F

Answers

9. C

10. 0.16 or $\frac{4}{25}$

11. G

12. B

Item Analysis (continued)

9. **A.** The student does not square the radius.

B. The student multiplies by the diameter instead of the radius squared.

C. Correct answer

D. The student multiplies by the diameter squared instead of the radius squared.

10. **Gridded Response:** Correct answer: 0.16 or $\frac{4}{25}$

Common Error: The student writes the answer as a negative number.

11. **F.** The student mistakes the roles of slope and intercept, thinking same intercept means what same slope means.

G. Correct answer

H. The student chooses a conclusion for lines that have the same intercept and the same slope, overlooking that these lines have different slopes.

I. The student thinks that because the two lines have the same y-intercept, they must have the same slope.

12. **A.** The student uses part of the formula for the volume of a sphere.

B. Correct answer

C. The student uses part of the formula for the volume of a sphere.

D. The student thinks the volume of a sphere is πr.

9. A cylinder and its dimensions are shown below.

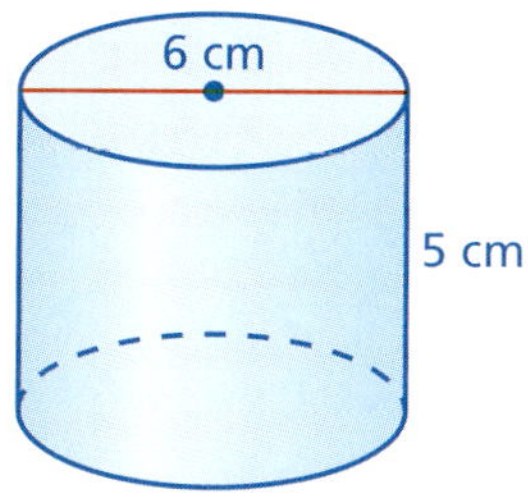

What is the volume of the cylinder? (Use 3.14 for π.)

A. 47.1 cm^3

B. 94.2 cm^3

C. 141.3 cm^3

D. 565.2 cm^3

10. Find $(-2.5)^{-2}$.

11. Two lines have the same y-intercept. The slope of one line is 1, and the slope of the other line is -1. What can you conclude?

F. The lines are parallel.

G. The lines are perpendicular.

H. The lines are neither parallel nor perpendicular.

I. The situation described is impossible.

12. The volume of a sphere is $4\frac{1}{2}\pi$ cubic centimeters. What is the radius of the sphere?

A. $\frac{4}{3}$ cm

B. $1\frac{1}{2}$ cm

C. π cm

D. $4\frac{1}{2}$ cm

Additional Topics

For Your Information

The following topics are taught with Chapter 3–Expressions and Equations. These topics build on students' knowledge of equation solving.

Pacing Guide for Chapter 3

Chapter Opener	
Regular	1 Day
Accelerated	1 Day
Section 1	
Regular	2 Days
Accelerated	0.5 Day
Section 2	
Regular	3 Days
Accelerated	1.5 Days
Study Help / Quiz	
Regular	1 Day
Accelerated	0 Days
Section 3	
Regular	2 Days
Accelerated	1 Day
Section 4	
Regular	2 Days
Accelerated	1 Day
Section 5	
Regular	2 Days
Accelerated	1 Day
Topic 1	
Regular	0 Days
Accelerated	1 Day
Topic 2	
Regular	0 Days
Accelerated	2 Days
Topic 3	
Regular	0 Days
Accelerated	1 Day
Chapter Review/ Chapter Tests	
Regular	2 Days
Accelerated	2 Days
Total Chapter 3	
Regular	15 Days
Accelerated	12 Days
Year-to-Date	
Regular	42 Days
Accelerated	22 Days

Technology for the Teacher

BigIdeasMath.com
Chapter at a Glance
Complete Materials List
Parent Letters: English and Spanish

What Your Students Have Learned

- Use parentheses, brackets, or braces in numerical expressions, and evaluate expressions with these symbols.
- Write and interpret numerical expressions without evaluating them.
- Use and interpret simple equations.
- Write and evaluate numerical expressions involving whole-number exponents.
- Read, write, and evaluate algebraic expressions.
- Apply the properties of operations to generate equivalent expressions.
- Factor out the greatest common factor (GCF) in algebraic and numerical expressions.
- Identify equivalent expressions.
- Determine if values are solutions of equations.
- Add, subtract, factor, and expand linear expressions with rational coefficients.
- Understand that writing expressions in different forms can show how the quantities are related.
- Write, graph, and solve one-step equations (includes negative numbers).
- Solve two-step equations.
- Compare algebraic solutions to arithmetic solutions.

What Your Students Will Learn

- Solve linear equations with rational number coefficients, including equations whose solutions require expanding expressions using the Distributive Property and collecting like terms.
- Show that linear equations in one variable have one solution, infinitely many solutions, or no solution by transforming the equations into simpler forms.

Additional Topics

Topic 1 Solving Multi-Step Equations

Topic 2 Solving Equations with Variables on Both Sides

Topic 3 Rewriting Equations and Formulas

Topic 1 Solving Multi-Step Equations

Check It Out
Lesson Tutorials
BigIdeasMath.com

Solving Equations

In this lesson, you will

- use inverse operations to solve multi-step equations.
- use the Distributive Property to solve multi-step equations.

Key Idea

Solving Multi-Step Equations

To solve multi-step equations, use inverse operations to isolate the variable.

EXAMPLE 1 Solving a Two-Step Equation

The height (in feet) of a tree after x years is $1.5x + 15$. After how many years is the tree 24 feet tall?

$1.5x + 15 = 24$ Write an equation.

Undo the addition. → $-15 \quad -15$ Subtraction Property of Equality

$1.5x = 9$ Simplify.

Undo the multiplication. → $\frac{1.5x}{1.5} = \frac{9}{1.5}$ Division Property of Equality

$x = 6$ Simplify.

So, the tree is 24 feet tall after 6 years.

EXAMPLE 2 Combining Like Terms to Solve an Equation

Solve $8x - 6x - 25 = -35$.

$8x - 6x - 25 = -35$ Write the equation.

$2x - 25 = -35$ Combine like terms.

Undo the subtraction. → $+25 \quad +25$ Addition Property of Equality

$2x = -10$ Simplify.

Undo the multiplication. → $\frac{2x}{2} = \frac{-10}{2}$ Division Property of Equality

$x = -5$ Simplify.

The solution is $x = -5$.

On Your Own

Now You're Ready
Exercises 6–9

Solve the equation. Check your solution.

1. $-3z + 1 = 7$
2. $\frac{1}{2}x - 9 = -25$
3. $-4n - 8n + 17 = 23$

Laurie's Notes

Introduction

Applying Mathematical Practices

- **Make Sense of Problems and Persevere in Solving Them:** Mathematically proficient students look for entry points in a problem. In solving a multi-step equation, they analyze the given information and determine how they can begin the solution.

Connect

- **Yesterday:** Students solved two-step equations by undoing the operations.
- **Today:** Students will solve multi-step equations by using inverse operations to isolate the variable.

Motivate

- Make a card for each student in your class. Write a variable term on each card. Students will walk around to find others with a card containing a *like term* to the one they are holding.

  Samples: $5x$, $-13x$, $5y$, $6xy$, x, $3.8x$, $\frac{1}{2}y$, $-3.8y$
- Ask students to explain what it means for terms to be *like* terms.

Lesson Notes

Key Idea

- **Connection:** When you evaluate an expression, you follow the order of operations. Solving an equation undoes the evaluating, in reverse order. The goal is to isolate the variable term and then solve for the variable.

Example 1

- One way to explain the equation is to think of the tree as being 15 feet tall when being planted. It then grows at a rate of 1.5 feet each year.
- **Model with Mathematics:** Make a table to show the height of the tree from the first year to the sixth year.

Example 2

? "Why is $8x - 6x = 2x$?" Use the Distributive Property to subtract the terms; $8x - 6x = (8 - 6)x = 2x$.

On Your Own

- In Question 2, students may divide both sides by $\frac{1}{2}$ and get $x = -8$. Remind students that dividing by $\frac{1}{2}$ is the same as multiplying by 2.

What Your Students Will Learn

- Use inverse operations, combining like terms, and the Distributive Property to solve multi-step equations.

Goal Today's lesson is solving multi-step equations.

Technology for the Teacher

Dynamic Classroom

Lesson Tutorials
Lesson Plans
Answer Presentation Tool

Extra Example 1

The height (in inches) of a plant after t days is $\frac{1}{2}t + 6$. After how many days is the plant 21 inches tall? 30 days

Extra Example 2

Solve $-2m + 4m + 5 = -3$. $m = -4$

On Your Own

1. $z = -2$
2. $x = -32$
3. $n = -0.5$

Laurie's Notes

Extra Example 3

Solve $-4(3g - 5) + 10g = 19$. 0.5

Example 3

- Ask students to identify the operations involved in this equation. from left to right: multiplication (by 2), subtraction, multiplication (5x), addition
- **Note:** Combining like terms in the third step is not obvious to students. When the like terms are not adjacent, students are unsure of how to combine them. Rewrite the left side of the equation as $2 + (-10)x + 4$.

Words of Wisdom

- Take time to work through the Study Tip and discuss the steps. Instead of using the Distributive Property, both sides of the equation are divided by 2 in the third step. This will not be obvious to students, nor will they know why it is okay to do this.
- Explain to students that the left side of the equation is 2 times an expression. When the expression $2(1 - 5x)$ is divided by 2, it leaves the expression $1 - 5x$. In the next step, students want to add 1 to each side because of the subtraction operation shown. Again, it is helpful to write $1 - 5x$ as $1 + (-5)x$ so that it makes sense to students why 1 is subtracted from each side.

Extra Example 4

You have scored 7, 10, 8, and 9 on four quizzes. Write and solve an equation to find the score you need on the fifth quiz so that your mean score is 8.
$\frac{x + 7 + 10 + 8 + 9}{5} = 8$; 6

Example 4

- You may need to review *mean* with the students.
- Discuss the information displayed in the table and write the equation.

? "Is it equivalent to write $\frac{x + 3.5}{5} = 1.5$ instead of $\frac{3.5 + x}{5} = 1.5$? Explain." yes; Commutative Property of Addition

- **FYI:** It may be helpful to write the third step with parentheses: $5\left(\frac{3.5 + x}{5}\right)$.
- **Reason Abstractly and Quantitatively:** Ask students to explain the impact of trying to achieve a mean of 1.5 miles run per day when you ran 0 miles on two of the days.
- **Note:** This is a classic question. When all of the data are known except for one, what is needed in order to achieve a particular average? Students often ask this in the context of wanting to know what they have to score on a test in order to achieve a certain average.

On Your Own

4. $x = -1.5$
5. $d = -1$
6. $\frac{88 + 92 + 87 + x}{4} = 90$; $x = 93$

On Your Own

- Encourage students to work in pairs. Students need to be careful with multi-step equations, and it is helpful to have a partner check each step.

English Language Learners

Vocabulary

English learners will benefit from understanding that a *term* is a number, a variable, or the product of a number and variable. *Like terms* are terms that have identical variable parts.

3 and 16 are like terms because they contain no variable.

$4x$ and $7x$ are like terms because they have the same variable x.

$5a$ and $5b$ are *not* like terms because they have different variables.

Closure

- **Exit Ticket:** Solve $8x + 9 - 4x = 25$. Check your solution. $x = 4$

EXAMPLE 3 Using the Distributive Property to Solve an Equation

Solve $2(1 - 5x) + 4 = -8$.

> **Study Tip**
>
> Here is another way to solve the equation in Example 3.
>
> $2(1 - 5x) + 4 = -8$
>
> $2(1 - 5x) = -12$
>
> $1 - 5x = -6$
>
> $-5x = -7$
>
> $x = 1.4$

Step	Reason
$2(1 - 5x) + 4 = -8$	Write the equation.
$2(1) - 2(5x) + 4 = -8$	Distributive Property
$2 - 10x + 4 = -8$	Multiply.
$-10x + 6 = -8$	Combine like terms.
$\underline{-6} \quad \underline{-6}$	Subtraction Property of Equality
$-10x = -14$	Simplify.
$\dfrac{-10x}{-10} = \dfrac{-14}{-10}$	Division Property of Equality
$x = 1.4$	Simplify.

EXAMPLE 4 Real-Life Application

Use the table to find the number of miles x you need to run on Friday so that the mean number of miles run per day is 1.5.

Day	Miles
Monday	2
Tuesday	0
Wednesday	1.5
Thursday	0
Friday	x

Write an equation using the definition of *mean*.

(sum of the data → numerator; number of values → denominator)

Step	Reason
$\dfrac{2 + 0 + 1.5 + 0 + x}{5} = 1.5$	Write the equation.
$\dfrac{3.5 + x}{5} = 1.5$	Combine like terms.
Undo the division. → $5 \cdot \dfrac{3.5 + x}{5} = 5 \cdot 1.5$	Multiplication Property of Equality
$3.5 + x = 7.5$	Simplify.
Undo the addition. → $\underline{-3.5} \quad \underline{-3.5}$	Subtraction Property of Equality
$x = 4$	Simplify.

∴ So, you need to run 4 miles on Friday.

On Your Own

Now You're Ready
Exercises 10 and 11

Solve the equation. Check your solution.

4. $-3(x + 2) + 5x = -9$
5. $5 + 1.5(2d - 1) = 0.5$
6. You scored 88, 92, and 87 on three tests. Write and solve an equation to find the score you need on the fourth test so that your mean test score is 90.

Topic 1 Exercises

Check It Out
Help with Homework
BigIdeasMath.com

Vocabulary and Concept Check

1. **WRITING** Write the verbal statement as an equation. Then solve.

   2 more than 3 times a number is 17.

2. **OPEN-ENDED** Explain how to solve the equation $2(4x - 11) + 9 = 19$.

Practice and Problem Solving

CHOOSE TOOLS Find the value of the variable. Then find the angle measures of the polygon. Use a protractor to check the reasonableness of your answer.

3. 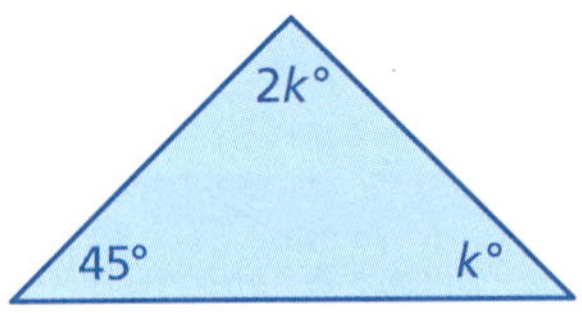

Sum of angle measures: 180°

4. 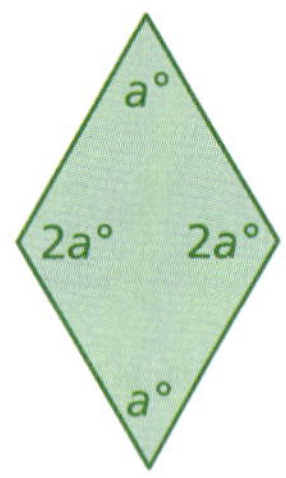

Sum of angle measures: 360°

5. 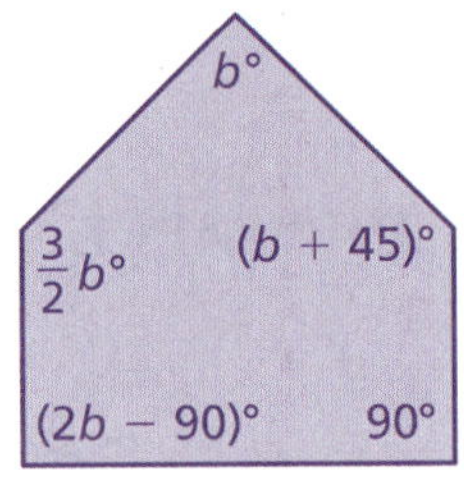

Sum of angle measures: 540°

Solve the equation. Check your solution.

1 2 6. $10x + 2 = 32$

7. $19 - 4c = 17$

8. $1.1x + 1.2x - 5.4 = -10$

9. $\frac{2}{3}h - \frac{1}{3}h + 11 = 8$

3 10. $6(5 - 8v) + 12 = -54$

11. $21(2 - x) + 12x = 44$

12. **ERROR ANALYSIS** Describe and correct the error in solving the equation.

   ✗
   $$-2(7 - y) + 4 = -4$$
   $$-14 - 2y + 4 = -4$$
   $$-10 - 2y = -4$$
   $$-2y = 6$$
   $$y = -3$$

13. **WATCHES** The cost C (in dollars) of making n watches is represented by $C = 15n + 85$. How many watches are made when the cost is \$385?

14. **HOUSE** The height of the house is 26 feet. What is the height x of each story?

Assignment Guide and Homework Check

Level	Assignment	Homework Check
Accelerated	1–5, 6–14 even, 15–22	10, 12, 14, 16

For Your Information

- **Exercises 3–5** Students will learn about sums of interior angles of polygons in Chapter 12. Here, the sums are given.
- **Exercise 16** Remind students about percents. They will need to remember how to do calculations with percents to find the solution.

Common Errors

- **Exercises 8 and 9** When combining like terms, students may square the variable. Remind them that $x^2 = x \cdot x$, and in these exercises they are not multiplying the variables. Remind them that when adding and subtracting variables, they perform the addition or subtraction on the coefficient of the variable.
- **Exercises 10 and 11** When using the Distributive Property, students may forget to distribute to all the values within the parentheses. Remind them that they need to distribute to all the values and encourage them to draw arrows showing the distribution, if needed.
- **Exercise 16** Students may struggle with writing the equation for this problem because of the tip that is added to the total. Encourage them to write an expression for the cost of the food and then add on the tip.

Topic 1 Record and Practice Journal

Solve the equation. Check your solution.

1. $3x - 11 = 22$

   $x = 11$

2. $24 - 10b = 9$

   $b = 1.5$

3. $2.4z + 1.2z - 6.5 = 0.7$

   $z = 2$

4. $\frac{3}{4}w - \frac{1}{2}w - 4 = 12$

   $w = 64$

5. $2(a + 7) - 7 = 9$

   $a = 1$

6. $20 + 8(q - 11) = -12$

   $q = 7$

7. Find the width of the rectangular prism when the surface area is 208 square centimeters.

   $w = 4$ cm

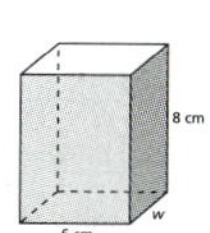

8. The amount A of money in your savings account after m months is represented by $A = 135m + 225$. After how many months do you have $765 in your savings account?

   4 months

9. A rectangular garden has a length of $(x + 8)$ feet and a width of 20 feet. The perimeter of the garden is 110 feet. How much greater is the length than the width?

   15 feet

10. A sponsor of a youth baseball team buys new hats and jerseys for the players on the team. Each hat costs $10 and each jersey costs $20. The sponsor also donates $200 to help pay for umpires. The sponsor spends a total of $650. How many players are on the team?

    15

Vocabulary and Concept Check

1. $2 + 3x = 17; x = 5$
2. *Sample answer:* Subtract 9 from each side. Divide each side by 2. Add 11 to each side. Divide each side by 4.

Practice and Problem Solving

3. $k = 45$; 45°, 45°, 90°
4. $a = 60$; 60°, 120°, 60°, 120°
5. $b = 90$; 90°, 135°, 90°, 90°, 135°
6. $x = 3$
7. $c = 0.5$
8. $x = -2$
9. $h = -9$
10. $v = 2$
11. $x = -\frac{2}{9}$
12. They did not distribute the -2 properly.

$$\begin{aligned} -2(7 - y) + 4 &= -4 \\ -14 + 2y + 4 &= -4 \\ 2y - 10 &= -4 \\ 2y &= 6 \\ y &= 3 \end{aligned}$$

13. 20 watches
14. 10 ft

Practice and Problem Solving

15. $4(b + 3) = 24$; 3 in.

16. $1.15(2p + 1.5) = 11.5$; \$4.25

17. $\dfrac{2580 + 2920 + x}{3} = 3000$; 3500 people

18. See *Taking Math Deeper*.

Fair Game Review

19. < **20.** =

21. > **22.** D

Mini-Assessment

Solve the equation. Check your solution.

1. $18 = 5a - 2a + 3$ $a = 5$
2. $2(4 - 2w) - 8 = -4$ $w = 1$
3. $2.3y + 4.4y - 3.7 = 16.4$ $y = 3$
4. $\frac{3}{4}z + \frac{1}{4}z - 6 = -5$ $z = 1$
5. The perimeter of the picture is 36 inches. What is the height of the picture? 10 in.

Taking Math Deeper

Exercise 18

This problem is an example of how algebra can be used in fields outside of mathematics.

1 Begin by translating the scoring system into a mathematical formula.

Final score = 0.6(degree of difficulty)(sum of countries' scores) — sum of countries' scores: not including the highest and lowest

2 Substitute the given information.

Let x = the degree of difficulty.

$$77.7 = 0.6(x)(7.5 + 8.0 + 7.0 + 7.5 + 7.0)$$
$$77.7 = 0.6x(37)$$
$$77.7 = 22.2x$$
$$3.5 = x$$

a. The degree of difficulty is 3.5.

3 This question has many answers.

Let x = sum of the five countries' scores.

$$97.2 = 0.6(4)(x)$$
$$97.2 = 2.4x$$
$$40.5 = x$$

One possibility is the following:

b. 8.0 + 8.0 + 8.0 + 8.0 + 8.5 with a low score of 7.5 and a high score of 9.0

Project

Use the Internet or school library to find all the different dives that are scored in a diving competition. Find the degree of difficulty that goes with each dive.

Reteaching and Enrichment Strategies

If students need help...	If students got it...
Resources by Chapter • Practice A and Practice B • Puzzle Time Record and Practice Journal Practice Differentiating the Lesson Lesson Tutorials Skills Review Handbook	Resources by Chapter • Enrichment and Extension • Technology Connection Start the next section

In Exercises 15 – 17, write and solve an equation to answer the question.

15. POSTCARD The area of the postcard is 24 square inches. What is the width b of the message (in inches)?

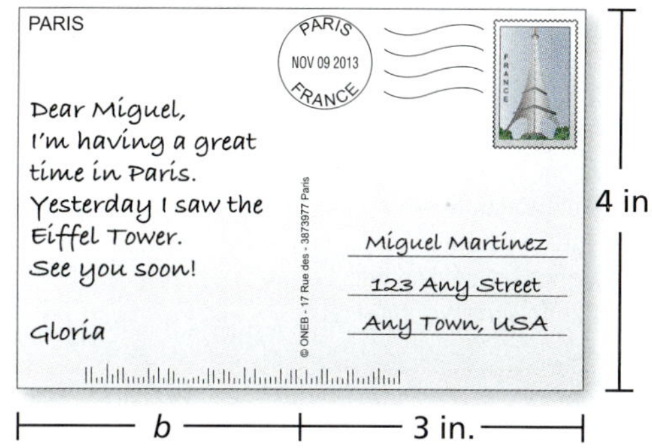

16. BREAKFAST You order two servings of pancakes and a fruit cup. The cost of the fruit cup is \$1.50. You leave a 15% tip. Your total bill is \$11.50. How much does one serving of pancakes cost?

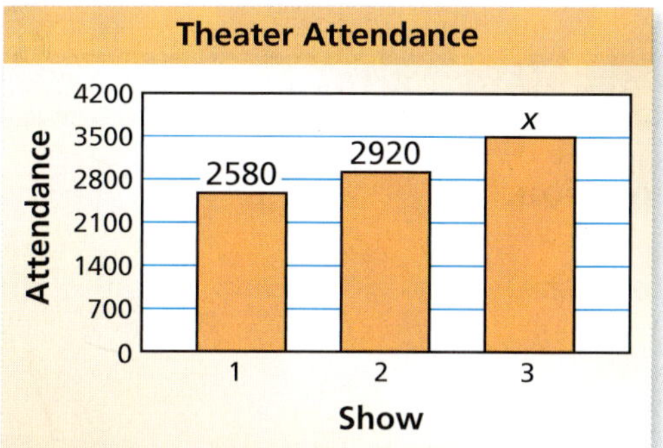

17. THEATER How many people must attend the third show so that the average attendance per show is 3000?

18. DIVING Divers in a competition are scored by an international panel of judges. The highest and the lowest scores are dropped. The total of the remaining scores is multiplied by the degree of difficulty of the dive. This product is multiplied by 0.6 to determine the final score.

a. A diver's final score is 77.7. What is the degree of difficulty of the dive?

Judge	Russia	China	Mexico	Germany	Italy	Japan	Brazil
Score	7.5	8.0	6.5	8.5	7.0	7.5	7.0

b. Critical Thinking The degree of difficulty of a dive is 4.0. The diver's final score is 97.2. Judges award half or whole points from 0 to 10. What scores could the judges have given the diver?

Fair Game Review What you learned in previous grades & lessons

Let $a = 3$ and $b = -2$. Copy and complete the statement using <, >, or =. *(Section 1.2, Section 1.3, and Section 1.4)*

19. $-5a$ ☐ 4 **20.** 5 ☐ $b + 7$ **21.** $a - 4$ ☐ $10b + 8$

22. MULTIPLE CHOICE What value of x makes the equation $x + 5 = 2x$ true? *(Skills Review Handbook)*

Ⓐ -1 Ⓑ 0 Ⓒ 3 Ⓓ 5

Topic 2 Solving Equations with Variables on Both Sides

Check It Out
Lesson Tutorials
BigIdeasMath.com

Solving Equations

In this lesson, you will

- solve equations with variables on both sides.
- determine whether equations have no solution or infinitely many solutions.

Key Idea

Solving Equations with Variables on Both Sides

To solve equations with variables on both sides, collect the variable terms on one side and the constant terms on the other side.

EXAMPLE 1 Solving an Equation with Variables on Both Sides

Solve $15 - 2x = -7x$. Check your solution.

Step	Work	Reason
	$15 - 2x = -7x$	Write the equation.
Undo the subtraction.	$+2x \quad +2x$	Addition Property of Equality
	$15 = -5x$	Simplify.
Undo the multiplication.	$\dfrac{15}{-5} = \dfrac{-5x}{-5}$	Division Property of Equality
	$-3 = x$	Simplify.

The solution is $x = -3$.

Check

$15 - 2x = -7x$

$15 - 2(-3) \stackrel{?}{=} -7(-3)$

$21 = 21$ ✓

EXAMPLE 2 Using the Distributive Property to Solve an Equation

Solve $-2(x - 5) = 6\left(2 - \frac{1}{2}x\right)$.

Step	Work	Reason
	$-2(x - 5) = 6\left(2 - \frac{1}{2}x\right)$	Write the equation.
	$-2x + 10 = 12 - 3x$	Distributive Property
Undo the subtraction.	$+3x \qquad +3x$	Addition Property of Equality
	$x + 10 = 12$	Simplify.
Undo the addition.	$-10 \quad -10$	Subtraction Property of Equality
	$x = 2$	Simplify.

The solution is $x = 2$.

On Your Own

Now You're Ready
Exercises 6–14

Solve the equation. Check your solution.

1. $-3x = 2x + 19$
2. $2.5y + 6 = 4.5y - 1$
3. $6(4 - z) = 2z$

Laurie's Notes

Introduction

Applying Mathematical Practices

- **Attend to Precision:** Mathematically proficient students try to communicate precisely with others. In today's lesson, students will work with vocabulary related to solving equations.

Connect

- **Yesterday:** Students solved multi-step equations by using inverse operations to isolate the variable.
- **Today:** Students will solve equations with variables on both sides by using Properties of Equality and collecting variable terms on one side. The equations could have one solution, no solution, or infinitely many solutions.

Motivate

- Share some information about the Mississippi River. It is the third longest river in North America, flowing 2350 miles from Lake Itasca in Minnesota to the Gulf of Mexico. The width of the river ranges from 20–30 feet at its narrowest to more than 11 miles at its widest.
- Explain that one of the examples in today's lesson involves a boat traveling upstream on the Mississippi River.

Lesson Notes

Key Idea

? "In the expression $5x - 2 - 9x + y + 4$, what are the variable terms? Constant terms?" $5x$, $-9x$, and y; -2 and 4

- **Common Error:** Students may forget to include the sign of the variable term.

Example 1

- Notice that a constant is on the left side only. For this reason, it makes sense to solve for the variable term on the right side of the equation. It is possible to solve for the variable term on the left side of the equation, but finding the solution involves an extra step. Show students this approach as well, and point out that it gives the same solution.

Example 2

- **Teaching Tip:** Before distributing on the left side, rewrite the inside expression as $x + (-5)$ (add the opposite). Students are likely to recognize that they are multiplying $(-2)(-5)$ to get a product of 10.

? "How do you multiply a whole number and a fraction like $6\left(\frac{1}{2}\right)$?" Listen for language such as, write the whole number over a denominator of 1 and multiply straight across.

- **FYI:** Solve for the variable on the side of the equation where the coefficient is the greatest. The coefficient of the variable term will be positive, a condition that generally renders fewer mistakes.

On Your Own

- **Neighbor Check:** Have students work independently and then have their neighbors check their work. Have students discuss any discrepancies.

What Your Students Will Learn

- Solve equations with variables on both sides using properties of equality and the Distributive Property.
- Determine whether equations have one solution, no solution, or infinitely many solutions.
- Write and solve equations with variables on both sides.

Goal Today's lesson is solving equations with variables on both sides.

Technology for the Teacher

Dynamic Classroom

Lesson Tutorials
Lesson Plans
Answer Presentation Tool

Extra Example 1

Solve $r = -5r + 18$. Check your solution. 3

Extra Example 2

Solve $6\left(1 + \frac{1}{2}x\right) = 2(x + 1)$. -4

On Your Own

1. $x = -3.8$
2. $y = 3.5$
3. $z = 3$

Laurie's Notes

Discuss

- Ask a few questions about equation solving:
  - "Does every equation have a solution?" no
  - "Does every equation have just one solution?" no
  - "Is it possible for an equation to have two solutions?" yes Students may say no, but using the third example below will convince them otherwise. They will encounter these types of equations in Chapter 14.
- Share some common equations and discuss the number of solutions:

  $x + 2 = 7$ one solution, 5

  $x + 2 = x + 7$ no solution

  $x^2 = 4$ two solutions, 2 and -2

  $x + 2 = x + 2$ infinitely many solutions
- Explain that in Examples 3 and 4, students will investigate equations that have no solution or infinitely many solutions. Assure students that they will use the same techniques for solving equations as before.

Extra Example 3

Solve $3x - 5 = 7 + 3x$. no solution

Example 3

- **Teaching Tip:** Instead of telling students when an equation has no solution, work through the example and ask students about the "solution" $3 = -7$.
- Work through the problem in two ways, as shown and by collecting the constant terms on one side of the equation as the first step. Show that the solution is the same both ways.
- For the final step, write $3 \neq -7$ to emphasize that there is no solution.
- **Reason Abstractly and Quantitatively** and **Look for and Make Use of Structure:** Students should notice that the same quantity, $4x$, is being subtracted from different numbers, 3 and -7. They should reason that the two sides of the equation can never be equal, so there is no solution.
- "How do you know when an equation has no solution?" Solve the equation normally and if you end up with a false statement, the equation has no solution.

Extra Example 4

Solve $\frac{1}{4}(8x - 12) = 2x - 3$ infinitely many solutions

On Your Own

4. no solution
5. infinitely many solutions
6. infinitely many solutions
7. no solution

Example 4

- **Teaching Tip:** Instead of telling students when an equation has infinitely many solutions, work through the example and ask students about the "solution" $4 = 4$.
- To check the solution, ask volunteers to choose several values for x. Substitute these values into the original equation and show that all result in true equations.
- **Reason Abstractly and Quantitatively** and **Look for and Make Use of Structure:** Students should notice that in the second step, the expressions on both sides of the equal sign are the same. They should reason that both sides will be equal for any value of x, so there are infinitely many solutions.

On Your Own

- **Common Error:** In Exercises 5–7, students may forget to distribute the factor to the second term within the parentheses.
- **Neighbor Check:** Have students work independently and then have their neighbors check their work. Have students discuss any discrepancies.

English Language Learners

Vocabulary

Remind English learners that *like terms* are terms with the same variables raised to the same exponents. As the number of terms in an equation increases, an important skill is to identify and combine like terms.

Some equations do not have one solution. Equations can also have no solution or infinitely many solutions.

When solving an equation that has no solution, you will obtain an equivalent equation that is not true for any value of the variable, such as $0 = 2$.

EXAMPLE 3 Solving Equations with No Solution

Solve $3 - 4x = -7 - 4x$.

$3 - 4x = -7 - 4x$ — Write the equation.

Undo the subtraction. → $+ 4x \quad + 4x$ — Addition Property of Equality

$3 = -7$ ✗ — Simplify.

The equation $3 = -7$ is never true. So, the equation has no solution.

When solving an equation that has infinitely many solutions, you will obtain an equivalent equation that is true for all values of the variable, such as $-5 = -5$.

EXAMPLE 4 Solving Equations with Infinitely Many Solutions

Solve $6x + 4 = 4\left(\frac{3}{2}x + 1\right)$.

$6x + 4 = 4\left(\frac{3}{2}x + 1\right)$ — Write the equation.

$6x + 4 = 6x + 4$ — Distributive Property

Undo the addition. → $- 6x \quad - 6x$ — Subtraction Property of Equality

$4 = 4$ — Simplify.

The equation $4 = 4$ is always true. So, the equation has infinitely many solutions.

On Your Own

Now You're Ready
Exercises 18–29

Solve the equation.

4. $2x + 1 = 2x - 1$

5. $\frac{1}{2}(6t - 4) = 3t - 2$

6. $\frac{1}{3}(2b + 9) = \frac{2}{3}\left(b + \frac{9}{2}\right)$

7. $6(5 - 2v) = -4(3v + 1)$

EXAMPLE 5 Writing and Solving an Equation

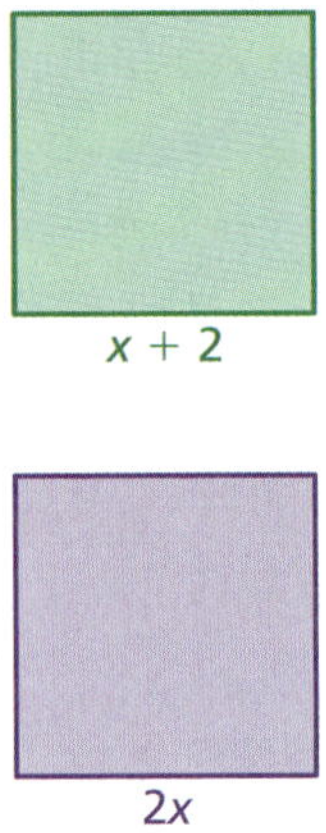

The squares are identical. What is the area of each square?

Ⓐ 2 Ⓑ 4 Ⓒ 16 Ⓓ 32

The squares are identical, so the side length of each square is the same.

$x + 2 = 2x$	Write an equation.
$\underline{-x} \quad \underline{-x}$	Subtraction Property of Equality
$2 = x$	Simplify.

Because the side length of each square is 4, the area of each square is $4^2 = 16$.

So, the correct answer is Ⓒ.

EXAMPLE 6 Real-Life Application

A boat travels x miles per hour upstream on the Mississippi River. On the return trip, the boat travels 2 miles per hour faster. How far does the boat travel upstream?

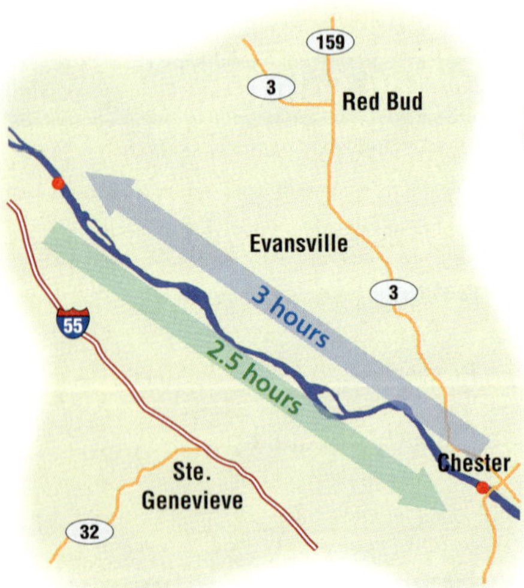

The speed of the boat on the return trip is $(x + 2)$ miles per hour.

Distance upstream = Distance of return trip

$3x = 2.5(x + 2)$	Write an equation.
$3x = 2.5x + 5$	Distributive Property
$\underline{-2.5x} \quad \underline{-2.5x}$	Subtraction Property of Equality
$0.5x = 5$	Simplify.
$\dfrac{0.5x}{0.5} = \dfrac{5}{0.5}$	Division Property of Equality
$x = 10$	Simplify.

The boat travels 10 miles per hour for 3 hours upstream. So, it travels 30 miles upstream.

On Your Own

8. **WHAT IF?** In Example 5, the side length of the purple square is $3x$. What is the area of each square?

9. A boat travels x miles per hour from one island to another island in 2.5 hours. The boat travels 5 miles per hour faster on the return trip of 2 hours. What is the distance between the islands?

Laurie's Notes

Example 5

- Solve the equation as shown.
- This example reviews the concept of the area of a square. You might also ask about the formula for the perimeter of a square. $P = 4s$

Example 6

? "How do you find the distance traveled d when you know the rate r and time t?" multiply; $d = rt$

- Discuss with students that the distance both ways is the same, a simple but not obvious fact.

? "If you travel 40 miles per hour for 2 hours, how far will you go? How about 40 miles per hour for a half hour?" 80 mi; 20 mi

? "How far do you travel at x miles per hour for 3 hours?" $3x$ mi

- Students need to read the time from the illustration. The rates for each direction are x and $(x + 2)$.
- Write and solve the equation as shown.

On Your Own

- **Think-Pair-Share:** Students should read each question independently and then work in pairs to answer the questions. When they have answered the questions, the pair should compare their answers with another group and discuss any discrepancies.

Closure

- Give an example of an equation that has
  a. no solution. *Sample answer:* $x + 5 = x + 2$
  b. one solution. *Sample answer:* $x + 5 = 2x + 3$
  c. infinitely many solutions. *Sample answer:* $4x + 12 = 2(2x + 6)$

Extra Example 5

The legs of the right triangle have the same length. What is the area of the triangle? $1\frac{1}{8}$ square units

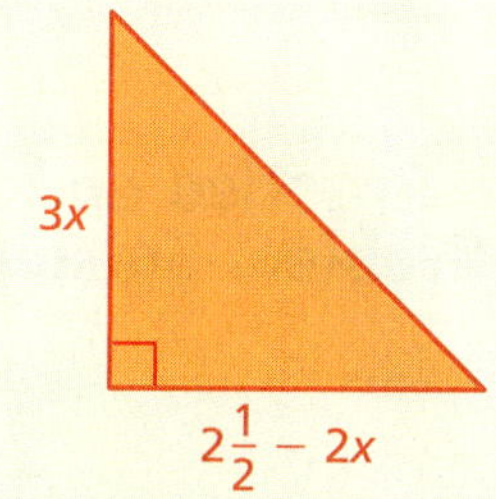

Extra Example 6

A boat travels 3 hours downstream at r miles per hour. On the return trip, the boat travels 5 miles per hour slower and takes 4 hours. What is the distance the boat travels each way? 60 mi

On Your Own

8. 9

9. 50 mi

Vocabulary and Concept Check

1. no; When 3 is substituted for x, the left side simplifies to 4 and the right side simplifies to 3.
2. *Sample answer:* $4x + 1 = 3x - 2$

Practice and Problem Solving

3. $x = 13.2$ in.
4. $x = 7.2$ in.
5. $x = 7.5$ in.
6. $m = -4$
7. $k = -0.75$
8. $x = 8.2$
9. $p = -48$
10. $w = 2$
11. $n = -3.5$
12. $z = 1.6$
13. $x = -4$
14. $d = 14$
15. The 4 should have been added to the right side.

$$\begin{aligned} 3x - 4 &= 2x + 1 \\ 3x - 2x - 4 &= 2x + 1 - 2x \\ x - 4 &= 1 \\ x - 4 + 4 &= 1 + 4 \\ x &= 5 \end{aligned}$$

16. 2 lb
17. $15 + 0.5m = 25 + 0.25m$; 40 mi

Assignment Guide and Homework Check

Level	Assignment	Homework Check
Accelerated	1–5, 15, 16–42 even, 43–47	20, 26, 34, 36, 40

For Your Information

- **Exercises 3–5** Students will need to use nets to write an expression for surface area.
- **Exercise 16** The equation represents a mixture problem in which peanuts are added to other ingredients, making trail mix. The equation shows that p pounds of peanuts that cost $4.05 per pound are added to other ingredients that cost a total of $14.40. This mixture creates $(p + 3)$ pounds of trail mix that costs $4.50 per pound.

Common Errors

- **Exercises 6–14** Students may perform the same operation that they are trying to undo instead of the opposite operation when trying to get the variable or constant terms on the same side. Remind them that whenever a variable or constant term is moved from one side of the equal sign to the other, the opposite operation is used.
- **Exercises 10–14** When using the Distributive Property, students may forget to distribute to all the values within the parentheses. Remind them that they need to distribute to all the values and encourage them to draw arrows showing the distribution, if needed.

Topic 2 Record and Practice Journal

Solve the equation. Check your solution.

1. $x + 16 = 9x$
   $x = 2$
2. $4y - 70 = 12y + 2$
   $y = -9$
3. $5(p + 6) = 8p$
   $p = 10$
4. $3(g - 7) = 2(10 + g)$
   $g = 41$
5. $1.8 + 7n = 9.5 - 4n$
   $n = 0.7$
6. $\frac{3}{7}w - 11 = -\frac{4}{7}w$
   $w = 11$
7. One movie club charges a $100 membership fee and $10 for each movie. Another club charges no membership fee but movies cost $15 each. Write and solve an equation to find the number of movies you need to buy for the cost of each movie club to be the same.
   $100 + 10x = 15x$; $x = 20$

Solve the equation. Check your solution, if possible.

8. $4x + 6 = 4x - 6$
   no solution
9. $5y - 7 = 7 - 5y$
   $y = 1.4$
10. $3(p + 3) = 3p + 9$
    infinitely many solutions
11. $g - 2 = \frac{1}{4}(4g - 2)$
    no solution
12. $12 - 4n = -12 - 4n$
    no solution
13. $\frac{1}{3}(3w - 15) = -\frac{1}{5}(25 - 5w)$
    infinitely many solutions
14. Thirty percent of all the students in a school are in a play. All students except for 140 are in the play. How many students are in the school?
    200

Topic 2 Exercises

Check It Out
Help with Homework
BigIdeasMath.com

Vocabulary and Concept Check

1. **WRITING** Is $x = 3$ a solution of the equation $3x - 5 = 4x - 9$? Explain.
2. **OPEN-ENDED** Write an equation that has variables on both sides and has a solution of -3.

Practice and Problem Solving

The value of the solid's surface area is equal to the value of the solid's volume. Find the value of x.

3. 

4. 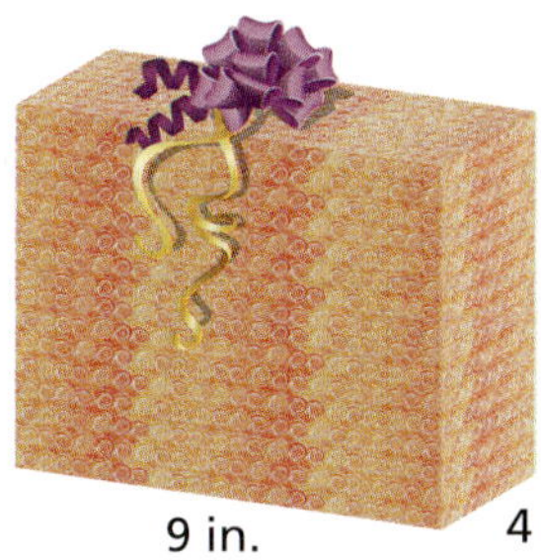

5. 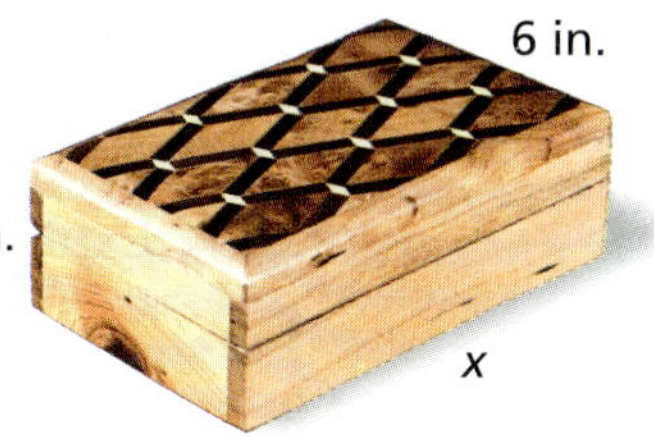

Solve the equation. Check your solution.

1 2 6. $m - 4 = 2m$

7. $3k - 1 = 7k + 2$

8. $6.7x = 5.2x + 12.3$

9. $-24 - \frac{1}{8}p = \frac{3}{8}p$

10. $12(2w - 3) = 6w$

11. $2(n - 3) = 4n + 1$

12. $2(4z - 1) = 3(z + 2)$

13. $0.1x = 0.2(x + 2)$

14. $\frac{1}{6}d + \frac{2}{3} = \frac{1}{4}(d - 2)$

15. **ERROR ANALYSIS** Describe and correct the error in solving the equation.

$$3x - 4 = 2x + 1$$
$$3x - 4 - 2x = 2x + 1 - 2x$$
$$x - 4 = 1$$
$$x - 4 + 4 = 1 - 4$$
$$x = -3$$

16. **TRAIL MIX** The equation $4.05p + 14.40 = 4.50(p + 3)$ represents the number p of pounds of peanuts you need to make trail mix. How many pounds of peanuts do you need for the trail mix?

17. **CARS** Write and solve an equation to find the number of miles you must drive to have the same cost for each of the car rentals.

$15 plus $0.50 per mile

$25 plus $0.25 per mile

Solve the equation. Check your solution, if possible.

3 4 **18.** $x + 6 = x$

19. $3x - 1 = 1 - 3x$

20. $4x - 9 = 3.5x - 9$

21. $\frac{1}{2}x + \frac{1}{2}x = x + 1$

22. $3x + 15 = 3(x + 5)$

23. $\frac{1}{3}(9x + 3) = 3x + 1$

24. $5x - 7 = 4x - 1$

25. $2x + 4 = -(-7x + 6)$

26. $5.5 - x = -4.5 - x$

27. $10x - \frac{8}{3} - 4x = 6x$

28. $-3(2x - 3) = -6x + 9$

29. $6(7x + 7) = 7(6x + 6)$

30. ERROR ANALYSIS Describe and correct the error in solving the equation.

✗
$$-4(2n - 3) = 12 - 8n$$
$$-8n + 12 = 12 - 8n$$
$$-8n = -8n$$
$$0 = 0$$
The solution is $n = 0$.

31. OPEN-ENDED Write an equation with variables on both sides that has no solution. Explain why it has no solution.

32. GEOMETRY Are there any values of x for which the areas of the figures are the same? Explain.

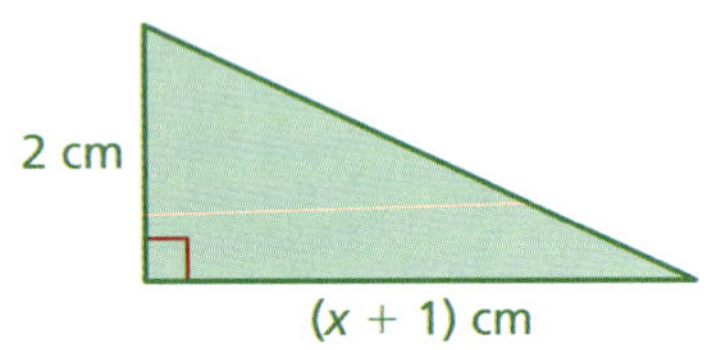

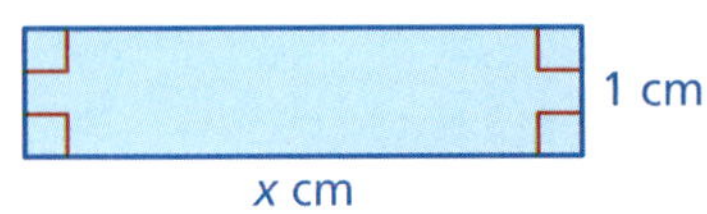

33. SATELLITE TV Provider A charges \$75 for installation and charges \$39.95 per month for the basic package. Provider B offers free installation and charges \$39.95 per month for the basic package. Your neighbor subscribes to Provider A the same month you subscribe to Provider B. After how many months is your neighbor's total cost the same as your total cost for satellite TV?

34. PIZZA CRUST Pepe's Pizza makes 52 pizza crusts the first week and 180 pizza crusts each subsequent week. Dianne's Delicatessen makes 26 pizza crusts the first week and 90 pizza crusts each subsequent week. In how many weeks will the total number of pizza crusts made by Pepe's Pizza equal twice the total number of pizza crusts made by Dianne's Delicatessen?

35. PRECISION Is the triangle an equilateral triangle? Explain.

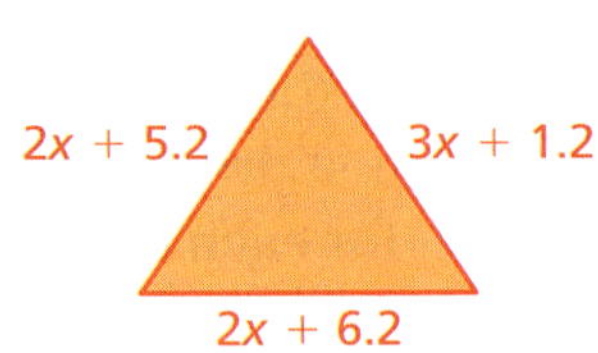

For Your Information

- **Exercises 41 and 42** Remind students about percents. They will need to remember how to do calculations with percents to find the solution.

Common Errors

- **Exercises 18–29** Students may end up with an equivalent equation such as $6 = 0$ or $3x + 15 = 3x + 15$ and get confused about how to state the final answer. Remind students that an equation can have no solution or infinitely many solutions.
- **Exercises 18–29** When using the Distributive Property, students may forget to distribute to all the values within the parentheses. Remind them that they need to distribute to all the values and encourage them to draw arrows showing the distribution, if needed.
- **Exercise 20** Students may end up with an equivalent equation such as $0.5x = 0$ and not know how to proceed to a final answer. Remind them that dividing 0 by a nonzero number is permissible and gives a result of 0. Encourage students to check their solutions. The solution $x = 0$ checks in this exercise.
- **Exercise 25** Students may be confused about the meaning of the minus sign outside the parentheses. Remind students that it means -1 times the quantity in parentheses, and they can make this substitution in a solution step if it helps them see how to arrive at a solution.
- **Exercise 27** Students may combine like terms on the left side of the equation incorrectly, ending up with $14x$ instead of $6x$. Remind them to consider the signs of the variable terms.

Practice and Problem Solving

18. no solution

19. $x = \frac{1}{3}$

20. $x = 0$

21. no solution

22. infinitely many solutions

23. infinitely many solutions

24. $x = 6$ **25.** $x = 2$

26. no solution

27. no solution

28. infinitely many solutions

29. infinitely many solutions

30. When the equation is $0 = 0$, it means it is true for all values of n, not just 0; The equation has infinitely many solutions.

31. *Sample answer:* $8x + 2 = 8x$; The number $8x$ cannot be equal to 2 more than itself.

32. no; There is no solution to the equation stating the areas are equal, $x + 1 = x$.

33. It's never the same. Your neighbor's total cost will always be $75 more than your total cost.

34. The total number of crusts made by Pepe's Pizza is always twice the total number of crusts made by Dianne's Delicatessen.

35. no; $2x + 5.2$ can never equal $2x + 6.2$.

Differentiated Instruction

Kinesthetic

Some students may benefit from using algebra tiles to model and solve the equations. Check that the students have correctly modeled the equation before attempting to solve.

Practice and Problem Solving

36. 3 units **37.** 7.5 units

38. 232 units

39. See *Taking Math Deeper.*

40. fractions; Because $\frac{1}{3}$ written as a decimal is repeating.

41. 10 mL **42.** 25 grams

43. See Additional Answers.

Fair Game Review

44. 27 cm^3

45. 15.75 cm^3

46. 24 $in.^3$

47. C

Mini-Assessment

Solve the equation.

1. $n - 4 = 3n + 6$ $n = -5$
2. $0.3(w + 10) = 1.8w$ $w = 2$
3. $-3x + 15 = 3(5 - x)$ infinitely many solutions
4. $\frac{1}{2}(4x + 14) = 2(x - 7)$ no solution
5. The perimeter of the rectangle is equal to the perimeter of the square. What are the side lengths of each figure? rectangle: 4 units by 10 units; square: 7 units by 7 units

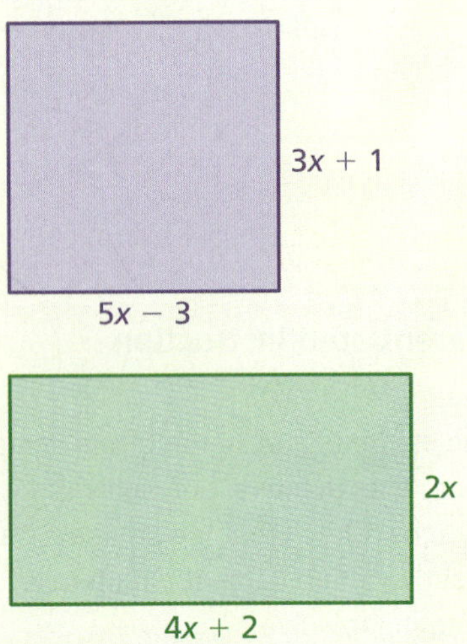

Taking Math Deeper

Exercise 39

This problem seems like it is easy, but it can actually be quite challenging.

1 Identify the key information in the table.

	Packing Material	Priority	Express
Box	\$2.25	\$2.50/lb	\$8.50/lb
Envelope	\$1.10	\$2.50/lb	\$8.50/lb

2 Write and solve an equation.

Let $x =$ the weight of the DVD and packing material.

Cost of Mailing Box: $2.25 + 2.5x$

Cost of Mailing Envelope: $1.10 + 8.5x$

$$2.25 + 2.5x = 1.10 + 8.5x$$
$$1.15 = 6x$$
$$0.19 \approx x$$

3 Answer the question.

The weight of the DVD and packing material is about 0.19 pound, or about 3 ounces.

Project

Postage for special types of mail, such as Priority Mail®, is determined by the weight of the package and the distance it needs to travel. Find the cost of sending a 15-ounce package from your house to Los Angeles, Washington, D.C., and Albuquerque.

Reteaching and Enrichment Strategies

If students need help. . .	If students got it. . .
Resources by Chapter • Practice A and Practice B • Puzzle Time Record and Practice Journal Practice Differentiating the Lesson Lesson Tutorials Skills Review Handbook	Resources by Chapter • Enrichment and Extension • Technology Connection Start the next section

A polygon is *regular* if each of its sides has the same length. Find the perimeter of the regular polygon.

36.

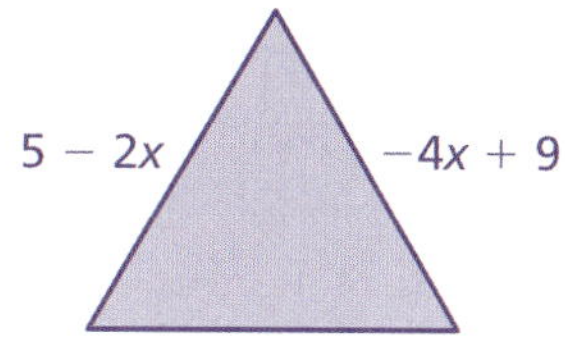

37.

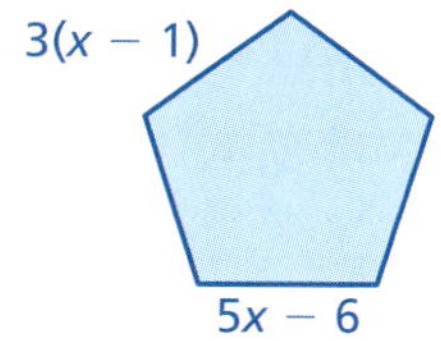

38.

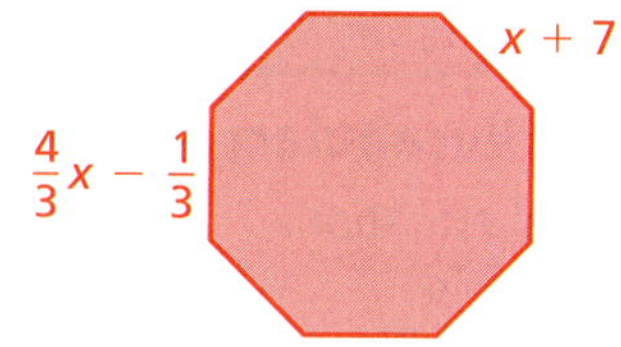

39. PRECISION The cost of mailing a DVD in an envelope by Express Mail® is equal to the cost of mailing a DVD in a box by Priority Mail®. What is the weight of the DVD with its packing material? Round your answer to the nearest hundredth.

	Packing Material	Priority Mail®	Express Mail®
Box	\$2.25	\$2.50 per lb	\$8.50 per lb
Envelope	\$1.10	\$2.50 per lb	\$8.50 per lb

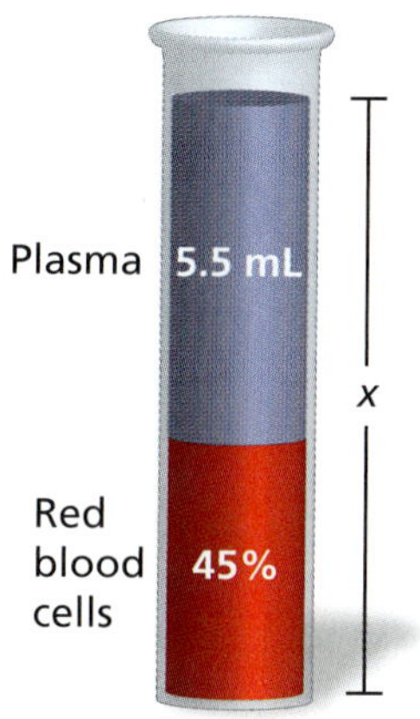

40. PROBLEM SOLVING Would you solve the equation $0.25x + 7 = \frac{1}{3}x - 8$ using fractions or decimals? Explain.

41. BLOOD SAMPLE The amount of red blood cells in a blood sample is equal to the total amount in the sample minus the amount of plasma. What is the total amount x of blood drawn?

42. NUTRITION One serving of oatmeal provides 16% of the fiber you need daily. You must get the remaining 21 grams of fiber from other sources. How many grams of fiber should you consume daily?

43. Geometry A 6-foot-wide hallway is painted as shown, using equal amounts of white and black paint.

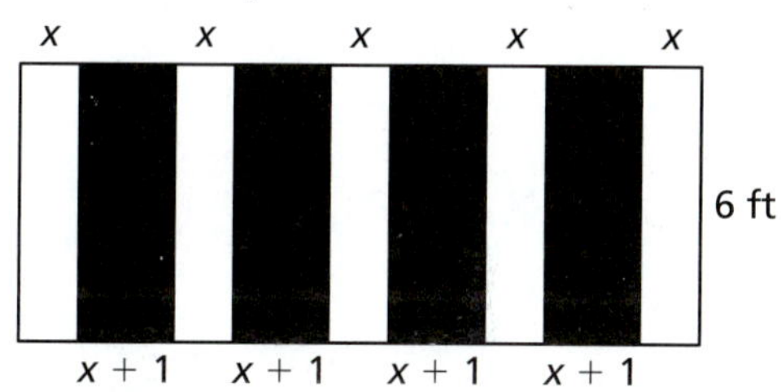

a. How long is the hallway?

b. Can this same hallway be painted with the same pattern, but using twice as much black paint as white paint? Explain.

Fair Game Review What you learned in previous grades & lessons

Find the volume of the prism. *(Skills Review Handbook)*

44.

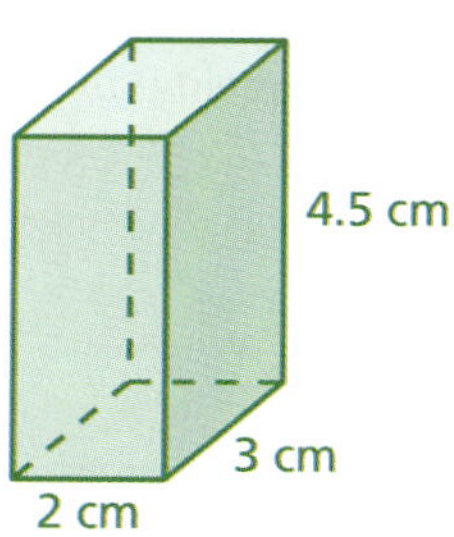

45.

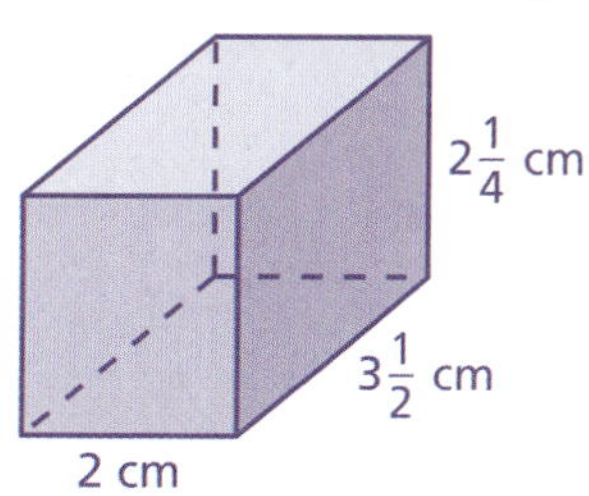

46.

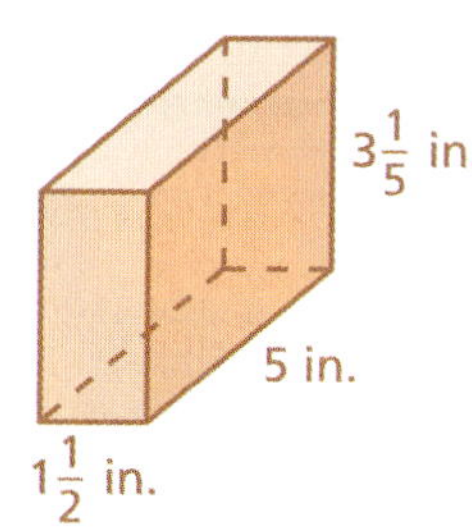

47. MULTIPLE CHOICE A car travels 480 miles on 15 gallons of gasoline. How many miles does the car travel per gallon? *(Skills Review Handbook)*

Ⓐ 28 mi/gal　Ⓑ 30 mi/gal　Ⓒ 32 mi/gal　Ⓓ 35 mi/gal

Topic 3 Rewriting Equations and Formulas

Check It Out
Lesson Tutorials
BigIdeasMath.com

Key Vocabulary
literal equation, *p. 774*

An equation that has two or more variables is called a **literal equation**. To rewrite a literal equation, solve for one variable in terms of the other variable(s).

EXAMPLE 1 Rewriting an Equation

Solve the equation $2y + 5x = 6$ for y.

$2y + 5x = 6$ — Write the equation.

Undo the addition. → $2y + 5x - 5x = 6 - 5x$ — Subtraction Property of Equality

$2y = 6 - 5x$ — Simplify.

Undo the multiplication. → $\frac{2y}{2} = \frac{6 - 5x}{2}$ — Division Property of Equality

$y = 3 - \frac{5}{2}x$ — Simplify.

On Your Own

Now You're Ready
Exercises 5–10

Solve the equation for y.

1. $5y - x = 10$ **2.** $4x - 4y = 1$ **3.** $12 = 6x + 3y$

EXAMPLE 2 Rewriting a Formula

Remember
A *formula* shows how one variable is related to one or more other variables. A formula is a type of literal equation.

The formula for the perimeter P of a rectangle is $P = 2\ell + 2w$. Solve the formula for the length ℓ.

$P = 2\ell + 2w$ — Write the formula.

$P - 2w = 2\ell + 2w - 2w$ — Subtraction Property of Equality

$P - 2w = 2\ell$ — Simplify.

$\frac{P - 2w}{2} = \frac{2\ell}{2}$ — Division Property of Equality

$\frac{P - 2w}{2} = \ell$ — Simplify.

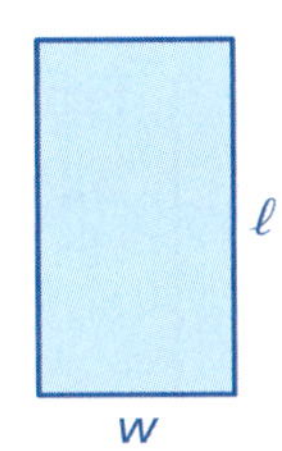

On Your Own

Now You're Ready
Exercises 14–19

Solve the formula for the red variable.

4. Area of rectangle: $A = bh$ **5.** Simple interest: $I = Prt$

6. Height of a dropped object: $h = -16t^2 + s$

Multi-Language Glossary at BigIdeasMath.com

Laurie's Notes

Introduction

Applying Mathematical Practices

- **Look for and Express Regularity in Repeated Reasoning:** Mathematically proficient students look for shortcuts for solving problems, such as solving a literal equation for a specified variable.

Connect

- **Yesterday:** Students solved equations with variables on both sides by using Properties of Equality and collecting variable terms on one side.
- **Today:** Students will rewrite literal equations.

Motivate

- To pique interest, share with students the highest and lowest recorded temperatures, in degrees Fahrenheit and degrees Celsius, in New Mexico.

Highest Recorded Temperature		Lowest Recorded Temperature	
122°F	50°C	−50°F	−46°C

Lesson Notes

Example 1

? "Can 6 and 5*x* be combined? Explain." no; They are not like terms.

- **Look for and Make Use of Structure:** Simplifying the last step is not obvious to all students. The expression $\frac{(6-5x)}{2}$ is a fraction with two terms in the numerator. You subtract the numerators and keep the same denominator. For example:

$$\frac{5-3}{7} = \frac{5}{7} - \frac{3}{7} \quad \text{and} \quad \frac{6-5x}{2} = \frac{6}{2} - \frac{5x}{2} = 3 - \frac{5}{2}x.$$

On Your Own

- Spend sufficient time on these problems. Students will encounter many equations in which they need to rewrite *y* in terms of *x*.
- Notice in Question 2 that the coefficient of *y* is −4. Suggest students rewrite the equation as $4x + (-4)y = 1$.

Example 2

- **Teaching Tip:** Highlight the variable ℓ in red as shown in the textbook. Discuss the idea that everything except the variable ℓ must be moved to the left side of the equation using Properties of Equality.
- **Look for and Make Use of Structure:** Structurally, the formula has the form $A = Bx + C$. It can be solved by subtracting *C* from both sides and then dividing both sides by *B*.

? "The term $2w$ is added to the term 2ℓ. How do you move it to the left side of the equation?" Subtract $2w$ from each side of the equation.

On Your Own

- **Think-Pair-Share:** Students should read each question independently and then work in pairs to answer the questions. When they have answered the questions, the pair should compare their answers with another group and discuss any discrepancies.

What Your Students Will Learn

- Rewrite equations and formulas to solve for one variable in terms of the other variable(s).
- Convert temperatures by rewriting formulas and solving for the unknown variable.

Goal Today's lesson is solving **literal equations**.

Technology for the Teacher

Dynamic Classroom

Lesson Tutorials
Lesson Plans
Answer Presentation Tool

Extra Example 1

Solve the equation $-2x - 3y = 6$ for *y*.

$y = -\frac{2}{3}x - 2$

On Your Own

1. $y = 2 + \frac{1}{5}x$
2. $y = x - \frac{1}{4}$
3. $y = 4 - 2x$

Extra Example 2

The formula for the surface area of a square pyramid is $S = x^2 + 2x\ell$. Solve the formula for the slant height ℓ.

$\ell = \frac{S - x^2}{2x}$

On Your Own

4. $b = \frac{A}{h}$
5. $P = \frac{I}{rt}$
6. $s = 16t^2 + h$

English Language Learners

Vocabulary

Have students start a *Formula* page in their notebooks with the formulas used in this section. Each formula should be accompanied by a description of what each of the variables represents and an example. In the case of area formulas, units of measure should be included with the description (e.g., units and square units). As students progress throughout the year, additional formulas can be added to the *Formula* notebook page.

Extra Example 3

Solve the temperature formula $F = \frac{9}{5}C + 32$ for C. $C = \frac{5}{9}(F - 32)$

Extra Example 4

Which temperature is greater, 400°F or 200°C? 400°F

On Your Own

7. greater than

Laurie's Notes

Key Idea

- Write the formula for converting from degrees Fahrenheit to degrees Celsius.
- Use this formula if you know the temperature in degrees Fahrenheit and you want to find the temperature in degrees Celsius.

? "You are traveling abroad, and the temperature is always stated in degrees Celsius. How can you figure out the temperature in degrees Fahrenheit, with which you are more familiar?" Students may recognize that you will want to have a different conversion formula that allows you to substitute for C and calculate F.

Example 3

? "What is the reciprocal of $\frac{5}{9}$?" $\frac{9}{5}$

- Remind students that multiplying by the reciprocal $\frac{9}{5}$ is more efficient than dividing by the fraction $\frac{5}{9}$.

Example 4

- **FYI:** The graphic on the left provides information about the temperature of a lightning bolt and the temperature of the surface of the sun. The two temperatures use different scales.

? "How can you compare two temperatures that are in different scales?" Listen for understanding that one of the temperatures must be converted.

? "How do you multiply $\frac{9}{5}$ times 30,000?" Students may recall that you can simplify before multiplying. Five divides into 30,000 six thousand times, so $6000 \times 9 = 54{,}000$.

? "Approximately how many times hotter is a lightning bolt than the surface of the sun?" 5 times This is a *cool* fact for students to know!

On Your Own

- **Neighbor Check:** Have students work independently and then have their neighbors check their work. Have students discuss any discrepancies.

Closure

- **Exit Ticket:** Solve $2x + 4y = 11$ for y. $y = -\frac{1}{2}x + \frac{11}{4}$

Solving Equations

In this lesson, you will

- rewrite equations to solve for one variable in terms of the other variable(s).

Key Idea

Temperature Conversion

A formula for converting from degrees Fahrenheit F to degrees Celsius C is

$$C = \frac{5}{9}(F - 32).$$

EXAMPLE 3 Rewriting the Temperature Formula

Solve the temperature formula for F.

$C = \frac{5}{9}(F - 32)$ — Write the temperature formula.

Use the reciprocal. → $\frac{9}{5} \cdot C = \frac{9}{5} \cdot \frac{5}{9}(F - 32)$ — Multiplication Property of Equality

$\frac{9}{5}C = F - 32$ — Simplify.

Undo the subtraction. → $\frac{9}{5}C + 32 = F - 32 + 32$ — Addition Property of Equality

$\frac{9}{5}C + 32 = F$ — Simplify.

The rewritten formula is $F = \frac{9}{5}C + 32$.

EXAMPLE 4 Real-Life Application

Which has the greater temperature?

Convert the Celsius temperature of lightning to Fahrenheit.

$F = \frac{9}{5}C + 32$ — Write the rewritten formula from Example 3.

$= \frac{9}{5}(30{,}000) + 32$ — Substitute 30,000 for C.

$= 54{,}032$ — Simplify.

Because 54,032 °F is greater than 11,000 °F, lightning has the greater temperature.

On Your Own

7. Room temperature is considered to be 70 °F. Suppose the temperature is 23 °C. Is this greater than or less than room temperature?

Topic 3 Exercises

Check It Out
Help with Homework
BigIdeasMath.com

Vocabulary and Concept Check

1. **VOCABULARY** Is $-2x = \frac{3}{8}$ a literal equation? Explain.

2. **DIFFERENT WORDS, SAME QUESTION** Which is different? Find "both" answers.

Solve $4x - 2y = 6$ for y.	Solve $6 = 4x - 2y$ for y.
Solve $4x - 2y = 6$ for y in terms of x.	Solve $4x - 2y = 6$ for x in terms of y.

Practice and Problem Solving

3. **a.** Write a formula for the area A of a triangle.
   b. Solve the formula for b.
   c. Use the new formula to find the base of the triangle.

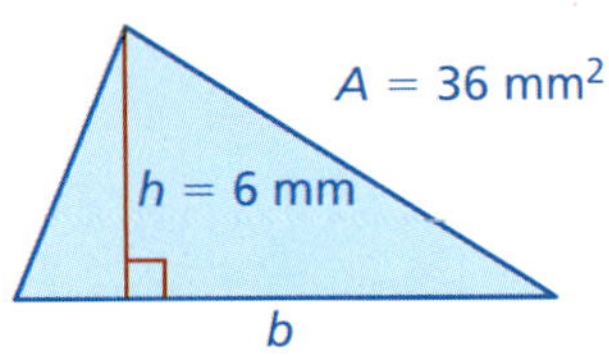

4. **a.** Write a formula for the volume V of a prism.
   b. Solve the formula for B.
   c. Use the new formula to find the area of the base of the prism.

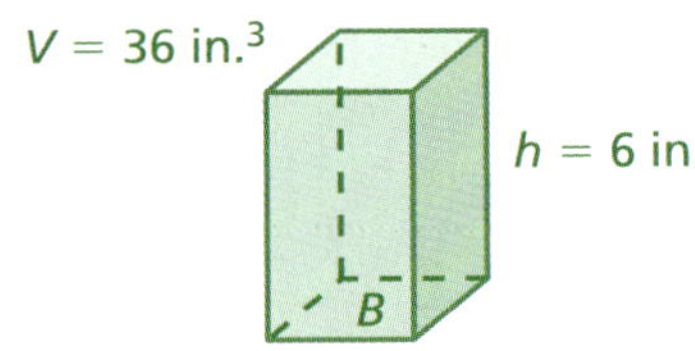

Solve the equation for y.

① 5. $\frac{1}{3}x + y = 4$

6. $3x + \frac{1}{5}y = 7$

7. $6 = 4x + 9y$

8. $3 = 7x - 2y$

9. $4.2x - 1.4y = 2.1$

10. $6y - 1.5x = 8$

11. **ERROR ANALYSIS** Describe and correct the error in rewriting the equation.

✗ $2x - y = 5$
$y = -2x + 5$

12. **TEMPERATURE** The formula $K = C + 273.15$ converts temperatures from Celsius C to Kelvin K.
   a. Solve the formula for C.
   b. Convert 300 Kelvin to Celsius.

13. **INTEREST** The formula for simple interest is $I = Prt$.
   a. Solve the formula for t.
   b. Use the new formula to find the value of t in the table.

I	\$75
P	\$500
r	5%
t	

Assignment Guide and Homework Check

Level	Assignment	Homework Check
Accelerated	1–4, 6–10 even, 11, 12–22 even, 23–28	6, 10, 18, 20, 22

For Your Information

- **Exercise 2** In the *Different Words, Same Question* exercise, three of the four choices pose the same question using different words. The remaining choice poses a different question. So there are two answers.
- **Exercise 22** This exercise is a preview of circles that will be taught in Chapter 8.
- **Exercise 22** Remind students about percents. They will need to remember how to do calculations with percents to find the solution.

Common Errors

- **Exercises 5–10** Students may solve the equation for the wrong variable. Remind them that they are solving the equation for *y*. Encourage them to make *y* a different color when solving so that it is easy to remember that they are solving for *y*.
- **Exercises 14–19** In each exercise, different steps are required to solve for the red variable, and some of these steps could confuse students. Remind them to take their time and think about the process for solving for a variable.

Topic 3 Record and Practice Journal

Solve the equation for *y*.

1. $2x + y = -9$

   $y = -2x - 9$

2. $4x - 10y = 12$

   $y = \frac{2}{5}x - \frac{6}{5}$

3. $13 = \frac{1}{6}y + 2x$

   $y = -12x + 78$

4. The formula for the perimeter of a triangle is $P = a + b + c$.

   a. Solve the formula for c.

   $c = P - a - b$

   b. Use the new formula to find the value of c.

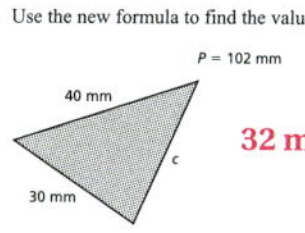

   32 mm

Solve the formula for the bold variable.

5. $V = \ell \mathbf{w} h$

   $w = \frac{V}{\ell h}$

6. $f = \frac{1}{2}(\mathbf{r} + 6.5)$

   $r = 2f - 6.5$

7. $V = \frac{1}{3}B\mathbf{h}$

   $h = \frac{3V}{B}$

8. The formula for the area of a triangle is $A = \frac{1}{2}bh$.

   a. Solve the formula for h.

   $h = \frac{2A}{b}$

   b. Use the new formula to find the value of h.

   $h = 9$ in.

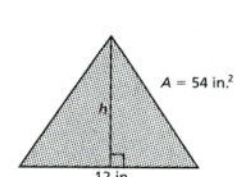

Vocabulary and Concept Check

1. no; The equation only contains one variable.
2. Solve $4x - 2y = 6$ for x in terms of y.;

   $x = \frac{3}{2} + \frac{1}{2}y$; $y = -3 + 2x$

Practice and Problem Solving

3. **a.** $A = \frac{1}{2}bh$

   b. $b = \frac{2A}{h}$

   c. $b = 12$ mm

4. **a.** $V = Bh$

   b. $B = \frac{V}{h}$

   c. $B = 6 \text{ in.}^2$

5. $y = 4 - \frac{1}{3}x$
6. $y = 35 - 15x$
7. $y = \frac{2}{3} - \frac{4}{9}x$
8. $y = \frac{7}{2}x - \frac{3}{2}$
9. $y = 3x - 1.5$
10. $y = \frac{4}{3} + \frac{1}{4}x$
11. The *y* should have a negative sign in front of it.

    $$\begin{aligned} 2x - y &= 5 \\ -y &= -2x + 5 \\ y &= 2x - 5 \end{aligned}$$

12. **a.** $C = K - 273.15$

    b. 26.85°C

13. **a.** $t = \frac{I}{Pr}$

    b. $t = 3$ yr

Practice and Problem Solving

14. $t = \frac{d}{r}$

15. $m = \frac{e}{c^2}$

16. $C = R - P$

17. $h = \frac{2A}{b_1 + b_2}$

18. $V = \frac{Bh}{3}$

19. $w = 6g - 40$

20. The rewritten formula is a general solution that can be reused.

21. **a.** $F = 32 + \frac{9}{5}(K - 273.15)$

b. 32°F

c. liquid nitrogen

22. See *Taking Math Deeper*.

23. $r^3 = \frac{3V}{4\pi}$; $r = 4.5$ in.

Fair Game Review

24. $3\frac{3}{4}$

25. $-5\frac{1}{3}$

26. $\frac{1}{3}$

27. $1\frac{1}{4}$

28. D

Mini-Assessment

Solve the formula for the red variable.

1. Distance Formula: $d = rt$ $r = \frac{d}{t}$
2. Area of a triangle: $A = \frac{1}{2}bh$ $h = \frac{2A}{b}$
3. Circumference of a circle: $C = 2\pi r$ $r = \frac{C}{2\pi}$
4. The temperature in Portland, Oregon, is 37°F. The temperature in Mobile, Alabama, is 22°C. In which city is the temperature higher? Mobile, Alabama

Taking Math Deeper

Exercise 22

This problem reviews circles and percents, as well as distance, rate, and time. It also has a bit of history related to George Ferris, who designed the first Ferris wheel for the 1893 World's Fair in Chicago.

1 Organize the given information.
Circumference (Navy Pier Ferris Wheel): $C = 439.6$ ft
Circumference (first Ferris wheel): x ft
Relationship: $439.6 = 0.56x$

2 Find the radius of each wheel.

a. Radius (Navy Pier Ferris Wheel):

$$C \approx 6.28r$$
$$439.6 = 6.28r$$
$$70 = r$$

Circumference (first Ferris wheel):

$$439.6 = 0.56x$$
$$785 = x$$

b. Radius (first Ferris wheel):

$$C \approx 6.28R$$
$$785 = 6.28R$$
$$125 = R$$

3 **c.** The first Ferris wheel made 1 revolution in 9 minutes. How fast was the wheel moving?

$$\text{rate} = \frac{785 \text{ ft}}{9 \text{ min}} \approx 87.2 \text{ ft per min}$$

It might be interesting for students to know that the first Ferris wheel had 36 cars, each of which held 60 people!

Project

Use your school's library or the Internet to find how long one revolution takes for the Ferris wheel on the Navy Pier in Chicago and the one in London, England. Which one has the greater circumference? Which one travels faster? How do you know?

Reteaching and Enrichment Strategies

If students need help...	If students got it...
Resources by Chapter • Practice A and Practice B • Puzzle Time Record and Practice Journal Practice Differentiating the Lesson Lesson Tutorials Skills Review Handbook	Resources by Chapter • Enrichment and Extension • Technology Connection Start the next section

Solve the equation for the red variable.

2 **14.** $d = rt$ **15.** $e = mc^2$ **16.** $R - C = P$

17. $A = \frac{1}{2}h(b_1 + b_2)$ **18.** $B = 3\frac{V}{h}$ **19.** $g = \frac{1}{6}(w + 40)$

20. **LOGIC** Why is it useful to rewrite a formula in terms of another variable?

21. **REASONING** The formula $K = \frac{5}{9}(F - 32) + 273.15$ converts temperatures from Fahrenheit F to Kelvin K.

a. Solve the formula for F.

b. The freezing point of water is 273.15 Kelvin. What is this temperature in Fahrenheit?

c. The temperature of dry ice is −78.5 °C. Which is colder, dry ice or liquid nitrogen?

Navy Pier Ferris Wheel

22. **FERRIS WHEEL** The distance around a circle is called the *circumference*. The Navy Pier Ferris Wheel in Chicago has a circumference C that is 56% of the circumference of the first Ferris wheel built in 1893.

a. The *radius* of a circle is the distance from the center to any point on the circle. The circumference of a circle is about 6.28 times the radius. What is the radius of the Navy Pier Ferris Wheel?

b. What was the radius of the first Ferris wheel?

c. The first Ferris wheel took 9 minutes to make a complete revolution. How fast was the wheel moving?

23. Repeated Reasoning The *radius* of a sphere is the distance from the center to any point on the sphere. The formula for the volume V of a sphere is $V = \frac{4}{3}\pi r^3$, where r is the radius and π is a constant whose value is about 3.14. Solve the formula for r^3. Use Guess, Check, and Revise to find the radius of the sphere.

Fair Game Review What you learned in previous grades & lessons

Multiply. *(Skills Review Handbook)*

24. $5 \times \frac{3}{4}$ **25.** $-2 \times \frac{8}{3}$ **26.** $\frac{1}{4} \times \frac{3}{2} \times \frac{8}{9}$ **27.** $25 \times \frac{3}{5} \times \frac{1}{12}$

28. **MULTIPLE CHOICE** Which of the following is not equivalent to $\frac{3}{4}$? *(Skills Review Handbook)*

Ⓐ 0.75 Ⓑ 3:4 Ⓒ 75% Ⓓ 4:3

Key Vocabulary Index

Mathematical terms are best understood when you see them used and defined *in context*. This index lists where you will find key vocabulary. A full glossary is available in your Record and Practice Journal and at *BigIdeasMath.com*.

Student Index

This student-friendly index will help you find vocabulary, key ideas, and concepts. It is easily accessible and designed to be a reference for you whether you are looking for a definition, real-life application, or help with avoiding common errors.

A

B

C

D

E

Student Index

F

G

H

I

Student Index

K

L

M

Student Index

Student Index

S

W

X

Y

Student Index

Additional Answers

Chapter 11

Try It Yourself

5. a. $(0, -1)$ b. $(0, 1)$

6. a. $(-5, 0)$ b. $(5, 0)$

7. a. $(4, 6.5)$ b. $(-4, -6.5)$

8. a. $\left(-3\frac{1}{2}, 4\right)$ b. $\left(3\frac{1}{2}, -4\right)$

9. 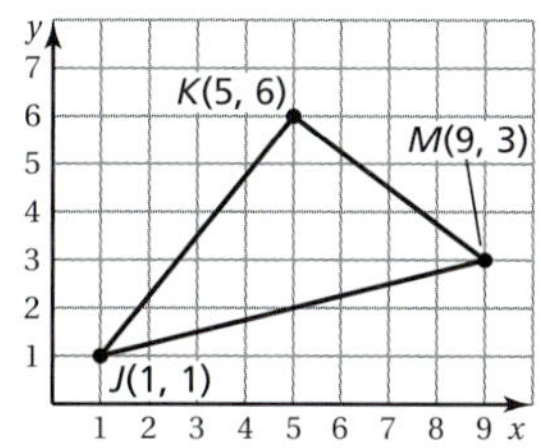

10. 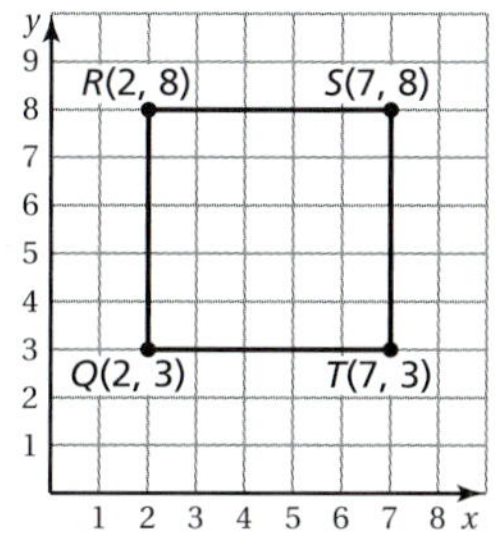

Record and Practice Journal Fair Game Review

5. $(-1, -2)$; $(1, 2)$

6. $(3, -2)$; $(-3, 2)$

7. 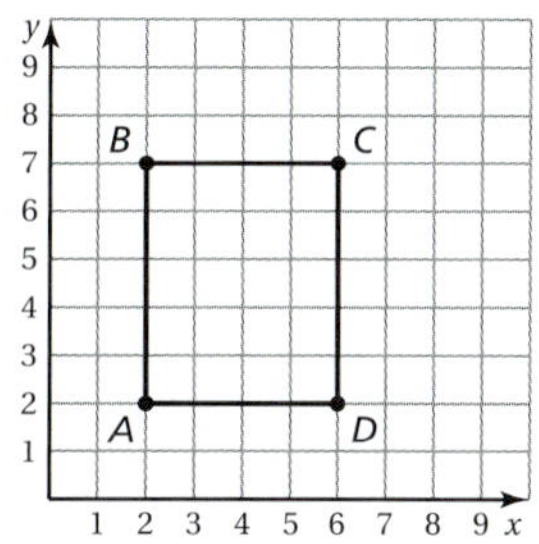

8. 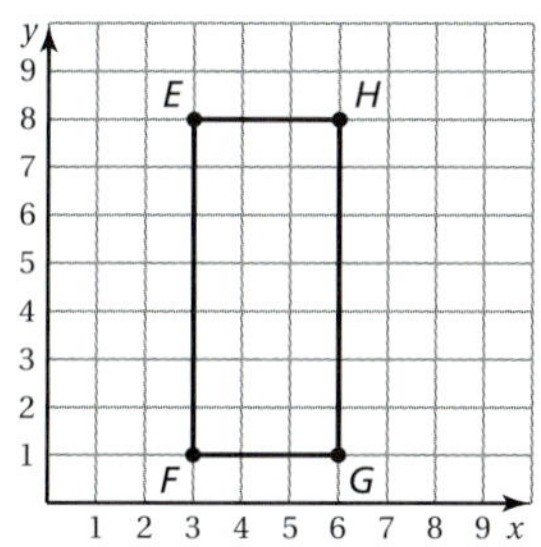

9. 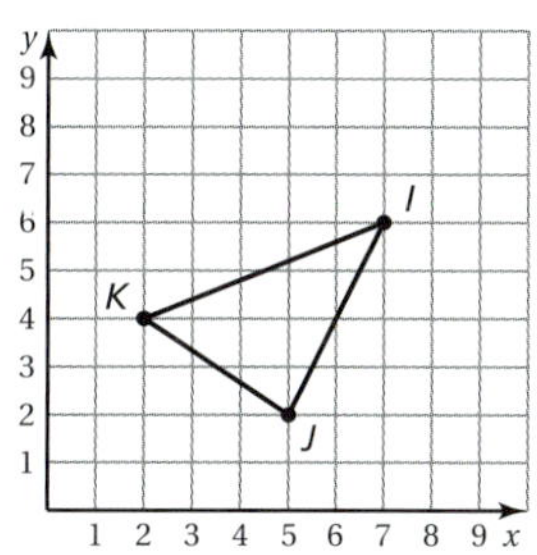

10. 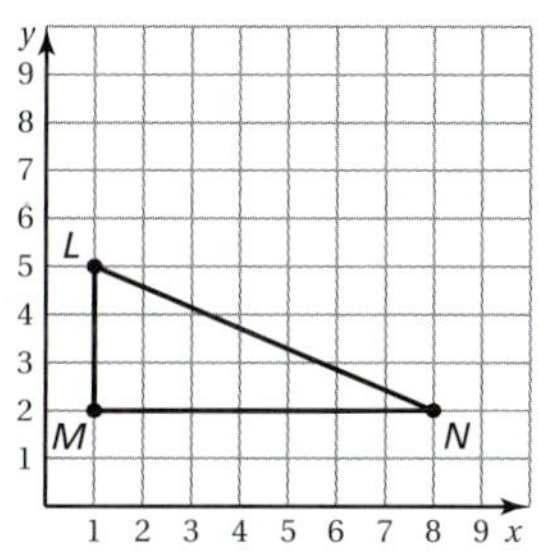

11. 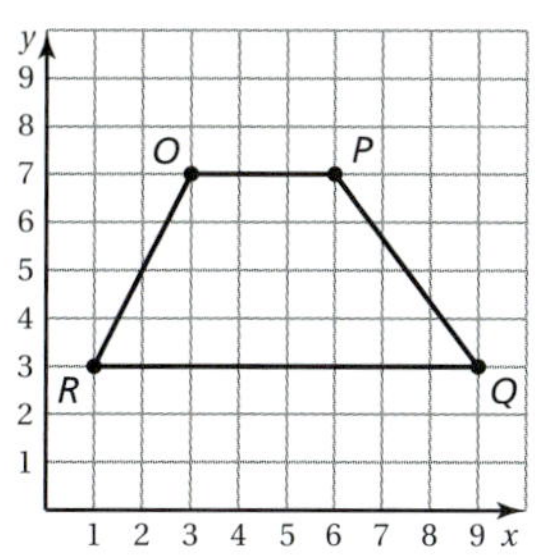

12. 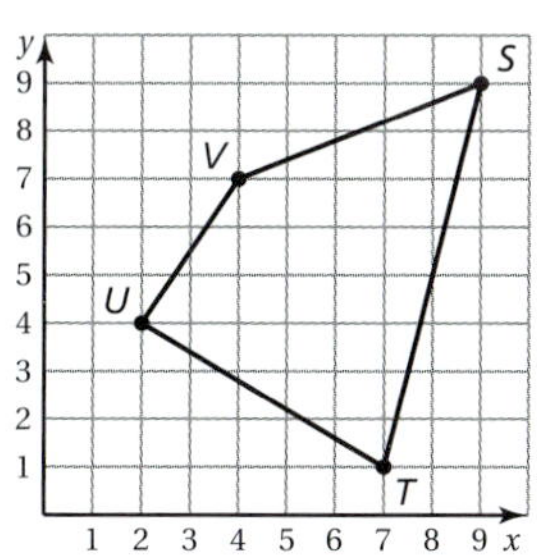

Section 11.1

Record and Practice Journal

2. a. 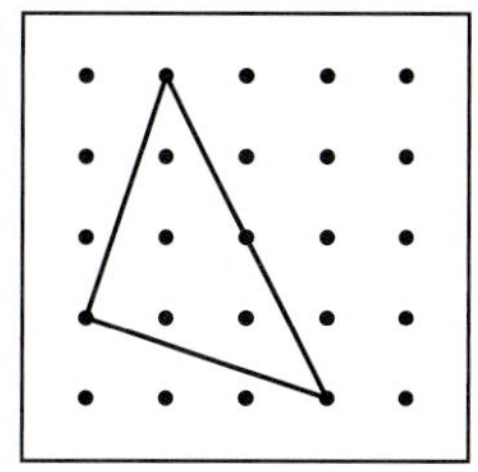

c. 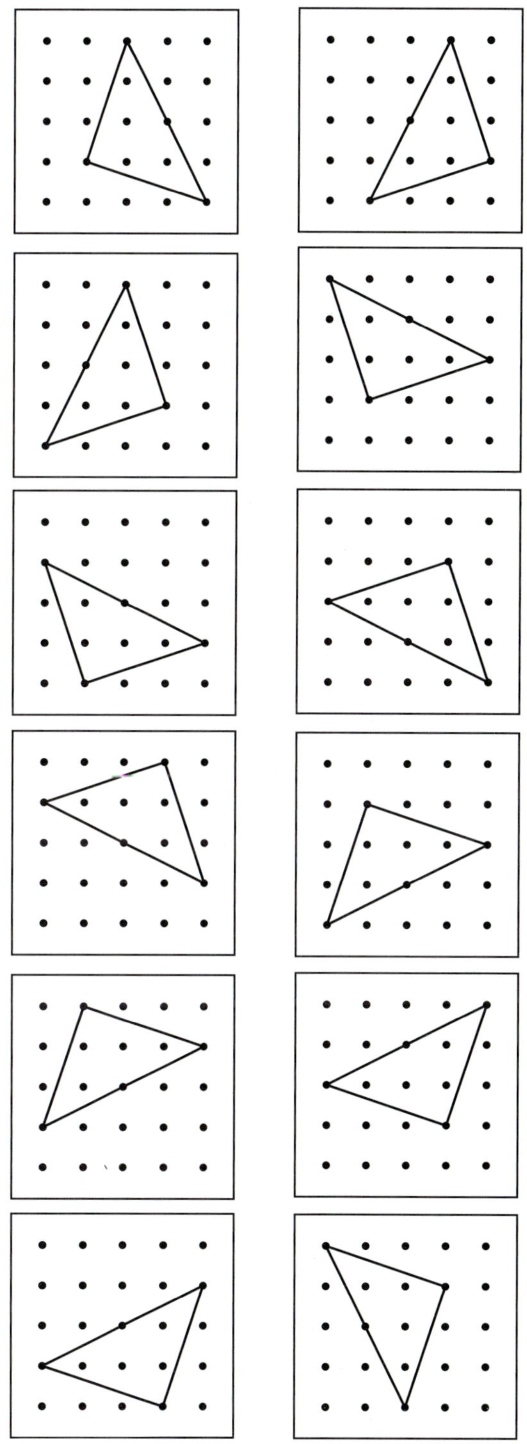

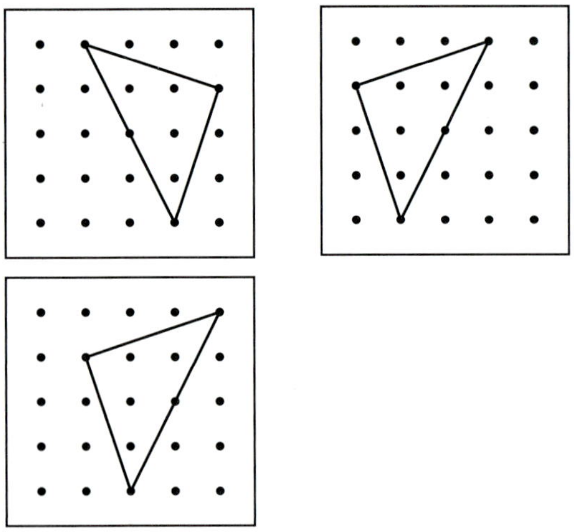

Practice and Problem Solving

12. **a.** 32 ft **b.** $\angle M$

c. 20 ft **d.** 96 ft

14. yes; The dimensions of congruent figures are equal, so the areas of the figures are equal.

15. **a.** true; Side *AB* corresponds to Side *YZ*.

b. true; $\angle A$ and $\angle X$ have the same measure.

c. false; $\angle A$ corresponds to $\angle Y$.

d. true; The measure of $\angle A$ is 90°, the measure of $\angle B$ is 140°, the measure of $\angle C$ is 40°, and the measure of $\angle D$ is 90°. So, the sum of the angle measures of *ABCD* is $90° + 140° + 40° + 90° = 360°$.

Fair Game Review

16–19.

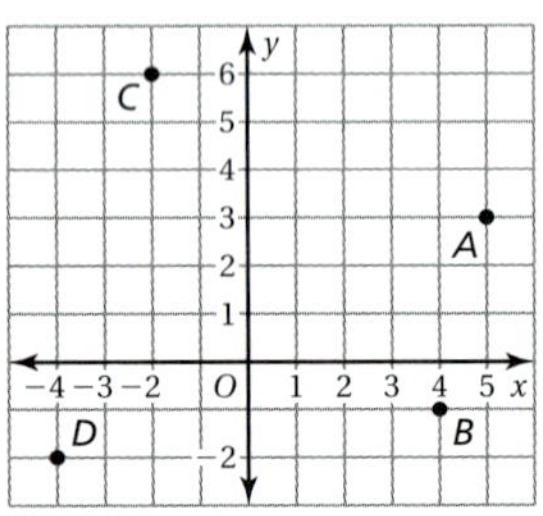

20. B

Record and Practice Journal Practice

1. Corresponding angles: $\angle A$ and $\angle J$, $\angle B$ and $\angle K$, $\angle C$ and $\angle L$, $\angle D$ and $\angle M$

Corresponding sides: Side *AB* and Side *JK*, Side *BC* and Side *KL*, Side *CD* and Side *LM*, Side *DA* and Side *MJ*

2. Corresponding angles: $\angle J$ and $\angle Q$, $\angle K$ and $\angle P$, $\angle L$ and $\angle T$, $\angle M$ and $\angle S$, $\angle N$ and $\angle R$

Corresponding sides: Side *JK* and Side *QP*, Side *KL* and Side *PT*, Side *LM* and Side *TS*, Side *MN* and Side *SR*, Side *NJ* and Side *RQ*

Section 11.2

On Your Own

5.

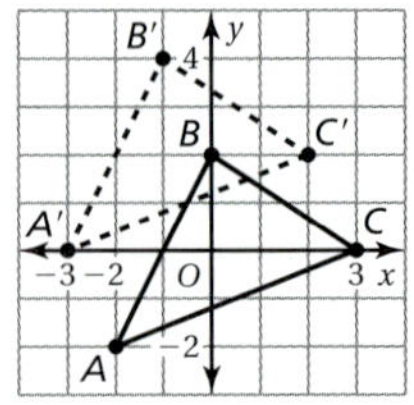

Practice and Problem Solving

15.

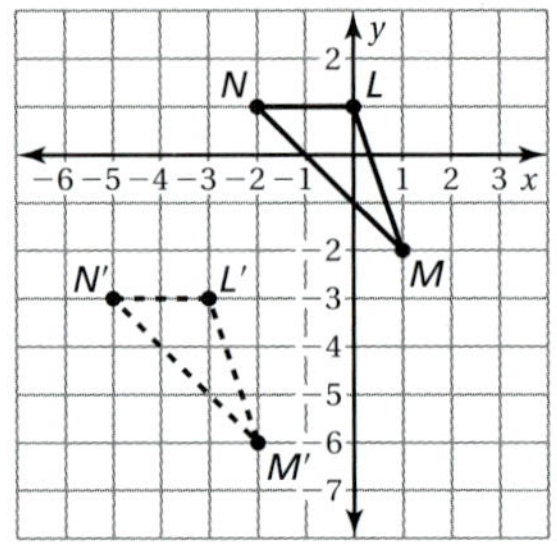

19. 6 units right and 3 units down

20. 5 units left and 2 units down

21. **a.** 5 units right and 1 unit up

b. no; It would hit the island.

c. 4 units up and 4 units right

22. yes; You can write one translation to get from the original triangle to the final triangle, which is $(x + 2, y - 10)$. So, the triangles are congruent. You can also measure the sides and angles to determine that the triangles are congruent.

Fair Game Review

24. yes **25.** no

26. no **27.** yes

28. B

Section 11.3

On Your Own

4. **a.**

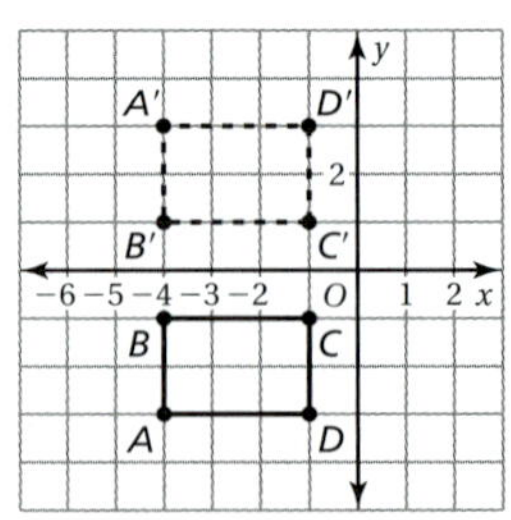

b.

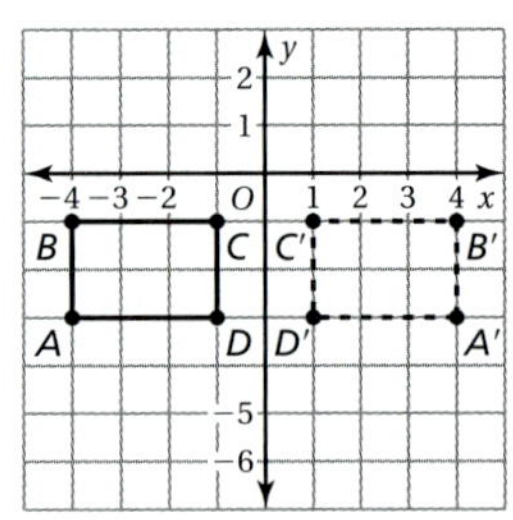

c. yes; They are all rectangles of the same size and shape.

Practice and Problem Solving

13.

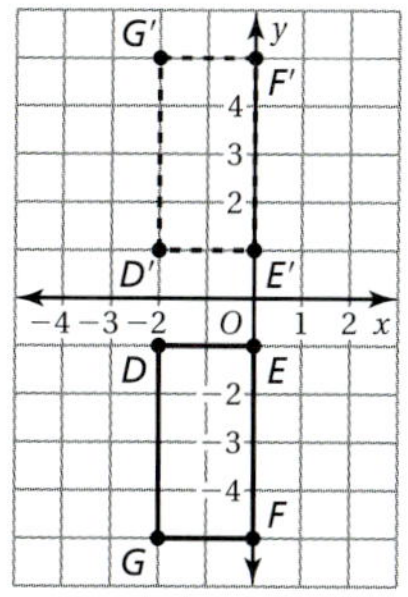

$D'(-2, 1)$, $E'(0, 1)$, $F'(0, 5)$, $G'(-2, 5)$

14.

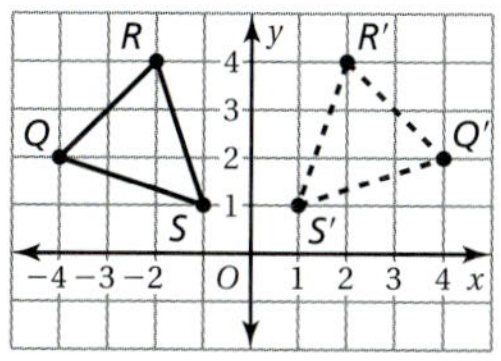

$Q'(4, 2)$, $R'(2, 4)$, $S'(1, 1)$

15.

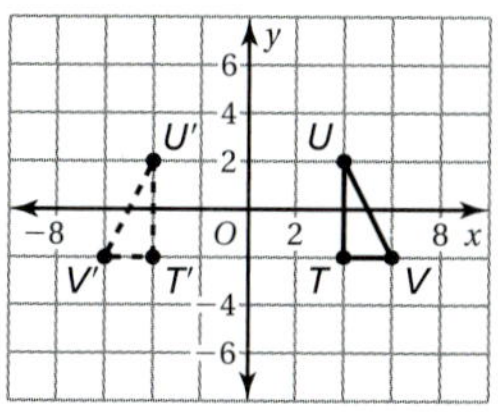

$T'(-4, -2)$, $U'(-4, 2)$, $V'(-6, -2)$

16.

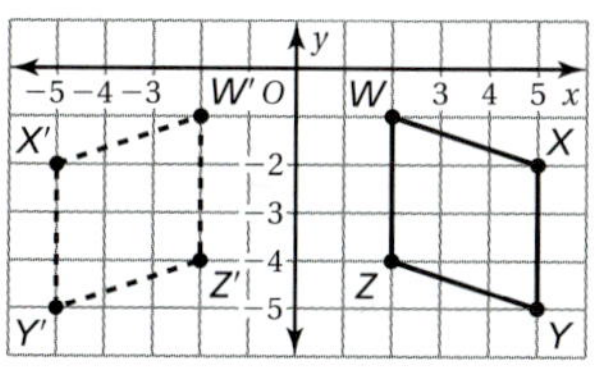

$W'(-2, -1)$, $X'(-5, -2)$, $Y'(-5, -5)$, $Z'(-2, -4)$

17.

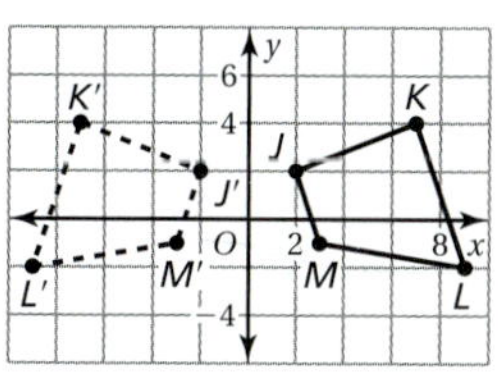

$J'(-2, 2)$, $K'(-7, 4)$, $L'(-9, -2)$, $M'(-3, -1)$

28.

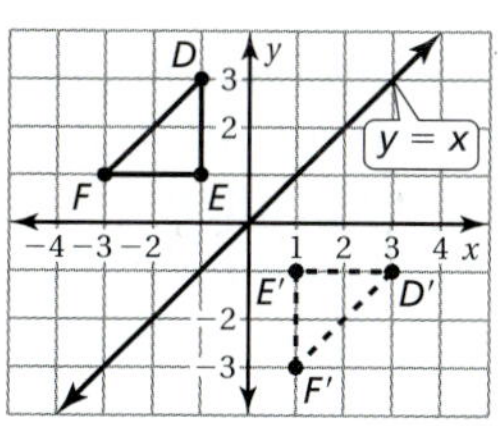

The x-coordinate and y-coordinate for each point are switched in the image.

Section 11.4

On Your Own

3. a.

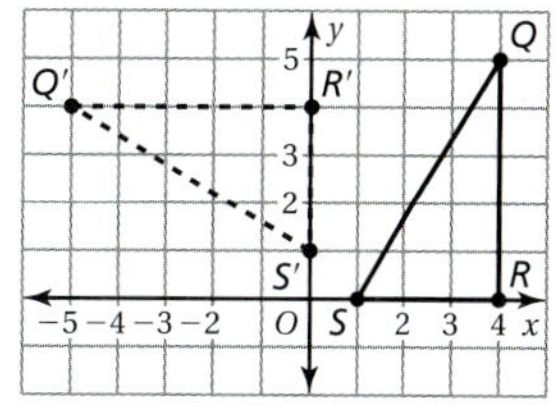

b.

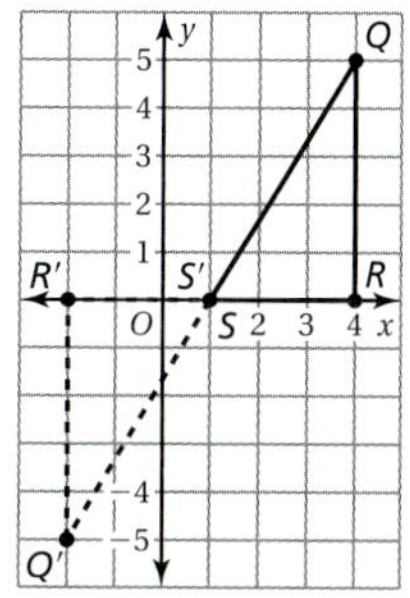

c. yes; In a rotation, the original figure and its image are congruent. Because both images are congruent to the same (original) figure, they are congruent to each other.

Practice and Problem Solving

28. a.

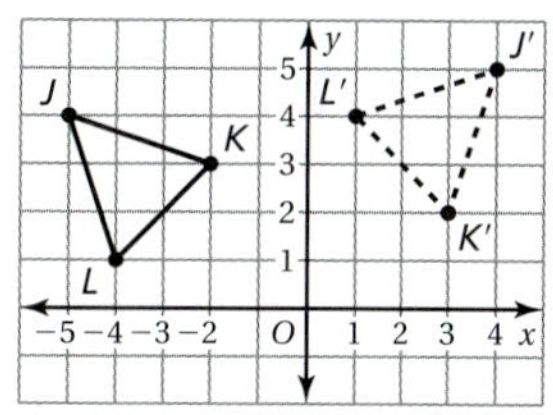

The x-coordinates of $\triangle J'K'L'$ are the same as the y-coordinates of $\triangle JKL$. The y-coordinates of $\triangle J'K'L'$ are the opposite of the x-coordinates of $\triangle JKL$.

b.

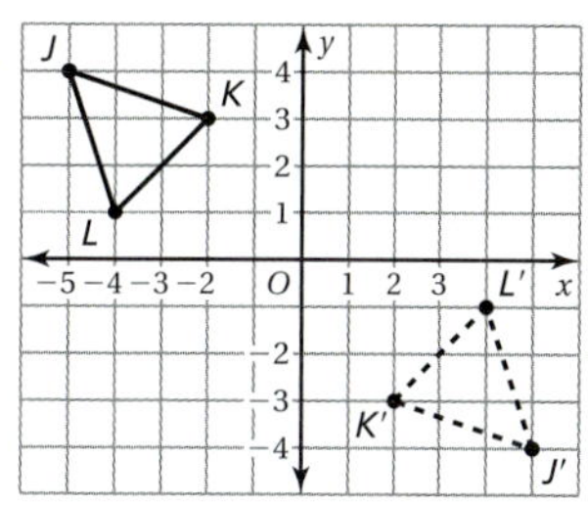

The x-coordinates of $\triangle J'K'L'$ are the opposite of the x-coordinates of $\triangle JKL$. The y-coordinates of $\triangle J'K'L'$ are the opposite of the y-coordinates of $\triangle JKL$.

c. yes; Explanations will vary.

Section 11.5

Practice and Problem Solving

6.

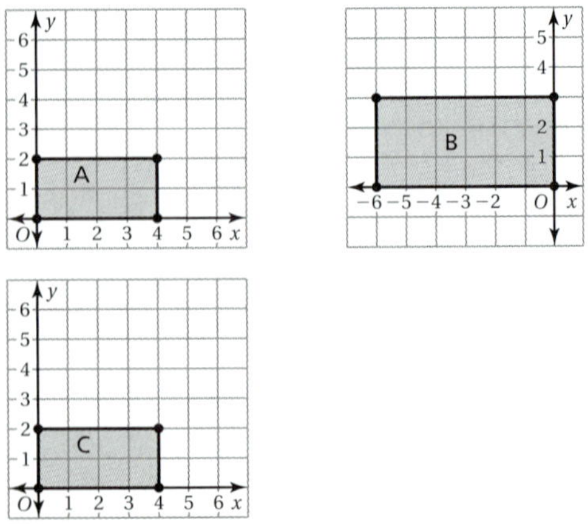

A, B and C; Corresponding side lengths are proportional and corresponding angles are congruent.

7.

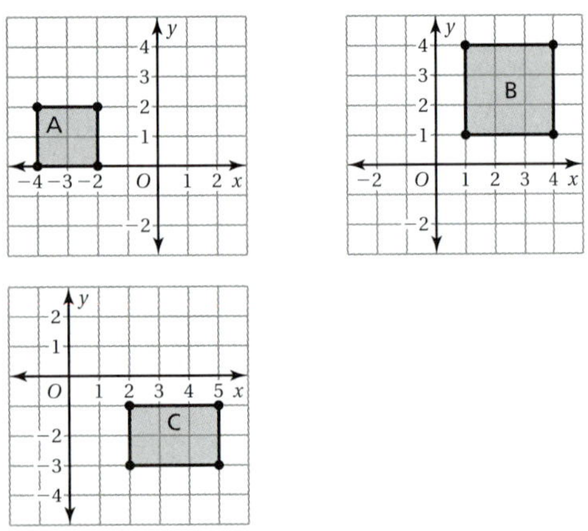

A and B; Corresponding side lengths are proportional and corresponding angles are congruent.

19. **a.** yes

b. yes; It represents the fact that the sides are proportional because you can split the isosceles triangles into smaller right triangles that will be similar.

20. yes; *Sample answer:*

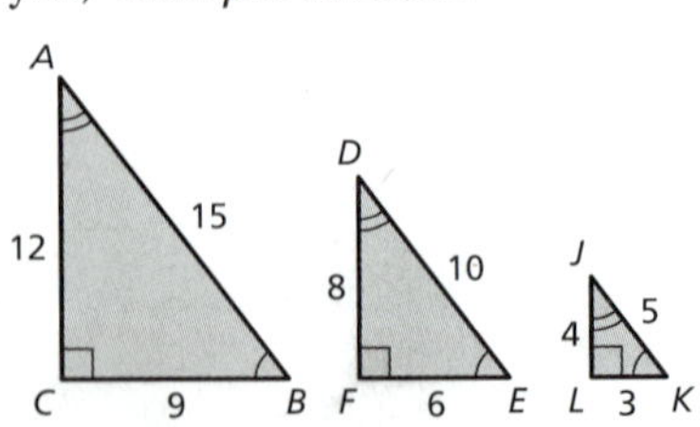

Because corresponding angles are congruent and corresponding side lengths are proportional, $\triangle ABC$ is similar to $\triangle JKL$.

Section 11.6

Record and Practice Journal

1. **a.**

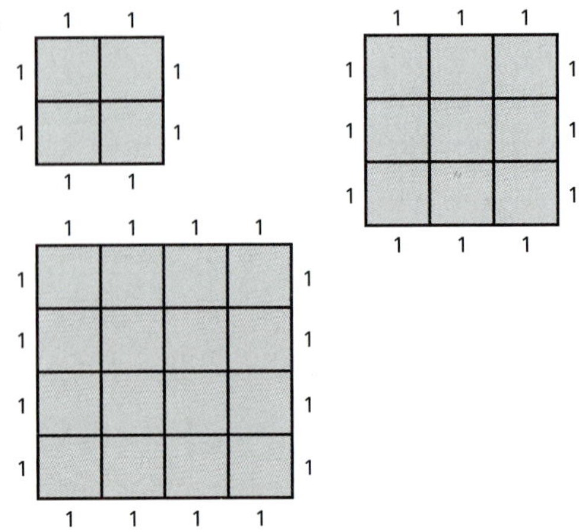

b.

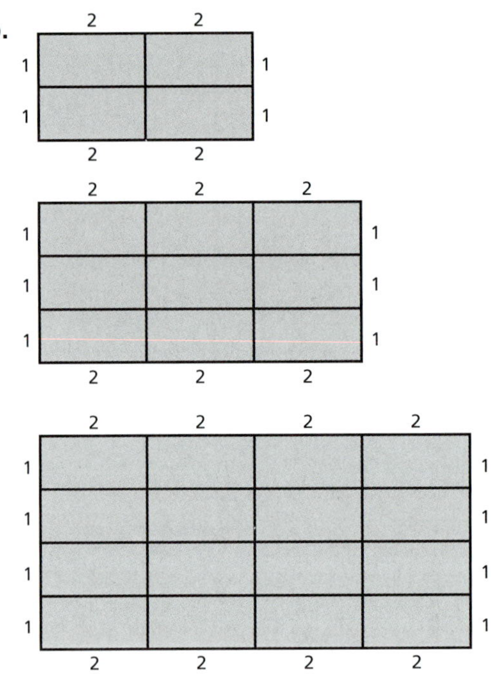

4. **b.** perimeter of shaded rectangle: 18 units; area of shaded rectangle: 18 units^2;

perimeter of unshaded rectangle: 36 units; area of unshaded rectangle: 72 units;

yes; The dimensions are doubled. So, the perimeter of the unshaded rectangle is twice the perimeter of the shaded rectangle and the area of the unshaded rectangle is 4 times the area of the shaded rectangle.

c.

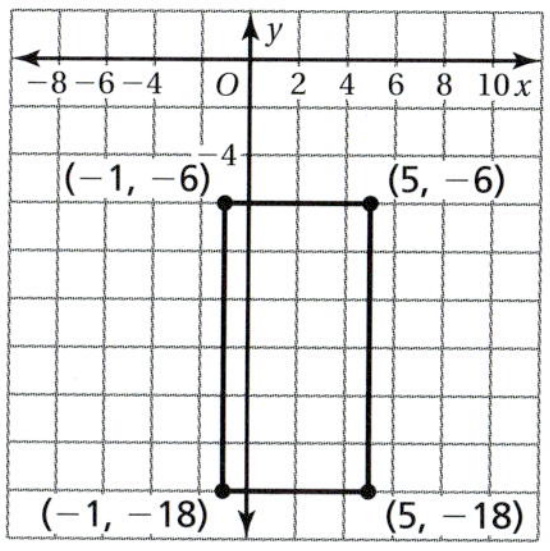

$$\frac{\text{Shaded Length}}{\text{Unshaded Length}} \stackrel{?}{=} \frac{\text{Shaded Width}}{\text{Unshaded Width}}$$

$$\frac{6}{12} \stackrel{?}{=} \frac{3}{6}$$

$$\frac{1}{2} = \frac{1}{2}$$

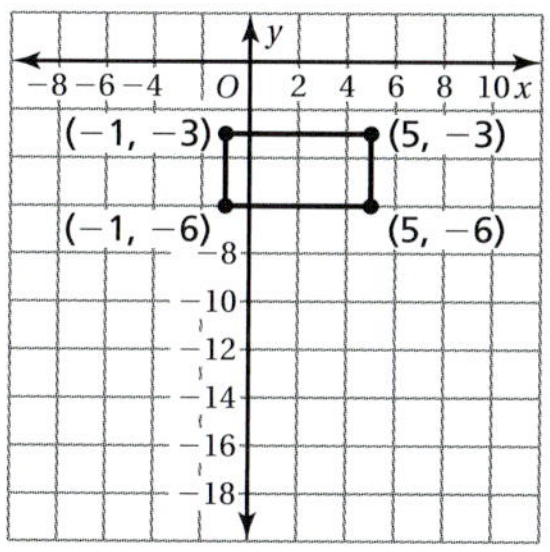

$$\frac{\text{Shaded Length}}{\text{Unshaded Length}} \stackrel{?}{=} \frac{\text{Shaded Width}}{\text{Unshaded Width}}$$

$$\frac{6}{6} \stackrel{?}{=} \frac{3}{3}$$

$$1 = 1$$

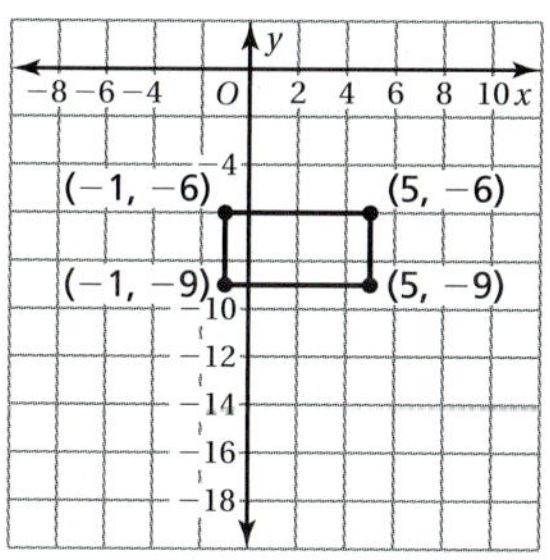

$$\frac{\text{Shaded Length}}{\text{Unshaded Length}} \stackrel{?}{=} \frac{\text{Shaded Width}}{\text{Unshaded Width}}$$

$$\frac{6}{6} \stackrel{?}{=} \frac{3}{3}$$

$$1 = 1$$

Practice and Problem Solving

18. a. $\frac{1}{4}$; $\frac{1}{4}$; $\frac{1}{16}$

b. The ratio of the circumferences is equal to the ratio of the radii. The ratio of the square of the radii is equal to the ratio of the areas. These are the same proportions that are used for similar figures.

Section 11.7

Practice and Problem Solving

7. yes

8. yes

9. no

10. yes

11. yes

12. no

13.

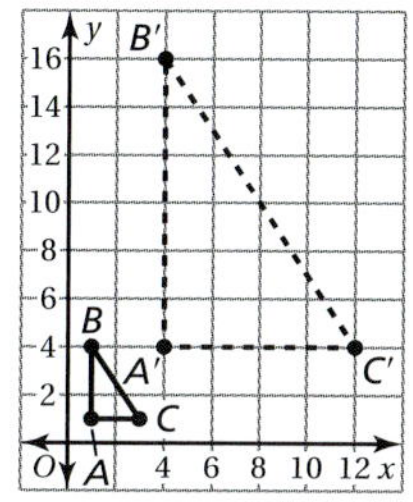

enlargement

14.

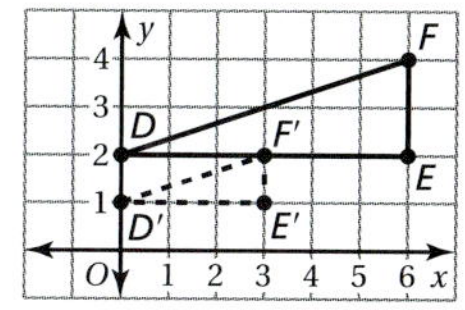

reduction

15.

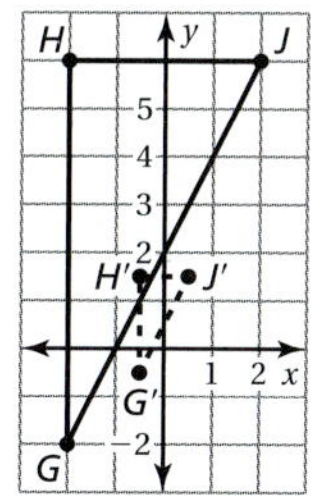

reduction

16.

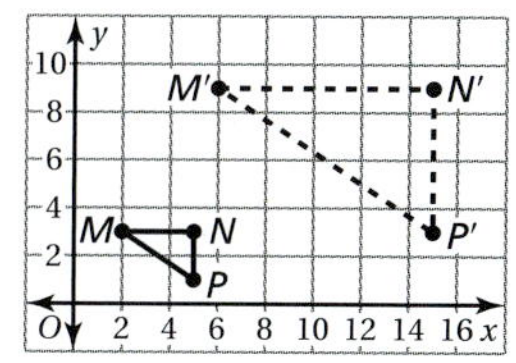

enlargement

17.

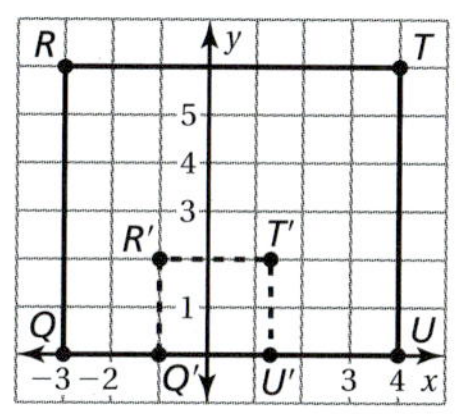

reduction

18.

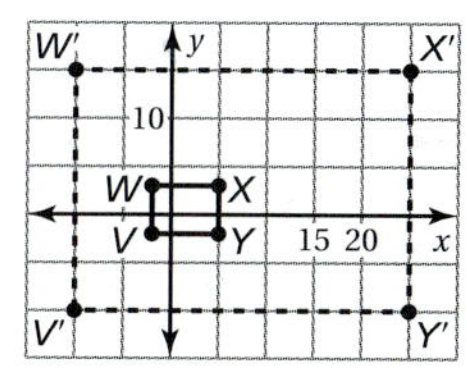

enlargement

31. a. enlargement

b. center of dilation

c. $\frac{4}{3}$

d. The shadow on the wall becomes larger. The scale factor will become larger.

32. $\frac{3}{2}$; Multiply the two scale factors.

33. The transformations are a dilation using a scale factor of 2 and then a translation of 4 units right and 3 units down; similar; A dilation produces a similar figure and a translation produces a congruent figure, so the final image is similar.

34. The transformations are a reflection in the y-axis and then a translation of 1 unit left and two units down; congruent; A reflection produces a congruent figure and a translation produces a congruent figure, so the final image is congruent.

Additional Answers

35. The transformations are a dilation using a scale factor of $\frac{1}{3}$ and then a reflection in the x-axis; similar; A dilation produces a similar figure and a reflection produces a congruent figure, so the final image is similar.

36. $(2x + 3, 2y - 1)$ is a dilation using a scale factor of 2 followed by a translation 3 units right and 1 unit down. $(2(x + 3), 2(y + 1))$ is a translation 3 units right and 1 unit down followed by a dilation using a scale factor of 2.

37. $A'(-2, 3)$, $B'(6, 3)$, $C'(12, -7)$, $D'(-2, -7)$; Methods will vary.

Chapter 12

Section 12.1

Record and Practice Journal

4. *Sample answer:* When two parallel lines are intersected by a transversal, eight angles are formed. In the figure $\angle 1$, $\angle 3$, $\angle 5$, and $\angle 7$ are congruent and $\angle 2$, $\angle 4$, $\angle 6$, and $\angle 8$ are congruent.

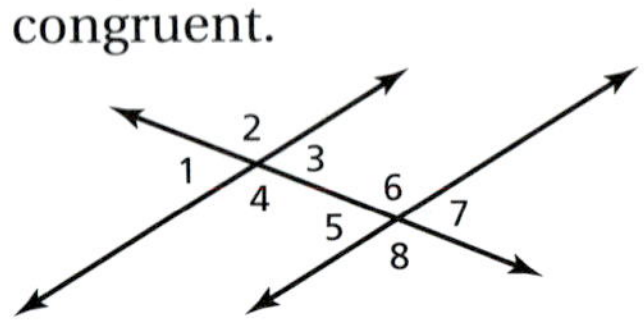

5.

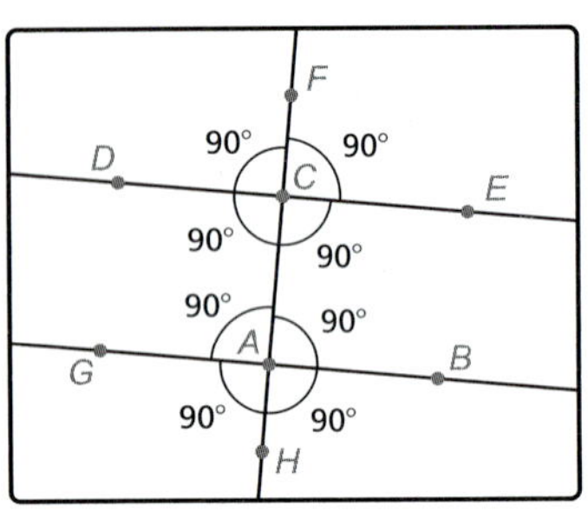

Practice and Problem Solving

17. $\angle 2 = 90°$; $\angle 2$ and the given angle are vertical angles.
$\angle 1 = 90°$ and $\angle 3 = 90°$; $\angle 1$ and $\angle 3$ are supplementary to the given angle.
$\angle 4 = 90°$; $\angle 4$ and the given angle are corresponding angles.
$\angle 6 = 90°$; $\angle 4$ and $\angle 6$ are vertical angles.
$\angle 5 = 90°$ and $\angle 7 = 90°$; $\angle 5$ and $\angle 7$ are supplementary to $\angle 4$.

18. 56°; *Sample answer:* $\angle 1$ and $\angle 8$ are corresponding angles and $\angle 8$ and $\angle 4$ are supplementary.

19. 132°; *Sample answer:* $\angle 2$ and $\angle 4$ are alternate interior angles and $\angle 4$ and $\angle 3$ are supplementary.

20. 55°; *Sample answer:* $\angle 4$ and $\angle 2$ are alternate interior angles.

21. 120°; *Sample answer:* $\angle 6$ and $\angle 8$ are alternate exterior angles.

22. 129.5°; *Sample answer:* $\angle 7$ and $\angle 5$ are alternate exterior angles and $\angle 5$ and $\angle 6$ are supplementary.

23. 61.3°; *Sample answer:* $\angle 3$ and $\angle 1$ are alternate interior angles and $\angle 1$ and $\angle 2$ are supplementary.

24. 40°

25. They are all right angles because perpendicular lines form 90° angles.

26. *Sample answer:* 1) $\angle 1$ and $\angle 7$ are congruent because they are alternate exterior angles.
2) $\angle 1$ and $\angle 5$ are corresponding angles and $\angle 5$ and $\angle 7$ are vertical angles. So, $\angle 1$ and $\angle 7$ are congruent.

Section 12.2

Record and Practice Journal

2. **b.** *Sample answer:* The sum of the measures of angle D, angle B, and angle E is 180° by definition of a straight line. Angle A is congruent to angle D and angle C is congruent to angle E because they are alternate interior angles of a transversal through parallel lines. Then by substitution, the sum of the measures of angle A, angle B, and angle C is 180°.

Practice and Problem Solving

19. sometimes; The sum of the angle measures must equal 180°.

20. always; Because the sum of the interior angle measures must equal 180° and one of the interior angles is 90°, the other two interior angles must sum to 90°.

21. never; If a triangle had more than one vertex with an acute exterior angle, then it would have to have more than one obtuse interior angle which is impossible.

22. You know that $x + y + w = 180$ and $w + z = 180$. Substitute $w + z$ for 180 in the first equation and you get $x + y + w = w + z$. Now subtract w from each side to get $x + y = z$.

Section 12.3

Practice and Problem Solving

25. 60°; The sum of the interior angle measures of a hexagon is 720°. Because it is regular, each angle has the same measure. So, each interior angle is $720° \div 6 = 120°$ and each exterior angle is 60°.

33. a. *Sample answer:*

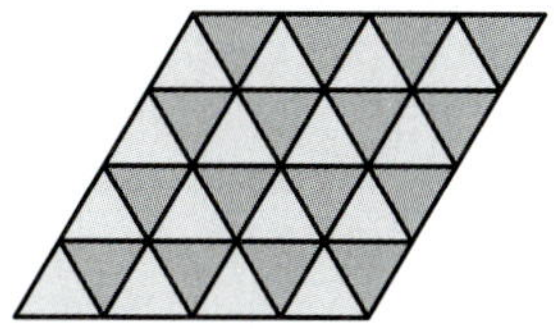

b. *Sample answer:* square, regular hexagon

c. *Sample answer:*

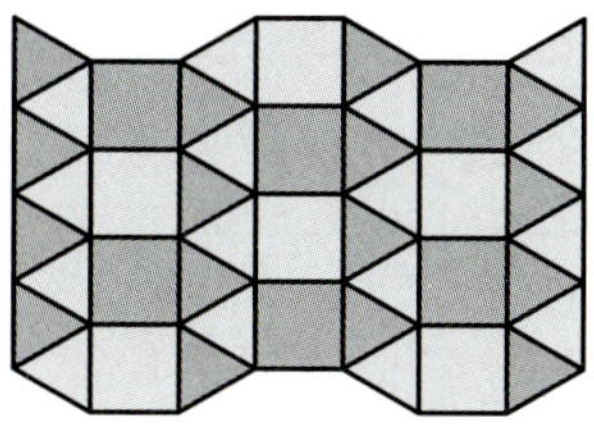

d. *Answer should include, but is not limited to:* a discussion of the interior and exterior angles of the polygons in the tessellation and how they add to 360° where the vertices meet.

Section 12.4

Record and Practice Journal

2. a. *Sample answer:*

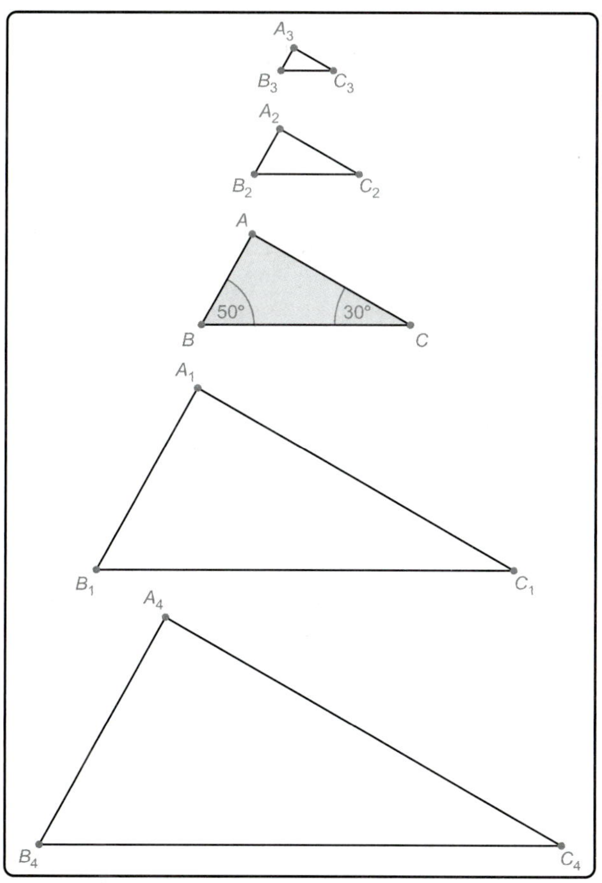

6. *Sample answer:* In the figure, you know that the streetlight forms a right angle with the ground and the person forms a right angle with the ground. Each triangle shares the same angle formed by the ground and the top of the streetlight. Because two angles in one triangle are congruent to two angles in another triangle, the third angles are also congruent. So, the triangles are similar.

Sample answer: Students can also use properties of parallel lines cut by a transversal to show that the two triangles are similar.

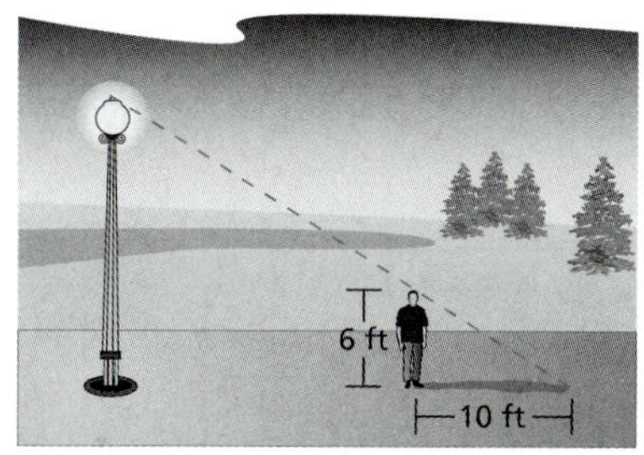

Practice and Problem Solving

14. no; Each side increases by 50%, so each side is multiplied by a factor of $\frac{3}{2}$. The area is $\frac{3}{2}\left(\frac{3}{2}\right) = \frac{9}{4}$ or 225% of the original area, which is a 125% increase.

15. 30 ft

16. *Sample answer:* 10 ft

Assume that you are 5 feet tall.

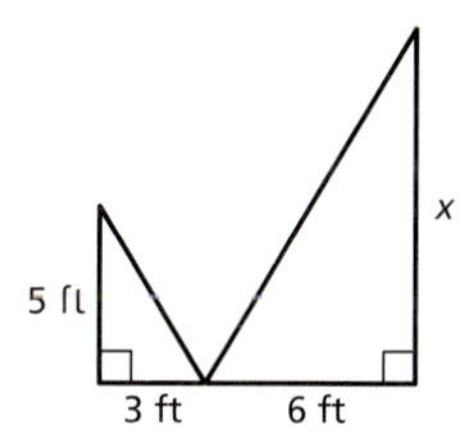

17. maybe; They are similar when both have measures of 30°, 60°, 90° or both have measures of 45°, 45°, 90°. They are not similar when one has measures of 30°, 60°, 90° and the other has measures of 45°, 45°, 90°.

18. $\triangle ABG \sim \triangle ACF$, $\triangle ABG \sim \triangle ADE$, $\triangle ACF \sim \triangle ADE$; 2 ft; 4 ft

Chapter 13

Record and Practice Journal Fair Game Review

16–20.

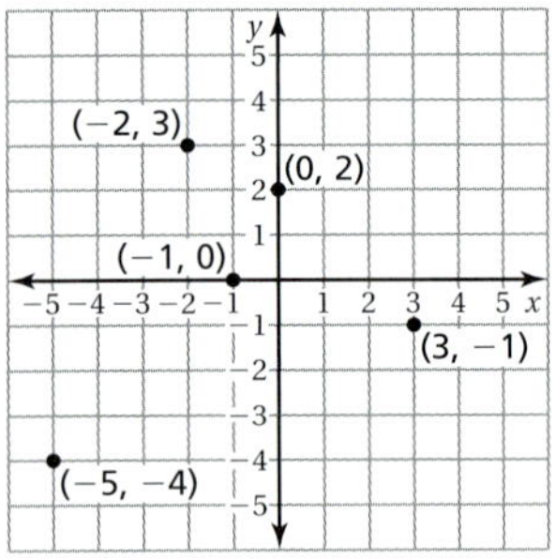

Section 13.1

On Your Own

1.

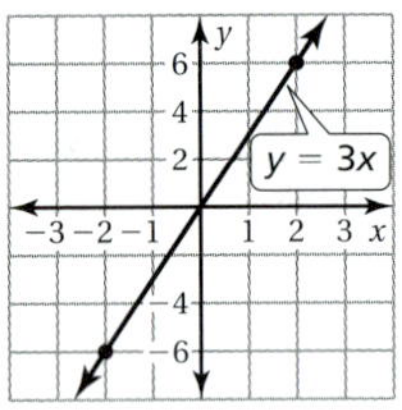

2.

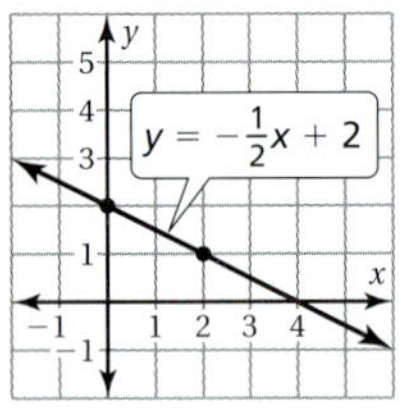

3.

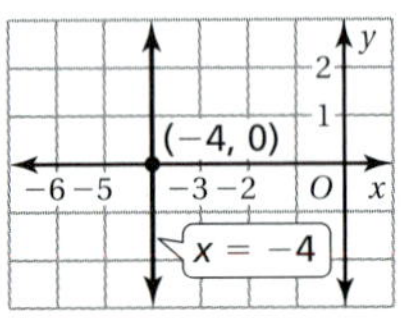

4.

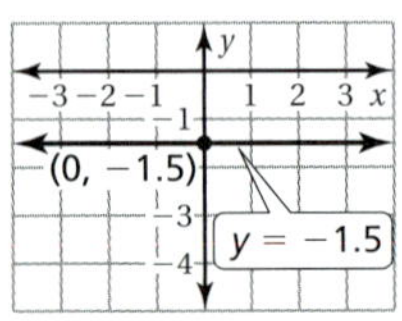

Practice and Problem Solving

7.

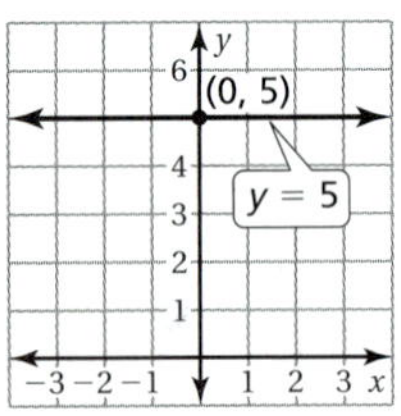

8.

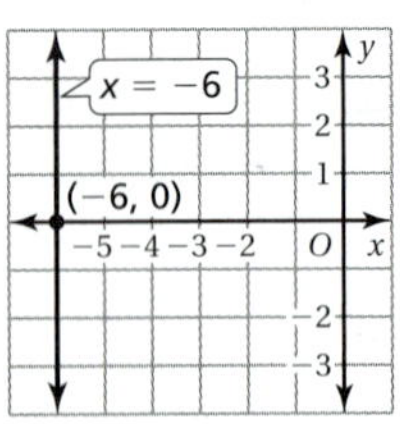

9.

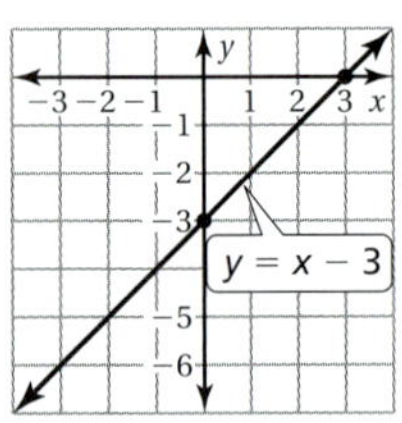

10.

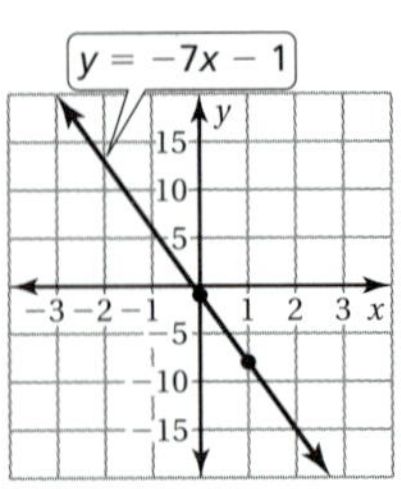

11.

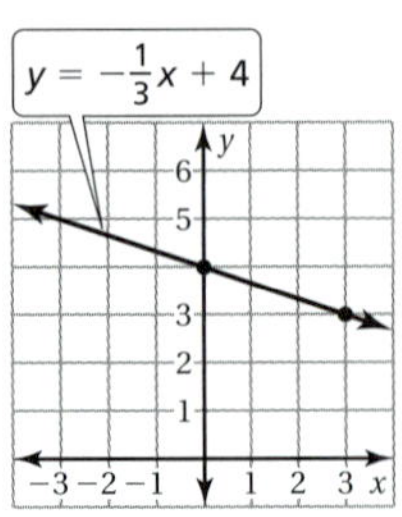

12.

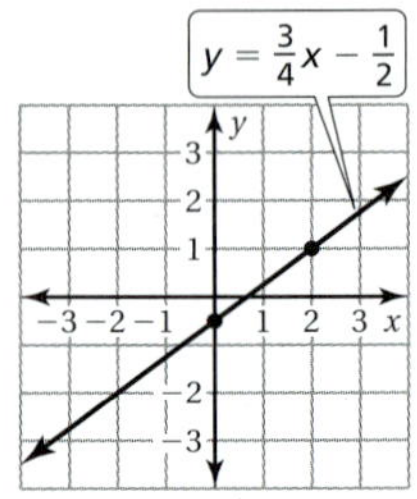

13.

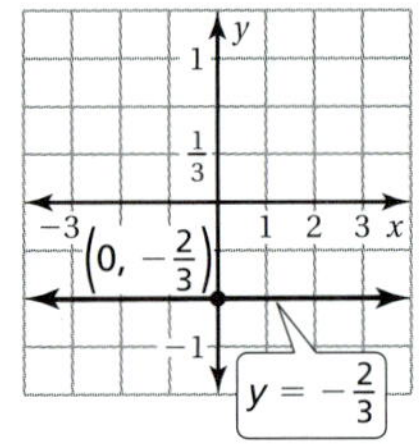

14.

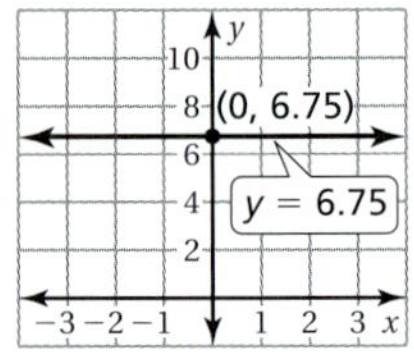

15.

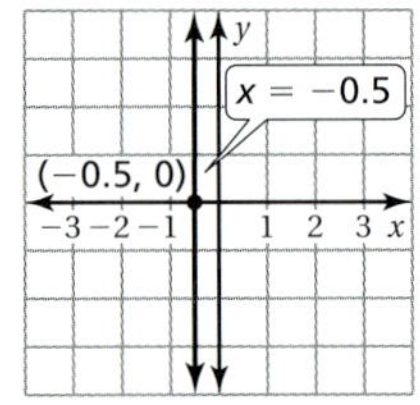

16.

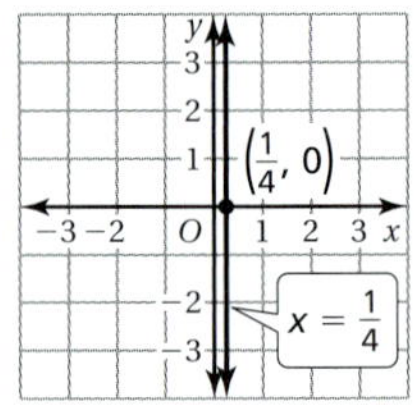

17. The equation $x = 4$ is graphed, not $y = 4$.

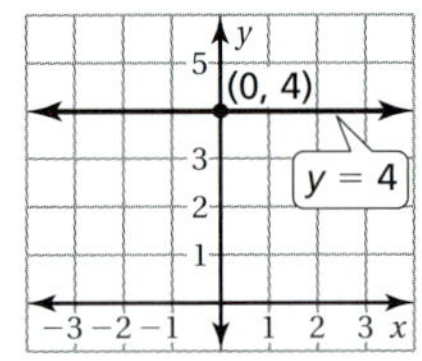

18.

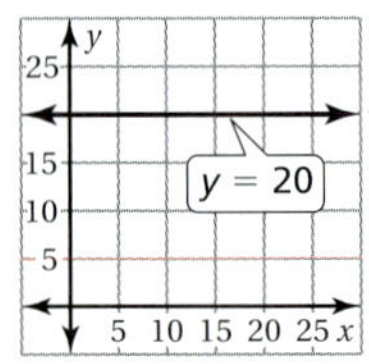

Sample answer: No matter how many text messages are sent, the cost is \$20.

19. a.

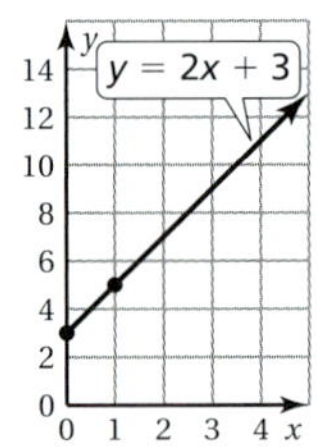

b. about \$5

c. \$5.25

21. $y = -\dfrac{5}{2}x + 2$

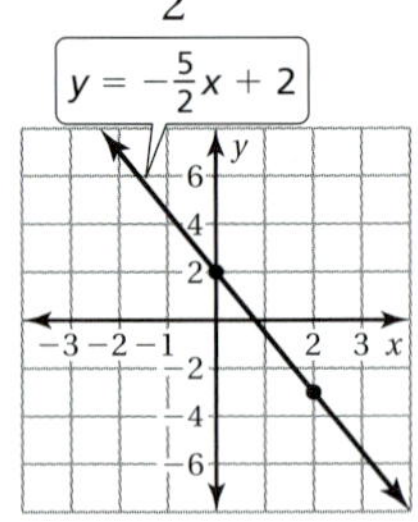

22. $y = 12x - 9$

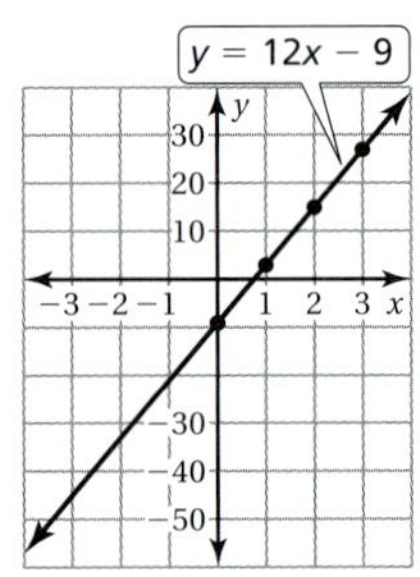

23. $y = -2x + 3$

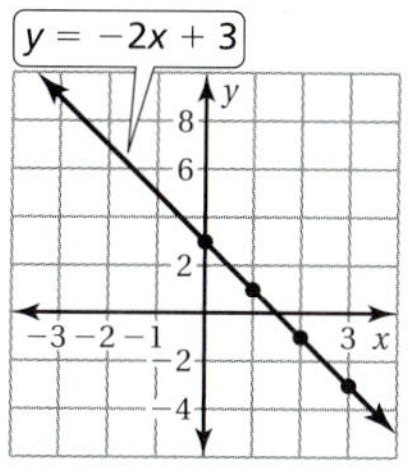

24. a.

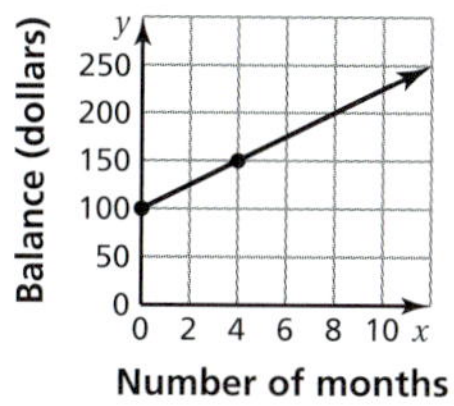

b. 6 mo

25. a. *Sample answer:*

yes; The graph of the equation is a line.

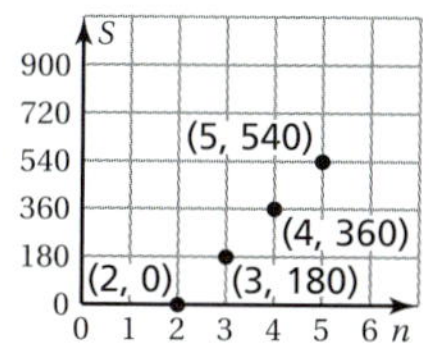

b. no; $n = 3.5$ does not make sense because a polygon cannot have half a side.

26. *Sample answer:* If you are 13 years old, the sea level has risen 26 millimeters since you were born.

$y = 2x$

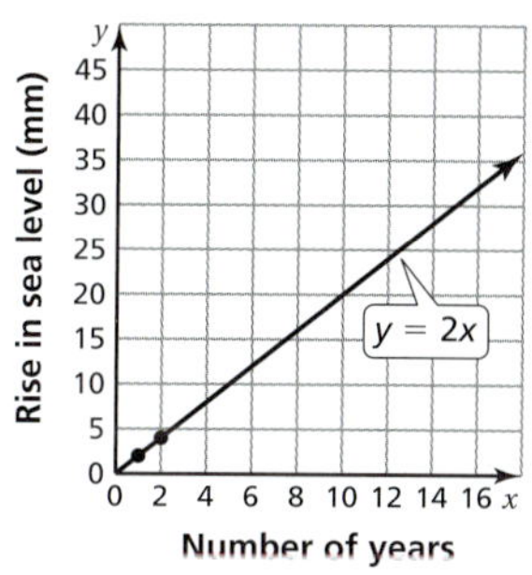

Section 13.2

Practice and Problem Solving

26. *Sample answer:*

a. Yes, it follows the guidelines.

b.

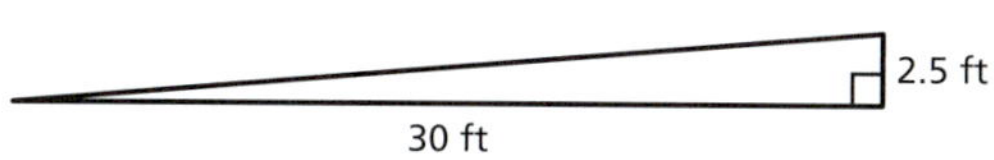

Extension 13.2

Extra Example 2

blue and green; The blue line has a slope of $-\frac{3}{2}$. The green line has a slope of $\frac{2}{3}$. The product of their slopes is $-\frac{3}{2} \cdot \frac{2}{3} = -1$.

Practice

3. yes; Both lines are horizontal and have a slope of 0.

4. no; $y = 0$ has a slope of 0 and $x = 0$ has an undefined slope.

5. yes; Both lines are vertical and have an undefined slope.

6. Use the vertices of the quadrilateral to find the slope of each side. When opposite sides are parallel (have the same slope), the quadrilateral is a parallelogram.

slope of $AB = -\frac{1}{7}$; slope of $BC = -\frac{5}{2}$;

slope of $CD = -\frac{1}{6}$; slope of $DA = -\frac{5}{3}$;

No, it is not a parallelogram. Because they have different slopes, opposite sides are not parallel.

7. blue and green; The blue line has a slope of 6. The green line has a slope of $-\frac{1}{6}$. The product of their slopes is $6 \cdot \left(-\frac{1}{6}\right) = -1$.

8. blue and green, red and green; The blue and red lines both have a slope of $-\frac{1}{2}$. The green line has a slope of 2. The product of their slopes is $2 \cdot \left(-\frac{1}{2}\right) = -1$.

9. yes; The line $x = -2$ is vertical. The line $y = 8$ is horizontal. A vertical line is perpendicular to a horizontal line.

10. no; Both lines are vertical and have undefined slopes.

11. yes; The line $x = 0$ is vertical. The line $y = 0$ is horizontal. A vertical line is perpendicular to a horizontal line.

12. Use the vertices of the quadrilateral to find the slope of each side. When adjacent sides are perpendicular, then the parallelogram is a rectangle. Note that the quadrilateral is a parallelogram, so you already know that opposite sides are parallel.

slope of $JK = \frac{2}{3}$; slope of $KL = -\frac{3}{2}$

slope of $LM = \frac{2}{3}$; slope of $MJ = -\frac{3}{2}$

Yes, it is a rectangle.

JK is perpendicular to KL because $\frac{2}{3} \cdot \left(-\frac{3}{2}\right) = -1$.

KL is perpendicular to LM because $-\frac{3}{2} \cdot \frac{2}{3} = -1$.

LM is perpendicular to MJ because

$\frac{2}{3} \cdot \left(-\frac{3}{2}\right) = -1$.

MJ is perpendicular to JK because $-\frac{3}{2} \cdot \frac{2}{3} = -1$.

Record and Practice Journal Practice

1. line B and line G; they both have a slope of $\frac{5}{3}$.
2. line B and line R; They both have a slope of 9.
3. yes; Both lines are vertical and have undefined slopes.
4. no; The line $x = 3$ has an undefined slope and the line $y = -3$ has a slope of 0.
5. yes; Because opposite sides have the same slope, they are parallel. Because opposite sides are parallel, the quadrilateral is a parallelogram.
6. line B and line R; Line B has a slope of 1. Line R has a slope of -1. the product of their slopes is $1 \cdot (-1) = -1$.
7. line R and line G; Line R has a slope of 4. Line G has a slope of $-\frac{1}{4}$. The product of their slopes is $4 \cdot \left(-\frac{1}{4}\right) = -1$.
8. yes; The line $x = 0$ is vertical. The line $y = 3$ is horizontal. A vertical line is perpendicular to a horizontal line.
9. no; Both lines are horizontal and have a slope of 0.
10. yes; Because the products of the slopes of intersecting sides are equal to -1, the parallelogram is a rectangle.

Section 13.3

Record and Practice Journal

2. The quantities in parts (a), (d), and (f) are in a proportional relationship.

   For part (a): slope = 10; The value of y for $(1, y)$ is 10.

   For part (d): slope = 6; the value of y for $(1, y)$ is 6.

   For part (f): slope = 2; The value of y for $(1, y)$ is 2.

   The value of y is equal to the slope of the line. The value of y represents the unit rate.

Extra Example 3

3. b.

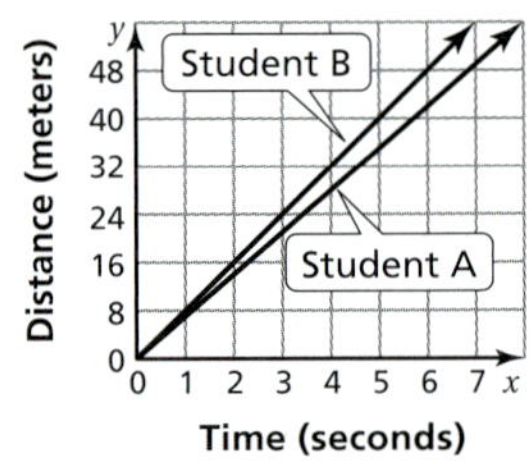

The graph that represents Student B is steeper than the graph that represents Student A. So, Student B is faster than Student A.

Practice and Problem Solving

10. a. fingernails; Fingernails grow about 0.7 millimeter per week and toenails grow about 0.25 millimeter per week.

b.

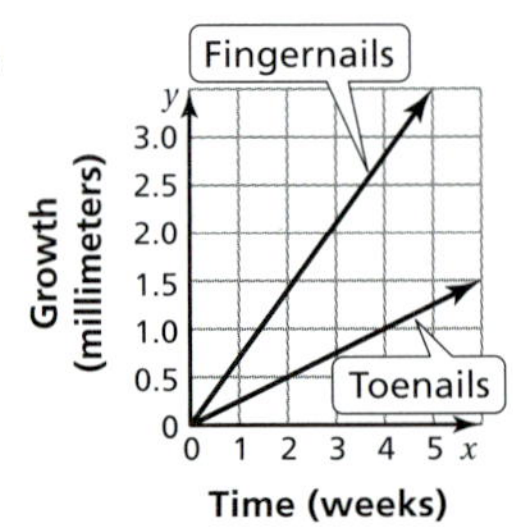

The graph that represents fingernails is steeper than the graph that represents toenails. So, fingernails grow faster than toenails.

13. a. yes; The equation is $d = 6t$, which represents a proportional relationship.

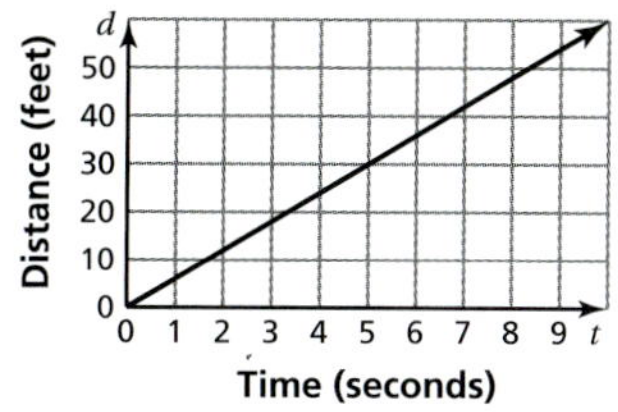

b. yes; The equation is $d = 50r$, which represents a proportional relationship.

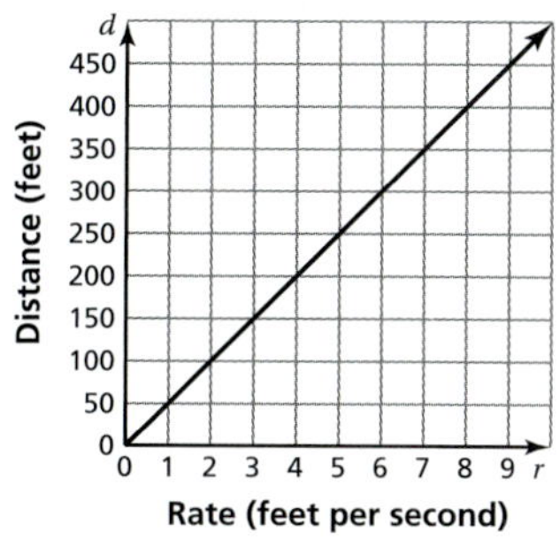

c. no; The equation is $t = \dfrac{300}{r}$, which does not represent a proportional relationship.

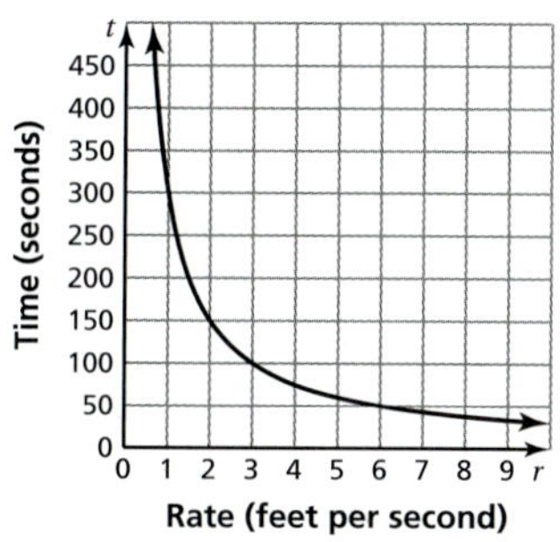

d. part c; It is called inverse variation because when the rate increases, the time decreases, and when the rate decreases, the time increases.

Fair Game Review

14.

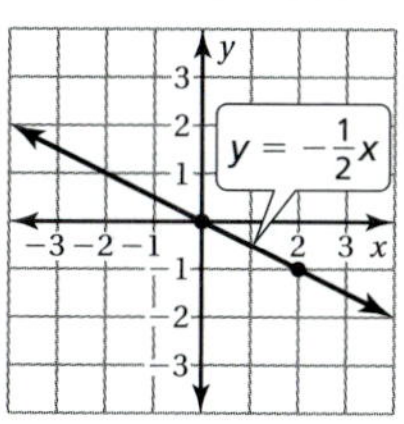

15.

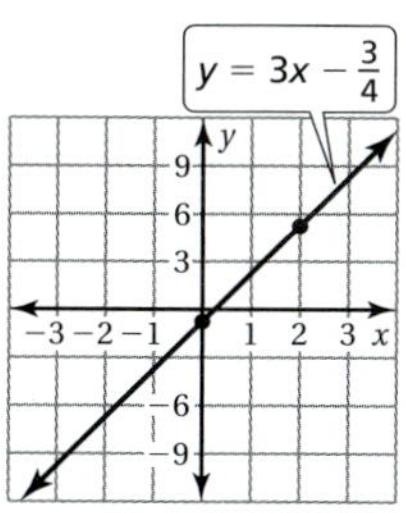

16.

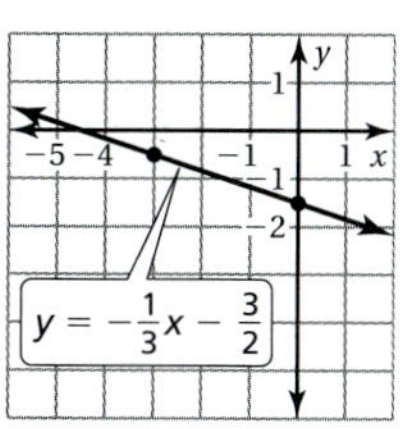

13.1–13.3 Quiz

1.

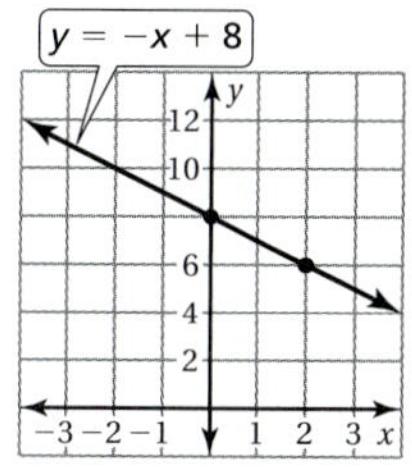

2.

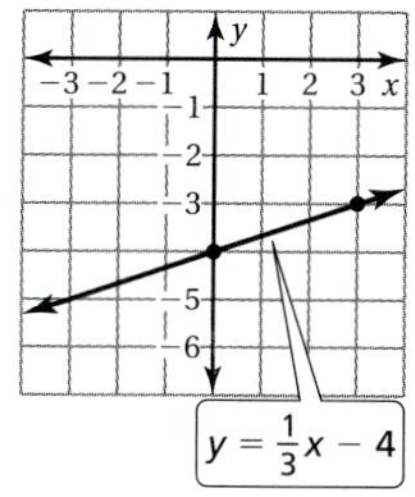

3.

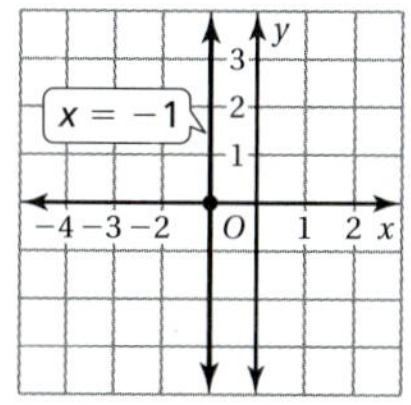

4.

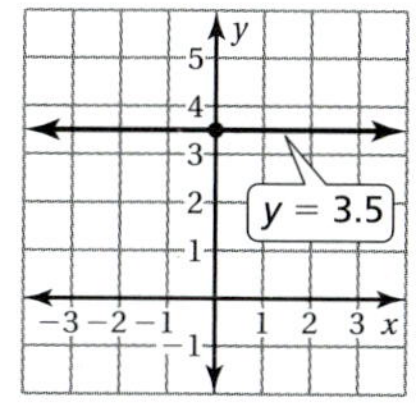

Section 13.4

Record and Practice Journal

1. a.

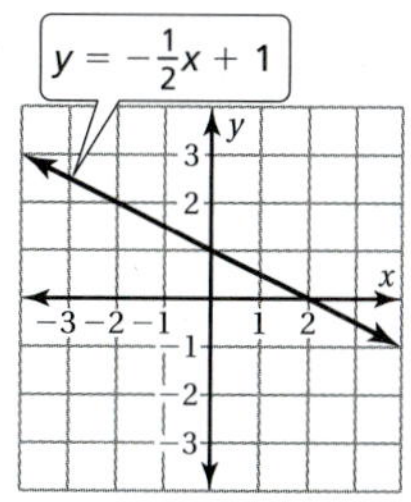

b.

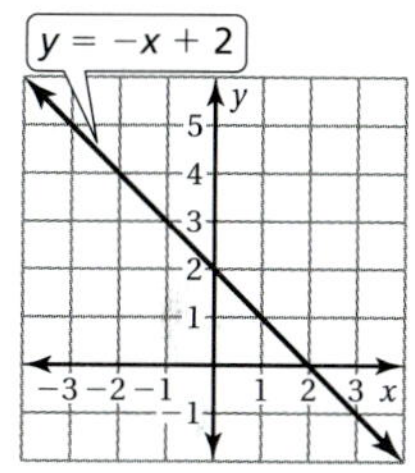

c.

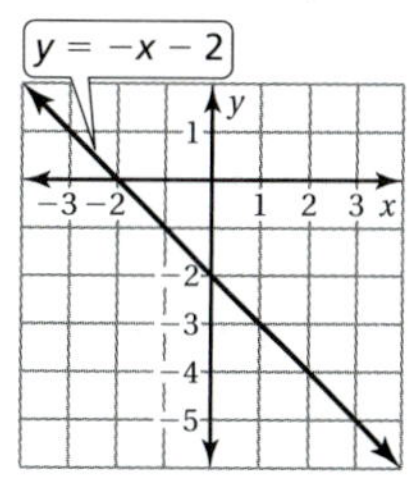

d.

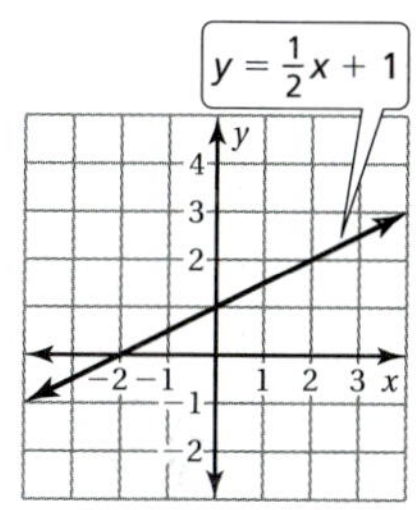

e.

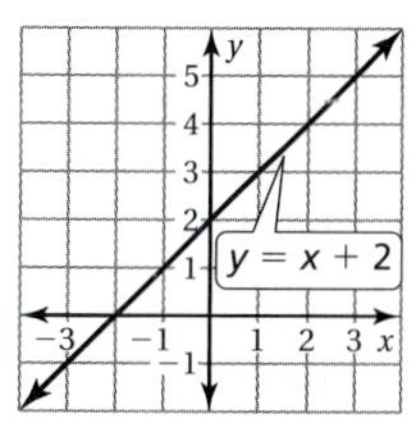

f.

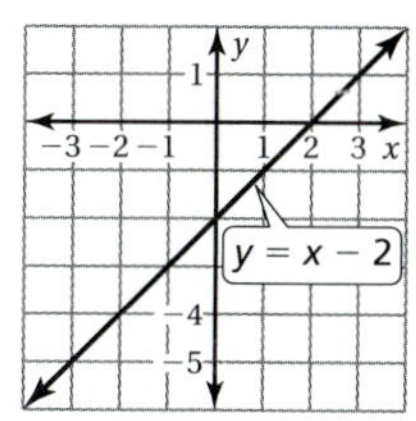

g.

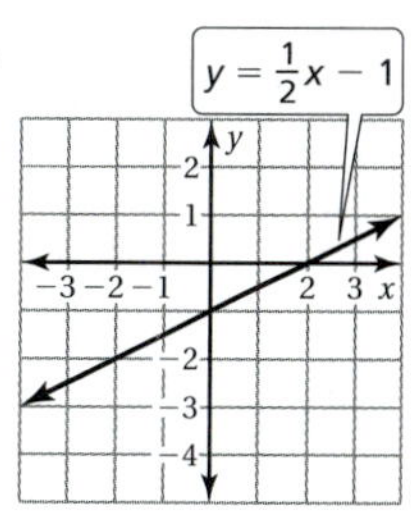

h.

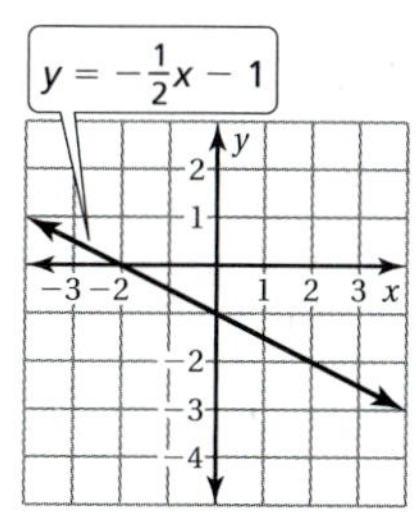

i.

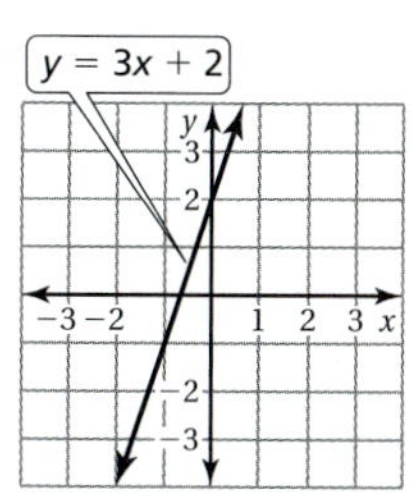

j.

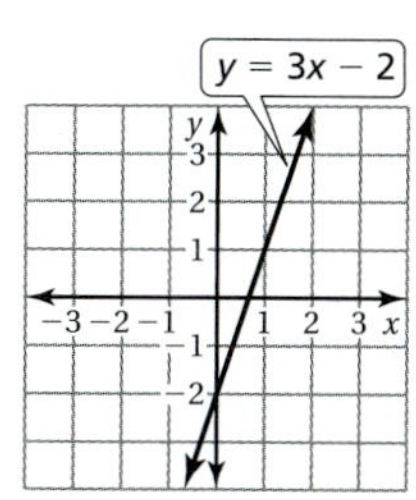

Additional Answers

On Your Own

3.

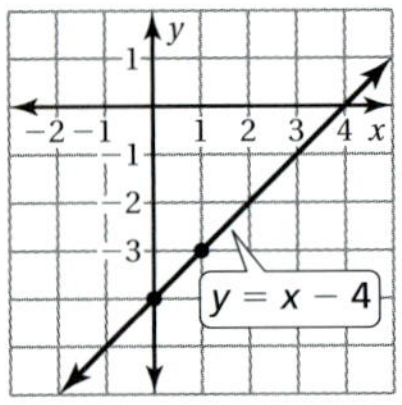

x-intercept: 4

4.

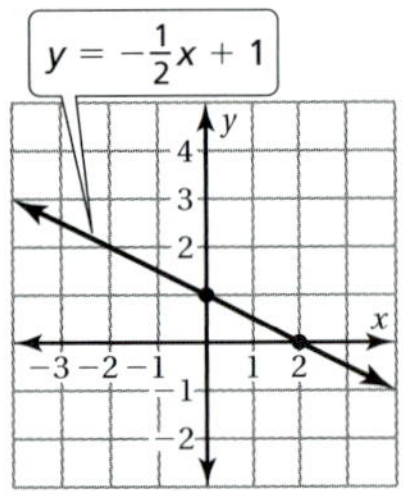

x-intercept: 2

5. The y-intercept means that the taxi has an initial fee of \$1.50. The slope means the taxi charges \$2 per mile.

Practice and Problem Solving

17. a.

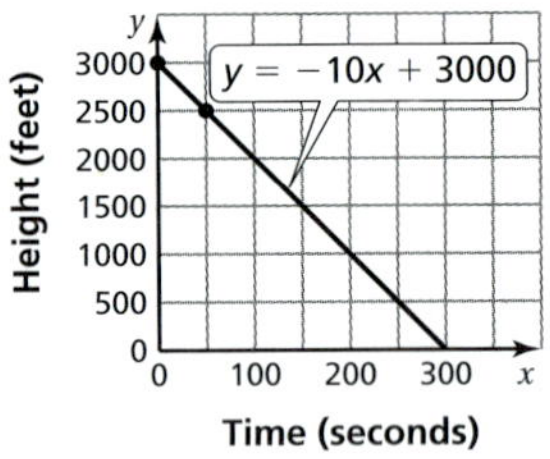

b. The x-intercept of 300 means the skydiver lands on the ground after 300 seconds. The slope of -10 means that the skydiver falls to the ground at a rate of 10 feet per second.

19.

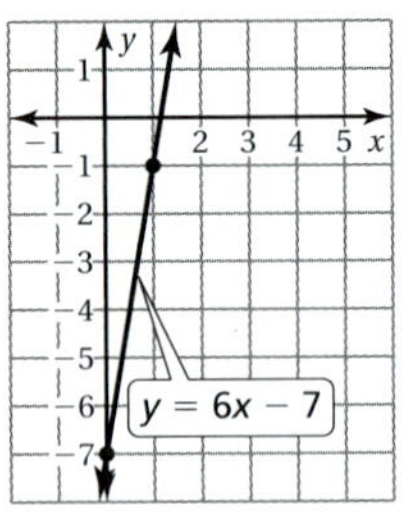

x-intercept: $\frac{7}{6}$

20.

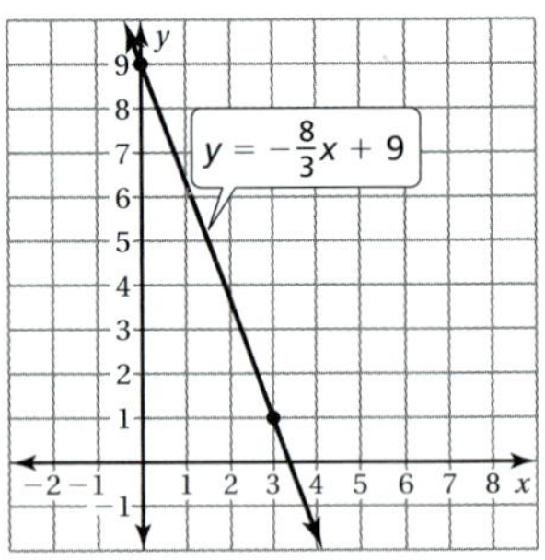

x-intercept: $\frac{27}{8}$

21.

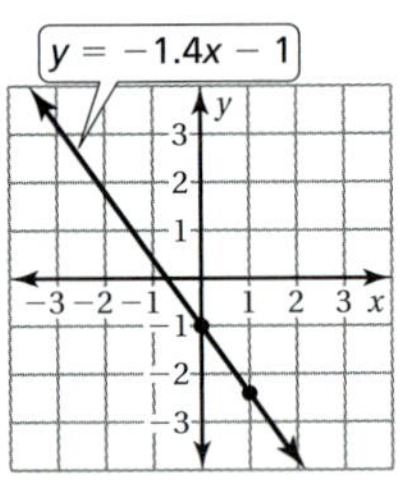

x-intercept: $-\frac{5}{7}$

22.

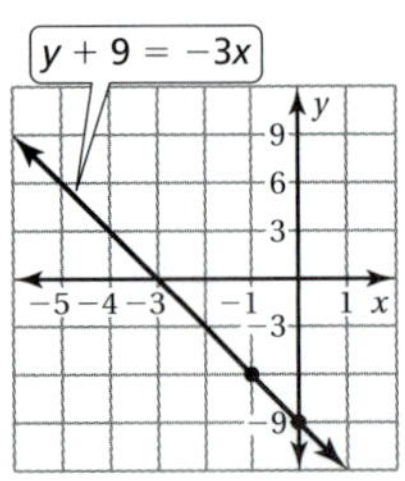

x-intercept: -3

23.

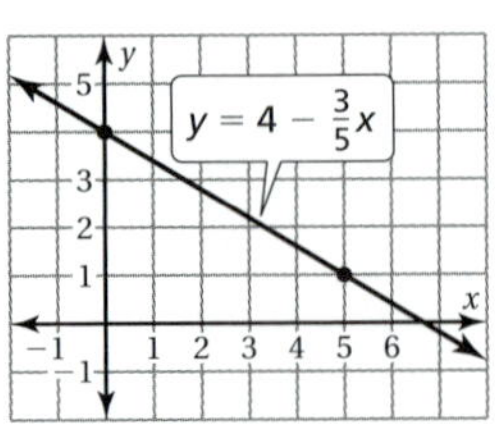

x-intercept: $\frac{20}{3}$

24. a. $y = 0.75x + 5$; The cost of going to the festival is the sum of the cost of picking x pounds of apples, $0.75x$, and the cost of admission, 5.

b.

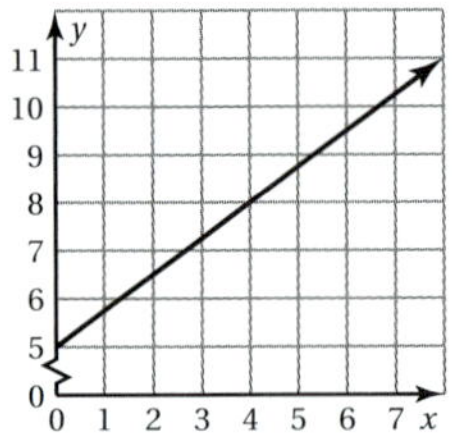

25. a. $y = 2x + 4$ and $y = 2x - 3$ are parallel because the slope of each line is 2; $y = -3x - 2$ and $y = -3x + 5$ are parallel because the slope of each line is -3.

b. $y = 2x + 4$ and $y = -\frac{1}{2}x + 2$ are perpendicular because the product of their slopes is -1; $y = 2x - 3$ and $y = -\frac{1}{2}x + 2$ are perpendicular because the product of their slopes is -1; $y = -\frac{1}{3}x - 1$ and $y = 3x + 3$ are perpendicular because the product of their slopes is -1.

Section 13.5

On Your Own

3.

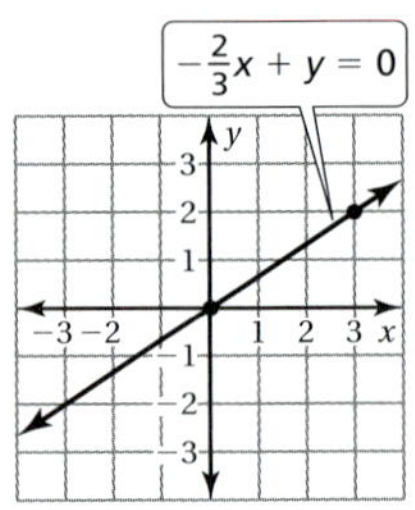

4.

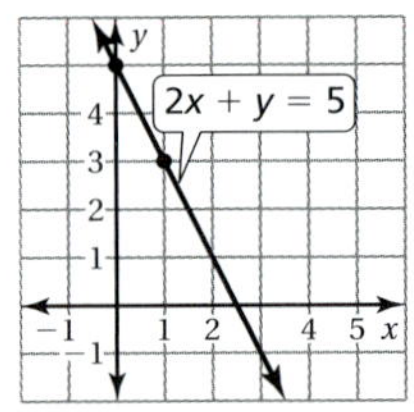

5.

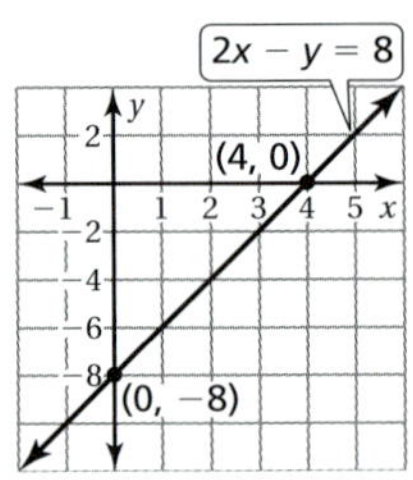

6.

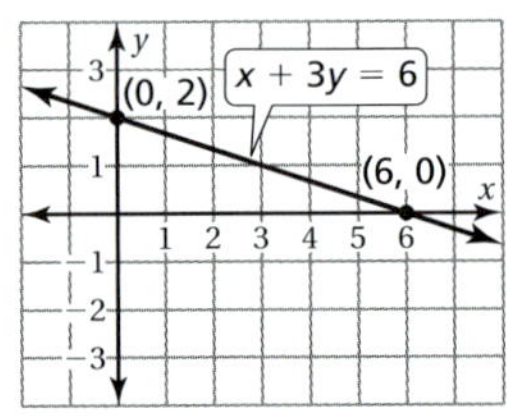

7.

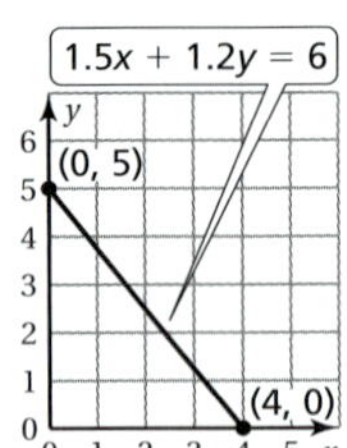

The x-intercept shows that you can buy 4 pounds of apples if you do not buy any oranges. The y-intercept shows that you can buy 5 pounds of oranges if you do not buy any apples.

Practice and Problem Solving

8.

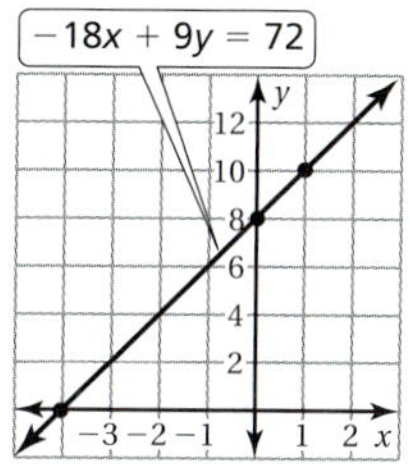

9.

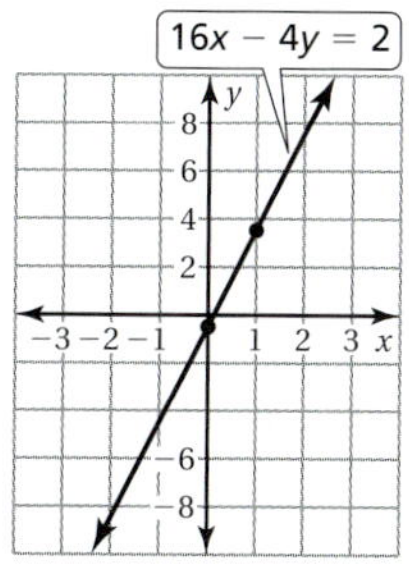

10.

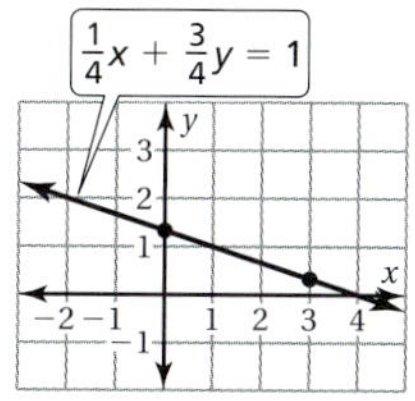

11. B

12. A

13. C

14. They should have let $y = 0$, not $x = 0$.

$$-2x + 3y = 12$$
$$-2x + 3(0) = 12$$
$$-2x = 12$$
$$x = -6$$

15. **a.**

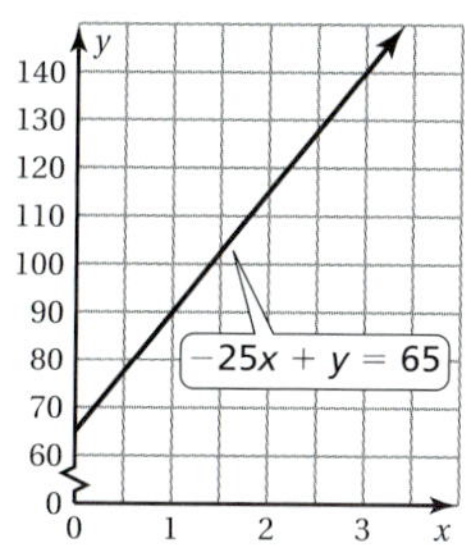

b. \$390

16.

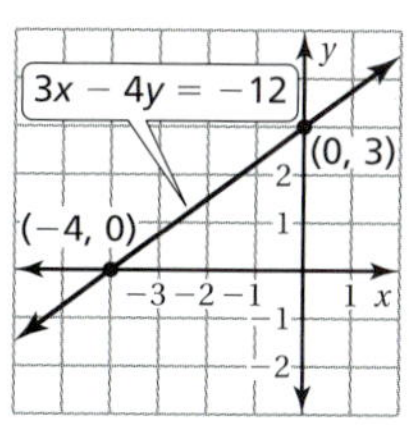

17.

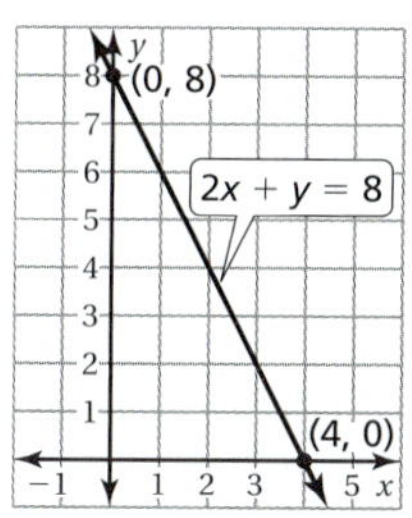

18.

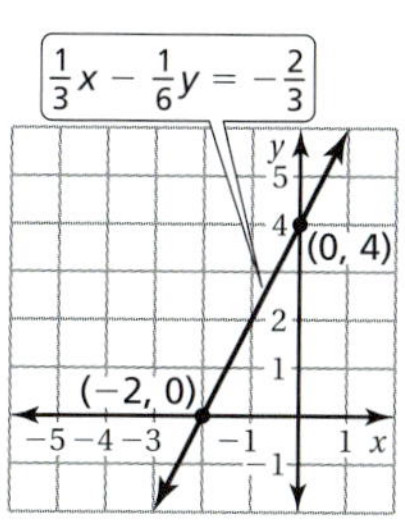

19. x-intercept: 9

y-intercept: 7

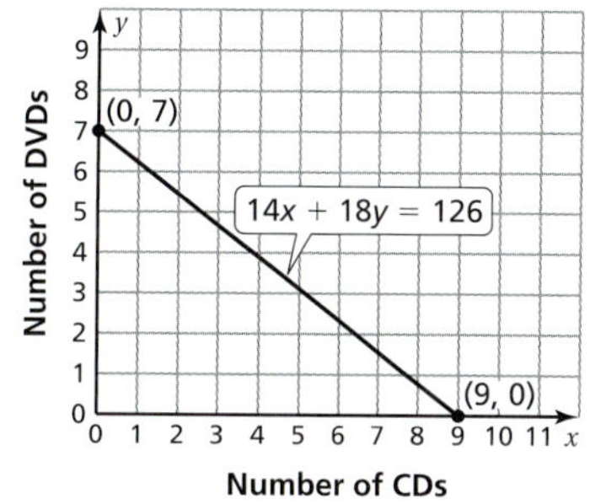

21. **a.** $9.45x + 7.65y = 160.65$

b.

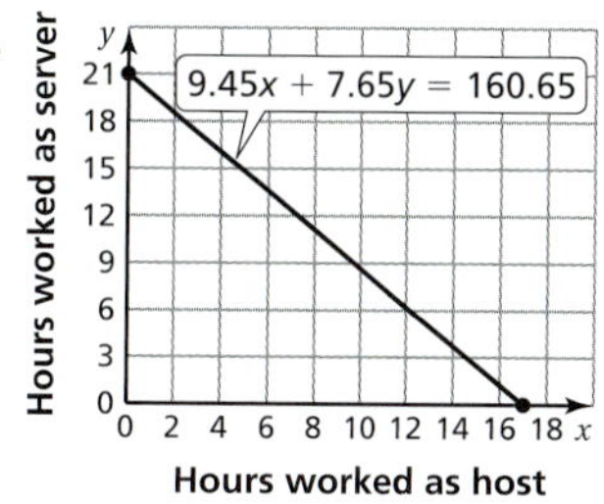

22. no; For example, $y = 5$ does not have an x-intercept, neither do any horizontal lines except $y = 0$.

23. **a.** $y = 40x + 70$

b. x-intercept: $-\frac{7}{4}$; no; You cannot have a negative time.

c.

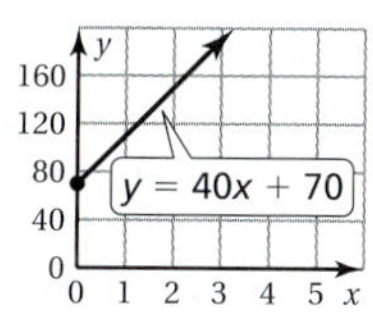

Section 13.6

Record and Practice Journal

1. **a.** top line: slope: $\frac{1}{2}$; y-intercept: 4; $y = \frac{1}{2}x + 4$

middle line: slope: $\frac{1}{2}$; y-intercept: 1; $y = \frac{1}{2}x + 1$

bottom line: slope: $\frac{1}{2}$; y-intercept: -2;

$$y = \frac{1}{2}x - 2$$

The lines are parallel.

b. right line: slope: -2; y-intercept: 3;
$y = -2x + 3$

middle line: slope: -2; y-intercept: -1;
$y = -2x - 1$

left line: slope: -2; y-intercept: -5;
$y = -2x - 5$

The lines are parallel.

c. line passing through (3, 2):

slope: $-\frac{1}{3}$; y-intercept: 3; $y = -\frac{1}{3}x + 3$

line passing through (3, 7):

slope: $\frac{4}{3}$; y-intercept: 3; $y = \frac{4}{3}x + 3$

line passing through (6, 4):

slope: $\frac{1}{6}$; y-intercept: 3; $y = \frac{1}{6}x + 3$

The lines have the same y-intercept.

d. line passing through (1, 2):

slope: 2; y-intercept: 0; $y = 2x$

line passing through (1, -1):

slope: -1; y-intercept: 0; $y = -x$

line passing through (3, 1):

slope: $\frac{1}{3}$; y-intercept: 0; $y = \frac{1}{3}x$

The lines have the same y-intercept.

2. a. $y = 4$; $y = -2$; $y = -2x + 8$; $y = -2x - 6$

b. $y = 5$; $y = -2$; $y = x + 5$; $y = x + 1$

Practice and Problem Solving

13. $y = 5$

14. $y = 0$

15. $y = -2$

16. $y = 0.7x + 10$

17. a–b.

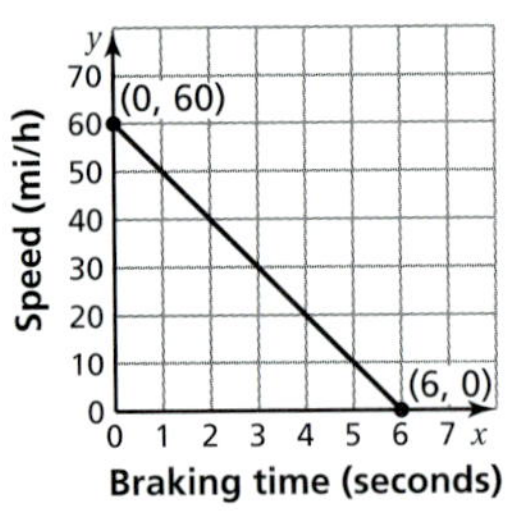

(0, 60) represents the speed of the automobile before braking. (6, 0) represents the amount of time it takes to stop. The line represents the speed y of the automobile after x seconds of braking.

c. $y = -10x + 60$

Section 13.7

Fair Game Review

24.

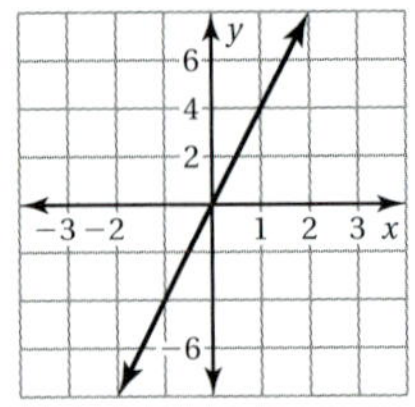

25.

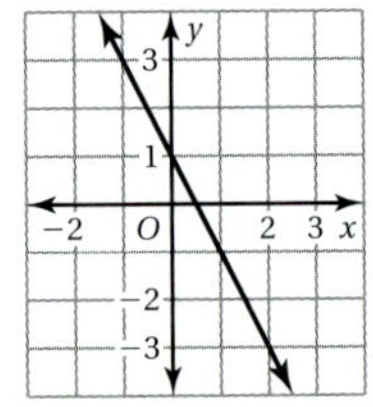

26.

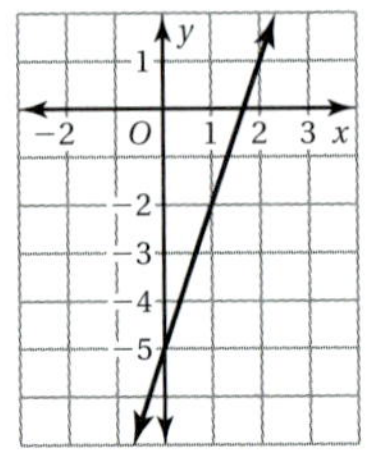

13.4–13.7 Quiz

15. a.

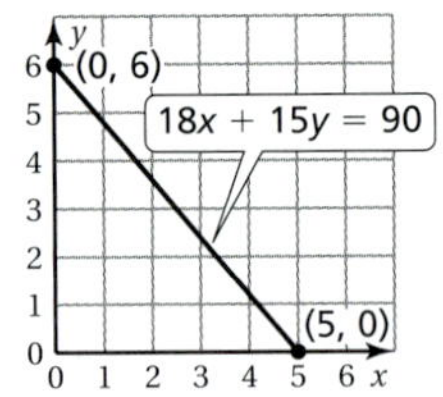

b. The x-intercept, 5, shows that you can buy 5 gallons of blue paint if you do not buy any white paint. The y-intercept, 6, shows that you can buy 6 gallons of white paint if you do not buy any blue paint.

Chapter 13 Test

7.

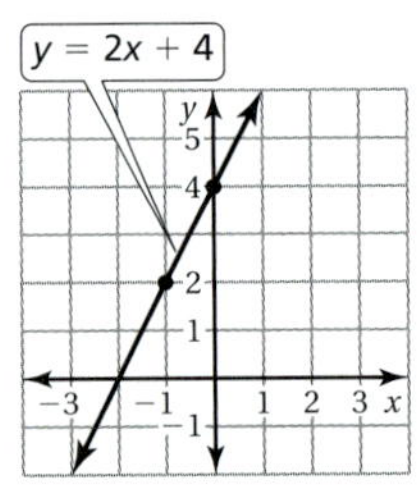

8.

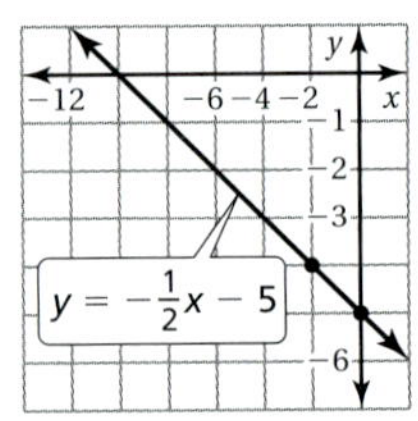

9.

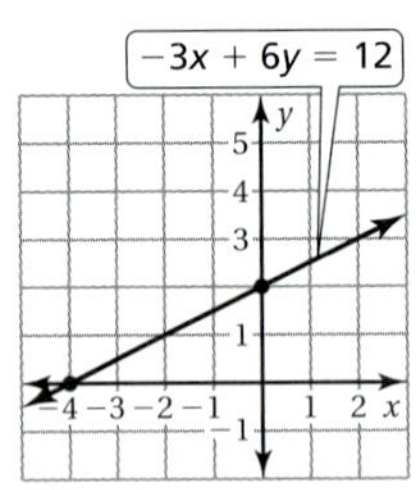

Chapter 14

Record and Practice Journal Fair Game Review

12. 63 **13.** 116

14. −51 **15.** 1

16. $\dfrac{24 + 32 + 30 + 28}{2}$; 57

Section 14.3

Practice and Problem Solving

17. a. *Sample answer:*

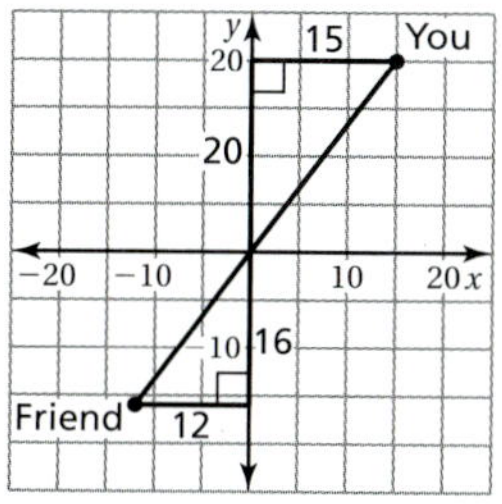

b. 45 ft

Section 14.4

Practice and Problem Solving

20. a. 7 **21. a.** 26

b. 6.8 **b.** 26.2

22. a. −8 **23. a.** −10

b. −7.8 **b.** −10.2

24. a. 3 **25. a.** −13

b. 2.6 **b.** −12.9

Extension 14.4

Record and Practice Journal Practice

1. $\frac{1}{3}$ **2.** $-\frac{2}{9}$

3. $1\frac{7}{9}$ **4.** $-2\frac{2}{3}$

5. $\frac{7}{15}$ **6.** $-1\frac{15}{18}$

7. $-\frac{11}{15}$ **8.** $\frac{2}{11}$

9. $-3\frac{8}{33}$ **10.** $1\frac{1}{11}$

11. $1\frac{17}{30}$ in.

Section 14.5

Practice and Problem Solving

26.

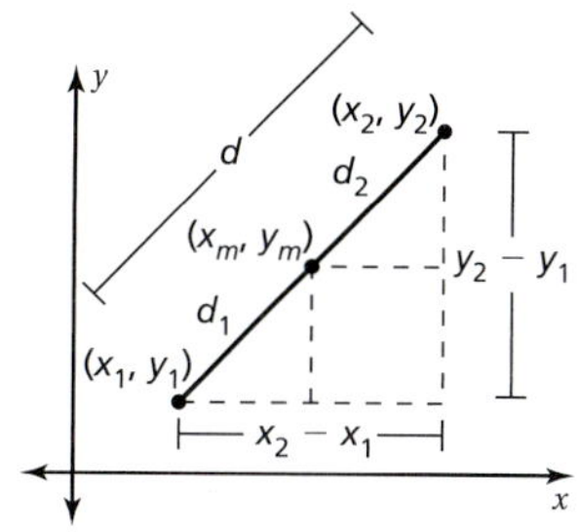

$$x_m = x_1 + \frac{1}{2}(x_2 - x_1) = \frac{2x_1 + x_2 - x_1}{2} = \frac{x_1 + x_2}{2}$$

Similarly, $y_m = \dfrac{y_1 + y_2}{2}$.

$$d_1 = \sqrt{\left(x_1 - \frac{x_1 + x_2}{2}\right)^2 + \left(y_1 - \frac{y_1 + y_2}{2}\right)^2}$$

$$= \sqrt{\left(\frac{x_1 - x_2}{2}\right)^2 + \left(\frac{y_1 - y_2}{2}\right)^2}$$

$$d_2 = \sqrt{\left(\frac{x_1 + x_2}{2} - x_2\right)^2 + \left(\frac{y_1 + y_2}{2} - y_2\right)^2}$$

$$= \sqrt{\left(\frac{x_1 - x_2}{2}\right)^2 + \left(\frac{y_1 - y_2}{2}\right)^2}$$

Because $d_1 = d_2$, the point (x_m, y_m) is located in the middle of the two given points.

Chapter 15

Section 15.4

Practice and Problem Solving

17. a. 9483 pounds; The ratio of the height of the original statue to the height of the small statue is 8.4 : 1. So, the ratio of the weights, or volumes is $\left(\frac{8.4}{1}\right)^3$.

b. 221,184 lb

Fair Game Review

20. 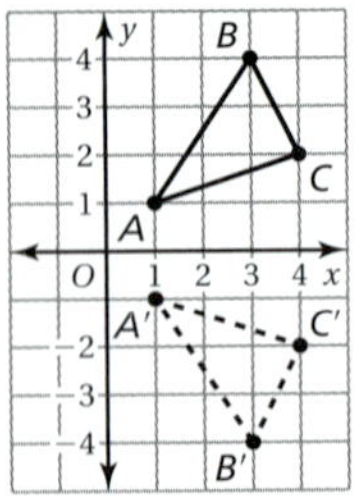

$A'(1, -1), B'(3, -4), C'(4, -2)$

21. 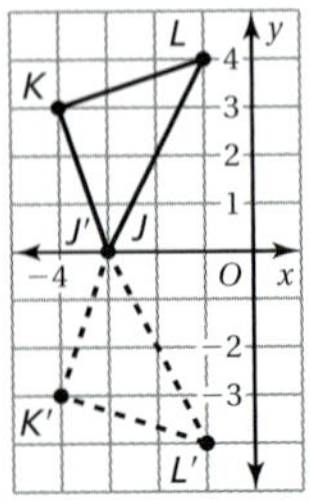

$J'(-3, 0), K'(-4, -3), L'(-1, -4)$

22. B

Chapter 16

Section 16.1

Practice and Problem Solving

27.

h	1	2	3	4	5
$2^h - 1$	1	3	7	15	31
2^{h-1}	1	2	4	8	16

$2^h - 1$; The option $2^h - 1$ pays you more money when $h > 1$.

Section 16.4

Record and Practice Journal

4. **b.**
$$\begin{aligned} 3452.867 &= 3 \cdot 10^3 + 4 \cdot 10^2 + 5 \cdot 10^1 + 2 \cdot 10^0 \\ &\quad + 8 \cdot 10^{-1} + 6 \cdot 10^{-2} + 7 \cdot 10^{-3} \\ &= 3 \cdot 1000 + 4 \cdot 100 + 5 \cdot 10 + 2 \cdot 1 \\ &\quad + 8 \cdot \frac{1}{10} + 6 \cdot \frac{1}{100} + 7 \cdot \frac{1}{1000} \\ &= 3000 + 400 + 50 + 2 \\ &\quad + 0.8 + 0.06 + 0.007 \end{aligned}$$

Section 16.6

Record and Practice Journal

4. *Sample answer:* Move the decimal point left or right so the number is at least 1 but less than 10. Then multiply by ten raised to the number of times you moved the decimal. If you moved the decimal point to the left, the exponent will be positive. If you moved the decimal point to the right, the exponent should be negative.

Section 16.7

Record and Practice Journal

5. *Sample answer:* To add or subtract numbers written in scientific notation, add or subtract the factors when the powers of 10 are the same. When the powers of 10 are different, first use properties of exponents to rewrite the numbers so the powers are the same.

   To multiply or divide numbers written in scientific notation, use the properties of multiplication to rewrite the expressions, then use the Product of Powers Property and the Quotient of Powers Property to simplify.

Additional Topics

Topic 2

Practice and Problem Solving

43. **a.** 40 ft

b. no;

$$\begin{aligned} 2(\text{white area}) &= \text{black area} \\ 2[5(6x)] &= 4[6(x + 1)] \\ 60x &= 24x + 24 \\ 36x &= 24 \\ x &= \frac{2}{3} \end{aligned}$$

$$\begin{aligned} 5x + 4(x + 1) &\stackrel{?}{=} 40 \\ \text{Length of hallway is } 5\left(\frac{2}{3}\right) + 4\left(\frac{2}{3} + 1\right) &\stackrel{?}{=} 40 \\ 10 &\neq 40 \end{aligned}$$

Photo Credits

Front matter
iv Big Ideas Learning, LLC; **iii** *top* Alexander Chaikin/Shutterstock.com, ©iStockphoto.com/Ann Marie Kurtz; *bottom* Evok20/Shutterstock.com; **ix** *top* ©iStockphoto.com/pagadesign, ©iStockphoto.com/Ann Marie Kurtz; *bottom* Sinisa Bobic/Shutterstock.com; **x** *top* ©iStockphoto.com/Alistair Cotton; *bottom* Artpose Adam Borkowski/Shutterstock.com; **xi** *top* ©iStockphoto/Michael Flippo, ©iStockphoto.com/Ann Marie Kurtz; *bottom* ©iStockphoto.com/Aldo Murillo; **xii** *top* ©iStockphoto.com/ALEAIMAGE, ©iStockphoto.com/Ann Marie Kurtz; *bottom* ©iStockphoto.com/kate_sept2004; **xiii** *top* Varina and Jay Patel/Shutterstock.com, ©iStockphoto.com/Ann Marie Kurtz; *bottom* ©iStockphoto.com/Ferli Achirulli

Chapter 11
464 Alexander Chaikin/Shutterstock.com, ©iStockphoto.com/Ann Marie Kurtz; **472** Azat1976/Shutterstock.com; **476** ©iStockphoto.com/Er Ten Hong; **T-477** ©iStockphoto.com/Tryfonov levgenii; **477** *center left* ©iStockphoto.com/Sergey Galushko; *center right* ©iStockphoto.com/Tryfonov levgenii; **478** ©iStockphoto.com/ingmar wesemann; **T-483** ©iStockphoto.com/Hazlan Abdul Hakim; **483** ©iStockphoto.com/Hazlan Abdul Hakim; **491** ©iStockphoto.com/Maksim Shmeljov; **494** *top* ©iStockphoto.com/Viatcheslav Dusaleev; *bottom left* ©iStockphoto.com/Jason Mooy; *bottom right* ©iStockphoto.com/Felix Möckel; **497** gary718/Shutterstock.com; **507** Diego Cervo/Shutterstock.com; **514** *center left* Antonio Jorge Nunes/Shutterstock.com, Tom C Amon/Shutterstock.com; *center right* ©iStockphoto.com/Alex Slobodkin

Chapter 12
524 ©iStockphoto.com/pagadesign, ©iStockphoto.com/Ann Marie Kurtz; **526** PILart/Shutterstock.com, Wildstyle/Shutterstock.com; **527** Estate Craft Homes, Inc.; **538** Marc Dietrich/Shutterstock.com; **544** *bottom left* ©iStockphoto.com/Evgeny Terentev; *bottom right* ©iStockphoto.com/Vadym Volodin; **545** NASA; **548** iStockphoto.com/Evelyn Peyton; **549** *top right* ©iStockphoto.com/Terraxplorer; *top left* ©iStockphoto.com/Lora Clark; *center right* ©iStockphoto.com/Jennifer Morgan

Chapter 13
564 ©iStockphoto.com/Alistair Cotton; **569** NASA; **570** ©iStockphoto.com/David Morgan; **571** *top right* NASA; *center left* ©iStockphoto.com/jsemeniuk; **578** ©iStockphoto.com/Amanda Rohde; **579** Julian Rovagnati/Shutterstock.com; **583** RyFlip/Shutterstock.com; **586** Luke Wein/Shutterstock.com; **589** AVAVA/Shutterstock.com; **594** ©iStockphoto.com/Dreamframer; **595** *top right* Jerry Horbert/Shutterstock.com; *center left* ©iStockphoto.com/Chris Schmidt; **597** ©iStockphoto.com/biffspandex; **600** ©iStockphoto.com/Stephen Pothier; **601** *top left* Gina Smith/Shutterstock.com; *center left* Dewayne Flowers/Shutterstock.com; **605** Herrenknecht AG; **606** ©iStockphoto.com/Adam Mattel; **T-607** ©iStockphoto.com/marcellus2070, ©iStockphoto.com/beetle8; **607** *top left* ©iStockphoto.com/Gene Chutka; *center right* ©iStockphoto.com/marcellus2070, ©iStockphoto.com/beetle8; **611** ©iStockphoto.com/Connie Maher; **612** ©iStockphoto.com/Jacom Stephens; **613** *top right* ©iStockphoto.com/Petr Podzemny; *bottom left* ©iStockphoto.com/adrian beesley; **614** Richard Goldberg/Shutterstock.com; **620** Thomas M Perkins/Shutterstock.com

Chapter 14
624 ©iStockphoto/Michael Flippo, ©iStockphoto.com/Ann Marie Kurtz; **629** Perfectblue97; **630** ©iStockphoto.com/Benjamin Lazare; **T-631** *top left* ©iStockphoto.com/Jill Chen; *top right* Oleksiy Mark/Shutterstock.com; **631** *top right* ©iStockphoto.com/iShootPhotos, LLC; *center left* ©iStockphoto.com/Jill Chen, Oleksiy Mark/Shutterstock.com; **636** Gary Whitton/Shutterstock.com; **637** Michael Stokes/Shutterstock.com; **638** ©Oxford Science Archive/Heritage Images/Imagestate; **642**©iStockphoto.com/Melissa Carroll; **645** *center left* ©iStockphoto.com/Yvan Dubé; *bottom right* Snvv/Shutterstock.com; **646** ©iStockphoto.com/Kais Tolmats; **650** *top left* ©iStockphoto.com/Don Bayley; *center left* ©iStockphoto.com/iLexx; **653** ©iStockphoto.com/Marcio Silva; **657** Monkey Business Images/Shutterstock.com; **665** LoopAll/Shutterstock.com; **666** CD Lanzen/Shutterstock.com

Chapter 15
670 ©iStockphoto.com/ALEAIMAGE, ©iStockphoto.com/Ann Marie Kurtz; **672** ©iStockphoto.com/Jill Chen; **675** ©iStockphoto.com/camilla wisbauer; **T-677** ©iStockphoto.com/Matthew Dixon; **677** *Exercises 13 and 14* ©iStockphoto.com/Prill Mediendesigns & Fotografie; *Exercise 15* ©iStockphoto.com/subjug; *center left* ©iStockphoto.com/Matthew Dixon; *center right* ©iStockphoto.com/nilgun bostanci; **683** ©iStockphoto.com/Stefano Tiraboschi; **689** Donald Joski/Shutterstock.com; **690** ©iStockphoto.com/Yury Kosourov; **691** Carlos Caetano/Shutterstock.com; **698** Courtesy of Green Light Collectibles; **T-699** ©iStockphoto.com/ivanastar **699** *top right* ©iStockphoto.com/wrangel; *center left* ©iStockphoto.com/ivanastar; *bottom left* ©iStockphoto.com/Daniel Cardiff; **700** Eric Isselée/Shutterstock.com; **704** ©iStockphoto.com/Daniel Loiselle

Chapter 16
708 Varina and Jay Patel/Shutterstock.com, ©iStockphoto.com/Ann Marie Kurtz; **710** ©iStockphoto.com/Franck Boston; **711** *Activity 3a* ©iStockphoto.com/Manfred Konrad; *Activity 3b* NASA/JPL-Caltech/R.Hurt (SSC); *Activity 3c and d* NASA; *bottom right* Stevyn Colgan; **713** ©iStockphoto.com/Philippa Banks; **714** ©iStockphoto.com/clotilde hulin; **T-715** ©iStockphoto.com/Boris Yankov; **715** ©iStockphoto.com/Boris Yankov; **720** ©iStockphoto.com/VIKTORIIA KULISH; **721** *top right* ©iStockphoto.com/Paul Tessier; *center left* ©iStockphoto.com/subjug, ©iStockphoto.com/Valerie Loiseleux, ©iStockphoto.com/Linda Steward; **726** ©iStockphoto.com/Petrovich9; **727** *top right* Dash/Shutterstock.com; *center left* NASA/JPL-Caltech/L.Cieza (UT Austin); **731** ©iStockphoto.com/Aliaksandr Autayeu; **732** EugeneF/Shutterstock.com; **733** ©iStockphoto.com/Nancy Louie; **735** ©iStockphoto.com/Dan Moore; **736** ©iStockphoto.com/Kais Tolmats; **737** *Activity 3a and d* Tom C Amon/Shutterstock.com; *Activity 3b* Olga Gabay/Shutterstock.com; *Activity 3c* NASA/MODIS Rapid Response/Jeff Schmaltz; *Activity 3f* HuHu/Shutterstock.com; *Activity 4a* PILart/Shutterstock.com; *Activity 4b* Matthew Cole/Shutterstock.com; *Activity 4c* Yanas/Shutterstock.com; *Activity 4e* unkreativ/Shutterstock.com; **739** *top left* ©iStockphoto.com/Mark Stay; *top center* ©iStockphoto.com/Frank Wright; *top right* ©iStockphoto.com/Evgeniy Ivanov; *bottom left* ©iStockphoto.com/Oliver Sun Kim; **740** ©iStockphoto.com/Christian Jasiuk; **741** Microgen/Shutterstock.com; **742** *Activity 1a* ©iStockphoto.com/Susan Trigg; *Activity 1b* ©iStockphoto.com/subjug; *Activity 1c* ©iStockphoto.com/camilla wisbauer; *Activity 1d* ©iStockphoto.com/Joe Belanger; *Activity 1e* ©iStockphoto.com/thumb; *Activity 1f* ©iStockphoto.com/David Freund; **743** NASA; **744** *center* Google and YouTube logos are registered trademarks of Google Inc., used with permission.; **745** *top left* Elaine Barker/Shutterstock.com; *center right* ©iStockphoto.com/breckeni; **746** *bottom left* ©iStockphoto.com/Max Delson Martins Santos; *bottom right* ©iStockphoto.com/Jan Rysavy; **747** *top right* BORTEL Pavel/Shutterstock.com; *center right* ©iStockphoto.com/breckeni; **751** *center left* Sebastian Kaulitzki/Shutterstock.com; *center right* ©iStockphoto.com/Jan Rysavy; **753** ©iStockphoto.com/Boris Yankov; **754** mmutlu/Shutterstock.com; **758** *bottom right* ©iStockphoto.com/Eric Holsinger; *bottom left* TranceDrumer/Shutterstock.com

Additional Topics
763 ©iStockphoto.com/ALEAIMAGE, ©iStockphoto.com/Ann Marie Kurtz; **764** ©iStockphoto.com/Harley McCabe; **765** ©iStockphoto.com/Jacom Stephens; **766** ©iStockphoto.com/Harry Hu; **767** ©iStockphoto.com/Ralf Hettler, Vibrant Image Studio/Shutterstock.com; **771** ©iStockphoto.com/Andrey Krasnov; **772** Shawn Hempel/Shutterstock.com; **777** *top right* ©iStockphoto.com/Alan Crawford; *center left* ©iStockphoto.com/Julio Yeste; *bottom right* ©iStockphoto.com/Mark Stay

Cartoon illustrations Tyler Stout

Learning Progression

Kindergarten

Counting and Cardinality	– Count to 100 by Ones and Tens; Compare Numbers
Operations and Algebraic Thinking	– Understand and Model Addition and Subtraction
Number and Operations in Base Ten	– Work with Numbers 11–19 to Gain Foundations for Place Value
Measurement and Data	– Describe and Compare Measurable Attributes; Classify Objects into Categories
Geometry	– Identify and Describe Shapes

Grade 1

Operations and Algebraic Thinking	– Represent and Solve Addition and Subtraction Problems
Number and Operations in Base Ten	– Understand Place Value for Two-Digit Numbers; Use Place Value and Properties to Add and Subtract
Measurement and Data	– Measure Lengths Indirectly; Write and Tell Time; Represent and Interpret Data
Geometry	– Draw Shapes; Partition Circles and Rectangles into Two and Four Equal Shares

Grade 2

Operations and Algebraic Thinking	– Solve One- and Two-Step Problems Involving Addition and Subtraction; Build a Foundation for Multiplication
Number and Operations in Base Ten	– Understand Place Value for Three-Digit Numbers; Use Place Value and Properties to Add and Subtract
Measurement and Data	– Measure and Estimate Lengths in Standard Units; Work with Time and Money
Geometry	– Draw and Identify Shapes; Partition Circles and Rectangles into Two, Three, and Four Equal Shares

Grade 3

Operations and Algebraic Thinking	– Represent and Solve Problems Involving Multiplication and Division; Solve Two-Step Problems Involving Four Operations
Number and Operations in Base Ten	– Round Whole Numbers; Add, Subtract, and Multiply Multi-Digit Whole Numbers
Number and Operations—Fractions	– Understand Fractions as Numbers
Measurement and Data	– Solve Time, Liquid Volume, and Mass Problems; Understand Perimeter and Area
Geometry	– Reason with Shapes and Their Attributes

Grade 4

Operations and Algebraic Thinking	– Use the Four Operations with Whole Numbers to Solve Problems; Understand Factors and Multiples
Number and Operations in Base Ten	– Generalize Place Value Understanding; Perform Multi-Digit Arithmetic
Number and Operations—Fractions	– Build Fractions from Unit Fractions; Understand Decimal Notation for Fractions
Measurement and Data	– Convert Measurements; Understand and Measure Angles
Geometry	– Draw and Identify Lines and Angles; Classify Shapes

Grade 5

Operations and Algebraic Thinking	– Write and Interpret Numerical Expressions
Number and Operations in Base Ten	– Perform Operations with Multi-Digit Numbers and Decimals to Hundredths
Number and Operations—Fractions	– Add, Subtract, Multiply, and Divide Fractions
Measurement and Data	– Convert Measurements within a Measurement System; Understand Volume
Geometry	– Graph Points in the First Quadrant of the Coordinate Plane; Classify Two-Dimensional Figures

Mathematics Reference Sheet

Conversions

U.S. Customary
1 foot = 12 inches
1 yard = 3 feet
1 mile = 5280 feet
1 acre = 43,560 square feet
1 cup = 8 fluid ounces
1 pint = 2 cups
1 quart = 2 pints
1 gallon = 4 quarts
1 gallon = 231 cubic inches
1 pound = 16 ounces
1 ton = 2000 pounds
1 cubic foot ≈ 7.5 gallons

U.S. Customary to Metric
1 inch = 2.54 centimeters
1 foot ≈ 0.3 meter
1 mile ≈ 1.61 kilometers
1 quart ≈ 0.95 liter
1 gallon ≈ 3.79 liters
1 cup ≈ 237 milliliters
1 pound ≈ 0.45 kilogram
1 ounce ≈ 28.3 grams
1 gallon ≈ 3785 cubic centimeters

Time
1 minute = 60 seconds
1 hour = 60 minutes
1 hour = 3600 seconds
1 year = 52 weeks

Temperature
$C = \frac{5}{9}(F - 32)$

$F = \frac{9}{5}C + 32$

Metric
1 centimeter = 10 millimeters
1 meter = 100 centimeters
1 kilometer = 1000 meters
1 liter = 1000 milliliters
1 kiloliter = 1000 liters
1 milliliter = 1 cubic centimeter
1 liter = 1000 cubic centimeters
1 cubic millimeter = 0.001 milliliter
1 gram = 1000 milligrams
1 kilogram = 1000 grams

Metric to U.S. Customary
1 centimeter ≈ 0.39 inch
1 meter ≈ 3.28 feet
1 kilometer ≈ 0.62 mile
1 liter ≈ 1.06 quarts
1 liter ≈ 0.26 gallon
1 kilogram ≈ 2.2 pounds
1 gram ≈ 0.035 ounce
1 cubic meter ≈ 264 gallons

Number Properties

Commutative Properties of Addition and Multiplication

$a + b = b + a$

$a \cdot b = b \cdot a$

Associative Properties of Addition and Multiplication

$(a + b) + c = a + (b + c)$

$(a \cdot b) \cdot c = a \cdot (b \cdot c)$

Addition Property of Zero

$a + 0 = a$

Multiplication Properties of Zero and One

$a \cdot 0 = 0$

$a \cdot 1 = a$

Distributive Property:

$a(b + c) = ab + ac$

$a(b - c) = ab - ac$

Properties of Equality

Addition Property of Equality

If $a = b$, then $a + c = b + c$.

Subtraction Property of Equality

If $a = b$, then $a - c = b - c$.

Multiplication Property of Equality

If $a = b$, then $a \cdot c = b \cdot c$.

Multiplicative Inverse Property

$n \cdot \frac{1}{n} = \frac{1}{n} \cdot n = 1, n \neq 0$

Division Property of Equality

If $a = b$, then $a \div c = b \div c, c \neq 0$.

Squaring both sides of an equation

If $a = b$, then $a^2 = b^2$.

Cubing both sides of an equation

If $a = b$, then $a^3 = b^3$.

Properties of Inequality

Addition Property of Inequality
If $a > b$, then $a + c > b + c$.

Subtraction Property of Inequality
If $a > b$, then $a - c > b - c$.

Multiplication Property of Inequality
If $a > b$ and c is positive, then $a \cdot c > b \cdot c$.
If $a > b$ and c is negative, then $a \cdot c < b \cdot c$.

Division Property of Inequality
If $a > b$ and c is positive, then $a \div c > b \div c$.
If $a > b$ and c is negative, then $a \div c < b \div c$.

Properties of Exponents

Product of Powers Property: $a^m \cdot a^n = a^{m+n}$

Quotient of Powers Property: $\frac{a^m}{a^n} = a^{m-n}, a \neq 0$

Power of a Power Property: $(a^m)^n = a^{mn}$

Power of a Product Property: $(ab)^m = a^m b^m$

Zero Exponents: $a^0 = 1, a \neq 0$

Negative Exponents: $a^{-n} = \frac{1}{a^n}, a \neq 0$

Slope

$$m = \frac{\text{rise}}{\text{run}} = \frac{\text{change in } y}{\text{change in } x} = \frac{y_2 - y_1}{x_2 - x_1}$$

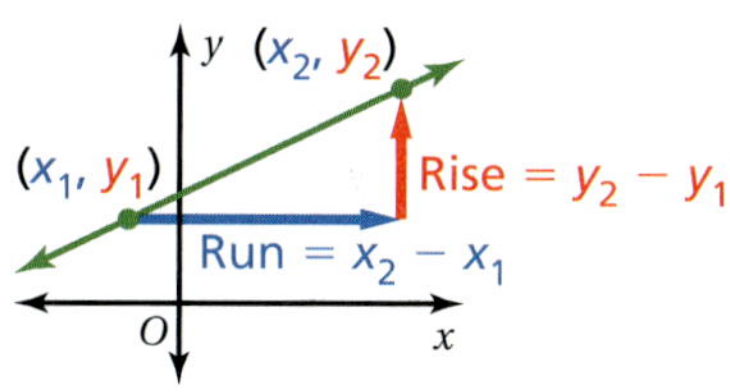

Equations of Lines

Slope-intercept form
$y = mx + b$

Standard form
$ax + by = c, a, b \neq 0$

Point-slope form
$y - y_1 = m(x - x_1)$

Angles of Polygons

Interior Angle Measures of a Triangle

$x + y + z = 180$

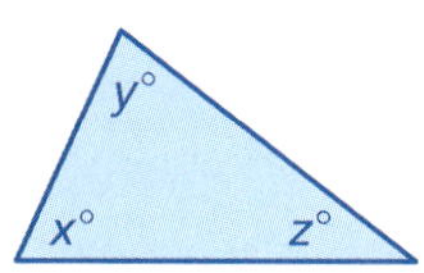

Interior Angle Measures of a Polygon

The sum S of the interior angle measures of a polygon with n sides is $S = (n - 2) \cdot 180°$.

Exterior Angle Measures of a Polygon

$w + x + y + z = 360$

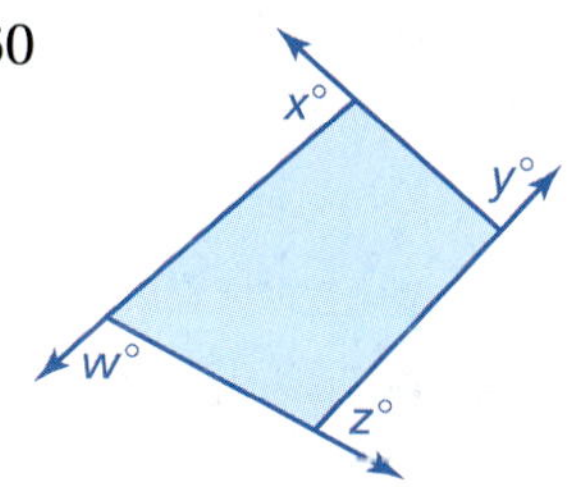

Formulas in Geometry

Prism

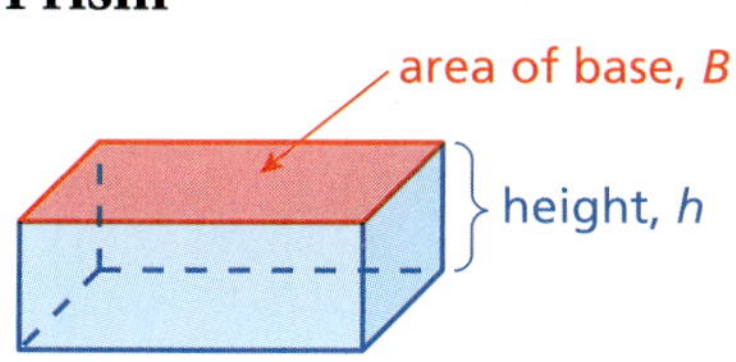

$S =$ areas of bases + areas of lateral faces
$V = Bh$

Pyramid

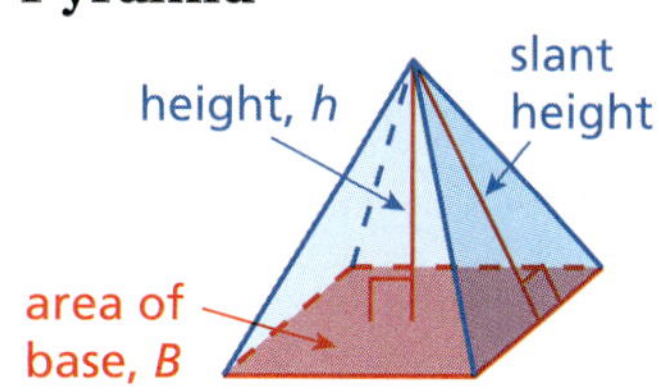

$S =$ area of base + areas of lateral faces
$V = \frac{1}{3}Bh$

Circle

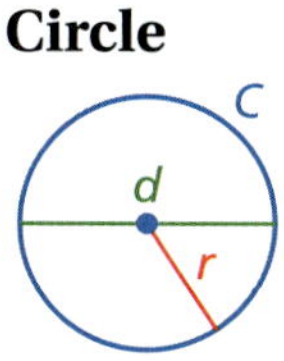

$C = \pi d$ or $C = 2\pi r$
$A = \pi r^2$

Cylinder

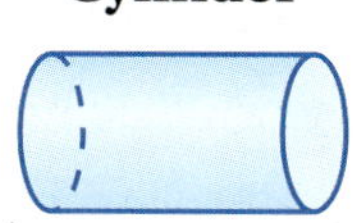

$V = Bh = \pi r^2 h$

Cone

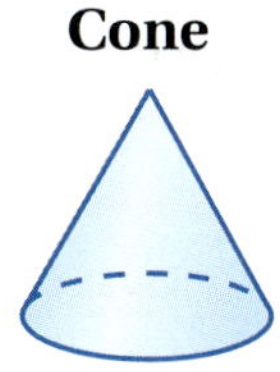

$V = \frac{1}{3}Bh = \frac{1}{3}\pi r^2 h$

Sphere

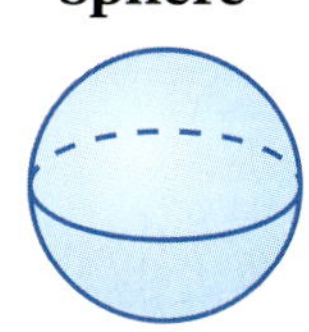

$V = \frac{4}{3}\pi r^3$